W9-BWM-871

ENCYCLOPEDIA OF HUMAN BIOLOGY

VOLUME 4 Go–Me

ENCYCLOPEDIA of HUMAN BIOLOGY

VOLUME 4 Go–Me

Editor–in–Chief
Renato Dulbecco
The Salk Institute
La Jolla, California

ACADEMIC PRESS, INC. *Harcourt Brace Jovanovich, Publishers*
San Diego   New York   Boston   London   Sydney   Tokyo   Toronto

Academic Press, Inc.
San Diego, California 92101

United Kingdom Edition published by
Academic Press Limited
24–28 Oval Road, London NW1 7DX

Library of Congress Cataloging-in-Publication Data

Encyclopedia of human biology / [edited by] Renato Dulbecco.
     p. cm.
   Includes index.
   ISBN 0-12-226751-6 (v. 1). -- ISBN 0-12-226752-4 (v. 2). -- ISBN
0-12-226753-2 (v. 3). -- ISBN 0-12-226754-0 (v. 4). -- ISBN
0-12-226755-9 (v. 5). -- ISBN 0-12-226756-7 (v. 6). -- ISBN
0-12-226757-5 (v. 7). -- ISBN 0-12-226758-3 (v. 8)
    1. Human biology--Encyclopedias.   I. Dulbecco, Renato, 1914-
    [DNLM: 1. Biology--encyclopedias.   2. Physiology--Encyclopedias.
QH 302.5 E56]
QP11.E53   1991
612'.003--dc20
DNLM/DLC
for Library of Congress   91-45538
                                               CIP

PRINTED IN THE UNITED STATES OF AMERICA
91 92 93 94   9 8 7 6 5 4 3 2 1

CONTENTS OF VOLUME 4

HOW TO USE THE ENCYCLOPEDIA

We have organized this encyclopedia in a manner that we believe will be the most useful to you and would like to acquaint you with some of its features.

The volumes are organized alphabetically as you would expect to find them in, for example, magazine articles. Thus, "Food Toxicology" is listed as such and would *not* be found under "Toxicology, Food." If the first words in a title are *not* the primary subject matter contained in an article, the main subject of the title is listed first: (e.g., "Sex Differences, Biocultural," "Sex Differences, Psychological," "Aging, Psychiatric Aspects," "Bone, Embryonic Development.") This is also true if the primary word of a title is too general (e.g., "Coenzymes, Biochemistry.") Here, the word "coenzymes" is listed first as "biochemistry" is a very broad topic. Titles are alphabetized letter-by-letter so that "Gangliosides" is followed by "Gangliosides and Neuronal Differentiation" and then "Ganglioside Transport."

Each article contains a brief introductory Glossary wherein terms that may be unfamiliar to you are defined *in the context of their use in the article.* Thus, a term may appear in another article defined in a slightly different manner or with a subtle pedagogic nuance that is specific to that particular article. For clarity, we have allowed these differences in definition to remain so that the terms are defined relative to the context of each article.

Articles about closely related subjects are identified in the Index of Related Articles at the end of the last volume (Volume 8.) The article titles that are cross-referenced within each article may be found in this index, along with other articles on related topics.

The Subject Index contains specific, detailed information about any subject discussed in the *Encyclopedia.* Entries appear with the source volume number in boldface followed by a colon and the page number in that volume where the information occurs (e.g., "Diuretics, **3:** 93"). Each article is also indexed by its title (or a shortened version thereof) and the page ranges of the article appear in boldface (e.g., "Abortion, **1: 1–10**" means that the primary coverage of the topic of abortion occurs on pages 1–10 of Volume 1).

If a topic is covered primarily under one heading but is occasionally referred to in a slightly different manner or by a related term, it is indexed under the term that is most commonly used and a cross-reference is given to the minor usage. For example, "B Lymphocytes" would contain all page entries where relevant information occurs, followed by *"see also* B Cells." In addition, "B Cells, *see* B Lymphocytes" would lead the reader to the primary usages of the more general term. Similarly, "*see under*" would mean that the subject is covered under a subheading of the more common term.

An additional feature of the Subject Index is the identification of Glossary terms. These appear in the index where the word "defined" (or the words "definition of") follows an entry. As we noted earlier, there may be more than one definition for a particular term and, as when using a dictionary, you will be able to choose among several different usages to find the particular meaning that is specifically of interest to you.

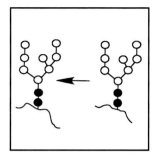

Golgi Apparatus

BECCA FLEISCHER, *Vanderbilt University*

Glossary

Glycoproteins Proteins that contain short, often branched oligosaccharide chains (2–10 monomers); the chains contain hexoses, hexosamines, and sometimes fucose, and they often terminate in sialic acid

Glycosylation Formation of an acetal or glycosidic bond between a hydroxyl or amino group in a molecule and a carbohydrate or oligosaccharide unit

Glycosyltransferases Enzymes that transfer carbohydrate moieties from a donor molecule (often a nucleotide sugar) to an acceptor molecule such as a protein or lipid

Proteoglycans Proteins that contain many long, unbranched chains of repeating disaccharide units, one of which is always an amino sugar, often sulfated, and the second is usually a uronic acid; thus the chains are highly negatively charged

THE GOLGI APPARATUS (also termed the Golgi complex) is a membrane-bound compartment or organelle found in most eukaryotic cells. When visualized by transmission electron microscopy, it has a characteristic structure consisting of a series of flattened cisternae and associated vesicles.

I. General Properties

A. Structure and Cytochemistry

The Golgi apparatus is named after Camillo Golgi, who first described it in neuronal cells almost a cen-

tury ago. It can be visualized by light microscopy after impregnation with heavy metals such as osmic acid (Fig. 1). It has a dispersed, perinuclear location in epithelial cells or ganglia but a compact localization between the nucleus and the apical pole in secretory cells. The organelle consists of at least three morphologically distinct features when viewed by transmission electron microscopy in stained, embedded, and thinly sectioned tissues (Fig. 2). The structure consists of a series of flattened, curved cisternae, many small vesicles peripheral to the ends of the cisternae, and large secretory vesicles clustered near one face of the stacked cisternae. In mammalian cells, there are usually four to eight cisternae with one side of the stack apposing the rough endoplasmic reticulum (RER) and the other the plasma membrane. The number of stacks and their general organization are highly dependent on the cell type under study.

The Golgi apparatus shows a marked polarity both in structure and in cytochemical staining (Fig. 3). This has resulted in one side being designated the *cis* face, and the opposite side the *trans* face. The *cis* side faces the RER, while the *trans* side faces the cell membrane. The first cisterna on the *cis* side is usually fenestrated, while on the *trans* side the apparatus includes a network of tubules and vesicles, termed the *trans* Golgi network (TGN). Small, uniformly sized transitional vesicles are seen between the endoplasmic reticulum and the *cis* side of the Golgi. Morphologically similar transport vesicles are present around the periphery of all the Golgi stacks.

Table I summarizes the reactivity of various parts of the structure to commonly used staining techniques. In some cases an intermediate, or *medial*, region can also be distinguished cytochemically from the *cis* and *trans* elements. The biochemical basis of the reactivity to the various staining techniques is indicated when known and is discussed

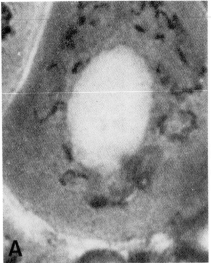

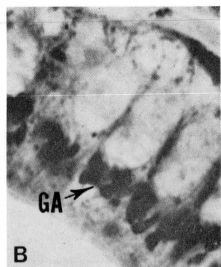

FIGURE 1 Golgi apparatus stained for light microscopy with heavy metal impregnation. A. Ganglion cell with perinuclear distribution of the Golgi apparatus. B. Epithelial cell with apical distribution of the Golgi apparatus. GA, Golgi apparatus. [Reproduced, with permission, from H. W. Beams and R. G. Kessel, 1968, The golgi apparatus: structure and function. *Int. Rev. Cytol.* **23,** 209, Academic Press, Orlando, Florida.]

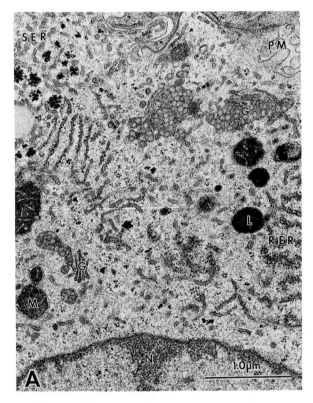

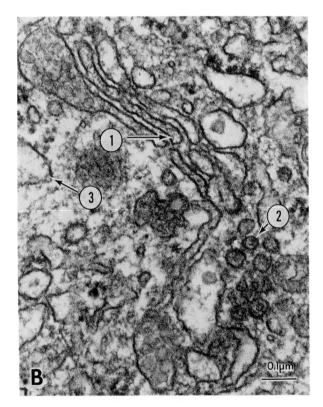

FIGURE 2 A. Electron micrograph of a section of a hepatocyte from rat liver. The organelles illustrated are nucleus (N), Golgi apparatus (G), mitochondrion (M), rough endoplasmic reticulum (RER), lysosome (L), and plasma membrane (PM). B. The Golgi apparatus in the rat hepatocyte. Characteristic features are (1) three or four stacks of flattened cisternae, (2) peripheral transport vesicles, and (3) large secretory vesicles on the *trans* side of the stacks. [Reproduced, with permission, from B. Fleischer and S. Fleischer, 1981, The golgi apparatus. *in* "Advanced Cell Biology" (L. M. Schwartz and M. M. Azar, eds.), Van Nostrand

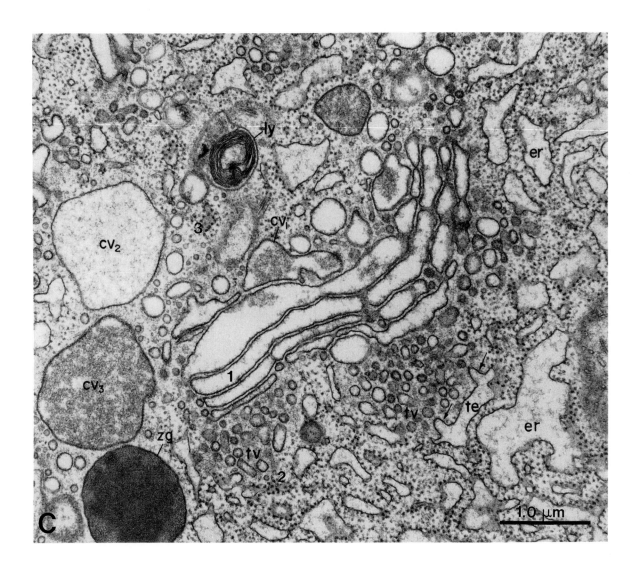

Reinhold Co., New York.] C. Golgi region of an exocrine pancreatic cell of the guinea pig. Characteristic features include four or five slightly dilated Golgi cisternae associated with condensing vacuoles (CV) and mature zymogen granules (ZG) on the *trans* side and peripheral transport vesicles (tv) on the *cis* side. These vesicles bud (arrows) from transitional elements (te) of the rough endoplasmic reticulum (er) carrying secretory products to the *cis* Golgi. Similar vesicles are involved in intracisternal traffic. The condensing vacuoles (CV$_{1-3}$) are in various stages of maturation to form zymogen granules. ly, lysosome. [Reproduced, with permission, from M. G. Farquhar and G. E. Palade (1981). The Golgi Apparatus (Complex)—(1954–1981)—from artifact to center stage. *J. Cell Biol.* **91**, 77s, Rockefeller University Press.]

further in Section II. The polarization of the structure can also be observed in the increased thickening of the cisternal membranes from the *cis* to the *trans* aspect of the stack.

B. Function

The Golgi apparatus is part of the secretory machinery of the cell. It modifies and sorts proteins and lipids, which it receives from the endoplasmic reticulum (ER), and thus helps direct them to their final destination. Secreted proteins and glycosaminoglycans, as well as plasma membrane glycoproteins and glycolipids, pass through the Golgi and are collected into secretory vesicles. These then fuse with the plasma membrane to release their contents into the extracellular space in a process termed exocytosis. Two types of secretion can be distinguished: constitutive secretion occurs continuously in all cells; regulated secretion occurs in exocrine or glandular cells. Regulated cells store their secretory products in granules that are released only upon stimulation of the cell by specific chemical signals.

TABLE I Cytochemical Staining of Golgi Cisternae

Procedure	Biochemical basis	Cis	Medial	Trans (+TGN)
OsO$_4$ impregnation	unknown	+	−	−
Periodic acid–silver methenamine	complex carbohydrate	+	+ +	+ + +
Thiamine pyrophosphatase	nucleoside diphosphatase	−	−	+
Acid phosphatase	unknown	−	−	+
Nicotinamide adenosine dinucleotide phosphatase[a]	unknown	−	+	+
DAMP[b]	acidic compartment	−	+	+[c]
Galactosyltransferase	enzyme immunolocalization	−	−	+
Sialyltransferase	enzyme immunolocalization	−	−	+
N-acetylglucosamine transferase	enzyme immunolocalization	−	+	−
Cation-independent mannose-6-phosphate receptor[a]	enzyme immunolocalization	+	+	+
Concanavalin A–peroxidase	lectin for Glc, Man	+	−	−
Wheat-germ agglutinin–peroxidase	lectin for (GlcNAc)$_2$, NAN	−	−	+
Ricinus 120 immunolocalization	lectin for galactose	−	+	+

[a] Localization is dependent on cell type.
[b] 3-(2,4-Dinitroanalino)-3-amino N-methylpropylamine.
[c] Particularly in fibroblasts and hepatocytes.
Glc, glucose; GlcNAc, N-acetylglucosamine; Man, mannose; NAN, N-acetylneuraminic acid; TGN, trans Golgi network.

Secreted proteins are first segregated into the lumen of the ER during their synthesis on ribosomes attached to the RER. The Golgi maintains this segregation throughout the transport process; therefore, the lumen of the Golgi is topologically continuous with the lumen of the ER, the external face of the plasma membrane, and the extracellular space.

The Golgi also takes part in the biosynthesis and transport of lysosomal enzymes. Lysosomes are organelles involved in cellular digestion, which contain many types of acid hydrolases (which break down various types of molecules).

A large variety of enzymatic modifications of transported proteins and lipids occurs in the Golgi apparatus. The earliest known and best characterized biochemically are the series of enzymes involved in the terminal glycosylation of N-asparagine-linked glycoproteins, a major class of secreted and membrane-bound glycoproteins. These proteins are first glycosylated in RER by the transfer of a core oligosaccharide structure from the carrier lipid dolichol pyrophosphate to an appropriate asparagine residue of the protein while the protein is synthesized and secreted either entirely or partially into the lumen. While in transit through the ER, the core oligosaccharide structure is trimmed by removal of terminal glucose and some mannose units. Glycoproteins destined for secretion or for residence in the plasma membrane or in lysosomes are transferred to the Golgi apparatus where they undergo further stepwise processing and terminal glycosylation by a series of specific glycosyltransferases uniquely present there.

The Golgi is believed to be the sole site for the formation of a number of other types of protein-bound glycan chains. These include the O-linked glycosides found in glycoproteins, in collagen, and in proteoglycans. In addition to glycoproteins, most of the glycosylations involved in the formation of glycolipids occur in the Golgi; however, the first glycosylation step in the formation of glycolipids (i.e., the conversion of ceramide to cerebroside) has not been localized exclusively in the Golgi.

A second type of modification of both secreted and membrane components generally carried out in the Golgi is sulfation. Sulfation of sugar moieties on glycoproteins, glycosaminoglycans, and glycolipids as well as sulfation of tyrosine residues of proteins also occurs there; however, not all sulfations carried out in the cell occur in the Golgi. Steroids, for example, are sulfated by cytoplasmic enzymes in hepatocytes. [See STEROIDS.]

The large variety of modifications of transported proteins that take place in the Golgi apparatus are believed to be related to the central role the Golgi plays in directing these constituents to their proper destination during their biosynthesis or in resorting membrane constituents from contents during recycling of endocytic vesicles. Details of these pathways are elaborated further in Section III.

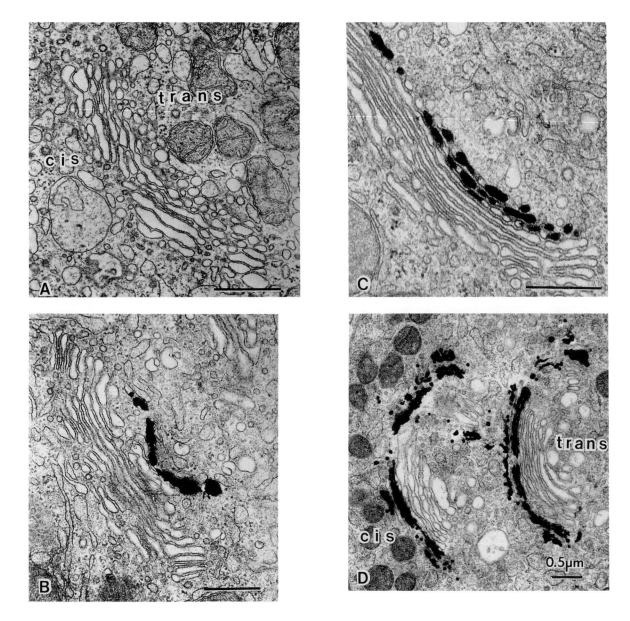

FIGURE 3 Golgi complex from rat epidydymis illustrating the polarity of the complex using cytochemical staining. In all cases, the Golgi is similarly oriented and the bars represent 0.5 μm. A. Unstained. B. Stained to reveal acid phosphatase activity. C. Stained for thiamine pyrophosphatase activity. D. Stained using heavy impregnation with OsO₄. [Micrographs courtesy of Daniel Friend.]

C. Isolation

The Golgi apparatus has been isolated in relatively intact form from a number of mammalian tissues and cell types (Fig. 4). In general, the methods are similar and involve gentle homogenization followed by differential centrifugation to obtain a Golgi-enriched fraction. The latter is then purified from contaminating intracellular organelles by centrifugation in a density gradient because Golgi membranes are generally less dense than other cell organelles. To maintain its characteristic structure, however, care must be taken to avoid excessive shear or osmotic shock during the isolation because, like most membranous organelles, the Golgi is easily disrupted to form small vesicles, or microsomes. The cellular origin of such vesicles is difficult to determine by morphological means. To evaluate the purity of any preparation containing smooth membrane vesicles,

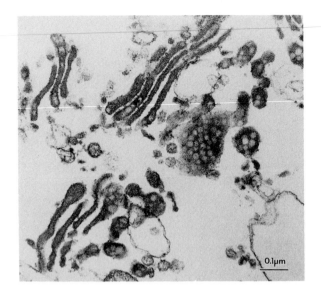

FIGURE 4 Purified Golgi apparatus isolated from rat liver. The characteristic structure of the Golgi has been retained during the isolation. Compare with rat liver Golgi *in situ* (Fig. 2B). [Reproduced, with permission, from B. Fleischer and S. Fleischer, 1981, *in* "Advanced Cell Biology" (L. M. Schwartz and M. M. Azar, eds.), Van Nostrand Reinhold Co.]

enzymological means are usually employed. This can be done because most organelles contain "marker" enzymes, i.e., membrane-bound enzymes ideally localized in only that organelle. A commonly used enzymatic activity for monitoring Golgi purification is galactosyltransferase, which adds the penultimate galactose to N-asparagine-linked glycoproteins. To use this approach most effectively, the Golgi apparatus must remain relatively intact because the enzyme is not uniformly distributed throughout the Golgi (Table I; see also Section II). Despite this, the enzyme has proven to be a useful marker for Golgi membranes because it is widely distributed, membrane-bound, stable, and relatively easy to assay. A few exceptions should be noted. In mammary gland, a soluble form of galactosyltransferase is produced and is secreted into milk complexed with α-lactalbumin to form lactose synthetase. A soluble form is also found at high levels in embryonic chicken brain and at low levels in human serum. In certain cases, low levels of galactosyltransferase are believed to be present on the cell membrane and play a role in cell recognition. For example, mouse sperm cells appear to bind to the egg zona pellucida via cell-surface galactosyltransferase. Embryonic mouse cell compaction at the eight-cell stage also appears to be mediated by the presence of galactosyltransferase on the cell surface.

II. Biochemistry

A. Composition

The Golgi apparatus has a typical biological membrane in that it consists mainly of lipids and proteins. In addition, it contains soluble secretory components and substrates in its lumen. Its membrane composition reflects its biological function and is therefore unique from other organelle membranes within the same cell. Composition of the Golgi isolated from various organs varies considerably, which reflects their functional differentiation.

1. Lipids

Detailed analysis of the lipids of the Golgi and comparison with other organelles from the same tissue has been carried out in only a few cases. The lipid content of Golgi isolated from several rat tissues is summarized in Table II. Golgi, like the plasma membrane, contains a high proportion of neutral lipid, mainly cholesterol. This fact partly accounts for its low buoyant density (1.12–1.14 g/ml). As in other organelles, the main constituent of Golgi lipids are phospholipids. In tissues rich in glycolipids (i.e., kidney and testes), the Golgi also contains significant levels of glycolipids. The phospholipid composition of the Golgi isolated from several different tissues is remarkably similar (Table III). [*See* Lipids.]

TABLE II Lipid Content of Golgi Apparatus Compared with Other Organelles of Rat Liver[a]

	Liver				Kidney	Testes[b]
	Mito	RER	Golgi	PM	Golgi	Golgi
Total lipid	0.202	0.425	1.01	0.994	0.859	1.26
% NL	13.4	13.6	33.3	32.4	18.9	39.4
% GL	—	—	—	—	17.1[c]	2.3[d]
Cholesterol	0.003	0.014	0.071	0.128	0.071	0.090

[a] All values expressed as mg/mg protein. Total lipid is the sum of the neutral lipid (NL), phospholipid, and glycolipid (GL) values. % NL is (neutral lipid/total lipid) × 100. % GL is (glycolipid/total lipid) × 100. Dashes indicate values not determined. Mito, mitochondria; PM, plasma membrane; RER, rough endoplasmic reticulum.
[b] Data of T. N. Keenan, S. E. Nyquist, and H. H. Mollenhauer, 1972, *Biochim. Biophys. Acta* **270**, 433.
[c] The major glycolipid in kidney Golgi is sulfatide.
[d] The major glycolipid in testes Golgi is a sulfated galactosyldiglyceride.

TABLE III Phospholipid Composition of Golgi Apparatus from Rat Tissues[a]

Phospholipid	Liver	Kidney	Testes[b]
PE	23.5	23.5	21.5
PC	54.0	52.1	56.8
Sph	7.8	6.7	7.0
DPG	1.0	0.0	0.0
PI	8.6	7.1	9.0
PS	3.0	5.4	3.1
LPE	0.3	0.3	1.1
LPC	0.4	3.6	1.4
PA	0.4	0.7	—
Unidentified	1.0	1.1	—

[a] All values expressed as % of total phospholipid phosphorus.
[b] Data of T. N. Keenan, S. E. Nyquist, and H. H. Mollenhauer, 1972, *Biochim. Biophys. Acta* **270**, 433.
DPG, diphosphatidylglycerol; LPC, lysophosphatidylcholine; LPE, lysophosphatidylethanolamine; PA, phosphatidic acid; PC, phosphatidylcholine; PE, phosphatidylethanolamine; PI, phosphatidylinositol; PS, phosphatidylserine; Sph, sphingomyelin.

2. Proteins

The most common analytical tool used to define the protein composition of isolated subcellular structures is dissociation in a detergent such as sodium dodecylsulfate followed by resolution of the proteins on the basis of particle size by polyacrylamide gel electrophoresis (SDS-PAGE). When analyzed in this manner, the Golgi apparatus shows a complex mixture of proteins, which is unique for that organelle in each tissue; however, Golgi from different cell types are very different in their overall protein profiles. This probably reflects their specialized function and the significant level of secretory products, which are processed through the Golgi and vary considerably in composition in different tissues and cell types. [*See* PROTEINS.]

B. Enzymes

1. Processing of N-Asparagine-Linked Glycoproteins

After initial formation and trimming of the relatively simple "core" carbohydrate chain of N-asparagine-linked glycoproteins in the RER, the proteins are transferred to the Golgi apparatus by means of transport vesicles. In the Golgi, a series of further sequential trimming and glycosylation steps are carried out by specific mannosidases and glycosyltransferases to form complex type chains on some of these proteins. The pathway is illustrated

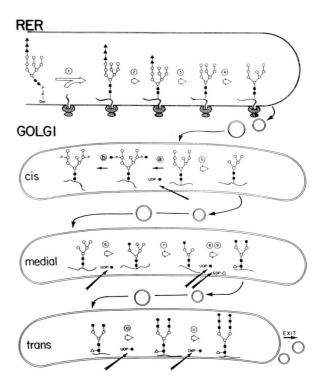

FIGURE 5 Schematic pathway of oligosaccharide processing on newly synthesized N-asparagine-linked glycoproteins. The reactions are catalyzed by the following enzymes: (1) oligosaccharyltransferase, (2) α-glucosidase I, (3) α-glucosidase II, (4) ER α1,2-mannosidase, (a) N-acetylglucosaminylphosphotransferase, (b) N-acetylglucosamine-1-phosphodiester α-N-acetylglucosaminidase, (5) Golgi α-mannosidase I, (6) N-acetylglucosaminyltransferase I, (7) Golgi α-mannosidase II, (8) N-acetylglucosaminyltransferase II, (9) fucosyltransferase, (10) galactosyltransferase, (11) sialyltransferase. ■, N-acetylglucosamine; ○, mannose; ▲, glucose; △, fucose; ●, galactose; ◆, sialic acid. [Reproduced, with permission, from R. Kornfeld and S. Kornfeld, the *Annual Review of Biochemistry* Volume 54, © 1985, by Annual Reviews Inc.]

schematically in Fig. 5. Many of the enzymes indicated in the sequence have been isolated and characterized. A few have been localized within regions of the Golgi stack in some cell types using immunohistochemistry (Table I and Fig. 6). The enzymes are membrane-bound and luminally oriented. The transferases use nucleotide sugars as cofactors. The various processing enzymes do not occur in a fixed ratio but vary in a tissue-specific and developmentally regulated manner.

2. Formation of O-Linked Glycoproteins

The initial step in the formation of O-linked glycoproteins is catalyzed by an N-acetylgalactosamine (GalNAc) transferase, which transfers

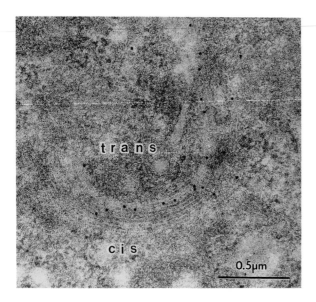

FIGURE 6 The Golgi apparatus in HeLa cells illustrating the localization of galactosyltransferase using immunohistochemistry. An antibody specific for the enzyme is conjugated to gold particles and used to localize the enzyme. Galactosyltransferase immunoreactivity appears concentrated in two to three *trans* cisternae. [Reproduced from J. Roth and E. G. Berger (1982). Immunocytochemical localization of galactosyltransferase in HeLa cells: Codistribution with thiamine pyrophosphatase in *trans*-golgi cisternae. *J. Cell Biol.* **92**, 223, by copyright permission of the Rockefeller University Press.]

GalNAc from uridine diphosphate (UDP)GalNAc to the hydroxyl group of a threonine or serine residue of a protein substrate. The preponderance of evidence is that this step occurs very close to the time when secreted proteins are transferred from the ER to the Golgi and probably is a *cis* Golgi function. Further elongation of the chain is carried out by glycosyltransferases in the Golgi in a similar manner to the formation of the complex carbohydrate chains present in N-asparagine-linked chains.

A special form of O-linked glycosylation is found in collagen consisting of disaccharide groups linked to hydroxylysine. These are formed in the Golgi by the sequential actions of a specific hydroxylysyl galactosyltransferase and a specific glucosyltransferase.

3. Formation of Sugar Chains of Proteoglycans

Proteoglycans differ from glycoproteins in the nature and number of their sugar side chains. The chains consist of polymers of disaccharides containing an amino sugar, often sulfated, and a uronic acid. The disaccharide chain is long and negatively charged. It is linked to the protein via the xylose portion of a trisaccharide (xylose-gal-gal) attached to serine residues of the core protein. The formation of the chains occurs primarily in the Golgi apparatus via specific glycosyltransferase activities acting in a stepwise manner. [*See* PROTEOGLYCANS.]

4. Formation of Glycolipids

Glycolipids are membrane lipids formed by the glycosylation of ceramide. The initial step in their formation is the transfer of glucose or galactose to the free hydroxyl group of ceramide to form cerebroside. This step can be carried out by Golgi but is not exclusively present there. However, the further elongation of cerebroside by the stepwise addition of galactose, *N*-acetylgalactosamine, and sialic acid has been shown to occur only in the Golgi membrane. These glycosyltransferases are membrane-bound enzymes, are luminally oriented, and utilize nucleotide sugars as cofactors. They are, however, distinct from the enzymes that form glycoproteins. The expression of these enzymes varies greatly from tissue to tissue. Liver, for example, contains very low levels of these lipids, whereas brain is very rich and kidney is intermediate. Testes and sperm contain a unique glycolipid that is a monoalkyl-monoacylglycerolgalactoside sulfate.

5. Formation of Sphingomyelin

The Golgi apparatus is the main cellular site for sphingomyelin biosynthesis in rat liver cells. However, the major phospholipids present in the Golgi, namely phosphatidylcholine and phosphatidylethanolamine, are primarily synthesized in the ER and later transferred to the Golgi. Sphingomyelin is formed by the transfer of phosphocholine from phosphatidylcholine to ceramide. The transferase activity is found predominantly in rat liver Golgi fractions enriched in cis and medial cisternae and is luminally oriented.

6. Modification of Secretion Products by Sulfation

A number of secretion products are sulfated by membrane-bound sulfotransferases uniquely localized in Golgi membranes. The enzymes utilize a nucleotide-bound sulfate (3'-phosphoadenosine-5'-phosphosulfate [PAPS]) as a cofactor. A number of such Golgi sulfotransferases have been characterized. Modifications carried out by Golgi include sulfation of galactosyl ceramide to form sulfatide, sulfation of the sugar moieties of glycosaminoglycans and of glycoproteins, and sulfation of tyrosine residues of proteins. In testes, galactosyl diglyceride is sulfated in the Golgi.

In general, the distribution of sulfotransferases within the Golgi stack has not been defined; however, sulfation of tyrosine groups of immunoglobulin M appears to coincide with the addition of galactose and sialic acid to the protein during its biosynthesis. This indicates that tyrosine sulfotransferase is a *trans* Golgi function.

7. Other Enzyme Activities Associated with Golgi Membranes

A number of other enzymatic processes not particularly unique to that organelle have been demonstrated to be present there. These include protein kinase activities, proton-translocating adenosine triphosphatase (ATPase) activity, phospholipase A_1 and A_2 activities, and protein acyltransferase activity.

C. Topology

As in the case for most cellular membranes, the Golgi membrane is asymmetric in its transmembrane organization. Most of the enzymes described in Section II.B act on secreted products, and many are luminally oriented. It is possible to study their orientation even in isolated Golgi vesicles because the latter largely retain their original orientation during isolation, even if vesiculation occurs. A major consequence of this orientation in the case of the various glycosyl and sulfotransferases present in the Golgi is that the nucleotide-bound sugars and sulfate, which act as donor substrates for the enzymes, must be present in the lumen of the Golgi. To accomplish this, various nucleotide sugars as well as PAPS are transported from the cytoplasm into the lumen of the Golgi by a series of specific transport proteins uniquely localized in the Golgi membrane. As yet, the transport proteins have not been identified.

In addition to the entrance of nucleotide substrates into the Golgi lumen, the nucleotide products of the transferases must not accumulate in the lumen or inhibition of the transferases would result. UDP, for example, is a potent inhibitor of galactosyltransferase. No accumulation of UDP occurs, however, during glycosylation, because the Golgi exhibits a high level of UDPase activity, which is luminally oriented and codistributes with galacto-

syltransferase (Table I). The product that accumulates is uridine monophosphate, which is not as inhibitory and is efficiently transported out of the Golgi lumen. For sialyl transferases, the product is cytidine monophosphate which is similarly transported out. During sulfation, adenosine monophosphate (AMP) has been shown to accumulate outside the Golgi. Presumably, the 3',5'-adenosine diphosphate formed by the enzyme is further hydrolyzed to AMP in the lumen, and AMP is transported out.

At least some of the lipids present in the Golgi membrane also appear to be asymmetrically distributed across the bilayer. In particular, sphingomyelin and glycosphingolipids are probably present almost entirely in the luminal side of the membrane. Phosphatidylcholine (PC), the major phospholipid present in the Golgi membrane, is about equally distributed between the two faces of the membrane. The two pools of PC do not equilibrate rapidly with each other. This is in contrast to the ER membrane, which exhibits a fast rate of exchange (or "flip-flop" rate) of PC from one side of the membrane to the other.

D. Functional Polarity and Dynamics

The Golgi apparatus is a highly dynamic organelle. It is constantly receiving newly synthesized proteins and lipids from the RER packaged in transitional vesicles. Exit from the RER is the rate-limiting step in most protein secretion. The vesicles fuse with the *cis* Golgi cisternae and eventually shuttle back to the RER without their contents. The secreted proteins then make their way unidirectionally through the various cisternae until they reach the *trans* Golgi network, where they are sorted for delivery to the plasma membrane, secretion granules, or lysosomes. This transit through the Golgi is rapid with a half-time of about 10 min. Movement between cisternae is accomplished by means of transport vesicles, which bud off from the ends of the cisternae and fuse with the next. Throughout this process, resident proteins of the RER membrane remain in the RER, and resident proteins of the different cisternae are not transported. The most likely explanation is that these proteins contain signals that cause their selective retention. Golgi intercisternal transport vesicles have been purified from rabbit liver (Fig. 7). They consist of nonclathrin-coated vesicles of about 75 nm in diameter (or 106 nm with their protein coat).

Both transitional vesicles and intercisternal trans-

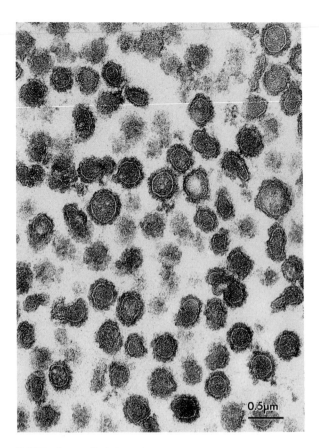

0.5μm

FIGURE 7 Purified intercisternal transport vesicles derived from Golgi apparatus isolated from rabbit liver. A distinct non-clathrin coat is visible on the vesicles. [Reproduced, with permission, from V. Malhotra, T. Serafini, L. Orci, J. C. Shepherd, and J. E. Rothman (1989). Purification of a novel class of coated vesicles mediating biosynthetic protein transport through the golgi stack. *Cell* **58,** 329, Cell Press, Cambridge, Massachusetts.]

port vesicles are formed by the interaction of the membranes with cytosolic proteins and ATP. Furthermore, both require hydrolysis of guanosine triphosphate and interaction with an *N*-ethyl-maleimide-sensitive fusion protein in the cytosol before fusion with the Golgi cisternae can occur. The same fusion machinery also appears to operate during fusion of endocytic vesicles. It may also be involved during mitosis, when the Golgi apparatus vesiculates and the vesicles disperse. They later re-assemble into recognizable Golgi structures in each daughter cell during telophase.

III. Role in Processing of Cellular Constituents

Secretory, lysosomal, and plasma membrane proteins share the same intracellular pathway from their site of synthesis in the RER through the Golgi apparatus to the *trans* Golgi network. There they are sorted for transport to their final destination. Although the mechanisms are not completely understood for this complex processing, this is at present an active area of research in cell biology, and current views will be outlined briefly in this section.

A. Constitutive Secretory Products

All cells exhibit unregulated, constitutive secretion, which takes place all the time. Proteins secreted via this pathway may be glycosylated or otherwise modified, or they may not. They are believed to enter this pathway by default; i.e., proteins that contain no specific sorting signals for segregation into secretory granules or lysosomes are secreted at a constant rate via a bulk route. In the TGN, such proteins appear to be largely excluded from clathrin-coated vesicles destined for secretory granules or for lysosomes. Instead, they are removed by smooth secretory vesicles that constantly bud off and fuse with the plasma membrane to discharge their contents. Serum protein secretion by hepatocytes, immunoglobulin secretion by plasma cells, and collagen and fibronectin secretion by fibroblasts are important examples of proteins secreted by this pathway. A special case of constitutive secretion occurs in epithelial-type cells, which are polarized (i.e., contain at least two different domains of plasma membrane). These cells maintain different constitutive secretion through each of the domains, although the products exit from the same Golgi structure. The targeting mechanism for this domain-specific segregation is not known.

B. Extracellular Matrix

The main constituents of the extracellular matrix are proteoglycans, fibrous proteins such as collagens, and adhesion glycoproteins such as fibronectin. Like soluble, secreted proteins, they are synthesized in the RER and individually processed and secreted by the Golgi apparatus. They are finally processed and assembled into a matrix extracellularly. These components are constitutively secreted. In epithelial-type cells, the components of the extracellular matrix are secreted in a polarized manner from the basal plasma membrane only to form the basement membrane. [*See* EXTRACELLULAR MATRIX.]

C. Plasma Membrane

The plasma membrane is highly enriched in glycoproteins, glycolipids and in sphingomyelin. The glycolipids, sphingomyelin and the carbohydrate moieties of the glycoproteins are extracellularly oriented. They are transported to the plasma membrane by secretory vesicles from the Golgi apparatus probably via the constitutive pathway. In some epithelial cells, the glycolipid content of the apical plasma membrane has been found to be twofold higher than the basolateral plasma membrane, indicating selective targeting of the glycolipids. The mechanism responsible for this enrichment is not known. Similarly, membrane glycoproteins destined for different domains of the plasma membrane travel through the Golgi apparatus together and are sorted into separate secretory vesicles upon exiting the Golgi by as yet unknown mechanisms. The liver cell may be an exception to this general scheme. Newly synthesized proteins destined for the apical surface have been shown to go first to the basolateral surface and then selectively move to the apical surface possibly via trancytotic vesicles.

D. Formation of Lysosomes

Acid hydrolases destined for transport to lysosomes are N-glycosylated in the RER as are many secreted proteins. They are specifically modified in the Golgi apparatus by a two-step enzymatic process, which introduces a lysosomal marker, mannose-6-phosphate (Man-6-P; see steps a and b in Fig. 5). The first step (a) involves *N*-acetylglucosamine phosphotransferase, a *cis* Golgi enzyme, which recognizes a common polypeptide feature of lysosomal enzymes. A genetic defect in this enzyme occurs in I-cell disease (mucopolydosis II) and in pseudo-Hurler polydystrophy (mucopolydosis III). The defects cause secretion of the hydrolases and nonfunctional lysosomes. In the second step (b), in the formation of the lysosomal marker, a specific *cis* Golgi phosphodiesterase removes the GlcNAc residue to form the Man-6-P residue. Proteins containing this marker bind to membrane-bound Man-6-P receptors present in *trans* Golgi. These receptors are responsible for segregating the hydrolases into clathrin-coated vesicles destined for transport to lysosomes. Man-6-P receptors are also present on the plasma membrane, where they pick up secreted lysosomal hydrolases by endocytosis to recycle them back to lysosomes.

Lysosomal membrane proteins also are segregated into the lumen of the RER during biosynthesis and are often N-glycosylated. They are then transferred to the Golgi apparatus and may be further glycosylated but do not carry the Man-6-P marker. In I-cell disease, they are correctly sorted into lysosomelike inclusion bodies. Nonglycosylated lysosomal membrane proteins are also correctly sorted in fibroblasts from I-cell disease patients, indicating that the recognition for sorting resides in the polypeptides themselves.

E. Formation of Secretory Granules

Secretion granules are secretory vesicles with a dense core. They accumulate in glandular cells and fuse with the plasma membrane only when the cell is triggered by a specific stimulus. Some cells, such as pancreatic acinar cells, store a variety of enzymes within a single type of secretory granule. Others, such as pancreatic islet cells, store a single hormone such as insulin in their granules. Pituitary gonadotrophs store two different glycosylated polypeptide hormones in separate and morphologically distinct granules. Immature secretory vesicles, or condensing vacuoles, appear to bud off from the *trans* Golgi or the TGN. Proteolytic processing of peptide hormones such as insulin probably occurs in the early secretory granules, which are clathrin-coated. Condensation of early granules, after removal of their clathrin coat, leads to formation of mature granules. Proteins destined for sorting into secretory granules are believed to carry positive signals for their sorting, although the nature of the signals is not known. The secretory granule membrane contains a limited, unique set of membrane proteins and is distinct from other membranes of the secretion pathway. After fusion with the targeted domain of the plasma membrane, the components are recycled back to the TGN where they can be reused to form secretory granules.

IV. Drug Effects on Golgi Function

A number of drugs that affect Golgi function are summarized in Table IV. Most exert their effects indirectly by lowering cell energy levels, causing microtubule disassembly or disturbing ionic equilibria between vesicles and cytoplasm. The molecular basis of the action of brefeldin A, a fungal metabolite, is not known. It appears to block the formation of transitional vesicles from the RER but not their

TABLE IV Drugs Affecting Golgi Function

Drug	Effect observed	Biochemical basis
Cyanide, azide	secretion products remain in RER	lowers cellular ATP and, thus, GTP levels, blocking vesicular transport from RER to Golgi
Ammonia, chloroquine	inhibits constitutive secretion from Golgi; causes vacuolization of *trans* Golgi	increases pH of endosomes and lysosomes, thus inhibiting receptor recycling and probably membrane recycling to *trans* golgi network
Colchicine	secretory products remain in Golgi	disassembles microtubules, thus preventing secretory vesicle movement?
Monensin	swollen, vesiculated Golgi cisternae; secretion blocked	neutralizes *cis* to *trans* Golgi pH gradient?
Swainsonine	neurological symptoms in some species; vacuolization in all cell types studied	inhibits mannosidase II
Tunicamycin[a]	blocks galactosylation of glycolipids and glycoproteins in isolated Golgi apparatus	inhibits UDPGal transport in Golgi membranes
Atractyloside[b]	blocks sulfation of endogenous acceptors in isolated Golgi	inhibits PAPS transport in Golgi membranes
Brefeldin A	blocks secretion; Golgi disassembled into vesicles; RER dilated	blocks RER to Golgi transport but not Golgi to RER recycling?

[a] This effect has been demonstrated *in vitro* at relatively high (mM) levels of tunicamycin. *In vivo,* tunicamycin at low levels (μM) blocks the enzyme that forms dolichol diphospho GlcNAc and, thus, the formation of all N-asparagine-linked glycoproteins.
[b] Atractyloside inhibits the mitochondrial ATP–ATP exchange transporter at significantly lower concentration.

return to the RER, thus disrupting the steady-state level of Golgi membrane. The recognizable structure disappears, but the effect is reversible when the drug is removed. Swainsonine, a toxic alkaloid found in plants of the genus *Swainsona* and in locoweed, inhibits both Golgi mannosidase II and lysosomal mannosidase *in vitro*. Ingestion of the toxin induces increased lysosomal vacuolization in both liver and kidney but no morphological changes in the Golgi structure. Atractyloside, a toxic plant glycoside, inhibits PAPS transport in Golgi membranes *in vitro*. *In vivo,* however, its toxicity is mainly due to its inhibition of mitochondrial ATP production because it is a potent inhibitor of the ATP–ADP exchange transporter of the mitochondrial inner membrane. Similarly, the main physiological effect of tunicamycin, a family of nucleoside antibiotics, is to block the formation of N-asparagine-linked glycoproteins because it is a potent inhibitor of the formation of dolichol diphospho-*N*-acetylglucosamine. *In vitro,* however, tunicamycin also inhibits UDPGal transport into the Golgi lumen.

Bibliography

Alberts, B., Bray, D., Lewis, J., Raff, M., Roberts, K., and Watson, J. D. (1989). ''Molecular Biology of the Cell,'' 2nd ed. Garland Publishing, Inc., New York.

Burgess, T. L., and Kelly, R. B. (1987). Constitutive and regulated secretion of proteins. *Ann. Rev. Cell Biol.* **3,** 243–293.

Farquhar, M. G. (1985). Progress in unraveling pathways of golgi traffic. *Ann. Rev. Cell Biol.* **1,** 447–458.

Fleischer, B. (1988). Functional topology of golgi membranes. *In* ''Protein Transfer and Organelle Biogenesis'' (R. C. Das and P. W. Robbins, eds.). Academic Press, New York.

Kornfeld, S. (1987). Trafficking of lysosomal enzymes. *FASEB J.* **1,** 462–468.

Kornfeld, R., and Kornfeld, S. (1985). Assembly of asparagine-linked oligosaccharides. *Ann. Rev. Biochem.* **54,** 631–664.

Pfeffer, S. R., and Rothman, J. E. (1987). Biosynthetic protein transport and sorting by the endoplasmic reticulum and Golgi. *Ann. Rev. Biochem.* **56,** 829–852.

Rodriguez-Boulan, E., and Nelson, W. J. (1989). Morphogenesis of the polarized epithelial cell phenotype. *Science* **245,** 718–725.

Sabatini, D. D., and Adesnik, M. B. (1989). The biogenesis of membranes and organelles. *In* ''The Metabolic Basis of Inherited Diseases,'' Vol. I, 6th ed. (C. R. Scriver, A. L. Beaudet, W. S. Sly, and D. Vallee, eds.). McGraw Hill, New York.

van Meer, G. (1989). Lipid traffic in animal cells. *Ann. Rev. Cell Biol.* **5,** 247–275.

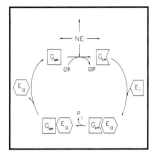

G Proteins

RAFAEL MATTERA, *Department of Physiology and Biophysics, School of Medicine, Case Western Reserve University*

Glossary

Agonists Synthetic analogs that interact with receptors, eliciting the biological effects normally caused by the physiological ligand (i.e., hormone or neurotransmitter)

Antagonists Synthetic analogs that interact with the receptors with high affinity and saturability, but do not produce the biological effects of the physiological ligands

cDNA Complementary DNA from an mRNA template, rather than a DNA template

Exon Gene sequence encoding amino acids present in the gene product

Hormones Molecules that are secreted by a particular set of cells and travel through the bloodstream to produce their biological effects in distant cells (i.e., effector or target cells)

Intron Noncoding (or intervening) sequence in a given gene

Neurotransmitters Molecules involved in the conduction of electrical impulses from one nerve cell to another or from one nerve cell to muscle cells

Patch-clamp Recordings Measurement of single ionic channels in small (1–10 um^2) patches of biological membranes tightly sealed to glass microelectrodes and subjected to adjustable electric fields

Receptors Target cell molecules that bind hormones or neurotransmitters, with high affinity and saturability, and mediate their actions

RNA Splicing Removal of the intron sequences present in the primary RNA transcript, carried out by a complex of RNA-processing enzymes

Signal Transduction Conversion of information carried to a cell by chemical or sensory stimuli, and recognized by specific receptors, into a new form, such as the change in the activity of cellular proteins and the intracellular concentration of ions and metabolites

Toxin-catalyzed ADP-ribosylation Covalent modification of a substrate protein produced by the enzymatic transference of ADP-ribose from NAD donor molecules to specific amino acids in the acceptor molecule

G PROTEINS are heterotrimeric molecules located in cell membranes which transduce and amplify the signal carried to cells by hormones, agonists, light, or odorants, inducing changes in the activity of cellular enzymes.

The G proteins possess quaternary structure, defined by the interaction of three different subunits: α, β, and γ, in order of decreasing molecular weights. The heterotrimers bind and hydrolyze GTP; this catalytic activity controls the deactivation of the protein.

The G proteins become activated upon interaction with specific hormone–receptor complexes, in the presence of Mg^{2+} and GTP. Nonhydrolyzable analogs of GTP, such as guanosine-5-O-(3-thiotriphosphate) (GTPγS) or guanosine-5'-yl imidodiphosphate, or ligands such as AlF_4^-, can induce stable activation of G proteins in the absence of hormone–receptor complexes. Activation by GTP analogs induces a conformational change which, at least in solution, causes dissociation of the oligomeric G protein into free α subunits, which interact with their effectors, and β–γ dimers. An important property of the G proteins is that they are substrates for ADP ribosylation catalyzed by bacterial toxins. ADP ribosylation of substrates by cholera toxin abolishes the GTPase activities associated with these molecules, causing their permanent activation. ADP ribosylation of substrates mediated by one of the exotoxins of *Bordetella pertussis* uncouples them from their specific hormone–receptor complexes, with impairment in the signal transduction.

The existence of these molecules was first evidenced in 1971 by Rodbell, Birnbaumer, and colleagues, who observed that GTP was required to obtain adequate activation of the enzyme adenylyl cyclase by glucagon.

I. Guanine Nucleotide-Binding Proteins in Biology

The G proteins constitute a subgroup of a family of proteins that can bind and hydrolyze GTP. Nature has assigned an important role to the GTPases, involving them in critical biological functions, such as initiation, elongation, and termination of protein synthesis; microtubule assembly; and regulation of enzyme activity (Table I). These molecules have different conformations and activities, depending on whether they have GDP or GTP bound to them. The GDP-bound forms can interact with other molecules that catalyze the exchange of GDP for GTP; the GTP-bound forms interact with, and change the activity of, a third class of molecules: effectors. The GTPase activity converts the protein to the GDP-bound form, which has a decreased ability to interact with the effector and is able to interact with the nucleotide exchanger in order to resume its effector active conformation (Fig. 1). In some cases the biological effects can occur without GTP hydrolysis, such as translocation in protein synthesis (mediated

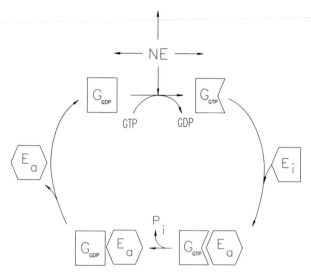

FIGURE 1 The interaction among a GTP-binding protein (G), a GDP/GTP exchanger (symbolized by NE, for nucleotide exchanger), and the inactive and active forms of an effector (E_i and E_a, respectively). One NE molecule (e.g., photolyzed rhodopsin in the activation of the retinal G protein transducin) catalyzes the nucleotide exchange of several hundreds of GTP-binding proteins (illustrated as multiple arrows emerging from NE). The GTP-binding proteins present two different conformations, depending on their interaction with either GDP or GTP. In this example the GTP-bound form is able to interact with E_i. As a consequence of this interaction, the effector changes tertiary structure, acquiring its active conformation. The activation of the effector by G_{GTP} is sometimes the consequence of a change in the quaternary structure of the effector (i.e., the number of subunits present in the oligomeric enzyme). The GTPase activity, which is either constitutively regulated by G or activated by the interaction with a separate molecule (e.g., ribosomal proteins or GTPase-activating proteins (see Table I)], shifts the G protein back into the G_{GDP} conformation. This form has low affinity for the effector, resulting in the release of E_a and the availability of G_{GDP} for new interaction with NE. Note the main features of this biological device: (1) the amplification of the NE signal and (2) the self-inactivation of the amplifier, which provides a mechanism for quick blockage of the response, unless the NE is still present to push the system into additional rounds. P_i, Inorganic phosphate.

by elongation factor G or 2), tubulin polymerization, and signal transduction by heterotrimeric G proteins. In others (e.g., termination of protein synthesis) GTP hydrolysis is required for manifestation of the biological effect.

In this article I focus only on the heterotrimeric GTP-binding proteins involved in signal transduction (referred to here as G proteins). The G proteins which have been cloned and/or purified are listed in Table II.

TABLE I GTP-Binding Proteins in Cell Biology

GTP-binding protein	GDP/GTP exchanger	Cellular process	Specific effectors	Signal for GTPase activity
Eukaryotic initiation factor 2 (eIF-2), prokaryotic initiation factor 2 (IF-2)	eIF-2B	Initiation of protein synthesis	Methionyl-tRNA, allowing binding to mRNA and small ribosomal subunit	Codon recognition large ribosomal subunit joins complex
Prokaryotic elongation factor Tu (EF-Tu)	Ef-Ts	Protein elongation	Amino acid end of aminoacyl tRNA, allowing tRNA to associate with ribosome	Binding of elongation factor complex to ribosome
Eukaryotic elongation factor 1	EF-1	Protein elongation	Catalyzes the binding of aminocyl tRNA to ribosome	Interaction with ribosome
Prokaryotic elongation factor G	Unknown	Translocation during protein synthesis	Deacyl-tRNA and ribosomal proteins	Interaction with ribosome
Eukaryotic elongation factor 2	Unknown	Translocation during protein synthesis	Deacyl-tRNA and ribosomal proteins	Interaction with ribosome[a]
Eukaryotic releasing factor	Unknown	Termination of protein synthesis	Stop codons and peptidyl transferases	[b]
Tubulin	Unknown	Microtubule formation	Ends of growing polymers (microtubules)	[c]
Heterotrimeric signal-transducing G proteins	Rhodopsin, hormone receptors	Signal transduction	Cellular enzymes and ionic channels	[d]
Family of low-molecular mass (i.e., 21–25 kDa) GTP-binding proteins (e.g., mammalian ras, rap, ral, rho, rab, and smgs, proteins)	Yeast, *CDC25* gene products, mammalian rGEF and rGEF-like proteins[e]	Functions of mammalian proteins unknown; probably involved in signal transduction and organelle traffic	Yeast ras proteins activate adenylyl cyclase; yeast YPT1 and SEC4 involved in organelle transport and exocytosis	Interaction with GTPase-activating proteins

[a] Translocation can occur without GTP hydrolysis. GTPase activity is required for the release of elongation factor G and the entry of ribosome in a new elongation step.
[b] GTP hydrolysis is required for the termination of protein synthesis.
[c] Delayed GTPase activity is responsible for dynamic instability.
[d] Some authors have postulated that the constant and relatively high GTPase activity is controlled by a built-in domain that mimics the function of GTPase-activating proteins.
[e] rGEF is a 100-kDa protein recently purified from bovine brain.

II. Signal Transduction in Biology: Concept of Amplification

The concept of signal transduction first appeared when studying the regulation of adenylyl cyclase by hormones and the biological effects of light in photoreceptor cells. The introduction of a third element, coupling the formation of a hormone–receptor complex to the change in the activity of an intracellular effector, has the potential of providing an amplification effect (see the legend to Fig. 1). In this way one molecule of agonist can interact with approximately 10 molecules of G_S, which, in turn, interact with the enzyme adenylyl cyclase. Also, each molecule of light-activated rhodopsin interacts with approximately 500 molecules of transducin. Each molecule of transducin activates one molecule of a phosphodiesterase enzyme that hydrolyzes cGMP into GMP. The initial amplification provided by the G proteins is subsequently magnified by the activated enzymes: Each molecule of cyclase or phosphodiesterase catalyzes the conversion of a large number of nucleotide molecules.

TABLE II G Proteins Characterized by Purification and/or Cloning[a]

G protein	Toxin specificity	Cellular distribution	Effectors
T_1, T_2[b]	CT, PT	T_1, retinal rods; T_2, retinal cones	cGMP-dependent phosphodiesterase (+)
G_s	CT	Ubiquitous	Adenylyl cyclase, voltage-gated Ca^{2+} channels (+)
G_{olf}	CT[c]	Olfactory tissue	Adenylyl cyclase (+)
G_z	—	Brain, retina, platelets and erythrocytes	Unknown
G_{i1} G_{i2} G_{i3}	PT	Ubiquitous	Atrial K^+ channels (+), adenylyl cyclase (−)[d]
G_o	PT	Brain, pituitary, heart, adipose tissue	K^+ channels in hippocampal pyramidal cells (+), opioid ligand-gated Ca^{2+} channels (−)[e], neuropeptide Y-gated Ca^{2+} channels in ganglion cells (−)[f], phosphoinositide-specific phosphodiesterase[f]

[a] CT, cholera toxin; PT, pertussis toxin; (+) and (−), activation and inhibition, respectively, of the effector.
[b] Only T_1, the transducin molecule present in retinal rods, has been purified and characterized as both a PT and a CT substrate. T_2 has been identified by molecular cloning and localized in retinal cones by immunocytochemistry.
[c] Inferred by homology with G_s and by the identification of an abundant CT substrate in olfactory tissue.
[d] Relatively weak inhibition of the adenylyl cyclase catalyst has been observed using G_i preparations. These observations have not been confirmed using recombinant proteins.
[e] Results were obtained using preparations of porcine brain G_o.
[f] Results were obtained using preparations of bovine brain G_o.

III. Cellular Functions Regulated by G Proteins

A. Phototransduction

Transducin interacts with the γ subunit of cGMP-dependent phosphodiesterase. This subunit is inhibitory to the catalytic activity of the α–β dimer of the phosphodiesterase. The dimer catalyzes the conversion of cGMP into GMP. The decrease in the levels of cGMP produces closing of the Na^+ channels; in this way a light signal is transduced into an electrical one.

B. Adenylyl Cyclase Activity

The enzyme adenylyl cyclase catalyzes the conversion of ATP into cAMP. This enzyme is subjected to dual regulation by stimulatory and inhibitory receptors. The signal carried by stimulatory receptors is transduced by G_s. The interaction of this protein with ligand-occupied stimulatory receptors induces dissociation of the α subunit from the β–γ dimers. The GTP-bound α subunit activates the catalytic subunit of adenylyl cyclase. The signal elicited by ligand-occupied inhibitory receptors is transduced by G_i. The activation of this G protein reduces the activity of the adenylyl cyclase, with effects opposite to those described above.

The cAMP, one of the most important intracellular messengers, binds and activates the cAMP-dependent protein kinase. This enzyme phosphorylates serine or threonine residues in specific protein substrates that change their activity as a result of the phosphate transfer. The phosphorylation of the cAMP-dependent protein kinases constitute the most important biological effect of cAMP. [See PROTEIN PHOSPHORYLATION.]

C. C-Type Phospholipases

A variety of Ca^{2+}-mobilizing receptors interact with G proteins called G_p's. The physical identity of these G proteins has not been defined and might vary from cell to cell. The activation of these G_p's increases the activity of phosphoinositide-specific phospholipase C (PLC). This enzyme, or group of enzymes (since several isoforms have been purified), catalyzes the rupture of the phosphodiester bond of phosphoinositides, minor lipid components of biological membranes. The reaction produces inositol phosphates (soluble compounds) and a membrane-bound lipid: diacylglycerol. These compounds have profound influences on the state of the cell: The inositol phosphates produce an increase in the concentration of cytosolic Ca^{2+}, due to the mobilization of the cation present in intracellular stores and also to the influx of extracellular cation. Ca^{2+} is another extraordinarily important messenger; changes in its levels can drive fundamental processes such as secretion and smooth muscle contraction. The diacylglycerol activates the Ca_2^+ and phospholipid-dependent protein kinase C (PKC). This kinase catalyzes the transfer of phosphates to specific serine and threonine residues on protein substrates. Among the PKC substrates in membranes, we can mention the Na^+/H^+ exchanger, Ca^{2+}-ATPase, glucose transporter, and several ion channels.

D. A₂-Type Phospholipases

Another physically undefined G protein couples the interaction of arachidonic acid-mobilizing receptors with the enzyme phospholipase A_2. This enzyme catalyzes the release of fatty acids esterified in the sn-2 position of phospholipids. The arachidonic acid, together with its metabolites (i.e., prostaglandins, thromboxanes, and leukotrienes), represents a key intracellular messenger controlling stimulation of secretion from anterior pituitary, pancreatic β, and mast cells; release of Ca^{2+} from intracellular stores (independent of inositol phosphate formation); activation of PLC, adenylyl cyclase, guanylyl cyclase, and PKC; increase in DNA synthesis; and regulation of K^+ channels. [*See* DNA SYNTHESIS.]

E. Ionic Channels

Evidence for direct interaction of ionic channels with receptor-activated G proteins has been obtained recently. This represents an effect independent from the activation of ionic channels resulting from the action of second messengers or protein kinases activated by them (e.g., cAMP-dependent protein kinase, PKC, and cGMP). Evidence for direct interaction was obtained from experiments in isolated membrane patches, using incubation conditions designed to minimize secondary effects. Examples of these direct interactions are the coupling of muscarinic agonist-gated K^+ channels in the heart by G_i-like proteins (see below), K^+ channels in hippocampal pyramidal neurons (G_o), the regulation of voltage-gated dihydropyridine-sensitive L-type Ca^{2+} channels (G_s), ATP-sensitive K^+ channels, cardiac Na^+ channels (G_s), and amiloride-sensitive Na^+ channels from renal epithelium. [*See* ION PUMPS.]

F. Regulation of Intracellular Vesicle Traffic

Experiments performed with permeabilized cells and patch–clamp recordings using dialyzed single cells indicate that exocytosis could be directly controlled by G proteins. This regulation seems to be independent of the changes in the levels of intracellular Ca^{2+} or PKC activity which are brought about by G protein regulation of phosphoinositide-specific PLC. It is still unclear whether this is indeed a direct coupling of a G protein to the fusion between secretory granules and plasma membranes, or whether it is due to the G protein-mediated release of another second messenger (e.g., arachidonic acid).

G. Olfaction

The signal carried by at least some odorants results in activation of adenylyl cyclase in olfactory cilia. This transduction is likely to be mediated by $G_{olf}\alpha$,[1] a G protein α subunit exclusively expressed in olfactory tissue, which shares 88% identity with $G_s\alpha$. The increased levels of cAMP gate a Na^+/K^+ conductance in the ciliary plasma membranes of the olfactory sensory neurons, resulting in membrane depolarization. [*See* OLFACTORY INFORMATION PROCESSING.]

IV. Primary Structure of G Protein Subunits

A. α Subunits

The primary structure of approximately 12 α subunits of G proteins have been determined in the past five years. Comparison of the analogous subunits among different species shows minimal differences.

---

1. The G protein subunits are indistinctly referred to by using either the subunit symbol with the appropriate subscript (e.g., α_{i1}) or "G" and the corresponding subscript, followed by the subunit symbol (e.g., $G_{i1}\alpha$).

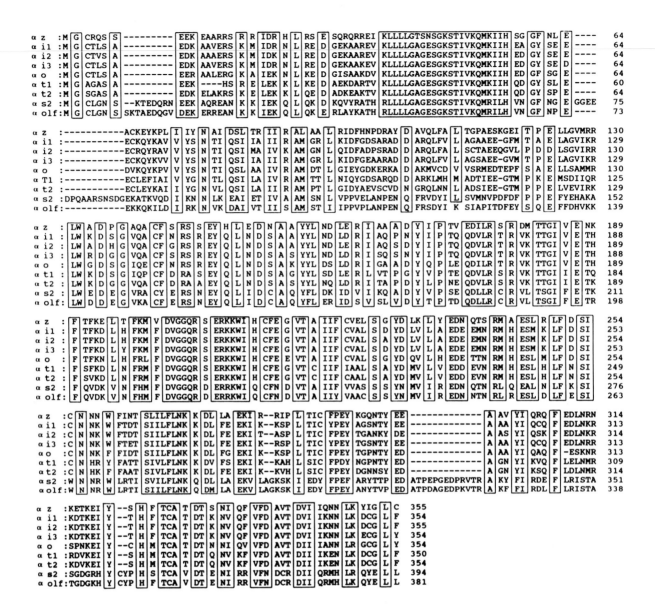

FIGURE 2 The sequences of the human α_z, α_s, α_{i2}, α_{i3}, and α_{o1}, bovine α_{t1}, α_{t2}, and α_{i1} and rat α_{olf} are given. Variations of the corresponding proteins between species is minimal: the human α_s is 99.7% identical to the corresponding bovine and rat proteins; comparison of the sequences of each of the different types of α_i in human bovine and rat also show that they are at least 98% identical. These proteins are the products of nine separate genes. The α sequence shown here (α_{s2a}, 394 amino acids) was deduced from one of the four splicing products of the α_s RNA primary transcript. Single-letter nomenclature for amino acids is as follows: A, alanine; C, cysteine; D, aspartic acid; E, glutamic acid; F, phenylalanine; G, glycine; H, histidine; I, isoleucine; K, lysine; L, leucine; M, methionine; N, asparagine; P, proline; Q, glutamine; R, arginine; S, serine; T, threonine; V, valine; W, tryptophan; and Y, tyrosine. Boxed areas are regions of identity or conservative substitutions. Sequences have been spaced to obtain the highest degree of homology. Groups of conservative substitutions are (1) D, N, E, and Q; (2) R, H, and K; (3) F, Y, and W; (4) S, T, P, A, and G; and (5) M, I, L, and V.

Four isoforms of α_s have been sequenced: α_{s1a}, α_{s1b}, α_{s2a}, and α_{s2b}, encoding proteins of 379, 380, 394, and 395 amino acids, respectively. The four isoforms are thought to be derived from the alternative splicing of a single RNA precursor. Two α subunits of transducin have been cloned; the two proteins (i.e., α_{t1} and α_{t2}) are 80% homologous and contain 350 and 354 amino acids, respectively.

The cDNAs encoding three different pertussis toxin substrates—named α_{i1}, α_{i2}, and α_{i3}—have also been cloned. Their designation ("α_i" for "inhibitory") was based on the known effects of pertussis toxin blocking hormonal inhibition of adenylyl cyclase. The exact functions of the proteins encoded by these cDNAs are the object of current investigation.

Two cDNAs encoding the α subunit of a protein abundant in brain (G_o) have been cloned in vertebrates. ADP ribosylation and protein purification experiments indicate the existence of further heterogeneity in this molecule. This is reinforced by the fact that two forms of α_o, differing in only seven residues at the amino terminus of the molecule, have been identified in the fly *Drosophila melanogaster*.

Figure 2 compares the sequences of human α_{s2a}, α_{t1}, α_{i1}, α_{i2}, α_{i3}, α_{o1}, α_{olf}, and α_z.

An additional set of five novel α subunits has been recently evidenced in the brain and the spermatids (i.e., spermatozoa predecessor cells). This finding indicates that we are still ignorant of the total complexity and size of the G protein family.

B. β Subunits

SDS-PAGE analysis of transducin purified from rod outer segments show an apparently unique β subunit (molecular mass 36 kDa). Heterotrimers other than transducin (G_i-like proteins, G_o, G_s's, etc.) contain at least two β species of 35 and 36 kDa. Molecular cloning experiments have evidenced three b subunit mRNAs: β_1, (encoding a 36 kDa protein), β_2 (encoding the 35 kDa form), and β_3. The proteins encoded by β_1 and β_2 are 90% homologous; β_3 has 83% and 81% homology with β_1 and β_2, respectively. The size of the translation product corresponding to the β_3 mRNA (cloned from a retinal cDNA library and present in several tissues) is still ignored; it might represent another 36 kDa form, or

a subunit of a different size that has not been detected in the transducin heterotrimers purified up to now. The comparison between the primary structures of the β_1 and β_2 subunits of G proteins is shown in Fig. 3.

C. γ Subunits

The γ subunits of transducin ($T\gamma$) and G_s, G_i, and G_o ($G\gamma$) have been cloned from retinal, brain, and adrenal tissues. The $T\gamma$ and $G\gamma$ subunits contain 74 and 71 amino acids, respectively, and differ in 44 residues (Fig. 4). Several studies indicate that the heterogeneity in the population of G_γ includes at least four subtypes.

The presence of multiple forms of β and γ subunits raises the possibility of preferential interaction of some β and γ subtypes with the different α subunits. However, initial experiments in this area show that G_s, G_{i2}, and G_{i3} purified from human erythrocytes contain similar ratios of β_{35} and β_{36} subunits (approximately 1:1).

V. Relationship Between Structure and Function of G Proteins

A. α Subunits

The three-dimensional structures of the G protein subunits are still unknown. Only the crystal structures of the GTP-binding proteins elongation factor Tu (EF-Tu) and c-H-ras (the product of the H-*ras* oncogene) have been elucidated. The effects of mutations on the GTPase and GTP binding activity of the α subunit of G_s have established that the inferred homology in the domains involved in guanine nucleotide interaction in G protein α subunits and ras proteins is indeed correct. The significant functional domains in α subunits correspond to the regions involved in the interaction with the β–γ subunits (β–γ site), with guanine nucleotides, with ef-

```
β1 - MSELDQLRQEAEQLKNQIRDARKACADATLSQITNNIDPVGRIQMRTRRT  -50
     |||  ||||||||||  ||||||||| ||  |  || |||   |||||||||||||
β2 - MSELEQLRQEAEQLKNQIRDARKACGDSTLTQITAGLDPVGRIQMRTRRT  -50

β1 - LRGHLAKIYAMHWGTDSRLLVSASQDGKLIIWDSYTTNKVHAIPLRSSWV  -100
     |||||||||||||||||||||||||||||||||||||||||||||||||||
β2 - LRGHLAKIYAMHWGTDSRLLVSASQDGKLIIWDSYTTNKVHAIPLRSSWV  -100

β1 - MTCAYAPSGNYVACGGLDNICSIYNLKTREGNVRVSRELAGHTGYLSCCR  -150
     |||||||||| |||||||||||||||| ||||||||||||  ||||||||||
β2 - MTCAYAPSGNFVACGGLDNICSIYSLKTREGNVRVSRELPGHTGYLSCCR  -150

β1 - FLDDNQIVTSSGDTTCALWDIETGQQTTTFTGHTGDVMSLSLAPDTRLFV  -200
     ||||||| ||||||||||||||||||||  ||| ||  ||||||||||  |||
β2 - FLDDNQIITSSGDTTCALWDIETGQQTVGFAGHSGDVMSLSLAPNGRTFV  -200

β1 - SGACDASAKLWDVREGMCRQTFTGHESDINAICFFPNGNAFATGSDDATC  -250
     ||||||| |||||||| ||||||||||||||| ||||||| ||||||||||
β2 - SGACDASIKLWDVRDSMCRQTFIGHESDINAVAFFPNGYAFTTGSDDATC  -250

β1 - RLFDLRADQELMTYSHDNIICGITSVSFSKSGRLLLAGYDDFNCNVWDAL  -300
     ||||||||||| ||||||||||||||| ||  |||||||||||||| |||||
β2 - RLFDLRADQELLMYSHDNIICGITSVAFSRSGRLLLAGYDDFNCNIWDAM  -300

β1 - KADRAGVLAGHDNRVSCLGVTDDGMAVATGSWDSFLKIWN  -340
     |  ||||||||||||||||||||||||||||||||||||||
β2 - KGDRAGVLAGHDNRVSCLGVTDDGMAVATGSWDSFLKIWN  -340
```

FIGURE 3 Comparison of the primary structures of the human β_1 and β_2 subunits. Vertical lines indicate identical residues.

```
Gγ - MASNNTASIAQARKL---VEQLKMEANIDRIKVSKAAADLMAYCEAHAKE  -47
     |           ||  | ||||  | ||| |    | |
Tγ - MPVINIEDLTEKDKLKMEVDQLKKEVTLERMLVSKCCEEFRDYVEERSGE  -50

Gγ - DPLLTPVPASENPFREKKFFCAIL  -71
     |||   |     || |||
Tγ - DPLVKGIPEDKNPFKELKGGCVIS  -74
```

FIGURE 4 Comparison of the primary structures of bovine T_γ and G_γ. The gap was introduced in order to obtain the highest degree of homology. Vertical lines show the position of identical residues.

FIGURE 5 Primary structure of human $G_{s2}\alpha$ showing the domains considered to be involved in the interaction with $\beta-\gamma$ subunits, phosphate anion (PO_4^{3-}), Mg^{2+}, guanine ring (gua), and hormone–receptor complexes (receptor), respectively. For details see text.

fectors, and with the hormone–receptor complexes (r site). The putative domains involved in the interaction of α_s (human α_{s2a}) with guanine nucleotides, $\beta-\gamma$ dimers, and receptors are depicted in Fig. 5. The assignment of the domains is based on the homology of the α subunits with c-H-ras and EF-Tu and in the effects of mutations.

The domains highlighted include the following.

1. Region 47–53 (GAGESGK)

This region is involved in the interaction with phosphates. This sequence is present in all of the G protein α subunits that have been sequenced and is conserved in other GTP-binding proteins as well (the sequence GXXXXGK is considered a consensus element for the interaction with phosphates). Substitution of a G residue for V in position 49 (i.e., *G49V* mutation) reduces GTPase activity in α_s. The K residue in position 53 is thought to neutralize the phosphate charge. This region is homologous to the region 10–16 in c-H-ras, included in loop 1 of the three-dimensional structure, closely juxtaposed to the α and β phosphates of the GDP molecule.

2. Region 223–227 (DVGGQ)

This region is also involved, or affects at a distance, the interaction with guanine nucleotides. Reciprocally, this region also seems to undergo conformational changes elicited by the interaction of the nucleotide molecule with the 47–53 region. The *Q227L* mutation (found in pituitary tumors and also introduced by site-directed mutagenesis) dramatically reduces GTPase activity. The *G226A* mutation in α_s, present in the *H21* mutant of the S49 lymphoma cell line, is able to couple to hormone–receptor complexes, but is unable to change the activity of the effector. Since this residue is conserved in

G protein α subunits that interact with other effectors (i.e., it is unlikely to be located in the effector site), it is possible that this region changes conformation after sensing the interaction of GTP with region 47–53. The D residue in position 223 is proposed to form a salt bridge with Mg^{2+} that links to the phosphoryl groups of the nucleotide. This region is adjacent to the R residue in position 201. This residue is the substrate for cholera toxin-mediated ADP ribosylation of G_s. This covalent modification, as well as the *R201C* found in pituitary tumors (see Section X), inhibits GTPAse activity, confirming the connection of this area with the GTP interaction domain.

3. Region 292–296 (NKQDL)

This region is proposed to be involved in the interaction with the guanine ring. The sequence NKXD has been defined as a consensus sequence involved in guanine nucleotide specificity. The three-dimensional structure of the c-H-ras shows that the homologous residues 116–117 and 119–120 form a side pocket for the guanine ring. Also, the homologous residues N135 and D138 in EF-Tu are proposed to interact with the keto and amino substituents of the ring, respectively.

4. Region 374–394

Several lines of evidence suggest that the 21 residues in the carboxy terminus of the protein are involved in the interaction with receptors (r site). The R389P α_s substitution, found in the *unc* mutant of the S49 lymphoma cell line (no coupling of receptors to G_s is detected in these cells), has provided experimental support for this hypothesis.

5. Region 1–25

This region is involved in the interaction with $\beta-\gamma$ dimers. This is inferred from studies measuring the interaction of $\beta-\gamma$ with the proteolytic fragments of the transducin α subunit.

The location of the effector site is still ambiguous. Chimeric proteins formed by the amino terminus of α_i (60%) and the carboxy terminus of α_s (40%) have shown that the effector domain is contained within the carboxy-terminal 40% of the molecule.

B. β and γ Subunits

No information is available on the relationship between structure and function of the β and γ subunits. The cysteine residues located in the fourth position from the carboxy terminus of γ subunits are conserved in α subunits of G proteins and in ras proteins. Acylation of this residue in some ras proteins is essential for membrane association. It is possible to speculate that the acylation of this cysteine might also be critical for the interaction of this subunit with the plasma membrane. It is also possible that the residues surrounding the cysteine residue favor the acylation of $G\gamma$ but not $T\gamma$, explaining the hydrophilic and hydrophobic characteristics, respectively, of these subunits.

More studies on the relationship between structure and function of the β and γ subunits will be obtained in the near future. Most of the functional assays will be based on the fact that retinal $\beta-\gamma$ dimers are less efficient than the corresponding brain or placental dimers in inhibiting G_s-mediated stimulation of adenylyl cyclase in membranes, as well as in reconstituted systems.

VI. Structure of G Protein Genes and Chromosomal Localization

The human α_s gene is approximately 20 kb long and is composed of 13 exons and 12 introns. The $G_{i2}\alpha$, $G_{i3}\alpha$, and $T_1\alpha$ genes are composed of eight exons and seven introns in their coding region. The positions of the splice junctions on the sequences of $G_{i2}\alpha$ and $G_{i3}\alpha$ are identical; this organization seems to be conserved in the $G_{i1}\alpha$ and $G_o\alpha$ genes. Also, three of the 12 splicing sites present in the $G_s\alpha$ gene are also present in the $G_i\alpha$ genes. This suggests that all of these genes evolved from a common ancestral precursor. [See GENES.]

The localization of genes encoding G protein subunits in human chromosomes has been reported. $T_1\alpha$ and $G_{i2}\alpha$ genes are located on chromosome 3. $T_2\alpha$, $G_{i3}\alpha$, and $G\beta_1$ genes map to chromosome 1. Genes encoding $G_{i1}\alpha$ and $G\beta_2$ are found on chromosome 7. $G_s\alpha$ and $G_z\alpha$ genes are distributed on chromosomes 20 and 22, respectively, It is interesting to note that highly homologous subunits, probably derived from common ancestral genes, are located on different chromosomes. [See CHROMOSOMES.]

VII. Regulation of G Protein Gene Expression

Chronic administration of corticosterone to normal and adrenalectomized rats shows that the steady-state levels of $G_s\alpha$ and $G_i\alpha$ mRNA in the cerebral cortex are subjected to positive and negative modulation, respectively, by glucocorticoids. Similar variations in the corresponding levels of proteins are observed. In contrast, $G_o\alpha$ and G protein β subunits do not appear to be regulated by the glucocorticoid. They could be regulated due to (1) direct regulation of subunit gene expression, (2) changes in the synthesis and/or catabolism rate of the mRNAs, or (3) an indirect effect elicited by glucocorticoid-sensitive mediators. The identification of consensus hexamer sequences for the interaction with steroid hormone receptors in G_{i1} supports the first possibility. The prolonged hormone treatments required for observation of the effects would be compatible with an indirect effect.

Thyroid hormones exert a negative regulation of the levels of β subunits and their corresponding mRNAs in fat cells, as measured in normal, hypothyroid, and hyperthyroid rats. Hypothyroid rat adipocytes also contain increased levels of pertussis toxin substrates and $G_{i1}\alpha/G_{i2}\alpha$-immunoreactive material, without significant changes in the levels of $G_{i2}\alpha$ mRNA. These results seem to be consistent with the enhanced inhibition of adenylyl cyclase activity produced by specific agonists in fat cells from hypothyroid rats. [See THYROID GLAND AND ITS HORMONES.]

Expression of β subunits in fat cells also seems to be positively regulated by glucocorticoids, as opposed to the lack of effects observed in the cerebral cortex.

VIII. Co- and Posttranslational Modifications of G Proteins

Myristoylation of human astrocytoma α_{41} and α_{40} (α subunits with molecular masses of 40 and 41 kDa, as determined by sodium dodecyl sulfate–polyacrylamide gel electrophoresis) has been reported. Purified brain α_{41} and α_{39}, but not retinal α_t, have also been shown to contain myristate residues. Overexpression of α_{i1}, α_{i2}, α_{i3}, α_o, α_z, α_t, and α_s in transfected COS monkey kidney cells demonstrated that all of the α subunits except α_s are able to undergo myristoylation. This modification is due to the cotranslational acylation of the amino-terminal glycine, as indicated by results obtained with α subunits mutated in this residue and by hydroxylamine treatment of the modified proteins. The absence of a serine residue in position 6 of the α_s-deduced primary structure, which is present in all of the other α subunits, is probably responsible for its lack of reaction with the myristoyl Co-N-myristyltransferase complex. This acylation seems to be critical (with the exception of α_z and α_t) for the attachment of the α subunits to the plasma membrane. The absence of acylation in the native α_t probably reflects the absence of enzyme–substrate interaction in the photoreceptor cell.

The proteins of the ras superfamily undergo posttranslational palmitylation of a conserved cysteine at the fourth position from the carboxy terminus. This residue is located in a sequence CAAX ("A" stands for amino acids with aliphatic chains; "X," for any amino acid), which is conserved in all ras proteins, in the γ subunits of G proteins, and in many of their α subunits. Recent experiments have shown that yeast *ras* gene products are cleaved at the above-mentioned cysteine residue, which subsequently undergoes methylesterification and palmitylation. This sequence of events has been proposed as a model for the membrane attachment of soluble proteins. Given the presence of the CAAX tail in many subunits of heterotrimeric G proteins, it is possible that they might undergo a similar processing. It has been shown that G protein γ subunits are cleaved at the fourth position from the carboxy terminus, followed by carboxyl methylation and isoprenylation of the cysteine residue.

Several lines of evidence indicate that G proteins undergo phosphorylation. The α subunits of G_i-like proteins purified from rabbit liver have been shown to be substrates for phosphorylation mediated by PKC. Also, the GDP-bound form of $T\alpha$, but not the GTP-associated form, is phosphorylated *in vitro* by PKC and by hormone-occupied insulin and insulin-like growth factor I receptors. Activation of PKC by phorbol esters in multiple cell lines impairs GTP-dependent hormonal inhibition of adenylyl cyclase and activation of phosphoinositide-specific phosphodiesterase. Some of these effects can be explained by the inhibition of activity of G_i-like proteins as a result of PKC-mediated phosphorylation. Other effects of the PKC activity on the adenylyl cyclase system include the phosphorylation of its catalytic component in some cell types.

Activation of PKC in platelets results in the phosphorylation of the α subunit of G_z or G_z-like proteins and the impairment of hormonal inhibition of adenylyl cyclase. These findings show a contradiction: hormonal inhibition of adenylyl cyclase is sensitive to pertussis toxin, but α_z lacks a consensus site for pertussis toxin-mediated ADP ribosylation. Future research is required to clarify the physiological relevance and identity of the phosphorylated G proteins.

Cholera toxin-mediated ADP ribosylation of G_s in GH_3 pituitary cells has been shown to increase the degradation of this protein, in spite of significantly increasing G_s-mediated stimulation of adenylyl cyclase. This observation suggests that ADP ribosylation could mark the G protein for increased degradation by cellular proteases. Three endogenous ADP-ribosyltransferases have been purified from eukaryotic cells; these enzymes (i.e., A, B, and C) can specifically transfer the ADP-ribose moiety to arginine, diphtamide, and cysteine residues, respectively. The ADP-ribosyltransferase C activity, purified from human erythrocytes, uses as specific substrate the α subunit of G_i-like proteins. This modification impairs the epinephrine-induced inhibition of adenylyl cyclase in platelets. It is then possible to speculate that ADP ribosylation of G proteins by endogenous ADP-ribosyltransferases might represent a physiological posttranslational modification, affecting the stability and coupling capability of these proteins.

IX. Cellular and Tissue Distribution of G Proteins

G_s and G_i-like proteins have been detected in almost all tissues studied. In fact, they are now considered to be encoded by "housekeeping" genes. Examples of G proteins with specific organ localization are T_1, T_2, and G_{olf}. Immunochemical studies have shown that the highest concentrations of $G_o\alpha$ are present in the central nervous system, followed by the pituitary gland and the sciatic nerve. Other peripheral organs containing G_o (less than 2% of the cerebrum concentration) are the urinary bladder, stomach, intestines, and heart atrium.

The mRNAs encoding for $G_o\alpha$, $G_s\alpha$, and $G\beta_1$ subunits have been mapped in the brain by *in situ* hybridization. The localization of these probes varies significantly. $G_s\alpha$ and $G\beta_1$ mRNAs show a similar extensive distribution throughout the brain, particularly in large neuronal cell bodies. On the other hand, $G_o\alpha$ mRNA is localized in fewer areas, including claustrum, habenula, hippocampal pyramidal cells, endopiriform nucleus, granule cells of the dentate gyrus, and the cerebellar Purkinje cells. The distribution of G_o mRNA in the brain is consistent with that of the corresponding protein, as determined by immunohistochemistry.

The localization of $G_s\alpha$ mRNA in the brain differs from the distribution of adenylyl cyclase, as determined by binding of labeled forskolin. This diterpene has been shown to interact with the catalytic subunit of the adenylyl cyclase. It is therefore possible that most of the brain G_s is involved in the coupling with effectors other than the adenylyl cyclase. This possibility has received experimental support, with the finding that G_s can interact with voltage-gated Ca^{2+} channels.

The intracelullar distribution of these mRNAs shows that the $G_s\alpha$ message is predominantly located in the cytoplasm, as opposed to the mainly nuclear signal detected for $G\beta_1$ mRNA. It is possible that the transport of β subunit mRNA from the nucleus to the cytoplasm can regulate the synthesis of β subunits and the function of the different heterotrimers.

The relative abundance of mRNA encoding G protein subunits in the brain is $\alpha_s > \alpha_o > \alpha_i$-like proteins. However, the steady-state levels of protein show a different proportion: α_o is approximately five times more abundant than α_i and 10 times more abundant than α_s. This would indicate differences in the translational efficiency of the mRNAs or in the turnover of the corresponding proteins.

Measurement of the concentration of G proteins in biological membranes has been performed by different techniques. Some examples of concentration measurements are:

1. G_s in S49 lymphoma cells: 20 pmol/mg (0.2% of the plasma membrane protein)
2. G_o in fibroblasts and adipocytes: 10 and 40 pmol/mg, respectively (0.1% and 0.4% of the membrane protein)
3. β_1 and β_2 subunits in undifferentiated and differentiated 3T3-L1 cells (fibroblastlike and adipocyte-like, respectively): 70 and 150 pmol/mg, respectively (0.25% and 0.5% of the membrane protein)
4. β_1 subunit in bovine cerebral cortex: 1.1 nmol/mg of protein (4% of the membrane protein)
5. β_1 subunit in human erythrocytes: 11 pmol/mg of protein (0.04% of the membrane protein)
6. Binding of GTPγS to bovine brain membranes and to fractions obtained in the purification of the G proteins shows that G_o constitutes approximately 1% of the brain membrane protein.

X. Diseases Involving Defects in G Proteins

A dominant inherited deficiency of α_s, due to the presence of an abnormal allele, is responsible for the resistance to parathyroid hormone and other hormones present in Type I pseudohypoparathyroidism. These patients show a 50% reduction in G_s activity in skin fibroblasts, lymphoblasts, and renal cells, together with olfactory dysfunction. A single base mutation in the α_s gene (A to G at the methionine initiation codon) has been detected in two related patients affected by this disorder. Other unrelated patients did not show this substitution, indicating that the type I pseudohypoparathyroidism phenotype is caused by more than one mutant form of the α_s gene. The recent identification of α_{olf} and the observation that bulbectomy decreases the levels of its mRNA (but not α_s mRNA) suggest that changes in the levels of this protein might be responsible for the impairment in olfaction in these patients. [*See* PARATHYROID GLAND AND HORMONE.]

The cholera toxin-mediated ADP ribosylation of G_s in intestinal cells causes a prolonged increase in intracellular cAMP, which, in turn, results in the large effluxes of salt and water in the gut of a patient with the diarrhea of cholera.

Two mutations in the G_s gene (*R201C* and *Q227R*) have been detected in growth hormone-secreting human pituitary tumors. These mutations block the GTPase activity of the α subunit, which explains the constitutive activation of adenylyl cyclase characteristic of the tumor cell phenotype. The tumor genome contains both mutant and nonmutant alleles, indicating the dominance of the genetic alteration. The conceivable demonstration that expression of the mutant proteins in pituitary somatotrophs induces their malignant transformation will define the G_s gene as a new oncogene. [*See* ONCOGENE AMPLIFICATION IN HUMAN CANCER.]

There is conflicting information regarding alterations in G proteins in diabetes mellitus. There is agreement, however, in that the disease is associated with an increase in the liver cAMP content. Some groups have reported that the expression of rat liver G_i is abolished in streptozotocin-induced experimental diabetes. Contrary to this, others have measured increased levels of liver G_s, with no significant changes in G_i, in similarly treated animals. The spontaneously diabetic BB/Wor rat, a model for insulin-dependent human diabetes, has increased levels of liver G_s and adenylyl cyclase activity. It has also been reported that the diabetic encephalopathy is associated with an altered balance between G_s and G_i/G_o activities, which is observed 14 weeks after the induction of the diabetic state. Clearly, more experimentation is required to determine the alterations in G proteins directly produced by the absence of circulating insulin. [*See* INSULIN AND GLUCAGON.]

Acknowledgments

The author wishes to thank Carlos Obejero-Paz, Paul Kungl, and Mark McKnight for their help in the preparation of the figures, and George Dubyak and Michael Maguire for critical reading of the manuscript. Work in the author's laboratory is supported in part by grants from the Biomedical Research Support Grant Program, National Institutes of Health (BRSG SO7 RR-05410-28), the Diabetes Association of Greater Cleveland (#326-89), the American Heart Association-Northeast Ohio Affiliate-(#4719), and the Ohio Board of Regents (#RIFOBR).

The format and purpose of this encyclopedia excludes the citation of references in the text. The author has selected a list of recent review articles, written by some of the leading scientists in the field, to guide the readers interested in the original research articles.

Bibliography

Birnbaumer, L., Abramowitz, J., Yatani, A., Okabe, K., Mattera, R., Graf, R., Sanford, J., Codina, J., and Brown, A. M. (1990). Roles of G proteins in coupling of receptors to ionic channels and other effector systems. *CRC Crit. Rev. Biochem.* **25,** 225–244.

Bourne, H. R. (1988). Do GTPases direct membrane traffic in secretion? *Cell* **53,** 669–671.

Casey, P. J., and Gilman, A. G. (1988). G protein involvement in receptor–effector coupling. *J. Biol. Chem.* **263,** 2577–2580.

Hughes, S. M. (1983). Are guanine nucleotide binding proteins a distinct class of regulatory proteins? *FEBS Lett.* **164,** 1–8.

Lochrie, M. A., and Simon, M. I. (1988). G protein multiplicity in eukaryotic signal transduction systems. *Biochemistry* **27,** 4957–4965.

Spiegel, A. M. (1987). Signal transduction by guanine nucleotide binding proteins. *Mol. Cell. Endocrinol.* **49,** 1–16.

Stryer, L., and Bourne, H. R. (1986). G proteins: A family of signal transducers. *Annu. Rev. Cell Biol.* **2,** 391–419.

Growth, Anatomical

STANLEY M. GARN, *University of Michigan*

Glossary

Aging Decreases primarily in size, functional capacity, or performance differing in onset and timing for different organs and tissues

Cusps Pointed projections on the molars and premolars that serve to grind food in mastication and are the earliest manifestations of odontogenesis

Development Changes in relative sizes of different body units caused by changes in localized growth rates; also increases in complexity of specific organs caused by changes at the cellular level

Environmental effects Effects of nutritional level and disease processes on the tempo of growth, maturity timing, and size and proportions; also includes effects of toxins, penetrating radiation, trace elements, and atmospheric and water-borne pollutants

Epiphyses Bony caps on the ends of long (tubular bones) that first appear as secondary ossification centers and ultimately unite (or "fuse"), bringing linear growth to an end. The iliac crest also has its own epiphysis

Growth Increase in size primarily caused by an increase in the number of cells, although growth in size may also be caused by an increase in size of individual cells (hypertrophy)

Hypoplastic markings On the enamel surfaces of the teeth, resulting from disturbances in enamel formation during dental development

Incremental or "growth" lines Seen in tooth cross sections, regularly spaced and attributed to cyclical changes in growth during dental development

Lean body mass (LBM) or lean body weight (LBW) "Lean" tissue mass as contrasted with the mass of fatty tissue (FM)

Life cycle Entire range of defined or definable stages in the life span including the period of the embryo and the fetal period (prenatally) through infancy, childhood, adolescence, adulthood, and senescence

Maturation Attainment of defined steps or events leading toward the specified end-points in size, complexity, or function (*vide* skeletal maturation, dental maturation, sexual maturation)

GROWTH, DEVELOPMENT, AND MATURATION are all simple concepts taken from the common language. *Growth* is by definition an increase in size caused by an increase in number of cells. *Development* involves changes in proportions caused by differential growth and an increase in complexity. *Maturation* includes the steps toward attainment of ultimate size, peak performance, or gametogenesis.

Growth in size may be gained through increases in the size of individual cells. Alternatively, a large number of smaller cells may come to occupy less space than a smaller number of larger cells. Development may include calcification of bony elements previously formed in cartilage, or simultaneous changes in dozens of organs. The maturation of the dentition, sexual maturation, and osseous maturation involve different and partially unrelated processes.

Growth of tissues may involve simultaneous gains and losses as tissues and discrete structures alter their form and "remodel." Development may include programmed cellular death. The maturation of different organs and tissues involves different growth rates and relative timing, so that neural, skeletal, and sexual units peak at different ages. Brain growth ends early but continues to increase in the complexity of neural connections and in storage of information. Skeletal units ("bones") may undergo involution (or loss) at some sites while still gaining at other sites.

I. Growth in the Embryo

Growth begins when one sperm successfully enters an ovum. This large fertilized cell then divides and subdivides into a larger number of smaller cells. There is a reduction in mass but an increase in cell number and complexity. Even at this early stage of embryogenesis, directionality is present. One end of the cell mass will become the head. Sidedness (laterality) is also established for the entire duration of life. [See FERTILIZATION.]

By the time the embryo is 25 mm long, a surprising number of body parts have attained a recognizable appearance. The individual segments of the bones of the hand, although still tiny in size, already exhibit adult proportions, one to the other. However, the teeth are but rudimentary dental "germs" at that time, and much of their development takes place in postnatal time.

During much of prenatal time the conceptus is only a tiny fraction of the mother's weight and body mass, with minimum demands on maternal tissues. During the entire 269 days of pregnancy, only 25 g of calcium is transferred to the conceptus, less than 4% of the mother's calcium stores. Most of this is added late in fetal development, and most of the gains in fetal weight are added late, at which time the demands on the mother and her nutrient and caloric reserves may then be heavy. Studies on neonates conceived and born during famines show the lasting effects of maternal undernutrition on the conceptus.

However, the fact that different organs and tissues form and complete at different times places great emphasis on the health of the mother and on her environment at these critical periods. Although the placenta provides a partial barrier for the protection of the fetus, it is only partial. Rubella (German measles) can affect the conceptus and bring about abnormalities of development, as can many viruses and toxins. Penetrating radiation at higher levels can seriously disturb embryogenesis. The workplace and the home now hold dangers for the conceptus not present in previous eras, yet improvements in food supply and distribution also minimize some of the hazards that existed in earlier hunting or agricultural communities. [See EMBRYO, BODY PATTERN FORMATION; FETUS.]

II. First Year of Growth

The human infant, and the infants of most other species, is best described as "an eating machine with a tremendous capacity for growth." So the human infant in Westernized countries quadruples or quintuples birth size in the course of 1 year, synthesizing an enormous amount of tissue. It does so by consuming food energy in vast amounts that may exceed 10% of its body weight per day.

Growth rates are highest in the first 3 months after birth, continuing the rapid growth of late prenatal time. Growth rates then decrease and decrease again at around the 9-month marker. By 9 months of age energy needs for growth are sufficiently lower, and the infant (and the parents) may sleep through the night without a nocturnal feeding.

During infancy there is a large increase in muscle mass, and as the muscles grow under neural control, activities increase along a predetermined schedule. Rolling over, then hitching, and then creeping and crawling are the behavioral markers of neuromuscular growth, leading to the ability to stand at the end of the first year of life. Fat stores markedly increase through the 9th month, then plateau. The one-year-old may be larger and leaner but only marginally heavier than at the 9-month horizon.

III. Growth During Childhood: The Period of Stretching Out

The period of growth we call "childhood" exemplifies the effect of changing relative growth rates on size and proportions. Brain growth (and neural growth in general) slows to a small annual pace, so the child no longer appears as large-headed and big-eyed as the infant. Changing relative growth rates of the axial and appendicular skeletal units increase the legs relative to the trunk so that the child is no longer so short-legged. Although the weight of fat and the percentage of fat both increase during childhood, it is the stretching out of the body that comes to our attention.

During childhood, the first of the permanent teeth emerges through the gums. Later, the roots of the deciduous teeth resorb, and they are shed and replaced by the successional teeth, which are larger for the most part and occupy more space in the jaws than their deciduous predecessors. With growth and remodeling of the facial skeleton, the face becomes longer and deeper, relative to the skull, and more adultiform in appearance.

Most of the postnatal ossification centers become radiographically visible during childhood, and their subsequent remodeling serves to provide useful measures of bone age ("skeletal age"). Individual tubular bones gain in widths and circumferences as well as in length, and the mineral mass gains volumetrically. However, blood-forming tissues continue to expand, so the medullary cavities also expand. Consistent with the increase in body mass and the synthesis of new tissues, caloric intakes also increase. So it is that undernutrition, diarrheas, and limiting protein may slow growth during childhood, increasing the size disparity between the poor and the affluent. Conversely, overnutrition is growth promoting so that obese boys and girls have more lean tissue as well: Obese children increase stature beyond the average and are usually advanced in postnatal ossification.

During childhood, weight is gained and stature increases at relatively constant rates, so that massed-data growth charts provide useful indications for clinical assessment. However, boys and girls reflect parental size in their growth progress, so that parent-specific growth charts rather than massed-data percentile values are more appropriate for the children of tall or short parents.

Unusual body proportions during childhood typify many of the congenital malformation syndromes and so merit clinical attention. Growth failure or slow growth during childhood may indicate malabsorption states, emotional-disturbance dwarfism, or abnormalities of the growth-regulating hormones. Intervention, as with growth hormone, is now practicable, and even the genetically short child may be accorded a larger adult size if therapy is begun in time.

IV. Sex Differences in Size and Developmental Timing

Long before birth the female is already more advanced than the male in many aspects of development and holds this advancement through to skeletal maturity. The female is more advanced than the male in ossification timing, the timing of tooth formation and emergence, the timing of epiphyseal union, and the age at which gametogenesis begins. For some ossification centers, the conception-corrected sex difference in ossification may be as much as 25%: for some teeth, the conception-corrected sex difference in eruption timing may exceed 7%. Female advancement in sexual maturation is especially evident at ages 12–14, and it is reflected in dating behavior and (later) in the difference in spousal ages.

Long before birth and throughout life, the male is the larger of the two sexes. The male is heavier, also, because of a larger lean body mass. In adulthood, the lean body mass of the male exceeds that of the female by 3:2; the skeletal mass and mineral mass of the adult male exceeds that of the female by 4:3. Because it requires more food energy to grow a larger body mass and to maintain it, the energy requirement of the male exceeds that of the female at all ages, and this is compounded (after sexual maturation) by higher voluntary and involuntary activity levels of the male and a higher basal oxygen consumption per unit of lean tissue. In fat weight, however, the female equals that of the male or may exceed it, so percent fat is higher in the female, providing an energy reserve against nutritional stresses.

Even at those ages when the sex difference in length and weight is small, males are larger than females of the same developmental or "skeletal" age. Comparing the two sexes at the same developmental level (i.e., the isodevelopmental level), boys average some 7% taller than girls, a value comparable to the adult sex difference.

Why the female is advanced over the male developmentally is not clear. The earlier notion that attributed the difference to the second X chromosome present in females and not in males is not in accordance with knowledge of the XXX ("superfemale") or the XYY or XYY "male." Although the developmental advancement of the female over the male for the first two decades of life might (in theory) be associated with a shorter life span, exactly the reverse is the case.

V. Genetic Determinants of Growth and Development

Bone lengths, muscle volumes, and tooth dimensions all show strong evidence of genetic control, as demonstrated by sibling and twin correlations. So do such discrete developmental events as ossification timing and formation and emergence timing of the 52 deciduous and permanent teeth. Even birth weights follow family line, as documented by mother–daughter comparisons. Mother–daughter correlations for the age of menarche (first menstruation) are systematically positive, approximately 0.25. The order of ossification of the postnatal ossification centers also follows family line, comparing parents and children at the same ages.

These family-line resemblances in dimensions, timing, sequence, and growth rates are equally demonstrable in the poor and the rich, in poorly nourished populations and in Westernized countries. Genetic control is superimposed on environmental determinants, so that taller boys and girls are disproportionately the progeny of taller parents even though the "tall" children in a famine area population may be shorter than the short boys and girls in an affluent society.

Genetic control of individual bone lengths and localized growth rates is also demonstrable in congenital malformation syndromes, including many types of dwarfism. Animal breeders select comparable genetically determined extremes to produce "breeds" or varieties of particular economic value or as household pets. Terriers and bulldogs and Pekinese are the result of intensive inbreeding. These animal examples serve as useful models for some human extremes, especially the achondroplasias. Genetically small animals may evidence unusual variants of growth-regulating hormones, again of value to the understanding of their human analogues.

VI. Nutritional Modification of Growth Rates, Size Attainment, and Body Proportions

From the fetal period through the third decade, growth rates and size attainment are affected by available energy, first supplied through the maternal circulation and (in the postnatal period) by food digested and absorbed. A small or abnormal placenta may limit fetal growth, as may maternal undernutrition, or a small material body size, or a low weight gain during pregnancy. Conversely, a large maternal size and a large weight gain during pregnancy favor prenatal growth. The "parity effect" (describing the tendency of later-born neonates to be larger in size) reflects the larger body mass and especially the larger fat weight of older mothers.

In the postnatal period, the caloric balance is a major determinant of growth rate, size attainment, maturity timing, and ultimate size. Obese infants, children, and adolescents are faster-growing and earlier to mature. The caloric density of the infant formula as well as the volume consumed directly affects growth rates. Undernourished boys and girls grow less rapidly, mature later, and are of smaller adult size. Food intolerances and malabsorption syndromes limit growth, as is evident in lactose intolerance, wheat-gluten sensitivity, and in numerous allergic states. Emotional-deprivation dwarfism may result from reluctance to eat, refusal to eat, or (even in subteen girls) fear of fatness.

Nutritionally deprived children not only grow less rapidly but attain final size with shorter legs relative to trunk. They are more "paedomorphic" (i.e., more childlike in proportions). Although better-nourished children do attain sexual maturity and cease growth earlier in the third decade, the larger growth rate (cm/year) during the growing period more than compensates for the earlier cessation of growth. Even before the 5th year, well-nourished boys and girls are 3–8 cm taller or longer than their poorly nourished Third World peers, and the dimensional advantages of a more positive caloric balance continues during the entire growing period.

VII. Ossification of the Round Bones and Epiphyses

The epiphyses at the ends of the tubular bones and the individual round bones are all fully developed during embryogenesis. However, most of them ossify in postnatal life, at which time they become radiographically apparent, and may be followed (in radiographs) through maturity. Postnatal ossification is earlier for every such epiphysis or for round bone in girls than in boys, and the sex difference in ossification timing exceeds 10% overall in girls.

This is consistent with the female developmental advancement in general.

Epiphyseal union (union of the epiphyses to the shafts of long bones) is also earlier in the female, and it is the operational reason why girls cease linear growth earlier. However, there is considerable individual variability in ossification timing and timing of epiphyseal union. This is genetic in part, as shown in twin, sibling, and parent comparison. The timing of union is also nutritionally mediated, as shown in the obese and the lean, and in malabsorption syndromes. There are also population differences in ossification timing that transcend socioeconomic differences.

The order or sequence of ossification of different bony nuclei also differs from family to family. The order of epiphyseal union also differs between families. Individuals may therefore differ much from the "textbook" order or sequence of ossification or union. Bone-to-bone differences in the timing of epiphyseal union and relative bone growth underlie ultimate differences in the relative length of different long (tubular) bones.

VIII. Dental Development

The 20 deciduous or "primary" teeth and the 32 permanent or "secondary" teeth of human beings and other anthropoids develop in quite different fashion from the bony units, reflecting their different historical origins and different embryogenesis. All the teeth form and calcify from top to bottom (crown to root) beginning with the enamel cusps and terminating with closure of the apices of the roots. Each tooth spends the first part of its history within the bony jaws, then emerging as the roots elongate and continuing "eruption" until it meets its opponent from the opposite jaw.

Statistically, the teeth are less variable in their developmental timing than are bony units. Teeth are also far less affected by extremes of nutrition and by such hormonal diseases as hypothyroidism or hypopituitarism. Nevertheless, teeth may bear permanent markings such as incremental lines in the dentin and hypoplastic markings in the enamel as a result of injurious prenatal events, birth trauma, postnatal disorders, or antibiotic therapies during their formations. Like the spicules or bony projections within the marrow cavities of tubular bones, such incremental lines and enamel defects are permanent historical records long preserved in the teeth.

Because the teeth are resistant to nutritional extremes, and with lower developmental variability, tooth formation and emergence timing provides a useful approach to age determination in forensic medicine. However, the teeth are quite variable as to number, and the last tooth of each morphological class (i.e., the second incisor, the second premolar, and the third molar) is most likely to be missing (agenesis) because of failure to form. The third molar ("wisdom" tooth) may be congenitally missing in the majority of individuals in some Asiatic populations, but it is rarely missing in ethnic Australians.

As with other calcified structures, there are consistent sex differences in dental developmental timing, females being more advanced than males in all teeth, except for the third molar. There are also consistent population differences in tooth formation and movement, later in those of European ancestry than in African, Amerindian, or Eastern Asiatic.

IX. Adolescent Growth

Just before the 10th year, especially in Westernized girls, a new set of growth-regulating mechanisms becomes operative, heralded by increased production of follicle-stimulating hormone (FSH), luteinizing hormone (LH), and several other hormones of pituitary origin that control the ovarian cycle and sexual development in both sexes. This is the onset of "adolescence" or of "puberty" (although the two terms have different meanings), and it reflects the responses of many different tissues and organs to hormones of gonadal (testicular and ovarian) and adrenal origin. Adolescence, for want of a better term, actually continues through the teens in human beings and is not complete until well into the third decade. [*See* PUBERTY.]

Functionally, the body is converted from an eating and growing organism to one capable of gametogenesis, sexual attraction, and (in the female) pregnancy and lactation. So the secondary sexual characteristics develop, in both sexes, including distinctive patterns of subcutaneous fat distribution and distinctively different patterns of body hair and facial hair distribution. There are also distinctively different patterns of sebaceous gland activity and the production of sexually attractant odorous substances (pheromes). Within the female body the uterus enlarges and its lining begins to undergo cyclic cellular losses and regeneration, as part of the menstrual cycle. In the male, the testes enlarge to produce more testosterone from the Leydig cells and to produce sperm from the Sertoli cells. The prostate enlarges to produce prostatic fluid.

There is, of course, a major increase in muscle mass in both sexes, more so in the male than in the female. However, all muscle groups are not equally affected; those of the legs are least differentiated during sexual development, and those of the back and neck far more. The supporting bones increase in length, until their linear growth is terminated by epiphyseal union. These tubular bones gain also in outer diameter (by subperiosteal apposition) and at their inner surfaces (by endosteal or inner surface apposition). So the skeletal mass increases and continues to increase for many years in both sexes, in human beings, in the primates, and in mammals in general.

Most boys and some girls are characterized by an adolescent "spurt" in linear growth, which may (at maximum) approximate 14 cm/yr. At this time, and in the rapidly growing adolescent, growth is almost visible, at 0.5 mm/day. This is the time of greatly increased food ingestion to fuel the rapid synthesis of tissue, and this is a time when the mineral intake may lag behind requirements for bone growth. However, many girls and some boys do not demonstrate a textbook adolescent spurt. As a general rule the spurt is greater for those early to mature and lesser for those who are late to mature. The magnitude of the adolescent spurt and its duration is also a familial characteristic; both may be attenuated by inadequate nutrition during adolescence.

During adolescence the skeleton of arms and legs grows more rapidly than the spine, so that the adolescent becomes more "leggy," like the adult. However, adolescents in undernourished populations retain a more childlike appearance, whereas those in Westernized nations are more old-looking (gerontomorphic) in proportions (i.e., with a long appendicular skeleton relative to the axial skeleton or trunk).

X. Secular Trend in Size and Maturational Timing

During the past century and a half, Westernized populations have shown increased growth rates, larger body size, and an earlier age at first menstruation. Such generational or *secular* changes have not yet appeared in some Third World populations, but the secular trends in size and maturational tim-

ing in Japan since 1946 have been of outstanding magnitude. Increases in size from one generation to the next are also known for colony-reared primates as compared with wild-captured specimens. The trend toward larger size is not restricted to human beings.

With increased body size at all ages and longer legs and arms, the spatial needs of human beings have also increased, not only for school desks, airplane seats, beds, and mattresses, but also for coffins and cemetery vaults. Bigger men, women, and children need more food to fuel them, more fiber to clothe them, and more energy to cool and warm them, and they excrete more bodily wastes. The trend toward larger body size is therefore of both economic and ecological importance.

Although various explanations have been proposed for the secular trends in size and timing, including out-mating (i.e., interbreeding between genetically unrelated groups) leading to increased gene diversity in the offsprings, the secular trend can be reversed under famine stress or wartime deprivations or when Third World economies falter. The secular trend was reversed in Germany after World War I, and after World War II, in Russian cities under wartime seige, and in Jerusalem during the seige. This shows that nutrition plays a predominant role.

Hands and feet and legs and arms have shown disproportionate secular increases, so that bodily proportions as well as body size has changed over time. Faces have become longer and deeper although slightly narrower, as shown in parent to adult-child comparisons.

Repeatedly, since 1942, investigators have suggested that the secular trend is at an end. However, additional data and larger samples confirm its continuation, particularly in the progeny of less affluent parents and in populations formerly characterized by small body size. Roman and Milanese boys and girls now exceed British stature norms, and formerly short peoples may emerge as the taller peoples of the world as economic conditions change.

XI. Continuing Growth in Adulthood

Although most growth charts end at age 18, linear growth continues for some years, through the mid-20s and often beyond. The lean body mass (LBM) continues to increase, through age 30. The skeletal mass increases through the end of the fourth decade. In most individuals in Westernized countries, fatty tissue (obesity tissue) increases through the seventh decade, through the increase in number and size of cells.

The skeletal volume, as apart from the skeletal mass, continues to expand lifelong. Tubular bones become larger, although not longer. Round bones (like the carpals in the hand) increase in size. The head grows in dimensions, so that the hat size increases. The facial skeleton continues to grow. Because the ears and the bony nose also grow throughout adulthood, there is a continuing change in appearance. The continued growth of skeletal volume can be advantageous, compensating in part for the diminution in the skeletal mass. The capacity for continued growth at the outer (subperiosteal) surface of tubular bone may be disadvantageous and possibly a contributing factor in osteoarthritis. The increase in fat mass or fat weight during adulthood may also be disadvantageous with respect to blood pressure and chronic diseases. Growth is not necessarily at an end at age 20 or 25. For some organs and tissues it may continue through life.

Bibliography

Daughaday, W. H., ed. (1979). ''Endocrine Control of Growth.'' Elsevier Press, New York.

Falkner, F., and Tanner, J. M., eds. (1986). ''Human Growth: A Comprehensive Treatise,'' 2nd ed. (3 vol.) Plenum Press, New York and London.

Goose, D. H., and Appleton, J. (1982). ''Human Dentofacial Growth.'' Pergamon Press, New York.

Prader, A. (1984). Biomedical and endocrinological aspects of normal growth and development. *In* ''Human Growth and Development—Proceedings from the Third International Congress on Auxology.'' (J. Borm, R. Hauspie, A. Sand, C. Susanne, and M. Hebbelinick, eds.). Plenum Publishing Corp., New York.

Tanner, J. M. (1981). ''A History of the Study of Human Growth.'' Cambridge University Press, Cambridge.

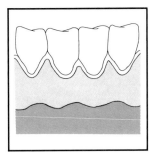

Gum (Gingiva)

ROSEMARY DZIAK, *State University of New York at Buffalo*

Glossary

Lamina propria Support layer of connective tissue
Alveolar bone Bone lining the sockets of the teeth
Periosteum Fibrous layer covering bone and intimately associated with collagen fibers of lamina propria
Keratinized Converted into keratin, a sulfur-containing fibrous protein that is insoluble and elastic; forms the chemical basis of epidermal tissue
Parakeratinized Irregular formation of keratin in the epidermal tissues
Glycosaminoglycan Substance characterized by a highly polymerized carbohydrate chain (from 150 to several thousand sugar units) having a regularly repeating sequence of usually two kinds of sugar units alternating along the whole length of the chain
QO_2 Rate of oxygen consumption, usually expressed as μl O_2/min/mg nitrogen

THE GUM (GINGIVA) is part of the investing and supporting structure of the *teeth*. It is composed of variable amounts of connective tissue and epithelium and surrounds the cervical (neck) region of the teeth and the surface of the *alveolar bone*. The gingival unit around a tooth along with the attachment apparatus consisting of the *cementum, periodontal ligament,* and the alveolar and supporting bone form a complex of tissues called the *periodontium*.

I. Introduction

The tissues of the periodontium have important physiological functions in the human body. They (1) attach the tooth to the underlying bone; (2) absorb forces exerted by speech, swallowing, and mastication; (3) assist in the maintenance of the oral cavity by compensating for structural changes associated with wear and aging by continuous remodeling and regeneration; and (4) act as a barrier to penetration of the body by microorganisms and noxious substances present in the mouth.

II. Classification of the Gum (Gingiva)

The gingiva extends from the cervical portion of the tooth to the mucogingival junction and is divided into the free (marginal) attached gingiva,* and the *interdental papilla* (Fig. 1). The free marginal gingiva and the interdental gingiva serve as the transitional zone between the soft tissue and the tooth surface. This is the site of initiation of the inflammation of gingival and *periodontal disease*. The interdental papilla is that part of the gingiva that fills the space between two adjacent teeth. The shape and size of the interdental gingiva are determined by the anatomy and the contact areas of the teeth. When viewed from the oral or vestibular aspect, the surface of the interdental papilla is triangular. In a three-dimensional view, the interdental papilla of the posterior teeth is tent-shaped and pyramidal between the anterior teeth. When the interdental papilla is tent-shaped, the oral and vestibular corners are high and the central part is like a valley. The central concave area fits below the contact point,

* At the International Conference on Research in the Biology of Periodontal Disease, Chicago, June 1977, it was voted to drop use of ''attached'' for simply ''gingiva.''

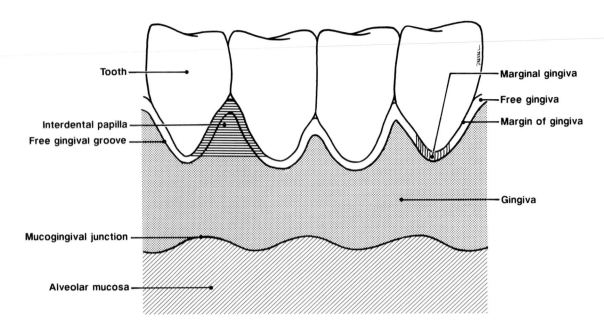

FIGURE 1 Diagram representing the gum (gingiva) in relation to teeth. Major areas of tissue are noted as described in the text.

and this depressed part of the interdental papilla is called the *col*. The col is covered by thin non-keratinized epithelium, and it has been suggested that the col is most vulnerable to periodontal disease.

In about 50% of the population, there is a dividing line between the free gingiva and the gingiva. It is called the *gingival groove* and runs parallel to the margin of the gingiva for 0.5 to 1.5 mm.

The surface of the gingiva has a stippled appearance. The pattern and extent of the stippling depends on the region of the mouth. It is less prominent on the lingual than facial surfaces and in some individuals there may be no stippling. The disappearance of stippling is often an indication of edema, a symptom of the development of progressing *gingivitis*.

III. Histology

A. General Features

A stratified squamous epithelium covers the surface of the free and attached gingiva. The gingiva is para-keratinized 75%, keratinized 15%, and non-keratinized 10% of the time. It appears that inflammation interferes with keratinization. The more highly keratinized the tissue, the whiter and less translucent it is. The epithelium is separated from the underlying connective tissues by a basal lamina and consists, like epidermis, of basal, spinous, granular, and cornified layers. The lamina propria is a dense connective tissue that does not contain large blood vessels. Small numbers of lymphocytes, plasma cells, and macrophages are present in the connective tissue of normal gingiva. These cells are involved in repair of the tissue and immune defense mechanisms. The connective tissue of the gingiva has characteristic numerous long, slender papillae. It contains few elastic fibers, confined in most part to the walls of the blood vessels. These characteristics make the lamina propria of the gingiva easily distinguishable histologically from that of the *alveolar mucosa* in which the papillae are low and the elastic fibers numerous. [*See* CONNECTIVE TISSUE.]

The gingiva is firmly attached to the teeth by the gingival fibers of the periodontal ligament and to the periosteum of the alveolar bone. The gingiva also contains the dense fibers of collagen. These gingival fibers are arranged in three groups: (1) The gingivo-dental group is embedded into the cementum just beneath the epithelium at the base of the gingival sulcus. The fibers spread in a fan-like fashion into the periosteum of the facial and lingual alveolar periosteum, crest of the interdental gingiva, and attached gingiva. (2) The circular group runs through the marginal and interdental gingival connective tissue to encircle the tooth. (3) The *trans*-septal group forms horizontal bundles between the cementum of approximating teeth and into which they are embedded.

B. Blood and Nerve Supply

The blood supply of the gingiva is mainly derived from the branches of the alveolar arteries that pass upward through the interdental septa. The interdental alveolar arteries perforate the alveolar crest in the interdental septa. The interdental alveolar arteries perforate the alveolar crest in the interdental space and end in the interdental papilla, supplying it and the adjacent areas of the buccal and lingual gingiva. In the gingiva these branches are connected with superficial branches of arteries that supply the oral and vestibular mucosa and marginal gingiva, including branches of the lingual, buccal, mental, and palatine arteries. The numerous lymph vessels of the gingiva lead to submental (under the chin) and submandibular lymph nodes.

The gingiva is well-innervated, by fibers arising from the periodontal ligament as well as those from the labial, buccal, and lingual nerves. These include Meissner or Krause corpuscles, end bulbs, loops, or fine fibers that enter the epithelium as ultraterminal fibers. These function as sensory systems responsive to pain, temperature, pressure, and tactile sensations.

C. Coloration

The gingiva is normally pink or coral in Caucasians but may sometimes have a grayish tint. The color depends in part on the surface (keratinized or not) and thickness and in part on pigmentation, with a great variation existing between individuals within a race and among different races. The surface may be translucent or transparent, permitting the color of the underlying tissues to be seen. The reddish or pinkish tint is due to the color of the underlying tissue imparted by the circulating blood in the vessels. In some cases the gingiva has a brown to black coloration, with the most intense color at the base of the interdental papilla. This coloration is due to the presence of melanin pigment elaborated by special cells, the *melanocytes,* in the basal layer and stored in basal cells in the epithelium.

IV. Biochemical Composition

A. General Considerations

In general, gingiva has a high water content with reported values from 74 to 82% of total tissue. The bulk of the tissue is formed by the connective tissue matrix composed of *collagen, proteoglycans, glycoproteins,* and *lipids.* The epithelium contains additional noncollagenous proteins and lipids as well as nucleic acids and basement membrane components.

B. Collagen

Collagen accounts for the bulk of gingiva with reports in the literature ranging from 37.7 to 45.0% of dry weight. The total noncollagenous proteins have been estimated at 47% of dry weight, with approximately one-half of that being characterized as noncollagenous structural glycoproteins. DNA accounts for less than 1% and RNA for approximately 2.0% of the total gingiva. *Glycosaminoglycans* account for about 1.4% and are relatively more abundant in gingiva than in other soft tissues.

The collagens of gingiva have been extensively studied and their polymorphism well demonstrated. Collagen molecules are synthesized by resident fibroblasts and are secreted into the extracellular compartment where they rapidly aggregate end to end and laterally to form collagen fibrils and fibers. The molecules are initially produced in a precursor form, which is converted into a *tropocollagen* molecule of approximately 300,000 daltons. Each tropocollagen molecule is composed of three polypeptide chains of approximately 100,000 daltons, with each chain coiled into a left-handed helix and all three into a right-handed helix. Based on the types of polypeptide chains, the following types of collagen molecules have been identified either in normal or inflamed gingiva: type I, type I-trimer, type III, type IV, and type V. Types I and III are the predominant types at approximately 80 and 20% of the total collagen, respectively. The concentration of type V chains (A and B) in normal gingiva has been approximated at 0.53 and 0.67%, respectively. The exact proportions of the different types of collagen in adult human gingiva are difficult to ascertain, however, because the relative concentrations of each type varies with inflammation. In inflamed tissue the percentages of type III appear to decrease (3.8%), with relatively higher amounts of type I (87.0%) and type V (8.0%) present. [See COLLAGEN, STRUCTURE AND FUNCTION.]

C. Glycosaminoglycans and Proteoglycans

There is a general agreement that in human gingival tissue, *chondroitin 4-sulfate* (C-4-S), *dermatan sul-*

fate (DS), and *hyaluronic acid* account for the bulk of glycosaminoglycans with relative percentages of 23.0, 33.0, and 43.0, respectively. Neither *chondroitin 6-sulfate* nor heparan sulfate appear to be present in preparations from human gingiva.

Proteoglycans have also been isolated from human gingiva. As are typical for proteoglycans in other tissues, they are composed of a protein core to which glycosaminoglycans are attached. The best-characterized human gingival proteoglycan contains two glycosaminoglycans, C-4-S and DS, with a total protein content of 46% and an equimolar ratio of uronic acid and hexosamine. It resembles a proteoglycan of the skin and is rich in aspartate, glutamate, serine, threonine, glycine, alanine, and leucine. [*See* PROTEOGLYCANS.]

Fibronectin [synonyms: large external transformation sensitive protein (LETS); cold insoluble protein; cell surface protein] is a component of the extracellular matrix of many cells and is also found in serum. It is synthesized in appreciate amounts by gingival fibroblasts in tissue culture and has been implicated in the attachment of cells to the collagen matrix. Fibronectin has been shown to be altered in several pathological conditions affecting the oral cavity and may also be important in the invasiveness of cancer as well as in periodontal reattachment. Laminin is a glycoprotein, a major constituent of basement membranes, comprising 30–50% of the total proteins. It binds to type IV collagen and to receptors on the surfaces of epithelial and endothelial cells, mediating attachment of these cells to basement membrane. Laminin has been shown to increase the proliferation and chemotaxis of gingival epithelial cells and has been postulated to play a critical role in the structure and function of these cells.

D. Lipids

The lipid content of human gingiva is approximately 1.5% on a wet weight basis or 5.8% on a dry weight basis, with approximately 70% of the total being neutral lipids and free fatty acids and phospholipids accounting for 30.0%. Gingiva contains large amounts of free fatty acids (17.0%) and cholesterol esters (12.0%) with triglycerides and diglycerides accounting for more than 20% of the total lipids. Total phospholipids account for approximately 30.0% of all lipids, with phosphocholine and phosphoethanolamine accounting for the majority of these lipids. There is a considerable amount of

arachidonic acid, the precursor of prostaglandins and leukotrienes, in the gingiva. The inflammatory compounds *prostaglandin* E_2 and $F_{2\alpha}$, as well as *thromboxane, prostacyclin,* and *leukotriene* B_4, have also been identified in human gingiva in varying amounts, dependent on the presence of inflammation. The phospholipids, free fatty acids, and cholesterol esters appear to be localized in largest amounts in the keratinized layer of the epithelium. *Arachidonic acid* and its metabolites are predominantly found in connective tissue. [*See* LIPIDS.]

E. Enzymes

Enzyme action in the gingiva has been extensively studied with particular emphasis on *collagenase*. Collagenase has been isolated and characterized in inflamed human gingival tissue. It is a protein of 40,000 mol wt that acts at neutral pH and requires calcium for its activity. It appears to be localized in fibroblasts subjacent to the epithelium and is present in an inactive form that can be activated by trypsin or by thiocyanate. Collagenase can be bound to an inhibitor, two of which, $\alpha2$ macroglobulin and β-anticollagenase, have been identified in serum. Collagenase can be strongly bound to collagen and obviously plays a major role in the metabolism of this protein and the integrity of the gingival tissue.

Other enzymes known to be important in the normal metabolism of gingiva and in the breakdown of tissue during inflammation are the *lysosomal enzymes*. The most widely studied of these enzymes include hyaluronidase, β-glucuronidase, alkaline phosphatase, acid phosphatase, and cathepsin B.

Enzymes of the *glycolytic* and *hexose monophosphate* shunt have also been localized in the gingiva. The activities of these enzymes are higher in the epithelium than in connective tissue portion of the gingiva. Hexokinase activity has been particularly implicated in the synthetic activity of the gingival epithelium.

V. Metabolism

A. Respiratory Rate

The biochemistry of gingiva is a dynamic system poised to preserve the structural integrity of the tissue as it is continually subjected to wear and aging. Measurements of respiratory rate have been used to gain insight into the overall biochemical activity of the tissue. The rate of oxygen consumption, QO_2 (expressed as μl O_2/min/mg nitrogen), reported in the literature for normal human gingiva is 1.49. This QO_2 value is within the range of QO_2 values for other human tissue and is only slightly lower than that of skin and higher than uterine tissue and gastric smooth muscle. The gingival QO_2 values are known to be affected by the degree of inflammation and the proliferative state of the tissue. In general, samples with light to moderate degree of inflammation have a higher QO_2 value, and those with moderate to heavy inflammation have lower values. Tissues in a high proliferative state have higher QO_2 values, whereas those in a state of degeneration have lower values. In accordance with this trend, the average endogenous QO_2 of human gingiva decreases with advancing age.

B. Turnover Rates

The biosynthetic rates of the tissue constituents of gingiva have also been evaluated. The protein turnover rate studies by the incorporation and release of radiolabeled proline is high in the gingiva, with this tissue having a relatively rapid protein turnover rate compared with the periodontal ligament, alveolar bone, or skin corium. The half-life times for gingival glycosaminoglycans suggest that this fraction is metabolized more quickly than in skin or cartilage.

C. Glucose Metabolism

Although there is evidence from the turnover rates of the components of the gingiva that it is a biochemically active tissue, there is relatively little information concerning the enzymatic sites of the synthetic processes except for the mechanisms involved in glucose metabolism. It appears that human gingiva possesses a relatively high glycolytic metabolism and a low aerobic metabolism, with possible involvement of both the Embden-Meyerhof and pentose phosphate sequences functioning.

VI. Gingival Diseases

Several diseases affect the gingiva and other elements of the periodontium. Some such as scurvy are systemic conditions that leak to gingival pathology. *Scurvy* results from a deficiency in ascorbic acid (vitamin C) leading to an aberration of collagen formation throughout the body. An early symptom of this disease is gingival bleeding. Other diseases such as gingivitis and periodontal disease are more specific to the gingiva and surrounding oral components.

Gingivitis is clinically characterized by inflammation of the gingiva with increased redness, swelling, ease of bleeding, and altered consistency. *Peridodontitis* involves inflammation of the supporting tissues of the teeth and usually leads to loss of the periodontal ligament and alveolar bone. Clinically, periodontitis is characterized by the formation of pathologically deepened gingival sulci known as *pockets*. Pocket formation and loss of alveolar bone lead to progressive loosening of the tooth and eventual loss. Gingivitis alone is more prevalent in children and adolescents, whereas periodontitis with pocket formation and bone loss occurs in a small number of individuals before age 20, with steady increases thereafter. The etiology of these diseases is multifactorial, but there is little doubt that bacteria present in the oral cavity leading to the production of plaque are the major etiological agents. Plaque consists of soft bacterial deposits that adhere to the teeth supragingivally or subgingivally. Extensive research in this area indicates that in humans, reduction of the microbial flora by brushing and flossing of the teeth, surgical procedures, or antibiotic therapy is effective in controlling gingivitis and periodontitis. [*See* ORAL PATHOLOGY.]

Bibliography

Bhasker, S. N., ed. (1980). "Oral Histology and Embryology," 6th ed. C. V. Mosby Co., St. Louis.

Engel, D., Schroeder, H. E., Gay, R., and Clodett, J. (1980). Fine structure of cultured human gingival fibroblasts and demonstration of simultaneous synthesis of Type I and III collagen. *Arch Oral Biol* **25,** 283–290.

Grant, D., Stern, I., and Listgarten, M., eds. (1988). "Periodontitis," pp. 216–251. C. V. Mosby Co., St. Louis.

Orlowski, W. A. (1983). Composition of gingiva. *In* "CRC Handbook of Experimental Aspects of Oral Biochemistry" (E. P. Lazzari, ed.), pp. 211–230. CRC Press, Boca Raton, Florida.

Simpson, J. W. (1983). Gingival metabolism. *In* "CRC Handbook of Experimental Aspects of Oral Biochemistry" (E. P. Lazzari, ed.), pp. 231–242. CRC Press, Boca Raton, Florida.

Terranova, V. P., and Wikesjö, U. M. (1987). Extracellular matrices and polypeptide growth factors as mediators of functions of cells of the periodontium—A review. *J Periodontol* **58,** 371–375.

Williams, R. C., and Howell, T. H. (1978). Periodontology. *In* "Textbook of Oral Biology" (J. H. Shaw, E. Z. Sweeney, C. C. Cappuccino, and S. M. Miller, eds.), pp. 1019–1031. W. B. Saunders, Philadelphia.

Hair

C. R. ROBBINS, *Colgate Palmolive Co.*

Glossary

Alopecia Loss of hair; may be localized or diffuse and over any part of the skin that contains terminal hair

Cortex Inner portion of a hair fiber; contains the major part of the fiber mass and consists of spindle-shaped cells bonded together by intercellular cement

Cuticle (scale) Outer covering of hair fibers; consists of several layers of flat overlapping scalelike structures that envelop each hair fiber

Dermal papilla Germinative tissue at the base of a hair bulb (in the follicle) that produces hair proteins

Hair follicle Microscopic canal leading from the skin surface to the inner dermis. One or more hair fibers grow out of a hair follicle

Hypertrichosis Excessive hairiness, which may occur on the limbs, trunk, face, or head. It may be endocrinologic in origin, induced by drugs, or of unknown origin

Keratin Group of highly specialized structural proteins containing the amino acid cystine found on the surface of mammals for protection and/or warmth (e.g., hair, nails, claws, horn, scales, and feathers are very rich in keratin)

Medulla Loosely packed region of cells containing vacuoles or air pockets, running along the axis of hairs, generally located near the center of thick, but not thin, animal hairs

Pilosebaceous unit Unit or system comprising a hair follicle and sebaceous gland

Terminal hair Long, thick hairs that are usually pigmented

Vellus hair Short (about 1-mm maximum length), fine hairs containing no pigment

HUMAN HAIR is a keratin-containing tissue that grows from large cavities called follicles, extending from the surface of the skin through the epidermis to the dermis. Hair is characteristic of all mammals and provides protective, sensory, and sexual attractiveness functions. In humans, it grows over a large percentage of the skin surface. Hair fibers grow in a cyclical manner consisting of three distinct stages called anagen, catagen, and telogen. Structurally, a fully formed hair fiber contains two, and sometimes three, different morphological units called cuticle, cortex, and medulla. Hair fibers are enveloped by layers of flat overlapping cuticle scales. Within the cuticle layers is the cortex, the major part of the fiber mass. The cortex consists of fibrous cell remnants containing highly oriented alpha-helical proteins embedded in a highly cross-linked, less-organized matrix. Thicker hairs often contain a third loosely packed porous region, the medulla, located near the center of the cortex.

I. Development, Types and Functions

A. Development of Hair and Types

Human hair contains keratins, highly specialized proteins produced in the bulb of the follicles. Hair keratins have several common features with other keratins (e.g., nails, claws, scales, and feathers). Keratin is not a single substance but a complex mixture of proteins characterized by a high concentration of the amino acid cystine.

Hairs grow from large cavities or sacs called *folli-*

cles that originate in the subcutaneous tissue (see Fig. 1). Rapid cell division occurs at the base of the *hair bulb,* beginning in the dermal papilla, located in the zone of differentiation and biological synthesis. These cells migrate upward as they move away from the base of the bulb; they elongate, dehydrate, and eventually die. Finally these cells move into the zone of *keratinization* where the fiber structure is stabilized by the formation of disulfide cross-links between cysteine residues. As the fiber continues to move upward toward the skin surface, it enters the region of the *permanent hair fiber.* The permanent hair fiber is a completely formed hair and consists of fully structured dehydrated cornified cells and intercellular binding matter.

In humans, prenatal hairs originate from the Malpighian layer (stratum germinativum) of the epidermis usually in the 3rd or 4th month of fetal life. Prenatal hairs are called *lanugo* and are either lightly pigmented or contain no pigment at all. Lanugo hairs are usually shed before birth or soon thereafter.

Soon after birth, hairs of the eyelashes, the eyebrows, and the scalp begin to grow at a more rapid rate, and the amount of pigment they contain usually increases. Smaller, finer hairs called *vellus hair* grow over the rest of the body. These fine, short (about 1 mm long) unpigmented hairs grow on those parts of the body that appear ''hairless,'' such as

the ''bald'' scalp and the forehead. *Terminal hairs* are long and thick, and generally finer and shorter on the scalps of children than on those of adults.

Hair in the axillary, pubic, and beard areas (for males) becomes longer and thicker at the onset of puberty. Those hairs that develop more fully at puberty are called *secondary terminal hair.* With further aging, hairs become less pigmented (graying) and thinning occurs.

In humans, either terminal or vellus hair generally grows over most of the skin. Those truly hairless areas include the palms of the hands, the soles of the feet, the undersurface of the fingers and toes, the margin of the lips, the areolae of the nipples, the umbilicus, the immediate vicinity of the urogenital and anal openings, the nail regions, and scar tissue.

Terminal hairs, at one stage or another of the life cycle, grow on the scalp, eyelash area, eyebrow area, axillary and pubic areas, the trunk and the limbs, and the beard and mustache areas of males. Vellus hairs grow on all other areas except for the truly hairless areas mentioned above.

B. Functions of Hair

Human hair provides limited protective, sensory, and sexual attraction or adornment functions and in an evolutionary sense may be becoming vestigal. *Scalp hair* provides warmth by insulating the scalp, and it can enhance sexual attractiveness. Hair also protects the scalp against light radiation and against mechanical impact. Hairs provide protective and/or adornment functions on other parts of the body as well. *Eyebrows* protect the bony ridges of the eye sockets and inhibit sweat and other liquids from running into the eyes. Eyebrows and eyelashes are both important as adornment. Eyelashes protect the eyes from sunlight and foreign objects and assist in communication. *Nasal hairs* help to filter air as it enters the respiratory system.

All hairs are attached to sensory nerve endings; therefore, hairs can function as *sensory receptors.* For example, one can sense an object by touching hair without even touching the skin.

II. Hair Growth

A. General Features of Hair Growth

The *dermal papilla,* at the center of the bulb (see Fig. 1), is responsible for controlling the growth

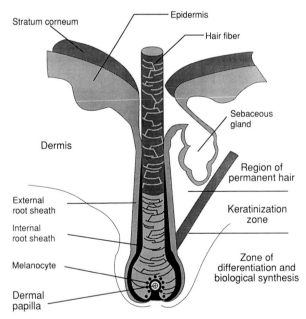

FIGURE 1 Schematic of an active hair follicle illustrating regions of synthesis.

cycle of hairs. *Basal layers,* which produce hair cells, nearly surround the bulb. *Melanocytes,* which produce hair pigment, are also within the bulb, and blood vessels feed and nourish the growing hair fiber near the base of the bulb.

All human hair fibers have three distinct stages in their life cycle (anagen, catagen, and telogen) (see Fig. 2). *Anagen* is the period during which the hair fiber grows. The growing period of hair not only varies among individuals but also depends on the region of the body. Anagen for scalp hair usually lasts longer than for hair growing on other body regions, usually 2–6 years, compared with less than 6 months for hair on the finger.

Catagen is a transition period when metabolic activity slows down in the dermal papilla. During catagen, the bulb shrinks and moves toward the epidermal surface. Catagen usually lasts only 2–3 weeks. *Telogen* is the resting stage when growth stops. At telogen, a new bulb begins to grow beneath the shrunken follicle. The old fiber is pushed out and is eventually shed and replaced by a new hair fiber. Therefore, the observation of hairs in the sink after shampooing is generally nothing to be alarmed about. It is usually normal hair fallout. Estimates of the length of telogen are in the range of 6–12 weeks. Because hair grows in a mosaic pattern on the scalp, in any given area hairs are in various stages of their life cycle.

Terminal human hair fibers grow at slightly different rates on different body regions. For example, hair grows at approximately 16 cm/year on the vertex or the crown area of the scalp but at a slightly slower rate (~14 cm/year) in the temporal area and

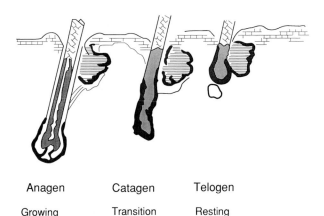

Anagen	Catagen	Telogen
Growing stage	Transition stage	Resting stage

FIGURE 2 Schematic illustrating the life cycle of a human hair fiber. [From C. R. Robbins (1988). Reprinted with permission of Springer-Verlag, New York.]

at a much slower rate in the beard region (~10 cm/year).

Because human scalp hair grows at a rate of approximately 16 cm/year and anagen generally lasts 2–6 years, scalp hair often grows to a length of 1 m before it is shed. In long hair contests, lengths greater than 1.5 m are frequently measured.

B. Abnormalities in Hair Growth

In rare cases, for unknown reasons, anagen of scalp hair persists for decades. Thus, human scalp hair greater than 3 m in length has been documented.

Hypertrichosis or excessive hairiness may be localized or diffuse. The most common type of hypertrichosis is called *essential hirsutism* or *idiopathic hypertrichosis* of women. This is the condition in which terminal hairs grow on women in those areas where hairiness is considered a secondary sex characteristic of males (e.g., the trunk, the limbs, or the beard or mustache areas). Such women generally do not show endocrinologic abnormality.

Endocrinopathic hirsutism is an uncommon condition resulting from excessive synthesis of hormones with androgenic properties. This condition produces masculinization of females, one symptom of which is excessive growth of terminal hairs in regions that are normally "hairless" in females. Classic examples of this disease are sometimes exhibited in circus side shows (e.g., Pastrana, the Mexican bearded lady, who appeared in London in the 1850s).

Hypertrichosis can also be induced by drug therapy. Streptomycin, estradiol, oxandrolone, or minoxidil, when taken internally, can produce an excessive growth of hair on the limbs, the trunk, or on the face.

C. Hair Loss

Alopecia (hair loss) may occur over any body region such as the scalp, face, trunk, or limbs. However, alopecia of the scalp, for cosmetic reasons, has received the most attention. Several types of alopecia are commonly seen on the scalp, all involving the papilla. The most common forms are usually linked to the endocrine system or to stress. Some of the different types of stress associated with hair loss are physical, psychological, disease or illness, or chemically induced stress (including drugs or the immune system).

Hair growth and hair loss are often linked to the

endocrine system. The most common form of hair loss is genetically involved, thus the term *androgenetic alopecia*. Androgenetic alopecia or *common baldness* is a normal aging phenomenon and occurs in both sexes. It occurs in about 4 of 10 men and in about 1 of 10 women. In women, the hair loss is generally more diffuse. Longer hair in women also tends to help cover up the hair loss. In males, hair loss occurs much faster in the vertex and frontal areas of the scalp, thus the term *male pattern baldness*.

Androgen production at puberty enhances hair growth over several parts of the body (e.g., limbs, pubic area, and axilla in males and females and chest and beard areas of males). *Testosterone* is somehow involved in androgenetic alopecia, because injections of this hormone induce the transformation of terminal hair to vellus-type hairs in the frontal scalp of stump-tailed macaques. However, the mechanism by which testosterone induces alopecia in the vertex and frontal areas of human males and at the same time induces hair growth in other regions (e.g., axillae and pubic areas) is unclear.

A normal scalp contains 175–300 terminal hairs per square centimeter, and about 85% of these are in anagen, 1–2% in catagen, and the remainder in telogen. In androgenetic alopecia, there is a progressive miniaturization of hair follicles producing smaller vellus-type hair from terminal hairs. As hair loss progresses, the percentage of hairs in anagen decreases while the percentage of hairs in telogen (at rest) increases. A small reduction in the number of follicles also occurs in the bald scalp.

Both during and after *pregnancy*, shedding rates normally decrease. However, some women during pregnancy report thinning of scalp hair. It is likely that these effects are related to endocrine changes.

Alopecia areata is believed to be related to the immune system (e.g., autoimmunity). This disease generally occurs as patchy baldness on an otherwise normal scalp, although sometimes hair from other body regions is affected. When the entire scalp is involved, the condition is called *alopecia totalis*. If terminal hair loss occurs over the entire body, which is rare, it is called *alopecia universalis*. Emotional stress has been shown to be an initiating cause of areata. Topical application of steroids is sometimes used to treat this condition.

Trichotillomania is a term used to denote alopecia induced by physical stress (i.e., physically pulling or twisting a localized area of hair until noticeable thinning develops). This type of hair loss sometimes occurs in children who unconsciously pull or twist a region of hair. A similar type of hair loss may also occur in adults.

Telogen effluvium indicates sudden but diffuse hair loss caused by an acute physical or psychological stress. This condition usually lasts only a few months and is reversible. Drugs used in chemotherapy often induce alopecia; however, this type of hair loss is usually reversible.

III. Structure of the Permanent Region

A. General Structural Features

For practical scientific reasons, terminal human hair fibers are assumed to have a circular cross section. However, some hairs (particularly kinky hair) deviate a great deal from circularity (see Fig. 3). In fact, the cross sections of hair fibers are better defined as ellipses; for scalp hair the major to minor axis ratios from three racial groups are defined in Fig. 3. With the approximation of circularity, the diameter of human scalp hair varies from approximately 30 to 120 μm, being finer in prepuberal children than in adults. The average diameters of hair fibers from the scalp, axilla, and thigh are similar. Whereas hair from the pubic region is about 30% thicker. Cross sections of beard hair are also more irregular than scalp hair. The degree of curl or axial shape of human hair also varies considerably. On average, scalp hair tends to be straightest for Orien-

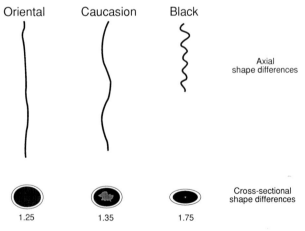

FIGURE 3 Some characteristics of scalp hairs of different racial groups. [From C. R. Robbins (1988). Reprinted with permission of Springer-Verlag, New York.]

tals and most curly for Blacks (see Fig. 3), although pubic hair and beard hair tend to be more curly than scalp hair.

B. The Cuticle

Cross sections of the permanent region of human hair fibers generally show three distinct regions (see Fig. 4). The outermost region, consisting of several layers of flat overlapping scalelike structures, is called the *cuticle*. The cuticle surrounds the major part of the fiber mass called the *cortex*. In thicker hairs, generally near the center of the cross section, a third region, the *medulla,* is sometimes present.

The cuticle is chemically resistant and serves as a protective covering for hair. Cuticle scales are translucent and for scalp hair do not contain pigment. They are attached at the root end and resemble shingles on a roof (see Fig. 5). Each scale is about 0.5–1 μm thick and about 45 μm long. The cuticle of human scalp hair is about 5–10 scale layers thick near the scalp. The number of scale layers often varies among different animal species and can sometimes assist in species identification in forensic studies (e.g., wool fiber generally contains 1–2 scale layers, human hair 5–10, but furs may contain up to 20 scale layers).

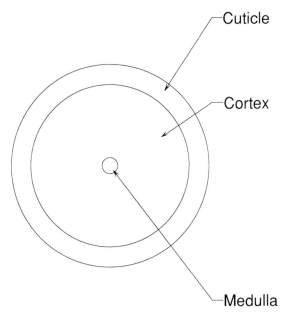

FIGURE 4 Schematic of a cross section of the permanent region of a hair fiber illustrating the three morphological regions.

A cuticle scale, the residue of a highly deformed, flattened cell, is still surrounded by a structure originating from the cell membrane. Internally, each scale has three layers (Fig. 6): the A-layer, the exocuticle, and the endocuticle. The A-layer is chemically resistant and contains a high concentration of the amino acid cystine (~30%), which cross-links neighboring polypeptide chains, through disulfide bonds, building a rigid structure. The exocuticle is also rich in cystine (~15%), whereas the endocuticle has little cystine (~3%).

As the permanent hair fiber emerges through the skin, it contains smooth unbroken scale edges (Fig. 5A). However, mechanical effects (e.g., combing and brushing) and chemical action from cosmetics (e.g., permanent waves, bleaches, and hair straighteners) and from sunlight degrade the cuticle (see Fig. 5B–D). In some extreme cases, severe mechanical damage can produce split ends (see Fig. 5D).

C. Intercellular Matter

This region sometimes called *cell membrane complex* consists of cell membranes together with adhesive matter that binds both cuticle and cortical cells. Cell membrane complex is primarily proteinaceous matter and structural lipid that is highly resistant to alkalinity and reducing agents.

D. The Cortex

The cortex constitutes the major part of the fiber mass. It is approximately 90% of the dry weight of hair and consists of spindle-shaped cells oriented parallel to the fiber axis (see Fig. 6). Each cortical cell is approximately 6 μm thick and 100 μm long. Cortical cells contain helical proteins embedded in a matrix rich in cystine cross-links. Each cortical cell is composed of smaller filamentous structures called *macrofibrils* (see Fig. 6), which in turn consist of highly organized filaments called *microfibrils* surrounded by a less-organized matrix that is highly cross-linked via cystine.

The microfibrils contain even smaller filaments called *protofibrils* that are composed of alpha-helical proteins. The alpha-helical proteins of the cortex of hair produce its well-known wide-angle X-ray diffraction pattern that provided the basis for the pioneering protein structure work of Pauling and Corey.

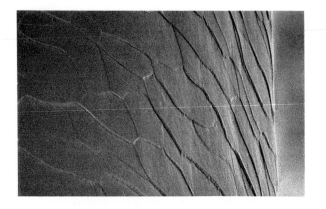

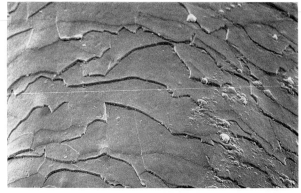

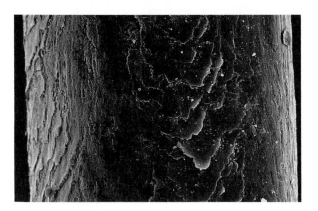

E. The Medulla

This morphological region is present only in thicker hairs (e.g., thick human hair, horse tail or mane, and porcupine quill) and is generally absent in most fine human hair. When present, it generally constitutes only a small percentage of the fiber mass. The medulla consists of a porous, loosely packed region of cells with air pockets. It may be continuous along the fiber axis, or it may be present or absent at different spots along the fiber axis. Some thicker animal hairs contain multiple medullas. Two medullas or a double medulla may be present in thicker human hair. The medulla enhances the insulation properties of hair fibers in fur-bearing animals but otherwise provides no known function in human hair.

IV. Chemistry of Hair

A. General Chemical Components

Morphologically, human hair contains two and sometimes three different types of cells and intercellular binding material. Chemically, human hair

FIGURE 5 Scanning electron micrographs illustrating changes in surface architecture of hair caused by weathering. *Upper left-hand corner:* near scalp, note smooth cuticle scale edges and scale surfaces; *upper right-hand corner:* 6 inches from scalp, note broken scale edges; *lower left-hand corner:* 12 inches from scalp, note broken scales and worn cuticle; *lower right-hand corner:* 18 inches from scalp, note split hair fragments.

contains four different types of components:

· minerals
· pigments
· lipid
· proteins

B. Minerals

A large number of elements have been reported in human hair. Some of the elements found in hair, other than C, H, O, N, and S, are Ca, Mg, Sr, B, Al, Na, K, Zn, Cu, Fe, Ag, Au, Hg, Pb, Sb, Ti, W, Mo, I, P, and Se. The primary sources of these elements in hair are from metabolic irregularities, sweat deposits, and the environment. The most important environmental sources of trace elements in hair are

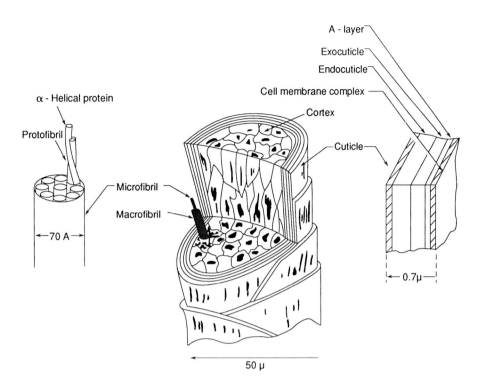

A - layer
Exocuticle
Endocuticle
Cell membrane complex
Cortex
Cuticle
α - Helical protein
Protofibril
Microfibril
Macrofibril
70 A
0.7μ
50 μ

FIGURE 6 Stereogram of the human hair fiber structure, illustrating substructures of the cuticle and cortex. [From C. R. Robbins (1988). Reprinted with permission of Springer-Verlag, New York.]

cosmetics, the water supply (bathing, washing, and pool water), and air pollution.

Certain heavy metals in the body tend to accumulate in hair at concentrations well above those present in blood or in urine. For example, levels of some toxic metals in hair (e.g., cadmium, arsenic, mercury, and lead) have been shown to correlate with the amounts of these metals in internal organs. Human hair has been used as a forensic tool in heavy metal poisoning and can be used as a diagnostic tool as well. High levels of cadmium have been reported in dyslexic children, and there is an unusually high ratio of potassium to sodium in hair of persons suffering from celiac disease.

Of metals derived from cosmetics or consumer products, lead has been demonstrated in hair treated with dyes based on lead acetate. Zinc and selenium can be detected in hair treated with antidandruff products containing these metals. Many shampoos or toilet soaps contain potassium, magnesium, or sodium ions that can deposit or bind to carboxylate groups in hair. Metals such as lead, cadmium, and copper in hair can derive from air pollution.

C. Hair Pigments

The cortex of human hair also contains remnants of cell nuclei and pigment granules. The pigments of human hair are the brown-black melanins (*eumelanins*) and the red (*pheomelanins*). In scalp hair the pigment granules exist only in the cortex and the medulla, but not in the transparent cuticle. Hair from other regions of the body such as the beard area sometimes contains pigment in the cuticle as well as the cortex and the medulla.

The formation of hair pigments takes place in the *melanocytes* (melanin-producing cells) starting with the amino acid tyrosine. The reaction is enzymatically controlled, and dopaquinone is believed to be a common intermediate for both the brown-black and the red pigments. Oxidative polymerization of dopaquinone in the absence of cysteine produces brown-black pigments, but in the presence of cysteine produces red pigment (see Fig. 7). *Melanosomes* or pigment granules are injected into newly differentiating cells near the base of the bulb. Thus the pigments are within the cells as they undergo keratinization.

The amount of pigment in the permanent hair fiber varies among racial groups. Generally, there is more pigment in the hair of Blacks and Orientals than in the hair of Caucasians. Melanocytes as well as keratinocytes are dormant during telogen (resting

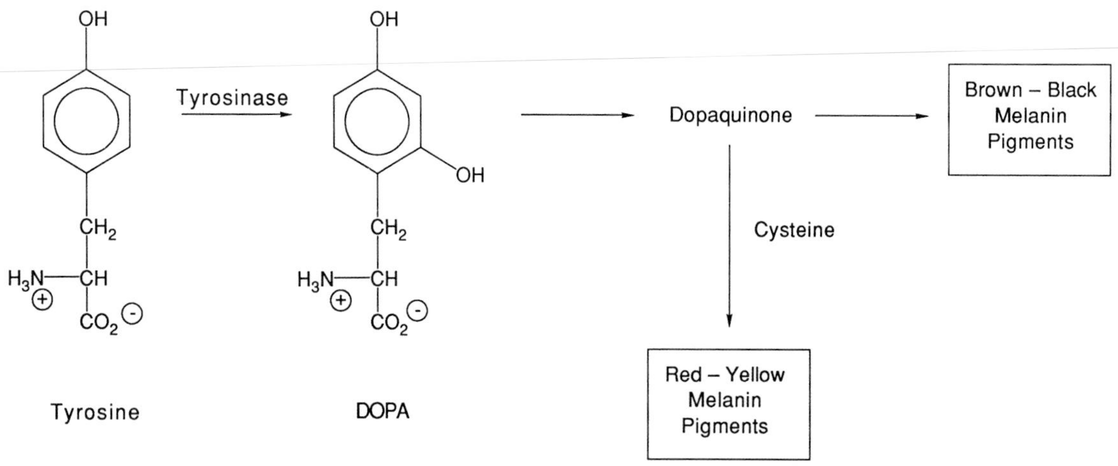

FIGURE 7 Schematic for formation of melanin pigments.

phase), but they resume activity during anagen. Graying results when melanocytes become less active during anagen.

D. Hair Lipids

Lipids exist in human hair as *free lipid* or as *structural lipid*. The structural lipids of human hair are generally associated with the cell membrane material and have not been fully characterized. The free lipid of human hair is primarily of sebaceous origin, and its concentration is governed by several factors including androgens (which control sebaceous gland activity), shampooing frequency, and by rubbing against objects such as combs, brushes, and pillows.

Free lipid exists both on the surface of hair and inside the fibers. Shampoos generally remove only surface lipids. The free lipid of human hair is chemically complex, and it is primarily responsible for the condition known as oily hair. It consists of about 50% free fatty acids (saturated and unsaturated) and neutral lipids including triglycerides, free cholesterol, wax esters, paraffins, and squalene. The free lipid of oily hair differs from that of dry hair in both amount and composition. Thus oily hair is more fluid and contains a higher percentage of wax esters, a higher ratio of unsaturated to saturated fatty acids, and a larger percentage of cholesterol esters than lipid from dry hair [*See* LIPIDS.]

Free lipid from children's hair generally contains higher concentrations of paraffinic hydrocarbons and lower concentrations of squalene and cholesterol than that from adult's hair.

E. Hair Proteins

At low humidities, when the water content is low, hair contains more than 90% proteins, mostly structural proteins with high concentrations of the amino acid cystine, up to 18% in human hair. Hair also contains hydrocarbon, hydroxyl, primary amide, basic amino acids, and carboxylic amino acid side chains, all at high frequencies or concentrations.

It must be remembered that hair fibers are a complex tissue composed of several different structural regions that are chemically and physically different. For instance, they greatly vary in the concentration of cystine. Therefore, they respond differently to cosmetic or chemical treatments.

Some of the proteins of hair are hygroscopic; therefore, hair responds to changes in relative humidity by either picking up water or losing it. The water content of hair varies from a theoretical zero at 0% relative humidity (RH) to about 32% of its own weight at 100% RH. The binding of water to hair and its content is critical to many hair properties (e.g., hair fibers swell with increasing water content, fiber friction increases and static charge decreases as the water content of human hair increases). The response of human hair to RH is so sensitive that hair has been used as a hygrometer.

All cuticle cells are rich in those amino acids that resist helix formation such as cystine and proline. The cortex contains microfibril and matrix regions. The microfibrils contain alpha-helical proteins and tend to be richer in the amino acids alanine, leucine, and glutamic acid, which favor helix formation. The matrix in contrast is richer in cystine and proline and other amino acids that tend to resist helical formation.

The medulla of porcupine quills has been analyzed, showing that medullary proteins are rich in basic and acidic amino acids and poor in cystine content.

When considering the reactivity of human hair to environmental factors and to cosmetic treatments, one may view its reactions in terms of the most reactive functional groups of its proteins. However, one must keep in mind that hair is not a homogeneous material. The best illustration of this fact is the difference in reactivity of the surface of hair to simple acids and bases. Hair proteins of the surface contain a greater frequency of acidic side chains relative to basic side chains as evidenced by its isoelectric point of pH 3.7. This fact demonstrates that the surface of hair has a greater capacity for combination with alkalies than with acids. However, when titrating whole fiber with acids and bases, we find that the entire hair has a greater capacity for acid than for alkalies.

F. Cosmetically Treated Hair

Several types of cosmetics react with functional groups of hair proteins, altering its functional groups as they react. For example, hair is generally bleached cosmetically using strong oxidizing agents consisting of alkaline hydrogen peroxide either alone or combined with persulfate salts. The function of a bleaching system is to lighten hair by degrading hair pigments, but the strong oxidizing agents used also oxidize cystine to cysteic acid, and other amino acid groupings of the hair proteins that are sensitive to oxidation are chemically modified. Changes in surface friction, wet tensile properties, and swelling behavior occur as a result of these changes of the fibers.

Permanent waving and chemical straightening also change the chemical behavior of hair. During permanent waving, hair is treated with a reducing agent that opens disulfide bonds, allowing molecular shirting to occur under stress.

$$\text{Hair-S-S-Hair} + 2 \text{ R-SH} \rightleftarrows \text{R-S-S-R} + 2 \text{ Hair-S-H}$$
$$\underset{\text{agent}}{\text{Reducing}} \qquad\qquad \text{Reduced hair}$$

The stress is induced by bending or curling the hair (permanent waving) or by combing it straight (chemical straightening), while it is in the chemically reduced state. The hair is then treated with a mild oxidizing agent to reform disulfide bonds after the hair has taken on a new shape.

$$2 \text{ Hair-SH} \rightarrow \text{Hair-S-S-Hair}$$

Another form of hair straightening involves treatment of hair with strong solutions of alkalies, which attack cystine and break open the disulfide cross-linkages to allow the hair to be straightened.

Hair dyes may be classified in various groups: permanent or oxidation dyes, semipermanent dyes, temporary dyes or color rinses, and metallic dyes. Each of these types of dyes has its specific chemistry that may be used to assist in identification in forensic analysis.

Shampoos, conditioners, and hair fixatives (hair sprays, gels, glazes, and mousses) primarily react at or near the fiber surface. These cosmetics can induce temporary and in some cases permanent changes to the surface properties of the hair.

Figure 5 illustrates physical changes that may be observed in some human scalp hairs with time. Because scalp hair grows at a rate of about 15 cm/year, the tip ends of 30-cm hair have been exposed to environmental effects including sunlight, wind, rain, and shampoos for about 2 years. We generally refer to these effects collectively as *weathering* of hair. Chemical changes have been demonstrated in weathered hair from the action of sunlight. Physical changes have also been demonstrated from combing and brushing of hair. The ultraviolet (uv) region of the spectrum oxidizes cystine to higher oxidation states. Changes to other amino acids have also been demonstrated in hair fibers exposed to uv radiation. These chemical weathering changes generally occur more readily near the fiber surface.

V. Some Abnormalities and Disorders

The louse is a blood-sucking, wingless insect that parasitizes human beings. One species generally attacks the head and another the body. *Pediculus capitis,* the parasite that is found on scalp hair, is the most common. The adult louse is a small gray-white insect, only about 3 mm long. It grasps hairs with its legs and lays eggs that are cemented to hairs. The eggs hatch into nymphs that grow into adults. The empty nit or egg sacs are left on the hair fiber surfaces, producing gross fiber distortions.

Monilethrix or beaded hair is a congenital, hereditary disease resulting in abnormal human scalp hair. This disease is easy to diagnose because individual hairs look beaded (i.e., composed of thin and thick segments along the length of the fiber axis). Hair length in monilethrix rarely exceeds a few centimeters because the fibers are fragile and dry.

Pili torti or twisted hair is also a rare congenital deformity. Pili torti is characterized by flattened fibers with multiple twists. Sometimes this condition produces hairs of normal length, but frequently the hairs are twisted and broken.

Pili annulati or ringed hair is a rare hereditary condition characterized by light (silvery) and dark bands or "rings" along the fiber. This visual effect is produced by regions with and without a medulla. Annulati generally does not present a medical problem because the hairs of this condition often appear otherwise normal.

Trichorrhexis nodosa occurs more often in facial hair than in scalp hair. This condition consists of bulbous nodes producing the appearance of thin and thick segments along the fiber axis resembling monilethrix. However, these nodes are actually partial fractures that crack under stress forming "broomlike" breaks. These fractures are actually fragmented cortex in which individual separated cortical cells appear like splintered wood.

Bibliography

Fraser, R. D. B., MacRae, T. P., and Rogers, G. E. (1972). "Keratins, Their Composition, Structure, and Biosynthesis." Charles C. Thomas, Springfield, Illinois.

Robbins, C. R. (1988). "Chemical and Physical Behavior of Human Hair," 2nd ed. Springer-Verlag, New York.

Zviak, C. (1986). "The Science of Hair Care." Marcel Dekker, New York.

Headache

OTTO APPENZELLER, *Oxygen Transport Program, Biomedical Research Division, Lovelace Medical Foundation*

Glossary

Aura Warning of impending attack (e.g., flashing lights, black spots appearing before the eyes, bright zigzag lines)
Biteplates Plastic covering inserted between upper and lower teeth
Extracranial vasodilatation Increase in diameter of blood vessels outside the bony cavity of the head
Focal neurologic deficits Failure of part of the nervous system to function
Hypoxia Lack of oxygen
Metabolism Chemical processes of living cells and organisms necessary for maintenance of function
Migraineur Person suffering from migraine
Neuron Nerve cell
Neurotransmitters Chemicals used for communication between nerve cells
Oxidative metabolism Where oxygen is used for metabolic processes
Pulsatile pain Wavelike increases in pain coinciding with heartbeats
Refractory period Time after an attack during which a new attack does not occur
Trigeminal perivascular nerves Nerves surrounding blood vessels that take origin in the trigeminal nerve cells, which also subserve sensory function
Trigeminal vascular system Nerves surrounding some blood vessels supplying blood to the brain

HEADACHE IS AMONG the most common afflictions of human beings, but it is seldom the most dangerous. It ranges from slight nondisabling pain in the head or neck to severe pulsating localized or generalized head pain. Most headaches, migraine and tension headache being the most common, are neither symptoms of serious disease nor disease themselves. Rather, recurring headaches, like migraine and tension headaches, are a manifestation of stress, a peculiarly human response to a perceived or actual threat to the headache sufferer's well-being. Only occasionally are headaches a sign of serious disease. These rare headache types include those associated with tumors of the brain, inflammation, or ruptured blood vessels within the cranial cavity.

Most people experience headache at some time in their lives, in part because humans are uniquely constructed to suffer headaches. Unlike other animals, humans have a denser innervation in the head, face, and neck. This suggests a possible evolutionary purpose for the malady, as it may protect the head and its content from damage. The ancestors of humans depended for their lives on the functions (sight, hearing, and oral communication) of the richly innervated head and its organs, also suggesting headache's evolutionary purpose, for when the head or its organs were threatened, life was threatened as well. The highest density of pain fibers (i.e., nerves that conduct impulses to the brain that are perceived as painful) is found in tissues of the head, such as the skin and the lining of the sinuses making up much of the foundation of the face. (Brain tissue itself is insensitive to pain.) Pain is perceived more acutely and more often in or around the head than in any part of the body. Historical evidence (e.g., ancient skulls containing holes drilled before death) suggests that early humans not only experienced headaches but sought remedies for them. It is thought that the holes served to relieve painful pressure within the head.

Modern humans are probably at greater risk for headache than their counterparts in earlier times because they are, arguably, exposed to more stress, a precipitating factor in most headaches. An estimated 40 million Americans suffer from headache; the numbers are similar in other Western countries but are lower in developing countries and in Japan. Headaches, including migraine, are more common in females of all ages but tend to decline in both sexes after the age of 65 yr. Estimates of the incidence of headaches (i.e., the percentage of individuals suffering from head pain) vary with age and range from 70% of men and 84% of women aged 18–24 yr to 26% of men and 50% of women aged 75–79 yr. Headache sufferers in each sex and age group consistently experience more depression and anxiety, particularly about their own health, which they monitor obsessively and frequently imagine is at risk.

I. Anatomy of Headache

Most of the information-processing necessary for the perception of external stimuli of any kind and of pain occurs within the head, i.e., at the site of the most highly developed sense organ—the brain. Because of this, the head has a very rich supply of nerves and is exquisitely sensitive to painful stimuli.

The brain is the most important and well-protected organ. It lies encased in several layers, the outermost being the skin. The skin itself consists of several distinct layers, all of which contain a large number of nerve endings. These endings receive and transmit the conscious sensations, including temperature, touch, pressure, and pain. When this rich nerve network is excessively stimulated, pain is perceived and protective measures are taken. [*See* BRAIN.]

The largest sensory nerve in this network is the trigeminal. Its sensitive endings are spread not only in the skin of the face and head but also within the muscles subserving chewing, in the lining of the mouth, and within the lining of the nasal cavity. Trigeminal nerve fibers also surround blood vessels within the cranial cavity that supply blood to the brain. These pain-sensitive fibers, which constitute a perivascular nerve fiber network known as the trigeminal vascular system, control the diameter of cranial blood vessels. They are important in headache because, by causing blood vessels to enlarge, they can cause pulsatile head pain.

Within the envelope of skeletal muscles is the boney brain cavity, which is rigid but pierced by numerous holes through which nerves and blood vessels enter and exit. The bones themselves are not sensitive to pain, but their covering of tough membrane and their numerous openings have pain-sensitive tissues. The bones of the cranial cavity and the face contain numerous air cavities, or sinuses, which reduce the weight of the skull without changing its strength. The sinuses are lined with membranes richly supplied with nerve endings and blood vessels that protect these cavities from air pollutants. When these membranes become irritated, they can give rise to pain.

Within the cavity of the skull, the brain rests on several membranes that absorb shocks and provide support. The membranes are composed of a tough outer membrane and finer inner membranes, the latter richly supplied with blood vessels and their accompanying nerves. Between the tough outer membrane and the one directly apposed to brain tissue, the so-called pia mater (the "tender mother"), a fluid layer provides further cushioning from undue displacement or blows to the head. The brain itself does not respond to pain because it lacks appropriate nerve endings or receptors sensitive to painful stimuli. Nevertheless, it can induce headache when it is displaced or swollen, the displacement or swelling thus exerting pressure on blood vessels, veins, and the brain's covering, all of which contain pain-sensitive nerve endings.

II. Classification

Painful symptoms that arise in or about the head are usually called headaches. The main groups recognized from an earlier system of classification (devised in 1962) are still valid in the revised classification (1988). They are vascular headaches, muscle contraction or tension headaches, traction and inflammatory headaches (due to displacement of pain-sensitive structures of the head), and psychogenic headaches. These have no definitive origin in any of the pain-sensitive structures but may be related to biochemical changes in the brains of victims. Other headache types may be lumped into a miscellaneous group (neuralgia, postconcussion headache, and other types). Many headache types respond to a slowdown, relaxation, or removal from the triggering situation, often exemplified by the relief sufferers get from isolation and lying down. Although each group is characterized by its symptoms, most, if not all, have symptoms in common, such as nausea, vomiting, and light and sound sensitivity. Fortunately, most headaches are mild and respond to simple measures, but some require interruption of activities and may last from 2 to 24 or more hours. Even though the pain is often intense and frightening, particularly if it is accompanied by signs of nervous system dysfunction, such as paralysis, numbness, or visual phenomena, as may occur in migraine attacks, these headaches are usually benign and transient. Only perhaps 5% are due to serious disease.

A. Vascular Headaches

A large number of pains in the head, ranging from those associated with hangover and hunger to migraine and cluster headache are in the vascular headache category. All of these pains can be traced to a transient abnormality of blood vessels that may not lie at the root of the disorder but that accompanies the headache. These headaches are by definition throbbing; their pain accompanies the pulsations of the heart. Although a given attack may be localized, at least initially, to one or the other side of the head, successive headaches may occur on different sides. These headaches often are triggered by specific stimuli, and in particular cold foods (ice cream headache) and drinks.

1. Migraine

Migraine and cluster headache, one subclass of vascular headaches, differ from other such headaches in that they are chronic and are usually lifelong afflictions. Unlike occasional toxic vascular headaches (such as hangovers), migraine has a congenital aspect, but exactly what the hereditary basis of these headaches is has not yet been elucidated. The changes in diameter of cerebral blood vessels resulting in migrainous attacks can occur in anybody, provided external and internal stressors have come together in the right amount and in the right sequence. Nevertheless, the physical and behavioral constitution of people suffering from frequent migraine is different from that of people who may have only one attack during a lifetime.

Each attack consists of at least two phases: the first is associated with a decrease in blood flow to specific areas of the brain, and the second with a marked increase in flow, both in the brain and scalp, and severe pulsatile pain. In some migraines, the phase of falling blood flow brings about dysfunction of the brain (neurologic deficits) resulting from the decrease in oxygenation of circumscribed areas of the brain. These attacks are known as migraine with aura (formerly classic migraine). Others, known as migraine without aura (formerly common migraine), lack neurologic deficits and consist only of pulsatile pain. The nature of a patient's migraine, the side of the pain, and the kind of focal neurologic deficits often change over a lifetime. Migrainous attacks are often accompanied by nausea, ushered in by sleepiness, excessive vigilance, or seemingly boundless energy. After the attack a refractory period lasts 1 or more days.

In childhood migraine sufferers, boys are more often affected by migraine than girls, but the proportion becomes reversed after the onset of puberty. In adults, women are by far the commonest victims of the disorder. Children may also have symptoms, such as colic, motion sickness, or periodic vomiting, without actual headache. Now classed as childhood periodic syndromes, they may be precursors to later migraine attacks.

a. Hypoxic, Neurogenic, and Vascular Theories of Migraine Causation
Numerous theories about the causation of migraine headaches have been proposed and discarded. None has taken into account and explained all the many triggers for headache, but with modern techniques for noninvasive assessment of oxygen supply and demand in the human brain, a comprehensive theory is likely to emerge. Existing theories (hypoxic, neurogenic, and vascular), as yet unproved, merit further investigation.

b. Hypoxic Theory of Causation
One hypothesis proposes that lack of sufficient oxygen supply to the brain is the cause of migrainous attacks. Oxygen is essential to brain function, which is dependent on neurotransmitter activity; when oxygen supply to the brain is diminished, brain function is impaired. This hypothesis does not state precisely why oxygen is lacking, but two possibilities exist: Some migraine sufferers may have a physical impairment that prevents adequate oxygen supply to their brains; others may have impairments that interfere with the energy supply to nerve cells and consequently with oxidative metabolism. To support this hypothesis, many similarities have been found between manifestations of migraine and those caused by abnormalities in oxidative metabolism. Thus, this theory proposes that triggering mechanisms for migraine may be either a reduction in the supply of oxygen or in the supply of the primary

fuel for brain metabolism, glucose. (Glucose is the only nutrient available for brain cell survival; it requires oxygen for its utilization by brain cells.) Situations that increase the demand for oxygen can also cause attacks, and an inappropriate matching between brain metabolism and blood flow, which brings oxygen and glucose to the brain, may be a trigger as well. Last, metabolic processes that result in biochemical abnormalities may also trigger attacks. Because the hypoxic theory examines several possible headache triggers, it deserves further clinical investigation.

Investigations outside the clinic (at high altitude) support the hypoxic theory. At high altitude, unless they acclimate first, people who do not ordinarily suffer from headaches invariably develop headaches, partly due to hypoxia. Their headaches have many of the same features of migraine, including pounding pain, accompanying nausea, and aggravation of the pain by exertion. Mountain-climbing migraine sufferers learn through experience the altitudes that trigger their migraines because they are doubly affected at high altitude, which contributes a predictable factor (hypoxia) in triggering their attacks.

The experience of patients whose migraine is triggered by eating certain foods supports the hypoxic theory as well. These foods change the availability of brain neurotransmitters by altering the electrical discharge of nerve cells. The neurotransmitters can change the requirements of oxygen of some cells by increasing the firing rate of their electrical impulses. This in turn can promote electrical conductance along certain pathways in the brain involved in the perception of pain.

The hypoxic theory is also supported by the experience of patients suffering from diseases (e.g., strokes) that affect the patency and distensibility of blood vessels in the brain, and, thus, oxygen supply to local areas of the brain. In these cases, the metabolic requirements of the brain arguably may not have been met with adequate oxygen, because the damaged blood vessels block the flow of blood (and oxygen) to the brain. Because many fits stop normal respiration and thus oxygen uptake by the blood, people who have disease of the brain (e.g., epilepsy) that manifests itself by fits may have associated headaches. [*See* EPILEPSY; STROKE.]

The hypoxic theory may also explain some aspects of tension headache, brought about through the intimate relationship of blood vessels and nerves within the muscles surrounding the head.

The contraction of muscles, which are initiated by nerve electrical impulses, together with chemical release of substances at the nerve–muscle junction, induces individual muscle fiber contraction and shortening. Muscle contraction initiated by stress and prolonged (without intermittent relaxation of the muscles), by the additional stress of irritated, contracted muscles, stimulates the perivascular nerve fibers, and reduces the flow of blood. In turn, the lack of oxygen produced by the decreased flow causes the release of a variety of chemicals that stimulate pain-sensitive endings in the muscles. Direct pressure on nerves by contracting muscles or by unusual positions of the head and neck bones also contributes to this pain.

c. Neuronal Theory of Causation The well-known relationship of migraines to stress, fatigue, and anxiety, all manifestations of brain nerve cell activity, and the frequency of attacks on weekends and holidays, when many drink alcoholic beverages (which impair brain cell function), suggest that abnormalities of nerve cells may be of primary importance in causing attacks. Similarly, the sensitivity of many sufferers to bright lights, which may trigger attacks within seconds, suggests a reflex, entirely neuronal causation.

Supporting evidence for a neuronal mechanism is the occurrence of deficits of function in discrete brain areas in some sufferers, especially patients who have had frequent attacks spread over many years. Abnormalities on brain imaging scans, only comparatively recently available, have also been found.

What is more, a weak association between migraine and epilepsy is well-documented. Because both of these disorders are common, however, most instances of the two occurring together are coincidental.

The postulate that migrainous attacks are the result of primarily neural dysfunction implies that, because attacks are paroxysmal, this abnormality in neural function occurs only in paroxysms, and that all vascular phenomena that accompany the attacks are secondary to the abnormality in nerve cell function. The most widely held and longest surviving explanation for the neuronal mechanism is that migraine is an episode of the ''spreading depression'' first described by Leao, a Brazilian physiologist. This spreading depression is a transient disruption of neuronal electrical activity, which can be elicited in experimental animals and spreads like a wave through some brain areas. The depression seen in experiments travels at a speed of 3 mm/min. The characteristic initial visual phenomena of migraine (bright zigzag lines with surrounding dark areas) have also been calculated to spread at the same rate. When cerebral blood flow was measured during attacks of migraine in some patients, an initial fall in blood flow was seen to spread, like a wave, at a rate of 3 mm/min. But when similar attacks were induced in patients whose migraine was not accompanied by focal neurologic deficits (e.g., visual phenomena), no such abnormalities in cortical blood flow were noted. In spontaneously occurring attacks similarly studied, the expected reduction of cerebral blood flow in those with focal neurologic deficit was found, but this decrease in flow extended well into the headache phase known to be accompanied by extracranial vasodilatation, and an increase in intracranial flow was recordable during the later stage of the attacks.

d. Vascular Theory of Causation The many vascular phenomena that accompany migraine, including increased pulsation of blood vessels supplying blood to the head and in the scalp and face, the transient improvement of pain when these blood vessels are compressed, and the effect of substances either causing blood vessel dilatation and increasing pain, or causing vasoconstriction and decreasing pain, have suggested that the primary cause of migraine is not in abnormal neuronal function but primarily in changes in blood vessel diameter. While the neural hypothesis does not deny the importance of vascular phenomena in the genesis of pain, this explanation itself does not account for peripheral effects outside the brain that accompany migraine attacks, but these, perhaps, could be activated through centers in the posterior part of the brain, the so-called brain stem, which does control to some extent the caliber of blood vessels throughout the body.

A migrainous attack may be considered an adaptive or reactive phenomenon. This phenomenon in turn depends on changes in the threshold of responsiveness of the brain to a number of stressful and precipitating factors. Such factors as hormonal balance, changes in the level of neurotransmitters of the brain, lack of oxygen, jet lag, changes in sleep patterns, and changes in blood vessel diameter are important triggers. If these triggers are acting in concert in the right sequence and in sufficient numbers, an attack of migraine occurs. Thereafter, ex-

haustion of the brain compels rest and recuperation until the setting becomes right for a subsequent attack. The threshold at which attacks occur varies from person to person and is probably genetically determined to some extent, but it also varies during various times of the life cycle and fluctuates with physical or mental stress, broadly defined. This concept allows for the occasional attack in otherwise ordinary people never subject to headache and accounts for those who have frequent and incapacitating recurrences of migraine and a family history of migraine.

2. Cluster Headache

Cluster headache, like migraine, is considered a "vascular" headache because pain during attacks is brought about by changes in cerebral blood flow and blood vessel diameter. The intensity of pain in cluster headache is unique. Unlike migraineurs, who lie still during their attacks, cluster patients pace back and forth in agony. Also different from migraine is the near certainty cluster patients have about the timing of their attacks during a cluster period. These attacks occur in series that may last for weeks or months (the cluster periods) and are separated by intervals of complete freedom that may last months or even years. Patients are equally uncertain about when a cluster period will end. The cyclic nature of this disease remains unexplained. The prevalence of attacks at night suggests a relationship to sleep phases and perhaps to biologic clocks, but cluster periods themselves do not seem to follow any patterns, although in some patients they are more likely in autumn. The attacks are much shorter in duration than migrainous attacks, lasting at the most 3 hr, but usually only 20–40 min. During a cluster period, attacks can occur just once every other day or as often as eight times daily. Characteristically, these attacks are associated with autonomic nervous system dysfunction ipsilateral to the side of the head pain, which is strictly unilateral. Conjunctival injection, a red eye, tearing, congestion of the nose, nasal discharge, and forehead and face sweating indicate the autonomic nervous system involvement, together with a smaller pupil and droopy eyelid and sometimes swelling of the eyelid ipsilateral to the pain.

Cluster males (the condition is far commoner in men than in women) are thought to have a characteristic appearance. They often have a ruddy complexion, thick facial skin, visibly dilated small blood vessels across the bridge of the nose and cheeks, and a prematurely deep-furrowed forehead, together with a broad chin and skull. They are often tall and have a rugged appearance and have a higher prevalence of hazel-colored eyes than other men. Few of them are blue-eyed. There is no pattern of inheritance for cluster headache, although these physical characteristics suggest that heredity may play some role. Personality factors are found to be similar to those of migraineurs: difficulty in handling stress is also characteristic of cluster patients. In addition, male sufferers smoke more and drink more coffee and alcohol (in between cluster periods only) than do ordinary men. During a cluster period, however, they all abstain from alcohol, because it invariably precipitates an attack.

Like migraine, cluster headache is treated with drugs that affect blood vessel diameter. One special type of cluster headache, mainly affecting women, strikes 10–18 times a day and may remain uninterrupted by pain-free periods for up to 20 yr, so that victims are unable to participate in ordinary life activities. This type of headache, called chronic paroxysmal hemicrania, responds to the anti-inflammatory drug indomethacin and may completely disappear with the use of this drug. This remains the only type of headache that can be "cured" permanently by the use of medication, even though it is not known how this type of pain is produced, and it remains the rarest type of the so-called benign headaches.

3. Toxic Vascular Headache

Anyone who has experienced fever has usually also experienced a toxic vascular headache, also known as fever headache. Infection by bacteria or viruses causes release of fever-producing substances, the so-called endogenous pyrogens. These substances, after they enter the blood and pass to certain areas of the brain, cause release of prostaglandins (vasodilators), which also affect the brain's thermostat and lead to an increase in body temperature by shivering and blood vessel constriction or other reactions, such as metabolic heat production by the liver. When the higher body temperature has finally been reached, an increase in oxygen demand and fuel supply, caused by the increased temperature, in turn demands an increased blood supply, which is accompanied by dilatation of blood vessels and headache. Such headache is dull, deep, often aching, generalized, and throbbing in its early stage, although it may occasionally be restricted to a specific area at the back of the head. As

it does with other vascular headache, any jolt, sudden movement, or strain aggravates fever headache.

B. Muscle Contraction or Tension Headaches

Tension headaches associated with continuous contraction of muscles around the head and neck are by far the commonest kind of headache. Typically, the pain is dull, lasts several hours, and recurs over periods of days, weeks, or years. It is often felt on both sides of the head, in the temple, and at the back of the head; the pain is sometimes described as a hatband around the crown. These headaches are often aggravated by physical or emotional stress or both, but in all cases severe disabling pain is real and not imagined. Excessive muscle contraction around the neck and scalp also occurs in other kinds of headaches.

C. Traction and Inflammatory Headaches

An important category of clinical pain is that caused by traction or displacement of pain-sensitive structures of the head or inflammation. The headache is due to disease or injury to the head, and if ignored or untreated may lead to permanent damage and even be life-threatening. While such headaches may initially respond to painkillers, delay is often dangerous, and sufferers should seek medical attention. Such headaches, grouped because of their ability to cause permanent damage, have diverse causes. The traction headaches are the result of mechanical forces that distort by pressure or traction pain-sensitive structures within the cranium. The forces are mostly due to brain tumors, blood clots, or collections of pus. If such lesions are sufficiently large, and particularly if they change size relatively rapidly, as blood clots and deposits of pus can, they cause pain. On the other hand, some slow-growing tumors can become large before pain becomes severe.

The second subgroup of these disease-related headaches is caused by inflammation, the site of the inflamed tissue usually giving the name to the underlying disease: meningitis, phlebitis, and arteritis. In meningitis, the meninges, or coverings of the brain, become inflamed. In phlebitis, from the Greek word for vein, the large veins draining blood from the brain are often involved; in arteritis, an inflammation of arteries, the vessels at the temple are often at fault. Arteritis is a disorder that most often strikes the elderly and may be associated with loss of sight or multiple strokes. It often causes an intense and deep-seated ache, with tender and rigid blood vessels on the surface of the skull.

1. Temporal-Mandibular-Joint Dysfunction

A widely touted cause of headache is temporal-mandibular-joint dysfunction (TMJ). The importance of abnormal jaw muscle contraction or arthritis in the jaw joints, causing imbalance of the bite and thus headache, is debated. TMJ typically is initially treated with bite plates prior to grinding teeth for better alignment. Some specialists, who do not generally believe in this type of causation, do not treat patients with headache by bite plates. Others use these devices to treat a variety of headaches and report frequent success. Bite plates are also used by athletes, who report improvement in performance when the plate is used and hard clenching of teeth during muscular exertion is prevented mechanically by the device.

2. Headache Caused by Degenerative Disease of the Cervical Spine

Degenerative disease of the cervical spine causing narrowing of joint spaces can also cause headache, because the disease causes imbalance in muscle contraction, which in turn may put pressure on the nerve roots exiting from the spinal cord. The pain these patients experience is indistinguishable from the pain of tension headaches, but, unlike tension headache pain, may respond to massage, traction on the neck, or surgery.

3. Inflammatory Headache Caused by Sinus Infection

Inflammation of the sinuses rarely cause chronic headaches but may cause pain in the face if acute infection is present. Most patients with long-standing sinusitis suffer from ordinary tension-type headache. That is not to say that such inflammation may not on occasion give rise to serious neurologic problems manifested by headache. A chronically blocked and inflamed sinus cavity, for example, can sometimes lead to inflammation of veins that drain blood from around the sinuses into the cranial cavity, and infection thus spreads, causing meningitis, which may be followed by brain abscess. [*See* INFLAMMATION.]

D. Psychologic/Psychogenic Headaches

1. Emotion, Stress and Mental Illness-Associated Headache

Chronic head or face pain is often associated with the muscle contractions of repeated and involuntary facial grimacing, nearly always a symptom of psychologic stress or mental illness. Patients suffering from such illnesses frown, squint, and grind their teeth (particularly at night) for prolonged periods of time. The contracted muscles become tired and release substances that cause chronic pain, which may be aggravated by malalignment of the jaw and improved by a bite plate.

2. Delusional or Conversion Headaches

Even though psychologic factors can aggravate the headache types described above, the pain is real: it has a physical basis. On the other hand, a complaint of headache can also express emotional disturbances, usually associated with sexual conflicts, repression, anxiety, or depression, and is often related to changes in an individual's life that are not wanted or cannot be controlled. Though patients suffering from these headaches are convinced of the organic nature of their illness, they often are rather indifferent to the pain they complain of so bitterly. Such people typically cannot establish normal social and situational contacts; they often experiment with drugs, including painkillers and tranquilizers, and usually present a complicated management problem because of their frequent associated drug-dependency. Such patients often require prolonged behavioral manipulations.

E. Miscellaneous Headache

1. Neuralgia

Several varieties of neuralgia, or pain arising from nerves around the head, are dreaded because of the intensity of the pain. While these disorders usually are not life-threatening, they make existence miserable. The most common is trigeminal neuralgia or tic douloureux, which is related to the trigeminal nerve system, the most widespread sensory system of the head. Its victims are usually 40 yr old or older. The pain is an intense aching, burning, or stabbing, lasting only a few seconds but coming in bursts, typically on one side of the head. Characteristically, the pain of trigeminal neuralgia is triggered by stimuli that affect the trigeminal sensory system, including cold air, touch, chewing, and sometimes clearing the nose. Individual sufferers become vividly aware of specific trigger zones and stimuli. Glossopharyngeal neuralgia is much rarer than tic douloureaux. Clinically, the pain is similar, but the sensitive trigger zones are located in the back of the throat or the tongue, from which the pain radiates toward the ear in terrifying, painful stabs. Swallowing is the most common trigger for this neuralgia.

2. Postconcussion/Post-Traumatic Headache

Occasionally after head injuries resulting in lacerations of the scalp, scar tissue forms around the small sensory nerve fibers that are abundant in the head region. Scarred nerves can produce occasional stabs of pain, initiated by touch or pressure, that later evolve into continuous contraction of scalp muscles. The result is indistinguishable from tension-type headaches. A clue to this headache mechanism is often the presence of an extremely sensitive spot related to a scar. The injection of a local anesthetic and sometimes steroids may interrupt the vicious cycle of abnormal and spontaneously arising nerve impulses in the constricted or scarred nerve, thereby relieving the scalp muscle contraction and pain.

a. Postconcussion Syndrome A constellation of symptoms following head injury, consisting of headache, dizziness, irritability, difficulty in concentration, and alcohol intolerance is labeled the postconcussion syndrome, headache being the most troublesome and prominent of the symptoms. This constellation of complaints is common after head injury, but its cause and treatment continue to arouse controversy.

b. Post-Traumatic Headache A brain concussion changes the levels of neurotransmitters (substances that are released or taken up by neurons for communications with other neurons) or neuromodulators (which have a similar function but take effect more slowly). The electrical brain activity is therefore affected by the impact, which changes levels of neurotransmitters, thus interfering with the regulatory capacity of the central autonomic nervous system. Subsequently the peripheral autonomic nervous system and particularly the trigeminal vascular systems are affected. The resultant abnormalities in cerebral blood flow following concussion are well demonstrated and may lead to very

serious consequences. Occasionally from this alteration in vasomotion arises primary traumatic brain damage and more diffuse secondary disorders of the brain. These secondary changes in the brain after trauma do not seem to be pathogenetically related to post-traumatic headache, because the brain itself is insensitive, but they point to a relationship between brain concussion and vascular disorders occurring immediately after the injury. The resulting decrease in blood flow to the brain and change in diameter of blood vessels, known to be sensitive to pain, can cause a headache that is clinically very similar to migraine. It is pulsatile and is often aggravated by exertion and by bending forward. Unlike typical migraine, however, the pain is not concentrated in one side of the head; it is either diffuse or frontal. Headache persisting for years after head injury, however, cannot easily be attributed to the original impact, and its causes are mysterious. Most authorities suggest that 30–50% of patients develop headache after injury, and because head injuries are common, the post-traumatic headache, persistent and incapacitating, forms the bulk of post-traumatic handicaps (serious neurologic deficits are less common in these patients). No clear correlation between the severity of the injury or damage to the brain and the incidence of post-traumatic headache has emerged. It has even been suggested that post-traumatic headache is less troublesome after major cerebral injury, and more disabling after minor concussion. Psychologic factors, such as anxiety or the desire for unemployment compensation, that are related to the trauma undoubtedly play a role in many post-traumatic headaches. Some post-traumatic headaches mimic ordinary tension headaches; others have many, if not all, of the features of migraine, including the accompanying nausea and light and sound sensitivity. A third, distinctive, type often follows injury to the neck. The cause is thought to be the autonomic innervation in the sheaths of major arteries to the head. Some rare patients experience pain and tenderness of the neck for weeks after the injury, later giving way to severe pain with sweating on the injured side in temple and frontal areas, accompanied by dilatation of the pupil, nausea, and light sensitivity. The pain may last for hours or days and responds to specific drugs.

3. Headache Associated with Drug Use or Drug Withdrawal

Drug-related headaches usually indicate that a chemical gained access to the body either through the skin or through swallowing or inhaling, causing a chemical imbalance. Such "toxins" may also give rise to classic attacks of migraine or cluster headache, but in many sufferers these headaches occur in the absence of focal neurologic deficits and in close relationship to the absorption of the offending substances. The pain is, as in other vascular headaches, pulsatile, and it may occur after ingestion of excessive amounts of alcohol or in particularly susceptible individuals after eating hot dogs, Chinese food, or frozen dinners, or after inhaling air pollutants. It also may occur in association with fever or infectious diseases such as influenza. Blood vessels are affected by triggers that cause relaxation of smooth muscle cells, thus causing dilatation of the vessels. The vessels dilate and, in turn, the pain-sensitive nerves surrounding them signal, in the form of pain, the abnormal diameter to the brain. Such headaches often respond to treatment with aspirin or similar drugs and often to physical manipulation, such as the application of ice bags to the painful area, a well-recognized treatment for hangover headaches. The application of ice bags tends to constrict the painful and swollen blood vessels, at least on the surface of the scalp, thus providing a nonchemical relief for the painful pulsatile vasodilation. In tension headaches, however, chemical treatment is sometimes warranted. Alcohol, the culprit in hangover headaches, can be used to advantage in treating tension headaches when it is consumed in moderation because alcohol temporarily blocks stress and allows the headache sufferer to relax the constricted muscles causing the pain.

Hangover headaches depend not only on the kind of alcohol consumed but also on how quickly and in what quantities it is drunk. Psychologic effects are also important because, characteristically, the pain does not start until blood alcohol levels decrease following breakdown in the liver and subsequent excretion. Moreover, because of intangible psychologic effects, the actual blood alcohol levels reached during any drinking bout have little relation to the subsequent hangover headache.

Most habitual coffee drinkers (those who routinely consume three or more cups of coffee a day) will suffer withdrawal headaches when coffee is unavailable. Caffeine constricts blood vessels, and in moderation may be useful for some migraine attacks, for example, or for hangover headaches. But, in those users who consume large amounts, the sudden withdrawal of the constricting effects of caffeine, as often occurs over weekends, may lead to

sudden and unaccustomed dilatation of blood vessels and headache. Such withdrawal headaches are rarely recognized for what they are and may sometimes account for what is known as the letdown headache of migraineurs, which also tends to occur at weekends at a time when many sufferers from caffeine withdrawal headaches are incapable of maintaining their blood vessels in a constricted state through ready access to the coffee urn.

4. Headache Associated with Substances Ingested or Inhaled

A Chinese restaurant syndrome has been traced to the liberal use of monosodium glutamate (MSG), which causes blood vessel dilatation, headache, and many other symptoms, such as chest pain. MSG is also liberally found in frozen TV dinners, canned meat, and powdered soups. In susceptible persons MSG, even in small quantities, may cause headache. If taken in large quantities, its vasodilatory capacity causes pain, even in subjects not usually prone to headache.

Inhalation of substances that trigger headache is not uncommon. These include nitrates and nitrites. Both compounds have a nitrogen atom and an oxygen atom and are used for the preservation of foods and for preservation of color in meats. Nitrates are part of the nitroglycerin molecule, an explosive sometimes used as a treatment of cardiovascular diseases. If used in medicine, it may be absorbed through the skin and cause vasodilation and complicating headache, a frequent affliction also of workers who handle nitrate in munition factories and on other jobs requiring the use of explosives. Such headaches are often accompanied by flushing of the face, an increase in heart rate, and lightheadedness.

Many air pollutants, and occasionally perfumes, may trigger headaches. Dyes and paints cause headache in some people; carbon monoxide from exhaust fumes or tobacco smoke in poorly ventilated rooms cause headache in others.

Solvents, paints, and paint removers are volatile vasoactive substances and may quickly cause severe headache because of their dilatory effect on blood vessel diameter.

5. Headache Associated with Infection

Infections of the eyes, ears, nose, mouth, teeth, and sinuses cause pain in the affected structures. A natural reaction to pain in these structures is stiffening of neck muscles and contraction of facial muscles, which in turn cause tension headache, making the infection-related pain doubly painful. Thus, pain in association with such inflammatory or other diseases of these parts of the head and neck have many of the associated features of tension headache, plus local pain in the primary structures affected.

6. Headache Associated with Metabolic Disorder

Patients having abnormal metabolism such as hypoglycemia (low blood sugar) or kidney disease may have a worsening of headache if they suffered before the onset of the metabolic disorder. Others have a new onset of headache in close relationship to the beginning of their metabolic disorder; the headache usually disappears within about 7 days after the abnormal metabolic state is corrected.

Headaches in this category include those associated with the hypoxia of altitude exposure, an altitude-induced metabolic disorder, or that occurring in association with pulmonary disease, which interferes with transfer of oxygen from the air to the blood. Some patients become hypoxic because of prolonged periods of apnea (lack of breathing) during sleep and also develop daytime headache. Occasionally, an increase in the arterial blood content of carbon dioxide without accompanying hypoxia can cause headache.

Those patients who become hypoglycemic, those on dialysis for kidney disease, and occasionally those without hypoglycemia who are fasting or who are undergoing plasma exchange may develop headache. These headaches usually disappear with correction of the metabolic abnormality induced by dialysis or plasma exchange.

7. Miscellaneous Headache without Structural Lesions

Tight head coverings can cause external compression headaches, also known as "goggle headaches."

Sexual activity can give rise to so-called coital headache, which may be dull or explosive and which occurs at the time of orgasm or immediately thereafter, when the headache sufferer resumes an upright posture.

8. Headache Associated with Exercise

Acute headache associated with exercise, straining, coughing, sneezing, laughing, or stooping can be a symptom of life-threatening intracranial disorders, but more often it results from poorly under-

stood benign causes related to exercise. Exertional headache must be distinguished from the more common aggravation of an established headache by exercise. Exercise aggravates vascular headaches of the migrainous type, but in some migraines and many cluster headaches it can be useful in aborting the attack. Benign exertional headache is more common in men than in women and is more frequent in older individuals. Benign exertional headache not associated with intracranial lesions is usually bilateral, frontal, occipital, or generalized. It is often abrupt in onset and severe; it may be sharp, stabbing, or pulsatile and last for several minutes after the exercise. The cause of benign exertional headache in nonathletes is obscure.

Disease-associated exertional headaches are rare, but they can accompany a number of serious disorders that increase the pressure within the skull. In headache associated with lesions within the head, headache may be induced by any kind of exercise, but its severity is related to the degree of effort. In some patients the headache is pulsatile and associated with visual disturbances; it may be mistaken for attacks of migraine induced by exertion.

Migrainous attacks can occur after athletic effort of any kind but tend to be more frequent at high altitude. Many such effort-induced migrainous attacks include only some of the symptoms. They may have just visual phenomena, usually occurring immediately after the exertion and remaining without subsequent pain. Nausea and severe pulsatile, unilateral pain indistinguishable from the classic variety of migraine may also occur. However, effort migraine is most common in poorly trained individuals, and dehydration, excessive heat, low blood sugar, and unaccustomed altitude place such individuals at particular risk.

III. Conclusion

Common headache is annoying but not life-threatening. It may cause disability if severe and frequent. Because most headache is aggravated by stress, the necessity to attend to the pain usually removes the sufferer temporarily from stressful situations. In this sense, headache may protect from prolonged and threatening situations that seriously jeopardize function. Most headache can be managed but not permanently cured. Only a few headache types have specific curative remedies.

Bibliography

Blau, J. N. (ed.) (1987). ''Migraine, Clinical, Therapeutic, Conceptual and Research Aspects.'' Chapman and Hall, London.

Dalessio, D. J. (ed.) (1987). ''Wolff's Headache and Other Head Pain,'' 5th ed. Oxford University Press, New York.

Rose, J. C. (ed.) (1989). ''New Advances in Headache Research.'' Smith-Gordon, London.

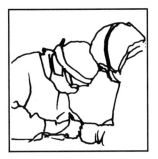

Health Care Technology

HAROLD C. SOX, JR., *Dartmouth Medical School*

Glossary

Effectiveness Results of using a technology under the usual conditions of medical practice
Efficacy Results of using a technology under ideal circumstances

HEALTH CARE TECHNOLOGY is a technology that health care professionals apply to their patients. This definition includes what physicians use in diagnosis, treatment, and prevention, as well as systems of health care. The Office of Technology Assessment of the United States Congress defines technology as "techniques, drugs, equipment, and procedures used by health care professionals in delivering medical care to individuals, and the systems within which such care is delivered."

I. Problems in Using Health Technology

The goal of using a technology is to apply it selectively to patients who will benefit most. The outcome of using a health technology is often unpredictable for several reasons. First, the individual's biological characteristics that determine the response to the technology are often unknown to the physician. Second, true diagnosis of the patient is often unknown at the time of a decision to use a technology. Consequently, some patients are treated unnecessarily, yet are subjected to the risks and costs of the technology.

Health technology is often costly. Whether or not this factor is important in the decision to use a technology depends on who must pay. The individual with comprehensive health insurance is likely to be indifferent, while the insurance company will be concerned, particularly in the case of extraordinarily costly technologies such as organ transplantation. In the aggregate, health technology is very costly in many countries. Some countries expend up to 11% of their gross national product on health care, while others expend only 7% of their gross national product or less. A crucial question is whether or not greater expenditure buys measurable differences in the quality of life.

The health effects of many health technologies have not been adequately characterized. Measures of outcome have been limited to the primary end point of the disease (e.g., blood pressure in studies of high blood pressure treatment; length of life in studies of cancer treatment). Studies have seldom included measures such as quality of life, functional status, and satisfaction with care. Many studies of new technology do not have a concurrent control group, which receives no intervention or receives the current standard intervention. This problem is particularly true of diagnostic tests and procedures. The advantage of using a very costly technology as compared with using a less costly technology is often unknown because the studies did not examine the relationship between cost and effectiveness. As a result of these deficiencies, physicians are unable to answer their patients' questions about the effect of a proposed treatment or test.

Physicians cannot always use what is known about a technology in their decision-making because other factors dominate their thinking in some patients. Patients sometimes request the use of a technology that is unlikely to benefit them, which creates a conflict between the physician's role as patient advocate and their responsibility to a society that is concerned about the costs of health care. Physicians sometimes are influenced by concern about being sued for malpractice when deciding whether or not to use a technology that has a small chance of benefitting the patient.

II. Methods for Evaluating Health Technology

Physicians and patients are concerned with using technology that is effective. The evidence on which the effectiveness of technology is based is derived from several different sources. These sources differ in the extent to which they conform to two requirements of an ideal study: validity and general application. A valid study is one whose conclusions are true. Validity is most likely to occur in a study of patients when any effects on clinical outcomes can be confidently ascribed to one and only one variable—the technology itself. To achieve this goal in a clinical study, one must control a number of extraneous factors that might affect the outcome of care. Studies of care under carefully controlled, ideal circumstances are called *efficacy studies*. The problem with such studies is that the conclusion, however valid, may apply only to the patient population that was used for the study. By restricting the study population to a narrowly defined group, one sacrifices general applicability in order to achieve validity. Ultimately, physicians need to know if a technology works in their patients. Studies of patient care under usual workaday circumstances are called *effectiveness studies*. In such studies, the study population is representative of usual practice, and the physicians are community physicians. Effectiveness studies are difficult to design so as to assure their validity. Thus, most studies of health technology compromise either validity or general application in order to achieve the other.

Several methods for evaluating health technology are described in the following sections.

A. Randomized Clinical Trial

In a randomized clinical trial of a new health technology, patients are assigned to receive the new technology or to receive an alternate technology. In a study of a new drug for high blood pressure, the patient receives either the new drug or a placebo that appears just like the new drug. Assignment to drug or placebo is random. If the study groups are large enough, random assignment assures that they are identical with respect to all variables, *known and unknown,* that might affect the outcome of care.

Blinding, by which the study participants are prevented from knowing the treatment assignment, is an important feature of the randomized trial. In a single-blind study, the patient does not know the treatment, but the doctor does know. In a double-blind study, neither the doctor nor the patient knows what the patient is receiving. Blinding should prevent the results from being foreordained by the patient's or the physician's preconceptions.

Two types of errors occur in interpreting the results of a randomized trial. The first type is called an alpha, or Type I, error. A Type I error means that a difference is assumed to exist when it does not exist. In other words, a Type I error is a false-positive trial result. The commonly reported "p value" for an observed difference between control and intervention groups is the probability that the observed difference is due simply to chance variation in two samples from the same universe. A low p value means that the observed differences are very unlikely to be due to chance and probably represent a true effect of the intervention. The second type of error is called a beta, or Type II, error. A Type II error is a false-negative trial result: There was a true difference, but the study failed to detect it. The reason for a Type II error is that the sample represented by the enrolled patients is too small to be representative of the universe of patients and has an atypical response to the intervention. Studies are designed to enroll a sufficient number of patients to reduce the probability of a Type II error to 5 or 10%.

The controlled clinical trial with randomized controls has several advantages. The most important is that, given a sufficiently large number of participants, the intervention is the only way in which the experimental and control groups can differ. This characteristic means that differences in outcome can be attributed to the intervention. The clinical trial is a well-accepted method for assessing a health technology; well-designed trials on topical clinical problems are, in the aggregate, quite influential. Clinical trials generate new data: A major clinical trial of coronary artery surgery led to nearly 100 journal articles in less than a decade.

Randomized clinical trials have a number of shortcomings. Clinical trials in chronic disease can be very costly, particularly if standardized patient care must be provided during many years of follow-up observations. If the number of outcome events is too small, a clinical trial may have a false-negative result, a Type II error. Thus, clinical trials are unsuited for the study of rare diseases, diseases with a low rate of outcome events, or interventions that are expected to have a small effect. Some studies take so long to accumulate sufficient trial end points that intervention has become obsolete long before the study has been completed. To maximize the chance of seeing an effect, a clinical trial may exclude patients that are likely to experience a trial end point for reasons unrelated to the intervention. For example, a study of coronary artery surgery excluded patients with known cancer because they might die of cancer before having a full opportunity to experience a trial end point. Such studies sacrifice general applicability to maximize the chance of observing an effect of the intervention. Finally, a randomized trial result favoring one treatment over another may not apply to all patients (e.g., some patients will do well with placebo treatment and experience fewer side effects than with an active treatment).

B. Meta-Analysis

Meta-analysis, or after-analysis, is a method for evaluating the weight of evidence as accumulated in many studies. The basic premise is that by combining several similar but small studies that showed no effect, you may enlarge the size of the pooled population to the point where a difference is seen or you can say with confidence that a difference does not exist.

Meta-analysis is performed in several steps:

1. Define a literature search strategy. Typically, one uses an electronically stored literature data base such as MEDLINE, which is maintained by the United States National Library of Medicine. Keywords from the title and abstract are stored in coded form in the MEDLINE data base. The choice of keywords specifies the search strategy. By publishing the search strategy, the author makes it possible for other investigators to duplicate the search.

2. Define criteria for excluding studies. The exclusion criteria define the population to which the conclusions of the meta-analysis will apply. Criteria may be broad (no articles in a foreign language; no studies of

animals; no review articles) or relatively narrow. All publications must state the exclusion criteria.

3. Identify poorly designed studies. The conclusions of poorly designed studies will be less valid than well-designed studies. One may exclude such studies or may analyze the results in subgroups defined on the basis of study design quality. Often, the best studies of a technology show a smaller effect on health outcomes than poorly designed studies.

4. Test for homogeneity of studies. In the ideal meta-analysis, all studies have as a study population a sample from the same universe of patients and use the same intervention in exactly the same way. While this ideal is unlikely to be achieved in practice, it is an unsound principle to combine studies that are very dissimilar. Such studies may be excluded by inspection or may be identified by tests for homogeneity of the pooled population. Typically, one uses a statistical test to see if the results of the separate studies differ significantly from one another. If the statistical test indicates differences, studies with discrepant results are eliminated one at a time until the test does not show a significant difference.

5. Pool the studies. The results of each study (the differences between each intervention group and control group) are combined using a weighted average. The studies may be weighted by sample size, study quality or any other characteristic. If the average is near zero, conclude that there is no difference between the two groups.

6. Perform a statistical test to confirm that the weighted average does not differ from zero just because of chance variation.

7. Perform sensitivity analysis. Sensitivity analysis is a test of the stability of one's conclusions. For example, one might pool all good quality studies and perform a statistical test. If the difference between the control and intervention group was small and no longer significant, one could conclude that the differences seen in the total population were due to poorly designed, perhaps invalid, studies.

C. Insurance Claims Analysis

Among the defects in prospective, controlled studies, two stand out: small sample size and highly selected study populations that sometimes bear little resemblance to the everyday practice of medicine. Recently, epidemiologists have turned to a new source of data: hospitals' claims for reimbursement for services rendered to a patient. When a patient who is enrolled in the United States government's Medicare system is discharged from the hospital, the hospital submits a claim for payment. This claim contains a code number for the patient's principal diagnoses and codes for major procedures that were performed during the hospitalization. These data are stored in electronic form in a Medicare computer file. Another Medicare file contains information on enrollees' vital status. One can assess the impact of an intervention by comparing survival at a specified interval after an admission for a particular disease for which one or another intervention was provided. For example, one can compare survival or incidence of stroke one year after a hospital admission in which narrowing of the blood supply to the brain was a coded discharge diagnosis and a patient either had an operation to relieve the narrowing or did not have the operation.

Analyzing insurance claims to measure the effectiveness of health technology has several advantages. The most important is the representativeness of the study population, which is the entire membership of the insurance program, and the providers, which include all participating hospitals and physicians. The breadth and size of this population is in sharp contrast to that for randomized trials, which are often small and narrowly selected. A second advantage is complete follow-up information. Randomized trials sometimes lose contact with patients and can say nothing about their clinical outcome; sometimes patients cannot be contacted because they have died, which is an important study end point. One can reliably discover if a patient is alive or dead from insurance claims files because the next of kin must report the death of a patient. A third advantage is the large number of patient records in a claims data base. In a very large data base, enough records are available from patients with a rare disease to study it. A final advantage of claims data bases is low cost: The data are collected for administrative purposes, and the additional cost of using them for research is slight.

Claims data bases have one very important disadvantage, which becomes evident when interpreting a finding such as longer survival in a group of patients who had a surgical procedure for narrowing of the arteries to the brain. Did the patients survive longer because of the procedure or because of other factors, such as overall good health, that led their surgeon to encourage them to have surgery? If many of the untreated patients had another disease, such as cancer, that limited the benefit that could be derived from surgery, their surgeon would not have encouraged them to have the operation. If many went on to die of cancer, survival would be shorter in the untreated patients even though the treatment

might actually be harmful. These extraneous factors can be taken into account by comparing only patients who did not have cancer, and these comparisons can often be done without losing statistical power because of the large number of patients in a claims data base. However, this method allows one to correct only for factors that are known and that are recorded in the data base. The advantage of randomization is that all confounding factors, known and unknown, may be taken into account. A second disadvantage of claims data bases is the relative paucity of data, which would allow one to examine clinically important relationships. This problem should resolve as technical advances lead to greater ease in storing complex clinical data electronically. A final disadvantage of current data bases is the lack of information on the patients' ability to function in daily life.

D. Consensus Conferences

A consensus conference is a panel of individuals who have been convened to render a joint opinion. For example, the panelists may be asked to read an article that describes the current state of the art. Often the article is comprehensive and is critical of the quality of the evidence. Many conferences proceed by means of several steps in which the panelists use a closed ballot to express an independent opinion, the results of the balloting are discussed, and a second vote is taken. This method, which is a variation of the Delphi technique, obtains an expression of group opinion while minimizing the risk that several vocal individuals will unduly influence the proceedings. This method identifies areas of controversy and areas of agreement.

Consensus conferences have several advantages. They are relatively inexpensive, as compared with studies in which new data are collected. They often lead to a description of the current practices of expert clinicians. Consensus conferences can be quite influential because the opinion of experts carries considerable weight with many clinicians.

Consensus conferences also pose some risks. Some formats place great emphasis on coming to a conclusion in a few days, which compromises critical discussion. The conclusions of consensus conferences may be biased by individuals whose opinion is influenced by factors other than scientific evidence. The validity of the conclusions are limited by the quality of the data. Unless the conferees refuse to give an opinion because of insufficient data, the conference may endorse a status quo that is based on poor-quality evidence. If the conferees are renowned and the conclusions are well publicized, the conference may retard the search for truth rather than advance it.

E. Decision Analysis and Cost-Effectiveness Analysis

Decision analysis is a method for simulating the events that follow a medical intervention. The method takes into account the uncertainty about when events will occur or whether they will occur at all. The method can also take into account patients' feelings about the outcomes that they might experience. Decision analysis can generate data, such as average length of life, that are similar to those that are obtained from randomized trials. Decision analysis uses data from published studies and is not a substitute for a randomized trial, which makes observations on patients' response to illness and treatment. Decision analysis does provide an answer in circumstances where a randomized trial cannot be performed or will take a long time to produce a result.

A model is a representation of reality. In a model of a decision problem, the events are typically represented by a tree structure. Events that are under the physician's control are represented by decision nodes. Unpredictable events (such as death following a surgical operation) are depicted by chance nodes, with a branch for each event that might occur. Each branch at a chance node is labelled with the probability that it will occur. The probabilities are obtained from published studies, from analysis of claims data bases, or from an expert clinician. Figure 1 is a decision tree for management of chronic pancreatitis, a disease in which inflammation of the pancreas causes abdominal pain. By calculating a weighted average value at each chance node starting with the tips of the tree on the right and working back to the left, one calculates an expected, or average, outcome for medical treatment and for pancreatic surgery. The average length of life spent in good health ("quality-adjusted life expectancy") is the measure of outcome in this model. The basic principle of expected value decision-making is to choose the decision alternative that has the largest expected, or average, value.

Cost-effectiveness analysis is a form of an expected value decision model in which both expected cost and a measure of clinical outcome, such as life

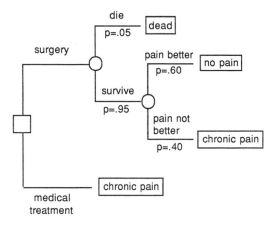

FIGURE 1 A decision tree for the management of chronic pancreatitis. The choices are a surgical procedure or medical treatment. The outcomes are perioperative death, resolution of pain, and continued pain. The □ is a decision node; the ○ is a chance node.

expectancy, are calculated. The results are expressed as cost per unit of clinical outcome. When comparing two decision alternatives, the difference in cost between the two alternatives (marginal cost) is divided by the difference in outcome (marginal outcome). Thus, an improved treatment might be compared with the usual treatment by the quotient of the additional cost of the new treatment and the additional length of life provided by it.

The advantages of decision analysis, as compared with randomized clinical trials, include low cost and the speed with which a conclusion may be reached. A decision analysis can be brought up to date by substituting new values for probabilities or changing the structure of the model as new data or treatment alternatives become available. A decision model may be used to develop a clinical policy that can be applied to many patients or can be individualized to a particular patient by substituting patient-specific values for the parameters of the model.

The role of decison analysis in the hierarchy of methods for assessing technology is to provide new insights about old data by casting them in the perspective of making a decision. However, the parameters of the model are obtained from prior observations of patients, and the validity of the conclusions are limited by the quality and applicability of these data. The problem of uncertainty about the true value of a variable in the model is approached by a technique called *sensitivity analysis*. The expected value of each decision alternative is recalculated as the value of the uncertain variable is given first one value and then another over the entire range of possible values. If the preferred decision alternative is the same in all cases, the variable is not an important determinant of the decision. Sensitivity analysis identifies the variables that are really important to decision-making and focuses the research agenda on obtaining better information about these variables.

III. Systems for Evaluating Health Technology in the United States

Systems for evaluating health technology are in a state of change. Professional organizations participate because they have a responsibility to educate physicians. They evaluate technology to provide their membership with current information about how to use technology effectively. The American College of Physicians is an example of a professional society with a very active program of technology assessment. The federal government and others who pay for health care are motivated to discover which medical practices are appropriate and should therefore be reimbursed. The Medicare system in the United States has a program to obtain advice about what practices should be reimbursed, as does a private insurer, the Blue Cross–Blue Shield Association.

IV. The Spread of New Technology

Several factors appear to promote the adoption of new technology. The theory underlying the technology should be consistent with prevailing medical theory. A technology that is easily learned, such as the use of new drug, may be more quickly adopted than a complex procedure that requires manual dexterity and long practice. A technology that deals successfully with a common and urgent clinical problem is especially likely to be adopted. A strong and influential advocate in the professional community speeds adoption of new technology. The practice setting can have an influence, especially if it is one in which physicians control the decision to acquire a new technology. In light of these complex influences, the effect of a careful evaluation of the technology may have less effect than might be expected, especially if the study results receive little publicity.

V. Future Directions

A. New Sources of Data

Technology evaluation has long been synonymous with carefully controlled studies of narrowly defined populations. This paradigm is being complemented by population-based studies that use data sets, which, although collected for administrative purposes, can serve the needs of clinical research. Patients are taking a greater role in deciding what care they will receive, and the study of technology now includes formal evaluation of patients' preferences for the outcome states they may experience.

B. New Measures of Effectiveness

The clinical outcomes that serve as the end points for a study are the measures of effectiveness. Past studies have used survival or a measure of disease activity as the sole study end point. As competing technologies become more powerful, they often differ little in their effects on disease and survival, and the choice between them will depend on outcomes, such as the ability to function in society physically, emotionally, and socially. Current studies of technology are starting to use these secondary measures.

C. New Methods to Disseminate Information about Technology to Physicians

The most important unsolved problem in technology evaluation is how to influence the physician to adopt a new practice style. Some evidence indicates that professional organizations can create the conditions in which change can occur. The practice setting is another impetus for adopting an appropriate practice style. More and more practice organizations provide comprehensive care in return for a single advance payment. Physicians in these "health maintenance organizations" often have a strong motivation to adopt clinically effective patient care policies that minimize expense.

D. The Role of the Computer

The computer-based administrative data sets are already playing a major role in technology evaluation. Soon, all of a patient's clinical data will be stored in a computer rather than in the traditional paper medical record. This development will mean a very rich data source for clinical studies of technology. The computer will probably soon play a role in guiding the selection of technology for a patient, because the computer can present the results of technology assessments to the physician at the time that a decision is required.

Bibliography

Bunker, J. P., Barnes, B. A., and Mosteller, F. (1977). "Costs, Risks, and Benefits of Surgery." Oxford University Press, New York.

Feeny, D., Guyatt, G., and Tugwell, P. (Eds.) (1986). "Health Care Technology: Effectiveness, Efficiency, and Public Policy." The Institute for Research on Public Policy, Montreal.

Fineberg, H. V., and Hiatt, H. H. (1980). Evaluation of medical practices: The case for technology assessment. *N. Engl. J. Med.* **303,** 1086–1091.

Kassirer, J. P., Moskowitz, A. J., Lau, J., and Pauker, S. G. (1987). Decision analysis: A progress report. *Ann. Int. Med.* **106,** 275–291.

Light, R. J., Pillemer, D. B. "Summing Up: The Science of Reviewing Research." Harvard University Press, Cambridge, Mass., 1984.

Sox, H. C. (Ed.) (1987). "Common Diagnostic Tests: Use and Interpretation." American College of Physicians, Philadelphia.

(1985). "Assessing Medical Technologies." National Academy Press, Washington, D.C.

(1989). "Medical Technology Assessment Directory: A Pilot Reference to Organizations, Assessments, and Information Resources." National Academy Press, Washington, D.C.

Health Psychology

GEORGE C. STONE, *University of California, San Francisco*

Glossary

Behavioral contracting Procedure in which provider and client mutually negotiate a set of goals for behavior change on the part of client and explicit rewarding or punishing consequences for fulfilling or failing to fulfill the contract

Cognitive processes Manipulation of information by the brain (i.e., interpreting, storing, retrieving, comparing, linking, etc.)

Conditioning Fundamental process, presumably in the central nervous system, whereby changed patterns of response to specific situations are brought about by temporally contingent relations between environmental events and behavior; occurs in many phyla of multicellular animals

Holism Idea, in this context, that animal organisms demonstrate emergent properties not predictable from knowledge of their component subsystems

Long-term versus short-term memory Symbol memories of humans and higher mammals. Long-term memory retains information for years, short-term, a few seconds, without rehearsal; long-term, vast amounts, short-term, about seven items

Pain behavior Behavior typically performed when painful stimuli are applied: verbal and nonverbal complaints, impaired functioning, and relief-seeking actions, such as self-medication

Psychoneuroimmunology Interdisciplinary field of research that demonstrates and seeks to explain influences on the immune system by cognitive and emotional processes

Psychosexual development Development of sex-related beliefs, attitudes, desires, behaviors. In psychologies deriving from the work of Freud, an elaborate sequence centering on parent–child relations has been proposed

Retrospective versus prospective design In psychological epidemiology, seeking to establish causal relations from recall or records of a group characterized by a consequent event, versus following a group possessing the antecedent attribute to see if the consequent occurs

IN 1980 the Division of Health Psychology of the American Psychological Association formally adopted a definition of a new field of specialization in psychology:

> Health psychology is the aggregate of the specific educational, scientific, and professional contributions of the discipline of psychology to the promotion and maintenance of health, the prevention and treatment of illness, the identification of etiologic and diagnostic correlates of health, illness and related dysfunction, and the analysis and improvement of the health care system and health policy formation.

In short, health psychology is the application of any aspect of the science and profession of psychology to problems of health and illness.

I. Antecedents

Although the term *health psychology* was first used in the 1970s, work that falls within the definition given above has been going on since the term *psychology* was coined in the 16th century and, in fact, since long before that.

A. Clinical Psychology

In the present century, the field of clinical psychology has developed to apply psychological knowledge and techniques to health problems. For a variety of reasons, until recently almost all the work of clinical psychologists was concentrated on "mental health problems," in which the behavior or emotional experience of the client/patient defined the issue to be diagnosed and treated. Although the first psychological clinic was established in the United States in 1896, the major growth of clinical psychology began after the Second World War, when a massive program was initiated and supported by the National Institute of Mental Health to prepare clinical psychologists to treat returning veterans and members of the armed forces. By 1970 there were 3,662 members of the Division of Clinical Psychology in the American Psychological Association, concentrating their effort on problems of psychotherapy, schizophrenia, mental retardation, and alcoholism. It was at about that time that note began to be taken of the neglect of the psychological aspects of all the other issues of health and illness and efforts began to redress the balance.

B. Medical Psychology

Unlike clinical psychology, "medical psychology" never developed into a regulated field of practice nor did it give rise to major professional organizations. Perhaps as a consequence, several different kinds of activity have been going by this name. In Great Britain and associated countries, medical psychology was almost synonomous with "psychiatry"—the treatment of mental illness. Throughout most of Europe, and to some extent in the United States, the term refers to the psychological aspects of medical care, emphasizing the experience of the patient and the "humanizing" of the doctor–patient relation. Medical psychologists of this kind mostly teach physicians-in-training to be attentive to these aspects of their work. Yet a third usage, found in the United States and elsewhere, refers to the practice of clinical psychology in medical settings or with patients suffering from problems of bodily disease. The distinction between this version of medical psychology and the clinical practice of health psychology or behavioral medicine is not precise.

C. Behavioral Medicine

The most widely accepted definition was an outcome of the Yale Conference on Behavioral Medicine, held in 1977, and a second conference sponsored by the Institute of Medicine in 1978:

> Behavioral medicine is the interdisciplinary field concerned with the development and integration of behavioral and biomedical science knowledge and techniques relevant to health and illness and the application of this knowledge and these techniques to prevention, diagnosis, treatment, and rehabilitation.

The scope encompassed by this definition is broader in some respects than that of health psychology, because it includes contributions from other behavioral sciences such as sociology and anthropology and also from biomedical sciences. Implicit in the definition, however, is a concentration on the biomedical aspects of health problems, an emphasis that has generally been characteristic of both research and practice in the field. Psychologists who refer to their practice as "behavioral medicine" usually employ behavioral techniques to address a medical problem *per se,* with a view to altering some aspect of bodily function either directly or indirectly. Those who call themselves "medical psychologists" may engage in such work also, but they are more likely to include concern for the individual patient's efforts to cope with the illness and treatment as a human problem—a problem of living—and perhaps to engage in psychological diagnostic procedures to help fit medical procedures to the particular person to whom they are to be applied. Conceptually, both psychologists who engage in behavioral medicine and those who go by the name medical psychologist are health psychologists, although they may not so identify themselves nor belong to the organizations of health psychologists.

D. Psychosomatic Medicine

Another line of development that began even earlier than clinical psychology was "psychosomatic medicine." The term was introduced in 1818 by a German physician, Heinroth, who was concerned with the role of internal conflict in causing mental disease. Active development of the psychosomatic concept began in the 1920s, when physicians, mostly of psychoanalytic persuasion, began to collect case histories and speculate about the relation of the emotions and symbolic representations of inner conflicts to bodily disease. To varying degrees, the ideas of three major theorists were interwoven to create sometimes fanciful explanations for several "psychosomatic diseases" (e.g., asthma, stomach ulcers, and rheumatoid arthritis). Sigmund Freud's concepts of psychosexual development were linked with Ivan Pavlov's discoveries about the processes of conditioning, and physiological mechanisms were proposed using Walter Cannon's "wisdom of the body" approach to physiology.

The concept of psychosomatic diseases reached its peak in the 1940s and early 1950s. Then increasing methodological rigor revealed that the relations were more general and less symbolically colorful than the early workers had portrayed them. Emphasis shifted from specific psychosomatic diseases to a recognition of the great importance of psychological phenomena in most disease processes. This idea was expressed as early as 1950 by George Engel, a quarter of a century before he coined the heuristic term *biopsychosocial model* to refer to it. This viewpoint now pervades behavioral medicine and guides the work of most health psychologists who work at the mind–body interface. The term *psychosomatic medicine* persists to refer to a special emphasis on holism and on symbolic influences within the body of behavioral medicine.

II. Origins of the Field of Health Psychology

Although work now seen as health psychology had been going on for many years previously, the recognition that it represented an appropriate field of psychological specialization emerged only in the 1970s. Psychologists, noting the lack of work on most aspects of health and illness, began to engage in new kinds of research and to develop new forms of intervention. From the learning laboratories came techniques, known as biofeedback, that allowed people to gain some voluntary control over behavior, such as the regulation of heart rate or skin temperature, that had traditionally been considered involuntary. Social psychologists began to study a phenomenon known as "noncompliance," i.e., the failure of patients to follow physicians' recommendations for which they had often paid dearly in time, trouble,

and money. As more and more psychologists began to work on the many problems they encountered in the new environments of the health system, they organized, first as a task force, then as a Section on Health Research in one of the divisions of the American Psychological Association, and finally, in 1978, as an independent Division of Health Psychology. Similar actions were taking shape in other parts of the world. A Cuban Sociedad de la Psicologia de la Salud was formed in 1974. A Task Force for Health Psychology was established in the Interamerican Society of Psychology in 1982 and a European Society of Health Psychology in 1987. A truly international enterprise has grown up, emphasizing health as much as illness and prevention as much as cure, based on a universal recognition of health as a primary human value.

III. Scope of the Field

Health psychology represents the overlay of psychological theory, data, and methods on the problems of the health system. These problems include not just the diagnosis and treatment of disease and injury, but identifying and responding to environmental hazards to health; fostering health-promoting behavior on the part of the public; supporting persons in their appraisals of their symptoms and their decisions as to what should be done about them; minimizing the discomfort and ardors of medical treatments that patients must undergo and helping them to cope with the irreducible stresses and to adhere to their treatment regimens; providing programs and facilities for those with residual disabilities after treatment; educating health professionals; and planning, financing, and staffing appropriate health care facilities and organizations.

Health psychology, like its parent discipline, psychology, encompasses both research and practice. In a recent comprehensive survey made by the American Psychological Association, the percentages of health psychologists affiliated with its various divisions paralleled quite closely the overall distribution of members. Two-thirds of the division members listed research as a major component of their work and about the same percentage said they were engaged in educational activities, while four-fifths said they were engaged in delivering health services. These figures indicate that most health psychologists are involved in more than one kind of activity. As primary activity, 39% listed faculty po-

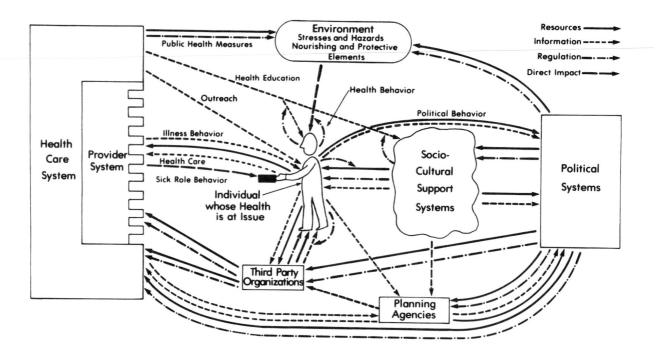

FIGURE 1 The health system.

sitions, which typically involve both research and teaching, 48% said they were engaged in the delivery or administration of health services, whereas only 6% reported research as their primary activity. These figures are also close to those for the Association as a whole, with a somewhat higher proportion of health psychologists in faculty positions.

Figure 1 diagrams some of the relations in the health system in which psychological processes influence health outcomes.

To recapitulate, health psychology, in principle, relates to all phases of the health process from hazard abatement and avoidance to rehabilitation after illness or trauma. In principle, it relates to all aspects of the health system (e.g., financing, training of personnel, design of organizations, and health care delivery). It is also concerned with all parties to health transactions, including persons at risk, their families, health care providers, organizational managers and support staff, and policy makers.

IV. Major Research Issues

Even though the domain of health psychology, as outlined here, is broad, most of the work actually undertaken in the early years has been concentrated in a relatively few areas. Examples of research are given in this section and examples of interventions in the next.

"Health" refers usually to some aspect of the physiological functioning of the body. "Psychology" is concerned with the behavior and experience of individuals. A fundamental issue in health psychology, therefore, concerns the mechanisms that link the physiological with the psychological. Many details of such linkages are mentioned in the sections that follow. A brief overview may help to introduce them.

Causal relations between health and behavior can be organized in either direction: Behavior can have health effects, and health can affect behavior. Pathways by which behavior can change the health status of an individual can be broadly grouped as follows:

1. Behavior can lead to physical trauma, broken bones, contusions, concussions, etc. In some cases, the relation is truly accidental, but there are also persons who deliberately expose themselves to situations of high risk for various psychological reasons. Some earn their livelihoods in hazardous occupations; some build traditional houses with unreinforced walls and heavy roofs; some crave excitement and a sense of danger. On the positive side, some persons take precautions against injury by wearing seat belts or protective helmets, making plans and preparations for natural disasters, and supporting regulations or legislation to remove or mitigate hazards in the environment.

2. Behavior can lead to the penetration of the body by toxic or pathogenic materials. The exposure may be deliberate, as in substance abuse and high-risk sexual behaviors, or unintended, as when someone buys a home on a site once used for the disposal of toxic wastes or in an area infested by insect vectors for serious diseases. Behavior taken to avoid exposure to or damage by such materials is properly considered preventive health behavior. Oral hygiene reduces the damage to teeth and gums of the bacteria that are naturally present in the mouth.

3. Behavior can affect the bodily condition of the individual through the choice of the amount and kinds of foods ingested, the amount of physical exercise engaged in, and the care taken in conjunction with such exercise to avoid injuries from excessive strain or improper postures and movements. Much of the current emphasis on "healthful life-styles" is directed toward behavior in this category.

4. Behavior can greatly influence the level of activity in the body's stress systems: the autonomic nervous system that regulate glands and organs, and the immune system that responds to pathogenic intrusions and novel substances that may appear in the body. Stress can be increased by life choices and by the ways in which one perceives or construes life events. Stress can also be reduced by certain practices, including various forms of meditation or contemplation. It may also be possible to reduce stress through the adoption of different frameworks for interpreting life circumstances.

In relation to each of these pathways, health psychologists attempt to understand the ways in which the risks are perceived and to devise and evaluate methods for communicating accurate understanding of the risks. They study decision-making processes and the determinants of risk-taking behavior. In relation to the stress pathways, they study the psychophysiological processes involved and the appraisal and coping mechanisms associated with greater or lesser physiological response to comparable stresses.

An individual's health status can also have profound effects on his or her behavior. Several psychologically meaningful attributes of such effects can be described.

1. Duration and mutability: Most conditions of disease, illness, or injury cause some degree of physical limitation for shorter or longer periods. In some cases it is of only a few days duration and in others it is present from birth and lasts for a lifetime. Closely related to this characteristic of the limitation is the degree to which it can be expected to change. Some problems are transitory and can be expected to pass without special

(psychological) intervention. Some are totally unchangeable, and some can be changed to varying degrees by varying expenditures of effort, time, and other resources.

Health psychologists study the impact of duration of the limitation and of its time of onset on the total reaction of the person and on the design of interventions to limit the degree of incapacitation.

2. Type of limitation: Limitations may involve sensory systems, as in blindness or deafness, paralysis of motor systems, amputation of limbs, lesions of the brain that impair mental functioning, or metabolic disturbances that reduce the capacity to engage in physical activity or, alternatively, that lead to hyperactivity that impairs concentration and focus. Seizure disorders may restrict the kinds of activities that can be undertaken because of the possibility of serious consequences that could follow from a seizure episode while driving, climbing, or performing highly skilled work. Some health conditions are disfiguring or otherwise stigmatizing; their effects may be primarily mediated through the reactions of others—actual or anticipated—to the health problem. Psychologists may be involved in the development of physical prosthetic devices to circumvent such limitations (e.g., sensory transducers that encode visual stimuli in a form so that the ears or the skin can receive information about them). They may, however, be concerned with emotional and attitudinal reactions to the limitations and with finding ways to minimize the secondary effects of such reactions.

3. Increasing versus decreasing limitations: Limitations on behavioral capacities from health causes vary greatly in their time courses. Some appear suddenly in full force, as in the case of strokes. Others, such as Alzheimer's disease, come on gradually over a period of years. Reactions of patients and of those around them are different as a result of variations in this attribute. There may also be substantial difficulties arising from the removal of limitations. Persons recovering from heart attack often retain psychologically based limitations that are not warranted by the degree of their recovery of cardiac function. Persons relieved of long-term obesity or facial disfigurement through surgical procedures may undergo catastrophic disturbances of their life adaptations and require significant psychological intervention to reestablish equilibrium.

Much more research has been focused on those issues in which psychological factors influence health than on those in which the health status of the individual has psychological sequelae. There are, however, a great many instances of each. In this article it is possible to present only a few representative examples.

A. Prevention of Injury and Illness

As costs of health care continue to grow, reaching more than 10% of the gross national product in recent years, recognition is increasingly being given to the idea that prevention costs less than curing—in monetary terms as well as in terms of human suffering. Most people agree with this viewpoint, but nevertheless, substantial difficulties need to be overcome in developing effective programs of prevention.

Of course, before one can develop methods for preventing health problems, one must understand how they come about. The pathogens, toxins, and other hazards of the environment must be causally linked to the injuries and diseases that we seek to prevent. Geneticists, epidemiologists, physiologists, structural engineers, and many others, including psychologists, contribute to our understanding of etiology (i.e., the causes of illness) and to the discovery of interventions that could mitigate the harmful effects of the hazards identified. Without exception, however, when a cause has been discovered and a potentially effective intervention has been designed, someone's behavior will have to be changed before prevention can be accomplished. In many cases, it is the behavior of the person whose health is at risk that must change. People must wear seat belts, stop smoking, change their sexual practices, and change the ways they clean their infants or build their houses. Changing behaviors of health care providers can also be difficult, even in an era characterized by technological progress. In other cases, it is the behavior of persons in corporate entities that must change, reducing the release of toxic substances into the environment, the sales of dangerous products, or the imposition of health-damaging working conditions. Sometimes changes involve the passage of legislation or the adoption of administrative regulations, in which case, the behavior of legislators and other policy makers must change, and perhaps so must the behavior of voters. In every case, however, the ultimate change process occurs in the behavior of individual humans, and the psychological processes of behavior change are involved.

A special characteristic of most preventive behaviors makes them difficult to establish and maintain. In most cases, it is necessary to replace behavior that provides more immediate gratification with behavior that may have little or no reward other than the knowledge (or belief or hope) that it may reduce the likelihood of some unfavorable health condition or event at some time in the future. Psychological research on prevention is mainly concerned with discovering ways to overcome barriers to behavior change so that well-established principles and procedures can be applied.

B. Stress and Illness, Psychoneuroimmunology

For more than 50 years there has been increasing understanding of the ways in which our individual appraisals of what happens to us give rise to responses in our nerves and glands that influence our bodies in many ways. Among other things, these responses alter our digestive processes; our metabolism of sugars, fats, and other substances; the functioning of our hearts and blood vessels; and the growth and distribution of the blood cells that help to protect us from invading bacteria and toxic materials such as plant pollens.

Between the conditions and events that occur in our environments, as they might be recorded by cameras and scientific instruments, and the physiological mechanisms of our body, there are psychological processes known as "stress appraisal." We all know that an event that may be devastating to one person may be handled with equanimity by another. Psychologists seek to characterize the properties of situations and of individuals that give rise to these differences. The psychological essence of the stressful situation is that it threatens the integrity of the individual, however that may be conceived. People attempt to cope with stressful situations in two major ways. These have been termed *emotion-focused* and *problem-focused* coping, depending on whether the individual tries to reduce the distressing emotions aroused by the threat or attempts to mitigate the source of the stress. Depending on the nature of the stressful situation (i.e., whether it can be altered by human effort), one or the other of these methods may be more effective in minimizing the impact on the person. Interactions between these psychological mechanisms and the physiological mechanisms of the stress response are extraordinarily complicated. In the new field of psychoimmunoneurology a trend in the data suggests that when coping mechanisms involve repression of (i.e., failure to express) emotion, suppression of the immune system may be greater. However, recent studies also suggest that hyperreactivity of the emotional system may have deleterious effects on the heart. [See PSYCHONEUROIMMUNOLOGY; STRESS-INDUCED ALTERATION OF GI FUNCTION.]

C. Personality and Illness

The idea that certain kinds of people were especially prone to develop particular diseases was, as previously noted, a major preoccupation of the field of psychosomatic medicine in the middle of this century. As more sophisticated kinds of statistics were applied and larger groups of patients studied with better sampling methods, the inductions that had been put forward proved difficult to substantiate. There are, however, continuing indications that differences in personality are related to health.

One of the best known examples of this kind of association concerns the relation between the "type A personality" and the incidence and course of coronary heart disease. According to the clinical observations of two cardiologists, Meyer Friedman and Ray Rosenman, first published in 1959, the type A, "coronary-prone" individual was competitive, achievement-oriented, time-urgent, and hostile and given to rapid speech and movements. A "structured interview" was developed as a means for identifying the type A individual, and decades of research were initiated by the finding of substantial differences between type A's and others in large epidemiological studies of initial and repeated heart attacks. Much of this research was directed to finding the critical components of the type A syndrome. It appears at this time that a critical factor is a disposition of "cynical hostility," although this variable does not exhaust the predictive value of the full type A measure. Two avenues of mediation are being investigated: One posits that hostile individuals show greater cardiovascular reactivity in conflict situations; the other is that they induce more hostility in others and thus provoke more negative life situations while at the same time they have less social support from relations with others.

Another personality characteristic that has been the subject of increasing research has to do with an overall outlook that is optimistic versus pessimistic. A preponderance of studies that have looked at this question has found that optimists are healthier. Martin Seligman proposed one of the most analytic approaches to this area and looked at three dimensions of optimism/pessimism: The pessimist is prone to see unfavorable events as resulting from *internal* causes arising within the person, from causes that are *global,* affecting wide reaches of behavior, and from causes that are *stable,* not likely to change much from time to time or situation to situation, whereas to the optimist the causes are more *external, specific* to the situation, and *changeable*.

D. Decisions in Health Care

Every phase of the health care process is fraught with decisions. Persons at risk and patients must decide about their health behaviors, what risks they will take, whether their symptoms require expert care and, if so, where to seek it, and whether to follow the recommendations made by the expert consulted. On the other side, a major component of the expertise of health professionals resides in their capacity to decide about the patient's diagnosis and what treatment to recommend.

For many years, psychologists have been studying the ways in which people make decisions, and health psychologists have been active in applying and extending their findings to the decisions of the health system. Much of this work has been based on a model that treats the person as rational—seeking to optimize outcomes under conditions of limited information. Within this model are two principal issues: (1) How one can best estimate the probabilities of various states of the world (e.g., that a person does or does not have a particular disease), and the probabilities of various outcomes given a particular state of the world and a particular course of action (e.g., the probability that a cardiac catheterization will yield important diagnostic information, given the patient's presenting symptoms, versus the probability that the patient will die during the procedure); and (2) how the values of various outcomes can be compared. How does one weigh the dollar cost of a treatment plan against the possibility that it will remove a particular symptom or reduce the chance of a serious health problem at some time in the future?

When there is sufficient time and ample resources, estimation of probabilities is the business of statisticians, although they require the input of psychologists as well as medical researchers whenever human behavior is a part of the picture. For example, it has been determined that the choice between a surgical and a pharmacological approach to a certain kind of hypertension depends on one's estimate of the probability that patients will faithfully follow their prescribed drug regimens. The problem of assigning numerical values to human values for purposes of formal decision making is one that psychologists and philosophers have struggled with without much success. In the end, it simply comes down to a careful measurement of human preferences. However, such measurement is no easy matter. Policy makers decide how the preferences of various persons and groups should be weighted, but psychologists are needed to deal with difficulties in getting accurate, stable, and consistent expressions of an individual's preferences.

For most of the day-to-day decisions that are made, however, patients and health professionals do not call on statisticians, philosophers, or policy makers for help. Sometimes they use approximations to the rational decision model, subject to limitations of the human mind in knowledge, long-term and short-term memory, and their understanding of the model itself. Often, irrational factors enter in. Important among these are the suppression of information that gives rise to discomfort when contemplated, and the rejection of mental calculations that are lengthy and effortful. Health psychologists study the operation of such tendencies in patients and professionals and seek ways to overcome them.

E. Interpersonal Issues in Health Care

As in every aspect of human life, the interactions among people play crucial roles in the health system. Three kinds of interaction may be considered here as examples. Most studied, because they are basic to all health care, are those between patients and the workers in the health care system. These represent a microcosm of the society at large, but there are many special characteristics. In the case of patient–provider communication, there is an agreed on difference in expertise, with clearly differentiated roles.

1. Provider Patient Interaction

One useful way of construing the situation is as a joint problem-solving situation. The patient brings the problem and presents it to the provider. The provider uses his or her expert knowledge to elicit further information, achieve a diagnosis, decide on possible forms of treatment, and present one or more options to the patient. The patient, in most situations, accepts or rejects the recommendation and follows or fails to follow the prescribed plan. The results of this sequence of actions are then evaluated, and if necessary, additional information is sought or new options considered. There are

many possible variations on this basic plan. A widely accepted classification, first proposed in 1956 by Szasz and Hollender, divides them into three models: (1) active provider–passive patient; (2) guidance–cooperation; and (3) mutual participation. These models differ in the degree to which there is continuing two-way communication throughout the transaction and the degree to which responsibility for decisions is shared. Although psychologists and other social scientists have tended to favor the mutual participation model on philosophical grounds, research has made it clear that different models are appropriate for different situations. One important determinant is the expectation by the patient and the preparation for assuming responsibility. Problems arise when provider and patient are applying different models to the transaction without knowledge that they are doing so.

2. Social Support

A second aspect of interpersonal relations that has been much studied by health psychologists is that of "social support." People do turn to each other for support. Some people are surrounded by networks of relatives and friends; others live alone and may have few or no significant persons in their lives. In general, people with strong networks of social support, particularly those with loving spouses, are healthier, live longer, adhere to treatment regimens better, and cope better with chronic or life-threatening illness. Early studies tended to focus on the quantity of support (i.e., the size of social networks). Increasingly, it is recognized that the quality of the support is at least as important; some families, typically counted as support, might be better considered as sources of stress. Quality of support is much more difficult to appraise than quantity.

3. Informed Consent

An ethical principle increasingly important in the health care system is that of "informed consent." In its strongest form, this doctrine holds that no procedure should be performed on persons until they have full information about the possible consequences and an opportunity to reflect on the implication of this information for their own lives and those of others. Only then should consent be given, and if it is not, the procedure should not be undertaken.

Several psychological problems arise in conjunction with this concept. How much information does one need to make the best possible decision? How can a person without medical training understand highly technical information? How can one anticipate what it would be like to be in a condition one has never experienced or known anyone to have experienced? What should be done when a patient asks the health care provider to make the decision? Should we insist on informing people who prefer not to have the information? What happens when a person with inadequate understanding of probability is informed about rare but serious side effects of a treatment—do they make better or worse decisions? For that matter, how do we compare the quality of decisions made? Psychologists have a good deal of information about "postdecisional cognitive processes" that point to major difficulties in assessing satisfaction or regret concerning the consequences of actions taken. Health psychologists cannot provide definitive answers to many of these questions, which mostly involve significant value issues, but they can help to determine the probabilities of alternative outcomes of the consent process based on analyses of the situations and the participants.

4. Cross-Cultural Issues in Health Psychology

Cultural beliefs and traditions are associated with a range of health-related behaviors, from diet to meditative practices. There is a wealth of evidence that certain religious groups, such as Mormons and Seventh Day Adventists, are generally healthier than the population of the United States as a whole. This difference can be readily explained by their lower use of alcohol and tobacco and, in the case of Seventh Day Adventists, their lower consumption of saturated animal fats. Japanese who live in Japan have lower incidence of cancer of the bowel and a higher incidence of cancer of the stomach than persons of Japanese ancestry in the United States who have adopted the American culture. Although bodily effects of dietary practices are not a subject for psychological research, they point to the possibility of other, more psychological differences between cultures that may be discovered through psychosocial research. Cultures differ, for example, in their "display rules" as to the degree to which emotions may be expressed in social intercourse. Some psychological theories have proposed that suppression of emotion can have deleterious health consequences. Thus, a culture's display rules for emotional effects might have health effects. Beliefs of different religious groups about the origins and

meaning of suffering may influence the seeking of health care and the experience of pain.

V. Major Forms of Intervention

In terms of activities performed by health psychologists, the field is at least as much a form of practice as it is a domain of research. The practice of psychology spans a wide range of approaches from precisely controlled techniques addressed to some specific component of an individual person's behavior to participation in the design of large-scale systems. In the broadest terms, psychological practice has one of two purposes: It may attempt to describe or predict—to *assess*—how individuals act or feel in specified situations, or it may attempt to *change* the ways in which they behave or feel. Although these two purposes are conceptually related, the procedures by which they are accomplished are altogether different.

A. Assessment in Health Psychology

Assessment, in the context of psychological practice, attempts to answer three kinds of questions: (1) what needs to be done to improve the condition of the client(s)? (2) which of the possible approaches to the problem thus defined is most likely to work with this particular person or group? and (3) are the selected interventions accomplishing the kinds of change that was intended? The fundamental problem of assessment in health psychology is that of describing psychologically relevant attributes of persons and environments that can predict what interventions will lead to improvements in the health of the target persons or populations. Because health psychology encompasses all aspects of psychology, the range of its assessment techniques is correspondingly broad. A few examples must suffice as illustrations.

1. Assessment of Health Status

Basic to many problems is the need to specify the "health status" of individuals. This has proven to be difficult. Many persons from many disciplines have tried their hand at defining health and developing numerical indices to represent it. The difficulty arises because health is not a uniquely valued state, but a collection of statuses on multiple dimensions. A person who is able to carry out the activities of daily living is healthier, all other things being equal, than the one who is not. Similarly, one who is free of pain or other troublesome symptoms is healthier. A person who can run far and fast, jump high, and lift heavy weights is generally considered healthier than one who cannot, even though neither may expect or care to engage in such activities. A body that can resist disease or injury is healthier. A community whose members engage in socially valued work is usually said to be healthier than one characterized by crime, violence, and vandalism.

Although there is general consensus as to the healthy direction on most of these dimensions, there are substantial differences in the weight assigned to each in an overall judgment of health. Health psychologists in collaboration with other social scientists have employed their techniques for measuring preferences to bring maximum clarity and precision to what remains a difficult and cloudy area.

2. Measurement of Pain

A persistent and difficult problem in health care has been that of evaluating patients' pain reports, particularly in those conditions in which the pain is chronic and cannot be correlated with any tissue damage. Based on their tradition of working with subjective experience, psychologists have been able to make some contribution to the recognition that such pain is "real" and deserving of professional attention. Behavioral methodologies developed to objectify the subjective have also provided some assistance in the understanding and treatment of pain conditions. In many pain clinics this has taken the form of shifting emphasis from the treatment of pain to the treatment of "pain behavior." [See PAIN.]

3. Psychological Risk Factors

Research demonstrating psychological differences between those who succumb to many different kinds of health problems, from relapse in smoking programs to myocardial infarctions, has given rise to the concept of "psychological risk factors." There is a logical progression from retrospective discoveries to prospective verification to the development of instruments to assess risk and ultimately to the attempt to modify the behavior that indexes risk and the psychological characteristics that give rise to such behavior. Two current areas of active interest are the investigations of the type A, coronary-prone personality, already described, and the "addictive personality." There are by now a few

studies suggesting that having identified a coronary-prone person, it may be possible to reduce his or her risk through psychological interventions. Characterization of the addictive personality is not yet sufficiently advanced to have given rise to assessment-based risk-reducing interventions based on assessment.

4. Assigning Patients to Treatments

An application just beginning to be exploited is the use of psychological testing to assign patients to medical treatments. For example, treatment outcomes for pain patients assigned to anesthesiological versus behavioral treatments have been predicted on the basis of scores from the Minnesota Multiphasic Personality Inventory, a widely used psychological test. Treatment programs for chronic, severe asthma patients have been designed on the basis of prognoses for rehospitalization, medication compliance, and other aspects of reaction to medical care based on a battery of psychological tests. Conclusive validation of such assignments is difficult, so that progress in this area can be expected to be slow.

5. Assessing the Quality of Life

Advances in medical technology have made it possible to sustain life in many situations and conditions in which death would have been almost certain in previous times. With this capacity has come, inevitably, the question of whether there are circumstances in which it is undesirable to do so. Are there times when the quality of life preserved is "worse than death"? Most people now agree that the answer to this question is "yes." Much more difficult to face is the question of whether there are qualities of life that are not worth the money it costs to sustain them. When the question is posed in this way, most people are reluctant to agree that there are. However, because the resources applied to health care inevitably compete with other applications of those resources, both within the health system and beyond it, choices must be made. Arguably, maximizing quality of life over a population is the best basis for allocating money for health care. Whether it is or not is a question that must be debated by philosophers, theologians, and people-at-large. But if it is, another arguable proposition, that the essence of quality of life is the integration of feeling states over time, becomes relevant. Psychological assessment includes the appraisal of feeling states (e.g., joy, pain, sadness, pleasure). Categories are still unstable and measurement is crude, but if human preference is the measure, psychologists are the ones to develop the instruments.

B. Behavior Change Applied in Health Psychology

Many people feel a certain ethical unease at the idea of interventions designed to change another's behavior. Since first there were humans, and even before that in the animal kingdom, individuals have tried to change the behavior of others to accomplish their own purposes. The socialization and education of children, the design of wage and salary systems, torture, and the writing of persuasive books are all examples of behavior change methods applied with little or no concern for the consent of the target of change. However, just as the surgeon's ethical standards are different from those of the swordsman, so the psychologists' should be different from those of the huckster or the law enforcement officer. Concepts of informed consent in health care should be as fully applied to behavior change as they are to any other form of intervention.

There are three major avenues to changing behavior, defined by the level of the person-in-environment system that is addressed. At the most basic level, the laws of conditioning, worked out by a long series of researchers from Pavlov to B. F. Skinner and their students, are adapted for use with human beings. The psychologist works directly or indirectly, through the agency of the individual whose behavior is to be changed, to control the association between behavior and events in the immediate environment. ''Biofeedback'' and ''behavior modification'' are names applied to techniques based mainly at this level. At the second level, verbal and other symbolic input to the individual is used in an effort to alter the internal, symbolic representations of the person to be changed. Examples of this approach are counseling, education, and persuasion. The third approach is through modifying social and physical contexts within which behavior occurs in the expectation that these changes will interact with the motivational and learning systems of individuals to change their behavior. Actual methods used in attempting to change behavior are rarely pure application of techniques from a single level. Examples of several applications in health psychology are given.

1. Biofeedback

For many years a distinction was made in psychology between voluntary and involuntary behavior. In general, actions of striated muscles were said to be voluntary and those of the smooth muscles and glands were involuntary (i.e., not subject to conscious control). Beginning in the 1960s, evidence began to accumulate that many bits of behavior previously thought to be involuntary could be changed if information about the ongoing state of the organ was provided to the person in visual or auditory form. By means of this biofeedback a person could gain some control over such things as heart rate, brain rhythm, and stomach acidity. In many cases, the pathways by which such control was achieved were obscure and sometimes varied from person to person. Attempts have been made to use biofeedback to alter many different bodily conditions. At this time there is substantial evidence for its use in controlling skin temperature of the hands, which provides relief in Reynaud's disease, a disturbance of peripheral circulation, and in migraine headache. Control of heart rate has proven clinically valuable in treating cardiac arrythmias, and feedback of information about blood pressure has been reported to permit hypertensives to gain some control over their conditions.

The specificity of the effect has been questioned, however, because in many situations relaxation training without biofeedback produces equally good results. The most clear cut evidence of the specific value of biofeedback is in conditions in which control of skeletal muscles has been inadequately learned or impaired through illness or injury. Well-documented effects have been found in torticollis (twisted neck), muscle spasms, scoliosis (crooked back), sphincter control to eliminate urinary or fecal incontinence, and rehabilitation of muscle control after stroke or injury. [*See* BIOFEEDBACK.]

2. Behavior Modification

The nervous system is organized in such a way that behavior leading to pleasurable or satisfying states of the organism becomes more likely and behavior that leads to pain or distress becomes less likely. The techniques known collectively as behavior modification make use of these fundamental properties of individuals to produce change. They may proceed by increasing clients' recognition of the situations in which an action is rewarded (produces pleasure or satisfaction), punished (produces pain or distress), or has no appreciable consequence of either kind. One of the techniques used is called "modeling" (i.e., essentially demonstrating the desired behavior and its rewarding consequences). Often the clarification is largely verbal, in which case this approach is not much different from the counseling techniques described below. The other major technique of behavior modification is to alter the rewarding or punishing consequences of health-related behaviors. In some cases a monetary reward can be linked to performing a desired action, such as appearing for an appointment on time, or to a result presumed to follow from a recommended action, such as losing weight or controlling blood sugar level (in the case of diabetics). Some programs have used punishments, usually in the form of forfeiting all or part of a deposit made at the beginning of a program such as weight control or smoking cessation, for failure to achieve specified goals.

Often a process of "behavioral contracting" is employed in which the health care provider and the client together work out a set of contingencies. Patients are often taught to design and apply reinforcing contingencies for themselves in "self-control" programs. In such programs, and in most others, even when behavior modification is not the designated mode of approach, the rewards and punishment of social approval and disapproval play an important part in changing behavior.

3. Counseling

Counselors are persons trained to facilitate behavior change primarily by talking with their clients. Most health psychologists who engage in counseling also subscribe to the view that behavior is controlled to a large extent by rewards and punishments (including such highly personal and subjective satisfactions as satisfaction with "knowledge of a job well done"), but they place much greater emphasis on the idea that it is not the actual relations of actions to consequences that determines behavior but the ways in which these relations are represented in clients' memories. Internal representations are limited in their accuracy by the limited experience of the individual and by the failure of the categories within which experience is recorded to correspond fully to the attributes of actions and situations that actually determine outcomes. Distortions arise, often serious ones, as a result of experiences in which strong emotions have occurred. Such traumatic experiences (or thoughts) can affect the handling of information received later

in such a way that errors of representation are perpetuated or aggravated.

Counselors work by encouraging their clients to express and explore the emotions associated with the health issues that confront them and then to evaluate their own representations in light of the consensual representations of others, revealed to them through the words of their counselors and such reading or other inputs as the counselor may recommend. Counseling is more useful the more emotionally laden the health problem being faced and the more perplexing the values that must be balanced against each other in choosing a course of action. Helping patients cope with severely disabling or terminal illness is one of the areas in which counseling is most often employed. Counseling can often be helpful when drastic changes in behavior are required to accomplish some health objective, although in such cases behavior modification techniques will usually be needed as well. (Many health psychologists are adept at both.)

The expertise of health counselors resides both in their skills in the counseling process and their knowledge of the probabilities and the value issues associated with the particular health problems their clients are confronting. Counseling skills, even without specific knowledge about the health condition, can help clients make use of the information and advice they are receiving from other sources. Knowledge of the health issues without the counseling skills is not sufficient for counseling. Considerable disagreement exists as to how much and what kind of education and training is required to produce a skilled counselor.

Some psychologists trained in this area believe that worthwhile levels of expertise can be imparted to health care providers in a course or two; others believe that counseling should only be undertaken by persons with full-fledged professional training. An intermediate position is represented in the training of subprofessionals trained for specific problem areas. Genetic counselors are examples of this type. A typical training program for such persons, who assist clients to reach decisions regarding their family planning on the basis of information provided by clinical geneticists, requires 2 years of graduate study with a faculty composed of both geneticists and psychologists.

4. Education and Persuasion

Education is a term with broad and varied meanings. In one sense it refers to any activities systematically conducted with a view to increasing the knowledge, skills, or competence of those educated. Most of the activities described above could be incorporated within such a broad definition. A common narrow definition refers to a change process involving verbal transmission of information that appeals primarily to human rationality. Often the educator disclaims any intent to produce *specific* changes in beliefs or behavior, such as might be the result of "propaganda," "indoctrination," or "coercion." "Persuasion" incorporates both the rational and irrational. In its most rational expression, persuasion seeks to organize the material for maximum effectiveness in accomplishing some specific change. At the other extreme it uses, covertly, knowledge of motivating factors to achieve change that could not be gained simply by the presentation of information. A striking example is the use of psychological "reactance," the tendency of humans to resist apparent efforts to constrain their freedom of thought, to back them into desired positions by suggesting a graded series of contrary views for them to resist. The statements, "You really have to be more concerned about your business than about your family," and "your family is probably more concerned about the income you produce than the effect of your work on your health" are reactance-inducing items that have been used in a series to induce a person to seek an appointment with his or her physician." [*See* PERSUASION.]

Most individual educational efforts in the health system are directed to patients and to health care personnel. However, they may be addressed to family members or other affiliates of patients and occasionally to individuals involved in the planning or administration of health-related activities. Education of persons in groups is more often directed to health care workers and professionals. For most of this century, psychologists have participated in teaching health care providers about the psychology of the patient and about provider–patient interactions.

The involvement of health psychologists in educative and persuasive interventions is more often than not as critics and evaluators of the efforts of others. Carefully controlled studies show that simply increasing the information available to individuals, even when they have incorporated it to the degree that they can reproduce it on demand, is not an effective way of changing behaviors deeply entrenched by habit, tradition, or preference. Educators, journalists, and media specialists are usually more artful at composing effective presentations of information, but psychologists can be useful in tailoring presentations for special audiences and in analyzing the details of what is working and what is not. Sometimes they can point to irrational barriers between knowledge and action and propose methods of penetrating them. "Health education," however, is often designed and conducted by health professionals who not only lack the expertise to do it effectively, but do not realize that such expertise exists. Health psychologists, working with them, can sometimes be the agents of educating the would-be educators.

5. Systems and Community Approaches

Some interventions by health psychologists are directed to groups of people who live or work together. In such cases interactive properties of "systems" of people are taken into account. The whole field of "systems analysis" has developed in the past 50 years in an attempt to describe, objectify, and quantify the idea that when one acts on one of a set of interconnected elements (i.e., on a system), there are compensating adjustments throughout the system. Systems enormously vary in size and complexity. Three levels of human systems (i.e., those in which the elements are humans) that have been addressed by health psychologists are families, health care organizations, and communities.

1. The work of health psychologists with families has developed largely in collaboration with the rebirth of the family doctor in the form of the specialist in family medicine. Another factor has been the importance recently placed on life-styles, obviously and demonstrably much influenced by family patterns. Yet another is based on the growing emphasis on social support as a major influence in the response to stress, the formation of health goals, and the adherence to intended health regimens. Families are often the primary source of social support. Simply recognizing the importance of families is only the beginning, however, of the approach to families being developed by psychologists and other social scientists. Drawing on general systems theory, group dynamics, dynamic psychology, and anthropology, a "family system theory" is growing up that takes families as a unit of analysis. The impact of stress on family units, their decision-making processes, and their mobilization to deal with crises of extended and life-threatening illness are studied to develop interventions directed to the family as whole.

2. Psychologists working in health care organizations have focused most of their attention on two issues: how to increase the satisfaction of clients/patients with the

care they receive, and how to match the utilization behavior of patients to their need for health care to make the best use of health care resources. Much of the effort in the area of patient satisfaction has been devoted to developing methods for assessing it. Many studies, particularly the earlier ones, have been limited in their discovery of correlates of patients' evaluation of their health care by the fact that patients tend to give high marks to their physicians and their services, even when other evidence suggests that substantial amounts of dissatisfaction exist. Furthermore, it has not proven easy to obtain differential ratings of the quality of different aspects of the care received. However, when health psychologists applied sophisticated techniques of questionnaire development to the problem it was possible to break through these barriers and to obtain information that could guide administrators and teachers in their efforts to improve the quality of care.

In early studies of factors influencing patient satisfaction it was often found that interpersonal aspects of the provider–patient encounter dominated, with technical quality taking a distinctly secondary place. Many contemporary discussions of the topic continue to rest on this generalization. However, careful analyses using better methods for gathering patients' appraisals reveal that patients are sensitive to providers' competence in performing the technical tasks of health care. Their own effective participation in the health care process is more affected by the latter factors than by purely interpersonal processes.

The organizational issue of utilization is complex. In essence, it is the problem of matching health care services to human needs in such a way as to make the best possible use of available resources. There are three aspects of the problem: underutilization, overutilization, and getting people to the right place, including the problem of referral. In general, poorer segments of the population have greater health needs and receive less health care services than the more affluent. This maldistribution may be reduced, but it is not eliminated even when the services are offered free of charge. However, a small proportion of persons uses a relatively large proportion of health services. For example, in a nationwide survey conducted in the United States in 1980, it was found that about 5% of the population made 26% of all outpatient visits and accounted for more than 22% of outpatient expenditures. The problem of misdirection of the initial request for service affects both under- and overutilization. In underutilizing groups, entry into the health care system will often result in a referral to another site, and such referrals result in a completed appointment at the proper facility less than half of the time. By contrast, persons who are thought to seek more from the health care system than it can properly deliver—the overutilizers—actively seek referrals and make many self-referrals to a great

variety of service providers in usually unsuccessful efforts to gain relief from symptoms.

Health psychologists have not been much involved in this set of problems as yet. Sociologists and demographers have tried to explain underutilization and psychiatrists have studied overutilizers, whereas economists have studied the impact of the pricing of services on these behaviors. Problems of making successful referrals are usually approached as instances of interaction and communication.

3. At the community level, health psychologists have focused on the use of community organizations and other resources to decrease health damaging behaviors and increase behaviors favorable to health. Examples of such programs are those conducted by Stanford University behavioral scientists in several communities in Northern California and by Finnish researchers in the county of North Karelia, Finland. With follow-up periods of 10 or more years, changes in health behaviors and in physiological and epidemiological indicators of cardiovascular risk and morbidity have been demonstrated. Large-scale projects of this kind have been conducted in Australia, South Africa, and Switzerland as well as several areas in the United States. Cardiovascular risk and substance abuse are the topics of most studies thus far conducted. These are multidisciplinary projects. Psychologists join with physicians and other health professionals, health educators, and others in their planning, design, implementation, and evaluation phases. Psychologists' expertise in individual and community dynamics can spell the difference between success and failure in attaining project goals.

6. Public Health Interventions

In practice, public health workers today often make use of educational approaches on a large scale, as described in the previous section. They support media campaigns to get people to wear seat belts or laws mandating warning messages on packages of tobacco products and in advertising for such products. In terms of a defining philosophy, however, the public health approach primarily relies on actions that do not require the active engagement of the public whose health is to be protected. A classic example comes from the work of a London physician, John Snow, during a cholera epidemic in the mid-19th century. Careful mapping of cases led him to conclude that a particular public water pump was the source of contamination, even though the mechanisms of cholera infection were unknown at the time. Snow did not try to convince people not to use water from the pump; he used his authority as a government physician to have the pump handle removed. Other examples of the "passive" approach that requires little or no participation by members of the public are draining swamps to control malaria, engineering highways and protective structures to reduce deaths and injuries from highway accidents, and promoting mandatory innoculation to provide life-long protection from infectious diseases.

The two major instruments for changing behavior via the public health approach are environmental change and regulation. The health psychologist can contribute to both of these by anticipating reactions to the measures taken. There is almost no intervention that does not require some degree of acquiescence or cooperation from individuals, and psychological knowledge applied in the selection, design, and implementation of environmental or regulative changes can increase their effectiveness.

VI. Psychology of the Addictions: An Example

The integration of the various concerns and activities of health psychologists as they relate to a major class of health problems can be exemplified by a consideration of the situations in which some behavior has become intractably excessive. Until recently, problems of alcohol abuse, drug addiction, smoking, obesity, pathological gambling, and pathological hypersexuality have been considered separately, using different theoretical approaches and independently developed bodies of intervention technique. Now there is a growing recognition that the underlying psychological mechanisms have much in common so that a coalescence of research and practice is underway.

In the past, there has been much argument over the meaning and use of such terms as *addiction, dependence,* and *vice,* and whether excesses of behavior were the result of disease, moral or psychological weakness, or faulty habits. There is now emerging a consensus that these conditions are complex biopsychosocial phenomena in which predisposing constitutional factors, environmental influences, psychological mechanisms of learning and defense, and moral and ethical beliefs all play important parts. In some patterns of excessive behavior, the physiological processes of tolerance (i.e., requiring increasing amounts of a substance to achieve the same bodily effects) and of withdrawal symptoms that are uncomfortable, painful, and even life-threatening assume much larger roles than

in others. These are the processes that have been strongly associated with the term *addiction.* Even in those conditions in which these physiological processes figure most strongly, in the case of the opiates, barbiturates, and alcohol, psychobiological processes of learning and the generalization of over-learned habits to ever-greater numbers of eliciting or discriminative stimuli give rise to the intractability of the conditions and form the basis for relapse. In the remainder of this section, the complexly determined pattern of intractably excessive performance of some behavior is referred to as an addiction.

The addictions have several essential features in common. First, some behavior that elicits a pleasurable response soon after the behavior has occurred becomes so frequent that it causes serious harmful consequences to the body of the addicted individual or in the social network of that individual. The forces that maintain the behavior are so strong that even though the affected person fully understands the harm that is being done, it is extremely difficult to change the behavior pattern. Commonly, sporadic efforts to change are increasingly overwhelmed by patterns of defensive denial, rationalization, and other behaviors that protect the addicted person from full recognition of the need for change. When addicted persons enroll in programs to change behavior, usually about a quarter of them succeed in eliminating or substantially reducing the unwanted behavior, but unless there is some continuing factor at work, most of these individuals will revert to the addicted pattern within a few months to a year.

A. Societal Aspects of Addiction

Specification of what constitutes an excess of an appetitive behavior is a social definition, although biological and psychological factors are deeply involved in its establishment. What is considered pathological excess in one culture or in a particular time period may be well within the range of the acceptable in others. When certain behaviors (e.g., the use of particular substances such as opium or cannabis) are totally proscribed by a society, there is no legally acceptable level of the behavior, although there may still be a socially tolerated level. The distribution of the amount of the behavior performed by the members of a population is skewed because most persons are moderate, with ever fewer engaging as the level of the behavior increases. Unless there is rigid enforcement of a legal or religious proscription of the behavior, the mode or peak is usually somewhat above total abstinence.

The costs to society of the addictions are enormous. Two kinds of costs are recognized: internal costs that are borne by the addict (including in some analyses, costs to family members), and external costs that are borne by employers, taxpayers, victims of accidents, purchasers of insurance policies, etc. Estimates made in the late 1980s of combined internal and external costs identify direct costs for facilities and treatment of alcohol and other drug abuse, exclusive of tobacco, at about $2 billion per year. In the same period, it was estimated that total costs, mainly arising from lost productivity and unemployment, amounted to nearly $200 billion per year. Direct costs of smoking amount to some $12–13 billion per year. These costs are offset to a greater or lesser degree by "savings" to society because substance abusers have shorter lives and therefore collect less in the way of pensions, make less use of nursing homes, etc. Some investigators have attempted to place a dollar value on expected years of life lost to the substance abuser. A common method of doing so is in terms of lost income as the substance abuser is withdrawn from the work force. A more psychological approach has drawn the estimate from data regarding the amount that people are willing to pay for a small change in life expectancy. One careful estimate of this amount was approximately $10 per hour. Using this figure and an equally careful estimate that a smoker loses 28 min of life expectancy for every pack of cigarettes smoked, $200 billion per year of internal costs would have to be added to the costs of smoking in the United States, including about $4 billion for passive smokers. Careful monetary cost estimates for obesity, pathological gambling, and hypersexuality are not available.

Analyses of this kind are mainly carried out by economists. Psychologists can contribute, however, through their knowledge of methods for assessing human preferences, thus refining subjective costs of pain, suffering, discomfort, lost life, etc. Applications of these analyses are found in legislation and regulations that set excise taxes, proscribe sale of certain commodities, or establish schedules of premiums or copayments in health insurance plans. Involvement of psychologists in planning such social interventions can lead to greater accuracy in anticipating public response to them.

In terms of health effects, smoking has been named as the single greatest preventable source of illness and mortality in the developed world. A major epidemiological study in a midwestern state concluded that alcohol abuse contributed to 12% of all years of life lost, mostly because of consequences of accidental injuries. Obesity that exceeds 130% of the average or recommended weight is associated with premature mortality that is 1.3 or more times that of persons with normal weight. The harm that results from excesses of gambling and sexual behavior are difficult to document in terms of physical illness and mortality figures. Often their deleterious effects have greater impact on families and other associates than on the addicted person. For these reasons, some critics claim that these last two conditions should not be considered in the category of illnesses, but rather as socially, perhaps arbitrarily, defined deviance. However, there are excessive rates of suicide and mental illness in these conditions. [See NONNARCOTIC DRUG USE AND ABUSE; OBESITY; TOBACCO SMOKING, IMPACT ON HEALTH.]

Health psychologists contribute to the scope and accuracy of the epidemiological studies that document these effects by studies of methods for phrasing questions and structuring questionnaires that appraise both the problematic behavior and the health status of respondents. They are also involved in making use of the results of such studies in developing educational and treatment programs for preventing and treating addictive behavior.

B. Individual Aspects of Addiction

The addictions are peculiarly psychological disorders, in that the problem arises specifically from a particular pattern of behavior. In almost every kind of health problem, behavioral factors are involved in the etiology, treatment, and recovery from the disease, but in the addictions behavior is the central aspect. This is not to say that physiological and social factors are unimportant. In every facet of addiction, biopsychosocial interaction is pervasive. These interactions are traced through the stages of addiction in the sections that follow.

1. Predisposing Factors

The occurrence of excessive appetitive behaviors is not uniform over populations but is associated with ethnicity, education, income, family membership, and other sociodemographic variables. Although the epidemiological evidence on which this statements rests is of variable quality for the several forms of the addictions, it has provided the basis for efforts to attribute predispositions based on genetic, familial (including psychodynamic), and socioenvironmental factors. By far the most extensively studied condition is that of excessive use of alcohol. It has been possible to accumulate conclusive evidence that risk factors for alcoholism can be associated with each of these domains of variables. Details of mechanisms are, for the most part, not well established. Genetic factors have also been sought with some claimed success in the case of excessive eating, and recent studies point to a possibly genetic abnormality of norepinephrine metabolism in pathological gamblers. Repeated proposals that endocrine abnormalities may be involved in hypersexuality have not been substantiated as yet. Postnatal predisposing factors have their impact through learned aspects of the behavior and are treated in subsequent sections.

2. Taking Up the Addictive Behavior

Addictive behavior is learned behavior. Therefore, opportunities to observe others engaging in such behavior usually precede the actual performance of the behavior. The observation may occur in real life, in the behavior of parents or peers, or it may be encountered first in television, films, or written materials. There is even evidence that experimentation with drugs is increased by educational programs intended to prevent drug use among adolescents. Studies of the initiation of these behaviors, which in their excess become problematic, have mostly been conducted with adolescents. In this group, the most strongly associated factor is almost always the degree to which same-aged associates, especially close friends, engage in the behavior. Parental behavior, attitudes, and permissiveness all make important contributions. A variable sometimes called "tolerance for deviance," which is closely related to the belief as to whether the behavior is wrong, is also important. Personality factors, such as rebelliousness, stress on independence, low sense of psychological well-being and self-esteem, and lower academic motivation and achievement are all associated with relatively early initiation of substance use, sexual behavior, and gambling. Investigation of these topics is one of the most active areas of research in health psychology.

3. Active Phase of the Addiction

Except for legally proscribed behaviors, the addictions consist of socially defined excesses of behavior exhibited by most members of a population to some degree. Even when a behavior is totally forbidden by law, sizable proportions of a population may engage in it on one or a few occasions. The processes by which persons change from occasional, or moderate, controlled degrees of the behavior to a level deemed excessive are multiple and complex. At the simplest level, basic principles of learning dictate that the more a behavior occurs, the more situations become occasions for the performance of the behavior. (In technical terms, by response generalization, more and more situations come to be discriminative stimuli in whose presence the behavior is highly probable.)

For most persons, both the rewarding effect and the generalization process are opposed and checked by social and physiological consequences of occasional excess or by the anticipation of such consequences, learned vicariously. A significant factor in establishing the equilibrium point at this first stage, as in the initiation process, is the behavior and attitudes of other persons in the settings that the individual frequents. Poorly understood differences among individuals in the positive, rewarding effects of the behavior, the propensity to inhibit behavior in anticipation of punishment or disapproval, and the impact of belief systems on behavior influence the progression into a second stage in which guilt, shame, and deceit become prominent. These new factors introduce individually characteristic degrees of anxiety arising from self-awareness of the problematic nature of the behavior. The rewarding effects of the addictive behavior may now be enhanced by their capacity to reduce anxiety and to distract attention from the awareness of transgression of social or internalized ethical norms. Thus a complicated positive feedback situation is created in which the knowledge of excess becomes a factor in continuing and increasing it.

When the addiction has reached this stage of self-perpetuation, it begins to transform the life patterns of the individual. Family and friends are likely to become alienated, and the addict becomes isolated or affiliated with other addicts in a way of life that is centered on and dominated by the excessive behavior pattern. The life course of the alcoholic or the person addicted to other drugs often ends prematurely as a result of drug-related deterioration of the body or from illness or injury brought on by intoxication and loss of domicile. Compulsive gamblers and those who are pathologically hypersexual may become drug addicts or alcoholics as the social structure of their lives is destroyed. Some of the eating disorders, notably, anorexia nervosa and bulemia, are serious health problems that can result in death.

Although the "milder" addictions of smoking and excessive eating leading to obesity are associated with significantly shortened life expectancy and often with a poor quality of life in later years, they typically do not progress into the stage of the addict life-style. The probable reason is that they do not invoke the guilt–shame–deceit complex.

4. Resolution Without Treatment

Many persons who are at one time in their lives actively addicted to some form of excessive behavior eventually return to a more balanced way of living. Authoritative estimates of the percentages who do so are hard to come by, and they vary substantially among the several forms of addiction. The quality of the data varies as a function of the covertness of the behavior and the social investment in its study. Probably the best data are available for the tobacco addiction. These data were almost all collected during the period after the intensification of public health efforts to get people to stop smoking in the 1960s. This fact reminds us that quantitative generalizations from one form of addiction to another, from one cultural context to another, and from one social era to another are not possible.

In the case of smoking, somewhat more than half the persons who report they were at one time regular smokers will also report that they have now been abstinent for 1 year or more. Other data indicate that relapse is relatively rare after such a period of abstinence. Elimination of the addiction is not often achieved on the first effort, however. Perhaps 4–6% of those who attempt to quit smoking succeed on any given attempt. Of those who do succeed in quitting, it appears that more than 9 of 10 quit on their own without having been involved in any formal treatment programs. Recent, methodologically sound research on persons with serious alcohol addictions indicates that about half of them are abstinent or are free of drinking problems 4 years after initial contact even if they have received no treatment in the interim. (A third or more are free of alcohol problems if they have received treatment.) Even in the case of addiction to heroin, which is certainly one of the most intractable of all the addic-

tions, it has been reported that during a 10-year period as many as 40% of those previously addicted will become abstinent. Accurate figures for the number of heroin addicts who become abstinent without participating in an effective treatment program at some time are not available.

Although the mechanisms for freeing oneself from an addiction considerably vary in detail, there are some patterns in common. The shifting equilibrium that prevails between the positive and negative consequences of the addictive behavior during active addiction is disrupted through some event or series of events that bring the addict to a "turning point." Death of a close associate, loss of job, disruption of family, or sometimes just the realization that one is living in total subjugation to a dependency leads to a crystalized intention to change. In the traditions of alcoholism, it is said that one must reach "rock bottom" before the turn to recovery can be made. However, data indicate that the turning point can occur at any degree of addiction and that recovery is more likely and quicker the less intense the addiction.

Many devices are tried in the course of the difficult process of change. Some persons are successful with a process of "tapering off" (i.e., gradually reducing the amount of the addictive behavior until its strength is lessened). They may restrict the times and settings when they engage in it and thereby gradually reverse the process of generalization referred to earlier. Others, influenced more by the perceived hazards of relapse, prefer abrupt and total change to abstinence—"cold turkey." Addicts committed to change often adopt patterns of avoiding the situations in which the urge to engage in the addictive behavior is particularly strong. Smokers may give up a morning cup of coffee or an afternoon cocktail, because these are strongly associated with the pleasure of cigarettes. Particularly important is finding new interests, activities, and desires to fill the void created by the removal of all the behaviors associated with the addiction. Episodes of relapse are almost universal, and the manner in which they are handled by the addict is critical for the ultimate success of the recovery process.

Although present understanding of the origins of, development of, and independent recovery from the addictions is substantial, continuing psychological research is adding further knowledge of the mechanisms that underlie the phenomena and is providing details of the variations to be found among the several kinds of addictions and as a result of variations in the social settings and individual characteristics of the addicts. This knowledge is used to design treatment programs for interrupting the patterns and restoring addicts to more normal functioning.

5. Treatment of the Addictions

There is an enormous literature on the treatment of addiction. In the case of obesity and hypersexuality, surgical and hormonal approaches can be effective in some cases. However, in the vast majority of cases, some method for producing behavior change must be used. Incarceration in a prison or hospital is usually effective for its duration, but relapse is virtually certain unless other change techniques are employed during the period of confinement. Psychological programs for change are offered by proponents of all the major theoretical approaches to the understanding of human behavior. Behaviorists emphasize the elimination of the habitual aspects of addictive behavior and the substitution of new habits. Cognitivists concentrate on clarification of the consequences of the addiction, the clarification of values, the formation of new intentions, and reappraisal of the addict's potential for change ("self-efficacy" beliefs). Psychodynamicists focus on the underlying conflicts that have contributed to the susceptibility to addiction preparatory to subsequent re-educative processes. Socially oriented therapists may try to help the addict establish supportive social networks made up of nonaddicted persons, and they may involve family members in the treatment process. In fact, most therapies involve elements from all these approaches.

Each of these levels of address can contribute to the change process in individuals who have reached the turning point. All except the most stringently habit-focused can aid the ambivalent addict to come to the turning point. Claims of success in treatment vary widely. Short-term change may be achieved in 50–80% of those who complete a treatment, but the long-term success rate is much lower because of relapse. Careful studies of long-term (1 year or more) outcomes usually find 20–30% abstinence in the case of substance abuse. Many more addicts demonstrate significant reduction in the amount of the addictive behavior in which they engage, often with significant lessening of adverse consequences. These accomplishments are better than those achieved by addicts on any given independent effort to change. Extensive studies of heroin and alcohol treatment programs indicate that they can be cost effective, saving more in terms of identifiable costs to society than is spent on them. Success rates are higher the more broadly the problem is addressed, and no one of the approaches mentioned above appears to be substantially better than the others. The most basic conclusion regarding recovery from addiction is that it must be regarded as a complex process extended in time rather than a single event.

6. Postrecovery Condition

Relapse after short-term success in eliminating an addiction is likely. Research emphasis has therefore shifted to discovering methods for preventing relapse. Investigation of the circumstances under which it occurs provides a basis for warnings to ex-addicts. A common finding is that high stress is frequently involved, so training in stress management is often offered. Formal programs of aftercare can substantially reduce the relapse rate. Even an agreed on schedule of follow-up contacts can be a significant factor in maintaining recovery.

A major controversy in the psychology of the addictions has been the question of whether it is possible for a former addict to continue in the controlled performance of the formerly excessive behavior. Of course, in the case of excessive eating, it is essential to do so. This fact has been pointed to as an explanation of the belief that excessive eating is among the most difficult of the addictions to eliminate. Data on long-term outcomes of weight reduction programs are scarce, however. When the problem behavior is illicit or illegal, controlled continuation is an unstable behavior pattern, because engaging in the behavior tends to be incompatible with normal life patterns. Alcohol and tobacco are two substances that are not essential to life, but whose use is legal and socially acceptable. Some persons are able to develop long-term patterns of well-controlled use of these substances after periods of strong addiction. The preponderance of evidence indicates that most of those who attempt to sustain such a pattern will revert to the pattern of addiction at some stressful time in their lives. This question has not, however, been adequately studied.

VII. Conclusion

In every aspect of the health system, problems arise that involve human reactions to complex situations that affect health outcomes. Human emotions, beliefs, decisions, intentions, communications, and interactions influence the processes of abating and avoiding hazards in the environment, interpreting symptoms, seeking and participating in health care, and recovering maximum ability to function effectively after injury or illness. These psychological factors also determine the actions of those who provide care, develop and manage health organizations and public health programs, and formulate health policies, regulations, and legislation. They influence the behavior of those who conduct, interpret, and support research on health problems.

Contemporary psychology, in conjunction with adjacent fields of information science, neuroscience, and social science, has much to offer in the understanding of the role of such factors and in the application of this understanding to the design of effective interventions. To realize the potential contributions, psychologists must immerse themselves in the problems and settings of the health system. Those who do so are properly called health psychologists. Their number is rapidly increasing throughout the developed and developing world.

Bibliography

Baum, A., and Singer, J. E. (eds.) (1987). "Stress." L. Erlbaum Associates, Hillsdale, New Jersey.

Baum, A., Taylor, S. E., and Singer, J. E. (eds.) (1984). "Social Psychological Aspects of Health." L. Erlbaum Associates, Hillsdale, New Jersey.

Belar, C. D., Deardorff, W. W., and Kelly, K. E. (1987). "The Practice of Clinical Health Psychology." Pergamon Press, New York.

Dasen, P. R., Berry, J. W., and Sartorius, N. (eds.) (1988). "Health and Cross-cultural Psychology: Toward Applications." Sage Publications, Beverly Hills, California.

Feuerstein, M., Labbe, E. E., and Kuczmierczyk, A. R. (1986). "Health Psychology: A Psychobiological Perspective." Plenum Press, New York.

Fisher, S., and Reason, J. (eds.) (1988). "Handbook of Life Stress, Cognition and Health." John Wiley and Sons, New York.

Gentry, D. (ed.) (1984). "Handbook of Behavioral Medicine." Guilford Press, New York.

Karoly, P. (ed.) (1985). "Measurement Strategies in Health Psychology." Wiley, New York.

Matarazzo, J. D., Weiss, S. M., Herd, J. A., Miller, N. E., Weiss, S. M. (eds.) (1984). "Behavioral Health: A Handbook of Health Enhancement and Disease Prevention." Wiley, New York.

Melamed, B. G., Matthews, K. A., Routh, D. K., Stabler, B., Schneiderman, N. (eds.) (1988). "Child Health Psychology." L. Erlbaum Associates, Hillsdale, New Jersey.

Orford, J. (1985). "Excessive Appetites: A Psychological View of Addictions." John Wiley and Sons, Chichester, England, and New York.

Stone, G. C., Weiss, S. M., Matarazzo, J. D., Miller, N. E., Rodin, R., Belar, C. D., Follick, M. J., Singer, J. E. (eds.) (1987). "Health Psychology: A Discipline and a Profession." University of Chicago Press, Chicago.

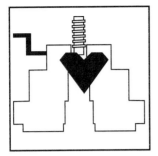

Heart–Lung Machine

HANS H. J. ZWART, *Wright State University*

Glossary

Bubble trap Specialized blood filter that can also trap a limited amount of air bubbles from the blood.
Bypass flow Amount of blood pumped per minute.
Cannula Short piece of tubing inserted into major blood vessel.
Cardioplegia Production of cardiac arrest by infusing the coronary arteries with potassium solution, often combined with cooling.
Cardiopulmonary bypass Procedure to substitute the function of heart and lungs.
Cell saver Device to collect and purify blood spilled in the operative field.
Crystalloid solution Solution of simple electrolytes such as sodium chloride.
Doppler sound probe Device that emits sound waves and senses echoes of those waves. Air bubbles distort normal echoes from blood and can thus be detected.
Emboli Foreign particles in the blood, traveling through the circulation and lodging somewhere in the tissues.
Heat exchanger Device to control the temperature of the blood.
Left heart bypass Bypass from left heart to aorta.
Oxygenator Device to add oxygen and to remove carbon dioxide from the blood.
Right heart bypass Bypass from right heart to pulmonary artery.

THE HEART–LUNG MACHINE is an apparatus that is used to temporarily replace the function of both the heart and the lungs. It was designed primarily for use in heart surgery since most heart operations can only be carried out on the arrested heart. The heart–lung machine is made up of an artificial heart and a lung-part, connected with each other and with the patient by a system of cannulas and tubes. The system also incorporates filters and equipment to control the temperature of the blood. The blood is drained from the right side of the heart to the lung-part of the machine where oxygen is added and carbon dioxide is removed. The pump part thus transports the blood back to the aorta or one of its branches.

I. Components of the Heart–Lung Machine

A. Overview

The patient's blood flows to the heart–lung machine via a $\frac{1}{2}$-inch diameter plastic tube called the venous line. The line is connected on the patient's side, to one or two cannulas inserted into the right atrium of the heart, and on the machine side, to a venous reservoir that is usually open to air. The air blood level is 50–100 cm below the patient's heart and the sole propellant for blood flow through the venous line is gravity. One-quarter-inch plastic suction tubes are also connected to the reservoir to return to the heart–lung machine blood suctioned out of the surgical field. One of these suction tubes may be temporarily used to decompress the heart. Medications and fluids supplied to the patient via the heart–lung machine are introduced into the blood circuit via the venous reservoir as well. All fluids and substances drain from the reservoir to an oxygenator either by gravity or by means of a pump, depending on the type of oxygenator. The oxygenator usually incorporates a heat exchanger to help keep the blood temperature at the desired level. Finally, the blood is pumped back to the patient via a $\frac{3}{8}$-inch inside diameter tube called the arterial line. On the patient's side the line is connected to the blood return cannula, which is usually inserted into the ascending aorta with the tip pointing downstream. Still another set of tubes serves to perfuse and cool the heart itself. This so called cardioplegia system consists of $\frac{3}{16}$-inch plastic tubing connected on the patient's side to a needle inserted into the ascending aorta just downstream from the aortic valve. On the machine side, the cooling tube is connected to a separate apparatus specifically designed for cardiac cooling.

B. Artificial Heart Components; Blood Pumps

The pump in the heart–lung machine circuit must be capable of propelling blood at a rate of up to 8 l/min against a pressure as high as 150 mm Hg. During a typical run, the flow rate is 4–5 l/min at a blood pressure of 60–80 mm Hg. Depending on the type of oxygenator used, the arterial pump is located either upstream or downstream from the oxygenator. In circuits with a high resistance artificial oxygenator, two pumps may have to be used, one up and one downstream from the artificial lung.

The pump most commonly used is of the *roller type*, in which one, mostly two, or sometimes three rollers squeeze a flexible tube, containing the blood, within a rigid housing. The rollers rotate, isolating a certain volume of blood within the tube, and transporting it from the inlet to the outlet. It is of simple construction, easy to operate, inexpensive, and reliable. In case of power failure, it is easy to rotate the rollers by hand using a special handle fitted on the axis of the pump. The tube containing the blood is disposable, so the cleaning process is simple. Extensive laboratory testing has provided evidence that the pump causes little damage to the blood, provided it is not used for periods of longer than 4–6 hours at flow rates of about 5 l/min. The disadvantage of the pump is the occurrence of spellation, that is, dislodgment of small plastic particles from the wall of the tubing, which occurs mainly after prolonged use. These particles can travel into the arterial circulation and lodge in any part of the body. Other members of the group of rotating pumps are the Archimedean screw pump and the finger pump. The Archimedean pump consists of a rotor, in the form of a helical screw, which revolves within a housing with a double internal helical thread. During pumping, the fluid pockets between rotor and housing steadily progress along the thread toward the outlet. Finger pumps transport the blood by sequential compression of a straight rubber tube by multiple fingers. These two types of pumps were only of historical interest until recently, when the Archimedean screw was skillfully engineered into a promising mechanical circulatory device to assist in closed-chest left heart bypass (see Section III, C).

During many years the roller pump has served reliably in millions of operations lasting 2–5 hours. More recently though, the need for prolonged application of bypass led to further refinements of the heart–lung machine, and a less traumatic pump type was introduced. This *centrifugal* or *impeller pump* imparts velocity to blood with a propeller contained in a disposable plastic translucent chamber, without touching it, and rotates at a high rpm. The chamber is mounted on a pump house and the propeller rotation is effected by an electromagnetic field. This type of pump increases output as more volume is presented at the inlet; it is very sensitive to pressure at the outlet, decreasing output as the pressure rises. The pump cannot generate much suction at the inlet. These characteristics make this pump type well suited for prolonged use, such as days or weeks. There is less damage to the blood than in roller pumps, and no spellation. At present, centrifugal pumps are expensive, and are not therefore used widely for routine heart surgeries. As production costs will decrease, the pumps will probably be used for short-term applications as well.

Finally, reciprocating pumps comprise the last family. These pumps imitate the natural heart: the blood enters the chamber, and is ejected by a decrease of the chamber volume. They typically need an inlet and an outlet valve. These pumps are used primarily in artificial heart research. To expel the blood the volume of the pump chamber is decreased either by a rigid plate connected to a plunger or by circumferential compression within a rigid outer chamber by gas or by a liquid. The most commonly used artificial heart pump so far was of the reciprocating type.

C. Artificial Lung

The purpose of the artificial lung (oxygenator) is to transfer oxygen into the patient's blood and to eliminate carbon dioxide from it. This is a formidable task and it was the major limiting factor of the original heart–lung machines in 1950s, because the gas transfer has to be carried out reliably and gently.

Oxygen and carbon dioxide are carried by the hemoglobin molecule, which is present within the red blood cell. The ventilating gasses have to transfer through the blood plasma to the red cell, must cross the cell membrane, and diffuse through the cell fluid to the hemoglobin molecules. Diffusion of oxygen through the layers is much slower than that of carbon dioxide. To speed up the gas exchanges the blood exposed to the ventilating gas should be in a thin layer, the blood transit time through the oxygenator should be long, and mixing of the blood efficient. The artificial lung should be able to oxygenate venous blood fully at a flow rate of up to 6 l/min. To reach this goal, the oxygenator has to transfer 250–500 ml of oxygen per minute into that volume of blood and remove an equal amount of carbon dioxide.

The earliest clinical oxygenator model was of the vertical screen type. Venous blood from the patient was delivered to the top of a vertical screen. The blood flowed down the screen in a thin film and was exposed to a high flow of oxygen. The oxygenation capacity could be varied by increasing the number or the surface area of the screens. A large volume (3.5 l) was necessary to fill the device (priming volume). This was all very problematic because blood was used to prime the machine.

A less bulky oxygenator of the same type was developed in which blood was distributed as a film on a number of rotating disks mounted in a cylinder. Blood filled only the bottom of the cylinder; the rest was ventilated with oxygen. While rotating, the disks picked up a thin layer of blood and exposed it to the oxygen. The capacity of these oxygenators could be adjusted by adding more disks and by varying their speed of rotation.

In still another type of oxygenator, oxygen was simply blown into the blood via a showerhead-like inlet. As a result, the blood starts to foam and it is in the foam bubbles that gas transfer takes place. The initial problem of these so called bubble oxygenators was the inability to defoam the blood before it was returned to the patient. It should be noted that even small gas particles in blood can cause major damage to all organs of our bodies by stopping flow through small blood vessels. Effective debubbling was eventually achieved by exposing the foam to surfaces sprayed with silicone. Bubble oxygenators were accepted because they were easy to regulate and had a small priming volume. Once they were produced as disposable units, they became very popular. However, even the late models cause too much blood damage, especially when used over prolonged periods of time.

The oxygenator type that reproduces most closely the normal lung is the membrane oxygenator. In this artificial lung, the blood and gas phases are separated by a membrane. Its development depended on membrane technology. Early oxygenators were plagued by membrane rupture and inefficient gas transfer, but as the membrane quality improved, oxygenators of this type have become the artificial lung of choice even in short-term perfusions.

Membrane oxygenators have many other advantages. The primary volume is low and since there is little direct contact between gas and blood, blood damage is minimal. The entire heart–lung machine blood circuit is closed to air, promoting sterility and safety, and further decreasing blood damage. A late variety of the membrane oxygenator is the capillary oxygenator. The blood is contained within or around multiple small rubber tubes rather than between sheets of membrane, reducing the priming volume even further. The modern membrane oxygenators are all disposable.

D. Other Parts

Since in the heart–lung machine circuit blood is exposed to room temperature, its temperature must be controlled. Once the patient is on the machine, the normal blood temperature of 37°C is allowed to drop to around 30°C, decreasing the oxygen need of the body. This gives an extra margin of safety to the use of the heart–lung machine. Certain complicated operations necessitate cessation of all blood perfusion through the body, which can be done safely only if the body temperature is decreased to around 10°C. Decreasing the temperature is easier and quicker than increasing it. Modern heat exchangers can increase blood temperature by about 0.5–1.0°C/min at a blood flow rate of 5 l/min. Thus, the heart surgeon has to notify the perfusionist to start rewarming the patient some 20–30 minutes ahead of terminating the surgery. The heat exchangers are usually built into the artificial oxygenator and are thus disposable as well.

Filters are incorporated to ensure that no foreign particulate matter is being injected into the patient. They are essential in suction lines through which excess blood is removed from the operative field and returned to the venous reservoir. This blood may contain small tissue fragments, dissolved fat, blood clots, and even small pieces of sutures. Filters are also frequently employed in the blood return tube. The pore size of the filters varies from 25 μ in the venous filters to 40 μ in the arterial lines. A bypass circuit is always provided around the arterial line filter in case the latter clogs up. The modern arterial filters also serve as bubble traps, but only some 100–200 ml of gas can thus be prevented from entering the patient. Larger amounts would escape the bubble trap and enter the patient.

One or two tubes are provided in the heart–lung machine circuit to enable the surgeon to collect excess blood from the chest wound and to return it to the blood circuit. This so-called *cardiotomy suction* is of great importance. Sudden accidents occurring during surgery could lead to dangerous blood loss if it were not promptly returned to the patient. The suctioning, however, causes most of the blood damage: it is the primary port of entry through which foreign particles and infectious sources enter the blood circuit of the heart–lung machine. The surgeon must realize these dangers and use the cardiotomy suction sparingly and wisely. The suction lines are mounted in one or two roller pumps, which return the blood to the venous reservoir. While the aorta is clamped and the heart is stopped, one of the suction lines is often used to decompress the heart by connecting it with the perfusion cannula in the aorta. Any blood accumulating within the heart is thus returned to the heart–lung machine circuit.

The roller pump produces a near continuous outflow into the arterial blood circulation of the patient, whereas the natural heart produces a pulsatile flow with a difference of 20–50 mm Hg between highest and lowest pressure (pulse pressure). Whether the pulsatile flow or the continuous flow is better for the patient has been a matter of dispute ever since heart surgery has been practiced. For the 2–4 hour duration of routine heart surgery, advantages of pulsatile over nonpulsatile flow cannot be demonstrated. It is difficult to produce pulsatile flow with the roller pump, but several pulsating devices have been designed and are used in surgery of any duration; centrifugal pumps are especially suitable. Most surgeons prefer pulsatile flow if it is readily available.

Knowledge of the rate of bypass flow is of vital importance to the heart–lung machine operator. When roller pumps are used, the flow is simply a function of the rpm of the pump. If a centrifugal pump is used, the flow is usually measured by a Doppler sound probe in the arterial line or with an electromagnetic probe.

Several chemical constituents of the blood must also be measured frequently. They include hemoglobin, potassium, glucose, activated clotting time (ACT), and arterial blood gasses. In most hospitals, blood samples from the patient are taken to the laboratory every 20 minutes to obtain these parameters. The laboratory returns the values by telephone. Depending on the quality of the lab, it takes 5–20 minutes to obtain these vital measurements. Equipment is now available that can provide on-line measurements of many of these values. A small tube delivers blood from the heart-lung machine to the lab machine. There it is exposed to all manner of analysis without being contaminated, so that it can be returned to the heart–lung machine. These laboratory machines are very expensive and therefore are not widely in use.

E. Ancillary Equipment

Modern heart surgery requires additional equipment used in concert with the heart–lung machine. A fairly recent addition is an apparatus to cool the heart. Once the patient is on cardiopulmonary bypass, it is safe to stop the heart. Cardiac arrest is needed to facilitate the miniature vascular operations on the coronary arteries and/or to open the heart to repair or replace diseased valves.

In the past, the aorta was simply clamped, causing anoxic heart arrest. A much better method is to cool the heart. Cooling alone, however, often leads to arrhythmias including ventricular fibrillation, during which the heart contracts with short, uncoordinated movements. In this state, the oxygen need of the heart increases sharply. To avoid this complication, potassium is added to the cooling solution, resulting in immediate arrest. Over the years, special equipment has been designed to facilitate this so-called hypothermic-cardioplegic arrest. A cooling unit contains a roller pump and a reservoir for cold water. Blood is drained from the arterial line of the heart–lung machine and passed through a coil suspended in the cold water. Once the blood temperature has decreased to some 5–10°C, a crystalloid solution containing a high potassium concentration is added, resulting in a final potassium level of some 20 mEq/l. The mixture is pumped into a needle inserted into the ascending aorta. Some 300–600 ml of cold mixture are delivered to the heart every 20 minutes.

A newer apparatus is the so-called cell saver. It provides a special suction line separate from the ones connected to the heart–lung machine, which is used before and after heart–lung machine bypass. The blood is first mixed with heparin, which makes it unclottable, then is passed through a filter and emptied into a reservoir. From there, it is moved to a centrifuge where the cellular elements of the blood are isolated. These cells are returned to the patient. This machine has led to a further decrease in the use of donated blood in heart surgery.

In patients with poor renal function, an artificial kidney may be combined with the heart–lung machine. The artificial kidney is generally used just before heart surgery. However, an artificial kidney-like apparatus called an *ultra filtration device* is connected in series within the arterial line. This apparatus can be used advantageously during cardiopulmonary bypass. The blood flows between membranes that retain the cells and the larger molecules of the proteins, allowing the removal of large amounts of excess fluid, including small molecules and waste products.

II. The Heart–Lung Machine in Use

A. Hemodynamics during Bypass

Before entering the operating room, the heart–lung machine is assembled and filled (primed with fluid). Nowadays blood is not used to prime the circuit but rather crystalloid solutions that closely resemble normal blood plasma in electrolyte contents. Albumin (30–50 g) is added to increase osmotic pressure and to coat the inside of all plastic surfaces, in the hope of decreasing blood damage. The total priming volume is 2000–2500 ml. The fluid is then circulated through the machine by short circuiting the arterial with the venous line. At this point, a $0.2-\mu$ filter is temporarily inserted into the circuit to sift out all plastic particles that could have been left behind in the devices by the manufacturer. After testing the function of all moving parts the heart–lung machine is wheeled into the operating room and positioned next to the operating table.

The surgeon is usually well under way with the operation. The tubes that will connect the patient with the machine are handed over to the surgical assistant. The patient is given some 300 units of heparin per kilogram of body weight intravenously. The surgeon places so-called pursestring sutures into the front wall of the aorta and in the right atrial wall. Pursestring sutures, as the name implies, are circle-shaped stitches that can be pulled up with a resulting narrowing of the tissue captured within the circle. (The aortic and venous cannulas are inserted into the bloodstream via the center of the pursestrings.) About 5 minutes after the heparin has been given to the patient, a blood sample is analyzed for clotting, usually by measuring the ACT and comparing it to the control value. The heparinized values should be some three times the control value of 80–150 sec. During complete bypass the ACT level should be at least 500 sec. At that point the surgeon inserts a short tube (cannula) via the pursestring into the aorta with the tip pointing downstream. The cannula is connected to the arterial line of the heart–lung machine. Taking extreme care not to insert any air bubbles in the system, the surgeon inserts another cannula via a pursestring into the right atrium. This cannula is then connected to the venous line of the heart–lung machine. Two suction lines are provided by the machine. Up to this point, the surgeon has been using the suction tube provided by the cell saver machine. Once the patient is heparinized, this suction is not used any more until the heparin is neutralized. Still another tube system is provided by the apparatus to cool the heart.

With all the tubes in place, the heart–lung perfusion or cardiopulmonary bypass can be started. The venous line is opened first and as venous blood flows by gravity from the right atrium to the heart–lung machine, the perfusionist starts the arterial pump and returns the blood to the patient via the arterial line. The arterial blood pressure of the patient nearly always falls, even though the bypass flow is the same as or higher than the amount of blood that the patient's heart pumped prior to the bypass. The lowering of blood pressure is caused by a reflex causing a marked drop of the vascular tone, and thus decreasing vascular resistance. The arterial blood pressure waveform changes from pulsatile to straight (continuous flow pattern). Often, the vascular resistance recovers spontaneously but, if it does not, the perfusionist will inject drugs to contract the arteries.

During bypass, the arterial blood pressure depends on the bypass flow times the resistance of the arteries. The flow has to be at least two liters per meter square body surface area per minute (that is, 4–6 l/min) for an average adult. If the flow is less, the metabolism of the body suffers and the patient will sustain damage. Thus, the primary requirement in bypass is adequate flow. Too low a flow is considered an emergency that must be corrected immediately. With an adequate blood flow, the blood pressure is controlled by changing the arterial resistance. Powerful drugs are available to either increase or decrease it. Most operators keep arterial blood pressure between 60 and 90 mm Hg.

During bypass, the circulation is very sensitive to the blood volume. Loss in blood volume is reflected in the immediate decrease in blood flow. Volume can be lost directly into the operative field and it is the surgeon's responsibility to frequently check for accumulation of blood and to return it to the heart–lung machine via the suction. Urine production, which can be as high as 1000 cc/30 min, is another possible source of volume loss. It is replaced by adding crystalloid solution to the heart–lung machine.

As soon as the patient is on the heart–lung machine, the blood temperature is allowed to drift down to about 30°C in most routine heart operations. At the end of the operation, the surgeon initiates the discontinuation of the bypass. The bypass flow is slowly decreased, taking care not to accumulate more volume in the heart–lung machine. Thus, the natural heart gets back more and more of its original blood supply, and, as a result, pumps out more blood. Once it is certain that the heart increases its output, the bypass flow is lowered further, until finally it is stopped altogether. At this point, the heart should have taken over the pump function and put out at least as much as the last amount of blood pumped by the heart–lung machine. If the heart fails to take over the circulation, a rare event, the patient has to be put back on the heart–lung machine. Once the heart–lung machine has been switched off, the surgeon will try to return as much of the priming volume to the patient as is tolerated. The aortic cannula is removed and connected to the venous cannula. The patient is given protamine sulfate to counteract heparin, and the proper level of blood coagulability is established by remeasuring the ACT. Once the priming volume of the heart–lung machine is returned to the patient completely, the venous line is removed as well. At this point, the patient is on his or her own.

B. Metabolism during Bypass: Cooling the Body

Metabolism is evaluated by measuring the rate of oxygen usage by the tissues. At rest, it is approximately 120–150 ml/m²/min. The heart pumps 3.0–3.5 l/m²/min. The metabolism is said to be adequate if sufficient amounts of oxygen are delivered to tissues and carbon dioxide and other metabolites are removed. A number of blood parameters determine this ideal state: the hemoglobin content, the oxygen content, the viscosity, the flow, and pressure and resistance relationships, in addition to the conscious level of the patient and the body temperature.

During bypass the hemoglobin is decreased from a normal of 12–14 to 6–8 mg/100 ml due to dilution of the blood by the nonblood priming volume, which does not contain hemoglobin. The hemoglobin can be increased by supplying the patient with blood transfusions. The oxygen content of the blood is easily optimized with modern oxygenators, resulting in complete saturation of the hemoglobin. The hemodynamic parameters of flow, pressure, and resistance can be controlled readily in routine operations (see Section II, A). The blood viscosity decreases with hemodilution and increases with cooling, but in practice only at the very low temperatures of 10–15°C does it significantly affect bypass flow and tissue perfusion. At the lower temperatures, low flow is compensated by decreased demand for oxygen. Anesthesia alone decreases oxygen demand by 10–25%, depending on the depth. When patients start to wake up during surgery, the oxygen consumption rises immediately. The depth of anesthesia is actually evaluated by measuring the oxygen content of venous blood leaving the patient for the heart–lung machine. If all other factors are satisfactory, a decrease of the venous oxygen level signals the need for more anesthetic.

Probably the most effective way to decrease oxygen demand is to lower the body temperature: the demand is 50% of normal at 30°C; 25% at 22°C; and 12% at 16°C. As the temperature lowers, heart action decreases, and eventually the heart stops. Since blood circulation is controlled while on cardiopulmonary bypass, these effects are not worrisome, and are even encouraged. Thus, from the early days of heart surgery, body cooling usually to about 30°C has been used to create a margin of safety against low bypass flow conditions.

C. Effects of Heart–Lung Machine on Organs: Complications

During cardiopulmonary bypass, all organs are at the mercy of the artificial circulation. If inadequate, one or all vital organs may be damaged, sometimes irreversibly. The injection of foreign particles or air bubbles may also cause severe damage. The heart is probably the strongest organ to resist bypass-induced damage, and the most amenable to treatment. The brain is the most vulnerable. Damage to the brain, kidneys, liver, and lungs is difficult or impossible to influence therapeutically.

As cardiopulmonary bypass is started, the heart receives less blood and consequently pumps less, but, if its function is normal, the intracardiac pressures do not change. The metabolic need of the heart does not change much either even though the heart pumps less blood. The oxygen consumption of the heart depends more on the pressure against which it pumps than on the volume load. If the heart is failing, for instance in valvular heart disease, the start of the heart–lung machine may actually cause heart damage by increasing the arterial pressure. The increasing afterload creates a greater demand for oxygen, which the heart cannot always satisfy. This state of overloading lasts only a short time in the everyday practice of heart surgery, but it may become a problem when prolonged closed chest cardiopulmonary bypass techniques are used.

Once the patient is on full bypass, nearly all the venous blood returned to the heart is diverted and the cardiac output drops low enough that the heart only occasionally empties and causes a pressure wave in the aorta. At this point the heart is cooled. A tube from the cooling apparatus is inserted via a pursestring suture into the aorta close to the aortic valve. The aorta is clamped just downstream from the cooling cannula so that the arterial blood cannot reach the coronary arteries. The cooling machine pumps diluted, cold blood with a high potassium content at a rate of 200–300 ml/min into the coronary arteries (see Section I, E). A total amount of 300–400 ml is delivered every 20 minutes. The result is cardiac arrest and cooling of the heart down to 10–15°C. Under these circumstances, the heart can safely be without blood supply for 20–30 minutes. Once the operation on the heart proper is completed, the aortic clamp is removed and warm aortic blood is allowed to reperfuse the coronary arteries. This causes the heart to resume its function. Malfunction may cause mild functional changes, and, rarely, massive cardiac failure and death.

The lungs frequently suffer some decrease in function during bypass. Good measurements on pulmonary function are hard to obtain in the immediate postoperative period because the patient does not put forth much of a breathing effort. Blood gas levels usually indicate a defect in oxygen diffusion in the lung, resulting in low partial oxygen tension in the arterial blood. Generally these abnormalities subside in 2–5 days, but in some cases the pulmonary dysfunction progresses to intractable failure. This condition, called *acute respiratory distress syndrome* (ARDS) is life threatening, and requires the use of the respirator for prolonged periods of time. The mortality rate can be as high as 50%. It is of great importance that the patients be frequently encouraged to breath deeply in the early postoperative period. Another complication is *atelectasis,* that is, collapse of air space in the lungs, which arises when chest pain makes the patient suppress deep breathing efforts. It leads to decrease of arterial oxygen content, cardiac arrhythmias, and pulmonary infection. [*See* PULMONARY PATHOPHYSIOLOGY.]

The brain is very sensitive to inadequate perfusion. If the blood supply to the brain stops for only a few minutes, irreversible damage occurs and the patient never wakes up, although all of the other vital organs function properly. This complication is rare, especially because the decrease of body temperature down to 30°C gives some margin of safety. *Emboli*, consisting of foreign particles when lodged in the brain, may cause strokes, the severity of which depends on the size of the embolus and its location in the brain. Emboli may originate from the heart–lung machine, from the aortic wall at the site of the arterial cannula, and from other areas of the vascular system, notably ulcerations in the carotid arteries. Heart–lung machine emboli used to be a major problem, but not now. Particles such as air bubbles, plastic fragments, and antifoam coagulates have been located in peripheral tissues. By careful selection of the aortic cannulation site, the second source of embolization can be kept to a minimum. The effect of carotid artery disease is also minimized by screening patients before heart surgery, and cleaning out these vessels if necessary. Since such preventive measures have been taken, the incidence of postoperative stroke has dropped to some 0.5%. Psychiatric abnormalities occur at fairly high frequency but subside a few months after the surgery. Depression is quite common; fully developed psychosis is rare. Efforts to study the brain function during surgery by recording the EEG (electroencephalogram) have not given useful results.

The function of kidneys depends on a blood pressure of at least 60 mm Hg. Below that pressure, renal perfusion is still adequate for tissue survival but not for urine production. Kidney dysfunction and even failure may be produced in the immediate postoperative period by medications or by insufficient circulation. Renal failure calls for the use of a temporary artificial kidney. Complete recovery may occur.

Cardiopulmonary bypass causes damage to the blood owing to the blood's contact with foreign surfaces, turbulence, squeezing in the tubes, foaming in the bubble oxygenator, or mixture with tissue debris in the wound.

Red blood cells are destroyed during cardiopulmonary bypass as shown by an increase of hemoglobin in the plasma. Most of the damage is caused by suctioning of the blood from the operating area. If it is extensive, as during multiple valve replacements, the urine must be alkalinized by keeping the partial pressure of the carbon dioxide in the blood low and urine production high. A proportion of red cells are not destroyed but are weakened to the point that they are captured and eliminated by the body itself in the first few days postoperatively, resulting in a drop of hemoglobin without discernable blood loss.

After bypass, white cells are less effective against foreign intruders. In the postoperative period, the white blood cell count frequently increases and remains high for 5–10 days, raising the question of the presence of infections. The platelet count drops approximately 40% during bypass and increases to slightly below normal levels in a few hours afterwards. Blood proteins are denatured and destroyed but not to a dangerous extent. The clotting factors are altered significantly, primarily as a result of the administration of heparin before and protamine sulfate after bypass.

III. Applications of the Heart–Lung Machine

A. Heart Surgery

During heart surgery the machine is needed to exclude the heart from the blood circulation, stop, and open it. The most frequent operation is the coronary artery bypass operation, in which cardiopulmonary bypass is needed for 1–3 hours.

In valve surgery, cardiopulmonary bypass must last for a longer time, up to 6 hours. The wound area suction is used extensively in this operation, and

thus more blood damage is incurred. In heart transplantation, the heart–lung machine is used for about 3 hours.

During heart surgery for congenital defects, a heart–lung machine miniaturized to adapt to babies is used. Whole body cooling down to 10°C is generally applied.

B. Other Applications: Lung Surgery and Aortic Surgery

The heart–lung machine has been used in the past for complicated pulmonary operations, but this use is now discontinued. Cardiopulmonary bypass is used in operations on the first part of the aorta from the aortic valve down to the origin of the artery to the left arm. The rest of the intrathoracic aorta can be operated on with or without bypass. If bypass is used, it does not have to include the heart and the lungs. A simple tube from the aortic arch to the descending aorta suffices, although some surgeons use complete heart-lung bypass. The simple bypass has the advantage that generalized heparinization is not needed, an important consideration since descending thoracic aortic surgery is frequently done on patients shortly after major accidents.

C. The Heart–Lung Machine as an Assist Device for Heart and Lung Failure

Early in the development of the heart–lung machine, it was recognized that cardiopulmonary bypass could be used for assistance to the failing heart. However, prolonged bypass periods are needed, causing considerable damage to blood and its cellular elements, even with the advanced equipment presently available. Uncontrollable bleeding is also a problem, which becomes particularly severe if the patients are heparinized.

Cardiopulmonary bypass can be used to help patients who become very unstable just before surgery, especially during complicated procedures in the heart catheterization laboratory. Cannulas are inserted into a femoral artery and a femoral vein, and are connected to the heart–lung machine. Bypass flows of 2–4 l/min can be obtained and with that, circulation is usually sufficiently restored. If the increased arterial pressure puts a heavier workload on the heart, threatening heart failure, the left side of the heart must be decompressed. This result can be obtained by applying short periods of external cardiac massage, or, more effectively, by draining the left ventricle. The Archimedean screw pump proved very effective in pumping blood from a cannula inserted in the ventricle via the aorta.

Cardiopulmonary bypass using peripheral cannulation can be used for the treatment of cardiogenic shock after myocardial infarction. Since the refinement of equipment and the recognition of the need to decompress the left ventricle, the application of the bypass in this area has been revived during the last few years. Given the high incidence of myocardial infarction this field of application may grow significantly.

Another area of application of the heart–lung machine is treatment of severe pulmonary failure. The bypass can be lifesaving in some 50–90% of newborn babies; in adults, it is applied in patients who have acute respiratory distress syndrome. In these cases, only the lungs are bypassed, not the heart. The technique is called extracorporal membrane oxygenation (ECMO). Blood can be drained from and returned to either a femoral vein or an artery. In adults, the blood circuit is usually from vein to vein. The bypass is operated at a low flow of some 500–1000 ml/min, sufficient to normalize the blood gasses even though the patient only breathes two to four times per minute. Oxygen transfer is affected primarily in the lungs, whereas the elimination of carbon dioxide occurs in the artificial lung. Such a bypass can be carried on for many days until recovery of pulmonary function occurs. The major complication is profuse bleeding. Equipment improvement may control this complication in the future. Using materials coated with heparin makes heparin-

ization of blood unnecessary, reducing bleeding danger.

Variants of the cardiopulmonary bypass are the *left heart bypass* and *right heart bypass,* in which the patient's own lungs are used for oxygenation. Left heart bypass is of interest because it effectively decompresses the left ventricle; it has therefore been studied extensively for assisting the failing heart. In heart surgery it is used as a last resort in patients who cannot be weaned off the heart–lung machine. The mortality of such patients, however, is understandably high. Implantable left heart bypass pumps have been used in these cases for prolonged assistance, up to 10 days. The ultimate permanent application of both the left and right heart implantable pump systems is the artificial heart.

Bibliography

Akutsu, Tetsuzo, ed. (1986). "Artificial Heart I." Springer-Verlag, Tokyo, Berlin, Heidelberg, and New York.

Friedman, Eli A., ed. (1988). "ASAIO Transactions," vol. 34, no. 3. J. B. Lippincott Co., Philadelphia.

Galletti, Pierre M., and Brecher, Gerhard A. (1966). "Heart–Lung Bypass: Principles and Techniques of Extracorporeal Circulation." Grune and Stratton, New York.

Ionescu, M. F., ed. (1982). "Techniques in Extracorporeal Circulation," 2d ed. Butterworths, London.

Reed, C. C. (1985). "Cardiopulmonary Bypass." Texas Medical Press, Stafford.

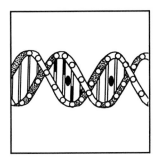

Heat Shock

MILTON J. SCHLESINGER, NANCY C. COLLIER, *Washington University School of Medicine*

Glossary

BiP Immunoglobulin heavy-chain binding protein

Chaperonin Family of proteins whose proposed role is to ensure that the folding of certain polypeptides and their assembly into oligomeric structures occur correctly

Glucose regulated protein (GRP) Number following GRP refers to the protein's subunit molecular weight; thus, GRP94 is the glucose-regulated protein with a subunit molecular weight of 94,000

GRO EL Protein in *Escherichia coli* required for cell division and for bacteriophage replication

Heat shock protein (HSP) Number following HSP refers to the protein's subunit molecular weight (cf. above, GRP)

Heat shock promoter element Sequence in the DNA, usually in the 5′ nontranscribed region of a heat shock gene, that controls expression of that gene

Heat shock transcription factor Protein that binds to the heat shock promoter element and affects expression of a heat shock gene after a heat shock

Rubisco Ribulose bisphosphate carboxylase-oxygenase

HEAT SHOCK refers to the activation of a set of evolutionarily conserved genes and the subsequent synthesis of their encoded proteins as a result of an increase in temperature above that normally experienced by a cell. The temperature range required to initiate a heat shock generally does not exceed a 10% change from the physiological temperature; thus, heat shock can occur in organisms that normally live in temperatures ranging from 4°C to >60°C. Many heat shock genes are also activated by other agents that stress the cell such as organic solvents, heavy metals, nutrient deprivation, oxidants, and viral infection. Some heat shock genes are activated during normal growth and development of the organism.

I. Background

The internal temperature of the human body can fluctuate only within a very narrow range without leading to death of the individual. Even within this range, severe damage can occur; however, evolution has built within the cells of the human as well as those of virtually all species in the living world—including the bacteria—a set of biochemical activities that protects the cell during a thermal stress and allows for its recovery later when temperatures return to normal. Not all of the mechanisms employed by a cell for thermal protection are known, but one that has been studied in considerable detail is the activation of a small set of genes almost immediately after the cell senses the temperature shock.

The first observation of a heat-inducible gene activation was the appearance of new puffs in the chromosome of *Drosophila* embryos that had been moved from their normal temperature of 25°C to 30°C. Puffs are sites of intense RNA transcription, and the RNAs from the heat shock puffs are selectively translated to produce high levels of proteins called heat shock proteins. Similar proteins were subsequently found in heat-shocked cells obtained from many different organisms including human, chicken, plant, yeast, and bacteria (see Fig. 1), although the temperatures required to activate their genes were vastly different. The major proteins fell into four groups based on their size: small proteins of 15,000–30,000 molecular weight, intermediate proteins of 50,000–60,000 and 70,000–80,000, and large proteins of 90,000–100,000. Many features of the structure and function of these proteins are described below; however, it is the formation of these proteins and the factors controlling the expression of their genes that provide the operational definition for the heat shock response.

II. Heat Shock Genes and Their Expression

A. Features of a Heat Shock Gene

Heat shock genes can be distinguished from other protein-encoding genes by a short sequence of DNA, noted as the heat shock promoter element, which is located in the 5′ nontranscribed portion of the gene positioned about 80–200 nucleotide base pairs from the start site of messenger RNA (mRNA) transcription. In eukaryotic cells, the promoter consists of contiguous arrays of an inverted repeat of five nucleotide base pairs containing nGAAn, where n can be any of the four bases (see Fig. 2).

Another feature of many heat shock genes is the lack of introns (noncoding sequences) within the protein-coding region of the gene—a property that enables heat shock mRNAs to bypass the normal pre-mRNA splicing activities that are inhibited in a heat-shocked cell. Also, some sequences in the 5′ and 3′ transcribed but noncoding domains of heat shock genes stabilize the mRNA during heat shock and confer a selectivity for its translation by the cell's protein synthetic machinery. [*See* GENES.]

B. Expression of Heat Shock Genes

Activation of all eukaryotic heat shock genes occurs when the heat shock promoter element is occu-

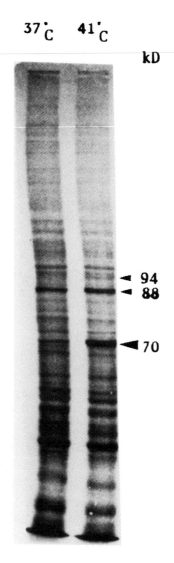

FIGURE 1 Synthesis of heat shock proteins in human endometrium incubated at 41°C. Tissue samples were given radioactive amino acids and incubated for 2 hr at either normal or heat-shocked temperatures. Proteins were extracted and separated by electrophoresis in polyacrylamide gels. The pattern shown is an autoradiogram; heat shock proteins are indicated. [Reproduced, with permission, from A. Ron and A. Birkenfeld, 1987, Stress proteins in the human endometrium and decidua. *Hum. Reprod.* **2**, 277–280, IRL Press, McLean, Virginia.]

pied by a DNA-binding protein called the heat shock transcription factor. The transcription factor is a protein of about 90,000 molecular weight and is present in the nonstressed cell. Almost immediately upon heat shock, this protein is phosphorylated and binds to the promoter element, where it interacts with other proteins bound to adjacent regions of the

DNA. The complex leads to transcription of the heat shock gene. In the *Drosophila* chromosome, an RNA polymerase is already bound to and transcribes a short region of the heat shock gene in the nonstressed cell, but it is blocked in its ability to proceed beyond a specific region of the gene. Binding of the heat shock factor relieves this inhibition and allows for the complete transcription of the gene. [*See* DNA AND GENE TRANSCRIPTION.]

The transcription factors are different in bacteria. In *Escherichia coli,* the heat shock transcription factor is an isoform of the sigma factor, the regulatory subunit of the DNA-dependent RNA polymerase. The heat shock sigma factor is made during normal cell growth but is degraded very rapidly and its steady-state level is low. Heat shock blocks degradation of this protein, resulting in higher levels of the factor, which then replaces the normal sigma factor in the RNA polymerase complex. This new complex binds selectively to heat shock promoter elements, thereby transcribing only heat shock genes.

The biochemical mechanisms that sense heat shock and produce the specific alterations in heat shock transcription factors are unknown. In bacteria, some evidence indicates that a thermal unfolding of proteins is the trigger. In the eukaryotic cell, the thermal-sensitive element could be the heat shock transcription factor itself or a protein bound to it. The latter might be a heat shock protein.

III. Heat Shock Proteins and Their Functions

A. General

The precise number of heat shock proteins in an organism is not known, but in the human it is unlikely to exceed 50. Only a few have been studied in detail, but several features have emerged that are common to the major heat shock proteins of many species. Perhaps the most striking characteristic is the high degree of conservation in protein structure that exists between the human proteins and those of highly divergent species. For example, about 50% of the amino acid sequences in the human 70,000-molecular weight heat shock protein is found in a heat shock protein of similar size in *Escherichia coli,* and antibodies raised against a chicken heat shock protein of this size recognize the human and the yeast protein.

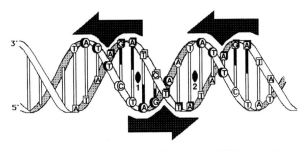

FIGURE 2 A segment of double-stranded DNA containing a heat shock promoter element (*n*GAA*n*). Shaded large arrows are heat shock transcription factors bound to DNA. [Based on a model, with permission, presented by O. Persic, H. Xiao, and J. T. Lis, 1989, Stable binding of drosophila heat shock factor to head-to-head and tail-to-tail repeats of a conserved 5 bp recognition unit. *Cell* **59,** 797–806, Cell Press.]

A second important property of the major heat shock proteins is that they can be grouped into distinct protein families. Individual members of a family are closely related in structure but are distinct with regard to their location in a cell as well as by the conditions controlling their synthesis and turnover. Features of several major heat shock protein families are noted below.

B. The HSP70 Protein Family

The protein whose synthesis appears most rapidly and most intensely in the human cell after heat shock has a molecular weight of 70,000. There is, however, a basal level of this protein in nonstressed cells, and the protein also appears at specific stages during the normal cell division cycle as well as after infection of human cells with viruses. Thus, expression of the human hsp70 gene is regulated by multiple promoter elements including one responding to the heat shock transcription factor. HSP70 is usually present as a dimer in the cytoplasm, but most of it migrates to the nucleus during a heat stress where it binds tightly to proteins in the nucleolus—the site of ribosome assembly.

The human HSP70 family contains four other proteins that are slightly larger in molecular weight. One of these, HSP72, is found only after a heat shock. Another of similar size is present at high levels in normal cells and is identical to a protein that disassembles clathrin-coated vesicles. The third and fourth members of the family are localized to the endoplasmic reticulum and the mitochondria of the cell. The former is a glycoprotein called glucose-regulated protein 78, GRP78, and is identical to BiP, a protein that was first identified by its binding to the immunoglobulin heavy chain in preB-cells that did not make the immunoglobulin light chain. In the absence of light chains, immunoglobulin heavy chains denature; however, BiP prevents this inactivation. Later, when light chains are produced, the heavy chains are released from BiP and complex with light chains to form active immunoglobulins. The mitochondrial protein is termed GRP75 and is closely related to a similar protein found in yeast mitochondria where its function is essential for survival of the organism. Synthesis of these GRPs is induced by agents that affect protein glycosylation or increase intracellular calcium concentrations.

The functions of HSP70 have been surmised from a variety of activities detected for several of the HSP70 family members from different organisms. Some of these have been described above. In yeast, an HSP70-like protein is required to unfold certain cytoplasmic proteins destined for import into mitochondria or for transport into the endoplasmic reticulum and ultimately secretion from the cell. In *Escherichia coli,* the HSP70 homologue (DNA K) participates in the initiation of DNA replication for bacteriophages and for the host cell at high temperatures.

In all of these examples, the HSP70 protein first forms a tight complex with a partially folded polypeptide or a complex of polypeptides and then—in a reaction driven by adenosine triphosphate (ATP)—modifies the bound protein so that it either unfolds or the complex dissociates (see Table I). In fact, HSP70 has a weak ATPase activity and the

amino-terminal domain of the protein contains an ATP-binding site. During heat shock, HSP70 is postulated to bind to protein complexes that are particularly sensitive to temperature denaturation, to nascent polypeptide chains, or to polypeptide subunits that have not yet folded to a thermostable conformation. In these complexes, the proteins are protected from denaturation, but when the cell returns to its normal physiological state, HSP70 is released, thereby allowing the protein to resume its normal structure and function.

C. The HSP60 Protein Family

This group of proteins is the most highly conserved in structure among all of the heat shock proteins. Tetradecamers of the 60,000-molecular weight subunit with a sevenfold axis of symmetry are localized to the inner compartment of the cell's mitochondria where they catalyze the refolding of polypeptides newly imported across the mitochondrial membrane. These polypeptides are encoded by nuclear genes and are synthesized on cytoplasmic polysomes. They contain amino acid sequences that target them to receptors on mitochondrial membranes, but to be transported through the mitochondrial membrane, they must first be unfolded—a reaction catalyzed by an HSP70-family protein. Upon emerging into the lumen of the mitochondria, a refolding occurs, catalyzed by the HSP60-type protein. The term chaperonin has been invoked to identify this kind of interaction. In plants, the chaperonin protein is the rubisco-binding protein. It is abundant in the chloroplast, where it participates in formation

TABLE I Heat Shock Protein Complexes

Heat shock protein	Target protein	Result after ATP action
HSP70 family member	Clathrin-coated vesicles	Dissociation
	Premitochondrial and presecreted proteins	Unfolding
	DNA replication–initiation complex	Dissociation
	Newly imported, unfolded proteins in the lumen of the endoplasmic reticulum	Dissociation
	Unfolded forms of tumor-suppressor proteins	?
	Nucleolar proteins (pre-mRNA splicosomes; pre-rRNA ribosomes)	Dissociation
	Steroid hormone receptors	Dissociation
HSP60 family member	Large rubisco subunit	Folding
	Bacteriophage structural proteins	Assembly
HSP90 family member[a]	Steroid hormone receptors	Activation of the receptor
	Phosphokinases (oncogene products)	Activation of the kinase
	Eukaryotic initiation factor 2 kinase	Activation of the kinase
	Actin, tubulin	?

[a] Requires binding of ligands other than ATP.

of the multimeric complex of enzymes that fix carbon dioxide into organic material. The bacterial protein, called GRO EL, is needed for assembly of bacteriophages and for normal cell growth. Antibodies directed against the protein isolated from tetrahymena recognize homologous proteins in *Escherichia coli* and in humans.

D. The HSP90 Protein Family

Members of this group are highly phosphorylated, abundant cytoplasmic proteins. About 50% of their sequences are conserved among yeast, trypanosomes, fruit flies, chickens, and mammals. A bacterial form is 40% related in sequence to that of other species. One member of the family, noted GRP94, is localized to the eukaryotic cell's endoplasmic reticulum.

A number of other cellular proteins form a stable complex with HSP90. These include viral and cellular protein kinases, steroid hormone receptors, tubulin, and actin. When complexed with HSP90, the kinases and receptors are inactive. Binding of hormone releases the HSP90 from the receptor complex and unmasks the DNA-binding domain of the receptor.

The HSP70, HSP60, and HSP90 protein families share a common function in that all form dissociable complexes with a variety of cellular polypeptides (summarized in Table I).

E. The HSP15–30 Protein Family

Proteins in the 15,000–30,000-molecular weight range are synthesized in almost all cells after heat shock and are the most abundant heat shock proteins in plants. Their function is unknown, but they are essential for survival and recovery of many organisms from stress. Many of these proteins form very large polymers, which appear as heat shock granules in the cytoplasm. Some of these granules, particularly in stressed plant cells, are highly regular in structure and contain specific small RNAs. Upon recovery from stress, the granules dissociate. A portion of the amino acid sequence for many of the small heat shock proteins is closely related to a sequence in bovine lens α-crystallin, and this may be the region responsible for polymerization.

Some of the genes encoding these proteins in fruit flies are activated during late larval and early pupal stages of development, possibly by the hormone ecdysone. In yeast, the protein is synthesized when

cells stop growing and when they initiate sporulation.

F. The Ubiquitin System

All cells contain enzymes or enzymatic pathways for degrading intracellular polypeptides, proteins whose activities must be carefully regulated during normal cell function as well as proteins that become misfolded and denatured as the result of stress such as a heat shock. In the eukaryotic cell, a universal system for degrading intracellular proteins to small peptides and amino acids employs a heat stable cofactor called ubiquitin, a polypeptide of 76 amino acids whose sequence is almost invariant from plants to humans. A set of enzymes activates ubiquitin in the presence of ATP and covalently attach it to lysine groups on the protein destined for degradation. Multiple ubiquitin molecules are subsequently added and the polyubiquitinated protein is recognized by a large multicomponent complex containing proteases. Proteins targeted for ubiquitination have either charged or very bulky hydrophobic amino acids at their amino terminus or have an abnormal, misfolded domain within the body of the protein. The latter would be most likely present during a heat shock.

Genes encoding several of the enzymes involved in the ubiquitin pathway are activated by heat shock. These include a gene-encoding ubiquitin itself and two of the enzymes that transfer the activated ubiquitin to the target protein. Other cellular components utilize ubiquitin, and some of these are also affected by a heat shock (see below).

Ubiquitin is not present in prokaryotes; however, the gene (lon) that encodes the ATP-dependent protease La in *Escherichia coli* is activated by heat shock. La probably functions like the ubiquitin system to remove aberrant proteins formed during stress in bacteria.

Thus, proteins damaged (i.e., unfolded) by heat shock have two alternative fates: they may be rescued and repaired by a heat shock protein, or they may be proteolytically degraded.

G. Other Proteins

In addition to those proteins cited above, the synthesis of the glycolytic enzymes enolase, glyceraldehyde-3-phosphate dehydrogenase, and phospho-glycerate kinase are stimulated by a heat shock. Thermal stress also increases the formation of some extracellular and cell-surface proteins such as thrombospondin, C-reactive protein, and a collagen-binding protein.

IV. Events in a Heat-Shocked Cell

The activation of heat shock genes is only one of several changes that occur in the cell in response to a stress. Many of the cell's metabolic activities are affected, and virtually every organelle in the cell shows some modification. There are also changes in the cell's morphology and cytoskeleton. As expected, the extent of these alterations is closely related to the intensity and duration of the stress. Below are cited some of the more profound changes.

Among the cytoskeletal elements—microfilaments, microtubules, and intermediate filaments—it is the latter that is most sensitive to stress. Upon a mild heat shock, the intermediate filament network collapses to a perinuclear cytoskeletal ring (see Fig. 3) but reappears in its extended form shortly after removal of the stress. Formation of heat shock proteins appears to be necessary for this recovery. With more severe stress, actin bundles are found in the nucleus and cytoplasmic microtubules are destroyed.

Metabolic activities in the nucleus are among the most sensitive to stress. DNA synthesis slows dramatically, ribosomal RNA precursors accumulate, and splicing of mRNA is inhibited. Histones are modified by phosphorylation and deubiquitination—changes that are normally found during condensation of the chromatin.

The protein synthetic machinery is not as sensitive but polysomes dissociate immediately after a brief severe heat shock. Most proteins are stable to temperatures in the physiological range of a heat shock; however, an increase in the level of ubiquitin-conjugated proteins indicates increasing numbers of denatured proteins in the heat-shocked cell.

At the plasma membrane, there is a decrease in sodium–potassium ATPase activity. With severe stress, internal cell membranes are affected: the Golgi stack and other cytoplasmic vesicles are disrupted, and the endoplasmic reticulum is swollen and fragmented. In addition, mitochondria move to the perinuclear space. Subsequently, they swell, the normal cristae organization is altered, and aerobic ATP generation decreases. The cell shifts to the glycolytic pathway for ATP generation and accu-

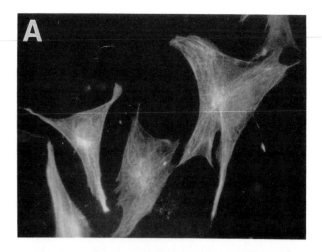

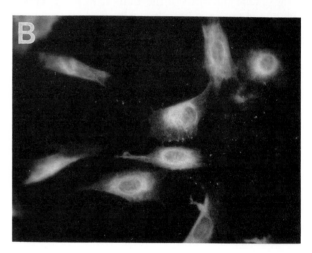

FIGURE 3 Effect of heat shock on the intermediate filament network in chicken embryo fibroblasts. The normal pattern (A), revealed by immunofluorescence staining with antibodies to vimentin, is immediately altered after a heat shock (B). [Reproduced from N. C. Collier and N. Schlesinger (1986). The dynamic state of heat shock proteins in chicken embryo fibroblasts. *Journal of Cell Biology* **103**, 1495–1507 by copyright permission of the Rockefeller University Press.]

mulates lactic acid. Levels of calcium also increase in the heat-shocked cell.

Proteins are modified in a variety of ways after a heat shock (e.g., the ribosomal subunit S6 is dephosphorylated).

V. Stressors Other Than Heat Shock

In addition to heat shock, a variety of other kinds of environmental abuses reprogram the cell's protein synthetic machinery to produce heat shock proteins or related family proteins. Table II lists many of these together with the types of stress proteins elicited. Whether or not these different stressors act through a common mechanism to signal the activation of heat shock genes is unknown; however, heat shock genes have multiple promoter elements, and some of these are used for factors specific to a particular stressor. Even where a heat shock element is the effective promoter, activation of the heat shock transcription factor could be different for distinct stressors.

VI. Thermotolerance

Damage caused by a severe heat shock can be reduced if the cell or organism is first subjected to a milder heat stress. This phenomenon is called thermotolerance, but it is not limited to either thermal stress or thermal protection. For example, a mild stress by ethanol will protect a cell from heat shock damage, and a mild heat shock protects cells from ethanol.

The molecular mechanisms leading to stress tolerance have not been completely defined; however, a strong correlation exists between synthesis and persistence of heat shock proteins and thermotolerance. Agents that induce tolerance also induce the heat shock proteins, and some tissue culture cell lines, selected for resistance to heat shock damage, constitutively synthesize high levels of certain heat shock proteins such as HSP70. Furthermore, tolerance cannot be induced during stages of tissue development where heat shock protein induction is blocked, nor can tolerance occur when cells are stressed by growth in the presence of analogues of amino acids. In the latter case, the analogues interfere with the formation of active forms of the heat shock proteins. There are, however, examples where thermotolerance is achieved in the absence of heat shock protein synthesis and other cases where thermotolerance does not occur despite high levels of heat shock proteins. Thus, while heat shock proteins may be necessary for thermotolerance, they may not be sufficient. [See THERMO-TOLERANCE IN MAMMALIAN DEVELOPMENT.]

VII. Heat Shock Proteins and the Immune Response

The immune system of higher vertebrates consists of a complex, interactive set of cells whose activities lead to secretion of circulating proteins (the immunoglobulins) and to activation of specific types of cells (the T-lymphocytes), which together protect the organism from a wide variety of potentially toxic materials and from microorganisms whose growth can cause severe damage and death. Among the latter are various parasitic and nonparasitic protozoa, fungi, bacteria, and viruses whose life cycles involve growth and development at widely different temperatures. When many of these invade the human body (e.g., by transmission from insects), the microorganism experiences an immediate heat shock and responds by producing heat shock proteins. The immune response to infection by many microorganisms elicits immunoglobulins and T-lymphocytes that recognize heat shock proteins as the major foreign antigens. As a result, the organisms are destroyed and the infection is limited. Because the structure of heat shock proteins is so widely conserved, a broad cross-protection to many different types of invasive microorganisms could occur

TABLE II Nonheat Shock Inducers of Heat Shock Protein Synthesis

Type	Class	Examples
Stressors that induce mainly the heat shock proteins	Oxidizing agents and drugs affecting energy metabolism	arsenite, peroxide
	Transition series metals	mercury, copper
	Chelating drugs	salicylate
	Amino acid analogues	canavanine, azetidine
	Inhibitors of gene expression	puromycin
	Steroid hormones	dexamethasone
	Virus infection	herpes simplex
	Others	ethanol, wounding, recovery from anoxia
Stressors that induce mainly the glucose regulated proteins	Glucose deprivation	deoxyglucose, tunicamycin
	Membrane permeants	calcium ionophores

early in the life of the individual. These same proteins exist in normal human cells; however, they are present at low levels and confined inside a cell and, thus, are hidden from the immune recognition system. In abnormal, stressed, or damaged cells, however, there are higher levels of these proteins, and some may be present at the cell surface. This situation would lead to immune-mediated destruction of these cells—a phenomenon referred to as immune surveillance. In fact, a major mouse tumor antigen is identical to HSP86, a member of the HSP90 family of heat shock proteins. Immune recognition of human heat shock proteins that are induced by a localized inflammation or a systemic febrile response could also lead to autoimmune diseases. In fact, individuals suffering from the latter possess antibodies and T-lymphocytes recognizing heat shock proteins. [*See* IMMUNE SURVEILLANCE, IMMUNE SYSTEM.]

Bibliography

Bond, U., and Schlesinger, M. J. (1987). Heat shock proteins and development. *Adv. Genet.* **24,** 1–29.

Lindquist, S., and Craig, E. A. (1988). The heat shock proteins. *Ann. Rev. Genet.* **22,** 631–677.

Neidhardt, F., and VanBogelen, R. (1987). Heat shock response *In* "*Escherichia coli* and *Salmonella typhimurium:* Cellular and Molecular Biology" (F. C. Neidhardt, ed.). American Society of Microbiology, Washington, D.C.

Nover, L. (1984). "Heat Shock Response of Eukaryotic Cells." Springer-Verlag, Berlin.

Pelham, H. R. B. (1989). Heat shock and the sorting of luminal ER proteins. *EMBO J.* **8,** 3171–3176.

Rechsteiner, M. (1988). "Ubiquitin." Plenum Press, New York.

Rothman, J. E. (1989). Polypeptide chain binding proteins: Catalysts of protein folding and related processes in cells. *Cell* **59,** 591–601.

Schlesinger, M. J., Ashburner, M., and Tissieres, A. (1982). "Heat Shock: From Bacteria to Man." Cold Spring Harbor Press, Cold Spring Harbor, New York.

Young, R. A., and Elliott, T. J. (1989). Stress proteins, infection, and immune surveillance. *Cell* **59,** 5–8.

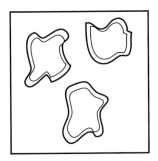

HeLa Cells

SHIN-ICHI AKIYAMA, *Kagoshima University*

Glossary

Hybrid Fusion product of somatic cell hybridization
Multidrug-resistant Simultaneously resistant to many drugs that are structurally and functionally unrelated to the selecting drug

THE HeLa CELL was isolated from a biopsy of the cervical carcinoma of a 31-yr-old black female, Henrietta Lacks (HeLa), in February 1951. Researchers suggest that HeLa is an adenocarcinoma rather than an epidermoid carcinoma. Currently, there are three original HeLa lines in the American Type Culture Collection (ATCC): HeLa (CCL2), HeLa 229 (CCL 2.1), and HeLa S3 (CCL 2.2). HeLa 229 and HeLa S3 are derivatives of HeLa. As the first established human cell line, the HeLa cell has contributed for over three decades to the development of biology, virology, and somatic cell genetics.

I. History of HeLa Cell Lines

On February 1, 1951, a 31-yr-old woman complaining of intermenstrual spotting appeared at the gynecology outpatients' department of Johns Hopkins University Hospital. A biopsy revealed carcinoma. On February 8, 1951, the HeLa cells were established from a biopsy of the cervical carcinoma. The original biopsy of the cervical carcinoma had been misinterpreted as epidermoid carcinoma. Twenty years after the first diagnosis, the original slide was re-examined and diagnosed as a very aggressive adenocarcinoma of the cervix, an unusual type of cervical cancer. Although it was diagnosed to be in an early stage and was treated with radium therapy, the patient died in August 1951.

For about 1 yr, the HeLa cells were maintained in a plasma clot culture in roller tubes. In 1952, the cells were maintained serially in monolayer cultures in a medium composed of either human placental serum, human adult serum, or human ascitic fluid (50%); chicken embryonic extract (2 or 5%); and balanced salt solution (Hanks', 48 or 45%).

II. Characteristics of HeLa Cell Lines

The three original HeLa cell lines have been retained in the ATCC repository. HeLa (CCL2) is considered most similar in characteristics to the cells described in classic studies. Both HeLa 229 and HeLa S3 are derived from HeLa (CCL2) but differ slightly from each other in certain respects. Some characteristics of these cell lines are listed in Table I.

The genetic characteristics of HeLa cells have been extensively investigated. HeLa cells express M, N, S, s, Tj[a], and HLA (A28, A3, BW 35) antigens and the enzyme phenotypes listed in Table II. In Caucasians, the glucose-6-phosphate dehydrogenase(G6PD)(A) frequency is essentially zero with the exception of the Mediterranean area; this characterizes the difference between HeLa and Caucasoid-derived lines.

Marker chromosomes characteristic of HeLa cells in culture are revealed by chromosome banding. There are four HeLa markers (1–4; Table III)

TABLE I Some Characteristics of HeLa Cell Lines[a]

Characteristic	HeLa (CCL2)	HeLa 229	HeLa S3
Number of serial subcultures from tissue of origin	Unknown; 90–102 from culture received by W. F. Scherer, 1952	Unknown; 78–88 from culture sent to W. F. Scherer in 1952	Unknown
Morphology	Epitheliallike	Epitheliallike	Epitheliallike
Virus susceptibility	Susceptible to poliovirus type 1 and adenovirus type 3	Susceptible to adenovirus type 3 and vesicular stomatitis (Indiana strain) virus; this strain is 2–3 logs less sensitive to poliovirus types 1–3 than HeLa CCL2	Susceptible to poliovirus type 1, adenovirus type 5, and vesicular stomatitis virus
Chromosome number	Mode 84 Range 58–179		Mode 68 Range 51–74
HeLa marker chromosomes	1 copy of No. 1; 1 copy of No. 2; 4–5 copies of No. 3; 2 copies of No. 4		1 copy of No. 1; 1 copy of No. 2; 2 copies of No. 3; 1 copy of No. 4
Tumorigenic	in nude mouse		in cheek pauch of hamster
Submitted to ATCC by:	W. F. Scherer	J. T. Syverton	T. T. Puck

[a] From American Type Culture Collection (ATCC).

with marker 3 characteristically present in more than one copy per cell. HeLa cells lack the Y chromosome. The determination of stable genetic markers on cultured cells is a powerful tool to estimate cellular identity or contamination. [*See* DNA MARKERS AS DIAGNOSTIC TOOLS.]

Many cell lines subsequently isolated are suspected of being HeLa contaminants because they contain genetic markers of HeLa. These lines exhibit (1) G6PD(A), (2) phosphoglucomutase (PGM(1-1) and PGM(1-3)), (3) absence of a Y chromosome by fluorescent staining, and (4) possession of a complex of HeLa chromosome markers. According to these criteria, >40 established human cell lines seem to be contaminated by the HeLa cell line. A list of cell lines with the characteristic peculiar to HeLa cells has been reported (Table IV). Many investigators, however, have reported con-

siderable differences in the chromosome composition of these cell lines. Karyotypes of a heteroploid cell line reportedly change only slightly in a limited number of chromosomes even after years of continued culture; thus, whether all the cell lines indicated as HeLa contaminants should be classified as HeLa cells because of their possession of those HeLa-specific markers or each should be considered a distinct line bearing significant HeLa cell characteristics is still unclear.

TABLE II Enzyme Phenotypes in HeLa Cells

Enzyme	Phenotype
Adenylate kinase	AK(1-1)
Adenosine deaminase	ADA(1-1)
Acid phosphatase	AcP(AB)
Phosphoglucomutase	PGM(1-1)
6-Phosphogluconate dehydrogenase	6PGD(A)
Glucose-6-phosphate dehydrogenase	G6PD(A)
Esterase D	ESD(1)
Peptidase D	PEP-D(1)

TABLE III HeLa Chromosome Markers[a]

Designation	Description
Number 1 marker	Short arm and centromere of No. 1 chromosome and arm of No. 3 chromosome
Number 2 marker	Probably short arm of No. 3 chromosome and long arm of No. 5 chromosome
Number 3 marker	Small isochromosome in two or more copies
Number 4 marker	"Dull" short arm and long arm of No. 9 chromosome or No. 18 chromosome with bright fluorescence (weakly staining short arm of No. 22 chromosome and long arm and centromere of No. 12 chromosome [W. A. Nelson-Rees *et al.*])

[a] From Miller *et al.*

TABLE IV Cell Lines with Characteristics Peculiar to HeLa Cells[a]

Designation	Source	Designation	Source
HeLa (adenocarcinoma cervix)	ATCC	SA4 (TxS-HuSa₁) (liposarcoma)	C. Pfizer, Inc.
HeLa (=CCL2)	ATCC via A. Deitch	SA4	D. Morton
	A. Mukerjee	RT4 (carcinoma, bladder)	J. Leighton via N. Abaza
	V. Klement from Flow Labs., Inc.	Detroit 30A (carcinoma, ascitic fluid)	W. D. Peterson, Jr.
	G. Gey		
	Unlisted	Detroit 98 (=CCL18) (sternal marrow)	ATCC
	Four individuals, unlisted	Detroit 98s (=CCL18.1)	ATCC
	Grand Island Biological Co.		
	N. Differante	Detroit 98/AG (CCL18.2)	ATCC
HeLa 229 (=CCL2.1)	ATCC	Detroit 98AT-2 (=CCL18.3)	ATCC
HeLa S3	G. Nette from E. Robbins		
	L. Levintow	Detroit 98/AHR (=CCL18.4)	ATCC
HeLa S3g	M. Griffin via G. Melnykovych	FL (=CCL62) (amnion)	ATCC
HeLa S3k	K. Kajievara via G. Melnykovych	CaOV (carcinoma, ovary)	N. P. Mazurenko
KB (carcinoma, oral)	Unlisted	J96 (leukemic blood)	T. A. Bektemiov
KB (=CCL 17)	ATCC	JIII (monocytic leukemia) (=CCL24)	Commercial, unlisted
	S. Mak		Unlisted
	V. Klement from MBA		ATCC
	H. Sussman	T-9 (transformed normal diploid)	O. G. Andzaparidze
	E. Priori		
	Commercial, unlisted	DAPT (astrocytoma, piloid)	A. O. Bykovsky
H.Ep.-2 (carcinoma, larynx)	Unlisted	AO (amnion)	A. O. Bykovsky
H.Ep.-2 (=CCL 23)	ATCC	KP-P₁ (carcinoma, prostate)	P. Lee via M. Glovsky
	Individual, unlisted		
	P. Dent	ElCo (carcinoma, breast)	R. Patillo
	V. Klement from MBA	HCE (carcinoma, cervix)	D. Brown
	M. Webber		
	K. McCormick	CMP (adenocarcinoma, rectum)	Unlisted
H.Ep.-2 (clone)	K. V. Ilyin		
AV3 (amnion)	Unlisted	CMPII C2	D. Rounds via J. Kim
AV3 (=CCL21)	ATCC	JHT (placenta)	J. Cho via J. W.-Peng
AV3 (103)	I. Keydar from ATCC	OE (endometrium)	The originators
AV3 (F-49-1)	P. Peebles from ATCC		P. Di Saia via L. Milewich
L132 (=CCL5) (lung)	ATCC	SH-2 (carcinoma, breast)	The originators
L132 (G-38-7)	P. Peebles from ATCC		G. Seman via R. Miller
Chang liver (liver)	Unlisted	SH-3 (carcinoma, breast)	The originators
Chang liver (=CCL13)	ATCC		G. Seman via R. Miller
	R. Chang	ESP₁ (Burkitt lymphoma, American)	P. Price, from E. Priori
	Individual, unlisted		E. Priori
HBT3 (carcinoma, breast)	P. Arnstein from R. Bassin	EB33 (carcinoma, prostate)	F. Schroeder
HBT-E (16c, clone of HBT-3)	R. Bassin	D18T (synovial cell)	D. A. Peterson
		M10T (synovial cell)	D. A. Peterson
HBT-39b (carcinoma, breast) (clone 6)	P. Arnstein from E. Plata	Detroit 6 (sternal marrow)	Unlisted
			Commercial, unlisted
HEK (kidney)	Commercial, unlisted	Detroit 6 (=CCL3)	ATCC
	J. Rhim from C. Pfizer, Inc.	Detroit 6 (clone 12) (=CCL3.1)	ATCC
	C. Pfizer, Inc.		
HEK/HRV (HEK, virus transformed)	S. Aaronson		
MA160 (prostate)	The originators		
MA160	P. Price, MBA		
	M. Vincent, MBA		
Prostate (=MA160)	Unlisted		

(continued)

TABLE IV (*continued*)

Designation	Source
Detroit 6 (=CCL3)	Child Research Center of Michigan or ATCC
Minnesota EE (esophageal epithelium)	Individual, unlisted
Minnesota EE (=CCL4)	ATCC
Intestine 407 (jejunum, ileum) (=CCL6)	ATCC
Intestine 407	Commercial, unlisted
Intestine 407 (=HEI = CCL6)	G. Spahm from ATCC
NCTC2544 (=CCL19) (skin) (epithelium)	ATCC
NCTC3075 (=CCL19.1)	ATCC
WISH (amnion)	Individual, unlisted
WISH (=CCL25)	ATCC
Girardi heart (heart) (=CCL27)	ATCC
TuWi (=CCL31)	ATCC
Wong-Kilbourne (conjunctiva) (=CCL20.2)	ATCC

[a] From Nelson-Rees and Flandermeyer.

III. Growth of HeLa Cells

Many kinds of media with different components have been used for the culture of HeLa cells. Usually Eagle's minimal essential medium or Dulbecco's modified Eagle's medium with 10% newborn or fetal calf serum, and L-glutamine (2 mM; pH 7.3), is used as a monolayer culture at 37°C in an atmosphere of 5% CO_2–95% air with >98% humidity. Penicillin (50–100 units/ml) and streptomycin (50–100 μg/ml) are also added to the medium to eliminate or suppress microbial contaminants unless they affect any ultimate use intended for the cells. Glutamine is reported to be the major energy source for cultured HeLa cells. It provides energy by aerobic oxidation from citric acid cycle metabolism.

HeLa cells also grow in serum-free medium (Ham's F12) supplemented with insulin, transferrin, hydrocortisone (aldosterone), fibroblast growth factor, and epidermal growth factor. The omission of any single hormone results in less than maximum cell growth. The growth rate and clonal growth of HeLa cells in the serum-free medium is equal to that in the serum-supplemented medium. Clone S3

is readily adaptable to growth in spinner culture using a medium deficient in Ca^{2+} or Ca^{2+} and Mg^{2+}.

IV. Storage of HeLa Cells

HeLa cells can be stored frozen indefinitely and retain high viability. The cells are frozen in culture medium containing 10% dimethyl sulfoxide to a concentration $>1 \times 10^6$ cells/ml.

Into each sterile 2-ml polypropylene freezing tube, 1 ml of the cell suspension is dispensed. If the cells are to be kept in liquid nitrogen, step-wise freezing is required. The simplest way to accomplish this is to wrap the cells in cotton and place them in a small box. After covering the box, it is placed in a −80°C freezer overnight. At any convenient time thereafter the vials can be transferred to liquid nitrogen for storage.

V. HeLa Cell Variants

The HeLa cell has also contributed to the study of somatic cell genetics such as genetic control mechanisms and mutation at the cellular level.

Many variant cell lines have been isolated from HeLa cells. Some interesting variant cell lines have recently been listed (Table V). In particular, ouabain resistant mutants of hypoxanthine phosphoribosyl transferase (HPRT)-deficient HeLa cells (D98-OR) have been useful as universal hybridizers for the production of intraspecific human hybrids. The HPRT-deficient D98 cells named D98AH2 are tumorigenic in nude mice. Tumorigenicity assays of intraspecies hybrids between D98AH2 and normal human diploid cells have revealed that most are nontumorigenic. Cytogenetic analyses of the hybrid cell populations suggest that genes that map to normal chromosome 11 may be involved in suppressing the tumorigenic potential of the D98AH2 cells.

Multidrug-resistant sublines KB-8-5, KB-C1, -C1.5, -C2, -C2.5, -C3, -C3.5, and -C4 were recently isolated from KB cells that are now considered to be HeLa cells. They have been very useful in elucidating the molecular basis for multidrug resistance, especially for the function of P-glycoprotein as an energy-dependent efflux pump.

TABLE V Variant Cell Lines Derived from HeLa and KB Cells

Character	Cell lines	Derivation
Nutritional variants		
Glutamine requiring	HeLa I-11a	HeLa I-11
Glutamine independent	HeLa I-11	HeLa S3-1
Carbohydrate	Ribose variants	
	Xylose variants	
	Lactate variants	
Grow in protein- and lipid-free synthetic media	HeLa-P3	HeLa
Quantitatively different in their requirement of serum	HeLa S1	HeLa
Drug-resistant		
Actinomycin D	HeLa-R	HeLa
1-β-D-arabinofuranosylcytosine	KB/araC	KB
8-Azaguanine	S3AG1	HeLa S3
Bromodeoxyuridine (Brd Urd)	HeLa BU-10	HeLa S3
	HeLa BU-15	HeLa S3
	HeLa BU-25	HeLa S3
	HeLa BU-50	HeLa S3
	HeLa BU-100	HeLa S3
Chloramphenicel	296-1	HeLa S3
		HeLa S3
Colchicine, multidrug-resistant	KB-8-5	KB
	KB-Cl ~ -C4	KB
Adriamycin, multidrug-resistant	KB-Al	KB
Vinblastine, multidrug-resistant	KB-V1	KB
Erythromycin	ERY2301	HeLa
Ethylmethane sulfonate	HeLa A6	HeLa S3
Ouabain	D98-OR	D98/AH-2
6-Thioguanine	H23	HeLa
	D98/AH-2	HeLa
Toxin-resistant		
Diphtheria toxin	KB-R2	KB
	KB-R2A	KB
Epidermal growth factor–*Pseudomonas* toxin	ET	KB
Altered virus susceptibility		
Poliovirus-sensitive	HeLa I-3	HeLa S3
Poliovirus-resistant	HeLa S3-1C	HeLa S3
Poliovirus-resistant	"R"	HeLa S3
Others		
Alkaline phosphatase lacking	A clonal line of giant HeLa cells	HeLa
Ultraviolet sensitive	S-1M	HeLa S3
	S-2M	HeLa S3

Bibliography

Akiyama, S. (1987). HeLa cell lines. *In* "Molecular Genetics of Mammalian Cells" (M. M. Gottesman, ed.). "Methods in Enzymology" **151**, 38. Academic Press, San Diego.

Akiyama, S., Fojo, A., Hanover, J. A., Pastan, I., and Gottesman, M. M. (1985). Isolation and genetic characterization of human KB cell lines resistant to multiple drugs. *Somat. Cell Genet.* **11**, 117.

Chen, C.-J., Chin, J., Ueda, K., Clark, D., Pastan, I., Gottesman, M. M., and Roninson, I. (1986). Internal duplication and homology with bacterial transport proteins in the mdr 1 (P-glycoprotein) gene from multidrug-resistant human cell. *Cell* **47**, 381.

Chen, T. R. (1988). Re-evaluation of HeLa, HeLa S3, and HEp-2 karyotypes. *Cytogenet. Cell Genet.* **48**, 19.

Kaelbling, M., and Klinger, H. P. (1986). Suppression of tumorigenicity in somatic cell hybrids. *Cytogenet. Cell Genet.* **41**, 65.

Srivatsan, E. S., Benedict, W. F., and Stanbridge, E. J. (1986). Implication of chromosome 11 in the suppression of neoplastic expression in human cell hybrids. *Cancer Res.* **46**, 6174.

Helminth Infections, Impact on Human Nutrition

CELIA HOLLAND, *Trinity College, Dublin, Ireland*

Glossary

Anorexia Loss of appetite

Anthropometric measurements Assessments of protein–energy malnutrition which include weight, height, weight-for-height, mid/upper arm circumference, and skinfold thickness. Assessments are made relative to a standard or reference value from a well-nourished population of the same age. The reference values most commonly used are the Harvard standards or the more recent NCHS–WHO reference values. The NCHS references have been recommended by WHO for use in developing-country surveys.

Anemia A reduction in the hemoglobin concentration, the hematocrit, or the number of red blood cells to a level below that which is normal for a given individual. Anemia has been defined as the hemoglobin level below -2 S.D of the mean value in the population. WHO recommend characterizing the hematologic status of populations by determining the frequency distribution of hemoglobin or hematocrit values.

Intensity of infection Number of individuals (determined directly or indirectly) of a particular helminth species in each infected host in a sample

Malabsorption Failure to assimilate essential ingredients from the diet leading to malnutrition. In this context, failure of the intestine to absorb the products of digestion as a consequence of changes in the intestine

PEM Protein–energy malnutrition. Clinical signs and anthropometric measurements are used to diagnose serious protein–energy malnutrition.

Steattorhea Steattorhea is characterized by loose, smelly, fatty stools and is a symptom of fat maldigestion which is accompanied by malabsorption of more than just fat. It is generally defined as fecal fat excretion greater than 6 g per day.

D-xylose absorption This is used specifically as a measure of carbohydrate malabsorption but is often used generally as an indicator of malabsorption. A measurement of urinary xylose excretion is usually made 5 h after a 5-g oral dose of D-xylose. It is used because it is rapidly absorbed by the small intestine and excreted unmetabolized in the urine.

THE IMPACT of four major soil-transmitted helminth nematode infections on human nutrition is considerable. The soil-transmitted nematodes are defined as those parasites in which part of life cycle development occurs in the soil (i.e., the development of eggs or larvae prior to their ingestion by or penetration of the definitive host). As far as humans are concerned, the most prevalent species are the human roundworm, *Ascaris lumbricoides*, the human hookworm, *Ancylostoma duodenale* and *Necator americanus*, the human whipworm, *Trichuris trichiura* and *Strongyloides stercoralis*. All four species are nematode parasites which inhabit the human intestine as adult worms and produce eggs or larvae which pass out in the feces. These species are discussed both because of their high prevalence worldwide and the fact that their distributions frequently overlap with those of malnutrition. In addition, they have tended to be neglected and somewhat underestimated in public health terms.

I. Background

In many parts of the world, particularly the subtropics and tropics, human beings suffer the double insult of parasitic infection and malnutrition. Recent global estimates indicate that the soil-transmitted helminths are some of the most common infections in the world (Table I). Nutrient deficiency diseases are equally prevalent (Table II), and intestinal nematode infections have been implicated in the aggravation of protein energy malnutrition, iron deficiency anemia, and to a lesser extent Vitamin A deficiency. Actively growing children who are often heavily infected with several parasite species are particularly at risk of developing malnutrition. [*See* MALNUTRITION.]

Both gastrointestinal nematodes and malnutrition flourish in an environment dominated by poverty, insufficient sanitation, crowded living conditions, and inadequate health care facilities. Morbidity associated with such parasitic infections is likely to be underestimated, and little systematic study of the symptomatology associated with them has been undertaken. Certainly the fact that they inhabit the intestinal tract makes it highly likely that they will influence nutritional status, but there is a need to establish the importance of their role and, there-

TABLE I Prevalence of Major Infections in
Africa, Asia, and Latin America, 1977–1978[a,b]

Infection	Rate of incidence in millions/yr
Diarrheas	3000–5000
Tuberculosis	1000
Ascariasis[c]	800–1000
Hookworm infections	700–900
Malnutrition	500–800
Malaria	800
Trichuriasis	500
Amebiasis	400
Filariasis	250
Giardiasis	200
Schistosomiasis	200

[a] The infections are arranged in order of their frequency. Adapted from Pawlowski Z. S. (1984). Implications of parasite–nutrition interactions from a world perspective. *Fed. Proc.* **43**, 256–260. Reproduced with permission.
[b] Data on Strongyloidiasis is not available from this source. *Strongyloides stercoralis* has been estimated to infect 56 million people worldwide but this is likely to be an underestimate.
[c] Major soil-transmitted helminths are shown in boldface type.

TABLE II Gross Estimates of the Total Number of Persons Affected by Three Dietary Deficiency Diseases Worldwide[a]

Deficiency	Prevalence[b]	Age (years)
Protein and energy[c]	500 million	0–6
Iron	350 million	18–45 (women)
Vitamin A[d]	6 million	All

[a] Adapted from Latham, M. C. (1984). Strategies for the control of malnutrition and the influence of the nutritional sciences. *Food and Nutrition* **10**, 5–31. Reproduced with permission. Also adapted from Stephenson (1987).
[b] Estimates are gross and do not express the significant variations within and between countries.
[c] Those suffering from stunted growth defined by weight below the 2.5 m percentile of the WHO growth standards.
[d] This is an underestimate because it includes cases from Southeast Asia only.

fore, a realistic picture of their public health significance.

The intestinal nematode infections are transmitted as a result of inadequate disposal of human feces, thereby disseminating the infective eggs or larvae into the environment. Infection by ingestion of infective eggs (*Ascaris lumbricoides, Trichuris trichiura*) or penetration by infective larvae (hookworm species, *Strongyloides stercoralis*) can occur. Direct ingestion of the larvae, in the case of *Ancylostoma duodenale* can result in complete development. The development of *Necator americanus* appears to require percutaneous entry with a phase of larval development in the lungs. The eggs or larvae require a certain period of development in the environment under appropriate conditions in order to become infective. All four species can produce large numbers of infective stages, particularly *A. lumbricoides,* which has been estimated to produce an average of 200,000 eggs per day per female worm. On ingestion or penetration, the larval stages undergo a form of migration and eventually end up as adult worms in the small intestine. Their location varies, with *A. lumbricoides* inhabiting the jejunum, *T. trichiura* the cecum, hookworm species, the upper small intestine, and *S. stercoralis* the duodenum and the upper jejunum. In heavy infections the parasites may extend their location.

The intensity of a helminth infection is a particularly important parameter for the accurate assessment of the dynamics of infection and the morbidity and mortality associated with that infection. Intensity of infection has been found to have a strong positive correlation with symptomatology and nutritional morbidity. This applies to the eggs and worms of *Ascaris, Trichuris,* and hookworm and the larvae of *Strongyloides.* It is therefore important to know the intensity of infection for each individual in a nutrition–parasite study.

In terms of population dynamics, the distribution of worm intensity is characteristically aggregated or overdispersed within a host community or population. This means that most infected hosts harbor few or no parasites while a small proportion harbor heavy parasite burdens. This distribution has been described for *Ascaris, Trichuris,* and hookworm from a variety of countries worldwide. Further evidence has shown that individuals exhibit a predisposition to heavy or light infection with these three parasite species, although the precise explanations or mechanisms for this predisposition are still unknown. One factor may be host nutritional status, indicating not only that heavy infection can contribute to nutritional morbidity but poor nutritional status may enhance the risk of harboring a heavy infection.

Strongyloides must be considered differently from the other three species. The process of autoinfection means that the dynamics of infection are not governed purely by the processes of immigration and emigration. Despite this, heavy infections of *Strongyloides* do occur, described as the hyperinfection syndrome, and those individuals who suffer it are predisposed by a number of factors (see Section III on strongyloidiasis).

The epidemiology of these infections is an important consideration when assessing the relationship between intestinal parasitic infection and malnutrition.

Some attention has been paid to the design of nutrition–parasite studies, and certain recommendations were made by L. S. Stephenson in 1987. Three types of studies have been commonly employed to look at the interaction between host nutrition and parasitism. These are studies in laboratory animals, clinical studies in a hospital setting, and community-based field studies. Community-based field studies, which are often the most useful, are limited by ethical considerations. The choice of population, the study design, and the form of intervention must be evaluated with care. The aggregated distribution of the helminths in a community needs to be taken into account, and selective sampling may be necessary to identify a number of heavily infected individuals. Polyparasitism can also make it difficult to assess the impact of a single parasite species. Host dietary intake is another important consideration, as this will influence the impact of an infection and the ability of a host to respond to losses incurred.

II. Hookworm

Adult hookworms live in the upper small intestine, attached to the mucosa by their buccal capsules. Worms feed by drawing a plug of mucosal tissue into their buccal capsules. Blood loss occurs because of the sucking of the worms and from lesions in the mucosa.

The clinical severity of the disease is closely related to the intensity of infection and the condition of the host. Host factors include iron intake and bioavailability, iron resources and needs, and general state of health. The most important nutritional impact of infection is the blood and iron loss as a consequence of the feeding of the adult worms in the intestine. In acute infections, nausea, vomiting, and diarrhea and abdominal pain can occur. In chronic infections, most of the symptoms are those of iron deficiency anemia. In a classic study Layrisse and Roche demonstrated a highly significant relationship between circulating hemoglobin levels and hookworm egg counts in a rural Venezuelan setting. Hemoglobin levels were significantly lower in subjects passing more than 2000 eggs per gram of feces for women and children and more than 5000 eggs for adult men. Mean hemoglobin level decreased linearly as egg counts increased.

A number of estimates of intestinal blood loss and iron loss of hookworm-infected people are shown in Table III. A person who passes 2000 eggs per gram of feces loses an estimated 1.3 mg of iron for *Necator* and 2.7 mg of iron for *Ancylostoma.* In tropical countries only about 10% of the iron ingested is absorbed. The type as well as the amount of dietary iron consumed has a major influence on the bioavailability and hence the amount of iron a person actually absorbs daily. In most areas of the world where hookworm is prevalent iron intake is of predominantly vegetable origin. The nonhaem iron present in foods of vegetable origin is more poorly absorbed than the haem iron in foods of animal origin. Another source of iron is contaminant and insoluble organic iron which is relatively unavailable for absorption but probably accounts for some of the reports of high iron content of meals from some countries where nutritional iron deficiency is endemic.

TABLE III Intestinal and Fecal Blood and Iron Losses in Hookworm Infection in Humans[a]

Parameter[b]	Necator americanus	Ancylostoma duodenale
Intestinal blood loss in ml per worm per day mean	0.03	0.15
Range	0.01–0.04	0.05–0.30
Intestinal blood loss in ml per 2000 epg mean ± SD	4.3 ± 2.02	8.9 ± 2.32
Intestinal iron loss in mg per 2000 epg	2.0	4.2
Fecal iron loss in mg per 2000 epg[c]	1.3	2.7

[a] From Stephenson (1987) pp. 128–160 with permission from Taylor and Francis.
[b] Epg = eggs per gram of feces. An infection of 2000 epg, or about 80 worms, is considered a moderate subclinical infection by public health standards.
[c] These values assume a hemoglobin level of 14 g/dl. Fecal iron loss is less than intestinal iron loss because some of the iron lost is reabsorbed and reutilized and is not excreted in the feces. Average fecal iron loss per person per day has been estimated to be on the order of 5–9 mg.

Studies of the absorption of fat, carbohydrate, vitamin A, vitamin B_{12}, and folic acid have been undertaken in hookworm-infected people. The majority of these investigations have provided little evidence for a malabsorption syndrome associated with hookworm infection and indicated that the structural and functional abnormalities found were more likely to be due to PEM or tropical sprue.

One study that did provide evidence for a link between hookworm infection and malabsorption involved 14 anemic infected patients from Costa Rica. Malabsorption of carbohydrate, fat, and Vitamin A was detected in the majority of the cases, and absorption improved after treatment. Biopsies showed abnormal villous architecture which receded after treatment. Six lightly infected patients and eight people with other parasites showed no evidence of malabsorption. In contrast to these findings, 10 studies from the Indian, African, Asian, and Latin American continents found little or no evidence of malabsorption of fat or carbohydrate in patients with hookworm infection of varying intensity.

Further evidence for the lack of a malabsorption syndrome associated with hookworm infection was provided by a study of 15 Colombian subjects. Patients with severe protein energy malnutrition but no hookworm infection showed abnormal small bowel absorptive function with flattened villi and a decreased crypt–villous ratio. This returned to normal after protein repletion. Patients with protein energy malnutrition and anemia secondary to hookworm infection showed similar functional and histological findings which normalized after protein repletion, despite the persistence of their infection and associated anemia.

Patients without protein energy malnutrition and with severe hookworm infection and anemia failed to show as pronounced biochemical or histological evidence of malabsorption. Findings like these, similar to those described for *Strongyloides,* further emphasize the need for meticulous evaluation of the nutritional status of infected patients.

Hookworm infection has been shown to be associated with reduced food intake and anorexia, although there have been no specific studies undertaken to examine this. Another important, yet ill-studied, aspect of hookworm is its possible association with growth stunting. A significant increase in growth was reported in children treated for hookworm and *S. hematobium.* The degree of growth improvement correlated well with a decrease in hookworm eggs after treatment. Possible causes could include reduced food intake or protein loss as a result of blood loss caused by worms feeding or because of a protein-losing enteropathy in an inflamed intestine. Some studies have shown albumin or protein loss to increase in line with increasing worm burden. It is likely that protein loss does occur but will only cause protein deficiency in severe chronic infections and/or extremely low intakes.

In terms of food intake and its effect on growth, an ingenious "natural experiment" took place involving 26 well-nourished Indian men who played a game on a previously used but recently ploughed defecation field. They manifested intense itchiness of the skin after the game and went on to develop heavy hookworm infection with abdominal pain and vomiting. They lost an average of 7 kg in weight. None of the men had steatorrhea or villous atrophy, and only six of them had abnormal D-xylose values. The authors attributed the men's weight loss to deficient caloric intake due to fear of food precipitating abdominal pain.

Another important aspect of hookworm-induced iron deficiency anemia is the functional outcome that can affect worker productivity, reproduction, and cognitive performance. A number of studies have demonstrated reduced work capacity among anemic workers and increased work output after iron supplementation. A study among latex tappers in Indonesia demonstrated a correlation between work output and hemoglobin concentration. The work output was significantly less (19%) in anemic tappers than in nonanemic tappers. After treatment, the output of the anemic group increased to that of their nonanemic colleagues.

Severe anemia is associated with increased risk of premature delivery, increased maternal and fetal morbidity, and mortality. Even relatively mild anemia has been found to be associated with premature delivery and a correlation between maternal hemoglobin levels and fetal birth weight has been demonstrated.

There are many indications (but little firm evidence) of a relationship between impaired cognitive and behavioral function and helminth disease, including hookworm infection. Iron deficiency anemia has been found to have deleterious effects on cognitive performance and this outcome of hookworm infection is obviously in need of further study.

III. Ascariasis

Ascariasis is a prevalent infection worldwide, particularly in children who manifest the heaviest infections. The parasite produces many resistant eggs which can survive for long periods of time in the environment.

Most nutritionally significant symptomatology is associated with the presence of adult worms in the small intestine and includes abdominal pain, nausea, diarrhea, anorexia, and nutrient malabsorption. Complications resulting from the aggregation of adult worms in the intestine and the migration of worms to other parts of the body have been described. These can be fatal and are expensive and difficult to treat, particularly in inadequate health care systems. Pathology associated with the larval migration has not received much systematic attention, although cough, asthma, substernal pain, fever, a skin rash, and eosinophilia have been noted.

Extensive studies of *Ascaris suum* in the pig have been undertaken which can act as a useful laboratory model for *Ascaris lumbricoides* infection in children. Evidence has accumulated that *Ascaris*-infected pigs fed a low protein diet exhibit reductions in food intake, in weight gain, and in weight gain per amount of food consumed. This reduction can be correlated with worm burden. Decreased nitrogen retention and decreased absorption of lactose and fat have also been described in infected

animals. Reduced mucosal lactase activity and histological abnormalities of the jejunal mucosa have been demonstrated. These findings have stimulated an interest in the relationship between nutritional morbidity and ascariasis in growing children who may be malnourished.

A number of clinical studies have shown that *Ascaris* infection is associated with decreased absorption of nitrogen, steattorhea, and abnormal D-xylose absorption. Jejunal biopsies of infected children also showed abnormal villous architecture, including broadening and shortening of the villi, elongation of the crypts, and cellular infiltration of the lamina propria. After treatment there was evidence of reduced fecal nitrogen, and steattorhea and D-xylose malabsorption regressed. Mucosal damage also decreased after treatment.

Lactose maldigestion has also been shown to be associated with *Ascaris* infection. *Ascaris*-infected Panamanian children exhibited a marked decrease in lactose digestion, measured by breath H_2 concentration compared to age- and sex-matched controls. This decrease in lactose digestion was linearly related to worm burden and returned to normal 3 weeks after treatment. Symptoms associated with milk intolerance have been described from infected children and include abdominal pain, diarrhea, and flatulence after drinking milk. Uninfected controls did not manifest these conditions after lactose intake. An assessment of milk intake among *Ascaris*-infected children revealed a significant reduction in lactose consumption compared to controls using the 24 hr recall method. The food weighing method failed to detect this difference. [*See* LACTOSE MALABSORPTION AND INTOLERANCE.]

Accelerated mouth-to-cecum transit time has also been recorded from infected Panamanian children. This was significantly negatively correlated with worm burden. This may be a possible mechanism for decreased nutrient absorption. Evidence has also been provided that ascariasis interferes with Vitamin A absorption, and in one study there was a clear improvement in the absorption of Vitamin A after deworming.

A number of community-based longitudinal studies have concluded that ascariasis can contribute to growth retardation in preschool children and that periodic deworming can improve growth.

One study based in the Machakos district in Kenya examined children aged 12 to 72 months three times at 3.5 month intervals. Infected and uninfected children did not differ in terms of socioeconomic status or anthropometric measurements at the start of the study. However, there was a decrease in skinfold thickness of *Ascaris*-infected children compared with the controls. Three and a half months after treatment, previously infected children showed significantly higher weight gain and percent expected weight gain and an increase in skinfold thickness compared with the controls. This represented a 33% increase in expected growth rate over the 3.5 month period. Multiple regression analysis revealed that *Ascaris* infection was the most important of 37 possible health, nutritional, and socioeconomic variables in explaining the decrease in skinfold thickness before and the increase in skinfold thickness after treatment.

In Tanzania, improved growth rates were also reported amongst pre-school children who received levamisole at three monthly intervals over a one year period compared to a control group who received a placebo.

Recently, Malaysian children treated for ascariasis every 3 months for a year gained significantly more weight and showed greater skinfold thickness compared with untreated infected children. The difference in weight gain between the treated and untreated children was about 16% of the 1 year increment.

Furthermore, a recent Burmese study demonstrated a significant increment in height for age amongst children 12, 18 and 24 months after treatment with levamisole compared to a group of children who did not receive treatment.

Several other studies which have examined the relationship between ascariasis and child growth have failed to demonstrate similar findings. In a recent book on the impact of helminth infection on human nutrition, Stephenson reviewed these studies in detail and offered some explanations for the variations in the findings concerning child growth. Aspects of experimental design, including the choice of population, sample size, successful drug treatment, and data analysis are considered. Successful drug treatment is particularly important, and in a number of studies *Ascaris* had not been completely eradicated from the infected children at the time that growth assessment took place.

Another important factor is the underlying socioeconomic status of the individuals or community under study. In a cross-sectional survey of the relationship between ascariasis and percent weight for age in three Balinese hamlets, the interaction between infection and nutritional status was influenced by other socioeconomic factors. This underlines the importance of evaluating the dietary and socioeconomic variables in a community which may in turn influence the relationship between infection and malnutrition.

IV. Strongyloidiasis

Strongyloides stercoralis is unusual in that it possesses both free living and parasitic larval and adult stages. Under certain conditions, not yet fully understood, free living larval forms can give rise to infective larvae which can penetrate and infect a human host. Unlike the other species of parasites described here, it is the larval form which is used for diagnosis. Female worms, which live in the intestinal mucosa, lay eggs which hatch into larvae in the gut lumen. They are then passed out in the feces. These larvae can be difficult to detect in a routine smear and require the use of a special technique, which is not widely used, and therefore the prevalence of this parasite is likely an underestimation.

Two forms of autoinfection can occur and play an important role in the public health significance of the disease. Some larvae may metamorphose and molt into infective larvae en route down the bowel. These larvae can then directly penetrate the wall of the gut and enter the bloodstream; this is termed *internal autoinfection. External autoinfection* can occur when fecal matter containing infective larvae makes contact with the perianal region and larvae invade the local tissue. These processes are responsible for the perpetration of the infection long after an individual has left an endemic area and can contribute to the severe form of the disease.

Strongyloides infections can be divided into asymptomatic, chronic, and severe. In both chronic and severe forms of the disease, abdominal pain, nausea, vomiting, diarrhea, and weight loss of varying severity occur. These symptoms can lead to a decrease in food intake and an increase in nutrient excretion.

If allowed to persist, strongyloidiasis can cause malnutrition and even death as a consequence of hypovolemic shock due to extensive diarrhea and vomiting. A malabsorption syndrome has been found to be associated with *Strongyloides* infection, including lesions in the duodenum and jejunum. Fibrosis as a consequence of bacterial contamination of ulcers in the small intestine may also contribute to this malabsorption.

Infection with *Strongyloides* can persist for many years as a consequence of autoinfection and occasionally progress to the so-called hyperinfection syndrome or disseminated strongyloidiasis. Acute strongyloidiasis involves a massive build up of both adult and larval worms in the intestinal mucosa. Parasites can also invade other organs. This condition has a mortality rate of 50–70%. Cause of death includes malnutrition but bacterial sepsis in individuals immunocompromised by therapy or disease often contributes to the fatal outcome. A number of conditions are thought to predispose individuals to the development of these heavy infections, including malignant neoplastic disease, corticosteroid or immunosuppressive therapy, and malnutrition. Recently disseminated strongyloidiasis has been reported from a number of AIDS-infected patients. It is particularly difficult to establish if malnutrition was present before infection and therefore contributed to it or arose as a consequence of the disease. Depressed cell-mediated immunity, as a consequence of PEM, has been described as a causative agent of the hyperinfection syndrome. But a recent case description of an immunologically competent patient developing disseminated strongyloidiasis indicates that even in uncompromised hosts *Strongyloides* should not be allowed to persist.

Evaluation of the relationship between strongyloidiasis and malnutrition is complicated by the fact that most of the information available comes from case studies of one or more patients, suffering from the hyperinfection syndrome, in a hospital setting. This type of data has obvious limitations, particularly in terms of the evaluation of morbidity at the community level.

A number of studies have focused on the issue of a malabsorption syndrome associated with strongyloidiasis. The findings tend to be somewhat contradictory, and in general the sample sizes were low, no uninfected individuals were included as controls, follow-up after treatment was not always described, and the biochemical, radiological, and histological investigations undertaken varied from study to study. Five studies provided evidence for a malabsorption syndrome associated with *Strongyloides* infection and reported steattorhea, abnormal D-xylose and folic acid absorption, and abnormal jejunal and duodenal biopsies. Some patients manifested diarrhea, vomiting, and weight loss. Several studies which undertook follow-up after treatment reported a regression of steattorhea and that intestinal biopsies returned to normal.

In contrast, a Costa Rican study which included infected and uninfected controls found no evidence of a malabsorption syndrome with normal fat and carbohydrate absorption, despite pathological changes in the jejunum and duodenum. In response to these somewhat contradictory findings, a carefully controlled study of individuals of known nutritional status was undertaken, the premise being that the malabsorption and intestinal changes recorded in previous studies might be secondary to malnutrition and not directly due to the parasitic infection. In malnourished patients with *Strongyloides,* malabsorption persisted in spite of total parasite eradication and disappeared in patients fed a low protein diet in spite of a continued parasite presence. These findings indicate how important it is to evaluate the nutritional status of an infected host to untangle the relative contributions of infection and malnutrition.

V. Trichuriasis

Adult worms of the human whipworm, *Trichuris trichuira,* inhabit the mucosa of the cecum, into which they burrow by means of their anterior stylet. In heavy infections the worms can be found in the wall of the appendix, the colon, and the most posterior section of the ileum.

Children harbor the heaviest infections; intensity declines with adulthood. This decline has been attributed to behavioral changes in adulthood which result in a reduction in exposure or acquired resistance which may lead to diminished establishment. Severity of infection is dependent on parasite intensity but also on such host-related factors as age, general health status, iron resources, and experience of past infections.

Infections can be classified as light, moderate, and heavy. Light infections tend to be regarded as asymptomatic. Moderate infections produce abdominal pain, diarrhea, vomiting, weight loss, and anemia in malnourished children. Heavy infections, which tend to be confined to endemic areas, are accompanied by bloody diarrhea, abdominal pain, pronounced weight loss, severe anemia, rectal prolapse, and finger clubbing. These symptoms are likely to contribute to a reduction in food intake and an increase in nutrient excretion. The diarrhea described is not of small intestinal origin and is likely to result from inflammation of the colon.

Despite high prevalence worldwide, there has been relatively little measurement of the morbidity associated with trichuriasis at the community level. Heavy trichuriasis has been associated with anemia, chronic diarrhea, and dysentery and protein energy malnutrition.

It has not been conclusively proved that blood is an important food of *Trichuris,* but several workers have found blood and blood products inside worms at autopsy. Certainly the anterior stylet is responsible for the laceration of tissue and blood vessels, causing bleeding from the mucosa. A number of studies have reported an association between trichuriasis and anemia and an inverse relationship between hemoglobin level and intensity of infection. Only one study has concluded that trichuriasis is a cause of anemia. An estimated blood loss of 0.005 ml per day per worm and a blood loss per child of 0.8–8.6 ml per day has been reported. This is higher than that attributed to losses from uninfected persons (0.2–1.5 ml per day) but considerably less than that described for hookworm. Inconsistencies in the findings of a number of the studies may be explained by differences in the intensity of infection and the selection of study subjects. Individual variation in iron nutritional status and iron requirements may also be important. More community-based studies are needed to examine the relationship between trichuriasis, blood loss, and anemia in children with heavy and light infections, some of whom are anemic.

Evidence that trichuriasis can contribute to PEM has been provided from Malaysia, South Africa, and St. Lucia. Heavily infected Malaysian children and controls of similar socioeconomic status showed striking differences in nutritional status and clinical signs. All the infected children had chronic dysentery, and many manifested rectal prolapse, edema, anemia, and finger clubbing in contrast to the uninfected children (Table IV). Their growth was also significantly poorer and they had lower hematocrit and serum albumin levels. After treatment, nutritional index, number of red blood cells, and serum albumin significantly increased, and the frequency of rectal prolapse, diarrhea, edema, and clubbing was reduced in the infected children. A South African study revealed a significant increase in weight gain after treatment. The authors concluded that this weight gain was likely to be a consequence of the cessation of diarrhea, an improved appetite after treatment for anemia, and the improved hospital diet.

TABLE IV Features of *T. trichiura* Infection in Heavily Infected Malaysian Children (*n* = 67)[a,b]

Feature	Incidence (percent)
Blood in stools	73
Edema	18
Clubbing	12
Rectal prolapse	51
Refeeding syndrome	26
Pica	67
Previous attendance as in-patients for diarrhea	44

[a] Adapted from Gilman, R. H., Chong, Y. H., Davis, C., Greenburg, G., Virik, M. K., and Dixon, H. B. (1983). The adverse consequences of heavy *Trichuris* infection. *Trans. Royal Soc. Trop. Med. and Hyg.* **77,** 432–438. From Stephenson (1987) pp. 161–201 with permission from Taylor and Francis.
[b] Visualization of *T. trichiura* worms in the rectum by anoscopy was used to diagnose heavy infection. The control group of children (*n* = 73) had 0% for all categories.

Finally, heavily infected children from an endemic community in St. Lucia showed greater growth deficits than the rest of the population. These deficits were attributed to growth stunting because low height-for-age was more significantly associated with heavy trichuriasis than low weight-for-age or arm circumference. The authors concluded that trichuriasis is associated with growth stunting as a consequence of dysentery. These provocative observations indicate the need for further investigations using a longitudinal design. Some workers are of the opinion that the contribution of the whipworm to childhood morbidity, and in particular malnutrition, has been significantly underestimated.

VI. Conclusions

In general further investigations are needed of the relationship between helminth infections and malnutrition, with a particular emphasis on carefully controlled, longitudinal, community-based studies.

Evidence is plentiful that hookworm is an important cause of anemia in many countries. It now seems unlikely that hookworm infection causes a malabsorption syndrome. The contribution of hookworm infection to reduced food intake, anorexia, and possible growth stunting merits further attention.

A number of researchers have now demonstrated convincingly that successful treatment for ascariasis can improve growth and macronutrient absorption in malnourished individuals in an environment where ascariasis and malnutrition are endemic. The first evidence of an improvement in height for age among children treated for ascariasis over a 2-yr period has been provided from Burma.

Careful attention needs to be paid to the socioeconomic background of the communities under study. Reductions in appetite and food intake may be one of the explanations for the impact of ascariasis on child growth. As in the case of hookworm, there is a lack of systematic studies on the influence of ascariasis on food intake and anorexia. Reduced food intake has been demonstrated in *Ascaris*-infected pigs. Precise measurement of food intake in individuals living within their own communities is notoriously difficult to undertake, but an attempt to do so is urgently needed.

Community-based observations on the epidemiology of strongyloidiasis are generally lacking, and this makes an evaluation of the influence of this parasite on nutritional status difficult. Some evidence of a malabsorption syndrome associated with *Strongyloides* has been found, but more studies are needed employing larger sample sizes, individuals of known nutritional status, and controls.

Evidence presented on trichuriasis indicates that the impact of this parasite on the nutritional status of children has not received adequate attention. The syndrome of chronic dysentery, rectal prolapse, anemia, poor growth, and clubbing associated with heavy *Trichuris* infection is of obvious public health significance and indicates that the identification of heavily infected individuals is important. The observation that heavy *Trichuris* infection is associated with growth stunting is also significant and needs further investigation using a longitudinal design.

In terms of control, some effective anthelmintic drugs are now available for *Ascaris, Trichuris,* and hookworm. Despite this, the prevalence of these infections remains staggeringly high in the countries of the subtropics and tropics. Mass chemotherapeutic programs have proved to be expensive and difficult to implement in such countries. In addition, these diseases have not always received the attention they deserve in public health terms. [*See* CHEMOTHERAPY, ANTIPARASITIC AGENTS.]

The concept of targeted or selected chemotherapy has been discussed, given the aggregated distribution of the parasites concerned. Selection for treatment of the most heavily infected individuals or those perceived to be at particular risk, such as children, could reduce transmission and the number of infective stages in the environment. Limited resources would also be used more cost effectively. The identification and characterization of these heavily infected individuals is therefore an important priority in epidemiological research.

In addition, the provision of improved sanitary facilities and public health education can only assist in the reduction of the soil-transmitted helminths. A purely biomedical approach to parasite control will not be adequate. There is a need to take account of behavioral and cultural attitudes in the design of control programs. *Ascaris* control programs have been found to act as a useful tool in the introduction of other primary health care programs and priorities.

In the case of *Strongyloides,* mass treatment is not appropriate because of the lower worldwide prevalence. Despite this, because of autoinfection and the hyperinfection syndrome, vigilance is needed. Diagnostic procedures need to be improved, and epidemiological information is generally lacking.

Bibliography

Anderson, R. M. (1986). The population dynamics and epidemiology of intestinal nematode infections. Transactions of the Royal Society of Tropical Medicine and Hygiene **80,** 686–696.

Carvalho Filho, E. (1978). Strongyloidiasis. *Clin. Gastroenterol.* **7,** 179–200.

Cooper, E. S. and Bundy, D. A. P. (1988). *Trichuris* is not trivial. *Parasitology Today,* **4,** 301–306.

Crompton, D. W. T. (1985). Chronic ascariasis and malnutrition. *Parasitology Today,* **2,** 47–52.

Crompton, D. W. T. (1986). Nutritional aspects of infection. *Trans. Royal Society of Tropical Medicine and Hygiene* 80, 697–705.

Grove, D. I. (1984). Strongyloidiasis. *In* "Tropical and Geographical Medicine" (K. S. Warren and A. A. F. Mahmoud, eds.), McGraw-Hill, New York.

Holland, C. V. (1989). An assessment of the impact of four intestinal nematode infections on human nutrition. *Clin. Nutr.* 8**(6),** 239–250.

Nesheim, M. C. (1989). Ascariasis and human nutrition. *In* "Ascariasis and Its Prevention and Control" (D. W. T. Crompton, M. C. Nesheim, and Z. S. Pawlowski, eds.). Taylor and Francis, Ltd., London and Philadelphia.

Pawlowski, Z. S. (1984). Trichuriasis. *In* "Tropical and

Geographical Medicine'' (K. S. Warren and A. A. F. Mahmoud, eds.). McGraw-Hill, New York.

Pollitt, E., Soemantri, A. G., Yunis, F. and Scrimshaw, N. S. (1985). Cognitive effects of iron deficiency anemia. *Lancet* **1,** 158.

Schad, G. A., and Banwell, J. G. (1984). Hookworms. *In*
''Tropical and Geographical Medicine'' (K. S. Warren and A. A. F. Mahmoud, eds.). McGraw-Hill, New York.

Stephenson, L. S. (1987). ''Impact of Helminth Infections on Human Nutrition.'' Taylor and Francis, Ltd., London and Philadelphia.

Hematology and Immunology in Space

MEHDI TAVASSOLI, *University of Mississippi School of Medicine*

Glossary

Anemia Reduction from the normal red blood cell mass

Erythropoiesis Production of red cells in the bone marrow from differentiation, proliferation, and maturation of pluripotential stem cells. Daily production is 1/120th of red cell mass. This reflects the survival of red cells in the circulation, which is 120 days

Erythropoietin Hormone produced by the kidney in response to low pO_2. It acts on erythroid-committed progenitor cells in the bond marrow to stimulate erythropoiesis

Oxygen-carrying capacity Capacity of blood to transport oxygen from lungs to tissues. This capacity can be altered by a large number of physiological and pathological factors including the concentration of hemoglobin, molecular alterations in hemoglobin, rhealogy of blood, affinity of hemoglobin for O_2, red cell metabolism, etc.

Packed cell volume (hematocrit) Ratio of red cell mass to whole blood measured in a small sample. In practice it serves as an estimation of red cell mass.

Red cell mass Mass of circulating red cells within the vascular space, measured directly by dilutional study of a ^{51}Cr-red cell sample or derived indirectly from the measurement of ^{125}I plasma space and hematocrit. It is expressed as mg/kg body weight

Red cell survival Survival of red cell in the circulation from the time of its delivery into the circulation (in the form of reticulocytes) to its removal from the circulation in about 120 days. Because the removal of aged red cells follows an exponential curve, the survival is usually expressed as "half-life," which is 60 days. However, because the measurement is usually determined by labeling of cells with ^{51}Cr and the radioactivity is subject to physical decay as well as physical elution, normal ^{51}Cr half-life is considered 28–30 days

Reticulocytes Youngest form of red cell that can enter blood; lacks the nucleus, but still contains certain cytoplasmic elements such as may be necessary for hemoglobin synthesis that, on staining, give the cell a reticulated appearance (thus the name). These elements are lost within 24 hr which is the life span of the cell. The cell then takes the appearance of an ordinary red cell. The number of reticulocytes as a proportion of total circulating red cells serves as an index of bone marrow activity in erythropoiesis.

SMEAT Ground-based control study designed to mimic a full 56-day Skylab flight except for the conditions of microgravity and possible cosmic rays. SMEAT served as a control for the subsequent physiological alterations noted during the spaceflights

THE MOST CONSISTENT IMMUNOHEMATOLOGIC finding during orbital flights has been a substantial loss of red cell mass, often known as "anemia of spaceflight." However, anemia, as usually defined, is based on the concentration of red cells or hemoglobin in a given volume of whole blood. Total blood volume is comprised of red cell mass and plasma volume. In clinical and experimental conditions, anemia is recognized by the measurement of packed cell volume (PCV or hematocrit) or hemoglobin concentration. In these situations a loss of red cell mass occurs usually without or with a negligible loss of plasma volume. Thus, oxygen-carrying capacity of blood (PCV or hemoglobin concentration) falls, leading to anemia. During orbital flights, the loss of red cell mass is often associated with a decrease in plasma volume. This is the result of fluid shift from the lower part to the upper part of the body consequent to the lack of gravitational pull. Volume receptors then act to reduce plasma volume. Parallel decrease in plasma volume renders concentration measurements unrepresentative of the true red cell mass: hematocrit and hemoglobin concentration remain "normal," while the total red cell mass is reduced. Nonetheless, because of its currency, the term *anemia* of spaceflight is used here interchangeably with the reduction in red cell mass. [*See* HEMOGLOBIN.]

I. Methodology

A. Method of Detection

Because in anemia of spaceflight, the normally used measurements of PCV and hemoglobin concentration are not representative, the reduction of red cell mass is generally detected by isotope dilution study. A sample of blood is obtained, and red cells are labeled with radioactive ^{51}Cr. Labeled cells are then reinfused in the subject's circulation. After some 20 min when dilution equilibrium has been attained, another blood sample is removed and its radioactivity is compared with that of the original reference sample. Simple calculations then permit the determination of red cell mass. Determinations are made at several points before flights (to ensure the stability of readings) and repeatedly after flights (to determine the postflight pattern). Comparisons are then made between immediate preflight and postflight readings. By drawing a few other blood samples over a period and without the use of additional ^{51}Cr, the red cell mass study may also be used to gain information on red cell life span. This may otherwise be determined by incorporation of ^{14}C-labeled glycine into hemoglobin. Inflight determinations can also be done. But because most flights are of short duration and because of concern that excess blood-drawing may in itself lead to the loss of red cell mass, inflight determinations are not the rule. The rate of blood-drawing has been kept to a necessary minimum, and it has not been a factor in the loss of red cell mass.

B. Controls

Before the initiation of orbital flights, it was thought that physiological alterations as a result of exposure to microgravity can be predicted from bed rest studies: In its lack of requirement for muscular exercise, microgravity can indeed mimic bed rest. There was no indication, however, of loss of red cell mass from bed rest studies, although actual measurement of red cell mass was not done. To provide for a baseline, a definitive experiment was initiated during the Skylab program (1973–1974). Known as Skylab Medical Experiment Altitude Test (SMEAT), this was a ground-based experiment that simulated in all aspects a full 56-day Skylab mission. The physical facility, atmospheric pressure and composition, crew activity, diet, and the timetable of events were all representative of actual flight. Only the gravitational effects, effect of mental stress of being in space and possibly the effect of cosmic radiations undetectable by current methods, differed between SMEAT and the actual Skylab flight. Immunohematologic findings of SMEAT are listed here:

- No significant changes in the red cell mass (2.7% ± 0.4, which is not significant)
- No significant shortening of red cell life span
- Slight but insignificant (1.6%) increase in plasma volume (a decrease has been reported in most orbital flights)
- No significant changes in red cell glycolytic enzymes
- Essentially normal routine blood cell counts
- Essentially no alterations of chromosomes and genetic materials
- Essentially no immunologic alterations

These findings provide a firm control basis for the interpretation of actual flight data, so that the immunohematologic alterations observed during actual flights could likely be attributed to the effect of microgravity.

II. Magnitude of the Loss

In all orbital flights where red cell mass has been determined, a loss has been noted. The consistency is seen in both the American and Russian programs, which are similar in their essential features. Table I summarizes the results obtained. The most profound deficit is seen in Gemini flights where the atmosphere of Spacecraft cabins consisted of 100%

TABLE I Red Cell Mass Losses During American and Russian Flights in Which Measurements Are Made Under Comparable Methodologies and Conditions[a]

Flight	Duration of flight	No. subjects	Percent decrease
American Flights			
Gemini 4	4	2	13
Gemini 5	8	2	21
Gemini 7	14	2	14
Apollo 7	11	3	3
Apollo 8	7	3	1
Apollo 9	10	3	7
Apollo 14	10	3	4.7
Apollo 15	12	3	10.1
Appollo 16	12	3	14.2
Appollo 17	13	3	11.2
Skylab 2	28	3	14
Skylab 3	59	3	12
Skylab 4	84	3	7
Spacelab 1	10	4	9.3
Russian Flights			
Soyuz 13	8	—	3
Salyut 3	16	—	14
	5	18	14
	4	30	16
	5	49	31
	4	63	20
	6	96	26
	6	40	14
	6	175	18

[a] All data obtained directly by ^{51}Cr labeling except those from Gemini 4 where the data were calculated from plasma volume and hematocrit.

O_2 at a partial pressure of 258 mm Hg (far greater than the normal 160). Such hyperoxia can lead to a loss of red cell mass via two different mechanisms: (1) peroxidation of red cell membrane, leading to shortening of red cell life span (hemolysis), and (2) suppression of the production of erythropoietin hormone, and thus a decrease in red cell production. Both mechanisms were documented. The ^{51}Cr survival of red cells in the Gemini crews was significantly reduced (half-life 20 days, normal 28–30 days). This was associated with reduced plasma α-tocopherol levels and red cell membrane lipids, indicating oxidative injury and subsequent hemolysis. Moreover, no compensatory reticulocytosis was observed, suggesting suppression of red cell formation. When, during subsequent flights, the atmosphere of cabins was changed to a mixture of O_2 and N_2 to provide a pO_2 of 150–185 mm Hg, the shortening of red cell survival was alleviated. Table II compares red cell survival in Gemini and subsequent flights. In Spacelab 1, postflight survival of red cells was entirely normal.

III. Pattern

Almost all the red cell mass deficit occurs during the first 2 weeks of flight as a linear function of flight length. Red cell mass appears then to stabilize as physiological adjustment to the new conditions occurs. The cause of the initial loss and the subsequent adjustment may be the reduced requirement for O_2 as a result of a decrease in muscular work under microgravity conditions. Thus, the loss of red

TABLE II Red Cell Survival[a]

	Preflight	Flight	Postflight
Gemini 4, 7	—	20	—
Apollo 7–8	25	28	25
Apollo 14–17	24	23	27
Skylab 2–4	26	(123)[b]	24

[a] Means of ^{51}Cr half-life in days; normal 28–30 days. Most studies were done in two members of the crew.
[b] C-glycine red cell life span in days.

cell mass, is not progressive and does not interfere with flights of very long duration. This is in contrast to the loss of calcium by the kidney that also occurs and is progressive. Calcium loss is thought to be the limiting factor for the duration of man's exposure to microgravity.

IV. Mechanisms

Reduction in the circulating red cell mass is usually attributed to either a more rapid elimination of red cells from the circulation or a decrease in their input into the circulation. The rapid appearance of the deficit, within 2 weeks, indicates a loss of red cells from the circulation. The life span of red cells in the circulation is relatively long (normally 120 days). Although, with normalization of cabin atmospheric O_2, hemolysis caused by hyperoxia has now been eliminated, there is evidence that some random hemolysis may occur. This can be caused by exposure to cosmic rays, particularly HZE particles (particles with high charge Z and high energy E). However, probably a major factor is sequestration and subsequent destruction of red cells in times, and can lead to reduction in red cells within the body. This retention could be due to reduced requirement for O_2 as a result of reduced muscular work in microgravity. Consistent with this interpretation is an increase in serum ferritin (reflecting the iron released from hemoglobin in the red cells and stored by the body to be used for further synthesis of hemoglobin) parallel to the reduction in red cell mass, indicating the release of iron from sequestered hemoglobin and its storage in the form of ferritin (Table III). Other parameters of iron kinetics are also consistent with changes of ferritin levels.

Evidence for erythropoietic suppression is derived from reticulocyte studies and marrow differential cell counts. Reticulocytopenia, or at least reticulocyte counts not high enough to compensate for the degree of red cell deficit, has been a consistent finding during spaceflights. Table IV shows serial reticulocyte counts in the crew members of the

TABLE IV Mean Reticulocyte Counts ($\times 10^9$/liter)

Mission	Duration	Preflight	0	1	7	14
				Day after landing		
Skylab 2	28	34	15	18	27	29
Skylab 3	59	36	23	—		
					66	82
Skylab 4	84	45	40	53	64	83
Spacelab 1	10	64	24	54	48	64

three Skylab and Spacelab 1 missions. In every case there is a fall in total circulating reticulocytes on the day of spacecraft recovery as compared with preflight. Morphological study of the marrow shows a reduced erythropoietic activity as well as a 20% reduction in bone marrow cellularity. Russians have also reported a decrease in the concentration of red cell progenitors and the contraction of the pool of younger cells in the bone marrow, indicating a suppression of red cell production. Erythropoietin levels are also reduced. There is no evidence that ineffective erythropoiesis or nutritional deficiencies contribute to the loss of red cell mass.

Hence, it appears that a reduction in muscular work caused by microgravity leads to rapid sequestration of a proportion of red cell mass. This reduces the circulating red cell mass and also leads to an increase in the body and serum ferritin. Random hemolysis may also contribute to the loss of red cell mass. Because of the lack of demand for oxygen-carrying capacity, production of erythropoietin hormone is decreased, leading to the suppression of bone marrow erythropoiesis and consequent reticulocytopenia. This scheme is diagrammatically shown in Fig. 1, but it must be considered tentative.

TABLE III Serum Ferritin during Spacelab 1 Flight (Mission Duration: 10 Days)[a]

Flight day	1	7	10	11	18	22
Flight	103	134	133	145	96	84
Control (Simulation)	99	82	86	84	61	67

[a] Data given as % of preflight mean value.

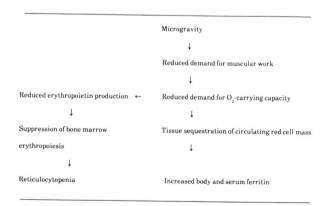

FIGURE 1 Possible mechanism of how microgravity can lead to the suppression of erythropoiesis in bone marrow and consequently to red cell mass deficit.

As more data become available, this scheme could be modified.

V. Recovery

Recovery of red cell mass invariably occurs after landing, indicating lack of damage to the physiological control mechanisms for red cell production. As expected, the recovery of red cell mass is heralded by reticulocytosis (Table IV), indicating the resetting of the regulatory mechanisms for erythropoiesis. Curiously, in flights of short duration, reticulocytosis is somewhat delayed, and appropriate response is not seen for several weeks. This might be attributable to the dormancy of bone marrow

TABLE V Changes in Hematologic, Immunologic, and Serum Parameters Immediately After Landing as Percent of Changes from Preflight Measurements in the Four Members of the Crew of Spacelab 1

Parameter	% Change on landing (± SEM)	
Blood volume		
Red cell mass, ml/kg	−9.30	1.60
Plasma volume, ml/kg	−5.98	4.30
Blood volume, ml/kg	−10.50	0.87
Erythrocyte hematology		
Erythrocyte count, $\times 10^{12}$/liter	−5.83	1.03
Hemoglobin, g/dl	−3.4	0.3
Hematocrit, liter/liter	−3.2	0.8
Mean corpuscular volume, fl	2.0	1.0
Mean corpuscular hemoglobin, pg/cell	8.3	0.9
Mean corpuscular hemoglobin concentration, g/dl	5.1	1.4
Erythrocyte production		
Reticulocyte number, $\times 10^9$/liter	−61	8
Reticulocyte production index	−60	8
Reticulocyte RNA, % of cytoplasm	24.7	3.1
Erythropoietin, units/ml	−72	10
Iron		
Transferrin, mg/dl	−10.7	7.5
Serum iron, μg/dl	−32	6.6
Unbound iron-binding capacity, μg/dl	−13.7	5.9
Total iron-binding capacity, μg/dl	−17.5	6.1
Saturation of transferrin, %	−17	4
Platelets and leukocytes		
Platelet count, $\times 10^9$/liter	12.0	10.7
Leukocyte count, $\times 10^9$/liter	17	16
Neutrophils, $\times 10^9$/liter	34	14
Lymphocytes, $\times 10^9$/liter	−0.4	13
Monocytes, $\times 10^9$/liter	40	50
Eosinophils, $\times 10^9$/liter	80	100
Serum chemistry		
Osmolality, mOsm/liter	0.6	0.3
Sodium, mEq/liter	−0.4	0.2
Potassium, mEq/liter	−1.5	6.5
2,3-DPG, mol/g Hb	−9.3	6.9
ATP, mol/g Hb	−6.9	14.2
Serum proteins		
Total serum protein, g/dl	−3.0	1.7
Albumin, g/dl	−0.7	3.9
Alpha-1 globulin, g/dl	10	10
Alpha-2 globulin, g/dl	10	10
Beta globulin, g/dl	−3	8
Gamma globulin, g/dl	−30	5
Haptoglobulin, mg/dl	12	8
Ferritin, ng/ml	62	36

stroma, which is needed for regulation of erythropoiesis. In flights of longer duration, reticulocytosis occurs within 1 week after landing, and the recovery of red cell mass is swift and uncomplicated.

VI. Functional Capacity of Space-Born Red Cells

Salyut 6 was an orbiting space station and the home for several successive missions of increasing duration. These missions provided the unusual opportunity to study the functional capacity of space-born red cells. Some of the crewmen in these missions remained aboard the Salyut for periods of 140 and 175 days. Because the life span of red cells is approximately 120 days, circulating red cells in these crew members were entirely replaced by red cells produced in space. The oxygen-carrying capacity of these cells was completely normal, suggesting that the cellular proliferation in the weightless state leads to the production of normally functioning cells.

Similarly, no major alterations have been reported in other proliferating cells systems such as the skin or the mucosa of gastrointestinal tract. However, subtle changes cannot be excluded.

VII. Other Hematologic and Immunologic Findings

Table V summarizes the findings in the hematologic, immunologic, and serum parameters that were determined in the four members of the crew of Spacelab 1. Alterations are either generally related to the loss of red cell mass, or insignificant on the level of our present understanding. Blood lymphopenia is seen early during the flight and has been attributed to the flight stress and the consequent hormonal changes. Consistent immunological changes have not been seen with routine methodology used. Hyporeactivity of lymphocytes to such external stimuli as lectins have occasionally been reported, but this is not a consistent finding. No consistent chromosomal alterations occur. [*See* LYMPHOCYTES.]

Bibliography

Leach, C. S., Chen, J. P., Crosby, W., Johnson, P. C. Lange, R. D., Larkin, E., and Tavassoli, M. (1988). Hematology and biochemical findings of Spacelab 1 flight. *In* "Regulation of Erythropoiesis" (E. D. Zanjani, M. Tavassoli, and J. L. Ascensao, eds.), pp. 415–453. PMA Publishing Corp., New York.

Talbot, J. M., and Fisher, K. D. (1986). Influence of spaceflight on red blood cells. *Fed. Proc.* **45,** 2285–2290.

Tavassoli, M. (1982). Anemia of spaceflight. *Blood* **60,** 1059–1067.

Tavassoli, M. (1986). Medical problems of space flight. *Am. J. Med.* **81,** 851–854.

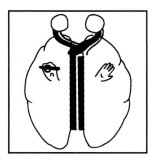

Hemispheric Interactions

MALCOLM JEEVES, *University of St. Andrews*

Glossary

Cerebral specialization Neuropsychological studies of brain-damaged people, patients with the forebrain commissures surgically divided, and normals have shown that each cerebral hemisphere is selectively superior for processing and analyzing certain kinds of sensory input and for performing motor output; for example, in most right-handed people, while the left hemisphere is better at handling speech, writing, language, and calculation, the right hemisphere is better at spatial construction, singing, and playing musical instruments

Forebrain commissures Axons of some cortical neurons cross from one cerebral hemisphere to the other; commissural fibers are distinct bundles of these axons clearly identifiable using autoradiographic techniques; they cross the midline in one of the forebrain commissures, namely the anterior commissure, corpus callosum, and dorsal and ventral hippocampal commissures

Splenium Commissural fibers from a given cortical region tend to occupy a distinct location in one of the forebrain commissures. The splenium is in the posterior part of the largest of the forebrain commissures, the corpus callosum, and cross connects areas of the cortex involved in visual processing.

GIVEN THAT WE have two cerebral hemispheres, it has long been debated whether or not this implies the possibility of two minds within one brain. That two hemispheres acting together are better than one is in accord not only with our intuition but with empirical data. How the two hemispheres work together to ensure the unified activity of the mind has for the past four decades been the subject of intense experimental investigation. Questions arise readily. What sorts of interactions take place to ensure the moment-by-moment, efficient, unified functioning of a system made up of two partially independent modules? What mechanisms make possible the efficient sharing of raw sensory data, yet also, as required, ensure sensory isolation and response inhibition? We consider in turn hemispheric integration, hemispheric inhibition, hemispheric facilitation, and the developmental aspects of each of these questions. Normally, the brain acts as a unified whole despite the fact that the two cerebral hemispheres in most people are functionally different and anatomically distinct. Moreover, each hemisphere appears to represent the world differently. Although not a clear-cut dichotomy, most left hemispheres emphasize a digital linguistic representation, whereas most right hemispheres specialize in analogue perceptual representations. These complementary functional systems were revealed most clearly following the now classical split-brain studies of Roger Sperry and his colleagues.

I. Interhemispheric Connections

The two cerebral hemispheres are intimately cross-connected through three transverse commissures, referred to collectively as the forebrain commissures (Table I). The axial or midline parts of the body tend to have the most numerous commissural connections, whereas those that deal with sensory and motor functions of extreme distal parts of the extremities have relatively few. The higher-order association areas, in the frontal lobe and at the parieto-temporo-occipital junction, have variable densities of contralateral connections. After conception, if development proceeds normally, the first crossing is of the fibers that will form the anterior commissure at about the 50th day. The second crossing, the hippocampal commissure, appears soon thereafter at the end of the second gestational month. Finally, the corpus callosum begins its development at about the 12th gestational wk. The corpus callosum is formed in all its parts 18–20 wk after gestation but does not reach full functional maturity until myelination is completed at around 10 yr of age. [See BRAIN.]

Different topographical regions of the forebrain commissures contain fibers relating to different functional specializations (see Fig. 1). While somatic sensory functions are localized to the central portion of the body of the callosum, the splenium deals mainly with visual function. This topographical organization is confirmed and illustrated by clinical and behavioral evidence from studies of patients with discrete lesions of the forebrain commissures. The anterior commissure cross-connects the olfactory areas on the two sides of the brain as well as parts of the temporal lobes.

The cross-connection of the two hemispheres

TABLE I Hemispheric Interactions Mediated through the Forebrain Commissures

Integration	ensures sharing of lateralized sensory input and coordination of bilaterally controlled motor output
Inhibition	(1) ensures the development of topographic sensation and precise motor control by suppressing the contribution from uncrossed ipsilateral pathways
	(2) ensures hemispheric independence, thus allowing implementation of modularity of representation and processing in the normal brain
Facilitation	(1) in the intact cortex, the corpus callosum exerts a facilitatory, or modulating, influence on the neural activity of both hemispheres
	(2) this modulatory action may actively participate in the functional reorganization that takes place after brain injury

through the commissures has, perhaps naturally, focused on the likely information transmission and information transfer functions of the system. However, the fact (see below) that patients in whom the commissures have been cut are indistinguishable from normals in activities of daily living warns against overemphasizing this role for the commissures to the exclusion of other possibilities. And why such a massive structure as the corpus callosum should be present if other significant roles are not carried by it is hard to conceive.

At birth, there is exuberant growth of the fibers in the commissural systems. The extent of this is indicated by the elimination of 70% of these fibers in the mature callosal system.

II. Hemispheric Integration

When the forebrain commissures are cut in humans, a dramatic disconnection syndrome occurs. Detailed laboratory studies of such split-brain patients indicate the coexistence of two seemingly independent cognitive systems within the same brain; thus, one hemisphere is unable to communicate with the other hemisphere. The split-brain syndrome, however, has relatively few effects on everyday life. Thus, normally, the forebrain commissures and particularly the corpus callosum ensure that the processing operations of the two hemispheres can be integrated. For example, in the tactile modality, if a familiar object is placed in the right hand of a split-brain patient, he has no difficulty in naming it; however, if the same task is given but now with the object in the left hand, the patient is usually unable to name it, because the information from the left hand goes to the right hemisphere but due to the commissure section is then denied access to the brain's left hemisphere where in most right-handed people the language system is found. Similarly, in the visual modality, if the picture of an object is flashed briefly to the right side of where the patient is fixating, the information goes to the left side of the brain. A split-brain patient has no difficulty in naming an object thus presented. If, however, a picture is flashed to the left side of the visual field so that the information goes to the right side of the brain, the split-brain patient is unable to name correctly what he has seen. The same is true if letters of the alphabet or words are presented to the right or left visual fields. Thus, it was concluded that the two hemispheres communicated and passed infor-

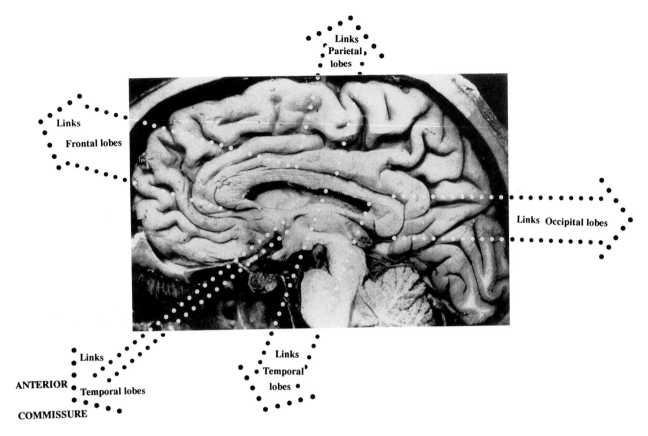

Links Parietal lobes

Links Frontal lobes

Links Occipital lobes

Links Temporal lobes

ANTERIOR COMMISSURE

Links Temporal lobes

FIGURE 1 The human corpus callosum and the anterior commissure seen *in situ* contain the majority of the fibers cross-linking the two cerebral hemispheres. Different topographical regions of the forebrain commissures contain fibers relating to different functional specializations. Somatic sensory and motor functions are localized to the central portion of the body of the corpus callosum; the splenium deals mainly with visual-related functions.

mation for specialized processing back and forth through the corpus callosum (Fig. 2).

In recent years, the necessity of the forebrain commissures for hemisphere integration, as outlined above, has been modified somewhat. Studies have shown that some rudimentary forms of visual information pass from one hemisphere to the other even though the corpus callosum, the anterior commissure, and the hippocampal commissure have all been cut. Researchers conclude that subcortical structures may have the functional capacity to subserve the integration of information at a visuospatial and also higher-order cognitive level; however, they noted that such integration as achieved was far from perfect.

The possibility for two hemispheres to share in-

formation even though the neocortical commissures have been cut gains support from earlier studies of people who were born without the corpus callosum and, in some instances, also without the anterior commissure. Extensive studies on a small group of these patients showed that in everyday living they manifest no obvious difficulties; like the surgical split-brain patients, their deficits only became evident with careful laboratory testing. So long as the anterior commissure is present, they seem able to share information between the two cerebral hemispheres although not as efficiently as a person with a normal intact brain. The compensatory mechanisms making possible their seemingly normal performance probably vary depending on the sense modality being tested. Thus, visual transfer may be effected through the anterior commissure, whereas tactile transfer may be made possible through the enhanced elaboration and development of uncrossed sensory pathways.

The importance of the integrative action of the neocortical commissures in ensuring efficient hemispheric interaction is further underlined by the many reports demonstrating the so-called bilateral processing advantage. This refers to the finding that

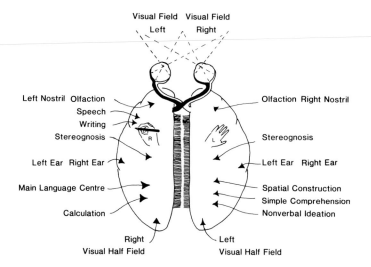

FIGURE 2 A schematic representation of some of the specialized functions of the cerebral hemispheres. The corpus callosum is shown as sectioned down the midline, thus abolishing most routes for hemispheric interactions as described in the text.

when two stimuli, normally visual, must be compared and a judgment made as to whether they are the same or different, if one stimulus is sent separately to each hemisphere, the task is accomplished faster and with fewer errors than if both stimuli go to one hemisphere. To a point, this finding is counter-intuitive, because in the bilateral presentation condition one stimulus must be transferred through the forebrain commissures to the other hemisphere for comparison to be made. From a physiological point of view, such transfer involves information transmission with the consequent possibility of degradation of the signals, which in turn might be expected to result in poorer performance than when both stimuli arrive within the same hemisphere and no such interhemispheric transfer is called for. Because the net effect in the bihemisphere condition is improved performance, it further underlines the efficiency of the integrative processes through the forebrain commissures.

III. Reciprocal Hemispheric Inhibition

When clear differences in hemispheric functioning were first reported, physiologists pointed out that such cerebral specializations most likely could only come about as a result of inhibitory processes mediated through the forebrain commissures. This view was subsequently taken up by psychologists who argued that, in the course of normal development, the inhibitory processes mediated by the forebrain commissures ensured the specializations of function of the two cerebral hemispheres observed at maturity (Fig. 2) (e.g., special linguistic capacities

develop in the left hemisphere so it inhibits and prevents similar developments in the right hemisphere). The anatomical and physiological substrate for this process is assumed to be the forebrain commissures and primarily the corpus callosum; however, this view has been challenged. The results of detailed studies of people born without the corpus callosum failed to find the predicted bilateral representation of function, which should, according to the inhibitory hypothesis, occur in the absence of the corpus callosum from birth. Thus, while it seems unlikely that the normal role of the corpus callosum is to bring about hemispheric specialization, fine-tuning of hemispheric specialization still may be afforded by the forebrain commissures. If this is true, then while the brains of acallosals are lateralized, the extent of lateralization may be less than that found in the normal population. [*See* CEREBRAL SPECIALIZATION.]

Because hemispheric interaction in the form of inhibition may not be necessary to ensure normal hemispheric specialization, it does not mean that other important inhibitory roles are not mediated through the corpus callosum. Researchers have suggested that inhibition, at the cognitive level, is called for to enable component mental operations elaborated concurrently in both hemispheres to be protected from interfering cross-talk between the hemispheres until the operations are fully developed and ready to adopt specific relationships with one another within the total action program. In a somewhat similar vein, others have suggested that hemispheric independence is a common and ubiquitous state in the normal brain and that a process of callosal inhibition could ensure such independence. It was argued that in the normal person, the uncrossed ipsilateral pathways for sensory input and motor control are not allowed to compete with the crossed pathways, and this occurs through the suppressive action or the inhibition through the corpus callosum. Because this is absent in an acallosal patient, the competition between crossed and uncrossed pathways remains, resulting in reduced sensory and motor performance. These ideas were based on the reduced ability of adult acallosals to make fine sensory discriminations and to exercise precise motor control. These findings have been confirmed and extended. The inhibitory role can account for the development of dominance of the contralateral over the ipsilateral pathways in the sensory and motor systems, which appear to be essential for finely tuned sensation and motor control. [See MOTOR CONTROL.]

IV. Hemispheric Facilitation

Evidence indicates an important facilitative role of the corpus callosum. Studies of patients born without the corpus callosum, as well as of children and adults who had undergone partial or total callosotomy for control of intractable epilepsy, demonstrated deficits that were not limited to *inter*hemispheric processing but were evident also in *intra*hemispheric processing; therefore, the corpus callosum seems to exert a facilitatory action modulating neural activity in *both* hemispheres. In the absence of the corpus callosum, this is assumed to be reduced or absent and, thus, the deficits in intrahemispheric as well as interhemispheric processing occur. One implication of this hypothesis on the modulating role of the corpus callosum is that in the absence of the callosum or following damage to it, neither hemisphere will likely achieve its full potential. Furthermore, clinical reports indicate that the corpus callosum, through its modulating action, may actively participate in the functional reorganization that takes place after brain injury. Restitution of language functions only occur when the forebrain commissures remain intact. Thus, this type of hemispheric interaction, labelled modulatory action, may be one important way in which an intact cerebral hemisphere can help to compensate for loss or impairment of functions when the other hemisphere is damaged.

V. Developmental Aspects of Hemispheric Interactions

As noted above, the corpus callosum is one of the last paths of the nervous system to mature. It is generally accepted that the callosal fibers are not fully myelinated (and therefore mature) until around 10 yr of age. Therefore, as a system, the forebrain commissures are among the last to achieve full functional capacity. It has been further suggested that hemispheric function during development, and presumably up until about 10 yr of age, reflects the increasing interhemispheric influences discussed earlier, because these are mediated largely through the maturing corpus callosum. Attempts to demonstrate this developmental progression physiologically and behaviorally have met with limited success. One study reported that relay time across the corpus callosum decreased linearly with logarithm of age from 3.5 yr to puberty. Others have given children the task of judging whether the textures of two pieces of cloth, lightly touched by a finger, are the same or different. The judgments were made by either two different fingers on one hand or two different fingers on different hands. They found that whereas all the children showed an expected improvement in overall performance with increasing age, the younger preschool children had greater difficulty when the two judgements were made by different hands than when they were made by the same hand. With the two-handed condition, information must cross the forebrain commissures, whereas with the one-hand condition, no such crossing is involved. Other researchers have studied the ability of children of different ages to indicate correctly when the tip of a finger has been lightly touched. They show the experimenter which finger was touched, by touching it themselves with the thumb of that hand (this is the uncrossed condition). Alternatively, they may be asked to indicate in the same manner, on the other hand, the finger that was touched on the first hand (this is the crossed condition). In the latter case, a callosal crossing is involved, in the former it is not. Younger children found this crossed condition much harder relative to the same-hand, uncrossed condition when compared with older children. The importance of the development of fully efficient interhemispheric integration is evident when one remembers that complex tasks demand that the human brain enlists and coordinates the special abilities of both hemispheres. The question naturally arises of what happens if, for whatever reason, the corpus callosum is not fully functional in a developing child. A study of the efficiency of children of different ages in transmitting information from one hemisphere to the other showed that dyslexic children resemble normal children younger than them both in overall level of correct response and in the ratio of crossed (same hand) to uncrossed (two hands) errors in a task such as that described above. Typically, the reading-disabled children approximate normal children 4 to 6 yr younger than themselves. However, the callosal deficit may simply be a particular instance of a more general deficit, which is prominent in the studies described above, because the corpus callosum is simply the largest of an extensive set of cortico-cortical connections. Pictures of brains using magnetic resonance imagery link underdevelopment of the corpus callosum with ill-defined learning difficulties in children.

Thus, hemispheric interactions through the forebrain commissures seemingly play a variety of important roles to ensure the development to full capacity of the potential for unified functioning of the human brain.

VI. Conclusions

Hemispheric interaction is not a simple matter. Several distinguishable interhemispheric functions are mediated through the forebrain commissures. Of these, hemispheric integration is crucial for the unified functioning of the differently specialized cerebral hemispheres. An inhibitory role of the commissures is equally important, not only during development, but also at maturity to make possible the independent functioning as and when required

of a particular cerebral hemisphere without interference from the other hemisphere. A facilitatory function has been suggested, which ensures the fine-tuning of what is happening in both hemispheres and guarantees that both hemispheres achieve their full potential. Regarding development, there are two distinguishable functions, both of which depend on a properly functioning neocommissural system. First, there is the sharing function, alternatively called information transmission or integration, which increasingly guarantees the sharing of information as and when appropriate between what is happening in each cerebral hemisphere. The shielding function is equally important. It is the way in which one hemisphere may, through the callosum, inhibit the activity of the other hemisphere so that it does not interfere with the ongoing activity of the first hemisphere.

Bibliography

Jeeves, M. A. (1986). Callosal agenesis: Neuronal and developmental adaptations. *In* "Two Hemispheres—One Brain" (F. Lepore, M. Ptito, and H. H. Jasper, eds.). Alan R. Liss, Inc., New York.

Jeeves, M. A., Silver, P. H., and Milne, A. B. (1988). Role of the corpus callosum in the development of a bimanual motor skill. *Dev. Neuropsychol.* **4(4)**, 305–323.

Lassonde, M. (1986). The facilitatory influence of the corpus callosum on intrahemispheric processing. *In* "Two Hemispheres—One Brain" (F. Lepore, M. Ptito, and H. H. Jasper, eds.). Alan R. Liss, Inc., New York.

Njiokiktjien, C. (1988). "Pediatric Behavioural Neurology," Ch. 35 Developmental Interhemispheric Disconnection, Suyi, Amsterdam.

Quinn, K., and Geffen, G. (1986). The development of tactile transfer of information. *Neuropsychologia* **24**, 793–804.

Zaidel, E., Clarke, J. M., and Suyenobu, B. (1990). Hemispheric independence: A paradigm case for cognitive neuroscience. *In* "Neurobiological Foundations of Higher Cognitive Function" (A. Scheibel and A. Wechsler, eds.). Guildford, New York.

ologists therefore called it heme–heme interaction. The term "cooperativity" is now preferred.

The equilibrium curve of hemoglobin begins with a straight line at 45° to the axes, because at first oxygen molecules are so scarce that only one heme in each hemoglobin molecule has a chance of catching one of them, and all of the hemes therefore react independently, as in myoglobin. As more oxygen flows in, the four hemes in each molecule begin to cooperate and the curve steepens. The tangent to its maximum slope is known as Hill's coefficient (n), after the English physiologist A. V. Hill, who first attempted a mathematical analysis of the oxygen equilibrium.

The normal value of Hill's coefficient in the blood of healthy human subjects under standard physiological conditions is 3.0, which means that in this part of the curve the fractional saturation of hemoglobin with oxygen increases with the third power of the partial pressure of oxygen. The curve ends with another line at 45° to the axes, because oxygen has now become so abundant that only a single heme in each molecule is likely to be free. In this situation there can be no cooperativity, and all hemes in the solution once more combine with oxygen independently, as in myoglobin.

II. Cooperative Effects: Physiological Purpose

Hill's coefficient and the oxygen affinity of hemoglobin depend on the concentrations of several chemical factors in the red blood cell: protons (hydrogen ions; i.e., hydrogen atoms without electrons, whose concentration, measured as pH, reflects the acidity of the solution), carbon dioxide, chloride ions, and an ester of glyceric acid and phosphate, called 2,3-diphosphoglycerate (DPG). These are known as heterotropic ligands, as opposed to oxygen or carbon monoxide, which are called homotropic ligands. Increasing the concentration of any of these heterotropic ligands shifts the oxygen equilibrium curve to the right, toward lower oxygen affinity, and makes it more sigmoidal (Fig. 2). The cooperative binding of oxygen and the influence of the heterotropic ligands on the oxygen equilibrium curve are known collectively as the cooperative effects of hemoglobin. Strangely, none of the heterotropic ligands influences the oxygen equilibrium curve of myoglobin, even though the chemistry and the structure of myoglobin are related closely to those of the individual chains of hemo-

globin. The explanation emerged ony recently (see below).

What is the purpose of these cooperative effects? Why is it not good enough for the red blood cell to contain a simple oxygen carrier such as myoglobin? The answer is that such a carrier would not allow enough of the oxygen in the red blood cell to be unloaded to the tissues, nor would it allow enough carbon dioxide to be carried to the lungs by the blood plasma.

The cooperativity of oxygen binding and release, the effects of the heterotropic ligands, and the effect of temperature conspire to maximize the difference in fractional saturation with oxygen between arterial and venous blood. The partial pressure of oxygen in arterial blood is normally 90–100 mm Hg; in mixed venous blood, 35–40 mm Hg. Under standard conditions (i.e., pH 7.4, pCO_2 of 40 mm Hg, a DPG concentration of 5 mM/liter of packed red blood cells, carbon monoxyhemoglobin of 1%, and a temperature of 37°) in whole blood from apparently healthy subjects, these partial pressures correspond to oxygen saturations of 98–100% in arterial and 66–73% in venous blood. Under these circumstances only one-quarter to one-third of the total oxygen carried is released in the tissues; the fractional saturation with oxygen in muscular veins during heavy exercise is believed to be lower, so that more oxygen would be delivered to the muscles.

The more pronounced the sigmoidal shape of the equilibrium curve, the greater the fraction of oxygen that can be released. Several factors cooperate to that purpose. In the tissues oxidation of nutrients by the tissues liberates lactic acid and carbonic acid; these acids, in turn, liberate protons, which shift the curve to the right, toward lower oxygen affinity, and make it more sigmoidal, which allows more oxygen to be released. DPG has the same effect. The number of DPG molecules in the red cell is about the same as the number of hemoglobin molecules, 280 million, and probably remains fairly constant during circulation; a shortage of oxygen, however, causes more DPG to be made, which further lowers the oxygen affinity and therefore helps to release more oxygen. The human fetus has a hemoglobin with the same α chains as the hemoglobin of an adult human, but different β chains that have a lower affinity for DPG. This gives fetal hemoglobin a higher oxygen affinity and facilitates the transfer of oxygen from the maternal to the fetal circulation.

If protons lower the affinity of hemoglobin for oxygen, then the laws of action and reaction de-

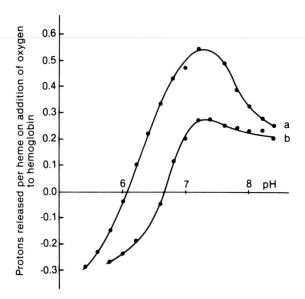

FIGURE 3 Discharge of protons upon uptake of oxygen by hemoglobin, known as the Bohr effect, after Danish physiologist Christian Bohr. (a) Native human hemoglobin. (b) Human hemoglobin from which the C-terminal histidines of the β-chains, which are two of the principal residues responsible for the Bohr effect, has been cleaved.

mand that oxygen lowers the affinity of hemoglobin for protons. Hence, the liberation of oxygen causes hemoglobin to combine with protons and vice versa; at physiological pH about two protons are taken up for every four molecules of oxygen released, and two protons are liberated again when four molecules of oxygen are taken up (Fig. 3). This reciprocal action is known as the Bohr effect and is the key to the mechanism of carbon dioxide transport. The carbon dioxide released by respiring tissues is too insoluble to be transported as such, but it can be rendered more soluble by combining with water to form a bicarbonate ion and a proton. The chemical reaction is written

$$CO_2 + H_2O \rightarrow HCO_3^- + H^+$$

In the absence of hemoglobin this reaction would soon be brought to a halt by the excess of protons produced, like a fire going out when the chimney is blocked. Deoxyhemoglobin acts as a buffer, mopping up the protons released in the tissues and thus tipping the balance toward the formation of soluble bicarbonate. In the lungs the process is reversed. There, as oxygen binds to hemoglobin, protons are cast off, driving carbon dioxide out of the solution so that it can be exhaled.

III. Allosteric Proteins

Many proteins besides hemoglobin exhibit cooperative effects. Most of these are enzymes that catalyze (i.e., speed up) chemical reactions in the living cell. Such enzymes consist of several subunits, each containing a site that catalyzes the same reaction. The combination of substrate (i.e., the molecule that undergoes the reaction) with one of the subunits raises the substrate affinities of all of the subunits, just as the combination with oxygen of one of the four subunits of hemoglobin raises the oxygen affinity of all of them. Chemical compounds that bear no resemblance to the substrates themselves often regulate the substrate affinity of such enzymes, just as hydrogen ions or DPG regulate the oxygen affinity of hemoglobin.

In most of these proteins, the cooperative effects arise through a transition between two or more alternative structures with different substrate affini-

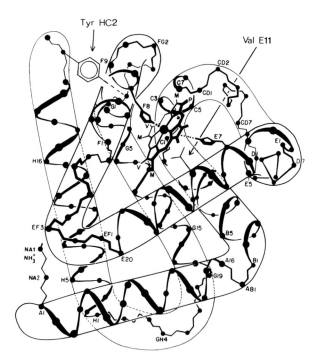

FIGURE 4 Secondary and tertiary structure of the hemoglobins showing α carbons and coordination of the hemes. Shown are the proximal histidine F8, which links the heme iron to the globin; the distal histidine E7 and valine E11, which make contact with the bound oxygen; and tyrosine HC2, which links the carboxyl terminus of the chain to helix F by a hydrogen bond. The exact numbers of residues in the different segments are the same in all mammals, but vary in other vertebrates and especially invertebrates. The lettering of the helical and nonhelical segments is explained in the text. M, V, and P, Methyl, vinyl, and proportionate side chains, respectively, of the heme.

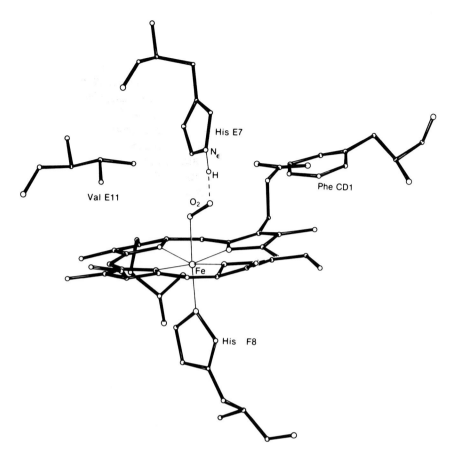

FIGURE 5 Arrangement of proximal and distal histidines in oxy-myoglobin, showing the hydrogen bond between N_ϵ of the distal histidine and the bound oxygen. His, Histidine; Val, valine; Phe, phenylalanine.

ties. These structures are distinguished by the arrangement of the subunits and the number and kinds of bonds between them. If there are only two alternative structures, the one with fewer and weaker bonds between the subunits would be free to develop its full catalytic activity or oxygen affinity. It is therefore called R, for "relaxed." The activity would be damped in the structure with more and stronger bonds between the subunits; this form is called T, for "tense." In the absence of oxygen or substrate, nearly all of the protein molecules have the T structure; if the protein molecules are saturated with oxygen or substrate, nearly all of them have the R structure. The greater the proportion of molecules in the R structure, the greater the affinity of the solution for oxygen or substrate.

Cooperatively arises by a change in the relative population of protein molecules in the T and R structures as the reaction progresses. For example,

in hemoglobin the steep slope of the oxygen equilibrium curve near half-saturation of the solution with oxygen arises when the change in the relative population of the two forms as a function of oxygen saturation is greatest. Proteins that exhibit cooperatively and change their structures in response to chemical stimuli are called allosteric. Compounds other than oxygen or substrates that change the equilibrium between T and R states, such as DPG in red blood cells, are called allosteric effectors. In hemoglobin the heterotropic ligands all act as effectors that shift the allosteric equilibrium towards the T structure. This is the reason why they shift the oxygen equilibrium curve of hemoglobin to the right, towards lower oxygen affinity. They do not affect the oxygen equilibrium curve of myoglobin because it is a monomer, and allosteric effects arise only if several monomers combine, as in hemoglobin.

IV. Three-Dimensional Structure

It has become customary to speak of the primary, secondary, tertiary, and quaternary structure of proteins. Primary refers to the amino acid sequence; secondary, to the local conformation of the polypeptide chain, such as the α helix or pleated sheet; tertiary, to the fold of a single polypeptide chain, as in myoglobin; and quaternary, to the assembly of several chains or subunits (e.g., the hemoglobin tetramer).

The fold of the polypeptide chain is the same in myoglobin and in the α and β chains of hemoglobin. It is made up of seven or eight α-helical segments and an equal number of nonhelical ones placed at the corners between them and at the ends of the chain (Fig. 4). The helices are named A–H, starting from the amino terminus, and the nonhelical segments that lie between helices are named AB, BC, CD, etc. The nonhelical segments at the ends of the chain are called NA at the amino terminus and HC at the carboxyl terminus. Residues within each segment are numbered from the amino terminus, A1, A2, CD1, CD2, etc. Evolution has conserved this fold of the chain, despite great divergence of the sequences: The only residues common to all hemo-

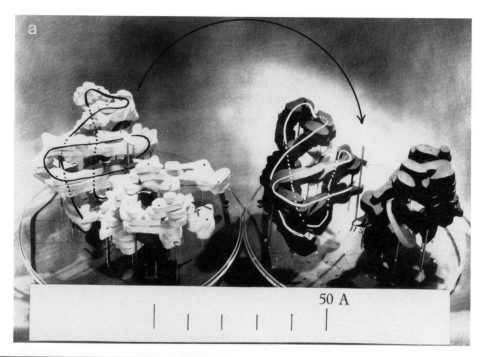

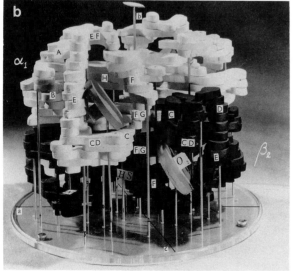

FIGURE 6 Assembly of hemoglobin tetramer. (a) A pair of α subunits (white) and a pair of β subunits (black) are placed on either side of the twofold symmetry axis (central rod with white sign). The pair of α subunits is then inverted and placed on top of the β subunits to form the complete tetramer (b). Note the hemes in separate pockets. Rotation by 180° about the twofold symmetry axis brings the molecule to congruence with itself. The letters refer to the helical and nonhelical segments shown in Fig. 4 and explained in the text.

globins are the proximal histidine F8, which attaches the heme to the globin, and phenylalanine CD1, which wedges the heme into its pocket (Fig. 5). Most globins also have a histidine on the distal (i.e., oxygen) side of the heme. In myoglobin and in the α subunits this histidine forms a strong hydrogen bond with the bound oxygen, but in the β subunits the bond is either weaker or absent.

Myoglobin has ionized side chains distributed all over its surface, but the surfaces of the α and β globin chains have nonpolar patches that allow them to combine to form the $\alpha_2\beta_2$ tetramer. To make a model of the tetramer, each of the subunits must first be joined to its partner around a twofold symmetry axis, which brings one subunit into congruence with the other by a rotation of 180°. One pair of chains is then inverted and placed on top of the other to make a tetramer in which the four subunits are arranged at the corners of a tetrahedron (Fig. 6). The twofold symmetry axis that relates the pairs of α and β chains runs through a water-filled cavity at the center of the molecule. This cavity widens upon transition from the R structure to the T structure to form a receptor site for the allosteric effector DPG between the two β chains.

The heme is made up of one atom of ferrous iron at the center of protoporphyrin IX, which is synthesized in the bone marrow from glycine and acetate (see Fig. 1). It is wedged into a pocket of the globin with its hydrocarbon side chains interior and its polar propionate side chains exterior; it is in contact with about 20 side chains of the globin, all hydrophobic apart from the two histidines.

V. Transition from the Deoxy Structure to the Oxy Structure

Upon transmission from the deoxy (T) structure to the oxy (R) structure , one α–β dimer rotates relative to the other by 12–15° (Fig. 7). The rotation is accompanied by a shift of ~1 Å along the rotation axis and makes the two α–β dimers move relative to each other along the α_1–β_2 and $\alpha_2\beta_1$ contacts. The two contacts form a two-way switch which ensures that the α–β dimers click back and forth between no more than two stable positions, so that any stable intermediates in the reaction of hemoglobin with ligands must have either the quaternary R or the T structure. The T structure is constrained by additional bonds between the subunits, which oppose the changes in tertiary structure needed to flatten

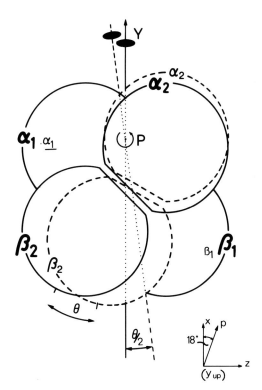

FIGURE 7 The change in quaternary structure that accompanies the ligation of hemoglobin. Bold symbols and plain lines in the diagram refer to deoxyhemoglobin; light symbols and broken lines refer to oxyhemoglobn. The binding or release of oxygen causes little movement of the α_1 and β_1, or α_2 and β_2, subunits relative to each other. The oxygenated and deoxygenated α_1–β_1 dimers have been superimposed. The position of the oxygenated α_2–β_2 corresponds to that obtained by moving the deoxygenated α_2–β_2 dimer as follows: rotating it about an axis P (which is perpendicular to the twofold symmetry axes, marked Y, of both the oxygenated and deoxygenated molecules, and to the picture plane) by an angle $\theta = 12$ to 15° and shifting it along the axis P by 1 Å into the page. [From J. Baldwin, and C. Chothia, Haemoglobin: The structural changes related to ligand binding and its allosteric mechanism. *J. Mol. Biol.* **129,** 175 (1979).]

the hemes upon combination with oxygen. These bonds take mainly the form of salt bridges (i.e., hydrogen bonds between oppositely charged ions), such as

$$-NH_3^+ \cdots \begin{matrix} -O \\ C- \\ O \end{matrix} \quad \text{or} \quad -NH^+ \cdots \begin{matrix} -O \\ C- \\ O \end{matrix} \quad \text{or} \quad -NH_3^+ \cdots Cl^-$$

One pair of salt bridges is formed by the carboxyl-terminal arginines of the α chains, another by the carboxyl-terminal histidines of the β chains, and another by chloride ions (Cl^-) at the amino-terminal valines of the α chains. In addition, four pairs are formed by DPG with cationic groups of the β chains (Fig. 8).

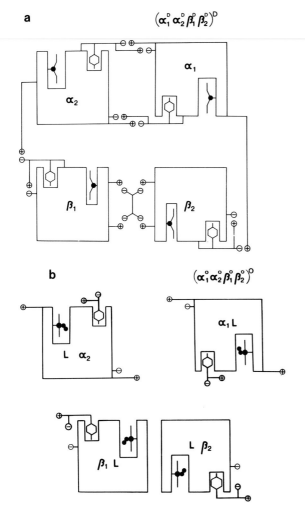

a $(\alpha_1^D \alpha_2^D \beta_1^D \beta_2^D)^D$

α_1

α_2

β_1

β_2

b $(\alpha_1^o \alpha_2^o \beta_1^o \beta_2^o)^o$

L α_2

α_1 L

β_1 L

L β_2

FIGURE 8 (a) Salt bridges by carboxyl-terminal residues, marked by juxtaposed positive and negative charges, in the deoxyhemoglobin (T) structure. The molecule with the four negative charges between the β chains of the T structure is 2,3-diphosphoglycerate. It must be released before transition to the R structure (b), where the gap between the β chains becomes too narrow to accommodate it, and the salt bridges are broken. The black circles in (a) represent the iron atoms; in (b) the iron atoms and oxygen molecules attached to them. The lines extending from the black circles represent the porphyrins. The hexagons represent the tyrosines HC2 that precede the carboxyl-terminal amino acid residues.

Hydrogen ions, DPG, chloride ions, and carbon dioxide all stabilize the T structure. They lower its oxygen affinity and retard the T → R transition. For example, at low concentrations of the allosteric effectors, most of the hemoglobin molecules in a solution may click from the R structure to the T structure when the second oxygen molecule is bound. At high effector concentration they may click over only when the third oxygen molecule is bound.

Every one of the effectors acts by fortifying the salt bridges of the T structure. In this way they promote the release of oxygen to the tissues. The allosteric effectors leave the oxygen affinity of the R structure unchanged.

The transition from the T structure to the R structure is triggered by stereochemical changes at the hemes. In deoxyhemoglobin the porphyrins are domed, and the iron atoms are displaced by 0.4 Å from the plane of the porphyrin nitrogens, regardless of the quaternary structure of the globin. Upon

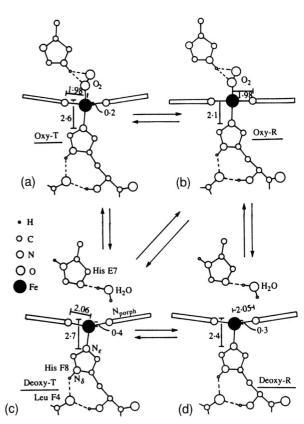

O_2 — 1·98 — 0·2 — 2·6 — Oxy-T (a)

O_2 — 1·98 — 2·1 — Oxy-R (b)

• H
○ C
○ N
○ O
● Fe

His E7 — H_2O

H_2O

2.06 — N_{porph} — 0·4 — 2·7 — N_ε — His F8 — Deoxy-T — N_δ — Leu F4 (c)

2·05 — 0·3 — 2·4 — Deoxy-R (d)

FIGURE 9 Changes in the heme stereochemistry on binding of oxygen by the α subunits in the R and T structures (a) Upon uptake of oxygen by the T structure, the heme remains domed and the iron remains displaced from the porphyrin plane, due to the constraints by the tightly packed side chains of the globin around the heme; on the other hand, upon dissociation of oxygen from the R structure (d), the porphyrin becomes doomed, because R structure, being relaxed, does not force the heme to remain flat. (c) In the oxygen-free T structure the porphyrin is domed and the iron atom is displaced from the porphyrin plane. (b) In the oxygenated R structure the porphyrin is flat and the iron lies in the porphyrin plane. The numbers give distances in Ångstrom units. [Reprinted with permission from M. F. Perutz, G. Fermi, B. Luisi, B. Shaanan, and R. C. Liddington, Stereochemistry of cooperative effects in hemoglobin. *Acc. Chem. Res.* **20**, 309 (1987). Copyright 1987 American Chemical Society.]

binding oxygen the iron atoms move toward the porphyrin. These remain domed in the T structure, but flatten upon transition to the R structure. As a result, upon transition from deoxyhemoglobin (oxygen free) in the quaternary T structure to oxyhemoglobin (hemoglobin saturated with oxygen) in the quaternary R structure, the iron atoms and the proximal histidines move toward the mean planes of the porphyrins by 0.5–0.6 Å (Fig. 9). In deoxyhemoglobin in the quaternary T structure, combination of the β hemes with oxygen is blocked by the distal valine; in the R structure this block is removed by shift of the β hemes relative to helix E. The doming and undoming of the porphyrins, the movements of the irons and the proximal histidines, and, in the β subunits, the movement of the distal residues relative to the heme are seen as the only perturbations that could set in motion the changes in quaternary structure.

VI. Embryonic and Fetal Hemoglobins

The transport of oxygen differs in the embryonic, fetal, and adult stages of development. The early embryo obtains oxygen from the maternal interstitial fluid and uses a hemoglobin known as ζ_2–ε_2. The developing fetus obtains its oxygen via the placenta, using a hemoglobin known as $F_{II(\alpha_2-\gamma_2)}$. This has the same α chain as the adult form, but its β chain, known as γ, differs from the adult form in 39 positions. In fact, two different kinds of γ chain are made; they differ only in position H14(136), where one chain has a glycine and the other has alanine. There is also a minor component, known as hemoglobin F_I, in which the amino-terminal valines of the γ chains are acetylated. The oxygen affinities of hemoglobins F_{II} and F_I are higher than those of the two adult forms [hemoglobin A (α_2–β_2) and A2 (α_2–δ_2)], which facilitates the transfer of oxygen across the placenta from the adult to the fetal circulation. Their higher oxygen affinity is due to their lower affinity for the allosteric effector, DPG; that is, at equal concentrations of hemoglobin and DPG, the latter lowers the oxygen affinity of fetal hemoglobin less than that of adult hemoglobin.

VII. Hemoglobin Diseases

Hemoglobin genes are subject to mutations that alter the structure of the globin. Several hundred variants of human hemoglobins have been isolated and chemically characterized. In most of them the abnormality consists of the replacement of one pair of identical amino acid residues by another. All of these replacements are consistent with single-base substitutions in the DNA coding for the globin chains. Some variant hemoglobins have residues deleted or inserted; in others the chains are cut short or elongated, and yet others contain hybrids of β and δ or β and γ chains. While many of the abnormal hemoglobins are without effect on their carriers' health, others give rise to symptoms. In over 30 abnormal hemoglobins the stereochemical causes of these symptoms are now known.

Among the hemoglobin diseases caused by single-amino acid substitutions, sickle cell anemia is the most serious, affecting the largest number of people. Many other substitutions give rise to hemolytic anemia when they affect the stability of hemoglobin, causing unfolding of the globin chains and the formation of clumps of unfolded hemoglobin molecules in red blood cells. These anemias are generally less serious, first, because they do not lead to the blocking of blood vessels that causes the crises characteristic of sickle cell anemia, and second, because each of these substitutions is sufficiently rare to exclude its occurrence in the father and the mother of the same child. Many substitutions raise the oxygen affinity of hemoglobin, which diminishes the amount of oxygen delivered to the tissues. A lack of oxygen is registered by a sensor in the kidneys that responds by releasing a hormone called erythropoietin. This hormone stimulates the synthesis of red blood cells in the bone marrow, with the consequence that such patients tend to suffer from polycythemia (i.e., an excess of red blood cells), which could be severe enough to cause discomfort and occasional symptoms. On the other hand, their high erythrocyte count and high oxygen affinity make them better adapted than other people to life at high altitudes. There are far fewer substitutions that lower the oxygen affinity. Those that do occur often manifest themselves by cyanosis (i.e., bluish color of the skin, due to insufficient oxygenation of the blood) and sometimes lead to diminished red blood cell synthesis. [See SICKLE CELL HEMOGLOBIN.]

Another group of hemoglobin diseases, known as thalassemias, arises from failures to synthesize either the α or the β globin chains. Individuals with only one sickle cell or thalassemia gene and one normal hemoglobin gene are generally healthy, because their red blood cells contain enough normal hemoglobin to supply them with oxygen, but people with two defective hemoglobin genes tend to be severely crippled. In 1949 the British geneticist J. B. S. Haldane first spotted that these diseases are most frequent in areas where malaria is prevalent. This has been confirmed by studies in many parts of the world. For reasons that we do not yet understand, the altered hemoglobin inhibits the multiplication of the malaria parasite in the red blood cells of infants carrying one sickle cell or one thalassemia gene. These infants are therefore more resistant to malaria than are normal infants and stand a better chance of surviving to adult age. [*See* MALARIA.]

It seems that the mutations causing either sickle cell anemia or thalassemia arise spontaneously in human populations. In the absence of malaria, selective pressure penalizes the carriers of the altered hemoglobins and they produce fewer children, but if malaria is present, fewer children with normal hemoglobin than children who carry one defective hemoglobin gene survive to reproductive age. The high incidence of thalassemia in malarial islands of Melanesia and its rarity in malaria-free islands are particularly impressive, since people have inhabited these islands for no more than 3000 years; Darwinian selection must therefore have operated in histor-ical times. It is the best example of evolution by natural selection in humans.

Bibliography

Antonini, E., and Brunori, M. (1971). "Hemoglobin and Myoglobin and Their Reactions with Ligands." North-Holland, Amsterdam.

Bunn, H. F., and Forget, B. G. (1986). "Hemoglobin: Molecular, Genetic and Clinical Aspects." Saunders, Philadelphia, Pennsylvania.

Dickerson, R. E., and Geiss, I. (1983). "Hemoglobin." Cummings, Menlo Park, California.

Fermi, G., and Perutz, M. F. (1981). "Haemoglobin and Myoglobin: Atlas of Biological Structures" (D. C. Phillips and F. M. Richards, eds.). Oxford Univ. Press (Clarendon), Oxford, England.

Imai, K. (1982). "Allosteric Effects in Haemoglobin." Cambridge Univ. Press, Cambridge, England.

Perutz, M. F. (1979). Regulation of oxygen affinity of hemoglobin. *Annu. Rev. Biochem.* **48,** 327.

Perutz, M. F. (1990). Mechanisms regulating the reactions of human hemoglobin with oxygen and carbon monoxide. *Annu. Rev. Physiol.* **52,** 1–25.

Perutz, M. F., Fermi, G., Luisi, B., Shaanan, B., and Liddington, R. C. (1987). Stereochemistry of cooperative effects in hemoglobin. *Acc. Chem. Res.* **20,** 309.

Hemoglobin, Molecular Genetics

DAVID WEATHERALL, *Institute of Molecular Medicine,
University of Oxford, John Radcliffe Hospital*

$$\alpha_2\beta_2 \quad \alpha_2\delta_2$$

Glossary

Enhancer DNA sequence that enhances the transcriptional activity of genes and may regulate their expression in particular tissues

Exon Segment of the coding sequence of a gene

Intron Intervening sequence within a gene that contains noncoding sequences, which do not appear in the processed messenger RNA

Promoter Region on the DNA molecule to which RNA polymerase binds and initiates transcription

Pseudogene Sequence with homology to a particular gene but that contains one or more mutations that prevent its normal function

***Trans*-activating factor** DNA-binding protein that interacts with specific regulatory regions within or close to structural genes

HEMOGLOBIN IS THE oxygen-carrying protein of the red cell. Like all mammalian hemoglobins, the human hemoglobins vary in structure during different periods of development, an adaptive process designed to meet differing oxygen-transport requirements. All human hemoglobins have a tetromeric structure, consisting of two pairs of different globin chains, each associated with one heme molecule. Hemoglobin is an allosteric protein; i.e., its configuration undergoes alterations that are essential for its normal function. Its oxygen-binding properties are reflected in a sigmoid oxygen dissociation curve, which means that it can bind oxygen tightly in the lungs and release it rapidly when it encounters a low partial pressure of oxygen in the tissues. Furthermore, the oxygen affinity of hemoglobin can be modified according to physiological needs, the curve shifting to the left or right in response to pH, temperature, and carbon dioxide levels. Some of these adaptive changes are the result of binding small molecules, notably 2,3-diphosphoglycerate. All these allosteric functions require the interaction of two unlike pairs of globin chains. Thus, the red blood cell precursors must synthesize these globin subunits in a synchronous manner, and a regulatory system has evolved, which ensures that different structural hemoglobins are produced at appropriate times during fetal and adult life; fetal hemoglobin has a higher oxygen affinity than adult hemoglobin, an adaptive response to the oxygen requirements of the fetus. Apparently, by a series of gene duplications followed by mutations that have modified the function of particular forms of hemoglobin, we have arrived at our present state of evolution, in which we have different hemoglobins adapted specifically to varying physiological needs at particular phases of development.

I. Structure and Heterogeneity of Human Hemoglobin

Human adult hemoglobin is a heterogeneous mixture of proteins consisting of a major component, hemoglobin A, and a minor component, hemoglobin A_2, constituting about 2.5% of the total. In intrauterine life, the main hemoglobin is hemoglobin F. The structure of these hemoglobins is similar. Each consists of two different pairs of identical globin chains. Except for some of the embryonic hemoglobins (see below), all normal human hemoglobins have one pair of α-chains: in hemoglobin A these are combined with β-chains ($\alpha_2\beta_2$), in hemoglobin A_2 with δ-chains ($\alpha_2\delta_2$), and in hemoglobin F with γ-chains ($\alpha_2\gamma_2$).

Human hemoglobin shows further heterogeneity, particularly in fetal life. Hemoglobin F is a mixture of two molecular species, which differ by only one amino acid residue, either glycine or alanine at position 136 in their γ-chains; they are designated $\alpha_2\gamma_2^{136\,Gly}$ and $\alpha_2\gamma_2^{136\,Ala}$. The γ-chains containing glycine at position 136 are called $^G\gamma$-chains; those that contain alanine are called $^A\gamma$-chains. At birth, the ratio of molecules containing $^G\gamma$-chains to those containing $^A\gamma$-chains is about 3 : 1; this ratio varies widely in the trace amounts of hemoglobin F present in normal adults. The fetal–adult hemoglobin switch, which relates γ- to β-chain production, starts before birth and, except in some pathological states, is complete by the end of the first year of life.

Before the eighth week of intrauterine life, there are three embryonic hemoglobins: Gower 1 ($\zeta_2\varepsilon_2$), Gower 2 ($\alpha_2\varepsilon_2$), and Portland ($\zeta_2\gamma_2$). The ζ- and ε-chains are the embryonic counterparts of the adult α- and β-chains and γ- and δ-chains, respectively. ζ-chain synthesis persists beyond the embryonic stage of development in some of the genetic disorders of hemoglobin production; so far, persistent ε-chain production has not been observed. During fetal development, there is an orderly switch from ζ- to α-chain and ε- to γ-chain production, followed by β- and δ-chain production after birth. In some inherited hemoglobin disorders, γ-chain production is persistent into childhood and adult life. [*See* HEMOGLOBIN.]

II. Globin Gene Clusters

Each different globin chain is the product of a specific gene. The α and ζ genes form a cluster on chromosome 16, while the γ, β, and δ gene cluster is on chromosome 11. The different human hemoglobins together with the arrangement of these gene clusters is shown in Fig. 1.

Although there is some individual variability, the α gene cluster usually contains one functional ζ gene and two α genes, designated $\alpha2$ and $\alpha1$. It also contains four pseudogenes; $\varphi\zeta$, $\varphi\alpha1$, $\varphi\alpha2$, and θ; i.e., gene loci with homology to the α or ζ genes but with mutations that render them functionless. They

are thought to be evolutionary remnants of once-active genes. The θ gene has only been discovered recently and is remarkably conserved among different species. Although it appears to be expressed in early fetal life, its function is unknown; it seems unlikely that it can produce a viable globin chain.

Each α gene is located in a region of homology approximately 4 kb (kb = 1,000 nucleotide bases) long, interrupted by two small nonhomologous regions. It is thought that the homologous regions have resulted from gene duplication and that the nonhomologous segments may have arisen subsequently by insertion of DNA into the noncoding regions around one of the two genes. As is the case for most mammalian genes, the α genes are divided into coding regions (exons) separated by intervening sequences (IVS), or introns. All the globin genes have two introns and three exons. The exons of the two α-globin genes have identical sequences. The first intron in each gene is identical, but the second intron of $\alpha1$ is 9 bases longer and differs by 3 bases from that in the $\alpha2$ gene. Despite their high degree of homology, the sequences of the two α-globin genes diverge in their 3' untranslated regions 13 bases beyond the TAA stop codon (see below). These differences provide an opportunity to assess the relative output of the two α genes. Apparently, the production of $\alpha2$ messenger RNA exceeds that of $\alpha1$ by a factor of 1.5–3.

The $\zeta1$ and $\zeta2$ genes are also highly homologous. The introns are much larger than those of the α-globin genes, and, in contrast to the latter, IVS 1 is larger than IVS 2. In each ζ gene, IVS 1 contains several copies of a simple repeated 14-bp sequence, which is similar to sequences located between the two ζ genes and near the human insulin gene. There are three base changes in the coding sequence of the first exon of $\zeta1$, one of which gives rise to a premature stop codon, thus turning it into an inactive pseudogene.

The region separating and surrounding the α and α-like structural genes has been analyzed in detail. This gene cluster is highly polymorphic. There are five so-called hypervariable regions in the cluster, one downstream from the $\alpha1$ gene, one between the $\zeta2$ and $\zeta1$ genes, one in the first intron of both ζ genes, and one 5' to the cluster. These regions have been sequenced and found to consist of varying numbers of tandem repeats of nucleotide sequences. Taken together with numerous single-base restriction fragment-length polymorphisms (RFLPs; i.e., nucleotide variants that alter the pat-

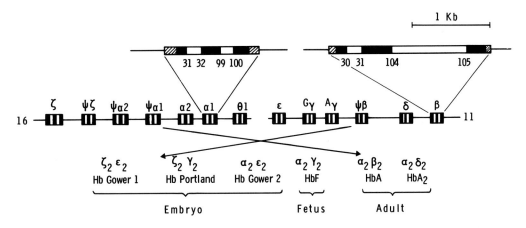

FIGURE 1 The human hemoglobins and the globin gene clusters on chromosomes 11 and 16. The extended figures of the α1 and β genes show the exons in dark shading, the introns in open boxes, and the flanking regions in lined shading. 1 Kb = 1,000 nucleotide bases. Hb, hemoglobin.

terns of DNA fragments after treatment with restriction enzymes), the genetic variability of the α gene cluster reaches a heterozygosity level of approximately 0.95. Thus, identifying each parental α-globin gene cluster in the majority of persons is possible. This heterogeneity has important implications for tracing the evolutionary history of the α genes.

The arrangement of the β-globin gene cluster on the short arm of chromosome 11 is ε, Gγ, Aγ, ψβ, δ, β. Each of the individual genes and their flanking regions have been sequenced. Like the α1 and α2 gene pairs, the Gγ and Aγ genes share a similar sequence. In fact, the Gγ and Aγ genes on one chromosome are identical in the region 5′ to the center of the second intron, yet they show some divergence 3′ to that position. At the boundary between the conserved and divergent regions, there is a block of simple nucleotide sequence, which may be a "hot spot" for the initiation of recombination events that have led to unidirectional gene conversion (i.e., matching of the two genes) during evolution.

Like the α-globin genes, the β gene cluster contains a series of single-point RFLPs, although in this case no hypervariable regions have been identified. The arrangements of RFLPs, or haplotypes, in the β-globin gene cluster falls into two domains. On the 5′ side of the β gene, spanning about 32 kb from the ε gene to the 3′ end of the ψβ gene, three common patterns of RFLPs exist. In the region encompassing about 18 kb to the 3′ side of the β-globin gene, three common patterns also exist in different popu-

lations. Between these regions, there is a sequence of about 11 kb in which randomization of the 5′ and 3′ domains occurs, and, hence, where a relatively higher frequency of recombination may occur. Recent studies indicate that β-globin gene haplotypes are similar in most populations, although they differ markedly in individuals of African origin; these results suggest that these haplotype arrangements were laid down very early during evolution and are consistent with data obtained from mitochondrial DNA polymorphisms, which point to the early emergence of a relatively small population from Africa, with subsequent divergence into other racial groups.

III. Structure of the Noncoding Regions

The regions flanking the coding regions of the globin genes contain a number of conserved sequences that are essential for their expression. The first is the ATA box, which serves to accurately locate the site of transcription initiation at the CAP site, usually about 30 bases downstream, and which also appears to influence the rate of transcription. In addition, there are two so-called upstream promoter elements; 70 or 80 bp upstream is a second conserved sequence, the CCAAT box, and further 5′, approximately 80–100 bp from the CAP site, is a GC-rich region with a sequence that can be either inverted or duplicated. These promoter sequences are also required for optimal transcription; mutations in this region of the β-globin gene cause its defective expression. The globin genes have also conserved sequences in their 5′ flanking regions, notably AATAAA, which is the polyadenylation signal site.

IV. Expression and Regulation of the Globin Genes

The primary transcript of the globin genes is a large messenger RNA precursor containing both intron and exon sequences. During its stay in the nucleus, it undergoes a good deal of processing, which entails modifying the 5' end and polyadenylation of the 3' end, both of which probably serve to stabilize the transcript. The introns are removed from the messenger RNA precursor in a complex two-stage process, which depends on certain critical sequences at the intron–exon junctions.

Very little is known about the regulation of globin gene transcription. The methylation state of the genes clearly plays an important role in their ability to be expressed; in human and other animal tissues, the globin genes are extensively methylated in non-erythroid organs and are relatively undermethylated in hemopoietic tissues. A change occurs in the methylation pattern and chromatin configuration around the globin genes at different stages of human development. Increasing evidence indicates that these genes come under the influence of so-called *trans*-acting factors, DNA-binding proteins that may be developmental stage-specific. The promoter sites are involved in efficient transcription of the globin genes. In addition, however, other sequences apparently are involved, particularly in tissue-specific expression. For example, evidence indicates the existence of so-called enhancer sequences, which may act by coming into spatial apposition with the promoter sequences to increase the efficiency of transcription of particular genes. Several enhancer sequences for the human globin genes have been defined, although their precise role in the regulation and specificity of expression remains to be determined. The most important sequences, called dominant control regions, lie many kilobases upstream from the α and β gene clusters.

Globin chain synthesis is directed in the cytoplasm of red cell precursors by the processed messenger RNA. Translation is initiated by the binding of a specific initiator transfer RNA, which carries the amino acid methionine to the initiation codon AUG. The initiation complex includes the two subunits that make up the ribosomes together with a number of proteins called initiation factors and with guanosine triphosphate and adenosine triphosphate. The first amino acid incorporated into globin chains is methionine, which is later cleaved. The globin messenger RNA is then translated in a stepwise fashion during which time amino acids are added to the growing chain until the ribosomes reach the termination codon UAA. At this time, the ribosomal subunits drop off the messenger RNA and are reutilized for further protein synthesis. The completed peptide chain, which starts to fold into its complex secondary and tertiary configuration and which probably binds heme while still on the ribosomes, then associates with its partner chains to form the definitive hemoglobin molecule. α- and β-chain synthesis is almost synchronous, although a slight excess of α-chains is degraded by proteolytic enzymes in the red cell precursor.

V. Developmental Changes in Globin Gene Expression

Knowledge about the developmental regulation of the globin genes is equally incomplete. β-globin synthesis commences early during fetal life, at approximately 8–10 wk gestation. Subsequently, it continues at a low level (ca. 10% of the total non-α-globin chain production) up to about 36 wk gestation, after which it is considerably augmented. At the same time, γ-globin chain synthesis starts to decline, so that at birth approximately equal amounts of γ- and β-globin chains are produced. Over the first year of life, γ-chain synthesis gradually declines, and by the end of the first year, this amounts to <1% of the total non-α-globin chain output. In adults, the small amount of hemoglobin F is confined to an erythrocyte population called F cells.

How this series of developmental switches is regulated is unknown. They are not organ-specific but are synchronized throughout the developing hemopoietic tissues. Although environmental factors may be involved, some form of "time clock" is built into the hemopoietic stem cell. At the chromosomal level, regulation apparently occurs in a complex manner involving the interaction of developmental stage-specific *trans*-activating factors with sequences in the $\gamma\delta\beta$-globin gene cluster.

VI. Molecular Pathology

The gene disorders of hemoglobin are the most common single gene disorders in the world population and affect millions of individuals. They result from either mutations that alter the structure of a globin chain or those that cause a drastic reduction in the rate of production of one or more globin chains.

Most of the structural hemoglobin variants result from a single-base change in one or another of the globin genes, causing the production of a hemoglobin variant with a single amino acid substitution. Structural variants with shortened or elongated subunits are more rare. While many hemoglobin variants are harmless, some give rise to diseases of varying severity. The best-known example of a disease due to a structural hemoglobin variant is sickle-cell anemia. Sickle-cell hemoglobin results from the substitution of glutamic acid by valine in the sixth position of the β-globin chain. This causes abnormal aggregation of hemoglobin molecules and, hence, a sickling deformity of the red blood cell. This in turn leads to shortening of the red cell survival and aggregation of sickled erythrocytes in small blood vessels with subsequent death of tissue due to a reduced oxygen supply. Some hemoglobin variants cause molecular instability; the hemoglobin molecule precipitates in the red blood cell, causing its premature damage in the circulation. Others interfere with oxygen transport.

The genetic disorders of hemoglobin that are characterized by a reduced rate of production of the α- or β-globin chains are called thalassemias. There are many forms of thalassemia, the most common being α- and β-thalassemias, which are due to a reduced rate of production of the α- or β-globin chains. This leads to imbalanced globin chain synthesis with the precipitation of the chain that is produced in excess, damage to the developing red blood cell, and, hence, a varying degree of anemia.

Over 100 different mutations have been described in the globin genes of patients with thalassemia. They include partial or complete deletions of the globin genes, point mutations that produce premature termination codons within exons, mutations that cause a shift in the reading frame of the genetic code and, hence, premature termination of chain synthesis, a wide variety of different mutations that involve the critical splice junctions and, hence, cause abnormal splicing of messenger RNA, and point mutations that lie in or near to the promoter boxes and cause defective transcription of the globin genes. In addition to their medical importance, the characterization of these mutations has produced invaluable information about the important regulatory regions within and flanking the globin genes and, because of their worldwide distribution, has thrown considerable light on the population genetics and evolution of human hemoglobin.

Bibliography

Bunn, H. F., and Forget, B. G. (1986). "Hemoglobin: Molecular, Genetic and Clinical Aspects." W. B. Saunders Company, Philadelphia and London.
Stamatoyannopoulos, G., Nienhuis, A. W., Leder, P., and Majerus, P. W. (1987). "The Molecular Basis of Blood Diseases." W. B. Saunders Company, Philadelphia and London.
Weatherall, D. J., Clegg, J. B., Higgs, D. R., and Wood, W. G. (1989). The hemoglobinopathies. *In* "The Metabolic Basis of Inherited Disease," 6th ed. (C. R. Scriver, A. L. Beaudet, W. S. Sly, and D. Valle, eds.). McGraw Hill, New York.

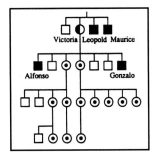

Hemophilia, Molecular Genetics

EDWARD G. D. TUDDENHAM, *Clinical Research Centre, United Kingdom*

Glossary

Coagulation Process whereby blood is transformed into a solid clot

Exon Protein coding region of gene

Factor VIII Protein cofactor of coagulation, deficient in hemophilia A

Factor IX Inactive zymogen precursor of factor IXa; deficient in hemophilia B

Factor IXa Activated form of factor IX, which activates factor X to factor Xa

Factor Xa Enzyme that activates prothrombin

Factor XI Inactive zymogen precursor of factor XIa; deficient in hemophilia C

Factor XIa Activated form of factor XI, which activates factor IX

Fibrinogen Protein that gives rise to blood clot when acted upon by thrombin

Hemophilia Literally, "love of blood"; medically, any bleeding disorder due to coagulation factor deficiency, usually inherited

Intron Intervening sequence of gene between exons (i.e., spliced out during messenger RNA processing)

Platelet Blood cell involved in coagulation

Prothrombin Inactive zymogen precursor of thrombin

Splice junction Sequences of DNA at junction of intron and exon that are essential for correct splicing

Thrombin Terminal enzyme of the coagulation process

von Willebrand factor Protein that carries factor VIII in the blood and promotes platelet adhesion; deficient in von Willebrand's disease

HEMOPHILIA COMPRISES a diverse group of disorders in which deficiency of a clotting factor leads to an inheritable tendency to bleed. The bleeding tendency may be mild, occurring only after injury, or severe, occurring spontaneously depending on the factor and its level of deficiency. Diagnosis of hemophilia depends on specific testing of the coagulation system for deficiency of particular coagulation factors. Treatment is by means of replacement of the relevant factor. Inheritance varies for different factors according to their chromosomal location and physiological behavior. Mutations in the genes for the 10 clotting factors involved in hemophilia are now being discovered and are highly diverse, including all of the known types of mutation affecting gene function. Treatment of hemophilia with blood-derived clotting factor concentrates has produced a high rate of infection with viruses and is likely to be replaced with bioengineered recombinant DNA-derived clotting factors during the present decade.

I. History of Hemophilia

Failure of the clotting mechanism produces very dramatic clinical problems, which have impressed medical observers from early times. A village in which many men and boys suffered bleeding after injury was described by Khalif Ibn Abbas, the great Moorish surgical writer in the tenth century. In 1803, John Otto described in the Medical Reposi-

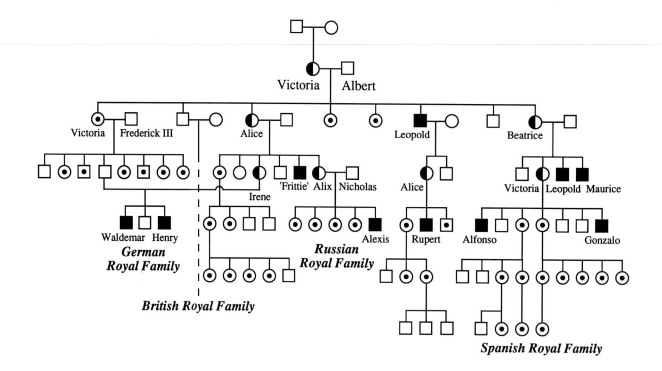

FIGURE 1 Queen Victoria's antecedents and descendents show that she was a carrier of sex-linked hemophilia, most probably arising as a new mutation in the gamete she received from her elderly father. Two of her daughters were definite carriers and passed the hemophilic gene to three other European royal families. Her son Leopold was a hemophiliac. Whether the royal hemophilia was due to factor VIII or factor IX deficiency is not known and will probably never be determined because the trait died out in this family before the availability of specific tests. The present British royal family descends through a normal male and, therefore, cannot carry the gene. □, Normal; ■, Haemophiliac; ▣, Possible haemophiliac; ○, Normal; ◐, Carrier; ⊙, Possible carrier.

tory of New York a family with hemophilia affecting males and transmitted by apparently unaffected carrier females. Many other cases and family reports appeared during the nineteenth century and the clinical features and patterns of inheritance of sex-linked hemophilia became well known. Queen Victoria was a carrier of hemophilia, which she transmitted through her daughters to three of the royal households of Europe (Russian, German, and Spanish; Fig. 1). Delayed coagulation of hemophilic blood was demonstrated toward the end of the nineteenth century, but the prevailing theory of blood coagulation at that time was inadequate to account for the deficiency in hemophilia. In 1936, researchers partially isolated a principal from normal blood able to correct the clotting time of sex-linked hemophilic blood. Around 1950, two kinds of sex-linked hemophilia emerged (then called hemophilia A and B), and they are now known to be due to deficiency of coagulation factor VIII and factor IX, respectively. Treatment of these two prevalent bleeding disorders with concentrates prepared from normal blood began in the 1960s, transforming the lives of hemophiliacs, who up to then had suffered early onset of crippling due to bleeding in the joints and a very short life expectancy due to catastrophic bleeding in the brain or after injury. Factor VIII and factor IX were purified to homogeneity in the 1970s and cloned and sequenced in the early 1980s, lead-

ing to discovery of the underlying mutations in hemophilia A and B.

II. Normal Blood Coagulation

Figure 2 diagrams the processes whereby exposure of blood to surfaces outside the circulation, as occurs whenever a blood vessel is breached, leads through a linked series of enzymic interactions to the production of fibrin clot. The blood clotting factors circulate as inactive precursors, either of enzymes (called zymogens) or of cofactors (called procofactors). Both the so-called intrinsic pathway, starting with factor XI, and the extrinsic pathway, starting with plasma factor VII, are required for

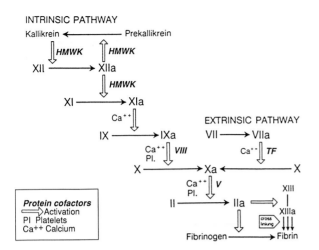

FIGURE 2 Blood coagulation cascade. The intrinsic pathway is initiated when blood contacts extravascular surfaces such as collagen. The extrinsic pathway is initiated by contact with tissues bearing the cell-surface receptor tissue factor (*TF*). Subscript a indicates the activated form of a clotting factor.

the visible and structurally solid fibrin clot. The clot is stabilized by another enzyme that cross-links fibrin monomers covalently. This factor is designated factor XIII and, like factors V and VIII, requires to be activated by thrombin.

III. Defects of Coagulation—The Hemophilias

A. Factor VIII Deficiency—Hemophilia A (Classic Hemophilia)

Hemophilia A is the most common severe bleeding disorder in man, affecting approximately 1 in 5,000 male births, or 1 in 10,000 of the general population. The gene for factor VIII is situated at the tip of the long arm of the X-chromosome, not far from the gene for factor IX and close to the genes for color vision and for glucose-6-phosphate dehydrogenase (Fig. 3). The severity of bleeding in hemophilia A

normal blood clotting. The way in which factor XI or factor VII is activated is unclear and remains controversial. In test-tube experiments, factor XI is activated by factor XII in concert with two other proteins, but all three of these contact factor proteins (factor XII, high-molecular weight kininogen [HMWK] prekallikrein) are apparently dispensable for normal coagulation, because individuals who lack them have no problem with bleeding. Factor XIa activates factor IX. Tissue factor is present at the surface of many cells in the body and is a highly efficient cofactor for the activation of factor X by factor VIIa. No individual lacking tissue factor has been described, and such a deficiency would probably not be compatible with life. Factor VIII is a procofactor, which is itself activated by thrombin and then becomes efficient in promoting activation of factor X by activated factor IX, a step that occurs on the surface of platelets. Thus, the two processes of initiation/activation converge on factor X generating factor Xa. Factor Xa activates prothrombin, and for this it requires the assistance of a cofactor V and platelet surface. Factor V circulates as a procofactor requiring to be activated by thrombin. (From where the initial thrombin for activation of factors V and VIII is derived is uncertain.) Finally, thrombin removes short peptides from fibrinogen (fibrinopeptides A and B), yielding fibrin monomers. Fibrin monomers spontaneously polymerize by end-to-end and staggered side-to-side association, producing

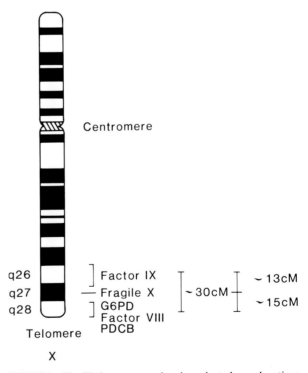

FIGURE 3 The X-chromosome showing selected gene locations near the tip of the long arm. Fragile X is associated with mental retardation, glucose-6-phosphate dehydrogenase (G-6PD) with hemolytic anemia, and protan deutan color blindness (PDCB) with color blindness. Factor VIII and factor IX genes are mutated in hemophilia A and B, respectively. The scale of distances is given in centimorgans (cM) q.v. One cM represents approximately 1 million base pairs of DNA.

depends on the residual amount of clotting factor in the patient's blood. If none can be measured, the patient has a severe tendency to bleed with apparently spontaneous episodes of bleeding into his joints and muscles as often as three or four times a week. Paradoxically, the healing of small cuts and scratches is normal in hemophilia A because blood platelet function is normal in these patients and able to plug small holes in the smaller blood vessels. If untreated, bleeding into the joints causes severe pain and immobility but spontaneously resolves in a few days to weeks. The consequence of such bleeding is rapid and progressive articular damage leading to deformity and fixation of the joints most affected, the large load-bearing joints—knees, ankles, elbows. Any joint in the body can be affected and bleeding into the muscles can give rise to obstruction of blood flow, leading to death of tissue and contraction (see Fig. 4). Mild hemophilia occurs in patients with residual factor VIII >5% of the normal level. Episodes of bleeding into joints and tissues only affects these patients after significant injury (Table I). Hemophilia A follows the typical inheritance of an X-linked recessive disorder. Females who carry a defective factor VIII gene are themselves protected from bleeding by its normal counterpart on their other X-chromosome (Fig. 5). Occasionally, due to the process of random inactivation of one X-chromosome per cell, which occurs early in female embryogenesis, most of the normal X-chromosomes are inactivated and, thus, a female carrier may have mild or moderate hemophilia. On average, half of the sons of a carrier female are affected by hemophilia A and half of her daughters are carriers. A male with classic hemophilia or hemophilia B has only normal sons and only carrier daughters.

FIGURE 4 Hemophiliac who grew up before treatment with factor VIII was available. He suffered repeated joint and muscle bleeding, which led to deformity and fusion of the joints and wasting of the muscles.

TABLE I Clinical Severity of Hemophilia[a]

Level of factor VIII or IX (% of normal)	Bleeding tendency
<2	Severe: Frequent apparently spontaneous bleeds into joints and other internal organs.
2–10	Moderately severe: Bleeding after minor trauma, with occasional spontaneous bleeds.
10–30	Mild: Bleeding only after major trauma or surgery.

[a] Applies to hemophilia A and B, and not to von Willebrand's disease nor to factor XI deficiency (see text).

B. Factor IX Deficiency—Hemophilia B

Hemophilia B cannot be distinguished from hemophilia A by examining the patient or taking his history, and the inheritance of the two disorders is the same. For example, which disorder was transmitted by Queen Victoria (Fig. 1) is unknown. Specific blood tests are necessary to reveal a deficiency of clotting factor IX. The frequency of this disorder is somewhat less than that of hemophilia A affecting about 1 in 30,000 males.

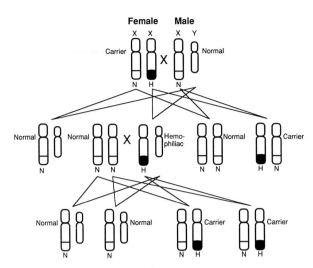

FIGURE 5 Inheritance of X-linked hemophilia. Females who carry a defective factor VIII or IX gene are protected from its effects by the normal gene on the paired chromosome. Males have only one X-chromosome and, therefore, suffer from the defect.

C. von Willebrand's Disease

von Willebrand's disease resembles hemophilia A in that there is deficiency of factor VIII, but an additional defect is found in failure of platelets to adhere to surfaces such as collagen that are exposed after blood vessel injury. This leads to prolonged bleeding from cuts because platelet plugs do not form in small capillaries after injury. The explanation for this combined defect of coagulation and platelet function puzzled researchers for many decades. It is now clear that the deficient blood protein in this disorder—von Willebrand factor—has two functions: (1) to support and protect factor VIII in the circulation, preventing its rapid destruction, and (2) to promote adhesion of platelets under conditions of high shear rate. Patients with von Willebrand's disease lack von Willebrand factor due to a defect of the von Willebrand factor gene located on chromosome 12. Most often, the disorder is transmitted as an autosomal dominant with a moderate reduction of both factor VIII and von Willebrand factor levels. These patients suffer bleeding from small vessels—the vessels where shear rate is high and von Willebrand factor platelet-adhesion promotion is most important. This gives rise to bruising, nose bleeds, bleeding from other mucous membranes of the intestinal tract, and heavy menstrual bleeding in women. Bleeding in the joints and muscles is rare because the factor VIII levels are usually above

10%. Some surveys indicate that the incidence of mild von Willebrand's disease may be as high as 1% of the general population. Because the bleeding is mild, diagnosis is often delayed into the second or third decade of life. Rare individuals have inherited a recessive gene for von Willebrand factor deficiency from both parents and completely lack the factor from their circulation and have very low levels of factor VIII. These patients suffer both from the type of bleeding seen in hemophilia A and from the superficial bleeding characteristic of platelet-adhesion defect.

D. Factor XI Deficiency—Hemophilia C

Hemophilia C has a most curious racial distribution, being common among Ashkenazi Jews and rare among all other racial groups, including the Sephardic Jews. Up to 8% of Ashkenazis (Jews of the Eastern Diaspora) have a defect in synthesis of factor XI with moderately reduced levels of this contact factor. They generally have little or no clinical problem from bleeding except after severe trauma, and many such individuals have no problems with bleeding at all. Because the gene frequency is so high, homozygous cases are quite common, and these individuals have little or no factor XI in their circulation but still only suffer with a mild bleeding tendency, quite unlike the situation for factor VIII and factor IX deficiencies.

E. Deficiency of Other Clotting Factors

Most of the other clotting factors represented in Fig. 2 (factors I, II, V, X, VII, and XIII) are rarely found to be deficient in individuals with a hemophilialike bleeding tendency. Because the inheritance in each case is autosomal recessive, usually only one generation is affected. Curiously, deficiency of factor XII or of prekallikrein, or HMWK, is not associated with hemophilic bleeding.

IV. Mutations That Cause Hemophilia

A. Factor VIII Gene Defects in Hemophilia A

The factor VIII gene spans a large region of the X-chromosome being about 190,000 base pairs in length. The protein-coding regions of the gene are divided into 26 exons (shown as shaded bars in Fig. 6). The processed messenger RNA for factor VIII is

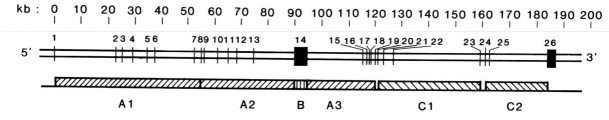

FIGURE 6 The factor VIII gene. Top line: scale in kilobase pairs of DNA. Middle line: diagram of gene; exons shaded dark, introns unshaded. Bottom line: factor VIII protein. A1–A3 domains homologous to the copper carrying protein ceruloplasmin. The B domain is removed when factor VIII is activated by thrombin. C1 and C2 domains homologous to the slime mold protein discoidin.

about 9,000 base pairs in length and specifies a protein of 2,351 amino acids. The gene is expressed mainly in liver cells, which produce the protein and modify it in various ways before releasing it into the circulation. Various kinds of mutation can affect the functioning of genes, and with such a very large gene an extremely large number of small or large mutations that would affect its function can be predicted. Furthermore, because hemophilia A is often lethal to affected males and greatly reduces a sufferer's chances of having children, mutations that cause it will be rapidly lost from the population and the incidence of the disorder must be maintained by a constant input of new mutations. This gives rise to the prediction made in 1935: a high proportion of cases of hemophilia would be due to recent mutation and, hence, these mutations are likely to be diverse. Due to the extremely large size of the gene, there are technical problems of localization of point mutations, but those that have been found are indeed highly diverse. The simplest category of mutation to identify is a large deletion of the gene, and this has been found in about 5% of cases. Table II lists the size and extent of various deletions of the factor VIII gene that have been found in hemophilia

A patients. All except one of the patients with these deletions have severe hemophilia A. The exception is a case in which exon 22 has been deleted. It so happens that the splice junctions between exons 21 and 22 and exons 22 and 23 are in frame, so that removal of the exon does not interrupt the reading frame of the messenger RNA. It simply removes 52 amino acids from near the carboxy terminal of the protein. Apparently, the shortened protein retains about 5% activity in the circulation. Some, but not all, of these individuals have made antibodies to factor VIII, which was given to them as treatment for bleeding. Another type of mutation is the substitution of a single base creating either a stop signal or the substitution of an amino acid that affects the function of the protein. Note that substitution of an amino acid that had no effect on factor VIII function would not be detected clinically but simply be part of the natural variability as found, for example, in blood groups. Point mutations that create a premature stop codon have been found to occur preferentially at certain sites where the nucleotides cytosine and guanine occur together in the order CG followed by an A coding for arginine. Table IV lists such mutations identified in patients with severe hemophilia A. The proposed mechanism for these mutation "hot spots" is based on the fact that mammalian DNA is often methylated at cytosines followed by a guanine. Spontaneous deamination of methyl cytosine would convert the base to a thymidine, which being a normal constituent of DNA is not recognized by repair mechanisms. Where CG is followed by an A coding for arginine, the conversion will create a TGA codon for termination of translation. For example, the finding of four independent mutations at codon 2209 in the factor VIII coding

TABLE II Deletions in the Factor VIII Gene Causing Hemophilia A

Exons deleted	Factor VIII level (% of normal)	Antibodies to factor VIII
1–26	<1	No
1–22	<1	Yes
1–5	<1	No
3	<1	No
6	<1	No
14[a]	<1	No
22	5	No
26[b]	<1	No
15–18	<1	Yes
23–25	<1	No
23–26	<1	Yes
24 and 25	<1	No

[a] Two independent cases.
[b] Four independent cases.

sequence converting an arginine to stop is far too frequent to have happened as a result of chance alone given the numbers of patients screened. A similar mechanism operating on the opposite strand of DNA will convert the G to an A (its pair having been converted to a T on the lower or antisense strand). Such mutations have indeed been found associated with either severe or moderate hemophilia depending presumably on the effect of the substitution of a glutamine (CAA) for an arginine (CGA) or a cysteine for an arginine at various sites. These examples are also listed in Table V. The mutations in which arginine occurs at a site where thrombin activates factor VIII by cleaving the primary chain of amino acids are of great theoretical interest for the function of factor VIII in coagulation. In some hemophilia A patients, either one of two critical thrombin-sensitive arginine residues has been replaced by a cysteine due to the mutation of a C to T in the first base of codon 372 or 1689. Factor VIII protein with 1689 cysteine in place of arginine has a light chain, which is resistant to thrombin, and no functional activity in coagulation. A similar mutation at codon 372 has been found in a patient with moderately severe hemophilia A. This man's factor VIII with 372 cysteine in place of arginine has a thrombin-resistant heavy chain and low residual activity.

All these point mutations have been discovered by screening, specifically at the CG codons, either with DNA cutting enzymes, which cut at such sequences, or by using synthetic short stretches of DNA to probe these regions of the factor VIII gene. As other regions of the factor VIII gene are searched for defects in hemophilia, other mutations are expected to be found affecting the function of the factor VIII protein or the processing of the factor VIII message. The factor VIII gene represents a large target area of the human genome, and the accumulation of knowledge on the random pattern of mutations across this region causing hemophilia should help us to understand more of the mechanisms by which such errors in DNA arise. The discovery of the mutation hot spots at CG dinucleotides is an example of a result from this natural laboratory.

B. Factor IX Gene and Hemophilia B

The factor IX gene is located on the X-chromosome on the centromere side of the factor VIII gene at a probable distance of 30 million base pairs. As

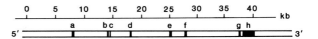

FIGURE 7 The factor IX gene. Top line: scale in kilobase. Middle line: diagram of gene; exons shaded, introns unshaded.

shown in Fig. 7, eight exons are within a gene spanning 33,500 base pairs. The protein for which this gene codes—factor IX—is shown in Figure 8, a diagram using the single-letter amino acid code (Table III). The primary amino acid sequence consists of 461 amino acids as the nascent chain emerges from the ribosome into the endoplasmic reticulum. This undergoes several modifications before it is released into the circulation from the liver cells in which it is synthesized. Certain glutamic acid residues are modified by the addition of an extra COOH group to their gamma carbon, and these are indicated in the figure as a gamma symbol (γ). This step requires the presence of vitamin K and explains why antagonists of vitamin K, such as warfarin, produce a bleeding tendency. The signal for carboxylation to occur is contained within the prepro leader sequence, which itself is cleaved from the main protein by a processing protease during secretion from the cell. Furthermore, an aspartic acid residue at position 64 is hydroxylated in the β position (indicated by a β). Two asparagine residues have sugar molecules added.

TABLE III Notation for Amino Acids

Amino acid	Three letter	One letter
Alanine	Ala	A
Cysteine	Cys	C
Aspartic acid	Asp	D
Glutamic acid	Glu	E
Phenylalanine	Phe	F
Glycine	Gly	G
Histidine	His	H
Isoleucine	Ile	I
Lysine	Lys	K
Leucine	Leu	L
Methionine	Met	M
Asparagine	Asn	N
Proline	Pro	P
Glutamine	Gln	Q
Arginine	Arg	R
Serine	Ser	S
Threonine	Thr	T
Valine	Val	V
Tryptophan	Trp	W
Tyrosine	Tyr	Y

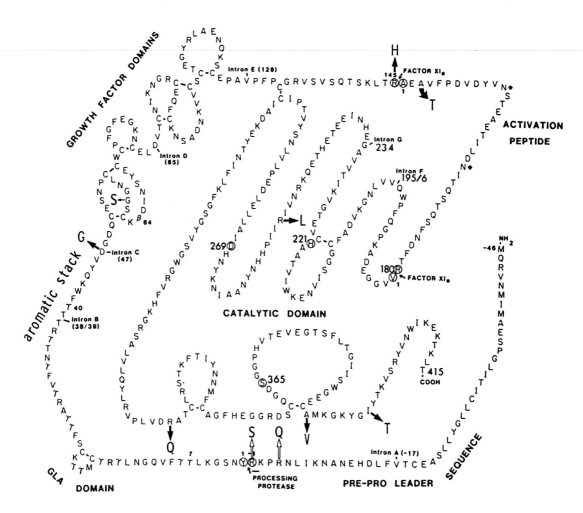

FIGURE 8 Factor IX protein (amino acid sequence single-letter codes as in Table III). Point mutations with substitution of a single amino acid producing hemophilia B; →Q, →S, etc. A ↘ T Protein Polymorphism. [Modified, with permission, from Yoshitakes *et al.* (1985). Nucleotide sequence of the gene for human factor X (Antihaemophilic factor B). *Biochem.* **24**, 3736. Courtesy of the American Chemical Society.]

By comparing the sequence of factor IX with that of other proteins, some fascinating similarities have emerged. The overall structure of factor IX very closely resembles that of three other proteins of blood coagulation: factors VII, factor X, and the anticoagulant protein C. The region labeled catalytic domain resembles other protein cutting enzymes on the general pattern of trypsin. A similar domain is found in all the proteolytic enzymes of coagulation. Quite surprisingly, two domains resemble epidermal growth factor, and a domain labeled aromatic stack is thought to be involved in protein–protein interaction in coagulation. Activation of factor IX by factor XIa the preceding enzyme in the intrinsic clotting pathway, occurs through cleavage between residues 145 and 146, and 180 and 181, releasing the activation peptide, on which the two asparagine-linked sugar chains are located (marked with lozenges on Fig. 8).

As with hemophilia A, hemophilia B mutations are rapidly lost from the population due to their highly adverse effect on reproductive fitness; there-

fore, new mutations account for a large proportion of cases and are diverse at the molecular level. About one-third of patients with hemophilia B have a non- or poorly functional factor IX molecule in their circulation. Some of these were the first mutations to be solved at the molecular level in hemophilia. One such factor is factor IX Chapel Hill (named after the location of the laboratory that carried out the research) and is caused by the substitution of histidine for arginine at residue 145. This interferes with activation by factor XIa, and factor IX Chapel Hill has low activity. The mutation responsible for the amino acid substitution affects the middle residue of codon 145 converting CGT to CAT. It is an example of the type of mutation dis-

TABLE IV Point Mutations in the Factor VIII Gene Causing Hemophilia A

Codon[a]	Mutation	Substitution	Factor VIII level in patients blood (%)	Number of unrelated cases
272	GAA → GGA	Glu → Gly	2	1
336	CGA → TGA	Arg → Stop	<1[b]	4
372	CGC → TGC	Arg → Cys	4[b]	2
1689	CGC → TGC	Arg → Cys	<1 or 4[b,c]	3
1941	CGA → TGA	Arg → Stop	<1	2
2116	CGA → TGA	Arg → Stop	<1	1
2116	CGA → CCA	Arg → Pro	<1	1
2147	CGA → TGA	Arg → Stop	<1	2
2209	CGA → TGA	Arg → Stop	<1	4
2209	CGA → CAA	Arg → Gln	<1	3
2307	CGA → TGA	Arg → Stop	<1	1
2307	CGA → CAA	Arg → Gln	9	1

[a] Codons numbered 1–2332 in the sequence of mature factor VIII.
[b] Nonfunctional factor VIII protein present in these cases with failure to activate due to mutation at thrombin cleavage site.
[c] Unexplained variation in residual activity of different cases.

cussed above, occurring at CG dinucleotides thought to be caused by deamination of methyl cytosine. Also shown in Figure 8 are nine other substitution (termed missense) mutations producing moderate or severe hemophilia B. The substitution of arginine by serine at position −1 in the prepro sequence led to failure of release of the N-terminal peptide and the abnormal factor IX molecule, which is nonfunctional. Table V lists more exhaustively other point mutations found in the factor IX gene in hemophilia B. Among these are examples of mutations affecting gene regulation, often referred to as hemophilia B Leyden. This class of mutations affects the way in which factor IX gene is controlled and leads to a condition characterized by low levels of factor IX in childhood that rise gradually toward normal levels around puberty. It seems that the mutations affect the response of the gene to testosterone, such that the gene is poorly expressed until the testosterone levels rise at puberty when the gene is turned on. Other point mutations affect the critical junctions between exon and intron, such that there is failure of correct processing of messenger RNA as seen in hemophilia B Oxford 1 and 2.

Deletion of part or all of the factor IX gene has been found in other patients, many of whom have made antibodies to factor IX after treatment for bleeding. Some of these are listed in Table VI. Disorders of the factor IX gene in hemophilia B are proving to be a natural laboratory for study of the mechanisms of mutation and the correlation between structure and function of the factor IX protein.

C. von Willebrand Factor and von Willebrand's Disease

von Willebrand factor circulates in plasma as a series of multimers (Fig. 9) of a basic protomer with molecular weight approximately 240,000. During biosynthesis in endothelial cells lining the blood vessel, the proteins are assembled first into dimers through carboxy-terminal disulfide interchange and then into multimers of higher order through N-terminal disulfide interchange in a step that depends on the presence of a very long propeptide. This peptide is excised from von Willebrand factor multimers during assembly and secreted separately but has no known function other than to promote multimer assembly. The importance of multimer assembly is attested by the fact that some individuals with von Willebrand's disease have von Willebrand factor in their circulation that is not properly multimerized and of low molecular weight and has very poor function in platelet adhesion although able to support factor VIII. The complete amino acid sequence of the von Willebrand factor monomer has been established and consists of a single chain of 2,050 residues. The sequence of the propeptide was obtained from complementary DNA sequence and consists of 763 amino acids that are cleaved from the protomer during synthesis and assembly. The gene for

TABLE V Mutations in the Factor IX Gene Causing Hemophilia B

(Location[a]) codon	Mutation	Substitution/effect	Factor IX level in patients blood (%)	Region of gene or protein	Number of unrelated cases
(−20)	T → A	Alters gene regulation	Increases at puberty	Promoter	1
(−6)	G → A	Alters gene regulation	Increases at puberty	Promoter	1
(13)	A → G	Alters gene regulation	Increases at puberty	Promoter	2
(6704)	GTA → GGA	Destroys exon C splice donor	<1	Promoter	1
−4	CGG → CAG	Arg → Gln	<1	Propeptide release	7
−4	CGG → TGA	Arg → Trp	<1	Propeptide release	1
−1	AGG → AGC	Arg → Ser	<1	Polypeptide release	1
7	GAA → GAC	Glu → Asp	<1	Gla domain	1
11	CAA → TAA	Gln → Stop	<1	Gla domain	1
27	GAA → AAA	Glu → Lys	<1	Gla domain	1
29	CGA → TGA	Arg → Stop	<1	Gla domain	2
29	CGA → CAA	Arg → Glu	5	Gla domain	1
33	GAA → GAC	Glu → Asp	10	Gla domain	1
47	GAT → GGT	Asp → Gly	10	Aromatic stack	1
50	CAG → CCG	Glu → Pro	<1	Aromatic stack	1
55	CCA → GCA	Pro → Ala	10	EGF[b]	2
60	GGC → AGC	Gly → Ser	10	EGF	3
114	GGA → GCA	Gly → Ala	?	EGF	1
120	AAC → TAC	Asn → Tyr	<1	EGF	1
145	CGT → TGT	Arg → Cys	10	Activation site	3
145	CGT → CAT	Arg → His	10	Activation site	8
173	CAA → TAA	Gln → Stop	<1	Activation peptide	1
180	CGG → CAG	Arg → Gln	<1	Activation site	1
180	CGG → CAG	Arg → Trp	?	Activation site	2
182	GTT → TTT	Val → Phe	<1	Protease	1
(20,566)	GGT → GTT	Destroys exon f splice donor	<1	Protease	1
194	TGG → TGA	Trp → Stop	<1	Protease	1
222	TGT → TGG	Cys → Trp	<1	Protease	1
248	CGA → TGA	Arg → Stop	<1	Protease	3
248	CGA → CAA	Arg → Gln	5	Protease	1
252	CGA → CTA	Arg → Leu	?	Protease	1
252	CGA → TGA	Arg → Stop	<1	Protease	2
257	CAC → TAC	His → Tyr	?	Protease	1
260	AAT → ACT	Asn → Thr	10	Protease	1
291	GCT → ACT	Ala → Thr	10	Protease	1
291	GCT → CCT	Ala → Pro	?	Protease	1
296	ACG → ACT	Thr → Met	<1	Protease	1
307	GTA → GCA	Val → Ala	10	Protease	1
311	GGA → CGA	Gly → Arg	5	Protease	1
333	CGA → CAA	Arg → Gln	1–5	Protease	4
336	TGT → CGT	Cys → Arg	5	Protease	1
338	CGA → TGA	Arg → Stop	<1	Protease	4
363	GGA → GTA	Gly → Val	?	Protease	1
390	GGA → GTA	Ala → Val	1–5	Protease	2
396	GGA → AGA	Gly → Arg	<1	Protease	1
397	ATA → ACA	Ile → Thr	5–10	Protease	7
407	TGG → AGG	Trp → Arg	<1	Protease	1
411	AAA → TAA	Lys → Stop	<1	Protease	1

[a] (N) Nucleotide position from capsite; −m Codon number of prepropeptide; −m, Codon number of mature protein.
[b] EGF = Epidermal Growth Factor like domain.

TABLE VI Deletions in the Factor IX Gene Causing Hemophilia B

Exons deleted	Antibodies to factor IX	Number of unrelated cases
a–h	Yes	12
a–h	No	2
a–c	Yes	1
e, g, h	Yes	1
f–h	Yes	1
e and f	No	1
d	No	1
d	Yes	1

von Willebrand factor encoded on chromosome 12 has only recently been cloned and mapped and consists of 52 exons. von Willebrand's disease is highly diverse in terms of clinical expression and the presence of different types of structural defects in von Willebrand factor.

D. Mutations Causing von Willebrand's Disease

To date, only a few of the mutations in the von Willebrand factor gene associated with von Willebrand's disease have been identified, although genetic linkage studies clearly show that the mutations responsible for the disease are indeed associated with the von Willebrand factor gene. A small subgroup of patients with homozygous recessive von Willebrand's disease develop antibodies to the protein after infusion, and these individuals have deletions of all or part of their von Willebrand factor gene. The mutation in a patient with von Willebrand's disease in which there was failure of proper multimer formation has been localized to substitution of valine by asparagine at residue 844. It was shown by mutagenizing a normal von Willebrand factor clone at this precise position that the resulting von Willebrand factor made *in vitro* in a tissue culture system failed to form multimers. Thus, proving that this mutation was indeed the cause of the clinical condition, and coincidentally that residue 844 must be involved in multimer assembly.

E. Factor XI Gene and Hemophilia C

Factor XI protein circulates as a dimer of two identical subunits, each consisting of a single chain of 607 amino acids. The gene for factor XI is 23,000 base pairs in length containing 15 exons. The chro-

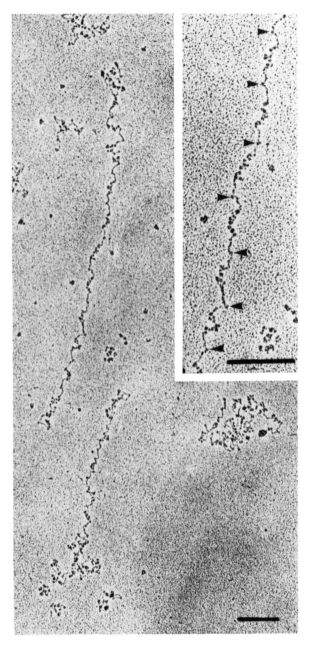

FIGURE 9 Electron micrograph of von Willebrand factor produced by glycerol spraying and rotary shadowing. Note the long extended forms with a repeating feature corresponding to the dimer. Folded up ''fuzzy balls'' are a relaxed form of the protein. A symmetrical IgM molecule is visible above the scale bar in the inset. [Reprinted, with permission, from W. E. Fowler and L. J. Fretto, 1989, Electron microscopy of von Willebrand factor, *in* ''Coagulation and Bleeding Disorders,'' courtesy of Marcel Dekker, Inc., New York.]

mosomal location of the factor XI gene is 49, 35. Jewish patients with factor XI deficiency invariably have reduced amounts of circulating factor XI protein in parallel with the reduction of coagulant activity. Three mutations responsible for factor XI deficiency in Jews have now been identified, as shown in Table VII. This information raises a very interesting question in evolutionary genetics. Prior to the discovery of this molecular heterogeneity, it had always been proposed that factor XI deficiency in this racial group was due to the founder effect and subsequent genetic drift in a closed breeding population. Because the polymorphism clearly is due to at least three separate mutational events, some heterozygote advantage to partial factor XI deficiency is much more likely; however, the nature of this advantage can only be conjectural at present. One possibility is that some environmental factors, perhaps dietary and/or cultural, would promote a high incidence of thrombosis in this group, protection from which is provided by a partial coagulation factor deficiency.

F. Mutations Associated with Other Coagulation Factor Defects

Abnormal fibrinogen, which clots poorly (dysfibrinogenemia), has been shown in some cases to be due to mutation at or close to the thrombin cleavage site of the fibrinogen α-chain. Some cases of factor X deficiency have been associated with deletions of the factor X gene. So far, the molecular basis of factor VII deficiency and of factor V deficiency has not been reported, but it is to be expected that a range of mutations will be found in different families with these various coagulation factor deficiencies.

V. Treatment of Hemophilia

The mainstay of treatment for hemophilia A and B for the past 20 years has been with concentrates derived from pooled normal human plasma. Such concentrates were relatively impure and, until fairly recently, were not treated in any way to inactivate viruses. As a consequence, the population of pa-

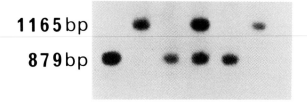

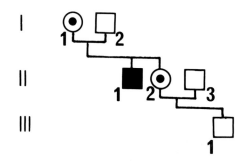

FIGURE 10 Use of gene probe to diagnose hemophilia from chorion villus biopsy sample at 8 wk gestation. In this family, hemophilia A is associated with the lower band on the Southern blot (q.v.). Because the male fetus (III 1) has the upper band, he cannot have hemophilia A, as proven 7 mo later. (Reprinted, with permission, from Gitschier *et al.* (1985) Antenatal diagnosis and carrier detection of haemophilia A using a factor VIII gene probe, *Lancet.* **1,** 1093)

tients with hemophilia A and B became successively infected with hepatitis B, hepatitis non A non B (one form of which has recently been identified as hepatitis C virus), and human immunodeficiency virus. The drastic consequences of these infections are now being seen in end stage liver failure and acquired immunodeficiency syndrome. In the past 3 years, severely heat-treated or detergent-treated concentrates, which are probably free of infective risk, have become available, and very recently highly purified clotting factor proteins have become available based on monoclonal antibody purification schedules. Simultaneously genetically engineered factor VIII is undergoing clinical trials and is proving efficacious. Over the next decade, treatment of hemophilia A probably will depend more and more on genetically engineered replacement therapy.

TABLE VII Mutations in the Factor XI Gene Causing Hemophilia C

Type	Site	Mutation	Effect on expression	Frequency[a]
I	Junction of exon 14 and intron N	GT → AT	Destroys splice donor of exon 14	$\frac{1}{12}$
II	Exon 5, codon 117	GAA → TAA	Glu → Stop	$\frac{5}{12}$
III	Exon 9, codon 283	TTC → CTC	Phe → Leu	$\frac{6}{12}$

[a] Results of preliminary survey amongst Ashkenazi Jews.

Eventually, it is hoped that gene therapy by introduction of suitably modified coagulation factor genes into somatic tissues will provide an effective cure or amelioration for the bleeding tendency in hemophilia.

VI. Genetic Counseling

Every male patient with hemophilia A or B has a number of female relatives who are definite or possible carriers of his gene. These females request information on their carrier status and advice on risk of producing hemophilic offspring. Based on the new knowledge of the genetic loci involved, it is now possible to provide definitive carrier status information in a high proportion of cases. For hemophilia B, now that the precise mutation can be localized in all cases, carrier prediction can be provided with certainty for all female relatives of a patient with this disease. For hemophilia A, genetic linkage analysis can provide such diagnosis in about 50% of cases, but in other cases, due to an unknown level of a new mutation linkage, analysis is only partially informative. Antenatal diagnosis using gene probes can be performed on biopsy samples from chorion villus material as early as 8 wk gestation (Fig. 10). Thus, carriers of hemophilia can make the choice to bear only normal sons. [*See* GENETIC COUNSELING.]

Bibliography

Colman, R. W., Hirsh, J., Marder, V. J., and Salzman, E. W. (eds.) (1987). "Hemostasis and Thrombosis. Basic Principles and Clinical Practice," 2nd ed. J. B. Lippincott Co., Philadelphia.

Kane, W. H., and Davie, E. W. (1988). Blood coagulation factors V and VIII: Structural similarities and their relationship to hemorrhagic and thrombotic disorders. *Blood* **71**, 539–555.

Tuddenham, E. G. D. (ed.) (1989). "The Molecular Biology of Coagulation. Baillière's Clinical Haematology," Vol. 2, No. 4. Baillière Tindall, London.

Tuddenham, E. G. D. (1989). von Willebrand factor and its disorders: An overview of recent molecular studies. *Blood Reviews* Vol. 3, No. 4, pp. 1–12.

Tuddenham, E. G. D. (1989). Inherited bleeding disorders, chap. 23. *In* "Postgraduate Haematology," 3rd ed. (A. V. Hoffbrand and S. M. Lewis, eds.). Heinmann Medical Books, Oxford.

Vehar, G. A., Lawn, R. M., Tuddenham, E. G. D., and Wood, W. I. (1989). Factor VIII and factor V: Biochemistry and pathophysiology, chap. 86. *In* "The Metabolic Basis of Inherited Disease" (C. R. Scriver, A. L. Beaudet, W. S. Sly, and D. Valle, eds.). McGraw Hill, New York.

White, G. C., and Shoemaker, C. B. (1989). Factor VIII gene and hemophilia A. *Blood* **73**, 1–12.

Zimmerman, T. S., and Ruggeri, Z. M. (eds.) (1989). "Coagulation and Bleeding Disorders." Marcel Dekker Inc., New York.

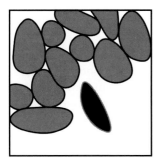

Hemopoietic System

KENNETH DORSHKIND, *University of California*

I. Blood Cell Types and Their Functions
II. Hemopoietic Hierarchy
III. Organization of Hemopoietic Tissues
IV. Regulation of Blood Cell Production
V. Interface between Basic and Clinical Studies

Glossary

Chemotaxis Orientation of a cell along a chemical concentration gradient or movement in the direction of the gradient

Cytokine Cell product, usually a glycoprotein, that has effects on the growth, differentiation, activation, and/or survival of another cell population

Hemopoiesis Process of blood cell formation that takes place in the bone marrow or medullary cavity

Pluripotent hemopoietic stem cell Immature cell that can self-renew and generate additional stem cells and has the potential to differentiate and generate progeny of all hemopoietic lineages

Progenitor cell Cell with limited proliferative capacity and a restricted differentiative potential

Retrovirus RNA virus; after infecting a cell, its RNA is copied into DNA by the enzyme reverse transcriptase, and the DNA becomes part of the genome

THE MAJORITY OF blood cells have a relatively short life span, and new cells must be continually produced to replace them. This process depends on a highly regulated series of events in which immature precursors progress through several stages of growth and differentiation, resulting in the formation of mature blood cells of a particular type. Blood cell development is known as hemopoiesis, and this occurs in cavities within the bones. The soft tissue that occupies the cavities is the bone marrow, and it is composed of the various hemopoietic populations and the supporting cells that form the environment in which blood cell production occurs. These supporting tissues are collectively referred to as the stroma, and they play an important role in the regulation of blood cell production.

I. Blood Cell Types and Their Functions

A. Classification of Blood Cells

Hemopoietic cells are classified into two broad categories: myeloid cells and lymphoid cells. Myeloid populations include red blood cells, or erythrocytes, as well as members of the granulocytic, monocytic, and megakaryocytic lineages. The two principle lymphoid cells are B and T lymphocytes. Additional terms, such as leukocytes or white blood cells, are also used and refer only to granulocytes, monocytes, and lymphocytes. In any case, mature myeloid and lymphoid cells are easily detected upon examination of the marrow, and some of their distinguishing characteristics are described below. In addition to the fully differentiated cells, immature progenitors of each of the different blood cell types are also present.

B. Lymphoid Cells

Lymphocytes are found in the blood and various lymphoid tissues such as the spleen and lymph nodes. These mononuclear, nonphagocytic white blood cells are key components of the body's immunological defense system and are classified as T cells or B cells. [*See* LYMPHOCYTES.]

Several subpopulations of T cells have now been identified based on functional criteria and their expression of cell-surface antigenic determinants;

these include helper T cells and cytotoxic T cells. Helper cells are important in the regulation of the immune response and can affect the function of other T cells or B lymphocytes through direct contact with those cells or via the release of soluble factors. Cytotoxic T cells function to eliminate tumor cells or virus-infected cells. Some investigators believe that a third type of suppressor T cell exists, but this is controversial. T cells are the exception to the rule that all blood cells are produced in the bone marrow. T-cell development occurs in the thymus, and precursors present in the marrow are believed to migrate to that tissue and provide a continual source of immature cells from which thymocytes are generated. The maturational state of these marrow emigrants is unknown, but they could be a prothymocyte committed to T cell development or the pluripotent hemopoietic stem cell itself.

The other main type of lymphocyte is the B cell. B lymphocytes are produced in the bone marrow and are the class of lymphocyte responsible for the production of immunoglobulin (Ig) or antibody molecules. Each newly produced B cell expresses an immunoglobulin on its cell surface that can specifically recognize a particular foreign material known as an antigen. Following binding of the antigenic substance to the Ig molecule, the B cell is stimulated to synthesize identical Igs, which are secreted. All of the immunoglobulin molecules secreted by any one B cell or its progeny have the same antigenic specificity. Whether or not, analogous to T cells, subpopulations of B cells exist is an active area of interest. Some investigators believe that those murine B cells that express the Ly1 cell-surface antigen or human B cells that express the CD5 cell-surface antigen constitute a separate B lymphocyte lineage. [*See* B-CELL ACTIVATION, IMMUNOLOGY.]

Natural killer (NK) cells are an additional population best considered with the lymphocytes. These cells are present in the spleen, lymph nodes, peripheral blood, and peritoneal cavity and are capable of killing a target, usually a tumor or virus-infected cell, without prior sensitization. NK cells are also known as large granular lymphocytes due to the presence of granules in their cytoplasm. To what lineage NK cells belong is unclear, and this has been an area of considerable controversy. NK cells originate in the bone marrow from hemopoietic precursors. [*See* NATURAL KILLER AND OTHER EFFECTOR CELLS.]

C. Myeloid Cells

The majority of blood cells can be classified as myeloid cells and include monocytes–macrophages, several different types of granulocytes, megakaryocytes, and erythrocytes. Macrophages are mononuclear cells derived from monocytes and are specialized for phagocytosis. These cells can ingest particulate antigens or microorganisms and digest these materials with hydrolytic enzymes they produce. Macrophages vary in size and shape depending on their state of activation. In general, however, they have a cell surface from which projects numerous membrane processes, a spherical or oval nucleus, and a cytoplasm with granules that contain numerous hydrolytic enzymes. Macrophages play a key role in the immune response by presenting antigen to lymphocytes and through the secretion of various factors that can modulate the function of lymphocytes and other cells in the body. [*See* MACROPHAGES.]

The granulocytes are a second category of myeloid cells. These cells receive their name because of the presence in their cytoplasm of granules that contain various enzymes or active proteins. Based on their appearance after hematologic staining, granulocytes are classified as neutrophils, eosinophils, and basophils. Neutrophils can be recognized by their multilobed nucleus and numerous cytoplasmic granules. The latter are classified as primary granules, which develop first during neutrophil differentiation and contain hydrolases, lysozymes, myeloperoxidase, and cationic proteins, and secondary granules, which constitute up to three-quarters of the granules and contain lysozymes and lactoferrin. These agents function to destroy or break down substances phagocytized by the cells. Neutrophils are sensitive to chemotactic agents and can migrate to the site of an inflammatory response at which they phagocytize and kill bacteria and other infectious agents. Eosinophils are bilobed cells that are characterized by the presence of large reddish orange staining granules that contain peroxidase and other enzymes. Because their numbers increase during parasitic infections, they may be involved in defense against those organisms. Eosinophils may also play a role in various allergic reactions. Basophils have a constricted nucleus and a cytoplasm that contains coarse bluish black staining granules. The granules in these cells contain heparin, histamine, and serotonin. These substances are released following stimulation of the cells and contribute to such allergic responses as smooth muscle contraction and increased blood vessel permeability. Basophils have also been implicated in defense against parasites. [*See* NEUTROPHILS.]

Megakaryocytes or thrombocytes are huge cells that can have large nuclei that form as a result of polyploidy. The function of megakaryocytes is to produce platelets, and this occurs as a result of budding of cytoplasmic fragments from the cells. Platelets function in coagulation, hemostasis, and thrombus formation.

Erythrocytes, or red blood cells, are non-nucleated biconcave disks that are able to transport oxygen to the tissues. This is because the cells contain hemoglobin, a protein that combines readily with oxygen.

II. Hemopoietic Hierarchy

A. Introduction

If a sample of bone marrow is smeared on a slide and stained with an appropriate hematologic reagent, a marked heterogeneity of cell types can be observed. The mature cells described above can be easily recognized as well as other populations not readily identifiable. These are hemopoietic cells of various lineages and at different stages of development. The maturational state of these cells can be determined using a variety of criteria that include their histologic appearance, expression of defined cell-surface or cytoplasmic antigenic determinants, responsiveness to growth and differentiation factors, and proliferative and differentiative potential. All of these parameters have provided useful information that has permitted a scheme of hemopoiesis to be formulated.

B. The Pluripotent Hemopoietic Stem Cell

At the head of the hemopoietic hierarchy is the pluripotent hemopoietic stem cell (PHSC). This is defined as a cell that has the ability to divide and generate additional stem cells (a property known as self-renewal) as well as to differentiate and give rise to the blood cell types described above. The PHSC is a rare cell and may account for <0.1% of the nucleated cells in the bone marrow. Their incidence

is even lower in other tissues such as the spleen. Despite this, these cells clearly exist.

One study that provided evidence for the existence of a multipotential stem cell analyzed blood cells in women who were heterozygous for the enzyme glucose-6-phosphate dehydrogenase (G-6-PD). Each cell and all of its progeny, in these individuals expresses only one of the two forms of the enzyme. Therefore, all blood cells that derive from any one pluripotent hemopoietic stem cell would express one or the other G-6-PD type. When cells from G-6-PD heterozygous females who had developed chronic myelogenous leukemia (a disease that originates in a single cell) were analyzed, cells from most of the hemopoietic lineages were shown to express the same G-6-PD type. This provided strong evidence that these populations were the progeny of a common stem cell that exhibited a marked differentiative capability. Another classic study that demonstrated the existence of the PHSC involved the induction of chromosomal markers in bone marrow cells by subjecting them to a low, sublethal dose of ionizing irradiation. This induces random chromosomal translocations in a high proportion of the surviving cells, and because each is unique, it can be used as a cytogenetic marker to detect cells derived from the marked cells. This cell population (called clonal) was used to repopulate X-irradiated mice, in which the bone marrow is destroyed. At various times following reconstitution, cytogenetic analysis was performed on the myeloid, B, and T cells which derived from the cells used to repopulate the animal. The study showed that in some of the mice myeloid, B, and T cells all bore the same unique chromosomal marker, providing very strong evidence that they were all progeny of a single pluripotent stem cell in which the marker had been created.

Another series of experiments that demonstrated the existence of the PHSC repopulated mice with bone marrow cells that were infected with a retroviral vector carrying the bacterial neomycin resistance gene (which allows for the selection of the cells carrying the gene). The retrovirus genome integrates randomly into the genome of the marrow cells, and the viral integration site creates in each cell a unique change (called restriction fragment-length polymorphism), which can be recognized by preparing DNA from the cells and cutting it with an appropriate restriction enzyme. These fragments can be visualized by a suitable laboratory technique. At appropriate periods after reconstitution of irradiated mice, myeloid, B, and T cells were isolated from the animals, and the cells' DNA was analyzed. These studies demonstrated that neomycin resistant B cells, T cells, and myeloid cells could be detected in which the retroviral vector had integrated at the same site. The most straightforward explanation for these results was that these cells were the progeny of a pluripotent hemopoietic stem cell that had been infected by the retrovirus.

A clonal analysis of hemopoietic cells in the above mice was also performed at various times after reconstitution. This investigation showed that some hemopoietic stem cells stably contributed to all hemopoietic lineages in a reconstituted mouse over time, while others underwent temporal changes. For example, in some animals, myeloid and lymphoid progenies of a particular stem cell were observed at the initial sampling time, but only one or the other cell type was present at a later point. In additional animals, new stem cell clones were observed to have emerged, based on the observation that lymphoid and myeloid cells that bore the same unique viral integration site were detected at the time of a second sampling but not an earlier point following reconstitution. These studies are of importance, because they document that the contribution of a pluripotent stem cell to the various lineages can vary with time.

The above experiments have provided evidence for the existence of the pluripotent hemopoietic stem cell, but they do not allow it to be directly studied in isolation. A major goal of experimental hematologists is to isolate and grow pure stem cell populations. This would present great advantages for studying how the growth and differentiation of that population is regulated and for identifying and characterizing genes that are temporally expressed during blood cell development. A variety of approaches have been used to purify stem cells from the marrow based on their physical characteristics and expression of cell-surface determinants. Recent work has indicated that murine stem cells have special surface characteristics by which they can be isolated. Efforts are now underway to identify markers expressed on human pluripotent hemopoietic stem cells.

C. Restricted Stem Cells

A restricted stem cell is a cycling population that has extensive differentiative potential, but more limited than that of a pluripotent stem cell. Therefore, some investigators prefer to designate these cells as multipotential precursors to clearly distinguish them from the pluripotent stem cell. Two types of multipotential precursors have been described. A multipotential myeloid-precursor is defined as a cell that can generate all hemopoietic cells except lymphocytes. The evidence that these restricted populations exist is based on several criteria. First, in the above experiments using retrovirally marked cells, patterns of repopulation were observed in which only myeloid but not lymphoid cells bore the same viral integration site. A second piece of evidence is based on classic experiments that demonstrated that bone marrow cells will form macroscopic colonies on the spleen 8–13 days following their injection into lethally irradiated mice. The cell that initiates these colonies is known as the spleen colony-forming unit (CFU-S; S stands for spleen) and the colonies are called spleen colonies.

Based on the observation that lymphoid cells are rarely, if ever, detected as progeny of the CFU-S and the colonies primarily contain myeloid cells, the assay is generally accepted as detecting a multipotential myeloid-restricted precursor. Spleen colonies form at various times after injection of bone marrow cells into the irradiated mice. Those that appear on day 13 are generally thought to be derived from a multipotential myeloid-restricted precursor. Spleen colonies also appear at earlier times, but they may be derived from more differentiated progenitor cells described below.

A common lymphoid precursor is a second type of restricted stem cell. By definition, such a cell would have the potential to generate both B and T lymphocytes but not myeloid cells. Several laboratories have provided indirect evidence for the existence of a common lymphoid precursor, but to date, a pure population of such cells has not been isolated. Whether lymphocytes are the progeny of a common lymphoid precursor committed to the B- and T-cell differentiative pathway or are direct descendants of the PHSC remains an unanswered question.

D. Progenitor Cells

Progenitor cells have a limited proliferative capability and a restricted differentiative potential. For each of the blood cell populations described above, a progenitor cell committed to that particular lineage exists. There are two principle means by which progenitor cells can be detected. The classical method takes advantage of the ability of progenitors to form lineage-restricted hemopoietic colonies in semisolid medium. The growth of these colonies depends on the presence of a growth factor that potentiates their formation. Progenitor cells can also be identified by their expression of lineage-specific antigens.

1. Myeloid Progenitors

Investigators in Israel and Australia were the first to demonstrate that bone marrow or spleen cells could form colonies in a semisolid medium such as agar following culture under appropriate conditions. The colonies that grew contained granulocytes and/or macrophages, and subsequent analyses indicated they had arisen from a single progenitor cell. Distinct progenitor cells for most of the myeloid lineages have now been identified, and their formation of colonies in semisolid medium depends on the presence of a particular lineage specific growth and/or differentiation factor. For example, as shown in Fig. 1, it is now known that a bipotential granulocyte–macrophage progenitor cell exists that can differentiate into cells of one or both of these lineages.

Progenitor cells are not necessarily of the same precise maturational state, and multiple cells may be included in a progenitor cell compartment. This point is best illustrated by analysis of cells of the erythroid lineage. Two stages of erythropoiesis have been defined based on the ability of erythroid progenitors to form colonies that appear at different times following initiation of cultures. An erythroid progenitor known as the erythroid burst-forming unit (BFU-E) is an immature erythroid progenitor and the erythroid colony-forming unit (CFU-E) is a more differentiated cell. Both the BFU-E and the CFU-E meet the definition of progenitor cells, because they are lineage-restricted in their differentiative potential. Figure 1 also shows the other myeloid progenitors that have been identified, and heterogeneity within these compartments also likely exists.

2. Lymphoid Progenitors

The bone marrow contains precursors that can generate B and T cells, but the development of these two types of lymphocytes occurs in different sites. B lymphopoiesis proceeds in the marrow of adult mammals, while primary T lymphocyte development occurs in the thymus.

Many researchers now believe that T-cell production depends on the migration of a bone marrow-derived T-cell precursor to the thymus. Whether this is a lymphoid stem cell, a progenitor committed to the T-cell developmental pathway (the prothymocyte) and/or the pluripotent stem cell itself is not entirely clear. In any case, once the T-cell precursor enters that tissue, it interacts with the thymic stroma and a series of events occur that result in the expression of the T cell-receptor and T cell-surface antigens. Major efforts are underway to identify the anatomical sites within the thymus in which various differentiative events occur, to define the lineage relationships, which result in the production of the various T-cell subpopulations, and to further clarify the molecular events that result in the expression of the genes that encode the T cell receptor. It is generally accepted that a cellular intermediate that expresses both the CD4 and CD8 antigens develops during the intrathymic stage of differentiation, and this gives rise to T-cell subpopulations that include the CD4+ T helper cells and the CD8+ cytotoxic T cells. These newly produced T cells express the heterodimeric T-cell receptor, which most commonly consists of an α- and a β-chain (Fig. 1), although it may also be formed by a γ- and a δ-chain. These chains are expressed during T-cell differentiation as a result of a complex process in which genes that encode segments of these molecules rearrange and become expressed in developing thymocytes. The T-cell receptor on each newly produced T lymphocyte has a unique site at its amino terminus, known as the variable region, that recognizes antigen in association with cell-surface determinants encoded by the major histocompatibility locus.

B-cell development occurs in the bone marrow, and maturation of cells in this lineage results in the expression of the Ig molecule as well as other non-immunoglobulin cell-surface antigens. Distinct stages of B lymphopoiesis have been defined through the use of antibodies that recognize these molecules (Fig. 1). For example, the B220 molecule is a 220,000-molecular weight glycoprotein present on murine B-cell progenitors. Similar molecules may be present on human B-cell progenitors.

The Ig or antibody molecule is formed by the assembly of two identical heavy-chain and two identical light-chain glycoproteins. The heavy-chain genes are expressed first, and following translation

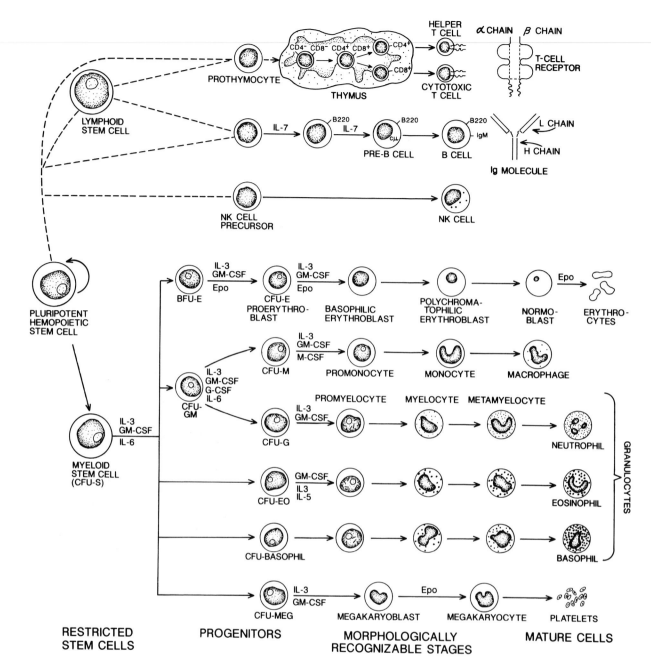

RESTRICTED
STEM CELLS   PROGENITORS   MORPHOLOGICALLY
RECOGNIZABLE STAGES   MATURE CELLS

of heavy-chain messenger RNA, heavy chain of the immunoglobulin M class is expressed in the cytoplasm of a cell population termed a pre-B cell. Light-chain genes are subsequently expressed, and as light-chain protein appears, the complete Ig molecule is assembled in the cytoplasm of the pre-B cell and immediately appears on the surface of what is now termed a newly produced B lymphocyte. As with the T-cell receptor, expression of the Ig molecule depends on the rearrangement of gene seg-

FIGURE 1 A scheme of the cells of the hemopoietic system showing the various lineages and maturational states that have been defined. Broken lines indicate pathways in which lineage relationships remain unresolved. The arrow on the pluripotent hemopoietic stem cell indicates that the cell is capable of self-renewal. Some of the cytokines that regulate the growth and/or differentiation of blood cells are indicated. Their placement in the diagram is intended to indicate the point at which they act, but this is only an approximation. The reader must consult primary literature for the precise, up to date effects of these and other cytokines.

ments that encode the heavy- and light-chain molecules. The amino terminus of the heavy and light chains form a unique site specialized for recognition of an antigen.

Colony assays for B-cell progenitors, analogous to those for erythroid and myeloid populations, are a recent development that provides an additional means to identify B cell precursors.

E. Morphologic Identification of End Cell Populations

Before the development of colony assays or the availability of antibodies that recognize antigens expressed by hemopoietic precursors, morphology was the principle means by which immature cells in the various lineages were identified. However, by the time a cell of a particular lineage is histologically recognizable, it is more mature than progenitors that are detected in colony assays. For example, it is possible to distinguish immature erythroid cells such as the erythroblast. This cell, which is thought to correspond to a mature CFU-E, has a basophilic cytoplasm and a nucleus containing multiple nucleoi. More differentiated progeny of the erythroblast can also be defined based on cell size, the nuclear chromatin pattern, and cytoplasmic basophilia (Fig. 1). Similarly, various stages of myelopoiesis can be recognized under the light microscope. One of the earliest recognizable precursors of granulocytes is the promyelocyte. This cell, which is certainly more differentiated than the progenitors that initiate colonies in semisolid medium has a large nucleus with a nucleolus. Peroxidase-positive granules are present in the cytoplasm, and a well-developed Golgi apparatus is present. From the promyelocyte stage on, cells in the granulocytic series are distinguished based on the formation of specific granules in their cytoplasm, and once these appear, the cells are classified as myelocytes. Cells are classified as metamyelocytes following the development of the distinguishing nuclear characteristics noted in Section I,C.

Some appreciation of the appearance of the various hemopoietic cells remains important. This is particularly true for the clinician, because diagnosis of certain types of hematologic dyscrasias such as leukemia is still based on the morphological appearance of the cells present in the marrow or peripheral blood.

III. Organization of Hemopoietic Tissues

Blood cell development takes place in association with a fixed framework of supporting elements known as stromal cells. An understanding of the location of these stromal populations and the architecture of the marrow cavity is best appreciated in the context of blood cell circulation to the bone.

The principle blood supply to a typical long bone is the nutrient artery. This enters at midshaft through the nutrient foramen and travels in the central portion of the medullary cavity in parallel with the longitudinal axis of the bone. This centrally running artery sends off branches, which radiate toward the inner or endosteal surface of the medullary cavity, and at that point these branches anastomose with tributaries from other arteries that supply the bone. These subendosteal vessels become continuous with a series of venous sinusoids that deliver their contents toward a central sinus in the center of the marrow cavity. Blood exits the bone from the central sinus via the nutrient foramen and enters the systemic circulation.

Stromal cells are present in the intersinusoidal spaces and form the framework on which hemopoiesis proceeds. These cells, which have been referred to as reticular cells by morphologists, have numerous cytoplasmic projections that form extensive connections with one another and the processes of stromal cells in the adjacent intersinusoidal spaces. Stromal cell processes also have extensive contact with the endothelial cells, which form the lining of the sinuses. Developing blood cells in the intersinusoidal spaces are in close contact with the stromal cells, and this association led early morphologists to propose that the stroma can regulate blood cell production. As blood cells mature, they must ultimately exit the medullary cavity and enter the circulation. They do so by passing from the intersinusoidal spaces and through the endothelial cells to enter the venous sinusoids.

Under normal, steady-state conditions, the bone marrow is the only site in which the simultaneous production of myeloid cells and B lymphocytes occurs. This suggests that local tissue influences critical for that process are operative in the medullary cavity. It is generally accepted that the various stromal cell types and their products are one source of these regulatory signals, and the term hemopoietic microenvironment has been used to describe the localized sites in the medullary cavity in which blood cell production proceeds. Some investigators consider that components of the hemopoietic microenvironment are able to induce a pluripotent hemopoietic stem cell to differentiate along a particular lineage. However, this is not universally accepted, and it remains unclear if the microenvironment acts directly to induce the commitment and differentiation of multipotential stem cells or, instead, provides the permissive signals that allow the further development of a precursor that has become randomly committed to a particular lineage through its own intrinsic genetic programming. These controversies aside, it is evident that a number of growth and differentiation signals exist that affect hemopoiesis.

IV. Regulation of Blood Cell Production

A. Introduction

The identification of the environmental signals that regulate blood cell production and the characterization of their effects on hemopoietic targets remain challenging areas of investigation. Such knowledge is of relevance to the basic understanding of a fundamental developmental process and has clinical relevance as well. When regulatory controls do not function properly, a number of hemopoietic abnormalities can occur. Much of what is known about the regulation of hemopoiesis is based on studies of murine tissues and cells, and it is reasonable to expect that similar regulatory controls are operative in human marrow.

The environmental cues a differentiating hemopoietic cell receives can be delivered via interactions with other cells, through the association of the cell with components of the extracellular matrix, or by the binding of a soluble mediator to a receptor on the cell membrane.

B. Cell–Cell Interactions

Morphologic examination of marrow sections reveals that hemopoietic cells are tightly clustered in the intersinusoidal spaces in association with each other and the supporting stromal elements. This obviously creates the potential for direct cell–cell interactions between these populations and has suggested that hemopoietic and stromal cells might

associate with one another via recognition molecules that permit direct membrane binding. Such molecules are now being identified by several laboratories that have prepared monoclonal antibodies to stromal cells. These antibodies apparently block the interactions between stromal cells and blood cells. These reagents will be of particular value in functional studies and in purifying and characterizing cell interaction molecule(s).

The endothelial cells that line the marrow sinusoids also play a role in regulating the passage of cells from the marrow into the systemic circulation. A number of elegant ultramicroscopic analyses have demonstrated that blood cells actually pass through the endothelial cell wall on their way into the sinusoids. What signals may be transmitted from the endothelial cell to the migrating blood cells remain to be defined, but the potential for this exists, because endothelial cells are known to secrete various factors with effects on hemopoiesis.

An important means of regulating passage of cells from the marrow may be mediated by the reticular and endothelial cells. The processes of reticular cells are associated with the abluminal surface of the endothelial cells and the degree of association can be variable. It has been hypothesized that an extensive reticular-endothelial cell association may serve to retain cells in the marrow, while a retraction of reticular cell process from the endothelium may facilitate blood cell emigration.

C. Cell–Matrix Interactions

The extracellular matrix (ECM) is a meshwork of polysaccharides and proteins that form an organized framework in the tissues. Fibroblasts in the tissues secrete much of ECM. The two main ECM components include polysaccharide glycosaminoglycans that are linked to a protein backbone and fibrous proteins. The latter include structural components such as collagen and elastin and adhesive proteins such as fibronectin and laminin. The matrix was initially believed to serve as a packing material in the tissues, but studies now indicate that it plays an important role in the development, migration, proliferation, shape, and metabolic function of cells. [*See* EXTRACELLULAR MATRIX.]

Hemopoietic cells have been reported to have receptors for various ECM components to which they may bind. What particular events may be signalled by such binding remains to be determined. Also, evidence indicates that hemopoietic growth factors can bind to ECM components and that the ECM thereby compartmentalizes these cytokines and helps to present them to developing blood cells.

D. Soluble Mediators

Experiments in which a conditioned medium was used to stimulate the formation of hemopoietic colonies in agar demonstrated that some soluble factors could regulate the growth and/or differentiation of hemopoietic cells. It is now known that the molecules that potentiate the formation of the granulocyte and macrophage colonies are members of a family of cytokines known as the colony-stimulating factors. Additional factors that can affect the growth, differentiation, survival, and activation of hemopoietic cells have now been identified and include erythropoietin, which acts on erythroid cells and, more recently, a variety of interleukins with effects on lymphoid and myeloid cells (Table I). The considerable progress that has been made in the identification and characterization of the actions of these multiple cytokines is due in part to the application of molecular approaches to cytokine analysis. Techniques for the cloning of the genes that encode these mediators has become relatively routine, and because of this, high concentrations of essentially pure recombinant factors are now available for *in vitro* and *in vivo* experiments. [*See* Cytokines in the Immune Response.]

Among the best characterized factors are the family of glycoproteins known as the colony-stimulating factors (CSFs). Three of the four CSFs were named according to their initially described effects on the growth and differentiation of a particular progenitor cell population and include macrophage (M-CSF), granulocyte (G-CSF), and granulocyte–macrophage (GM-CSF). While the primary targets of these mediators are indicated by their name, their actions are not restricted to those lineages. For example, GM-CSF targets include eosinophils, megakaryocytes, and multipotential myeloid precursors. A fourth CSF is known as multi-CSF or interleukin-3 (IL-3). IL-3 targets include granulocytes, macrophages, eosinophils, megakaryocytes, and multipotential precursors. The various hemopoietic cells noted above each express several hundred CSF receptors, and occupancy of as few as 10% of them can stimulate a biological response by the cells. The CSFs have a high specific activity and are active at concentrations of 10^{-10}–10^{-12} molar. While a major action of CSFs is the stimulation of cell proliferation, they also have effects on the survival, differentiation, maturation, and activation of their targets.

Red blood cell production depends on the hormone erythropoietin. This is produced in the kidney, and its level of production is sensitive to the oxygen levels in the blood. A lack of oxygen or shortage of red blood cells results in an increase in erythropoietin synthesis and a stimulation of red blood cell production in the bone marrow.

Another class of molecules with effects on the

TABLE I Selected Cytokines and Their Effects on Hemopoietic Cells[a]

Cytokine	Source	Effects
IL-1	Macrophages, monocytes, fibroblasts, endothelial cells	Synergy with CSFs, stimulation of CSF and IL-6 secretion by stromal cells
IL-2	T cells	Stimulates thymocyte and T-cell growth
IL-3	T cells	Stimulation of proliferation–differentiation of myeloid-restricted stem cells and erythroid, monocytic, granulocytic, and megakaryocytic progenitors
IL-4	T cells, stromal cells	Pre-B-cell maturation, potentiates responsiveness of myeloid progenitors to CSFs
IL-5	T cells	Stimulation of eosinophil progenitors
IL-6	T cells, macrophages, stromal cells	Acts as CSF to stimulate proliferation of myeloid progenitors
IL-7	Stromal cells	Stimulation of B cell progenitor–proliferation; stimulates thymocyte proliferation; augments proliferation of mature T cells to other cytokines
M-CSF	Macrophages, endothelial cells, stromal cells	Stimulates proliferation–differentiation of macrophage progenitors
G-CSF	T cells, stromal cells	Stimulates proliferation–differentiation of granulocyte progenitors
GM-CSF	T cells, stromal cells	Stimulates proliferation–differentiation of granulocyte–macrophage progenitors

[a] The cellular sources of a particular factor and the effects are not restricted to the parameters noted. This is only a selected tabulation of sources and actions of these cytokines. In addition, these are not the only cytokines with effects on hemopoietic progenitors.

growth and differentiation of hemopoietic cells is the interleukins. This name was chosen to indicate that the source of these cytokines is leukocytes. In addition to IL-3 described above, at least nine other interleukin molecules have now been described. Interleukin-1 (IL-1) is primarily a macrophage–monocyte product, although other tissue cells such as fibroblasts and endothelial cells can produce it. IL-1 exhibits potent synergistic effects with other CSFs. Other interleukins with hemopoietic effects include interleukins-4, -5, -6, and -7. For example, IL-5 acts as a differentiation factor for eosinophils. IL-6 is a recently described interleukin that was originally defined by its ability to promote the growth of Ig-secreting B cells. This molecule is now recognized to act as a CSF to stimulate granulocyte progenitors and multipotent myeloid precursors. IL-7 is the first cytokine available in recombinant form that has been shown to affect B-cell precursors in the bone marrow. The gene for IL-7 was cloned from a bone marrow stromal cell line and encodes a 14.9-kD molecule. IL-7 can stimulate the proliferation of B-cell progenitors. Its effects are not restricted to B-lineage cells, as both thymocytes and peripheral T cells are sensitive to its actions. IL-7 apparently does not have any differentiative function, and this would suggest that additional molecules with such actions on B-lineage cells exist.

Cytokine biology is a dynamic and complex area, and it is difficult to discuss all of the subtleties of this field. However, several important facts regarding the actions of these regulatory molecules must be appreciated and can be summarized as follows.

1. The effects of any one cytokine are pleiotropic (i.e., multiple). For example, the interleukins are critical to the mediation of the immune response and many were initially described by their effects on mature lymphocytes. The fact that most of these factors have potent effects on the growth and differentiation of hemopoietic cells of various lineages was a secondary discovery.

2. The same factor may exhibit differential effects on cells within the same lineage. IL-4 has been proposed to stimulate or potentiate the differentiation of mature B cells while acting as an inhibitor of the proliferation–differentiation of early B-cell precursors. These dichotomous effects may be due to different concentrations of the factor.

3. The potential exists for additive and synergistic interactions between various cytokines. IL-1 and the CSFs are known to synergize and IL-4 has been shown to potentiate the response of myeloid progenitor cells to the actions of CSFs. Precisely how the interactions between two cytokines result in an augmented cellular response remains to be defined. If a cell expresses receptors for both cytokines, then the result of dual binding to these could easily explain an additive response. The explanation for a synergistic response is not as straightforward. In the case of IL-1–CSF synergy, it is possible that one molecule, such as IL-1, could stimulate an increase in the number of CSF receptors expressed by the cell and, thus, potentiate the response to that cytokine.

4. The effects of any particular hemopoietic factor are not necessarily mediated through a direct action on the hemopoietic target. Many of these molecules indirectly stimulate bystander cells to secrete a second molecule, which in turn acts on the hemopoietic progenitor. Numerous cells such as fibroblasts, bone marrow stromal cells, and endothelial cells are known to express IL-1 receptors. One effect of IL-1 binding to them is that they are stimulated to secrete a number of other mediators that include the CSFs. These in turn may then act on the hemopoietic target alone or synergistically with the IL-1 molecules that originally induced their production.

5. Certain hemopoietic cells can secrete cytokines to which they respond. For example, macrophages are sensitive to the effects of M-CSF and may also produce that mediator. This raises the possibility that autocrine (self-regulating) and paracrine (regulating neighboring cells) regulatory mechanisms are operative.

6. It is becoming clear that control of cell production in the hemopoietic system is also governed by various negative regulatory molecules that can have inhibitory effects on the growth and differentiation of cells. One such molecule, transforming growth factor-β, has been shown to affect the production of lymphoid and myeloid lineage cells. There have also been suggestive reports that the positive effects of many cytokines may be modulated via negative feedback circuits. In this case, a cytokine could stimulate a particular hemopoietic cell while inducing the secretion of a negative regulator by another cell.

It is important to note that virtually nothing is known about factors that regulate the self-renewal and differentiation of pluripotent hemopoietic stem cells. Some of the factors described above and in Table I might act on stem cells, but it is difficult to obtain a pure population on which to test this. A factor known as leukemia inhibitory factor (LIF) has been shown to inhibit the differentiation and promote the growth in culture of a cell known as an embryonic stem cell. These cells are isolated from the inner cell mass and have the capability to generate all cells and tissues of an embryo. Whether or not LIF acts to promote the self-renewal and expansion of the hemopoietic stem-cell compartment is unknown. Nevertheless, the fact that such molecules with effects on undifferentiated embryonic cells exist suggests that comparable mediators with actions on the hemopoietic stem cell will be described.

V. Interface between Basic and Clinical Studies

Because hemopoietic cells are easily obtained and manipulated, the hemopoietic system is an excellent model for investigating the parameters that regulate cell growth and differentiation. The development of culture systems that facilitate these studies has been a major technological advance, and systems that allow long-term, stromal cell-dependent murine myelopoiesis or lymphopoiesis have been described and are in widespread use. These cultures are being used to study the growth of human hemopoietic cells as well.

The accessibility of bone marrow, the relative ease in manipulating it, and the development of techniques for its transplantation have contributed to great interest in applying the basic knowledge obtained about hemopoietic cells to clinical medicine. For example, the experiments in which retroviral vectors were used to infect stem cells have potential use far beyond simply establishing lineage relationships. There is now considerable interest in using this approach to correct genetic defects that affect blood cell development or function. Thus, hemopoietic stem cells could be harvested from an individual and infected with a retrovirus that contains a normal copy of the defective gene. These cells would then be used to repopulate that individual's hemopoietic system with "normal" cells. This therapy depends on perfecting techniques for retroviral infection of stem cells and for ensuring that the exogenous gene is expressed in a regulated manner.

Another example of how the basic studies of the hemopoietic system have rapidly overlapped with clinical applications is evident from the rapid progress in cloning the genes for the various cytokines. Virtually unlimited quantities of factors such as erythropoietin, GM-CSF, and G-CSF are now available and are being administered to patients with promising results. While considerably more basic research in the mechanisms of action of these factors remains to be conducted, information regarding their actions is being obtained from these clinical trials. Thus, not only can basic information about the hemopoietic system be applied to human medicine, but the results obtained from these applications can contribute to the fundamental understanding of blood cell production.

Bibliography

Abramson, S., Miller, R. G., and Phillips, R. A. (1977). Identification of pluripotent and restricted stem cells of the myeloid and lymphoid systems. *J. Exp. Med.* **145,** 1567.

Dorshkind, K. (1990). Regulation of hemopoiesis by bone marrow stromal cells and their products. *Annu. Rev. Immunol.* **8,** 111.

Keller, G., Paige, C., Gilboa, E., and Wagner, E. F. (1985). Expression of a foreign gene in myeloid and lymphoid cells derived from multipotent haematopoietic precursors. *Nature* **318,** 149.

Kincade, P. W., Lee, G., Pietrangeli, C. E., Hayashi, S. I., and Gimble, J. M. (1989). Cells and molecules that regulate B lymphopoiesis in bone marrow. *Annu. Rev. Immunol.* **7,** 111.

Klein, J. (1982). "Immunology: The Science of Self–Non-self discrimination." John Wiley & Sons, New York.

Jordon, C. T., McKearn, J. P., and Lemischka, I. R. (1990). Cellular and developmental properties of fetal hematopoietic stem cells. *Cell* **61,** 953.

Metcalf, D. (1988). "The Molecular Control of Blood Cells." Harvard University Press, Cambridge.

Metcalf, D., and Moore, M. A. S. (1971). "Haemopoietic Cells." North-Holland Publishing Company, Amsterdam.

Nicola, N. A. (1989). Hemopoietic cell growth factors and their receptors. *Annu. Rev. Biochem.* **58,** 45.

Spangrude, G. J. (1989). Enrichment of haemopoietic stem cells: Diverging roads. *Immun. Today* **10,** 344.

Hemostasis and Blood Coagulation

CRAIG M. JACKSON, *American Red Cross*

DIANE L. LUCAS, *National Heart, Lung and Blood Institute*

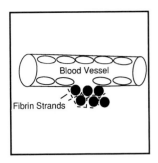

Glossary

α_2-Antiplasmin Inhibitor of plasmin

Antithrombin III Inhibitor of the proteinases of blood clotting

Cl inactivator Inhibitor of Factor XIIa

Factor V (proaccelerin) Protein cofactor to Factor Xa during activation of prothrombin

Factor VII (proconvertin) Proteinase which, with tissue factor, activates Factor X

Factor VIII (antihemophilic factor) Protein cofactor to Factor IXa during activation of Factor X

Factor IX (antihemophilic factor B, Christmas factor) Precursor of proteinase which, with Factor VIIIa, activates Factor X

Factor X (Stuart–Prower factor) Precursor of proteinase which, with Factor Va, activates prothrombin

Factor XI (plasma thromboplastin antecedent) Precursor of proteinase which activates Factor IX (intrinsic pathway)

Factor XII (Hageman factor) Precursor of proteinase which activates Factor XI and prekallikrein

Factor XIII (fibrin-stabilizing factor) Precursor to transamidase which stabilizes fibrin by cross-linking

Fibrinogen (Factor I) Soluble protein which, after proteolytic cleavage by thrombin, polymerizes to form the fibrin clot

Heparin cofactor II Inhibitor of thrombin

High-molecular-weight kininogen Protein cofactor of Factor XIIa and kallikrein in the activation of Factor XII and prekallikrein

α_2-Macroglobulin General inhibitor of hemostatic and fibrinolytic proteinases

Plasminogen Precursor of proteinase which dissolves the fibrin clot

Plasminogen activator inhibitor, Types I, II, and III Inhibitors of tissue plasminogen activators in plasma; Type III also inhibits activated protein C

Platelet Cell (i.e., megakaryocyte) fragment which adheres to tissue exposed as a result of injury in order to form the primary hemostatic plug

Prekallikrein Precursor of kallikrein, the proteinase which activates Factor XII

Proenzyme Inactive precursor to the fully functional form of a proteolytic enzyme

Protein C Precursor of proteinase which inactivates Factors Va and VIIIa

Protein S Protein cofactor to activated protein C

Prothrombin (Factor II) Circulating precursor (proenzyme) to the proteinase thrombin

Thrombomodulin Membrane protein which acts as a protein cofactor for activation of protein C by thrombin

Tissue factor (thromboplastin, Factor III) Cellular protein responsible for initiation of clotting on ruptured cells; in mixture with phospholipids, called thromboplastin

Tissue-type plasminogen activator Proteinase of endothelial cell origin which activates plasminogen

Urokinase-type plasminogen activator (prourokinase) Proteinase of cellular origin which activates plasminogen

von Willebrand factor Adhesive protein required for platelet adhesion to cells and a carrier for Factor VIII

HEMOSTASIS IS THE spontaneous arrest of bleeding from ruptured blood vessels. Normal hemostatic response to vessel injury includes blood vessel constriction, adhesion of platelets to cells not normally exposed to blood, and a complex series of proteolytic reactions, which culminate in the formation of a fibrin network to reinforce and consolidate the platelet plug at the injury site. The hemostatic response is finely balanced under normal physiological conditions. Through the antagonistic actions of the components that promote the proteolytic reactions of the blood clotting system and the inactivators of these components, the risk of excessive bleeding (i.e., hemorrhage) is balanced against the risk of formation of unwanted blood clots (i.e., thrombosis). Decreased numbers of functional platelets, or deficiencies of clotting factors, predisposes to hemorrhage; deficiencies in inhibitors of the clotting proteinases, or in the enzymes that digest the blood clot after tissue repair is completed, predisposes to thrombosis.

The principal reactions of the blood clotting cascade are transformations of proenzymes of proteinases into active enzymes. Physiologically adequate proenzyme activation occurs when an activating proteinase, a protein cofactor, and the proenzymes bind to a cell membrane surface to form a complex. Proenzyme activation is several hundred thousand times faster in the complex than it is when only the activating proteinase is present. The formation of the complex at the site of the exposure of damaged cell membranes localizes rapid activation to the site of injury. Activator complexes are characteristic of the proenzyme activation reactions in blood clotting.

I. Physiology of Hemostasis and Blood Clotting

The rupture of blood vessels initiates the hemostatic response. The ruptured blood vessels, particularly arteries, constrict, thus narrowing the opening through which blood may escape. When blood proteins and platelets come into contact with cells or cell components other than the intact endothelial cells which normally line the lumen of the blood vessel, they adhere to them, forming a platelet plug, which physically occludes the opening in the ruptured vessel. Adhesion of platelets to exposed smooth muscle cells and exposed collagen fibrils requires the plasma protein von Willebrand factor, which binds to specific receptors on the outer membrane of the platelet. von Willebrand factor thus anchors the platelet to the subendothelial cells and the extracellular matrix. Deficiency in von Willebrand factor is among the more common of the

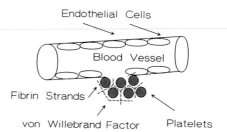

FIGURE 1 An endothelial cell-lined blood vessel with platelets released into the extravascular space via a rupture in the vessel wall.

blood clotting disorders and leads to prolonged bleeding in affected individuals. The primary hemostatic response is shown in Fig. 1.

As platelet adhesion begins, a sequence of the reactions described as blood clotting commences with the formation of a complex between a membrane protein, tissue factor, and the clotting protein, Factor VII.[1] The formation of the tissue factor–Factor VII complex initiates the sequence of blood clotting reactions, which culminates in the formation of the proteolytic enzyme thrombin. Thrombin then catalyzes the conversion of soluble fibrinogen into a form which polymerizes to produce the fibrin meshwork of the hemostatic plug, or blood clot. Thrombin also stimulates platelets to release several clotting factors and aggregating agents (e.g., ADP) from storage granules within the platelet. These platelet factors synergistically aid in the further development of the platelet plug. The hemostatic response which is initiated by tissue injury, or *in vitro* by the addition of a tissue hemogenate to the blood sample, is called the extrinsic pathway of blood clotting. The alternate pathway of blood clotting, which does not begin with tissue factor, is called the intrinsic pathway (Fig. 2). [*See* TISSUE FACTOR.]

II. Biochemistry of the Hemostatic System Components

The blood clotting system, or cascade, consists of a series of proteolytic reactions in which the circulating inactive blood clotting factors are converted into catalytically active forms. The proenzymes are

---

1. The Roman numeral designations are those given in 1958 [I. Wright, *J. Am. Med. Assoc.* **170,** 135–138 (1959)] to eliminate the use of patient names for coagulation factors. The letter "a" following the numeral indicates the activated derivative.

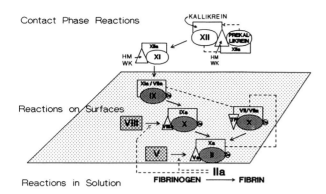

Contact Phase Reactions

Reactions on Surfaces

Reactions in Solution

FIBRINOGEN ⟶ FIBRIN

FIGURE 2 The blood clotting cascade. Components are enclosed by geometric figures which indicate the type of component as follows: Proenzymes are shown as ellipses; proteinases are shown as polygons, protein cofactors are shown as rectangles; activated protein cofactors are shown as triangles. The tissue factor triangle (TF) is inverted to indicate that it "sticks into" the membrane. The contact phase reactions are shown away from the cell surface, because the surfaces to which they bind are only known *in vitro* and are not of biological origin. The pathway which is initiated through exposure of tissue factor is called the extrinsic pathway; that which is initiated through contact of Factor XII and prekallikrein with an artificial surface is called the intrinsic pathway. HMWK, high molecular weight kininogen. (Reproduced, with permission, from the *Annual Review of Biochemistry*, Vol. 49. © 1980 by Annual Reviews Inc.)

converted to active proteolytic enzymes. The protein cofactors have latent binding sites for a proteinase and a proenzyme, which are unmasked by proteolysis. The transformations of proenzymes into active proteinases are the elementary reactions of the blood clotting cascade; they occur rapidly only when a proteinase is assisted by an activated protein cofactor and when all components, including the proenzyme substrate, are bound to a membrane surface. Such assembled activator complexes catalyze activation of their respective proenzymes at more than 100,000 times the rate observed with a proteinase alone.

A. Clot Formation—The Polymerization of Fibrin

Understanding the blood clotting and fibrinolytic processes is aided by classifying the components into the following categories on the basis of their structures and functions: (1) precursors of proteolytic enzymes and active proteinases; (2) precursors of protein cofactors, which act catalytically, but are not directly involved in the splitting of peptide bonds (two of the cofactors, Factors V and VIII, are structurally similar; the other, high-molecular-

weight kininogen, is structurally different); (3) inhibitors of proteolytic enzymes; (4) the precursor of the transamidase that, after proteolytic activation, catalyzes the formation of cross-links in the fibrin clot; and (5) fibrinogen and its proteolysis product, fibrin, which is the substance of the blood clot. In addition, fibrin functions almost like a protein cofactor in the activation of the fibrinolytic enzyme precursor, plasminogen, by plasminogen activators of the tissue and urokinase types.

The only substance necessary to form the blood clot is the colorless protein fibrin, but a blood clot formed in a test tube, or a thrombus formed in a vein in which there is stagnant flow, is red because it contains entrapped red blood cells. In a hemostatic plug fibrin fibers are attached to platelets via specific receptors on the platelet surface, forming a mesh which envelopes the platelets and the trapped cells. Fibrin is not present in blood in an appreciable concentration prior to injury and initiation of the reactions of blood clotting; fibrinogen, the precursor of fibrin, is the normally circulating form. The presence of fibrin without an injury could indicate a pathological thrombus formation.

Fibrinogen molecules, because they are large, can be seen by electron microscopy. They consist of three globular domains (Fig. 3), two of which are

A

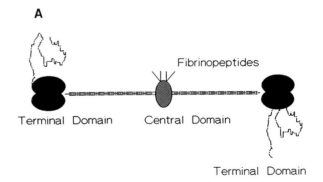

Fibrinopeptides

Terminal Domain   Central Domain

Terminal Domain

B

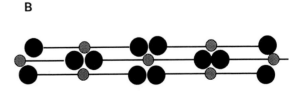

FIGURE 3 (A) Fibrinogen is shown with its central domain, attached fibrinopeptides, and the two terminal domains with the free additional polypeptide chains, which are not folded into the terminal domains. (B) Polymerized fibrin monomers are shown, with central and terminal domains of adjacent molecules making contact. The free polypeptide chain portions of the terminal domains have been omitted.

at the ends of the molecule and are attached to a central domain by long thin protein strands. Thrombin action on fibrinogen to convert it to fibrin removes four short polypeptides from the central domain. This small change, however, exposes sites that enables noncovalent association of individual fibrin monomers, leading to the formation of macroscopic fibrin strands. By electron microscopy the fibrin stands appear to be similar to a frayed rope, with small fibrils extending from the larger fibers.

A fibrin clot formed only by noncovalent association of fibrin monomers is not stable in flowing blood. The dilution caused by the blood flow permits fibrin monomer molecules to dissociate from the polymer; thus, the gel dissolves and bleeding recurs. Such dissolution is prevented by the plasma transamidase, Factor XIIIa, which catalyzes the formation of covalent cross-links between fibrin monomer molecules within the polymer. Factor XIII is a necessary component in the blood clotting system, and when it is defective or deficient, poor wound healing may occur. The most dramatic example of the importance of Factor XIII is that of a woman who suffered many spontaneously aborted pregnancies prior to the diagnosis of Factor XIII deficiency. After transfusion to replenish the Factor XIII, she delivered a normal infant without significant complications.

B. Activation of the Circulating Proenzymes

1. Initiation of Blood Clotting and the Extrinsic Activator of Factor X

The blood clotting factors circulate as inactive precursors, unless injury to the blood vessel occurs. After injury clotting factors at the injury site are converted to their functionally active forms by proteolysis at specific amino acid residues. The single exception to this generalization is Factor VII, which expresses proteolytic activity without proteolytic cleavage. However, its efficiency as a proteolytic enzyme is extremely low and becomes physiologically significant only when it is associated with its protein cofactor, tissue factor.

It seems likely—although it is not proven—that the exposure of tissue factor, an integral membrane protein, as a result of endothelial injury initiates the extrinsic pathway of blood clotting. The exposed tissue factor then binds Factor VII in the presence of membrane lipids with negatively charged functional groups, which are exposed to the blood as a result of the cell rupture. The binding of Factor VII

to the lipid molecules requires calcium ions; this is one of the ways in which calcium is involved in blood clotting. Factor VII is over 100,000 times more efficient as a proteinase in the complex with tissue factor than in its absence; its proteolytic action is therefore restricted to the site at which tissue factor is exposed. The complexed Factor VII catalyzes the cleavage of a single peptide bond in Factor X, transforming it from proenzyme to active proteolytic enzyme (called Factor Xa). In turn, Factor VII can be proteolytically activated by Factor Xa, a reaction that converts it to an even more active proteinase. However, this feedback reaction, which further enhances the activation of Factor X, is not obligatory. Physiologically, this described mechanism of activity enhancement links the initiation of clotting and subsequent reactions to the site of injury. Factor VII also activates Factor IX.

An *in vitro* test of the function of the clotting system uses a mixture of tissue factor and lipid vesicles in the form of an extract of the lung or the brain, which is referred to as thromboplastin. Because this mixture contains materials not normally present in the circulating blood, the resulting clotting reactions are described as occurring by the extrinsic pathway of blood clotting.

2. Prothrombinase and the Activation of Prothrombin

The prototype activator complex of the coagulation system is prothrombinase (Fig. 4). This complex contains the substrate prothrombin, the proteinase Factor Xa, the activated protein cofactor Factor Va, the requisite lipids from cell or platelet membranes, and calcium ions. Activation of prothrombin to form thrombin can be catalyzed by Fac-

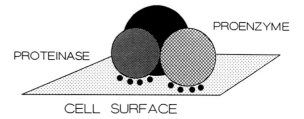

FIGURE 4 The association of Factor Va (the activated protein cofactor), Factor Xa (the proteinase), and prothrombin (the proenzyme substrate) and calcium ions to create the prothrombinase complex on a membrane surface is illustrated. The overlapping circles represent the contacts made among all of the individual components of the complex.

tor Xa alone, but in the absence of Factor Va and a membrane surface, it is exceedingly slow. The complex converts prothrombin to thrombin more than 300,000 times faster than does Factor Xa alone. Expressed in terms of the time required to form a given amount of thrombin, a 300,000-fold increase in the reaction rate is equivalent to reducing the time for thrombin formation from 6 months to a little more than 1 minute.

Binding of prothrombin and Factor Xa to a cell membrane increases the concentrations of these components, thereby increasing the rate of prothrombin activation about 100 times. Such activation of prothrombin in the absence of the activated cofactor is probably sufficient to produce enough thrombin for the activation of Factor V and thus unmask its binding sites for Factor Xa and prothrombin. This enables the formation of the complete complex, which optimally converts prothrombin into thrombin. Because Factors V and Va bind to the cell membrane surface and to prothrombin and Factor Xa, activation of prothrombin, like activation of Factor X by Factor VII, is localized to the injury site. When the membrane surface is provided by platelets stimulated by thrombin, Factor Va may bind to a specific platelet receptor molecule. The physiological importance of complex formation is thus both localization and the rapidity of response.

Prothrombin and Factor Xa binding to Factor Va and to the membrane phospholipids in the presence of calcium ions depends on specialized domains present at the amino-terminal ends of the protein molecules. Structural and functional properties of the calcium-binding domains are related to vitamin K action in blood clotting (see Section II, F,3).

The prothrombinase complex, and similar complexes as well, differ in one important respect from the Factor VII–tissue factor complex. Whereas the latter remains anchored to the cell membrane, the individual components of prothrombinase can spontaneously dissociate, because they are held together by weaker noncovalent interactions. Dissociation of Factor Xa and prothrombin from the complex leaves Factor Va susceptible to inactivation by activated protein C, a component of the anticoagulant pathway. Furthermore, Factor Xa separated from Factor Va and the membrane surface is readily inactivated by the proteinase inhibitor antithrombin III. These inactivation reactions are discussed in Section II,D.

C. Other Activator Complexes of the Clotting Cascade

1. The Intrinsic Activator of Factor X

In addition to being activated by the tissue factor–Factor VII complex, Factor X is activated by a complex activator in which Factor IXa is the proteinase, Factor VIIIa is the activated protein cofactor, and either damaged cells or platelets provide the membrane surface (see Fig. 2). Similar to the activation of prothrombin, the complex activator increases the rate of activation of Factor X by nearly 500,000 times. Factor VIII, the protein cofactor precursor, is activated by thrombin, as in the activation of Factor V. Factor VIII is the missing or defective clotting factor in classical hemophilia, a relatively common life-threatening disease characterized by bleeding into joints. Hemophilia is perhaps best known because of its prevalence within the royal families of Europe. Factor VIII circulates in the blood bound to von Willebrand factor, although von Willebrand factor is not involved in Factor X activation.

Factor IX deficiency gives rise to a bleeding disorder similar to classical hemophilia, not surprisingly, because both of these factors are components of the complex activator of Factor X. Because the genes for both of these clotting factors are on the X chromosome, hemophilias are diseases almost exclusively of males. For historical reasons this activator of Factor X is sometimes referred to as the intrinsic Factor X activator. [*See* HEMOPHILIA, MOLECULAR GENETICS.]

2. The Contact Phase of Blood Clotting: An *in Vitro* Phenomenon

Several reactions which occur when blood clots in a test tube might not be important for *in vivo* hemostasis. These reactions are labeled in Fig. 2 as the contact phase reactions of blood clotting. These apparently nonphysiological reactions take place in contact with the surface of the test tube or, in one of the *in vitro* clotting tests, with kaolin, a claylike powder added to the blood plasma. As in the previous cases, they also occur most efficiently in complexes containing a proteinase and a protein cofactor. The contact phase clotting factors are Factor XII, prekallikrein, and high-molecular-weight kininogen. In marked contrast to the situations which exist with deficiencies of the other clotting factors, individuals deficient in these factors do not have an evident bleeding disorder, implying that these factors are not important for hemostasis.

Another contact phase factor, Factor XI, is, however, important for hemostasis, because deficient individuals suffer from excessive bleeding under some circumstances. The only known activator of this factor is Factor XIIa, suggesting that another factor might remain to be discovered.

D. Regulation of Blood Clotting Reactions

Two distinctly different mechanisms exist for terminating the reactions of the clotting cascade. One is proteolytic inactivation of the protein cofactors Factors Va and VIIIa by activated protein C. As with the activators, activated protein C is only optimally effective when assisted by protein S as a cofactor and in the presence of a membrane surface and calcium ions. The reactions that lead to Factors Va and VIIIa inactivation have been designated the anticoagulant pathway of hemostasis. Protein C is activated by thrombin and, like the other proenzyme activations, requires a protein cofactor, the integral membrane protein thrombomodulin, for an optimal activation rate.

The second mechanism by which clotting proteinase action is stopped is through reaction with irreversible inhibitors, most importantly antithrombin III. This inhibitor reacts with all of the proteinases, except Factors VII and VIIa, to form a covalent complex which renders the proteinase inactive. Reaction of these proteinases is catalyzed by heparin, a sulfated polysaccharide, or glycosaminoglycan. Heparin increases the rate of inactivation of thrombin nearly 1 million times, and that of Factor Xa more than 1000 times. Heparin is found in small amounts on the surface of endothelial cells, where it may have a role in preventing clotting on the normal uninjured vessel lining; it is primarily present in blood only when added as a therapeutic agent to reduce the formation of thrombi. Other inhibitors of clotting proteinases are α_1-proteinase inhibitor, α_2-macroglobulin, Cl inactivator, and heparin cofactor II, but these inhibitors are believed to be less important than antithrombin III. Thrombomodulin on the endothelial cell surface, because of its capacity to bind thrombin, could act as a sink for thrombin prior to inactivation by antithrombin III.

The regulation of thrombin action is critical because of its multiple actions in the hemostatic process. Thrombin produces clotting by its action on fibrinogen and promotes its own formation through activation of Factors V and VIII and through the stimulation of platelets. On the other hand, thrombin tends to stop its own formation by activating Protein C and, as a consequence, inactivating Factors Va and VIIIa. The existence of these antagonistic processes suggests that thrombin must be closely regulated. The optimal hemostatic response must minimize the risk of bleeding when injury occurs, while also minimizing the risk of thrombosis.

E. Fibrinolysis

The fibrin strands which are intertwined within the hemostatic plug are transient structures and are digested by the fibrinolytic enzyme, plasmin. The proenzyme, plasminogen, is activated by proteinases that are released from cells into the circulating blood. Two types of plasminogen activator are known: tissue type and urokinase type. Activation of plasminogen and the action of plasmin on fibrin are facilitated by their binding to the fibrin strands. Fibrin, therefore, acts like a protein cofactor for plasminogen activation. The association of the tissue-type plasminogen activators and plasminogen with fibrin for maximally efficient activation of plasminogen and fibrinolytic activity provides the rationale for the development of recombinant tissue-type plasminogen activators as therapeutic agents for dissolving thrombi in coronary arteries and other blood vessels.

Irreversible inhibitors of plasmin and plasminogen activators present in the blood act to restrict the action of these proteinases to the fibrin clot, similar to the inhibitors of the clotting proteinases. Deficiencies of these inhibitors, α_2-antiplasmin or the plasminogen activator inhibitors, produce excessive fibrinolytic action and result in bleeding analogous to that associated with clotting factor deficiencies.

F. Protein Structural Domains and the Functional Properties of the Coagulation and Fibrinolytic System Proenzymes.

The ability of the clotting proteins to form specific activator complexes with high catalytic efficiencies is a direct consequence of unique structural and functional domains within these molecules. The unique structures of the amino-terminal ends of the proenzymes account for approximately one-half of the total mass. The proteinase domains account for the remaining half. Whereas the proteinase domains are structurally similar, the amino-terminal regions of the individual proenzyme molecules are differ-

A

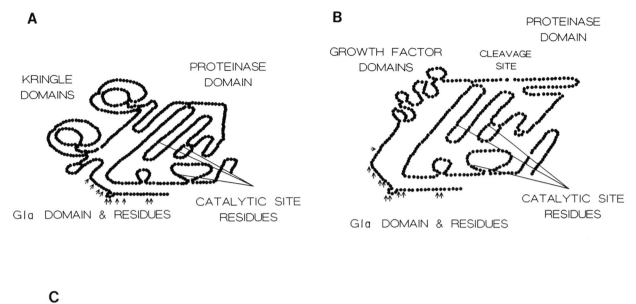

B

C

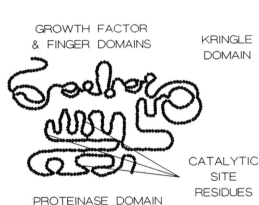

D

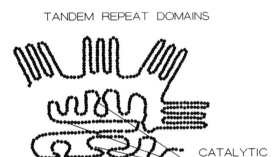

FIGURE 5 Schematic structures for four proenzyme molecules: (A) prothrombin, (B) Factor X, (C) Factor XII, and (D) prekallikrein. The various domain structures, as represented in two dimensions, are shown to illustrate the varieties of structure that make up the proenzyme molecules. Gla, γ-Carboxyglutamic acid. [Modified from E. W. Davie, (1987). *In* ''Hemostasis and Thrombosis: Basic Principles and Clinical Practice'' (R. W. Colman, J. Hirsh, V. J. Marder, and E. W. Salzman, eds.), 2nd ed. pp. 242–267. Lippincott, Philadelphia, Pennsylvania.]

ent. They contain several common structural domains, but in different combinations, which determine their specific functions. Based on their amino acid sequences and other structural similarities, particularly the arrangement of disulfide bridges within the molecules, the schematic structures shown in Fig. 5 can be described as being constructed from six distinct domain structures, discussed below.

1. The Proteinase Domains: The Carboxyl-Terminal Regions

The proteinase domains of the proenzymes, identified by the similarity of their amino acid sequence with those of trypsin and the other pancreatic serine proteinases, possess all of the functional groups necessary for proteolytic activity. Several particular amino acid residues, common to all of the proteinase domains, are responsible for the hydrolysis of the peptide bond. The catalytic triad, or active site residues, consist of one histidine, one serine, and one aspartic acid residue. All of the proteinases preferentially hydrolyze peptide bonds at the carboxyl group of arginyl or lysyl residues. The complementarity between the structures of the proenzyme substrates and the active sites of the individual proteinases is determined by the amino acid residues that precede the arginine or lysine residues in the proenzyme molecules. Specificity of the proteinase with respect to a particular protein substrate, which is exhibited by each of the individual clotting and fibrinolytic proteinases, is determined further by the amino acid residues that either surround or are adjacent to the active site residues in the three-dimensional structure.

In all of the coagulation and fibrinolytic proenzymes, except prothrombin, the proteinase domain is covalently attached to the domains of the amino-terminal region through a single disulfide bond. This simple structural feature maintains all proteinases, except thrombin, covalently bound to their own amino-terminal regions in the activator complex. Thrombin, in contrast, by not being so bound, is able to diffuse to the more than five different proteins of the clotting system on which it acts (see Fig. 2).

2. The Amino-Terminal Regions of the Clotting Proenzymes

In contrast to the great similarities in the structure and function of the proteinase domains, the amino-terminal regions possess a variety of structural domains with specific functions. The domain structure called a kringle (after a form of Danish pretzel) is illustrated in Fig. 5A. This structure is found in at least five of the clotting or fibrinolytic system proenzymes. In three dimensions this domain is folded into a fat disk, or oblate ellipsoid. Prothrombin and the tissue-type plasminogen activator each contain two kringle domains; Factor XII and the urokinase-type plasminogen activator each contain one kringle domain, and plasminogen contains five kringle domains.

Specific functions are attributed to the kringle domains in prothrombin and plasminogen. In prothrombin the first kringle has the γ-carboxyglutamic acid (Gla) domain attached to it (see below). The second kringle domain in prothrombin is necessary for the binding of prothrombin to activated Factor V. Sites in the kringles of plasminogen and plasmin are responsible for binding to fibrin.

3. Gla Domains in the Vitamin K-Dependent Proteins of Clotting

The clotting factor precursors and active proteinases that depend on the action of vitamin K for full activity contain several residues of Gla. The Gla residues are located in a 33- to 45-amino acid domain at the amino-terminal end of the proenzymes. Gla domains are found in prothrombin (Fig. 5A);

Factors VII, IX, and X (Fig. 5B); and proteins C and S. Ten to eleven Gla residues are present in each of these human proteins and occur in three pairs plus four or five separate residues. The presence of Gla residues confers unique calcium ion binding properties on the vitamin K-dependent proteins.

Vitamin K Action and the Formation and Function of Gla Residues The vitamin K-dependent clotting proteins are synthesized on the ribosome, but subsequently undergo modification to add a carboxyl group to the 10 or 11 Glu residues within the Gla domain, changing them to Gla. The addition directly depends on vitamin K. The proteins are synthesized and carboxylated in the liver and then secreted into the blood. Calcium ions bind specifically to the Gla residues, enabling the vitamin K-dependent proteins to bind to the negatively charged lipids that are exposed on cell membranes at the site of injury.

Several drugs (i.e., oral anticoagulants) interfere with the formation of Gla residues in the liver and are used therapeutically to reduce the occurrence of unwanted blood clotting, particularly after heart attacks. This anticoagulation is achieved by eliminating the ability of the vitamin K-dependent proteins to bind to the membranes of platelets and at injury sites. This consequently abolishes the large reaction rate enhancements which normally occur as a result of activator complex formation; and the unbound proteinases are more readily inactivated by antithrombin III.

4. Growth Factor Domains

A domain first identified in polypeptide growth factors is found in several clotting and fibrinolytic system proteins. In the vitamin K-dependent proteins two growth factor domains are found in Factors VII, IX, X (Fig. 5B), and XII (Fig. 5D) and protein C. The tissue-type and urokinase-type plasminogen activators each have a single growth factor domain. A second modified amino acid, β-hydroxyaspartic acid, is found in the growth factor domains and is involved in calcium ion binding distinct from that of the Gla residues. Specific functions for these domains have not been identified.

5. "Finger" Domains

Another structural domain, the finger domain, is found in Factor XII, which contains two such domains, separated by a growth factor domain (Fig.

5C). The second finger domain is followed by a second growth factor domain, which, in turn, is followed by a kringle domain. Factor XII is thus the most varied of the proteins with respect to structural domains within the amino-terminal region.

6. Tandem Repeat Domains

Two of the clotting factors, prekallikrein (Fig. 5D) and Factor XI, possess unique domain structures, which, in two dimensions, resemble tandem repeated amino acid sequences. No specific function has been identified for these domains.

G. Protein Cofactor Structural and Functional Domains

Amino acid sequences have been established from the cDNAs for the protein cofactors Factors V and VIII. Again, as discovered with the proenzymes, the structures of the two proteins show extensive homology. The cofactor activity of Factors Va and VIIIa is derived from portions of the precursor molecules located at the two ends of the precursor polypeptide chains. A calcium or other divalent ion links the two polypeptide chains of the activated protein cofactors. A large internal region of the protein cofactor molecules has no known function in the clotting reactions. The carboxyl-terminal polypeptides of the activated cofactors are involved in their binding to membranes and, in Factor VIIIa, to von Willebrand factor.

III. Summary and Conclusions

Although the hemostatic system is composed of many components and reactions, the structural and functional similarities among the members of a few categories make it possible, although not simple, to understand how this system works. Similar structures express similar functions. Interactions among the various components form activator complexes. Extremely large increases in activation reaction rates occur in the activation complexes, which normally form only at the site of an injury, where blood loss is probable. Irreversible inhibitors inactivate the proteinases after completion of the hemostatic response, thus ensuring the fluidity of the blood and its ability to transport nutrients and other substances throughout the body.

Bibliography

Bloom, A. L., and Thomas, D. P. (eds.) (1987). "Haemostasis and Thrombosis," 2nd ed. Churchill-Livingstone, New York.

Colman, R. W., Hirsh, J., Marder, V. J., and Salzman, E. W. (eds.) (1987). "Hemostasis and Thrombosis: Basic Principles and Clinical Practice," 2nd ed. Lippincott, Philadelphia, Pennsylvania.

Jackson, C. M., and Nemerson, Y. (1980). Blood coagulation. *Annu. Rev. Biochem.* **49,** 765.

Suttie, J. W., and Jackson, C. M. (1977). Prothrombin structure, activation and biosynthesis. *Physiol. Rev.* **57,** 1.

Herpesviruses

BERNARD ROIZMAN, *University of Chicago*

HERPESVIRUSES share two fundamental properties that differentiate them from other viruses. All herpesvirus particles are architecturally identical even though they may differ considerably with respect to their biological properties and the size, structure and composition of their genomes. In addition, all herpesviruses investigated to date are able to remain latent, i.e., in a nonmultiplying form, in their natural hosts.

Most of the herpesviruses infecting humans and the diseases they cause have been known for many years. Herpes simplex viruses (HSV) were first isolated during the second decade of this century, but only in 1962 did it become known that there are two distinct serotypes, type 1 (HSV-1) and type 2 (HSV-2). The latest human herpesvirus was discovered in 1989. Recent interest in the biology of these agents was spurred by a general resurgence of interest in sexually transmitted agents.

A. Definition and Taxonomy of Herpesviruses

Herpesviruses are recognized on the basis of the structure of their virus particle, the *virion* (Fig. 1). From the center out, the typical herpesvirus consists of a *core* consisting of the viral chromosome coiled in the form of a ring (toroid); a shell of proteins called the *capsid* and consisting of 162 protein complexes or *capsomeres;* a variable, asymmetric layer of proteins designated as the *tegument;* and a membrane called the *envelope* to differentiate it from cellular membranes from which it is derived (Fig. 1). The infectious particle contains approximately 30–40 species of viral proteins.

The herpesviruses constitute the family Herpesviridae, which comprises more than 100 different species. Members of this family have been isolated from oysters, turtles, fish, marsupials, and just about every mammal in which they were sought. Although in the past they have been named after the disease they produce (e.g., varicella zoster virus, VZV) or their discoverers (e.g., Epstein-Barr virus, EBV), the rules adopted in 1981 require that with few exceptions, herpesviruses must be named after the family of the host in ascending numerical order (e.g., canid herpesvirus No. 1, the only canine herpesvirus known, and human herpesvirus No. 7 (HHV7), the latest herpesvirus isolated from humans).

The herpesviruses differ from each other considerably with respect to several characteristics (i.e., (1) the animal host range, (2) the cell host range, (3) the rate of multiplication, (4) the tissue in which the virus remains latent in its native or experimental host, and (5) the size, average base composition, and sequence arrangement of their genomes. Herpesviruses have been provisionally classified into three subfamilies on the basis of their biological properties, largely because these are easily determined. Thus, *alphaherpesvirinae,* exemplified by HSV-1 and HSV-2 and VZV, usually multiply in a variety of cells from different species and tissues, spread rapidly from cell to cell in culture, and are associated with neuronal tissues during latency. The major characteristics of *betaherpesvirinae,* exemplified by human cytomegalovirus (CMV), are a restricted host range and a sluggish spread from cell to cell. The *gammaherpesvirinae* (e.g., EBV) grow in B or T lymphocytes and remain latent in these

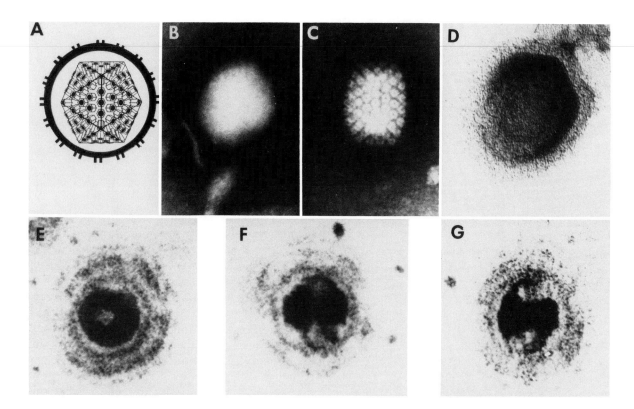

cells. Unambiguous differentiation of herpesviruses into subfamilies based on gene arrangement and conservation, and specific characteristics of nucleotide sequences of their genes is likely to supplant current criteria for classification.

B. Enumeration of Herpesviruses Infecting Humans

Humans become infected with eight known herpesviruses. Of this number, seven are human pathogens transmitted from person to person mostly by physical contact. These viruses and their taxonomic designations are HSV-1 and HSV-2 (human herpesviruses No. 1 and No. 2), VZV (human herpesvirus No. 3), human CMV (human herpesvirus No. 4), EBV (human herpesvirus No. 5), and HHV6.

Another herpesvirus that occasionally finds its way into the human host is the simian B virus. Although its biology in its native hosts (i.e., the old world rhesus and maccacus monkeys) is similar to that of HSV-1 in humans, transmission of the virus to humans by bites can cause severe disease and death. Herpesviruses infecting domestic animals do not cross into humans. Conversely, human herpesviruses are not as a rule transmitted to domestic

FIGURE 1 Structure of herpesviruses. (A) Schematic representation of a herpes virion. The outer ring represents the viral envelope with glycoproteins (spikes) projecting from its surface. The shell, capsid, has a diameter of 105 nm and the shape of an icosadeltahedron consisting of 150 hexameric and 12 pentameric protein subunits (capsomeres). A hole runs through each capsomere along its long axis. (B) and (C) Electron micrographs of an intact virion and capsid, respectively, stained with an electron dense stain. The envelope prevents stain from penetrating the virion, and therefore the virion is electron translucent. The electron dense stain between the capsomeres and in the holes of the capsomeres outlines the structure of the capsid. (D) Stained capsid showing the outlines of the DNA coiled in its interior. DNA is coiled in the form of a toroid seen as a ring when viewed from the top (E) or bar when viewed from the side (F) and (G). The stain used on preparations shown in E, F, and G reacted primarily with the DNA and with a lower intensity with proteins in the capsid and envelope.

animals although rare transmission of HSV-2 to horses can cause severe disease.

I. Biology of Lytic and Latent Human Herpesvirus Infections

A. Definition of Lytic and Latent Infections

Herpesviruses interact with their hosts in two different ways. In *lytic* infections, the herpesviruses multiply in *permissive* cells and these cells are destroyed. The diseases caused by herpesviruses are the direct consequences of cell destruction that obligatorily accompanies virus multiplication, although the immune response to viral multiplication may also contribute to tissue destruction. In *latent* infections the viral genome is retained in an *episomal state* in the nucleus of the cell harboring the virus. The virus may express a small fraction of its genome designed primarily to keep the cell alive with the viral genome contained in that cell, but virus multiplication and the ensuing destruction of the infected cell does not take place. The virus, however, may become activated, and, as a result, multiply and destroy the cell. [*See* VIROLOGY, MEDICAL.]

B. Infection and Interaction of Herpesviruses with Their Human Hosts

1. Herpes Simplex Virus 1 and 2

These two viruses are transmitted by contact of healthy mucosal cells with infected saliva, genital secretions, or tissues from an individual with virus-containing lesions. HSV-1 is more commonly transmitted by oral contact, whereas HSV-2 is more commonly acquired by sexual contact. Both viruses are readily transmitted to the newborn during birth by contact with infected tissues. Individuals exposed for the first time to either HSV-1 or HSV-2 may experience a mild to severe infection depending on their immune state: The infection is much more severe in the newborn and in immunologically compromised individuals than in healthy adults. The virus multiplies in cells at the portal of entry and spreads through nerve endings to the sensory nerves innervating that site. Once in the nerve, the virus is transported through the axon to the nucleus of the nerve cell, where it remains latent, sometimes for the life of the host. While in the latent state, the virus is shielded from the immune system, and as long as it expresses a small number or no viral gene products, it is not affected by drugs that can suppress viral multiplication during lytic infection. In a fraction of individuals harboring the virus in a latent state, the virus is periodically activated, multiplies, and is then transported through the axon to a site at or near the portal of entry, most frequently to the mucocutaneous junction of the mouth (i.e., fever blister at the junction of the lips and skin), the cornea, or genital organs. The reacti-

vated virus may spread to cells innervated by the infected nerve and cause a blister, which ultimately is cleared by the immune system. The molecular mechanisms for activation of viral multiplication are not known. However, termination of the latent state and virus multiplication follow trauma (e.g., ultraviolet light damage, burns) of tissues innervated by sensory nerves harboring latent virus, emotional stress, menstruation, intake of hormones, etc. Most initial infections are self-limited, but rare cases, especially in neonates, result in encephalitis, which can be fatal. Encephalitis in adults is rare and usually the result of reactivated latent virus.

2. Varicella-Zoster Virus

Varicella-zoster virus causes chicken pox, a childhood disease rapidly transmitted from virus shedders to nonimmune contacts. In the nonimmune individual, the infecting virus multiplies at the portal of entry (lung and pharynx) and is disseminated by white blood cells to the skin where it multiplies and causes erythematous skin vesicles characteristic of chicken pox. In a latent form, the virus remains harbored in sensory ganglia. Activation of the virus, which occurs usually in adults and only once, results in shingles, a painful disease resulting from the radial spread of infection to cells and tissues innervated by a ganglion harboring the virus. In immunocompromised individuals, primary infection or reactivation of the virus may result in extensive, life-threatening lesions, which are difficult to treat. As in the case of HSV-1 and HSV-2 infections, the latent virus is shielded from the immune response of the host and plays a key role in the maintenance of the virus in the human population. The child recovering from chicken pox becomes the reservoir of the virus and maintains it in the largely immune population. This virus becomes available for transmission to the children of the next generation when it is reactivated in the adult shingles patient.

3. Human Cytomegalovirus Virus

Human cytomegalovirus rarely causes significant disease in healthy, immunologically competent adults although virus can remain in latent form. Severe and often fatal disease can result from reactivation of latent virus or first infection in severely immunocompromised adults. Infection of a pregnant woman can result in transplacental transmission of the virus and moderate to extensive tissue damage to the fetus evident post partum in the form of organ malformations, mental retardation, etc. Newborns infected during birth as well as infants infected during gestation may secrete virus in urine and saliva for long periods of time and show moderate to severe sequelae of disease including impairment of hearing, etc. Not much is known of the mechanisms of latency, reactivation of virus multiplication or of the cells harboring the virus.

4. Epstein–Barr Virus

Epstein-Barr virus is associated with a number of clinical syndromes. In Europe and North America it is the causative agent of infectious mononucleosis; a self-limiting polyclonal lymphoproliferative disease of young adults resulting from transmission of the virus contained in infected saliva by kissing. The virus multiplies in nasopharyngeal tissues; it is readily isolated from throat washings and from B lymphocytes circulating in the blood. The immune system of most healthy individuals can effectively cope with infectious mononucleosis. The virus persists, however, in a latent state in B lymphocytes. In many communities infection occurs at an early age, and except in immunocompromised individuals, the infection can be inapparent. Some children are genetically unable to respond to the EBV infections and develop a fatal polyclonal lymphoma after infection. The presence of latent EBV is a common feature of a monoclonal B cell lymphoma described first by Burkitt and occurring in some areas of Africa and New Guinea known for a high rate of infection with the malarial parasite. The B cell lymphoma is most likely a sequel of chromosomal translocation, and the role of EBV is not clear. EBV is also a marker of cancer cells constituting a nasopharyngeal carcinoma prevalent in some genetically predisposed Chinese populations in South East Asia. Although the presence of secretory antibody to EBV structural proteins is an important marker of nasopharyngeal carcinoma, the role of the virus in the etiology of the cancer is not known. [See EPSTEIN-BARR VIRUS.]

5. Human Herpesvirus No. 6 and 7

Human herpesvirus No. 6 virus was initially isolated from lymphocytes of healthy adults as well as from individuals infected with the human immunodeficiency virus 1. The virus grows best in T lymphocytes. Studies on the prevalence and distribution of antibodies to the virus in the human population indicate that infection with this virus oc-

curs early in life and is widespread. HHV6 appears to be the causative agents of roseola, a childhood exanthem that occurs in afebrile children after a short febrile episode. Like other herpesviruses, HHV6 likely remains latent in the human host. The tissue in which it remains latent, the mode of transmission, and the long-term effects of infection with this virus are not yet known. HHV7 is distantly related to HHV6 and like HHV6 it multiplies in T lymphocytes. HHV7 was discovered in 1989 and little is known of its biology. [See LYMPHOCYTES.]

II. Molecular Biology of Herpesvirus

A. Structure and Sequence Arrangements of Viral Genomes

The herpesvirus genomes vary in length, from approximately 120,000 to approximately 240,000 base pairs, and in base composition, from 32 to 76 moles of guanine + cytosine per cent. The variability in both size and base composition is extraordinary; it reflects the evolutionary divergence of viruses well integrated in their hosts and evolving with them through a large portion of the evolution of life on Earth. With the exception of HSV-1 and HSV-2, the human herpesviruses identified to date are no more related to each other than they are to viruses isolated from other species. Thus HSV and EBV are each much more closely related to other herpesviruses isolated from lower animals than they are to each other. These observations suggest that each virus evolved a niche in its human host independently of the other. Notwithstanding the difference in the sequence arrangements and base composition of their genomes, a significant fraction of genes are conserved. Structurally, the virus particles that harbor and protect the genomes of different viruses cannot be differentiated from each other visually although they differ significantly with respect to immunological specificity, the exact size and number of viral proteins.

The herpesvirus genomes studied most extensively are those of the HSV, EBV, CMV, and VZV. The significant features of the genomes are as follows:

1. In the virus particle, the genomes are made up of linear double-stranded DNA. A single nucleotide protrudes at each end (3′ extension).
2. In all viral genomes, *reiterated DNA sequences* are interspersed among unique, nonreiterated se-

quences. Large repeats divide the herpesvirus genomes into major domains. In the case of HSV genomes, it is convenient to describe the genome as consisting of two covalently linked components designated as large (L) and small (S). Each component consists of unique sequences flanked by inverted repeats. The inverted repeats of the L component are approximately 9 Kbp long and can be further subdivided into the terminal *a* sequence and the internal *b* sequence; the sequence *ab* and its inverted repeat, *b'a'*, flank a long stretch of quasi unique sequence designated as U_l. The inverted repeats of the S component, each 6.6 Kbp long and designated as *a'c'* and *ca*, flank the quasi unique sequence of this component designated as U_s. The size of terminal *a* sequence varies from approximately 200 to 500 bp depending on strain and is the only shared sequence by the two components. The *a* sequences at the termini of the L component can be present in several copies that are directly repeated and adjacent to each other.

The two components L and S can invert relative to each other. The frequency of inversion is so high that viral DNA extracted from viruses produced in a few thousand cells and representing the progeny of a single infectious virus particle consists of *four equimolar populations* of viral genomes differing solely in the relative orientation of the L and S components. HSV-1 and HSV-2 genomes are identical in size and colinear with respect to the arrangement of their genes. The CMV genome (225 Kbp) is significantly larger than the genomes of the other herpesviruses, but the arrangement of its reiterated sequences is similar to those of HSV-1 and HSV-2 genomes. The genomes of VZV and of other herpesvirus differ from those described above in that only one set of unique sequences (U_s) is flanked by inverted repeats and only that unique stretch of DNA inverts relative to the other, longer (U_l) stretch of unique sequences. DNA extracted from these viruses or from infected cells consists of only two isomeric populations. The EBV genome is representative of still another group of herpesviruses. Some representatives of this group contain a sequence at the termini, which is repeated in tandem (directly) many times. Other, internal stretches of viral DNA may also be repeated in tandem numerous times. In the absence of large stretches of sequences that are repeated in an inverted orientation, such as are present in HSV-1 or HSV-2, the rearrangement of viral genomes does not occur.

3. A significant feature of the herpesvirus genomes is their *polymorphism*. The variability in the number of repeats of a particular reiterated sequence probably reflects both intragenomic and intergenomic recombinational events. They are of lesser interest than polymorphism in the *restriction endonuclease cleavage sites*. The observation that restriction endonuclease cleavage sites are stable and do not change on passage in cell culture and from one individual to his or her contact has established the *restriction endonuclease fingerprinting* as an important tool for tracing the transmission of virus from person to person. The significant conclusion drawn from the application of this epidemiologic tool is that the increase in the rate of genital herpesvirus infections seen in the past two decades reflects a "frenzy of transmission" (i.e., an increase in the number of sexual contacts between infected and uninfected individuals rather than the consequence of an epidemic caused by a new virus sweeping the continent). [*See* POLYMORPHISM, GENES.]

B. The Herpes Virion: Architecture and Composition

The herpesvirus envelope is derived from cellular membranes but contains viral rather than cellular proteins. The number of *viral proteins* varies. HSV-1 and HSV-2 contain seven proteins on the surface (i.e., glycoproteins B, C, D, E, G, H, I) and an unknown number of proteins on the underside of the envelope. The glycoproteins D, B, and H are the only ones shown to be essential for infection and multiplication in cells in culture. Of these proteins, B and H appear to be conserved among several herpesviruses. The capsid contains approximately 10 proteins. Another 10–15 proteins are located in the tegument, between the envelope and the capsid.

III. Reproductive Cycle in Productive Infection of Permissive Cells

A. The Reproductive Cycle

The sequence of events in the replication of HSV-1 and HSV-2 is depicted in Fig. 2, and with minor variations it is applicable to all herpesviruses. The key feature of the herpesvirus reproductive cycle is a highly ordered, tightly regulated sequence of viral gene expression that lasts 18 (HSV) to 72 (CMV) hr. The process may be summarized as follows:

1. Infection is initiated by the attachment of viral glycoproteins (most likely B or C) to as yet unidentified cell surface proteins. Viral proteins (most likely glycoprotein D with B and H) fuse the envelope with the plasma membrane. The capsids are then transported along the cytoskeleton to nuclear pores. Factors at the environment of the nuclear pore act on viral proteins and trigger the release of viral DNA into the nucleus. The DNA circularizes immediately on entry into the

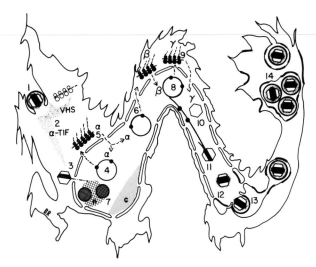

FIGURE 2 Schematic representation of replication of herpes simplex viruses in susceptible cells. (1) The virus initiates infection by the fusion of the viral envelope with the plasma membrane after attachment to the cell surface. (2) Fusion of the membranes releases two proteins from the virion. *VHS* shuts off protein synthesis (broken RNA in open polyribosomes). *α-TIF* (the α gene *trans*-inducing factor) is transported to the nucleus. (3) The capsid is transported to the nuclear pore where viral DNA is released into the nucleus and immediately circularizes. (4) The transcription of α genes by cellular enzymes is induced by α-TIF. (5) The 5 αmRNAs are transported into the cytoplasm and translated (filled polyribosome); proteins are transported into the nucleus. (6) A new round of transcription results in the synthesis of β proteins. (7) At this stage in infection the chromatin (c) is degraded and displaced toward the nuclear membrane whereas the nucleoli (round hatched structures) become disaggregated. (8) Viral DNA is replicated by a rolling circle mechanism, which yields heat-to-tail concatemers of unit length viral DNA. (9) A new round of transcription/translation yields γ proteins consisting primarily of structural proteins of the virus. (10) Capsid proteins form empty capsids. (11) Unit length viral DNA is cleaved from concatemers and packaged into preformed capsids. (12) Capsids containing viral DNA acquire a new protein. (13) Viral glycoproteins and tegument proteins accumulate and form patches in cellular membranes. Capsids containing DNA and the additional protein attach to the underside of membrane patches containing viral proteins and are enveloped. (14) Enveloped proteins accumulate in the endoplasmic reticulum and are transported into extracellular space.

nucleus. [*See* NUCLEAR PORE, STRUCTURE AND FUNCTION.]

2. Fusion of the envelope with the plasma membrane results in the release of several viral proteins along with the capsid. These include a protein that turns off the synthesis of host proteins and DNA (virion host shut off, VHS) and a protein (α *trans* induction factor, αTIF) that is transported into the nucleus and *trans*-activates the transcription of the 5-α genes, the first set

to be expressed after infection. *Trans*-activation results from the concurrent binding of αTIF and a host transcriptional factor (variously designated as OTF-1, αH1, and NFIII) to a specific site on the viral DNA. [*See* DNA AND GENE TRANSCRIPTION.]

3. The 5 α proteins perform regulatory functions. One, infected cell protein No. 4 (ICP4), is required for the *trans*-activation of all genes expressed later in infection. It directly binds to viral DNA at specific sites in both 5' nontranscribed and transcribed noncoding domains. This protein also autoregulates the transcription of its own gene and that of another *trans*-activator (the α gene specifying ICP0). Two other α genes, α22 and α27, regulate the expression of genes transcribed late in infection. The first set of genes induced by the α proteins designated as β specify proteins that replicate viral DNA by a rolling circle mechanism. β proteins include a DNA polymerase, primase, helicase, topoisomerase, thymidine kinase, DNase, dUTPase, ribonucleotide reductase, and both single-strand and double-strand specific DNA binding proteins.

4. Most virion structural proteins are specified by the late or γ genes, a large, heterogeneous group whose expression requires functional α proteins and viral DNA synthesis. The γ proteins are transported into appropriate compartments. Thus, the capsid proteins assemble into empty capsids in the nucleus whereas some tegument and envelope proteins form patches in membranes of the infected cells. Viral DNA is cleaved from concatemers at the *a* sequence and packaged into the preformed capsids. In HSV-infected cells the capsids containing viral DNA have been shown to bind an additional protein that enables the capsid to bind to the patches containing the tegument and envelope proteins. After envelopment at the inner lamellae of the nuclear membrane, viral particles are transported in the endoplasmic reticulum through the Golgi to the extracellular space. In cultured cells, a large number of particles may become de-enveloped and some may be enveloped *de novo* in the membranes of the endoplasmic reticulum. Recent studies suggest that glycoprotein D made in the infected cells blocks reinfection of the cells with newly released virus.

5. In the case of the HSV-1 genome, at least 17 genes can be deleted without affecting the ability of the virus to grow in cells in culture. These genes include α0, α22, and α47, the genes specifying thymidine kinase, dUTPase, ribonucleotide reductase, and glycoproteins C, E, G, and I. It is highly unlikely that all these viral genes are truly redundant. At least some of the deleted genes are dispensable only because cells in culture express cellular functions that complement the deletion mutants and allow them to multiply. Other viral genes may be required for alternative modes of entry or replication in cells that lack certain host factors.

B. Fate of the Infected Cells

Cells productively infected with any of the known herpesviruses invariably die. In addition to the early shutoff of host macromolecular metabolism characteristic of some herpesviruses, the infected cells usually exhibit other important changes (e.g., irreversible disaggregation of the nucleoli, margination and pulverization of chromatin, and major alterations in the structure of cellular membranes). Much of the pathology associated with herpesvirus infections is caused by destruction of the productively infected cells.

IV. Latent Infection

The hallmark of herpesviruses is their ability to establish latent infections in a specific population of host cells (i.e., in neurons of sensory or autonomic ganglia in case of HSV-1 and HSV-2, in cells associated with dorsal root ganglia in case of VZV, and at least in B lymphocytes in the case of EBV). In latently infected cells, the viral DNA is maintained in an episomal (closed circular) form, only a fraction of viral genes is expressed, and the function of these genes is to restrict viral gene expression and to maintain both the copy number (~10–200 copies per cell) and the state (episomal, not integrated into host chromosomes) of the viral DNA. Latent infections of cells capable of dividing or with a limited life span (e.g., the latent infection of B lymphocytes with EBV) is different from that of neuronal cells with HSV-1 or HSV-2. The EBV genome expresses several genes in latently infected B lymphocytes. The products of some of these genes act on a specific site on the EBV genome, the *ori P,* to initiate the replication of the DNA to the desired copy level per cell and to maintain the genome in a closed circular form whereas other genes appear to modify the physiology and, possibly, the longevity of the cell. The sensory neurons do not divide, and although the genome copy number appears to be elevated, there is no evidence for the synthesis of viral proteins that perform similar maintenance functions as those seen in EBV-infected B lymphocytes.

The factors that activate the latent virus to multiply appear to vary. Only approximately half of those harboring latent HSV reactivate the virus. Reactivation follows physical or emotional trauma, hormonal imbalance, or injury of tissues innervated by the neurons harboring latent virus. Experimentally, activation has been associated with increased prostaglandin synthesis. Activation may occur with such frequency as to appear to be a chronic smoldering infection or as infrequently as once in a lifetime. Activation of latent VZV (shingles) rarely occurs more than once.

V. Prospects for Control of Human Herpesvirus Infections: Chemotherapy and Prophylaxis

A. Chemotherapy

Effective chemotherapy of herpesvirus infections is based on differences in substrate specificities and structure of viral and cellular enzymes. The most significant drug currently licensed for use against HSV infections, *acyclovir,* a nucleoside that is activated by phosphorylation largely, but not exclusively, by viral thymidine kinase. The triphosphate form is used preferentially by the viral DNA polymerase, and it acts as a DNA chain terminator. Because its phosphorylation in the absence of the viral thymidine kinase is drastically reduced, its toxicity to uninfected cells is low. Its virtue is also its weakness. Acyclovir and drugs that act in a similar fashion are effective only in the presence of enzymes that can phosphorylate it and if they are administered while viral DNA synthesis takes place. They are not effective against latent viruses. Although daily administration blocks the appearance of symptoms of reactivated virus, the long-term effects of intake of drugs, which could also act as human DNA chain terminators of the cellular DNA, remain to be established. Drugs similar to acyclovir are being tested for other human herpesvirus infections. There are no drugs or strategies as yet to rid humans of latent herpesviruses. [*See* CHEMOTHERAPY, ANTIVIRAL AGENTS.]

B. Prophylaxis

It is far more desirable to prevent disease by prior immunization than to treat it with drugs. In practice, human herpesviruses present unique challenges that have not been met as yet. Central to the practical aspects of disease prevention is that each virus has a specific *portal of entry* (e.g., mouth and gastrointestinal tracts for poliovirus, respiratory tract for influenza virus) and a specific *target organ* in which cell destruction caused by virus multiplication is the cause of the illness. In principle, when the target organ and portal of entry differ, the immunity induced by vaccination will be rapidly stimulated by the virus multiplying at the portal of entry, blocking the dissemination of the virus to the target organ. Vaccination does not, as a rule, block the infecting virus from multiplying at the portal of entry. Furthermore, in the case of herpesviruses, immunity to the virus does not affect latent virus or the activation of its multiplication in the neuron although the severity of the lesions and the time it takes for the lesions to heal are determined by the immune system. For HSV, the target organ is the portal of entry, because the virus multiplying at the portal of entry (i.e., cell of the mucous membranes of the mouth or genitals) colonizes the sensory ganglia and is responsible for the recurrent lesions. To block establishment of latency, the immunity must be so high as to preclude or at least reduce viral multiplication at the portal of entry. In essence, the immunity induced by vaccination must be as high and as sustained as that induced by natural infection and maintained by periodic reactivation of latent virus. None of the many anti-HSV vaccines tested to date have met this objective.

The situation appears to be different in the case of VZV. Here the portal of entry and target organ are sufficiently different to render vaccination useful in the case of highly susceptible populations. A vaccine developed by Professor Michiaki Takahashi of Osaka University in Japan has been tested for several years in the United States. There are no vaccines for prevention of human EBV, CMV, or HHV6 infections.

Bibliography

Corey, L., and Spear, P. G. (1986). Infections with herpes simplex viruses. *N. Engl. J. Med.* **314,** 686–691, 749–757.

Gelb, L. D. (1990). Varicell-Zoster virus. *In* "Fields' Virology" 2nd Edition. (B. N. Fields, D. M. Knipe, R. M. Chanock, M. S. Hirsch, J. L. Melnick, T. P. Monath, and B. Roizman, eds.) pp. 2011–2054. Raven Press, New York.

Kieff, E., and Leibowitz, D. Epstein-Barr Virus and its replication. (1990). *In* "Fields' Virology" 2nd Edition. (B. N. Fields, D. M. Knipe, R. M. Chanock, M. S. Hirsch, J. L. Melnick, T. P. Monath, and B. Roizman, eds.) pp. 1889–1920. Raven Press, New York.

Lopez C., and Honess R. W. Human Herpesvirus—6. (1990). *In* "Fields' Virology" 2nd Edition. (B. N. Fields, D. M. Knipe, R. M. Chanock, M. S. Hirsch, J. L. Melnick, T. P. Monath, and B. Roizman, eds.) pp. 2055–2062. Raven Press, New York.

Meignier, B. (1985). Vaccination against herpes simplex virus infections. *In* "The Herpesviruses," vol 4 (B. Roizman and C. Lopez, eds.) pp. 265–296. Plenum Press, New York.

Roizman, B., and Sears, A. E. (1990). Herpes simplex viruses and their replication. *In* "Fields' Virology" 2nd Edition. (B. N. Fields, D. M. Knipe, R. M. Chanock, M. S. Hirsch, J. L. Melnick, T. P. Monath, and B. Roizman, eds.) pp. 1795–1894. Raven Press, New York.

Roizman, B., and Tognon, M. (1983). Restriction endonuclease patterns of herpes simplex virus DNA: Application to diagnosis and molecular epidemiology. *Current Topics Microbiol. Immunol.* **104,** 275–286.

Stinski, M. F. (1990). Cytomegalovirus and its replication. *In* "Fields' Virology" 2nd Edition. (B. N. Fields, D. M. Knipe, R. M. Chanock, M. S. Hirsch, J. L. Melnick, T. P. Monath, and B. Roizman, eds.) pp. 1959–1980. Raven Press, New York.

Whitley, R. J., and Hutto, S. C. (1987). Therapy of viral infections in children. *Adv. Pediatr. Infect. Dis.* **2,** 35–56.

Specialized Reviews

Roizman, B., ed. (1985). "Herpesviruses," vol. 3. Plenum Press, New York.

Roizman, B., and Lopez, C., eds. (1985). "Herpesviruses," vol. 4. Plenum Press, New York.

Lopez, C., and Roizman, B., eds. (1986). "Human Herpesvirus Infections: Pathogenesis, Diagnosis and Treatment." Raven Press, New York.

Higher Brain Function

STEVEN E. PETERSEN, *Washington University School of Medicine*

I. Introduction
II. Fundamental Concepts in the Study of Higher Brain Function
III. Example Studies of Higher Brain Function

Glossary

Cytoarchitecture Arrangement of cells in the brain, used in defining functional areas

Equiluminance When two (or more) visual stimuli have the same perceived brightness but can vary in other characteristics (such as color)

Functional area Brain region that is specialized to perform a set of related information-processing operations

MT Primate brain region that appears to be specialized for the processing of visual motion

Myeloarchitecture Distribution of myelin in the brain, used to define functional areas

Neuron Specialized cells in the nervous system that are electrically excitable and perform information processing in the brain

Parallel distributed processing models Computer simulations that use many simple elements arranged in parallel (working at the same time) to perform complex computations, based on simplified concepts of how the brain processes information

Positron emission tomography (PET) Method for measuring the distribution of low-level positron radiation in tissues, used to measure activity in the brain related to how hard different regions are working

THE DOMAIN OF neuroscience extends from the biological basis of individuality to the molecular biology of nerve cells; higher brain function focuses on a subdiscipline within this larger field. The study of higher brain function in its largest sense is the study of the brain as an information-processing device. Studies of higher brain function generally emphasize aspects of information processing that are neither simple input processing (sensation and sensory transduction) nor output processing (final implementation of movement), and also do not focus on the implementation of organism maintenance such as breathing and thermoregulation. Examples of the information-processing problems that are commonly approached include studies of language, attention, perception, learning, and memory. While higher brain functions have been studied for >100 years, recent technological, methodological, and conceptual advances allow for a useful interface among several disciplines including neurobiology, experimental psychology, and computer science.

I. Introduction

Because studies of higher brain function are at the interface of several disciplines, these studies produce many different kinds of information from studies in another discipline. For example, a study in computer vision, designed to simulate the way that a person recognizes a particular shape, produces very different information from neurobiological studies in which electrical signals are recorded from single nerve cells in the parts of a monkey's brain that are related to vision. Because of these differences, debates are often waged over which type of information is most appropriate for understanding human visual perception. Most profitably, however, the information from these different studies should be viewed cooperatively rather than competitively; bits of information from both computer science and neurobiology can provide clues as to how the brain implements the complex function of visual perception.

To understand how kinds of information could profitably be integrated, perhaps it would be useful to think of how a reverse engineer (i.e., an engineer trying to understand an existing machine's design instead of trying to design a currently nonexistent machine) might approach any information-processing machine (machine X).

The engineer might begin by taking an outside-in approach, observing *what* machine X is capable of doing when given many different types of input, under many different output demands, and with different settings of any controls on the machine. Our engineer might well be able to make preliminary guesses about several ways that machine X might have been designed.

Armed with the knowledge of what machine X's capabilities are, the reverse engineer might take the cover off and attempt to get a general view of the *where* some of the possible design components are. He or she might attempt to trace different input and output lines from the outside-in and to determine how many different subcomponents are used by the machine. By understanding the general physical layout of the machine's components, the engineer can begin to test ideas about how the machine works. Measurements of the electrical activities might be made in different regions of the machine when the machine is performing in different conditions. He or she might remove some components to see how they affect the performance of the machine during particular performance tests.

The reverse engineer might now turn to the question of exactly *how* machine X was designed to perform its various functions. As any engineer knows, many different sets of chips, circuit boards, and physical layouts of other electrical components can be used to realize a similar end product. If our interest is in how machine X works in particular, our reverse engineer must now explore how the design is implemented in the special case of this machine,

looking at exactly what components are used, and in what exact configurations.

A reverse engineering approach can be directed at answering three fundamental questions about a machine. *What* are the performance characteristics of the machine with different inputs, or output demands, or in different states? *Where* in the machine are components located that contribute to the different capabilities of the machine (i.e., what is the general physical layout of the machine)? *How* does the machine implement the performance of its different capabilities in terms of particular circuit designs and problem-solving techniques?

This same reverse engineering approach can be applied to the study of the brain. The many disciplines that study aspects of higher brain function tend to produce information that will give partial answers to one or the other of these questions. From these partial answers, fundamental concepts about how the brain functions as an information processor have emerged. The next section will describe some of these fundamental concepts. In the final section, two very different functions—the processing of visual motion and the processing of visually presented words—will be used as examples of how higher brain function is studied. Hopefully, these examples will show how integrating information from very different approaches can advance our understanding of how the brain supports human performance.

II. Fundamental Concepts in the Study of Higher Brain Function

A. Elementary Mental Operations (What)

When a human performs a task such as answering the question "Is the red object moving to the right?", the task can be thought of as consisting of many component subtasks: understanding the question, looking at the object, appreciating its color, determining its direction of movement, deciding if the object is red, if it is moving to the right, programming a response, and making the response. Each of these component jobs may or may not be broken into further components. Each of the component jobs is a candidate for an elementary mental operation.

An elementary mental operation is difficult to define precisely, but such an operation has several characteristics: like a subroutine in a computer program, it is composed of a set of related computations; it is not easily divided into more component parts; and when one of these operations is performed, information is present in a form that was not available before the process took place.

Many elementary mental operations will be orchestrated in the performance of any task. Different processing operations will be used to deal with the input presented, the instructions or training needed to perform the task, in organizing further processing, and in programming output.

The methods used to determine elementary operations measure the performance characteristics of the brain (what the brain is capable of doing). The studies usually measure reaction time (how long it takes a person to perform a task), percent correct (how often a person can correctly perform a task), and other, more specialized measurements of behavior. The subdisciplines that most often address performance studies include psychophysics and cognitive psychology.

Psychophysics is the study of human performance characteristics when stimuli are manipulated along different physical dimensions such as color, size, or speed of movement in vision, and pitch or loudness in hearing. By carefully observing the responses to different stimuli, insight might be gained into the nature of perception. For example, color psychophysics studies the ability of people to distinguish between different colors, or how bright different colored objects must be before they can be seen. The manipulations of the different stimuli can get quite complicated, so that the effect of color changes in one part of the visual field on the person's judgment of color in another part of the visual field might be tested. [*See* VISION (PSYSIOLOGY).]

Not only the physical nature of stimuli can be manipulated in performance experiments, but other processes such as the ability to remember, to learn, to assign meaning, and to attend to different stimuli when instructed can also be studied. The study of these processes is the turf of the cognitive psychologist.

Section III includes one example from each of the above subdisciplines.

B. Functional Area

Where in the brain are the different elementary mental operations taking place? Perhaps each elementary mental operation is distributed all over the brain, so that studying the physical layout becomes a useless level of analysis. Many lines of evidence argue, however, for functional localization; i.e., different parts of the brain perform different functions.

Strong evidence that different functions are localized in the brain comes from the study of deficits following brain injury. Injuries to the back of the brain produce disturbances in vision. Injuries to other regions of the brain may produce disturbances in touch or hearing, or some more complex function (see Fig. 1).

Given the observation that different locations of brain injury causes different deficits, can different functional units, or functional areas, of the brain be described in more precise terms? A brief history about this question may prove illuminating.

In animals, the precise anatomy and physiology of the brain can be studied with a variety of techniques that have shaped the understanding of functional areas. In the nineteenth century, it was observed that movements could be elicited by electrically stimulating a region of a dog's cerebral cortex. If movements of the paw were elicited at one site, then an adjacent site of stimulation would produce movements in the same leg. It was as if a little "map" of one-half of the dog's body was on this small piece of cortex. The left half of the dog was mapped to the right hemisphere, and the right half of the dog was mapped to the left hemisphere.

Around the same time, people were beginning to look at the microscopic anatomy of slices of the brain in several species, including humans. Six layers of cells were observed in mammalian cerebral

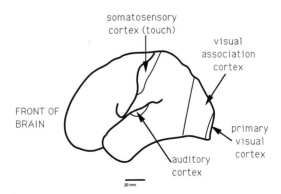

FIGURE 1 A diagrammatic, lateral (side) view of a human brain showing regions related to touch, vision, and hearing. Extrastriate or nonprimary visual areas such as MT (see text) are located in visual association cortex.

cortex, and the thicknesses of the layers, the types of cells in the layers, and the arrangement of cells within the layers were not uniform across the cortex but seemed to organize into a series of distinct zones. Agreement as to the number of zones or to their exact boundaries was unclear, but the presence of regional differences in what was called the architectonics of the brain was well accepted. [*See* CORTEX.]

In the twentieth century, these two observations were pursued. Topographic mapping of the cortex, not just for electrically elicited movements but for responses to different kinds of sensory stimulation, was done for several different species of animals. For example, microscopically fine electrodes could be placed in cortex of an anesthetized animal, and different kinds of visual stimuli could be presented in different parts of the visual field, or different types of sounds could be played to the ears. When this was done, for each of the sensory modalities, as well as for movements, there was not just one map, but several separate and complete maps. For the most completely studied system, the visual system, it was shown that for cats, several species of monkeys, and presumably for humans as well, that there are some two dozen or more visually responsive 'maps'.

The twentieth century also saw much progress in anatomical techniques. New staining techniques made archictectonic regional analysis clearer. Several techniques were also developed that allowed the tracing of connections between these different zones, both in the forward direction (so that all of the regions to which one region sends messages could be labeled) and in the backward direction (so that all of the regions from which a region receives messages could be labeled).

Finally, in the second half of the twentieth century, the exact responses of the cells in these different areas began to be studied. As mentioned earlier, while there were several electrically defined maps of the visual field, the cells in these different regions do not all seem to be doing the same type of processing. One visual region has cells that signal the direction and speed of moving visual stimuli, and other regions carry more information about the color and shape of visual stimuli. These analyses of the information carried by different cells are studies of the functional properties of neurons. [*See* VISUAL SYSTEM.]

As the studies of these different methodologies progressed, particularly in the visual system, the different types of studies began to agree with one another. For example, when the borders of an electrically recorded map of the visual field was compared with the cytoarchitectonic or myeloarchitectonic borders, there was agreement. An architectonic area seemed also to have a coherent set of connections with other areas (i.e., in most cases, if one part of area A connects with areas B and C, but not with area D, then all parts of area A will connect with areas B and C, and not D).

Because the distinctions between different areas on any one measure can be very subtle, the current way of defining functional areas in animals is to find agreement between these different methodologies:

1. distribution of functional properties of cells
2. distributions of anatomical cell and fiber distributions
3. uniformity of connections with other areas
4. presence of an electrically recorded (or electrically stimulated) topographic map

Only a small number of areas to date have been studied with all four methodologies, but those that have produce very encouraging results; the study of visual motion processing area MT, presented later, is an example of such a study.

While the use of these four criteria is not possible for directly studying human functional areas, the animal studies provide a firm conceptual basis for understanding functional localization in humans.

C. Neural Implementation

A third concept important in understanding the brain as an information processor is in the way a functional area actually might implement the elementary operation, i.e., the way in which the processing is specified on the hardware of the brain. At this point, there is no complete description of this level for any elementary operation, but there is information about how the nervous system carries and encodes information and, in some specific cases, how processing is carried out in quite specific ways at the level of single nerve cells.

To understand the idea of a neural implementation, it is necessary to know something about how nerve cells work. Neurons (nerve cells) are usually made up of a cell body, dendrites, and an axon. The cell body, which contains the nucleus, performs many of the same functions that all other types of cells must perform in terms of cell maintenance, etc. Mostly, the axon and dendrites distinguish neurons from many other cells. Neurons are electrically excitable cells and can produce spikes of electrical activity. These spikes often travel along the length of the axon, a long hairlike process that extends away from the cell body. The electrical message usually travels from the cell body along the axon, where it makes connections to other cells. These connections, or synapses, are often made on the dendrites of the receiving cell, the dendrites being a network of processes extending from the cell body. The connections between an axon and its receiving cell are not electrical, however, but chemical. When the electrical message comes to the end of the axon, it induces the release of a minute amount of a chemical (called a neurotransmitter) that moves across the synapse to find a home (receptor) on the receiving cell. This in turn induces electrical changes in the receiving cell. These changes can either encourage the cell to send an electrical message further on in the system, in which case we call the synapse an excitatory synapse, or they can discourage a message, in which case we call the connection inhibitory. Each cell of the brain receives messages from many others and must total up all of its inputs in deciding whether or not to send a message in terms of electrical spikes, and how many spikes to send. [*See* NEUROTRANSMITTER AND NEUROPEPTIDE RECEPTORS IN THE BRAIN.]

The electrochemical nature of the synapse makes the transmission of information from one neuron to the next slower than if information was transmitted by direct electrical connections. The time course of these steps in information processing contribute to our ability to measure reaction time differences in different tasks. If each step of a task took only microseconds (instead of the milliseconds that it normally takes), our reaction time in any task would be so quick that measuring differences would be very unlikely. Fortunately, for generations of psychophysicists and cognitive psychologists, this is not the case.

Most often, studies of neural implementation take advantage of the spiking nature of the electrical message that neurons use to communicate with one another. Neurobiologists can measure the spikes that a cell is producing by placing microscopically fine wires (electrodes) next to a cell body. The functional properties of a cell are most often assessed by comparing the number of spikes in one condition with the spikes produced under another set of conditions. If the number of spikes is affected by some

manipulation, the cell is assumed to be coding information important to that manipulation. In the example presented in the next section, this idea is used to assess the information encoded in the directionally selective cells of visual area MT.

A second approach to the study of neural implementation is to take what is known about how neural systems are organized and to make computer simulation using, in essence, very simplified model versions of little pieces of the brain. Sometimes computers are used to simulate human performance, and different types of simulations include control systems, or connectionist (parallel distributed processing [PDP]) models. The hope is that such simplifying models will help give insight into how the complicated system of the brain might be organized. The most explicitly neural of these types of models have been called neural network, connectionist, or PDP models.

Both of these approaches to neural implementation, direct measures of the information-processing capacities of neurons, and computer simulation are present in the following examples.

III. Example Studies of Higher Brain Function

The way in which information about elementary operations, functional areas, and neural implementation can help in understanding higher brain function can be seen in a very simple framework. Any task performance consists of the orchestration of several elementary operations. Each of these operations might be localized in a separate functional area of the brain. Within that area, the computations are implemented through the specific actions of single neurons, which carry and transform the information necessary to compute the elementary operation. The examples below are intended to clarify these ideas.

A. Visual Motion Processing

In the early 1970s, several workers described a region in several different primate species that had a high concentration of cells that coded for the direction of movement of a visual stimulus. Remember that the way neuronal coding is assessed is by the frequency of spikes they make. In the case of direction-selective cells, many more spikes are counted when an object is moved in one direction (e.g., upward) than when it is moved in the opposite direction (downward). This area is called MT.

MT has been the focus of studies utilizing several criteria for the definition of a functional area. Area MT had a distinctive architectonic appearance in brain sections; the area was heavily stained by myelin-staining techniques. The area defined by this heavily myelinated zone contained a complete representation of the visual half-field as measured by electrical mapping. The heavy concentration of directionally selective cells agreed with borders defined by the myelin and by the representation of the visual field assessed by electrical mapping.

MT also has a unique set of connections when compared with nearby cortex, particularly a set of two-way connections with primary visual cortex, the first way station for visual information in the cerebral cortex.

Along with the functional property of direction-selectivity, further evidence that MT related to processing of visual motion information came from a series of microlesion experiments. In these studies, a monkey was trained to find and track with its eyes a moving spot of light that was presented in different locations of the visual field. After training, a microscopically small amount of a chemical that inactivates brain cells was applied to a small part of MT. Because MT has a map of the visual field, inactivating part of MT inactivates processing of only a part of the visual field. When this was done, the animal could find and move its eyes to a stationary spot with the same accuracy as prior to the inactivation. But when the stimulus was moving, the accuracy of the eye movement to the stimulus and the ability to track the movement of the stimulus was poorer than before the inactivation. The animal could still see a stimulus and could still move its eyes accurately to it if it was stationary, but it acted as if it could not appreciate the motion and speed of the stimulus. Thus, the monkey was inaccurate in movements to a moving stimulus and in matching the velocity of its eye to the velocity of the stimulus. The sum of evidence at this point was that area MT was part of a pathway processing visual motion information.

Further study of the functional properties of these cells also showed velocity tuning; the cells fired much more vigorously at some velocities than at others. These cells were also less sensitive to changes in the color of visual stimuli than they were sensitive to changes in brightness, irrespective of the color of the stimulus.

Another study further delved into the neural implementation of motion processing in MT. In these studies, monkeys were trained to perform a discrimination of the motion of a field of randomly placed moving dots. The field contained a large number of dots, and the direction of movement of each dot could be independently determined. The direction information in the field could be totally random, or a percentage of the dots could be moved in the same direction, while the remaining dots were randomized. The percentage of dots moving in the same direction varied, and the monkey was to report one of two directions that the correlated dots were moving. When there was near 0% correlation (each dot moving in a random direction), the monkey would guess right about 50% of the time by chance, and when all of the dots were moving in the same direction (100% correlation), the monkeys rarely, if ever, reported the incorrect direction. As the amount of correlation increased from 0 to 100%, the number of correct responses also increased.

Single cells were recorded in area MT during the performance of this task. If the cell was directionally selective for movement in the upward direction, the two directions chosen for the monkey's task were up and down. When all of the dots were moving in an upward (preferred) direction, the cell's response was very vigorous, and the animal's response was about 100% "up." When the dots were all moving down, the cell's response was very low, and the animal's response was about 100% "down." For these extremes, the response of the cells would be a good predictor of the response of the animal: When the cell fires vigorously, the animal's response will be "up"; when the cell fires a little, the response will be "down." But this was known before the test was performed. What about for the intermediate correlations for up and down dot percentages? Would the intermediate firing rates of the cells predict the intermediate percentages of up and down responses. The answer was a startling yes. Single cells in MT seemed to carry enough information to predict the percentage of up and down responses of the whole monkey. In other words, if the decision-making apparatus of the monkey could "look" at only this one cell in area MT, the cell would be just as good at this task as the whole monkey. It appears as if the information for *perceiving* the coherent motion of dots in a noisy background is precisely coded at the level of single MT neurons. Much of the raw information used for visual motion *perception* itself is probably encoded in the visual area MT.

How does knowledge about this motion processing area map onto studies of human performance? One observation about MT was that while much information about visual movement is computed, MT cells do not appear to carry much information about the color of the stimulus. MT cells are much more sensitive to levels of brightness than to changes in color. It is as if the motion-processing pathway in the brain is somewhat color-blind.

If this were the case, then what would happen if a person was asked to judge the movement of objects that were different only in color, not in the level of brightness? There is a long history of studies on this question, but a recent review of these studies illuminate a very interesting aspect of this question. When colors are equiluminant (same brightness), a person's appreciation of the velocity of the stimulus is much less accurate.

For example, imagine that a movie is made of a black cylinder on a black background. Nothing can be seen. Now paste small white circles to both the cylinder and the background. A pattern of white spots can be seen, but the cylinder blends into the background; it is effectively camouflaged. If the cylinder is rotated, the cylindrical surface becomes apparent, the only clue being the relative motions of the dots of light with one another. Even if the dots in the background are moved randomly, the rotating cylinder will still stand out.

A computer can be made to simulate this effect, which is called three-dimensional structure from motion. In this case, rather than white dots on a black background, red dots can be placed on a green background. If the red dots are much brighter than the green background, and are moved with the same relative motions as the lights glued to the can, again a cylinder is "seen" by an observer. Now the observer turns down the brightness on the red dots. When the red dots are the same perceived brightness as the green background, the cylinder disappears! The red dots are still visible against the green background, but there is no perception of the cylinder. The observer continues to turn down the brightness of the red dots. When the red dots are sufficiently dimmer than the green background, the cylinder reappears.

When the moving dots have a brightness signal, the motion pathway can "see" them and can compute the cylinder from the relative motion of the dots. When the only difference between the dots and the background is the color, then the color-blind motion pathway cannot see them, but some other color-sensitive pathway can. So in both the same- and different-brightness conditions, the dots can be seen, but only in the different-brightness condition can the cylinder be seen. This psychophysical observation concurs with the neurobiological evidence that motion and color information are computed by separate pathways at some point in the visual system.

B. Visual Processing of Words

Similar approaches can be made to a different kind of visual "understanding," how visual words are processed during reading.

Many different elementary operations must be performed when reading. The lines and curves that make up the words must be visually processed, recognized as words, the meaning of the word must be "looked up," and a series of words must be integrated to understand the full meaning of a sentence. This example focuses on the visual processing of single words. [*See* READING PROCESSES AND COMPREHENSION.]

The visual processing of a word is often thought to consist of three levels of processing: (1) visual features, the lines and squiggles that make up letters; (2) letters, the single characters that make up words; and (3) words, a group of letters that make up a functional unit.

Studies in cognitive psychology have supported the idea that there are complex computations made on words and wordlike strings of letters that unify the visual input into a chunk. For example, each letter of a word can be perceived at dimmer contrasts than when that letter is presented alone or as part of a nonsense string of letters. This effect—that letters within words are more visible than the same letters in random letter strings—is known as the word superiority effect. The word superiority effect extends to meaningless letter strings that are visually similar to words (e.g., POLT), suggesting that this effect does not involve the meaning of the string but its regularity (i.e., similarity to strings of letters that could be words). It seems reasonable from these results to conclude that some level of visual processing involves coding at the level of visual words.

These results have led to some models of visual word processing that implement separate levels of feature, letter, and word form analysis. One of these models takes the form of a connectionist, or neural network, model run on a computer.

Each level is made up of a set of elements. Each of the elements is thought of as a neuron or set of neurons. Each element at the feature level represents one of the lines or squiggles that make up a letter; each element at the letter level represents a letter; each element at the word level, a single word. Elements at a single level inhibit one another, and elements between levels excite one another. All of the connections between elements are reciprocal (they go both ways). As visual information enters at the feature level, a particular set of features might be activated that make up the letters DOG. Elements at this letter level coding these letters become active and inhibit other letters at its level. Elements at this letter level in turn excite elements at the word level containing the letters DOG. The excited elements also excite appropriate features at the feature level through backward connections. The same thing takes place at the word level with the element representing the word DOG inhibiting the other words and adding further excitation to the constituent letters at the letter level. A random collection of letters does not get this extra feedback from the word level, so the input of a word or near word produces a word superiority effect. For sets of letters that are close to words, several elements at the word level are partially activated and still produce added activation at the letter level. In terms of a quasineuronal model of implementation, a multilayer processing network is a possible implementation that simulates human processing of visual words.

Two different types of studies provide evidence that separate functional areas in the human brain are related to these different levels of processing.

Studies of patients with damage to particular parts of level visual association cortex provide evidence that visual word form codes might be present in areas near the back of the cortex. Such lesions sometimes cause pure alexia (i.e., the inability to read words without other language deficits). Some people with pure alexia can read the letters that make up words and, by saying the letters aloud, can assemble the words, thus allowing them to read in a very laborious way. Letter-by-letter reading could be attributed to a deficit in the visual word form processing, but it could also involve earlier stages of the network. These results are certainly consistent with an important role of the extrastriate cortex in processing visually presented words. [See SPEECH AND LANGUAGE PATHOLOGY.]

A second set of results that reinforces this localization comes from positron emission tomographic (PET) studies of normal individuals. Methods have been developed that allow the PET scanner to localize areas of the human brain that are active during the performance of a task. Very low doses of a short-lived radioisotope, oxygen-15, are administered to a subject in the form of radioactive water to trace changes in blood flow during task performance. Forty seconds are needed to collect information to develop a scan of the activity that takes place. The scan is made up of pictures of the brain at seven horizontal levels. By placing these pictures in exact registry, it is possible to subtract the activity in one task from another, or to average the activity in the same task when performed by different people. Use of a standard brain atlas allows identification of the actual brain regions activated.

When subjects were presented with visual words, areas in visual association cortex were activated. Some of these areas were activated only by words but were not activated by other types of visual stimuli. The area activated in the left hemisphere is at a similar location to that found damaged in pure alexics.

From the examples presented in the previous section, it can be seen that very different kinds of information can influence the understanding of how the brain supports different mental processes. Only through the integration of these different kinds of information can progress be made toward understanding the mysteries of how we perceive, think, and communicate.

Bibliography

Damasio, H., and Damasio, A. R. (1989). "Lesion Analysis in Neuropsychology." Oxford University Press, New York.

Livingstone, M. S., and Hubel, D. H. (1987). Psychophysical evidence for separate channels for the perception of form, color, movement, and depth. *J. Neurosci.* **7(11),** 3416–3468.

Newsome, W. T., Britten, K. H., Movshon, J. A., and Schadlen, M. (1989). "Single Neurons and the Perception of Visual Motion." Retina Research Foundation Symposium **2**.

Petersen, S. E., Fox, P. T., Posner, M. I., and Raichle, M. E. (1989). Positron emission tomographic studies of the processing of single words. *J. Cog. Neurosci.* **1(2),** 153–170.

Posner, M. I., Petersen, S. E., Fox, P. T., and Raichle, M. E. (1988). Localization of cognitive functions in the human brain. *Science* **240,** 1627–1631.

Rumelhart, D. E., and McClelland, J. L. (1986). "Parallel Distributed Processing," Vols. 1 and 2. MIT Press, Cambridge.

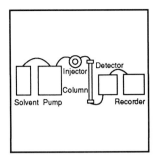

High Performance Liquid Chromatography

ROBERT MACRAE, *University of Hull*

Glossary

Chromatography Process that allows the separation of compounds by differential partition between a mobile and a stationary phase

Column Cylindrical tube that contains the stationary phase, usually in particulate form

Resolution Ability of the system to separate compounds of differing structure

Band-broadening Separation between molecules of the same type, leading to elution profile (peak) of finite width, ideally very small

Capacity factor Measure of the degree of retention of a compound

Plate number Measure of the efficiency of a column

HIGH PERFORMANCE LIQUID CHROMATOGRAPHY (HPLC) is one of a series of liquid chromatographic techniques in which separation between components is achieved by differential partition between a stationary phase and a liquid mobile phase flowing over it. The mechanisms involved and the underlying theory are exactly the same as those active in classical open column chromatography. HPLC, as all chromatographic techniques, is separative in nature, and any subsequent identification and quantification relies on the ability of the method to separate structurally similar compounds, often in complex mixtures. Thus, the factors that affect the resolving power of a particular system are of paramount importance.

I. Basic Theory

An ideal chromatographic separation would involve separation between molecules in the sample with different structures while molecules of the same type stayed together. In the case of column chromatography (including HPLC) if the sample is applied as a narrow band to the head of the column and passes down the column under the influence of the mobile phase, the components in the mixture would separate into individual narrow bands no wider than the initial sample. In practice this ideal situation does not arise, and there is significant spreading within the bands. The degree of retention of the compounds on the column depends on the relative affinity of the components for the mobile and stationary phases. This process is essentially under thermodynamic control, whereas the degree of band spreading is dependent on a number of kinetic factors.

II. Chromatographic Retention

The degree of retention of a compound on a column packed with a stationary phase under the influence of flowing mobile phase depends on the partition (or distribution) of that compound between the two phases. Partition is characterized by the distribution coefficient (K) for the compound, which is simply the ratio of its concentration in the stationary phase (C_s) to that in the mobile phase (C_m):

$$K = C_s/C_m$$

In practice this distribution is more commonly characterized by the capacity factor (k'), which is the ratio of the amount of sample in the stationary phase (a_s) to that in the mobile phase (a_m).

$$k' = a_s/a_m$$

This parameter is extensively used to quantify retention of peaks and can be shown to be equal to the expression

$$k' = \frac{t - t_0}{t_0}$$

where t is the retention time of the peak of interest and t_0 is the time in which an unretained compound will emerge from the column. The capacity factor can be used for theoretical studies of the chromatographic process as a thermodynamic model [e.g., $\log k'$ is found to be proportional to T^{-1} (temperature in kelvin)]. More practically k' is not altered by changes in column geometry, whereas t would be. The chromatographic process can only be considered by the thermodynamic model under ideal conditions; deviations are to be expected once the distribution coefficient is no longer independent of sample concentration. The most frequent deviations can be found at high sample concentrations, in which peak broadening occurs together with a reduction in retention time.

III. Band-Broadening

The ideal chromatographic peak would have the same width as that corresponding to the sample as applied to the head of the column. In practice this does not occur, and band-broadening takes place as the sample passes down the column. The smaller the extent of this broadening, the more efficient the column and hence the greater its ability to resolve similar compounds. Thus, in the example shown in Fig. 1 the resolution has been increased by decreasing the bandwidths. Alternatively the resolution could be increased by increasing the difference in retention times of the two compounds, but this would also lead to an increase in analysis time.

The effect of these two changes can readily be seen from the mathematical definition of the term resolution (R_s):

$$R_s = \frac{t_2 - t_1}{\frac{1}{2}(w_2 + w_1)}$$

where t_1, t_2 are the retention times and w_1, w_2 are the peak widths.

A further significant advantage in reducing bandwidths is that for a given amount of material a narrower band will yield a taller peak and hence greater sensitivity.

Narrow peaks will only result in practice where

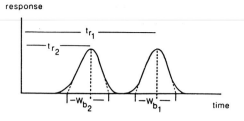
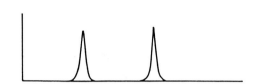

FIGURE 1 Increase in resolution caused by reduction in bandwidths. t_r, retention times; w_b, bandwidths.

there is efficient exchange of sample molecules between the mobile and stationary phases, in other words there is good mass transfer. Additionally there will be contributions to band-broadening as a result of eddy diffusion (multiple paths through the stationary phase, each with differing velocities) and mobile phase mass transfer (gradation of solvent velocities across a flowing stream). Pools of stagnant mobile phase held within pores can also lead to broadening as sample molecules held within them will be further delayed. The factors affecting these individual processes are complex, but the most important are the particle size of the stationary phase material and the velocity of the mobile phase over it. These two factors are shown in Fig. 2, where it will be noted that the material with smaller particles gives narrower bands but also that these can be achieved at a higher solvent velocity, so that this will allow more rapid analyses. Indeed, it is the ability of HPLC to form narrow peaks with short analysis times that makes it such an important technique.

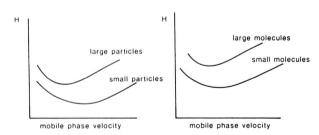

FIGURE 2 Effect of particle size and mobile phase velocity on bandwidth. Low values of H represent narrow bands.

Another important factor that affects the transfer of sample molecules between the phases is the ease with which they can diffuse in the mobile phase, which in turn will depend on the diffusion coefficient of the solvents involved, which will also be influenced by temperature. Solvents with low viscosity will tend to produce narrow peaks, and an increase in column temperature will, generally, result in reduced band spreading.

The efficiency of a chromatographic column is expressed by its plate number (by analogy with fractional distillation) and is given by the expression:

$$N = 5.54(t_r/w_h)^2$$

where t_r is the retention time of the peak and w_h is the peak width at half height. Modern microparticulate phases (5μ) will have plate numbers approaching 100,000 per meter.

IV. Modes of Chromatography

The mode of chromatography refers to the mechanism by means of which the solute is retained on the stationary phase. The most commonly encountered modes are shown in Table I. The choice of mode of chromatography for a particular group of compounds is far from simple, and although some types of compounds are not compatible with a given mode, there are often two or three modes that could be used.

A. Adsorption Chromatography

Silica gel is by far the most common material used for this mode of chromatography, in which solute molecules partition between solution in the mobile phase and adsorption onto the polar surface of the silica. The activity of the material depends strongly on the degree of hydration of the surface. In the

TABLE I Modes of Chromatography

Mode	Basis of Retention
Adsorption	Polarity
Partition	Solubility
Ion exchange	Charge
Size exclusion	Molecular size
Chiral	Optical activity
Affinity	Biological activity

presence of polar solvents, such as alcohols, which are often used as polar modifiers in a less polar solvent, the silica surface becomes solvated and the solute (sample) molecules then interact with this solvated surface. The elution power of the mobile phase is related to the solvent polarity and can readily be calculated from a weighted average for solvent mixtures. However, the solvent strength alone does not adequately predict its chromatographic behavior as different solvents show differing selectivities in relation to their ability to interact with solutes (samples) of particular types. Solvents are therefore further characterized with respect to their ability to interact *via* hydrogen bonding or dipole effects. This classification of solvents, together with their solvent strength values, then allows significant predictions as to the most suitable solvent for a particular separation. Adsorption chromatography is not extensively used with HPLC as the columns are slow to equilibrate and can readily be contaminated with highly polar compounds due to irreversible binding.

B. Partition Chromatography

This mode of chromatography involves partition between the liquid mobile phase and a stationary liquid phase held on a support material. Originally the stationary liquid was simply adsorbed onto an inert material, but the resulting phases were unstable, the liquid coating being stripped from the column. The logical solution to this problem is to bond the liquid stationary phase covalently to the support material. This may be achieved by esterification of the surface silanol groups of silica, for example, but the resulting silicate esters can only be used with nonaqueous mobile phases. A more permanent bonded phase is achieved by silylation of the surface hydroxy groups with chlorosilanes to form siloxanes. Monochlorosilanes give rise to monomeric phases whereas di- and tri-chlorosilanes lead to polymeric phases. The "art" of producing good bonded phase materials is to have sufficient liquid stationary phase present for good chromatographic retention yet at the same time have phases with good mass transfer characteristics, which demands relatively thin bonded layers. It may also prove necessary to remove any residual unreacted silanol groups by "capping" with a highly reactive reagent such as trimethylchlorosilane.

Bonded phases may be polar in nature (e.g., an aminopropyl phase), or they may be hydrophobic (e.g., an octadecyl bonded material). Polar phases behave in a similar manner to adsorption materials: more polar compounds are more strongly retained, and an increase in mobile phase polarity results in decreased retention times. Hydrophobic materials, however, form the basis of reversed phase chromatography, in which compounds of lower polarity are more strongly retained and mobile phases of lower polarity have increased eluting power. Octadecylsilane phases are often used with aqueous methanol (or acetonitrile) as mobile phases. Reversed phase materials are extensively used with gradient elution, as they equilibrate rapidly with changes in solvent composition, and in the example cited here, a gradient would be run with increasing methanol concentration to yield increasing eluting power. Modern reversed phase materials are extremely robust and not prone to contamination with polar compounds in complex mixtures (these are not retained). The materials are therefore widely used with complex samples of food or of physiological origins. The versatility of reversed phase chromatography can be further extended to ionic compounds by the technique of ion-pair chromatography. Here a hydrophobic ionic species of opposite charge to that of the analyte is added to the mobile phase. The resulting ion pair is far more hydrophobic than the initial ionic species and can then be chromatographed under reversed phase conditions, rather than with ion exchange.

C. Ion Exchange

In ion-exchange chromatography the mechanism of retention is simply electrostatic between opposite charges on the analyte and the stationary phase. The stationary phase may carry a net positive charge and is therefore able to retain anions (anion exchange) or it may be negatively charged and retains cations (cation exchange). Ion exchangers can be classified further as strong or weak exchangers, by analogy with strong or weak bases, depending on the charged groups. Thus, a sulfonic acid group will behave as a strong cation exchanger and a carboxymethyl group as a weak cation exchanger. Similarly quaternary ammonium groups will form strong anion exchangers and tertiary amino groups will act as weak exchangers. The prerequisite for ion exchange is that the stationary phase, and determinand must be charged. The charge on weak exchangers will alter gradually over wide pH ranges, whereas strong exchangers will only lose their surface charge at extremes of pH.

Ion-exchange chromatography is a two-stage process: adsorption where the determinand (analyte of interest) displaces the counterion on the exchanger and desorption where the sample is displaced by other counterions. Separation between compounds will only be effected where there is selectivity, in either the adsorption or desorption stages, for the particular compounds. The desorption stage is often carried out with gradient elution, either by gradually changing the pH of the mobile phase or by altering the concentration of competing counterions. The most common material for ion-exchange stationary phases is chemically modified cross-linked polymers of styrene and divinylbenzene. The degree of cross-linking affects the porosity of the material and hence the accessibility to the charged sites, particularly for macromolecules.

D. Size-Exclusion Chromatography

This mode of chromatography differs from those already discussed in that there are no direct interactions between the solute molecules and the stationary phase. Separation is effected between molecules of different sizes due to differential migration into porous material. The molecular size range of compounds that can be separated by this procedure depends on the pore size range of the stationary phase. If the solute molecules are all too large, in relation to the pores present, total exclusion will take place with no separation. On the other extreme, if all the solute molecules are small, allowing complete penetration into the pores, they will all be retained to the same extent, also resulting in no separation. Each phase is therefore characterized by a fractionation range between complete penetration and complete exclusion, and a wide range of such phases is commercially available.

Porous soft gels can only be used under low pressure conditions (e.g., Sephadex or Biogel), whereas porous glass or cross-linked polymers can be used as small particles under high pressure conditions, allowing higher mobile phase velocities and shorter analysis times. In practice, most size-exclusion materials also exhibit secondary interactions, such as hydrophobic bonding or anionic exclusion, so that pure size exclusion is rarely encountered. Stationary phases are available that are compatible with both organic or aqueous mobile phases.

E. Chiral Chromatography

The enantiomeric form of optically active compounds is of vital importance to their biological activity. There is considerable interest in being able to separate these forms to determine their optical or enantiomeric, purity (e.g., with drugs). There are two methods to achieve these separations by HPLC. In the first a chiral phase is used to distinguish the spatial differences between enantiomeric forms, and in the second, or indirect, method, diastereomeric derivatives are formed using optically pure derivatizing reagents.

In the direct method the stationary phase may contain the chiral component or a chiral additive may be included in the mobile phase. Both of these methods are used, but where the chiral agent is included in the stationary phase, it is preferable to use a bonded phase, in which the chiral agent is covalently bonded to the base silica. These are exemplified by Pirkle phases.

F. Affinity Chromatography

Affinity chromatography is based on the specific interaction of the determinand with a complementary compound immobilized on the stationary phase. Such specific interactions are commonly encountered in biological systems (e.g., hormone and binding protein), and it is in this field that affinity chromatography is widely used. A prerequisite is that it must be possible to bond covalently the complementary compound to the support material without reducing its biological activity. Also the specific binding must be reversible so that the determinand can be subsequently eluted, for example, with a pH shift or change in ionic strength. It may also be important that the elution conditions do not cause loss of biological activity if the material is to be isolated for further studies. For analytical purposes only, it is possible to elute the bound compound by using denaturing conditions.

V. Instrumentation

The basic instrumentation required for an isocratic (constant solvent composition) HPLC system is shown in Fig. 3 and consists essentially of a solvent reservoir, a pump, an injector, columns, and a detection system. The options available for these ba-

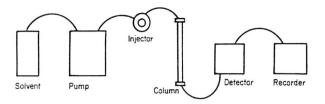

FIGURE 3 Basic high performance liquid chromatograph (isocratic).

sic components are numerous, and this fact provides the basis of the versatility of HPLC as an analytical technique.

A. Solvent Reservoir

Any convenient container may be used for a solvent reservoir, provided there is some means of degassing the solvent before entering the inlet of the pump. This may be effected by purging with an inert gas or by applying ultrasound, heat, or vacuum. The nature of the solvent will depend on the mode of chromatography used, but it must be free from particulate material and other contaminants that may mask the determinand. In particular, very high purity solvents are needed when sensitive detectors, such as fluorescence, are being used.

B. Pumps

The pumping requirements depend on the column being used. For traditional 4–5-mm i.d. columns of 10–25-cm length flow rates will be 1–2 ml min^{-1} with pressures as much as 400 bar. If microcolumns are to be used (1 or 2 mm i.d.), the flow rates will need to be as low as 10–20 μl min^{-1}, and this places severe demands on the pump. Most modern HPLC pumps are single- or dual-piston reciprocating types and, provided they are well maintained, are now significantly more reliable than earlier models.

C. Injection Devices

Several types of injector have been used for HPLC, but the injection valve is now standard among most chromatographers. The sample is injected with a microsyringe into a holding loop, and when the value is turned, the mobile phase is diverted via the loop to push the sample onto the head of the column. With careful attention to the connections between the valve and the head of the column, it is possible to introduce the sample with minimal dilution, which would otherwise result in band-broadening. Injection valves may be incorporated into autosamplers to allow automatic sample introduction under the control of the system programmer.

D. Columns

Most published analytical HPLC separations still employ stainless steel columns (10–25 cm long, 4–5 mm i.d.) packed with 5 or 10 μm diameter stationary phases. Trends to use narrower columns (1–2 mm i.d.) or shorter columns (*ca* 5 cm) packed with 3-μm material have been limited, despite the significant advantages that can theoretically be realized in terms of reduced analysis time and reduced solvent consumption. Many cartridge systems are available in which the stationary phase is packed in cartridges, which are then placed in a steel column. When the column fails, it is then necessary to replace the cartridges only, rather than the entire column, with a significant cost saving.

For the analysis of complex samples, it is advisable to use a short guard column between the injector and the head of the analytical column to protect the latter from impurities in the sample. With cartridge systems the guard column is usually a short (1 cm) separate cartridge placed on the top of the first analytical column.

E. Detectors

The wide range of detectors available for HPLC allows the detection of any compound likely to be encountered in a column eluate, although there are considerable differences in sensitivity and selectivity between the various detectors. In some applications a universal detector is required, so that all the components in a mixture will be detected; in others detectors with high selectivity are desirable so that the determinand of interest can be distinguished from the other compounds present. The weakness in detector systems currently available is the lack of a universal detector with good sensitivity. In addition to known selectivity and sensitivity, an ideal detector should be stable, possess a wide linear range, not adversely affect the chromatographic separation achieved, and be nondestructive so that the separated compounds can be further studied by additional techniques.

1. Ultraviolet Visible Detection

Ultraviolet (UV) visible detection is by far the most important of all the detection options available. Instruments may be based on interference filters providing a limited selection of wavelengths or, more importantly, a monochromator to provide any wavelength throughout the UV range. The wavelength range may be extended to cover the visible region by the inclusion of a tungsten filament lamp in addition to the deuterium radiation source used for the UV region. Spectral information from the separated compounds may be obtained either by stopping the mobile phase when the component of interest is in the flow cell and then scanning through the wavelength range or, more conveniently with the use of diode array technology, where the spectral information is obtained directly without the need for stopping the flow of the mobile phase to obtain a spectral span.

The detection wavelength should be chosen to provide good sensitivity for the compound of interest, ideally at the λ_{max} position. However, it may also be important to select a wavelength that provides good selectivity over other potentially interfering compounds. The final choice may therefore be a compromise between sensitivity and selectivity. Low wavelength UV detection, approximately 200 nm, is subject to considerable interference from many compounds in biological extracts and also from solvent impurities and should only be used when alternative wavelengths are not suitable. This is the situation for simple sugars and saturated lipids.

2. Fluorescence Detection

This method of detection approaches the ideal for many applications in the biological field, providing high sensitivity and selectivity with good stability. Relatively few compounds fluoresce in relation to the number that adsorb UV radiation, and the selectivity is further improved by having the facility to choose both excitation and emission wavelengths. Cheaper instruments may be based on interference filters for both excitation and emission wavelength selection, whereas more sophisticated equipment will make use of monochromators for emission characterization and also possibly for excitation wavelength selection. The sensitivity of fluorescence detection may be enhanced by the use of a laser radiation source for excitation and such instruments have allowed the determination of extremely low levels of compounds with high biological activity (e.g., mycotoxins).

3. Refractive Index Detection
Determination of the refractive index of the column eluate and comparison of this value with that of the pure mobile phase provides, at least in theory, an ideal universal detection system. In practice, although refractive index detectors of considerable sensitivity are available, the instruments are also very sensitive to changes in pressure (including pulsing), temperature, and even levels of dissolved gases. Also if the temperature of the HPLC column changes, there will be a shift in the equilibrium of mobile phase components with the stationary phase and hence a change in eluate composition, which will result in a refractive index response. The final limitation of these detectors is, of course, that they cannot be used with gradient elution. Despite all these problems, refractive index detection is still used for the determination of sugars and lipids (including phospholipids) if there are no better alternatives available.

4. Electrochemical Detection
Compounds, organic or inorganic, that can be oxidized or reduced at a charged electrode surface may be electrochemically detected. This provides the basis for an extremely sensitive and selective means of detection, with the selectivity being enhanced by the ability to select the potential on the charged electrode in relation to the redox properties of the determinand. Electrochemical detectors are sensitive to pressure changes (and pulsing) and mobile phases should be degassed to avoid complications caused by dissolved oxygen. Several important groups of biologically active compounds (e.g., catecholamines) can be determined at low levels using electrochemical detection.

5. Derivatization
For some compounds in which there is no suitable means of detection, the situation may be improved by derivatization to form a colored compound or, more commonly, a compound with good UV absorption or fluorescence properties. This may be performed postcolumn, where the derivatizing reagent is mixed with the column eluent before passing through the detector, or the derivatization may be carried out precolumn, before the sample is injected. There are more restrictions with post-

column derivatization, as the reaction conditions must be compatible with the presence of the mobile phase and the reaction must be rapid, although holding coils and elevated temperatures can be used to accommodate a wide range of reaction conditions. With precolumn derivatization, in theory, there are no such restrictions, but reactions are generally selected that can be carried out under mild conditions, possibly in an autosampler. The derivatives must be sufficiently stable, although with automation it is possible to arrange for all the samples and standards in a batch to be derivatized with identical time lapses before injection. Precolumn derivatization will also alter the chromatographic properties of the compounds involved, and this is often associated with a reduction in polarity as functional groups, such as hydroxyl or amino, are derivatized. In practice this may result in a change of mode of chromatography (e.g., sugars that are normally determined on bonded phase columns can be chromatographed on silica as less polar hydrazone derivatives).

6. Other Detectors
Several other detection systems have been employed for specialized applications where the determinand, or matrix, presents particular problems. The field of lipid analysis has generated interest in several other types of detection including infrared detection, flame ionization detection, and the mass detector.

Infrared detection is limited by the strong absorption of many common HPLC solvents, and absorption coefficients are much lower than in the UV region, leading to poorer sensitivity.

The earliest flame ionization detectors made use of a moving wire to transport the column eluate to the ionization flame via an evaporator to remove the solvent. Only a small proportion (1%) of the eluate was retained on the wire, and thus the resulting sensitivity was poor, although the instrument was compatible with gradient elution. A more modern detector uses a quartz belt for transportation, which allows retention of all the column eluate. Good sensitivity is claimed for both lipids and carbohydrates.

The mass detector is an evaporative instrument in which the mobile phase is removed by nebulization and evaporation before any nonvolatile solute in the column eluate is detected by light scattering. The detector has been extensively used for lipid and carbohydrate determinations and provides high sensitivity with gradient compatibility.

The discriminating power of HPLC can be greatly increased when the detector used provides additional characterization, as already noted with UV detectors. A particularly powerful combination of techniques is HPLC coupled to mass spectrometry (MS), but there are significant problems of interfacing as solvent cannot be introduced into the mass spectrometer ion source in appreciable amounts. earlier devices were based on belts to transfer the column eluate to the ion source via suitable evaporators, but these proved unreliable and showed "memory effects." The most important HPLC–MS interface currently in use is the thermospray. The column eluate, containing a volatile ionic additive, is passed through a heated capillary tube. A supersonic jet of vapor is formed containing a mist of fine droplets and ions from the solute present. The jet then enters a sampling core where about 7% of the ions are abstracted and only 1% of the solvent vapor. The residual gas and ions are pumped away using a rotary pump. The resulting spectra obtained from a simple thermospray usually show pseudomolecular ions together with a small degree of fragmentation. An additional filament may be introduced in the ion source to produce spectra more similar to those obtained by chemical ionization.

F. Gradient Elution and Automation

The basic isocratic liquid chromatograph shown in Fig. 3 can be extended in a number of ways to increase its versatility and efficiency. Gradient elution allows the composition of the mobile phase to be altered during the course of the analysis, and automation allows unattended operation of the chromatograph for many samples.

The use of gradient elution is particularly beneficial where complex mixtures are to be analyzed and where there are significant differences in properties between the components in the mixture. Gradient elution is most commonly used with bonded phases (polar, reversed phase, and ion exchange) as these materials rapidly equilibrate with changes in solvent composition. This is important if reproducible chromatography is to be obtained between separate chromatographic runs. Silica gel, used in adsorption chromatography, is slow to equilibrate and is therefore not ideally suitable for use under gradient conditions. Gradient profiles may be obtained with high pressure mixing, in which the outputs from two or

more HPLC pumps are mixed and the resulting composition is varied by altering the relative flow rates of the pumps. Alternatively, the solvents may be mixed via a proportioning value on the low-pressure side of a single HPLC pump. The composition is varied by altering the relative amounts of time the valve is open to a particular solvent reservoir. These configurations are shown in Fig. 4. Although gradient elution will often provide improved resolution of complex mixtures, the technique should only be used when an adequate isocratic separation cannot be achieved. Gradient elution is of necessity more complex and hence more prone to mechanical problems, and quantitation may not be precise because of slight differences in the elution position of the determinand.

An isocratic liquid chromatograph can be readily automated by the simple addition of an autosampler, which allows injection of samples at preset time intervals. The autosampler may then be linked to the microprocessor controlling the gradient to allow automated gradient elution. A significant part of any automation is the data handling, which is carried out on-line, and this is one area in which microcomputers have had a vast impact in recent years. The computer will carry out automated calibration from "samples" designated as standards and can then report the data in any desired format. Peak areas (or heights) can be converted into concentrations by a number of methods of quantifica-

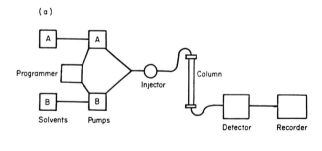

(a)

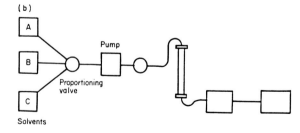

(b)

FIGURE 4 Gradient high performance liquid chromatograph systems: **(a)** high pressure mixing; **(b)** low pressure mixing.

tion. The simplest method is based on external standards where the peak assigned to the determinand in the sample is compared with a calibration curve prepared from known concentrations of the compound in question. Precise results will only be obtained with precise injection volumes, and this is most readily achieved by using an injection valve and completely filling the loop each time. The dependency on injection volume may be overcome by using an internal standard. Here a compound not present naturally in the sample is added in known amounts to both the sample and standards. The ratio (peak areas or heights) of the determinand to the internal standard in the sample is compared with the same ratio for the standard solution. This procedure overcomes any variation in injection volume and, if the internal standard is added before the extraction and clean-up stages, will partially compensate for losses during these stages. The internal standard must clearly be well resolved from the determinand and other compounds present and, as far as possible, should be similar chromatographically and in terms of detectability. In mixtures in which all the components have a similar detector response, it is possible to quantify the determinand by percentage area. In HPLC this is not a common situation, although carbohydrates with similar structures, as found in partially hydrolyzed starch, do have similar refractive index responses.

HPLC is now well established as an essential technique for the separation, identification, and quantification of nonvolatile compounds in complex mixtures, and this is nowhere more true than in the biological sciences. Here the low levels of deter-

minands and the complexity of the sample matrices demand a sensitive technique with high resolving power, and these are the characteristics of HPLC.

However, it must be emphasized that HPLC is fundamentally a separative technique, and any identification of compounds based on retention similarity with standards alone must be considered tentative. The ability of HPLC to characterize compounds more fully is greatly enhanced by interfacing the separative ability of the HPLC system with a detector that can provide both qualitative and quantitative information. Thus, the recent developments of interfaces for HPLC and MS and detectors such as photodiode array UV spectrometers and FT infrared instruments allow peak assignments to be made with much greater confidence. Recent developments in stationary phases (e.g., chiral materials or porous graphite), together with improvements in detector performance, have also greatly extended the versatility of the technique.

Bibliography

Macrae, R. (1988). "HPLC in Food Analysis," 2nd ed. Academic Press, London.
Poole, C. F., and Schuette, S. A. (1984). "Contemporary Practice of Chromatography." Elsevier Science Publishers, Amsterdam.
Ravindranath, B. (1989). "Principles and Practice of Chromatography." Ellis Horwood, Chichester, U.K.
Snyder, L. R., and Kirkland, J. J. (1979). "Introduction to Modern Liquid Chromatography," 2nd ed. John Wiley, New York.

High Performance Liquid Chromatography in Food Analysis

ROBERT MACRAE, *University of Hull*

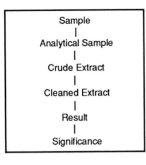

Sample
|
Analytical Sample
|
Crude Extract
|
Cleaned Extract
|
Result
|
Significance

Glossary

Additives Compounds deliberately added to foods to alter natural flavor, stability, color, or other property. The use of some of these compounds has generated much controversy

Contaminants Minor components that have inadvertently been incorporated in foods (e.g., processing or agricultural residues)

Determinand Compound of interest in the sample to be determined quantitatively

HPLC High-performance liquid chromatography. An instrumental liquid chromatographic technique using small particle–size stationary phases leading to high sensitivity and resolution

Macronutrients Major components of foods (i.e., protein, carbohydrate, lipid, water, and minerals). Proteins, carbohydrates, and lipids can be characterized in detail by chromatographic methods but are commonly determined by routine chemical methods

Micronutrients Minor food components with an important physiological role, especially vitamins and trace metals

THE ANALYSIS OF FOOD materials is carried out for a number of reasons, including nutritional value, safety and toxicology, authenticity and adulteration, identification of processing losses, and deter-mination of additives and contaminants. Thus, there is a wide range of components to be determined, some of which are present in large amounts and some at very low levels. The situation is made more difficult by the complex nature of many food substances, especially after cooking or processing. The use of food additives has increased substantially in recent years, and legislation has been introduced to control these compounds and the levels to which they can be added. This has greatly increased the demands made of the food analyst.

Chromatographic techniques, including high-performance liquid chromatography (HPLC), have played an important part in this field of food analysis.

I. Characteristics of HPLC

The background theory and basic instrumentation of HPLC have been outlined in a separate entry. The characteristics of HPLC make it a suitable analytical technique for the determination of many components in foods, and these are the same characteristics that have made it so important in many other biological areas. [*See* HIGH PERFORMANCE LIQUID CHROMATOGRAPHY.]

A. Resolution

The resolving power of HPLC is superior to most other liquid chromatographic techniques, and the majority of determinations in the food area requires this capability. This is required both to resolve compounds structurally related to the determinand in question (e.g., resolving a group of related compounds with the same biochemical function, as in vitamers) or, perhaps more importantly, to resolve the determinand from components of the food matrix. Food materials, particularly after cooking, are extremely complex both in terms of high and low molecular weight components present. In many instances HPLC is not adequate alone to deal with such complex mixtures, and sample preparation (clean-up) becomes crucial.

B. Sensitivity

The sensitivity of HPLC is a reflection of both the low on-column dispersion that is achieved and the inherent sensitivity of modern detectors. Most detectors are concentration dependent, so the narrower and more concentrated the chromatographic bands, the greater the sensitivity. In some food analyses, sensitivity is not a major problem because high levels of the determinand are present (e.g., simple sugars). However, even in these cases, the ability to handle dilute solutions minimizes the sample clean-up that is required. For many other compounds, high sensitivity is essential (e.g., drug residues or vitamins) and there are a number of determinations, particularly for highly toxic compounds, in which it is not possible to say what an acceptable level of detection would be. The major shortfall in instrumentation available for HPLC is still a sensitive universal detector, able to detect all substances.

C. Versatility

HPLC can be used to determine a wide range of food components because a number of different types (modes) of chromatography can be used. Thus, low molecular weight hydrophobic compounds can be analyzed with adsorption columns, whereas more polar compounds are more suitable for partition chromatography. Ionic compounds can be conventionally handled with ion-exchange stationary phases or via ion-pair formation and reversed phase chromatography. Higher molecular weight components, polysaccharides, and proteins can also be separated on the basis of molecular size on size-exclusion phases. The versatility of HPLC is further enhanced by having a wide selection of detectors available so that the required degree of selectivity over potential interferents can often be achieved. The detector may also be able to provide qualitative information about the compounds that have been separated (e.g., with diode array detection or with HPLC coupled to mass spectrometry). This qualitative information greatly enhances the significance of any peak assignments made in addi-

tion to chromatographic similarity between standards and unknowns.

The HPLC stage is the final part of analytical methods using this technique, but it is often the previous stages that are more difficult and also where large errors may be introduced into the method. A typical analytical scheme is shown in Fig. 1.

Sampling is a particular problem in food analysis as the produce concerned is often heterogeneous

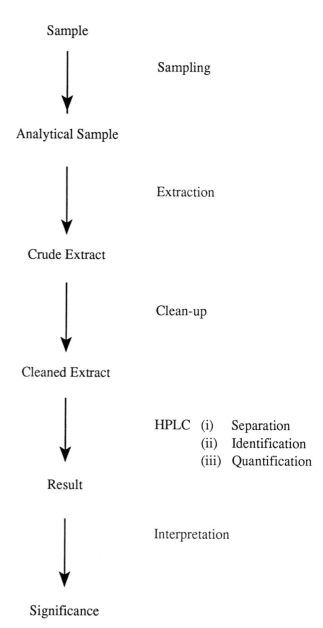

FIGURE 1 Typical analytical scheme. HPLC, high-performance liquid chromatography.

(e.g., meat and vegetables). The determination of contaminants is probably one of the more difficult areas of analysis in this respect. The contamination may be localized as a result of exposure (e.g., pesticide residues on the surface of produce) or the contaminants may be concentrated in a small area if they are formed by fungal growth (e.g., aflatoxins from *Aspergillus flavus*). Representative sampling is only possible once the degree of heterogeneity is known, and in absence of this knowledge extreme caution must be adopted. Where possible large samples should be homogenized before sampling, which may need to be carried out in several stages.

The ideal extractant would remove all the determinand while leaving all other compounds behind in the food matrix. In practice there is always a conflict between efficiency and selectivity. For example, in the determination of sugars, water would be an efficient extractant but would not be very selective as it would also remove some proteins, polysaccharides, and salts. However, 80% aqueous alcohol would be a less powerful extractant, but it would yield cleaner extracts from most foods. The recent trend has been to use the most efficient extractant available and rely on improved methods of clean-up to remove potential interfering compounds.

The nature of the clean-up employed will depend on the characteristics of the determinand with respect to those of the interfering compounds present in the extract. Traditional methods, such as ion exchange, solvent extraction, or precipitation, are still widely used. However, the advent of solid phase extraction, in which the determinand or the interfering compounds are removed by retention on a small chromatographic cartridge, has greatly increased the ease with which the clean-up of complex samples may be achieved. It is essential that the recovery of both the extraction and the clean-up stages is determined, otherwise substantial errors will have been introduced even before HPLC.

The HPLC stage *per se* must achieve separation, identification, and quantification if the determination of which it is part is to be valid. It is therefore necessary to ensure that interfering compounds do not coelute with the determinand, that the peak assigned to the determinand in the sample is in fact correct, and also that the basis of the quantification is justified (e.g., linearity of detector).

The final stage of the HPLC determination, as with any other method, is the interpretation of the data obtained as to its significance, and this should take into account the precision and accuracy of the overall method.

II. Carbohydrates—Sugars

The low molecular weight sugars, consisting of one to four or five saccharide units, can be extracted from foods most efficiently with water. Ideally the extraction should be at low temperature (ambient) to avoid the risk of hydrolysis of oligosaccharides, especially the inversion of sucrose. Acidic foods should be neutralized before extraction, again to reduce the risks of hydrolysis. In foods likely to contain active hydrolytic enzymes, the incorporation of some ethanol, or other denaturing agent, in the extractant is advisable.

The simplest method to clean-up crude aqueous extracts of sugars is to add an equal volume of a solvent such as acetonitrile. Not only will this precipitate out high molecular weight material, it will also reduce the polarity of the sample solution, making it more compatible with the conditions used subsequently for partition HPLC. However, solid phase cartridges are increasingly being used for clean-up of aqueous solutions such as these. With a reversed phase cartridge, the sugars pass straight through, and the more hydrophobic contaminants are retained. A more selective clean-up may be achieved with phases such as those containing phenyl boronic acid groups, which retain the sugars by interaction with the hydroxy groups present on adjacent carbon atoms.

Size-exclusion chromatography has been applied to the separation of sugars with some success, particularly where the mixture contains sugars with different numbers of saccharide units. Thus, the oligosaccharides of the raffinose family found in soya products (sucrose, raffinose, stachyose, and verbascose) can be separated on soft gels such as BioGel or on rigid polymeric size-exclusion phases under high-pressure conditions. Another relevant example would be the separation of glucose oligomers in partially hydrolyzed starch, in which all the components will significantly differ in molecular weight.

Bonded phase partition columns, such as those containing aminopropyl functional groups, have been widely used for the separation of groups of monosaccharides as well as oligosaccharides. Aqueous acetonitrile is employed as the mobile phase, with the amount of water used depending on the application, 12–20% for monosaccharides and as much as 50% for higher molecular weight oligo-saccharides. Most of the monosaccharides may be resolved, but there are "difficult pairs" (e.g., glucose and galactose). Refractive index detection is normally used for such sugar separations and does not allow gradient elution. Refractive index detectors have considerably improved in recent years with respect to sensitivity, although temperature and flow-rate control are still critical. Sugars may be detected using low wavelength ultraviolet (UV) absorption (ca. 200 nm) but solvents of high purity (and cost) are required. Limited gradients are possible with this means of detection. The mass detector, an evaporative analyzer, can be used with aqueous acetonitrile under gradient conditions and has been successfully used to resolve all the components of a glucose syrup from one saccharide unit (glucose) up to oligomers with more than 20 units (Fig. 2). The amino partition phase can be formed *in situ* by using a straight phase silica column and a suitable amine modifier in the mobile phase, which would again be aqueous acetonitrile. Attempts to replace the toxic solvent acetonitrile with other organic solvents (e.g., acetone/ethyl acetate/water) have met with some success but have not been widely adopted.

Reversed phase chromatography is not usually associated with carbohydrate analyses. However, by use of a C-18 column with water as the mobile phase, it is possible to separate sugars with different numbers of saccharide units. So here again the technique has found application to the separation of

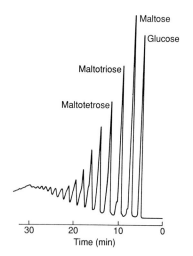

FIGURE 2 Separation of glucose syrup with gradient elution. Column, Spherisorb S-5 NH₂ 250 × 4.6 mm; solvent, water (32–60%) in acetonitrile; detection, mass detector.

components of glucose syrups, where some resolution of anomeric forms is also observed.

Ion-exchange materials are currently used in two quite distinct ways for the chromatography of sugars. Aminex resins contain sulfonic acid groups attached to a styrene–divinylbenzene matrix and are used in the sodium, potassium, calcium, lead, or silver forms for sugar analyses, in which the mechanism involved is a combination of ligand exchange and size exclusion. For resins with a low degree of cross-linking and a porous structure, size exclusion will predominate, especially for oligosaccharides, whereas for monosaccharide separations, ligand exchange will be more important. A significant advantage of this type of chromatography is that pure water can be used as the mobile phase, although good peak shapes are only realized at elevated temperatures. Detection is achieved by refractive index or postcolumn derivatization.

Sugars can also be separated with ion-exchange resins by true ion exchange under alkaline conditions, in which anions are formed. Sugars are surprisingly stable even in 0.1 M sodium hydroxide, and problems with degradation and hydrolysis are not observed, provided the column temperature is held below 50°C. This form of ion chromatography is often combined with the pulsed amperometric detector, which allows electrochemical detection of sugars, with cleaning pulses to prevent irreversible adsorption of oxidation products onto the electrode surfaces. The combination of anion exchange and electrochemical detection provides a sensitive method for the determination of sugars and forms the basis of the systems produced by Dionex.

The applications of HPLC to the characterization of polysaccharides are mainly restricted to size-exclusion chromatography, which provides relatively limited resolution based on differences in molecular size. HPLC can, however, be used to provide compositional data on the sugars present in polysaccharides after hydrolysis with acid or enzymes.

III. Vitamins

The nutritional role of vitamins is crucial to the maintenance of health, and considerable research effort has been aimed at methods for their determination. However, despite this effort the methods available are not ideal, and although HPLC methods play an important part, they do not produce satisfactory results for all vitamins in all foods. The situation is somewhat different for the fat-soluble and water-soluble vitamins, and these groups may be considered separately.

A. Fat-Soluble Vitamins

The most important members of this group of vitamins are retinol (vitamin A), tocopherol (vitamin E), cholecalciferol (vitamin D_3), and phylloquinone (vitamin K), all of which may be present in a number of closely related forms.

1. Vitamin A

Unfortified foods contain vitamin A mainly as retinol and retinyl esters, while fortified foods will predominantly contain vitamin A as its acetate or palmitate. Unless knowledge of the nature of the ester is of importance, it is usual to saponify the food, or a suitable extract, to liberate free retinol for analysis. The saponification process also removes the triglycerides, which are the major lipid components. Retinol is readily oxidized in the atmosphere, and it is usual practice to incorporate an antioxidant in the extracting solvent. [*See* VITAMIN A.]

Retinol can be chromatographed on normal (silica) or reversed phase columns; the latter is more suitable when the sample extract contains only small amounts of lipid (i.e., after saponification). Retinol is naturally fluorescent, and this mode of detection provides the greatest selectivity and sensitivity. It also has a strong UV absorption at 328 nm, and this will prove adequate for detection for many samples.

2. Provitamin A

A large number of carotenoids are present in nature, and those containing the retinol structure can be converted to retinol *in vivo*. A complete analysis of a food product for vitamin A activity should therefore include both vitamin A and provitamin A. β-Carotene is the major provitamin A carotenoid, and many HPLC analyses concentrate on this compound alone. Extraction is achieved with a range of organic solvents, with or without saponification. Care is necessary to avoid oxidation, as in the case of vitamin A itself. Normal or reversed phase materials can be used for HPLC, although the latter have proved to be more robust and less prone to contamination. Detection in the visible region (ca. 440 nm) provides a high degree of selectivity. Confirmation of identity can be obtained with the use of more

sophisticated detection systems such as photodiode arrays or mass spectrometers.

3. Vitamin D

Vitamin D (D_2 or D_3) is only present in foods at low levels, on account of its high biopotency. Its chemical/physical properties are similar to other lipid components present at much higher levels, and extensive sample preparation is therefore required before HPLC. Saponification is usually the first stage, followed by extraction into an organic solvent. This crude extract may then be purified by a number of routes: precipitation to remove sterols, alumina column chromatography, thin-layer chromatography (TLC), solid phase extraction (reversed phase cartridges), or preparative HPLC (normal or reversed phase). Vitamins D_2 and D_3 can only be separated on reversed phase columns, often with aqueous methanol as mobile phase. Detection is achieved at 264 nm with adequate sensitivity for fortified products and for these food products naturally rich in vitamin D. [*See* VITAMIN D.]

4. Vitamin E

Vitamin E is present in vegetable oils and other food products in eight forms, four with saturated (tocopherols) and four with unsaturated (tocotrienols) side chains. The various forms (α-, β-, γ-, and δ-) differ in the degree and position of methylation of the aromatic ring. The individual compounds have differing biological and antioxidant activities, and therefore it is necessary to determine these compounds separately to provide an assessment of the total activity. Oil samples can be directly analyzed by simply dissolving the oil in the mobile phase and injecting directly into the HPLC. If tocopheryl esters are present, these are usually hydrolyzed by saponification as they are only weakly fluorescent, which makes detection more difficult. Tocopherols are extremely sensitive to oxidation, particularly under alkaline conditions, and care must be taken to remove oxygen during the saponification stage. [*See* VITAMIN E.]

This is one application of HPLC in which it is essential to use straight phase columns (silica) if all the eight vitamin forms are to be separated, Fig. 3. However, for some food products that do not contain all the forms in appreciable amounts, reversed phase may be used, which has advantages in terms of robustness. Fluorescence detection is employed wherever possible to use the strong natural fluorescence of the free tocopherols and tocotrienols. UV

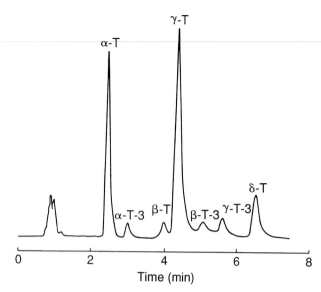

FIGURE 3 Tocopherols in oil seeds. Column, LiChrosorb Si60 250 × 4.6 mm; solvent, 4% dioxane in hexane; detection, fluorescence. T, tocopherol; T-3, tocotrienol.

detection at 280 nm can still provide adequate sensitivity, particularly in fortified foods.

5. Vitamin K

The importance of vitamin K as an antihemorrhagic agent has only been fully realized in recent years, and this has led to an increase in methods for its analysis. Phylloquinone (K_1) and the menaquinones (K_2 group) are unstable in alkali, and enzymic hydrolysis is required to breakdown lipid components before extraction. Further clean-up is carried out before HPLC, which may be achieved on open silica gel columns or by preparative HPLC. Normal and reversed phase columns can be used to determine vitamin K, usually in conjunction with UV detection. Fluorescence can only be used for vitamin K, after postcolumn conversion to the corresponding hydroquinone.

B. Water-soluble Vitamins

In general the HPLC methods available for the water-soluble vitamins are less successful than those described above for the fat-soluble vitamins. In most instances this is not due to problems with the chromatography *per se* but more related to extraction and sample purification before HPLC. Many of the water-soluble vitamins are present in bound forms, and this must be taken into account when considering the extraction procedure.

1. Vitamin C

L-Ascorbic acid and its oxidation product dehydroascorbic acid show vitamin C activity, but D-ascorbic acid (erythorbic or isoascorbic acid), which may be added to foods as an antioxidant, shows greatly reduced activity. Ascorbic acid is very susceptible to oxidation, which is catalyzed by trace metals (e.g., copper and extractants such as metaphosphoric acid are used to complex with these ions). Total vitamin C levels are usually determined by reducing dehydroascorbic acid with homocysteine before measuring total ascorbic acid. [*See* Ascorbic Acid.]

Ascorbic acid is very polar and therefore elutes near to the void volume with reversed phase columns, even when water alone is used as the mobile phase. The degree of retention may be increased by the addition of ion-pair reagents or by the use of low pH to suppress ionization. Detection is achieved by UV absorption at 254 nm, although electrochemical detection can provide greater selectivity. Sensitivity is rarely a problem in foods, as ascorbic acid is present at high levels, but in physiological samples, electrochemical detection can provide improved sensitivity. Where necessary it is possible to separate dehydroascorbic acid from ascorbic acid on reversed phase columns. However, the former no longer contains a strong chromophore, and detection is only possible at low wavelengths (210 nm). Separation of ascorbic acid from isoascorbic acid by chiral chromatography is rarely carried out on food samples.

2. Vitamin B_1

Thiamin (vitamin B_1) is found naturally both in free form and as the pyrophosphate ester, the latter predominating in animal products. The extraction process involves acid hydrolysis with dilute hydrochloric acid to release the vitamin from association with proteins. Additionally an enzymic hydrolysis is required to break down any ester form to yield free thiamin. Traditionally this is realized with a takadiastase, which possesses phosphatase activity, although the latter is only present as an impurity. If further clean-up is necessary, this may be achieved by solid phase extraction on reversed phase cartridges. Thiamin is an ionic compound and is therefore only adequately retained on a reversed phase column with ion pairing, which may be as simple as acetate. Alternatively, C_5, C_6, or C_7 sulfonates may be used. Unfortunately, thiamin often produces a poor peak shape with tailing caused by adsorption. Detection at around 250 nm provides adequate sensitivity for fortified products, but natural levels may require the additional sensitivity of fluorescence, which is only applicable after pre- or postcolumn oxidation of thiamin to thiochrome. Oxidation of thiamin to thiochrome also forms the basis of a method for the separation of thiamin esters. After oxidation with alkaline cyanogen bromide, the thiochrome phosphate esters can be separated on an amino column. Such methods are of more relevance to neurological studies than in food analysis.

3. Vitamin B₂

Like thiamin, riboflavin (vitamin B_2) is found naturally in a number of forms, mainly as the 5'-phosphate and flavin adenine dinucleotide. These are protein bound, and extraction follows the same procedure as for thiamin, again with an enzyme with phosphatase activity to hydrolyze the phosphate. Any additional clean-up may be achieved by solid phase extraction. Riboflavin is naturally fluorescent and can be detected by this means with high sensitivity after reversed phase ion-pair chromatography. Riboflavin is very photolabile, and exposure to light should be kept to a minimum throughout the analysis. The combination of acidic and enzymic hydrolysis used in the extraction of foods ensures that all the forms are converted to riboflavin, and so separation of the various natural forms is not usually attempted in food analysis.

4. Vitamin B₆

Pyridoxine (or pyridoxol) is the main form of vitamin B_6, although pyridoxal and pyridoxamine are also found. In their natural form they are predominantly present in phosphate forms and therefore require sample treatment similar to vitamins B_1 and B_2. Clean-up may be carried out with ion-exchange or reversed phase materials, and indeed both of these modes have been successfully employed for the HPLC stage as well. Mobile phases of low pH (2.2) are used sometimes in conjunction with alkane sulfonates as ion pairs. Such conditions allow the separation of pyridoxol, pyridoxal, and pyridoxamine. All three forms of vitamin B_6 are fluorescent, and this is the method of choice for detection, although UV absorption at 280 nm has been used where the vitamin is present at reasonably high levels.

5. Niacin

Niacin (nicotinic acid) is one of the most stable vitamins in foods and can be extracted into acid or alkali. Hydrolysis under either of these conditions will also convert any niacinamide present to free niacin. Some differences have been observed between analytical results obtained under differing conditions of extraction. Nicotinic acid has been determined using ion exchange, although the more usual procedure is to use reversed phase ion-pair chromatography together with UV detection.

6. Folates

Folic acid (pteroylglutamic acid) is not found naturally but rather as conjugates with as many as 12 glutamic acid residues as tetrahydrofolates. Assays are usually simplified by deconjugation with a suitable enzyme to remove the glutamic acid residues. Extraction and clean-up from foods may be extremely complex. HPLC is carried out with ion-exchange or reversed phase ion-pair chromatography. UV detection is only sufficiently sensitive for fortified foods, and fluorescence is only applicable after postcolumn oxidation to a pterin derivative.

IV. Amino Acids

Amino acid analysis has traditionally been carried out by ion-exchange chromatography with postcolumn derivatization. Indeed, amino acid analyzers were the first chromatographs to be run under high resolution conditions (i.e., with small particle size stationary phases and high inlet pressures). Resolution of the 20 or so naturally occurring amino acids required ionic strength and/or pH gradients using a cation-exchange resin. Detection was achieved with postcolumn derivatization with ninhydrin, which necessitated a high-temperature reaction bath. This traditional methodology can be transferred unchanged to modern HPLC instrumentation, but alternative methods that allow more rapid separations have been developed. The most widely used of these techniques is that based on precolumn derivatization with o-phthalaldehyde in the presence of a thiol. The reaction is rapid and can readily be carried out in a robotic autosampler. The fluorescent derivatives formed are more hydrophobic than the parent amino acids and can be separated by gradient elution on a reversed phase column. Physiological samples, which contain a wider range of amino acids, can also be handled by these methods. Analysis times are significantly shorter than with ion-exchange-based instruments, but controversy still exists as to the relative precision/accuracy of the methods, with many analysts still preferring traditional methods.

The use of amino acid analysis to determine the composition of proteins involves prior acid hydrolysis to break down the protein to its constituent amino acids.

V. Lipids

The lipid fraction of foods is complex, and HPLC has found a number of applications in this area. The simplest of such analyses is the separation into lipid classes (e.g., sterol esters, triglycerides, diglycerides, monoglycerides, and free fatty acids). These components widely differ in polarity and may be separated on silica columns using an alkane mobile phase with an alcohol such as 2-propanol as a polar modifier. An acid, acetic or formic, is often also included to suppress the ionization of free fatty acids and therefore allow their elution with reasonable peak shapes. Complex mixtures may require gradient elution although adequate group separations can often be achieved under isocratic conditions when refractive index detection can be employed. [*See* LIPIDS.]

The determination of free fatty acids is used as an indicator of quality, as the fatty acids predominantly result from lypolysis of triglycerides. The fatty acids may be directly determined by HPLC using refractive index detection, but it is more usual to form derivatives, e.g., phenacyl esters, which assist with both resolution and detectability. In practice, however, free fatty acids are more commonly determined by gas-liquid chromatography (GLC) after esterification to form volatile derivatives. This also applies to the determination of the fatty acid composition of triglycerides, which is readily achieved by GLC after transesterification, again to form volatile derivatives such as methyl esters.

The most important application of HPLC to lipid analysis is with respect to triglycerides. High-temperature capillary GLC can provide separations based on chain length (or carbon number), although there are problems with some components at the elevated temperature. HPLC on the other hand is a far milder technique, and so problems of deterioration during the analysis are not encountered. Ini-

tially, separation was achieved on reversed phase columns with mobile phases of methanol/water. However, this restricted applications to relatively small triglycerides, and much greater success was achieved with nonaqueous reversed phase systems using acetonitrile/acetone as mobile phase. Under these conditions it was noted that a double bond had the same effect on retention as a reduction in chain length of two carbon atoms. This finding introduced the concept of equivalent carbon number, defined as the number of carbon atoms minus twice the number of double bonds. Triglycerides with the same equivalent carbon number would then have similar retention characteristics, and their separation often proves difficult. Complex mixtures of triglycerides, especially those containing components with the same equivalent carbon number, require gradient elution to achieve adequate resolution. This method excludes the possibility of using refractive index detection, and UV absorption is weak for these compounds. The recently introduced mass detector and flame ionization detector have found excellent applications in this area, and an example of the resolution that can be realized is shown in Fig. 4.

Argentation chromatography has been well established for TLC for many years, and the technique has successfully been transferred to HPLC. Silica columns loaded with 10% silver nitrate have been used to separate triglycerides with differing degrees of unsaturation and also positions of the double bonds. This technique can be used in conjunction with reversed phase HPLC to provide additional information. It may also be used as a preliminary stage for HPLC.

The nontriglyceride fraction of lipids is mainly made up of polar components, of which the phospholipids are an important group. The major com-

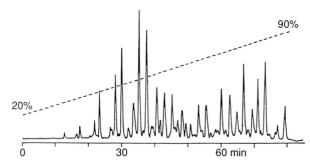

FIGURE 4 Butter triglycerides. Column, Spherisorb ODS 250 × 4.6 mm; solvent, acetone (20–90%) in acetonitrile; detection, mass detector. Elution in order of equivalent carbon number.

pounds (e.g., phosphatidyl choline, phosphatidyl ethanolamine, and sphingomyelin) can be separated on silica or reversed phase columns without any major problems. However, these compounds do not absorb in the UV and are not fluorescent, and so detection remains a problem, which has been partially solved by the advent of the mass detector and flame ionization detector. Similar detection problems also apply to the determination of cholesterol and other sterols, and for this reason their analysis still relies heavily on GLC.

VI. Food Additives

The range of food additives is enormous, and HPLC has found application for many of these compounds, with the possible exception of volatile flavor compounds for which GLC is a far more suitable technique. Applications to some of the more important areas illustrate the power and versatility of HPLC in these types of analyses. However, it is often the case that the HPLC stage *per se* is not a problem but rather the extraction from complex, and often processed, foods.

A. Acidulants

Organic acids are naturally present in many foods, e.g., dairy products, and they are also added to other products, particularly soft drinks. In some cases natural products are adulterated by the addition of acids to mask poor quality or even to disguise origin. The common organic acids may be separated on ion-exchange materials with simple aqueous buffers as mobile phases and refractive index detection. Some food extracts lead to rapid column deterioration, and alternative modes of chromatography have been employed. The polarity of the acids in solution can be reduced by ion suppression (low pH) or by the addition of an ion-pair reagent to the mobile phase, and then separation may be achieved on reversed phase columns. Sample clean-up required depends on the foods involved; removal of sugars may be essential especially when a nonselective detector such as refractive index is used. This may be achieved by a simple ion-exchange isolation.

B. Synthetic Dyes

Food colorants (synthetic) are mainly water-soluble azo dyes containing sulfonic acid groups. They may

be isolated from foods by aqueous extraction, preceded by enzymic digestion in appropriate cases, and then purified on polyamide columns before HPLC. Being ionic in nature they may be separated on ion-exchange columns, but more recently the tendency has been to use reversed phase columns with ion-pair reagents in the mobile phase. This does not always lead to reproducible chromatography, and simpler systems using ammonium acetate or high ionic strength mobile phases have been introduced. Maximum selectivity for detection is obtained by using a wavelength in the visible region corresponding to absorption by the dye (i.e., its complementary color). The same techniques may be employed to study dye degradation within food systems, and it has been shown that the stability of azo dyes depends heavily on the presence of other additives. Ascorbic acid has a particularly strong reducing effect on the azo group within the dyes, leading to loss of color.

Public concern over the safety of synthetic dyes has encouraged the use of natural colorants. These are generally multicomponent mixtures and are difficult to determine in food systems. The most commonly encountered colors are derived from anthocyanins (red/purple), carotenoids (red/yellow), and chlorophyll (green). HPLC can be used to characterize mixtures of natural pigments by providing profiles or to quantify specific compounds, where reference standards are available. These compounds are not as stable as their synthetic counterparts and degrade substantially during processing and cooking.

C. Antioxidants

Many foods contain natural antioxidants, such as tocopherols and ascorbic acid, but it is common practice to add synthetic antioxidants to increase storage stability of fat-rich foods. The most commonly encountered compounds of this type are butylated hydroxyanisole (BHA), butylated hydroxytoluene (BHT), and propyl gallate (PG). These may be extracted from fats directly into acetonitrile or aqueous alcohols or from hexane extracts of the foods. Steam distillation has also been used. Synthetic antioxidants may be readily separated on reversed phase columns with aqueous acetonitrile as mobile phase. Detection is straightforward with UV absorption at 280 nm, although enhanced selectivity can be realized with fluorescence or even electrochemical detection.

D. Preservatives

These compounds are added to foods to prevent, or inhibit, microbial growth. Sorbic, benzoic, and propionic acids, together with the esters of hydroxybenzoic acid, are the most commonly used preservatives. Sulfite is also extensively used but is not usually determined by HPLC, although the sulfite anion can be quantified by ion chromatography. Preservatives in liquid foods (e.g., wine) can often be directly determined with little sample preparation, whereas solid foods must first be extracted with a suitable organic solvent and the extract cleaned by solid phase extraction, if necessary. The acid preservatives may be analyzed on reversed phase columns with ion-pair modified mobile phases. Detection using UV absorption is adequate for most of the preservatives with the obvious exception of propionic acid, which only absorbs at low wavelength. The alkyl esters of hydroxybenzoic acid (parabens) can be separated on reversed phase columns with aqueous acetonitrile as mobile phase.

VII. Food Contaminants

Our food, apparently, is becoming more and more contaminated with an ever wider range of organic compounds. It is also true that our ability to determine a greater range of compounds at lower and lower levels is also increasing. These facts may not be unrelated. HPLC with its high sensitivity and ability to separate and hence to identify single compounds in complex mixtures is being used extensively in this area of analysis. The applications selected below give some idea of the versatility of this technique.

A. Polynuclear Aromatic Hydrocarbons

A wide range of polynuclear aromatic hydrocarbons (PAH) is formed during combustion and roasting, and they are therefore ubiquitous in nature; wood smoke is a particularly rich source. Certain members of this group (e.g., benzo(a)pyrene) are extremely carcinogenic, and it is debatable whether there is a "safe level" for these compounds in foods. However, some countries have adopted a maximum allowed level of 1 μg kg^{-1} in products such as smoked meat for benzo(a)pyrene. At these levels extraction from foods is not straightforward; most methods involve initial saponification of the foods followed by extraction into hexane. This crude extract is then further purified by re-extraction and open column chromatography on silica and/or alumina. The large number of PAHs likely to be encountered in a food extract, or an environmental sample, may require gradient elution for adequate resolution. Surprisingly, aqueous methanol, or acetonitrile, appears to provide sufficient solubility for use as mobile phase, albeit at low concentrations. Most of the PAHs are naturally fluorescent, and this method of detection is mandatory to obtain adequate sensitivity.

B. Nitrosamines

These compounds also appear to be widespread as contaminants of many industrial chemicals, including pesticides and cutting and hydraulic fluids. In foods the major sources of contamination appear to be from direct heating processes in which products from high-temperature oxidation of nitrogen come in contact with the food (e.g., kilning of malt in the past). Another major potential source of nitrosamines is from curing brines that contain nitrate

TABLE I

Compounds	Sample-type	Sample preparation
Glucose, fructose, sucrose	Fruit, drinks, processed foods	Water extraction-precipitation or solid phase extraction
Sugars-sugar alcohols	As above	As above
Triglycerides	Oils, fats	Solution in mobile phase
Tocopherols	Fatty foods, oils	Extraction, saponification, isolation of nonsaponifiables
Riboflavin	Fortified cereals	Extraction with dilute HCl, solid phase extraction
Amino acids	Protein hydrolysates	Acid hydrolysis, dissolve in buffer
Amino acids	Protein hydrolysates	Acid hydrolysis
Synthetic dyes	Beverages	Polyamide column
Aflatoxins	Groundnuts	Methanol or acetone, solid phase extraction
Polynuclear aromatics	Vegetable oils	Saponification, extraction, silica chromatography

and nitrite. The volatile nitrosamines such as *N*-nitrosodimethylamine can be determined by GLC or HPLC, and indeed the high resolution capability of capillary GLC and the ease with which this can be coupled to mass spectrometry probably make this the method of choice. However, in foods there is often a mixture of volatile and nonvolatile nitrosamines (e.g., *N*-nitrosopyrolidine from proline), and in these instances HPLC has considerable advantages in terms of compound stability. Reversed phase chromatography appears to separate most of the commonly encountered nonvolatile nitrosamines. The thermal energy analyzer provides a selective means of detecting nitrosamines, which can be used with GLC or HPLC. The separated compounds are pyrolyzed to form NO radicals, which react with ozone to form excited NO_2^*, and as this returns to its ground state, light is emitted. This detector is sensitive and selective; unfortunately it is not compatible with aqueous mobile phases.

C. Pesticides

Maximum residue levels have now been established for many pesticides in foods, based on the known, or predicted, toxicity of the compounds involved. These have raised stringent analytical requirements for residue analysis, and most of the methods are centered around GLC, with electron capture detection for organochlorine compounds, the flame photometric detector for phosphorous and sulfur compounds, and the thermionic detector for nitrogen compounds. HPLC has really only found application for those compounds that are thermally unstable (e.g., the carbamates). A wide range of carbamates are used both as insecticides and herbicides,

and many of these can be separated on silica columns with an alkane mobile phase containing 2-propanol as a polar modifier. A number of these compounds do not contain strong UV chromophores, and so detection must be carried out at low wavelength, which is not ideal for residue analysis.

D. Mycotoxins

Mycotoxins are secondary metabolites produced by certain molds, the most important of which are produced by the *Aspergillus, Fusarium,* and *Penicillium* genera. Several hundred of such toxins have been identified, but only a relatively small number of these have been found naturally in food systems; of these the aflatoxins have been most widely studied. The aflatoxins are strongly carcinogenic, and methods are required for their analysis at ppb levels in foods. This demands both efficient extraction and extensive clean-up before HPLC. The former may be achieved with a variety of solvents singly or in combination (acetone, methanol, or acetonitrile), while clean-up now relies heavily on the use of minicolumns or solid phase cartridges. Aflatoxins can be determined using either normal (Fig. 5) or reversed phase columns, the latter being preferred as they are more robust. Aflatoxins and many of the other mycotoxins are fluorescent, and the sensitivity of fluorescence detection is required to obtain the necessary limits of detection. [*See* MYCOTOXINS.]

In addition to the aflatoxins, established HPLC methods exist for ochratoxin, patulin, zearalenone sterigmatocystin, and the ergot alkaloids. The tricothecenes are normally determined by GLC, as they do not have natural fluorescence and only weak UV absorption.

Examples of High Performance Liquid Chromatography Food Analyses

Column	Mobile phase	Detector
Aminopropyl silica	Aqueous acetonitrile	Refractive index
Anion exchange	Sodium hydroxide	Pulsed amperometric
C₁₈ Reversed phase	Acetonitrile/acetone	Refractive index or mass detector
Silica	Dioxane/hexane	Fluorescence
C₁₈ Reversed phase	Methanol/water, heptane sulfonate	Fluorescence
C₁₈ Reversed phase	Acetonitrile/buffer gradient	Fluorescence (precolumn *o*-phthalaldehyde derivatization)
Cation exchange	Citrate buffers gradient	Postcolumn ninhydrin derivatization
C₁₈ Reversed phase	Ammonium acetate/methanol	Visible absorption
C₁₈ Reversed phase	Water/acetonitrile, methanol	Fluorescence
C₁₈ Reversed phase	Water/acetonitrile	Fluorescence

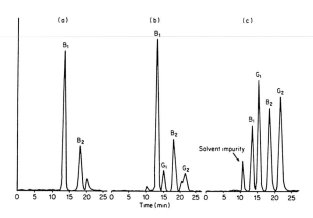

FIGURE 5 Aflatoxins in groundnuts. Column, silica A/10 250 × 2.6 mm; solvent, chloroform/dichloromethane (1 : 1, vol/vol); detection, fluorescence. (a) Naturally contaminated extract; (b) spiked extract; (c) standards, aflatoxins B_1, B_2, G_1, and G_2.

E. Other Contaminants

In addition to the above classes of compounds many other contaminants in foods have been studied by HPLC, including

- antibiotic and steroid residues in meats
- plasticizers from packaging materials
- taints (e.g., chlorophenols)

VIII. Conclusion

The versatility of HPLC allows many macro- and microconstituents of foods to be determined. In many instances recent developments in clean-up techniques have allowed HPLC to be applied to a wider range of compounds in more complex food matrices. Cooked, especially roasted, products tend to provide the food analyst with the most difficult task in terms of producing adequately clean extracts. This wide range of applications is illustrated by some selected methods shown in Table I. Improvements in column design, detector sensitivity, and automated sample preparation will further increase the applications of HPLC to this area of analysis.

Bibliography

De Leenheer, A. P., Lambert, W. E., and De Ruyter, M. G. M., eds. (1985). "Modern Chromatographic Analysis of the Vitamins." Marcel Dekker, New York.

Gilbert, J. (1984). "Analysis of Food Contaminants." Elsevier Applied Science Publishers, Barking, U.K.

Lawrence, J. F. (1984). "Food Constituents and Food Residues, Their Chromatographic Determination." Marcel Dekker, New York.

Macrae, R. (1988). "HPLC in Food Analysis," 2nd ed. Academic Press, London.

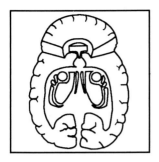

Hippocampal Formation

DAVID G. AMARAL, *The Salk Institute*

Glossary

Anterior Toward the front of the brain

Anterograde amnesia Loss of the ability to acquire and/or recall new information

Axon Output process of neurons which forms connections between brain regions

Cytoarchitectonics Characteristic organization and packing density of neurons. A major criterion used in defining boundaries of brain regions

Dendrites Treelike processes of neurons which form their receptive surface

Lateral Toward the side of the brain

Medial Toward the middle of the brain

Posterior Toward the back of the brain

Temporal lobe One of the four lobes of the human brain, the others being the frontal, parietal, and occipital. The temporal lobe is located on the side of the brain in the region below and behind the temple.

Ventricles System of cavities in the brain that are filled with cerebrospinal fluid

THE HIPPOCAMPAL FORMATION is a brain region located in the inner, or medial, portion of the temporal lobe. It comprises four subregions: the dentate gyrus, the hippocampus, the subicular complex, and the entorhinal cortex. These four subregions are linked by prominent connections which tend to unite them as a functional entity. The major behavioral function associated with the hippocampal formation is memory. Damage to the human hippocampal formation results in anterograde amnesia. The hippocampal formation is particularly sensitive to a variety of traumas and disease states and is often damaged in anoxia/ischemia, epilepsy, and Alzheimer's disease.

I. Location and Structure of the Hippocampal Formation

The most distinctive subregion of the hippocampal formation, and the one from which it takes its name, is the hippocampus. The term *hippocampus* (or sea horse) was first applied in the 16th century by the anatomist Arantius, who considered the three-dimensional form of the grossly dissected human hippocampus to be reminiscent of this sea creature (Fig. 1). Others likened the hippocampus to a ram's horn, and De Garengeot named the hippocampus "Ammon's horn" after the mythological Egyptian god. The terms *hippocampus* and *Ammon's horn* (or Cornu Ammonis) are now used synonymously.

The hippocampal formation is located in the medial portion of the temporal lobe, and the hippocampus and dentate gyrus form a prominent bulge in the floor of the lateral ventricle (Fig. 1). The hippocampal formation is widest at its anterior extent, where it bends toward the medial surface of the brain. The subtle bumps (or gyri) formed in this region give it a footlike appearance, and the name *pes* (foot) *hippocampi* has classically been applied to this area. The main portion, or "body," of the hippocampus becomes progressively thinner as it bends posteriorly and upward toward the corpus callosum (Fig. 1).

The medial surface of the hippocampus contains a flattened bundle of axons called the fimbria (Fig. 1 and 2). Axons originating from neurons in the hip-

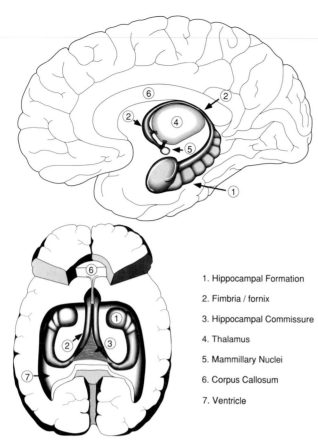

FIGURE 1 The position of the hippocampal formation and related structures in the human brain. The top image represents the medial surface of the human brain. The front of the brain is to the left. The position of the hippocampal formation in the medial portion of the temporal lobe is illustrated. The subcortically directed output bundle of axons, here labeled the fimbria/fornix, can be seen to arc over the thalamus and to ultimately descend into the diencephalon. Connections are made both with the mammillary nuclei and with the thalamus. The drawing in the lower left of the illustration is a cutaway of the brain as viewed from above and behind. The front of the brain is at the top of the image. The C-shaped structure of the hippocampal formation can be seen to occupy the floor of the lateral ventricles.

1. Hippocampal Formation

2. Fimbria / fornix

3. Hippocampal Commissure

4. Thalamus

5. Mammillary Nuclei

6. Corpus Callosum

7. Ventricle

pocampus and subicular complex travel in a thin layer called the *alveus* that sheaths the hippocampus and coalesce in the medially situated fimbria. At rostral levels, the fimbria is thin and flat but it becomes progressively thicker caudally as fibers are continually added to it. As the fimbria leaves the posterior extent of the hippocampus, it fuses with the ventral surface of the corpus callosum and travels anteriorly in the lateral ventricle. The major portion of the rostrally directed fiber bundle is called the body of the fornix. At the end of its anterior trajectory, the body of the fornix descends and

is called the column of the fornix. The fornix then divides around the anterior commissure to form the precommissural fornix, which enters the basal forebrain and the postcommissural fornix, which terminates in the diencephalon. Near the point where the fimbria fuses with the posterior portion of the corpus callosum, some of its fibers extend across the midline of the brain to form the hippocampal commissure. A variety of gross anatomical terms have been applied to the commissural fibers but the term *psalterium* (alluding to a harplike stringed instrument) is most common.

The four subregions of the human hippocampal formation can be differentiated cytoarchitectonically in neuroanatomical preparations in which thin sections of the brain have been stained to show the distribution of neuronal cell bodies (Fig. 2). The dentate gyrus is the simplest of the subregions and has a trilaminate appearance. The principal cell layer of the dentate gyrus (the granule cell layer) is populated primarily by one class of neuron, the granule cell. The human dentate gyrus contains approximately 9 million granule cells on each side of the brain. The dendrites of the granule cells extend into the overlying molecular layer, where they receive their main input from the entorhindal cortex.

The human hippocampus can be divided into three distinct fields labeled CA3, CA2, and CA1 (Fig. 2). The distinction of the three fields relates primarily to differences in their connections and more subtle differences in neuronal size and shape. All of the hippocampus has essentially one cellular layer, the pyramidal cell layer, which is populated by neurons with triangularly shaped cell bodies (the pyramidal cells). Pyramidal cells have dendrites emanating from their top (or apical) and bottom (or basal) surfaces. On each side of the brain, there are about 2 million neurons in the CA3 region, 220,000 in CA2, and almost 5 million neurons in field CA1. The relatively acellular layers above and below the pyramidal cell layer all have distinctive names. The outside limiting surface of the hippocampus located deep to the pyramidal cell layer is formed by axons of the pyramidal cells and is called the alveus (Fig. 2). Between it and the pyramidal cell layer is stratum oriens, which contains the basal dendrites of the pyramidal cells. The region above the pyramidal cell layer contains the apical dendrites of the pyramidal cells and is divided into several strata (stratum lucidum, stratum radiatum, stratum lacunosum-moleculare). Different connections are formed in each of these strata.

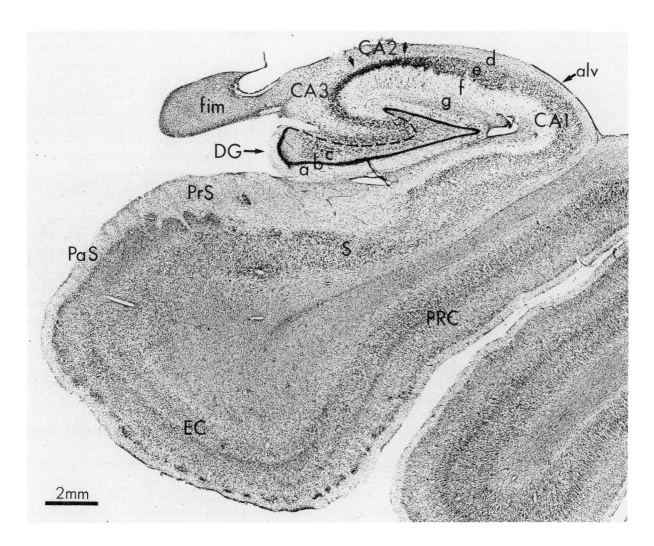

FIGURE 2 A thin section cut through the hippocampal formation of the human brain that is stained for the visualization of neurons. Each dark spot indicates the location of one neuronal cell body. The four major fields of the hippocampal formation can be differentiated in this type of preparation. The dentate gyrus (DG) has three layers: the molecular layer (a), granule cell layer (b), and polymorphic cell layer (c). The hippocampus can be divided into three fields, CA3, CA2, and CA1. The covering of the hippocampus is composed of axons originating from the pyramidal cells and is called the alveus (alv). The next layer, stratum oriens (d) contains the basal dendrites of the pyramidal cells. The cell bodies of the pyramidal neurons are contained in the pyramidal cell layer (e). The apical dendrites of the pyramidal cells extend into the overlying stratum radiatum (f) and stratum lacunosum-moleculare (g). The subicular complex comprises three distinct regions: the subiculum (S), the presubiculum (PrS), and the parasubiculum (PaS). The last subregion of the hippocampal formation is the entorhinal cortex, which is a multilaminate cortical area that resembles the neocortex. The adjacent perirhinal cortex (PRC) is one source of sensory information to the hippocampal formation.

The subicular complex can be subdivided into three distinct fields, the subiculum proper, the presubiculum, and the parasubiculum (Fig. 2). As in the hippocampus, one principal cell layer in the subiculum is populated by about 2.5 million neurons on each side. Details concerning the laminar organization of the other components of the subicular complex, the presubiculum and parasubiculum, are complex and not yet well understood. The subicular complex is an important region of the hippocampal formation, however, because it originates the major connection with subcortical regions such as the thalamus and hypothalamus. [*See* HYPOTHALAMUS; THALAMUS.]

The term *entorhinal cortex* was coined by the early neuroanatomist Korbinian Brodmann to name the region that lies adjacent to the shallow rhinal sulcus in nonhuman brains. More than any other subregion of the hippocampal formation, the entorhinal cortex has undergone substantial regional and laminar differentiation in the human brain. Unlike the other hippocampal subregions, the entorhinal cortex is a multilaminate cortical region that resembles the cytoarchitectonic appearance found in the neocortex. It is distinguished from the neocortex by the presence of clusters of darkly stained neurons that constitute the first cell layer (layer II) located just below the surface of the brain. It is also distinct in lacking a layer of small granular cells that typically forms the fourth layer in neocortex.

II. Connections of the Hippocampal Formation

The connections between the four subregions of the hippocampal formation, i.e. the intrinsic connections, and between the hippocampal formation and other portions of the brain, i.e. the extrinsic connections, have been the topic of extensive neuroanatomical study for more than a century. The fundamental intrinsic circuitry of the hippocampal formation and its major inputs and outputs are illustrated in Fig. 3.

The four subregions of the hippocampal formation are connected by unique and largely unidirectional connections (Fig. 3). The entorhinal cortex can, for convenience, be considered the first step in the intrinsic hippocampal circuit. Cells located primarily in layers II and III of the entorhinal cortex project to the molecular layer of the dentate gyrus. The connection between the entorhinal cortex and the dentate gyrus is called the *perforant pathway*. Some of the entorhinal projections also terminate in the subiculum and in the CA1 and CA3 fields of the hippocampus.

The dentate granule cells give rise to axons, the mossy fibers, that form connections with pyramidal cells of the CA3 region of the hippocampus. The other main constituent of the granule cell layer is the dentate basket pyramidal cell that gives rise to a dense plexus of fibers and terminals that surround the granule cell bodies. These basket cells are known to use the inhibitory neurotransmitter, γ-aminobutyric acid (GABA). Other classes of neurons in the dentate gyrus form a variety of feedback and feedforward circuits within the dentate gyrus.

The pyramidal cells of CA3 give rise to axons that project to other levels of CA3 (associational connections) and to subcortical regions, especially the septal nuclei. CA3 cells also contribute the major

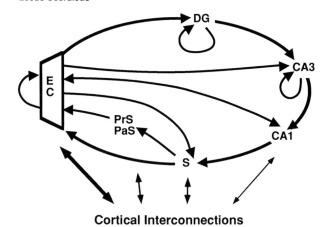

Subcortical Inputs

Amygdala
Claustrum
Septal Nuclei
Basal Nucleus (Meynert)
Supramammillary Nucleus
Anterior Thalamus
Midline Thalamus
Ventral Tegmental Area
Raphe Nuclei
Locus Coeruleus

Subcortical Outputs

Olfactory Regions
Claustrum
Amygdala
Septal Nuclei
Nucleus Accumbens
Caudate and Putamen
Hypothalamus
Mammillary Nuclei

Cortical Interconnections

Perirhinal Cortex (Areas 35 and 36)
Parahippocampal Cortex (Areas TF and TH)
Cingulate Cortex
Piriform Cortex
Insular Cortex
Orbitofrontal Cortex
Superior Temporal Gyrus

FIGURE 3 Summary of the major intrinsic and extrinsic neural connections of the hippocampal formation. The oval in the center of the illustration represents the hippocampal formation. Arrows indicate the direction of the major intrinsic hippocampal connections. The major subcortical inputs and outputs of the hippocampal formation are listed at the top of the illustration. Several cortical areas that are interconnected with the hippocampal formation are listed on the bottom of the illustration. The cortical interconnections with the hippocampal formation are primarily made with the entorhinal cortex.

input system to hippocampal field CA1 (the Schaffer collaterals). The CA3 field of the hippocampus also contains a number of nonpyramidal cells, many of which form local circuits within CA3. The pyramidal cells in the CA1 field have a pattern of connections that is quite distinct from the one in CA3. What is perhaps most striking is that CA1 pyramidal cells do not project significantly to other levels of CA1, i.e. there are virtually no associational connections within CA1. Rather, CA1 pyramidal cells project predominantly to the subiculum. The subiculum, in turn, projects to the presubiculum and parasubiculum, and all three components of the subicular complex project to the entorhinal cortex.

One of the more striking features of the intrinsic circuitry of the hippocampal formation is that it is largely unidirectional. The CA3 field does not project back to the granule cells of the dentate gyrus, for example. Nor do CA1 pyramidal cells project back to CA3. Thus, aside from the initial entorhinal input that reaches all of the hippocampal fields in parallel, information flow from the dentate gyrus through the other fields follows an obligatory serial and largely unidirectional pathway. This is in marked distinction to the situation in most other cortical regions, where connections are usually reciprocated.

A variety of chemical substances (neurotransmitters) mediate the transfer of information within the hippocampal circuitry. As in all other cortical regions, some of these substances excite the neurons onto which they are released and some are inhibitory. The hippocampal formation is particularly rich in the glutamate family of transmitters and receptors. One class of receptor, the *N*-methyl *D*-aspartate (NMDA) receptor, has been implicated both in the modulation of neural activity that accompanies learning and memory and with the pathological activation of neurons that accompanies ischemia-induced neuronal cell death.

The fimbria and fornix form the classical efferent, or output, system of the hippocampal formation, and the human fornix is said to contain about 1.2 million axons. The precommissural fornix, which primarily innervates the septal nucleus and other basal forebrain structures, arises mainly from neurons of the hippocampus and, to a lesser extent, from the subiculum and entorhinal cortex. Axons originating in the subicular complex are mainly connected to the diencephalon, i.e., the thalamus and hypothalamus, particularly the mammillary nuclei (Fig. 1).

Connections from one brain region to the same region on the other side of the brain are called *commissural connections*. In nonprimate brains, the hippocampal formations of both sides are heavily interconnected. However, in the primate brain, including humans, the side-to-side interconnections of the hippocampal formation appear to be rather meager. This anatomical observation is consistent with the behavioral finding that damage to the hippocampal formation on one side of the human brain preferentially impairs memory for either verbal (on the left) or spatial (on the right) types of information. This indicates that memory function may be lateralized in the human hippocampal formation.

One fact that has been uncovered about the human hippocampal formation comes from the analysis of patients who have had bilateral damage to the hippocampal formation. These people are unable to learn or retrieve new information about their day-to-day existence. Their ability to recall old memories, however, remains largely preserved. This implies that memory is not stored in the hippocampal formation but perhaps in one of the brain areas that the hippocampal formation communicates with. Thus, it is important to know from what brain regions the hippocampal formation receives its information about ongoing events (sensory information) and to what brain regions the hippocampal formation delivers its processed information. These latter regions would be strong candidates for memory storage sites.

For many years it was thought that the hippocampal formation received little input from the higher processing centers of the neocortex. This view was based largely on neuroanatomical studies conducted in rodents. However, with neuroanatomical studies of the organization of the nonhuman primate hippocampal formation, it has become clear that this area is in receipt of substantial sensory information from the highest levels of the neocortex. Several cortical regions, located mainly in the frontal and temporal lobes, are connected to the entorhinal cortex. Information is then relayed to the other hippocampal fields via the intrinsic connections. It is also quite clear that the entorhinal cortex sends return projections back to the neocortex. Thus, the hippocampal formation is privy to high level sensory information processing that takes place in the cortex and has the connections necessary to send its own processed information back to potential storage sites in the neocortex. [*See* NEOCORTEX.]

III. Behavioral Functions of the Hippocampal Formation

The hippocampal formation has historically been implicated in a variety of functions. In the latter part of the 19th century, for example, the neurologist Wilhelm Sommer considered the hippocampal formation to be a component of the motor system because he found that damage to the hippocampal formation correlated with the seizure disorder associated with temporal lobe epilepsy. For much of the first half of this century, the hippocampal formation was thought to be primarily related to olfactory function and was consider to be a prominent component of what was called the rhinencephalon, or olfactory brain. But evidence that indicated that the hippocampal formation was as prominent in anosmic species as in species which rely heavily on the sense of smell put this notion to rest. In the 1930s the neurologist James Papez considered the hippocampal formation to be a central component in a system for emotional expression. In his view, the hippocampus was something of a conduit by which perceptions could be collected and channeled to the hypothalamus to recruit appropriate emotions. Unfortunately, there has been little substantiation of Papez's theory, and the role of modulator of emotional expression is now more closely linked with another prominent medial temporal lobe structure, the amygdaloid complex.

Perhaps the most widely accepted and long-lived proposal of hippocampal function relates to its role in memory. It has been known for nearly a century that damage to certain brain regions can result in an enduring amnesic syndrome that is characterized by a complete, or near complete, anterograde amnesia. Affected patients are incapable of recreating a record of day-to-day events although most past memories remain largely intact. It is now clear that damage isolated to the human hippocampal formation is sufficient to produce this form of memory impairment.

While fairly convincing evidence now exists that the hippocampal formation plays a prominent role in the formation of enduring memories, the mechanism by which it exerts its influence is far from clear. Since damage to the hippocampal formation does not cause a loss of distant or well-established memories, it appears that the hippocampal formation cannot be the final repository for stored information. Rather, it appears that the hippocampal formation must interact with storage sites in other, presumably cortical, regions to consolidate ephemeral sensory experiences into long-term memory.

Electrophysiological studies conducted primarily in rodents have demonstrated that neurons in the hippocampus are preferentially activated by certain aspects of the environment. If one records the neural activity of a single hippocampal cell while a rat is running around in a maze, for example, the cell might be activated only when the rat travels through a certain location of the maze. Data of this type have prompted the suggestion that the hippocampus can form a "cognitive map" of the outside world. In a more general sense, it might be though that the neurons of the hippocampal formation, acting as an assembly of differentially activated units, can form

a representation of ongoing experience. Perhaps the interaction of this hippocampal representation of experience with the more detailed information of the experience located in the neocortex is the route through which long-term memories are formed. One implication of the electrophysiological data is that neurons in the hippocampal formation are not uniquely sensitive to certain types of information. Rather, neurons in the hippocampal formation may act more like random-access memory (RAM) in a computer and are therefore potentially activated by all types of information. Since it would be difficult for evolution to anticipate all the various forms of information that might need to be stored as memory, a generalized memory buffer system would be highly adaptive.

IV. Sensitivity of the Hippocampal Formation of Illness

Various clinical conditions result in morphological alterations of the human hippocampal formation. While the causative factors are not yet known for most of these disease states, it is clear that each of the different hippocampal cytoarchitectonic fields are more or less vulnerable to damage. In ischemia and temporal lobe epilepsy, for example, the field CA1 of the hippocampus (the so-called Sommer's sector) suffers the greatest neuronal cell loss. In other neuropathological conditions, such as Alzheimer's disease, the entorhinal cortex may suffer greater pathology. Several neuroscientists have suggested that the physiological "plasticity" inherent in the hippocampal formation as a biological memory device may predispose it to selective vulnerability to a variety of environmental and biological stressors.

Among the many conditions that produce pathological changes in the hippocampal formation, Alzheimer's disease is probably the most devastating. Alzheimer's disease is an age-related neurodegenerative illness that results in profound memory impairment and dementia. A number of pathological profiles, including senile (or neuritic) plaques and neurofibrillary tangles are consistently seen in some of the hippocampal fields, especially the entorhinal cortex. Ultimately, Alzheimer's disease leads to a massive death of neurons in the hippocampal formation and other brain regions. There is good reason to suspect that the memory impairment associated with Alzheimer's disease is due, in large part, to the devastation of the hippocampal formation by the disease. [*See* ALZHEIMER'S DISEASE.]

Temporal lobe or complex partial epilepsy is another neurological disorder in which the hippocampal formation is severely affected. This most common form of epilepsy was first associated with damage to the hippocampal formation in the late 1800s by Sommer who conducted the first postmortem microscopic examination of a brain from a long-term epileptic patient. Sommer noted a dramatic loss of neurons in the hippocampal formation that was relatively selective and involved a region that in modern terminology would encompass CA1 and part of the subiculum. In approximately two-thirds of the cases of temporal lobe epilepsy, the hippocampal formation is the only structure that shows pathological modifications. During the first part of this century, it was generally believed that hippocampal pathology was a consequence of the epileptic seizures rather than their cause. Currently, however, increasing emphasis is being placed on the idea that disruption of normal hippocampal function

may be an initiating factor in temporal lobe seizures. [*See* EPILEPSY.]

In a number of other pathological conditions the hippocampal formation is preferentially damaged. Among these is the loss of neurons, primarily in CA1, consequent to the ischemia associated with cardiorespiratory arrest. As noted above, these patients often demonstrate an anterograde memory impairment apparently resulting from the hippocampal damage.

Bibliography

Amaral, D. G., and Insausti, R. (1990). The hippocampal formation. *in* "The Human Nervous System" (G. Paxinos, ed.). Academic Press, San Diego.

Squire, L. R., Shimamura, A. P., and Amaral, D. G. (1989). Memory and the hippocampus. *in* "Neural Models of Plasticity" (J. Byrne and W. Berry, eds.). (pp. 208–239) Academic Press, San Diego.

Chan-Palay, V., and Kohler, C. (eds.) (1989). "Hippocampus New Vistas." Alan R. Liss, New York.

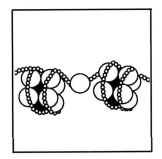

Histones and Histone Genes

GARY S. STEIN, JANET L. STEIN, *University of Massachusetts Medical Center*

Glossary

Cell cycle Interval between the completion of mitosis in the parent cell and the completion of the next mitosis in one or both progeny cells; the periods of the cell cycle are sequentially defined as mitosis (prophase, metaphase, anaphase, and telophase), G_1 (the period between the completion of mitosis and the onset of DNA replication), S phase (the period of the cycle during which DNA replication occurs), and G_2 (the period between the completion of DNA replication and the onset of mitosis)

Histone proteins Five principal species of basic chromosomal proteins designated H2a, H2b, H3, H4, and H1, which range in size from 11,000 to 25,000 kDa; histone proteins complex with DNA to form the primary unit of chromatin structure, the nucleosome

Mitosis Period of the cell cycle during which cell division occurs: prophase, metaphase, anaphase and telophase

Nucleosome Primary unit of chromatin structure in eukaryotic cells consisting of approximately 200 nucleotide base pairs of DNA and two each of the core histone proteins (H2a, H2b, H3, and H4).

Post-transcriptional control Components of gene expression involving regulation mediated at the level of messenger RNA processing within the nucleus and/or cytoplasm, the translatability and/or stability of mRNA, and the assembly or post-translational modifications of polypeptides

Post-translational modification Modifications of protein primary structure that occur subsequent to

protein synthesis by the ribosomes (e.g., phosphorylation [covalent addition of phosphate groups], glycosylation [addition of carbohydrate residues])

Promoter regulatory elements DNA sequences, generally but not necessarily 5' (upstream) from the messenger RNA transcription initiation site, which modulate the specificity and/or level of transcription

Transcriptional control Component of gene expression involving the synthesis of RNA, utilizing DNA as a template

HISTONE PROTEINS ARE the molecules responsible for packaging DNA into chromatin, the protein–DNA complex that constitutes the eukaryotic genome. Histone–DNA complexes form the primary unit of chromatin structure, the nucleosome. Equally important, modifications in histone–DNA interactions occur in association with modifications in the expression of specific genes. Human histone genes have been cloned and characterized with respect to the regulation of expression that occurs in proliferating cells at the time when DNA is replicated, providing histone proteins to package newly replicated DNA into chromatin. Regulatory sequences of the histone genes, which determine the specificity and levels of transcription, as well as factors that bind to regulatory elements to mediate histone gene expression, have been identified.

I. Histone Proteins

A. General Properties

There are five principal species of histone proteins designated H2a, H2b, H3, H4, and H1. These low-molecular weight chromosomal polypeptides (Fig. 1) range in size from 11,000 to 25,000 daltons. They

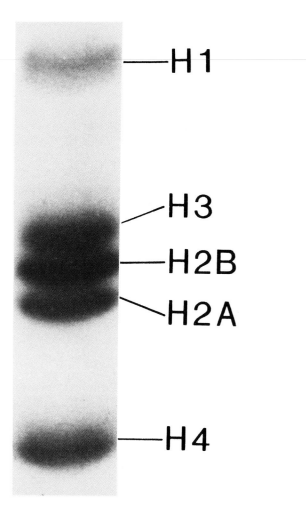

FIGURE 1 The five principal classes of histone polypeptides (H1, H3, H2b, H2a, and H4) fractionated electrophoretically in acetic acid–urea polyacrylamide gels.

share the common feature of a net positive charge due to a high representation of the basic amino acids arginine, lysine, and histidine, which facilitates the interactions of histones with negatively charged DNA molecules. The amino acid sequences of the histone proteins have been highly conserved during evolution, as illustrated by only limited amino acid substitutions in the histones of organisms separated as far phylogenetically as plants and mammals. This retention of the histone amino acid sequences reflects the conserved role of these proteins in chromatin structure and the apparently stringent requirement to support the similarly conserved primary unit of chromatin structure, the nucleosome (see below).

Despite the conserved nature of the histone proteins, these chromosomal polypeptides are encoded in a multigene family with approximately 20 non-

identical copies of each core (H2a, H2b, H3, and H4) and H1 gene. As a result, the five principal classes of histone polypeptides can be separated into several groups: (1) those that are represented in most cells and tissues and synthesized only in proliferating cells at the time of DNA synthesis (>90%); (2) those that are found in many cells and tissues but are expressed independently of proliferation, either constitutively during the cell cycle or following the completion of proliferation at the onset of tissue-specific gene expression associated with differentiation; and (3) those that are expressed solely in specialized cell types, such as sperm, in which there are highly specific requirements for modifications in the packaging of DNA into chromatin. Additional heterogeneity of the histone proteins is reflected by a series of post-translational modifications that include acetylation, methylation, phosphorylation, and adenosine diphosphate (ADP)-ribosylation. Such modifications alter the distribution of charge in specific domains of the histone proteins and may influence histone–DNA, as well as histone–histone, interactions. In the case of acetylation and phosphorylation, nuclear deacetylases and phosphatases permit the removal of acetate and phosphate moieties from histone polypeptides. These post-translational modifications are involved in the incorporation of newly synthesized histones into chromatin (e.g., histone–DNA binding or histone–histone interactions) and may provide a basis for changes in the interactions of histones with DNA for remodeling chromatin architecture. An example of a dramatic reorganization in chromatin structure and organization is the condensation of chromatin into discrete and identifiable chromosomes at the onset of mitosis, which is accompanied by changes in histone phosphorylation. A more subtle, yet functionally important, modification in chromatin structure, which is reflected by changes in histone–DNA interactions occurs when the expression of specific genes is activated or repressed. [*See* CHROMATIN FOLDING.]

B. Chromatin Structure

The magnitude of the problem associated with chromatin structure is illustrated by a requirement for the ordered packaging of 2.5 yards of DNA within the confines of the cell nucleus in a manner that supports the selective expression of specific genes required for the biogenesis and maintenance of cellular phenotypes. An equivalent amount of histone and DNA are present in cell nuclei, forming complexes known as nucleosomes, which are the primary unit of chromatin structure (Fig. 2). Each nucleosome consists of a core particle of approximately 140 base pairs of DNA wound around a complex consisting of two H2a, H2b, H3, and H4 molecules and a linker DNA region of approximately 40–60 base pairs. Under the electron microscope, the nucleosomes appear as a series of beads (protein–DNA complexes) on a string (linker DNA joining the nucleosomes). The H1 histones bind to the linker region and are involved with nucleosome–nucleosome interactions. While the interactions of histones with DNA and the H3–H4 and H2a–H2b histone interactions within the nucleosomes have been firmly established, this organization only accounts for a packing ratio of seven. Clearly, higher-order structural organization of chromatin is required for accommodating the genome within the cell nucleus, and here our understanding of chromatin structure is minimal. While additional insight into the interactions of histone with DNA is required, it is also necessary to further define the role of ''nonhistone,'' sequence-specific DNA-binding proteins in directly mediating DNA conformation and in modulating histone–DNA interactions within the nucleus. [*See* DNA IN THE NUCLEOSOME.]

II. Histone Gene Expression

Historically, the observations that the cellular content of histone proteins doubles during the S phase of the cell cycle (the period of the cell cycle when DNA replication occurs) and that histone protein

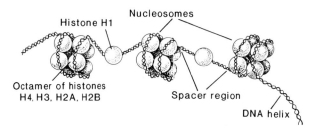

FIGURE 2 Schematic representation of nucleosomes, the primary unit of chromatin structure. As indicated, each nucleosome consists of approximately 140 base pairs of DNA wound around a complex consisting of two H2a, H2B, H3, and H4 histone protein molecules and 40–60 base pairs of linker DNA. The H1 histone proteins are involved with nucleosome–nucleosome interactions.

synthesis occurs concomitantly with DNA synthesis provided the first example of gene expression functionally related to cell growth. As such, investigations in several laboratories over the past two decades have focused on the regulation of histone gene expression within the context of understanding the control of gene expression as it relates to the complex and interdependent series of events required for cell proliferation. The selective synthesis of histone proteins during the S phase of the cell cycle is mediated by the regulation of histone gene expression at both the transcriptional and post-transcriptional levels. At the transcriptional level, the extent to which messenger RNA (mRNA) is transcribed from histone genes is regulated. Post-transcriptional regulation involves the control of histone mRNA processing and/or stability.

Multiple levels of control have been established by addressing experimentally the relationship of DNA replication to histone protein synthesis, cellular histone mRNA levels and histone gene transcription. The rationale for this approach is that a stoichiometric relationship among DNA replication, histone protein synthesis, and histone mRNA levels would be indicative of transcriptionally mediated expression, while the transcription or presence of histone mRNAs in cells when histone proteins are not synthesized would be a direct indication of post-transcriptional control. Differences, and particularly fluctuations, in the relationship between the principal parameters of histone gene expression—mRNA levels, mRNA synthesis, and protein synthesis—would suggest that both transcriptional and post-transcriptional regulation are operative.

A functional, as well as temporal, relationship between DNA replication and the expression of human core and H1 histone genes was initially indicated by the constant histone–DNA ratio (1:1) observed in a broad spectrum of cells, tissues, and organs and the doubling of cellular levels of histone protein during the S phase of the cell cycle. Direct measurements then confirmed that histone protein synthesis is largely confined to S phase and that inhibition of DNA replication results in a rapid cessation of histone protein synthesis. Cellular levels of histone mRNAs have been measured throughout the cell cycle in continuously dividing cell populations and after stimulation of quiescent cells to proliferate. In all cases, the cellular levels of human histone mRNA accumulation reflect cellular levels of both histone protein synthesis and DNA replication. Similarly, inhibition of DNA replication brings about a dose-dependent loss (selective destabilization) of histone mRNAs, which parallels decreases in DNA and histone synthesis. Measurements of histone gene transcription indicate enhanced synthesis of histone mRNAs early during the S phase of the cell cycle. A summary of the principal biochemical events associated with histone gene expression during the cell cycle is presented in Fig. 3. [*See* DNA SYNTHESIS.]

A viable model for regulation of human histone gene expression must therefore account for (1) an increased transcription of histone genes at the time DNA replication is initiated; (2) elevated cellular levels of histone mRNA and histone protein synthesis in conjunction with DNA replication; and (3) a rapid and selective destabilization of histone mRNAs concomitant with the termination of DNA replication at the natural end of S phase or following inhibition of DNA synthesis.

The increased transcription of histone genes early during S phase and the coordinate accumulation of mRNAs for core and H1 histone proteins that closely parallels the initiation of DNA and histone protein synthesis suggest that the onset of histone gene expression is at least in part transcriptionally mediated. In fact, it is reasonable to postulate that throughout S phase the synthesis of histone proteins is modulated by the availability of histone mRNAs. The stabilization of histone mRNAs throughout S phase and the destabilization of histone mRNAs when DNA replication is completed or inhibited are highly selective and largely post-transcriptionally controlled. The selectivity of histone mRNA destabilization is suggested because both the natural end of S phase and inhibition of DNA replication are associated with a rapid loss of histone mRNAs, while only minimal fluctuations, qualitative or quantitative, are observed in other cellular mRNA populations under these conditions. The extremely tight coupling of histone gene expression with the extent of ongoing DNA replication is supported by the coordinate and stoichiometric decreases of histone mRNAs and histone protein synthesis after inhibition of DNA replication. [*See* DNA AND GENE TRANSCRIPTION.]

In summary, the preferential expression of histone genes during the S phase of the cell cycle apparently involves control at both transcriptional and post-transcriptional levels. The initiation of DNA replication in human cells is associated with a three- to fivefold increase in histone gene transcription, which after several hours returns to a basal level that is maintained outside of S phase. This increase represents enhancement rather than an activation of histone gene transcription and, hence, the mechanisms that regulate histone gene transcription can be expected to differ from those observed when activation of nonexpressed genes is initiated. In contrast, the selective destabilization of histone mRNAs at the end of S phase or after inhibition of DNA synthesis is post-transcriptionally mediated.

III. Organization and Regulation of Human Histone Genes

A. General Organization

The human histone genes are organized into clusters of core alone (H2a, H2b, H3, and H4) or core together with H1 histone coding sequences. Several such segments of human genomic DNA containing histone coding sequences are illustrated schematically in Fig. 4. Chromosome mapping studies based on *in situ* hybridization of metaphase chromosomes with radiolabeled human histone gene probes, Southern blot analysis of DNAs from panels of mouse–human somatic cell hybrids, and hybridization with DNAs from flow-sorted human chromosomes have indicated that the human histone gene clusters are represented on at least chromosomes 1 and 6. Within these clusters, there is generally a pairing of H2a with H2b genes and H3 with H4 genes. This organization of human histone genes is similar to that observed for mouse and chicken, but somewhat more complex than the simple, tandemly repeated clusters found in lower eukaryotes such as sea urchin and *Drosophila*.

B. Organization of a Cell Cycle-Regulated Human Histone Gene

Despite the clustering of human histone genes, each histone coding sequence is an independent transcription unit. All amino acids of the histone protein are encoded in contiguous nucleotides because these genes lack introns. Also noteworthy are the absence of a polyadenylation site in the 3' region and the nontranslated leader and trailer segments of the mRNA that are <50 nucleotides. This simple organization of the mRNA coding region of the histone gene appears to be optimal for rapid processing of the mRNA transcript and export to the cyto-

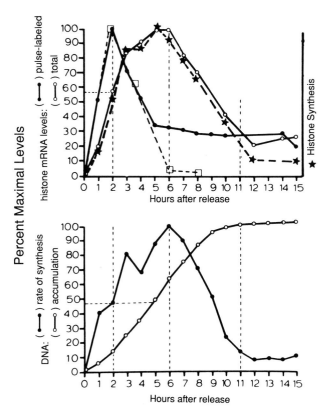

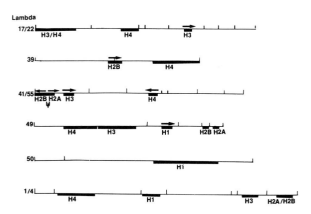

FIGURE 4 Organization of cloned human histone genes. Six cloned segments of human DNA containing histone genes are schematically illustrated. The locations of core (H2a, H2b, H3, and H4) and H1 histone coding sequences are designated. Horizontal arrows indicate the direction of RNA transcription.

FIGURE 3 Expression and regulation of histone genes during the cell cycle. The upper panel shows cellular levels of histone mRNAs determined by Northern blot analysis using ^{32}P-labeled cloned human histone genes as hybridization probes (O——O); histone mRNA synthesis measured by incorporation of ^{3}H-uridine into histone mRNA in intact cells (●——●); histone protein synthesis (★---★); and replication of core and H1 histone genes measured by hybridization of BUDR-substituted DNA with cloned human histone genes (□---□). The lower panel shows relative rates of DNA synthesis monitored by pulse labeling with ^{14}C-thymidine (●——●) and the accumulation of DNA calculated from rates of DNA synthesis (O——O).

plasm for immediate use as a template for histone protein synthesis.

The 5′ flanking regions of histone genes contain consensus regulatory sequences of many genes transcribed by RNA polymerase II (Fig. 5). Proximal regulatory elements include a TATA box approximately 30 base pairs upstream from the transcription start site and a CAAT box further upstream but within the initial 100 base pairs of 5′ sequences. As has been observed for all genes that have been analyzed for the localization of promoter regulatory elements, an extensive series of sequences that influence both specificity and level of transcription have been identified in the initial 1,000 nucleotides of the 5′ promoter region.

C. Regulatory Factors That Control Expression of Human Histone Genes

Two approaches have been pursued to identify and characterize proteins that bind in the 5′ flanking regions of human histone genes at regulatory elements that influence transcription. The first experimental approach has been to establish the sites of protein–DNA interactions in the histone gene promoter in intact cells. The second approach has been to assay the binding of factors in fractionated nuclear extracts to isolated sequences upstream from cell cycle-regulated histone genes.

In vivo protein–DNA interactions were established by a technique known as "genomic sequencing," in which the guanine sequencing reaction of Maxam and Gilbert is carried out in intact cells. Cells are treated with dimethylsulfate, which methylates the N7 position of guanine residues; after isolation of the DNA, it is treated with base to cause breakage of the strand at the modified guanine residue. The extent to which each guanine residue is methylated, compared with its reactivity in isolated DNA, reflects sites of specific protein–DNA interactions. After fractionation of the DNA by electrophoresis in sequencing gels and subsequent transfer to a nylon membrane, the sequences are visualized by hybridization with radiolabeled probes followed by autoradiography. Sites of protein–DNA interactions can be detected at single nucleotide resolution. Schematically diagrammed in Fig. 5 are a series of protein–DNA interactions that have been established both in intact cells and by *in vitro* pro-

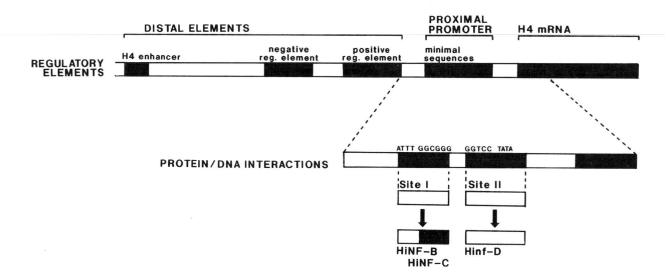

FIGURE 5 Diagram showing the organization of the regulatory elements (distal elements and proximal promoter) that regulate expression of an H4 histone gene. The modular organization of the regulatory sequences is reflected by enhancer, negative regulatory elements, and positive regulatory elements. Protein–DNA interactions in the proximal promoter of the gene at two primary regulatory elements in the proximal promoter are designated (HiNF-B and HiNF-C factors at site I and HiNF-D factor at site II).

tein-binding studies using isolated promoter elements for sequence-specific protein binding.

Taken together with the contribution of multiple regulatory elements to the specificity and level of histone gene transcription, this series of promoter–factor interactions reflects the modular organization of the cell cycle-regulated histone genes. A single, rate-limiting step to the transcription of the cell cycle-regulated histone genes has not been established; rather a series of sequences comprising both positive and negative regulatory elements contribute individually as well as synergistically to determine the extent to which the gene is expressed.

Bibliography

Heintz, N., Sive, H. L., and Roeder, R. G. (1983). Regulation of human histone gene expression: Kinetics of accumulation and changes in the rate of synthesis and in the half-lives of individual histone mRNAs during the HeLa cell cycle. *Mol. Cell. Biol.* **3,** 539–550.

van Holde, K. E. (1988). "Chromatin." Springer-Verlag, New York.

Marzluff, W. F., and Pandey, N. B. (1988). Multiple regulatory steps control histone mRNA concentrations. *Trends. Biochem. Sci.* **13,** 49–52.

Pauli, U., Chrysolgelos, S., Stein, G., Stein, J., and Nick, H. (1987). Protein–DNA interactions in vivo upstream of a cell cycle regulated human H4 histone gene. *Science* **236,** 1308–1311.

Sierra, F., Lichtler, A., Marashi, F., Rickles, R., Van Dyke, T., Clark, S., Wells, J., Stein, G., and Stein, J. (1982). Organization of human histone genes. *Proc. Natl. Acad. Sci. USA* **79,** 1795–1799.

Stein, G., Lian, J., Stein, J., Briggs, R., Shalhoub, V., Wright, K., Pauli, U., and van Wijnen, A. J. (1989). Altered binding of human histone gene transcription factors during the shutdown of proliferation and onset of differentiation in HL-60 cells. *Proc. Natl. Acad. Sci. USA* **86,** 1865–1869.

Stein, G. S., Stein, J. L., and Marzluff, W. F. (eds.) (1984). "Histone Genes." John Wiley and Sons, Inc., New York.

Hormonal Influences on Behavior

EDWARD P. MONAGHAN, S. MARC BREEDLOVE, *University of California, Berkeley*

Glossary

Androgen Class of steroid hormones, the primary source of which is the testes. The major circulating androgen is testosterone

Endogenous Originating from within the organism

Estrogen Class of steroid hormones, the primary source of which is the ovaries. The major circulating estrogen is estradiol

Exogenous Originating from outside the organism

Gonads The sex organs; testes in the male, ovaries in the female

Hormone Chemical secreted by one cell and capable of affecting other cells throughout the organism

Perinatal Around the time of birth

Perineum Genital region and structures occupying the pelvic floor

Sexual dimorphism Structural or behavioral difference between males and females of the same species

Steroids Class of hormones with cholesterol as the common precursor. The primary sources of steroids are the gonads, placenta, and adrenal cortex

HORMONES, without doubt, can influence the behavior of animals, including humans. However, the causes of behavior are seldom simple, and it is often difficult to conclude that hormone X causes behavior Y. Rather, through controlled study and manipulation of hormone levels in various situations, we can demonstrate that a given hormone is involved in certain behaviors and whether it increases or decreases the likelihood that the behavior will appear.

I. Introduction

Before dealing with particular hormones and behaviors, we need to discuss briefly the mechanisms and timing of hormone actions and the scientific approaches to investigating hormonal influences on behavior. Classically speaking, a hormone is a chemical substance that is secreted into the bloodstream by an endocrine gland and that travels to and influences particular target tissues. To produce an effect in the target tissue, the hormone must bind to a specific receptor that is itself a protein. Much of this chapter will deal with steroid hormones, such as androgens and estrogens, that readily diffuse into cells. Steroid receptors are contained within the target cells and, when bound to the appropriate hormone, interact with the cell's DNA to modulate gene activity. The other class of hormones, protein or peptide hormones, usually bind to receptors on the external surface of the cell membrane and trigger a series of internal events that greatly vary from hormone to hormone and even from tissue to tissue. Two points should be emphasized: first, if a tissue or an individual lacks receptors, the hormone cannot have its effect, and second, the hormone does not completely change the target, but instead acts by adjusting ongoing processes. [*See* ENDOCRINE SYSTEM; STEROID HORMONE SYNTHESIS.]

When can hormones act to influence behavior? The organizational/activational model for hormone action proposes that during early development hormones permanently organize neural substrates such that the individual will be capable of displaying particular behavioral patterns in response to later hormone exposure. As we will discuss later, this model is best illustrated by sexually dimorphic behaviors (i.e., those behavior patterns more common in one sex than the other). For example, the comparatively high testosterone (T) level in male rats around the time of birth organizes their central nervous system (CNS) (i.e., the brain and spinal cord) such that the rise in T levels during puberty activates male-specific behavior patterns. Treatment of adult female rats with T produces little masculine behavior. Thus early exposure to some hormones can regulate neural development, which will then determine hormone responsiveness later in life.

The study of hormonal involvement in a particular behavior is a three-step process. We start by observing and recording behavior and plasma hormone levels in an intact subject. Next we remove the endocrine gland, which produces an implicated hormone, and note the effect on the behavior. Finally, if removal of the hormonal source affects behavior, we supply, via injection, exogenous hormone and see if the behavior is restored to normal levels. Obviously, such controlled surgical manipulations are not carried out on human subjects. Most basic biomedical research is conducted on animals, and it is important to note that although there are species differences, the information gained in animal studies provides the basic framework to explain human phenomena. In this chapter we will focus on human behavior, but references will be made to animal research, which has provided particular insight.

II. Sexual Behavior

A. Introduction

It has often been stated that the further advanced along the "evolutionary ladder" an animal is, the less influence gonadal hormones have on its behavior. Thus hormones seem to play a greater role in sexual behavior of rats than of primates. This may be an oversimplification. The effects of hormonal alterations may be easier to see in rodents than in primates, partly because of the greater control we have over extraneous variables when dealing with rodents. It is difficult to conduct well-controlled experiments involving primates (especially humans), but recent findings have suggested that gonadal hormones do play an important role in the behavior of "higher animals."

Hormones are essential for the expression of sex-ual behavior in vertebrates. They are involved in everything from the fetal development of sexual structures to the production of copulatory behavior. It may not be immediately obvious why we mentioned the development of sexual structures when we are concerned with behavior, but looking at behavior without looking at morphology can lead to confusion. Whenever a behavior involves two or more individuals, whether they are rats or humans, the physical appearance of each individual can profoundly affect the resultant behavior. Furthermore, research on animals has shown that the same signals that direct masculinization of the body also direct masculinization of the nervous system and therefore behavior. With this in mind, we will begin by briefly describing sexual differentiation, the process by which a fetus develops as either male or female.

Early during fetal development, the individual's genetic sex, XX or XY, determines the differentiation of the gonads: a genetic female, XX, will develop ovaries, and a genetic male, XY, will develop testes. From that point on, further development of phenotypic sex (the male or female body type) depends not on the sex chromosomes directly, but on hormones secreted by the gonads. In the normal case, the testes of an XY individual release hormones that masculinize bodily structures: T acts to masculinize the external genitalia, and another testicular factor works in concert with T to differentiate the internal structures. It is the absence of these testicular secretions in the XX individual that results in the feminine phenotype. In addition to determining these somatic characters, hormones also act on the CNS of developing embryos to affect the likelihood that the individual will engage in certain behaviors later in life; this is referred to as the organizational role of hormones in behavior of animals. The early exposure to androgen seems to organize irreversibly the CNS in a masculine fashion. The challenging question, which is largely unanswered, is whether this is also true for humans.

Because it is our assumption that any differences in behavior must be reflected by differences in the nervous system, it is worth while to look at some of the sexual dimorphisms demonstrated in the CNS, especially those found in both human and nonhuman animals. Firstly we will consider a sexual dimorphic area in the hypothalamus: that region of the brain involved in homeostasis and appetitive behaviors. The preoptic area (POA) of the hypothalamus appears to be important in sexual behavior because lesions there selectively abolish male sexual behavior in a variety of species. Located within this region is the sexually dimorphic nucleus of the POA (SDN-POA). The SDN-POA of both rats and humans is larger in males than in females. Hormonal manipulation in rats have shown that it is the presence of high levels of perinatal T that cause males to have a larger SDN-POA. Although we know that the POA is involved in sexual behavior, the behavioral relevance of the SDN portion of the POA has not been established. [See HYPOTHALAMUS.]

A second sexual dimorphism of the CNS is located in the spinal cord of both rats and humans. Male rats have a group of large motoneurons called the spinal nucleus of the bulbocavernosus (SNB) that innervate and control muscles of the perineum, specifically those attached to the penis. Both the motoneurons and their muscles are absent in adult female rats. At birth, both sexes have SNB cells and the perineal muscles, but perinatal T levels prevent the death of these cells in males. Adult T levels regulate activity of this neuromuscular system, and as might be expected, it is important in male copulatory behavior. These same perineal muscles in humans are innervated by an homologous spinal area called Onuf's nucleus, which has more motoneurons in males than in females. Continued research involving sexual dimorphisms will lead to a greater understanding of how early androgens influence neural development and therefore how differences in behavior arise in both rats and humans. [See SPINAL CORD.]

The final aspect of sexual differentiation in humans is the formation of gender identity, which is believed to be completed early in childhood and refers to how an individual perceives oneself (i.e., as either male or female). There are conflicting opinions as to what determines gender identity, but a reasonable hypothesis is that prenatal hormone levels and postnatal environmental factors (experience and learning) interact in the process of psychosexual development. As we are about to see, these three aspects of sexual identity (genotypic sex, phenotypic sex, and gender identity) are not always in agreement.

B. Syndromes Affecting Sexual Development

1. Androgen Insensitivity

Several syndromes alter the course of sexual differentiation. The most striking disagreement between genetic and phenotypic sex is the androgen insensitivity syndrome (AIS). AIS individuals are genetically male (XY) but, because of a genetic mutation, lack androgen receptors. During development the Y chromosome causes their gonads to differentiate as testes that release T, but because there are no receptors present, body tissues fail to respond and therefore a feminine phenotype results. In cases of complete insensitivity, the external appearance of such individuals is quite feminine, and they are often undetected and are raised and behave in a normal female pattern, except that at puberty menstruation fails to appear.

Animal studies have indicated that it is a metabolite of T, estradiol (see Fig. 1), that is responsible for much of the sexual differentiation of brain tissue and that this neural masculinization ultimately affects behavior. Because AIS individuals have receptors for estradiol, we might predict that masculinization of the neural structures would occur despite the feminine appearance. In that case, we would expect problems in the development of gender identity and sexual orientation, but no such problems are reported. The lack of difficulty in psychosexual development of AIS individuals suggests that either these early hormones do not influence the formation of gender identity and sexual orientation or that consistent environmental factors can override hormonal cues. The difficulty in deciding between these two possibilities becomes especially evident when we consider the conflicting conclusions reached in the study of other syndromes.

2. 5 Alpha-Reductase Deficiency

As the name suggests, this syndrome is due to a deficiency in the enzyme, 5 alpha-reductase, which, as illustrated in Fig. 1, plays an essential role in allowing T to masculinize the external genitalia during development. Individuals born with this syndrome are genetically male (XY) but because of the enzyme deficiency, they do not convert T to the more active androgen, DHT; therefore they are born with external genitalia that are feminine or at least ambiguous: the scrotum resembles labia and the size of the phallus is intermediate between that of a clitoris and a penis. This disorder is common in a small village in the Dominican Republic where it is referred to by the villagers as "guevedoces"—penis at twelve. Reportedly, these individuals are raised as girls until puberty, when testicular T secretions trigger development of the male phenotype (i.e., deepening of the voice, increase in muscle

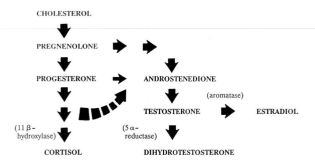

CHOLESTEROL

PREGNENOLONE

PROGESTERONE → ANDROSTENEDIONE

(aromatase)

TESTOSTERONE → ESTRADIOL

(11β-hydroxylase) (5α-reductase)

CORTISOL DIHYDROTESTOSTERONE

FIGURE 1 Simplified schematic diagram of synthesis of some steroid hormones. Cholesterol is the precursor for all steroid hormones. In gonads, enzymes facilitate synthesis in the direction of androstenedione, a precursor to testosterone. Testosterone can then be converted to dihydrotestosterone (DHT) by 5-alpha reductase, which is present in many peripheral tissues, or to estradiol by aromatase, which is present in the ovaries and some areas of the central nervous system. One of the major steroids produced by the adrenals is cortisol, and the enzyme 11 beta-hydroxylase is important in its production. In the absence of 11 beta-hydroxylase (congenital adrenal hyperplasia syndrome), steroid production in the adrenals is shifted in the direction of androgens.

mass, enlargement of the phallus, and descension of testes). After puberty, their gender identity and sexual orientation becomes masculine. Some researchers have concluded from these cases that the environment (i.e., being treated as a girl throughout childhood) has little effect on adult gender identity and that the presence of T either early in development, at puberty, or both, completely determines psychosexual development as a male. An alternative explanation of the guevedoces phenomenon is that although the affected individuals look somewhat feminine, the villagers are clearly aware of some differences and treat them subtly differently than normal boys or girls. This enables a smooth assumption of a male gender identity around puberty. It is also possible that hormones have little influence on human gender development, and in a secure setting where such gender shifting is accepted, humans are so adaptable that they can accept a pubertal sex change, especially when the change brings the many social advantages accorded males in that culture.

3. Congenital Adrenal Hyperplasia

The adrenal glands are a source of steroid hormones, the corticosteroids. As can be seen in Fig. 1, the synthesis of adrenal and gonadal steroids is closely related, but normally only small amounts of androgen are released from the adrenals. However, in individuals with congenital adrenal hyperplasia (CAH), a deficiency in 11 beta-hydroxylase results in an abnormally high production of androgen. If this occurs in an XX individual, depending on the extent of the deficiency and the time it arises during development, the genitalia can be masculinized to varying degrees. The appearance of ambiguous genitalia at birth leads physicians to a diagnosis of CAH. The genitalia can be surgically corrected, and maintenance treatment with exogenous steroids can halt excess androgen production.

Because of differences in the time of initiation of hormone treatment and surgical correction and the possibility that the parents communicated ambiguities about the sex of their child, it is quite difficult to draw definitive conclusions about whether this early exposure to T itself affected the later behavior of these females. Despite problems finding appropriate comparison groups, some studies have reported a consistently higher level of early tomboyish behavior, and as adults, a higher incidence of bisexual fantasy and experience in CAH women than in AIS or normal women. However, the majority of the CAH females have exclusively heterosexual relations. Thus the studies of CAH women leave open the possibility of hormonal influences on gender development but do not provide evidence for an overpowering role for hormones.

C. Female Sexual Behavior

Thus far we have dealt exclusively with the effect of T on development and behavior. Although some T is secreted by the ovaries of adult females, estrogen (E) and progesterone (P) are the major steroid products of the ovaries. Hormone levels vary throughout a female's menstrual cycle (referred to as "estrous cycle" in nonhumans). E levels are high in the early to midcycle (i.e., before and around the time of ovulation). P levels are high late in the cycle, after ovulation. T levels, although always much lower than E, also peak around midcycle.

With these normal, predictable changes in hormone levels, one might expect experiments on hormonal effects on female sexual behavior to be among the first studies conducted in hormones and behavior. This is true when we consider the "lower" mammals in which the females will mate only when they are in hormonally activated estrus (heat), but human and many other primate females were considered to be sexually receptive throughout their cycle. Thus it was long concluded that their sexual behavior was independent of proximate hormonal stimulation, but scientists have recently taken a closer look at this proposal.

Female sexual behavior can be divided into three categories. Whether or not a female will copulate if approached by a male is referred to as receptivity. Proceptivity describes the actions a female will take to attract a male or to initiate a sexual encounter. Attractivity refers to how well an individual female serves as a sexual stimulus for a male. For example, male monkeys show preferences for sexual partners who are at a particular stage of their estrus cycle; therefore we could say that these females are more attractive. In terms of stimulus value, a given rat, monkey, or woman can have a different level of attractivity depending on her hormonal state.

Although rhesus females will allow males to mount (i.e., are receptive) at any point during their estrus cycle, it was found that around midcycle they were much more likely to work to gain access to a male, indicating that proceptive behavior peaks around the time of ovulation. Rhesus males are more attracted to females (i.e., are more likely to mount and continue until ejaculation) when the females are in midcycle and least likely to show extended mounting late in the female's cycle. We can thus see that although rhesus females can show receptive behavior independent of specific hormone level, other important components of sexual behavior *are* influenced by hormones. Animal studies such as these have led to a re-evaluation of hormonal effects on the sexual behavior of women.

Obviously, collecting data with human subjects is more difficult. Scientists usually rely on self-reports, which can be inaccurate and prejudiced. To determine how hormones might influence a woman's proceptive behavior, some researchers have asked that in addition to keeping track of sexual activity the couples record who initiates the encounter; some such studies have reported an increase in female-initiated encounters around midcycle. Another approach is to look at frequency of masturbation, and again there are reports of a midcycle peak. What hormone is responsible for the increased interest in sexual activity around midcycle? Both E and T levels peak at this time, but several lines of evidence implicate T as the effective hormone.

Most human studies that demonstrate a relation between hormones and female sexual behavior have correlated peak or average hormone level with overall sexual gratification, interest, and activity. No correlation was found between E levels and a variety of sexual behavior/interest measures. However, T levels positively correlated with sexual gratification, arousal, desire, and masturbation frequency. Further implication of T in female sexual behavior comes from hormone replacement therapy in women who have undergone hysterectomy. T, but not E, treatment increased sexual desire and fantasies, although E treatment did augment vaginal lubrication. However, neither hormone had an effect on frequency of copulation. Thus in women, hormones may not affect receptivity, but T seems to be involved in the proceptive or cognitive aspects of sexual behavior.

A discussion of female sexual behavior must address the premenstrual syndrome (PMS). During the last decade, PMS has become a popular topic not only in the general press and scientific literature, but also in the courtroom, where attorneys have used PMS as a defense in criminal cases. Despite all this attention there is little scientific understanding of the syndrome.

PMS encompasses a wide variety of disturbances including somatic complaints of bloating and breast tenderness, mood changes related to tension, anxiety, and irritability, and behavior changes such as crying, social withdrawal, and spousal fighting. Few clinicians or researchers doubt that many women suffer to varying degrees from some form of PMS, but there is much disagreement as to its treatment and cause. Early reports concluded that PMS was a result of abnormally high levels of prolactin, a peptide hormone released by the pituitary, and claimed that bromocriptine, a prolactin antagonist, alleviated symptoms in a group of patients. Later studies demonstrated that a placebo was equally effective in treating all the symptoms of PMS except breast pain.

Recent well-controlled studies measured a variety of hormonal levels throughout the menstrual cycle and compared women who suffer from PMS and those who show no symptoms; no differences in hormone levels were found. Thus there appears to be no simple hormonal basis for PMS. A possible explanation for the reported effectiveness of various treatments is that acknowledgment of the symptoms and attempts to provide relief serve to alleviate some of the suffering in women who previously had been told nothing was wrong. At present the best explanation for the cause of PMS involves differences in sensitivity or dynamic response to normal hormonal fluctuations between PMS sufferers and other women.

D. Male Sexual Behavior

When looking at the activational effects of T on male sexual behavior one general rule seems to emerge regardless of the species being considered—there is plenty of T. In normal males, whether rodents or men, individual differences in sexual behavior cannot be predicted from differences in T levels. Although levels vary, even those male rats with relatively low T concentrations have 2–3 times the amount of hormone necessary to maintain normal copulatory behavior. Behavioral differences between males with intact, functional testes are most likely a result of differences in their neural substrate, i.e., the quantity or dynamics of the hormone receptors, or neural connections, or nonhormonal influences.

Although low versus high levels of T do not explain differences in sexual behavior in male rats or humans, there is considerable evidence that having *no* T influences sexual behavior in both species. Hypogonadal men have *extremely* low T levels and show varying degrees of hyposexuality. Treatment with exogenous T restores normal levels of sexual activity and interest. To determine how T acts to affect male sexual behavior, researchers compared hypogonadal and normal men in their erectile responses. No differences were found in their ability to maintain erections in response to an erotic film. However, unlike normal males, hypogonadal men were unable to maintain erections when sexual fantasy was the only stimulus. These individuals also showed significantly fewer and less intense spontaneous erections. Taken together, these data suggest that the mechanical or physiological component of sexual arousal is not drastically influenced by T but that the hormone works at a motivational or cognitive level. In other words, although T is not necessary to produce erections, it may act to lower the threshold for sexual arousal.

Early in our discussion of hormones and behavior, we mentioned the organizational/activational hypothesis of hormone action. From the review of both male and female sexual behavior, it is evident that hormones do have an activational effect (i.e., the presence of hormones in the adult influences the expression of sexual behavior). Although humans are not totally dependent on the presence of gonadal hormones for the expression of sexual behav-

ior, there is evidence that the presence or absence of gonadal hormones can affect many aspects of sexual interactions in both men and women. However, in contrast to most other mammals, it has been difficult to demonstrate an organizational effect of early hormones on later sexual behavior in humans. The various human syndromes resulting in prenatal hormonal abnormalities do not always produce the alterations in sexual behavior predicted by animal models, and the problems that are observed could be explained by nonhormonal mechanisms. Presently, it is unclear whether humans differ from other mammals in the early action of hormones on the CNS or whether social factors are much more effective in overriding early hormonal organization in humans. [*See* SEXUAL BEHAVIOR, HUMAN.]

III. Aggressive Behavior

In almost all mammalian species, hamsters and hyenas being among the few exceptions, spontaneous intraspecies aggression is primarily a male phenomenon. This is one reason why it is widely accepted that in animals, aggression is very much dependent on T levels. Unfortunately, it is somewhat easier to find exceptions to this rule than support for it. In naive animals (those who were never before exposed to an aggressive situation), the presence or absence of T will affect whether they will fight or merely submit. But once an animal has experienced aggressive encounters, his T level has little effect on subsequent responses. On the contrary, success or failure in an aggressive encounter tends to affect T levels more than do T levels affect the probability or outcome of an aggressive encounter. Considering the difficulty in seeing a simple relation between hormones and aggression in animals, it is not surprising that the study of human aggression has proven exceedingly difficult. Many of the antagonistic encounters between people occur on the verbal level: Is this aggression? Animal researchers use tissue damage as the criterion for an aggressive act: Is corporate ''back-stabbing'' the modern socialized form of intrahuman aggression? The difficulty in answering such questions, and the tendency for many researchers in the area to ignore them, has resulted in a less than satisfying literature. [*See* AGGRESSION.]

Clinical research involving hormone manipulations has been conducted on men convicted of violent crimes. Most of the participants in these studies were sexual offenders, making interpretation difficult because of the combination of sexual and aggressive components of the act. Nonetheless, some interesting and consistent conclusions have been reached. In several studies participants were treated with drugs that act as antiandrogens either by reducing T levels or by interfering with hormone receptor dynamics. Recidivism in subjects treated with antiandrogens was significantly lower than in those individuals who did not receive drug treatment. While on the drug, the men reported a reduction in the frequency and intensity of fantasies and urges related to sexually deviant/aggressive activities. Researchers have concluded that the antiandrogens act to reduce an individual's motivation to engage in deviant behaviors, thus allowing for the adaptation of socially acceptable alternatives. However, men so motivated to reform that they will accept antiandrogen treatment may be less prone to recidivism even without the drugs.

In normal adult males, no clear correlation has been found between plasma T levels and nonsexually related aggression, although some evidence exists for such a relation in adolescents. Studies of high school-aged males have reported a greater incidence of hostile acts among individuals with high T levels. Such studies show a correlation (i.e., there is some relation between T and aggression in adolescents), but they do not demonstrate that T causes an increase in aggressive behavior. T may increase the motivation to engage in aggressive behaviors or permit the expression of such behaviors in individuals predisposed to aggression due to social factors. Another possible explanation involves the peripheral effects of the hormone, (i.e., the appearance of the individual). Peripheral effects of T include increased muscle mass and body hair and this more "mature" or "intimidating" appearance may influence the way peers or authorities respond to that individual.

The importance of considering the appearance of an individual when assessing behavioral interactions is often neglected by researchers in behavioral endocrinology, who are often eager to jump to conclusions about neural substrates and motivation. For example, reports indicate a higher than expected proportion of XYY individuals involved in violent crimes. These individuals tend to develop faster and have higher T levels than their adolescent peers. Genetic and hormonal explanations have been offered to explain this high level of aggression. An alternate explanation might take into account the appearance of XYY adolescents. Actions that might be interpreted as adolescent pranks when committed by a normal 15-year-old may be viewed as violent when committed by an XYY individual if he looks more like an adult. Thus in humans, if we exclude sexually related actions, it is difficult to see a direct effect of hormones on aggressive behavior.

A great deal of attention has recently been focused on steroids in sports. A controversy has arisen about the role androgens may play in physical performance. As for endogenous levels, androgens reportedly rise before competition and remain high in the winners relative to the losers. However, the hormone levels both before and after a match appear to correlate more closely with the individual's mood than with his or her performance (i.e., the contestants with the highest spirits before a match had higher T levels, and among the winners, those who were most satisfied with their performance had the higher postmatch T levels). As for exogenous steroid use, much clinical research suggests that it does little to improve fitness or performance, despite widespread assumptions to the contrary. Why then is its illicit use so widespread in sports? In addition to a possible placebo effect, some researchers suggest that the dosage used by many bodybuilders is 10–100 times that used in clinical studies and that, at these levels, a significant effect on training may be seen. At such high doses, steroid use is exceedingly dangerous: in addition to the cardiovascular and liver damage that may result, many individuals show affective dysfunctions (e.g., depression, mania, hallucinations, and paranoid delusions). In a recent report, more than 20% of the subjects showed significant behavioral problems while on steroids. Even here we must consider the possibility that the behavioral problems may be caused by the internal conflict such users must feel about knowingly risking their health to achieve recognition or acceptance. Considering that the relation between endogenous steroids and behavior is complicated and often not well understood, the use of exogenous steroids may or may not be useful in producing the desired effects but definitely can have deleterious effects.

IV. Cognitive/Motor Tasks

There are a variety of tasks in which performance differs between men and women; tests involving dichotic listening, verbal, spatial, or complex manual skills are the most often reported. In addition to developmental and environmental factors that may contribute to these differences, recent reports suggest that hormonal levels at the time of testing can also affect such performance. [See SEX DIFFERENCES, PSYCHOLOGICAL.]

We will consider two tasks, one involving spatial skills in which men tend to score higher, the other a test of complex manual ability in which women excel. Women's performance on both tasks varies throughout the menstrual cycle and is correlated with the level of E and P. Scores on the manual dexterity test were higher at the midluteal phase when E and P are high than during menstruation when E and P levels are low. In contrast, women were more accurate in spatial judgments during menstruation than during the midluteal phase. Another study examined androgen levels and performance on spatial tasks and found that females with high androgen levels scored higher than those with low levels. The opposite trend was observed in men, those with high androgen levels scoring lower than those with low levels. Given that E and P are present in much higher concentrations in females and much more androgen is present in males, it is interesting to note that among females the performance on the manual dexterity task (female dominated) was highest when E and P levels were high and that females with high androgen levels performed best on spatial tests (male dominated). However, the data on men point out once again that simple relations between hormones and behavior are rare. Taken together, both studies suggest that observed differences in nonsexual and nonaggressive behaviors may also be subtly influenced by hormones.

V. Ingestive Behavior

Both the hormones and some of the behaviors we have discussed thus far are important in the survival of the species but are not necessary for the survival of an individual. This enabled us to study how drastic hormonal manipulations affected various aspects of behavior. Unlike gonadal hormones, an individual is dependent on adrenal, thyroid, pancreatic, and pituitary hormones for survival, and it is obvious that behaviors such as drinking and eating cannot be drastically altered without deleterious effects to the individual. As we will see, when considering such behaviors, it is easy to demonstrate simple relations with some hormones, but because the be-

haviors are so essential, auxiliary mechanisms of control have also evolved.

Two hormones are known to play important roles in water retention and excretion: aldosterone and vasopressin (also known as antidiuretic hormone). Aldosterone is a steroid hormone secreted by the adrenal cortex; it acts on the kidneys to reduce salt excretion. Vasopressin is a peptide hormone secreted by the posterior pituitary; it stimulates the reabsorption of water by the kidneys. The release of vasopressin is under neural control, but aldosterone secretion is dependent on the plasma levels of another hormone, angiotensin. Both vasopressin and aldosterone act to conserve water, and loss of function of either of these hormones leads to increased thirst and therefore increased water consumption in both humans and other mammals.

An important factor in the control of feeding behavior is the level of glucose in the blood. Insulin, a peptide hormone secreted by the pancreas, is a major determinant of blood glucose levels. The presence of insulin in the bloodstream allows muscles and organs to take up and use circulating glucose; in the absence of insulin, only the brain and liver can use this nutrient. An injection of insulin will deplete blood glucose levels and therefore stimulate hunger, which will result in feeding behavior. High blood glucose levels, such as after a meal, stimulate the release of insulin, which will allow its uptake and storage by peripheral cells. However, a small release of insulin can be stimulated by the taste or smell of food, or by some conditioned signal that a meal is eminent. This is why light hors d'oeuvres may increase our appetite—their taste and smell can trigger a pulse of insulin that depletes blood glucose and therefore increases hunger. In individuals afflicted with diabetes mellitus, the pancreas does not produce insulin; if left untreated, these people will increase their food intake and maintain high blood glucose levels, but because their cells cannot use the energy, they grow thin and eventually quite ill. [See INSULIN AND GLUCAGON; PEPTIDE HORMONES OF THE GUT.]

Several neural and hormonal factors appear to play roles in cessation of feeding behavior. The presence of food in the stomach both stimulates stretch receptors, which directly signal the brain to stop eating, and causes the gut to release peptide hormones, glucagon and cholecystokinin. The precise role these and other hormones play in producing satiety is not yet clear.

In addition to providing necessary energy, feeding behavior serves to maintain and, in some cases, increase body energy stores of fat. The question of how the body monitors and maintains a particular mass of adipose tissue has intrigued researchers as well as the weight-conscious public. A recent theory suggests that the brain monitors the average blood insulin levels over relatively long periods. The amount of food intake, adipose tissue, and activity interact to affect plasma insulin. Average insulin levels may indicate the level of fat storage and may in turn influence long-term feeding behavior.

VI. Conclusion

Hormonal effects on behavior are obviously extensive. In this chapter we touched on some of the areas that have been vigorously researched and in which a fair level of understanding has been attained. There are several other areas in which hormones are known to play a role. For instance, the gonadal, adrenal, thyroid, and pituitary hormones have all been implicated in depression, but as is often the case, it is not known whether it is dysfunction in the endocrine glands that contributes to the development of depression or whether the depressive disorder causes abnormal hormonal activity. Thus our discussion leads to a general conclusion that the role of hormones in human behavior is important and extensive but rarely simple.

Acknowledgments

We thank Nancy Forger and John Dark for their comments on the manuscript.

Bibliography

Davidson, J. M., and Myers, L. S. (1988). Endocrine factors in sexual psychophysiology. *In* "Patterns of Sexual Arousal" (R. C. Rosen and J. G. Beck, eds.). The Guilford Press, New York.

DeVries, G. J., DeBruin, J. P. C., Uylins, H. B. M., and Corner, M. A., eds. (1984). "Progress in Brain Research, vol 61: Sex Differences in the Brain." Elsevier, New York.

Fishman, R. B., and Breedlove, S. M. (1988). Sexual dimorphism in the developing nervous system. *In* "Handbook of Human Growth and Developmental Biology, vol 1, Part C: Factors Influencing Brain Development." (E. Meisami and P. S. Timeras, eds.). CRC Press, Boca Raton, Florida.

Hines, M. (1982). Prenatal gonadal hormones and sex differences in human behavior. *Psychol. Bull.* **92**(1), 56–80.

Reinish, J. M., Rosenblum, L. A., and Sanders, S. A., eds. (1988). "Masculinity/Femininity: Basic Perspectives." Oxford University Press, New York.

Rosenzweig, M. R., and Leiman, A. L. (1982). "Physiological Psychology." D. C. Heath and Company, Lexington, Massachusetts.

Sherwin, B. B. (1988). A comparative analysis of the role of androgen in human male and female sexual behavior: Behavioral specificity, critical thresholds, and sensitivity. *Psychobiology* **16**(4), 416–425.

Strunkard, A. J., and Stellar, E., eds. (1984). "Eating and Its Disorders." Raven Press, New York.

Svare, B. B., ed. (1983). "Hormones and Aggressive Behavior." Plenum Press, New York.

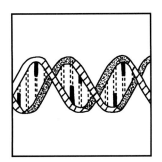

Human Genome and Its Evolutionary Origin

GIORGIO BERNARDI, *Institut Jacques Monod*

Glossary

Recombination Any process that gives rise to cells or individuals (recombinants) associating in new ways two or more hereditary determinants (genes) by which their parents differed

Sister chromatids Two chromatids derived from one and the same chromosome during its replication in interphase

Translocation Chromosomal structural change characterized by the change in position of chromosome segments (and the genes they contain)

THE BASIC QUESTION this article addresses is of the organization of nucleotide sequences in the human genome and the evolutionary origin of such organization. It shows that, far from being a "bean bag," the genome is highly ordered from the chromosomes down to the nucleotide level. Nucleotide sequences, whether in genes or in the very abundant nongenic segments, obey precise compositional rules. On the other hand, compositional rules also are evident in chromosome structure in that DNA composition is responsible for chromosome bands. These structural properties are associated with functional properties and shed a new light on genome evolution.

I. Introduction

A. The Genome

Every living organism contains, in its genome, all the genetic information that is required to produce its proteins and that is transmitted to its progeny. The genome consists of deoxyribonucleic acid (DNA), which is made up of two complementary strands wound around each other to form a double helix. The building blocks of each DNA strand are deoxyribonucleotides. These are formed by a phosphate ester of deoxyribose (a sugar), linked to one of four bases: two purines (adenine [A] and guanine [G]) and two pyrimidines (thymine [T] and cytosine [C]). In the DNA double helix, purines pair with pyrimidines (A with T, G with C) and the phosphates bridge the paired building blocks of the two strands to form the double helix. [*See* DNA AND GENE TRANSCRIPTION; GENOME, HUMAN.]

During cell replication, the two strands of the double helix are unwound, and a complementary copy of each is made, producing two identical copies (except for rare mistakes, or mutations) of the parental double helix. The two strands are also unwound at the time when one strand, the sense-strand carrying the genetic information, is copied into a complementary ribonucleic acid (RNA), which differs from the DNA master copy in having ribose instead of deoxyribose and uracil instead of thymine. RNA transcripts of genes are used as templates for the synthesis of proteins. The translation of each RNA transcript into the corresponding protein involves a very complex machinery that makes use of ribosomes (particles made up of two subunits, each containing a ribosomal RNA) and transfer RNAs specific for different amino acids. It will suffice here to mention that subsequent sets of three adjacent nucleotides (or triplets, also referred to as codons) of the transcript specify amino acids that follow each other in the protein chain. Because there are 64 triplets (three of which are termination codons, marking the end of translation) and only 20 amino acids, all amino acids, except for methionine and tryptophan, are encoded by more than one codon. In other words, several synonymous codons may be used to specify the same amino acid; therefore, the genetic code is said to be degenerate, which means that alternative possibilities (synonymous codons) exist for the same amino acid. Differences among synonymous codons are mainly at the third codon positions.

The genome of living organisms greatly differs in size, from 4.2 Mb (megabases, or millions of base pairs [bp]) for a typical bacterium such as *Escherichia coli* to 3,000 Mb or 3 Gb (gigabases, or billions of bp) for a eukaryote such as humans. While prokaryotes (bacteria) are characterized by small genome sizes, clustering around the value given above for *E. coli,* eukaryotes exhibit larger genome sizes that cover a wide range—from 20 Mb for the yeast *Saccharomyces cerevisiae* to 3 Gb for mammals.

The much larger genome size of higher eukaryotes is not only due to the presence of a larger number of genes (see below). In fact, the increase in size is mainly due to noncoding sequences that are both intergenic and intragenic. The latter sequences, called introns, separate different coding stretches, or exons, of most eukaryotic genes. The intron part of the primary transcript is eliminated by splicing; leaving the mature transcript or messenger RNA that encodes a given protein.

Eukaryotes differ from prokaryotes in other respects as well. They have a nucleus that is separated from the cytoplasm by a nuclear membrane. In addition to the nuclear genome, eukaryotic cells also have organelle genomes, which are located in mitochondria and, in the case of plants, also in chloroplasts. Organelle genomes contain a very limited yet an essential amount of genetic information (genes) encoding organelle-specific proteins and RNAs. Organelle genomes apparently originated from symbiotic bacteria, which entered proeukaryotic cells. Like the bacterial genomes, organelle genomes are physically organized in a rather simple way. In contrast, in the nuclear genome of eukaryotes DNA is wrapped around histone octamers to form nucleosomes, which are packaged into chromatin fibers. In turn, these fibers are folded into chromatin loops, consisting of 30–100 Kb (kilobases, thousands of bp) of DNA, which are, in turn, packaged into chromosomes.

B. The Human Genome

The nuclear genome of humans consists of about 3 billion base pairs, whereas the mitochondrial genome is only 16,000 bp. Estimates of the number of (nuclear) human genes range from 30,000 to 100,000. If coding sequences average 1,000 bp, they would represent 1–3% of the human genome; 97–99% of which, therefore, is made up of noncoding sequences. It should be noted that the larger number of genes in humans compared with bacteria is mainly due to the fact that many human genes exist as multigene families, the result of gene duplications during evolution.

Our present knowledge of human coding sequences, in terms of primary structures (or nucleotide sequences), is limited to about 1,000 of them. In other words, we only know about 1–3% of our coding sequences, or 0.03% of our genome. Our knowledge of noncoding sequences in terms of primary structure is even more limited. Nevertheless, we know that a sizable part of intergenic noncoding sequences are formed by repeated sequences that belong in several families. Two important families are called LINES and SINES (the long and short interspersed sequences), which are present in about 100,000 and 900,000 copies, respectively. These families of repeated sequences have been largely studied by reassociation kinetics. This experimental approach is based on the fact that, if small DNA fragments (having a size of 300–400 bp) are separated into their complementary strands, the reassociation of the latter proceeds at a rate that depends on the frequency in the genome of the sequences present in the fragments. Single strands from sequences that are present very many times in the genomes will find their complementary strands faster than single strands from genes that are present a few times or even only once in the genome. This technique allows estimating the relative amounts of repetitive sequences and single-copy sequences, the latter being present only once or a small number of times in the genome.

The other level of knowledge of the human genome concerns chromosomes. Each human germ cell (sperm or oocyte) contains 23 chromosomes. In haploid cells, 22 chromosomes (1–22 in order of decreasing size) are autosomes, which are identical in both sexes. The 23rd chromosome, the sex chromosome, is an X chromosome in females and a Y chromosome in males. Somatic cells are diploid; they have two haploid chromosome sets. Female diploid cells have two X chromosomes (one of which is inactive), whereas males have one X and one Y chromosome. During mitosis, chromosomes condense and, at metaphase, they are characterized by specific staining properties. G-bands (Giemsa-positive, or Giemsa dark bands; equivalent to Q-bands, or Quinacrine bands) and R-bands (reverse bands; equivalent to Giemsa-negative or Giemsa light bands) are produced by treating metaphase chromosomes with fluorescent dyes, proteolytic digestion, or differential denaturing conditions. [*See* CHROMOSOMES.]

Under standard conditions, Giemsa staining produces a total of about 400 bands that comprise, on the average, 7.5 Mb of DNA. If staining is applied to prophase chromosomes, which are more elongated, up to 2,000 bands can be visualized. At this high resolution, one chromosomal band contains, on the average, 1.5 Mb of DNA. Prophase chromosomes can also be studied at meiosis, the process leading to haploid germ cells. Staining meiotic chromosomes (usually at pachytene) produces results that are similar to those just mentioned. Indeed, a pattern of chromatin condensations, the chromomeres, are visualized. Chromomeres and the interchromomeres separating them correspond to high-resolution Giemsa-positive and Giemsa-negative bands, respectively. A number of approaches, ranging from purely genetical to molecular ones have allowed assigning genes not only to individual chromosomes but also to chromosome bands.

II. Isochores and Genome Organization

Our present knowledge of the human genome is clustered around two levels of organization: genes, which are in a DNA size range of a few Kb, and chromosome bands, which are in a size range of a few Mb. We know very little about the intermediate size range.

Recent discoveries, however, not only have linked the gene level with the chromosome level of genome organization but have also shed light on the functional and evolutionary implications of genome organization. Indeed, the human genome is made

up of isochores, large DNA domains (>300 Kb) that are compositionally homogeneous and belong to a small number of families ranging from 35 to 55% GC (see Fig. 1; GC is the molar percentage of the two complementary bases G and C in DNA). This point was demonstrated in two ways. (1) The relative amounts of isochore families, as judged by fractionating human DNA fragments according to their base composition, is independent of molecular size between 3 and >300 Kb. The fact that breaking down DNA fragments to small sizes does not change the relative ratios of the compositional families obviously indicates compositional homogeneity in the larger fragments. (2) Hybridization of single-copy sequences to compositional fractions of DNA fragments about 100 Kb in size occurs within a very narrow GC range, about 1%. This means that all DNA fragments that carry a given gene are extremely close in composition despite the fact that they were derived from intact chromosomal DNA by a random breakdown due to the unavoidable physical and enzymatic degradation that occurs during DNA preparation.

The discovery of an isochore organization in the human genome (and in the eukaryotic genomes in

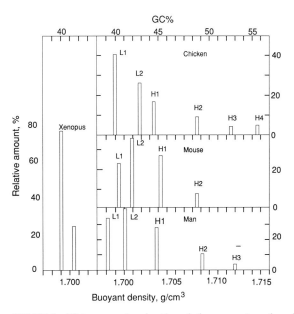

FIGURE 1 Histograms showing the relative amounts and modal buoyant densities in CsCl and GC levels of the major DNA components (L1, L2, H1, H2, H3, H4) from *Xenopus,* chicken, mouse, and man. Satellite (clustered repeated sequences) and minor components (like ribosomal DNA) are not shown. [Reproduced, with permission, from G. Bernardi (1989). The isochore organization of the human genome. *Annu. Rev. Genet.* **23,** 637–661.]

general) was accompanied by the discovery of compositional correlations (1) between genes and the isochores containing them, and (2) between chromosomal bands and isochores. These correlations will be discussed in the two following sections.

A. Isochores and Genes

The localization of a number of genes in compositional fractions of human DNA has shown a very important point, namely that a linear relationship exists between GC levels of coding sequences (and of their first, second, and third codon positions) and GC levels of the DNA fragments (about 100 Kb in size) containing them (see Fig. 2). Because the DNA fragments are mainly composed of noncoding intergenic sequences, the compositional correlation just described is in fact a correlation between coding sequences and the noncoding sequences that embed them. This suggests that both coding sequences representing <5% of the genome and the flanking noncoding sequences representing >95% of the genome are subject to the same compositional constraints.

The existence of this compositional relationship is also important from another viewpoint. Indeed, it

allows us to know the genomic distribution of genes over isochores having a different GC content. This can be deduced from histograms showing the compositional distribution of coding sequences or of their different codon positions. In fact, because the coding sequences are compositionally correlated with the vast DNA regions surrounding them, the composition of coding sequences themselves indicates the location of these genes in different isochore families. For instance, a coding sequence that is GC-rich will be present in a GC-rich isochore, whereas a gene that is GC-poor is located in a GC-poor isochore. If one examines the composition of all human coding sequences studied so far (or, even better, of their third codon positions, which are free to change with less alteration of the corresponding amino acids), one discovers (Fig. 3) a strong predominance of GC-rich coding sequences, which are mainly located in the GC-richest isochores. These results point to an extremely nonuniform distribution of genes in the human genome because the GC-richest isochores are the least represented in the human genome.

Very interestingly, this gene distribution is mimicked (1) by the distribution of CpG doublets, the only potential sites of methylation (CpG doublets in GC-poor genes are underrepresented relative to statistical expectations, whereas in GC-rich genes, CpG doublets have the statistical frequency), and (2) by the distribution of CpG islands, which are sequences >0.5 Kb in size and characterized by high GC levels, by clustered unmethylated CpG's, by G/C boxes (e.g., GGGGCGGGGC or closely related sequences), and by clustered sites for rare-cutting restriction enzymes (which recognize GC-rich sequences comprising one or two unmethylated CpG doublets). CpG islands are associated with the 5' flanking sequences, exons, and introns of all housekeeping genes (namely of the genes whose activity is required by every cell) and many tissue-specific genes (namely of genes that are expressed only in some specific tissues, such as liver cells, and with the 3' exons of some tissue-specific genes.

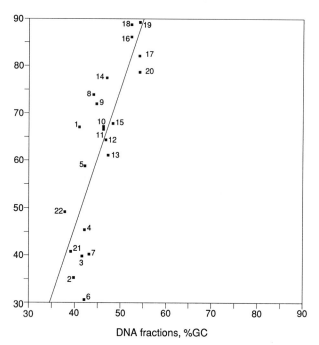

FIGURE 2 Plot of GC levels of third codon positions against the GC levels of compositional DNA fractions in which they were localized. The numbers indicate different coding sequences.

B. Isochores and Chromosome Bands

A number of lines of evidence show that GC-poor and GC-rich isochores largely correspond to the DNA of G- and R-bands, respectively. However, G- and R-bands not only differ in their overall isochore makeup, but also in their internal isochore

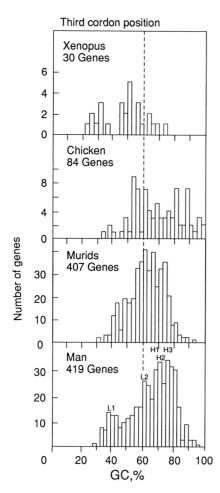

FIGURE 3 Compositional distribution of third codon position from vertebrate genes. (This distribution is the most informative because of its wider spread compared with coding sequences and first or second codon positions.) The number of genes under consideration is indicated. A 2.5% GC window was used. The vertical broken line at 60% GC is shown to provide a reference. Approximate identifications of different compositional classes of coding sequences corresponding (see Fig. 1) to the major components of the human genome (L1, L2, H1, H2, and H3) are indicated. The borders between L1-L2, H1-H2, and H3 can be tentatively estimated as 67.5% and 77.5% GC, respectively. [Reproduced, with permission, from G. Bernardi (1989). The isochore organization of the human genome. *Ann. Rev. Genet.* **23**, 637–661.]

structure as indicated by several lines of evidence, and, more recently, by the compositional mapping of the long arm of chromosome 21. Compositional maps may be constructed whenever long-range physical maps are available by assessing GC levels around landmarks (localized on the physical map) that can be probed. This simply requires the hybridization of the probes on DNA fractionated according to base composition, because this establishes the GC level of vast regions ($\geq$200 Kb) around the sequence probed. This approach has provided a direct demonstration for the compositional homogeneity of G-bands, for the compositional heterogeneity of R-bands, and for the highest GC levels and highest gene concentrations in the telomeric region of the long arm of chromosome 21 (a similar situation is possibly present in the telomeric regions of many other chromosomes).

III. Isochores and Genome Functions

While the functional correlates of the isochore organization of the human genome are still largely open problems, a number of points, as outlined below, are already well established.

A. Isochores and Integration of Mobile and Viral DNA Sequences

Stable integration of mobile and viral DNA sequences is mostly found in isochores of matching composition. Mobile sequences that have been amplified by retrotranscription and translocated to numerous loci of the human genome during mammalian evolution, such as LINES and SINES (see Section I) are predominantly located in isochores of matching GC levels. This indicates that reinsertion is targeted to matching genome environments and/ or that integration is more stable within such environments. Incidentally, such reinsertion may be a cause of mutation, if it occurs in genes that are thereby disrupted.

Needless to say, these observations are of interest in connection with the integration of foreign DNA into the genome of transgenic mammals. In this case, the important, yet unresolved, question concerns the effect of genomic compositional context on the expression of integrated sequences.

B. Isochores, Translocation Breakpoints, and Fragile Sites

Translocation breakpoints are not randomly located on chromosomes. R-bands and G/R borders are the predominant sites of exchange processes, including spontaneous translocations, spontaneous or induced sister-chromatid exchanges, and the chromosomal abnormalities seen after X-ray and chemical damage. Likewise, fragile sites tend to be more frequent in R-bands or near the border of R- and G-bands. Moreover, cancer-associated chromosomal aberrations are also nonrandom, because a limited number of genomic sites are consistently involved and frequently associated with cellular oncogenes or fragile sites. These observations indicate that R-bands and G/R borders are particularly prone to recombination. They also raise the question of the role played in these phenomena by the compositional discontinuities at G/R borders and within R bands as well as by the genomic distribution of SINES, CpG islands, and other recombinogenic sequences, such as minisatellites, which are predominantly located in R-bands.

Chromosomal rearrangements have two important consequences: the activation of oncogenes by strong promotors that could be put upstream of them by the rearrangement, and the formation, in evolutionary time, of reproductive barriers and speciation.

C. Other Functional Aspects of Isochores

Isochores are related to replication and condensation timing in the cell cycle. G-bands and genes located in GC-poor isochores replicate late in the cell cycle, whereas R-bands and genes located in GC-rich isochores replicate early. In contrast, condensation timing during mitosis occurs early in the cell cycle for G-bands and late for R-bands.

The distribution of genes over the isochores also has some correlation with gene function. Housekeeping genes (including oncogenes) are preferentially distributed in GC-rich isochores and R-bands, whereas tissue-specific genes are more abundant in GC-poor isochores and G-bands.

The range of GC values (30–100%) in codon third positions of human genes practically is as wide as that exhibited by the genes of all prokaryotes. This very extended range implies very large differences in codon usage (namely in the differential use of synonymous codons for the same amino acid) for GC-poor and GC-rich genes present in the same genome. In particular, at the high-GC end of the range, an increasing number (up to 50%) of codons are simply absent. In turn, GC values in codon third positions are paralleled (although, expectedly, to a more limited extent) by the GC values in first and second positions. The range of the latter values leads to very significant differences in the frequency of certain amino acids in GC-rich and GC-poor genes. For instance, the ratio of alanine+arginine to serine+lysine (namely, of the two amino acids that contribute most to the thermodynamic stability of proteins over the two that do so the least) increases by a factor of four between proteins encoded by the GC-poorest and the GC-richest coding sequences in the human genome.

The distributions of CpG doublets and of CpG islands suggests that the distribution of methylation in the genome of humans and other vertebrates is highly nonuniform, a point of interest in view of the role of DNA methylation in gene function and of the distribution of housekeeping and tissue-specific genes.

The results on CpG islands have an additional functional relevance. In GC-poor isochores, genes are usually endowed with a TATA or a CCAAT box and an upstream control region, whereas in the GC-rich isochores there is no TATA box but promotors containing properly positioned "G/C boxes" that bind transcription factor Spl, a protein that activates RNA polymerase II transcription. These GC-rich promotors apparently are associated with all genes located in GC-rich isochores.

IV. Isochores and Genome Evolution

As shown in Figures 1 and 3, the compositional patterns of the human genome are characterized by a wide GC range of both isochore families and coding sequences and by a predominance of GC-rich genes. These patterns are found in all mammals explored so far, with only slightly narrower distributions in three families of myomorpha (a suborder of rodents that includes mouse, rat, and hamster). Very similar, yet not identical, patterns are found in birds, a vertebrate class that arose from reptiles independently of mammals. In the case of birds, the compositional distributions of both DNA fragments and coding sequences attain, however, higher GC values than in mammals (see Figs. 1 and 3). In sharp contrast, the genomes of the vast majority of cold-blooded vertebrates exhibit compositional patterns that are narrower and do not reach the GC levels of the GC-rich DNA fragments of warm-blooded vertebrates or of the coding sequences contained in them (see Figs. 1 and 3).

The compositional patterns of vertebrate genomes define two modes of genome evolution. In the conservative mode, which prevails in mammals (and probably in birds), the composition of DNA fragments and coding sequences is maintained during evolution despite a very high degree of nucleotide divergence (which may be >50% in third codon positions). This compositional conservation appears to require negative selection, operating at the isochore level, to eliminate any strong deviation from a presumably functionally optimal composition. It is clear that if, during mammalian evolution, a number of genes retain a GC level of >90% in their third codon position, the mutations that would have decreased this level must somehow have been eliminated.

In contrast, in the transitional or shifting mode, parallel compositional changes are seen in both isochores and coding sequences. The two major compositional transitions found in vertebrate genomes are the GC increases that occurred between the genomes of reptiles on the one hand and those of birds and mammals on the other. (Apparently, these increases were accompanied by the replacement of TATA and CCAAT boxes by GC-rich promotors.) These compositional changes are due to a directional fixation of point mutations largely caused by both negative and positive selection at isochore levels. Selection appears to be for the higher thermal stabilities of proteins, RNA and DNA, which are required by the higher body temperature of warm-blooded vertebrates. Of course, selection implies functional differences and, therefore, supports the idea that isochores are functionally relevant structures. Moreover, the compositional relationships between coding and noncoding (particularly intergenic) sequences indicate that the same compositional constraints apply to both kinds of sequences. The selection pressures underlying such constraints cannot be understood if noncoding sequences are "junk DNA," with no biological functions.

V. Conclusions

Isochores represent a new structural level in the organization of the genome of warm-blooded vertebrates that bridges the enormous size gap between the gene level, including both their exon–intron systems and the corresponding regulatory sequences, and the chromosome level, with its banding patterns. These three levels are correlated with each other, because genes match compositionally the isochores in which they are harbored, while GC-poor and GC-rich isochores are DNA segments located in G- and R-bands, respectively.

The investigations that led to the discovery of isochores and of these two correlations have firmly established the existence of differences in the base composition and gene concentration of DNA segments present in G- and R-bands. Moreover, they have revealed that these segments are characterized by strikingly different complexities. The isochores present in G-bands are GC-poor, very close in composition, and characterized by a low gene concentration, whereas the isochores present in R-bands belong to different GC-rich compositional families, including those of the GC-richest family that have the highest concentration of genes and CpG islands. R-bands also comprise, however, a number of GC-poor isochores (as shown by the compositional mapping of chromosome 21) that may correspond to internal "thin" G-bands only seen at high resolution. In conclusion, isochores correspond to a chromosome organization level lower than standard chromosomal bands, possibly to chromomeres and interchromomeres.

While isochores from the genomes of warm-blooded vertebrates belong to a number of families characterized by large differences in base composition, this is not true for cold-blooded vertebrates, in which case isochores are characterized by much smaller differences in composition, which correspond to much weaker banding patterns in metaphase chromosomes.

Isochores are, however, not only structural units but also appear to play functional roles. Some of these, like the integration of mobile and viral sequences or recombination and chromosome rearrangements, are well established. The observations on the gene distribution of the genome, the relationships of such distribution with gene functions (housekeeping, tissue-specific), with codon usage, and with different kinds of regulatory sequences are also indicative of functional roles for isochores. In contrast, DNA replication timing and chromosome condensation timing at mitosis seem to be rather correlated with the chromomere–interchromomere organization of chromosomes, independently of the composition of the corresponding DNA stretches, because they are also found in cold-blooded vertebrates.

Isochores are evolutionary units of vertebrate genomes. Their composition may be conserved in spite of enormous numbers of point mutations or may undergo dramatic changes after more modest numbers of point mutations. In the case of the two independent evolutionary transitions from cold-

TABLE I   The Human Genome

Paleogenome (G-bands, GC-poor isochores)	Neogenome (R-bands, (GC-rich isochores)
Chromomeres	Interchromomeres
Late replication	Early replication
Early condensation	Late condensation
Abundance of LINES	Abundance of SINES
Compositional homogeneity	Compositional heterogeneity
Scarcity of genes	Abundance of genes (esp. in H3)
GC-poor gene (esp. tissue-specific)	GC-rich genes (esp. house-keeping)
Scarcity of CpG islands	Abundance of CpG islands
TATA box promotors	G/C box promotors
Less frequent recombination	More frequent recombination

Modified from G. Bernardi (1989). The isochore organization of the human genome. *Ann. Rev. Genet.* **23**, 637–661.

blooded vertebrates to mammals and birds, the compositional transitions occurring in the genome seem to be largely associated with the optimization of genome functions following environmental body temperature changes. Interestingly, these transitions appear to be accompanied by very conspicuous changes in promotor sequences.

In conclusion, two large compositional compartments can be distinguished in the human genome and, more generally, in the genomes of warm-blooded vertebrates (see Table I). The first compartment, the paleogenome, is characterized by its similarity to what it was, and still is, in cold-blooded vertebrates: the late-replicating, compositionally homogeneous, GC-poor isochores of early-condensing chromomeres contain relatively rare, GC-poor (largely tissue-specific) genes having TATA box promotors (CpG islands are scarce). The second compartment, the neogenome, is characterized by the fact that it changed its compositional features compared with what it was in cold-blooded vertebrates. In the neogenome, the ancestral, early-replicating, GC-poor isochores of late-condensing interchromomeres were changed into compositionally heterogeneous, GC-rich isochores that contain abundant genes (perhaps including most housekeeping genes) having G/C box promotors (genes and CpG islands are particularly abundant in the GC-richest isochores).

Bibliography

Bernardi, G. (1989). The isochore organization of the human genome. *Annu. Rev. Genet.* **23**, 637–661.

Bernardi, G., Mouchiroud, D., Gautier, C., and Bernardi, G. (1988). Compositional patterns in vertebrate genomes: Conservation and change in evolution. *J. Mol. Evol.* **28**, 7–18.

Gardiner, K., Aissani, B., and Bernardi, G. (1990). A compositional map of human chromosome 21. *EMBO J.* (in press).

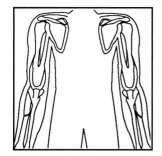

Human Muscle, Anatomy

ANTHONY J. GAUDIN, *California State University, Northridge*

Glossary

Agonists Muscles that produce the same movement of a body part when they contract
Antagonists Muscles that produce opposing actions when they contract
Fixator Muscle that immobilizes a joint
Insertion Place of attachment of a muscle that moves when the muscle contracts
Origin Point of attachment of a muscle that does not move when the muscle contracts
Synergist Stabilizing muscle that helps another muscle accomplish some desired action

THE HUMAN BODY contains over 600 voluntary muscles attached to the bones of the skeleton. Contraction of these muscles moves the bones and accomplishes such movements as walking, running, eating, smiling, reading, and writing. In this article the anatomy of the major groups of skeletal muscles is reviewed (see Color Plates 1 and 2).

I. Muscles and Their Movements

A. Naming the Muscles

The names of most muscles refer to some prominent feature of the muscle. For example, many muscles are named with reference to their size, such as the pectoralis major (large chest muscle)

and the pectoralis minor (small chest muscle). Some muscles are named for their position in the body, such as the tibialis anterior (anterior tibial surface) and the tibialis posterior (posterior tibial surface). Muscles can be named for movements they accomplish, such as the flexor digitorum (flexing of the fingers) and the levator scapulae (elevation of the scapula).

Muscles are also named for their shape—for example, the trapezius (trapezoidal shape) and the orbicularis oculi (circular muscle that surrounds the eye). Some muscles have been named for bones to which they attach—for example, the zygomaticus (which attaches to the zygomatic bone) and the temporalis (which attaches to the temporal bone). Muscles have been named for their number of divisions, such as the biceps brachii (a two-headed muscle of the arm) and the quadriceps femoris (a four-headed muscle on the femur). Finally, the name of a muscle might contain two or more references to prominent features; for example, the name "flexor carpi ulnaris" not only indicates that the muscle is a flexor of the wrist (carpus), but it also indicates its position adjacent to the ulna.

B. Shapes of Skeletal Muscles

The arrangement of fascicles within the whole muscle reveals four distinct patterns: parallel, convergent, pennate, and circular. Parallel fascicles lie parallel to one another along the longitudinal axis of the muscle, forming wide flat muscles. A convergent pattern is found in triangular-shaped muscles, in which fascicles attach along a broad tendon at one end of the muscle and converge to a narrow tendon at the other. Fascicles in a pennate pattern are attached like plumes of a feather to an elongate tendon that is nearly as long as the entire muscle. The fascicles may be attached only on one side of the tendon (unipenniform) or to both sides (bipenniform). A circular pattern of fascicles is characteristic of sphincter muscles that surround openings (e.g., the mouth or the anus). [*See* SKELETAL MUSCLE.]

Because of the different ways in which fascicles attach to tendons, skeletal muscles exist in a variety of sizes and shapes. Some common ones are illustrated in Color Plate 3. Fusiform muscles are columnar in shape, because the muscle fibers tend to run in the same direction. The tendons that attach fusiform muscles to bones are restricted to the ends of the muscle. The thickest part of the muscle, its belly, the fleshy portion in a fusiform muscle, is usually near the middle of the muscle. Penniform muscles are flattened, and either or both of the tendons will extend for some distance along the length of the belly. The tendon in penniform muscles may extend along one side of the belly (unipenniform), or it may course along the center of the belly (bipenniform). Radiate muscles are roughly triangular in shape, with the muscle fibers converging from a wide tendinous attachment at one end into a narrow tendinous connection at the opposite end. In addition to these generalized shapes, muscles can also be rectangular, triangular, rhomboid, and circular in shape; other shapes defy simple description.

C. Origin, Insertion, and Action

A muscle produces skeletal movements because its fibers or tendons pass across a movable joint connecting two different bones. Skeletal muscles can also move soft tissues that are not strictly skeletal parts (e.g., the eyelids, eyeballs, and lips). In most cases of movement, the joint connecting the two bones is a synovial joint, which permits significant movement between the two bones involved. When a muscle contracts and shortens, it moves one of the two bones attached to it. Usually, only one of the two bones moves. As discussed below, this is accomplished by the stabilizing contractions of other muscles attached to the nonmoving bone. [*See* ARTICULATIONS, JOINTS BETWEEN BONES.]

The site of attachment that does not move, or moves least, when the muscle contracts is the origin of the muscle. The place of attachment that moves is the insertion. For example, flexing the fingers to form a clenched fist is accomplished by muscles located on the front of the forearm, that are attached to the phalanges of the fingers at one end and to bones of the arm near the elbow at the other. When the muscle contracts to flex the fingers, only

the fingers move, not the entire hand. In this case the origin of the muscle is near the elbow, and the insertion is on the phalanges of the fingers. The specific movement caused by a muscle is its action. In the example just given, the muscle's action is flexion.

Muscles are capable of only one activity, namely, contraction, and only muscles contract. This limits the action of any given muscle to one specific movement or sometimes a few closely related movements. Thus, to move a bone in a variety of directions, differently oriented muscles are required. For example, in the fist-clenching action a muscle on the front of the forearm is used to flex the fingers. To unclench the fist, muscles on the back of the forearm are used.

A muscle of primary importance in any specific action is a prime mover, or agonist. In some cases only one muscle is used in certain actions, as in abducting the arm away from the body at the shoulder joint. At other times a group of agonists is used, as in adducting the arm toward the body at the shoulder joint. Most muscles of the body are arranged in opposing pairs on opposite sides of a joint. Each muscle produces opposing movements. When an agonist contracts, its opposing muscle relaxes. For example, when the fingers are flexed in a clenched fist, the opposing muscle—in this case the extensor on the other side of the forearm—relaxes. Similarly, when the fingers are extended in unclenching the fist, the opposing flexors relax. Any muscle that performs an opposing action to an agonist (e.g., in flexing and extending, or abducting and adducting) is referred to as an antagonist. [*See* MOVEMENT.]

In addition to producing significant movements, muscles can also function as synergists, stabilizing muscles that help another agonist accomplish some desired action. For example, two relatively powerful muscles positioned on the back of the forearm act as prime movers in extending the wrist joint, as in hitting a backhand shot in Ping-Pong.™ The same muscles act as synergists to the flexors of the fingers during the clenching of the fist, by contracting just enough to keep the wrist joint from flexing simultaneously with the fingers. Most muscles act as agonists, antagonists, and synergists at different times. Two muscles at a joint can be either agonistic or antagonistic to each other, but they cannot act in both ways simultaneously.

When a muscle is acting to immobilize a joint, it is called a fixator. A fixator can act as a synergist, such as the wrist muscles described above. Fixators can also stabilize a body part without specifically helping an agonist. For example, several groups of muscles lie in the posterior region of the trunk, attached to the vertebrae, pelvic girdle, and ribs. The chief function of these muscles is to keep the vertebral column erect. Whenever a person stands or sits upright on a backless stool or bench, these muscles remain contracted, maintaining proper postural position. The cooperation of several kinds and groups of muscles is necessary to maintain proper posture and to execute coordinated movements of body parts.

II. Muscles of the Head and the Neck

Muscles of the head can be divided into four major groups: facial expression, extrinsic muscles of the eye, mastication, and muscles of the tongue (see Color Plate 4).

A. Muscles of Facial Expression

Larger muscles of the facial region move the soft fleshy parts of the face and produce expressions of surprise, joy, sadness, and other nuances of body language. The epicranius fits over the cranium like a skull cap. The part that lies under the scalp is the galea aponeurotica. Its fleshy part is divided into two regions: the frontalis, which covers much of the forehead, and the occipitalis, at the rear of the skull. Contraction of these muscles either moves the scalp or elevates the eyebrows, producing, for example, horizontal ridges in the forehead associated with a surprised look. The orbicularis oculi muscles surround the eye and "flesh out" the eyelids. Contraction of these muscles can involve only superficial fibers that close the eyes in a blink or while sleeping. A forceful contraction of the deep fibers of this muscle produces squinting expressions and winks. The vertical furrows in the forehead between the eyebrows associated with a frown, sorrow, and crying are produced by the small corrugator supercilii muscle.

The orbicularis oris brings the lips together and purses, or puckers, them. Several muscles acting together—the zygomaticus major and minor, the levator labii superioris, and the buccinator—elevate and retract the corners of the mouth in a smile. The buccinator is used by musicians who play brass instruments, to force air out through the mouth. It is commonly known as the "trumpeter's muscle." The platysma and the depressor labii inferioris retract and depress the skin of the neck and the lower jaw in a snarl and bare the lower teeth.

B. Muscles That Move the Eyeball

Each eyeball has six extrinsic (i.e., attached to the outer surface) muscles that produce coordinated movements necessary for proper vision (see Color Plate 5). Four rectus muscles are concerned primarily with moving each eyeball up and down, and from side to side, and two oblique muscles rotate the eyeball.

C. Muscles of Mastication

The muscles of mastication are involved in biting and chewing. The large masseter and temporalis are powerful elevators of the mandible, assisted by the pterygoid medialis (Fig. 1). The pterygoid lateralis, assisted by the digastric (discussed in Section E) opens the mouth by depressing and protracting the mandible. The two pairs of pterygoid muscles may also be contracted one side at a time, producing the side-to-side rocking motion of the mandible associated with chewing and grinding food.

D. Muscles of the Tongue

The tongue is a muscular organ used to manipulate food, to speak, and to swallow. It is composed of both intrinsic and extrinsic muscles. Intrinsic muscles within the tongue consist of three groups of muscle fibers oriented in longitudinal, transverse, and vertical bundles. Most people possess all three groups of muscles and are able to twist and turn the tongue in a variety of actions. However, some people lack one or more of the bundles. In a genetically determined variation, the transverse bundles may be missing, making it impossible for the individual who lacks them to roll the tongue. An even rarer occurrence is the individual who lacks the longitudinal group, making it impossible to poke the tongue out and lift it up at the tip. [See TONGUE AND TASTE.]

Extrinsic muscles (i.e., the styloglossus, hyoglossus, and genioglossus) are at the rear and along the sides of the tongue, and the latter extend into the tongue (Fig. 2). They move the tongue up and down in the mouth as well as protract and retract it.

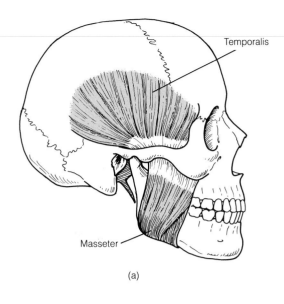

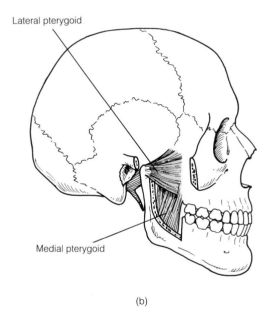

FIGURE 1 Muscles of mastication. (a) Superficial view. (b) Deep view. [From A. J. Gaudin and K. C. Jones (1989). "Human Anatomy and Physiology," p. 218. Harcourt Brace Jovanovich, San Diego.]

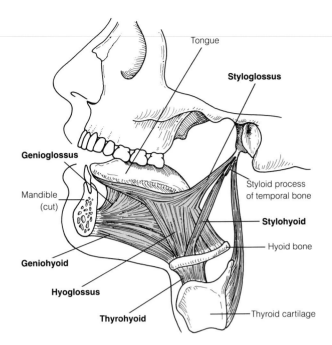

FIGURE 2 Extrinsic muscles of the tongue and the hyoid bone. [From A. J. Gaudin and K. C. Jones (1989). "Human Anatomy and Physiology," p. 219. Harcourt Brace Jovanovich, San Diego.]

E. Muscles of the Neck

If one looks at the side of the neck, the most prominent muscle is the sternocleidomastoid (see Color Plate 6). This muscle passes from the sternum and the clavicle to the mastoid process of the temporal bone. Acting together, this pair of muscles flexes the head and the neck forward; acting alone, each rotates the head in the opposite direction. The suprahyoid muscles elevate the hyoid bone when the mandible is fixed in place, and lower the mandible when the hyoid is held firm by muscles beneath it. One of the suprahyoid muscles, the digastric muscle, is divided lengthwise by its central tendon into an anterior belly that extends forward to the mandible and a posterior belly that extends backward to the mastoid process of the temporal bone. The infrahyoid muscles act antagonistically to the suprahyoid muscles, depressing the hyoid bone. One can easily feel these muscles moving by placing the fingers on the larynx (i.e., the Adam's apple) and then swallowing.

Several muscles located on the posterior side of the neck are concerned with additional movements of the head and the neck (see Color Plate 7). The semispinalis, splenius capitis, and longissimus capitis act as antagonists to the sternocleidomastoid by extending the head and the neck backward when acting together, or by rotating the head and the neck when the muscles on only one side of the neck are used.

III. Muscles of the Trunk

Trunk muscles are classed in four functional groups: those of the vertebral column, thoracic wall, abdominal wall, and the pelvic floor.

A. Muscles of the Vertebral Column

Muscles of the vertebral column, often referred to as the erector spinae group, are separated into bundles that stabilize the spine and extend it (i.e., arching the back) when used together, or move it from side to side when used singly (see Color Plate 7). Some smaller muscles rotate the spine. Several are relatively short, extending over only one to three vertebrae. Others are long, covering from the sacrum and the ilium to the rib cage or from the thoracic vertebrae to the occipital bone.

In the cervical region several small muscles (the rectus capitis, longus colli, splenius capitis, and splenius cervicis) flex, extend, and rotate the vertebrae when the head is being turned. Three muscles—the quadratus lumborum, psoas major, and psoas minor—in the lumbar region rotate and bend the vertebral column laterally.

B. Muscles of the Thoracic Wall

The thoracic wall includes muscles concerned primarily with breathing movements (see Color Plate 8). The most important of these muscles are the intercostales externi, intercostales interni, serratus posterior, traversus thoracis, and diaphragm. The intercostals fill spaces between the ribs. Their fibers are oriented at right angles. The external intercostals are positioned closer to the spine than the internal intercostals. They extend from the vertebrae to the costal cartilages, while the internals extend from the angles to the ribs to the sternum. The external intercostals overlie the internals. The latter move the ribs during breathing.

The diaphragm separates the thoracic and abdominal cavities. The fibers of this dome-shaped muscle pass radially from the lower ribs, costal cartilages, and vertebrae and insert on a thick central tendon. Contraction of the muscle fibers of the diaphragm pulls on the central tendon, which lowers the dome. This action expands the thoracic cavity and draws air into the lungs. When the diaphragm relaxes, the dome rises and elastic recoil of the lung tissue forces air out. Several large organs (i.e., the esophagus, aorta, and vena cava) pierce the diaphragm. In all other respects the diaphragm is a typical skeletal muscle.

C. Muscles of the Abdominal Wall

The chief function of abdominal wall muscles is to maintain physical support for organs in the abdomen. This act is accomplished by maintaining muscle tone, not by forcible contractions. The obliquus externus abdominis, obliquus internus abdominis, and transversus abdominis stretch across the lateral abdominal wall (see Color Plate 9). Fibers of these muscles course in different directions, resulting in three flat layers of muscular tissue that reinforce one another. The rectus abdominis extends from the sternal xiphoid process, lower ribs, and costal cartilages inferiorly to the pubes.

In addition to forming the abdominal wall, these four types of muscles can be voluntarily contracted to compress the abdomen during urination, defecation, and childbirth, or in anticipation of a trauma to the abdomen. The rectus abdominis, obliquus externus, and obliquus internus also help in flexing and rotating the trunk and are used in forceful exhalation. The quadratus lumborum lies on either side of the lumbar vertebrae just anterior to the transversus abdominis. It helps form the posterior abdominal wall and also rotates the vertebral column or depresses the lower ribs during forced breathing.

D. Muscles of the Pelvic Floor

Muscles of the pelvic floor support the organs that project into the pelvis from the abdominal cavity (see Color Plate 10). Two muscles are involved in this task: the levator ani, which surrounds the rectum, urethra, and vagina (in females), and the coccygeus. In addition to providing support, these muscles can be voluntarily contracted during defecation, urination, and sexual intercourse. Several smaller muscles are associated specifically with the reproductive organs, and two anal sphincters, one under voluntary control, surround the lower (i.e., distal) end of the intestinal tract.

IV. Muscles of the Pectoral Girdle and the Upper Extremity

For convenience the muscles of the pectoral (i.e., shoulder) girdle and the upper extremity are divided here into six groups. These include muscles that move the scapula, humerus, forearm, hand, and fingers.

A. Muscles That Move the Scapula

The scapula, a triangular bone, articulates solidly only with the small clavicle. Because the scapula is not held firmly in place, it is free to move in several directions. The trapezius is a large flat muscle that covers much of the upper back region (see Color Plate 11). Its fibers extend in several directions, and it rotates the scapula freely. The rhomboideus muscles and the levator scapulae retract and elevate the scapula. Acting agonistically to these muscles, the pectoralis minor and the serratus anterior protract the scapula forward, as when a person reaches for something. The subclavius muscle in this region attaches to, and acts indirectly through, the clavicle, which firmly articulates with the scapula to pull the scapula toward the sternum.

B. Muscles That Move the Humerus

Two groups of muscles move the humerus (see Color Plate 12). One group inserts close to the shoulder joint, produces small movements, and is used primarily in maintaining the articulation of the humerus with the glenoid fossa and in rotating the humerus. This group includes the supraspinatus, infraspinatus, subscapularis, and teres minor. The term ''rotator cuff'' is often applied to an area where tendons of these muscles fuse with tissues of the shoulder joint and rotate the humerus. This area is subject to trauma and pain in athletes who make extensive use of their arms (e.g., baseball pitchers and tennis players).

The second group of humeral muscles includes the deltoid, pectoralis major, latissimus dorsi, teres major, and coracobrachialis. The deltoid raises the arm and contributes to movements of the humerus associated with such activities as climbing. The other four oppose the deltoid by lowering the humerus.

C. Muscles That Move the Forearm

The forearm includes two large bones: the ulna and the radius. Two opposing sets of muscles move the forearm in two different directions at the elbow (Fig. 3). The biceps brachii, brachialis, and brachioradialis flex the forearm. The biceps brachii also is a powerful supinator. Extension of the forearm at the elbow is accomplished by the large triceps brachii and the smaller anconeus.

D. Forearm Flexors of the Wrist and the Hand

The anterior surface of the forearm contains three layers of muscles, most of which are concerned with flexing the wrist and the fingers (see Color Plate 13). The superficial layer includes the pronator teres, which rotates the radius, turning the palm down, and three flexors of the wrist: the flexor carpi ulnaris, flexor carpi radialis, and palmaris longus.

The flexor digitorum superficialis muscle is the middle muscle on the anterior side of the forearm. Its tendon of insertion divides at the wrist and attaches to the middle phalanges of fingers 2–5. The thumb has its own flexors. Along with the flexor digitorum profundus, which lies beneath it, the flexor digitorum superficialis flexes fingers 2–5, as when making a fist.

The deep layer of these forearm muscles includes the flexor digitorum profundus, pronator quadratus, and flexor pollicis longus. The flexor digitorum profundus inserts on the terminal phalanges of fingers 2–5 and assists its superficial counterpart in flexing the fingers. The pronator quadratus pronates the hand, and the flexor pollicis longus flexes the thumb into the palm.

E. Forearm Extensors of the Wrist and the Hand

The posterior aspect of the forearm has two layers of muscles concerned with extending the wrist and the fingers (see Color Plate 14). An outer layer consists of five muscles, including three wrist extensors (i.e., the extensor carpi radialis longus, extensor carpi radialis brevis, and extensor carpi ulnaris) and two finger extensors (i.e., the extensor digitorum communis, whose tendons are seen fanning over the back of the hand and the knuckles, and the extensor digiti minimi). The deep layer moves the fingers and includes the abductor pollicis longus, extensor pollicis longus, extensor policis brevis, and extensor indicis.

F. Intrinsic Muscles of the Hand

In addition to tendons from forearm muscles, the hand is operated by three groups of muscles on the palmar surface (see Color Plate 15), which permit a

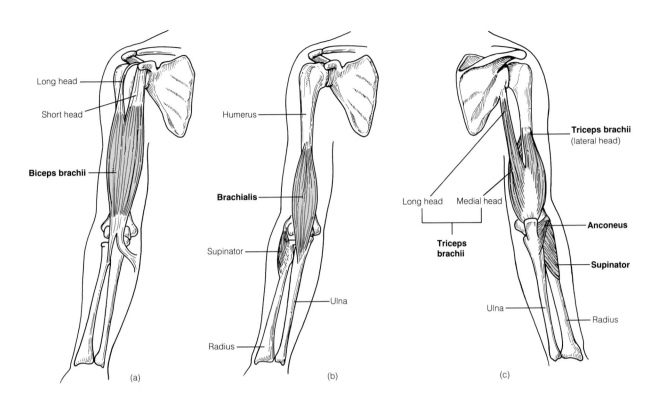

FIGURE 3 Muscles that move the forearm. (a) Superficial anterior view. (b) Deep anterior view. (c) Posterior view. (From A. J. Gaudin and K. C. Jones, "Human Anatomy and Physiology," p. 235. Harcourt Brace Jovanovich, San Diego, California, 1989.)

versatile range of movements of the thumb and the fingers. The first group, which moves the thumb, forms a fleshy pad, the thenar eminence, at the base of the thumb. On the opposite side of the hand is the narrower hypothenar eminence, which extends from the base of the little finger. It consists of four muscles, three of which move the little finger. Between these two eminences is a V-shaped depression, with the three sets of midpalmer muscles that move fingers 2–5.

V. Muscles of the Lower Extremity

The lower extremity has a different function than the upper extremity. While arms are used to manipulate objects, legs act in locomotion. This difference in function is reflected in the organizational complexity of the muscles in the two appendages. Muscles that move the lower ones are involved more with strength and stability rather than with complex manipulation. The muscles of the lower

extremity are divided here into functional groups based on their actions.

A. Muscles That Move the Thigh

Because the femur articulates with the pelvic bone through a ball-and-socket joint, a variety of movements are possible. Flexion of the thigh is accomplished by three muscles (i.e., the iliacus, psoas major, and psoas minor), which lie anterior to the vertebral column and the pelvic bone (see Color Plate 16). The psoas minor is another genetically variable muscle, being absent in about 40% of the population. Because the tendons of insertion of these three muscles fuse on the lesser trochanter of the femur, the three muscles are often referred to as the iliopsoas muscle group. In addition to flexing the thigh, they also help in the lateral rotation of the thigh. Extension of the thigh is accomplished chiefly by the gluteus maximus muscle that forms most of the fleshy buttocks.

Abduction of the femur is accomplished by the tensor fascia latae, gluteus medius, and gluteus minimus. These three muscles also rotate the femur medially. The tensor fasciae latae attaches the iliotibial tract (i.e., fascia lata, or iliotibial band) to a broad superficial sheet of connective tissue. It also attaches to the gluteus maximus and several other muscles on the anterolateral surface of the femur. The iliotibial tract extends to the lateral condyle of the tibia. Contraction of the several muscles that attach to it serves to strengthen the extended knee joint during walking and running.

Five major adductors of the femur are the gracilis, pectineus, adductor longus, adductor brevis, and adductor magnus. The gracilis is longer than the other four, extending from the pubic bone to the tibia, just below the medial condyle. Consequently, it also flexes the leg. The other four muscles in this group assist in the lateral rotation of the thigh. A group of six smaller muscles close to the hip joint rotate the femur laterally. They are the piriformis, obturatorius internus, obturatorius externus, gemellus superior, gemellus inferior, and quadratus femoris.

B. Muscles That Move the Leg

Leg muscles are organized into anterior and posterior functional groups on the thigh (see Color Plate 16). Some anatomists recognize a third group of muscles, consisting of the five adductors just discussed. The anterior group includes the sartorius, the longest muscle in the body, and the quadriceps femoris group, which consists of four separate muscles: the rectus femoris, vastus lateralis, vastus medialis, and vastus intermedius. The sartorius extends from the ilium diagonally over the anterior surface of the thigh, eventually passing behind the medial side of the knee. Its relationship to these two joints allows it to flex both the thigh and the leg. The massive quadriceps muscles fuse into a common tendon of insertion, the patellar tendon, which surrounds the patella and continues with the patellar ligament to insert on the tibial tuberosity. The quadriceps group, a powerful extensor of the leg, is used in walking, running, and specifically in kicking. The rectus femoris originates on the ilium and also assists in flexing the thigh.

Antagonists to these muscles form the posterior group of thigh muscles. They are usually referred to as the hamstring group, because tendons of these muscles in pigs are used by butchers to attach curing hams to meat hooks. The hamstring consists of three muscles—the biceps femoris, semitendinosus, and semimembranosus—which originate on the ischial tuberosity. Their tendons of insertion are easily felt posterior to the knee joint, the biceps laterally, the other two medially. All three ham-

string muscles extend the thigh and flex the leg. The popliteus also flexes the leg (see Section V,C).

C. Muscles That Move the Foot and the Toes

The foot attaches to the leg at the ankle joint, a diarthrosis that allows movement in four directions: up, down, and side to side. Because of their positions, the muscles that extend the toes also help lift the foot and muscles that flex the toes help lower the foot.

A group of four muscles lies on the anterior surface of the leg anterior to the interosseous membrane between the tibia and the fibula (see Color Plate 17). The tibialis anterior lifts and inverts the foot, while the smaller peroneus tertius lifts and everts it. The extensor hallucis longus and the extensor digitorum longus extend the toes and permit dorsiflexion of the foot.

A lateral pair of leg muscles, the peroneus longus and the peroneus brevis, extend along the lateral surface of the fibula. Both muscles lower and evert the foot.

A posterior group of muscles in the calf of the leg is composed of seven muscles divided into superficial and deep groups (see Color Plate 18). The thick fleshy gastrocnemius, soleus, and plantaris belong to the superficial group. The tendons of these muscles fuse at their lower ends, forming the calcaneal tendon (i.e., Achilles's tendon), which inserts on the calcaneus. All three muscles lower the foot. The gastrocnemius and the plantaris also help in flexing the knee joint. The deep group includes four muscles. The relatively short popliteus flexes the leg. The tibialis posterior lowers and inverts the foot. The flexor hallucis longus and the flexor digitorum longus flex the toes and also help lower the foot.

D. Intrinsic Muscles of the Foot

The hand and the foot are composed of three similar groups of bones and possess similar sets of muscles. However, because of very different functions of the two structures, significant differences exist in the muscles. The foot lacks the flexibility of the hand, but possesses more strength for support and locomotion.

The foot possesses a dorsal muscle, the extensor digitorum brevis, lacking in the hand. The remaining 10 muscles lie in four overlapping layers on the bottom surface of the foot (see Color Plate 19).

These muscles belong to three functional groups; one group moves only the great toe, another moves only the fifth toe, and a third group moves toes 2–5.

Bibliography

Basmajian, J. V. (1978). ''Muscles Alive,'' 4th ed. Williams & Wilkins, Baltimore, Maryland.

Brody, D. M. (1987). Running injuries: Prevention and management. *Clin. Symp.* **39**(3), 1–36.

Crelin, E. S. (1981). Development of the musculoskeletal system. *Clin. Symp.* **33**(1), 1–36.

Koivisto, V. A. (1986). The physiology of marathon running. *Sci. Prog.* **70,** 109–127.

McMinn, R. M., and Hutchings, R. T. (1988). ''Color Atlas of Human Anatomy,'' 2nd ed. New York Med. Publ., Chicago.

Melloni, J. L., *et al.* (1988). ''Melloni's Illustrated Review of Human Anatomy.'' Lippincott, Philadelphia.

Rosse, C., and Clawson, D. K. (1980). ''The Musculoskeletal System in Health and Disease.'' Harper & Row, New York.

Huntington's Disease

EDITH G. McGEER, *University of British Columbia*

I. General Description
II. Pathology
III. Theories as to the Cause
IV. Animal Models
V. Treatment

Glossary

Caudate One of two nuclei making up the striatum (see putamen)

Cholinergic Describes neurons that use acetylcholine as a neurotransmitter

Chorea Ceaseless occurrence of a wide variety of rapid, highly complex, jerky movements that appear to be well coordinated but are involuntary

Choreiform Having the characteristics of chorea

Dementia Loss of mental faculties, particularly memory

Excitotoxin Excitatory amino acid that destroys neurons in brain regions into which it is injected hypodermically. The exact mechanism is unknown but these materials appear to overstimulate or excite neurons to such a degree that they die. Most known excitotoxins are not found normally in brain although they are chemically related to glutamate which is.

Extrapyramidal system Group of subcortical nuclei, also called the basal ganglia, having to do with movement control

γ-aminobutyric acid Major central inhibitory neurotransmitter

Glutamate Excitatory amino acid that is major central excitatory neurotransmitter

Kainic acid Chemical extracted from a Japanese seaweed which acts as a very powerful excitotoxin

Neurotransmitter One of a group of specific chemicals used by the nervous system to carry messages from one neuron to another

Peptide Compound made up of a chain (polymer) of amino acids

Putamen One of two nuclei making up the striatum (see caudate)

Quinolinic acid An amino acid normally found in small amounts of brain that, at higher concentrations, acts as an excitotoxin.

Striatum Forebrain structure that forms a major part of the extrapyramidal system; used here as a synonym for caudate–putamen

HUNTINGTON'S DISEASE, formerly called Huntington's chorea, is a rare, autosomal dominant neurodegenerative disorder characterized by involuntary movements, cognitive impairment, and personality change. The choreiform movements cease during sleep. Huntington's disease generally manifests itself in affected individuals in about the fourth decade of life and progresses within 10–15 yr to death. Only palliative treatment is available. The brain is often shrunken, with extensive cortical atrophy, but the most characteristic and consistent pathology is loss of neurons in the caudate and putamen. The mutant gene has been localized to the short arm of chromosome 4, but the gene product is not yet known. In many families, however, DNA probes can now be used to estimate the risk of a fetus carrying the gene. Because injections of excitotoxins into the neostriatum in animals can reproduce much of the striatal pathology of the disease, it has been hypothesized that the genetic defect may be in some aspect of the endogenous excitatory amino acid systems and that neuronal death might be slowed by treatment with an inhibitor of excitation. Both hypotheses await future testing.

269

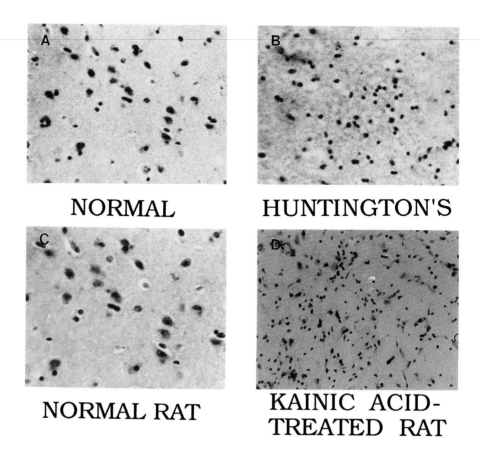

NORMAL

HUNTINGTON'S

NORMAL RAT

KAINIC ACID-
TREATED RAT

FIGURE 1 Histological sections of the caudate of a neurologically normal human (A), a case of Huntington's disease (B), a control rat (C), and a rat after an intrastriatal injection of kainic acid (D). Note the loss of the larger cells (neurons) in B and D, with preservation or increase in the numbers of very small cells (glia).

I. General Description

Huntington's disease (HD) occurs equally among men and women, and the abnormal movements usually begin between the ages of 37 and 47 yr. They may be preceded by as much as a decade by personality changes. Curiously, the mental symptoms can mimic schizophrenia in the early stages, and some victims have been kept in mental hospitals for several years before onset of the choreiform movements permits the correct diagnosis. In the later phases, however, the severe dementia is unmistakable. A few cases have been reported where choreiform movements are not seen; this is why the name Huntington's disease rather than Huntington's chorea is now preferred. The disease is relentlessly progressive, and death usually ensues within 10–15 yr.

In approximately 10% of HD cases, the onset may occur before the age of 15 yr, and the onset in a few cases may be delayed until the sixth or seventh decade. The progress of the disease is generally more rapid in affected children than in young adults and is unusually slow in the older onset cases.

HD is a rare disorder that affects all races, but its incidence can vary widely depending on the inheritance. The incidence in Europe, North America, and Australia is around 4–7/100,000 but is less than 1/100,000 in Japan and China. Pockets with a very high incidence exist, as in North Sweden, around Lake Maracaibo in Venezuela, and in Tasmania. In many of these instances, all of the cases can be traced back to a common ancestor.

The extent of suffering is increased because the onset of the disease commonly occurs after the birth of children who perpetuate it; both the unaffected spouse and the children suffer from the knowledge that, statistically, 50% of the children will develop this devastating, presently untreatable disorder. Moreover, it is difficult for such children to decide whether or not they should in turn procreate, knowing that, if they are not themselves vic-

TABLE I Examples of Some Biochemical Measures on Presynaptic Neurotransmitter Indices in the Striatum and Substantia Nigra in Huntington's Disease and Excitotoxic "Models"

Neurotransmitter	Huntington's disease	Excitotoxic "models"
In the striatum		
Neurotransmitter in striatal neurons		
GABA indices	Decreased	Decreased
Acetylcholine indices	Decreased	Decreased
Substance P	Decreased	Decreased
Enkephalin	Decreased	Decreased
Dynorphin	Decreased	Decreased
Somatostatin	Normal or increased	Normal or decreased[a]
Neurotransmitters in projections to the striatum		
Dopamine indices	Normal or increased	Normal or increased
Noradrenaline indices	Normal or increased	Normal or increased
Serotonin indices	Normal or increased	Normal or increased
In the substantia nigra		
Neurotransmitters in projections from the striatum		
GABA indices	Decreased	Decreased
Substance P	Decreased	Decreased
Dynorphin	Decreased	Decreased

[a] This is an example of a neurotransmitter system where a quinolinic acid lesion mimics HD better than does a kainic acid lesion.

tims they will not transmit the disorder. The existence of DNA probes that can predict, in many families, the likelihood of any individual carrying the gene could theoretically be helpful in genetic counseling, but how justified such testing is at this time, when no help can be offered to those who test positive, is still being debated.

II. Pathology

A. Histological Pathology

There is a general atrophy throughout the brain, but the greatest and most typical atrophy occurs in the caudate–putamen. The weight of the total brain is usually reduced by about 20%, while that of the caudate–putamen is reduced by more than 50%. Neuronal degeneration occurs in many brain regions, but the neuronal loss in the caudate–putamen is the characteristic pathological feature of the disease (Fig. 1). Many types of neurons with their cell bodies in the caudate–putamen are affected, although the large neurons tend to be better preserved than the small ones.

B. Chemical Pathology

Attention has been focused on the chemistry of the caudate–putamen because this is the site of the heaviest and most consistent pathology. The first reported chemical pathology was loss of cholinergic and γ-aminobutyric acid (GABA) indices in the striatum and of GABA indices in the substantia nigra. This led to clinical trial of compounds designed to raise the level of either GABA or acetylcholine, but these trials have been ineffective. This is not surprising because it turns out that this is only part of the biochemical pathology (Table I). All, or almost all, neuronal types with cell bodies in the striatum seem to be affected, while the same kinds of neurons are alright elsewhere in the brain. Thus, for example, GABA and cholinergic indices are preserved in the cortex, thalamus, and many other regions. Many of the identified biochemical losses in other regions such as the substantia nigra can be traced to loss of projections from the caudate–putamen (Fig. 2). Hence, HD has been called a region-specific disease as contrasted with a neuron-specific disease such as Parkinsonism, where the critical pathology seems to be loss of dopamine neurons throughout the brain. It is worth noting that these nigrostriatal dopamine neurons are preserved in HD.

Positron emission tomography studies of glucose metabolism in HD show clear, progressive decreases in the caudate–putamen. Whether such decreases in local glucose metabolic rate precede or follow the onset of clinical symptoms remains a matter of some controversy.

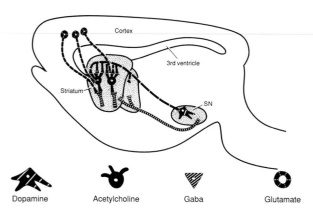

Dopamine   Acetylcholine   Gaba   Glutamate

FIGURE 2 Diagrammatic representation of a few of the projections to and from the striatum as well as a few of the striatal interneurons in a rat brain. Many other neurons using other transmitters exist in this system. For example, descending neurons to the substantia nigra (SN) also use various peptide neurotransmitters such as substance P and dynorphin.

III. Theories as to the Cause

HD is a genetic disorder dependent on inheritance of an autosomal dominant gene with close to 100% penetrance. Identification of the product of the affected gene will indicate the root cause of the disorder and, hopefully, suggest methods of effective treatment. Intensive research is being done with this objective, and one hopes that the gene product will be defined within a few years. In the meantime, considerable interest has been focused on the excitatory amino acid systems in brain because the kainic acid and quinolinic acid animal "models" of HD suggest that excitatory amino acids may be involved in the mechanism of cell loss (Section IV). According to this hypothesis, some genetic defect in an excitatory amino acid system of the brain results in progressive neuronal destruction through an excitotoxic mechanism. The caudate and putamen might be particularly vulnerable because of the massive excitatory amino acid projection that they receive from the cortex (Fig. 2). Although unproven, this hypothesis has led to suggestions of as yet untested therapeutic measures to slow the progression of the disorder (Section V). [*See* Genetic Diseases.]

IV. Animal Models

Injections of excitotoxins such as kainic acid or quinolinic acid into the striatum of rats produces local neuronal loss histologically similar to that seen in HD (Fig. 1), and extensive biochemical studies have indicated that the lesion-induced changes in many neurotransmitter systems in the basal ganglia are similar to those found in HD (Table I). Bilaterally lesioned rats also show some behavioral abnormalities and pharmacological responses similar to those seen in the human disorder. Thus, for example, the rats show enhanced activity during a rat's normal waking period (Fig. 3A), abnormal locomotion, learning problems, and weight losses reminiscent of those seen in HD. They also show markedly enhanced responses to amphetamine (Fig. 3B) and scopolamine but an attenuated cataleptic response to haloperidol and sedative effects with apomorphine. The hyperactivity in the rats is decreased on treatment with haloperidol or physostigmine. These findings have been interpreted as akin to the pharmacological responses seen in HD.

Rats sacrificed several months after such lesions show considerable striatal atrophy and some neuronal loss in the cortex and thalamus, which make the pathology even more similar to that in HD than the changes seen shortly after lesioning. However, the time course of neurotoxicity in such "models" clearly differs from that seen in the human disease, and the genetic factor is missing.

The quinolinic acid "model" seems biochemically to resemble HD more closely than the kainic acid "model." Moreover, quinolinic acid is a normal mammalian metabolite of tryptophan and is found in brain; kainic acid is derived from a seaweed and does not occur in animals. Hence, much attention has been focused on the quinolinic acid system, but the general hypothesis, based on these models, merely suggests some abnormality in excitatory amino acid systems in HD brain.

V. Treatment

No form of treatment is known to arrest the mental deterioration. Dopamine blockers (such as haloperidol or chlorpromazine) or dopamine depleters (such as reserpine or tetrabenazine) may help to control the involuntary choreiform movements in some cases; such drugs are better known for their tranquilizing action in schizophrenia.

As mentioned (Section II), attempts to replace the GABA or cholinergic deficiencies were without beneficial effect. In view of the complex nature of the biochemical losses in HD (Table I), it seems

A

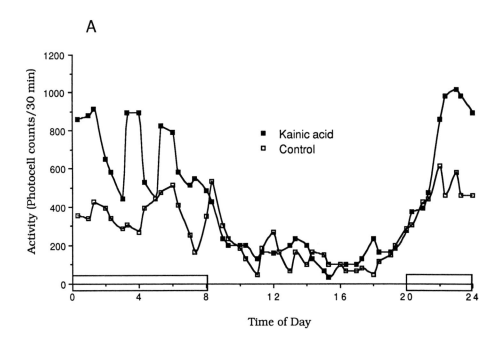

B

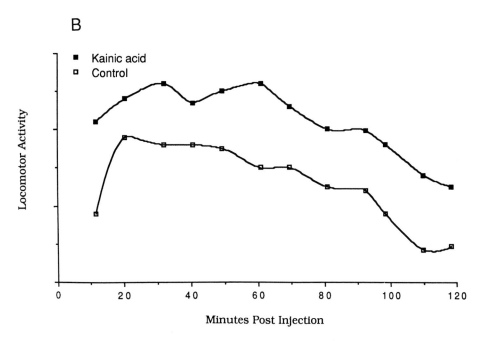

FIGURE 3 Mean activity measurements in groups of kainic acid-lesioned rats as compared with groups of controls. (A) Over the course of 1 day, the period of light being from 8 to 20 hr. (B) After injections of amphetamine.

highly unlikely that an effective replacement therapy can be devised. Side effects would also be large because of the existence elsewhere in the brain of healthy neurons of the affected types. Attention is therefore being focused on therapies designed to prevent or inhibit the cellular destruction. The existence of the excitotoxic "models" of HD (Section IV) has led to the suggestion that agents that inhibit either the release of neurotransmitter glutamate or the action of excitatory amino acids at postsynaptic receptors, particularly those of the N-methyl-D-aspartate (NMDA) subtype, might slow progression of the disease. Other agents such as taurine and naloxone, which antagonize excitotoxicity in animal models, have also been suggested. In theory,

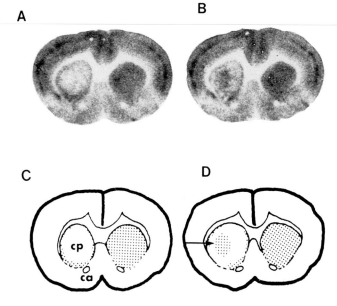

FIGURE 4 Autoradiographic studies of glucose metabolism in brain sections from rats that had been given injections of kainic acid into the left striatum about 11 weeks before sacrificing. Extent of the normal striatal area is dotted in the diagrams. Rat (B, B') received a transplant of striatal cells from a neonatal rat three weeks after the kainic acid injection. The darker the shading, the greater the rate of glucose metabolism. Note the lower glucose metabolism in the striatum on the injected side of the rats as compared to the uninjected side. HD cases show similarly low glucose metabolism in the striatum when studied by PET. The transplanted cells (arrow) appear to have nearly normal glucose metabolism.

none of these agents would have any effect on established symptomatology, but they might be particularly useful in delaying the onset of clinical symptoms in persons carrying the gene. The only drug of this type so far tested is the relatively weak and nonspecific GABA$_B$ agonist Baclofen; it did not appear to slow the progression in established cases. The present great interest in the development of NMDA antagonists and glutamate release inhibitors for the possible treatment of epilepsy and prevention of neuronal loss in ischemia and hypoxia should lead to the rapid identification of agents worthy of trial.

Experimental transplants have been done in the excitotoxic rat "models" of HD. Neonatal or fetal rat striatal cells are injected into the lesioned area some days after striatal injections of kainic or quinolinic acid. The cells can be shown histologically to live and develop a blood supply. The transplanted tissue also seems to show near-normal glucose metabolism (Fig. 4). Biochemical measurements indicate some recovery of GABA and cholinergic indices and no loss in weight of the caudate–putamen. Some normalization of the behavior and response to drugs has also been reported after such transplants. A limited number of such transplants have been done in baboons with excitotoxic lesions of the striatum, but analogous procedures have not been tried in human victims of HD.

Bibliography

Harper, P. J., Quarrell, O. W. J., and Youngman, S. (1988). Huntington's disease: Prediction and prevention. *Philos. Trans. R. Soc. Lond. [Biol.]* **319,** 285–298.

Menkes, J. H. (1988). Huntington's disease: Finding the gene and after. *Pediatr. Neurol.* **4,** 73–78.

Sanberg, P. R., and Coyle, J. T. (1984). Scientific approaches to Huntington's disease. *CRC Crit. Rev. Clin. Neurobiol.* **1,** 1–44.

Schwarcz, R., Foster, A. C., French, E. D., Whetsell, W. O., Jr., and Kohler, C. (1984). Excitotoxic models for degenerative disorders. *Life Sci.* **35,** 19–32.

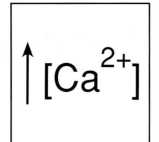

Hypercalcemia of Malignancy

GREGORY R. MUNDY, ASHLEY J. P. YATES, AND CHARLES A. REASNER, *University of Texas Health Science Center at San Antonio*

Glossary

Humoral hypercalcemia of malignancy Hypercalcemic syndrome occurring in patients with cancer that is due to circulating factors produced by the tumor cells

Hypercalcemia Increase in the calcium concentration in the blood

Myeloma Neoplastic disorder of plasma cells universally associated with bone destruction, which may be either localized or generalized and often complicated by hypercalcemia

Osteolytic metastasis Localized bone destruction occurring adjacent to cancer cells that have spread to bone

Primary hyperparathyroidism Overproduction of parathyroid hormone by a primary disorder of the parathyroid glands; usually due to a single benign parathyroid gland adenoma

HYPERCALCEMIA OF MALIGNANCY is one of the most common endocrine syndromes associated with cancer. These endocrine syndromes are caused by humoral factors produced by the tumors and are unrelated to direct tumor cell invasion of tissues or to tumor metastases (spread of the tumor to distant sites). These syndromes have also been called paraneoplastic syndromes or ectopic hormone syndromes. They frequently mimic overproduction of hormones by benign tumors of endocrine glands. In the case of hypercalcemia of malignancy, the hypercalcemic syndrome is similar to that seen in patients with benign adenomas of one of the four parathyroid glands that are associated with excess production of parathyroid hormone (PTH). The endocrine syndromes associated with malignancy are responsible for considerable morbidity and even mortality. They are important for physicians to recognize because they can be frequently treated or prevented by appropriate therapy. Other examples of these endocrine syndromes include Cushing syndrome, inappropriate antidiuretic hormone secretion, hypoglycemia, cachexia, fever, and other nonmetastatic hematopoietic, neurologic, and dermatologic syndromes.

Tumor cells have an amazing capacity to synthesize and secrete peptides that cause systemic effects. Tumors have been described that have produced up to nine ectopic proteins. It is not surprising therefore that so many syndromes caused by the systemic effects of tumor peptides have been described.

Hypercalcemia is one of the most serious of these syndromes. It always indicates increased calcium resorption from the skeleton, although there are also usually effects of tumor products on the other major organs of calcium homeostasis, the kidney and the gut. Tumor involvement of the skeleton also causes bone pain and fracture after trivial injury as well as hypercalcemia. This is a particularly unfortunate complication of cancer, because once tumors involve the skeleton, they are usually incurable, and the only therapy possible is palliative.

I. Frequency of Hypercalcemia in Patients with Malignancy

With the introduction of routine serum calcium measurements by the autoanalyzer in the mid 1970s, hypercalcemia has become recognized as a common electrolyte abnormality in the general population, as well as in hospitalized patients. Estimates of the incidence and prevalence of primary hyperparathyroidism have dramatically risen since that time. In addition, hypercalcemia has also been more frequently noted in patients with malignant disease. Epidemiologic surveys performed in England and the United States suggest there are approximately 270 new cases of primary hyperparathyroidism per million population per year and approximately 150 new cases of hypercalcemia occurring in patients with malignant disease per million population per year. Primary hyperparathyroidism and malignant disease are the cause of hypercalcemia in the vast majority of hypercalcemic patients. Other causes of hypercalcemia listed in Table I are much less common, comprising about 10% of all patients.

A relatively small number of specific malignancies are responsible for hypercalcemia in most cancer patients (Table II). Malignant hypercalcemia is most often caused by squamous cell carcinoma of the lung, whereas anaplastic carcinoma or oat cell carcinoma of the lung is rarely associated with hypercalcemia. Carcinoma of the breast of all types is often responsible for hypercalcemia. Twenty to forty percent of patients with hematologic malignancies, such as myeloma, develops hypercalcemia. Squamous cell tumors of the head and neck and the upper end of the esophagus are also often associated with hypercalcemia. Many common cancers are almost never associated with hypercalcemia (e.g., carcinoma of the colon and carcinoma of the female genital tract). In contrast, some rare malignancies are frequently associated with hypercalcemia (e.g., cholangiocarcinoma and VIPomas).

TABLE I Causes of Hypercalcemia

Primary hyperparathyroidism
    Sporadic (80% adenoma)
    Familial (hyperplasia)
Malignant disease
Hyperthyroidism
Immobilization
Vitamin D intoxication
Vitamin A intoxication
Familial hypocalciuric hypercalcemia (FHH)
Diuretic phase of acute renal failure
Chronic renal failure
Thiazide diuretics
Sarcoidosis and other granulomatous diseases
Milk-alkali syndrome
Addison's disease

TABLE II Relative Frequency of Causes of Hypercalcemia

	No. of patients	Percentage of cases	
Primary hyperparathyroidism	111	54	
Malignant disease	72	35	
Lung	25		35
Breast	18		25
Hematologic (myeloma 5, lymphoma 4)	10		14
Head and neck	4		5
Renal	2		3
Prostate	2		3
Unknown primary	5		7
Others (gastrointestinal 4)	6		8

II. Pathophysiology

In most patients with cancer, the hypercalcemia is due to a combination of increased calcium released from bone (increased bone resorption) and decreased excretion of calcium by the kidneys (increased renal tubular calcium reabsorption). In a few patients, impairment of glomerular filtration contributes to the decreased renal excretion of calcium. The combined effects of increased bone resorption and decreased renal excretion of calcium offset the decreased calcium absorption from the gut seen in most patients. The pathophysiology of hypercalcemia of malignancy is best considered in terms of three distinct syndromes: (1) humoral hypercalcemia of malignancy, (2) hypercalcemia associated with osteolytic disease, and (3) hematologic malignancies.

A. Humoral Hypercalcemia of Malignancy

This hypercalcemic syndrome is caused by tumor products that circulate and cause hypercalcemia because of their systemic effects. It is unrelated to local osteolytic bone destruction. Solid tumors are those most often associated with the humoral hypercalcemia of malignancy (HHM) syndrome (e.g., carcinomas of the lung, head and neck, kidney, pancreas, and ovary). Occasionally, patients with malignant lymphomas develop the HHM syndrome. Most patients with HHM have many of the biochemical features of primary hyperparathyroidism including hypercalcemia, and increased urinary excretion of phosphate, cyclic AMP. However, other characteristics of the HHM syndrome, including decreased calcium absorption from the gut, decreased bone formation, and metabolic alkalosis, are not seen in primary hyperparathyroidism. The variations that occur in different patients with the syndrome of HHM are most likely caused by the variety of factors produced by the tumor and the variable response of the host immune cells. Specific factors that have been implicated on the pathophysiology of HHM are discussed below.

1. Parathyroid Hormone-Related Protein

Albright first suggested that the hypercalcemia of malignancy is caused by a parathyroid hormone-(PTH) like substance produced by the tumor; the phrase ''pseudohyperparathyroidism'' is used to refer to this syndrome. Evidence supporting the presence of a PTH-related protein includes the presence of a factor in tumor tissue that stimulates adenylate cyclase activity in renal tubular membranes and osteoblast cells that can be inhibited by synthetic antagonists to parathyroid hormone. The factor responsible for this activity has been identified, and the gene molecularly cloned. Eight of the first 13 amino acids of this PTH-related protein (PTH-rP) are identical to PTH. The gene encoding PTH-rP is clearly different from the PTH gene (it is on a different chromosome) but probably had a similar evolutionary origin. There is more than one PTH-rP due to differential use of the information of the gene during formation of messenger RNA (alternate processing) and possibly due to proteolytic degradation of the protein after secretion. The PTH-rP is able to bind and activate the PTH receptor.

The PTH-rP seems to mimic all the biologic effects of PTH and in addition has its own unique effects. Like PTH, PTH-rP stimulates bone resorption both *in vitro* and *in vivo,* increases renal tubular calcium reabsorption, increases cyclic AMP generation, increases renal phosphate excretion, increases 1,25 dihydroxyvitamin D production, and causes hypercalcemia when given by injection. Despite the similarities between PTH-rp and PTH, the syndromes of HHM and primary hyperparathyroidism are clearly different. These differences are likely due to other factors that are produced by the tumor or host immune cells, which have unique effects on target tissues to modulate the effects of PTH-rP. [*See* PARATHYROID GLAND AND HORMONE.]

2. Transforming Growth Factor α

Solid tumors frequently produce polypeptide growth factors that may be responsible for maintaining the transformed phenotype in tumor cells. These autocrine factors have a similar conformation to that of epidermal growth factor and are able to bind to and activate the epidermal growth factor receptor. The best described of these factors is transforming growth factor alpha (TGFα). *In vitro* TGFα is a potent stimulator of osteoclasts, whose function is to solubilize bone. Its effects on bone are even more potent than those of PTH or PTH-rP. It also causes hypercalcemia when injected into normal mice. There is now considerable evidence that some tumors produce both TGFα and PTH-rP. This is clearly true in the rat Leydig tumor (a testicular tumor), a well-studied animal model of HHM. Because the effect of TGFα and PTH-rP on bone are synergistic *in vitro,* it is not surprising that these tumors are associated with profound increases in osteoclastic bone resorption. However, the relation between these factors may be complex. In some circumstances, TGFα opposes the effects of PTH-rP. For example, TGFα decreases the adenylate cyclase response to PTH-rP (and PTH) in both renal tubular cells and cells with the osteoblast phenotype (which cause bone formation). Moreover, TGFα stimulates replication of osteoblastic cells but inhibits their differential function. [*See* TRANSFORMING GROWTH FACTOR-α.]

3. Tumor Necrosis Factor

Tumor necrosis factor (TNF) is a potent bone resorbing cytokine (locally acting factor) that is secreted by normal activated macrophages. It stimulates the formation of osteoclasts from progenitor cells by enhancing both proliferation of progenitors and differentiation of committed progenitors and also activates mature osteoclasts. Like TGFα and PTH-rP, TNF causes hypercalcemia when infused or injected *in vivo* or whenever cells carrying the gene for TNF are introduced into animals. Although there is no current evidence that solid tumors themselves produce TNF directly, a number of tumors have now been identified that have the capacity to stimulate TNF production by normal host cells. In some circumstances, TNF production is due to tumor production of granulocyte-macrophage colony stimulating factor (GM-CSF), which in turn acts on host immune cells to provoke TNF release. In nude mice carrying tumors, the serum calcium can be lowered by injecting antibodies to TNF. In these circumstances, it appears likely that tumor cells produce bone-resorbing factors such as PTH-rP and TGFα and that host cells produce bone-resorbing cytokines such as TNF. Hypercalcemia is presumably due to the effects of these factors acting in concert. The hypercalcemia caused by the introduction of cells carrying the TNF gene is associated with markedly increased osteoclastic bone resorption, indicating that increased circulating TNF, produced either by tumor cells or by host cells, can cause the syndrome of hypercalcemia of malignancy. [*See* CYTOKINES IN THE IMMUNE RESPONSE.]

4. Interleukin-1

Interleukin-1 (IL-1) is a multifunctional protein that is present in supernatant media harvested from activated leukocyte cultures. It causes hypercalcemia when injected into normal mice. Both IL-1α and IL-1β appear to do so equally by stimulating osteoclastic bone resorption. An initial transient fall in the serum calcium may be mediated via prostaglandins because it can be inhibited by indomethacin, which blocks prostaglandin formation.

IL-1 causes marked changes in bone resorption. There are increases in osteoclast numbers, active bone resorption surfaces, and marked expansion of the marrow cavity. IL-1 causes increased osteoclast formation in a marrow culture system, which detects the effects of factors on the generation of osteoclasts from human mononuclear precursors. In addition, IL-1 stimulates differentiation of committed progenitors into mature osteoclasts and increases the capacity of the mature osteoclasts to resorb bone. Whether this effect is mediated directly or indirectly through other cells is still not clear.

Interleukin-1 is produced by several solid tumors and at least one lymphoma associated with the HHM. Surprisingly, these tumors produce IL-1α, which is the less predominant form produced by normal activated leukocytes.

It is important to note that both IL-1 and TNF act synergistically on bone with PTH, PTH-rP, and TGFα. Like TGFα, IL-1 has been shown to modify the effects of PTH or PTH-rP on target bone cells. Preincubation of osteoblastic cells with IL-1 decreases the subsequent adenylate cyclase response to PTH or PTH-rP.

5. Other Factors

It is possible that other factors produced by tumors can cause the HHM. Recently, a factor capable of producing hypercalcemia in nude mice was identified in a culture of human melanoma cells. It differs biologically from both PTH-rP and TGFα. Another factor is a form of vitamin D, probably 1,24 R dihydroxyvitamin D, produced by a small cell carcinoma of the lung. This particular vitamin D metabolite competes with 1,25 dihydroxyvitamin D for receptor binding blocking its activities. The result is bone resorption that is observed in organ cultures. [*See* VITAMIN D.]

A number of years ago, there was great interest in the potential role of prostaglandins as humoral mediators of hypercalcemia of malignancy. It seems unlikely that they might play an important role as systemic mediators, although it is still possible that local production of prostaglandin in bone may be important. Although prostaglandin production by several animal tumor models has been convincingly demonstrated, prostaglandins have to be infused or injected in enormous amounts to cause hypercalcemia *in vivo,* and treatment with drugs that effectively inhibit prostaglandin synthesis has been ineffective in the treatment of hypercalcemia of malignancy.

B. Osteolytic Metastases

Hypercalcemia often occurs in patients with widespread bone metastases associated with extensive osteolytic bone destruction. In these cases it is likely that the tumor cells either directly or indirectly stimulate local osteoclast activity, which is responsible for the bone destruction and hypercalcemia. The most common tumor in this category is breast cancer, in which hypercalcemia is almost always limited to patients with extensive metastatic bone disease. Renal tubular calcium reabsorption is increased in some of these patients due to mechanisms that are not clear. Although occasional patients with breast cancer and hypercalcemia have been shown to produce PTH-rP, in the great majority, nephrogenous cyclic AMP is not increased and calcium reabsorption is likely produced by other mechanisms. Hypercalcemia is estimated to occur in approximately one-third of patients with breast cancer. Because breast cancer cells have the capacity to cause the release of ^{45}Ca from devitalized bone *in vitro*, the tumor cells may have a direct effect in the development of hypercalcemia. However, examination by scanning electron microscopy of bone surfaces suggests that osteoclast stimulation by tumor factors is the major mechanism.

Breast cancer cells produce a number of factors capable of stimulating osteoclast activity in culture, in particular, TGFα. It is usually present in greater amounts in patients with hormone-independent, estrogen receptor negative cell lines. In the estrogen receptor positive cell lines, TGFα production can be increased by incubating the cells with estrogen. Prostaglandins are also produced by many of these breast cancer cells, and also their production is enhanced by estrogens. This may be important in the development of acute hypercalcemia seen in patients with breast cancer and widespread metastatic disease who have been treated with antiestrogens. Breast cancer cells also produce the zymogen precursor form of the lysosomal enzyme procathepsin D, which has no enzymatic activity but directly stimulates osteoclasts. This may be an important mechanism for local bone destruction. Because the mechanism of hypercalcemia in this important tumor is relatively unexplored, other factors produced by tumor cells could also be involved in the bone destruction. [*See* BREAST CANCER BIOLOGY.]

C. Myeloma and Other Hematologic Malignancies

In myeloma, hypercalcemia occurs in approximately 20–40% of patients and is almost always associated with extensive osteolytic bone disease. Local factors produced by myeloma cells activate adjacent osteoclasts, causing the osteolytic bone lesions and, in some patients, hypercalcemia. Hypercalcemia is far more common in patients with impaired glomerular filtration. Because impairment of glomerular filtration in myeloma frequently occurs from a variety of causes, it is not surprising that hypercalcemia occurs so frequently.

Concerning the mediators produced by myeloma cells that are responsible for increasing osteoclast activity, the major one appears to be lymphotoxin, a polypeptide cytokine released by normal activated T-lymphocytes. Human or murine lymphotoxin stimulates osteoclastic bone resorption in organ culture and causes increased bone resorption and hypercalcemia when injected into normal intact mice. Other factors may be also involved because not all the bone resorbing activity can be blocked by antibodies to lymphotoxin. One possibility is the cytokine IL-6, which has the capacity to increase osteoclast formation from committed progenitors *in vitro*. Possibly, production of IL-6 in combination with lymphotoxin could be responsible for the effects observed on bone in patients with myeloma.

Patients with lymphomas occasionally develop hypercalcemia. This can be seen with T-cell lymphomas, B-cell lymphomas, or histiocytic lymphomas. The mechanisms here are clearly heterogeneous. In some of these patients, increased circulating concentrations of 1,25 dihydroxyvitamin D have been found, probably produced by the tumor cells from 25 hydroxyvitamin D. However, this is a rare event because increased serum concentrations of this vitamin D metabolite are not present in the majority of patients. Other lymphomas have been shown to produce IL-1α and PTH-rP. [*See* LYMPHOMA.]

III. Treatment

A. General

Medical therapy for hypercalcemia of malignancy is usually successful. Many agents will lower the serum calcium satisfactorily. However, no therapy is ideal, and the treatment choice should be individualized for each patient. The selection of an individual form of therapy depends on a number of factors including the rate of rise of the serum calcium, specific contraindications to a particular form of therapy (such as the presence of renal failure), the extent of the increase in serum calcium, the presence of symptoms, and the pathophysiology of the hypercalcemia. The general principles of therapy for hypercalcemia are (1) effective treatment of the tumor where possible, (2) treatment of any exacerbating cause that has precipitated hypercalcemia, and (3) treatment of dehydration, which is common in patients with the hypercalcemia of malignancy and may lead to a spiral of disequilibrium hypercalcemia and progressive worsening of the serum calcium as fluid losses provoke renal tubular calcium reabsorption.

In patients with malignant disease, hypercalcemia is usually progressive, and the serum calcium may rise rapidly over a few weeks. This occurs because of progressive increases in bone resorption and overall bone destruction. Because the symptoms of hypercalcemia are unpleasant and can be prevented, we feel all patients with the hypercalcemia of malignancy should be treated actively.

B. Nonurgent Therapy for Hypercalcemia

The majority of patients with hypercalcemia of malignancy will be asymptomatic with a serum calcium less than 13 mg/dl at the time of diagnosis. A number of agents are satisfactory for this situation. These agents may also be used in patients previously treated for more severe hypercalcemia who have mild residual hypercalcemia.

1. Bisphosphonates

These analogues of pyrophosphate bind to mineralized surfaces and inhibit mineral deposition and osteoclastic bone resorption. Although their precise mechanisms of action are still unclear, it seems most likely that they do in fact have a direct cellular effect to inhibit the bone resorption process. In the United States, etidronate is the only biphosphonate approved for human use. Unfortunately, etidronate is not effective when taken only orally—it must first be used intravenously before it can be given orally. Intravenous therapy should be given in association with approximately 3 liters/day of intravenous saline. This therapy is relatively safe and effective in approximately 70% of patients. In most patients the serum calcium will return to the normal level within 3–5 days.

Of the other oral bisphosphonates, the most promising is amino-hydroxypropylidene bisphosphonate (APD), which is extremely effective and acts relatively quickly, lowering the serum calcium within 48–72 hours. In addition to inhibiting mature osteoclasts, it may also inhibit the recruitment of osteoclasts and their progenitors. A bisphosphate currently widely used in Europe is clodronate. In some of the earlier studies, several patients being treated with this agent developed acute leukemia, but it seems unlikely that this was related to the drug. Neither clodronate nor APD significantly impairs bone mineralization when used in doses effective at lowering the serum calcium. Their other toxicities are relatively minor.

2. Plicamycin

Plicamycin was developed as a cytotoxic drug during the 1960s and is an effective antitumor agent in certain rare embryonal tumors. It was found by serendipity that plicamycin also lowered the serum calcium, and it has been widely used since then for the treatment of hypercalcemia of malignancy. It inhibits DNA-dependent RNA synthesis, and it is probably cytotoxic for osteoclasts. In organ culture its effects are long-lasting and presumably irreversible. It is extremely effective in the treatment of hypercalcemia of malignancy, lowering the serum calcium in greater than 80% of patients. However, it must be given by intravenous infusion over several

hours. The infusion should not be repeated until the serum calcium rises again. It is potentially dangerous in patients who have impaired renal function because it has direct nephrotoxic effects and is cleared by the kidney. Other side effects include bone marrow suppression, a nonthrombocytopenic bleeding diathesis, and hepatotoxicity. For these reasons, plicamycin should be reserved for patients with normal renal function when other drugs have been unsuccessful.

3. Calcitonin and Glucocorticoids

Glucocorticoids have been used for many years in the treatment of hypercalcemia of malignancy. When used alone they reduce serum calcium in approximately 30% of all patients but rarely to the normal range. They are most likely to be effective in patients with hematologic malignancies such as myeloma or the lymphomas. They are occasionally effective in patients with breast cancer or lung cancer, but it is difficult to predict which patients will respond. *In vitro,* they are effective inhibitors of bone resorption stimulated by agents such as the cytokines but are less effective against PTH or PTH-rP. They are a convenient form of treatment for the ambulant patient, because the many chronic side effects of glucocorticoids are not seen in most patients who have no more than 3–6 months of remaining life owing to the progression of the tumor.

Calcitonin alone in many patients produces a transient fall of serum calcium, which is not maintained. This "escape phenomenon" is seen both *in vitro* and *in vivo*. However, the escape phenomenon can be blocked by glucocorticoids. The combination of the two drugs is extremely effective in patients who require a rapid fall in serum calcium, particularly in patients with hematologic malignancies. It is an attractive form of treatment in patients with fixed impairment of renal function, in whom most other forms of treatment are contraindicated. In many patients, it may be necessary to withdraw calcitonin for 48 hours every week to avoid escape. There are no significant side effects when calcitonin is used in these doses.

4. Phosphate

Oral phosphate has been used for many years in the treatment of hypercalcemia. It is only like to be effective in those patients in whom the serum phosphorus is less than 4 mg/dl. The mechanism of action is complex. Phosphate inhibits osteoclastic bone resorption, decreases calcium absorption from the gut, and promotes soft tissue calcium deposition. Oral phosphate should not be used in patients who have impaired renal function or who have serum phosphorus levels greater than 4 mg/dl. The major side effect of long-term phosphate therapy is troublesome diarrhea, which often limits its usefulness.

5. Gallium Nitrate

This drug, which was originally used as a cytotoxic agent, inhibits bone resorption *in vitro* and produces prolonged lowering of the serum calcium *in vivo*. It seems to be effective in most forms of hypercalcemia of malignancy. It has not had widespread application as yet, although preliminary results are encouraging and side effects are few.

6. Indomethacin and Aspirin

These very effective inhibitors of prostaglandin synthesis were widely used when prostaglandins were thought to be the major cause of hypercalcemia of malignancy. However, patients rarely respond to them.

C. Emergency Treatment of Hypercalcemia

Emergency treatment for hypercalcemia is necessary when patients are symptomatic. This usually occurs when the serum calcium is greater than 13 mg/dl. This level of serum calcium is potentially life-threatening in patients with malignant disease in whom the serum calcium may rapidly rise further, particularly if patients become dehydrated. Other situations that can precipitate severe hypercalcemia include thiazide therapy, treatment with estrogens or antiestrogens if the patient has breast cancer, or vomiting with associated dehydration and prerenal azotemia.

1. Fluids

Patients with severe hypercalcemia are often dehydrated and may require repletion with 6–10 liters of normal saline over 24 hours. Because severe hypercalcemia is associated with fluid depletion, and calcium and sodium handling by the renal tubules are linked, normal saline should be given to replace fluid losses. Normal saline repletion will lead to a sodium and calcium diuresis, which will help lower the serum calcium. Occasionally, patients will become hypernatremic (excess sodium); and under these circumstances half normal saline should be substituted for normal saline.

2. Loop Diuretics

Furosemide has been frequently used in the United States for the emergency treatment of hypercalcemia because it promotes a calcium diuresis. However, in the doses used by most clinicians it is not clear that it produces a significant effect beyond that of saline therapy alone. Moreover, loop diuretics at the required doses may be dangerous and lead to other electrolyte problems, dehydration, and worsening of hypercalcemia. [*See* DIURETICS.]

3. Bisphosphonates

The most effective current agents in the emergency treatment of hypercalcemia are the bisphosphonates. Etidronate is effective in approximately 70–80% of patients and will usually lower the serum calcium to the normal range after 48–72 hours, without serious toxicity. The major side effect is impairment of bone mineralization, which is not a concern in the short term but may be more important in patients receiving large doses over prolonged periods of time. The newer bisphosphonates APD and clodronate are probably more effective and act more rapidly.

4. Calcitonin and Glucocorticoids

The combination of calcitonin and glucocorticoids is often effective in the emergency treatment of hypercalcemia. It is particularly effective in patients with hematologic malignancies. It is probably the most rapid and safest form of therapy in patients who have hypercalcemia associated with cardiac failure or renal failure. Unfortunately, it is not universally effective. The use of calcitonin together with a bisphosphonate may cause a more rapid lowering of the serum calcium than the use of the bisphosphate alone.

5. Plicamycin

Plicamycin (mithramycin) is often efficacious in the treatment of severe hypercalcemia but has to be given by infusion and usually takes 24–48 hours to produce a beneficial effect. It should be avoided in patients who have impaired renal function because of its nephrotoxicity.

6. Dialysis

Dialysis has been used occasionally in the emergency treatment of hypercalcemia in patients with renal failure. It is useful only as a transient remedy where other therapies are contraindicated or ineffective.

Acknowledgments

The authors are grateful to Nancy Garrett, Thelma Barrios, Kim Samaniego, and Joanna Kelley for expert secretarial assistance in the preparation of this manuscript. Part of the work described here was supported by grants CA-40036, RR-1346, AR-28149, AR-39357, and DEO-8526-02.

Bibliography

Case Records of the Massachusetts General Hospital (Case 15-1971). (1971). *N. Engl. J. Med.* **284,** 839–847.

Garrett, I. R., Durie, B. G. M., Nedwin, G. E., Gillepsie, A., Bringman, T., Sabatini, M., Bertolini, D. R., and Mundy, G. R. (1987). Production of the bone resorbing cytokine lymphotoxin by cultured human myeloma cells. *N. Engl. J. Med.* **317,** 526–532.

Moseley, J. M., Kubota, M., Diefenbach-Jagger, H., Wettenhall, R. E. H., Kemp, B. E., Suva, L. J., Rodda, C. P., Ebeling, P. R., Hudson, P. J., Zajac, J. D., and Martin, T. J. (1987). Parathyroid hormone-related protein purified from a human lung cancer cell line. *Proc. Natl. Acad. Sci.* **84,** 5048–5052.

Mundy, G. R. (1988). The hypercalcemia of malignancy revisited. *J. Clin. Invest.* **82,** 1–6.

Mundy, G. R., Ibbotson, K. J., and D'Souza, S. M. (1985). Tumor products and the hypercalcemia of malignancy. *J. Clin. Invest.* **76,** 391–395.

Mundy, G. R., Ibbotson, K. J., D'Souza, S. M., Simpson, E. L., Jacobs, J. W., and Martin, T. J. (1984). The hypercalcemia of malignancy: Clinical and pathogenic mechanisms. *N. Engl. J. Med.* **310,** 1718–1727.

Mundy, G. R., and Martin, T. J. (1982). Hypercalcemia of malignancy—Pathogenesis treatment. *Metabolism* **31,** 1247–1277.

Myers, W. P. L. (1960). Hypercalcemia in neoplastic disease. *Arch. Surg.* **80,** 308–318.

Powell, D., Singer, F. R., Murray, T. M., Minkin, C., Potts, J. T. (1973). Non-parathyroid humoral hypercalcemia in patients with neoplastic disease. *N. Engl. J. Med.* **289,** 176–181.

Sporn, M. B., and Todaro, G. J. (1980). Autocrine secretion and malignant transformation of cells. *N. Engl. J. Med.* **303,** 878–880.

Suva, L. J., Winslow, G. A., Wettenhall, R. E. H., Hammonds, R. G., Moseley, J. M., Diefenbach-Jagger, H., Rodda, C. P., Kemp, B. E., Rodriguez, H., Chen, E. Y., Hudson, P. H., Martin, T. J., and Wood, W. I. (1987). A parathyroid hormone-related protein implicated in malignant hypercalcemia: Cloning and expression. *Science* **237,** 893–896.

Yates, A. J. P., Gutierrez, G. E., Smolens, P., Travis, P. S., Katz, M. S., Aufdemorte, T. B., Boyce, B. F., Hymer, T. K., Poser, J. W., and Mundy, G. R. (1988). Effects of a synthetic peptide of a parathyroid hormone-related protein on calcium homeostasis, renal tubular calcium reabsorption and bone metabolism. *J. Clin. Invest.* **81,** 932–938.

Hypertension

MORTON P. PRINTZ, *University of California at San Diego*

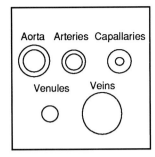

I. Blood Pressure—Definition and Characterization
II. Hypertension as a Disorder in Humans
III. Pathophysiology of Human Essential Hypertension
IV. Hypertension as a Disorder

Glossary

Autonomic nervous system Nervous system that carries information from the brain to all the internal organs *except* skeletal muscle and bones; this system operates with minimal, if any, conscious effort and is an automatic responding system; two parts of the autonomic system are recognized: a sympathetic branch and a parasympathetic branch; almost all organs receive nerve fibers of both parts, and optimum organ system function requires both to be operational

Cardiac output Amount of blood pumped by the left ventricle of the heart in 1 min; this is a measure of the pumping action of the heart

Cardiovascular system Refers to the organ system, which includes the heart and all the blood vessels, arteries, veins, and capillaries

Diastolic pressure Minimal arterial blood pressure, which occurs during the rest phase (or diastole) of the heart

Endothelial cell Cells that line the inner face of blood vessels and are in immediate contact with blood

Heart left ventricle The heart consists of four chambers: right and left atria and right and left ventricles; the atria serve to pump the blood into the ventricles, which are the main pumping chambers; blood returns from the body and enters the right atrium and is pumped to the right ventricle and then through the lungs; it returns to the left atrium, which pumps it into the left ventricle; the left ventricle is the pumping chamber that forces the blood out into the arteries toward all the tissues

Neurotransmitter Chemicals that are made and stored in nerve cells and that, when released by a process termed exocytosis, carry information from one nerve cell to another cell; some neurotransmitters excite the target cell, whereas others inhibit; however, all are carries of information

Systolic pressure Maximal arterial blood pressure, which occurs when the left ventricle of the heart pumps blood into the arteries

Vascular resistance Resistance to the flow of blood through the arteries and into the capillary blood vessels in the organs

HYPERTENSION, or high blood pressure, refers to an elevation of systemic arterial blood pressure that is *greater than expected* based on population measurements of systemic arterial pressure for individuals of comparable age and sex. Blood pressure and volume is controlled by many organ systems including the heart, nervous system, blood vessels, and the kidney. At the present time, hypertension is considered to be a disorder of regulation rather than a simple disease process. Essential or primary hypertension is the most common form of the disorder and its cause is still unknown. A clear diagnosis of hypertension is a first step followed by classification of patients according to the severity of the hypertension. Therapy of hypertension is then individualized and consists of both drug and nondrug approaches. There are several types of antihypertensive drugs, each type designed to influence one or more of the mechanisms that control blood pressure of volume. The high occurrence of this disorder throughout the world combined with the evidence that untreated hypertension leads to increased mortality from heart disease, stroke, kidney failure, and atherosclerosis makes hypertension a major medical problem.

I. Blood Pressure—Definition and Characterization

A. Definition

Blood pressure, in its simplest concept, is the pressure of the liquid within blood vessels; however, hypertension refers to elevated pressure in the arteries, the blood vessels that serve as the pathway for blood from the heart to all the tissues. The blood pressure in the arteries is determined by (1) the force provided the blood by the heart during its phase of contraction and ejection of blood into the arterial compartment, (2) the rate of flow of blood out of the arterial compartment and into the tissues via the capillaries, (3) the volume of blood within the vascular compartment, and (4) the tension generated by the walls of the blood vessel resisting the blood pushed into the arteries by the heart. To understand how hypertension could develop by a variety of possible mechanisms, a brief discussion of the physiology of the cardiovascular system is warranted.

B. Vascular Compartments

Humans, like all animals, have a closed and continuous compartment through which blood flows carrying oxygen and nutrients to tissues and carbon dioxide and waste products from tissues to sites of metabolism and elimination. This compartment, referred to as the vascular compartment, actually consists of three separate but connected compartments: arterial (arteries), venous (veins), and capillary. [See CARDIOVASCULAR SYSTEM, ANATOMY.]

The arterial compartment commences at the outflow from the left ventricle of the heart and extends, in an ever-branching manner, to all the oxygen systems and tissues. The large vessels of the arterial system are termed conduit vessels, because they serve mainly to be feeder vessels to the larger number of smaller vessels ramifying to the tissues. The arterial system is a high-pressure circuit, and the large conduit vessels receive the full pressure and flow of the blood pumped from the left ventricle of the heart. As the vessels ramify to smaller and smaller vessels, the pressure within these vessels declines and is lowest (in the arterial circuit) at the junction with the capillary network of the tissues.

Arterial blood vessels are complex and consist of multiple cell types and three discrete layers. In immediate contact with blood is a single-cell-thick lining consisting of endothelial cells with underlying connective tissue protein and termed the intima of the vessel. The middle of the vessel (termed the media) consists of multiple layers of contractile smooth muscle cells. The outer layer (termed the adventitia) consists of connective tissue cells (fibroblasts), elastic and connective tissue proteins, sympathetic nerve endings, and other cell types, which vary according to the size of the vessel and organ system. The media smooth muscle layer has probably the most important role in determining the level of arterial blood pressure in this vascular compartment. This will be discussed later.

The capillary compartment consists of the smallest blood vessels in all three compartments of the vascular system. These vessels consist of a single cell layer of endothelial cells with associated connective tissue proteins. The structure of the capillary system serves to permit rapid movement of materials, in both directions, between the blood contained within the vessel and the adjoining tissue. This vascular compartment lacks both the outer adventitial and medial smooth muscle layers found in the arterial system and, therefore, is primarily a passive circuit for exchange.

The venous compartment commences at the outflow from the capillary compartment and extends to the right atrium of the heart. This vascular system serves both to collect the blood and return it to the heart as well as a reservoir of blood in the body. The venous compartment is essentially a mirror-image of the arterial compartment with small veins, highly ramified, joining into successively larger vessels. Veins are complex vessels just like arteries and consist of an endothelial cell lining, medial smooth muscle, nerves, and multiple types of cells; however, the venous system is a low-pressure circuit. The pressure in the large veins is approximately 3–5 mm of mercury (abbreviated as mm Hg see below). Accordingly, the smooth muscle layer is much thinner than that found in the arterial compartment for an equivalent-sized vessel. The veins are referred to as capacitance vessels because they have the capacity to expand and hold large amounts of blood.

When needed, this blood can be recruited into the arterial and capillary compartments through the activity of the sympathetic nervous system. [*See* VASCULAR CAPACITANCE.]

The vessels connecting the heart and lung constitute a special compartment of the vasculature. While arteries from the left ventricle of the heart to all the organ systems (but the lungs) carry highly oxygenated blood and veins returning to the right atrium of the heart carry low-oxygenated blood from the tissues, the converse exists in the blood vessels of the lungs (called pulmonary vessels). The pulmonary artery, from the right ventricle of the heart to the lungs, carries venous blood of low-oxygen content, while the pulmonary vein, from the lungs to the left atrium of the heart, carries highly oxygenated blood. [*See* CARDIOVASCULAR SYSTEM, PHYSIOLOGY AND BIOCHEMISTRY.]

Pulmonary hypertension is excess blood pressure in the pulmonary artery from the heart to the lung. It is not equivalent to systemic hypertension, the subject of this chapter, and has, in general, other causes and treatment. For this reason, we will not deal with this special form of hypertension.

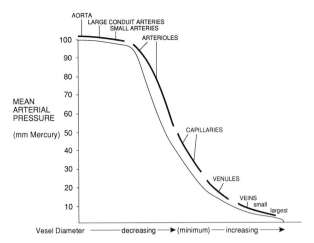

FIGURE 1 Schematic of the distribution of mean arterial pressure through the vascular compartment of man. As discussed in the text, the maximum decrease in arterial pressure occurs in the resistance vessels, the arterioles, with a further decrease in pressure through the capillaries. Arterial pressure continues to decrease through the venous side of the circulation, reaching a minimum at the entry to the right heart receiving chamber, the right atrium. In contrast to a continuous drop in arterial pressure, blood vessel diameter first decreases to a minimum in the capillary network and then increases again through the venous capacitance vessels.

C. Levels of Blood Pressure in Vascular Compartments

Blood pressure is highest in the large conduit arteries leaving the left ventricle of the heart (aorta and carotid arteries) and decreases steadily through the arterial, capillary, and venous compartments; however, the largest drop in pressure occurs on the arterial side (Fig. 1). Maintenance of this decreasing gradient of pressure is essential to ensure adequate blood delivery to all tissues, optimum fluid, and material exchange in the capillary network and to deliver the blood back to the right side of the heart to be pumped through the lungs and recycled through the body. Disease processes that interfere with any of these three functions will invariably result in a form of hypertension.

D. Determinants of Arterial Blood Pressure

Arterial blood pressure is determined by the pumping action of the heart, by the resistance to flow of the blood out from the arterial vascular bed and into the capillaries, and by the compliance (distensibility) of the walls of the arteries. The arterial pressure varies during the contraction phase (termed systole) and rest phase (termed diastole) of the left ventricle of the heart. When blood pressure is measured, both an upper and lower value is quoted (e.g., 120 over 80). The upper value is the maximum pressure measured (systolic pressure), while the lower value is the lowest (diastolic pressure). The magnitude of the difference between systolic pressure and diastolic pressure is pulse pressure.

During contraction of the left ventricle of the heart (i.e., systole), blood is ejected into the large arterial conduit vessels and arterial pressure builds rapidly to the maximum systolic pressure (Fig. 2). The magnitude of systolic pressure is determined by the rate and force of ejection of blood into the arterial vessels by the left ventricle of the heart, by the degree of compliance of the large conduit arteries, and by the rate of outflow from the arterial compartment into the capillary compartment. During diastole, as the left ventricle relaxes permitting a refilling of the left ventricle by blood from the left atrium, arterial pressure declines to the minimum diastolic pressure (Fig. 2). The rate of decline is

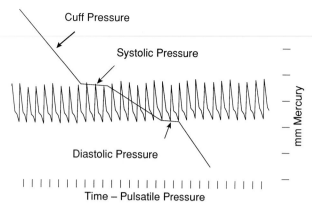

Cuff Pressure

Systolic Pressure

mm Mercury

Diastolic Pressure

Time – Pulsatile Pressure

FIGURE 2 Illustrated schematically is the pulsatile pressure profile in a large conduit artery. As discussed in the text, the maximum arterial pressure is termed systolic pressure while the minimum is termed diastolic pressure. Illustrated also is the relationship between the sphygmomanometer (cuff) pressure and a patient's actual arterial pressure. When the cuff pressure is maximum, blood flow is impeded. To measure systolic and diastolic pressures, the cuff pressure is slowly decreased. When the cuff pressure equals the maximum arterial pressure, flow begins through the artery and this marks systolic pressure. The sound of blood flowing through a partially restricted blood vessel is detected by a stethoscope. Diastolic pressure is then equal to that cuff pressure associated with a loss of the characteristic sound of flow through the partially restricted vessel.

governed both by the rate of outflow of blood from the arterial compartment and by the elasticity of the conduit arteries. The latter results in what is termed a windkessel effect, namely a rebound by the distended vessel wall, which results in pressure being exerted on the blood in the arteries. The windkessel process is very much like the stretched rubberband of a slingshot rebounding and exerting force on the object being propelled. The pressure does not drop to zero during diastole because the heart contracts, repeating the cycle. Under normal conditions, if the heart rate is slowed, the minimum (diastolic) pressure reached would be expected to be lowered.

A critical determinant of both systolic and, especially, diastolic pressure is the rate of outflow from the arterial compartment. This rate with which blood leaves the arterial compartment and enters the capillary compartment is determined by the degree of constriction of resistance elements in small arteries. These resistance elements, which are equivalent to valves regulating the flow of water from a water hose, are small arterial blood vessels (approximately 30–50 μm in diameter). They are termed resistance elements because by varying the degree of constriction of the vessel they set the re-

sistance to flow the blood out of the arterial compartment and, thereby, the level of arterial pressure. Constricting the resistance elements leads to increased arterial pressure, whereas relaxing (dilating) the elements leads to decreased arterial pressure. Note that both systolic and diastolic pressures are influenced by these resistance elements.

The smooth muscle of the media of the resistance elements controls the degree of constriction or relaxation of the vessel. Smooth muscle cells within the media layer are spindle-shaped and oriented side-by-side and at an angle to the direction of the vessel. The small resistance vessels and, specifically, their smooth muscle cells are richly supplied by nerves under the control of the sympathetic branch of the autonomic nervous system. The nerves controlling the resistance vessels release a chemical (neurotransmitter) called norepinephrine, which upon binding to a specific protein on the smooth muscle cell (termed the α_1-adrenergic receptor) causes the smooth muscle cells to constrict. Through coordinated contraction of the many smooth muscle cells within the wall of the vessel, the inner diameter (termed lumen) of the vessel is reduced and constricted. In this manner, blood flow through the lumen is impeded, resistance to flow is increased, and arterial pressure rises. [See SMOOTH MUSCLE.]

While systolic and diastolic pressures are important in determining whether or not a patient has true hypertension, to understand the possible causes and treatment of high blood pressure, we must also consider the interrelationships among the heart, blood vessels, and autonomic nervous system. Rather than distinguish the extremes of pressure (systolic or diastolic pressure), we speak about the arterial compartment as a whole and use an "averaged" arterial pressure, namely mean arterial pressure (MAP). Mean arterial pressure, when averaged over time, is defined by the following relationship involving cardiac output (CO) and total peripheral vascular resistance (PVR): MAP = CO × PVR; or mean arterial pressure is the product of cardiac output and total peripheral vascular resistance. Total peripheral vascular resistance is the total resistance offered by the arterial vascular compartment to the flow of blood out from the arterial compartment. Cardiac output is the amount of blood (in liters) pumped by the left ventricle of the heart over a full minute. This volume of blood pumped in 1 min is determined by the force of contraction of the left ventricle, by the rate of contraction of the heart (i.e., the heart rate), and by the amount of blood contained within the left ventricle chamber during each contraction. The latter is controlled partly by the amount of blood that returns to the heart from the venous compartment, termed venous return, and by the resistance encountered when the heart pumps the blood into the arterial circuit. Because the veins function largely as reservoirs for the blood, changes in both blood volume and the state of constriction of venous smooth muscle influence the low blood pressure in the veins and the amount of blood returned to the heart. Cardiac output is defined by the following relation involving stroke volume (SV), i.e., the volume of blood ejected by the left ventricle with each beat and heart rate (HR): $CO = SV \times HR$. Therefore, mean arterial pressure is determined by the stroke volume, the heart rate, and the total peripheral vascular resistance of the arteries, or $MAP = SV \times HR \times PVR$.

Because arterial blood pressure is governed by the action of the heart, by the state of resistance to flow by the vessels, by the state of constriction of veins, and by the total volume of the system, high blood pressure could be caused by several different factors; however, numerous control mechanisms operate to compensate for changes in one or more of the above factors so as to keep arterial blood pressure constant. For example, if peripheral vascular resistance increases, cardiac output is decreased through homeostatic mechanisms by decreasing heart rate. Likewise, if cardiac output drops, as, for example, might occur if there is sudden loss of blood due to hemorrhage, there is a homeostatic increase in sympathetic nervous system discharge to blood vessels resulting in constriction of smooth muscle in resistance elements of the arterial bed and a concomitant increase in total peripheral vascular resistance and arterial pressure. This homeostatic regulation of arterial pressure is dynamic and continuously operational. It responds very rapidly to changes in pressure. For example, changes may be sensed by the system within two heart beats and reflex responses initiated.

Hypertension is considered a disorder of regulation, in part because in hypertensives these homeostatic mechanisms appear to continue to operate but regulate at a higher level of arterial pressure. A "resetting" of the level of homeostasis permits sustained elevated arterial pressure and, ultimately, the development of hypertension. With chronic elevation of arterial pressure, there is a thickening of the smooth muscle layer of the arteries (possibly so as to resist the increased pressure within the blood vessel). This thickening causes the lumen of the vessel to become smaller, which raises the resistance to flow (increases vascular resistance). Over time, the changes in the blood vessels also lead to a decrease in the distensibility or compliance of the vessel. As it becomes less distensible, there is less "give" with each pressure pulse from the heart's contraction, and this also will result in a marked elevation of systolic and diastolic pressures.

II. Hypertension as a Disorder in Humans

A. Normative Levels of Arterial Pressure in Humans

There is not a single value for "normal" arterial blood pressure because there are age, sex, and weight influences on resting blood pressure. Blood pressure increases with age and generally is lower in women than in men who are the same age. This tendency toward lower blood pressures in women may be changing in Western societies due to the increasing involvement of women in professional and business life. The reasons for such a change are unclear but may range from increased stress or altered life-style, increased consumption of alcohol or cigarette smoking, or dietary changes. More epidemiological studies are needed to determine if a trend toward more hypertension in women is really developing. Blood pressure is also influenced by body mass or weight and by extent of physical conditioning (exercise, etc.). Therefore, what is normal in one person may be abnormal in another.

Through measurements of large numbers of adult subjects, all in apparent good health, the well-known values of 120 (systolic) and 80 (diastolic) have entered our vocabulary to imply "normal blood pressure." In general, these values are good approximations *for the population of such subjects as a whole*; however, biological variability results in a range of "normal" blood pressures with some being on the extreme low or extreme high end. Such individuals are not necessarily hypertensive; therefore, a complete history and physical examination with appropriate tests are essential to define whether or not a patient has the disorder of hypertension.

No single blood pressure value defines when a patient becomes classified as having hypertension. Because of the variability in blood pressure values among the population as a whole, as well as a result of the many factors that go into setting the level of arterial pressure, classification as to when an individual is hypertensive and requires therapy has been controversial. This question is especially important for patients who fall in the borderline hypertensive category (i.e., have diastolic pressures between 90 and 95 mm Hg) because many of the antihypertensive drugs used for treatment may, with long-term usage, have undesirable side effects, which must be monitored and prevented. Studies of the adult population have resulted in guidelines. For adults within the ages of 21–60 yr and with a weight within the normal range for their age and body build, a diastolic pressure >90 mm Hg, taken in an appropriate manner and with a complete evaluation (see below), is considered as evidence for a *possible* classification of hypertension. In patients *not* under any antihypertensive medications, diastolic pressures between 90 and 104 mm Hg defines the category of mild hypertension with 90–94 mm Hg constituting borderline hypertension. Likewise, patients not receiving therapy with diastolic pressures between 105 and 120 mm Hg are placed into the moderate to severe hypertension range. Because drug therapy is judged optimally effective when the diastolic pressures are brought into the normal range (i.e., <90 mm Hg), the above categorization only applies when the patients are free of any antihypertensive medications.

Why is diastolic pressure used to identify and categorize hypertensives? While elevated systolic pressures (160 mm Hg or more) generally alert health care professionals to the possibility of hypertension and in hypertensive patients may provide the physician information about the long-term effects of the disorder in that patient, systolic blood pressure is very sensitive to stress or anxiety. The systolic pressure, as discussed earlier, is determined largely by the action of the heart pumping blood into the arterial vascular bed. On the other hand, diastolic pressure is determined largely by the peripheral vascular resistance and blood vessel compliance. Transient and reversible episodes of elevated arterial pressure would not necessarily constitute a diagnosis of hypertension; however, sustained elevation of arterial pressure would be an indication either of changes in the blood vessel structure (as discussed earlier) or alterations in homeostatic regulation. These changes would be most evident when the heart is in diastole (rest phase). For this reason, diastolic pressure is critical to the diagnosis and decision as to when to begin therapy and the type of therapy. When the diastolic pressure rises >95–100 there is no question but that hypertension, or some other cardiovascular disorder, must be considered and appropriate treatment instituted.

B. How is Hypertension Recognized?

While many people believe that headaches or rapid heart beat may indicate high blood pressure, these symptoms are not most commonly associated with hypertension. Hypertension has been termed the "silent killer," because a patient may be hypertensive for many years without any evident symptoms. Not infrequently, prior to a greater effort among health professionals to identify hypertensive patients, ophthalmologists or even opticians would discover hypertensive patients when examining the retina (the back inner surface of the eye). The retina is highly vascularized and with sustained, chronic, untreated hypertension the vascular bed becomes damaged, which may be readily observed on examination. Atherosclerosis and diabetes may also lead to changes in retina blood vessels, and retinal examination by itself will not provide a diagnosis of hypertension.

Another common time for diagnosis of hypertension is during routine physical examinations. However, the procedures followed by the physician or health care professional must be rigorous to make a diagnosis. The institution of drug therapy *without* a thorough examination and analysis (see below) is not advisable. Why is diagnosis so difficult? The reason is that the measurement of arterial pressure using devices external to the body is not trivial! Almost all methods of measurement employ a form of the sphygmomanometer technique, an indirect method to measure pressure. This involves measurement of the *external* pressure required to occlude, totally or in part, blood flow in a readily accessible artery, such as the brachial artery of the upper arm. The procedure involves wrapping an appropriately sized rubber cuff (which is an inflatable air bag like a balloon) around the upper arm and pumping air into the cuff to inflate it, thereby impeding the flow of blood through the major artery (basilar) of the arm. A stethoscope is placed in a location on the side of the cuff away from the heart so as to listen to characteristic sounds (termed Korotkoff sounds) formed by the resumption of blood flow as the cuff air pressure is slowly lowered. In the absence of flow, no sounds are heard. Two sounds are generally listened for. When the first sound is heard, which is the start of blood flow in the vessel as the cuff pressure is lowered, the cuff air pressure at that point defines systolic (maximal) pressure. This air pressure, and therefore blood pressure, is expressed in terms of millimeters of mercury (mm Hg) because the first, and still most accurate, sphygmomanometers use a column of liquid mercury to determine air pressure. When flow sounds are lost (indicating no further cuff restriction on flow in the artery), the cuff pressure is taken as the diastolic (minimal) pressure. Automated devices, which are gaining wider usage, also use an occluding technique as described for the sphygmomanometer.

While the cuff blood pressure measuring devices found in pharmacies and food stores, which use the finger rather than the arm, may be quite reliable, they cannot be used to exclude or to diagnose hypertension; rather, to make the diagnosis, a systematic and rigorous protocol must be followed and blood pressures should be taken on both arms while the subject is sitting quietly. Most importantly, pressure cuffs must be used that are appropriate for the size and body mass of the individual. The use of too small a cuff may give an anomalously high pressure (more cuff pressure is needed to occlude the arterial blood flow), whereas too large a cuff may give a falsely low value. If a high reading is obtained, *before* the diagnosis of hypertension can be made, three separate blood pressure measurements should be taken and averaged, and this should be done on at least two separate visits at least 1 wk apart. Often, blood pressure is measured both when the patient is upright (sitting or even standing) and in a lying-down position. There is a well-described phenomenon known as "white-coat hypertension," which is an elevated blood pressure whenever the pressure is taken in a doctor's office by the physician or even the nurse. This is attributed to a stress-induced transient elevation of blood pressure and may disappear when the pressure is taken in a less forbidding environment. For this reason, it is essential that measurements be taken over several visits and in a calming environment so as to ensure an accurate assessment of the patient's blood pressure. If hypertension is suspected, a complete examination must be conducted to exclude secondary forms of hypertension (see below), including a complete family history to determine if there is a genetic predisposition to be hypertensive (see below).

C. Types of Hypertension in Humans

The most common form of human hypertension, estimated to be responsible for >80% of all hypertension, is of unknown cause and, therefore, is called primary, or essential, hypertension. Some experts feel that human essential hypertension actually reflects a variety of different forms of hypertension, each attributable to a different abnormality in blood pressure and/or volume regulation. In contrast, other experts believe that essential hypertension is one syndrome that arises from a mix of different abnormalities in regulation. Essential or primary hypertension may also include patients who exhibit a salt-dependent form of hypertension as well as obesity-dependent hypertensives, if there is no evident explanation for their hypertension.

A much smaller percentage (estimated to be approximately 10–15% of all hypertensives) have hypertension that is due to a known or identifiable cause. This form of hypertension is termed secondary hypertension. A common form of secondary hypertension is due to diminished or restricted arterial blood flow into the kidney. This results in a form of hypertension that is caused predominantly by activation of the renin-angiotensin system (see below) and is termed renovascular hypertension. Estimates are that approximately 5–8% of all human hypertension is renovascular in origin. Hypertension may also be due to other causes, some reversible and some irreversible. For example, secondary hypertension may be due to selected pharmacological (i.e., drug-induced) exposures. While the patients with these secondary forms of hypertension are much fewer than those with essential or renovascular hypertension, because they may reflect a particular group of the population, by age or sex, they may be the cause of a predominant form of hypertension in that group. For example, hypertension was found in young women who were taking selected estrogen-containing oral contraceptives. Because all the subjects were women of comparably young age, this form of hypertension comprised a significant fraction of the types of hypertension in this age group. When the association with the pill was established, modifications in the formula used in those oral contraceptives that had the highest frequency of causing the disorder resulted in a major drop in the incidence of hypertension in this group of patients. (It should be noted that some investigators believe that women who exhibited this form of hypertension are genetically prone to the disorder and may, as they get older, show signs of essential hypertension; however, this view is controversial and not yet proven or disproven. Clearly, many years are needed to determine if a correlation exists.)

Hypertension may also be caused by a relatively rare tumor of the adrenal medulla, termed a pheochromocytoma. Pheochromocytoma patients may exhibit episodes of very high arterial pressure. These periods of hypertension are attributable to the release of large quantities of neurohumoral substances (epinephrine and norepinephrine, among others) from the malignant tumor tissue at unpredictable times.

A form of hypertension called "isolated systolic hypertension" is found in the elderly and reflects markedly elevated systolic pressure (>160 mm Hg) without a diastolic pressure >90 mm Hg. The cause of isolated systolic hypertension is not fully known; however, one explanation is that it may result from reduced compliance of the arterial blood vessel (with age). This would result in less of a windkessel effect and higher systolic pressure with the pulse of blood pumped by the heart's left ventricle. The treatment of isolated systolic hypertension is different than with the more common form of hypertension.

Malignant hypertension is a term used for patients who show high, and frequently irreversible, hypertension. This term in no way implies a cancerous origin to the hypertension; rather, patients with malignant hypertension are generally those who have had hypertension for long periods of time and who fail to be responsive to drug therapy. Often, malignant hypertension is associated with severe kidney failure and/or marked changes in the vascular compartments. Such hypertensive patients exhibit very high, sustained arterial pressure, both systolic and diastolic, and constitute a medical emergency necessitating bold drug treatment to lower the blood pressure and prevent further organ damage.

D. Genetic Predisposition to Hypertension

Strong evidence indicates that human essential hypertension has a hereditary component; i.e., the potential for developing hypertension is passed from one generation to another. This means that there are hypertensive-prone individuals; however, evidence is not yet sufficient to indicate whether or not hypertensive-prone patients will become hypertensive. In fact, many experts believe that hypertension can still be prevented (or at least significantly delayed in onset) by appropriate dietary and exercise regimens. As part of the original diagnosis, physicians and health care professionals must determine if there is any family history of hypertension or other cardiovascular disorder. (The latter information is important because hypertension may have gone undetected in earlier generations but the end result, heart disease or stroke, may have brought the patient to the attention of the physician.)

In contrast to essential hypertension, very little evidence indicates (or disproves) that secondary hypertension has a genetic component. Even drug-induced hypertension may reflect some hereditary aspects because, in the example of oral contraceptive hypertension, only a fraction of all the women on the medications in question became hypertensive. As a rule for good health, whenever a patient knows of the presence of hypertension in their family, they should be on guard as to a possible hereditary relationship.

III. Pathophysiology of Human Essential Hypertension

The possible cause or causes of essential hypertension are still unknown, and discovering the cause is complicated by the multiplicity of factors that determine the level of blood pressure. At the very least, disturbances in autonomic nervous activity, in blood volume, in the action of the heart as a pump, or in the distensibility of arterial blood vessels all may result in abnormal levels of arterial pressure. A complete discussion of these factors is beyond the scope of this chapter; however, a brief introduction is warranted because therapy is often directed at one or more of these possible causative mechanisms.

A. Elevated Activity of the Sympathetic Nervous System

The sympathetic branch of the autonomic nervous system is important in setting the level of activity of the body so as to meet all the needs of the organism, other than those of a purely vegetative nature. For example, the sympathetic nervous system may be activated by such innocent activities as standing or walking. Blockade of the sympathetic nervous system may cause postural *hypo*tension, which is a failure to maintain adequate blood pressure when going from a recumbent (lying or sitting) position to a standing position. Individuals with postural hypotension may faint on standing or getting out of bed too quickly. The sympathetic system is also stimulated by swallowing or other consummatory activity. The sympathetic nervous system normally is also involved in regulating metabolism (glucose, fat, etc.), in setting the level of heart rate, peripheral vascular resistance, and blood pressure in proportion to anticipated needs, and in the regulation of body temperature during high-activity periods. Stimulation of the sympathetic nervous system results in an increased heart rate, increased peripheral vascular resistance, and, generally, increased mean arterial pressure. These are but a few of the obvious functions of this nervous system; however, the sympathetic system also is an important mechanism regulating hormone systems important in salt (sodium) and water balance. At the other extreme of human behavior, stress, fear, and defense reactions all markedly elevate the activity of the sympathetic nervous system. These behavioral responses may result in either selected or global activation of the sympathetic nervous system.

In the early stages of human essential hypertension, evidence indicates an enhanced sympathetic nervous system activity and, often, an increased heart rate. Under normal circumstances, in response to elevated arterial pressure (with or without increased heart rate) reflex mechanisms are activated and should reduce the level of activity of the heart and lower blood pressure. These reflex mechanisms involve changes in the activity of the sympathetic nervous system *and* the parasympathetic nervous system. The latter slows heart rate. However, at some point, these reflex mechanisms appear to fail and the original level of blood pressure (and, possibly, sympathetic nervous system activity) is not regained. In fact, a new blood pressure baseline becomes the point of regulation. If this new baseline is at a higher blood pressure, a resetting of the "setpoint" for blood pressure has occurred. Blood pressure is then regulated at this new level.

The sympathetic nervous system may also be activated *because* of improper function of an organ system such as the kidney. For example, if there is insufficient filtration and urine formation, increas-ing the activity of the sympathetic nervous system could overcome the insufficiency. However, if this activation persists, then a new setpoint in function results. Often this is associated with an increased arterial pressure. The net effect is a circle of continuing resetting of arterial pressure and/or sympathetic activity to maintain function. The net effect of chronically elevated sympathetic nervous system activity, which remains unchecked by reflex activity, is elevated heart activity, increased peripheral vascular resistance, increased sodium retention by the kidney, and elevation of the activity of the renin-angiotensin system (see below) and other neurohormones.

B. Retention of Sodium and Fluid

There is no question that with established hypertension abnormalities in kidney function could lead to retention of sodium and, with it, water. Maintenance of salt and water homeostasis is the result of optimum arterial pressure and kidney function as well as under the control of various peptide hormones and steroids. One peptide hormone system, the renin-angiotensin system, is intimately involved in maintaining sodium and water homeostasis. Renin is an enzyme released from the kidney into the blood in response both to sympathetic nervous system activity and to the level of sodium intake (lowered sodium leads to increased renin release). Angiotensin comes from a large protein (prohormone) that is made in the liver and continuously released into the blood. Renin in the blood then breaks a chemical bond in the prohormone releasing a small, inactive peptide called angiotensin I. This peptide is converted to a smaller, but active, peptide called angiotensin II by an enzyme called angiotensin converting enzyme (ACE). Angiotensin II can constrict blood vessel smooth muscle and, thereby, influences peripheral vascular resistance and also stimulates the release by the adrenal gland of a steroid hormone called aldosterone. Aldosterone acts directly on the kidney to increase sodium (and water) retention by the body. Angiotensin also has important actions within the kidney and the intestines to regulate sodium and water conservation. This hormone system also has other functions, which include regulating and enhancing the functional activity of the sympathetic nervous system as well as stimulating heart activity.

Because of the importance of the renin-angiotensin system in affecting so many systems that control blood pressure, many studies look at the possible involvement of this hormone system in the cause of some forms of hypertension. In fact, renovascular hypertension is believed to be caused by excess activity of the renin–angiotensin system. The renin–angiotensin system has also become a major target in the design of antihypertensive drug therapy. While renin–angiotensin system activity is elevated in some forms of essential hypertension, little evidence indicates that it is the cause of the hypertension. Nevertheless, ACE inhibitors, which interfere with the formation of angiotensin II, are very effective in lowering blood pressure in many forms of hypertension. In addition, one mechanism whereby β-blockers (such as propranolol) lower blood pressure in hypertensive patients is believed to be through interference in sympathetic nervous system stimulation of renin release by the kidney.

Retention of sodium and water can occur in almost all forms of hypertension, even those that appear to be independent of the renin–angiotensin system. For example, one form of hypertension is characterized by a low-plasma renin activity but high-sodium retention and elevated (expanded) volume. This form of hypertension may reflect abnormalities in kidney function or in hormones that regulate kidney function and excretion. Restriction on sodium (salt) intake is advised to all hypertensive patients and, by many investigators, to almost everyone in Western society. The role of sodium intake as a factor in the development of hypertension remains controversial. There is no question that dietary salt intake is much higher then needed in Western societies, and many studies have demonstrated a positive relationship between the amount of salt taken in the diet and the incidence of hypertension; however, not everyone is hypertensive. Therefore, many investigators believe that some patients are "salt-sensitive." These individuals may have a hereditary predisposition to retain excess sodium (and water) and, as a consequence or independently, a predisposition to develop hypertension. A moderate salt intake is advisable to all adults. Salt-sensitivity may explain racial differences in the incidence and possible causes of hypertension. Blacks and Afro-Americans appear to have a higher incidence of hypertension, compared with other races, especially Caucasians. Blacks also exhibit a form of hypertension that many investigators believe reflects an abnormal mechanism for handling salt. Because the form of hypertension appears to be different, the appropriate antihypertensive treatment differs between races.

C. Relationship of Chronic Stress and the Development of Hypertension

One attractive, but as yet unproven, theory is that repeated exposure to stress in everyday life activities may result in abnormal cardiovascular responses in hypertensive-prone individuals. This is theorized to result in irreversible changes in the arterial blood vessels resulting in a sustained elevation of peripheral vascular resistance and, ultimately, hypertension. As yet no evidence exists to prove or disprove this thesis. Intuitively, repeated exposure to stress would be expected to result ultimately in some adverse effect on the cardiovascular system. No evidence supports a hypertensive "type-A" personality, as has been described for patients at risk to a myocardial infarction. However, *disproving* a stress–hypertension link has been as difficult as trying to establish a linkage. One reason for this difficulty may lie in the complex interplay of different mechanisms regulating blood pressure and volume.

IV. Hypertension as a Disorder

A. When Does Elevated Arterial Pressure Require Treatment?

Hypertension is classified according to the level of diastolic pressure and three categories are relevant to our discussion: borderline hypertension with a diastolic pressure between 90 and 94 mm Hg; mild hypertension with a diastolic pressure between 95 and 104 mm Hg; and moderate to severe hypertension with a diastolic pressure >104 mm Hg. These three categories are loosely defined because some patients are at greater risk of developing severe cardiovascular disorders with any degree of hypertension than are other patients. In addition, essential hypertension is a chronic disorder that extends over many years. It is generally accepted that without any treatment a patient will progress from mild to severe hypertension in a short time. Even with treatment, many patients will progress over years to a higher diastolic pressure. Therefore, the goal of therapy is to reduce the rate of that progression or prevent it entirely.

Physicians and investigators believe that some therapy must be initiated when a *definitive* diagnosis of hypertension has been made. That therapy should be antihypertensive drugs and/or nondrug approaches. If a secondary form of hypertension is considered, drug therapy is usually initiated immediately along with efforts to deal with the source of the hypertension. However, with essential hypertension, whether to treat or not depends on the classification of severity with moderate or severe forms of hypertension necessitating antihypertensive drug treatment along with nondrug therapy. In the case of borderline hypertension and even with mild forms of hypertension, there is an ongoing debate as to the type of therapy, if any. The primary goal of all therapeutic approaches must be to lower the diastolic (and systolic) pressures into the normal range or, if that cannot be reached, to lower pressure as much as possible. A diastolic pressure <90 mm Hg is usually the target of therapy. To achieve that goal, the therapeutic approach may be variable and individualized to the patient. For example, for patients with a family history of hypertension, many physicians will move more quickly to institute some form of antihypertensive drug therapy. Because so many drugs are now available to treat hypertension, the type of therapy employed generally reflects the experiences of the physician in treating similar cases of hypertension.

B. Types of Therapy Used to Lower Blood Pressure

There are two general approaches to lowering blood pressure back to the normal range. For essential hypertensive patients who are classified as borderline hypertensives, many physicians will initially recommend nondrug therapy. This frequently includes weight reduction, restriction of salt intake, stress-reduction and changes in life-style, exercise regimens, and discontinuance of alcohol intake or cessation of smoking. All these approaches constitute nondrug therapy of hypertension (and other cardiovascular disorders as well). Depending on the individual, these therapeutic approaches may be all that is needed. For example, a relatively small reduction in weight has been shown to markedly lower blood pressure in essential hypertensives, resulting in complete cessation of all antihypertensive drugs in some patients. Weight reduction achieves two immediate goals: reduction in the workload placed on the heart and reduction in total blood volume. Both effects are beneficial. Another nondrug approach is stress relaxation and/or modification in life-style. While not fully proven, many investigators, physicians, and patients feel that stress exacerbates high blood pressure and, therefore,

stress reduction has potential benefits. Stress-reduction would lower sympathetic drive, thereby lowering activation of neurohormones as well as directly decreasing stimulation of the heart. Nondrug therapies such as stress reduction may take several months before evidence of their effectiveness is clear. Many physicians prefer to start the patient on limited antihypertensive drug therapy to assist in lowering the blood pressure more quickly. Whether or not this is necessary depends on each individual patient.

It is generally considered that with time (years) patients with essential hypertension will show progression of the disorder and require the institution of some form of drug therapy. Because in most patients hypertension progresses eventually to the point where stronger drugs are necessary, a "stepped care approach" with drug therapy is followed. With stepped care, a first drug is used (along with nondrug therapy) to achieve the target goal of a diastolic pressure <90 mm Hg. If (or when) that drug fails to achieve the target goal, either a second drug is added or a substitution of another "type" of antihypertensive drug is tried. Ultimately, after many years in most patients, two or even three different drugs are used to lower the blood pressure. The choice of the first drug differs between the United States and Europe and undergoes continuing debate. In the United States, a diuretic (see below) is frequently the first drug used in hypertensives, whereas in Europe, a β-blocker is the drug of choice. In all cases, nondrug therapeutic approaches must be continued because, when effective, nondrug therapy reduces the need or amount of drug therapy.

C. Categorization of Antihypertensive Drugs According to Their Target

As discussed above, blood pressure may be influenced by a wide variety of systems in the body. Antihypertensive drugs have been developed to affect all of these systems. Table I lists the main groups of drugs in common use.

The ultimate goals of all the forms of therapy is to reduce the arterial pressure, to reduce the rate of progression to more severe forms of hypertension, and to reduce the rate of damage to those critical organs (heart, brain, kidney, and eyes) that are damaged by continued high blood pressure. But the drugs themselves are not harmless. Chronic use of thiazide diuretics has recently been linked to abnor-

TABLE I Categories of Antihypertensive Drugs

Type of agent	Reason for their use
Diuretic	Diuretics increase urine production, thereby increasing the loss of sodium and water. They have many secondary actions, some undesirable such as increasing loss of potassium.
Potassium-sparing diuretics	Diuretics that reduce the loss of potassium.
β-blocker	β-blockers partially block the sympathetic nervous system stimulation of the heart and release of renin from the kidney. Some may also work in the brain to reduce sympathetic nervous system activity and secretion of hormones, which may raise blood pressure. There are many types of β-blockers.
ACE inhibitors	A relatively new form of antihypertensive therapy that has become very popular. These agents interfere with the formation of angiotensin II. They also affect other systems that may control blood pressure and kidney function. However, their blockade of angiotensin is believed to be their primary mechanism.
α-blocker	These agents are designed to block the sympathetic nervous system constriction of blood vessels by blocking the action of norepinephrine and epinephrine on blood vessel smooth muscle at a protein called the α_1-adrenergic receptor. By interfering with this system, they can cause postural hypotension.
Centrally acting drugs	These drugs are designed to work primarily in the brain to reduce the activity of the sympathetic nervous system but may have additional mechanisms by which they lower blood pressure.
Calcium-channel antagonists	These drugs, relatively new in the treatment of hypertension, are designed to interfere with the entry of calcium into blood vessel smooth muscle cells. Because calcium entry is essential for contraction of the smooth muscle cell, they reduce peripheral vascular resistance. These agents consist of several different types, some of which are most active on the heart and reduce contractile force of the left ventricle of the heart. For this reason, they are used in treating heart failure. However, the primary target in the treatment of hypertension is the blood vessel, not the heart. They may also reduce the release of renin by the kidney.

malities in lipid metabolism and possible cases of coronary artery atherosclerosis. Similarly, many of the antihypertensive drugs, by lowering blood pressure, reduce the ability of the kidney to excrete sodium and water. These drugs require the simultaneous use of a diuretic to enhance urine production and sodium elimination from the body.

Hypertension is a disorder of regulation and any therapeutic approach must focus on restoring, if possible, the regulation of blood pressure. The treatment of hypertension is an ongoing series of decisions and evaluations by the physician based on each individual patient. Continued interaction and communication among patient, physician, and health professional is *essential* to the prevention of the damaging effects of hypertension. In the case of hypertension, as with many diseases or disorders, prevention through changes in dietary intake, exercise, stress reduction, and life-style remain the ultimate first step in treatment.

Bibliography

Genest, J., Kuchel, O., Hamet, P., and Cantin, M. (eds.) (1983). "Hypertension." McGraw-Hill, New York.
Gifford, R. W., Jr., Kirkendall, W., O'Connor, D. T., and Weidman, W. (1989). Office evaluation of hypertension. A statement for health professionals by a writing group of the Council for High Blood Pressure Research, American Heart Association. *Circulation* **79**, 721–731.
Kaplan, N. M. (1986). "Clinical Hypertension," 4th ed. Williams & Wilkins, Baltimore.
Page, I. H. (1987). "Hypertension Mechanisms." Grune & Stratton, Harcourt Brace Jovanovich, Orlando, Florida.
WHO/ISH (1989). 1989 Guidelines for the management of mild hypertension: Memorandum from a WHO/ISH meeting. *J. Hypertension* **7**, 689–693.

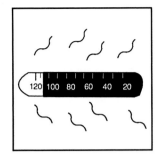

Hyperthermia and Cancer

GEORGE M. HAHN, *Stanford University School of Medicine*

I. Introduction
II. Hyperthermia as a Treatment Modality
III. Is Hyperthermia an Inducer of Cancer?
IV. Conclusion

Glossary

Carcinogen Drug or physical agent that induces cancer

Chemotherapy Treatment of disease by chemical agents

Mutagen Drug or physical agent that causes genetic changes in the offspring of an exposed cell or organism

Randomized prospective trial Clinical test comparing two or more treatments; neither the patient nor the physician is usually aware of the treatment assignment, which is made randomly

Transformation Cellular change that converts a normal cell to one that has the potential of growing as a cancer

WHEN THE TEMPERATURE of a human rises above its usual value (37°C), then that individual has had a hyperthermic episode. This episode may be due to one of several causes: fever, heat stroke, or intentionally induced hyperthermia for the treatment of cancer or other diseases. Fever and heatstroke necessarily involve increases in temperature of the whole body; intentionally induced hyperthermia may be whole body or localized. In the latter case, core body temperature may not change by more than a fraction of a degree. Both intentional and accidental hyperthermic episodes induce certain changes in cellular metabolism. Somewhat paradoxically, these may have beneficial or deleterious effects, particularly with respect to the treatment or the induction of cancer.

I. Introduction

The interior of the human body is designed to function within a limited temperature range. Under nonpathogenic conditions, the core temperature of the average person varies by only fractions of a degree around a set temperature. While this varies from individual to individual, the set temperature of the vast majority of people is between about 36.5° and 37°C. Throughout the body, temperature is maintained at surprisingly constant levels. Only in the liver and perhaps in other regions of high metabolic activity does the temperature increase slightly. Near the skin, particularly under conditions where convective temperature loss is considerable, a temperature gradient extends to a depth of perhaps 1 cm. Parts of the lung may be at lower temperature, largely because of local evaporation. In most of the rest of the body, however, temperature remains constant, within a fraction of a degree of the set point. Obviously, the body needs a variety of homeostatic systems to maintain such temperature uniformity. These systems operate on a local as well as a global level. It has been demonstrated in many experiments that the hypothalamus is a major controlling element that governs global control. Control on a local level is far less well understood, although it is effected primarily by changes in blood flow in the tissue of concern. [*See* BODY TEMPERATURE AND ITS REGULATION.]

Several situations, however, cause changes in the body temperature. We will consider here only increases in either local temperatures or increases at the level of the entire human. The global change that most of us are familiar with is that caused by fever. This form of hyperthermia apparently involves hypothalamus directed change in the body's set point. Homeostatic mechanisms are employed to maintain a core temperature around the new set point; this can be as high as 41°C. Possible beneficial functions of fevers have been discussed ever since physicians have attempted interventional medicine. For example, the Greek physician Parmenidis said that given the means to induce fevers, he would be able to cure all illnesses. In more modern times, physicians frequently thought fevers harmful and suppressed them with agents such as aspirin. In very recent years, the pendulum has begun to swing back, and many physicians now believe that fevers have, indeed, beneficial functions and should be suppressed only if they constitute a specific danger to the patient. Nevertheless, very few physicians would believe today that serious illnesses, particularly cancer, can be cured simply by means of fevers. This is because we can now satisfy Parmenidis' wish: We can induce fevers by injecting a variety of substances having pyrogenic properties. Perhaps of most interest among these substances are several bacterial toxins. These have indeed been tried as cancer remedies but only with limited success. Elevated temperatures induced by pyrogens behave very much like fevers; the body adjusts its set point and uses its homeostatic mechanisms to maintain the elevated temperature.

There are, however, other ways of raising the body's temperature. For example, in hot climates, particularly when the body is exposed excessively to sun, core temperature may rise. Mechanisms will then be set in motion to reduce body temperature. In the human, the primary mode to do this is sweating; because of the large amount of heat expended as heat of evaporation, skin temperature is cooled temporarily. Cardiac output is increased, so that the excess heat is rapidly carried to the skin and dissipated during evaporation. If all these maneuvers are insufficient, core temperature continues to rise and severe injury may result from "heatstroke." Exposure of the body to microwaves, either accidental or intentional, will increase core temperature. Again, homeostatic mechanisms will be brought into action to attempt to lower the body's temperature to its set point. Temperature can obviously also be increased locally. The most obvious means of doing so is conduction. If we touch a hot object, local skin temperature will be raised. Obviously, if the object is very hot, pain will quickly cause us to make every attempt to break contact. Here also, the body will attempt to maintain temperature by locally increasing blood flow to carry off the excess heat that has been introduced via conduction. The same phenomenon occurs when heating locally is carried out by absorption of energy emitted by infrared, focused, or locally applied microwaves or ultrasound radiations.

Parmenidis, when he mentioned his desire to cure all disease with heat, presumably included cancer. In his time, malignancies had been described in considerable detail. Is there really any reason to expect elevated temperatures to act as an anticancer treatment? Or is it perhaps the opposite: Does heat induce cancer? In this chapter, these two possibilities are discussed, and considerable experimental evidence is presented to suggest that both of these, at least in a limited sense, may be correct. As far as treatment is concerned, Parmenidis was obviously overoptimistic. We now know that heat by itself has only a limited influence on the progression of malignant disease. But many studies now indicate very strongly that in conjunction with more conventional treatments such as radiation therapy and chemotherapy, heat has much to offer. Thus, hyperthermia is an anticancer agent. In the fully developed organism, heat by itself may not be carcinogenic; it does, however, increase the carcinogenic efficacy of many agents, including ionizing radiations and many drugs, although this depends to some extent on the order of application. For example, if cells are heated and then radiated, the rate of transformation from a nonmalignant to a potentially malignant state is increased; if the order is inverted, it may be decreased. In the developing organisms, particularly in the fetus of rodents, heat clearly has deleterious, teratogenic effects. It is not unreasonable to suggest that in some cases developmental defects on the cellular level lead to cancer.

II. Hyperthermia as a Treatment Modality

Late in the nineteenth century, several attempts were made to treat malignancies by elevating body temperatures. The rationale for these attempts was based on well-documented occurrences of spontaneous remissions in cancer patients who had episodes of high fevers while suffering from malignant disease. This subject has been reviewed in detail. The reviewers found that one-third of 450 spontaneous remissions of histologically proven malignancies were known to be associated with the development of acute fevers that resulted from concurrent diseases such as malaria or typhus. Similarly, one-third of spontaneous remissions of lymphomas in children were also found to be associated with high fevers. The lymphoma remissions were of short duration, but several "cures" of carcinomas and sarcomas have been described. These observations certainly provided an impetus for physicians to attempt to treat incurable malignancies by elevating body temperatures. This was initially achieved primarily via the injection of bacterial agents or chemical pyrogens. The most prominent of these studies was performed by Coley late in the nineteenth century. He developed a bacterial toxin ("Coley's toxin") that, when injected into patients, induced fevers that ranged from 38° to 41°C. Some favorable results were obtained in these studies, particularly against osteosarcomas and soft tissue sarcomas. Current knowledge suggests that perhaps the effect of pyrogens and bacterial toxins relates not only to hyperthermia but perhaps also to stimulation of nonspecific immune responses. Following these early studies, progress in hyperthermic treatments of tumors was slow. Possibly, this was related to the development of X-ray therapy as a modality for treating cancers. In any case, it was not until the early 1930s that radiation was combined with hyperthermia. Because of the great difficulties involved in heating tumors, this work was not followed up by many investigators until the early 1960s when a series of experiments showed that hyperthermia by itself can have definitive antitumor effects, and, in combination with radiation therapy, it has curative potential. Both experiments involved localized heating. Regional and even whole-body heating have also been attempted. Because different physical means are employed with the two types of uses of hyperthermia in the clinic, they are described individually below.

A. Localized Hyperthermia

With localized hyperthermia, heating is achieved primarily via use of electromagnetic techniques or ultrasound, although for interstitial applications, heating via conduction of either water or electrically heated implants or catheters is also feasible. Each of these techniques has its own specific advantages and disadvantages. Unfortunately, the disadvantages are often very important. For example, electromagnetic energy is rapidly absorbed in tissue, particularly at the high frequencies where focusing and beam-shaping would permit selective heating of specific tissue volumes. Therefore, deep-seated tumors (e.g., cervical cancers, tumors near the intestines) do not receive enough microwave energy and do not heat readily. Ultrasound penetrates more readily, but cannot traverse tissue–air interfaces or penetrate bone. As a result, only a relatively small number of tumors can readily be heated by ultrasound. Interstitial techniques, while perhaps yielding the best temperature distributions, require surgery and, therefore, are primarily used in conjunction with brachytherapy (i.e., X-ray therapy delivered by implanted sources of radioactive materials).

1. Heat Alone

In one study, small human tumors (<4 cm in diameter, <3 cm deep) were heated by ultrasound. Tumor temperature varied typically ±1° across the tumor volumes. Near bone, even larger deviations were seen. Perhaps these temperature variations do not sound appreciable, but cell killing by heat varies in an exponential fashion, and a 1° change in temperature corresponds to a 50% change in effective dose. Temperatures in the region of 43°C were maintained for about 30 min, and treatments were repeated three times a week for a total of six treatments. Complete disappearance of the tumors was seen in about 10–15% of the patients, but those remissions were of short durations and lasted only about 6 wk. Very similar results were obtained by several other groups that employed a variety of means to heat the tumors. The conclusion from these studies was that heat alone was not able to eradicate enough tumor cells to cause long-term disappearance of the treated lesions. There were probably two reasons for this. First, uniform tumor heating proved to be difficult if not impossible, perhaps because of variations in localized blood flow. Blood acts as an efficient coolant of tissue. The other possibility is that a fraction of the cells in the treated tumors were heat-resistant and, therefore, exposed to an inadequate thermal "dose." Several of the studies did comment on one interesting finding, namely that heat seemed to do little, if any, damage to normal, nontumor tissue. The only problem seen was occasional skin blistering or minor necrosis of fatty tissue. Similar comments were made in many of the studies cited in the next paragraphs.

2. Hyperthermia Plus X-Irradiation

There have now been about 25–30 studies that in one way or another compared the antitumor efficacy of radiation only with that of radiation plus hyperthermia. Most of these studies used historical controls; i.e., the responses seen in patients treated with a combination of hyperthermia and radiation were compared with responses that had been previously seen in a more or less matched group of patients. Therefore, not all of these studies are truly randomized trials, and they have frequently been criticized for this reason; nevertheless, looking at the results is worthwhile. These have been surprisingly uniform. Complete response rates (i.e., clinical disappearance of the treated lesion) increased by about 50–75% when heat was added to X-irradiation. The typical target intratumor temperature that investigators attempted to reach was 43°–44°C. Some tumors proved extremely difficult to heat, and very likely only a small fraction of these tumors was heated to the target temperature. But even among the tumors that readily heated (and where such heating was documented by adequate thermometry), portions of the tumor remained well below the target temperature. It is reasonable to suggest that treatment failures were concentrated among poorly heated lesions. An interesting aspect of the studies was that the number of heat treatments seemingly were found irrelevant: tumors treated 3–4 times with hyperthermia did as well as those treated 10–15 times. Tumors treated included breast cancer, head and neck lesions, melanomas, and other surface-accessible lesions. Only one study examined the durations of local control and concluded these exceeded tumor-free intervals in historical controls.

Because prospective randomized studies are difficult and expensive to carry out and, furthermore, require a large patient number to yield statistically significant results, several investigators studied the effect of heat on matched pairs of lesions. In these studies, patients with multiple tumors had one of their tumors treated with radiation plus heat, another lesion matched in size with radiation alone, and in some cases a third with heat alone. Such studies, because of the internal controls, require relatively few patients. These studies also yielded uniform results. The tumors treated with a combination of the two modalities responded best; radiation-only treated tumors responded less well; and tumors heated but not irradiated rarely responded with complete remissions.

Only three randomized prospective studies have involved radiation and heat. Two of these showed results similar to those found in the earlier studies that had employed retrospective controls: approximately a doubling of the response rate when hyperthermia was added to radiation therapy. The third study, perhaps the largest of all clinical trials, was undertaken by a consortium of several radiation therapy centers. The results of this multi-institution trial purported to show that the addition of hyperthermia to radiation did not influence the response of treated tumors. Unfortunately, the study suffered from major defects. First, the majority of the tumors treated could not have been heated by available equipment; the study placed no restriction on tumor size. Second, because of the lack of adequate quality control, there was insufficient thermometry so that no estimate could be made on temperature distributions within most of the treated lesions. But even that study, when the tumors were stratified according to size, showed that in the smaller lesions (i.e., in those where adequate heating was possible), the addition of hyperthermia to radiation therapy indeed resulted in statistically significant benefits.

3. Hyperthermia Plus Chemotherapy

Cells in culture are much more sensitive to many drugs, if exposure is carried out, at elevated temperatures. Similarly, animal tumors are more sensitive to these drugs if chemotherapy is combined with local heating. For these reasons, physicians have also combined systemic drug treatment with localized hyperthermia. Most of the published reports originate in Japan. Most of the descriptions are anecdotal; few, if any, of the trials are randomized. Thus, although almost all the data suggest that hyperthermia increases response rates over those seen with chemotherapy alone, no conclusive proof for the added efficacy exists.

B. Regional and Whole-Body Hyperthermia

Most cancers are systemic. This may be because of the nature of the specific disease, such as the leukemias, or it may result from failure to treat localized disease sufficiently early or adequately so that metastases occur, and sites other than the primary one become involved. The use of whole-body hyperthermia is an attempt to deal with this problem. Regional heating, involving heating of entire limbs or other large volumes, is an intermediate application of hyperthermic therapy. Some major differences exist between the local and systemic hyperthermia treatments. First, for whole-body exposures, there is an absolute temperature limit of 42°C or less. Exceeding that temperature results in irreversible damage to liver and probably to brain.

Second, the physical implementation of systemic heating (and of regional heating) is far easier than that of localized heating.

There are several ways of inducing elevated temperatures throughout the body. The use of toxins and other pyrogens have already been discussed. While these do induce hyperthermia, the degree of hyperthermia induced tends to vary from patient to patient, and how long the elevated temperature will be maintained is difficult to predict in individual patients. Therefore, quality control becomes exceedingly difficult, and these agents have been abandoned in the clinic as inducers of hyperthermia.

Currently used techniques utilize conduction, absorption of electromagnetic energy, or external heating of blood and reinfusion via an arteriovenous shunt. The last technique is particularly useful for regional heating. Heating that relies on conduction alone is perhaps the slowest method of raising the patient's temperature. With hot water "blankets," for instance (i.e., thin layers of partially heat-conductive materials carrying conduits for rapid transport of heated water), the time required to do so is on the order of 3 hr. Such a long warm-up time may be undesirable for both practical and theoretical reasons. The arteriovenous shunt technique is perhaps the most rapid, but it requires a surgical procedure for the placement of the shunt and hospitalization during treatment. Perhaps the most convenient technique, and one that involves the least patient discomfort, combines heating by conduction with absorption of electromagnetic energy. With all the approaches, however, the most difficult and important aspect is the control of temperature to ensure that it does not rise above the target, usually 41.8°C. Exceeding that temperature for any appreciable length of time can lead to irreversible liver and brain damage.

1. Hyperthermia Alone

The results of exposing patients to whole-body hyperthermia without concomitant chemotherapy have not been particularly encouraging. For example, in one of the earliest studies reported, of 49 patients treated, 31 had either objective or subjective responses. Durations of responses were short; however, the disease progressed in most patients within a few weeks of completion of treatment. At the National Cancer Institute, out of 14 treated patients, 4 had objective responses. These responses lasted for 3–12 mo. These results are typical of clinical data obtained in a small percentage of responses (10–30%), and these only of short durations (typically 6 mo or less).

2. Heat Plus Chemotherapy

Better results were obtained when hyperthermia was combined with chemotherapy. Some studies have reported on treatments that combine drugs such as adriamycin, mitomycin C, and a nitrosourea with whole-body hyperthermia. A Texas group has used hyperthermia for many years in the perfusion of limbs affected with melanoma, apparently with considerable success. Recently, an East German group described the treatment of children with whole-body hyperthermia and chemotherapy and also claimed good results.

III. Is Hyperthermia an Inducer of Cancer?

No direct evidence, either from human or from animal data, suggests that heat by itself can cause cancer. Certainly, epidemiological studies show that in some warm countries, the incidence of skin cancers, particularly squamous cell carcinomas and melanomas, is very high; however, the causative agent here is certainly excessive exposure to the sun. Part of the sun-emitted spectrum of frequencies is in the ultraviolet range. Many studies have demonstrated that a portion of the ultraviolet radiation is preferentially absorbed by DNA, and the damage there, or its faulty repair, leads to changes in the cell (probably involving oncogenes) that finally result in neoplastic disease. The possibility that the increased temperature in the warm countries makes ultraviolet radiation a more efficient carcinogen has not as yet been examined. This concept—heat as a cocarcinogen—is discussed later. [See SKIN, EFFECTS OF ULTRAVIOLET RADIATION.]

What about indirect evidence? Here, two sets of data need to be considered. First, heat may be a mutagen. We know that most carcinogenic agents are also mutagenic, and vice versa. If heat can be demonstrated to be a mutagen for mammalian systems, then the thought would have to be entertained: It may also be a carcinogen. Second, if carcinogenesis is considered as an error in development, then data on heat-induced teratogenicity need to be examined.

A. Hyperthermia as a Mutagen or a Transformant

When *Drosophila* cells were exposed to 38°C for 1 hr (a hyperthermic exposure because the normal growth temperature of the fruit fly is 25°C), this treatment doubled the production of lethal mutations. In the only study on human cells, investigators measured 6-thioguanine resistance (a frequently used assay for mutagenesis) induced by heat in human lymphoblasts. A 10-min exposure to 45°C approximately doubled the mutant fraction, causing these workers to suggest that heat behaved like a strong chemical mutagen.

Transformation of mammalian cells *in vitro* has become a powerful technique used to examine the carcinogenicity of various treatments and drugs. The technique consists of exposing density-inhibited cells to the putative carcinogen and then, at various times later, examining the culture for foci of morphologically recognizable transformations. Cells from such transformed clones are then shown to be capable of producing tumors in appropriate hosts, whereas the untransformed cells are incapable of doing so. The process of acquiring this malignant potential is called transformation. Curiously, here several studies have shown that heat is not able to transform mammalian cells. Thus, an apparent discrepancy appears between the data on mutations and transformations. The explanation may well be that mutations usually involve only a single DNA modification, while it is known that transformation requires at least two such modifications. Thus, heat may be able to do the former, while unable or unlikely to do the latter.

B. Hyperthermia as a Teratogen

The teratogenic literature is much more definite. Heating cells clearly can cause modifications of development that can lead to abnormalities in the offspring. Heating of developing *Drosophila,* during the pupal stages, results in very specific, abnormal phenotypes of the emerging fruit flies. The type of abnormality relates to the time of heating and was surprisingly specific. For example, if a heat shock is given 38 hr after pupariation, this resulted in growth of abnormal hair in about 90% of the distal wing area, while on the proximal part of the wing, <10% of the hair was abnormal. Shifting the heat shock in time by only a few hours reversed this ratio. This curious phenomenon very likely is related to the induced synthesis of heat shock (or stress) proteins.

In other species, a great variety of heat-induced birth defects have been observed. Animals involved range from chickens to monkeys. These experiments have been reviewed recently. In chicken, defects involve the ventril body, head and feet, limbs, and eyes as well as other parts of the birds' bodies. In mice, the major abnormality that has been examined in detail is a neural tube defect, namely failure of the closure of the tube. This failure is frequently associated with the brain outside the cranium at birth.

These studies and many others also point toward the extreme heat sensitivity of the embryos' central nervous system. In guinea pigs, for example, heating at 43°C for 1 hr lead to offspring that were slow, clumsy, and frequently did not bond with the mother. Several of the newborn guinea pigs died within the first few days after birth. Autopsy showed that brain sizes were well below those of control animals, which had not been exposed to the elevated temperature. These findings are of considerable interest to medicine because heat, in many ways, acts like alcohol in inducing stress responses. Many of the characteristics found in the newborn guinea pigs delivered from mothers that had been exposed to hyperthermia resemble those found in children suffering from what is called the fetal alcohol syndrome, (associated with maternal intake of alcohol). Many biochemical modifications that are induced by heat are also induced by alcohol; it is not unreasonable to hypothesize that it is these biochemical changes that are responsible for malformations in the newborn.

Finally, in humans, several epidemiological studies have related maternal febrile episodes to births of malformed children. While these studies can hardly be considered as proof that the thermal episode was related to the malformation of the offspring, nevertheless the epidemiological studies on humans, when combined with the direct experimental evidence for many animal systems, are a strong indication that, indeed, hyperthermia is a teratogen in humans. [*See* THERMOTOLERANCE IN MAMMALIAN DEVELOPMENT.]

C. Hyperthermia as a Coinducer

In addition to the data cited, hyperthermia is definitely a cocarcinogen and a coteratogen. *In vitro* studies have shown that if carcinogens are applied to cells at elevated temperatures, the rate of transformation is usually increased. This is particularly true if the two treatments are combined. Similarly, if a teratogen is given in conjunction with an increase in body temperature, the probability of malformation in the newborn increases. Thus, heat acts as an agent able to increase the efficacy of many carcinogens and teratogens. In many ways, this is hardly surprising. Temperature increases the rate of chemical reactions and increases the probability of errors in error-prone DNA repair systems.

From these data, the following emerges. Heat by itself is not known to cause cancer; however, it is a mutagen and, as such, may contribute to its induction. In addition, many carcinogens, when applied at elevated temperatures, are more efficient than they are at 37°C. For these agents, heat acts as a cocarcinogen. The teratogenic effects of heat also suggest that elevated temperatures, if these occur at specific times during pregnancy, may contribute to induction of malignancies, although no data or experiments support this hypothesis.

IV. Conclusion

Hyperthermia appears to have a role in the treatment of cancer, particularly as part of a multimodality approach. It appears to be most effective against localized tumors. The major obstacle to its wider acceptance is the difficulty of heating lesions, particularly if these are not near the body's surface.

Somewhat paradoxically, hyperthermia may also influence induction of cancer. Again, its major importance may be in enhancing effects of other treatments, i.e., as a cocarcinogen or as a coteratogen.

Bibliography

Bull, J. M., Lees, D., Schuette, W., Whang-Pang, J., Smith, R., Bynum, G., Atkinson, E. R., Gottdiener, J. S., Gralnick, H. R., Shawker, T. H., and DeVita, V. T., Jr. (1979). Whole body hyperthermia: A phase I trial of a potential adjuvant to chemotherapy. *Ann. Intern. Med.* **90**, 317–323.

Coley, J. B. (1893). The treatment of malignant tumors by repeated inoculations of erysipelas, with a report of ten original cases. *Am. J. Med. Sci.* **105**, 488–511.

Crile, G., Jr. (1963). The effects of heat and radiation on cancers implanted in the feet of mice. *Cancer Res.* **23**, 372–380.

Edwards, M. J. (1986). Hyperthermia as a teratogen: A review of experimental studies and their clinical significance. *Teratogenesis, Carcinog. Mutagen.* **6**, 563–582.

Engelhardt, R. (1987). Hyperthermia and drugs. *In* "Hyperthermia and the Therapy of Malignant Tumors" (C. Streffer, ed.), pp. 136–203. Recent Results in Cancer Research, Vol. 104. Springer-Verlag, Berlin, Heidelberg, and New York.

Hahn, G. M. (1982). "Hyperthermia and Cancer." Plenum Press, New York.

Molls, M., and Scherer, E. (1987). The combination of hyperthermia and radiation: Clinical investigations. *In* "Hyperthermia and the Therapy of Malignant Tumors" (C. Streffer, ed.), pp. 110–135. Recent Results in Cancer Research, Vol. 104. Springer-Verlag, Berlin, Heidelberg, and New York.

Pettigrew, R. T. (1975). Cancer therapy by whole body heating. *In* "Proceedings of the International Symposium on Cancer Therapy by Hyperthermia and Radiation" (M. Wizenberg and J. E. Robinson, eds.), pp. 282–288. American College of Radiology Press, Baltimore.

Warkany, J. (1986). Teratogen update: Hyperthermia. *Teratology* **33**, 365–371.

Warren, S. L. (1935). Preliminary study of the effect of artificial fever upon hopeless tumor cases. *Am. J. Roentgenol.* **33**, 75–87.

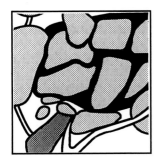

Hypothalamus

QUENTIN J. PITTMAN, *University of Calgary*

I. Background
II. Anatomy
III. Functions of the Hypothalamus

Glossary

Autonomic nervous system Part of the nervous system which is concerned with involuntary body functions through an action on glands, smooth muscle, and cardiac tissue

Baroreceptors Specialized stretch receptors located in the walls of some major blood vessels that are stimulated by elevations in blood pressure and transmit this information to the brain

Homeostasis Relatively stable state of equilibrium (of the internal body environment)

Neurohypophysial neuron Neuron with a cell body in the hypothalamus and an axon which extends into the posterior pituitary

Nucleus Aggregation of neuronal cell bodies

Tuberoinfundibular neuron Neuron in the medial basal hypothalamus which projects to the median eminence where it secretes either (a) hypothalamic release or (b) hypothalamic release–inhibiting hormones which are transported by the blood through capillaries to the anterior pituitary

THE HYPOTHALAMUS is composed of a number of groups of cells, lying close to the base of the brain and bordering on the third ventricle. An important area of the brain involved in the control of homeostatic functions as diverse as eating, drinking, sleep, thermoregulation, cardiovascular regulation, and hormone secretion, the hypothalamus receives a variety of inputs, both of a neural and of a humoral (blood-born) nature. Outputs from the hypothalamus result in hormonal, behavioral, and autonomic responses designed to maintain homeostasis. A

wide variety of neurotransmitters is utilized within hypothalamic neural networks to accomplish these integrative and control functions.

I. Background

Despite its small size (less than 4 g), the hypothalamus exerts enormous control over the body. An appreciation of this influence developed over the first few decades of this century as descriptions emerged of syndromes associated with hypothalamic lesions and dysfunction in humans. Much of our information about the hypothalamus, however, comes from animal experiments. It is such studies that have brought the hypothalamus out of the shadow of the overlying cortical areas to establish its important role in the control of homeostasis. As a result of these studies, the discipline of neuroendocrinology is virtually synonymous with that of the study of the hypothalamus. The identification and structural characterization of many of the chemicals important in brain function have arisen from analyses of extracts of hypothalami, where they are synthesized and stored in great abundance. Finally, the concept of neurosecretion, i.e., that a neuron could synthesize and release a hormone (peptide), which is now a commonly accepted fact in the neuropharmacological literature, had its infancy in studies of hypothalamic–pituitary relationships in lower vertebrates.

In order to understand hypothalamic function, it is necessary to have an appreciation of its anatomy.

II. Anatomy

A. Intrinsic Anatomy

The hypothalamus is an ill-defined area of neural tissue lying on each side of the midline and border-

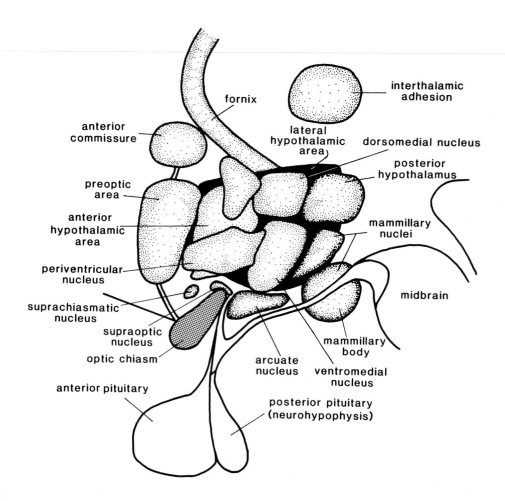

fornix

interthalamic adhesion

anterior commissure

lateral hypothalamic area

dorsomedial nucleus

posterior hypothalamus

preoptic area

anterior hypothalamic area

mammillary nuclei

periventricular nucleus

midbrain

suprachiasmatic nucleus

supraoptic nucleus

optic chiasm

mammillary body

arcuate nucleus

ventromedial nucleus

anterior pituitary

posterior pituitary (neurohypophysis)

FIGURE 1 Intrinsic hypothalamic nuclei and surrounding structures. This is a sagittal view from the medial aspect of the hypothalamus with anterior on the left.

ing the walls of the third ventricle. As its name indicates (hypothalamus), it is situated ventral to the thalamus at the base of the diencephalon. The pituitary stalk which joins the hypothalamus to the pituitary exits from the ventral surface; its anterior border is considered the plane of the anterior commissure and its posterior border the mammallary bodies which lie at the junction between the diencephalon and the midbrain. This small area of tissue, which in coronal section is barely the size of an adult's thumbnail, is made up of a number of nuclei. These nuclei are packed tightly together and have been named, as indicated in Fig. 1, on the basis of their location within the hypothalamus, their relationship to neighboring structures, and the organization of their afferent and efferent connections. In many (but not all) cases, a particular nucleus is associated with specific functions. At the anterior end of the hypothalamus is the preoptic area, which lies anterior to the optic chiasm. Because a well-defined group of neurons is not apparent from anatomical

studies, the term *preoptic area* has been given to this group of cells. Indeed, the preoptic area is often considered to be *anatomically* distinct from the hypothalamus but is so closely related *functionally* to the hypothalamus that it is generally included in any description of hypothalamic anatomy and physiology.

At its posterior edge, the preoptic area gives way to the anterior hypothalamic area, another ill-defined group of cells which often is functionally closely related to the preoptic area. On the medial aspect, near the walls of the third ventricle, the periventricular nucleus makes its appearance and continues posteriorly along the ventricle as a broad sheet of cells. At the anterior–posterior plane of the optic chiasm, small groups of cells (nuclei) with well-defined borders appear; these cells include the

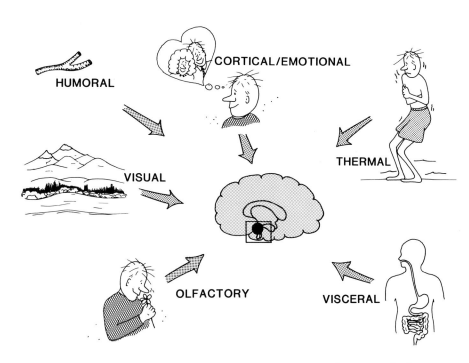

FIGURE 2 Inputs to the hypothalamus.

suprachiasmatic nucleus and, slightly more posteriorly and laterally, the supraoptic nucleus. The latter nucleus is particularly evident on histological sections because of its large magnocellular cell bodies; this nucleus is important in the control of posterior pituitary function. Dorsal and medial to the supraoptic nucleus and bordering the dorsal aspects of the third ventricle lies the functionally related paraventricular nucleus. Immediately posterior to the optic chiasm is a small nucleus lying immediately above the exit of the pituitary stalk; this nucleus, called the arcuate nucleus is important in endocrine control, as is the larger, ventromedial nucleus, which lies dorsal and slightly lateral to the arcuate nucleus. As can be seen in Fig. 1, the dorsal and posterior portions of the hypothalamus are occupied by the dorsal medial nucleus and posterior hypothalamus, respectively. A prominent landmark is provided by the ventral projection of the mammillary nuclei, which form a midline structure called the mammillary body. This protuberance (which is evident on the ventral surface) marks the posterior border of the hypothalamus. Running throughout the length of the hypothalamus and abutting laterally upon the subthalamic region and internal capsule is an anatomically indistinct area of tissue called the lateral hypothalamus.

B. Inputs to the Hypothalamus

In light of the predominant role played by the hypothalamus in a wide variety of internal body functions, it is hardly surprising to learn that the hypothalamus is extensively connected to numerous other parts of the nervous system. Some of these pathways are evident as large bundles of nerve fibers easily observed upon gross dissection; others are more diffuse, consisting of small nerve fibers which eluded anatomical description until the advent of more sensitive anatomical techniques. In addition to having neural inputs, the hypothalamus is also in the position to monitor both its local and distant environments through intrinsic sensory neurons which monitor both local conditions in the extracellular fluid and those of their blood supply. Through its afferent and intrinsic pathways, the hypothalamus receives a wide variety of information important to its control of the internal body environment (Fig. 2).

1. Visceral Inputs

The hypothalamus receives extensive information concerning the internal state of the body. Such visceral inputs are relayed to the hypothalamus from a series of ascending fiber tracts which arise in lower brain stem areas. For example, the nucleus tractus solitarius, a brain stem nucleus which receives much of the sensory information arising from the viscera, has a direct projection to various hypothalamic nuclei. A number of brain stem and pontine nuclei which synthesize and utilize noradrenaline as a neurotransmitter (for example, locus ceruleus and ventral lateral medullary nuclei) give rise to an ascending system of catecholamine fibers which innervate much of the hypothalamus. Many of these fibers ascend in a lateral area of the brain stem in the ventral noradrenergic tract of the medial forebrain bundle. Others ascend through a midline tract called the dorsal longitudinal fasciculus.

2. Somatic-Sensory, and Primary Sensory Inputs

Sensory information from the skin is relayed to the hypothalamus through ascending, polysynaptic pathways which arise in the dorsal horn of the spinal cord and cranial sensory nuclei. This information, which projects to the thalamus and higher cortical areas, probably reaches hypothalamic nuclei either from collaterals of ascending fibers or directly from the thalamus. Receptors in the skin transduce a wide variety of sensory signals; of particular importance to the hypothalamus are sensory fibers whose electrical activities change in response to temperature. These fibers carry thermal information to the hypothalamus. Other sensory inputs from the skin carry somatic-sensory information (e.g., touch, pain). [*See* THALAMUS.]

Although olfactory input is probably of relatively less importance in humans than it is in lower mammals, many of the connections of olfactory areas to hypothalamic nuclei have been retained in the human. Of particular importance is the input from olfactory areas of cortex to the amygdala, a nucleus lying in the temporal lobe. The amygdala extensively innervates a number of hypothalamic nuclei through two fiber pathways—the stria terminalis and the ventral amygdalo-fugal pathway. Visual sensory information reaches the hypothalamus directly through the retino-hypothalamic tract, which projects to the suprachiasmatic nuclei.

3. Cortical–Limbic Inputs

The limbic system consists of a number of structures which border on the ventricular system of the brain and are thought to be important in processing emotional behavior. The hypothalamus is one of the major destinations of limbic system information. Much of this information is relayed to the hypothalamus from a temporal lobe structure known as the hippocampus. The hippocampus gives rise to a large bundle of fibers called the fornix, which sweeps into the hypothalamus from its anterior dorsal aspect to innervate the posterior hypothalamic nuclei, in particular the mammillary complex. This white matter tract is easily identified on gross dissection of the brain, and its large myelinated fibers dominate histological sections of the hypothalamus. The septal nuclei, which lie anterior and dorsal to the hypothalamus, are also considered part of the limbic system, and they give rise to descending projections, which run through the medial forebrain bundle to innervate the lateral aspects of the hypothalamus and other descending structures. Cortical input to the hypothalamus arises from prefrontal cortex and reaches the hypothalamus either directly through cortical hypothalamic fibers or indirectly via relay through the thalamus. [*See* HIPPOCAMPAL FORMATION; LIMBIC MOTOR SYSTEM.]

4. Humoral (Blood-born) Inputs

In its position at the base of the brain, the hypothalamus is among the first of neural structures to be perfused by blood ascending in the carotid arteries. Indeed, the Circle of Willis, which distributes blood from the carotids to the various parts of the diencephalon, cortex, and subcortical areas, forms a ring at the ventral aspect of the hypothalamus. The hypothalamus contains sensory neurons of a variety of types which respond to local blood-born elements. This function is facilitated by a unique type of vascular epithelium in the medial basal aspects of the hypothalamus. Whereas most parts of the brain maintain a well demarcated functional barrier (the blood–brain barrier) between the blood and local extracellular fluid, the hypothalamus has a "leaky" blood–brain barrier due to the presence of fenestrated epithelia lining the capillaries. Thus, many substances which circulate (e.g., hormones, nutrients) diffuse freely into the hypothalamus, where specialized neurons can transduce substance levels into electrical activity. [*See* BLOOD–BRAIN BARRIER.]

It is also possible that the hypothalamus participates in a short, circulatory feedback loop from the pituitary. It has recently been established that blood may flow retrogradely from the pituitary to the brain; this provides a means by which pituitary secretions can quickly reach the hypothalamus, where their levels can be monitored via neurons having specialized receptors. [*See* PITUITARY.]

Other information of importance to the pituitary appears to be that of the metabolic state of the body. In addition to having neurons specialized for monitoring metabolically important substances (e.g., glucose), neurons also are present throughout the hypothalamus which respond to local temperature. As this temperature is largely a function of the blood perfusing this structure, these neurons are uniquely positioned to monitor core temperature as reflected by the temperature of blood arising from the thorax.

Because of its location bordering the walls of the third ventricle, the neurons of the hypothalamus are also in contact with the cerebrospinal fluid which fills the ventricular system and which can diffuse freely through the ependymal lining of the ventricles into hypothalamic tissue. Although evidence is still fragmentary, it is thought that specialized neurons may exist near the ventricular space which can respond to fluctuations in levels of neurotransmitters and ions which occur in the cerebrospinal fluid.

C. Outputs of Hypothalamus

The anatomical and humoral outputs of the hypothalamus provide this structure with the means to affect many behavioral and physiological functions. An overview of these various controls is given in the following section.

1. Endocrine Outputs

The identification and description of the control of pituitary function has provided one of the more fascinating stories of anatomical and physiological investigations of the past 60 years and has given rise to the entire discipline of neuroendocrinology. The intimate anatomical relationship between the hypothalamus and the pituitary is probably best appreciated when one realizes that the posterior pituitary (neurohypophysis) is actually a ventral growth of the hypothalamus. Magnocellular neurons of the paraventricular and supraoptic nuclei project via the hypothalamo–neurohypophysial tract through the infundibular stalk to end as free nerve endings in the neurohypophysis. The peptides arginine vasopressin and oxytocin (as well as a number of other peptides) are synthesized in these nuclei, transported down the axons of the neurohypophysis and released into the extracellular space from where they enter the bloodstream.

In contrast to the organization of the posterior pituitary, the glandular cells of the anterior pituitary receive no direct neural projections. Rather, a series of neurosecretory neurons (tuberoinfundibular neurons) scattered throughout the various hypothalamic nuclei (but concentrated in the arcuate and ventromedial nuclei) send their axons to the base of the brain (median eminence) and release their contents (releasing or release-inhibiting hormones) into a specialized vascular bed. These portal plexus capillaries transport the hypothalamic releasing or inhibiting hormones from the hypothalamus to the anterior pituitary, where they diffuse into this tissue to affect the release of the various anterior pituitary hormones. [See NEUROENDOCRINOLOGY.]

In some lower mammals, the pituitary contains an intermediate lobe, which secretes the hormone α-melanocyte-stimulating hormone and is controlled by arcuate nucleus neurons that project directly onto intermediate lobe cells. In humans, the intermediate lobe is present in the fetus but is largely absent, except for a few scattered cells, in the adult.

2. Cortical–Behavioral Outputs

The hypothalamus exerts its control over behavioral functions through mono- and polysynaptic pathways to limbic system structures. Among such projections are those to the septal nuclei, the amygdala, habenula, and medial dorsal nucleus of the thalamus. A particularly important tract providing a relay for hypothalamic information to the limbic system is the mammillo–thalamic tract, which, as its name suggests, arises in the mammillary body and projects to the anterior nucleus of the thalamus. The thalamus, in turn, relays this information to specific limbic structures, in particular, the cingulate gyrus.

3. Autonomic Outputs

Because of limitations in anatomical techniques, and possibly due to its well-described involvement in endocrine control of the pituitary, descending connections from the hypothalamus were originally thought to be few in nature. Anatomical studies on brains from patients succumbing to problems relating to hypothalamic lesions identified short descending pathways, including the mammillotegmental tract, the dorsal–longitudinal fasciculus, and caudal extensions of the medial forebrain bundle. With the advent of more sophisticated anatomical tracing methodology, however, studies in lower mammals have brought to light the fact that the hypothalamus possesses an extensive system of descending fiber tracts which innervate virtually all of the autonomic nuclei of the brain stem and even descend as uninterrupted axons to the most caudal aspects of the spinal cord, where they innervate the sympathetic and parasympathetic preganglionic neurons. A similar system of descending fibers is found in the human. Thus, in terms of size and distribution of their axonal projections, certain neurons of the hypothalamus rival that of the large upper motor neurons (Betz cells) of the cerebral cortex, which exert control over the somatic motor system.

4. Intrinsic Connections

In addition to widespread connections to and from the hypothalamus and other parts of the nervous system, the intrinsic hypothalamic nuclei are also extensively interconnected. Throughout the hypothalamus are many short axon interneurons, which receive information from adjacent nuclei, either directly or from collaterals of descending or ascending fibers.

D. Neurotransmitters

In the mammalian nervous system, neurons communicate with each other largely through the secretion of low molecular weight chemicals called *neurotransmitters,* which diffuse across specialized junctions between neurons called *synapses* to affect the electrical activity of neighboring cells. Virtually all the "classical" neurotransmitters (including acetylcholine, noradrenaline, serotonin, dopamine, adrenaline, γ-aminobutyric acid and glutamate) are well represented in the nerve fibers innervating hypothalamic nuclei. As a number of drugs of abuse and medications are known to act through their interference with such classical neurotransmitter systems, it is apparent that hypothalamic function can be altered by exposure to such agents. Of particular interest over recent years has been the realization that the hypothalamus is virtually a cornucopia of peptide hormones which act as neurotransmitters. Experiments describing the neurosecretory nature of hypothalamic neurons and subsequent electrophysiological studies of hypothalamic "neuroendocrine" neurons provided some of the first evidence that peptides could function as neurotransmitters. There have now been well in excess of 50 different peptides described as having putative neurotransmitter functions within the brain, and virtually all of these are represented within the hypothalamus. However, the task of identifying the roles of these peptides in specific aspects of hypothalamic anat-

omy and physiology has barely begun. [*See* PEPTIDES.]

III. Functions of the Hypothalamus

A. Homeostasis and Control Theory

The great French physician and scientist, Claude Bernard, first recognized the fact that the internal environment of the body, or "milieu interieur," is maintained in a constant state. Walter Cannon subsequently coined the term *homeostasis* to describe the mechanism by which animals maintain this relatively constant internal environment. The remarkable fact that the internal body environment can be regulated at a constant level in the face of widely varying external conditions of temperature, food availability, metabolic demands, and so on, is due, in part, to the hypothalamus. In order to achieve such regulation, the hypothalamus must be able to sense the various internal environments of the body, integrate this information and compare it to a theoretical "setpoint," and then respond to any perturbations with an appropriate effector signal. The means by which the hypothalamus obtains its afferent information and the pathways by which its output is directed to appropriate destinations have been outlined in the preceding sections. The following discussion examines the various functions controlled by the hypothalamus.

B. Specific Activities

The location of the hypothalamus deep within the brain has not facilitated investigation into the functions of this structure. From the dorsal aspect, one must pass through the entire cerebrum to reach the hypothalamus; from the ventral aspect, the base of the skull limits accessibility. In humans, tumors and cardiovascular accidents have provided the bulk of the experimental material revealing functional correlates to damage or stimulation of particular areas. Even under such circumstances, it has often been difficult to differentiate between a syndrome caused by *damage* to a specific area and that resulting from irritation or stimulation of an area adjacent to, for example, a small hemorrhage. Particularly with respect to the endocrine system, however, a number of diseases which exhibit classical sequelae can now be associated with specific hypothalamic deficits. The information which has been obtained about the human hypothalamus, however, is

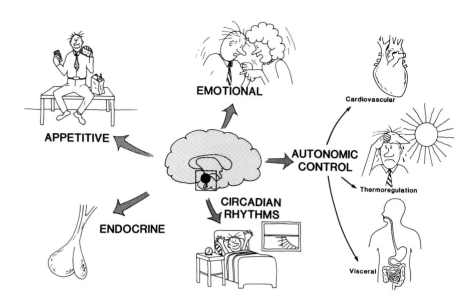

FIGURE 3 Activities under hypothalamic control.

strongly based on experimental studies in lower mammals. Fortunately, there appears to be remarkable preservation of form and function throughout the various mammalian genera, and experiments in mammals as diverse as the rat and a subhuman primate often give surprisingly congruent results. In these animals, it has been possible to carry out lesion, stimulation, and recording experiments which have revealed important aspects of hypothalamic function. However, even under controlled laboratory conditions, it has been difficult to achieve clear and easily interpretable results; the various hypothalamic nuclei are closely situated, and it has been difficult to affect one area of the hypothalamus discretely without impinging on adjacent areas. Nonetheless, with these caveats in mind, a description of hypothalamic functions emerges as outlined in Fig. 3.

1. Appetitive Behaviors

The hypothalamus plays an important role in the control of feeding behavior. It has been known for many years that stimulation of medial parts of the hypothalamus, in particular the ventromedial nucleus and adjacent tissue, will inhibit eating, whereas stimulation of the lateral hypothalamus will cause an increase in food intake in lower mammals. In keeping with these observations, lesions in medial hypothalamic areas occurring both naturally in humans and after experimental placement in lower mammals cause increased feeding; destruc-

tion of more lateral hypothalamic areas causes anorexia. Rather than a direct stimulation of the appropriate behavior (eating, chewing, food seeking), the evidence indicates that an actual setpoint of body weight regulation appears to be changed. These changes in food intake are associated with corresponding changes in metabolism, which suggest that they are part of an integrated response to the regulation of caloric balance. While the actual mechanisms responsible for the regulation of food intake are numerous, the hypothalamus plays an important role. Among other regulatory events is one in which neurons in the hypothalamus respond to changes in blood glucose level with changes in electrical activity. For example, cells in the ventromedial hypothalamus respond with increased activity to high blood glucose levels, whereas neurons throughout the lateral hypothalamus more often decrease their activity in response to elevations in glucose. Although it is undoubtedly simplistic to make the relationship, it may be noted that such changes are appropriate in terms of our understanding of the relative roles of medial and lateral hypothalamus in regulating food intake. Thus, as is expected for a controlled system, both the sensory mechanisms are present and the necessary connections exist from the hypothalamus to the limbic system to elicit the appropriate behaviors. [*See* FEEDING BEHAVIOR.]

The debate as to the exact cue for signalling thirst as well as the location of cells responsible remains vigorous to this day. It would appear that changes in blood osmolality and possibly sodium content activate neurons in areas near the hypothalamus (organum vasculosum of the lamina terminalis and subfornical organ), which in turn send strong projections to the hypothalamus. In addition, osmoreceptors are located in the anterior hypothalamus and preoptic area, which directly respond to changes in plasma osmotic pressure (concentration of solutes in the blood). [*See* THIRST.]

The hypothalamus appears to integrate signals from the various brain structures and activates drinking behavior via pathways to the limbic system in a manner similar to that described for eating behavior.

2. Autonomic Control

The pivotal role of the hypothalamus in the control of the autonomic nervous system has long been recognized, and it is often described as the "head ganglion" of the autonomic nervous system. The autonomic nervous system is divided on the basis of its anatomy, pharmacology, and to a certain extent, its function into two components: the sympathetic (involved in arousal, or "fight or flight," behavior) and parasympathetic (involved in vegetative function) systems. At one time, it was thought that there was a topographical relationship within the hypothalamus to these two systems, with the anterior hypothalamus considered the parasympathetic center and the posterior hypothalamus the sympathetic center. While it is true that in animals "sympathetic" effects can be activated by electrical stimulation of the posterior hypothalamic areas, and "parasympathetic" responses by stimulation of anterior hypothalamic areas, it is equally true that more discrete stimuli throughout a variety of areas of the hypothalamus can cause generalized autonomic effects. The strong interconnections between various hypothalamic nuclei and brain stem and spinal cord autonomic nuclei provide an anatomical basis for the hypothalamus to exert control of cardiovascular function, thermoregulation, and visceral function. [*See* AUTONOMIC NERVOUS SYSTEM.]

a. Cardiovascular System With respect to the cardiovascular system, it is known that information is relayed from baroreceptors synapsing in the medulla within nuclei which project directly or via polysynaptic pathways to the hypothalamus. Electrical or chemical stimulation of neurons in hypothalamic areas that receive baroreceptor information causes changes in both arterial pressure and in

heart rate; such effects are most likely mediated through descending projections to areas of the ventral lateral medulla and to the spinal cord, which control peripheral arterial muscles and cardiac muscle. In addition, a number of circulating hormones influence blood pressure, and their release can be influenced by such descending pathways or via hypothalamic connections to the pituitary (see Section III, B, 5).

b. Thermoregulation Classical experiments carried out in the 1930s established that animals with hypothalamic lesions were incapable of normal thermoregulation and gave rise to extensive studies that have unequivocally placed the hypothalamus as a pivotal structure involved in the neural control of body temperature. Thermal sensory information from the skin and other parts of the body is relayed via multisynaptic pathways to the hypothalamus, where it is integrated with information obtained from intrinsic receptors. Through mechanisms not yet well understood, this information is compared to a hypothetical "setpoint," and, where deviations from this temperature take place, appropriate effector mechanisms are brought into play to return temperature to normal. These effector mechanisms include the regulation of peripheral vasomotor tone, alterations in sweat secretion, changes in respiratory rate, changes in metabolism, and alterations of behavior and posture appropriate for changes in heat loss or heat gain. In addition to its control over normal thermoregulation, the hypothalamus, and particularly the preoptic/anterior hypothalamic areas, appears to be important in the development of fever. In lower mammals, application of minute quantities of pyrogens (agents that cause fever) directly to these areas causes prompt, experimental fevers; destruction of neurons in these same areas inhibits fever development in these animals. [*See* BODY TEMPERATURE AND ITS REGULATION.]

c. Visceral Function In accord with the hypothalamic control of appetitive behaviors described earlier, the hypothalamus also directly influences gut function to provide complimentary autonomic effects. For example, it has been shown in rats that electrical or chemical stimulation of a number of hypothalamic areas (e.g., paraventricular nucleus, ventromedial hypothalamus, dorsal hypothalamus) can influence gut muscle function as well as acid secretion from the stomach. It is thought that the pathways from the hypothalamus to brain-stem autonomic areas which control acid secretion may be of particular importance in the development of ulcers in response to stress, with the hypothalamus providing the intermediary role in relaying cortical and limbic information to the autonomic nervous system. [*See* Visceral Afferent Systems, Signaling and Integration.]

3. Emotional Expression

In lower mammals, electrical stimulation of several areas in the medial and posterior hypothalamic nuclei causes impressive displays of aggressive and vicious behavior; because it is thought that such behaviors are merely a behavioral output rather than the "thought" associated with anger, this phenomenon is known as sham rage. Destruction of other areas of the hypothalamus (for example, the ventromedial nucleus) will also lead to such behavioral alterations. In contrast to these aversive behaviors, it is known that other areas of the hypothalamus, in particular the lateral areas, appear to be associated with behaviors best described as "reinforcing" or "pleasurable." That is, animals will actually self-stimulate electrodes placed in the lateral hypothalamus, even to the detriment of other behaviors; presumably, stimulation of neurons in these areas activates pathways leading to cortical areas and are interpreted as pleasurable. [*See* Affective Responses.]

Subjective aspects of fear, rage, and pleasure are also associated with hypothalamic damage or irritation in humans. This evidence is consistent with the idea that the hypothalamus is one relay in the pathway involving limbic and cortical areas that are associated with emotional behavior.

4. Circadian Rhythms

Many animals exhibit behaviors which show a remarkable periodicity which approximates a 24 hr day. These circadian rhythms are thought to be generated by an oscillatory mechanism in the suprachiasmatic nucleus and entrain to the light–dark cycle via impulses arriving from retinal–hypothalamic tracts. In fact, destruction of the suprachiasmatic nucleus in such animals will cause these rhythms to become totally disrupted. Human beings also have a suprachiasmatic nucleus and have also been shown to display circadian rhythms in such functions as body temperature and endocrine functions. Of particular interest is the hypothalamic control of rhythms of sleep and wakefulness, which appear to be mediated by hypothalamic connections to the re-

ticular activating system of the midbrain. [*See* Circadian Rhythms and Periodic Processes.]

5. Endocrine Control

Through its anatomical connections to the posterior pituitary and to the median eminence, the hypothalamus represents the final common pathway for brain control of pituitary secretion. The availability of specific radioimmunoassays for various peptides and hormones has made it possible to measure these substances in the circulation and to identify the stimuli responsible for their release. With the advances in biochemical characterization of these peptide hormones, and the availability of pure synthetic peptides, it has also been possible to administer these substances and thereby identify target end organ responses.

The major hormones released from the posterior pituitary are the peptides arginine vasopressin and oxytocin. Arginine vasopressin is also known as antidiuretic hormone because of its action within the kidney to enhance water reuptake and concentrate the urine. In addition, this hormone also has other actions within the body to regulate blood pressure through its action on arterial smooth muscle and an action on metabolism through its enhancement of glycogen conversion to glucose in the liver. Oxytocin receptors are also found on smooth muscle, in this case on uterine smooth muscle and on mammary smooth muscle. Thus, its peripheral action is to cause milk letdown via contraction of the mammary smooth muscle and to cause uterine contractions (during the birth of a baby). Oxytocin receptors are also found on fat cells and on selected other peripheral tissues; oxytocin actions on these structures may explain its presence in males.

Regulation of the secretion of these hormones is brought about through specialized receptors in the hypothalamus and afferent stimuli arising from peripheral receptors. Neurons in the anterior hypothalamus, and possibly the vasopressin neurons themselves, are capable of sensing the osmolality of the blood and, in response to increases in osmolality, cause increased synthesis and release of vasopressin. Reductions in blood pressure and blood volume, signalled via specialized cardiovascular receptors in the periphery, are also potent stimuli for vasopressin release, as are some types of stressful stimuli (e.g., pain and nausea).

Oxytocin is released in response to tactile stimulation of the nipples (i.e., when a baby suckles) and after distention of the birth canal during delivery.

Cortical inputs are thought to provide a behavioral influence on oxytocin release; this is sometimes manifested by the spontaneous milk letdown experienced by a mother in response to the cries of her baby. Emotional stress, on the other hand, appears to inhibit oxytocin secretion, thereby making breast-feeding difficult for a mother under such stress.

Hormones responsible for the release or inhibition of release of anterior pituitary hormones are synthesized in tuberoinfundibular neurons scattered throughout the hypothalamus and terminating on the medium eminence. The secretory products of these neurons control glandular cells of the anterior pituitary, which synthesize and release luteinizing hormone, follicle-stimulating hormone, growth hormone, thyroid-stimulating hormone, adrenocorticotropic hormone, prolactin, and β-endorphin. The control of the tuberoinfundibular neurons is beyond the scope of this article. Nonetheless, in terms of general principles, there appear to be negative or positive feedback mechanisms which affect the release of their products. Thus, there are receptors in the hypothalamus which respond to circulating hormones produced both in distant parts of the body (e.g., steroids, such as estrogen produced in the ovary) as well as from the anterior pituitary. In addition, afferent stimuli from both viscera, as well as the cortex and limbic system, can affect release. Secretion of the anterior pituitary hormones is often synchronized to the light–dark cycle, and this circadian control is thought to be exerted via retinal signals entering the suprachiasmatic nucleus. [*See* POLYPEPTIDE HORMONES.]

C. Integrative Aspects

While the control of hypothalamic function has been discussed for each regulated function, it must be emphasized that response to perturbations in our internal body environment is accomplished through recruitment of many of the functions of the hypothalamus. For example, the response to a severe hemorrhage will include autonomic responses designed to elevate blood pressure (e.g., vasoconstriction of peripheral arteries and increased heart rate), hormonal responses mediated via descending autonomic fibers (e.g., adrenal secretion from the adrenal gland), and pituitary secretion of vasopressin and a number of anterior pituitary hormones. Similarly, the response to cold will include behavioral (e.g., warmth-seeking behavior, putting on warmer clothes), autonomic (e.g., vasoconstriction and shivering), and endocrine (e.g., increased secretion of thyrotropin-releasing hormone) responses.

The importance of these regulatory functions controlled by the hypothalamus cannot be understated. A lesion or deficit in the afferent circuitry, the central organization or the output of the hypothalamus for the control of these homeostatic functions, has serious consequences for the individual. As an "ancient" part of the brain present even in the lowest vertebrates, the hypothalamus has maintained its essential role in supporting bodily functions.

Bibliography

Appenzeller, O. (1982). "The Autonomic Nervous System." Elsevier, Amsterdam.

Ganten, D., and Pfaff, D. (eds.). (1986). "Morphology of the Hypothalamus and its Connections." Springer-Verlag, New York.

Gordon, C. J., and Heath, J. E. (1986). Integration and central processing in temperature regulation. *Ann. Rev. Physiol.* **48,** 595–612.

Lederis, K., and Veale, W. L. (eds.) (1978). "Current Studies of Hypothalamic Function. Vol. 1. Hormones." Karger, Basel.

Lederis, K., and Veale, W. L. (eds.) (1978). "Current Studies of Hypothalamic Function. Vol. 2. Metabolism and Behavior." Karger, Basel.

Nerozzi, D., Goodwin, F. K., and Costa, E. (eds.) (1987). Hypothalamic dysfunction in neuropsychiatric disorders. *In* "Advances in Biochemical Psychopharmacology" Vol. 43, Raven Press, New York.

Reichlin, S., Baldessarini, R. J., and Martin, J. B. (eds.) (1978). "The Hypothalamus." Raven Press, New York.

Smith, O. A., and DeVito, J. L. (1984). Central neural integration for the control of autonomic responses associated with emotion. *Ann. Rev. Neurosci.* **7,** 43–65.

Swanson, L. W., and Sawchenko, P. E. (1983). Hypothalamic integration: organization of the paraventricular and supraoptic nuclei. *Ann. Rev. Neurosci.* **6,** 269–324.

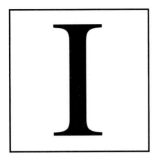

Idiotypes and Immune Networks

JOSE QUINTANS AND LINDA ZUCKERMAN, *University of Chicago*

Glossary

Ab1/Ab2 In Jerne's terminology, Ab1 is the idiotope produced in response to antigen, and Ab2 is the anti-idiotypic antibody that recognizes Ab1

Complementarity determining regions (CDR) or hypervariable regions Discrete portions of the V domains in which the variability in aminoacid sequence is most pronounced. Such variability creates the diverse paratopes found in T-cell receptors (TCR) and immunoglobulins (Ig). There are three in each immunoglobulin chain and possibly, in each chain of the TCR as well.

Epitope Antigenic determinant; the portion of an antigen complementary to the paratope

Idiotope Segment of a V domain recognized by other V domains said to be anti-idiotypic. A set of idiotopes in clonally distributed TCRs or immunoglobulins is called an *idiotype*. In general, anti-idiotopic reagents are monoclonal and anti-idiotypes polyclonal.

Immunoglobulin domain Structural building blocks consisting of a stretch of approximately 110 amino acids with a characteristic configuration called the *immunoglobulin fold*. There are two structural variants of Ig domains called V (variable) and C (constant). The V domains of TCRs and im-

munoglobulins contain variable amino acid sequences to create unique antigen-binding sites or paratopes

Internal image Idiotype or idiotope with epitope-like features

Paratope Antigen binding sites of V domains in TCRs and Igs. They are formed by the CDRs and happen to have physiochemical contours complementary to the epitope

IDIOTYPES ARE immunological markers characteristic of the antigen-binding portions of T-cell receptors (TCRs) and immunoglobulins (Igs). They are recognized by anti-idiotypic TCRs or Igs with complementary binding sites. The immune network is the operational term applied to the web of interconnected idiotypes and anti-idiotypes expressed by immunocompetent cells and their antigen-binding products.

I. Introduction: V Domains and Idiotypy

Lymphocytes are immunocompetent cells that recognize and respond to antigenic challenges. Antigen recognition is the property of specialized receptors exclusively made by lymphocytes and displayed on their membranes for interaction with antigens or their fragments. Both T and B lymphocytes express clonally distributed antigen-specific receptors (i.e., each lymphocyte and its clonal progeny carry the same receptors), approximately 10^4–10^5 per cell. This means that the coverage of the antigenic universe is partioned out among individual clonal specificities, and the immune repertoire is represented by the conglomerate of the different lymphocyte clones. Each lymphocyte clone synthesizes its unique receptors and advertises them on the cellular membrane to make them available for selection by antigen. The totality of individual specificities represents the repertoire. The human immune system consists of approximately 2×10^{12} lymphocytes expressing more than 10^7 different antigen-specific receptors. [*See* LYMPHOCYTES.]

Antigen-specific receptors are members of the Ig superfamily and consist of glycosylated polypeptide chains made of Ig domains. T lymphocytes express heterodimeric TCRs containing alpha/beta or gamma/delta chains. Each chain contains a variable (V) and a constant (C) domain, with the V domains of both chains making up the antigen-binding sites. B lymphocytes carry Igs, particularly IgM and IgD as antigen receptors. The receptors are cell surface antibody molecules with two identical heavy (H) and light (L) chains. The H chains contain one V_H and two, three or four C_H domains, whereas L chains are made of one V_L and one C_L. The antigen-binding sites of B-cell surface Igs are the same as those of their secreted forms (antibodies). They comprise one V_H and one V_L domain, so that each B-cell receptor or antibody molecule contains at least two antigen-binding sites. [*See* T-CELL RECEPTORS.]

As the name implies, V domains contain variable amino acid sequences that fold in different spatial configurations to create individual antigen-binding sites. In a restricted sense, the immune repertoire is the repertoire of V domains, because each unique amino acid sequence constitutes a different antigen-binding site(s). Certain structural features characteristic of individual V domains can be recognized by other V domains, a phenomenom known as *idiotypy*. Idiotypes are parts of V domains recognized by other V domains, which by definition are said to be anti-idiotypic. The concept revolves around the notion of V domains being both receptors and ligands and applies to cell-bound (T and B cells) receptors as well as secreted forms (antibodies). Operationally, the distinction between idiotypes and anti-idiotypes is arbitrary: An idiotype is by definition an anti-idiotype and vice versa, and there is no difference between recognizing and being recognized. Idiotopes are the individual markers of idiotypes; they can be found inside or outside the antigen-binding site. In the terminology introduced by Jerne, the antigen binding site is called the *paratope* and an antigenic determinant is the *epitope*

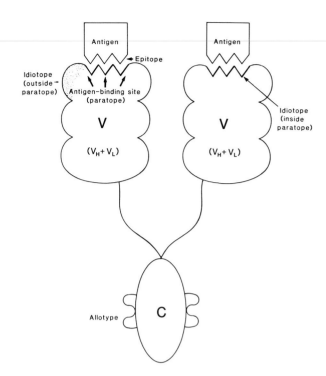

FIGURE 1 Illustration of an idealized antibody molecule to emphasize the features of V domains associated with idiotypy. V domains contain antigen-binding sites or paratopes with contours complementary to those of antigenic determinants or epitopes. Idiotopes can be found both inside and outside the paratope. An allotypic marker is depicted in the C domain.

(Fig. 1). A given V domain usually contains several idiotopes; some also found in other V domains (public idiotopes) and others that are exclusive (private idiotopes). An idiotope that happens to resemble an epitope is called an *internal image* of the antigenic determinant. Internal images can be thought of as "inside" representations of the antigenic universe that can function as surrogate antigens.

II. Historical Overview

The concept of idiotypy was first developed in the 1950s to refer to individual antigenic specificities found in antibacterial antibodies and myeloma proteins. Idiotypes were discovered in the course of experiments aimed at defining the phenomenom of antibody allotypy, a form of genetic polymorphism resulting in the presence of allelic forms of antibody molecules called *allotypes*. Allotypic markers were operationally defined by antiallotypic antisera that reacted with certain antibody classes of some but

FIGURE 2 Recognition of an allotype by the paratope of an antiallotypic antibody.

not all individuals within a species (Fig. 2). The sera containing antiantibodies recognized the slight changes in amino acid sequences of H or L chains responsible for the allotypic determinants. It was noted that certain antiantibodies lacked allotypic specificities but reacted with individual antibodies and appeared to recognize unique antigenic determinants. As noted above such determinants were called *idiotypes*. Thus the antigenic markers of antibodies fall in three major groups: isotypes characteristic of the antibody classes of all members of a species and defined by heterologous antisera; allotypes that are the allelic markers defined above detected by alloantibodies (intraspecies immunization); and idiotypes detected by antibodies lacking reactivities with iso- and/or allotypes.

Initially, it was thought that an idiotype was a unique marker of an individual antibody molecule but a broader notion of idiotypy evolved as a result of the detection of idiotypic cross-reactivity, the definition of public and private idiotopes with monoclonal anti-idiotypic reagents, and the finding of multiple idiotopic markers on antibody molecules. These concepts and the relevant terminology were extended to TCR, and their applicability to them was validated when their V + C domain structure was described in the 1980s. In contemporary

immunology, idiotopes are self-epitopes of V domains located both within and outside the paratopes of Igs and TCRs. They are operationally defined by anti-idiotypic reagents. The typical anti-idiotypic probes are antibodies because relatively little is known about idiotype-specific (anti-idiotypic) TCRs or how to use them to define idiotypic markers. Internal images are also defined operationally by their ability to act as surrogate antigens: Internal image anti-idiotypic antibodies are those that can be successfully used as immunogens to generate antiantigen immunity.

III. Structural Basis of Idiotypes and Internal Images

In general, more is known about the structure of Ig than TCR idiotypes. The structure of antibodies and their binding sites are essentially resolved, and excellent crystallographic analyses of antigen–antibody interactions have been published. In contrast, there is no detailed crystallographic data on idiotype–anti-idiotype interactions, and the molecular basis of most B-cell idiotypes remains to be elucidated. In practice, most antibody and B-cell idiotypes continue to be defined serologically.

A considerable amount of structural information on idiotypic markers has been generated in some murine idiotype systems. M104 and J558 are BALB/c myeloma proteins specific for dextran that express individual (private) (IDI) and cross-reactive (IdX) idiotopes. The IdIs have been tracked down to the third hypervariable region of their H chains: M104 and J558 have identical L chains, and their H chains differ only at the two amino acid residues (positions 100 and 101) encoded in the D region. The IdX of murine antidextran antibodies is dependent on two residues (H54 and H55) within the second hypervariable region. Another murine idiotypic system that has been extensively studied involves antiphosphorylcholine (PC) antibodies. PC is a highly immunogenic component of cell walls and membranes of many microorganisms. Most murine stains have protective anti-PC antibodies expressing the dominant T15 idiotype. T15 is an idiotype characteristic of the PC-binding myeloma TEPC 15 of BALB/c origin. In BALB/c mice, most anti-PC antibodies are T15$^+$. All T15$^+$ (as well as some T15$^-$) antibodies contain the V_H domain encoded by the V_{H1} germ line gene belonging to the V_{HS107} subfamily and a V_L domain encoded by the $V\kappa T15$ gene from the $V\kappa 22$ subfamily. Expression of the T15 idiotype is dependent on both V_H and V_L domains as it requires a dominant D region encoded sequence from V_{HS107} and the $V\kappa T15$ L chain. The exact contributions of V_H and V_L to the T15 idiotope have not yet been figured out.

Although the studies on the structural basis of internal images have not been as extensive as those on conventional idiotypes, some exciting findings have emerged. For instance, the molecular basis of antibody mimicry of the domain recognized by the platelet fibrinogen receptor was elucidated. The third hypervariable region (CDR 3) of the H chain of a monoclonal antibody specific for the fibrinogen receptor was found to contain an amino acid sequence that mimics the Arg-Gly-Asp (RGD) sequence within the fibrinogen A chain. A synthetic peptide with the 21-amino-acid-long sequence of the H chain CDR 3 specifically inhibited platelet aggregation and the binding of both the monoclonal antibody and fibrinogen to platelets. Thus the 21-amino-acid sequence comprising the CDR 3 of the H chain of a monoclonal antibody against the platelet fibrinogen receptor carries an internal image of the fibrinogen receptor–binding site. In another well-studied system, the molecular basis of idiotypic mimicry of

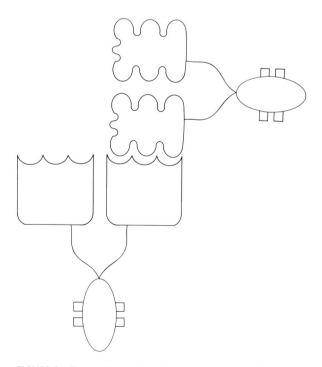

FIGURE 3   Recognition of nonbinding site–associated iditope by an anti-idiotypic antibody.

the mammalian reovirus receptor was elucidated. First a murine monoclonal antibody that recognized a viral epitope involved in the binding of type 3 reovirus to its cellular receptor was selected on the basis of its ability to inhibit the reaction between antiviral and anti(antiviral) antibodies. This particular antiviral antibody was used as an antigen to immunize mice with the expectation that it would stimulate the production of anti-idiotypic antibodies mimicking the receptor site used by the virus to infect cells. A monoclonal antibody with the desired viral mimicry was isolated and the amino acid sequences of its H and L chains determined and compared with those of reovirus type 3 hemagglutinin. Sequence homology was detected between the CDR 2 and a 15-amino-acid portion of the viral protein. A synthetic peptide based on the CDR2 was constructed and found to bind the viral receptor, down modulate its expression, and inhibit DNA synthesis in cells susceptible to reovirus infection. Thus this peptide represents an Ig internal image of a viral cell attachment site.

IV. Idiotype Network

Jerne proposed in 1974 that the immune system is a functional network resulting from paratope–idiotope interactions (Fig. 3). Initially postulated for B-cell idiotypes, the network theory is applicable to T cells and their receptors: TCR idiotypes have been successfully used to stimulate the production of anti-idiotypic T cells, and various types of idiotype-specific helper and cytotoxic effects have been reported.

Basically, the idiotype network is the closed system of self-recognizing V domains found in lymphocytes. The large repertoire of V domains contains representations of the antigenic universe in its internal images and has the intrinsic capacity to autoregulate itself through complementary anti-idiotypic interactions. Jerne's theory has been a strong stimulant of immunological research and generated much enthusiasm as well as some controversy. It is clear that anti-idiotypes with powerful immunoregulatory properties can be produced with relative ease. For instance, in the murine T15/PC system described above, we can readily observe that injections of anti-T15 idiotypic antibodies into neonatal mice cause deletion of T15 + B cells and long-lasting unresponsiveness to PC. The induction of idiotype suppression in neonates provides a striking demonstration of the immunoregulatory prowess of anti-idiotype reagents, but it is difficult to argue that it constitutes or represents a physiological procedure. It has also been reported that anti-T15 antibodies are produced naturally in the course of anti-PC responses giving credence to the notion that the antigen-induced rise in idiotypes (T15$^+$ anti-PC antibodies) stimulates the anti-idiotypic set and thus down-regulates the immune response. Also, anti-T15 antibodies expressing internal images of PC have been produced and shown to effect the expression of T15$^+$ B cells during development. Thus in experimental models such as the murine PC/T15 system, there is ample evidence supporting the existence and operation of an immunoregulatory idiotype network.

Other aspects of network regulation are more controversial. It is not clear to what extent the T- and B-cell networks interact and whether the generation of the T- and B-cell idiotypic repertoires is interdependent. Although it is well-documented that anti-idiotypic antibodies affect T cells and the presence of B cells is important for T-cell ontogeny, the physiological dependence of T cells on B-cell idiotypes is questionable. Similarly, the role of the idiotype network in autoimmunity and immunopathology is poorly defined. There are descriptions of autoanti-idiotypic antibodies in various clinical settings, but there is no conclusive evidence to suggest that dysfunctions of the idiotype network are primarily responsible for disease states.

V. Applications of Idiotypes and Internal Images

The number and variety of applications of idiotypic reagents in experimental immunology are great, and many examples can be found in Bona's book (see Bibliography). In this section we will describe some selected examples of applications relevant to clinical situations.

In principle, anti-idiotypes might be used immunotherapeutically to erradicate T- or B-cell cancers expressing receptors with unique idiotopes. The complete remission of one human B-cell lymphoma after treatment with rabbit anti-idiotype specific for the lymphoma's surface Ig has been reported, but unfortunately, other patients failed to respond to similar therapy. In part, the failure of the treatments was caused by the ability of the malig-

nant cells to modulate or modify the idiotypes expressed on their membranes.

The notion of an antibody or TCR posing as antigen has obvious applications in the development of vaccines. Anti-idiotypic antibodies expressing internal images of putative tumor-specific antigens have elicited antitumor immunity in experimental animals. Also, rodents have been successfully immunized against some bacteria and certain parasites by means of anti-idiotypes. Of greater significance for clinical trials is the vaccination of chimpanzees against hepatitis with anti-idiotype directed against antiviral antibodies. There is considerable interest in extending these studies in primates, particularly with regard to the possibility of developing a vaccine for AIDS based on anti-idiotypes.

In the area of T-cell immunology, there is potential for the application of idiotypic vaccines to prevent autoimmune diseases. There have been reports on the prevention of graft versus host disease in rats by idiotype-specific T cells. Rats have also been vaccinated against experimental allergic encephalomyelitis, an autoimmune disease mediated by T cells specific for myelin basic protein. The rats were immunized with TCR peptides corresponding to idiotypic determinants found in disease-inducing T cells specific for the encephalitogenic epitopes of myelin basic protein. The TCR idiotypes presumably stimulate the production of anti-idiotypic T cells that mediate the resistance to the induction of the disease. [*See* AUTOIMMUNE DISEASE.]

Other applications of internal images are in the areas of immune regulation and endocrinology. The field of hormone receptor mimicry contains a variety of examples of antibodies posing as hormones, neutrotransmitters, and in some instances, lymphokines. Anti-idiotopes can even mimic steroid hormones: Mice immunized with rabbit anti-idiotypic antibodies specific for mouse monoclonal antiprogesterone antibody made antiprogesterone antibodies that reduce their fertility. Presumably, the anti-idiotypic antibody contains a progesterone-like epitope (i.e., an internal image of the steroid hormone that induces the production of antibodies specific for progesterone that block pregnancy).

VI. Future Prospects

Several areas pertaining to idiotypy hold considerable promise for continuing study. At the basic level, immunologists would like to know the structural basis of idiotypy, the precise roles of T- and B-cell idiotypes in the generation of the immune repertoire, and how the immune network affects immunoregulatory pathways. The acquisition of such basic knowledge will provide the means to develop internal image vaccines rationally and to predict and possibly control the effects of anti-idiotype immunotherapy. As noted above, encouraging findings with experimental animals have already been reported. However, there still remain considerable obstacles to the development of internal image vaccines for clinical use, and most experts in network theory would probably not predict that the cures for AIDS and cancer will turn out to be anti-idiotypic vaccines.

Bibliography

Bona, C. A., ed. (1988). Biological applications of anti-idiotypes. CRC Press, Boca Raton, Florida.

Burdette, S., and Schwartz, R. S. (1987). Current concepts: Idiotypes and idiotypic networks. *N. Engl. J. Med.* **317,** 219.

Davie, J. M., Seiden, M. V., Greenspan, N. S., Lutz, Ch. T., Bartholow, T. L., and Clevinger, L. B. L. (1986). Structural correlates of idiotypes. *Annu. Rev. Immunol.* **4,** 147.

Gaulton, G. N., and Greene, M. I. (1986). Idiotypic mimicry of biological receptors. *Annu. Rev. Immunol.* **4,** 253.

Jerne, N. K. (1984). Idiotypic networks and other preconceived ideas. *Immunol. Rev.* **79,** 5.

Lee, V. K., and Hellstrom, K. E. (1988). Anti-idiotypes and cancer therapy. *Adv. Drug Delivery Rev.* **2,** 271.

Quintans, J. (1989). Cellular competition and the promotion of T15 to idiotypic dominance. *Immunol. Rev.* **110,** 119.

Rajewski, K., and Takemori, T. (1983). Genetics, expression and function of idiotypes. *Annu. Rev. Immunol.* **1,** 569.

Immune Surveillance

HANS WIGZELL, *Karolinska Institute, Stockholm*

Glossary

Homograft reaction Rejection reaction against transplanted tissue from a foreign individual where T lymphocytes are the dominating aggressive cells
Immunogenicity Capacity of a substance or cell to induce a specific immune reaction against it
Natural killer cells Aggressive cells with an inborn tendency to kill malignant cells more readily than normal cells
T lymphocytes White blood cells requiring the thymus to mature properly; they constitute a major specific protective force against many infectious diseases

IMMUNE SURVEILLANCE (IS) is here defined as a mechanism through which arising "potentially dangerous mutant cells" are constantly eliminated and where the mechanism(s) is "of immunological character." The need for IS was considered to constitute the major reason behind the development of the thymus-dependent part of the immune system, with the capacity to eliminate *in situ* malignant cells before they had developed into minitumors. IS would thus have the form of a natural, spontaneous immune reactivity already pre-existing before the appearance of the tumor cells. This should then be put in contrast to other, more conventional forms of specific immune reactions, which may become induced against an already established, growing tumor. Of course, in the latter situation, IS clearly has already shown itself as a failure with regard to function.

I. Concept of IS

The word immune means exempt. It is derived from past observations of a severe infectious disease that swept through a district and caused many people to become sick, with several even dying; however, when the same disease returned at a later time people who had survived the first epidemic were now exempt, or immune. When the term immune surveillance is used, however, it is not used in the context of protection against infectious diseases but rather as a constant guard against invaders from within (i.e., against cancer cells). Several scientists have suggested that a primary function of the immune system is to protect the multicellular society of the individual against its "asocial" members—the malignant tumor cells.

The first theory along these lines was probably formulated by the famous German immunologist Paul Ehrlich in 1909. He believed that malignant cells did arise throughout the life of the individual but that the vast majority of such cells would be eliminated by the immune system constantly surveilling the tissues against newly arising tumor cells. Accordingly, the appearance of tumors would more or less be a mistake, by which the cancer cells somehow managed to "sneak" through this constant watch-guard mechanism of IS. [*See* IMMUNE SYSTEM.]

More recent formulations of IS stem from researchers who, in the later part of the 1950s, put IS into a more modern concept of cellular immunology. As the strength of the homograft reaction at that time was just being cognized, this particular form of immune reactivity was proposed to have as its prime biological basis the elimination of malignant cells. When T lymphocytes later became known as the major force for the specific, cell-mediated immune reactions causing rejection of foreign grafts, T cells were considered the most likely candidates for the effector cells of IS. Still later, when natural killer (NK) cells were discovered, with their significant tendency to preferentially react against malignant as compared with normal targets, they also were indicated as potential effectors of IS. [*See* LYMPHOCYTES; NATURAL KILLER AND OTHER EFFECTOR CELLS.]

II. Attempts to Validate IS

The above concept of IS against cancer is intellectually highly attractive; however, a number of situations disprove the general validity of the hypothesis. A major assumption when considering the existence of IS is that the existence of a deficiency in the proposed immune effector mechanism for IS would automatically lead to an increased risk of tumor development. Superficially, significant support would seem to exist for IS using the above reasoning. If one assumes that T lymphocytes are the performers of IS, literature abounds with articles showing that immunodeficient individuals do indeed express a significantly increased risk for tumor development. It was initially shown for individuals with genetically defined immunodeficiency disorders being followed by similar observations in transplantation patients with drug-induced immunosuppression. When the acquired immunodeficiency syndrome epidemic began, the HIV-infected individuals, upon entering into immunodeficiency, displayed a most dramatic increase in malignant disorders, reaching cancer frequencies in the order of 25% or more in some cohorts.

These data would initially seem to support the existence of IS. However, a major problem with the above findings in relation to IS is the type of cancers arising. If IS is of general significance, one would expect that all forms of tumors would be increased in immunosuppressed conditions; however, this is not the case (see Table I). The most common forms of cancer in humans, such as mammary, pulmonary, gastrointestinal, and prostatic tumors a.s.o. do not appear at increased frequencies in immunodeficient individuals. In contrast, only a limited number malignancy types dominate among the tumors appearing in immunosuppressed humans. Among these tumors are lymphomas, skin tumors, and a tumor type normally very rare in Caucasian groups, namely Kaposi's sarcoma. Undoubtedly, the immunosuppression is intimately linked to the

TABLE I Increased Appearance of Certain Tumor Types[a] in Immunosuppressed Individuals

	Relative tumor incidence (control = 1)[b]		
	Kaposi's sarcoma	B lymphomas	Skin tumors
Organ-grafted	450 (23)	39 (36)	29
Acquired immunodeficiency syndrome	>10,000	>40	increased

[a] Tumors such as pulmonary, mammary, prostatic, and gastrointestinal show no increase in immunosuppressed individuals.
[b] Figures within parentheses indicate in months of tumor appearance after induction of immunosuppression.

increase of these tumor types. For instance, the depth of immunosuppression induced by drug treatment in organ-grafted individuals is directly positively correlated with the increase in percentage of individuals developing malignancies. However, these normally rare tumors occurring at high frequency in immunosuppressed individuals may not be due to a faulty IS. The lymphomas arise, for instance, within an immune system afflicted by damage. Many other systems show that chronic inflammation and repair per se in a tissue may serve as promoters for tumor development. Likewise, the lymphomas that appear are of a type called Burkitt's lymphomas, a tumor intimately associated with the Epstein–Barr virus, known to start replicating in immunodeficient individuals. For Kaposi's sarcoma, a similar situation does prevail. Evidence suggests that the immunodeficiency situations linked with this type of tumor cause the production of certain growth factors, one or more of which act as efficient promoters of the cells composing Kaposi's sarcoma. Again, this would suggest that the damage to the immune system is releasing tumor-promoting factors (viruses, growth factors) that are responsible for the increase in malignancies, with no evidence for IS playing a discernible role. [*See* LYMPHOMA.]

The third type of tumor with increased frequency in immunodeficient individuals, skin epithelial tumors of squamous or basal cell type, is the only one for which a plausible case for IS containing the carcinoma cells exists. Ultraviolet (UV) radiation at high doses is known to be immunosuppressive and can, at certain wavelengths, also be shown to induce tumors with high immunogenicity in inbred mice. UV radiation in humans likely will act in a similar manner. The fact that UV-induced tumors in mice are unusually immunogenic can be used to make a case that the ubiquitous UV radiation from the sun creates a vital need for efficient protection against mutagenized premalignant cells in the skin, perhaps via IS; however, this argument needs more proof to become a fact.

In animals, the results with regard to consequences of T-cell deficiency are similar to those in humans. In most circumstances, mice lacking thymus and T lymphocytes have normal frequencies of tumors, unless the mouse colony is infected with tumor-inducing viruses. In such a case, the T cell-deficient mice develop tumors in a very high percentage; however, this does not prove the existence of IS but, rather, a lack of conventional immune protection against a tumorigenic virus, allowing it to cause a more severe infection, and, thus, a higher probability to cause cancer.

In summary, immunodeficiencies involving T lymphocytes thus fail to reveal the existence of IS against tumor cells. But the more recently discovered cell type NK cells—with their tendency toward "spontaneous" killing of preferentially malignant cells, could serve as an alternative candidate taking care of IS. Such cells can clearly be shown in experimental animal systems to act *in vivo* against transplanted tumor cells. However, in human beings with relatively select NK cell deficiencies such as the rare Chediak–Higashi's disease no convincing evidence indicates a general increase in malignancies. Some increase in lymphoid malignancies have been recorded, but as this disease involves a damage within the bone marrow cells, this again may have secondary reasons. Likewise, in the animal systems no significant support for IS via NK cells has been forthcoming. Therefore, one can conclude that immunodeficiency disorders involving either T or NK cells have so far failed to provide evidence for the existence of any significant IS against tumor cells.

III. Does IS against Tumors Exist?

From the above information, one can make certain comparatively firm conclusions. All severe immunosuppression induced by various ways (genetic, drug-induced, acquired via viral infection) result in an increase in cancer disease in the afflicted individuals. However, the identical restricted types of tumors appear (lymphomas, skin tumors, Kaposi's sarcoma) at higher frequencies, whereas all common types of cancer appear at normal rates. The tumors that do arise at enhanced rates in these immunosuppressed individuals may all do so for reasons unrelated to a failure of IS.

At face value, the above data would seem like the final elimination of the existence of an IS mechanism constantly protecting the individual against the appearance of tumors. It may be wise to postpone this final disposal, at least for some time. It is normally considered that the immune system is composed of cells from the bone marrow with a division of labor between subsets of cells. This definition may be too narrow if one returns to the original meaning of the word immune (i.e., exempt). Thus, cell functions may exist in the body that are yet to be defined with "immune" consequences and where such cells are not necessarily derived from the bone marrow. For instance, in the body a situation exists where tumor cells may reside for prolonged periods of time—years—without starting to grow. This is most easily shown in the situations of metastasis coming up a decade or more after the primary tumor was removed. One tumor type comparatively frequently acting in such a manner is the melanoma. Whether normal cells in the vicinity of the "dormant" tumor cells keep the cancer cells from growing or there are inborn restraints in the tumor cells that may be released with time is unknown. From the clinic, a significant number of reports indicate the same message. A patient operated, for example, in the leg for a primary melanoma some 15 yr later develops mammary carcinoma. After surgery of the carcinoma, local X-irradiation is given. A few months after the X-ray hundreds of melanoma metastases can be seen growing locally in the skin of the irradiated area. The X-irradiation has undoubtedly activated dormant melanoma cells and/or alternatively eliminated some restraining element(s) provided for by the local tissue cells. If the latter is true, is it justified to call this IS despite the fact that it would not cause the death of the tumor cells but merely a failure to grow?

Studies on the possible existence of IS against tumors has been a fruitful way to explore how the immune system is functioning in general, despite the failure to provide its existence in a more general manner. New findings or concepts as to how the immune system may be functioning will surely be tested in the future researching the question of IS.

Bibliography

Burnet, F. M. (1957). Cancer—A biological approach. *Br. Med. J.* **1,** 779.

Burnet, F. M. (1970). The concept of immunological surveillance. *Progr. Exp. Tumor Res.* **13,** 1.

Ehrlich, P. (1957). "The collected papers of Paul Ehrlich". (F. Himmelweit, ed.). Pergamon Press, Oxford.

Old, L. J. (1989). Chapter IV Defense: Structural basis for tumor cell recognition by the immune system. *In* "Progress in Immunology VII" (F. Melcher, ed.). Springer Verlag, Berlin.

Penn, I. (1986). Cancer is a complication of severe immunosuppression. *Surg. Gynecol. Obstet.* **162,** 603.

Thomas, L. (1959). Reactions to homologous tissue antigens in relation to hypersensitivity. *In* "Cellular and Humoral Aspects of the Hypersensitive States" (F. S. Lawrence, ed.), p. 529. Harper, New York.

Immune System

ANTONIO COUTINHO, *Institut Pasteur, Paris*

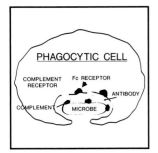

Glossary

Immune response Bodily response to an antigen that involves the interaction with lymphocytes to induce the formation of antibodies and lymphocytes capable of reacting with it and rendering it harmless

Immunoglobulins Any of the vertebrate serum proteins that are made up of light chains and heavy chains usually linked by disulfide bonds and include all known antibodies

Lymphocytes Agranular leukocyte formed primarily in lymphoid tissue; occurs as the principal cell type of lymph and composes 20–30% of the blood leukocytes

I. Biology and Paradigms

THE IMMUNE SYSTEM (IS) provides vertebrates with the ability to identify molecular shapes and react to variations in their concentrations, by modifying the production of its own components. The system is thus homeostatic for the "shape–space" in the organism and exquisitely adaptive. It shows properties usually considered cognitive, for it is capable of learning and of keeping a life-long "memory" of previous molecular experiences of the individual. These are also described as the ability to discriminate the treatment to give to different molecular shapes: coexistence at equilibrium (tolerance) or elimination from the body (immunity). Specific molecular recognition is not unique to the IS, but an essential property throughout the biological world. Molecular identification in all other systems, however, is limited to a narrow range of shapes (e.g., hormones, growth factors and respective receptors, or cell adhesion molecules), whereas the IS can principally identify any molecular shape, whatsoever. Three levels of organization can be distinguished in the IS, which seem necessary to account for its characteristics: (1) structural, a very large diversity in the shapes of its components; (2) dynamic, the rates of production of each of those shapes can be varied by many orders of magnitude; and (3) metadynamic, novel shapes are produced throughout life, which become available to be selected into the system's dynamics.

The IS is composed by a set of soluble molecules (immunoglobulins, Igs) and cells (lymphocytes) that circulate in blood and lymph through most tissues in the body, and, thus, have access to most molecular species in the individual. These, collectively named "antigens", are either produced by the organism itself or penetrated into the internal environment from the outside. The first thing the IS has to learn, therefore, is the molecular environment where it exists; the very large number of shapes that constitute the organisms itself. Interestingly, many of these shapes are components of the system itself, that is, the IS significantly contributes to the composition of the molecular environment where it exists and determines its own structure and dynamics. This is the reason why modern immunology ascribes some degree of autonomy to the immune activities, and the basis for the recursive properties of the IS that may well contribute to its ability to learn and remember. Thus, immune activities necessarily alter the composition of self by producing novel molecular shapes or by drastically altering their concentrations in the organism. Because lymphocytes react to alterations in their molecular environment, it follows that immune activities contribute to determine further dynamics in the system, thereby accounting for some stability in its composition.

Recursivity might be particularly relevant in the context of self–nonself discrimination, an expression that underlines the ability of normal IS to be tolerant to the sets of shapes in which it has developed, while eliminating from the organism other shapes falling outside that domain. The immunological self, however, is not a genetic or chemical listing of molecules, each existing within a fixed range of concentrations, but rather all shapes that are available in the organism to the IS as it develops to competence, even if encoded by the genome of a virus or that of an heterozygotic twin sharing the blood circulation. Shapes present when the system's organization is emerging, impinge in that organization, leading to a structure and dynamics of equilibrium, that is characteristic of tolerance. In contrast, shapes that do not participate in building the self-defining organization of the system elicit, when recognized, the typical desequilibrium dynamics of immune responses. Self and foreign, therefore, are not fixed categories, but qualitatively and quantitatively defined by the IS in its organization and ongoing operation. (Abnormally operating IS indeed treat the genetic or chemical self as foreign, and foreign materials as self.) Self–nonself discrimination is the manifestation of a global behavior of the IS that requires learning and thus reflects the molecular conditions existing in development, as well as the point of view of the system.

If the process of self-learning is defective or the corresponding memory is not accurate, immune activities can provoke disfunction of other body systems, by damaging tissues or organs, by eliminating self molecules, or by abnormally neutralizing or enhancing their biological activities. Paramount examples of such autoimmune diseases are juvenile diabetes, thyroid disfunctions, many forms of rheumatisms and arthritis, multiple sclerosis, some forms of anemia and hemophilia, hepatitis, and carditis or kidney destruction. Some of the so-called autoimmune diseases, however, may not always result from a primary malfunction of the IS, but represent its normal reaction to alterations in the molecular environment, or the loss of its adaptive

capacities in the presence of extreme conditions. Moreover, it is expected that the IS will adapt to perturbations in any other biological system in the organism that are accompanied by significant alterations in the qualitative or quantitative molecular composition of self. If this notion brings an immune component to most diseases, it may also open the possibility of directing immune activities to the correction of many disfunctions that are not immunological at all. Thus, the unique wealth of shapes that can be produced in the IS and the flexibility of their rates of production are such, that it is in principal possible to neutralize or mimic the active site of any other molecule which serves relevant functions in intercellular communications.

Once the IS reaches maturity and the molecular self's identity has been established, novel shapes are usually not tolerated, and learning essentially becomes the ability to identify and eliminate foreign shapes. Immune responses are reactions to the presence of immunologically foreign materials. Microorganisms which infect vertebrates, self-replicate and actively produce (over short periods of time) large amounts of molecules that are new to the host. Infections by virus, bacteria, and various types of parasites provide, therefore, for drastic and rapid alterations in the composition of the molecular environment, thereby precipitating adaptive reactions from the IS with similar characteristics of kinetics and magnitude. Typically, immune responses develop rapidly (within a few days), and predominantly represent the selective amplification of lymphocyte receptors and Igs with shapes that are complementary (specific) to those identified in the foreign antigen. Their characteristics of specificity and high magnitude can be assessed by measuring circulating concentrations of antigen-binding Igs (antibodies), or the numbers of lymphocytes in blood that recognize and react with the antigen. In both cases, little or no increases are found if a different antigen is used in those assays. This provides the basis for immunodiagnosis, for it becomes possible to infer, with some degree of certainty, whether or not an individual has been infected by (and made an immune response to) a given microorganism, or which microorganism is actually causing a given infection. Once the concentration of the new (foreign) shapes has been compensated by the rapid rise in concentrations of complementary antibody shapes, immune responses start a rapid drop. Typical immune responses in which a small subset of lymphocytes and Igs are amplified by many orders of magnitude over a few days or weeks, account for desequilibrium dynamics in the IS.

If the antigen is actually eliminated from the organism, the specific immune response decays and, with time, the serum concentrations of specific antibodies will not be much higher than before immunization. Individuals that have once responded to a foreign antigen, however, are "primed" for future contacts with the same microorganisms. Immune responses induced by reexposure to microorganisms or their products, are in most cases more prompt, of higher magnitude and precision (specificity) than the corresponding primary reactions. These characteristics of secondary responses, that also represent some form of immunological memory, explain the specific immunity of individuals surviving a primary infection, and provide the basis for vaccination. The great wealth of shapes in the IS—making it possible to vaccinate and protect from diseases caused by prevalent pathogens—also allows for a novel vaccination strategy, which dispenses of the microorganisms and of their products. Thus, the unique shape of an antibody that is complementary to a given microbial antigen, can be used to provoke the production of other immunoglobulins that are complementary to it. In broad terms, the shapes of the latter are "equivalent" to that of the antigen, for they are both complementary to the same antibody. Such immunoglobulins can, therefore, be used as surrogates of the antigen to vaccinate against microorganisms.

Other than pathogenic microorganisms, many foreign materials penetrate the organism, most often through mucosal surfaces primarily by the digestive tract. Most external molecular shapes to which vertebrates are exposed are actually necessary to their survival, and not potentially pathogenic. Many of these, however, only enter the body (circulation) after being appropriately processed at sites that may be specially designed to reduce their impact in the IS. Furthermore, most of these materials are encountered often and they are not self-replicating shapes. These conditions of dose, time kinetics, and repetition of exposure, are likely to be closer to the equilibrium dynamics characteristic of immune tolerance. There is indeed evidence for the establishment of "oral tolerance" in adult animals, but other findings show the induction of protective immunity by the same routes of contact. Unfortunately, experimental Immunology has predominantly used protocols of immunization that do not address physiological antigenic contacts with the external environment in normal animals, and this area would require more attention in the future.

Identification of shapes requires interactions between the immune components and motifs on the molecules to be identified, be these a hormone, a bacterial toxin, or a viral protein—permitting virus entry to the host cells. *Per se*, such interactions may lead to neutralization of the target molecules, sequestering them from other potential interactions (their active sites may be occupied, their accessibility to the respective receptors reduced, or their circulation and tissue diffusion impaired). In addition, immune interactions with target molecules or living microbes and cells, may lead to their destruction and elimination or rejection from the body. This is possible because the molecular basis of shape identification is coupled to a number of other molecular and cellular "effector" mechanisms, often recruited from evolutionarily more primitive systems (e.g., phagocytosis and digestion, complement and other acute phase reactants, cell-associated proteases). Some of these are autocatalytic chains of molecular processes that lead to greater amplification of the initial reactions. Specialized mechanisms have also been evolved to couple some immune effector functions (e.g., hormonelike peptides secreted by lymphocytes once activated) to the process of inflammation, again facilitating recruitment of other cell types to the site of specific interaction and, ultimately, the removal of the specifically identified materials. Often, such effector mechanisms need no Igs or lymphocytes to react directly to materials that are very foreign to the vertebrate composition (e.g., endotoxins from bacteria), demonstrating the preservation in vertebrates of more primitive forms of protective molecular identification.

Loss of shape–space homeostatic functions can also be manifested in the reaction of the IS to foreign materials, either by failing to promote their elimination from the organism or by exaggerated reactivity to them, even if present in minute concentrations. Both such immunodeficiencies and allergies can be life-threatening, in the first case by colonization of the organism with pathogenic microorganisms and in the second, by local reactions that compromise vital functions (e.g., asthma).

The property of the IS to neutralize foreign toxins and participate in the elimination of infectious agents is at the origin of its very designation, and historically central to immunology. The discipline was born as a branch of microbiology and owes much of its development to the medical interest in antiinfectious therapy and prophylaxis. The early, spectacular success of vaccines in protecting against some viruses, bacterias, or their products has profoundly marked the field. Over the recent decades, however, few good vaccines have been produced (none available against parasites), and the effectiveness of some that are widely used (e.g., BCG in tuberculosis) is not universally accepted. If, in most cases, the solution to the prevalence of such diseases is socioeconomic, they continue to afflict more than half of the world's population. Naturally, therefore, this continues to represent an active area of immunological research. Only recently, has the center of interests in immunology started shifting to immunity–to–self, as many diseases of unknown ethiopathogeny turned out to be autoimmune. The reduction of all immune activities to the defence against pathogenic microorganisms continues, nevertheless, to be paradigmatic. This view is justified by the observation that immunodeficient individuals die of infection, but it should be recalled that immunodeficiencies are also often accompanied by autoimmune syndromes. There is no doubt that the IS provides vertebrates with a particularly elaborated and effective antiinfectious protection, but it may be asked whether this is the only function for such an extremely complex system. Other forms of immunity that do not involve the IS (that is, Igs and lymphocytes) are known in vertebrates, which actually live at equilibrium with a variety of microorganisms that contribute to their general economy. Furthermore, the variety and effectiveness of antimicrobial defence systems already deployed in invertebrates (many of which were retained in evolution), could suggest that shape–space homeostasis has afforded to vertebrates major evolutionary advantages in functions other than protection from infections. The fact that the Ig gene superfamily has evolved from systems of recognition and communication among self-cellular components is also used in support of these notions. They suggest that immune recognition, used in some other function, has been redeployed for defence purposes, by coupling to such an universal process of molecular identification previously existing "effector" mechanisms of elimination. The ability of IS to ensure homeostasis in the space of shapes, together with the notion that tolerance–to–self represents a physiological equilibrium, has lead to the speculation that immune activities contribute to integrate (and possibly regulate) other biological systems into a coherent organism, by continuously ascertaining the normality of the self's molecular composition. These alternative teleologic proposals, however, have received little attention and no empirical support. Putative roles of immune activities in development, in the removal of normally decaying molecules and cells, or in the physiology of other body systems, are not ascertained by observations in severely immunodefective individuals. Such functions of the IS, if existing at all, would not be essential but simply contribute a higher robustness to those systems. If speculative, these ideas are, nevertheless, worth considering because they could open wide perspectives in biology and medicine. In summary, the IS contributes to establish, define, and maintain the molecular identity of the individual.

II. Diversity

The properties of the IS owe much to the structural diversity of its components, which is unique in the biological world, and provides a demarcation for what to consider as an IS: the set of diverse molecules carrying variable(V)-regions and the cells that produce them. As their designation already indicates, V-regions are the segments of the polypeptide chains composing Igs and lymphocyte receptors which differ among such molecules and, thus, provide for the specificity in the interactions with ligands. They constitute the wealth of shapes in the IS, the "constant" regions of the same molecules or any other structural component of lymphocytes showing only a handful of variants, as in any other system. The IS contains the greatest structural diversity in the world of biology. The genome of a vertebrate encodes some 10^4–10^5 different proteins, whereas the IS alone can potentially produce different V-regions up to 10^{12} or even higher orders of magnitude (it actually produces hundreds or thousands of millions). As discussed below, many such different sequences are generated by random nucleotide excisions and additions, making the potential upper limit of the structural immune diversity very high indeed. If this is not surprising for a system that has to do with identification of all possible molecular shapes, it is interesting to note that most of these are actually the system's own components. The genetic mechanisms generating this large diversity have been at the heart of immunology for many years, and were essentially solved by the end of the 1970s.

The families of genes encoding V-regions of Igs and lymphocyte receptors, as well as the mechanisms leading to their expression seem to have evolved for generating diversity; for randomly producing novel, evolutionarily unselected V-region shapes obviously within the possibilities of a conserved overall structure of the molecules. The basic strategy adopted in mammals has been compared to language, which again is "openended" because its structure allows for the generation of an infinite number of sentences with a limited number of words. Mammals such as man, keep in the germline relatively large groups of homologous gene segments (from a few to several hundred genes in each group), but none of these can be expressed alone. The structure of these loci is such that expression of a V-region requires somatic recombination in a single cell of gene segments from five different groups of genes. Since the process of assorting one segment from each group seems to be random, a large number of combinations is possible from a much smaller number of building blocks (e.g., mouse Igs are encoded by combinatorial assortment of gene segments from groups containing, respectively, 300, 12, 4, 200, and 4 members, a total of 500 genes; the number of different V-regions created by random assortment of those genes is in the order of $300 \times 12 \times 4 \times 200 \times 4$, i.e., over 10^7). Another set of enzymatic mechanisms superimposed on the former genetic strategy, contribute to a much larger diversity of unique V-regions, and reinforce the argument for an evolutionary drive for novel, unselected shapes. Thus, in contrast with most other biological systems, which the rule is conservation in the transmission of germ-line sequences and fidelity in their somatic expression, the genetic processes leading to the expression of V-regions are error-prone. The mechanisms of DNA rearrangement that assemble those gene segments utilize enzymes that amputate germ-line DNA sequences at the recombining ends, or generate new sequences without templates at the sites of recombination. Such joining diversity can, therefore, create an almost unlimited number of new random sequences. Such somatic processes of generation of diversity appear as relatively more important than the evolution of germ-lines, in particular the polygenism at each of those gene clusters. Thus, ISs seem to do reasonably well even when considerable sections of germ-line V-genes are deleted (as in mutants found in nature) or their

expression prevented (as can be done in the laboratory). This observation makes it difficult to consider evolutionary pressures for the selection of particular sets of germ-line genes, each of which can only be possibly selected together with another four, randomly chosen genes, most often after being altered by random processes of nucleotide excision and addition. Certainly, polygenism contributes to ensure immune diversity, and polymorphism of those genes to enlarge it in the species. On the other hand, such forms of genetic amplification and variability are to be expected from a neutral drift of homologous gene clusters. It seems as though evolution has driven at the generation of as many V-regions as possible, rather much independently of the precise shape of each one. For this view, evolutionary selection provided the mechanisms for potential completeness in shape–space, but this comes forth through the predominance of an ontogenetic process of production and selection of newly made shapes, in the construction of individual ISs. In other words, the construction of immune repertoires is essentially epigenetic; once those mechanisms are derived in evolution, each individual can, immunologically speaking, cope with any environment or universe of shapes, for its IS can start evolution a new with a complete wealth of possibilities. There are other current views on this problem, based on the fact that, for each V-region, two of those 5 gene segments encode subregions that constitute the larger part of the molecule. It is argued that some reactivities are ensured by such areas, regardless of the rest of the V-region. It is, therefore, possible that the corresponding sets of genes, precisely the most numerous, have been evolutionarily selected on the basis of important reactivities, be those directed to microbial pathogens or to self structures, irrespective of whatever else can be constructed with them.

A fundamental consequence of the strategy evolved for the generation of diversity, is the fact that the process of genetic recombination necessary for expressing a V-region, is irreversible and, for each haploid genome, unique. It follows that, in addition to ensuring the low probability that different cells produce precisely the same V-region, those genetic mechanisms also exclude that each lymphocyte expresses the whole variety of possibilities. That is, each lymphocyte expresses only one V-region and is essentially unique.

To all these "anticipatory" mechanisms directed at generating diversity without an apparent purpose of fitting evolutionarily selected shapes, we must add another source of diversity brought about by somatic mutation of antibody V-region genes. Exceptionally high rates of mutation (up to 10^{-3}/base pair/cell generation) have indeed been estimated to occur in rearranged, already expressed Ig V-genes, if the corresponding lymphocyte engages in immune responses in specialized sites of the secondary lymphoid tissues. While such rates of mutation will lead to loss of expression or specificity in most of the mutants, this mechanism can contribute, and it apparently does, to provide a multiplicity of variants from the original specificity which can be selected for the fine tuning of the responses.

The genesis of immune diversity shows powerful evolutionary driving forces to produce the largest possible multiplicity of different shapes, even if the vast majority of them will never be useful to the individual. Thus, each lymphocyte can only express a single V-region, and the number of lymphocytes being limited, an individual can express but a very small fraction of the full diversity potential. Furthermore, many (most?) of the lymphocytes in an individual are actually never used, for the V-regions they express do not happen to be necessary in self assertion or in immune responses. Correspondingly, these notions suggest an enormous amount of cellular waste, which is in agreement with the population dynamics of lymphocytes, and impose either the requirement for continuous selection of V-region repertoires or that the expressed diversity is simply the product of random events of gene recombination.

III. Degeneracy

In addition to this extreme wealth of shapes, the ability of IS to identify molecules also derives from the nature of the physicochemical reactions mediated by its elements: they are reversible, based on noncovalent weak forces, allowing for some degree of imprecision. Generally, these reactions are not biunivocal. The same V-region may bind, with a wide range of molecular fitness or precision (affinity), to a variety of structurally different molecules. Conversely, the same molecular shape may interact with a large number of distinct V-regions. The interacting areas on the surface of V-regions and of the ligands that they actually contact (usually called paratope or combining site, and epitope or antigenic determinant, respectively) are not fixed for each

molecule, but vary for each specific interaction. Combining sites and antigenic sites are mutually and circularly defined for each case (i.e., they are attributes defined for the interacting molecules by the interaction itself). The larger the antigenic molecule is, the more possibilities for interactions (epitopes) it contains. In contrast to the initial models derived for antibody interactions with small molecules, which suggested "key and lock" structures, it has become clear that antibody reactions with proteins involve large areas of contact (more than a dozen aminoacid residues in each molecule, and equally large numbers of hydrogen bonds). Clearly, this accounts for a great precision (affinity) in some interactions, but many small variants of such large interacting sites in either the antigen or the antibody, will still allow for specific recognition. There is, therefore, a great degeneracy and redundancy in immune recognition which enables the IS to recognize all molecular shapes, even those that have never existed in nature. This property, known as "completeness" or "openendedness", is essential to the biology of the IS and owes to diversity and degeneracy alike. Thus, IS with limited diversity, such as those of small animals (a tadpole can only produce a maximum of 10^4–10^5 different antibodies), is nevertheless complete and can "recognize" any antigen whatsoever, just like the IS containing a much larger diversity (a man can produce 10^{10}–10^{12} different antibodies). The smaller the immune diversity, however, the less precise recognition will be, because of limitation in the choices of the best fit for any given shape. Large immune systems in higher vertebrates can produce antibodies with extremely high affinity to nearly any antigen, if one allows for appropriate selection of rare V-regions and time for their amplification. Specificity is indeed often considered in the definition of Immunology. For each random shape, there will always be more low-than high-affinity V-regions, but the latter can be selected for. The fundamental problem here is that we have little information on the thresholds of affinity that are compatible with function in physiological conditions. Moreover, these are likely to vary with the nature of the function considered (binding and neutralization, complement activation, fixation of complexes on cellular receptors), as well as with the physical nature of the antigen (molecular flexibility, multiplicity of determinants) and of the V-region (valence, cell-bound or free).

Completeness of immune recognition poses interesting evolutionary questions. Thus, the marked difference in generation times between vertebrate hosts and microorganisms that infect them has led to the argument that, to be protective, ISs had to be prometheic. In some way, the system must be aware of future possibilities for novel microbial shapes that would necessarily arise from a much faster evolution. Every influenza epidemic is a demonstration of the argument: mutant virus strike even individuals previously immunized by another type of the same virus, making it *a fortiori* useless to select into the species germ-line those genes that helped good responders to survive infection with a different mutant a few decades before. Obviously, completeness accounts for these prometheic characteristics because no shapes fall outside the domain of the recognizable. The conservation of a complete repertoire in evolution is another problem, briefly discussed above.

IV. Individuality

A recent development adds another twist to this question on the universe of recognizable shapes. Thus, interactions of protein "antigens" with immunoglobulins are obviously limited to their solvent-exposed surfaces. Actually, the epitopes on a protein molecule are often not a contiguous sequence of amino acids, but are instead composed after the tertiary or quaternary structure of the protein, by stretches lying far apart in the polypeptide chain(s). These are called conformational epitopes and are lost when the protein is denatured. The interior of the native molecule, therefore, remains unaccessible to Ig recognition. Over the last ten years, however, it has become clear that all nucleated cells in the organism can break down or process their own proteins as well as those endocytosed from the outside, and present on their surfaces relatively small peptide fragments of the original proteins. These can then be recognized by a class of lymphocytes (the T cells), which do not produce Igs but express instead a different type of membrane bound receptor with V-regions. T-cell receptors (TCR) are structurally very similar to Igs, and the respective structural genes, although distinct, show essentially the same organization such that an equivalent large diversity of V-regions exists in both TCR and Ig. The ability of TCR to recognize linear short sequences of protein fragments provides the IS with a different "language" for molecular identification. Moreover, the T-cell recognition system gives it access, not only to shapes inside of molecules, but also to the molecular composition of all cells in the body. This process seems to be fundamental in the establishment of self identity, and thus in determining the ability of individuals to discriminate the treatment to give to different shapes, namely an immune response and elimination or an equilibrium dynamics and tolerance. It deserves a few comments akin to both immune degeneracy and learning.

Proteins hidden inside cells can be seen by the IS, because short peptides thereof are presented on the cell surface. A single type of molecules in the cell seems to be competent to carry such fragments from the inside to the surface, and to present them to T cells. These are proteins, encoded by a cluster of genes that evolved together from the most primitive vertebrates to man, which are expressed by all nucleated cells in the body, and which show the highest polymorphism of all loci in the genome: in a given species, there are dozens of structural variants of each gene in this complex. Since the complex contains up to ten or so genes that are codominantly expressed, that enormous polymorphism makes it very unlikely that two unrelated individuals carry exactly the same combination of alleles (in humans this probability is $<10^{-6}$). The shapes of such molecules expressed by every cell in the organism provide, therefore, the most exquisite marker for the uniqueness of the individual in the universe of shapes (i.e., they can serve as unique components in the definition of the self's molecular identity). This is the reason why those genes and respective products were first identified by their major influence in determining immune reactivities to tissue or organ grafts exchanged between individuals in the same species: they are known as the major histocompatibility complex (MHC). As these molecules seem to be the only ones that can bind and present fragments of other proteins to the IS, it follows that each individual IS can only see the set of peptides that can actually bind to the MHC proteins it expresses. As these show no somatic variability, each individual disposes of only a limited number to cope with the whole world of protein fragments, and, consequently, T cells in any IS ignore all but a small fraction of all possible peptides. Degeneracy of molecular interactions plays, here again, a significant role in the IS's biology, for the fraction of the universe of peptide shapes that exists for each individual varies inversely with the precision of peptide-

MHC molecule interactions. The most interesting aspect of these processes, however, concern the structural allelism of the presenting MHC molecules, which implies that each variant in the species will be specifically capable of presenting a limited and unique set of peptides. For any given protein, different MHC alleles present different fragments, and the T cells in each of those individuals will have distinct "points of view" of the same original antigenic shape, sometimes none at all because no peptide fragment of that protein can bind to any of the MHC molecular in the individual. This is actually the reason why individuals in the same species differ in their ability to recognize proteins, particularly those with relatively monotonous sequences (from which few different peptides can be made), a phenomenon first ascribed to Immune response (Ir)-genes, later recognized to be no other than the MHC. In other words, the fact that IS fail to see many (most?) proteins with T cells (in current nomenclature, these are refered to as "nonresponder" individuals) does not argue against the completeness of immune recognition. First, because the same IS contain Igs capable of recognizing any shape whatsoever; second, because the nonresponder IS seems to contain T cells that can recognize peptides of the same protein, provided these are presented in MHC molecules. If this is the case, T-cell reactivities are primarily limited by MHC-restriction, rather than by incompleteness in receptor repertoires. On the other hand, as seen below, T cells that react to peptides on the individuals MHC molecules are selected for in ontogeny, such that T-cell repertoires may be essentially limited to self-restricted cells.

The implication of these mechanisms for the establishment of the individual's molecular identity is clear. The areas in shape–space that are occupied by self components are greatly expanded by the (possibly very large) variety of shapes composed by complexes of MHC molecules and peptides from all proteins in the body. That is, to the immune definition of a three-dimensional molecular self existing in the body fluids and cell surfaces, we have to add another "one-dimensional" linear self, reconstructed from all proteins, even if intracellular. Furthermore, each differentiated tissue constitutes a local and unique compartment of self molecular shapes, available only to Igs and lymphocytes that transit through that tissue, making it unescapable that some form of specific intercellular "transfer of information" exists in the IS if the whole

system has to learn the whole of self. Given MHC polymorphisms, and even disregarding structural allelism at all other loci in the genome, the molecular self is unique for each individual, and each IS exists in a space of shapes that is different from all other IS, with a "point of view" that is also necessarily unique. If these constraints are advantageous for the evolution of the species and many hypotheses can be constructed to explain why, they also require that IS learn self accurately, from both the perspective of Igs and that of TCR. As T cells can only or predominantly interact with peptides bound to MHC molecules, and these contribute a great deal to the three-dimensional shape interacting with TCR V-regions, only a limited number of all T cells can actually interact with each MHC variant. It would seem necessary, therefore, to ensure that T cells in an individual are selected to predominantly interact with those MHC variants expressed by the individual. This process of "positive selection" or "education" of newly produced T cells to self MHC-restriction indeed occurs, primarily in the thymus (the organ where T lymphocytes are generated from precursors), and it constitutes today one of the most active areas of research in Immunology.

Positive selection of T cells is one example of the general process of selecting and adjusting V-region repertoires that are available to the individual at any point in life. This process continues throughout life in the normal IS, for new V-regions are generated all the time and can thus be either recruited into the system, let die, or be actively eliminated, in accordance with their coherence in the context of the everchanging molecular environment. Many of these changes result from exposure to external antigens, but the recruitment (and expansion) of new V-regions must also contribute to generate new shapes in the lymphocyte environment. The complexity of this process can readily be appreciated if one considers the fact that V-regions cannot only interact with each other as three-dimensional shapes, but also be "processed" into peptides and be presented to T cell (and antibody?) recognition together with MHC molecules, generating a possibly very large number of novel shapes. There is indeed some evidence for the selection of Ig repertoires, depending on T-cell V-regions and MHC genes, as well as for the modulation of the selected T-cell diversity by Ig V-regions. That is, the repertoires that are used (the V-region structure), and the manner how they are used (the dynamics of immune activities) contribute to the selection of new V-regions (metadynamics) and IS's evolution in time. It is clear, therefore, that the "starting repertoires," expressed at the initial stages of the IS's ontogeny will have marked effects in the future evolution of the system, and in the establishment of its uniqueness. This is the reason why maternally transmitted antibodies, present in the egg or crossing the placenta, and constituting most of the V-regions available in the embryo as it develops (in mammals, maternal antibodies are quite significant in the neonatal periods, due to their high concentrations in the colostrum and milk and specific mechanisms to absorb them in the gut), have marked effects in the repertoires expressed later in the progenies' life. This process provides vertebrates with a nongenetic mechanism for transferring to progenies information about the environment, that makes sense with the predominantly epigenetic and somatic (individual) construction of immune repertoires, and the evolution of the genetic bases of immune diversity.

The importance of the initial V-region repertoires for the evolution of each IS, is also manifested by a series of genetic mechanisms determining expression of V-regions in the embryo, which all seem directed at ensuring the presence of some reactivities and a relatively conserved overall structure of the original repertoires. Thus, for reasons that seem to relate to the relative position of each V-gene segment in the respective locus, the first lymphocytes to be produced predominantly or exclusively express only a minor subset of all V-region possibilities. Furthermore, the enzymatic mechanisms that generate junctional diversity are underdeveloped or absent at this stage of development, such that the first V-regions expressed are essentially encoded by germ-line sequences with little or no random diversity. Because the embryo is secluded from all antigenic contacts with the external environment, it has been proposed that such primordial, evolutionarily selected reactivities, must be directed at self structures. This suggestion, which has recently received considerable experimental support, would ensure conditions for driving the selection and expansion in the developing IS. Interestingly, it has been established that embryonically expressed Ig V-regions show a marked preference for reactivities with other Ig V-regions in the same subset, that is, they seem to constitute a "network" with relatively high connectivity. The existence of such a germ-line encoded, developmentally controlled V-region network, could provide the IS with structural conditions for dynamic autonomy in development and repertoire selection, and it brings support to theories ascribing to the IS a network organization.

V. Amplification and Renewal

Immunocompetent lymphocytes are produced at high rates throughout life, at least in higher vertebrates. Some of these new cells arise by division of previously existing lymphocytes (see below), but most of them are actually produced anew. Thus, many new lymphocytes continuously differentiate from uncommitted precursor cells in the bone marrow and in the thymus. As much as a quarter or more of the cellular content in the adult mammalian bone marrow, which also produces all other cells in the blood, is occupied by cells differentiating to become B lymphocytes. As expected from the error-prone mechanisms involved in DNA recombination in the construction of V-region–genes, many of the differentiating cells never "make it" to express Ig and become a B lymphocyte. Large numbers, however, do: if the determinations in rodents are transposed to human beings, it is estimated that we produce up to one million new B cells per second. Many of these are exported to the periphery, while others are not accepted in the system and seem to be destroyed locally, contributing to the considerable cellular waste in the organ. A similar process takes place in the thymus which, before its natural involution in adulthood, generates a large number of T lymphocytes. Here the process of local selection is even more stringent: More than 90% of all the newly formed T cells never leave the organ and die *in situ*.

Because most of the newly formed lymphocytes carry V-regions that did not exist before in the system, these conditions of continuous overproduction provide unique possibilities of somatic selection for the appropriate repertoire of specificities available in the individual at each point in time. This process is metadynamic, in the sense that the IS not only can vary in time the rates of production of its components (dynamics), but it can also recruit into its activity components that are altogether new and continuously provided by a never-ending ontogeny. Because the genetic mechanisms generating V-regions are part of the process producing new lymphocytes throughout life, the structure of the IS is not fixed but can be modified and adjusted to the ongoing activity. The total number of lymphocytes is obviously kept constant in normal individuals, indicating that an equally large number of lymphocytes is continually being lost from the organism. In other words, many lymphocytes have short life spans and do not persist in the organism for more than a few hours or days. It follows that available V-region repertoires in the individual are always much smaller than those produced in a given time span. This multiplicity of V-regions must compete for the limited size of lymphocyte populations, that is, available immune repertoires are continuously selected. Selection must ensure the identity of the molecular self (a structure and dynamics of repertoires at equilibrium with self), but also the possibility of responding appropriately to foreign before it is ever introduced. Obviously, such selection of "pre-immune" repertoires, which brings forth the "prometheic" characteristics discussed above, must be based on shapes that are part of the molecular self. As proposed by Jerne, a solution to this problem is to consider that the multiplicity of V-regions in the system provide internal images of all possible foreign antigenic shapes, accounting for yet another aspect of the ISs autonomy.

These notions of an ongoing endogenous selection of lymphocyte repertoires, are supported by the findings that persistence or life span of lymphocytes varies with their V-region specificity. Of all cells completing differentiation and expressing V-regions, some will die very rapidly, others will survive a few days or weeks, while others persist for years, or perhaps for a lifetime. As seen below, persistence of a given V-region can be achieved by activating and amplifying the corresponding lymphocyte. Conversely, a few lymphocytes decay as a consequence of terminal differentiation to effector functions. The largest part of lymphocyte populations that turnover, however, are resting cells. This mode of V-region selection could be based on rescuing for persistence (positive selection) cells otherwise submitted to programmed death (apoptosis), or else on eliminating newly formed lymphocytes after identification of their V-regions (negative selection). Both mechanisms seem to occur in normal IS, and contribute significantly in the process of adopting equilibrium dynamics and tolerance, as well as in keeping resting lymphocyte repertoires that can appropriately respond to external challenges.

Metadynamic selection of repertoires must reflect previous immune activities, that is, the identity of self, the ontogeny of the IS and foreign antigenic experiences, such that it provides repertoires ac-

counting for both "completeness" and memory, be that of self or of foreign. The control of total lymphocyte numbers constitutes an essential part of this process and, therefore, one of the most interesting questions in immunology. Unfortunately, the mechanisms that count lymphocytes, and pertain to the very organization of the system are completely unknown. Normal lymphocyte numbers are ensured not only in the presence of a continuous overproduction, but also in the opposite situations, namely if loss is greatly enhanced by repeated bleeding, even after natural involution of the thymus in adult individuals. In this case, existing lymphocytes in the periphery must stop decaying and expand, to compensate for losses occurring in the absence of central production. Again, the driving forces stimulating such expansions are not known, but are likely to be related to those leading to the accumulation of large numbers of lymphocytes in the neonatal period, where no decays are observed until normal adult numbers are reached.

The plasticity of the IS also owes much to the cell biology of lymphocytes. These are the immunocompetent cells, for they are the only cell type expressing diverse, V-region-bearing receptor molecules. As postulated by Burnet in the clonal selection theory, all receptors on each lymphocyte are identical and therefore distinct from the overwhelming majority of all other lymphocytes. It follows that each lymphocyte specificity covers only a very limited area of shape space in terms of the molecular species it can interact with. Each lymphocyte expresses 10^4–10^5 identical receptor molecules on its surface membrane, a concentration of the unique V-region that is probably too small to significantly contribute in the overall composition of the molecular self. Lymphocytes differentiate from uncommitted precursors to become receptor-bearing resting cells, with a dense nucleus and a thin rim of cytoplasm. These cells maintain expression of house-keeping genes, but the only differentiated function they perform is to advertise those surface receptors, together with several other molecules that seem to either serve accessory roles in their interaction with ligands, or else determine their ability to home to specialized areas of the lymphoid tissue, or to migrate out of the blood vessles into the tissues. Lymphocytes are disconnected from the mitotic cycle and serve no known function other than existing. The majority of lymphocytes in a normal individual exist and eventually decay without ever being engaged in any activity, for no complementary shapes to their receptor V-regions are available in the molecular environment. In contrast with most other cell types in the organism, however, lymphocytes if stimulated can divide and produce progenies of many cells. Doubling times of lymphocytes measured *in vitro* are rather constant, between 16 and 20 hours. It follows that a single lymphocyte, if optimally stimulated, will produce more than 1,000 progeny in one week. The proliferative potential of mature lymphocytes is strikingly large. Both B- and T lymphocytes were shown capable of undergoing up to 50 or 60 doublings *in vivo,* before "clonal senescence" ensues. If all progeny cells were to keep optimal proliferative rates, a single lymphocyte should potentially be capable of producing some 10^{15} or 10^{16} identical cells, which are many orders of magnitude more than the whole IS of a mouse or even a man. Obviously such amplifications do not ever occur in a normal IS, where clonal amplifications are comparatively very reduced, perhaps not exceeding a handful of doublings. These questions have been little addressed, and bring us to the problem of the regulation of clonal amplification, and the fate of the proliferating lymphocytes.

As the second postulate of the clonal selection theory had predicted, all members in the progeny of a lymphocyte are identical, as to the V-regions they express. Clonal expansion, therefore, contributes a fundmental strategy in the amplification of a given molecular shape, particularly because the selected V-region continues to function as a cellular receptor connected to the induction of effector functions in the amplified clone. Thus, there is another strategy for V-region amplification which is based on selectively upregulating the expression of genes encoding Igs and the necessary cellular machinery for protein synthesis. A fully differentiated plasma cell, which represents the stage of terminal maturation of a B cell, secretes antibodies at rates that are 1,000-fold higher than in the resting lymphocyte. It follows that, by concerting lymphocyte proliferation with terminal differentiation, a given antibody V-region specificity can be amplified more than 10^6-fold in a week. Interestingly, these two modes of amplification are dichotomic in time, that is, proliferating lymphocytes are not fully differentiated and terminally mature cells are uncapable of further proliferation. This delicate balance imposes some constraints to the multitude of possibilities in regulating lymphocyte performances, particularly because terminally differentiated cells are generally short-lived and do not perform for more than a few hours or days. Furthermore, because fully differentiated cells can no longer proliferate and expand, the process may also serve as a mechanism to eliminate unwanted V-region specificities, or at least keep their level at concentrations that are compatible with the normal dynamics in the system.

Differentiation to effector functions is not limited to B cells and Ig secretion. T lymphocytes as well, only perform after activation, by selective induction of expression of genes that are silent in the resting state. These encode effector function-associated proteins displaying a wide variety of activities on other cell types. Some are hormonelike polypeptides that serve as growth or differentiation factors for other lymphocytes, collectively known as "lymphokines." Others function as chemotactic and/or activating factors for other circulating or tissue associated cell types, regulating expression of various genes and functions. Yet some other T-cell products are proteolytic enzymes that promote the lysis of (target) cells. Interestingly, expression of the multiple genes associated with effector functions segregates in different T lymphocytes, defining their functional heterogeneity and the existence of T-cell subclasses. These were first named after the biological activity that can be recorded in functional assays, but are now extensively characterized, both with respect to the differential expression of surface molecules (that can be used as "markers" to identify and enumerate them) and to their differentiative pathways. T helper cells participate in the activation of B lymphocytes and, by the secretion of B-cell-directed growth and differentiation lymphokines, promote their proliferation and plasma cell maturation; they are instrumental in the amplification of antibody V-regions. T inflammatory cells produce factors that mobilize and activate blood-borne or tissue myelomonocytic, mast cells and other T lymphocytes, which all participate in the process of local inflammation. T cytotoxic cells are capable of producing the lysis of target cells, and also produce interferons, such that they appear particularly suited to deal with infections by budding viruses. T suppressor cells negatively regulate the activities of other lymphocytes and thus limit and control immune activities. The considerable variety of effector molecules and lymphokines produced by the different T-cell subsets, often displaying opposite effects on target cells depending on the time of their delivery, regulated by "feed back" or "feed forward" loops, contribute a great complexity to these processes and make it difficult to analyze and predict them. A helper factor inducing terminal differentiation is obviously suppressive of proliferation. Furthermore, some of these mediators regulate the expression of others in intricated patterns. For example, interferons secreted by inflammatory T cells inhibit growth of helper cells, whereas a product of the latter inhibits the secretion of those interferons and of the growth factor utilized by the former. Helper cells are thus suppressors for those of the inflammatory type and vice versa. The final result of a particular immune activity will represent the predominance of a given cell type, and the interplay of many such effects. Because the repertoire of specificities in either lymphocyte subset is probably comparable, these mechanisms are related to the class determination of immune responses (e.g., antibody production versus inflammation), particularly important in the context of localized activities (mucosae, lymphoid organs, parenchymal or connective tissues) and thus of antigen localization and route of entry into the organism.

All T-cell activities, including those mediated by soluble factors, are short-ranged and exclusively active in the immediate proximity of the effector cell, often upon membrane contacts with target cells. In contrast with Igs (the effector molecules of B cells), T-cell effector molecules carry no V-regions and are thus not competent in antigen recognition. Their short-range activity is therefore important to ensure that T-cell functions, even if nonspecifically mediated, are always triggered and limited by V-region specificity. (If lymphokines are administered at pharmacological concentrations, however, systemic effects are also obtained, opening therapeutic possibilities that are under intense investigation at present.) The processes of lymphocyte activation and control of performance are essentially based on intercellular communication and membrane contacts, a strict requirement for T cells which only recognize shapes on other cell surfaces, presented by MHC molecules. Accordingly, lymphocytes express several types of cell adhesion molecules common to other cell types in the organism, as well as lymphocyte-specific surface proteins that facilitate interactions with structures on target cell membranes, MHC in particular. The privileged sites for lymphocyte cooperation are the lymphoid organs with specialized presenting cells, a local architecture that ensures the feasibility of such contacts, and a lymphoid traffic that allows for the screening of many cellular reactivities.

The management of the plasticity potential, based on the selection of resting lymphocytes into activation, amplification and/or differentiation to effector functions, is very complex indeed, and it offers a great variety of possibilities which all account for increased adaptability to the molecular environment. It suffices to say that selection from the resting state involves interactions with V-regions of surface receptors, be these Igs or TCRs, such that selection is clonally specific. This is possible because V-region receptors and those accessory molecules are expressed in a membrane complex together with proteins that are coupled to mechanisms of intracellular signaling (protein kinases, G-proteins). Lymphocyte activity is thus regulated by ligand occupancy of V-region receptors. Most important, the dose-response curves of lymphocyte responses to specific ligands are bell-shaped, in other words, overstimulation results in paralysis. Lymphocyte performance, therefore, is only observed for values of receptor occupancy that fall within two thresholds: below the lower threshold of induction, the cell remains resting, while above the upper threshold of paralysis no response is induced and the cell rapidly decays. Although clonally specific, lymphocyte activation follows general rules of cell biology. As seen above, the shapes recognized by V-regions (the antigens) are not lymphocyte growth factors, as it was formerly believed. These are cell-type specific, but carry no V-regions. The process of lymphocyte activation represents, therefore, the coordinate induction of selective growth receptor expression (absent from the resting cell) by a V-region-dependent mechanism, and the production of growth factors by other lymphocytes, again in a specific manner. Clonal performance is also specifically regulated, that is, activation is not a once and for all decision: lymphocytes will only proliferate (or mature for effector functions) as long as they are stimulated to do so, by the maintenance of the conditions leading to the initial activation. Furthermore, mechanisms exist which actually turnoff or suppress ongoing lymphocyte activities, such that immune responses can be limited by other immune responses, and the overall levels of immune activities can be regulated. It should perhaps be underlined that the high levels of clonal amplification which are characteristic of immune responses, are by no means prototypic of all dynamics in the IS. Immune activities ongoing in normal, healthy individuals, particularly those associated with tolerance states, seem to rely on a much lower level of clonal amplification, and to depend instead upon the relatively slow recruitment and/or amplification of the appropriate lymphocyte specificities.

VI. The System At Work

A normal adult human being contains about 10^{12} B lymphocytes and a roughly equal number of T cells, that corresponds to a total of around 10^{17} cell-bound V-region molecules. Most of these lymphocytes are resting, their unique V-regions finding no complementarity in the environment where they exist. Given the diversity of V-region shapes, however, it is probable that some other lymphocytes in the body actually express a surface receptor that would fit anyone V-region existing in the same individual. These complementary cell-bound receptors, however, have little chance of meeting and engaging in productive interactions. In other words, even if "completeness" of V-region shapes imposes the circularity of a complete "network," this does not seem to come forth for the majority of lymphocytes. From a functional point of view at least, most lymphocytes are disconnected from a putative network, simply because they are resting cells. The repertoires of such lymphocytes have been selected from the moment each was produced. Those expressing V-regions with self complementarities, such that they suffered a degree of receptor occupancy exceeding the upper threshold of activation, have been eliminated (clonal deletion or negative selection). Those expressing V-regions with no complementarities to the self molecular environment, and thus no receptor occupancy at all, have probably died rapidly and let the place to others with some intermediate level of self affinity, sufficient for survival but insufficient for activation (positive selection). Finally, a minority of lymphocytes expressed V-region receptors with an affinity to shapes in their environment that are in concentration high enough to result in activation but not in paralysis. These were activated, and accordingly, a normal individual, even if secluded from all contacts with external antigens (as it can be done with laboratory animals), contain some 10% or so of all lymphocytes that are blasts and perform effector functions. We must now make a distinction between T and B lymphocytes. The former, only capable of interacting with peptide/MHC complexes on other cells, cannot be exposed to many different ligands simultaneously.

Their activation, inactivation or positive selection is therefore precisely determined by their local immediate environment, and it proceeds within a closed compartment (the thymus) from where the cells are not let to emigrate before maturity (resting state). B cells, in contrast, bind soluble three-dimensional shapes and are thus exposed to all circulating self shapes, plus those in the immediate tissue surroundings. It follows that activated B cells will be those expressing relatively low affinity complementarities to self molecules, and preferentially those with V-regions capable to bind many different shapes. Accordingly, circulating Igs in normal individuals (often called natural antibodies) show degenerate reactivities with many self antigens. As helper T cells with equivalent specificities do not exist, lymphocyte cooperation leading to large clonal expansions does not come about, the level of such autoantibodies is low, and they cannot be selected for high affinity by somatic mutation of Ig-genes accompanying clonal expansions.

The serum of the same normal individual contains, therefore, up to 10 g/l of Ig, what corresponds to 10^{16}–10^{17} molecules of Ig per milliliter, and a total of some 10^{20} free circulating V-regions. Even accounting for a large diversity, serum Ig must contain a considerable number of identical V-region shapes. These have access to essentially all other lymphocytes in the body, and they are likely, therefore, to find the V-region complementarities with enough fitness to result in functional activation of the corresponding cells. Accordingly, the repertoire of circulating Ig V-regions is likely to constitute a network, and to recruit into the same dynamics connected T-cell receptors as well, because antibodies to TCR are competent ligands in T-lymphocyte activation. Because T cells with both positive (helper) and negative (suppressor or cytotoxic) effects on B cell performances are equally likely to be engaged, this internal activity is regulated and self sustained. Normal individuals do contain equivalent numbers of activated T lymphocytes, capable of activating or suppressing B cell activities. Moreover, since interactions between soluble Ig molecules inhibit the access of either partner to the corresponding lymphocyte receptors, and this is necessary for cell activation, it necessarily follows that immune activities within such a network are at equilibrium. Accordingly, the variations in time of natural antibody concentrations show oscillatory dynamics, within ranges not exceeding one order of magnitude. It follows that natural lymphocyte activities, in the absence of external stimulation, are likely to compose a selfreferential network at equilibrium with, and therefore tolerant to, the self molecular environment. This internal immune activity is thought to represent the physiology of autoimmunity, an area of research that only recently has attracted some attention. It is likely that better knowledge of these physiological processes will help in understanding their pathological alterations.

All lymphocytes that exist resting in the individual, are obviously functionally excluded from the domain in the space of shapes that is characteristic of self and of the self-defining immune activities. If, however, a novel (foreign) shape is introduced in the organism, the lymphocytes with the appropriate V-region receptor complementarities will be selected into amplification. As they do not integrate a V-region network, they are free to expand as long as that shape persists, clonal amplification being essentially limited by the mechanisms discussed above, relating to class determination of immune responses. The characteristic dynamics of immune responses is thus obtained, and their protective value in anti-infectious defence justified. Both types of repertoires (those involved in responding to foreign as well as those participating in the physiological definition of the molecular self), however, contribute to the same general function, namely to maintain the identity of the individual in the space of shapes.

Bibliography

Coutinho, A. (1989). Beyond clonal selection and network. *Immunol. Rev.* **110,** 63–87.

Jerne, N. K. (1984). Idiotypic networks and other preconceived ideas. *Immunol. Rev.* **79,** 5.

Jerne, N. K. (1971). What precedes clonal selection? Ciba Foundation Symposium: Ontogeny of acquired immunity. Elsevier.

Jerne, N. K. (1974). Towards a network theroy of the immune system. *Ann. Immunol. (Inst. Pasteur).* **125C,** 373–389.

Schwartz, R. H. (1984). The role of gene products of the major histocompatibility complex in T cell activation and cellular activation. *In* "Fundamental Immunology." (W. F. Paul, ed.). pp. 379–438. Raven Press.

Tonegawa, S. (1983). Somatic generation of antibody diversity. *Nature (London).* **302,** 575–581.

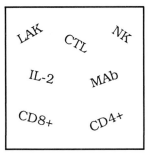

Immunity to Cancer

ARNOLD E. REIF, *Mallory Institute of Pathology*

Glossary

Antibody Protein secreted by an immune cell that can specifically bind to an antigen

Antigen Substance that elicits a specific immune response when introduced into the body

BCG Bacillus Calmette-Guérin, a viable mycobacterium used for immunotherapy

Biomodulators Abbreviated term for biological response modifiers: agents that improve the biological response of the host to diseases such as cancer

C. parvum Corynebacterium parvum, a killed bacterium used for immunotherapy

Cytokines Tissue activators secreted by cells of the lymphoid system

Immune system Organs, cells, and secretions that participate in immune reactions

Immunogenic Capable of inducing an immune response

Immunotherapy Treatment by stimulation of the immune system or by transfer of immune cells or factors produced extraneously

Immunosurveillance Surveillance (guarding) of the body by the immune system

Lymphocytes Spherical cells that are the chief constituents of the lymphoid tissues

Monoclonal antibody Antibody produced by immortalized cells derived from one clone

Natural killer (NK) cells Lymphoid cells that can attack tumor cells in a host that has not been immunized, but that do not attack normal cells

Tumor specific antigen Antigen found on tumor cells, not on normal tissues

CANCER IMMUNOLOGY is the subject that deals with all aspects of the interaction between cancer cells and cells of the immune system. The title "Immunity to Cancer" was chosen as an alternative, because it hints at a therapeutic benefit. It is also the title I chose for a series of conferences on cancer immunology, which so far have been held in 1984, 1987, and 1989. Tumor immunology is a young science, in which some of the most relevant discoveries may still lie ahead. The body maintains a precise control over the numbers and types of its cells. Depending on the kind and concentration of the agent that causes cancer, the body is able to mount a defense against the cancer cells that ranges from very weak to strong. This defense employs cells of the immune system and their secretions, namely, antibodies and lymphokines. Uncertainties still underlie such central problems as the potentialities and limitations of immunosurveillance against cancer in human beings and the degree to which the various types of human cancer are immunogenic. The answer to this latter question holds the key to successful immunotherapy of cancer, because no strong, active (ongoing) immune response is possible unless a cancer is immunogenic.

Immunotherapy uses a variety of strategies to intensify the natural defenses against cancer. However, the key to potent treatment of cancer with immunotherapy has proven elusive. Even in work in experimental animals, we know only the immunogenic strengths of tumors induced by high doses of a carcinogen (cancer-causing agent) but do not understand the extent to which immunogenicity is reduced as the dose of carcinogen is reduced. Nor do we understand as yet how to boost the immune response maximally, by injecting appropriate types and doses of biomodulators during appropriate phases of the immune response against tumors. On the bright side, the development of the use of monoclonal antibodies makes it possible to produce antibodies that react with specific antigenic portions of cancer cells. Trials are presently in progress on vaccination of patients with a minimal load of cancer cells against their very own tumor. Also under trial is vaccination against hepatitis B virus for prevention of liver cancer in Southeast Asia. This may be the prototype for the ultimate answer to those types of human cancer caused by viruses: prevention through vaccination. Unfortunately, most of such viruses have yet to be identified and purified.

I. Destruction of "Foreign" Cells

A. Cells from a Different Individual

In 1883, the Russian scientist Metchnikoff performed the experiment that launched immunology. Intrigued with the idea that host defense cells might attack a foreign body, he stuck rose thorns under the skin of some transparent starfish larvae. When he viewed them under the microscope the next day, the thorns were surrounded by mobile host defense cells, which he later called "macrophages"—scavenger cells. Subsequent work showed that cells from a genetically different organism are destroyed (rejected).

Cancer cells are also foreign intruders, in the sense that they have constituents not specified in the genetic blueprint for the body, or else appear in the body at a time far removed from their normal appearance—for instance, during embryogenesis or during regeneration of injured cells. Research focused on the transplantation of tissues has shown that the rejection of normal tissues depends on the presence of genetic disparity between donor of the tissue and the host injected with it. In general, the greater the disparity, the stronger the rejection reaction, which is immunological in nature. When the tissue in question is cancer, then there is the additional problem for the host defense cells that they must marshal against an invader whose numbers are increasing through cell division. If the rate of growth of the cancer cells overwhelms the rate at which the defending cells can be produced, the cancer cells will win the battle, even if they are antigenically quite distinct. In addition, cancer cells can elaborate immunosuppressive factors that may disrupt host defenses. Nevertheless, work in animal systems has shown that if there is a considerable genetic (and therefore, antigenic) difference between the donor of the cancer cells and the host, most frequently the cancer cells are completely eradicated. [*See* IMMUNOBIOLOGY OF TRANSPLANTATION.]

B. Cells from a Genetically Identical Person

Transplantation research has shown that tissues grafted between identical twins are successful: the tissues are not rejected. For this reason, experiments directed at detecting a tumor-specific antigen are carried out in strains in which all animals are genetically as similar as identical twins. These strains are produced by continual "inbreeding"—matings between brothers and sisters of a single litter. After 21 successive generations of inbreeding, uniformity in genetic makeup is attained, and animals can be called "isogeneic" (Greek "iso" = equal, "−geneic" = genetically). Grafts of normal tissues performed between isogeneic animals are accepted, just as grafts between identical twins.

C. How Does Tumor Rejection Depend on the Cancer Causing Agent?

When a known oncogenic (cancer-causing) agent is used in an animal of an inbred strain at appropriate dose and application, a tumor characteristic of that oncogenic agent is produced. Tumors induced by different types of oncogenic agents have widely varying immunogenicities (Table I). With the exception of some of the UV-induced tumors, all tumors are accepted if injected into an isogeneic "naive" animal—one that has not been immunized (vaccinated) before this "challenge" by injection of viable tumor cells. To immunize a naive animal, the tumor cells must be inactivated by a method that prevents further cell division, such as irradiation with a high dose (12,000 rad) of X-rays. Only when the animal to be challenged with viable tumor cells has first been immunized by injection of inactivated tumor cells can that challenge be rejected.

The results of such immunization experiments are recorded in Table I. Thus, "++++" for UV-

TABLE I Relation Between Etiology of a Tumor and Its Immunogenicity

Etiological agent	Dose of agent	Approximate immunogenicity	Cross-reactivity with other tumors caused by that agent[a]
None: tumors are spontaneous	−	−, ±, (+)	−, ±, (+)
X-rays, radioisotopes	high	−, ±, (+)	−, ±, (+)
Chemical carcinogens	high	++, +	±
	medium	+, ±	(±)
Oncogenic viruses			
Injected at birth	high	+++, ++	+++, ++
Vertical transmission[b]	−	+, −	+, −
UV radiation	high	++++, +++	±

[a] Immune cells or antibody obtained by immunization with one tumor also react positively with other tumors initiated by the same carcinogenic agent.

[b] Vertical transmission means passage from the mother to the fetus in her uterus.

induced tumors (Table I) indicates that some of the tumors are so highly immunogenic that they may be rejected even without prior immunization. The notation "+++" indicates that the tumor will be rejected if only a single vaccination is given, whereas "+" shows that three or four vaccinations may be necessary before a challenge with viable tumor cells can be rejected. In contrast, several injections of "inactivated" normal tissues from isogeneic animals will not prevent acceptance of a graft of the corresponding viable normal tissue. Thus, rejection of the graft of a tumor by an isogeneic animal preimmunized with inactivated tumor indicates the presence of an antigen or antigens not found in normal tissue, which we call "tumor-specific."

Animal experiments (Table I) indicate that several classes of tumors have very low immunogenicities: those that appear spontaneously, those induced by ionizing agents (X-rays, radioisotopes), and perhaps also those induced by very low doses of chemical carcinogens or by vertical transmission of some types of oncogenic viruses. Unfortunately, we lack adequate data for these latter two categories. In contrast, some classes of tumors are highly immunogenic: tumors induced by high doses of chemical carcinogens, or by injection at birth of oncogenic viruses, or by UV radiation.

Two tumors are said to be cross-reacting when an animal immunized against one tumor exhibits changes (usually, increases) in immunity to the other tumor. The data on cross-reactivity (Table I) are important for work with oncogenic viruses, by showing that only tumor cells produced by a particular oncogenic virus possess the same tumor-specific antigen that is present in *all* tumors induced by

that virus—even those induced in different individuals. Thus, it is possible to immunize one animal with inactivated cells of a given virally induced tumor taken from another animal. In contrast, for the other agents producing strongly immunogenic tumors, namely, chemical carcinogens and UV irradiation, only a very slight degree (±) of cross-reactivity exists. Thus, to protect against challenge with viable tumor cells, it is necessary to immunize with the identical tumor to which immunity is desired. This requirement makes the process of vaccination against a tumor difficult, because the vaccine must be prepared from the patient's own tumor cells. Still, the cross-reactivity of virally induced tumors in animal systems suggests that human tumors caused by known oncogenic viruses can be successfully prevented by vaccination; indeed this is the assumption behind the ongoing trial of vaccination with hepatitis B virus to prevent liver cancer.

II. Cells Implicated in Tumor Rejection

Many different types of immune cells have been described that take part in a positive or negative way in the rejection of tumor cells in naive or immunized animals.

A. Natural Killer Cells

Natural killer (NK) cells possess *spontaneous* cytotoxicity (cell-directed toxicity) against tumor cells and against a limited number of normal cell subpopulations, for instance, virus-infected fibroblasts.

Unlike T (thymus-derived) cells, NK cells require neither sensitization nor the presence of major histo-compatibility (tissue-compatibility) antigens, without which T cells cannot act. NK cells can kill target cells directly or through antibody-dependent cellular cytotoxicity (ADCC), which consists of killing by NK cells mediated by antibody coating of the target cells. In human beings, NK cells represent about 5–8% of circulating white cells. They can be distinguished from T cells because they do not express T-cell-antigen receptors nor have rearranged their T-cell-receptor genes. In animal experiments, NK cells have been shown to play an important role in immunosurveillance against cancer. [*See* NATURAL KILLER AND OTHER EFFECTOR CELLS.]

B. Lymphokine Activated Killer Cells

The cytolytic (cell-killing) activity of NK cells can be increased by interleukin-2 (IL-2) both *in vivo* and *in vitro*. IL-2 can greatly potentiate the natural antitumor activity of NK cells and recruit cytotoxic cells of T-cell origin. NK cells treated with IL-2 can kill types of tumor cells that untreated NK cells cannot kill. Clinical trials with oft-repeated large doses of IL-2, supplemented by infusing cells obtained from the patient's blood and expanded through tissue culture in the presence of IL-2 [which turns these cells into lymphokine activated killer (LAK) cells], have produced durable complete responses in 15–30% of patients with renal-cell carcinoma or melanoma. These responses sometimes include disappearance of the tumor. However, IL-2 is highly toxic at the doses employed, and the results are far less positive for more common types of cancer. A possible explanation is that renal-cell carcinoma and melanoma are highly immunogenic and that the positive results for these tumors result because LAK cell-killing of tumor cells is followed by conventional T-cell-mediated cytotoxicity (see Section II,D) that requires the presence of strongly immunogenic tumor antigens. Most recently, Rosenberg has separated tumor-infiltrating lymphocytes (TILs) and expanded these cells *in vitro* before injection into melanoma patients: much improved results have been obtained by this technique. [*See* INTERLEUKIN-2 AND THE IL-2 RECEPTOR.]

C. Macrophages

Macrophages are named for their ability to *swallow* bacteria or foreign particles. They are mononuclear (single-nucleus) cells present in both blood and body tissues. Macrophages can kill tumor cells in two ways. They can be "activated" nonspecifically, to become directly cytotoxic to tumor cells. Alternatively, they can kill tumor cells coated with specific antitumor antibody by means of ADCC (described in Section II,A). Macrophages also initiate the response to a strongly immunogenic tumor by presenting tumor antigen to T-helper lymphocytes (see Section II,D). Further, macrophages can secrete many different factors involved in host defense. In animal models, macrophages have been shown to prevent or cure cancer both by nonspecific and by specific mechanisms. Despite their impressive potentialities and positive animal experiments, the evidence for the involvement of macrophages in immune surveillance and in the suppression of metastases is suggestive rather than definitive. Indeed, macrophages also have suppressive functions. The identification of macrophages is made possible because they cling to surfaces and possess specific cell surface markers. Work in human beings has focused mainly on probing their involvement with primary tumors and metastases (extension of cancer to distant organs). [*See* MACROPHAGES.]

D. B and T Lymphocytes

As mentioned above, strongly immunogenic tumors can elicit a classical immune response in which macrophages act as antigen-presenting cells for helper/inducer (CD4-positive) T lymphocytes. These activated helper cells begin a sequence of cellular interactions that results either in B (bursa-derived)-cell activation, and/or activation of cytotoxic (CD8-positive) T lymphocytes to cytolytic cells (CTLs). The B-cells activation can lead to the secretion of specific antibodies to the tumor antigen involved, whereas the cytotoxic T cells can kill tumor cells bearing the antigen to which they have been specifically sensitized. Indeed, serum antibodies have been found in persons with certain types of tumors, and specifically cytotoxic T lymphocytes have been described in patients with melanomas, with UV-induced skin tumors, and with certain other types of solid tumors. While these findings suggest that the respective tumors are highly immunogenic, the presence of antibodies or of cytotoxic T cells does not imply that a given tumor will regress. [*See* LYMPHOCYTES.]

E. Suppressor Cells and Factors

Suppressor cells down-regulate both humoral (B-cell-mediated) and cellular (T-cell-regulated) immunity. In human beings, T suppressor cells (Ts) bear the CD8 marker (also displayed by cytotoxic T cells) and also the CD11b marker, which distinguishes them from cytotoxic T cells. Suppressor cells act through soluble factors, which directly or indirectly inhibit the activity of effector cells. Also, antigen-nonspecific suppressor T cells exist, which have a similar effect. Unfortunately, many types of immunogenic tumors stimulate host production of Ts, which depress the immune response to those tumors. To inhibit Ts, low doses of certain biologic agents as well as X-rays have been used. At high doses, cyclophosphamide inhibits the growth of rapidly dividing cells, but at low doses it inhibits Ts in cancer patients and thereby potentiates tumor immunity. Therefore, in current clinical immunotherapy trials, low doses of cyclophosphamide are frequently used to inhibit Ts cells.

Several suppressive factors have been found in cancer patients. Some of these factors represent tumor antigens produced by (or through breakdown of) the tumor. Others represent secretions of suppressor T cells or of suppressor macrophages, whereas others still are of unknown origin; all depress the immune response to immunogenic tumors.

III. Tumor-Associated and Tumor-Specific Antigens

Tumors often arise from the basement layer of cells that—even in adult life—maintain active cell division and provide replacements for cells that have died. As has been illustrated in the white cell series, such replacement cells may go through a series of differentiation steps (changes to more specialized function), at each of which tumors can arise that correspond to their normal counterparts at that particular step of differentiation. Because the number of normal cells seen at that step may be quite small, the corresponding tumor cells that are often present in large numbers may be judged incorrectly to possess a new "tumor-specific" antigen. In reality, the antigen will be a normal tissue antigen previously unrecognized, because it is confined to a small subpopulation of normal cells. Such antigens are "tumor-associated" antigens—normal antigens associated with the tumor.

A. Oncofetal Antigens

Carcinoembryonic antigen (CEA) was named thus because it is found in large amounts only in human embryonic tissues and on the part of the tumor cell surface (the glycocalix) from which it is shed. Another such antigen is α-fetoprotein (AFP), which is secreted by human fetal liver and by most human liver tumors. Because small amounts of both CEA and AFP are present in normal adults, it is possible that both CEA- and AFP-secreting tumors developed from small subpopulations of embryonic-type stem cells retained even in adulthood to serve as basement layers for cell replacement. Because oncofetal antigens are normal antigens, if found on tumors they are "tumor-associated."

B. Oncogene Product Antigens

During the past decade, a virtual revolution has taken place in our understanding of the molecular mechanism of carcinogenesis. Today we believe that many types of cancer cells develop because one or two normal genes—called "oncogenes" because they can endow a cell with cancer-like properties—are switched on at a time inconsistent with the time sequence of their activation set out in the genetic blueprint of the body. The most vital cancer-like property is the ability to continue cell division, irrespective of host attempts to limit it or to destroy the cells. Perhaps the body's ability to limit the proliferation of embryonic-type cells may be impaired by the time adulthood is reached. To date, approximately 50 human "oncogenes" have been discovered. Oncogenes encode molecules that are normal body constituents and would therefore be only "tumor-associated" antigens if located at the cell surface—unless mutations occurred in the oncogenes, leading to the encoding of tumor-specific antigens. Because oncogene-encoded molecules may be present in large amounts, they—like oncofetal antigens—may serve a valuable role as tumor markers and, to a lesser extent, also as potential targets for immunotherapy. [See ONCOGENE AMPLIFICATION IN HUMAN CANCER.]

C. Virally Induced Antigens

Hepatitis B virus (HBV), herpes viruses, retroviruses, and papovaviruses all can cause tumors under natural conditions. They can cause tumors not only in some species of mammals, but also in nonhuman primates and in human beings. In terms of the oncogene theory, oncogenic viruses can produce tumors in several ways. Most directly, once a DNA virus has inserted its DNA into the cellular genome, the viral DNA can function as an activated oncogene. For retroviruses, the viral RNA first must be transcribed into DNA before that DNA can be inserted. Other oncogenic viruses act by switching on the host cell's own oncogenes or else by inactivating oncogene repressor genes.

Oncogenic viruses can generate tumor-specific cell surface antigens by several different mechanisms. Most simply, and probably most frequently, they can produce copious new virus particles or viral components; these appear on the cell surface, thereby conferring the antigenicity of the oncogenic virus onto its virally induced tumor cell. Less directly, the viral DNA within the genome of the transformed cell can specify novel cell surface antigenicities that are distinct from those on the native virus. Finally, genetic recombinations between two or more viruses can occur that may specify novel cell surface antigens. As already noted, most tumor-specific antigenicities induced by one particular oncogenic virus are identical, irrespective of the individual or even of the species in which the tumor has occurred. This means that such tumors cross-react and that vaccination against such tumors—most of which are highly immunogenic—is entirely feasible.

Liver cancer is far more common in hot, wet areas of the world than in Western nations or in the United States, were it ranks 25th. Presently, experts believe that 80–90% of cases of liver carcinoma are caused by HBV. The mode of HBV carcinogenesis is unusual: the surface antigen of HBV acts as a chemical carcinogen for liver, whereas integration of the viral DNA into the cellular genome is merely a random event that has no effect on induction of liver cancer. Principal among the other agents that cause liver carcinoma or act as cofactors for HBV are aflatoxin B, a toxin produced by a fungus that thrives on cereals and nuts stored under hot and moist conditions, and long-continued heavy use of alcohol, which produces a 10% incidence of cirrhosis. Chronic HBV infection results in a relentless pathologic process that, barring death from other causes, leads to eventual death either from cirrhosis and/or from liver cancer. Effective vaccines against HBV are now available. In China, where liver cancer is the most common form of cancer, a program of vaccinating all newborn children has begun. In the United States, a cancer preven-tion center based in Philadelphia provides hepatitis B vaccinations for high-risk groups (i.e., the families of HBV carriers, drug addicts, and others). In the great majority of HBV-negative persons who have been vaccinated, complete protection against HBV infection—and therefore against HBV-induced liver cancer—is expected.

Many types of leukemogenic viruses have been discovered in animals. Among the most interesting is feline leukemia virus (FeLV), a retrovirus (RNA virus) that infects about 3% of cats in the United States. Of cats exposed to FeLV through contact with infected cats, about one-third receive insufficient virus to become either infected or immune, one-third become immune to the virus, and one-third become infected. Within 4 years of infection, more than 80% of infected cats die of leukemia. A vaccine is now commercially available to immunize cats against FeLV. The vaccine contains the FeLV envelope gp70 protein, which is produced by recombinant DNA techniques. [*See* RETROVIRAL VACCINES.]

In the case of human leukemias, viral causation is certain only for one rare type: T-cell leukemias caused by the human T-cell leukemia virus-I (HTLV-I). This virus is gradually spreading around the world. Although it shares only 2% of its genes with the AIDS virus, like that virus, HTLV-I is spread by blood transfusions, sex, and mothers' milk. Although the virus has been discovered in a significant proportion of people in Japan, the Caribbean, and the southernmost United States, no efforts to prepare a vaccine against it seem to have been made as yet. [*See* LEUKEMIA.]

A study by Margaret Davis published recently in the *Journal of the National Cancer Institute* shows that children who are breast-fed for more than six months are partially protected against childhood lymphoma, which is a rare disease. These children develop lymphomas only one-half as frequently as children who were breast-fed for a shorter period. Because breast-feeding allows the baby to absorb from the mother's milk her antibodies that protect against many types of bacteria and viruses, this finding suggests that some types of childhood lymphomas are caused by one or more cancer-causing viruses. When these have been identified, vaccination of babies against lymphomas will be possible. We have explored such vaccination in a model system in our laboratory. We found that vaccination of newborn mice against a known mouse leukemia virus is very effective in one but not in another strain of mice. This suggests that a future human leukemia virus vaccine will be effective in many but not in all infants. [*See* LYMPHOMA.]

The most common childhood tumor in hot and wet areas of Africa and Southeast Asia is Burkitt's lymphoma (BL). In the same areas, nasopharyngeal carcinoma (NC) is also common. In those areas in Africa, around 99% of patients with BL and a high percentage of patients with NC have tumor cells bearing markers of the Epstein-Barr virus (EBV), a herpes virus. (In contrast, BL is rare in the United States, and only 10% of U.S. patients with BL have cells that bear the EBV antigen.) All patients have a translocation involving the oncogene *myc*. Possible cofactors include immune depression through injury by diseases such as malaria or schistosomiasis, whereas possible causes other than EBV include carcinogens produced by fungi common in the same hot and humid climates and other, as yet unrecognized viruses. Thus, three different factors combine to cause BL: EBV, the *myc* oncogene, and additional cofactors. [*See* EPSTEIN-BARR VIRUS.]

Although the precise role of EBV is uncertain, it may not be premature to use a genetically engineered vaccine to EBV in areas where BL and NC are common, to determine whether immunity to EBV protects against these tumors. Because more than 80% of adults the world over have been exposed to EBV, immunization would be done in infants.

Potentially oncogenic venereal infections include herpes simplex virus type 2, cytomegalovirus, human papillomavirus (HPV), and the AIDS virus (HIV-I). Recently, interest has focused on HPV. A strong correlation exists between a silent HPV infection of the lower genital tract of women and the occurrence of cervical cancer. Analysis of tissue culture lines of human cervical cancer indicate the presence of the viral genome and suggest a functional role for HPV in malignant (cancer) transformation. Further, one-third of the male partners of women with cervical cancer develop penile cancer containing DNA sequences of HPV, contrasted with only 1% of male partners of women with normal cervical tissues. Because as many as one-third of German women who are sexually active are infected with HPV virus, and a similar situation may prevail in other Western nations, it is to be hoped that a genetically engineered vaccine against HPV will be prepared and tested soon. [*See* ACQUIRED IMMUNODEFICIENCY SYNDROME (VIROLOGY); HERPESVIRUSES; PAPILLOMAVIRUSES AND NEOPLASTIC TRANSFORMATION.]

Proof of the viral causes of human cancers is difficult to obtain, because it is impossible to infect human beings deliberately with suspected viral agents. Most of our present information comes from epidemiological studies and depends on questionable correlations between the incidence of a certain type of tumor, and the presence of a particular virus (or viruses) in that tumor, as compared with the incidence of that virus in the general population. Therefore, for all but a very few types of human cancers, a viral etiology presently is highly questionable. However, the new techniques of molecular biology are greatly helping to speed this quest. Such work is of great importance, because success in the identification of a particular virus as the cause of one type of human cancer brings with it the probability that a vaccine against the virus could be engineered, which could be eventually used to eradicate that type of cancer.

D. Chemically Induced Antigens

Tumors induced by chemical carcinogens in laboratory animals tend to be immunogenic, although often not as highly as virally induced tumors. Chemically induced tumors possess cell surface antigens that are stable and heritable. These antigens can be used to immunize animals both prophylactically (i.e., animals are first immunized, then challenged with viable tumor cells) or else immunotherapeutically (i.e., the immunizations are performed only *after* challenge with viable tumor cells).

Work with tumors induced by different types of chemical carcinogens has shown that for each individual tumor, its tumor-specific antigen is unique—even in the situation in which two tumors arise on the same animal. Tumors induced by a single chemical may show cross-reactivity, but this is often of such low degree that it is difficult to detect. This lack of cross-reactivity between different tumors induced by the same chemical carcinogen poses a nearly unsurmountable problem for use of such tumor-specific antigens for diagnosis, vaccination, or immunotherapy, except in the situation in which these are personalized to deal with a single patient's tumor. However, if a given chemical carcinogen produces a type of tumor cell that is the neoplastic (cancer) counterpart of a small, antigenically distinct subclass of normal cells, its tumor-associated (normal) antigens can be used for all tumors of that type produced by that carcinogen to accomplish the same purpose, although perhaps not as easily: for diagnosis, the antigenically identical subclass of normal cells must pose only a surmountable problem; for therapy, eradication of this subclass of normal cells must not threaten the patient's life.

The dose of the chemical carcinogen used to produce tumors in experimental animals is often orders of magnitude higher than the doses of carcinogens to which human beings are exposed. What little information we have suggests that the immunogenicity of chemically induced tumors decreases as the dose of the carcinogen is decreased. We completely lack information about the immunogenicity of tumors induced at the very low doses at which we encounter chemical carcinogens in our daily lives.

E. Radiation-Induced Tumors

Brilliant experimental work by radiologist Henry Kaplan proved that radiation-induced leukemias of mice seldom were caused directly, due to radiation-induced mutations in the DNA of lymphocytes from which they developed. More commonly, the effect of the radiation was permissive rather than direct: it was to kill host immune cells, thereby permitting the outgrowth of leukemia cells initiated by leukemogenic viruses being carried by the mice.

A similar situation may well obtain in mice for carcinogenesis by radioisotopes. For instance, the bone-seeking radioisotope strontium-90 appears to act primarily in a permissive role for the initiation of bone sarcomas. The primary oncogenic agents are C-type oncogenic viruses carried by the host, viruses that remain latent and only occasionally produce tumors in unirradiated hosts.

To answer the question "How immunogenic are radiation-induced tumors?" we need to take into account the effect of passaging and concentrating oncogenic viruses in the laboratory before injecting them into newborn animals. In direct contrast, oncogenic viruses are present in their native form and at low concentration when activated by the permissive action of radiation. Thus, we might expect radiation-induced tumors to be far less immunogenic than those produced by virus injection into newborn, and this seems to be the situation (Table I).

F. Tumors Induced by Other Causes—or None

Tumors may also arise by "physical" carcinogenesis: as when a piece of plastic film is introduced into the peritoneal cavity; or when asbestos fibers, too long to be swallowed and disposed of by macrophages, are inspired into the lung. As in the case of radiation carcinogenesis, physical carcinogens may merely permit latent oncogenic viruses to initiate tumor cells—which are initially out of reach of attack by host immune cells, being physically protected by film or fibers. Such tumors should have immunogenicities similar to those of radiation-induced tumors.

When tumors appear spontaneously, they may do so because they are genetically fated to arise, because of promotion by hormones or other cofactors, or because of the action of various carcinogens—viruses, chemicals, radiations, and fibers—on cells. Even when the primary oncogenic agent is a virus, the same dose concentration effects mentioned above may apply.

In human beings, much work has been done on melanomas, because they appear to be highly immunogenic. Unfortunately, animal studies suggest that UV-induced tumors (which include some melanomas) possess an extraordinary escape mechanism. A single tumor induced in a mouse by UV light may display a number of antigens on its surface; we shall call three of these A, B, and C. The host responds with a rejection reaction directed at the strongest antigen, A. Meanwhile, clones that possess only antigens B and C can grow unrestrained. Then the host eradicates all tumor cells displaying the next strongest antigen, B. However, in the meanwhile, clones displaying only C have grown unmolested; when the host finally switches to react against C, it finds itself overwhelmed by the large number of cells bearing antigen C. At least some human melanomas behave in a way that suggests that they use this escape mechanism. [*See* MELANOMA ANTIGENS AND ANTIBODIES.]

Other types of human tumors worthy of special attention because of suggestive evidence of immunogenicity include transitional cell carcinomas of the bladder, renal carcinoma, skin carcinomas, Burkitt's lymphoma, neuroblastoma-type IVS, childhood lymphomas and leukemias, bone sarcoma, and choriocarcinoma.

The cell surfaces of human tumors contain complex carbohydrate molecules that sometimes display deletions or changes at their ends. Such changes are responsible for the appearance of new antigenic specificities. For instance, a gastric adenocarcinoma arising in a patient whose blood type is O may display the "illegitimate" blood group antigen A. Further, tumor cells sometimes display the embryonic rather than the adult form characteristic of a given cell surface glycoprotein; this suggests that they originated from embryonic-type stem cells or else were converted to such cell types through activation of an oncogene. If identical antigenic changes were present on *all* tumor cells, their antigenic uniqueness could be used for immunoprevention or immunotherapy.

IV. Immunosurveillance in Light of AIDS

The basic thesis of immunosurveillance is that the immune system prevents the emergence of many more tumors than actually appear. Unfortunately, this thesis sometimes has been debated as by a lawyer seeking to make a winning case for a client, rather than as by a scientist seeking a fair exposition of pros and cons, to arrive at present truth dependent on present evidence. Here are some of the pros and cons:

1. *Pro:* The age-related incidence for some human cancers shows a peak for children 2–3 years old, expected from oncogenic viral infections when immunity is immature.
   Con: The cause of these tumors may be genetic, not viral.
2. *Pro:* The incidence of most human cancers increases greatly with age. This increase corresponds to a gradual decrease in immune competence beyond early adulthood.
   Con: The increase in tumor incidence is caused by the cumulative effects of exposure to carcinogens and promotors and reflects the long latent period of cancer.
3. *Pro:* In the many types of human immune deficiency diseases, the incidence of leukemias and lymphomas averages 10%, of epithelial tumors and sarcomas about 1%.
   Con: Survival of persons with these diseases is not as brief as is claimed; if immunosurveillance was a reality, more carcinomas and sarcomas should appear.
4. *Pro:* Transplant recipients and others receiving immunosuppressive treatment have a greatly increased incidence of leukemias and lymphomas; epithelial tumors and sarcomas are increased 50-fold, and sun-related skin cancer in sunny areas, 100-fold.
   Con: The immunosuppressive agents used are carcinogenic.
5. *Pro:* Athymic (nude) mice that have severe deficiencies in rejecting grafted tissues do develop malignancies when prevented from dying when young.
   Con: Many more malignancies might be expected if immunosurveillance is real.

6. *Pro:* Natural antibodies directed against tumor are found in some types of cancer.
   Con: Because the tumor continues to grow, these antibodies must be ineffective.
7. *Pro:* Specifically sensitized host cells have been demonstrated in many types of tumor.
   Con: Some such cells have been shown to enhance rather than inhibit the tumor.
8. *Pro:* Failure to reject an immunogenic tumor is due to presence of suppressor cells.
   Con: This merely explains why immunosurveillance is ineffective.
9. *Pro:* Tumor cells but not normal cells are attacked by NK cells and by macrophages.
   Con: On occasion, macrophages can stimulate rather than retard growth.

Vaccination for prevention of cancer and cancer immunotherapy are often perceived as potentiating the natural propensity for surveillance against tumors—and as impractical if surveillance does not exist. This perception is partly wrong, for artificial means for inducing immunity can be more potent than natural immunity. Because cancer strikes more than 20% of the U.S. population, it is obvious that immunosurveillance frequently fails. No wonder that it is hotly debated. [*See* IMMUNE SURVEILLANCE.]

Because the AIDS virus (HIV-I) specifically destroys human helper/inducer (CD4-positive) T lymphocytes, the spread of this virus allows us to see the outcome of this insult to the human immune system. Because helper cells are implicated in the T-cell response that only takes place against immunogenic tumors, we should expect AIDS patients to develop a higher than normal incidence of the type of tumors induced by oncogenic viruses, UV radiation, or high doses of chemical carcinogens (see Table I). Once the deficiency of helper T cells is sufficient for AIDS to develop, at present survival is rarely more than 3 years. Given this short survival, it is all the more remarkable that the incidence of Kaposi's sarcoma in male homosexuals has been reported as 2,000 times higher than in age-matched controls and that the incidence of B-cell and other types of lymphomas is increased several-fold. This limited and specific outgrowth of tumors after destruction of helper T cells suggests that immunosurveillance by the host employs many different cell types and mechanisms and that the destruction of only one type of host defense cells will lead only to a limited outgrowth of tumors. However, transgenic mouse studies suggest that Kaposi's sarcoma is directly induced by HIV-I and not the result of immunodeficiency. In fact, it usually begins before much immunodeficiency is evident.

There are many other arguments in favor of immunosurveillance, for instance, the experimental success in initiating tumors by injecting oncogenic viruses into newborn animals (which are immunologically immature) and the failure of most such viruses to induce tumors in adults. However, most arguments can be explained alternatively in other ways. The fact that the body obeys nearly all the time the dictates of its genetic blueprint suggests that there are precise mechanisms for maintaining the genetically designed integrity of each organ, which of necessity implies destruction of cells not designed to appear. Although we understand little about how this control is maintained, it argues for surveillance.

V. Immunodiagnosis of Cancer

Part of immunodiagnosis is the evaluation of the state of patients to determine whether the tumor has harmed their general immune competence and whether they possess specific reactivity (either humoral or cellular) against their tumor (see Section II). A second part of immunodiagnosis is related to the detection of tumor antigens (see Section III). Here, we shall deal with this second task.

A. Monoclonal Antibodies

Monoclonal antibodies have revolutionized the diagnosis of cancer at the cellular level. Their name reflects their secretion by cells expanded from a single clone of cells. This clone is constructed by fusing a cell secreting the desired antibody (a splenic lymphocyte from an animal immunized with tumor cells or purified tumor antigen) with a myeloma cell (a malignant tumor cell derived from the type of normal lymphocytes that secrete antibodies). The myelomas used for fusion are selected because they do not secrete a complete immunoglobulin. These hybrid cells (hybridomas) can be grown indefinitely to produce large amounts of the antibodies made by the splenic lymphocyte used for the fusion. Their main advantage over conventional antibodies is their specificity. Because many hybridomas can be screened for specificity, the investigator has a good chance of finding one hybridoma secreting an antibody of (or close to) the desired specificity and binding strength. The usual quest is for an antibody that reacts only with one type of cancer but not with its benign precursor cells, nor with other types of tumors or normal tissues.

It was only in 1966 that the first diagnosis of leukemic lymphocytes using a conventional antibody was performed in the writer's laboratory. Since the advent of monoclonal antibodies, the precise differentiation between different subsets of neoplastic cells of a given type is now possible for leukemias and lymphomas. Even for routine histological diagnosis using tissue sections, staining techniques employing monoclonal antibodies with precise specificities are now a valuable adjunct to conventional staining techniques, many of which may soon be obsolete.

Despite commercial availability, many investigators need to prepare their own monoclonals for special projects. For instance, monoclonals can reveal variations in antigenicity within a given tumor, information that is valuable if immunotherapy is planned. Other uses are to determine differences between antigens of the same histologic type taken from different patients and changes in antigenicity as tumors of a single type progress through stages from benign to malignant. [See MONOCLONAL ANTIBODY TECHNOLOGY.]

VI. Immunotherapy of Cancer

In "active" immunotherapy, the patient's own immune system is stimulated to become the "effector"—the active agent—for rejection of the patient's tumor. In "passive" immunotherapy, injected substances or cells are the effectors. The distinction between these two kinds of immunotherapy sometimes blurs, because effector materials may stimulate the immune system as a secondary effect.

A. Active Immunotherapy

Early in this century, Coley's toxin, a mixture of killed bacilli that causes fever rather like a bacterial infection, was used to obtain partial or complete regression of cancer in a number of patients. At least part of the explanation for its effects lies in the nonspecific "recruitment" (mobilization) of host immune cells as part of the fever process, with immune cells attacking the tumor cells as if they were "innocent bystanders," without specific immunization. However, later work in my laboratory showed that if this attack was to be powerful, it had to be followed by an immune response specific for tumor antigens, and that this specific response could occur only if the tumor was immunogenic. As a result of work in many laboratories on BCG, on *C. parvum* (a killed bacillus), and on many other biomodulators (biological response modifiers), the following precepts have emerged:

1. Genetic background greatly affects the host's response to any type of antigen—including tumor antigens.

2. Life history, including the extent of exposure to and immunity against different antigens, age, and current status of immune responsiveness are also important.

3. Restriction of immunotherapy to minimal tumor loads is an essential requirement for success, as illustrated by animal experiments. Thus, the best time for immunotherapy is soon after the primary tumor and accessible metastatic tissue have been surgically removed and any remaining tumor further reduced in mass by chemotherapy or radiotherapy.

4. A multifaceted approach should be used, in which immune stimulators (tumor vaccines and biomodulators) and inhibitors (cyclophosphamide, X-rays) are used sequentially at times chosen to maximize the tumor rejection response and to minimize the emergence of suppressors.

5. Immunotherapy with BCG or *C. parvum* in animals is ineffective unless the tumor is strongly immunogenic. A similar conclusion is suggested by recent trials of immunotherapy with IL-2 or cancer vaccines in human beings.

6. Lacking a method for determining the immunogenicity of human tumors, a rough guide can be obtained from indirect indications (last paragraph, Section III,C).

7. Because some animal models employ tumors with strong immunogenicities—often without quantifying them—the need for conservative extrapolation from the results of immunotherapy in animals to immunotherapy in patients is evident.

8. Methods that increase the immunogenicity of tumor cells should be explored further.

9. When indicated, restorative immunotherapy—to redress the immune imbalance created by the tumor in the patient—might well be chosen to precede immunotherapy.

The time has not yet arrived when immunotherapy will take its place beside surgery, radiation therapy, and chemotherapy as a fourth method of cancer treatment. The startling advances in chemotherapy during the past 30 years were based on a consensus of what constituted meaningful animal model systems and how the results of experiments should be expressed: namely, in terms of the increased tumor load (expressed in logarithms of tumor cell kill) that was rejected in animals receiving therapy as compared with untreated controls. A similar consensus must be reached in immunotherapy, to allow a comparison of the work performed by different investigators. Lacking such a consensus, meaningful extrapolation from animal experiments to the treatment of human tumors is difficult.

In immunotherapy, one stand-out success is the use of "adjuvant contact immunotherapy" in the case of bladder cancer. In this approach, BCG is injected directly into the tumor, where it can induce an intense delayed-type hypersensitivity reaction (a rejection reaction performed by immune cells). In a recent clinical trial that compared BCG immunotherapy with doxorubin chemotherapy for the treatment of bladder cancer, immunotherapy succeeded in producing 1.5-fold fewer recurrences and 8-fold longer durations of response than did chemotherapy. Bladder cancer has an incidence of almost 50,000 cases per year and comprises 2.5% of all cancer deaths in the United States.

Use of the lymphokine IL-2 deservedly has attracted much attention for immunotherapy of patients with renal-cell carcinoma or melanoma (see Section II,B). Equally promising—and far less toxic—is immunotherapy with tumor antigen vaccines, which is underway mostly in patients with the same types of tumors. Many different types of vaccine are being investigated. The protocol of one of the most exciting ongoing trials is based on extensive animal experiments. In this trial, the tumor of a patient with colon or rectal cancer is surgically removed (see 3 above), then immediately dissociated and stored frozen. When needed, tumor cells are thawed, irradiated to make them nontumorigenic, admixed with BCG, and used to vaccinate the patient. The initial result has been a 50% reduction in recurrences. It is still too early to know whether 5-year survival rates (the gold standard for meaningful cancer therapy) will be improved.

Many other types of immunotherapy are presently in different phases of clinical trial. Particularly interesting is work with tumor necrosis factor (TNF) and with interferon (INF). TNF is a product of stimulated macrophages that can induce hemorrhagic tumor necrosis directly or can activate macrophages to destroy tumor cells. Because TNF is toxic at the high doses required for effectiveness, clinical trials are being conducted in which TNF at lower total doses is introduced directly into the tumor. Endotoxin (a bacterial toxin that is also the active principal in Coley's toxin) is highly effective in stimulating macrophages to produce TNF and also induces secretion of IL-1 and INF. This explains the efficacy of Coley's toxin.

Interferons (there are several kinds) have many biological activities, including the enhancement of tumor-associated antigens such as CEA; they can act in synergy with TNF. INF-alpha can induce the remission and alter the natural history of hairy-cell leukemia and chronic myelogenous leukemia. INFs have shown lesser activity against several other types of tumors. [See INTERFERONS.]

Hematopoietic growth factors (HGFs) constitute another family of cytokines that promises to show clinical usefulness. They act on the bone marrow to stimulate rapid growth and maturation of bone marrow cells. HGFs may prove capable of countering bone marrow failure resulting from chemotherapy.

B. Passive Immunotherapy

Because cytokines such as TNF and HGF have *direct* as well as indirect effects on tumor cells, they could be classified equally well as passive immunotherapy agents. Modes of passive immunotherapy include the expansion in tissue culture of host lymphocytes in the presence of IL-2, followed by reinfusion of a large number of these expanded cells (LAK or TIL cells) into the tumor-bearing host. Several different clinical trials employing either LAK or TIL cells are in progress presently (see Section II,B).

Monoclonal antibodies possess the greatest potentiality for tumor therapy. Their production and use for diagnosis have already been described (Section V,A). Because substances from a different species provoke a powerful immune response, repeated use of mouse monoclonals in human beings leads to their rapid elimination. Instead, human monoclonals are coming into use. Monoclonals have been used for therapy mainly in three different ways.

First, they can act as cytotoxic agents: either for direct killing of tumor cells with the help of host complement (a family of proteins that facilitates cell killing); or indirectly, by acting as binding agents between host effector and tumor cells, thereby mediating the destruction of tumor cells. Presently,

this type of therapy has benefited only a few of the patients enrolled in clinical trials.

Second, monoclonals can act as carriers for cytotoxic drugs, natural toxins, or radioactive isotopes. Their mission is to act as bullets, delivering their attached lethal charge to each tumor cell in the body. The difficulty lies not in attaching a lethal substance to an antibody molecule, but in achieving antibody localization in a tumor. There are many problems: presence of tumor antigens in the circulation; lack of sufficient strength in the binding of antibodies to tumor cells; slow and uneven uptake of antibody by the tumor due to variations in blood supply and in antigen content between different portions of the tumor; loss of antibodies through specific or nonspecific binding to cells of the reticulo-endothelial system; and immunization of the patient against injections of the monoclonal antibody, which is a foreign protein. This last problem can be solved by the use of human rather than mouse monoclonals. Despite such problems, radiolabeled antitumor monoclonals already have attained some limited success in the diagnosis of metastases.

Third, anti-idiotypic antibodies (antibodies against the idiotype, the active binding portion of another antibody) have been used for therapy in two different ways. The first type of therapy takes the same approach as the body seems to choose, to regulate the numbers of cells in its normal B-cell clones: it produces anti-idiotypic antibodies against the idiotype of the antibodies attached to the cell surface of a particular B-cell clone, thereby suppressing it. Similarly, such antibodies can be used to suppress B-cell tumors. In a clinical trial with 11 patients with B-cell tumors, each was given monoclonal antibodies tailor-made to react with the idiotype of the immunoglobulin molecules on the cell surface of that patient's tumor. Only a single patient developed a long-lasting remission: the other 10 patients relapsed, due to the development of tumor cells with changed idiotypes. [See IDIOTYPES AND IMMUNE NETWORKS.]

A very different approach is to use idiotypes to mimic tumor antigens. If the tumor idiotype is shaped like a hand and the anti-idiotypic antibody as a glove that fits over the hand, a second antibody prepared against the first anti-idiotypic antibody (the glove) again assumes the shape of part of a hand—thereby mimicking the original tumor idiotype, which can act as a tumor-specific antigen. Thus, it is possible to use such second anti-idiotypic antibodies instead of the actual tumor antigen for immunization. So far, this approach has proven only mildly effective.

C. Tumors: Resourceful Adversaries

Finally, it must not be forgotten tumors have at their disposal a variety of mechanisms that can frustrate or even co-opt host defenses. One such mechanism is "antigenic modulation," the outgrowth of tumor cells with antigenicities different from those of earlier generations of the same tumor. A second mechanism is the release of blocking factors that can be either specifically or nonspecifically immunosuppressive. A third mechanism is the copious shedding of tumor-specific or tumor-associated cell surface antigens, thereby providing sham targets for host effector cells. Because of the existence of these mechanisms, immunotherapy is most effective when the tumor is as yet small. By the same token, the most effective strategy is to mobilize immunity at a time when the tumor load is zero, by vaccination against likely future tumors.

VII. Summary

Although the body can strongly reject cells that are genetically foreign, the mechanisms for regulating the numbers of its own cells are delicate. Tumors are the body's own cells but have escaped its finely tuned mechanism that impels normal cells to conform to the genetic blueprint. Depending on the tumor's immunogenicity, different types of cells respond to it. NK cells can kill tumor cells spontaneously and can be activated by IL-2 to kill more effectively. Macrophages can be activated to kill tumor cells both nonspecifically and specifically. Strongly immunogenic tumors induce a host response of cytotoxic T lymphocytes, but this can be inhibited by suppressor T cells and by suppressor factors. Some tumors express normal antigens in abnormal amounts, whereas others display tumor-specific antigens on their cell surface. Tumors induced by UV light, by viruses, and by high doses of chemical carcinogens fall into the latter class. Because tumors caused by a single type of virus share common antigens, vaccination against oncogenic viruses eventually should greatly reduce the incidence of virus-caused cancers in humans.

Immunosurveillance predicts that the immune system prevents the emergence of many more tumors than actually appear. For virally induced tumors, the evidence in favor of immunosurveillance seems overwhelming. Immunodiagnosis of cancer has been revolutionized by the introduction of monoclonal antibodies. Their use for therapy of human cancer—whether as cytotoxic agents, as carriers of toxic molecules, or as anti-idiotypic antibodies—is only in its beginning stages. In "active" immunotherapy, the patient's own immune system is to be stimulated to reject the tumor. For success, immunotherapy requires attention to many conditions, which include restriction of this therapy to strongly immunogenic tumors, reduction to minimal tumor mass before therapy, and inhibition of the emergence of suppressor cells. Some promising results have been obtained with tumor vaccines containing the patient's own tumor antigens and with lymphokine-activated killer cells. Work with other lymphokines, such as tumor necrosis factor and interferon, suggests higher effectiveness for multimodal therapy.

Bibliography

Awwad, M., and North, R. J. (1989). Cyclophosphamide-induced immunologically mediated regression of a cyclophosphamide-resistant murine tumor: A consequence of eliminating precursor L3T4+ suppressor T-cells. *Cancer Res.* **49,** 1649.

Badger, C. C., and Bernstein, I. D. (1989). Treatment of murine lymphomas with anti-Thy-1.1 antibodies. *In* "Cell Surface Antigen Thy-1" (A. E. Reif and M. Schlesinger, eds.). Marcel Dekker, New York.

Kagan, J. M., and Fahey, J. L. (1987). Tumor immunology. *JAMA* **258,** 2988.

Law, L. W., ed. (1985). Tumour antigens in experimental and human systems. *Cancer Surveys* **4,** number 1.

Metzgar, R. S., and Mitchell, M. S. (1989). "Human Tumor Antigens and Specific Tumor Therapy." Alan R. Liss, New York.

Mitchell, M. S., ed. (1989). "Immunity to Cancer. II." Alan R. Liss, New York.

Reif, A. E. (1978). Evidence for organ specificity of defenses against tumors. *In* "The Handbook of Cancer Immunology," vol. 1 (H. Waters, ed.). Garland STPM Press, New York.

Reif, A. E. (1982). Antigenicity of tumors: A comprehensive system of measurement. *Methods Cancer Res.* **20,** 3.

Reif, A. E. (1985). Vaccination of adult and newborn mice of a resistant strain (C57BL/6J) against challenge with leukemias induced by Moloney murine leukemia virus. *Cancer Research* **45,** 25.

Reif, A. E. (1986). Relationship of success in classical immunotherapy to the relative immunorejective strength of the tumor. *J. Natl. Cancer Inst.* **77,** 899.

Reif, A. E. (1987). Immunosurveillance reevaluated in light of AIDS. *In* "Lectures and Symposia of the 14th International Cancer Congress," vol. 5. Akademiai Kiado, Budapest.

Reif, A. E., and Mitchell, M. S., eds. (1985). "Immunity to Cancer." Academic Press, Orlando.

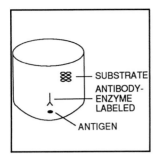

SUBSTRATE
ANTIBODY-
ENZYME
LABELED
ANTIGEN

Immunoassays, Nonradionucleotide

KENNETH D. BURMAN, *Walter Reed Army Medical Center, Washington, D.C.; Uniformed Services University of the Health Sciences, Bethesda, Maryland*

Glossary

Chemiluminescent assay Technique of measurement that uses compounds with emitted light as a detection system; in most assays, the greater the unknown concentration, the more light is emitted
Enzyme-linked immunosorbent assay (ELISA) Generally uses antibodies to detect unknown substances; an enzyme may be attached to an antibody, which when activated will change color; the color density will correlate with the concentration of the unknown substance
Fluorescent assay Technique of measurement that uses emitted fluorescence intensity as a detection system
Fluorescent polarization assay Technique of measurement that uses the deflection or degree of polarization of transmitted light in its detection system
Immunoassay Method of measurement that uses antibodies to detect concentrations of unknown hormones, drugs, or substances
Nonradionuclide assay Any technique or method of measurement of hormones, drugs, or other substances that does not use isotope (e.g., enzyme-linked immunosorbent and chemiluminescent assays)
Radionuclide assay Any technique or method of measurement of hormones, drugs, or other substances that does use isotope (e.g., radioimmuno- and radioimmunometric assays)
Thyrotrophin Secreted by the pituitary gland, a hormone that stimulates function of the thyroid

gland and production of the active thyroid hormones triiodothyronine (T_3) and thyroxine (T_4); specific thyrotrophin receptor molecules are on thyroid gland cells

NONRADIONUCLIDE IMMUNOASSAYS[1] allow the measurement of unknown concentration of drugs, hormones, or other substances; these assays do not employ isotopes. The major types of these non-nuclide techniques are enzyme-linked immunosorbent assay (ELISA), chemiluminescence, fluorescence, and fluorescence polarization. Non-nuclide assays can be highly sensitive and, in most cases, easy to perform. Compared with assays employing radionuclides (e.g., radioimmunoassays (RIAs), radioimmunometric), the reagents used for non-nuclide assays are, in general, inexpensive, easy to obtain, have a long shelf-life, and do not require special radioactive disposal or administration. [*See* RADIOIMMUNOASSAYS.]

I. General Comments

A nonradionuclide immunoassay refers to any assay that uses an antibody for detection or separation but does not employ any type of radionuclide. Several points should be emphasized at the onset. First, no inherent fundamental differences exist in principle between immunoassays that employ isotope and those that do not. The basic chemical interactions are usually quite similar; however, the detection system in these two assay systems are quite different. Second, RIAs and nonradionuclide

1. The opinions or assertions contained herein are the private views of the authors and are not to be construed as official or as reflecting the views of the Department of the Army or the Department of Defense. Mention or description of a particular assay or reaction does not imply endorsement of that product.

assays each have their own set of advantages and disadvantages, which may lead one laboratory to choose one system over another. For example, sensitivity or specificity usually depends on other factors in addition to the detection system, and, as a result, for a particular hormone or drug, an RIA may be more sensitive, whereas in another circumstance a non-nuclide assay may have advantages. Therefore, the detection system per se (e.g., isotope or nonisotope) is usually only one determinant of an assay's characteristics, and most often practical laboratory or cost considerations will dictate which assay is most useful in a given circumstance.

As a third general comment, one aspect of immunoassays that frequently enhances sensitivity and specificity relates to the use of two or more antibodies in a system, resulting in a "sandwich"-type assay. Usually, these type of assays are more sensitive and specific than classic single antibody assays, irrespective of whether or not isotope is employed in the detection system. In the following analysis, ELISA assays are used as a prototype for non-nuclide assays, and general and specific principles of these techniques will be considered; then, other important types of nonradionuclide assays will be described.

II. General Principles

A. Customary Assay

In a customary assay, the antigen is allowed to react with an antibody that would be typically labelled with an enzyme (e.g., alkaline phosphatase). Under suitable conditions, the antigen and antibody would be allowed to react, and then a substrate (in the case mentioned here, p-nitropenyl phosphate) would be added. This reaction results in a change in the solution color from clear to yellow. Furthermore, the intensity of the yellow color would be directly correlated with the number of antibody molecules that are bound to the antigen (Fig. 1). Control samples are always included, and time at which the substrate is allowed to react is constant in an individual assay. In the preliminary analysis of an assay, the appropriate control antigen and antibody reaction are analyzed such that it is known when the enzyme reaction has plateaued. An assay in which the reaction was still progressing rapidly would be unusual; comparison of one tube to another would be difficult

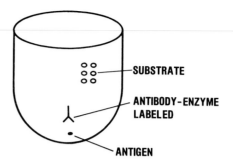

FIGURE 1 Schematic depiction of a customary ELISA assay.

in such a situation. We shall discuss further details of such a reaction later. In such a customary assay, the antigen concentration could be determined by analysis of the color change or by optical density of the solution. A standard curve, usually with several sets of known control samples, must also be run (Fig. 2). In this manner, an unknown sample concentration could be determined by comparison with known concentrations in the standard curve. In non-nuclide immunoassays, as in RIAs, a great deal of preliminary analysis must be performed. The concentration of antibodies, proper enzyme label, and substrate concentration as well as time of incubation, type of optimal density reader, and type of plate, for example, must all be taken into consideration before a non-nuclide assay can be used reproducibly. This type of assay can be modified by using two antibodies for detection, usually enhancing sensitivity (Fig. 3).

B. Single Sandwich ELISA

In single sandwich-type assay, an antibody is allowed to adhere to the bottom of an ELISA well, and the antigen, perhaps in serum or other biologic fluid, is then layered onto the antibody. After thorough washing, the second antibody, which is enzyme-labelled, is allowed to react with the solution (Fig. 4). Finally, a substrate is added and color change of the solution is quantitated, and the concentration of the antigen in the biological sample is determined by comparison to a standard curve. Of course, each assay contains several known standards, as well. Obviously, the concentration of antigen in the biologic sample is directly proportional to the amount of enzyme-labelled antibody bound to the antigen, and, as a result, optical density readings are directly related to the color change (Table I). The antibody plated on the bottom of the microti-

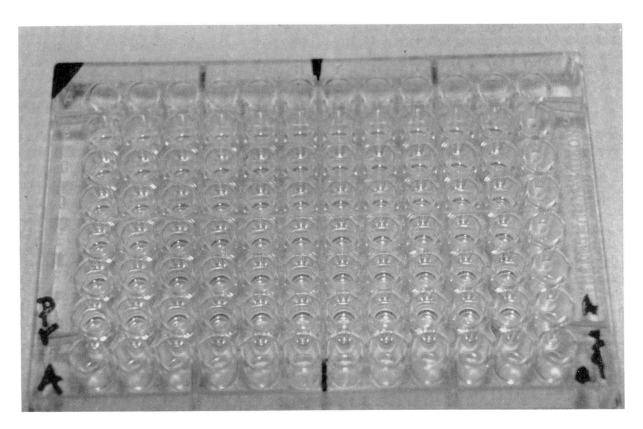

FIGURE 2 Typical ELISA plate using an alkaline phosphatase labelled antibody. The intensity of the yellow color after the addition of substrate is correlated with the unknown samples' concentration. Note that the standard curve is in columns 3–6 and that the unknown samples are in columns 7–11. The outside lanes are not used for sample determination.

specificity and sensitivity is high, largely because of the two different antisera used. Again, careful attention to each component of the assay is required.

ter plate usually recognizes a different portion, or epitope, of the antigen than does the enzyme-labelled antibody. The use of those two antibodies, which recognize different portions of the antigen, represents the most advantageous part of this assay (Fig. 5). The background is low and the degree of

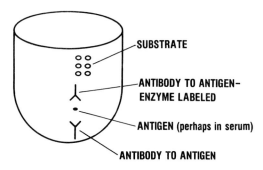

FIGURE 4 Schematic depiction of a single sandwich ELISA.

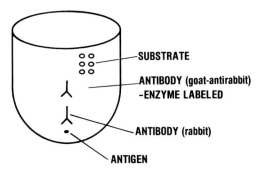

FIGURE 3 Schematic depiction of a modified ELISA assay.

TABLE I Examples of Different Enzymatic Systems in ELISA Assays

Enzyme	Substrate	Color
Alkaline phosphatase	p-nitrophenyl phosphate	Yellow
Horseradish peroxidase	5-Amino salicylic acid	Red
Biotin	Avidin	Blue

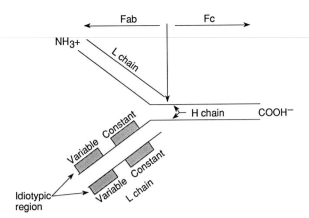

FIGURE 5 Schematic depiction of the structure of an immunoglobulin. There are two heavy and two light chains. The Fc portion of the molecule allows binding relatively nonspecifically, whereas the Fab portion contains a highly specific binding region, the most relevant of which is the variable, or hypervariable, region of the idiotypic region. This region is important in binding to antigens.

C. Double Sandwich ELISA

The double-sandwich-type assay is similar to those noted above but employs a total of three antibodies. The first antibody is directed against one site on the antigen; this antibody is used on the bottom of the microtiter well. The unknown concentration antigen, located in a biologic fluid, is then layered onto the antibody. After thorough washing, a second antibody, which recognizes a different epitope or binding site of the antigen, is then added; this second antibody is not enzyme-labelled. A third antibody (in this case, an enzyme-labelled antibody that recognizes the second antibody) is added, and then a substrate is added. In general, the first antibody is raised in one species (e.g., goat), whereas the second antibody is generated in a different species (e.g., a rabbit). Finally, the third antibody does not recognize the unknown antigen but, rather, is directed against the nonbinding protein (Fc portion) of the second antibody; i.e., the third antibody is an anti-rabbit antibody directed against the Fc portion of rabbit IgG. These types of assays are very specific and have high sensitivity. Because the first (goat) antibody may also react with the third (anti-Fc) antibody, increased background interference may result. Thus, frequently, the first (goat) antibody will be processed to leave the antigen-binding region intact, but the Fc fraction will be removed. Many variations of this type of assay can be used (e.g., several different rabbit antibodies, each di-

rected against different portions of the antigen molecule, can be used and may help to enhance sensitivity and specificity). Semantically, it must also be noted that the descriptive term "double sandwich" may be euphemistically pleasing, but it is a nonspecific term that can be applied to various types of assay; thus, it may be misleading.

D. Basic Principles

Several basic principles inherent in the development of an ELISA assay merit reiteration.

1. When an antigen or antibody is chemically linked to an enzyme, the resulting conjugated substance does not lose any of its original antigenicity and antigen-binding capacity. As a result, the characteristics of antigen–antibody binding are not affected.
2. Many substrate molecules can react with only one enzyme molecule. This assay characteristic leads to an enhancement of the detection system and, thus, leads to a highly sensitive assay.
3. Molecules unbound to microtiter wells or unattached to antigen or antibody can be effectively removed from the assay well by vigorous washing. Obviously, this results in minimal background interference and the relationship between optical density reading and assay conjugates becomes direct and generally linear. Usually, albumin or other large proteins are included into the wash buffers to decrease nonspecific binding.
4. The process of adsorption of an antigen or antibody onto the microtiter plates occurs without loss of biologic or immunologic activity.

III. Specific Constituents

A specific ELISA assay is capable of detecting thyrotrophin (TSH) receptor antibodies. The TSH receptor is a cellular membrane protein of thyroid cells, which binds TSH, a pituitary hormone, and initiates the complex and largely unknown process of thyroid gland stimulation. In Graves' disease, antibodies are formed against the TSH receptor; these antibodies act like TSH and are quite effective in causing prolonged thyroidal stimulation through the TSH receptor. Normally, TSH production and action is under negative feedback inhibition by the circulating thyroid hormones; however, stimulation by antibodies in Graves' disease is not under normal feedback homeostasis and results, in time, in a clinical disorder characterized by excessive secretion and action of the thyroid hormones triiodothyronine and thyroxine. In the ELISA assay, human tissue containing TSH receptors (e.g., thyroid or guinea pig fat cell membranes) is used to isolate a solubilized membrane-enriched fraction. About 2 μg protein in 100 lambda solution is layered onto flat-bottomed polystyrene microtiter plates at pH 7.2 and allowed to remain at 4°C for about 18 hr. This procedure allows the membranes to adhere to the plates. The precise mechanism by which this occurs is unknown, but it probably relates to charge attraction, because generally proteins are positively charged and plastics are negatively charged.

After application of the antigen, the wells are washed three or four times with phosphate-buffered saline (PBS) (pH 7.2), and then 100 lambda of a 1% human serum albumin solution (PBS, pH 7.2) is added and allowed to remain in the wells for 1 hr at 37°C. This step is extremely important because it reduces nonspecific binding in the assay.

Following washing with a PBS solution (pH 7.2) containing 0.05% Tween 20 (a detergent) and 0.1% albumin, the wells are ready for the biologic substance to be measured, for instance serum. One hundred lambda of serum is added and incubated for 1 hr at 37°C. Various dilutions of the serum are used, depending on the assay used and the concentration of the antigens in the biologic substance to be tested, as well as on the level of nonspecific binding or interference found in a particular fluid. Again, the wells are washed diligently and then the detection system is added. To detect TSH receptor antibodies, an anti-human IgG conjugated to the enzyme alkaline phosphatase (100 lambda in a 1 : 1,000 titer) is added and allowed to sit in the wells for 1 hr at 37°C. After more washes, 150 lambda of a solution containing p-nitrophenyl phosphate (1 mg/ml) in diethanolamine buffer (pH 9.8) (with 0.01 M MgCl$_2$) is added and incubated for a sufficient time to obtain good reading and low background as judged from positive and negative controls included in each plate. The incubation time of the activated enzyme in this last step is the most crucial and is based on preliminary studies aimed at optimizing the concentrations and times of each reagent.

The control samples are very important. They must include controls of (1) antibody conjugate alone, with buffer added instead of an unknown solution in an empty well (without antigen); (2) biologic fluid with antibody conjugate (in this instance, the wells also do not contain antigen; and (3) antigen with antibody conjugate. These controls should give very low readings. Background values also must be low.

IV. Various Types of Non-Nuclide Assays

In the last several years, there has been a virtual explosion in technology resulting in a large variety of other non-nuclide assays. Several of these will be discussed to give an overview of their basic characteristics.

In many areas, chemiluminescent assays are becoming very popular. Their main advantage is that they can exhibit tremendous sensitivity. In chemiluminescent assays, a specific chemiluminescent tracer or compound is used (either linked to antigen or antibody) in the detection system. Chemiluminescence or bioluminescence refers to the property of light emission of certain compounds, chemical or biological, respectively, especially after they are in an excited state. This luminescence may be used in assays in a manner analogous to radioisotopic compounds. These luminescent labels are stable in their basal state for long periods of time, thereby representing an advantage over isotopes and even over various enzyme labels. The greatest advantage of these compounds is the sensitivity they impart to an assay, mainly because of their decreased background "noise." As an example, acridinium esters are used in a commercially available kit as compounds that emit light when treated with hydrogen peroxide and alkali. In an assay, similar to the ELISA assays discussed above, the second antibody is conjugated with the acridinium ester. The amount of light emitted is in proportion to the amount of antibody bound. A chemiluminescent assay generally may detect 0.005 μU/ml of TSH, whereas a good ELISA and RIA will detect 0.10 μU/ml and 0.3 μU/ml, respectively.

Another new type of non-nuclide assay is designated as time-resolved fluorescence. In a commercially available assay, the detection system uses a Europium-chelate, which has a prolonged fluorescence delay time, conjugated to the second antibody. After thorough washing and addition of an enhancement solution, the Europium is converted into cations that dissociate, form chelates with portions of the solutions, and give rise to fluorescence, which is measured in a time-resolved fluorometer. A major advantage of this type of system is that it has low background interference, because of the prolonged fluorescent decay time, and, thus, increased sensitivity.

Fluorescent polarization assays also represent an innovative, non-nuclide, commercially available method of assessing the concentrations of various substances. These assays are based on the principle that molecules bound to a fluorophor will emit polarized light in a different manner, depending on whether or not the fluorophor can rotate freely; i.e., if the fluorophor is bound to the antibody, the complex is free to rotate; the polarized light directed through this solution will be deflected and becomes less polarized as it exits. In contrast, if the antibody is bound to the antigen, the resultant emitted light will be more polarized. This assay depends on competition between the fluorophor-labeled antigen and unlabelled antigen in biologic fluid. It is widely used in a variety of hormone or drug assays.

V. Conclusion

Several different types of non-nuclide radioassays have been examined. The basic principles in an assay that utilizes isotopes (e.g., RIA, immunoradiometric assay) are the same as those involved in a non-nuclide assay. There may be inherent advantages in using one assay over another, but the decision to use a particular assay must often involve practical laboratory considerations. Such issues as primary objective, finances, experience in a given area, and the laboratory environment are all important (Table II). Further, we must not forget that a comparison of isotopic versus nonisotopic assays must consider such nonreadily apparent issues as time involved in radiation safety administration, radiation disposal, equipment costs, and stability of reagents.

In this context, however, non-nuclide assays certainly must be considered a tremendous advance over nuclide assays, and we await with expectation further innovative assay developments in the future.

Acknowledgment

The author thanks representatives from London Diagnostics, Eden Prairie, Minnesota; Pharmacia Diagnostics, Fairfield, New Jersey; and Abbott Laboratories, North Chicago, Illinois, for helpful discussions and information regarding non-nuclide assays. Portions of this report have been presented in several immunoassay technique courses that have been sponsored by The Endocrine Society, Bethesda. The author also thanks Ms. Estelle Coleman for editorial assistance.

Bibliography

Baker, J. R., Lukes, Y. G., Smallridge, R. C., Berger, M., and Burman, K. D. (1983). Partial characterization and clinical correlation of circulating immunogloblins directed against thyrotrophin binding sites in guinea pig fat cell membranes. *J. Clin. Invest.* **72,** 1487.

D'Avis, J. R. E., Black, E. G., and Sheppard, M. C. (1987). Evaluation of a sensitive chemiluminescent assay for TSH in the follow-up of treated thyrotoxicosis. *Clin. Endocrinol.* **27,** 563.

Engvall, E., and Perlmann, P. (1972). Enzyme-linked immunosorbent assay (ELISA). *J. Imunol.* **109,** 129.

Soini, E., and Kojola, H. (1983). Time-resolved fluorometer for Lanthanide chelates—A new generation of non-isotopic immunoassays. *Clin. Chem.* **29,** 65.

Sturgess, M. L., Weeks, I., Evans, P. J., Mpoko, C. N., Laing, I., and Woodhead, J. S. (1987). An immunochemiluminometric assay for serum free thyroxine. *Clin. Endocrinol.* **27,** 383.

Symposium on Ultrasensitive TSH Assays (October–December 1988). *Mayo Clin. Proc.* **63,** Parts I–III.

Tseng, Y.-C., Burman, K. D., Baker, Jr., J. R., and Wartofsky, L. (1985). A rapid, sensitive enzyme-linked immunoassay for human thyrotropin. *Clin. Chem.* **31,** 1131.

Voller, A., Bidwell, D., and Bartlett, A. (1980). Enzyme-linked immunoassay. *In* "Manual of Clinical Immunology," 2nd ed., Chapter 45 (N. Rose and H. Friedman, eds.), p. 359. American Society for Microbiology, Washington, D.C.

Weeks, I., Sturgess, M., Siddle, K., Jones, M. K., and Woodhead, J. S. (1984). A high sensitivity immunochemiluminometric assay for human thyrotrophin. *Clin. Endocrinol.* **20,** 489.

Yolken, R. H. (1978). Enzyme-linked immnosorbent assay. *Hosp. Pract.* **December,** 121.

TABLE II Comparison of Different Types of Assays[a]

	RIA	ELISA	Luminescent
Isotope	++++	0	0
Convenience	+	+++	++
Cost	++	+	+++
Sensitivity	++	+++	++++

[a] These assessments are subjective and apply to each type of assay only in general terms. Individual exceptions exist.

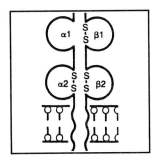

Immunobiology of Transplantation

MEGAN SYKES AND DAVID H. SACHS, *National Cancer Institute*

Glossary

Allogeneic Derived from another member of the same species and expressing one or more foreign histocompatibility molecules (alloantigens)

Allograft Graft obtained from an allogeneic donor

Anti-idiotypic antibodies Antibodies directed against idiotopes of other antibodies; Idiotopes are those antigenic determinants which are unique to a particular immunoglobulin molecule and which are therefore components of the portion of the immunoglobulin molecule which confers specificity.

B lymphocyte Lymphocyte expressing a clonally distributed immunoglobulin receptor on its cell surface. B lymphocytes differentiate into antibody-producing plasma cells. Antibody mediates the "humoral" component of an immune response.

CD3 Cluster of differentiation antigen 3; complex of five proteins expressed on the surface of T cells in close physical association with the T cell receptor

CD4 Cluster of differentiation antigen 4; surface molecule which defines one of two major subsets of mature T cells; CD4-positive T cells usually recognize or are restricted by a class II MHC molecule.

CD8 Cluster of differentiation antigen 8; surface molecule which defines one of two major subsets of mature T cells; CD8-positive T cells usually recognize or are restricted by a class I MHC molecule.

Polymorphism Genetic variability from one member of a species to another member of the same species

MHC restriction Phenomenon whereby antigen is recognized by a T cell receptor only when presented in association with a particular "self" major histocompatibility complex (MHC) molecule. Restriction results from a selective process taking place in the thymus during ontogeny.

T lymphocyte Lymphocyte expressing a clonally distributed T cell receptor complex on its cell surface. T lymphocytes are responsible for the "cell-mediated" component of an immune response.

Xenogeneic Derived from a member of a different species

Xenograft Graft obtained from a xenogeneic donor

THE FIELD OF TRANSPLANTATION represents a unique interface between basic immunology and clinical medicine. This article covers the major principles of transplantation biology and relates them to the field of clinical transplantation. The mechanisms whereby immune cells of an individual recognize and reject foreign tissues, as well as some approaches to avoiding such rejection, are discussed.

I. Goals of Research in Transplantation Immunology

In 1945 Dr. Ray Owen reported a surprising observation he had made in cattle twins born from a common placenta, known as Freemartin cattle. His observation was that these animals, when they reach adult life, carry in their circulation cells with surface antigens which could only be explained by the mixing of blood cell components between the cattle twins in utero. Although these findings were reported predominantly as an explanation for the unusual blood typing of Freemartin cattle, they also had important immunologic implications. The persistence of blood cells from a nonidentical twin into adult life suggested that exposure of the immune system to the antigens on such cells early in development must have led to a state of immunologic tolerance.

Several years later, Billingham, Brent, and Medawar repeated this experiment of nature by injecting neonatal mice (less than 24 hr old) with allogeneic bone marrow cells. They found that such animals matured into adults which were specifically tolerant to skin grafts from the same allogeneic donor strain. The animals were otherwise normal immunologically, suggesting again that early exposure to the allogeneic cells had fooled the immune system into considering this specific set of alloantigens as self, rather than as nonself.

These early experiments demonstrated that the phenomenon of specific immunologic tolerance to transplants is feasible, but they obviously did not provide a reasonable clinical methodology, since clinical transplantation is usually required for adults. Thus, the goal of inducing specific transplantation tolerance in adult animals has remained the major challenge to transplantation immunologists.

II. Histocompatibility Systems

The cell surface molecules which lead to rejection of a graft have been termed "histocompatibility antigens" or "transplantation antigens." These antigens can be defined as any cell surface molecule which, when present on cells of a graft but not on cells of the recipient, can lead to an immunologic rejection reaction. Such antigens have been found in all mammalian species which have been studied, and some of these antigens have been highly conserved throughout speciation. Obviously then, the names "histocompatibility antigens" and "transplantation antigens" are misnomers, since to be conserved in evolution, these molecules must have reasons for existing other than to thwart the field of clinical transplantation.

A. Genetics of Histocompatibility Antigens

Using inbred strains of mice, it is possible to determine how many different genetic loci code for histocompatibility antigens. One can breed two inbred strains of mice to produce an F_1 generation and then breed the F_1 animals to each other to produce an F_2

generation, in which all genetic loci will distribute according to the binomial distribution $1:2:1$. If one then places a tissue graft from one of the parental strains onto all of the F_2 animals, one can calculate from the outcome how many histocompatibility antigen differences exist between the two original parental strains. If only one locus encoded an antigen capable of causing rejection, then one would expect three-fourths of the grafts to be accepted. However, if there were two independently segregating loci, each capable of causing rejection, then three-fourths of three-fourths ($[\frac{3}{4}]^2$) would be the fraction of animals which would accept tissue grafts, or for n histocompatibility loci, $(\frac{3}{4})^n$ would be the fraction accepting. Using inbred mice, such an experiment is feasible and has actually been performed on numerous occasions. The value for n has been found to be as large as 30–40 for the strain combination C57BL/6 and BALB/c.

Therefore, it is clear that there are many loci in the genome capable of causing graft rejection. Fortunately, from the point of view of transplantation biology, one of these loci is of overwhelming importance in determining the outcome of allografts, and this locus has been called the major histocompatibility complex (MHC). It is now clear that the molecules encoded by the MHC are of great importance to the normal function of the immune system. In fact, immune T cells see foreign antigens not alone, but in the context of self MHC antigens, and this is undoubtedly the major reason for the existence of MHC molecules. T lymphocytes are probably selected during development for weak interactions with self MHC products; such T cells, when mature, exhibit strong interactions with self MHC molecules which are complexed to a foreign "nominal antigen." It is believed that allogeneic MHC molecules might resemble such complexes of self MHC plus nominal antigen, thus explaining the high frequency of T cells recognizing such alloantigens, which therefore behave as strong transplantation antigens. [See LYMPHOCYTES.]

What, then, are all the other 30–40 loci which have also been implicated as encoding transplantation antigens? These other antigens are often referred to as minor histocompatibility antigens in contrast to major histocompatibility antigens. Minor histocompatibility antigens may be molecules which happen to be polymorphic in a species but do not necessarily have any particular role in the immune system. Because the immune system is capable of recognizing foreign antigens to which it has not been "tolerized" during development, such molecules can be seen as antigenic upon immunization and can cause rejection. For any single minor histocompatibility antigen, however, the rejection seen is generally rather weak and easily overcome by low doses of immunosuppressive medications. Rejection due to the MHC antigens, on the other hand, is generally as vigorous as is physiologically possible. Recent molecular data suggest that minor histocompatibility antigens are probably peptides of other polymorphic molecules presented to the immune system in a groove of an MHC molecule on the cell surface.

B. Molecular Basis of Histocompatibility Antigens

The MHC, as its name implies, is not a single locus but rather a complex of several genetic loci. The loci of greatest importance to transplantation are known as class I and class II loci, which encode class I and class II antigens, respectively. As shown schematically in Fig. 1, class I antigens are 45,000

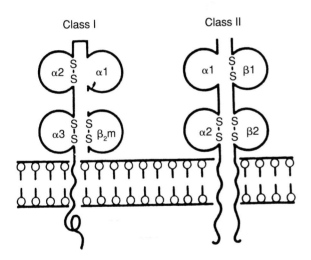

FIGURE 1 Structures of class I and class II MHC molecules. The class I molecule consists of three α domains (α1 through α3), noncovalently linked to the single domain β2 microglobulin (β2m) molecule. The class II molecule consists of an α chain containing two external domains (α1 and α2) noncovalently linked to a β chain also containing two external domains (β1 and β2). Both α chains and the class II β chain possess a transmembrane region and intracellular tail, whereas β2m consists only of an external domain. S–S indicates disulfide linkages, which occur within the external domains of class I and class II molecules. [Reprinted with permission from M. Steinmetz, "Genes of the Major Histocompatibility Complex in Mouse and Man," Science **222**, November 18, 1983, pp. 727–733. Copyright 1983 by the AAAS.]

MW transmembrane glycoproteins. They are found ubiquitously on the surface of all cells of the organism, with the possible exception of red blood cells, and are associated noncovalently with a 12,000 MW glycoprotein known as β2-microgloblin, which is nonpolymorphic and is encoded elsewhere in the genome. The class I heavy chain is among the most polymorphic of all known loci.

Class II antigens consist of two transmembrane glycoproteins, each of approximately 30,000 MW (Fig. 1), and both encoded within the MHC. These molecules are found predominantly on the surface of cells which present antigen to the immune system, such as the surfaces of B cells and macrophages. Both chains of a class II molecule, alpha and beta, can be polymorphic.

In every mammalian species studied so far, the MHC contains at least two class I and at least two class II loci. The linear relationship between these loci on respective chromosomes in different species is variable, but the class I and class II loci have remained closely linked throughout evolution. The class I and class II genes of several species have now been isolated and characterized.

C. The Role of Histocompatibility Antigens in Transplantation

The overwhelming importance of the MHC to the outcome of transplants is clear from an analysis of clinical transplant data in which the results of kidney transplantation between HLA (Human Leukocyte Antigen; refers to human MHC molecules) identical siblings are compared to the results of similar transplants between non-HLA identical siblings and between unrelated individuals (Fig. 2). On a statistical basis, HLA identical siblings or non-HLA identical siblings would be expected to share approximately the same number of minor histocompatibility antigens. Thus, the difference in kidney survival statistics between these two groups must depend on MHC matching.

Matching of MHC antigens for cadaver donor transplants, on the other hand, has not been found to be a reliable means of improving graft survival. Presumably, this is because the MHC antigens are extremely polymorphic. Because of the tight genetic linkage of MHC loci, when one matches MHC antigens within families, one can be assured that matching implies identity of the products of all MHC loci. Matching between unrelated individuals implies only that the determinants for which typing

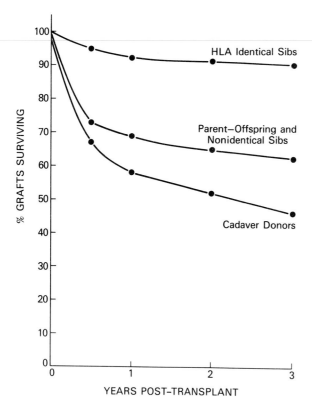

FIGURE 2 Influence of MHC matching on kidney graft survival. These data were compiled before the advent of cyclosporin, after which the survival curves for unmatched related and unrelated cadaveric transplants have improved significantly. [Reprinted with permission from Paul, W. E. (ed.) (1984). Fundamental immunology, Chapter 13 "The Major Histocompatibility Complex" David H. Sachs, p. 341.]

reagents exist are the same but not that the entire molecule is actually identical or that untyped MHC loci are the same.

Class I antigens are thought to be the major targets of the rejection reaction, while class II antigens appear to stimulate T helper cells and thus are predominantly responsible for stimulating the rejection reaction. There is, however, much overlap between the functions of the products of class I and class II genes in transplantation. When minor histoincompatibilities are also present, either class I or class II antigenic differences are sufficient to cause vigorous skin graft rejection. For vascular allografts, however, class II antigenic differences may be especially important in inducing prompt rejection reactions. In the absence of class II disparities, grafts bearing class I alloantigens or class I plus minor alloantigens may lead to the induction of tolerance rather than to rejection.

III. Graft Rejection

In order to reject a graft, lymphocytes of the recipient must first encounter and recognize a foreign histocompatibility antigen; this sensitization process represents the afferent arm of the allograft response. Sensitized cells function as either effector cells or as helper cells which help other lymphocytes to proliferate and differentiate into effector cells. T lymphocytes play a critical role in graft rejection, but other cell types appear also to be involved. Effector cells, which mediate the efferent arm of the rejection response, include T lymphocytes which are cytotoxic (cytotoxic T lymphocytes, or CTL) or which produce toxic soluble lymphokines, and B lymphocytes, which differentiate into antibody-producing plasma cells. Cytotoxic effector cells such as activated macrophages and natural killer (NK) cells may also play a role in rejection. The mechanisms by which sensitization and rejection occur are discussed in more detail below.

A. Sensitization Pathways

Several events must occur before lymphocytes become sensitized *in vivo*. First, circulating T cells must localize to a site where alloantigen is encountered. Such encounters can occur in the graft itself, in local lymph nodes draining the graft, or in distant lymphoid organs such as the spleen, to which passenger leukocytes from the allograft can migrate. Many mature lymphocytes circulate continuously between the blood and lymph, optimizing the opportunity for encounter with antigen. In the case of skin graft rejection, the earliest sensitization events occur in the lymph nodes which drain the graft bed. Migration of T cells into lymph nodes involves a class of lymphocyte "homing" receptors for ligands expressed on specialized endothelial cells, as well as other adhesion molecules. Sensitization can also occur within allografts lacking lymphatic connections through lymphocyte contact with vascular endothelial cells and/or "trapping" of lymphocytes from the circulation. The exact mechanisms leading to initial trapping are unclear but, in addition to recognition of antigen, may involve inflammatory mediators released by the healing-in process. Once T cells are sensitized, they demonstrate selective homing to allografts bearing their target antigens. Nevertheless, a large number of nonactivated, nondonor-specific CTL precursors can also be found among the lymphocytes which infiltrate allografts;

these cells probably migrate to the graft in response to chemotactic cytokines released during the inflammatory process.

Sensitization of T cells requires the presentation of antigen on a specialized antigen presenting cell, or APC. Much of what is known about the cellular requirements for sensitization of T lymphocytes has been learned from *in vitro* culture systems. One function of APCs is to endocytose exogenous antigen, break it down into fragments and present the fragments on its cell surface in association with class II MHC molecules. Most APCs express class II MHC molecules, either constitutively, as in the case of B lymphocytes and some macrophages, or in response to lymphokines such as γ-interferon, as in the case of macrophages and endothelial cells in some species. Specialized dendritic cells in tissues, such as Langerhans cells in the skin, also function as APCs. In addition to expression of class II antigens, however, APCs have other specialized functions, including the ability to take up and process exogenous antigen, as well as the secretion of interleukin 1 and possibly of other cytokines. Such factors may act as "costimulatory" signals, in addition to antigen, for the activation of previously unsensitized T lymphocytes. If certain costimulatory signals are not provided, T cell activation will not occur, and T cells may even become refractory to subsequent activation.

The class II-restricted pathway of antigen presentation discussed above is necessary for the activation of alloantigen-specific CD4-positive T cells, which, once activated, may function as lymphokine-secreting helper cells or, in some cases, as cytotoxic effector cells. In each case, the structure recognized by CD4-positive T lymphocytes includes a class II MHC molecule, which can be either an allogeneic class II molecule on a graft APC, or a host class II molecule on a host APC. Such host APCs can pick up exogenous minor or major histocompatibility antigens from the graft, process them, and present them on the cell surface in association with host class II molecules. Recognition of native allogeneic MHC molecules on allogeneic APC is referred to as "direct" recognition, whereas recognition of alloantigen re-presented on host APC is referred to as "indirect." [*See* CD8 AND CD4: STRUCTURE, FUNCTION, AND MOLECULAR BIOLOGY.]

In contrast to the recognition of class II molecules by CD4-positive T cells, direct recognition of allogeneic class I MHC molecules is mediated by CD8-positive T cells. Nevertheless, specialized APC are also required in order to sensitize class I-specific T cells. Class I-specific, CD8-positive T cells include cells with both helper and cytotoxic effector functions. Unlike responses to class II MHC or minor histocompatibility antigens, naive class I-specific T cells can be primed without the participation of a CD4-positive T cell.

Not all pathways that lead to T cell activation *in vitro* necessarily lead to graft rejection *in vivo*. Variations in immune responses to allografts according to the density of specialized APCs in each type of graft suggest that APCs, or "passenger leukocytes," may play an important role in determining the magnitude of a rejection response. Experimental results in animals have demonstrated that depletion of APCs can lead to marked prolongation or even permanent survival of some endocrine tissue allografts. Nevertheless, provision of an antigenic challenge containing the same alloantigens in association with donor APCs leads to sensitization and rejection of the original APC-depleted allograft. Even though host APCs eventually repopulate APC-depleted allografts, presentation of alloantigen on these cells does not lead to rejection of these grafts, suggesting that re-presentation of donor alloantigen on host APC (i.e., the indirect pathway) alone may be insufficient to cause graft rejection. Apparently, in order to generate appropriate effector cells, the same complex of antigen and MHC recognized by the effector cell must be found on the APC; local secretion of lymphokines by helper T cells does not appear to be sufficient to activate nearby effector cells specific for an MHC–antigen complex expressed only on parenchymal cells of the graft. The reason for this discrepancy with *in vitro* results is unclear, but it is apparent that *in vitro* results do not always mimic the *in vivo* situation. It is also unclear whether or not the same principles apply to grafts other than endocrine tissues.

Sensitization of helper T cells assists the differentiation, activation, and proliferation of cytotoxic effector cells and B cells, mainly through the secretion of soluble lymphokines. One such lymphokine, interleukin 2 (IL-2), plays a pivotal role in the proliferation of cytotoxic T lymphocytes (CTL), but other lymphokines may also be involved. The dichotomy of helper and cytotoxic T cell functions is not absolute, since some T cells, known as "helper-independent T," or "HIT," cells can both secrete IL-2 and mediate cytotoxic activity. [*See* INTERLEUKIN-2 AND THE IL-2 RECEPTOR.]

IMMUNOBIOLOGY OF TRANSPLANTATION

362

B. Effector Pathways

Both humoral and cellular components of the immune system can contribute to rejection processes. Antibodies, which confer specific antigen recognition for humoral rejection, may be either natural (i.e., present in the absence of known prior sensitization) as in the case of antibodies against blood group antigens or against xenogeneic antigens, or may result from prior sensitization due to pregnancy, blood transfusions, or previous graft rejections. The binding to graft endothelial cells of natural antibody, which is usually of the IgM class, leads to activation of the complement and clotting pathways, resulting in the dramatic phenomenon known as "hyperacute rejection," in which vascular thrombosis and rejection occur within minutes. Because of the existence of natural antibody against blood group antigens, ABO blood group matching between organ allograft donors and recipients is routinely performed, although successful transplants have recently been performed across ABO barriers after extracorporeal absorption of preformed antibodies. The problem of natural antibody represents a formidable barrier to the use of xenogeneic organ donors.

In order to avoid rejection by preformed antibody in previously sensitized patients, cross-matches are routinely performed before allogeneic organ transplantation, and a prospective donor is only considered acceptable if his or her lymphocytes are cross-match-negative with a recent serum sample from the recipient. Such donors can be extremely difficult to find for highly sensitized recipients. Antibody has also been strongly implicated in the phenomenon of chronic rejection of renal allografts in which cellular infiltrates are not prominent.

Virtually all cell types of the immune system can be found in acutely rejecting allografts, including macrophages, NK cells, B cells and their fully differentiated antibody-producing progeny, plasma cells, and T cells. Both CD4-positive and CD8-positive T cells are found, although the latter predominate. The many cell types which infiltrate rejecting allografts could potentially implicate multiple mechanisms of graft rejection, including complement-mediated cytotoxicity, antibody-dependent cell-mediated cytotoxicity (ADCC), release of toxic cytokines from helper T cells and macrophages, and specific and nonspecific cell-mediated cytotoxicity, mediated by CTL, NK cells, and macrophages, respectively. A consensus has not yet been achieved as to the relative importance of each of these components, which may vary according to the graft type and species under study. Several groups have attempted to define the T cell subsets necessary for rejection of skin grafts by adoptively transferring normal T cell populations into T cell–deficient mice and rats. While such models have been informative, they can only define the minimal components required for rejection, rather than demonstrate what actually occurs in the usual rejecting graft. In general, the results of such studies are consistent with a requirement for a source of lymphokines (helper T cells) as well as cytotoxic T cells for rejection to occur. Results of experiments involving skin and islet grafts containing mixtures of syngeneic and allogeneic cells indicate that destruction of allogeneic cells in a graft does not lead to destruction of neighboring host-type cells. These results suggest that specific CTL, rather than nonspecific inflammatory mediators released by helper cells, may be the actual effectors of graft rejection. Nevertheless, these results do not necessarily apply to primarily vascularized grafts, and evidence exists to implicate important roles for nonspecific cytotoxic cells and delayed-type hypersensitivity reactions in the rejection of human renal allografts. The role that cytokines such as tumor necrosis factor (TNF) and γ-interferon play in mediating graft injury is currently a subject of active investigation. One of the major effects of γ-interferon appears to be the upregulation of MHC expression on cells of the graft, which may facilitate antigen presentation and graft rejection. [See CYTOKINES IN THE IMMUNE RESPONSE.]

IV. Interfering with the Rejection Process

Graft rejection can be prevented in three major ways. (1) by modifying the graft in a way which renders it nonimmunogenic; (2) by globally suppressing host immunity; and (3) by specifically suppressing the response of the host against donor alloantigen. The first method has been discussed above and at present is applicable only to certain endocrine allografts. Most currently accepted clinical transplantation protocols utilize the second method, but the third approach would be most desirable, since it would bypass the need for lifelong immunosuppressive therapy and the concomitant risk of infection. We will discuss nonspecific and specific immunosuppression separately:

A. Nonspecific Immunosuppression

Most current nonspecific immunosuppressive regimens used for the prevention of graft rejection include cyclosporin A (CSA), a lipophilic, cyclic 1.2 kDa undecapeptide produced by the fungal species *Tolypocladium inflatum* Gams. In addition to its potent immunosuppressive activity, CSA has some desirable properties which distinguish it from other nonspecific immunosuppressive agents. Although the mechanisms of action of this drug are complex and incompletely understood, it is generally agreed that CSA diminishes the proliferation of cytotoxic T lymphocytes by inhibiting the secretion of IL-2 by helper T cells, while possibly sparing the activity of suppressor cells. The use of CSA has had an enormous impact on the field of transplantation, leading to marked improvements in the survival rates of organ allografts, as well as reductions in the cost and duration of hospitalization after transplantation. CSA is also used in standard regiments for the prevention of graft-versus-host disease (GVHD) in bone marrow transplant recipients. Unlike most other nonspecific immunosuppressive drugs, cyclosporin is not toxic to the bone marrow. In addition, cyclosporin is associated with a lower incidence of opportunistic infections compared with other immunosuppressive agents, making it a safer, more acceptable form of long-term immunosuppression. Nevertheless, cyclosporin does have significant toxic effects, including nephrotoxicity, which can be difficult to distinguish from graft rejection in recipients of renal transplants. Additional toxic effects include neurotoxicity, hepatotoxicity, hypertension, gingival hyperplasia, opportunistic infections, and malignancies. Since some of these toxic effects are dose related, most current immunosuppressive regimens include lower doses of cyclosporin, modified according to trough serum levels, in combination with other agents, such as azathioprine and steroids. FK506, a newer and possibly more potent immunosuppressive agent than CSA, is currently undergoing clinical evaluation.

Polyclonal and monoclonal anti-T cell antibodies have also found a place in standard immunosuppressive regimens, and the use of the monoclonal antibody OKT3 (anti-CD3) has proved to be particularly effective in the reversal of rejection episodes. Antibody against IL-2 receptor also shows promise in both clinical and experimental models and may prove to be an effective, nontoxic adjunct to cyclosporin therapy. The theoretical advantage of antibody against IL-2 receptor is that only T cells which are activated, and therefore expressing IL-2 receptor at the time of antibody administration, should be inhibited from functioning, so depletion of T cells with specificity for pathogens to which the patient is not concurrently exposed should not occur. Such an approach might lead to donor-specific tolerance if T cell clones reacting to donor alloantigen were specifically eliminated by antibody, whereas the use of antibodies which do not permanently deplete T cells would lead only to temporary dysfunction of alloreactive T cells. A major limitation of monoclonal antibody therapy is the tendency of recipients to develop antibody against determinants of the xenogeneic murine monoclonal antibodies, rendering subsequent antibody therapy ineffective. This problem could be partially resolved by the development of human monoclonals, but reliable methods for producing such antibodies are not yet available.

An additional form of immunosuppression, blood transfusion, deserves mention, although its mechanism of action remains controversial. The use of blood transfusions before cardiac or renal transplantation has been documented to improve the outcome of clinical transplantation. This observation is somewhat paradoxical, considering the capacity of blood transfusions to induce sensitization and alloantibody formation. In fact, some of the apparent benefit of transfusions may be due to such sensitization, which is reflected in the development of a positive cross-match, leading to the "selecting out" of donors against whom sensitization is most likely to occur. In addition to this selective process, however, blood transfusion does appear to have an immunosuppressive effect on recipients. One of the hypotheses which has been proposed to explain this immunosuppression is that alloantibodies "enhance" graft survival, perhaps by blocking immunogenic determinants on the graft or, alternatively, through the production of anti-idiotypic antibodies which specifically recognize the unique determinants of alloantibodies or T cell receptors reactive against the graft. An alternative mechanism proposed in some experimental systems is the "veto" phenomenon, wherein cells of the original transfusion sharing alloantigens with the subsequent graft donor are able to eliminate T cells which react against antigens expressed on the surface of the veto cell itself (and hence the allograft). Consistent with the specificity implied by these proposed mechanisms for the beneficial effect of blood transfusions, results of some clinical trials have indicated that pretransplant transfusions are most effective when the blood donor and organ donor are the same individual, so that the transfusions are "donor-specific." Nevertheless, donor-nonspecific transfusions, which are the only kind possible in the cadaveric organ allograft situation, are also beneficial, suggesting that nonspecific suppressive mechanisms may also be involved. As the drug regimens used for nonspecific immunosuppression and organ allograft survival have improved over the years, it has become less clear that the benefits of transfusion outweigh its attendant risks of sensitization and transmission of viruses.

Another potent nonspecific immunosuppressive regimen involves the use of total lymphoid irradiation (TLI). This approach has been successful in permitting successful kidney transplantation in patients at particularly high risk for rejection, and less cumbersome modifications of the original protocol are currently being evaluated.

B. Specific Immunosuppression

Until a clinically applicable method for inducing permanent, donor-specific tolerance is developed, transplant recipients will continue to be subjected to the risks of infection and malignancy associated with long-term nonspecific immunosuppression. Many experimental protocols have successfully achieved permanent donor-specific tolerance in animals; some of these regimens, such as a short course of peri-transplant cyclosporin in rodents, may be species-, organ-, or even strain-specific, and may not apply in humans. The development of donor-specific tolerance in some of these systems has been associated with the presence of specific suppressor T cells. One regimen which has been documented to induce donor-specific tolerance in humans, with successful withdrawal of immunosuppressive drugs, is the use of TLI. Unfortunately, the TLI protocol required before transplantation of a cadaveric kidney is somewhat impractical, and there are only a few documented cases in which tolerance was successfully induced using this regimen.

Another approach to the induction of specific tolerance involves the production of lymphohematopoietic chimerism in recipients which are immunoincompetent due either to early age (e.g., neonatal animals) or to ablation of the immune system by lethal whole body irradiation (WBI) or TLI.

Although the induction of chimerism may be the most potent and durable method for inducing donor-specific tolerance, regimens required to ablate host resistance and permit the development of such chimerism in adults are extremely toxic. It would not be justifiable, for example, to use lethal WBI and allogeneic bone marrow transplantation (BMT), with its high risk of graft-versus-host disease and mortality in the post-BMT period, to induce chimerism and donor-specific tolerance, when 80% 1-year survival of cadaveric kidneys can be achieved with the use of chronic immunosuppressive therapy. Less toxic preparative regimens are therefore required before the induction of chimerism can be considered a practical approach to the development of donor-specific tolerance. Such regimens, involving the specific ablation of host T lymphocytes using monoclonal antibodies, have recently been developed in animal models. For example, permanent chimerism and specific tolerance can routinely be induced across complete MHC barriers in mice using a low, sublethal dose of whole body irradiation in conjunction with high-dose local irradiation to the thymus and injection of monoclonal antibodies specific for CD4-positive and CD8-positive T cells. Much remains to be learned about the requirements for induction of tolerance and the applicability of such regimens in large animal models before they can be applied clinically.

The mechanism whereby induction of chimerism leads to donor-specific tolerance is incompletely understood. In neonatally tolerized animals, both specific suppression by suppressor T cells and clonal deletion of donor-reactive T cells have been implicated. In contrast, clonal deletion may be the primary mechanism of tolerance induction in lethal WBI-treated recipients of T cell-depleted allogeneic bone marrow transplants. Attempts to induce tolerance by lethal irradiation followed by infusion of exhaustively T cell–depleted autologous marrow have not been successful in inducing tolerance to MHC mismatched organs transplanted at the time of BMT. Thus, donor alloantigen seems to be required within the lymphomeatopoietic system for the most consistent induction of permanent donor-specific tolerance. It is likely that such tolerance induction occurs primarily within the thymus, where clonal abortion of self-reactive T cells is achieved. Considerable evidence has now been accumulated to indicate that bone marrow–derived cells within the thymus induce clonal abortion of T cells with specificity for histocompatibility antigens expressed by those bone marrow–derived cells. The constant presence of such cells in the thymus in permanent lymphohematopoietic chimeras might therefore ensure the continuous abortion of newly developing T cell clones with specificity for donor alloantigen. Thymic epithelial cells can also induce tolerance partly by inducing "clonal anergy."

V. Future Directions of Research

Several approaches currently under study may overcome the problems now associated with attempts to induce donor-specific tolerance. An alternative approach to providing alloantigens on allogeneic lymphohematopoietic cells might involve the use of molecular biologic techniques to introduce allogeneic MHC genes into autologous hematopoietic stem cells. Infusion of T cell–depleted autologous marrow carrying such allogeneic gene products into lethally irradiated recipients might permit the induction of donor-specific tolerance while circumventing the risks associated with allogeneic bone marrow transplantation. A better understanding of the mechanisms of specific tolerance induction may permit targeting of host lymphocyte subpopulations for elimination using monoclonal antibodies. In addition, the use of monoclonal antibodies to target cytokine receptors and adhesion molecules involved in graft rejection represents a more refined strategy for attenuating allograft responses than the use of nonspecific immunosuppressive drugs. The development of human monoclonal antibodies would represent a major advance in the ability to use such reagents in patients without risking the development of antibody against xenogeneic immunoglobulin constant region determinants. Finally, the problem of donor availability, which already limits the utilization of organ transplantation as a therapeutic modality, is unlikely to be resolved unless an alternative source of donors can be found. Thus, a greater understanding of the mechanisms responsible for xenogeneic graft rejection, and the development of methods of overcoming such rejection, is essential. Such new approaches could potentially revolutionize the field of transplantation.

Bibliography

Ascher, N. L., Hoffman, R. A., Hanto, D. W., and Simmons, R. L. (1984). Cellular basis of allograft rejection. *Immunol. Rev.* **77,** 216–230.

Auchincloss, H., Jr., and Sachs, D. H. (1989). Transplantation and graft rejection. *In* "Fundamental Immunology," 2nd Ed. (W. E. Paul, ed.). Chapter 33, pages 889–922. Raven Press, Ltd., New York.

Hancock, W. W. (1984). Analysis of intragraft effector mechanisms associated with human renal allograft rejection: Immunohistological studies with monoclonal antibodies. *Immunol. Rev.* **77,** 60–81.

Hansen, T. H., and Sachs, D. H. (1989). The major histocompatibility complex. *In* "Fundamental Immunology," 2nd ed. (W. E. Paul, ed.). Chapter 16, pages 445–487. Raven Press, Ltd., New York.

Kupiec-Weglinski, J. W., and Tilney, N. L. (1989). Lymphocyte migration patterns in organ allograft recipients. *Immunol. Rev.* **108,** 63–82.

Pullen, A. M., Kappler, J. W., and Marrack, P. (1989). Tolerance to self antigens shapes the T-cell repertoire. *Immunol. Rev.* **107,** 125–139.

Roser, B. J. (1989). Cellular mechanisms in neonatal and adult tolerance. *Immunol. Rev.* **107,** 179–202.

Unanue, E. R. (1989). Macrophages, antigen-presenting cells, and the phenomena of antigen handling and presentation. *In* "Fundamental Immunology," (W. E. Paul, ed.). Chapter 5, pages 95–115. Raven Press, Ltd., New York.

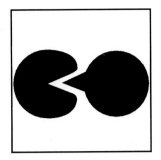

Immunoconjugates

CARL-WILHELM VOGEL, *Georgetown University School of Medicine and University of Hamburg*

I. Effector Molecules Used in Immunoconjugates
II. Targets For Immunoconjugates
III. Targeting Moieties Used in Immunoconjugates
IV. Synthesis of Immunoconjugates
V. Shortcomings and Future Challenges

Glossary

Heterobifunctional cross-linking reagent Chemical reagent with two different reactive groups, designed to react with two protein molecules for cross-linking

Monoclonal antibody Antibody with defined specificity derived from a single B-cell clone

Nude mouse Strain of hairless mice without the thymus. These animals are immunocompromised and cannot reject foreign tissue

IMMUNOCONJUGATES are a novel class of semisynthetic biomolecules. They consist of two different biomolecules linked together to create a hybrid molecule exhibiting the biological activities of the two parent compounds. The classical examples are antibody conjugates in which one component is a monoclonal antibody with specificity for a cell surface antigen of a tumor cell and the other component is a toxic effector molecule (e.g., a protein toxin or a toxic drug). In this example, the binding function of the antibody molecule is combined with the cytotoxic function of the toxic effector molecule to create a selective cytotoxic agent for antigen-bearing tumor cells. Although in principle any two biomolecules could be linked together to form a dual-function hybrid, the conjugates in the context of this article invariably consist of a binding moiety and an effector moiety. The binding moiety serves to target the conjugate to a defined molecular structure, usually on the cell surface of a cell, and is therefore also referred to as targeting moiety. The effector moiety commonly exerts cytotoxic activity or serves to detect the target cells. The binding moiety of most conjugates prepared so far is an antibody, and hence the term *immunoconjugates*. Although, strictly speaking, conjugates with non-antibody binding moieties are not immunoconjugates, these conjugates conceptually belong into the immunoconjugate category and are included in this review article. Fig. 1 shows the structure of a model immunoconjugate.

In the following, immunoconjugates will be reviewed and grouped according to their effector moieties, their targeting moieties, their cell targets, and their methods of synthesis. A final paragraph will outline current shortcomings and problems encountered with immunoconjugates for therapeutic applications and the existing research challenges.

I. Effector Molecules Used in Immunoconjugates

Several categories of effector molecules have been linked to antibodies and other targeting moieties to construct immunoconjugates (Table I). These effector molecules include protein toxins, low-molecular-weight drugs, biological response modifiers (BRMs), radionuclides, enzymes, and several other substances.

A. Antibody Conjugates with Protein Toxins (Immunotoxins)

Several protein toxins of bacterial or plant origin have been coupled to antibodies and other targeting moieties. Frequently coupled protein toxins include

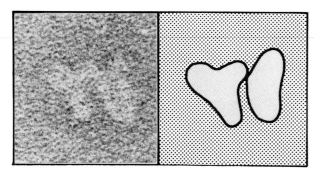

FIGURE 1 Structure of an immunoconjugate. Shown is an electron micrograph of an antibody conjugate with cobra venom factor (*left panel*) and its graphic reproduction (*right panel*). The Y-shaped IgG antibody molecule and the irregularly oval cobra venom factor molecule can easily be distinguished. The cobra venom factor molecule is linked to one of the Fab arms of the antibody molecule.

ricin, diphtheria toxin, and *Pseudomonas* exotoxin. Additional protein toxins used for the construction of immunoconjugates are gelonin, pokeweed antiviral protein, abrin, and several others. Immunoconjugates constructed with these protein toxins are frequently referred to as immunotoxins.

Several of these toxins (e.g., ricin, abrin) consist of two polypeptide chains attached through a disulfide bond. One polypeptide chain is responsible for the toxic activity of inhibiting cellular protein synthesis and is referred to as the *toxic A chain*. The toxic A chain is an enzyme that irreversibly modifies the ribosomal binding site for elongation factor 2. Because of the enzymatic nature of their toxic activity, these toxins are very potent once they have entered the cell. As a matter of fact, a single toxin molecule can be sufficient to cause cell death. The biological function of the nontoxic B chain is to allow the toxins to reach the cytosol. The B chain mediates binding to cell surface structures and facilitates in a not very well understood fashion the transfer of the toxin molecule into the cytosol. An-

other group of plant toxins, which includes, for example, the pokeweed antiviral protein and gelonin, is referred to as hemitoxins. They consist only of a toxic A chain, which is structurally similar to the A chain of ricin and other holotoxins; lacking a B chain, they enter the cell with much greater difficulty.

The two bacteria toxins diphtheria toxin and *Pseudomonas* exotoxin structurally differ from each other and from the above-mentioned plant toxins, but functionally resemble them. They are synthesized as a single polypeptide chain and contain three different domains responsible for cell binding, membrane translocation, and toxic activity. The toxic activity of diphtheria toxin and *Pseudomonas* exotoxin is the ADP-ribosylation of elongation factor 2, which inhibits protein synthesis.

Because of their high toxicity, toxins such as ricin have been coupled to many monoclonal antibodies with specificity for tumor antigens. These immunotoxins showed good cytotoxic activity for antigen-bearing tumor cells. However, the nonspecific binding function of the B chain, which in the case of ricin recognizes terminal galactose residues on cells, would severely limit the selectivity of the cytotoxic activity of the immunotoxins.

To overcome the unwanted toxic activity of immunotoxins for bystander cells caused by the binding domain of the toxin, several approaches have been employed. For *in vitro* applications, the binding site could be blocked by the addition of lactose to the incubation medium. The lactose competes for the B-chain binding site with cell surface galactose so that the binding specificity of the antibody determines the selectivity of the immunotoxin. Successful *in vitro* applications using immunotoxins with holotoxins (A chain + B chain) include the purging of T lymphocytes from bone marrow for allogeneic bone marrow transplantation to reduce the incidence and severity of graft-versus-host disease.

TABLE I Effectors Used in Immunoconjugates

Effector	Immunoconjugate	Examples of effectors
Protein toxin	immunotoxin	ricin, diphtheria toxin
Low-molecular-weight drug	chemoimmunoconjugate	doxorubicin, methotrexate
Biological response modifier (BRM)	BRM immunoconjugate	interferon, cobra venom factor
Radionuclide	radioimmunoconjugate	^{131}I, ^{111}In, ^{90}Y, ^{212}Bi
Enzyme	enzymoimmunoconjugate	glucose oxidase, alkaline phosphatase
Others	—	radiosensitizers, magnetic microspheres

Similarly, such immunotoxins have been used to purge leukemia cells from bone marrow before autologous bone marrow transplantation.

Another approach to eliminate nonspecific immunotoxin binding was the use of hemitoxins, which lack a binding domain, resulting in selective immunotoxins for antigen-positive target cells, in which the immunotoxin binding is only mediated through the antibody portion. More frequently, however, investigators coupled toxin fragments lacking the binding domain to antibodies. The most commonly employed procedure is the separation of the toxic A chain from the binding domain-bearing B chain of ricin and coupling of the ricin A chain to antibodies. In the case of the bacterial toxins diphtheria toxin and *Pseudomonas* exotoxin, toxin fragments lacking the binding domain were used for immunotoxin generation. These toxic fragments were obtained from naturally occurring mutants or by genetic engineering, deleting the binding domain from the cloned gene.

Many *in vitro* systems have been described showing selective cytotoxic activity of immunotoxins containing toxic A chains or truncated toxins for their target cells (mostly tumor cells). In most systems, the toxic activity for antigen-bearing target cells was several orders of magnitude greater than for antigen-negative control cells. In the case of the A-chain immunotoxins, it was found, however, that their potency was significantly lower than the potency of corresponding immunotoxins made with holotoxins containing both the A and B chain. This is a consequence of the fact that the B chain mediates toxin entry into the cell. Accordingly, a smaller fraction of the A-chain immunotoxins reaches the cytosol, a prerequisite for their cytotoxic activity.

Several attempts to overcome the disadvantage of decreased internalization of A-chain immunotoxins have been reported. One such approach is the use of holotoxins with "blocked" B chains (i.e., B chains where the binding site is not functional). The inactivation of the binding site was achieved by sterical hindrance caused by the proximity of the coupled antibody or by specific chemical inactivation. Another approach was to generate A-chain immunotoxins and B-chain immunotoxins separately, which act synergistically when bound to the same target cell.

Another concept to enhance the cytotoxic activity of A-chain immunotoxins was the simultaneous administration of pharmacological agents such as lysosomotropic amines or carboxylic ionophores. Lysosomotropic amines such as ammonium chloride, methylamine, and chloroquine and carboxylic ionophores such as monensin increase lysosomal pH, thereby inhibiting the degradation of A-chain immunotoxins, which enter secondary lysosomes rather rapidly. This approach enhanced the cytotoxic activity of A-chain immunotoxins *in vitro*. However, *in vivo* application of these enhancer substances proved rather difficult because sufficiently high concentrations cannot be reached because of toxicity (e.g., ammonium chloride) or rapid plasma clearance (e.g., monensin). A prolonged plasma half-life of monensin could be achieved by coupling it to serum albumin, resulting in better *in vivo* antitumor activity of A-chain immunotoxins.

Animal studies using immunotoxins as anticancer agents have been performed by many investigators. Model animal systems include guinea pigs, mice, and, predominantly, nude mice with human tumors. It was generally found that the immunotoxins were well-tolerated and that localization of immunotoxins into the tumor occurred. However, significant amounts of the immunotoxins were taken up by nontarget organs, and an antibody response to both the antibody moiety and the toxin moiety was usually observed. Nevertheless, antitumor activities were clearly observed. Several general conclusions can be made. By and large the antitumor activity was better the more artificial the tumor animal model was. For example, tumors could be completely suppressed if tumor cell inoculation and immunotoxin therapy were performed within a short period of each other. The larger and more established the tumors were, the lesser was the therapeutic effect of immunotoxin treatment. Generally, the therapeutic effect was better if immunotoxin treatment was at the same site as the tumor (e.g., intratumoral injections into subcutaneous tumors). Systemic therapy with immunotoxins administered by the intraperitoneal or intravenous route was usually less effective.

Several phase I clinical trials of ricin A-chain immunotoxins have been performed in patients with solid tumors such as melanoma or colon carcinoma as well as with leukemia. Similar to the animal studies, the immunotoxins were rather well tolerated, but a strong antibody response was usually noted. Therapeutic effects were limited, ranging from no effect and transient growth inhibition to an occasional disappearance of a metastasis.

B. Antibody Conjugates with Cytotoxic Drugs (Chemoimmunoconjugates)

Many studies have been reported in which low-molecular-weight toxins or chemotherapeutic drugs have been coupled to monoclonal antibodies. Among the substances coupled are many of the chemotherapeutic agents currently used in free form for cancer chemotherapy such as the anthracyclines doxorubicin and daunorubicin, the antimetabolite methotrexate, the alkylating agent melphalan, and many other agents including bleomycin, mitomycin, and vinca alkaloids. Low-molecular-weight mycotoxins such as α-amanitin or trichothecenes, which are not used as therapeutic agents, have also been coupled to antibodies.

Antibody-mediated delivery of toxic drugs is primarily intended to reduce uptake of the drug by nontarget tissues, thereby reducing general toxicity and widening the therapeutic window. The lower general toxicity might allow delivery of more of the cytotoxic drug to the tumor site. An alternative concept for the antibody-mediated delivery of drugs is to circumvent drug resistance in the case in which the resistance is due to decreased drug uptake. Antibody-bound drug might enter the cell by a different route and may reach different intracellular locations and may therefore circumvent or overcome drug resistance.

In contrast to antibody conjugates with protein toxins, which are enzymes and therefore require only a small number of toxin molecules to be internalized to become effective, the low-molecular-weight drugs are not enzymes and require a sufficiently high concentration in the tumor cell for their cytotoxic activity. Accordingly, it is desirable to deliver as many drug molecules in antibody-bound form to the cell as possible. The most severe limitation in derivatizing antibody molecules with numerous drug molecules is the inactivation of the binding function of the antibody. By and large, the derivatization of an IgG antibody with more than 10 drug molecules usually impairs the antigen-binding function of the antibody to a significant degree. One approach to increase the number of drug molecules delivered to a cell without interfering with the binding function of the antibody is the use of intermediate drug carriers. These are substances like serum albumin or dextran that can be derivatized with multiple drug molecules. The derivatized drug carrier can then be coupled to the antibody with comparatively minimal inactivation of its binding function, similar to the coupling of a protein toxin.

Similar to the immunotoxins, many *in vitro* studies have shown specific cytotoxic activity of chemoimmunoconjugates for antigen-bearing target cells. Many *in vivo* studies have also been performed, mainly in nude mice with human tumor transplants, with a various degree of success. Cure or increased survival time was found in those animal model systems in which the tumor burden was small and in which chemoimmunoconjugate treatment was started at an early stage. Some phase I clinical trials with chemoimmunoconjugates containing vinca alkaloids or methotrexate have been performed to gather information on general toxicity. No significant antitumor activity has so far been observed in these patients.

C. Antibody Conjugates with Biological Response Modifiers

The term *BRM* is used to describe a rather diverse group of biomolecules including synthetic analogues and derivatives that modify, direct, or augment the body's defense systems. This biological response is mediated by interaction with defined receptor molecules for the BRMs. Several well-defined BRMs have been coupled to monoclonal antibodies in an attempt to generate agents with antitumor activity. The common property of these antibody-BRM conjugates is that they exert their cytotoxic activity through activation or stimulation of one or several of the host's defense mechanisms. [*See* BIOLOGICAL RESPONSE MODIFIERS.]

Cobra venom factor (CVF) is the complement-activating protein in cobra venom. CVF activates the complement system analogously to the activated form of the third component of complement (C3b). However, the complement enzyme formed with CVF in the process of complement activation is not subject to the stringent control mechanisms that regulate the enzyme formed with C3b. Accordingly, CVF leads to exhaustive complement activation. By coupling CVF to monoclonal antibodies with specificity for tumor antigens, this complement-activating activity of CVF is targeted to the surface of the tumor cell where it will induce tumor cell killing by the host's complement system. Antibody conjugates with CVF have been shown to induce selective complement-depending killing of several tumors including melanoma, leukemia, and neuroblastoma. Although no extensive animal studies have been reported yet, CVF has two important properties that make it a promising therapeutic agent: First, CVF has no direct cytotoxic activities and second, because CVF is a structural analogue of mammalian C3b, it offers the theoretical possibility to develop a C3b derivative of significantly reduced or even absent immunogenicity.

Another BRM that has been coupled to monoclonal antibodies is the chemotactic tripeptide formyl-methionyl-leucyl-phenylalanine (f-Met-Leu-Phe). This tripeptide is chemotactic for inflammatory cells such as macrophages and polymorphonuclear leukocytes. The f-Met-Leu-Phe tripeptide has been coupled to antitumor antibodies and been shown to increase the number of macrophages in tumor-bearing animals.

Another BRM that has been coupled to monoclonal antibodies is α-interferon. Antitumor antibodies coupled with α-interferon have been shown to augment the killing of tumor cells in the presence of human peripheral blood mononuclear cells. This increased cytotoxic activity is believed to be caused by the binding of natural killer cells to their target tumor cells by the antibody conjugates and activation of the natural killer cells by the interferon. [*See* INTERFERONS; NATURAL KILLER AND OTHER EFFECTOR CELLS.]

D. Antibody Conjugates with Radionuclides (Radioimmunoconjugates)

Radioactive nuclides have been coupled to antibodies for two purposes: imaging and radiotherapy.

1. Radioimmunoimaging

Radioimmunoimaging or immunoscintigraphy is a new imaging modality for clinical diagnosis. A radiolabeled antitumor antibody binds to antigen-bearing tumor cells and thereby accumulates in the tumor. With the aid of a γ camera the patient will be scanned, and a tumor nodule can be imaged because of the γ radiation of the radiolabel. This method has been found useful to detect primary, recurrent, metastatic, and occult tumor nodules in cancer patients.

Many studies were performed with iodine isotopes, most frequently ^{131}I, because of well-established methods to radiolabel antibodies with iodine. However, ^{131}I has several drawbacks as an isotope for imaging purposes. Its high energy γ radiation is

not ideal for γ-camera imaging, its relatively long (8 days) radiation half-life and its β emission cause toxic effects, the oxidizing conditions necessary for its incorporation into the antibody damage the antigen-binding activity, and, most importantly, it exhibits low *in vivo* stability caused by deiodination of the iodine-labeled antibodies. ^{123}I has a shorter (13 hours) radiation half-life, and its photon energy allows more efficient γ-camera detection but still carries the other limitations of iodine.

For several metals such as technetium, indium, and gallium radioactive isotopes exist, which offer suitable γ-emission energies and radiation half-lives (^{99m}Tc, 6 hr; ^{111}In, 2.8 days; ^{67}Ga, 3.3 days). However, their use as radiolabels was only possible after methods for their stable binding to antibodies had been developed. These radioactive metals can now be bound to antibodies that have been derivatized with chelating agents. Using this chelation methodology, it has been found that these radioactive metal isotopes can be stably bound to antibodies, and radioimmunoimaging studies have been performed in experimental animals and tumor patients, particularly with ^{111}In-labeled antitumor antibodies. Tumors can be imaged with good specificity and sensitivity by radioimmunoimaging, and this procedure is a widely used experimental imaging method. However, the smallest tumor nodules imaged by radioimmunoimaging are approximately 1 cm^3 in size, and this is the reason why radioimmunoimaging has not replaced standard clinical imaging procedures [e.g., computed axial X-ray tomography (CAT scanning) or magnetic resonance imaging (MRI)]. The major problems are the background radioactivity caused by the blood pool and nonspecific uptake by nontarget organs and insufficient accumulation of the radiolabeled antibody in the tumor caused by limiting extravasation and tumor penetration. However, the intrinsic sensitivity of radioimmunoimaging is not a limiting factor, and it may well become a superior imaging tool for cancer patients if the pharmacological limitations can be overcome. One possibility to reduce those limitations is local application of the radiolabeled antibodies into body cavities or the lymphatics.

2. Radioimmunotherapy

Antitumor antibodies labeled with radioisotopes emitting high-energy β radiation or α radiation have been developed as cytotoxic agents. These reagents offer several advantages over antibodies conjugated with drugs or toxins. Radiolabeled antibodies do not need to be internalized by their target cells to become active. More importantly, they can exert cytotoxic activity for tumor cells with no or reduced surface antigen expression or with poor antibody access because the radiation has a tissue range of 100 to 1,000 cells in the case of β emitters and several cells in the case of α emitters.

The first β emitter used for radiotherapy was ^{131}I. This iodine isotope, however, carries all the disadvantages as outlined above. In addition, its γ emission, which is the basis for its application in radioimmunoimaging, adds to the nonspecific toxicity, which is dose-limiting because of its effect on the bone marrow. Nevertheless, therapeutic effects with ^{131}I-labeled antibodies have been described in various animal models and human trials. As with radiolabeled antibodies for imaging purposes, better results have been observed with local, intracavitary injections into the pleura, pericardium, and peritoneum.

The β-emitting isotopes of yttrium and rhenium with more favorable properties for radioimmunotherapy have been identified. ^{186}Re has a physical half-life of 3.7 days and little γ emission, whereas ^{90}Y has a physical half-life of 2.7 days with no γ emission. The physical half-life of these two β-emitting isotopes is long enough for tumor localization but short enough to minimize nonspecific toxicity. Both radioisotopes decay to stable products, and methods for their stable coupling to antibodies have been developed. Using ^{90}Y-labeled antibodies, many animal studies and human trials have been reported with promising therapeutic results. These studies have shown that higher dose rates and radiation doses may be delivered to tumor tissue with antibodies labeled with ^{90}Y compared with ^{131}I.

The α particles have a much shorter path length of only several cell diameters but exhibit an extremely powerful cytotoxic action. This is due to the high linear energy transfer and the limited ability of cells to repair the damage done to DNA by α particles. Two α-emitting isotopes of bismuth and astatine have been identified for the labeling of monoclonal antibodies for therapeutic purposes. ^{212}Bi has a physical half-life of 1 hr and can be coupled to monoclonal antibodies by chelation. ^{211}At has a half-life of 7.2 hr. Its coupling to monoclonal antibodies has been more problematic but was recently achieved using N-succinimidyl-astatobenzoate. ^{212}Bi-labeled antibodies have been shown to exhibit excellent *in vitro* cytotoxicity as well as anti-

tumor effects in animal studies, particularly after intraperitoneal application.

Another possibility to induce α-particle emission is the use of boron-labeled antibodies. ^{10}B absorbs thermal neutrons that release an α particle. Monoclonal antibodies have been labeled with ^{10}B. Because tissues not containing ^{10}B have a low neutron capture ability, selective tumor cytotoxicity may be achieved with ^{10}B-labeled antibodies bound to the tumor using subsequent irradiation with thermal neutrons.

E. Antibody Conjugates with Enzymes

Several enzymes with rather different catalytic activities have been coupled to monoclonal antibodies to achieve either cytotoxicity or other consequences of enzyme action. Strictly speaking, the ribosome-inactivating toxins such as ricin (see Section I,A) are enzymes and would also belong into this category. Other targeted enzymes to induce cytotoxicity include phospholipase C, glucose oxidase, and lactoperoxidase. Phospholipase C cleaves the phospholipids of plasma membranes and thereby induces cell death. Glucose oxidase and lactoperoxidase generate toxic metabolites such as hydrogen peroxide and oxidizing iodine species.

Another antibody-enzyme approach is to target an enzyme that will activate a prodrug at the tumor site. This approach offers several advantages over the coupling of a toxic drug directly to an antibody. First, many anticancer drugs are not potent enough to achieve a sufficiently high concentration at the tumor site by antibody-mediated delivery. In contrast, the prodrug approach allows enzymatical liberation of active drug at the tumor site, thereby increasing the local concentration. Second, the prodrug is released at the tumor site in free form. Consequently, it does not need to be released from the macromolecular antibody to become active. Third, the free drug liberated at the tumor site by the targeted enzyme will also be able to reach tumor cells that do not express many antigens or that cannot be reached by the macromolecular antibody-drug conjugate. Several model systems of antibody-enzyme conjugates and corresponding prodrugs have been reported. These include carboxypeptidase, which releases active benzoic acid mustard from its glutamic acid–conjugated prodrug form; alkaline phosphatase, which releases mitomycin or etoposide from its phosphorylated prodrug form; and penicillin amidase, which liberates doxorubicin from its phenoxyacetamide prodrug form.

Antibody conjugates for a different purpose have been synthesized with urokinase and tissue plasminogen activator. Both enzymes activate plasminogen to generate active plasmin, the major fibrinolytic enzyme. By coupling either of the two plasminogen activating enzymes to an antibody directed to fibrin, a significant enhancement of fibrinolysis could be achieved to dissolve blood clots.

F. Antibody Conjugates with Other Effectors

Several other effector molecules have been coupled to monoclonal antibodies. One example is miconidazole, a radiosensitizer that can enhance the lethality of ionizing radiation during radiotherapy of tumors. Another example of an effector molecule is hematoporphyrin. Hematoporphyrin is a photochemical that can be activated by visible light. Its cytotoxic effect is mediated through the production of an oxygen radical.

A different type of effector is polystyrene microspheres containing magnetite, which can be coupled to monoclonal antibodies directed against cell surface epitopes. The use of a flow system with a permanent magnet allows the removal of the magnetically tagged cells, a principle that has been used to purge tumor cells from bone marrow for autologous transplantation.

Liposomes are macromolecular phospholipid structures that have been used to encapsulate drugs such as anticancer agents. Such liposomes can also be targeted to specific tumor cells by covalent coupling to antitumor antibodies. For that purpose, a phospholipid derivative has been incorporated into the liposome structure, which allows the covalent binding of antibodies or their fragments to the surface of the liposome.

II. Targets for Immunoconjugates

The vast majority of immunoconjugates has been made to target an effector molecule to a defined cell population characterized by the expression of a specific surface structure (Table II). As is evident from the above discussions, most cell-directed immunoconjugates are directed to tumor cells. Obviously, tumor cells are an interesting target for immunoconjugates, which represent a novel modality of cancer treatment. The number of *in vitro* and *in vivo* stud-

TABLE II Targets for Immunoconjugates

Targets	Application
Tumor cells	tumor imaging and therapy
T lymphocytes	graft rejection, graft-versus-host disease, cancer therapy
B lymphocytes	autoimmune diseases, cancer therapy
Parasites	anti-infectious therapy
Blood clots	solubilization of clots (myocardial infarction), diagnosis of deep venous thrombosis, pulmonary emboli

ies is extensive, and accordingly, a great variety of tumors has been used as targets.

However, the selective killing of defined cell populations using immunoconjugates is certainly not restricted to tumor cells. Another group of cells that has been the target for immunoconjugates is lymphocyte subsets. These include T cells, which are responsible for graft rejection and graft-versus-host disease after organ transplantation. For example, antibodies directed against the CD5 antigen present on most peripheral blood T lymphocytes have been coupled to the ricin A chain. This immunotoxin has been shown to produce complete clinical responses in severe graft-versus-host disease after bone marrow transplantation. These rather promising results may point out that immunoconjugates directed against normal cells rather than tumor cells may be better models to study immunoconjugate therapy, because additional difficulties inherent to tumor models such as tumor cell heterogeneity and accessibility are absent. [*See* LYMPHOCYTES.]

Another group of immunoconjugates has been constructed to kill selectively B-cell clones producing antibodies against defined antigens. By coupling of the antigen to a cytotoxic agent such as the ricin A chain or a cytotoxic drug, selective immunosuppressive agents have been generated. Such immunosuppressive immunoconjugates have been produced with low-molecular-weight model haptens coupled to an ovalbumin carrier or model protein antigens such as tetanus toxoid. Antigen-toxin conjugates have also been generated with antigens involved in the pathogenesis of autoimmune diseases. For example, thyroglobulin has been conjugated to the ricin A chain. This thyroglobulin-ricin A-chain conjugate specifically suppressed the thyroglobulin autoantibody response of lymphocytes from patients with Hashimoto's thyroiditis. In another model system, the acetylcholine receptor had been coupled to ricin. This antigen-ricin conjugate caused reduced synthesis of antiacetylcholine receptor antibodies involved in the pathogenesis of myasthenia gravis.

Another potential class of target cells is infectious agents. However, surprisingly little work has been done in immunotoxin model systems using infectious agents as targets. Immunotoxins have been constructed with diphtheria toxin against a nonpathogenic amoeba, *Acanthamoeba castellani*. More recently, immunoconjugates have been constructed with the A chains of ricin or abrin and antibodies against *Trypanosoma cruzi*, the causative agent of Chagas' disease.

Antibody conjugates have also been constructed for noncellular targets such as the above-mentioned antibodies to fibrin coupled to a fibrinolytic agent. It should be mentioned in this context that a great variety of antibody-enzyme conjugates have been synthesized as developing reagents for enzyme immunoassays. These reagents usually consist of a secondary antibody directed against a primary antibody and an enzyme capable of producing a colored product such as horseradish peroxidase or alkaline phosphatase.

III. Targeting Moieties Used in Immunoconjugates

The vast majority of immunoconjugates has been constructed with antibodies or antibody fragments directed against surface structures of target cells (Table III). In the case of tumor cells, most of these antibodies were directed against tumor-associated antigens. The function and chemical nature of these antigens were usually not known; as a matter of fact, in many cases the antibodies were used to identify and chemically characterize the recognized surface antigens.

Most antibody conjugates were prepared with intact antibodies of the IgG subclasses. A few antibody conjugates were constructed with IgM, mainly using human monoclonal antibodies, which frequently are of the IgM class. In addition to intact antibodies, immunoconjugates were also constructed with antibody fragments such as $F(ab')_2$ and Fab. An even smaller antibody-derived binding domain used for the construction of immunoconjugates has been the Fv fragment by combining the light and heavy chain variable domains with a peptide linker using recombinant DNA technology.

In addition to monoclonal antibodies against newly discovered surface antigens of tumor cells, monoclonal antibodies to known surface receptors such as epidermal growth factor, hormones, and transferrin have been used to target cytotoxic agents to tumor cells. However, ligands to these known receptors on target cells have similarly been used as targeting moieties for immunoconjugates. In the case of hormone receptors, thyrotropin-releasing hormone (TRH) has been coupled to the toxic fragment of diphtheria toxin and shown to cause inhibition of protein synthesis in pituitary cells. Similarly, melanotropin (MSH) has been used as a targeting moiety for daunomycin or diphtheria toxin for the selective elimination of MSH receptor-binding melanoma cells.

Other molecules for which receptors exist on target cells have also been used as targeting moieties and include transferrin, α_2-macroglobulin, epidermal growth factor, and interleukins such as IL2, IL4, and IL6. Although the use of growth factors and interleukins as targeting moieties for cytotoxic agents such as *Pseudomonas* exotoxin generated selective cytotoxic agents *in vitro*, their *in vivo* application may be hampered by proliferative effects of the targeting moieties through binding to their respective receptors on the target cells.

Another molecule that has served as the targeting moiety for ricin A chain or *Pseudomonas* exotoxin is a soluble derivative of CD4, a surface protein found on T cells. CD4 has been identified as the receptor for the human immunodeficiency virus (HIV), the virus responsible for AIDS. CD4 binds to the gp120 glycoprotein that is expressed on the surface of the HIV virus. Soluble CD4 derivatives conjugated with a toxin have been shown to kill HIV-infected cells. [*see* CD8 AND CD4: STRUCTURE, FUNCTION, AND MOLECULAR BIOLOGY.]

Finally, a rather diverse group of targeting moieties has been used in antigen-toxin conjugates to eliminate selected antibody-secreting B cell clones. As discussed further above, thyroglobulin, the ace-

TABLE III Targeting Moieties of Immunoconjugates

Targeting moiety	Target site
Antibodies	various antigenic structures
Antibody fragments	various antigenic structures
Antigens	B-cell surface immunoglobulins
Hormones	specific receptors
Growth factors, interleukins	specific receptors
Transferrin, α_2-macroglobulin	specific receptors
CD4	gp 120 glycoprotein

tylcholine receptor, and several other model antigens have been used as targeting moieties in immunosuppressive antigen-toxin conjugates.

IV. Synthesis of Immunoconjugates

Immunoconjugates can be synthesized by two different procedures: chemical methods or recombinant DNA technology.

A. Chemical Synthesis of Immunoconjugates

The vast majority of immunoconjugates are synthesized by chemical methods. In proteins, available reactive groups most frequently used for chemical coupling are amino groups (ε-amino groups of lysine residues and amino-terminal amino groups) and free sulfhydryl groups. In the case of glycoproteins cis-diol groups in the carbohydrate portion have also been employed after oxidation to aldehyde functions. If no free sulfhydryl groups are available for coupling, they have been introduced by either the reduction of existing disulfide bonds or by chemical derivatization of amino groups with N-succinimidyl-3-(2-pyridyldithio) propionate (SPDP), which generates a free sulfhydryl group after reduction of its pyridyldithio moiety.

Low-molecular-weight drugs have usually been linked to antibodies using condensing reagents such as carbodiimides or after chemical activation of a group in the drug that allowed direct reaction with an antibody molecule. Several problems have been encountered with the synthesis of antibody-drug conjugates. Because the derivatization is random, the binding of the drugs usually impaired the binding function of the antibody, with a greater extent of inactivation occurring with an increasing average number of drug molecules bound per antibody molecule. However, the more drug molecules can be targeted by antibody-mediated delivery the greater is the cytotoxic potential. One approach to overcome this dilemma is the use of intermediate drug carriers such as dextran, polyaminoacids, or proteins such as albumin, which can be derivatized with multiple drug molecules and subsequently coupled to the antibody.

Another problem observed with antibody-drug conjugates is the requirement for the drug to be tightly bound to the antibody to prevent drug release before reaching the target site. However, in many instances it has been shown that release of the drug from the antibody is a prerequisite to exert its cytotoxic function in the target cell. Two approaches have been reported aiming at the release of the drug once it has reached the target cell. One such approach is the linkage of the drug to the antibody by an acid-labile linker, which releases the drug in the acidic environment of the lysosome of the target cell, an intracellular compartment that internalized antibody conjugates frequently reach. Another approach is the introduction of a short oligopeptide between the drug and the antibody, which is also cleaved by intralysosomal proteases in the target cell.

The synthesis of immunoconjugates consisting of two proteins, such as an antibody and a protein toxin, by chemical methods requires the use of a cross-linking reagent. With the exception of some early studies using rather nonspecific cross-linking reagents such as glutaraldehyde or homobifunctional reagents containing two identical reactive groups, the synthesis is performed with heterobifunctional cross-linking reagents, consisting of two differently reactive groups. One reactive group reacts with amino groups and invariably consists of an N-hydroxysuccinimide ester. The other reactive group usually reacts with a sulfhydryl group and can either be a pyridyldithio group, a maleimide group, or an aliphatic halogen such as iodine. The first step in the synthesis of an immunoconjugate consists of the derivatization of the first protein (e.g., an antibody) with a heterobifunctional reagent such as the above-mentioned SPDP. A second protein containing a free sulfhydryl group (e.g., ricin A chain) can then be coupled to the derivatized antibody, yielding the desired antibody-ricin A-chain conjugate. If the second protein to be coupled does not contain a free sulfhydryl group (e.g., CVF), free sulfhydryl groups can be introduced with SPDP as outlined above. Once free sulfhydryl groups have been incorporated, the derivatized protein can be coupled to the derivatized antibody.

The availability of heterobifunctional reagents has significantly contributed to the progress of immunoconjugate research during the past 10 years. The major advantage of these reagents is the fact that the conjugates generated by this procedure are always heteroconjugates, consisting of at least one molecule each of the two reactants. Because of the reaction sequence of heterobifunctional cross-linking reagents with proteins, the formation of homoconjugates (polymers consisting of only one of the two protein species) is excluded.

Nevertheless, immunoconjugate synthesis with heterobifunctional reagents has many inherent drawbacks. The derivatization of the protein to be coupled with the cross-linking reagent is random, which implies that individual protein molecules in a reaction mixture will be derivatized with different numbers of cross-linking molecules. In addition, a protein of the size of an IgG antibody has approximately 70 amino groups. Although they certainly differ in their individual reactivity, many amino groups will be able to react with the cross-linker. This implies that different antibody molecules derivatized with a small number of cross-linker molecules will be derivatized at different amino groups. This multiple and different derivatization of antibody molecules has two important consequences. One consequence is that chemical derivatization of amino groups leads to functional inactivation of the protein similar to derivatization of proteins with drugs, and increasingly with an increased average degree of derivatization. The other consequence is heterogeneity of the derivatized protein molecules caused by the derivatization at different amino groups.

Multiple derivatization of a single protein molecule has another disadvantage that becomes apparent after the coupling to the second protein. Each sulfhydryl-reactive group of the introduced cross-linking reagents can react with a free sulfhydryl group of the other reaction partner which yields immunoconjugates of heterogenous composition. If the second protein contains only one sulfhydryl group, the composition of the conjugates will be 1:1, 1:2, 1:3, etc. If the second protein contains multiple free sulfhydryl groups, the extent of compositional heterogeneity is even greater because the second reaction partner can also react with multiple molecules of the first reaction partner yielding a mixture of conjugates such as 1:1, 2:1, 1:2, 3:1, 2:2, 1:3, etc. Obviously, chemical inactivation and compositional (size) heterogeneity have significant effects on the biological activity of immunoconjugates.

Another disadvantage of the current heterobifunctional cross-linking reagents is that the proteins in the resulting immunoconjugates are rather close to each other so that they become functionally inactive because of steric hindrance by the other reaction partner. Further work will have to optimize the chemical procedures for immunoconjugative synthesis. An ideal process would result in the derivatization of only one defined reactive group in each protein, resulting in an immunoconjugate of predetermined composition (such as 1:1), with the linker being long enough not to interfere with the function of either protein. Several conceptual improvements in that direction have been published. Heterobifunctional cross-linking reagents have been developed that allow the coupling to carbohydrate moieties. In the case of antibodies, the use of these reagents will allow coupling of the effector molecule away from the antigen binding site of the Fab arms of the antibody molecule. Although it has clearly been shown that this approach will not interfere with the binding function of the antibody, the term *site-specific* may not be justified. Although the exact size of derivatization is confined to the carbohydrate side chains, it is certainly not site-specific. Furthermore, derivatization of the proteins at multiple oligosaccharide moieties occurs.

To overcome the steric hindrance of coupled proteins, several attempts to overcome this disadvantage by introducing a longer spacer between the two proteins have been published. In one example, a polypeptide spacer derived from the insulin B chain between the antibody and ricin resulted in an increased potency and efficacy of the immunoconjugate.

B. Synthesis of Immunoconjugates by Recombinant DNA Methods

More recently, several immunoconjugates consisting of two protein moieties have been synthesized by recombinant DNA methodology. A particular advantage inherent to such immunoconjugates is that they consist of the two components in a predetermined 1:1 ratio linked by a peptide linker between the amino terminus of one moiety and the carboxyl terminus of the other moiety. These recombinant immunoconjugates are therefore truly linked by a site-specific linkage. If, however, one of the terminal groups involved in the linkage is required for the activity of either of the moieties, such an immunotoxin will be inactive.

Several recombinant immunotoxins have been prepared. One consists of MSH spliced to a toxic fragment of diphtheria toxin. This recombinant fusion protein was toxic for MSH receptor-bearing melanoma cells. Several other recombinant immunotoxins have been generated using *Pseudomonas* exotoxin fused to various targeting proteins such as a single chain Fv fragment of an antibody or various growth factors and interleukins.

V. Shortcomings and Future Challenges

Immunoconjugates offer the potential to emerge into a new class of therapeutics, but the current use is still in the early experimental stage. Although it is possible to demonstrate good *in vitro* activity of virtually all immunoconjugates, their *in vivo* activity is low, inconsistent, or absent. To improve current immunoconjugates and to develop them into useful agents requires solving many complex problems.

The shortcomings of current immunoconjugates include the development of better methods for the synthesis allowing to generate more efficacious agents, but especially their lack of sufficiently specific accumulation in tumor tissue *in vivo* and their immunogenicity.

In fact in contrast to the misconception that antibodies and antibody conjugates home into their targets, they reach their targets by random distribution throughout the body through the bloodstream and by diffusive and convective transport through tissues. During their travel they distribute throughout the body and are nonspecifically taken up by tissues, and are subject to the various mechanisms of elimination from the body. These include excretion by the kidneys and uptake by the various cells of the reticuloendothelial system (RES), particularly in liver and spleen. In addition, cross-reaction of the antibody with normal tissues may cause specific uptake at an unwanted site.

Better tumor localization and increased plasma half-life can be achieved by reducing the uptake of immunoconjugates by the RES. One possibility is to preinject animals with a nonspecific monoclonal antibody that will subsequently block uptake of the specific immunoconjugate through Fc receptors on the cells of the RES. In the case of immunoconjugates with the ricin A chain it was found that the oligosaccharide portion of ricin mediated plasma removal of ricin-containing immunoconjugates through recognition of mannose-containing structures by receptors on Kupffer cells in the liver. The plasma half-life of immunoconjugates containing the ricin A chain and their localization into tumors could therefore be increased by injection of mannose-containing blocking reagents or by using chemically altered or deglycosylated ricin A chain.

The major obstacles in achieving sufficient tumor uptake, however, are the blood supply of the tumor, the extravasation of immunoconjugates in the tumor, and the penetration into the tumor tissue. Many factors will affect these parameters, including the vasculature of the tumor, the size of the immunoconjugate, and the affinity of the antibody portion for its antigen. The molecular weight of the immunoconjugate obviously will affect its extravasation at the tumor site and the penetration through the intracellular space. The binding affinity of the antibody for its surface antigen will also affect the distribution of the immunotoxin throughout the tumor. High antibody affinity will saturate antigens on tumor cells close to the capillaries, thereby decreasing the antibody penetration into the tumor and causing a more heterogeneous distribution.

To improve the efficacy of immunoconjugates, good accumulation of immunoconjugates at tumor sites must be achieved. Methods that will cause an easier penetration of immunoconjugates into the tumor will have to be developed, such as the use of vasoactive drugs.

The second major limitation of immunoconjugates is their immunogenicity. To date, the vast majority of immunoconjugates has been made with monoclonal antibodies of murine origin. As a protein from a different species, murine immunoglobulin elicits an immune response in humans. The generation of the antimurine antibody response restricts the use of monoclonal antibodies or immunoconjugates to a single course of therapy. Several attempts to reduce this immune response are being pursued. One obvious solution would be the use of human monoclonal antibodies. They have been generated, but for unknown reasons they are mainly of the IgM class and usually directed against intracellular antigens. A more promising approach therefore is to "humanize" murine monoclonal antibodies. This has been done by recombinant DNA technology by fusing the variable domains of a murine monoclonal antibody to the constant domains of a human antibody. Such chimeric antibodies have indeed been shown to elicit much less of an immune response and to exhibit longer plasma half-lives. In furthering this approach, humanized antibodies have been created by only inserting the hypervariable regions of the mouse antibody with the desired binding specificity.

More difficult, however, will be to overcome the immunogenicity of the effector moieties of immunoconjugates. Protein effectors, such as the ricin A chain, have invariably been found to elicit an immune response that severely affects the biodistribution and efficacy of ricin A chain containing immunoconjugates. For the most part, it appears unlikely to develop toxin analogues of lower or absent immunogenicity because there are no mammalian analogues to the ribosome-inactivating toxins. One possible exception is CVF, which is a structural analogue of mammalian C3; and it may be possible to generate a human C3 derivative with CVF-like functions but of significantly reduced or even absent immunogenicity.

Bibliography

Fujimori, K., Covell, D. G., Fletcher, J. E., and Weinstein, J. N. (1989). Modeling analysis of the global and microscopic distribution of immunoglobulin G, F(ab')$_2$, and Fab in tumors. *Cancer Res.* **49,** 5656.

Juliano, R. L., ed. (1987). Biological approaches to the controlled delivery of drugs. *Ann. N. Y. Acad. Sci.* **507.**

Petrella, E. C., Wilkie, S. D., Smith, C. A., Morgan, A. C., and Vogel, C.-W. (1987). Antibody conjugates with cobra venom factor. Synthesis and biochemical characterization. *J. Immunol. Methods* **104,** 159.

Rodwell, J. D., ed. (1988). "Antibody-Mediated Delivery Systems." Marcel Dekker, New York, Basel.

Sikora, K., Smedley, H., and Thorpe, P. (1987). Tumor imaging and drug targeting. *Br. Med. Bull.* **40,** 233.

Vitetta, E. S., Fulton, R. J., May, R. D., Till, M., and Uhr, J. W. (1987). Redesigning Nature's poisons to create anti-tumor reagents. *Science* **238,** 1098.

Vogel, C. W., ed. (1987). "Immunoconjugates." Oxford University Press, New York, Oxford.

Widder, K. J., and Green, R., ed. (1985). Drug and enzyme targeting. *Methods Enzymol.* **112.**

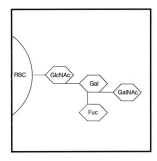

Immunogenetics

JAN KLEIN, *Max Planck Institut für Biologie*

I. Blood Group Antigens
II. Allotypes
III. Major Histocompatibility Complex Molecules
IV. Complement Components
V. Conclusion

Glossary

Antibody Soluble protein (i.e., immunoglobulin) produced by lymphocytes and capable of combining specifically with antigen

Antigen Substance capable of inducing an immune response

Antigenic determinant (epitope) Area of an antigenic molecule that determines the specificity of an antigen–antibody (T-cell receptor) interaction; site that fits into the combining site of an antibody or a T-cell receptor

Antiserum Fluid phase of blood containing antibodies against a specified antigen

Haplotype Particular combination of related genes borne by a single chromosome

Histocompatibility antigen Cell surface protein that determines the compatibility of transplanted tissue

Locus Region of a chromosome occupied by a gene

Polymorphism Occurrence in a population and at appreciable frequencies of two or more alleles at a given locus

IMMUNOGENETICS was originally conceived as a discipline in which researchers used immunological methods to study the inheritance of traits. In a typical situation scientists repeatedly injected human red blood cells (i.e., erythrocytes) into a rabbit or other animal until the recipient produced antibodies against antigens carried by the erythrocytes. The presence of the antibodies could be demonstrated, for example, by mixing the rabbit antiserum in a test tube with the human cells and observing their clumping. In this hemagglutination reaction, one of the two or more combining sites of the antibodies bind to one erythrocyte and another site binds to a second erythrocyte, connecting them. Because many antibodies can bind to the same erythrocyte, visible aggregates are formed.

The inheritance of the blood group antigens defined in this manner can then be studied by following their segregation in the progeny of parents, at least one of whom is heterozygous for the genes controlling these antigens (i.e., at least one of whom has two different forms, or alleles, of the same gene in the two chromosomes of a pair). Instead of erythrocytes, one can also inject other cells or body fluids into animals. Inoculation of blood serum, for example, produces antibodies against antigenic determinants borne by soluble proteins (i.e., allotypes).

More recently, the scope of immunogenetics has been widened to include the study of genes that control the immune response. Immunologists have learned that not all individuals of a species respond to the same degree when exposed to a given foreign substance. Some individuals respond to an antigen well, while others respond poorly, and this difference in the immune response often depends on the presence in the host of a particular immune response (*Ir*) gene. The best characterized of the *Ir* genes are those belonging to the major histocompatibility complex.

For many years the identification of antigens for the purpose of studying their inheritance was based on the use of antibodies produced by B lymphocytes. In the last few decades immunologists have realized, however, that the introduction of foreign substances into animal bodies also stimulates T lymphocytes that do not secrete antibodies, but combine with antigens via T-cell receptors, giving rise to the various forms of cell-mediated reactions.

Immunologists, therefore, distinguish two kinds of methods they use to study the inheritance of blood group antigens, allotypes, *Ir* genes, and other immunological characteristics: serological and cellular. Serological methods are based on the use of antibodies in the serum, whereas cellular methods rely on the functions of immune cells, particularly T lymphocytes. In many situations antisera are being replaced by monoclonal antibodies, which are antibodies of a single specificity produced in a culture of hybridomas, cells resulting from the fusion of an antibody-producing B lymphocyte with an immortal tumor cell.

I. Blood Group Antigens

Blood groups are any serologically detectable antigenic differences between blood components, particularly erythrocytes, of individuals belonging to the same species. The best-known human blood group antigens are those of the ABO and Rh systems. In addition to these, however, there are more than 20 other systems, as well as a large number of individual antigens.

The two main antigens of the ABO system are A and B, which divide humans into four groups: A, B, AB, and O. Group A individuals carry the A antigen on their erythrocytes and have B-reactive antibodies in their sera. Group B individuals express the B antigen on their erythrocytes and A-specific antibodies in their sera. Group AB individuals have both A and B antigens on their erythrocytes, but neither A- nor B-reactive antibodies in their sera. Finally, group O individuals lack both A and B antigens on their erythrocytes, but have A- and B-specific antibodies in their sera.

The origin of the A- and B-reactive antibodies in human sera is obscure, but, since structures similar to the A and B antigens are carried by certain microorganisms, researchers believe that infections early in human life cause their appearance. Because of the presence of A- or B-specific antibodies, the mixing of, for example, group A erythrocytes with group B serum or group B erythrocytes with group A serum leads to clumping of the blood cells. If clumping were to occur during a blood transfusion, it would have fatal consequences for the recipient.

The incompatibility of red blood cell antigens is the reason that donors and recipients must be matched in their blood groups before a transfusion.

The A and B antigens are borne by complex sugars consisting of several units (i.e., monosaccharides) strung together in a chain (Fig. 1). The two critical units are a galactose (Gal) and a glucosamine (GlcNAc) at the tips of the chain. To this Gal–GlcNAc doublet, an enzyme encoded in a gene on human chromosome 19 adds another sugar, fucose. The resulting trisaccharide is not recognized by either A- or B-specific antibodies, but it is recognized by other antibodies; the corresponding antigen is referred to as "H" (for heterogenetic). Individuals carrying two copies of a gene that codes for an inactive enzyme have no fucose attached to the disaccharide and do not react with H-specific antibodies. They possess the rare Bombay phenotype, named for the city in which it was first discovered.

Individuals carrying at least one copy of a gene coding for an active enzyme have fucose attached to the Gal–GlcNAc disaccharide and are H antigen positive. Once the fucose has been added, another sugar can be attached to the triplet thus generated. This addition is catalyzed by an enzyme encoded in a gene on chromosome 9. The gene exists in at least three forms (i.e., alleles).

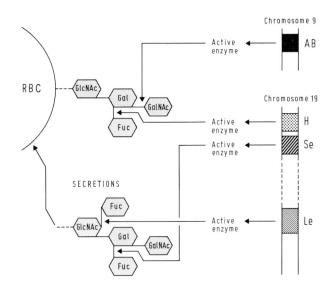

FIGURE 1 The ABH system of human blood groups. Genes on chromosomes 9 and 19 code for enzymes which catalyze the addition of different sugars to the basic disaccharide Gal–GlcNAc, either on the erythrocyte surface or in secretions (Se). Fuc, Fucose; Gal, galactose; GalNAc, galactosamine; GlcNAc, glucosamine; Le, Lewis antigen; RBC, red blood cell.

Some individuals carry two copies of a gene coding for an inactive enzyme which is incapable of attaching additional sugars; such individuals have the O group. Other individuals have at least one copy of a gene coding for an enzyme that adds a galactosamine to the basic trisaccharide and thus generates the group A antigen. A third category of individuals has at least one copy of a gene that encodes an enzyme capable of attaching a Gal to the trisaccharide, thus generating the group B antigen. Finally, a fourth category of individuals carries one copy of the A-encoding gene and one copy of the B-encoding gene, which together generate the AB group.

In addition to the gene responsible for the addition of fucose, chromosome 19 carries, at a different position, another gene that specifies whether the carbohydrate chain with ABH antigens will be released by cells of the digestive tract into saliva, stomach juice, bile, and other secretions. Those individuals who carry ABH antigens in their secretions are secretors, whereas those that do not are nonsecretors. Finally, chromosome 19 also bears genes (again at different positions) that specify, via corresponding enzymes, the addition of a second fucose to the gal–GlcNAc disaccharide, thus generating the Lewis antigen, named after a blood donor. Sugars with the Lewis antigen are present in the secretions of some individuals and can secondarily attach to erythrocyte surfaces.

The Rh system, so named because it was discovered with the aid of rabbit antibodies specific for rhesus monkey erythrocytes, is complex, consisting of many antigens. The Rh antigens are probably specified by a series of genes at three loci closely positioned on chromosome 1. The main antigens are D (or Rh1), C (or Rh2), E (or Rh3), c (or Rh4), and e (or Rh5), which occur in Caucasians with frequencies of 85%, 70%, 30%, 80%, and 90%, respectively. The most frequently occurring gene combinations are *Dce, DCe, DCE, dce, dCe, dcE,* and *dCE.* (The upper- and lower-case letters designate presumed alleles at the same locus; that is, *C* and *c* are alleles at one locus, *D* and *d* are alleles at the second locus, and *E* and *e* are alleles at the third locus.)

Individuals that possess the most prevalent RhD antigen are referred to as ''Rh positive''; those that lack it, ''Rh negative.'' In an Rh-negative mother carrying an Rh-positive fetus, a small number of fetal erythrocytes could pass into the mother's circulation, particularly during delivery, when the placenta separates from the uterine wall. The fetal erythrocytes might stimulate the production of antibodies reactive with the RhD antigen. If a high level of Rh antibodies builds up, some of them could pass through the placenta during subsequent pregnancies, enter the fetal circulation, and bind to the fetal Rh-positive erythrocytes. The antibody-coated erythrocytes are then retained and destroyed by macrophages in the spleen. The fetus attempts to compensate for the lost erythrocytes by pouring immature nucleated erythrocytes—the so-called erythroblasts—into the circulation, a condition known as erythroblastosis fetalis.

Portions of the hemoglobin molecules released from the destroyed erythrocytes are converted to bilirubin, the yellow pigment of jaundice, which is toxic at high concentrations. Some of the affected fetuses might be aborted, others might be delivered stillborn, and others still might be born alive, but with severe defects. The disease can be prevented by giving the mother RhD-specific antibodies to remove fetal cells from her circulation before they can immunize her and to shut off already ongoing antibody production via the regulatory pathway.

The discovery of blood groups has made blood transfusions much safer than before. It has also made it possible to obtain an average exclusion rate of 99% in disputed paternities, in cases in which the mother, the child, and the alleged father can be tested. Although paternity can be excluded with a high frequency, it cannot be proved by this approach, given the high frequency of all blood group combinations. The frequency of some blood group antigens is altered among individuals with duodenal and gastric ulcers, cancer of the stomach, and certain forms of anemia. The reasons for these alterations, as well as the function of most of the blood group antigens, however, are not known.

II. Allotypes

Antibodies not only can combine with antigen, they themselves can also serve as antigens and, when injected into another individual, can stimulate the production of antibodies that combine with them (Fig. 2). These antiantibodies thus define allotypes, which are antigenic determinants on antibody (i.e., immunoglobulin) molecules. (Allotypic differences can, however, also be defined between proteins

Antibody

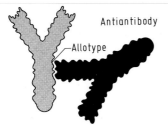

Antiantibody

Allotype

FIGURE 2 One antibody (solid area) recognizes, via its combining sites, antigenic determinants (i.e., allotypes) on another antibody (stippled area).

other than antibodies.) Allotypic determinants reside in the constant portion of the immunoglobulin molecule and arise because individuals have small differences in this region.

An individual X might have a particular amino acid residue at a certain position in the constant polypeptide, while Y might, because of a mutation in the corresponding gene, have a different amino acid at that position. When the immune system of individual X is exposed to immunoglobulins of individual Y, the immune cells recognize this small difference and begin to produce antibodies against it. No antibodies are produced against the parts that are identical. The exposure to immunoglobulins of other individuals could occur during blood transfusion or during pregnancy, when fetal immunoglobulins cross the placenta and immunize the mother.

Six allotype-encoding loci have been identified in humans and are designated *G1m, G2m, G3m, Mm, A2,* and *Km,* depending on the immunoglobulin chain (the $\gamma 1$, $\gamma 2$, $\gamma 3$, μ, and $\alpha 2$ heavy chains and the κ light chain) on which they occur. Each locus can be in one of two or more forms (i.e., alleles), each coding for a different antigenic determinant. Over 30 allotypic determinants have been identified. Five of the six loci form a cluster on chromosome 14; the *Km* locus is on chromosome 2. The frequency with which a given allele or a given combination of alleles occurs in a particular population is characteristic of that population.

Allotypes, therefore, can be used to investigate the origin of human populations. They are also used in paternity disputes, in the identification of blood stains, in the determination of twins' zygosity, and to solve other identification problems. Like some blood groups, some allotypes seem to occur more frequently in patients with certain diseases than among healthy individuals.

III. Major Histocompatibility Complex Molecules

The human leukocyte antigen (*HLA*) complex, the human version of the major histocompatibility complex, is a cluster of genes on chromosome 6 which codes for a unique set of proteins on the surfaces of most cells. These proteins have the ability to bind antigenic fragments of other proteins made in or entering the same cell and present them to the T-cell receptor of T lymphocytes. The peptides result from cleavage (i.e., processing) of the proteins. The presentation initiates the specific immune response in the human body. [*See* T-CELL RECEPTOR.]

Because different individuals have different variants of HLA molecules and because each variant can only bind some protein fragments, individuals differ in their ability to respond to different proteins. If a particular variant of the HLA molecule fails to bind a particular fragment, this individual's T lymphocytes ignore the corresponding antigen and fail to initiate a specific immune response to it, while responding normally to other antigens. The individual then appears to be a low responder, or even a nonresponder, to the particular antigenic determinant. Other individuals, carrying other variants of HLA molecules, respond normally to the particular determinant; they are high responders. The *HLA* complex thus functions as a set of *Ir* genes, deciding on the antigenic determinants to which a person can respond.

There are two classes of *HLA* genes on chromosome 6 (Fig. 3). Class I genes (*HLA-A, -B, -C,* and others) code for a long polypeptide which spans the plasma membrane of a cell and, with its extracellular part, associates with a shorter polypeptide, β_2-microglobulin (so named because it was originally discovered in a free form in the so-called β_2-globulin fraction of human serum, and because of its small size). The class I molecules specialize in binding fragments of viral proteins and presenting them to the killer version of immune lymphocytes (the cytotoxic T cells). [*See* LYMPHOCYTES.]

Class II genes are of two subclasses, A (α) and B (β), each being further divided into several families (*HLA-DR, -DP, -DQ, -DO,* and *-DN*). The α and β polypeptides are of approximately the same length and they associate in the plasma membrane into α–β dimers, according to their family relationships, (HLA-DR α chains normally associate only with HLA-DR β chains, HLA-DP α chains associate with HLA-DP β chains, and so on). The class II

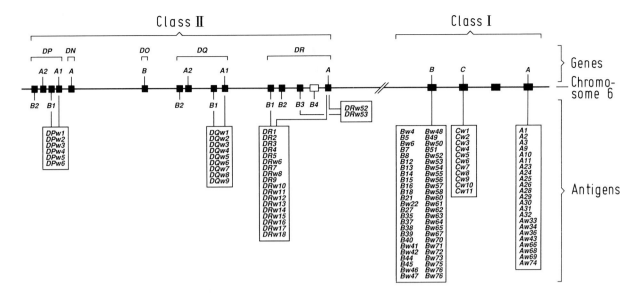

FIGURE 3 The HLA complex, the human version of the major histocompatibility complex. A short segment of human chromosome 6 (thin line) containing the class I and II *HLA* genes (solid boxes) is represented. Only some of the class I genes are shown; the total number is close to 20. The class II gene families are indicated by brackets. The full gene symbols are *HLA-A, HLA-B, HLA-C, HLA-DRA, HLA-DRB1, HLA-DRB2, HLA-DRB3, HLA-DRB4, HLA-DQA1,* and so on. The boxed symbols represent individual HLA antigens which have so far been identified. Each antigen is controlled by a different allele at the indicated locus.

molecules specialize in binding fragments of proteins brought into the cell from outside and presenting them to the helper T lymphocytes. While class I molecules have a wide tissue distribution, being present, albeit in different quantities, in most adult cells in the body, the class II molecules are preferentially expressed on B lymphocytes and a few other cells. The expression of class II molecules can, however, be induced on additional cells under abnormal conditions (e.g., under the effect of γ-interferon).

HLA-A, -B, -C, and some of the class II *HLA* loci are highly polymorphic, so that different individuals have different versions of the HLA proteins. In a randomly assembled sample of individuals, the chances are that no two individuals will have completely identical HLA molecules. This fact has grave consequences for attempts to transplant organs between individuals. The recipient's T lymphocytes recognize the different HLA molecules on the donor's organ as foreign and become strongly stimulated by them. They then initiate an immune response that can lead to the rejection (i.e., destruc-

tion) of the transplant. In an effort to minimize the antigenic stimulus to the recipient's lymphocytes, the donor and the recipient can be HLA typed and matched for at least some of the antigens. [*See* IMMUNOBIOLOGY OF TRANSPLANTATION.]

This matching can be relatively good between members of the same family, but complete matching between unrelated individuals is usually not possible. The HLA typing is carried out in specialized laboratories with batteries of antibodies with considerable specificity for the individual HLA antigens. Additional discrimination is achieved by cellular typing, in which the lymphocytes of the recipient are mixed in a test tube with lymphocytes from prospective donors (so-called mixed-lymphocyte reaction). The incompatible lymphocytes reveal themselves by their proliferation (i.e., cell division) or by their transformation into killer cells. So far, some 20 HLA-A, 50 HLA-B, 10 HLA-C, and 20 HLA-DR antigens have been identified by these methods.

Since these can occur in different combinations in different people, the reasons for the difficulty in matching unrelated individuals are apparent. Moreover, individuals also differ at numerous minor histocompatibility antigens, which can also stimulate incompatible lymphocytes, although less strongly than the HLA-controlled major histocompatibility antigens.

Several diseases have been found to be associated with certain *HLA* alleles. For example, all patients with narcolepsy, a sudden uncontrollable compulsion to sleep during waking hours, have the *HLA-DR2* allele. There are, of course, many *HLA-*

DR2-positive persons who do not suffer from narcolepsy. The association with other diseases is weaker, but statistically highly significant. For example, some 90% of Caucasians afflicted by ankylosing spondylitis (i.e., inflammation and stiffening of the spinal column) carry the *HLA-A27* allele, which is otherwise present in only 9% of healthy persons.

Significant associations of particular *HLA* alleles with insulin-dependent diabetes mellitus, rheumatoid arthritis, multiple sclerosis, systemic lupus erythematosus, myasthenia gravis, certain forms of thyroiditis, psoriasis, and other diseases have also been demonstrated. The reasons for most of these associations are not known. A reduction in the expression of HLA molecules is believed to be the cause of bare lymphocyte syndrome, in which the function of lymphocytes is greatly impeded; as a result the afflicted individuals might succumb to severe viral or other infections.

IV. Complement Components

Complement is a group of proteins in the blood plasma which, among other things, bind to antigen–antibody complexes causing lysis (i.e., destruction) of microorganisms and cells. There is at least one gene coding for each of the 20 or so complement components, and some of these genes occur in different versions—(i.e., they display genetic polymorphism). Some of the complement genes are located within the *HLA* complex (between the class I and II genes), while others are scattered over other human chromosomes. Defects at some of the complement loci can lead to frequent infections and an increased tendency to form antigen–antibody complexes and other kinds of immune complexes. A defect in the so-called C1 inhibitor leads to hereditary angioneurotic edema, characterized by sporadic attacks of swelling deep in the tissues, which is caused by increased blood vessel permeability. Defective complement function seems to contribute to a variety of diseases, such as systemic lupus erythematosus and certain forms of glomerulonephritis (i.e., inflammation of the kidney glomeruli). [*See* COMPLEMENT SYSTEM.]

V. Conclusion

Immunogenetics was conceived upon the union of two disciplines—immunology and genetics—at the turn of this century. Since then it has grown into a mature and healthy "offspring" that is now begetting its own subdisciplines. While immunogenetics has been dominated by the study of the blood group antigens, allotypes, and, in particular, major histocompatibility antigens, it can be expected to expand its realm even further in the future to encompass many more genetic systems, some of which might not yet have been discovered.

Bibliography

Fudenberg, H. H., Pink, J. R. L., Wang, A.-C., and Ferrara, G. B. (1984). "Basic Immunogenetics," 3rd ed. Oxford Univ. Press, Oxford, England.

Klein, J. (1986). "Natural History of the Major Histocompatibility Complex." Wiley Interscience, New York.

Race, R. R., and Sanger, R. (1968). "Blood Groups in Man." Blackwell, Oxford, England.

Williamson, A. R., and Turner, M. W. (1987). "Essential Immunogenetics." Blackwell, Oxford, England.

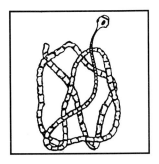

Immunology of Parasitism

G. V. BROWN AND G. F. MITCHELL, *The Walter and Eliza Hall Institute of Medical Research, Melbourne*

I. Host-Protective Immunity
II. Immune Evasion
III. Immunodiagnosis
IV. Immunopathology
V. Vaccine Design

Glossary

Antigen presentation Process by which antigens are taken up and presented to the immune system—in particular, T cells—by a variety of cells some of which have specificity for antigen—namely, macrophages, dendritic cells, and B cells

Cytokines Macromolecular products of cells; generally used in an immunological context, although, unlike antibodies, they do not have specificity for antigen. They affect the activities of other cell types. Lymphokines are from lymphocytes (e.g., interleukin-2) and monokines are from monocytes, although it is becoming clear that many cell types can produce cytokines of any particular type.

Hypersensitivity reaction Visible reaction to injection, ingestion, or inhalation of antigen that can manifest rapidly as wheezing, tissue swelling, and the like (i.e., immediate hypersensitivity) or after some hours (i.e., delayed-type hypersensitivity). Classically, the former is dependent on immunoglobulin E antibodies bound to mast cells (as in allergy to grass pollens), whereas the latter is initiated by inflammatory T cells.

Immunoglobulin E antibody A type of antibody that binds to mast cells in tissues. Other antibody types (known as immunoglobulin isotypes) include IgA, IgG, and IgM. Upon contact with antigen, the mast cell releases a variety of pharmacologically active molecules (e.g., histamine) that lead to the signs and symptoms of immediate hypersensitivity.

Monoclonal antibodies Antibody products of a perpetual cell line (called a hybridoma) derived from the fusion of a modified myeloma cell and an antibody-secreting cell.

T cell subpopulations CD4$^+$ and CD8$^+$ A division of the lymphocyte population derived from the thymus (i.e., T cells) into two broad categories according to surface molecules. In general, CD8$^+$ T cells participate in cytotoxic reactions (i.e., the killing of virus-infected cells), whereas CD4$^+$ T cells, as a result of the production of cytokines (actually lymphokines), help B cells in antibody production or engage in inflammatory responses (e.g., delayed-type hypersensitivity).

AS A CLASS OF organisms, parasites are noted for their ability to produce chronic infection, although they sometimes cause acute illness. These organisms have a global distribution, but are generally concentrated among the rural poor in the tropical less industrially developed parts of the world, where the insidious effects of parasitism by one or many species often go undetected in the vicious circle of poverty, malnutrition, and disease.

Because of the dependency of the parasite on the host and its propensity to cause chronic, even lifelong, infection, a successful parasite must not kill large numbers of individuals of the host species, at least in prereproductive life. A favorable balance must be established that ensures long-term survival and perpetuation of the parasite without endangering the host. To maintain this balance, partially effective host immune effector mechanisms against the parasite are balanced by partially effective parasite immune evasion mechanisms. Interactions of a single parasite with genetically diverse hosts, or of genetically diverse parasites with a single host, produce a spectrum from total resistance to life-threatening disease. Overreaction by the host to the offending organism can lead to immunopathological consequences which may be as threatening to life as the pathogen itself.

One important method for preventing life-threatening disease is for the parasite to induce immune responses that prevent successive infections and further "colonization" by the same species, a phenomenon called concomitant immunity, but also known as premunition or nonsterilizing immunity. Resident parasites develop immune evasion mechanisms, allowing their own survival, yet at the same time they induce or maintain responses that eliminate incoming parasites. A host can develop very efficient means for eliminating some protozoa in which case the parasite uses antigenic variation to thwart this host response. Age–prevalence and age–intensity curves obtained from epidemiological data in endemic areas strongly suggest that the development of substantial immunity to reinfection occurs in some human parasitic diseases in the face of continuing exposure. Even when apparent age-related resistance is not demonstrable, concomitant immunity can be operating very efficiently.

Another distinguishing feature of parasites, as a class of potentially pathogenic organisms in humans, is that none can be prevented by vaccination. There are, however, a few attenuated organisms that are used as vaccines in veterinary medicine. The task of developing prophylactic or therapeutic vaccines against these complex organisms is enormous: Parasite species are genetically diverse, and the genetic diversity of human hosts is, of course, extreme.

Topics discussed in this article are those of the most biological interest, namely, the nature of host-protective immune responses and the mechanisms of counteraction by immune evasion. Brief mention is also made of two other aspects of parasite immunology: immunodiagnosis and immunopathology.

I. Host-Protective Immunity

What immune effector mechanisms does the host use to limit parasite entry, proliferation, or continued residency? It is not surprising that practically all known immune effector mechanisms have been documented or implicated, often at every vulnerable stage of the life cycle. These mechanisms include complement-fixing antibodies for larval cestodes; opsonizing antibodies for many intracellular and extracellular protozoa; eosinophil-, macrophage-, and neutrophil-mediated antibody-dependent cellular cytotoxicity for schistosomules, microfilariae, and infective larvae; intracellular destruction following cytokine production by lymphocytes and/or macrophages for red blood cell-, macrophage-, and lymphocyte-dwelling protozoa; immunoglobulin A (IgA)-mediated effects (including maternal antibody effects) in *Giardia* infections; and recently, cytotoxic T cell activities (or at least CD8$^+$ T cell effects) for exoerythrocytic stages of plasmodia. [*See* CYTOKINES IN THE IMMUNE RESPONSE; LYMPHOCYTES; MACROPHAGES; NEUTROPHILS.]

Many proposed mechanisms of immunity are based on observations from *in vitro* or animal model systems that are of varying relevance to human parasitism. After identifying what can happen *in vitro* or in model systems, the laboratory scientist must proceed to the more difficult task of demonstrating that the proposed mechanism actually takes place in humans. The multiplicity of interactions between a mammalian host and a rapidly proliferating antigenically variable parasite (with a genome size of approximately 2×10^7 base pairs) that completes several life cycle stages in one host is several orders of magnitude more complicated than the interaction of a host and a relatively invariant virus (with a genome size of approximately 2×10^5 base pairs or less). Despite the striking advances of the past decade, there are great deficiencies in our understanding of the relative importance in natural host–parasite interactions of immunological mechanisms clearly demonstrated *in vitro,* and it is likely that the host mounts several different types of immune attack against every different stage of the parasite.

Most of the mechanisms described above reflect an interplay between various subpopulations of T cells reactive to specific antigens or B cells with the capacity to produce various kinds of antibodies and the activities of inflammatory cells (i.e., macrophages, eosinophils, neutrophils, mast cells, platelets, or fibroblasts) with no innate antigenic specificity. Of specific relevance to immunoparasitology is the method by which the host is exposed to the antigen, ranging from intravenous inoculation (e.g., hemoprotozoa) to direct introduction into lymph nodes (macrofilaria) or presentation by parasites dwelling in antigen-presenting macrophages (e.g., *Leishmania*). The route of antigen presentation can have profound effects on the immune response. For example, intravenously administered antigen is associated with low immunogenicity or tolerance, and for malaria it is clear that the immune response elicited in this way by repeated infections and drug cures might be different from the response to subcutaneous vaccination in the presence of adjuvant. [*See* MALARIA.]

Characteristically, initial parasitic infection occurs in a young host, often in the presence of maternally acquired antibody. A "trickle infection" with multiple challenge at low doses, in a semiimmune host might allow efficient antigen presentation and induction of effective immunity. A different result might occur with massive challenge in a nonimmune adult.

Parasites are such complex organisms that literally thousands of antigens may be released from living or dying parasites. Consequently, the immune system may be diverted toward thousands of foreign "housekeeping" or other intracellular molecules that induce responses irrelevant to host protection. Many of the responses demonstrable *in vitro* may fall into this category; but, as in other branches of the immunology of infectious disease, it is hoped that cytokine research will clarify the relevant signals derived from T cells and other cells that not only orchestrate the immune responses induced by pathogens, but may even have direct effects on the parasites themselves. In general, the *in vitro* tests of immune function can only view part of a phenomenal cascade of events initiated by the contact of host and parasites.

Medical textbooks state correctly that the hallmarks of helminth parasitic diseases as for immediate hypersensitivity reactions, are eosinophilia and IgE production. Precise definition of the functions of these host responses has been difficult. Model systems do not provide good examples in which IgE antibodies are indispensible for host resistance through, for example, elicitation of an immediate hypersensitivity response that creates a hostile local environment. Eosinophils participate in responses against schistosomules and eggs in *Schistosoma mansoni* infections but the range of positive and negative effects such cells may have on host resistance to helminths is not known. Monoclonal antibodies have been described that cause antibody-dependent cellular cytotoxicity *in vitro,* and other monoclonal antibodies of a different class can block this cytotoxic reaction.

II. Immune Evasion

A particularly interesting aspect of the immunology of parasitism is the study of the mechanisms used by parasites to evade expression or inhibit induction of host-protective immune responses (i.e., immune evasion). Two broad categories of mechanism are (1) an alteration of antigen display or immunogenicity of the parasite and (2) an alteration of immune responses by the host. Clearly, certain mechanisms within these two categories (see below) are closely related, and it is important to emphasize that it is highly unlikely that a successful parasite will have evolved only one mechanism of immune evasion.

Within the category of altered antigen display by the parasite are mechanisms such as antigenic variation (Section A below) and population heterogeneity with regard to antigen expression; sequestration in cells, cysts, and other immunologically privileged sites; masking and modulation of antigens; molecular mimicry; and reduced expression of associative recognition molecules (major histocompatibility complex-encoded molecules) for T cells.

A special situation applies to intramacrophage protozoa that thwart the hostility of their intracellular environment by inhibiting the fusion of phagosomes with lysosomes, escaping from the phagolysosome to lie in the cytoplasm, or inhibiting the action of digestive enzymes while resident in a phagolysosome within the macrophage. Some parasites at the time of entry may also alter the oxidative burst of phagocytes, which is a potent antiparasite mechanism.

Within the category of altered immune responses are specific and nonspecific immunosuppression (the latter through a multitude of possible mechanisms), immune deviation (section B below), elaboration of cellular toxins, antiinflammatory and anti-complementary factors, and digestion of bound antibodies. IgA, the immunoglobulin active at mucosal surfaces, can be cleaved by proteases secreted by some bacteria.

Two mechanisms are discussed in more detail: antigenic variation in trypanosomes and plasmodia and immune deviation in cutaneous leishmaniasis.

A. Antigenic Variation

African trypanosomes (the organisms responsible for sleeping sickness) have developed a sophisticated method for evading the host immune response. Parasites swimming freely in the blood are covered by a tough glycoprotein coat that envelops the entire plasma membrane of the organism and protects it against nonspecific host defenses. Furthermore, in the face of specific immune challenge, the parasite is able to change its surface coat, thereby rendering it invisible to the immune memory response. One hundred variant antigenic types have been documented to arise from a single cloned organism by the serial expression of a family of genes.

It is now clear that antigenic variation also occurs during chronic malaria infection. Once again, it is the erythrocytic form of the parasite that varies, but the variant antigens, rather than being expressed on the parasite, are found on the surface of the malaria-infected erythrocytes. This mechanism of immune evasion has been demonstrated in rodent and simian malaria and is generally believed to play a role during chronic human infection with *Plasmodium falciparum*. The mechanism for serial expression of antigenic variants is unknown, but these variants of cloned populations multiply the heterogeneity that already exists in antigenically diverse parasite populations.

B. Immune Deviation

Some parasites have the capacity to direct host immune responses along pathways unproductive for host resistance and the elimination of parasites. One of the best examples that has emerged from recent studies in the mouse concerns the causative organism of human cutaneous (Old World, zoonotic) leishmaniasis, the sandfly-transmitted protozoan, *Leishmania major*. It is now clear that T cells of one broad type, namely, CD4[+] T cells, can promote either host susceptibility in mice or, alternatively, host resistance, by activating macrophages to kill their intracellular parasites. Activation of the CD4[+] T cell type that secretes γ-interferon can aid in resistance, whereas activation of the CD4[+] T cell type that secretes interleukin-4 is largely counterproductive. If the parasite can divert the T cell response along the interleukin-4-secreting pathway (the class of T cells responsible being referred to as T_H2 CD4[+] T cells) rather than the γ-interferon-secreting type (i.e., T_H1 CD4[+] T cells), then the parasite will not be eliminated.

Consistent with this hypothesis is that the administration of antibody against γ-interferon exacerbates disease, whereas antibody to interleukin-4 is therapeutic. Relative dominance of the two types of CD4[+] T cell appears to determine the outcome in hosts of different genotypes. How the different subsets are induced is not known, but a difference in the modes of antigen presentation is the favored hypothesis.

Another mechanism of immune deviation may operate in malaria and Chagas' disease. These parasites all express protein antigens, with an unusual structure of blocks of tandemly repeated amino acid sequences. Immunodominant epitopes in these tandem repeat antigens may "tie up" B cells in a confusing network of irrelevant antibody production when host protection demands a response to a less dominant, but biologically more important, epitope. The response to a smokescreen of irrelevant antigens induces the gross splenomegaly (i.e., spleen enlargement) and hypergammaglobulinemia (i.e., high antibody content in the blood) typically associated with these infections, but the host remains unprotected.

Immune evasion mechanisms are of interest not only for the immunobiologist, but also for those attempting to develop vaccines. If the molecular bases of some of the immune evasion mechanisms mentioned in this section were known, opportunities for neutralizing these mechanisms immunologically (i.e., with a therapeutic vaccine) or with drugs could become apparent.

III. Immunodiagnosis

Detecting parasites in human hosts is often difficult, impractical, or tedious. For these reasons indirect methods are often used, the time-honored method being the detection of antiparasite antibody in the serum (e.g., immunodiffusion, agglutination, or direct reaction with parasites, such as circumsporozoite precipitation for antisporozoite immunity or the circumoval precipitin test for schistosomiasis involving precipitation around eggs). The most common assays are the enzyme-linked immunosorbent assay or the related radioimmunoassay. In these tests antigen anchored to a solid support (e.g., a plastic dish) is reacted with test serum, which could contain immunoglobulin specific for the antigen, and after washing, the complex of antigen and immunoglobulin is detected with antiimmunoglobulin labeled with enzyme or radioactivity. High radioactivity, or development of a color change after addition of the enzyme substrate, is indicative of the presence of antibody. [*See* RADIOIMMUNO-ASSAYS.]

Visualizing parasite eggs in feces or parasitized erythrocytes in a blood film provides direct evidence of current infection, but, in contrast, the presence of serum antibody is simply an indicator of past or present exposure to the parasite. Previously infected and/or highly resistant individuals might be positive for antibody, whereas individuals recently infected or responding weakly may be negative in an insensitive test. Skin tests for immediate (i.e., IgE-dependent) or delayed-type (i.e., T cell-dependent) hypersensitivity reactions also present problems in interpretation akin to those of serological tests. Moreover, cross-reactions with other parasites could reduce the specificity, although this problem can be reduced by the availability of non-cross-reacting epitopes synthesized chemically or produced using recombinant DNA techniques. [See RECOMBINANT DNA TECHNOLOGY IN DISEASE DIAGNOSIS.]

High specificity and sensitivity are required to increase the predictive value of a diagnostic test, and many immunodiagnostic tests for parasitic infection fall far short of the ideal. Much recent effort has been devoted to the development of immunoassays for the detection of parasite antigens or metabolites, generally in serum, but also in urine and in feces. The advent of monoclonal antibodies has provided the necessary tools to achieve this. One monoclonal antibody anchored to a solid support is reacted with serum containing circulating parasite molecules, and bound antigen is detected with a second labeled monoclonal antibody. Theoretically, the best parasite product for detection in serum is one that is poorly immunogenic so that in the infected host it is not masked or cleared by induced antibody. It is hoped that quantitative antigen detection methods (and gene amplification techniques such as the polymerase chain reaction) will provide some quantitative information on the *number* of parasites present in individuals, a critical aspect of many epidemiological studies in which the severity of pathology appears to correlate with the intensity of infection.

Monoclonal antibodies are being used for detection of parasites in vectors (e.g., malaria sporozoites in mosquitoes or *Leishmania* promastigotes in sandflies) and provide powerful tools for species differentiation (e.g., leishmaniasis) and for the detection of antigenic variants of parasite species. Monoclonal antibodies, of course, provide highly specific tools for the analysis of parasite epitopes recognized by B cells; we still lack such convenient tools for the study of T cell-stimulating epitopes. [See MONOCLONAL ANTIBODY TECHNOLOGY.]

IV. Immunopathology

Some antiparasite immune responses are beneficial for the host, but, as indicated in Section II, some are beneficial for the parasite. An exaggerated immune response to a parasite can be deleterious to the host, as shown by chronic schistosomiasis, which is the classical example of immunopathology in parasitic disease. In this snail-transmitted helminth disease, eggs laid by the female worm become impacted in organs and tissues. T cells of the inflammatory type respond to antigens emanating from eggs and produce lymphokines that result in granuloma formation and fibrosis around the eggs.

Some walling off the eggs is desirable for the host, in that adjoining cells (e.g., hepatocytes) are protected, but large granulomas and fibrosis, reflecting an exaggerated cellular response in organs such as the liver, ultimately impede blood flow, leading to the serious complication of raised pressure in the portal vein (i.e., portal hypertension). Other pathological consequences follow from this. Interestingly, exaggerated responsiveness decreases as the disease progresses, and the host benefits by producing a small controlled granuloma that is less likely to cause obstruction. The immunological basis of this phenomenon of granuloma modulation is of great interest, for, in theory, the same mechanism induced by vaccination could prevent severe disease in chronic schistosomiasis by a mechanism akin to desensitization. A vaccine that prevents infection would obviously be more desirable.

Onchocerciasis and Bancroft's filariasis (i.e., elephantiasis) are other human diseases in which exaggerated immune responses to antigens of adult worms or microfilariae are believed to be the major components of pathology. Immunopathology can be exacerbated by chemotherapy, which causes massive killing of microfilariae in patients with onchocerciasis. The peculiar tropical splenomegaly syndrome (i.e., hyperreactive malarious splenomegaly) seen in malaria-endemic areas is another example of an immunopathological consequence of infection in genetically predisposed individuals. It is clear that many of the manifestations of severe malaria, including respiratory distress, defective red blood cell production, hepatic dysfunction, fever, and hypoglycemia, could be produced by tumor necrosis factor, acting in concert with other cytokines. An accompaniment of concomitant immunity is the development of tolerance to the pathological effects of these cytokines, perhaps through masking of critical parasite epitopes by antibody or by "tolerance" with a diminution of symptoms during later infections. Other immunological responses that may have pathological consequences are hypergammaglobulinemia, the formation of complexes of antibodies with antigen in the blood and tissues, allergy, and autoantibody production.

Clearly, the analysis of immunopathology is a critical aspect of vaccine development, care being necessary to ensure that the vaccine sensitizes only for host protection in humans, but not for immunopathology. This could be particularly so in a vulnerable subset of vaccinees who might be genetically predisposed to respond differently from the majority of the population. [*See* IMMUNOPATHOLOGY.]

V. Vaccine Design

The first approach to vaccine design is to induce the response that occurs naturally in the proportion of individuals who develop high levels of resistance in response to repeated infection. The target antigens are called natural antigens. There are few examples of vaccines that induce resistance better than natural infection (tetanus toxoid perhaps being one). An alternative approach is to focus an immune response on molecules of the organism that are not recognized well during natural infection and that serve a critical function for the parasite, such as a receptor or an enzyme. These are often referred to as novel antigens. An attenuated organism, that is, one altered in such a way that it loses virulence yet remains immunogenic, could provide the best vehicle for presenting natural antigens; molecular vaccines, that is, emphasizing isolated molecules of the parasite, could theoretically be better as novel antigens. Information relevant to the first approach has been gained through the study of naturally induced immunity, by looking for responses correlating with immunity (e.g., antibody specificities or isotypes in serum of protected hosts that are absent in exposed but nonimmune individuals). Novel antigens are more likely to be discovered by finding specific receptors or enzymes involved in host–parasite interactions.

Vaccination to stimulate host resistance may be therapeutic for the treatment of current infection, or prophylactic to prevent initial infection. Successful immunotherapy has been reported in Venezuela against leishmaniasis using a mixture of killed organisms with bacille Calmette–Guérin (BCG) adjuvant, similar to the treatment of lepromatous leprosy using armadillo-derived *Mycobacterium leprae* with BCG. Immunotherapy is said to be as effective as chemotherapy in limited human trials performed to date. Molecular vaccines against human parasites that are at early stages of development include (1) vaccines against *P. falciparum* malaria, based on portions of the circumsporozoite protein of sporozoites and several merozoite antigens of blood stages of *P. falciparum*, (2) vaccines against schistosomes, based on integral membrane and other surface proteins and functional enzymes of the worms, and (3) both glycolipid and protein vaccines in leishmaniasis. Trials of a defined antigen vaccine against taeniasis have demonstrated protection of susceptible sheep.

Bibliography

Bloom, B. R. (1986). Learning from leprosy: A perspective on immunology and the Third World (Presidential address). J. Immunol. **137**, i–x.

Burnet, F. M., and White, D. O. (1972). "The Natural History of Infectious Diseases." Cambridge Univ. Press, Cambridge, England.

Cohen, S., and Warren, K. S. (1982). "Immunology of Parasitic Infection," 2nd ed. Blackwell, Oxford, England.

Mims, C. A. (1976). "The Pathogenesis of Infectious Disease." Academic Press, London.

Mitchell, G. F. (1986). Cellular and molecular aspects of host–parasite relationships. *Prog. Immunol.* **6**, 798–808.

Wakelin, D. (1984). "Immunity to Parasites: How Animals Control Parasitic Infection." Arnold, London.

Immunology, Regional

J. WAYNE STREILEIN, *University of Miami School of Medicine*

Glossary

Immune deviation Unique spectrum of immune effectors elicited by antigens under special immunizing circumstances; typically, one or another effector modality (e.g., delayed hypersensitivity, complement fixing antibodies) is selectively deleted from the systemic immune response to a particular antigen

Immune effectors Antibodies and T lymphocytes that possess effector functions associated with immune responses

Immunologic privilege Partial to complete failure of the immune system to respond to and eliminate antigenic tissues or grafts that are placed in special body sites (e.g., anterior chamber of the eye, brain, testis, cheek pouch of Syrian hamster)

Lymphocyte traffic Capacity of lymphocytes to travel from one site of the body to other sites, using the bloodstream and/or lymphatic channels as the medium for migration

Lymphoid organs and tissues Sites in the body wherein lymphocytes are normally generated and accumulate; these sites include lymph nodes, spleen, thymus, Peyer's patches, tonsils, appendix, and lymphocyte collections along mucosal surfaces

Mucosal immunity Specialized form of immune response to antigens encountered first through mucosal surfaces; in general, the most important feature of mucosal immunity is that the predominant immunoglobulin produced is IgA

Parenchymal cells Differentiated cell type that provides the unique function of an organ or tissue (e.g., keratinocytes in skin, hepatocytes in liver)

Regional specialization Modification of the generic systemic immune response to suit the special physiologic needs of various regions (organs and tissues) of the body

Selective immune deficiency Characteristic of individuals with immune deviation; one (or more) immune effector modality is not generated in response to an antigenic exposure

IN CARRYING OUT its primary function of providing protection against life-threatening pathogens, the immune system has developed an extraordinarily diverse array of antigen-specific effector modalities. Effector cells and molecules differ with respect to their ability to eliminate or eradicate various pathogens. Moreover, certain effectors are more appropriate for pathogens in one as opposed to other types of organs and tissues. As a consequence of these two considerations, the immune system has devised unique strategies for providing protection at individual tissue sites—protection that is appropriate for the pathogen and commensurate with preservation of the unique function of the individual tissue. The study of specialization in immune responsiveness according to different regions of the body is called regional immunology.

I. Introduction

Enormous advances have been made in the past few years in describing the cellular and molecular mechanisms responsible for the proper functioning of the immune system. As our knowledge of this extraordinary system has grown, we have begun to appreciate its diversity. The immune system possesses a broad range of strategies for dealing in a highly specific manner with a wide array of different antigens and pathogens. In this regard, not all immune effector modalities have comparable potential to neutralize, eliminate, or otherwise nullify specific pathogenic agents. In a sense, such a wide array of effector modalities addresses the evolutionary need to cope with the great diversity of pathogenic agents in our universe. [*See* IMMUNE SYSTEM.]

Beyond this type of diversity, there is another level of complexity that is not necessarily implied by these considerations. Experimental evidence indicates that the quality and quantity of immune responses to antigenic challenges in various organs and tissues of the body are not necessarily equivalent. One of the earliest reported examples of this phenomenon was the description of immunologic privilege in the anterior chamber of the eye. More than a century ago, investigators discovered that tumor tissues implanted into the anterior chamber of rabbit eyes often survived and even grew in that site, whereas similar tumor tissues implanted in the subcutaneous space did not grow or survive. This phenomenon was called immunologic privilege. Over the years, immunologic privilege has been described for several other unique sites of the body: brain, testis, ovary, cartilage, and cheek pouch of Syrian hamsters. Until very recently, immunologic privilege was regarded simply as a laboratory curiosity, rather than as an expression of an important, if enigmatic, aspect of systemic immune responsiveness.

The most clear-cut example of regional specialization in immune responsiveness is the presence of large amounts of IgA antibodies in external body secretions. Exposure to exogenous antigens via the gastrointestinal tract results in the production of IgA antibodies, which appear in secretions at mucosal surfaces but are poorly represented in the serum or in extravascular tissue compartments.

In this chapter, the conceptual framework upon which the idea of regional immunology is based is developed, specific examples of regional spheres of immunologic influence are described, and examples are presented of pathogenic consequences that may result from tissue-directed specializations in the immune response.

II. Conceptual Framework of Regional Immunology

A. Existence of Uniquely Differentiated Nonlymphoid Tissues and Organs

The body is comprised of different organs and tissues that in the aggregate work toward the common goal of sustaining and propagating life. While processes that are responsible for creating these different tissues from the zygote are beyond the scope of this chapter, certain features of this ontogenetic sequence are of particular interest to immunology.

In the eventual emergence of a fully differentiated organ or tissue (such as liver or skin), specificity is created via complex genetic paradigms. Unique genes are activated in cells that will become the specific (parenchymal) cells of a particular organ or tissue, and unique protein products are produced, proteins that are responsible in part for the fully differentiated state. Development and maintenance of the functional specificity of individual tissues and organs are also influenced by stromal elements. The skin affords a good example: the diverse properties of epidermis found on the sole of the foot, on the scalp, around the eyes, and covering the tongue appear to be directed by inducing factors released from underlying dermis. This example emphasizes that parenchymal cells of organs and tissues remain susceptible to modulation by external elements, even in the fully differentiated state—a point of importance in regional immunology.

In adult organisms, specific organs and tissues maintain their physiologic form and function by several different mechanisms: (1) endogenous genes encode specific proteins, which in turn endow individual cells with specific functions; (2) cells influence each other via cell-to-cell communication networks, such as cytokines and receptors, cell-adhesion molecules, and electrical coupling (gap junctions); and (3) stromal connective tissue cells, which also are capable of elaborating cytokines, undoubtedly serve as a further stabilizing influence.

Thus, fully differentiated, nonlymphoid somatic tissues carry out their unique physiologic functions and assume their unique microanatomical forms as a consequence of sustained activation of endogenous genes of parenchymal cells and genes acting via cells of the supporting connective tissue. The dynamic nature of these cellular communication devices is a crucial consideration for the concept of regional immunology.

B. Existence of Differentiated Lymphoid Organs and Tissues

The creation and development of the various organs and tissues of the lymphoid apparatus (lymph nodes, spleen, etc.) follow similar rules of ontogeny. The major themes of intercellular communication mediated by cytokines and their receptors, and of interacting cell-adhesion molecules, including specific receptors for ligands on the surfaces of other cells, are typical for both lymphoid and non-lymphoid tissues. However, throughout the life of the organism, the immune system (as well as the entire hematopoietic system) retains the capacity to generate fully differentiated cells from the most primitive pluripotent mesenchymal stem cells. Moreover, these ontogenetic processes are conducted in an almost ad hoc fashion in diverse regions of the body. The participating cells frequently migrate, via the blood, from site to site before completing their differentiation process. Perhaps as a consequence of this need to assemble and disassemble interacting multicellular units, gap junctions and electrical coupling are not utilized by lymphoreticular cells. Therefore, interacting lymphohematopoietic cells rely solely on cytokines and receptors and cell-adhesion molecules to exchange information and to influence each other. An extensive (if still incomplete) array of growth- and differentiation-promoting cytokines and receptors have now been described for the developing and intercommunicating cells of the immune and hematopoietic cell lineages. Our knowledge of the diversity of cell-adhesion molecules that also promote these processes is less advanced, although progress in this field of research is now moving at a rapid pace. It is important to bear in mind that not all intercellular communication is designed to promote or upregulate an effect; some cytokines can differentially downregulate (decrease) certain functional properties of target cells. [See HEMOPOIETIC SYSTEM.]

Over the past decade, it has been learned that cytokines used by nonlymphoid cells to exchange information are similar (often identical) to the cytokines used by interacting lymphoreticular cells. In fact, the receptors for these cytokines are often identical. Thus, the potential is great for signals derived from one type of cell to influence the functional activities of cells of a different tissue lineage and probably has physiological significance. [See CYTOKINES IN THE IMMUNE RESPONSE.]

C. The Perspective of the Immune System

Numerous immunological issues are raised when the immune system is called upon to respond to antigens in physiologically diverse regions of the body. Some of the most important issues are described below.

1. Unique Tissue-Specific Molecules (Antigens)

If differentiated tissues express one or more molecules (typically proteins) that are unique for that tissue alone, are these molecules incorporated into the immunological definition of "self"? Autologous proteins that come to be regarded as self by the immune system do so (largely, if not completely) because lymphocytes are exposed to these autologous molecules before their developmental program is complete. For T cells, this ontogenetic process is thought to occur within the thymus. If certain proteins are first expressed in a specific organ (e.g., surface molecules expressed on mature sperm) rather than in the thymus, and if this expression occurs after the full complement of mature T cells has disseminated from the thymus to peripheral tissue, then these proteins may not be regarded by the immune system as self. Other proteins may escape designation as self by the immune system because they are synthesized and subsequently remain "sequestered" only behind blood–tissue barriers (e.g., the S antigen of the retina). Molecules, which for these (and other) reasons never become incorporated into an immunologic definition of self, may serve as autoantigens. Therefore, tissue-restricted molecules can become the targets of autoimmune attack that can destroy the organs and tissues on which they are expressed.

2. Detection of Exogenous Antigens by the Immune System

Given that organs and tissues possess physicochemical barriers that limit the penetration of antigenic and pathogenic agents into the body, how do/ can cells of the immune system become aware of these exogenous agents? The epidermis is a good example of an important barrier. The barrier itself includes the stratum corneum and the basement membrane that separates the dermis from the epidermis. Langerhans cells, located within the epidermal compartment, appear to be specifically designed to pickup and transport foreign molecules to the draining lymph nodes where T cells normally reside. This mechanism maximizes the likelihood that pathogens will be detected by the immune system. Alternatively, in the case of the eye, and the brain, little if any effective lymphatic drainage pathways exist. This circumstance severely limits the possibility that T cells in regional lymph nodes could become aware of antigens in these sheltered organs.

3. Access of Immune Effectors to Regionally Located Antigens

To rid a local tissue of a specific antigen or pathogen, how do immune effector modalities gain access to that tissue? Effector modalities, such as specifically sensitized T cells and antibodies, are not chemotactically drawn to tissue sites by antigen. Instead, accumulation of immune effectors at tissue sites occurs by nonspecific means, relying first on random diffusion and migration of cells and molecules from peripheral blood into (virtually all) accessible sites in the body, and second on non-antigen-specific chemotactic gradients and by cell-adhesion molecules that are established by random encounters between effectors and antigens at tissue sites. Endothelial cells of the microvasculature are critical to these processes; for example, organs and tissues served by capillaries with endothelial cells united by tight junctions (such as the brain) represent formidable barriers to the infiltration of immune cells and molecules, whereas fenestrated capillaries, such as those found in skin, promote the accumulation of blood-borne cells and molecules. [See NATURAL KILLER AND OTHER EFFECTOR CELLS.]

4. The Problem of Immune Effector Response Amplification

In virtually all immune responses, successful eradication of antigens and pathogens relies on recruitment of nonspecific (nonadaptive) inflammatory defense mechanisms, which include macrophages, granulocytes, complement components, etc. The remarkable amplification potential of inflammation depends on recruitment of inflammatory cells, which is mediated via cytokines secreted by immune effector cells and is further amplified by cytokines released from the recruited cells themselves. Local factors, already present within specific organs and tissues (presumably subserving other physiologic purposes) may promote or interfere with the recruitment and amplification process. A good example of the latter is the presence of transforming growth factor-β (TGF$_\beta$) in normal aqueous humor of the anterior chamber of the eye. The relative inability of the anterior chamber to display and sustain delayed hypersensitivity responses can be ascribed in part to the profoundly inhibitory properties of TGF$_\beta$ on immunogenic inflammation. [See INFLAMMATION.]

D. The Perspective of Specific Organs and Tissues

Important issues are raised for parenchymal cells of specific organs and tissues by the possibility of expression of immunity within their midst.

1. Avoidance of Autoimmunity

The existence of microanatomical specializations that have the effect of isolating certain tissue-specific molecules and cells from the general circulation is probably not an accident. In the testis, numerous tight junctions encircle and bind together Sertoli cells. As a consequence, developing sperm, which are imbedded in Sertoli cells, are beyond the reach of potentially hazardous molecules and cells of the blood and of the interstitial spaces of the seminiferous tubules. In an analogous fashion, the continuous capillaries of the brain and retina form a barrier, ensuring that active mechanisms alone can permit entry of molecules and cells from the blood. Arguably, part of the reason for these anatomical specializations is the need to prevent the immune system from recognizing unique molecules (antigens) expressed on cells beyond these barriers (vide supra). Although simple mechanical barriers alone are not considered sufficient to account for the avoidance of autoimmunity in these tissues, barriers of this type certainly contribute.

2. Maintenance of Differentiated Tissue Function

In carrying out their unique physiologic functions, each organ and tissue must face the possibilities that (1) it may be destroyed by an intruding pathogen and (2) the pathogen may evoke tissue-localized immune responses that are themselves deleterious. In reference to the latter, immune effector cells secrete cytokines into an environment when they mediate protection. Therein lies a dilemma. Parenchymal cells often express, or can be induced to express, receptors that can bind these very same cytokines, and this may produce deleterious consequences, especially if the lymphocyte–macrophage-derived factors alter the physiologic properties of the parenchymal cells. In addition, recruitment of nonspecific host defense mechanisms may create amplified inflammation, which proves to be detrimental within a confined organ or tissue space. For example, accumulation of blood-borne cells and fluids within a restricted bodily compartment (such as the brain—confined by the bony skull) may compress parenchymal cells and produce tissue ischemia and death. Finally, the release of inflammatory mediators during an ongoing immune response may cause "innocent bystander" injury to parenchymal cells. In this regard, certain lymphokines (lymphotoxin) are directly toxic to nonlymphoid cells. These considerations reveal that expression of immunity within an organ or tissue may be deleterious to that tissue, even though the intent is one of protection.

III. The Thesis of Regional Immunology

Based on these considerations, the thesis is advanced that each region of the body (organ or tissue) has the potential to modify the quality of immune responses to antigens and pathogens in such a manner as to achieve a unique spectrum of immune effector modalities that are (1) specific for the antigen(s) in question, (2) sufficient to effect elimination or eradication of the pathogenic agent, and (3) unlikely to interfere with the proper physiologic function of that organ or tissue.

The thesis states that the most important force dictating the quality and quantity of the tissue-tailored immune response is the parenchymal cell, which provides the tissue its differentiated function. Thus, in skin, the epidermal keratinocyte is thought to assume the primary responsibility for setting the "tone" of local immune reactivity, whereas in the liver, the hepatocyte would be expected to play a similar role.

According to this thesis, the magnitude and type of immune protection afforded within each region of the body results from a compromise that must be struck between parenchymal cells and lymphocytes (chiefly T cells). This compromise takes into ac-

count such factors as (1) "ideal" immune effector modality for the pathogenic agent in question and (2) preservation of the functional properties of the differentiated organ or tissue. The compromise is mediated by intercommunications between immune cells (chiefly T cells and macrophages–dendritic cells) and parenchymal cells through cytokines, cell-adhesion molecules, and the extraordinary mobility and recirculation potential of lymphoreticular cells.

IV. Components of Regional Spheres of Immunological Influence

To accomplish regional specialization in immune responses, certain features appear to be essential ingredients in the process.

A. Tissue-Seeking Lymphocytes

Recirculating lymphocytes, especially T cells, possess the special capacity to migrate preferentially to one or another region of the body. Because migration of this type is *not* mediated by antigen, other mechanisms operate (vide supra). Tissue-tropism displayed by subpopulations of lymphocytes is mediated by different sets of tissue-specific homing receptors and their complementary ligands. As a consequence, some T cells migrate preferentially to lymphoid tissues associated with the gastrointestinal tract, whereas other T cells demonstrate a proclivity for trafficking to lymphoid tissues that drain the skin. Differential expression of homing receptors on T cells and of tissue-restricted ligands on high endothelial cells of postcapillary venules (called high endothelial venules [HEV]) in these disparate lymphoid compartments accounts for the existence of different migratory patterns. Within the past few years, our knowledge of the heterogeneity of lymphocyte migratory patterns has been expanded to include (1) homing receptors that direct certain T cells to the dermis of skin inflamed with active psoriasis or infiltrated with tumor cells and (2) homing receptors on other T cells that cause these cells to accumulate in the joints of patients with rheumatoid arthritis. Additional unique homing patterns (such as migrations to the brain, endocrine organs, and kidney) probably will be identified in the future. [*See* LYMPHOCYTES.]

The mechanisms responsible for creating and

maintaining diverse migratory pathways are essentially unknown at present. Homing receptors on lymphocytes seem to belong to classes of cell-surface molecules that function in cell adhesion. Beyond this, little is known about the factors that lead to the imprinting of recirculating lymphocytes in a manner that permits some to seek the skin while others seek the gastrointestinal tract.

B. Resident Antigen-Presenting Cells

The diversity of mechanisms by which antigens and pathogens gain access to different regions of the body implies that distinct strategies exist to promote recognition of antigens by lymphocytes. In some tissues, such as skin, specialized antigen-presenting cells (Langerhans cells) reside within the epidermis; by virtue of their functional properties, these cells confer upon this tissue site the specific ability to take up, process, and present antigens to lymphocytes. In addition, Langerhans cells that have captured antigen from the epidermal compartment can loosen their attachment to the epithelium, migrate across the dermal–epidermal junction, and flow with the afferent lymph to the draining lymph node. In other tissues, such as the gastrointestinal tract, parenchymal epithelial cells themselves are specialized (M cells) for the purpose of ensuring that antigenic materials in the lumen of the gut are transported across the epithelial surface into Peyer's patches. In these examples, evolution appears to have provided the organs and tissues with the capacity to maximize the likelihood that an antigen intruding into its environment will be detected by the immune system.

Alternatively, the eye and brain appear to have evolved toward a state that minimizes the possibility of local processing and presentation of antigens. Class II major histocompatibility complex positive dendritic cells and macrophages, which are usually expected to function in antigen presentation, are rare in the interior of the eye and within the brain. Moreover, cells harvested from the internal compartment of the eye are incapable of presenting antigens to primed T cells. The precise biologic meaning of this antigen-presenting cell deficiency is still open to question. It has been suggested that by minimizing *in situ* antigen presentation, the eye and brain may be able to avoid evoking those types of immune effectors that are particularly deleterious to these organs.

C. Uniquely Differentiated Parenchymal Cells

Each distinct organ and tissue contains parenchymal cells that carry out the differentiated functions unique to that organ or tissue. The keratinocytes of skin, the absorptive epithelial cells of the small and large intestine, the hepatocytes of the liver, and the follicular epithelial cells of the thyroid gland are but four examples. These cells are directly responsible for creating a regional microenvironment that is appropriate for their physiologic function. When the cells of the immune system enter these diverse microenvironments, the cells may be diverted into new patterns of behavior by the factors and mediators that provide the microenvironment with its distinctive features. Because even the immune system conducts its affairs in this way, the microenvironment of a lymph node is unique and dictated largely by the secretory products of resident lymphocytes, chiefly T cells.

The idea that each organ and tissue comprises a unique regional sphere of influence emanating from its definitive parenchymal cell is at the heart of the concept of regional immunology. According to this view, migrating lymphocytes regularly come under the temporary influence of diverse microenvironments, and as a consequence, their functions may be temporarily altered. Similarly, Langerhans cells (which are also migratory) may have their properties dictated by keratinocytes when they occupy the epidermal compartment. Alternatively, Langerhans cells that migrate to the draining lymph node may become functionally altered as they come under the influence of T cells in that site.

D. Specialized Endothelium of Postcapillary Venules

Certain T and B lymphocytes adopt at least two different migratory patterns: (1) to the lymphoid tissues associated with the gut and (2) to the lymphoid tissues draining the skin. Already mentioned is the fact that the pathway followed by any particular lymphocyte is governed by its own distinctive homing receptors and by complementary ligands expressed on specialized endothelial cells of postcapillary venules. The ability of endothelial cells to direct lymphocytes from the blood across the vessel wall into lymphoid organs depends, in part, on as yet unidentified factors elaborated by mature T cells. In addition, inductive influences also emanate from the parenchymal cells of tissues that drain into particular lymphoid organs, and these inductive factors help to modify the expression of appropriate ligands on postcapillary venules in these organs. Thus, specialization of stromal endothelial cells is an important component of a regionally distinct sphere of immune influence.

E. Draining Lymphoid Organs and Tissues

During primary immune responses to new antigens, the critical first step—antigen recognition by T cells—usually takes place within an organized peripheral lymphoid organ (lymph node, spleen), rather than in nonlymphoid tissues. The reasons for this requirement are beyond the scope of this chapter. Suffice it to say that the initial T cell-inductive event initiated by antigen must take place in the protected microenvironment of a peripheral (secondary) lymphoid organ. We presume that antigenic signals that arise from nonlymphoid tissues are delivered to secondary lymphoid organs where they can be transduced into appropriate effector modalities, but we know very little about this relationship. For example, excision of a lymph node prior to injection of antigen into skin can abort the expected immune response. A more dramatic example of a link between a tissue and its draining lymphoid tissue occurs with the eye. Antigens inoculated into the anterior chamber of the eye elicit a deviant systemic form of immunity, which selectively excludes delayed hypersensitivity and complement fixing antibodies, while other types of immune effectors are retained. If splenectomy precedes the anterior chamber injection of antigen, a conventional immune response develops, replete with both delayed hypersensitivity and complement-fixing antibodies. Thus, the spleen serves as the draining lymphoid organ for the eye, and this fact emphasizes that a meaningful relationship exists between a draining lymphoid organ and the specific region it serves.

In summary, a functional regional sphere of immunologic influence is comprised of five elements: tissue-tropic lymphocytes, specialized antigen-presenting cells, parenchymal cells (and the microenvironment they create), specialized endothelial cells of postcapillary venules, and a draining lymphoid organ. When antigens arise from an organ or tissue, these elements interact to create and mold immune responses that are protective yet appropriate; i.e., they do not interfere substantially with the physiologic function of the region in question.

V. Immune Consequences of Regional Spheres of Immunological Influence

The existence of specialized immune responses that are regionally determined has important ramifications for the systemic immune response as well as for the well-being of the entire organism.

A. Unique Spectra of Immune Effectors

The immune system can marshall a wide variety of functionally distinct, yet antigenically specific, effector cells and molecules. On the cellular side, we recognize cytotoxic T cells, delayed hypersensitivity T cells, and helper T cells—of more than one type. On the humoral side, the human immune system generates four different isotypes of IgG and two isotypes of IgA as well as IgM and IgE antibodies. Each of these isotypes has a special set of functional properties, typically related to molecular differences found within the Fc portion of each immunoglobulin molecule. As a further consideration, both cellular and humoral immune effectors can recruit macrophages, natural killer cells, granulocytes, complement components, clotting factors, etc. to assist in mediating their effects. In response to antigens first confronted in a distinct region of the body, a unique spectrum of immune effectors is selected, presumably because of the compromise described earlier. The spectra of elicited effectors may vary from tissue to tissue and organ to organ of initial antigenic confrontation. Because these effectors are generated centrally and are then disseminated systemically via the blood, the spectrum of immune effectors elicited becomes identical with the systemic immune response itself. Thus, when antigens are first presented to mucosal surfaces, the selective local production of IgA antibodies rather than IgG also characterizes the systemic immune response.

B. Selection of Immune Effectors Creates Selective Immune Deficiencies

If a regional immune response is comprised of some, *but not all,* possible effector modalities, apparently the ignored effectors will eventually comprise a selective immune deficiency, at least with regard to the antigen in question. This is an important consideration because elimination of pathogens sometimes depends solely on a single immune effector modality—and all other modalities are superfluous. As an example, antibodies to antigens associated with the tubercle bacillus confer essentially no protective immunity; only T cell-mediated immunity is able to eradicate this organism. Therefore, if a selective cell-mediated immune deficiency is created by regional exposure to the tubercle bacillus, the host would be deprived of a protective immune response, even though specific antibodies are produced. Thus, this immune deficiency would constitute a significant hazard to the host's survival.

The important conclusion from this set of statements is that while regional immune responses are created with the special needs of a specific tissue or organ in mind, the immune response that is generated is a systemic one. More specifically, a selective immune deficiency that is elicited by antigen in one region of the body will turn out to be a deficiency felt in all other regions of the body. This is the idea that brings home the importance of considering regional immune responses as "compromises."

C. When Regional and Systemic Immune Concerns Collide

Inevitably, the unique needs of a specific tissue are sometimes at odds with the needs of the entire organism. Not surprisingly, this conflict has been observed in experimental animal model systems in which regional immunity has been studied. For example, studies of experimental intraocular tumors have revealed that when tumor-associated antigens are relatively weak, the deviant systemic immune response that is generated protects the tumor and renders the host susceptible to death. By contrast, when intraocular tumors express strong antigens, a conventional immune response is generated, the host is spared, and the tumor is rejected, although frequently the eye itself is destroyed. Which factors determine whether regional concerns will dominate over systemic concerns or vice versa is not completely clear. Certainly, antigenic strength is an important variable, but much remains to be learned about other factors that influence this decision.

VI. Examples of Regional Spheres of Immunologic Influence

Several regions of the body are known to be subserved by specialized immune reactivities. These include the mucosal surfaces, the lower respiratory tract, the skin, the eye, the brain, certain endocrine organs, perhaps the liver, and the maternal–fetal interface. None are completely described or understood, but some are better characterized than others. As examples, three regional spheres of immunologic influence are described in Tables I and II.

In Table I, the five elements typically thought to contribute to regionally distinct responses have been indicated for the following regions: skin, mucosal surfaces, and eye. The parenchymal cells that dominate skin and mucosal surfaces are the respective epithelial cells; however, the dominant parenchymal force within the eye has yet to be clearly identified. Because this organ is comprised of several functionally distinct components, perhaps in the posterior compartment, the cells of the retina assume this role, whereas in the anterior compartment, the epithelial cells lining the ciliary processes and iris may be responsible.

TABLE I Comparison of Components of Selected Regional Spheres of Immunologic Interest

Component	Skin	Mucosa	Eye
Parenchymal cells	Keratinocytes	Absorptive epithelium	Retina and/or epithelia of iris/ciliary body
Antigen presentation	Promoted by Langerhans cells	Promoted by M cells of epithelia, and dendritic cells of Peyer's patches	Suppressed by Müller cells and bone marrow-derived cells of iris
Tissue-seeking lymphocytes	Have been described	Have been described	Not defined
Specialized endothelial cells	HEV of lymph nodes	HEV of Peyer's patches and lymph nodes	Not defined
Draining lymphoid organ	Lymph nodes	Lymph nodes	Spleen

HEV, high endothelial venules.

Local strategies for antigen presentation differ markedly for these three regions. Epidermal Langerhans cells provide the cutaneous surface with a pervasive network of dendritic processes that serve to trap antigens and pathogens crossing the stratum corneum and arising within the epidermis. By contrast, antigen uptake across the intestine and certain other mucosal surfaces is mediated by specialization of surface epithelial cells (M cells), which are strategically located on the lumenal surfaces overlying lymphocyte collections in the lamina propria (Peyer's Patches). Antigens that passes through M cells are picked up by dendritic cells of the subjacent lymphoid compartment, which then prepare the antigens for presentation to lymphocytes. Thus, the capacity for antigen presentation in the skin seems to be distributed evenly across the integument, whereas this function is promoted across mucosal surfaces at discrete sites that may be separated from each other by considerable distances. Antigen presentation takes on a completely different form in the eye. In fact, cells within the anterior chamber or the posterior compartment of the eye appear to be unable to carry out conventional antigen presentation. To the contrary, bone marrow-derived cells in the iris stroma and Müller cells in the retina have been shown to inhibit profoundly antigen presentation by conventional antigen-presenting cells. In part, this is accomplished by the release of suppressive cytokines, such as TGFβ.

Tissue-seeking lymphocytes have been demonstrated for the skin and for the mucosal surfaces. Whether these populations represent completely different subsets of recirculating cells or are partially overlapping is unclear. To date, the existence of lymphocytes that display tropic properties for the eye have yet to be identified, perhaps because such cells have not been looked for.

As might be expected, the existence of tissue-seeking lymphocytes implies the existence of tissue-specific ligands that are expressed uniquely on endothelial cells of the microvasculature supplying different regions. Ligands have been identified on high endothelial cells of mesenteric lymph nodes, and these ligands differ from similar molecules found on the high endothelial cells of lymph nodes that drain skin. The existence of lymphocytes that bind selectively to dermal microvessels indicates that a third set of molecules serves this function in the dermis. Ligands that direct lymphocytes selectively into the eye have yet to be defined.

Lymph nodes are situated along lymphatic vessels that drain skin. These organs trap antigen-bearing Langerhans cells escaping the skin and, therefore, serve as sites for antigen presentation to lymphocytes, the initiating event that leads to proliferation, differentiation, and the generation of immune effector modalities. Similarly, lymph nodes draining Peyer's patches and other mucosa-associated lymphoid collections function as regional sites for lymphocyte activation and differentiation. However, the internal compartments of the eye are not served by lymphatic vessels. As a consequence, molecules and cells that leave this organ are delivered directly into the bloodstream. Therefore, the spleen serves as the draining lymphoid organ for the eye. Considerable evidence indicates that the spleen transduces antigenic signals into effectors and regulators of the immune response that may differ considerably from those generated within lymph nodes. This may account for why immune responses to ocular antigens differ so remarkably from responses to antigens from the skin or the gastrointestinal tract.

The types of immune effector cells and molecules that are generated when antigens are introduced into the skin, the mucosa, and the eye are categorized in Table II. This table is incomplete, because several modalities are not even mentioned (IgM, IgE). The data are presented primarily to indicate that the spectrum of immune effectors elicited by antigens at these three regions are *not* identical.

Immune responses to cutaneous antigens are dominated by delayed hypersensitivity and complement-fixing antibodies, both of which depend on amplification through nonspecific inflammatory processes for successful elimination of the antigen or pathogen to which they are directed. The mucosal surfaces also generate delayed hypersensitivity responses, but the major immunoglobulin produced is IgA. Ocular immune responses are comprised of cytotoxic T cells and noncomplement-fixing antibodies, each of which can effect their specific function with minimal participation from the nonspecific mediators of inflammation.

Even this incomplete tabulation of effector responses is informative as to the teleology of regional specializations in immunity. By generating immune effectors that are unable to enlist nonspecific host defense mechanisms, the eye avoids the local accumulation of nonspecific inflammatory mediators, which have the potential to destroy vision by their tendency to disrupt the delicate microanat-

TABLE II Spectrum of Immune Effectors Generated in Response to Diverse Regional Exposures to Antigens

Region	T lymphocytes		Antibodies		
	Delayed hypersensitivity	Cytotoxic	Complement-fixing—IgG	Noncomplement-fixing—IgG	IgA
Skin	+++	+++	+++	+++	−
Mucosa	+++	(?)	+	+	+++
Eye	−	+++	−	+++	+

+ indicates that immune effector is present; +++ indicates that the effector is prominent; − indicates absence of an immune effector; ? unknown.

omy of the visual axis. The strategy of generating IgA antibodies ensures that mucosal surfaces will be bathed with antibodies that can withstand attack by proteolytic enzymes present in mucosal secretions, because protective antibodies must bind to their target antigen or pathogen in this environment. The intense inflammatory responses that attend delayed hypersensitivity and complement fixation can bring maximal defense mechanisms to bear within the skin. The logic of these vigorous responses apparently derives from the virulence of pathogens that gain access across the cutaneous barrier and from the skin's ability to withstand and recover its physiologic function after experiencing violent and destructive inflammatory responses that are necessary to rid the tissue of pathogens.

These few examples may serve to indicate the complexity and diversity that still awaits description in these and other regional spheres of immunologic interest. Only time and additional experiments will tell the extent to which other regions of the body will be similarly served. In addition, much remains to be learned about the mechanisms that operate to imprint regional patterns of reactivity on a "generic" systemic immune apparatus and response.

Bibliography

Alley, C. D., and Mestecky, J. (1988). The mucosal immune system. *In* "B Lymphocytes in Human Disease" (G. Bird and J. E. Calvert, eds.), pp. 222–254. Oxford University Press, Oxford, Great Britain.

Streilein, J. W. (1987). Immune regulation and the eye: A dangerous compromise. *FASEB J.* **1,** 199–208.

Streilein, J. W. (1988). Editorial—Regional spheres of immunologic influence. *Reg. Immunol.* **1,** 1.

Streilein, J. W. (1988). Skin associated lymphoid tissues. *In* "Immunologic Mechanisms in Cutaneous Disease" (D. A. Norris, ed.), pp. 73–96. Immunology Series Volume 46, Marcel Dekker, Inc., New York.

Streilein, J. W., Head, J. R., and Stein-Streilein, J. E. (1986). Regional specialization in antigen presentation. *In* "Reticulo-Endothelial System: A Comprehensive Treatise. Hypersensitivity States" (S. M. Phillips and P. Abramoff, eds.), pp. 37–93. Plenum Publishing Corporation, New York.

Streilein, J. W., and Wegmann, T. G. (1987). Immunologic privilege in the eye and the fetus. *Immunol. Today* **8,** 362–366.

Woodruff, J. J., Clarke, L. M., and Chin, Y.-H. (1987). Specific cell-adhesion mechanisms determining migration pathways of recirculating lymphocytes. *Annu. Rev. Immunol.* **5,** 201–222.

Immunopathology

STEWART SELL, *The University of Texas Medical School at Houston*

Glossary

Anaphylaxis Immediate hypersensitivity reaction (hive, shock) caused by release of mediators from mast cells as a result of reaction of antigen (allergen) with mast cell bound IgE antibody (Reagin)

Arachidonic acid 5,8,11,14-Eicosatetraenoic acid; an unsaturated fatty acid found in cell membranes that is a precursor in the synthesis of leukotrienes, prostaglandins, and thromboxanes, which are mediators of inflammation

Basement membrane Extracellular meshwork of collagen and reticular fibers in a proteoglycan matrix that supports epithelial cells and separates vessels from other tissues

Complement System of at least 13 serum proteins that are activated by enzymatic cleavages and aggregations to produce fragments or aggregates with biologic (inflammatory) activity. One method of activation is by reaction of the first component of complement with antibody–antigen complexes containing aggregated Fc portions of antibody

Epitope Antigenic determinant; the smallest structural area on an antigen that can be recognized by an antibody

Interleukin Substance produced by one white cell that can act on other white cells

Isoagglutinin Antibodies that act with antigens on the cells of different individuals of the same species. Anti-blood group A is an isohemagglutinin that reacts with type A red blood cells

Lymphokine Soluble substances, produced and secreted by lymphocytes, that act on other cells

Opsonin Something, usually an antibody, that enhances phagocytosis

Phagocytosis Engulfment of particles by cells

Prostaglandins Aliphatic acids derived from arachidonic acid that have a variety of biologic activities including increasing vascular permeability, causing smooth muscle contraction, and decreasing the threshold for pain. Originally found in prostatic fluid

Reagin Originally used for the complement fixing antibodies detected in syphilitic patients by the Wassermann reaction with cardiolipin; now used for skin-fixing antibodies of anaphylactic reactions

Reticuloendothelial system Complex of phagocytic cells of the body, primarily the sinusoidal cells of the liver (Kuppfer cells), spleen, and lymph nodes

Rheumatoid factor Autoantibody against slightly denatured immunoglobulin (usually IgM anti-IgG) found in patients with rheumatoid arthritis

Toxoid Altered form of a toxin that is immunogenic but not toxic

IMMUNOPATHOLOGY is a hybrid word, incorporating immunity and pathology. Immunity comes from the Latin word *immunitas* and means protection from; pathology from the Greek words *pathos*, meaning suffering or disease, and *-logy* indicates the "study of." Thus immunopathology is an oxymoron; a self-contradictory term meaning the study of how protective mechanisms cause disease. This emphasizes that the same immune mechanisms that protect us from disease, especially infectious diseases, may also produce disease. Immunity is in reality a double-edged sword, cutting down our enemies with one side and causing disease with the other side. *Immunology* is the study of immunity or the immune response. The statement that someone is "immune to measles" in common terms means that the person indicated has had measles once and will not get measles again. This specific protection against an infection has been established because of an immune response.

From the time of conception the human body must maintain its integrity in the face of a changing and often threatening environment. Physiologic mechanisms allow us to adjust to changes in temperature, nutrition, etc; immune mechanisms protect us against infectious agents. Immune mechanisms may be divided into two major groups: innate and adaptive (Table I). Innate mechanisms are present in all normal individuals and operate on different agents in the same manner regardless of whether the individual has been exposed to the infectious agent previously. The adaptive immune response requires previous exposure or immunization to become active, and the products of this response react specifically with the agent that stimulated the immunization and not with other unrelated infectious agents.

I. Immunization

Immunization (or vaccination) is the process of stimulating adaptive resistance by induction of a specific immune response. The immune response has three phases: afferent, central, and efferent. The afferent limb consists of the delivery of the foreign material to the reactive cellular components of the immune system. These cellular components are organized in specialized tissues known as lymphoid organs, such as lymph nodes and spleen. The central limb is composed of the interaction of the cells of the immune system within the lymphoid organs and the eventual production of specific products: antibody or sensitized cells (lymphocytes) that have acquired the capacity to react specifically with the immunizing agent. The efferent phase of the immune response consists of the delivery of the immune products to the site of antigen deposition or infection in the tissue and activation of immune effector mechanisms. Immunopathology deals with the effects of the immune effector mechanisms on tissues.

The state of immune reactivity is determined by cells of the immune system and their products. The cells that take part in the immune response are listed in Table II. Some of these cells have specific molecules (receptors) that recognize foreign material, and some do not. The molecules that are recognized as foreign are termed *antigens*. A complete antigen is able to both induce an immune response and react with the products of the immune response. The parts of the molecules that are recognized by antibody or sensitized lymphocytes are called *epitopes*. Specific antibody molecules or reactive T cells have a reactive site (paratope) that can recognize and combine with the epitope. An antigen may have more than one epitope. It is essential for the understanding of immunopathology that the role of both specifically reactive and nonspecific (accessory) cells in the inductive and expressive phases of the immune response be detailed. Foreign materials that are able to get past epithelial barriers and enter the body pass from the tissue spaces into lymphatic vessels and are delivered to the lymphoid organs. Two responding cellular areas in the lymphoid organs are the T-cell zone and the B-cell zone. T cells and B cells are the major populations of lymphocytes. Antigens are processed by a subpopulation of macrophages in the T-cell zones and presented to T cells. T effector cells specific for the antigen proliferate and are then delivered to the lymphatics that drain the lymphoid

TABLE I   Comparison of Innate and Adaptive Immunity

Characteristic	Innate resistance	Adaptive resistance
Specificity	Nonspecific, indiscriminate	Specific, discriminating
Mechanical	Epithelium	Immune induced reactive fibrosis (granuloma)
Humoral	pH, lysozyme, serum proteins	Antibody
Cellular	White blood cells	Specifically sensitized lymphocytes
Induction	Does not require immunization; constitutive	Requires immunization

TABLE II Features of Cells of Immunity and Inflammation

Cell type	Location	Distinguishing feature	Function
Polymorphonuclear neutrophil	Blood	Segmented nucleus, lysosomal granules	Nonspecific effector cell of acute inflammation
Lymphocyte			
T effector cell[a]	Blood	Small mononulcear, specific receptor for antigen, T-cell markers (CD 4+ or 8+)[b]	Specific effector cell for delayed hypersensitivity
T helper	Tissue	CD4+	Aids macrophages in sensitization of B cells
T suppressor	Tissue	CD8+	Controls sensitization process
B cell	Tissue	Surface Ig+	Precursor for antibody-producing plasma cells
Macrophage			
Monocyte	Blood	Large mononuclear, lysosomal granules	Late nonspecific effector cell in inflammation
Dendritic	Tissue	Follicles of lymphoid organs	Process antigen for antibody response (B cells)
Interdigitating	Tissue	Diffuse cortex of lymphoid organs	Process antigen for T cells
Mast cells	Tissue	Metachromatic granules	Contain mediators of immediate vascular + smooth muscular reactions
Fibroblasts	Tissue	Elongated connective tissue cells	Proliferate and fill in zones of tissue destruction (scarring)

[a] For further classification of T effector cells, see Table IV.
[b] CD, Cluster of Differentiation, an identifying site (epitope) detected by monoclonal antibodies.

organs into the blood (efferent lymphatic). The specifically sensitized T effector cells then circulate and localize in tissue sites containing the antigen. Two major subpopulations of T effector cells are important: T_D cells, which induce delayed hypersensitivity reactions, and T_K (killer) or T_{CTL} (cytotoxic) cells, which are able to recognize antigen on target cells and cause their destruction.

The second arm of the immune response is specific antibody. Antibody is a soluble protein molecule secreted by differentiated B cells. In the lymphoid organ, antigen is processed in B-cell zones by tissue-fixed macrophages (dendritic histocytes), in the presence of T helper cells. The T helper cells produce factors (interleukins), which serve to stimulate proliferation and differentiation of B cells that have reacted with antigen on the surface of the antigen-processing accessory cell. B cells differentiate into plasma cells that synthesize and secrete specific antibody. The secreted antibody is released into the efferent lymphatics and delivered to the systemic circulation. Antibodies in the blood are then able to reach tissue sites containing antigen. The effect of the reaction of specifically sensitized T cells or antibodies with antigens in tissue will be elaborated later in this chapter.

II. Inflammation

Inflammation is the process of delivery and activation of blood proteins and cells to tissues as a response to injury or infection. Cells in the blood take part in the inflammation induced by immune reactivity or may also be activated nonimmunologically by bacterial products or substances present in dying (necrotic) tissue. The hallmark cell of acute inflammation is the polymorphonuclear neutrophil (PMN). PMNs move into sites of acute inflammation, phagocytize dead tissue, or bacteria and may release enzymes from their cytoplasmic granules that cause further damage. Monocytes (blood macrophages) and lymphocytes are the hallmark cells of chronic inflammation. Lymphocytes reacting with antigen, or otherwise activated, release products (lymphokines) that attract and activate macrophages. Activated monocytes phagocytose and. "clean up" tissue debris resulting from the preceding chronic inflammation. [*See* INFLAMMATION.]

A. Antibody-Mediated Inflammation

The characteristics of the effector mechanisms mediated by antibody are largely determined by the nature of the antibody molecule. Antibody molecules are made up of units of four polypeptide chains joined together by disulfide bonds. The prototype antibody molecule, IgG, has a Y-shaped appearance. The stem of the Y contains the structures that determine the biological activity (i.e., placental transfer, ability to active complement). The arms of the Y contain the antigen binding sites (also called paratopes). Binding of antigen (epitope) to the paratopes causes alteration in the tertiary structure of the stem of the antibody molecule and activation of the complement binding site.

The activation of complement plays a critical role in antibody-mediated cell lysis and inflammation. Complement consists of a series of 11 serum proteins. Binding of the first component of complement produces activation of enzymes that cleave other components of the complement system, resulting in biologically active fragments of complement molecules. Some of these fragments attach to cell surface membranes (membrane attack complex) and cause destruction (lysis) of the cell. Other products of complement activation act to cause contraction of endothelial cells lining small blood vessels (anaphylatoxin) or attract PMNs (chemotaxis). Activation of complement in tissues results in opening up of small blood vessels so that components in the blood can enter the tissues and attracts activated polymorphonuclear leukocytes, the cells of acute inflammation.

Antibodies belong to a family of serum proteins termed *immunoglobulins*. The properties of the five major classes of immunoglobulins are listed in Table III. The most important from an immunopathologic standpoint are IgG, IgM, and IgE. IgG is the most prevalent class of immunoglobulin, and its reaction with soluble antigen or tissue antigens is most frequently the causative antibody in inactivation and immune complex reactions. IgM is composed of five antibody units joined together and is the first antibody formed after immunization. One molecule of IgM reacting on the surface of a cell is able to activate complement, whereas two or more IgG molecules reacting in close proximity are required. IgM is the most active antibody for cytolytic reactions. IgE has an affinity for mast cells. Mast cells are tissue-fixed leukocytes located near small blood vessels in tissue. They have cytoplasmic membrane-limited granules containing pharmacologically active agents that act on smooth muscle and endothelial cells. The resulting effect depends on the nature of the smooth muscle involved because of different types of receptors for the most active ingredient, histamine. Contraction of smooth muscle in the bronchi of the lung leads to constriction of the breathing tubes and symptoms of asthma. Dila-

TABLE III Some Properties of Human Immunoglobulins[a]

Property	Immunoglobulin class				
	IgG	IgA	IgM	IgD	IgE
Serum concentration (g/100 ml)	1.2	0.4	0.12	0.003	<0.0005
Molecular weight	140,000	160,000[b]	900,000	180,000	200,000
Electrophoretic mobility	γ	Slow β	Between γ and β	Between γ and β	Slow β
H-chains	γ	α	μ	δ	ε
L-chains	λ or κ	λ or κ	λ or κ	λ or κ	λ or κ
Complement fixation	Yes	No	Yes	No	No
Placental transfer	Yes	No	No	No	No
Percent intravascular	40	40	70	—	—
Half-life (days)	23	6	5	3	2.5
Percent carbohydrate	3	10	10	13	10
Antibody activity	Most Ab to infections; major part of secondary response; Rh isoagglutinins; LE factor	Present in external secretions	First Ab formed; ABO isoagglutinins; rheumatoid factor	Antibody activity rarely demonstrated, found on lymphocyte surface	Reagin sensitizes mast cells for anaphylaxis

[a] Modified from Fahey, J. L.: J.A.M.A. 194:183, 1966.
[b] Serum IgA 160,000 MW; secretory IgA 350,000 MW, may activate alternate pathway.

tion of the smooth muscles of small blood vessels, combined with contraction of the cells lining these vessels, results in increased blood flow and increased vascular permeability, the first manifestations of acute inflammation (see below). The operation of these IgE-mediated mechanisms (anaphylaxis) will be described in more detail below. [*See* SMOOTH MUSCLE.]

B. Cell-Mediated Inflammation

Cell-mediated inflammation is effected by one of a number of cellular mechanisms. Six general cell-mediated cytotoxic effector mechanisms have been recognized *in vitro* (Table IV). In addition, other cell types (e.g., polymorphonuclear leukocytes or macrophages) may function as killer cells via antibody that binds to cells (cytophilic antibody). Macrophages may also be activated by nonspecific stimulators such as phorbol esters or polynucleotides. Immune specific delayed hypersensitivity reactions are mediated by the first two cell types (i.e., T_D (delayed hypersensitivity) or T_K cells). T_K cells are also referred to as cytotoxic T lymphocytes (T_{CTL}). T_D and T_{CTL} are the major specific immune effector cells. [*See* NATURAL KILLER AND OTHER EFFECTOR CELLS.]

An infection with a pathogenic organism sets off a race between the infectious agent and the infected individual, the outcome of which will determine whether the infected person lives of dies. For this reason, the specific immune response must be rapid and effective but also be able to be deactivated when the infection has been cleared or neutralized, as continued production of immune products could lead to continued destruction of normal tissue or unrestricted proliferation of lymphoid cells (lymphoma or leukemia) (Table V). Rapid delivery of antigens is effected by afferent lymphatics, which deliver antigens to the lymph nodes. The lymph nodes contain the immune reactive cells, act as filters for the lymphatics, and provide a site where the immune products may be rapidly manufactured and delivered to the blood stream through efferent lymphatics, which drain into the systemic circulation. To be effective the immune products must be able to reach the sites of infection in tissues. This is accomplished by inflammation.

C. Acute Inflammation

Acute inflammation consists of the passage of fluid, proteins, and cells from the blood into the tissues. The cells taking part in inflammation are listed in Table II. During an inflammatory response, components of both the innate and the adaptive (specific) immune systems may be utilized. Initiation of an inflammatory response begins with increasing blood flow to infected tissues and contraction of the endothelial cells lining the blood vessels. The cardinal signs of the earliest stage of acute inflammation were first described by Celsus about 25 BC (Table VI). Increased blood flow causes redness (rubor) and increased temperature (calor). Passage of fluid into the tissue is seen as edema an swelling (tumor), and release of inflammatory mediators (e.g., prostaglandins) produces pain (dolor). In later stages of acute inflammation, infiltration of tissues with white blood cells (the cells of inflammation) produces a white color; leakage of red blood cells into tissue, a red color. If the inflammation is severe, activation and release of digestive enzymes from the inflammatory cells may produce death of the involved tissue, which is recognized as pus. Pus containing mainly white blood cells is white; if red cells are present the pus will be yellow or red, depending on the proportion of red cells.

D. Chronic Inflammation

Chronic inflammation features an infiltration of mononuclear cells (lymphocytes and monocytes) in

TABLE IV Mechanisms of Lymphoid Cell-Mediated Immunity

T_K cells (CTL)	Sensitized to target cell—direct killing
T_D cells	Sensitized to target cell—lymphokine release, indirect killing (accessory cells)
Null cells + specific antibody	Antibody-dependent cell-mediated cytotoxicity (ADCC)
NK cells	Natural killer cells
LAK cells	Lymphokine activated killer cells
Activated macrophages	Phagocytosis and digestion

1. *Recognition* of many different foreign invaders specifically
2. *Rapid synthesis* of immune products on contact with invaders
3. *Rapid delivery* of the immune products to the site of infection
4. *Diversity* of effector defensive mechanisms to combat infectious agents with different properties
5. *Direction* of the defensive mechanisms specifically to foreign invaders rather than one's own tissue
6. *Deactivation* mechanisms to turn off the system when the invader has been cleared

contrast to the polymorphonuclear cells characteristic of acute inflammation. Activated lymphocytes release products called lymphokines, which attract and activate monocytes. Activated monocytes (macrophages) clean up necrotic tissue and bacteria by engulfing and digesting the engulfed material in cytoplasmic vessels (phagocytosis). Macrophage means "big eater." If the tissue damage is not severe, the site of inflammation will be restored to normal (resolution); if more extensive, tissue fibroblasts will proliferate and fill in the necrotic area with fibrosis tissue (scarring).

E. Immune Effector Mechanisms

These are variations on the theme of inflammation. The specificity of antibody and specifically sensitized lymphocytes provides a means of applying the defensive action of the inflammation directly to the infectious agent. The expression of the specific immune response *in vivo* may be classified into six major mechanisms, with the understanding that more than one of these mechanisms may be active at the same time. The historical development of the classification of immune effector mechanisms is given in Table VII. These mechanisms have both

TABLE VI The Four Cardinal Signs of Acute Inflammation: Celsus (25 BC)[a]

Rubor	Redness
Tumor	Swelling
Calor	Heat
Dolor	Pain

[a] The fifth classic sign of acute inflammation, functio laesa (loss of function), was added by Virchow (1821–1902).

protective and destructive actions (Table VIII). For the remainder of this chapter, the destructive or disease-causing effects of these immune mechanisms will be stressed. The first four mechanisms are mediated by immunoglobulin antibodies. The characteristics of the reactions are determined by the properties of the immunoglobulin molecules, the nature and tissue location of the antigen, and on the accessory inflammatory systems that are activated. The fifth mechanism is mediated by sensitized lymphocytes (T cells): whereas the sixth mechanism may be initiated by reaction of either antibody or effector lymphocytes with antigen, the nature of the lesions observed is determined by the fact that the antigen is poorly degradable.

III. Autoimmunity

Tissue damage is caused by immune mechanisms when reaction to an infection is severe or inappropriate but may also occur when the immune response is directed not against foreign antigens but against self-antigens (autoimmunity). Autoimmunity occurs when an individual develops an immune response to his or her own tissue antigens. Normally during development, a person develops a condition that prevents him or her from recognizing his or her own tissue as foreign. This condition is known as tolerance. There are a number of possible mechanisms whereby tolerance is lost and autoimmunity may occur (Table IX). Once autoimmunity develops, diverse disease manifestations may occur depending on the location of self-antigen and on which of the six immunopathologic mechanisms are activated. [*See* AUTOIMMUNE DISEASE.]

IV. Immunopathologic Mechanisms

A. Inactivation or Activation

Inactivation or activation reactions occur when antibody reacts with an antigen that performs a vital function and inhibits that function. Inactivation may occur by reaction of antibody to soluble molecules (e.g., bacterial toxins) or by reaction of antibody with cell surface receptors (e.g., virus receptors). Reactions with soluble molecules may produce changes in the tertiary structure of the bio-

TABLE VII Classification of Immune Mechanisms

Gell and Coombs (1963)	Roitt (1971)	Sell (1972)
—	Stimulatory	Inactivation or activation
Type II	Cytotoxic	Cytotoxicity or cytolytic
Type III	Immune complex	Toxic complex (Arthus)
Type I	Atopic or anaphylactic	Atopic or anaphylactic
Type IV	Delayed hypersensitivity	Delayed hypersensitivity
—	—	Granulomatous

TABLE VIII The "Double-Edged Sword" of Immune Reactions[a]

Immune effector mechanism	Protective function: "Immunity"	Destructive reaction: "Allergy"
Neutralization	Diphtheria, tetanus, cholera, endotoxin neutralization, blockage of virus receptors	Insulin resistance, pernicious anemia, myasthenia gravis, hyperthyroidism
Cytotoxic	Bacteriolysis, opsonization	Hemolysis, leukopenia, thrombocytopenia
Immune complex	Acute inflammation, polymorphonuclear leukocyte activation	Vasculitis, glomerulonephritis, serum sickness, rheumatoid diseases
Anaphylactic	Focal inflammation, increased vascular permeability, expulsion of intestinal parasites	Asthma, urticaria, anaphylactic shock, hay fever
Delayed hypersensitivity	Destruction of virus-infected cells, tuberculosis, syphilis, immune surveillance of cancer	Contact dermatitis autoallergies, viral exanthems, postvaccinial encephalomyelitis
Granulomatous[b]	Leprosy, tuberculosis, helminths, fungi, isolation of organisms in granulomas	Beryllosis, sarcoidosis, tuberculosis, filariasis, schistosomiasis

[a] Modified from Sell, S. (1978). Introduction to symposium on immunopathology: Immune mechanisms in disease. *Hum Pathol* **9**, 24.

[b] Granulomatous reactions, like other inflammatory lesions, may result from nonimmune stimuli as well as from an immune reaction activated by antibody or by sensitized cells. The frequent association of granulomatous reactions with delayed hypersensitivity reactions has resulted in the inclusion of granulomatous reactions as a subset of delayed hypersensitivity.

TABLE IX Mechanisms of Autoimmunity

Immunizing event

1. Release of sequestered antigen: Tissue-specific antigens (e.g., myelin of central nervous system) normally not exposed to immune system are released into the circulation in amounts that stimulate an immune response.

2. Partial denaturation of antigen: Alteration of part of a molecule may lead to structures recognized as foreign by immune system. In responding to the denatured structures, an immune response to the nondenatured component may also take place.

3. Cross-reacting antigen: Molecules containing some "foreign" structures and some "self" structures are able to induce an immune response to both foreign and self epitopes.

4. Polyclonal activation: If the immune system is "turned on" by a strong stimulus, some of the products (antibodies or cells) may react with self-antigens.

Change in host

1. Regeneration of deleted clones by somatic mutation: Self-reactive lymphocytes that have been eliminated during development may reappear in the adult by mutation of previously unreactive lymphocytes.

2. Recovery from anergy: Self-reactive cells may be blocked from reacting to self-antigens by a failure to express receptors or to respond to antigen stimulation. A change in the status of the individual may remove this block.

3. Loss of suppressor T cells: T cells that hold self-reactive T or B cells in check may decline in numbers as a result of a failure to maintain their production, allowing the self-reactive cells to respond.

4. Loss of controlling antibody: Feedback control of antibody production by antiidiotype may become inactive if the level of the controlling antibody falls.

logically active molecule so that it no longer performs its biological function or is cleared from the circulation by the reticuloendothelial system as an immune complex. Reaction of antibody with cell surface receptors (i.e., disappearance from the cell surface). In some instances, antireceptor antibodies may mimic the action of the activating ligand for that receptor. Inactivation reactions to toxic agents (e.g., diphtheria or tetanus toxins) are beneficial, and antibodies to virus may prevent cellular infection. In fact, this is the goal of immunization to toxoids or viruses. However, when antibody reacts with something vital for normal function, the same reactions induce disease.

A list of some major diseases associated with inactivation effects of autoantibodies is given in Table X. Some examples include diabetes, myasthenia gravis, or hyperthyroidism. Although diabetes is actually a group of different diseases, the basic problem is a lack of insulin or an inability of cells of the body to respond to insulin. Diabetes may be mediated by at least three different antibodies as well as by nonimmune mechanisms. The antibodies include (1) antibodies to insulin, which block the effect of insulin, (2) antibodies to insulin receptors, which prevent the binding of insulin, and (3) antibodies to the cells that make insulin, resulting in a failure to produce insulin. Myasthenia gravis is a muscular unresponsiveness to neurogenic stimulation caused by autoantibodies that react with acetylcholine receptors and block the neuromuscular transmission effected by release of acetylcholine at the motor end-plate. Antireceptor activation is exemplified by some forms of hyperthyroidism caused by antibodies that react with receptors on thyroid cells that normally bind to thyroid-stimulating hormone. The

TABLE X Some Diseases Caused by Antibodies to Biologically Active Molecules

Disease	Target antigen
Diabetes	Insulin
	Insulin receptors
Juvenile diabetes	Insulin
Clotting deficiencies	Clotting factors
Pernicious anemia	Parietal cells
	Intrinsic factor
Myasthenia gravis	Acetylcholine receptor
Hyperthyroidism	Antithyroid-stimulating hormone receptor (LATS)[a]
Hypothyroidism	Antithyroid hormone
Asthma	Beta-adrenergic receptor

[a] LATS, long-acting thyroid stimulator.

antibody to the receptor mimics the activity of thyroid-stimulating hormone.

B. Cytotoxic or Cytolytic Reactions

Reaction of antibodies to cell surface antigens in the presence of a system of serum proteins called *complement* will sometimes cause lysis or destruction of the cells. Complement components can realign membrane structures and disrupt cell membrane integrity, or the cell can be coated with complement components that render the cell susceptible to phagocytosis by the reticuloendothelial system. Complement is activated by alterations in the tertiary structure of some antibodies when the react with antigen. One molecule of IgM antibody reacting with a cell surface antigen is capable of activating complement and of lysing a cell, whereas at least two molecules of IgG antibody must react in close proximity to one another to activate complement and accomplish lysis of a cell. For this reason, lytic reactions are much more effectively mediated by IgM than by IgG antibody. Other immunoglobulin classes do not fix complement in this way and thus do not lyse cells.

The sequence of complement activation is shown in Fig. 1. Complement consists of a series of 11 serum proteins that interact after activation. C1 through C9 are part of the activation sequence, whereas other components serve to inactivate the system. The first component of complement is activated by binding to aggregated antibody molecules in an immune complex of antibody and antigen. Activated C1 is a proteolytic enzyme and splits C4 and C2 into active products. On the cell surface, C4b and C2a join together and split C3 into C3a and C3b. C3b binds to the cell surface. C3b makes the cell more susceptible to phagocytosis (opsonizes) and also activates C5. C5 splits into C5a and C5b. C5b joins to the cell surface and activates C6-9 to form a complex of molecules that intercalates into the cell membrane and creates a channel in the cell that allows the cytoplasm to leak out (cell lysis). The activated C4a, C3a, and C5a products are released into the fluid phase. Together they are called anaphylatoxin and stimulate increased blood flow and increased vascular permeability. C5a is also chemotactic for PMNs. Thus the cell membrane bound components of complement are the killer molecules of cytolytic reactions; the activated soluble split products are critical for the inflammation of immune complex (Arthus) reactions (see below). When anti-

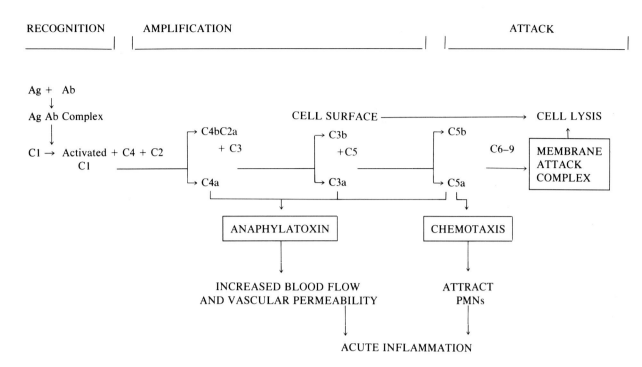

| RECOGNITION | AMPLIFICATION | ATTACK |

FIGURE 1 Role of complement activation of cytolytic and immune complex reactions.

body reacts with antigen on a target cell surface, cell lysis can result; if antibody reacts with soluble antigens forming an immune complex or with basement membrane antigens, complement-mediated inflammation will occur. Complement activation can directly inactivate viruses or bacteria via the alternate complement pathway (antibody independent) and can enhance phagocytosis of infectious agents, or contribute to an inflammatory response.

Cytotoxic or cytolytic reactions are often directed against cells in the blood. The disease caused by these antibodies reflects the loss of the function of the cell. Antibodies to red cells cause hemolytic anemia; to white blood cells, agranulocytosis and increased susceptibility to infection; to platelets, thrombocytopenia and bleeding disorders. Transfusion of red blood cells bearing blood group antigens to an individual who does not have the same antigens may result in a massive hemolysis (transfusion reaction) if the recipient has preformed antibody to the transfused cells. Erythroblastosis occurs during pregnancy when the mother has antibody to fetal red cell antigens contributed by the father and the antibody is able to cross the placenta. The maternal antibody destroys the red cells in the fetus. The fetus responds by increasing red cell production (erythroblastosis). Erythroblastosis fetalis usually

does not occur with the first incompatible pregnancy because immunization leading to antibodies to the fetal red cells occurs during delivery of an incompatible baby. This can be prevented in most instances by treating the mother with the appropriate antibodies at the time of delivery, thus preventing immunization and production of antibodies that might cause erythroblastosis during the next incompatible pregnancy.

C. Immune Complex (Arthus) Reactions

Immune complex reactions are due to the formation of complexes of antibody with soluble antigens in the circulation or by reaction of circulating antibody with basement membrane components. Activation of complement by antigen–antibody complexes (usually IgG antibody) in tissues is responsible for an inflammatory reaction mediated by activated components of complement. Activation of complement on the surface of the cell results in some components that are split off of larger molecules that bind to the cell surface, with additional components joining until a membrane lesion is produced and the cell is lysed (cytolytic reaction). In addition, biologically active split fragments of complement that do not bind to the cell surface are produced. The important activated fragments for inflammation are C3a, C4a and C5a. Activation of these products produces constriction of endothelial cells (anaphylatoxin), permitting blood components to pass into the tissue. C5a is also highly chemotactic (attractive) for polymorphonuclear neutrophilic leukocytes (PMNs). The result is the attraction and activation of these cells to tissues when antibody has reacted with antigen. PMNs are the hallmark of acute inflammation. They contain large numbers of lysosomes that are filled with digestive enzymes. Tissue damage is caused by digestion of tissue by enzymes released from the PMNs at the inflammatory site. Injection of antigen into the skin of an animal with circulating IgG antibody to the antigen produces an acute inflammatory reaction (Arthus reaction) that peaks at 6 hours and fades by 24 hours. It is characterized histologically by perivascular necrosis and PMN infiltration of arterioles and venules. Using immune labeling techniques (immunofluorescence), antigen–antibody complexes and complement components can be identified in the walls of the affected vessels. Some illustrative examples of immune complex reactions are serum sickness and some forms of glomerulonephritis. [*See* Antibody–Antigen Complexes: Biological Consequences.]

Serum sickness is a systemic inflammatory reaction following injection of foreign serum manifested by involvement of vessels throughout the body. This reaction was first noted at the turn of the century when horse antitoxin was used to treat individuals with tetanus. Ten days to 14 weeks after the injection of horse serum, inflammation of joints (arthritis) and kidneys (glomerulonephritis) and systemic inflammation of blood vessels (vasculitis) were observed. This reaction was due to the formation of soluble immune complexes in the circulation. The complexes deposit in vessel walls, renal glomerular basement membrane, and joints. On deposition the accumulation of these complexes leads to activation of complement and subsequent acute inflammation. If not fatal, the disease terminates when the immune complexes are cleared from the circulation and tissues. Clearance of circulating immune complexes occurs when more than one IgG antibody molecule reacts with a soluble antigen so that at least two antibody molecules are located close enough together to form a structure that is recognized by phagocytic cells, or activates complement. Thus early after an antibody response to foreign serum, antigen–antibody complexes will be found wherein antigen is in excess over antibody so that the complexes will essentially consist of one or two antigen molecules and one antibody. These complexes are not cleared by phagocytosis and deposit in tissues. When two or more such complexes deposit closely together in tissue, the antigen-bound antibody molecules can form the aggregates of IgG molecules required to activate complement and thus initiate inflammation in the tissue.

Glomerulonephritis is caused by deposition of soluble immune complexes in the basement membrane of the renal glomerulus or by reaction of antibodies specific for basement membrane antigens. Acute glomerulonephritis classically occurs after infection with certain strains of streptococcal bacilli. During resolution of the infection, the bacilli release antigens that have an affinity for the basement membrane. When the infected individual produces antibody to this protein, reaction with the antigen deposited in the membrane results in glomerulonephritis. A second, chronic phase of the glomerular inflammation results when, as a result of the release of basement membrane components, an autoantibody to the basement membrane is formed. This may result in progressive renal failure.

A variety of other human diseases are believed to be caused by the immune complex mechanisms. These include lupus erythematosus (DNA–Anti-DNA), rheumatoid arthritis (IgM–Anti-IgG, so-called "rheumatoid factor"), and polyarteritis nodosa (circulating immune complexes, e.g., hepatitis B antigen–anti-HBAg).

The protective function of immune complex reactions is to mobilize inflammatory cells at the site of infections. The activation of the chemotactic components of complement provides this signal, and the anaphylatoxic action of activated complement opens the vessel walls to permit the PMNs to pass into the tissue. The released lysosomal enzymes for the PMNs act on infectious organisms or their products and destroy or inactivate them.

D. Anaphylactic and Atopic Reactions

Anaphylactic reactions include hives (wheal and flare), systemic vascular shock, gastrointestinal allergy, and acute asthmatic attacks. Atopic reactions include hay fever and chronic asthma. The difference is most likely due to the amount of antigen (allergen) to which the susceptible individual is exposed in a given period. A large exposure producing anaphylactic symptoms; repeated small exposures during a longer period, atopic symptoms. The term *allergy* is commonly used for these reactions, and the antigen is termed *allergen*. [*See* Allergy.]

Anaphylactic or atopic reactions are initiated by pharmacologically active agents (histamine and serotonin) that are released from mast cells after antibody bound to mast cells reacts with antigen. In addition, arachidonic acid metabolites produced by macrophages from cell membrane phospholipids released from mast cell granules contribute to later inflammatory phases of the reaction. Injection of antigen into the skin of an anaphylactically sensitive individual results in an acute reaction composed of swelling (edema) surrounded by redness (erythema) that peaks within 15 to 30 minutes and fades in 2 or 3 hours (cutaneous anaphylaxis, wheal and flare, hives). A later cellular inflammatory phase of the reaction may occur after 10 to 20 hours. This later phase is more evident in the biphasic bronchospastic response of the lung, where production of arachidonic metabolites by lung tissue may be more pronounced. The antibody responsible is almost always of the IgE class, which has the capacity to bind to receptors on mast cells through the nonantigen binding part of the antibody molecule. Mast cells are located near smooth muscle (precapillary arteriole, bronchial smooth muscle, etc.) and contain granules filled with pharmacological agents that act directly on smooth muscle and endothelial cells. When antigen reacts with mast cellbound IgE antibody, the cell releases the contents of its granules by a nonlytic fusion mechanism (degranulation). The released pharmacologically active agents, primarily histamine and serotonin, act on end-organ target cells (e.g., vascular and bronchial smooth muscle) to cause immediate symptoms. In addition, the membrane phospholipids are metabolized by oxidative pathways to produce a series of longer acting biologically active compounds, leukotrienes and prostaglandins, that are responsible for later effects.

Anaphylactic or atopic reactions are elicited by reaction of antigens with IgE antibody on mast cells. The ability of IgE antibody to ''fix to skin'' is determined by its binding to mast cells in the skin (cytophilic antibody). The reaction can be passively transferred by injection of serum containing IgE antibody into the skin (passive cutaneous anaphylaxis). Antigen can be injected up to 45 days after injection of serum containing IgE antibody, as the antibody will remain ''fixed'' to mast cells in the skin. This is also referred to as the Prausnitz-Kustner reaction, in honor of the authors who first described the typical reaction. Passive transfer of an Arthus reaction is demonstrable only if the anti-

gen is injected within 24 to 48 hours after antibody, as the IgG antibody will diffuse away (it does not ''fix'' to mast cells). Skin fixing antibody is called reaginic antibody.

The protective function of anaphylactic reactions is not as well understood as in other immune effector mechanisms, but there are several possible functions. Histamine and serotonin are primary mediators of increased blood flow to sites of reaction. Anaphylactic reactions serve to open vascular endothelium, thus permitting blood-borne proteins and cells to enter inflammatory sites where they are needed. Anaphylactic reactivity to intestinal parasites often occurs, and increased intestinal motility and secretion may purge the gastrointestinal tract of them. The acute sneezing and coughing that is part of the acute asthmatic attack may help eliminate agents in the tracheobronchial tree. If nothing else, the acute reactivity certainly serves as a warning to avoid contact with exciting agents.

E. Delayed Hypersensitivity

These reactions are mediated by a population of specifically sensitized T lymphocytes that bear receptors for antigens. The lymphocytes that effect delayed hypersensitivity reactions belong to a subpopulation of the T lymphocyte class: T_D cells and/or T_K (T_{CTL}) cells. [See LYMPHOCYTES.]

The characteristic lesion of delayed hypersensitivity reactions is a perivascular mononuclear infiltrate. T_D lymphocytes reacting with antigen in tissues release lymphokines, which attract and activate macrophages. Most of the tissue damage is caused by macrophages, which have been attracted to the site of inflammation by lymphokines. T_K (killer or cytotoxic) lymphocytes may attack target cells directly. The term *delayed* is used because injection of antigen into the skin of an individual expressing delayed hypersensitivity induces an inflammatory reaction that peaks at 24 to 48 hours, in contrast to immune complex reactions, which peak at 6 hours (Arthus reactions), and cutaneous anaphylactic reactions, which peak at 15 to 30 minutes. Some examples of delayed hypersensitivity reactions are tuberculin skin test, contact dermatitis (i.e., poison ivy), tissue graft rejection, virus rash exanthems, such as measles, and viral-associated demyelinating diseases of the nervous system (e.g., postinfectious encephalomyelitis, Guillain-Barre, multiple sclerosis). [See MACROPHAGES.]

Delayed skin reactions become manifest when

sensitized lymphocytes react with antigen deposited into the skin. In normal skin, lymphocytes pass from venules through the dermis to lymphatics, which return these cells to the circulation. Recognition of antigen by sensitized lymphocytes (T cells) results in immobilization of lymphocytes at the site, production and release of lymphocyte mediators, and accumulation of macrophages with eventual destruction of antigen and resolution of the reaction. This results in an accumulation of cells seen at 24 to 48 hours after antigen injection. Macrophages may degrade the antigen. When the antigen is destroyed, the reactive cells return via the lymphatics to the bloodstream or draining lymph nodes. In this way,

specifically sensitized lymphocytes may be distributed throughout the lymphoid system after local stimulation with antigen.

The protective function of delayed hypersensitivity is mainly directed to intracellular agents such as viruses and to fungal and mycobacterial infections. Although delayed hypersensitivity reactions to infectious organisms may result in extensive tissue damage, the damage is a secondary effect of the immune attack on an infecting agent. It is incorrect to think, as in the past, that delayed hypersensitivity is a deleterious type of reaction. Delayed reactions are particularly important in eliminating intracellular virus infections when the viral antigens are

TABLE XI Characteristics of the Six Types of Hypersensitivity Reactions

Characteristic	Immunopathologic mechanism					
	Inactivation or activation	Cytotoxic or cytolytic	Toxic complex (Arthus)	Anaphylactic or atopic	Delayed cellular	Granulomatous
Immune reactant	IgG antibody	IgM > IgG	IgG antibody	IgE antibody	T_K and T_D cells	T_D cells
Accessory component		Complement, macrophages (RES)	Complement, polymorphonuclear leukocytes, coagulation	Mast cells, chemical mediators	Lymphokines, macrophages	Macrophages, (epitheloid and giant cells)
Skin reaction		Pemphigus, pemphigoid	Arthus: peaks 6 hr; fades by 24 hr	Wheal and flare: peaks 15–30 min; fades by 2–3 hr	Delayed (tuberculin): peaks 24–48 hr; fades by 3 days	Granuloma; takes weeks to develop and months to fade
Protective function	Inactivate toxins	Kill bacteria	Mobilize polys to site of infection	Open vessels, delivery of blood components to sites of inflammation	Kill organisms, virus-infected cells	Isolate infectious agents
Examples of protection	Tetanus, diphtheria, streptococci	Bacterial infections	Bacterial and fungal infections	Parasitic infections	Virus, fungal, mycobacterial, ?cancer	Leprosy, tuberculosis
Pathogenic mechanism	Inactivation of biologically active agents or cell surface receptors	Cell lysis or phagocytosis (opsonization)	Polymorphonuclear leukocyte infiltration, release of lysosomal enzymes	Bronchoconstriction edema, shock	Mononuclear cell infiltrate; target killing	Replacement of tissues by granulomas
Representative disease states	Insulin-resistant diabetes, myasthenia gravis, hyperthyroidism	Hemolytic anemias, vascular purpura, transfusion reactions, erythroblastosis fetalis	Glomerulonephritis vasculitis, arthritis, rheumatoid diseases	Anaphylactic shock, hives, asthma, hay fever, insect bites	Viral skin rashes, graft rejection demyelination	Sarcoidoses, berylliosis, tuberculosis

Modified from Sell, S. Immunology, Immunopathology and Immunity (4th ed) Elsevier, New York, 1987. Reproduced by permission.

expressed on the cell surface. It is unfortunate if the infected cell is one that performs a unique vital function (e.g., a neuron), because during elimination of the infected cells, irreversible tissue damage may result. T CTL is important in contact dermatitis, tissue graft rejection, tumor immunity, and some autoimmune lesions.

F. Granulomatous Reactions

A variant of delayed hypersensitivity or immune complex reactions, these reactions occur when the antigen is poorly broken down and remains as a chronic irritant. Indeed, the early lesions often resemble delayed hypersensitivity reactions or are associated with necrotizing vasculitis. However, instead of macrophages clearing away the antigen, there is a prolonged accumulation of macrophages and lymphocytes, which organize into granulomas. Typical granulomas consist of a ball-like mass of macrophages resembling epithelial cells (epithelioid cells) and multinucleated giant cells admixed with lymphocytes. Frequently the center of larger lesions is necrotic. Grossly this form of necrosis looks like cottage cheese and is described as caseous necrosis. The major destructive mechanism is simply the occupation of organ space. The lesions may become so extensive that the normal function of the organ is impaired. Granulomas serve to isolate the infectious agent by walling off infected areas of tissue. This operates when the delayed hypersensitivity mechanism is unable to eliminate the agent. Infectious diseases in which granulomatous reactions play a major role are tuberculosis, leprosy, parasite and fungus infections, and syphilis. Granulomas occur in children who have a deficiency in macrophage digestive function (granulomatous disease of children). In these individuals, granulomas form where macrophages ingest material that cannot be degraded and large numbers of macrophages collect in the tissue.

V. Summary

Specific antibody and sensitized T effector lymphocytes provide mechanisms for directing protective defense mechanisms to foreign invaders. These directed immune effector mechanisms serve to focus nonimmune specific inflammatory accessory systems on the specific infectious agent. Six mechanisms are recognized: inactivation, cytotoxic, immune complex (Arthus), anaphylactic (atopic), delayed hypersensitivity, and granulomatous reactions. A listing of the characteristics of each of these reactions is given in Table XI. Each of these mechanisms has important protective functions but may also cause tissue damage. Immunopathology is the study of how immune mechanisms cause disease.

Bibliography

Golub, E. S. (1987). "Immunology: A Synthesis." Sinauer Assoc., Sunderland, Massachusetts.
Hood, L. E., Weissman, I. L., Wood, W. B., and Wilson, J. H. (1984). "Immunology," 2nd ed. Benjamin-Cummings, Menlo Park, California.
Lackmann, P. J., and Peters, D. K., eds. "Clinical Aspects of Immunology," 4th ed. Blackwell, Oxford.
Paul, W. E., ed. (1984). "Fundamental Immunology." Raven, New York.
Roitt, I. (1984). "Essential Immunology," 5th ed. Blackwell, Oxford.
Sell, S. (1987). "Immunology, Immunopathology and Immunity," 4th ed. Elsevier, New York.

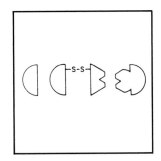

Immunotoxins

JON ROBERTUS, *University of Texas, Austin*

ARTHUR FRANKEL, *Florida Hospital Cancer and Leukemia Research Center, Orlando*

Glossary

Allogeneic Tissue with similar but not identical genetic makeup

CD5, CD19 Cell surface proteins that serve as markers for T and B cells, respectively

Monokines Soluble factors released from activated monocytes or macrophages, which can alter the behavior of neighboring cells

Neuropathy Disorder of nerve cells

Palliation Alleviation of symptoms without a genuine cure

Tumor burden Amount of tumor cells present in an individual, not precisely defined; in humans, tumor masses of 1 g, 1 kg, and 10 kg are referred to as small, moderate, and large, respectively

AN IMMUNOTOXIN is a conjugate of a cell-specific antibody and a toxic enzyme. The toxic enzymes are isolated from plant or bacterial sources and generally attack eucaryotic ribosomes or ancillary proteins. In theory, the antibody can guide the toxin to a tumor or other cell type and bind to it. After internalization, the toxin inhibits the protein-synthesis apparatus of the intoxicated cell, leading to cell death. Immunotoxins have been tested against cultured tumor cells and against tumors in animal systems to develop experience with their use. Recently, clinical trials have been undertaken showing mixed results. Although some cancers have been successfully treated with the conjugates, a number of troublesome side effects are usually manifest.

I. Theory of Cell-Selective Agents

Medical science has long been fascinated by the concept of a "magic bullet," which could be administered to a patient and specifically seek out infected or malignant cells and either destroy them or deliver a therapeutic agent to them.

In theory, a wide range of cell-specific molecules could be used to make magic bullets. In fact, protein hormones, such as human gonadotropin, have been used in this way. Certain cell types have receptors on their surfaces for gonadotropin. Conjugates made with the hormone and a given toxin are able to pick out those cells with gonadotropin receptors and kill them with reasonable selectivity. This type of targeting vehicle, however, is not general and can only be used in special cases. Immunotoxins (ITs) are another class of magic bullets, although currently they are not as selective as desired. The name "immunotoxin" describes the nature of this particular class of magic bullets: "immuno" refers to an antibody or immunoglobulin, which is used to select a particular cell type; "toxin" refers to a protein molecule, which, when delivered to and taken up by a target cell, is able to intoxicate or kill that cell. When these two proteins are chemically or genetically linked together by medical scientists, they form an IT. When introduced into the body, the IT typically circulates through the blood, making contact with a variety of cells. A chance collision with a cell of the type recognized by the antibody allows the IT to bind to it, and processes are initiated, which can intoxicate that cell. Although the IT is a magic bullet in the sense that it is selective, it is not a guided missile that homes in unerringly on the target.

II. Immunotoxin Structure and Function

A. Antibodies

In principle, antibodies can be used as targeting vehicles to a wide range of tissues. Antibodies are protein molecules produced by the B cells of all mammals. Antibodies of the G class are dimeric proteins shaped like the letter "Y." On each arm of the Y is a uniquely shaped pocket, which is complimentary in shape to some portion of a foreign molecule, called the antigen. By a complex selection mechanism, humans eliminate from circulation those B cells that produce antibodies capable of interacting with their own macromolecules, particularly their own proteins. Millions of antibody-producing cells remain, however, to recognize foreign proteins such as those on the surfaces of invading bacteria or virus particles. [*See* ANTIBODY-ANTIGEN COMPLEXES: BIOLOGICAL CONSEQUENCES.]

A common practice in biological science is to raise antibodies against a protein of interest. For example, one could isolate the coat protein of a common virus, which we can call "X." One could inject X into a rabbit and wait for the rabbit to produce a high yield of antibody that binds specifically to X and is called the "anti-X antibody." A large molecule (e.g., a protein) usually has many antigenic sites (i.e., capable of eliciting formation of an antibody) on its surface; in the experiment above, different antibody types would be raised against each such site. Methods now exist to raise antibodies against just a single antigenic site—these are called "monoclonal antibodies." [*See* MONOCLONAL ANTIBODY TECHNOLOGY.]

The ideal situation for constructing an IT would be to identify an antigenic marker that is unique to the cell type to be destroyed and is not present on healthy tissue. For example, if tumor cells had on their surfaces a protein or other molecular marker that existed on no other cell type, it would be possible to raise antibodies that bind only to the tumor cells bearing that marker. In such an ideal case, it would be possible to conjugate a toxin to the anti-tumor antibody and make very specific magic bul-

lets. Unfortunately, this ideal does not actually occur. Tumor cells arise from normal progenitor cells and survive only by hiding from surveillance by the host antibody system. If tumor cells bore different markers than normal cells, they would be attacked by our own immune systems and would not, in general, cause more trouble than any other invader. As a result, attempts to raise truly specific anti-tumor antibodies have met with considerable difficulty. In general, the best that can be hoped for is to raise an antibody that takes advantage of slight chance differences between a given tumor cell and normal tissue and that has a slightly higher affinity for the former than the latter.

In addition to cell specificity, another desirable property of an IT antibody is its ability to be taken into the target cell, where the toxin can act. The cell membrane surrounding normal or tumor cells is designed to keep extraneous matter outside the cell. Only certain molecules are allowed to enter, usually through specific channels. This uptake process is complex and is treated separately in this encyclopedia. Molecules taken inside a cell generally bind to specific cell-surface receptors, which are then taken up by endocytosis. An example is the uptake of iron by the transferrin system. Antibodies against the transferrin receptor will be readily taken into the cell by the natural endocytosis meant to take up iron; ITs targeted against the transferrin receptor are, thus, quite effective. These ITs are of little therapeutic value, however, because they are not tumor-specific and healthy tissue is targeted as well as cancer cells. Almost any antibody that binds a cell-surface marker will eventually be taken inside the cell by nonspecific endocytotic processes, but clearly there is a design advantage in identifying and using antibodies that are taken up by the target cell at a reasonable rate. [*See* CELL MEMBRANE TRANSPORT.]

B. Toxin Proteins

In principle, any agent toxic to the target cell could be used in the construction of ITs. Radioisotopes and cytotoxic chemicals have been carried to target cells by antibodies, but in general they lack the specificity and killing power of catalytic protein toxins. Protein toxins can be isolated from plants, fungi, and bacteria. Essentially all of these toxins act by disrupting protein synthesis in the target cell, although the modes of toxicity differ in detail.

1. Plant Toxins

Plant toxins fall into two broad classes: heterodimeric cytotoxins and single-chain ribosome-inhibiting proteins (RIPs) (Fig. 1). Heterodimeric toxins, such as ricin, abrin, and modeccin, possess two chains. One is the A chain, which enzymatically inhibits ribosomes by removing one specific adenine base from the ribosomal RNA. In the toxin, the A chain is linked by a disulfide bond to the B chain, which is a protein (lectin) capable of binding to certain galactose-containing carbohydrates located on cell-surface proteins. Binding of the B chain to cell surfaces triggers endocytosis and brings the enzymatic A chain into the cellular cytoplasm where it inhibits protein synthesis and kills the cell (Fig. 2). Ribosome-inhibiting proteins, such as pokeweed antiviral protein (PAP), trichosanthin, and gelonin, have only a single enzymatic chain, which is evolutionarily related to the A chain of the heterodimeric toxins and which catalyzes the same reaction. Because RIPs lack a surface recognition factor, they are only very weakly cytotoxic by themselves. If they are delivered to the cytoplasm however, they can be as lethal as the cytotoxic A chains.

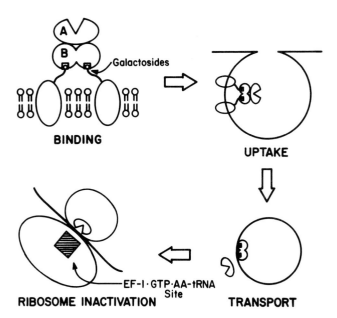

FIGURE 2 The action of cytotoxins. Ricin is the best-studied example of a true cytotoxin. It binds to cell surfaces through interaction of its lectin B chain with cell-surface galactose residues. This triggers endocytosis into a vesicle. By an unknown mechanism, the enzymatic A chain is transported into the cytoplasm where it attacks ribosomes. As a result of the N-glycosidation of a specific adenine base, the ribosomes are no longer able to utilize incoming aminoacyl tRNAs, and protein synthesis is disrupted.

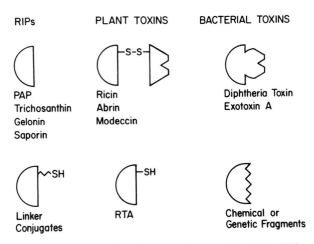

FIGURE 1 Three principle types of enzymatic toxins. Ribosome-inhibiting proteins (RIPs), also called hemitoxins, are generally isolated from higher plants and act as N-glycosidases on rRNA. A free thiol can be generated on the molecular surface by chemical modification or genetic engineering. True plant cytotoxins have an A chain similar to the RIPs and a lectin B chain. Chemical reduction of the toxin leaves a free-surface thiol on the A chain enzyme. Bacterial toxins often have an enzymatic domain, shown as a hemisphere, and a cell recognition domain linked as a single polypeptide chain. This chain can be cleaved into separate components by proteolytic digestion or genetic engineering.

The most thoroughly studied and commonly used plant cytotoxin is ricin, isolated from the seeds of *Ricinus communis,* the castor plant. The enzymatic A subunit and the lectin B subunits of ricin are called RTA and RTB, respectively. RTB is a glycoprotein of 262 amino acid residues. It binds the galactose moieties of cell-surface proteins (or perhaps glycolipids) and is carried into the cell by endocytosis. In vitro experiments show that free galactosides in the media can retard ricin-binding to cell surfaces by competing with surface-bound galactose. Cellular uptake may involve transfer to coated pits, but this is not required. The exact target or targets for ricin-binding are unknown, but as many as 3×10^7 potential binding sites on a cloned human HeLa cell exist. The half time for transport of ricin through the plasma membrane depends on cell type and conditions but is generally less than 3 hr. Once ricin is endocytosed, it travels through the Golgi and other portions of the secretory apparatus by a number of pathways. The majority of ricin molecules eventually recycle to the cell surface and are released; as little as 1% of the ricin initially endocytosed translocates into the cytoplasm. The mechanism of this translocation is unknown. RTB may aid RTA in its release to the cytoplasm, perhaps by forming some sort of permeable channel through which RTA moves. Alternatively, and probably more likely, RTB may act as a marker/flag for internal routing and sorting and may direct a fraction of the RTA into a relatively permeable vesicle from which escape is likely.

RTA is an N-glycosidase consisting of 267 amino acid residues. Initially, it is linked to RTB by a disulfide bond, but this bond is reduced inside the cell, allowing the chains to separate. RTA can efficiently recognize ribosomes at their intracellular concentration (roughly 1–10 μM). Under these conditions, RTA can inactivate ribosomes at a rate of 1,500/min. [See RIBOSOMES.]

Intact ricin is able to kill a wide variety of animal cells. A goal of IT design is to channel this toxicity to attack only those cells selected by the targeting antibody. RTA alone is roughly 10^3–10^5 times less toxic than ricin, and for this reason attempts have been made to conjugate only the enzyme chain with antibodies. By the same rationale, single-chain RIPs, such as PAP and gelonin, have been used in IT design to reduce the nonspecific toxicity caused by RTB's binding of cell surfaces.

Using recombinant DNA technology to alter natural toxins to make them more efficacious in cell-specific toxicity has prompted considerable interest. The three-dimensional structure of ricin is known, allowing a molecular description of the way in which RTB binds sugars and the way in which rRNA is recognized by RTA. This model allows functional roles to be assigned to certain amino acids. It is now possible to redesign the toxin and alter its action slightly. Experiments are underway to design toxins that will have reduced binding to normal cells but will be taken up rapidly as IT complexes by targeted cells.

2. Bacterial Toxins

In addition to plant toxins, bacterial toxins have also been used in the construction of ITs. Diphtheria toxin (DT), from *Corynebacterium diphtheriae,* is synthesized as a proenzyme of 535 amino acid residues. It is then cleaved to form an amino-terminal A chain of 193 residues linked to the B chain by a disulfide bond. The A chain is an enzyme that cleaves nicotinamide adenine dinucleotide (NAD) and inserts the adenosine diphosphate (ADP)–ri-

bose moiety into the active site of elongation factor 2 (EF2). This covalent modification of EF2 inhibits protein synthesis. The B chain binds to cell surfaces, probably by recognizing an ion-transport protein, and thereby appears to trigger endocytosis. The B chain also assists in the translocation of the enzyme A chain across the endosomal membrane into the cytoplasm, perhaps by inserting into the membrane to form an escape pore. The transport is facilitated by low pH inside the endosome. Although DT is a potent and well-understood toxin, its utility in the design of ITs is limited because many people are specifically immunized against DT by vaccination during childhood. As a result, conjugates injected into the body are met by a fully developed immune response. A second bacterial toxin, which holds some promise in IT construction, is exotoxin A, isolated from *Pseudomonas aeruginosa*. This 613-residue protein has three domains thought to function respectively in cell-surface recognition, membrane transport, and enzyme action, which, like DT, causes inhibition of EF2 by ADP-ribosylation.

C. Immunotoxin Linkage

As described above, antibodies and toxic proteins for ITs have generally been isolated separately and must then be joined to form an active agent. This is commonly accomplished by chemically attaching a reactive group, or "linker," to the antibody and allowing it to react with a toxin to form the derivatized antibody. We saw that RTA and RTB are joined by a reducible disulfide linkage, which allows the chains to separate after endocytosis. This reducible linkage is often mimicked in IT construction by using reagents, which form a disulfide bond between antibody and toxin. The most commonly used linker is *N*-succinimidyl-3-(2-pyridyldithio) propionate (SPDP), shown in Fig. 3. In this case, the reagent is allowed to interact with antibody under conditions designed to attach one or two linkers. Amine groups from lysine side chains act as nucleophiles, displacing the N-hydroxy succinimidyl moiety and forming a covalent bond to the linker. This process is random however, and the derivatized antibody population is heterogeneous. The 2-thiopyridine is now displaced in a disulfide exchange reaction by a free thiol on the toxin. If the toxin is isolated RTA, the thiol normally joined to RTB will carry out the displacement and form a cleavable covalent bond to the antibody. If intact

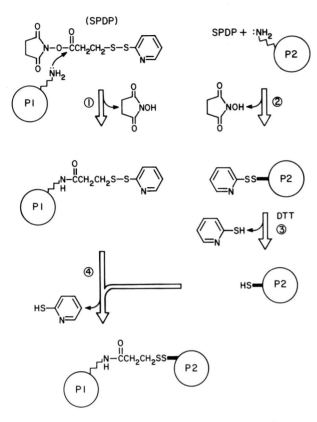

FIGURE 3 Chemical coupling of antibodies to toxins. The most commonly used reagent is N-succinimidyl-3-(2-pyridyldithio) propionate (SPDP). In step 1, a lysine side chain on the antibody (P1) attacks SPDP releasing N-hydroxy succinimide. This creates a reactive intermediate, which can exchange with a thiol on the toxin (P2). RTA has free thiol, as shown in Figure 1, and can react directly at step 4. This exchange reaction releases 2-thiopyridine and links P1 and P2. If the toxin of choice is an RIP, a thiol group must be added by reaction with SPDP (step 2). In step 3, an exchange reaction with dithiothreitol (DTT) creates a free thiol. The heavy bond on P2 is meant to suggest the thiol-containing linker may substitute for the natural thiol of a cysteine side chain.

ricin is to be used, or some single-chain toxin, then a thiol must be attached to that protein. SPDP could be attached randomly to lysine side chains on the toxin. A reducing agent, such as dithiothreitol, would then be used to remove 2-thiopyridine and create free thiols (Fig. 3).

In addition to these cleavable linkers, a wide range of noncleavable linkages can be synthesized, which irreversibly tether toxin and antibody in the IT construct. In general, isolated RTA and single-chain plant toxins must be linked to antibodies by a reducible bond, whereas whole toxins can be linked by either a reducible or nonreducible linker. In all cases, the exact choice of reagent and regimen of

chemical linkage have strong effects on the efficacy of the resulting conjugate. Some reagents may react preferentially with the antigen binding site of the antibody, reducing its cell-surface recognition ability; in other cases, conjugates might form in which the toxin physically blocks the antigen binding site. Alternatively, the reagent might selectively attack a group in the active site of the toxin, reducing its potency. Several agents are generally tested for each antibody–enzyme conjugate to optimize IT action.

Finally, it should be noted that it is now possible to use recombinant DNA technology to design ITS in which the cell-recognition protein and the toxin are fused into a single unit. In practice, light- and heavy-chain gene fragments coding for the antigen binding site are linked by genetic engineering methods; this unit is then fused to the gene or gene fragment coding for a toxic protein. Such a construct has several possible advantages: it is homogeneous and well defined; and it eliminates from the construct the antibody Fc fragment as well as the chemical linker, reducing the antigenicity of the IT and increasing its long-term effectiveness. Initial studies on human cell lines using this new technology suggest that in certain cases the recombinant ITs are at least as effective as those produced by chemical linkage.

D. Immunoconjugates

Once synthesized, the IT is purified to remove any free toxin or antibody and to isolate conjugates of a given size. For example, in analyzing IT action complexes of one antibody and one toxin molecule must be separated from those which might contain two or more toxins to one antibody. Size-exclusion chromatography and antibody or toxin affinity chromatography are used. The purity of the material is assessed by nonreducing sodium dodecyl sulfate polyacrylamide gel electrophoresis followed by dye staining and densitometric scanning. The pure IT should appear as a single molecular species on such a gel.

The derivatization and conjugation may alter the antibody affinity or the toxin enzymatic activity, and this must be assessed. The IT and unmodified antibody are compared in their abilities to block the binding of radiolabeled antibody to target cells. Alternatively, immunoperoxidase staining of tissue sections using ITs and peroxidase-conjugated antitoxin can be used to reveal changes in tissue speci-

TABLE I Immunotoxins Lethal to
Human Tumors

Tumor type	Toxins used[a]
Breast adenocarcinoma	RTA
Melanoma	RTA, DTA, PAP
Colorectal carcinoma	RTA, DTA
Hepatoma	Ricin
Ovarian carcinoma	RTA, PE
Graft-versus-host disease	RTA
Prostate carcinoma	RTA
B-lymphoma	RTA, PAP
Myeloma	RTA
Bladder carcinoma	RTA
Acute lymphoblastic leukemia	RTA

[a] Abbreviations used: RTA, ricin A chain; DTA, diphtheria toxin A chain; PAP, pokeweed antiviral protein; PE, *Pseudomonas* exotoxin.

ficity of the conjugated antibody. Toxin enzymatic potency is measured by testing the inhibition of rabbit reticulocyte protein synthesis (for RIP) or by testing the ADP-ribosylation of EF2 (for bacterial toxins). A large variety of ITs have been prepared with intact antibody and toxin functions and tested for selective cytotoxicity (Table I). [*See* IMMUNO-CONJUGATES.]

III. Immunotoxin Activity

A. *In Vitro* Characterization

Immunotoxins incubated separately with cells expressing the target antigen and with cells without the antigen should only kill the antigen-bearing cells. Varying concentrations of ITs are incubated with target cells for 12–36 hr, and the concentration required to kill 50% of the tumor cells (ID_{50}) and control cells is measured. In general, protein synthesis incorporation is measured. This assay measures the potency of the IT and should closely reflect both the antibody affinity and the rate of antigen internalization to intracellular compartments from which the toxin can escape to the cytosol. A number of ITs have shown potent selective cytotoxicity. ID_{50}'s (the dose that kills 50% of the cells) on sensitive cells range in concentration between 1 and 1,000 pM, with 100–10,000-fold higher ID_{50}'s on control antigen negative cells. Immunotoxins made with high-affinity antibodies to cell-surface antigens present in high copy number, which

are readily endocytosed, make active cytotoxins. Unfortunately, this assay does not measure better than 90% of cell kill. To accurately compare different active ITs, several tests of efficacy have been developed and appear to yield similar predictions. Saturating concentrations of IT are incubated with sensitive target cells. After prolonged incubation, the reduction in surviving cells, able to form a colony, is measured. Efficacious ITs can reduce the fraction of clonogenic cells to 1 in 1,000, or even 1 in 10,000. Kinetics of cell kill can be measured by incubating saturating concentrations of IT with target cells for varying lengths of time and measuring the cell kill, usually by a protein synthesis assay. Cell kill with a survival of 10–30%/hr correlates with more prolonged assays yielding 1 in 1,000 or 1 in 10,000 reduction in colony formation. Immunotoxins prepared with antibodies to different antigenic sites on the same cell-surface antigen have shown differing efficacy. Particular antigen–antibody complexes may internalize better, or the geometry of the complex may improve access of the toxin to membranes. When whole toxin and toxin fragments have been linked to the same antibody and compared, usually the whole toxin conjugates have shown greater efficacy. The augmented killing has been attributed to enhancement functions not present in the toxin fragment conjugates. Certain small molecular weight agents, such as ammonium chloride, chloroquine, and carboxylic acid ionophores, have been found to enhance both the potency and efficacy of ITs. The basis for this enhancement is unknown but may be related to weakening of intracellular vesicle membranes or prolonging transit time in certain intracellular vesicle compartments.

B. *In Vivo* Characterization

Immunotoxins have been administered to rodents, rabbits, and primates. After intravenous administration, the IT circulates in the bloodstream with a half-life of minutes to hours. This half-life is significantly shorter than the half-life of unmodified antibody ($t_{1/2}$ of days) and is due to either clearance by the reticuloendothelial system or liver cells, or metabolism with reduction of the disulfide bond between the toxin and antibody. Mannose-, fucose-, and xylose-terminated oligosaccharides on plant toxins are bound by specific receptors on Kupffer cells of the liver. The use of deglycosylated ricin A chain in the IT prolongs half-life twofold in mice and reduces isolated liver sinusoidal cell-binding by 50%. Unglycosylated toxins, however, are bound by undefined receptors on liver cells. New disulfide linkers have been constructed in which a methyl group and a benzene ring have been attached to the carbon atom adjacent to the disulfide bond to prevent attack by thiolate anions. Abrin A chain ITs prepared with the hindered disulfide linkers have a 10-fold longer half-life in mice. Penetration of IT into extravascular tissues, including tumors, has not been extensively studied. Intravenous administration of anti-T11-gelonin led to saturation of T cells in the lymph nodes and spleen of rhesus monkeys but required five times as much material/animal as antibody alone to achieve equal extravascular deposition. The large size of the ITs ($M_r \approx$ 180,000 daltons) may limit penetration of capillaries.

Toxicity studies of ITs in rodents (mice and rats) and monkeys (rhesus and cynomologous) suggests toxin- and antibody-specific organ damage as well as species-specific pathology. LD_{50}'s range from 0.05–0.1 mg/kg for whole toxin conjugates to 5–10 mg/kg for subunit ITs. The higher toxicity of whole toxin conjugates may be due to persistent, normal tissue binding sites on the toxin. Whereas most toxin conjugates show liver toxicity, *Pseudomonas* exotoxin and saporin conjugates show more dramatic liver damage. Anti-transferrin receptor antibody conjugates have much more toxicity than control antibody conjugates; presumably this is due to widespread presence of transferrin receptor in many normal cells.

Anti-tumor studies in mice have shown that local exposure of tumor cells to IT by regional therapies appears more efficacious than systemic treatment. The ineffectiveness of systemic therapy may be due to clearance or metabolism of the IT, or difficulty in capillary penetration by the large molecules. Tumor burden is important, because delaying the time between tumor implantation and the start of IT therapy decreases the chance of successful therapy. Greater tumor burden at the time of therapy may be associated with the presence of more resistant clones, and larger tumors may have more altered vascular beds, preventing adequate penetration of IT. Finally, several studies suggest that different antibody-toxin conjugates directed at the same tumor model can yield varying results. Many of the pharmacologic factors above may be responsible, or additional factors unique to tumor cells in vivo may play a role.

C. Clinical Applications

The first successful application of ITs has been the prevention and palliation of graft-versus-host disease in patients undergoing allogeneic bone marrow transplantation. Patients are treated with chemoradiotherapy to destroy malignant or abnormal hematopoietic stem cells. The treatment also destroys their normal stem cells so that a bone marrow graft is required from an identical twin or donor with a matched human leucocyte associated antigen. Unfortunately, while the donor marrow often survives in the recipient, the donor T cells recognize the recipient's tissues as foreign and attack the patient's tissues, producing graft-versus-host disease. The liver, gastrointestinal tract, immune system, and skin show the most damage. By depleting the donor graft of T cells using an anti-T cell IT, the patient suffers fewer incidences of graft-versus-host disease. Recently, patients with active graft-versus-host disease have been treated systemically with anti-T cell IT, and the severity and duration of the disease has been reduced. Some of the activated donor T cells in the patient are killed by the IT. Immunotoxin therapy may lead to successful HLA-mismatched allogeneic bone marrow transplantation. [See BONE MARROW TRANSPLANTATION.]

The same anti-CD5-ricin A chain IT has been used for the systemic treatment of chronic lymphocytic leukemia. No clinical responses were seen. The IT was much less cytotoxic to these chronic leukemia cells than to activated T cells. Clinical trials have begun using this IT for treatment of cutaneous T-cell lymphoma. Another, with an antibody to the IL-2 receptor conjugated to pseudomonas exotoxin, has been used on four patients with T-acute lymphoblastic leukemia (which displays IL-2 receptors). Again, no clinical benefit was observed, but doses were low due to liver toxicity, and the antigen density was low on patient cells. A trial has recently begun using anti-CD19-conjugated whole ricin for B-cell lymphoma and leukemia. Solid tumors that have been treated with ITs include metastatic melanoma, colorectal carcinoma, breast carcinoma, and ovarian carcinoma, but very few clinical responses have been seen. In each case, tumor penetration by IT has been difficult to document, and immune responses to the ITs have been observed. Toxicities have included fluid retention, edema, and hypoalbuminemia in melanoma and breast cancer trials and neurologic toxicities including diffuse encephalopathy in colorectal and ovarian cancer trials and peripheral neuropathy in the breast cancer trial. Reversible renal toxicity was seen in the colorectal cancer trial. A "capillary leak" syndrome was observed in all RTA conjugate trials but in no other conjugate trials. The neurotoxicities are unexplained, except in the breast cancer trial where antibody binding to Schwann cells was demonstrated. [*See* LEUKEMIA; LYMPHOMA.]

Future trials in leukemia and solid tumors will have to address the poor penetration and strong humoral immune response to IT. Current research has focused on the use of genetically engineered ITs, which should be smaller and yet retain the enhancing functions of the whole toxin conjugates. Several of these recombinant molecules will soon begin clinical study.

Bibliography

Blakey, D. C., Wawrzynczak, E. J., Wallace, P. M., and Thorpe, P. E. (1988). Antibody toxin conjugates: A perspective. *In* "Monoclonal Antibody Therapy. Prog. Allergy" (H. Waldmann, ed.) pp. 50–90. S. Karger, Basel

Frankel, A. E., ed. (1988). "Immunotoxins." Kluwer Academic Publishers, Boston.

Lord, J. M., Spooner, R. A., and Roberts, L. M. (1989). Immunotoxins: Monoclonal antibody-toxin conjugates—A new approach to cancer therapy. *In* "Monoclonal Antibodies: Production and Application." pp. 193–211. Alan R. Liss, Inc.

Pastan, I., Willingham, M. C., and FitzGerald, D. J. P. (1986). Immunotoxins. *Cell* **47,** 641–648.

Vitetta, E. S., Fullton, R. J., May, R. D., Till, M., and Uhr, J. W. (1987). Redesigning nature's poisons to create anti-tumor reagents. *Science* **238,** 1098.

Implantation (Embryology)

ALLEN C. ENDERS, *University of California–Davis, School of Medicine*

Glossary

Blastocyst Hollow spherical structure consisting of a shell of cells, the trophoblast (i.e., trophectoderm), and an inner cell mass located at one pole

Conceptus All of the derivatives of the blastocyst after implantation, including the embryo and the extraembryonic membranes

Endoderm Layer of cells that develops in the inner surface of the inner cell mass of the blastocyst

Endometrium Mucosal lining of the uterus, consisting of connective tissue stroma, uterine surface epithelium, and tubular glands

Eutherian mammal "True" placental mammals (i.e., mammals with efficient nutritive placentas); includes all mammals except monotremes and marsupials

Exocoelom Extraembryonic coelom; the cavity within the extraembryonic mesothelium

Inner cell mass Portion of the blastocyst giving rise to the embryo, as well as contributing to the extraembryonic membranes

Lacunae Spaces within the syncytial trophoblast, which fill with maternal blood and become intervillous spaces of the placenta

Placenta Specialized portion of the extraembryonic membranes and its associated maternal constituents, which form the major exchange area between the fetus and the maternal organism during most of pregnancy

Trophoblast Initially, the outer layer of the blastocyst; later, the outer (i.e., ectodermal) layer of the placenta

IMPLANTATION IN ANY eutherian mammal begins with the assumption of a relatively fixed position of the blastocyst within the uterus and initiates a stage during which the trophoblast of the blastocyst and the uterine endometrium become progressively more intimate. When we consider that the blastocyst is initially free to move about within the uterine cavity and that the placenta formed after implantation of the blastocyst not only is firmly embedded in the endometrium, but also has its surface, the trophoblast, bathed directly in maternal blood, a sequence of necessary events of implantation can be deduced. Implantation thus includes adhesion of the blastocyst to the uterine surface epithelium, penetration through this epithelium and its basement membrane, expansion into the underlying connective tissue stroma, tapping of the maternal blood vessels within the stroma, and the formation of blood-filled spaces within the enlarging trophoblast. These spaces are then formed around the entire circumference of the blastocyst.

I. Sequence of Events in Implantation

A. Time Sequence of Human Implantation

Much of the basic information concerning implantation in humans has been derived from carefully preserved specimens collected around the world, especially from the meticulously prepared specimens in the Carnegie collections. The classification of the stages of human development has recently been refined, placing the free blastocyst in stage 3, initial attachment of the blastocyst to the uterus in stage 4, and most of the events of implantation in stage 5 (Table I). Stage 4, the attaching blastocyst, has never been described in humans, but it has been deduced that attachment of the blastocyst probably occurs 6 days after ovulation. The events of stage 5 take place 7–14 days after ovulation, when placental villus formation has begun (i.e., the start of stage 6). By the end of the third postovulatory week, numerous branched placental villi surround the circumference of the conceptus, which is now called a chorionic vesicle. At approximately day 21 postovulation, the heart within the developing embryo begins to beat and red blood cells appear in the placental villi. The intra- and extraembryonic circulatory system, however, is not well developed for several more days.

B. Transition Stage from Free to Attached Blastocyst

A few human blastocyst stages have been examined by light microscopy, and blastocysts resulting from *in vitro* fertilization have been examined by electron microscopy, but neither the stage of adhesion of the blastocyst to the uterine surface nor the stage of penetration of the uterine epithelium by the implanting blastocyst has been described in humans.

Consequently, the nature of these crucial first events of implantation either must be extrapolated from other known events in humans or deduced from information gathered from other primates. Fortunately, both light- and electron-microscopic studies of preimplantation and early implantation stages of the rhesus monkey and the baboon are available to help interpret aspects of orientation and adhesion of the blastocyst and epithelial penetration stages. [*See* ELECTRON MICROSCOPY.]

The extracellular layer (i.e., the zona pellucida) that surrounds the blastocyst is shed from the blastocyst 1–2 days prior to implantation, permitting direct apposition of the trophoblast to the uterine surface. Although the blastocyst is "free" in the uterine cavity prior to implantation, there is little fluid within the cavity, and the blastocyst is probably in contact with the surface epithelium. After loss of the zona pellucida, the trophoblast cells near the inner cell mass of the blastocyst develop surface processes and fuse to form the syncytial trophoblast. It is only in this periembryonic region that the blastocyst adheres to the uterine surface.

C. Epithelial Penetration Stage

Following adhesion of the blastocyst to the uterine surface epithelium (i.e., the luminal epithelium), processes of the syncytial trophoblast intrude be-

TABLE I   Human Implantation Stages

Carnegie stage	Age (days)	Developmental event	Changing condition
3a	5	Blastocyst matures	
3b	5	Blastocyst loses zona pellucida	Develops the syncytial trophoblast that adheres to the uterine surface
4a	—[a]	Blastocyst attaches	Trophoblast intrudes between the uterine epithelial cells
4b	—[a]	Epithelial penetration	Trophoblast spreads along the basal lamina
5a	7.5–8	Trophoblastic plate formation	Trophoblast penetrates the basal lamina, invades the stroma, and taps the maternal vessels
5b	9–11	Lacuna formation	Trophoblast differentiation forms microvillus-lined clefts and lacunae
5c	12–13	Primary villus formation	Focal cytotrophoblast proliferations extend from the chorionic plate into the syncytium
6	13–15	Secondary placental villi, secondary yolk sac formation	Mesenchymal cells proliferate on the fetal side of the primary villi
7	16–18	Branching and anchoring villi formation	Cytotrophoblast extends through the syncytial trophoblast and spreads along the basal plate
8–9	18–22	Tertiary villi formation	Vessels differentiate *in situ* and are filled with blood when the heart beat links the yolk sac, the embryo, and the chorioallantoic placenta

[a] No adequate specimens were reported.

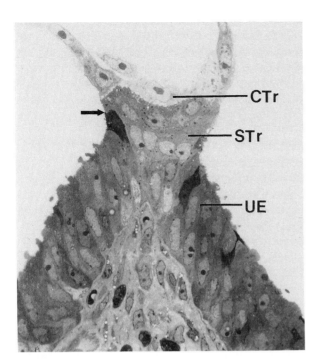

FIGURE 1 High-magnification light micrograph of a region where the trophoblast from a rhesus monkey blastocyst (above) is penetrating the uterine epithelium (below). A mass of syncytial trophoblast (STr) forms a wedge of cytoplasm extending between the uterine epithelial cells (UE). The arrow indicates the junction between the trophoblast and a darker uterine epithelial cell. This region of attachment was just to one side of the inner cell mass, (not shown). CTr, Cytotrophoblast. ×600.

tween cells of the epithelium, to the level of its basement membrane, which separates the epithelium from the underlying stroma (Fig. 1). The trophoblast adheres to lateral surfaces of apparently healthy uterine epithelial cells, especially in the region of the junctional complexes, which connect the epithelial cells to each other. Some of the uterine epithelial cells are surrounded by the trophoblast as it grows in the plane of the uterine epithelium and extends along the epithelial surface of the basement membrane and into adjacent glands. Although epithelial cells are phagocytized by the trophoblast, there is little evidence of necrosis of the uterine epithelium until after phagocytosis of the cells has been completed.

When the trophoblast has expanded on the uterine surface to form a broad region of attachment, the implantation site is said to be in the trophoblastic plate stage. During this stage, which in the rhesus monkey is reached 1 day after implantation, the trophoblast penetrates the basement membrane into the underlying stroma. The earliest human implan-

tation sites that have been examined (stage 5a) are in a late trophoblastic plate stage.

D. Human Previllous Stages

At early (i.e., stage 5a) human implantation sites the trophoblast near the inner cell mass has expanded into masses of cellular trophoblast (i.e., cytotrophoblast) and syncytial trophoblast (i.e., multinucleate trophoblast), whereas the rest of the trophoblast toward the uterine cavity remains as a simple epithelium of flattened cells (Fig. 2). Within the trophoblast many of the nuclei in both cytotrophoblast and the syncytium are large and are thought to be polyploid. At the same time the trophoblast adjacent to the endometrium of the uterus and that following the basement membrane into glands contain nuclei that are much smaller. The tissue of origin of these small nuclei is not yet clear: It may be trophoblast, or trophoblast fused with uterine epithelium. At early human implantation sites the inner cell mass is oriented toward the endometrial surface, indicating that the human blastocyst, like that of other primates, initially attaches through the trophoblast overlying or adjacent to the inner cell mass.

Since no studies of early human implantation stages have been performed using electron microscopy, it is not clear whether the trophoblast initially pauses at the basement membrane of the uterine surface epithelium, as it does in the rhesus monkey, nor is it clear how the trophoblast penetrates through this layer and into the uterine vessels. At the earliest implantation sites (i.e., stage 5a) the maternal vessels seem to be partially surrounded by the trophoblast (Fig. 3), but later (i.e., stages 5b and 5c) trophoblast shares part of the walls of the vessels, and maternal blood flows into spaces within the trophoblast, called lacunae (Figs. 4 and 5).

A single human implantation site in late stage 5c has been examined by electron microscopy. In this specimen the trophoblast intruding into the maternal vessels is adherent to the endothelial cells lining the vessels and shares with it endothelial cell junctional complexes.

While the trophoblast expands into the endometrium of the uterus, more areas of the wall of the blastocyst become differentiated into the syncytial trophoblast and cytotrophoblast masses. By stage 5c almost all of the wall of the blastocyst has been converted from flattened cells into syncytial trophoblast and cytotrophoblast, and the expanding conceptus is beneath the surface of the endometrium.

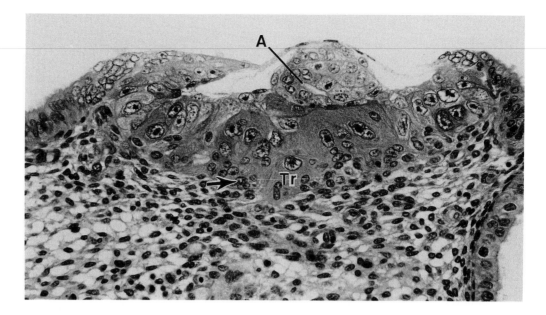

E. Trophoblast Differentiation

Shortly after trophoblast penetration of the maternal vessels in stage 5c, it can be seen that much of the trophoblast has changed its characteristics. Rather than the syncytium and the cytotrophoblast being intermixed, most of the cytotrophoblast cells now line the inner surface of the implantation site, facing the blastocyst cavity. Furthermore, the number of polyploid nuclei is reduced, and most of the nuclei within the syncytium are uniform in size and tend to be distributed singly along the syncytial

FIGURE 2 Stage 5a of human implantation. The inner cell mass already has a small amniotic cavity (A). The broad band of trophoblast (Tr) invading the endometrium has formed a trophoblastic plate, consisting of a mixture of pale cytotrophoblast cells and darker syncytial trophoblast, some of which contains large polyploid nuclei. Note that small nuclei (arrow) are present at the endometrial border of the trophoblast. Carnegie specimen No. 8020. ×350.

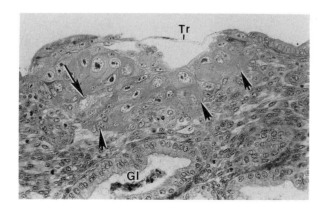

FIGURE 3 Margin of a human trophoblast plate stage (stage 5a). The arrow points to a nucleus in the wall of a maternal blood vessel that is largely surrounded by the trophoblast. Only a few cells of the inner cell mass are present, and the thin trophoblast (Tr) from the side of the blastocyst away from the implantation site has not yet begun to form the cytotrophoblast and the syncytial trophoblast. Arrowheads indicate the border of the trophoblastic plate with the endometrium. Gl, Uterine gland. Carnegie specimen No. 8225. ×210.

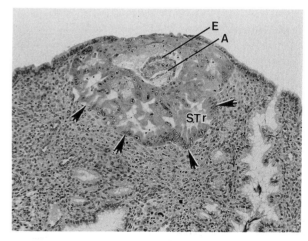

FIGURE 4 At stage 5b the endometrium has begun to undergo decidualization, which is seen as a richly cellular appearance of the endometrial stroma in this low-magnification micrograph. Many lacunae have developed within the syncytial trophoblast (STr). The endodermal layer (E) and the amnion (A) can be seen on either side of the embryonic disk. Arrowheads indicate the border between the trophoblast and the endometrium. Carnegie specimen No. 8004. ×108.

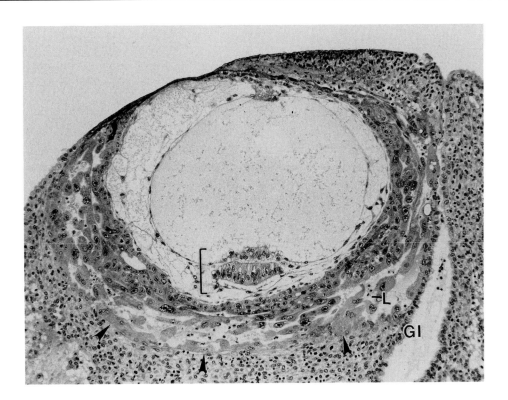

FIGURE 5 By stage 5c the regions of the cytotrophoblast and the syncytial trophoblast have spread around most of the circumference of the blastocyst. The border between the endometrium and the trophoblast is indicated by arrowheads. Note that there are now blood cells in the lacunae (L) within the trophoblast. The forming embryo (bracket) is in a disk stage. A delicate reticulum, the extraembryonic mesoderm, surrounds the embryonic disk and its amnion. Gl, Gland. Carnegie specimen No. 7700. ×150.

trophoblast lining the lacunae. Examination of the implantation sites in rhesus monkeys and baboons in the same period of development shows conversion of most of the syncytial trophoblast into a new type of syncytium that forms a thin polarized layer, with numerous microvilli lining the forming lacunae. This syncytium has a greatly increased surface area, does not cause clotting of the maternal blood, and is apparently less invasive than the previous multinucleate syncytial mass. The syncytial trophoblast, in its position lining lacunae filled with maternal blood, constitutes a layer without intercellular spaces, but with a large surface area that facilitates the transport of substances to the conceptus.

Since implantation is thought to begin approximately 6 days after fertilization, and lacunae are well formed by 11–12 days, it is clear that the stage in which the trophoblast is highly invasive lasts for only a few days. At the end of this time, the conceptus is only about 1 mm in diameter. It is farther

away from the muscle layers of the uterus than it was at the start of implantation, due to the increased thickness of the endometrium and the superficial position of the conceptus (Fig. 6).

F. Site of Implantation

Since the most common site of the placenta in successful pregnancies is the dorsal wall of the uterus (with the ventral wall second), it can be deduced that the blastocyst normally implants toward the midline of the dorsal or ventral wall. This is the position in which uterine muscular contraction would be expected to place the blastocyst. The abundance of other sites, including ectopic sites, indicates that the blastocyst attempts to implant at the place it occupies when development has advanced appropriately to the implantation stage.

II. Mechanisms of Cell Adhesion during Implantation

Although the nature of the changes in the cell surface that permit adhesion of the trophoblast to the uterus are of great interest, studies with laboratory animals have yet to determine the exact mechanisms by which this adhesion occurs in any species. It has been shown that there are at least three meth-

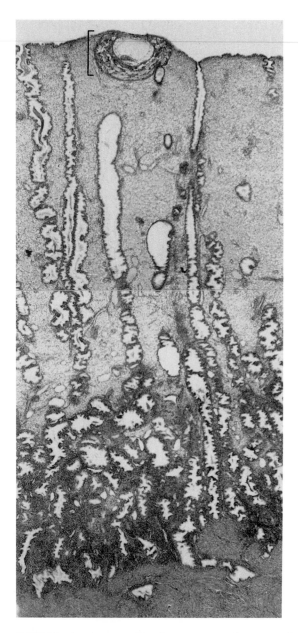

FIGURE 6 At a low magnification a conceptus from stage 5c is seen (bracket) near the surface of the endometrium. The coiled irregular uterine glands extend from the surface to the edge of the smooth muscle of the uterus (at the bottom of the micrograph). Note the small size and superficial position of the conceptus in relation to the total thickness of the endometrium. Carnegie specimen No. 7950. ×30.

ods of penetration of the uterine surface epithelium by the blastocyst in mammals: fusion of the trophoblast with uterine cells, intrusion of the trophoblast cell processes between epithelial cells, and cell death (i.e., apoptosis) of uterine epithelial cells, followed by phagocytosis of these cells by the trophoblast. Such critical aspects as to how the

uterine basement membrane is breached are not fully understood for any species. Studies of these crucial events in implantation in nonhuman primates and in laboratory mammals could lead not only to increased understanding of the events of implantation, but also to the ability to control these events.

III. Development of the Embryo during Implantation

At the time that the blastocyst first attaches to the uterus, the inner cell mass has differentiated into a layer of flattened cells adjacent to the cavity of the blastocyst, the primitive endoderm (i.e., the hypoblast), and a compact group of cells between the endoderm and the trophoblast, the ectoderm (i.e., the epiblast). In the youngest human implantation stage described, there is already a small cavity within the epiblast which constitutes the presumptive amniotic cavity (Fig. 2). The cells enclosing this space on the side toward the trophoblast are thus early amniotic epithelial cells, and the remainder of the epiblast gives rise to the embryonic disk. In rhesus monkeys and baboons it has been shown that formation of the amniotic cavity begins with a stage in which the cells of the epiblast develop apical junctions toward the center of the inner cell mass; the potential space thus formed becomes the amniotic cavity. It is consequently clear that in humans and similar primates the amnion forms by cavitation within the epiblast. A peculiarity of these species is that, during growth of the amniotic cavity, amniotic epithelial cells are adjacent to the trophoblast for a considerable period, and as the amniotic cavity enlarges a diverticulum of the amnion remains adherent to the trophoblast.

The primitive endoderm becomes irregular near the embryonic disk and also extends beyond it as a single layer, the parietal endoderm, around the cavity of the implanting blastocyst. The cells under the embryonic disk reorganize to form an irregular columnar layer, the visceral endoderm, and a group of thin cells separate from the rest of the endoderm to form a small secondary yolk sac adjacent to the embryonic disk (Fig. 7).

The fate of the parietal endoderm is particularly interesting and was, until recently, controversial. It has been shown that, in the rhesus monkey, the cells of this layer develop the cytological characteristics of mesenchymal cells and begin to form extracellular matrix constituents. Thus, parietal endo-

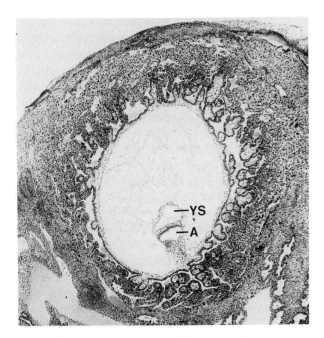

FIGURE 7 At stage 6 placental villi surround the conceptus, forming the chorionic vesicle. A secondary yolk sac (YS) has been formed, and the amnion (A) is beginning to enlarge. Although not apparent at this magnification, the embryo is in the primitive streak stage of development. Carnegie specimen No. 7802. ×32.

derm can differentiate into extraembryonic mesoderm and contributes the first cells to this layer and thus to the exocoelom. Since human endoderm forms a more extensive meshwork of cells (i.e., the endodermal reticulum) in the position of the forming extraembryonic mesoderm than does that of the rhesus monkey (Fig. 4), it is probable that parietal endoderm is also the source of the first extraembryonic mesoderm in humans. [*See* EXTRACELLULAR MATRIX.]

The primitive streak initially develops as an aggregation of cells at one end of the embryonic disk, thus establishing the caudal end of the disk. The primitive streak provides a source of cells that form the intraembryonic mesoderm and contribute to the extraembryonic mesoderm; the latter extends to the preexisting extraembryonic mesodermal layer by way of the forming embryonic stalk. The initial thickening of the embryonic disk is subtle and may be difficult to discern, but a clearly defined primitive streak and groove are present by stage 6. Whether or not the intraembryonic endoderm is also derived from the primitive streak in primates has not been established.

IV. Preparation of Endometrium for the Implanting Blastocyst

Under normal circumstances the blastocyst loses its zona pellucida prior to implantation, at a time when the lining of the uterus, the endometrium, is in an early progestational stage. That is, the endometrium shows the effects of progesterone deriving from the corpus luteum, including extensive glandular development, with coiling of the endometrial glands and good vascularity. The stroma of the endometrium tends to show fluid retention, or edema. The enlargement of the stromal fibroblasts that indicates the beginning of the decidual response does not begin until stage 5b or 5c.

As the implantation site grows the maternal blood vessels enlarge and some of the uterine glands are blocked and become cystic; thus, at stage 6 there is often leakage of blood into the glands and into the uterine lumen (which can produce the "spotting" that is occasionally confused with menstruation). In the normal menstrual cycle progesterone levels begin to diminish by 11–12 days after ovulation, as the corpus luteum declines. If implantation occurs, the corpus luteum continues to function under the stimulation of human chorionic gonadotropin (hCG) produced by the trophoblast. This hormone substitutes for the pituitary gonadotropins in maintaining the corpus luteum. Elevated levels of the hormone can normally be detected by 12 days after fertilization. Although undoubtedly some hCG is produced earlier, its detection is not practical before this time, and even the most sensitive methods cannot detect hCG consistently before day 8.

Other factors have been reported to be present early in pregnancy. One of these, the early pregnancy factor, could provide evidence of the presence of a viable embryo prior to implantation. Use of these factors to detect early pregnancy is not yet practical at the clinical level.

V. Implantation Rates and Fecundity

Fecundity, the chances of producing a full-term infant in a given menstrual cycle during which intercourse occurs, is surprisingly low in women. Estimates based on newlyweds in France at a time when the government was giving a bonus to increase the population and on newlywed Hutterites in the United States suggest a rate varying from 23% to 28%. This low rate includes failures of ovulation, fertilization, cleavage, implantation, and subsequent fetal development.

With the advent of *in vitro* fertilization, it has been possible to introduce into the uterus ova in which successful fertilization has already been judged to have occurred. Nevertheless, the success rate of infants produced by this procedure rarely reaches 17%. Although study of the cleavage stages indicates that some failures occur early, it is thought that much of the failure occurs during implantation. Preimplantation stages show high levels of chromosomal imbalance, but studies of normal *in vivo* pregnancies have shown that only a few, such as trisomy 21, which produces Down's syndrome, are viable.

One of the problems of determining the extent of pregnancy failure during implantation is the inability to determine how many blastocysts begin to implant. Studies of hormonal levels on days 12–18 after intrauterine introduction of an *in vitro* fertilized cleavage stage suggest that a number of pregnancies are initiated, indicated by elevated hCG on day 12, but soon fail, with menstrual flow occurring no later than day 18. In one study of *in vitro* fertilization and embryo transfer, 403 transfers to the uterus were made, with 26% pregnancies reported by normal methods, resulting in 17% viable pregnancies. However, in an additional 8% of the cases there appeared to be a transient hCG rise. Thus, the initial implantation rate may be considerably higher than the normal clinical data would indicate.

Whereas chromosome counts and chromosome banding techniques have revealed that a number of early abortions involve genetic abnormalities, a great many abortions are missed and never provide material for genetic studies. Certainly all of the embryo transfers with transient elevated hCG reported in the study described above would be implantation failures from which no material would be available for genetic study. Consequently, the extent to which the high failure rate during implantation is due to inappropriate endometrial conditions, poorly or improperly developed blastocysts, or genetic factors influencing the blastocysts has yet to be determined. [*See* ABORTION, SPONTANEOUS.]

VI. Models of Implantation *in Vitro*

A number of attempts have been made during *in vitro* fertilization to use fertilized ova to study as-

pects of implantation in an *in vitro* situation. In one study four human blastocysts were placed on a layer of flattened endometrial cells grown *in vitro*. Three of the four showed some outgrowth of trophoblast cells, which contacted the culture dish beneath the endometrial cells. However, most of the inner cell mass cells were lost, and the flattened cell layer was dissimilar from the normal lining of the human uterus. Provided an appropriate polarized and endocrine-sensitive endometrial layer can be established, such a model might be used to study adhesion of the human blastocyst to the epithelium and penetration between junctional complexes.

Another experimental procedure involved the use of blastocysts derived from *in vitro* fertilization and donor uteri maintained by the infusion of uterine vasculature *in vitro*. Although this study reported success in establishing an *in vitro* implantation, the implantation site illustrated was several days in advance of any stage that could have developed in the short time (just over 2 days) for which the uterus was perfused, indicating that the implantation had occurred in the donor uterus prior to surgical removal. This conclusion is reinforced by the observation that, unlike studies involving the ectopic placing of mammalian blastocysts, the implantation site was well defined and well organized.

Bibliography

Enders, A. C., and King, B. F. (1988). Formation and differentiation of extraembryonic mesoderm in the rhesus monkey. *Am. J. Anat.* **181,** 327–340.

Enders, A. C., Henderickx, A. G., and Schlafke, S. (1983). Implantation in the rhesus monkey: Initial penetration of endometrium. *Am. J. Anat.* **176,** 275–298.

Enders, A. C., Welsh, A. O., and Schlafke, S. (1985). Implantation in the rhesus monkey: Endometrial responses. *Am. J. Anat.* **173,** 147–169.

Enders, A. C., Schlafke, S., and Hendrickx, A. G. (1986). Differentiation of embryonic disc, amnion and yolk sac in the rhesus monkey. *Am. J. Anat.* **177,** 161–185.

Hertig, A. T., Rock, J., and Adams, E. C. (1956). A description of 34 human ova within the first 17 days of development. *Am. J. Anat.* **98,** 435–494.

Lindenberg, S., Hyttel, P., Lenz, S., and Holmes, P. V. (1987). Ultrastructure of the early human implantation *in vitro*. *Hum. Reprod.* **1,** 533–548.

O'Rahilly, R., and Müller, F. (1987). Developmental Stages in Human Embryos. *Carnegie Inst. Washington Publ.* **637.**

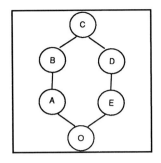

Inbreeding in Human Populations

L. B. JORDE, *University of Utah School of Medicine*

I. Historical Background
II. Basic Concepts of Inbreeding
III. Inbreeding Variation in Human Populations
IV. Causal Determinants of Inbreeding
V. Consequences of Inbreeding

Glossary

Autozygosity Type of homozygosity in which two genes at a locus are identical because of descent from an ancestor common to both parents

Allozygosity Type of homozygosity in which two genes at a locus are identical but are not descended from an ancestor common to both parents

Consanguinity Mating of two individuals who are related to one another

Inbreeding coefficient Probability that two genes at a locus are identical because of descent from an ancestor common to both parents

Inbreeding, random Amount of inbreeding that would occur in a strictly random-mating population of finite size

Incest Sibling or parent–offspring mating

Isonymy Marriage between persons having the same surname

Kinship coefficient Probability that a gene drawn randomly from a locus in one person is identical by descent to a gene drawn randomly from the same locus in another person

INBREEDING refers to the mating of two individuals who are related to one another. The mating of a man and woman who share a recent common ancestor is said to be "consanguineous" (from the Latin *consanguinitas,* "with blood"). The offspring of consanguineous unions are inbred. Because consanguineous couples share one or more common ancestors, their offspring are more likely to carry identical copies of the same gene at a locus. The probability of this identity is numerically measured by the inbreeding coefficient. Each human is thought to carry several recessive disease genes. By bringing identical genes together in individuals, inbreeding increases the frequency of recessive genetic diseases in populations.

I. Historical Background

For millennia, humans have given special recognition to marriage between relatives. The ancient Hebrews banned several types of consanguineous marriage (*Leviticus* 18:6–18). Meanwhile, the ancient Egyptians, perhaps in observance of the sibling marriage of the deities Isis and Osiris, appear to have encouraged brother–sister mating among the royalty. Cleopatra VII was the product of a brother–sister mating, and she in turn married her two younger brothers but produced no children by these marriages (her relations with Mark Antony and Julius Caesar were both fertile, however). Under Theodosius the Great (346–395 A.D.), a Roman could be put to death for marrying a first cousin. The Catholic Church forbade first-, second-, and third-cousin marriages for centuries, and most Western societies have traditionally had rather low rates of consanguineous marriage. Nevertheless, cousin marriages are seen in the pedigrees of a number of prominent Europeans. Charles Darwin, for example, married his first cousin and produced ten children (of whom four became scientists of note). Henri de Toulouse-Lautrec was the product of a first cousin marriage; his short stature is thought to be due to pyknodysostosis, a rare recessive disorder. Queen Elizabeth II of England and her husband, Prince Philip Mountbatten, are related approximately as third cousins.

Encyclopedia of Human Biology, Volume 4. Copyright © 1991 by Academic Press, Inc. All rights of reproduction in any form reserved.

II. Basic Concepts of Inbreeding

A. Definitions

1. Calculation of the Inbreeding Coefficient

Although the inbreeding coefficient was originally defined as the correlation between gametes uniting at fertilization, it is perhaps more convenient to think of inbreeding in terms of probabilities. To this effect, it should be remembered that at each locus (the site of a gene) in the human genome, there are two genes, which may or may not be identical, carried on the two homologous chromosomes. The probabilistic approach defines the inbreeding coefficient as the probability that two genes drawn from a locus in an individual are identical because both have originated from a single ancestor. An individual who carries two identical copies of a gene (or allele) at a locus is said to be a *homozygote*. One who carries two different alleles at a locus is a *heterozygote*.

Figure 1 is a schematic illustration of a consanguineous mating between two individuals, labeled A and E, who share one grandparent (they would be termed "half first cousins"). Individuals who are not related to both A and E are omitted from the diagram for simplicity. To calculate the inbreeding coefficient for the offspring (O) of this mating, we need to consider the "path" in this diagram that connects A and E. The path goes from A to B to C (the common ancestor) and then back down to D and finally to E (ABCDE). Our objective is to determine the probability that O has received the same gene from both A and E. The first event to consider is whether the gene that passed from A to the offspring, O, is identical to the gene transmitted from B to A. According to the rules of Mendelian genetics, an individual will transmit to his (or her) offspring either the gene derived from his mother or the one derived from his father, each with a probability of one-half. Thus, the chance that A transmitted the gene derived from B to his offspring is one-half. There is also a probability of one-half that the gene transmitted by A is derived from his other parent, who is not related to E and is therefore not part of the path. Next, we want to consider the probability that the gene transmitted from B to A is identical to the one transmitted from C to B. Using the same reasoning as before, this probability must also be one-half. Next, consider the probability that the gene transmitted from C to B is the same as the one transmitted from C to D. Suppose that C's two alleles at a locus are labeled X and Y. There are four possible combinations of alleles that could be transmitted to the two offspring: B and D could each receive Y; B and D could each receive X; B could receive X while D received Y; or B could receive Y while D received X. Because each of these combinations occurs with an equal probability, the overall probability that B and D receive the same allele is one-half. There are two final events to consider: the probability that the gene transmitted from C to D is identical to that transmitted from D to E, and the probability that the gene transmitted from E to O is the one derived from D. Again using the principles outlined above, each of these probabilities equals one-half.

We have evaluated five independent events, each of which occurs with a probability of one-half. The probability of all five of these events occurring together is obtained simply by multiplying the individual probabilities together: $\frac{1}{2} \times \frac{1}{2} \times \frac{1}{2} \times \frac{1}{2} \times \frac{1}{2}$, or $(\frac{1}{2})^5$. This is the probability that the offspring, O, received genes from each parent that are identical to one another because they both derived from the common ancestor, C. This defines the inbreeding coefficient, usually denoted as f. In this example, f is equal to $\frac{1}{32}$.

Full first cousins share two grandparents instead of one. Then there would be two paths to consider, one going through one grandparent and one going through the other grandparent. Because the transmission of genes through each of these paths would be independent events, one can add the probabilities resulting from each path together to obtain the inbreeding coefficient for full first cousins: $\frac{1}{32} + \frac{1}{32} = \frac{1}{16}$.

Sometimes the common ancestor is her- or him-

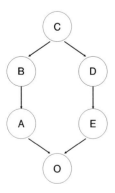

FIGURE 1 Abbreviated diagram of a first-cousin marriage. Only one grandparent is shown.

self the product of a consanguineous mating. This increases the probability that he or she would transmit the same gene to each of his or her offspring. The degree to which this probability is increased is measured by the ancestor's own inbreeding coefficient, denoted f_A (i.e., the probability that the ancestor carries two genes at a locus that are identical). To incorporate this effect, we multiply the inbreeding coefficient for O by the term $(1 + f_A)$.

In general, then, the inbreeding coefficient for an individual is given by

$$f = \Sigma(\tfrac{1}{2})^n(1 + f_A) \qquad [1]$$

where n is the number of individuals in the path connecting the parents of the individual (including the parents themselves), and the summation is taken over each path that goes through a common ancestor.

It is important to distinguish between genes that are *identical by descent* and those that are *identical in state*. The inbreeding coefficient, as we have defined it, measures identity by descent, also termed *autozygosity*. If, however, somebody who was *not* in the path shown in Fig. 1 (e.g., the spouse of individual E) happened to carry the same gene as the one transmitted by A to O, that person could also transmit the gene to O. The offspring would receive two identical genes, but they would not both be derived from the common ancestor. This identity in state is called *allozygosity*.

Another measure of interest is the probability that a gene drawn randomly from a locus in one parent is identical by descent to a gene drawn randomly from the same locus in the other parent. It is perhaps obvious that this probability is the same as the one defined by the inbreeding coefficient. A number of terms have been coined for this measure of relationship between mating couples, including *coefficient of kinship* (the term used here), *coefficient of consanguinity, coefficient of coancestry,* and *coefficient de parenté.*

The inbreeding level of a *population* is commonly assessed by averaging the inbreeding coefficients of all members of the population. The average inbreeding coefficient is denoted here as F (in other literature, it is sometimes denoted as α). The average kinship coefficient of a population is similarly estimated by averaging the kinship coefficients of all mating pairs.

Because all humans are probably related to one another to some extent, one might conclude that all matings would be somewhat consanguineous. This would clearly create confusion. In fact, consanguinity is defined *relative* to a reference population, often the ancestral founders of the current population. Thus, the inbreeding coefficient defined in Eq. (1) actually measures the extent to which the individual's inbreeding level *exceeds* the average inbreeding coefficient of the reference population.

2. Random and Nonrandom Inbreeding

Many models of population genetics assume that members of a population mate randomly. Given that human populations are necessarily finite in size, random mating implies that relatives would occasionally mate. This component of inbreeding in populations is referred to as *random inbreeding*. It is typically measured by computing the inbreeding coefficient for each individual assuming random mating in the parental generation: kinship coefficients are estimated for all possible pairs of individuals in the parental generation. These coefficients are then averaged. This coefficient can then be compared with the total average inbreeding coefficient for the population (F). A standardized measure of the difference between total and random inbreeding, termed *nonrandom inbreeding,* is given by

$$F_n = (F - F_r)/(1 - F_r) \qquad [2]$$

where F is the average inbreeding coefficient and F_r is the average random inbreeding coefficient. When $F_r = 0$, nonrandom inbreeding is equal to total inbreeding, F. When random inbreeding exceeds F, F_n is negative, reflecting avoidance of consanguineous matings. This pattern is frequently observed in human populations.

3. Close versus Remote Inbreeding

Population geneticists usually use an arbitrary rule to denote consanguinity. Matings between individuals more distantly related than third cousins (i.e., sharing a set of great-great-grandparents) are not commonly considered to be consanguineous. Matings between distant relatives are, however, sometimes said to reflect ''remote consanguinity'' or ''remote inbreeding.'' In a small, closed population, mating couples are often related through a large number of distant common ancestors. The amount of inbreeding due to remote consanguinity can often exceed the amount due to close consanguinity in these populations. It should be apparent that a significant amount of remote consanguinity will be observed in *any* finite population in which it is possible to trace ancestors back many generations.

4. Inbreeding in Subdivided Populations

Often, the population of interest consists of meaningful subdivisions (e.g., tribes, parishes). Average random kinship coefficients can be estimated for the total population as well as within and between subdivisions. Within-subdivision random kinship is estimated for each subdivision using the procedures outlined above. Between-subdivision random kinship is estimated by computing the average kinship coefficient between all members of one subdivision and all members of the other subdivision. For example, if subdivision i has 50 members and subdivision j has 100, a total of 5,000 kinship coefficients would be computed and then averaged to obtain the average random kinship coefficient for i and j.

B. Measurement of Inbreeding from Different Types of Data

1. Genealogies

The method for estimating inbreeding given in Eq. (1) is intended for use with pedigree or genealogical data (the terms *pedigree* and *genealogy* tend to be used interchangeably). Of the several methods of measuring inbreeding described here, genealogical data are thought to give potentially the most accurate description of inbreeding in populations. This is because they provide a direct description of the transmission of genes between individuals. However, a number of problems are also associated with estimating inbreeding from genealogical data. First, the approach is sensitive to data errors [e.g., "nonpaternity" (the reported father is not the actual father) and inaccuracies in archival sources]. Because the estimation of inbreeding is a multiplicative process (Eq. 1), a single error in a pedigree path will usually render the inbreeding coefficient grossly inaccurate. A second problem is that all pedigree information is truncated: few genealogies extend more than seven or eight generations into the past. Thus, much of the inbreeding measured in this way will be underestimated. However, it must be kept in mind that inbreeding coefficients are estimated relative to a founding population. If the genealogical data extend uniformly to a reasonably defined set of founders, this problem may not be serious. Finally, genealogical data are often difficult and expensive to obtain and manage.

During the past decade or so, several computerized genealogical databases have been developed to increase the accuracy and ease of manipulating genealogical data. Computers have made possible calculations of a magnitude that would have been impossible a generation ago. It is now feasible to estimate inbreeding coefficients for hundreds of thousands of individuals in a computerized database (and for millions of pairs of individuals when doing random inbreeding calculations). Some notable examples of large computerized genealogical databases include those of the Laredo, Texas, population, the Saguenay population of Quebec, and the Utah Mormon population.

2. Isonymy

In 1875 G. H. Darwin, a son of Charles Darwin, suggested that the frequency of marriages between individuals of the same last name might be a convenient measure of inbreeding in populations. Much later, in 1965, James Crow and Arthur Mange published a paper in which this approach, termed *isonymy* ("same name"), was formalized. They noted that in populations in which surnames are transmitted through the paternal line, there is a constant relationship between the kinship coefficient for a pair of individuals and the probability that the pair shares the same surname. For example, siblings share a surname with a probability of one; their kinship coefficient is $\frac{1}{4}$. First cousins share a surname with a probability of $\frac{1}{4}$ (i.e., when they are related through their fathers); their kinship coefficient is $\frac{1}{16}$. Second cousins share surnames $\frac{1}{16}$ of the time, and their kinship coefficient is $\frac{1}{64}$. This pattern leads to a useful generalization: if a fraction P of the marriages in a population occurs between individuals having the same surname (i.e., isonymous), the average kinship or inbreeding coefficient, F, is $P/4$.

This approach entails some important assumptions. First and foremost, it assumes that individuals who have the same surname obtained it from a common ancestor (e.g., all persons with the last name "Smith" would be related to one another). Obviously, this assumption is not always met, and failure to meet this particular assumption is one of the most important drawbacks to the isonymy method. A less obvious assumption is that in the founding population, each male must have a unique surname. Other assumptions of the method include monogamous mating, equal amounts of migration in the two sexes, a lack of inbreeding in common ancestors, and a random distribution of consanguineous relations occurring through male and female ancestors. Again, these assumptions are often violated. The primary advantage of the isonymy method is that it is relatively easy to obtain large amounts of usable data: one needs to know only the last names of married couples. In addition, isonymy in effect reflects mating patterns that have occurred many generations in the past. Thus, it can overcome the truncation problem that sometimes limits the use of pedigree data. Because of its cheapness and ease of application, the isonymy method has been widely applied in human population genetic studies.

It is straightforward to estimate random kinship (inbreeding) in a population using surname data. If p_i is the proportion of males in the population having surname i and q_i is the proportion of females having surname i, the average random kinship coefficient is given simply by multiplying these proportions together for each surname and then summing across surnames:

$$F_r = \tfrac{1}{4} \sum_i p_i q_i$$

Nonrandom inbreeding, F_n, can then be estimated using Eq. (2).

3. Migration Matrices

Two of the major determinants of random inbreeding are population size and gene flow between populations. In a subdivided population, gene flow between subdivisions decreases random inbreeding, whereas smaller population size increases random inbreeding. A number of models have been formulated in which these effects are used to predict the degree of random inbreeding in populations. Some of the most realistic models incorporate a matrix of migration probabilities among subdivisions. Gene flow among subdivisions is measured by ascertaining the birthplaces of parents and their offspring. From these data, a matrix, **S,** is formed in which each element, s_{ij}, specifies the probability that a gene now in subdivision j (the birthplace of a child) originated in subdivision i (the birthplace of the child's parent). Mathematical formulas are then applied to the migration matrix and to the population sizes of each subdivision to obtain a matrix of kinship coefficients, denoted as Φ. The diagonal elements of Φ predict average random inbreeding (or kinship) within each subdivision, and the off-diagonal elements of Φ represent average random kinship between each pair of subdivisions. Nonrandom in-

breeding cannot be estimated using the migration matrix technique.

As in the case of the isonymy approach, the migration matrix approach involves a number of assumptions. First, it is assumed that migration patterns do not change from one generation to the next. Also, it is assumed that migrants who enter the population from outside the study area are genetically homogeneous. Both of these assumptions are often violated in human populations.

4. Gene Frequencies

Consider a locus that has two alleles (two distinguishable forms of the gene), A and a, in a population. Denote the gene frequencies of A and a as p and q, respectively ($p + q = 1$). According to the Hardy–Weinberg law, the genotype frequencies for this system in a randomly mating population are given by

$$f(AA) = p^2$$
$$f(Aa) = 2pq$$
$$f(aa) = q^2$$

Inbreeding increases the proportion of homozygotes in the population in the following manner:

$$f(AA) = p^2(1 - F) + pF$$
$$f(Aa) = 2pq(1 - F) \qquad [3]$$
$$f(aa) = q^2(1 - F) + qF$$

where F is the average inbreeding coefficient, as defined above. The quantities pF and qF represent the frequencies of autozygous genotypes in the population, whereas the quantities multiplied by $(1 - F)$ represent the frequencies of allozygous genotypes.

Gene frequency data (from blood groups, electrophoretic systems, etc.) have been collected for many human populations. Sometimes departures from Hardy–Weinberg proportions are used to infer the level of inbreeding in these populations. The difficulty with this approach is that many factors other than inbreeding can cause deviation from Hardy–Weinberg proportions. These include natural selection, genetic drift, gene flow, and population subdivision. Thus, this approach to the estimation of inbreeding is particularly subject to error.

5. Comparison of Different Techniques for Measuring Inbreeding

Considering that there are several different approaches for estimating inbreeding in populations, it is natural to question how concordant the different approaches are. In general, the isonymy technique tends to overestimate inbreeding when compared with inbreeding based on pedigree data. This is usually attributed to the assumption that all individuals with the same surname are related to one another. The migration matrix approach has either overestimated or underestimated random inbreeding, depending on the way in which its assumptions are violated. The degree of inaccuracy can be large. When Hardy–Weinberg deviations have been compared with inbreeding estimated from genealogies, little concordance has been found. Most researchers have concluded that the Hardy–Weinberg approach is neither sensitive nor specific enough to detect the level of inbreeding usually seen in human populations.

A recent study compared the results of pedigree, isonymy, and migration matrix data in the large, geographically subdivided Utah Mormon population. This study showed that the isonymy approach overestimated within-subdivision inbreeding compared with the pedigree approach, while the migration matrix technique underestimated inbreeding. However, the overall patterns of between-subdivision kinship were quite concordant among all three approaches.

III. Inbreeding Variation in Human Populations

A. Interpopulation Variation

Table I summarizes average inbreeding coefficients for a number of human populations. The data were gathered from three principal sources: (1) Dispensation records kept by the Roman Catholic Church are commonly available for central and southern European and many American populations. Following the Council of Trent (1542–1563), parishioners were required to obtain dispensations for marriages between individuals related at any degree up to and including third cousins. In Sweden and Finland, dispensations for first-cousin marriages were required by the Crown until the mid 19th century. (2) Many other populations have extensive church or civil registers. It is often possible to reconstruct families and pedigrees from these registers. Computers are now commonly used to perform the arduous task of record linking. (3) A number of studies have employed the direct survey method: married

couples are simply asked about their degree of relationship.

The data in Table I show that large continental populations tend to have much lower inbreeding coefficients than do isolates (the latter are defined as populations that are nearly closed to immigration). In fact, most of the continental populations have average inbreeding coefficients less than 0.001, while all of the isolates exceed this value. The exceptions to this pattern among the continental populations are Japan, India, and several populations with large proportions of Moslems. Among the Japanese and Moslem populations, there has been a traditional preference for marriage between first cousins. In India, first-cousin marriage is also common, and uncle–niece marriage is preferred in some regions (particularly in South India). The highest inbreeding coefficients seen in this table, 0.02 to 0.04, are equivalent to a kinship coefficient less than that of first cousins but greater than that of second cousins. The lowest coefficient, 0.0001, is seen in two U.S. populations. This is consistent with the recent origins and high mobility rates of most populations in this country.

B. Temporal Trends in Inbreeding

Some of the most consistent temporal trends in inbreeding values have been observed in western European populations. First, a substantial increase in consanguineous marriage has been observed in many of these populations during the 19th century. This is commonly attributed to the combined effects of population pressure and the loss of primogeniture

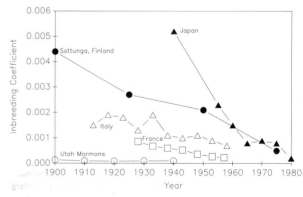

FIGURE 2 Decline of inbreeding during the 20th century in five selected populations.

TABLE I Average Inbreeding Coefficients in Selected Human Populations

Population	Time period	Sample size[a]	Data source	F
Large continental populations				
Belgium	1918–1959	2,040,027	Catholic dispensation	0.0005
Brazil	1956–1957	212,090	Catholic dispensation	0.0023
Canada	1959	51,729	Catholic dispensation	0.0005
Egypt	~1960–1980	26,574	Survey	0.0101
England and Wales	1940	49,315	Survey	0.0004
Finland	1810–1920	1,484,126	Royal dispensation[b]	0.0002
France (Loir-et-Cher)	1812–1954	212,837	Catholic dispensation	0.0011
Germany	1898–1953	1,002,175	Catholic dispensation	0.0004
India (Andhra Pradesh)	1957–1958	6,945	Survey	0.032
Italy	1956–1960	1,646,612	Catholic dispensation	0.0007
Japan	1900–1960	10,048	Survey	0.0049
Kuwait	1983	5,007	Survey	0.0219
Lebanon (Beirut Christian)	~1950–1980	1,001	Survey	0.0049
(Beirut Moslem)	~1950–1980	1,853	Survey	0.0109
Mexico	1956–1957	28,292	Catholic dispensation	0.0003
Netherlands	1906–1948	1,415,987	National archives	0.0002
Norway	1967–1972	336,818	Medical registration	0.0002
Pakistan (Lahore)	—	900	Survey	0.0250
Spain (Toledo)	1900–1979	25,061	Catholic dispensation	0.0018
Sweden	1750–1844	329,852	Royal dispensation[b]	0.0009
Turkey	1970–1987	55,175	Survey	0.0065
USA (Wisconsin Catholics)	1853–1981	920,461	Catholic dispensation	0.0001
(Utah Mormons)	1846–1945	435,777[c]	Genealogical database	0.0001
Isolates				
Amish (Old Order)	1800–1960s	3,107	Genealogies	0.0108
(other)	1800–1960s	5,056	Genealogies	0.0032
Arthez-d'Asson, France	1744–1975	5,196	Parish & civil records	0.005
British Pakistanis	~1960–1985	100	Survey	0.0375
Gypsies (Boston area)	—	21	Survey	0.017
Hutterites (S-leut)	1874–1960	667	Genealogies	0.0216
Khartoum, Sudan	—	4,833	Survey	0.0317
Outport Newfoundland	1960–1979	1,534[c]	Parish records	0.0081
Ramah Navajo	1820–1948	1,118[c]	Survey	0.0066
Saguenay, Quebec	1852–1911	7,607	Catholic dispensation	0.0017
Sottunga, Finland	1725–1975	3,030[c]	Parish & civil records	0.0031
Tristan da Cunha	1830–1959	456[c]	Survey	0.0289
Upper Bologna Apennine (Italy)	1565–1980	43,252	Catholic dispensation	0.0013

[a] Sample sizes are number of marriages unless specified otherwise.
[b] First cousins only.
[c] Individuals rather than marriages.

rights under the Napoleonic code. As population density increased and families were required to divide land among their sons, first cousins often married to maintain ever-diminishing land holdings in the same family. The other major trend that can be observed in most of these populations (as well as some others) is a decrease in inbreeding rates during the 20th century (Fig. 2). This trend appears to be due to the breakdown of genetic isolation as population mobility increased.

Another trend, the gradual buildup of remote consanguinity, is usually observed in the small isolated populations for which many generations of genealogical data have been collected. Even though the members of these populations may avoid close consanguineous marriages, they often marry somebody who is related through many distant common ancestors. Thus, their kinship coefficient can still be high. In small populations such as Tristan da Cunha and the Ramah Navajo, remote consanguinity accounts for *most* of the populations' high inbreeding levels. [*See* POPULATION GENETICS.]

IV. Causal Determinants of Inbreeding

A. Finite Population Size

As has already been discussed, a certain amount of random inbreeding will occur in any finite population. This effect becomes more pronounced as populations become smaller. Most of the isolates listed in Table I are small in size, often consisting of only a few hundred individuals.

B. Geographic Isolation

When a population is geographically removed from its neighbors, gene flow (and the availability of unrelated potential mates) will usually be reduced. A number of studies have demonstrated a positive correlation between geographic isolation, usually measured in terms of geographic distances, and inbreeding levels. The combined effect of migration rates and population size on the inbreeding coefficient can be summarized by a simple model:

$$F = \frac{1}{4N_e M + 1}$$

where N_e is the "effective population size" (i.e., the population size that would obtain under ideal conditions, including random mating, equal numbers of males and females, and equal numbers of progeny per couple) and M is the migration rate. This equation shows that as population sizes and migration rates get smaller, inbreeding increases.

C. Cultural Factors

Much of the population genetic theory regarding inbreeding assumes that populations behave in a mechanistic fashion. Although this may be true for some organisms, it is nearly always untrue for humans. Because of a variety of cultural influences on population dynamics, humans are the organisms least likely to conform to the expectations of population genetic models. The preferred uncle–niece and first-cousin marriage systems cited above are an example. The origins of such mating preferences are usually obscure, but they can sometimes be related to economic factors or religious mandates. In any case, as seen in India and in Moslem societies,

such mating systems can raise inbreeding levels substantially beyond the predictions based on simple population sizes and migration rates.

Inbreeding often varies by social class. Because members of the European nobility usually preferred (or were compelled) to marry within their own relatively small social class, inbreeding levels were often high in this stratum of society. Among the first-cousin dispensations issued in 19th-century Finland, for instance, 17% were issued to members of the nobility, even though the nobility accounted for only 2% of all marriages.

Another often-studied factor that leads to differences in inbreeding levels is urban versus rural residence. Most studies have found that rural populations experience substantially higher inbreeding than urban populations. This is explained partly by greater geographic isolation and lower population density in the rural setting. Rural residents, being "tied to the land," are also less likely to migrate than are city dwellers. As mentioned above, a need to maintain land holdings within families may further contribute to inbreeding in rural populations.

V. Consequences of Inbreeding

A. Inbreeding and Genetic Load

Because inbreeding increases homozygosity, it should increase the frequency of recessive diseases (which become manifest only when the two genes of the same locus are altered) among the offspring of consanguineous couples. Such offspring are more likely to die before reaching maturity ("prereproductive" mortality). This results in a reduction of genetic fitness, where fitness is defined in terms of the proportion of offspring who survive to reproductive age. The *genetic load* in a population is the relative amount of fitness that is lost because all members do not have the optimal genetic constitution. It is defined formally as $(w_{max} - w)/w_{max}$ where w_{max} is the fitness of the optimal genotype and w is the average fitness of the population. There are several types of genetic load. The type caused by expression of deleterious mutants has been termed the *mutational load*.

By comparing mortality rates at different levels of inbreeding (including $F = 0$), it is possible to estimate the average number of lethal recessive genes carried by each individual in a population. This statistical regression approach has been applied to a

TABLE II Mortality Levels Among Cousin and Unrelated Control Marriages in Selected Human Populations

Population	Mortality type	1.0 Cousin		1.5 Cousin[a]		2.0 Cousin		Unrelated	
		%	N	%	N	%	N	%	N
Amish (Old Older)	Prereproductive	14.4	1,218[b]	—	—	13.3	6,064	8.2	17,200
Bombay, India	Perinatal	4.8	3,309	2.8	176	0	30	2.8	35,620
France (Loir-et-Cher)	Prereproductive	17.7	282	6.7	105	11.7	240	8.6	1,117
Fukuoka, Japan	0–6 Years	10.0	3,442	8.3	1,048	9.2	1,066	6.4	5,224
Hirado, Japan	Prereproductive	18.9	2,301	15.3	764	14.7	1,209	14.3	28,569
Kerala, India	Prereproductive	18.6	391	—	—	11.8	34	8.7	770
Sweden	Prereproductive	14.1	185	13.7	227	11.4	79	8.6	625

[a] First cousins once removed.
[b] Includes 1.5 cousins.

large number of human populations. In general, these studies have shown that each human carries the equivalent of approximately two to five genes that would be lethal in the homozygous state (these are termed *lethal equivalents*). This conclusion must be regarded with some caution, however. It is often difficult to be certain that the environments of inbred and noninbred individuals are truly equivalent (or adequately controlled statistically). Also, this approach assumes that lethal genes act independently. There are a number of situations in which complex interactions may occur among genes, violating this assumption. Although this method of estimating lethal equivalents has its drawbacks, it serves an instructive purpose in showing that *all* humans are likely to carry several harmful recessive genes.

B. Inbreeding and Mortality: Empiric Data

Table II summarizes the results of several studies in which mortality was measured in several categories of inbred subjects. These studies show that mortality rates among the offspring of first-cousin marriages are substantially elevated above those of the offspring of unrelated couples. There is generally a decline in mortality rates with decreasing levels of consanguinity, as predicted by theory. It is usually impossible to detect a statistically significant increase in mortality beyond the second-cousin level of consanguinity. The largest comparative analysis of inbreeding and mortality published to date includes data from 31 studies of inbreeding and mortality. The analysis showed that the median relative risk for mortality among the offspring of first-cousin marriages was 1.4 (i.e., the offspring of these unions would be 1.4 times more likely to die before reach-

ing maturity than would the offspring of unrelated parents). The median relative risks for first cousins once removed[1] and second cousins were 1.21 and 1.28, respectively. Although most of the individual population comparisons involving first-cousin marriages yielded statistically significant results, most of the comparisons involving less closely related couples did not. This study also showed that even in societies with relatively high rates of inbreeding (up to 15% consanguineous marriages), the proportion of total mortality caused by inbreeding is small (generally less than 5%).

The studies summarized in Table II deal with postnatal mortality. A number of studies have compared the incidence of stillbirths between inbred and noninbred subjects. In general, these results have not been conclusive. This may be partly due to a lack of ability to detect early fetal and embryonic deaths. It has also been suggested that an *advantage* may be incurred by inbred fetuses: because they are genetically more similar to their mothers than noninbred fetuses would be, they may be less likely to be attacked by the mother's immune system. This hypothesis is still under investigation.

C. Inbreeding and Morbidity

Not all deleterious recessive genes are lethal. Therefore, it is also useful to examine the extent to which inbreeding contributes to *morbidity* (i.e., the disease rate in a population). Again, a large number of published studies have examined morbidity rates in relation to inbreeding. It is difficult to compare these studies because morbidity does not have an

---

1. First cousins once removed are the offspring of one's own first cousins.

exact definition. When reporting morbidity rates among newborns, different investigators have included diseases of varying levels of severity in their tabulations. Also, the degree of diagnostic accuracy varies considerably among studies, particularly in populations where medical care is not optimal. Keeping these difficulties in mind, it has often been stated that the rate of recognizable disorders among the newborn offspring of first-cousin marriages is approximately double that of the newborn offspring of unrelated couples (6–8% versus 3–4%, respectively).

For relatively common recessive diseases like cystic fibrosis and phenylketonuria, it is routine for unrelated heterozygous carriers (who may not be aware of being so) to mate and produce affected children (the carrier frequencies for cystic fibrosis and phenylketonuria in North America are about $\frac{1}{23}$ and $\frac{1}{50}$, respectively). Among rare recessive diseases, inbreeding is more frequently the cause of homozygous-affected offspring. In one published tabulation, only 5% of phenylketonuria cases were due to consanguineous marriage. In contrast, 50% of children born with Wilson's disease, a rare recessive disorder of copper metabolism, were the products of consanguineous marriage. [*See* CYSTIC FIBROSIS, MOLECULAR GENETICS; PHENYL-KETONURIA, MOLECULAR GENETICS.]

In addition to affecting disease rates, inbreeding has also been shown to have a negative effect on various measures of performance (this phenomenon is often referred to as *inbreeding depression*). One of the best-studied examples is the intelligence quotient (IQ). At least a dozen studies of inbreeding and IQ have been published, and all indicate that the average IQs of the offspring of first-cousin matings are several points lower than those of matched controls. Most of these studies matched the comparison groups for socioeconomic status.

D. Consequences of Incest

Incest, the mating of siblings or parents and off-spring, is universally prohibited in human societies. Recent studies indicate that it is even rare among other animals, with most bird and mammal species having fewer than 2% incestuous matings. Given the commonness of deleterious recessive genes, it is expected that incest would have serious consequences. This is indeed the case. Although data on the outcomes of incestuous human matings are understandably difficult to procure, several small studies have been published. In these studies, between one-fourth and one-half of the offspring of incestuous unions were born with diagnosable anomalies, many of which were severe. Mental retardation is especially common among these children. In part because of small sample sizes (on the order of one or two dozen per study), environmental factors were not controlled in these reports. Because many of the incestuous unions occurred in substandard home environments, nongenetic factors are doubtless partly responsible for the high morbidity rates.

E. Inbreeding and Natural Selection

Natural selection acting against a harmful recessive gene can only affect individuals with the homozygous genotype, *aa*. Because the great majority of recessive genes are "hidden" in the heterozygous state, selection against recessive genes is slow. This is particularly true when the frequency, q, of a becomes small. Recalling Eq. (3) and applying some algebra, it can be seen that inbreeding increases the frequency of the homozygous recessive genotype by the amount Fpq. In this way, inbreeding enables natural selection to work more quickly.

In a population in which inbreeding has been occurring for a long time, one might expect that natural selection would have eliminated a substantial proportion of deleterious recessive genes. It has been argued that this has taken place in South India, where close consanguinity has been common for centuries (possibly for as long as 2,000 years). One survey indicated that congenital malformation rates among inbred and noninbred newborns were not significantly different in this population, supporting the idea that a large share of deleterious recessive genes have been eliminated. However, most subsequent studies of the same population did show significant differences. These differences were comparable to those seen in populations such as that of Brazil, where inbreeding has not persisted for a long time. Thus, there is considerable doubt whether this phenomenon has yet been observed in a human population.

Acknowledgments

This work was supported by NIH grant HG-00347 and NSF grants BNS-8703841 and BNS-8720330. I am grateful for comments from Drs. Erich Kliewer, Kenneth Morgan, Derek Roberts, and Alan Rogers.

Bibliography

Cavalli-Sforza, L. L., and Bodmer, W. F. (1971). "The Genetics of Human Populations." W. H. Freeman, San Francisco.

Jorde, L. B. (1980). The genetic structure of subdivided human populations: a review. *In* "Current Developments in Anthropological Genetics, vol. 1" (J. H. Mielke and M. H. Crawford, eds.). Plenum Press, New York.

Jorde, L. B. (1989). Inbreeding in the Utah Mormons: An evaluation of estimates based on pedigrees, isonymy, and migration matrices. *Ann. Hum. Genet.* **53,** 339–355.

Khoury, M. J., Cohen, B. H., Chase, G. A., and Diamond, E. L. (1987). An epidemiologic approach to the evaluation of the effect of inbreeding on prereproductive mortality. *Am. J. Epidemiol.* **125,** 251–262.

McCullough, J. M., and O'Rourke, D. H. (1986). Geographic distribution of consanguinity in Europe. *Ann. Hum. Biol.* **13,** 359–367.

Schull, W. J., and Neel, J.V. (1965). "The Effects of Inbreeding on Japanese Children." Harper and Row, New York.

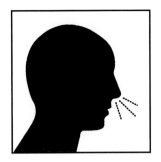

Infectious Diseases

ANNE CARTER, JOHN SPIKA, *Department of National Health and Welfare, Canada*

Glossary

Bacteria Unicellular microorganisms without a true nucleus that multiply by cell division; most have a cell wall and most are capable of living free of other cells, although some bacteria cannot carry on a major function without another cell to assist

Communicable disease Infectious disease in which the microorganism is transmitted from an infected host to an uninfected host; the transmission may be direct, by touching or through the air, or indirect via various vehicles such as water, food or fomites (articles contaminated by handling or use such as utensils, toothbrushes, and doorknobs), or vectors (living carriers of disease)

Helminths Multicelled parasites of three types: round worms (nematodes), flat worms (tapeworms or cestodes), and flukes (trematodes); they have a sexual reproductive cycle, usually in humans or other higher animals, and do not multiply within the definitive host; many roundworms are capable of living free in soil, but the remainder of the group of obligate parasites

Infectious disease Interaction between a host and a microorganism that usually causes signs and symptoms in the host; for this interaction to occur, the host must be susceptible to the microorganism, the microorganism must be present in sufficient numbers and capable of producing disease (pathogenic), and the environment must be supportive of the interaction

Parasites Organisms with a nucleus that live on or within another living organism and derive benefit from that organism; they may be single-cell (e.g., protozoa) or multicell (e.g., helminths)

Protozoa Single-celled organisms with a true nucleus; most are capable of living free of another cell; protozoan parasites can multiply within the host and many have several hosts incorporated into their live cycle; they usually multiply by asexual reproduction but some have a sexual reproductive phase in one of the hosts—the definitive host

Reservoir Ecological niche in nature in which an organism normally lives and multiplies between infectious episodes in humans

Viruses Minute microorganisms that lack ability to carry out essential metabolic processes such as respiration, excretion, and reproduction on their own; they replicate only within other living cells; however, they are true living organisms, because they reproduce with genetic continuity

Zoonosis Disease transmitted from animals to humans, which can be maintained in nature in the absence of humans

IN THE PAST, infectious diseases caused most of human ill health. Overall improvements in hygiene and sanitary facilities and in nutrition have been a major factor in infectious disease reduction in the developed world. With the development of modern medical technology, further advances have been made in controlling these diseases. The major advance has been the development of vaccines, which prevent disease, but the development of modern antimicrobial agents has contributed by making treatment possible. As a result, infectious diseases have become a much less important contributor to human illness in the areas of the world where these technologies are widely available. Because microorganisms often attack and even kill humans in the early years of life, the use of vaccines and antimicrobial agents has led to a longer life expectancy and an increase in the diseases of elderly people—chronic diseases such as cancer, heart disease, and dementia. In contrast to infectious diseases, the cause (etiology) of most chronic diseases is not well understood. Their etiology is probably multifactorial, and it is entirely possible that infectious agents may play an important role. In addition, newly discovered infectious diseases such as acquired immunodeficiency syndrome (AIDS) are becoming very important causes of death and disability. In the developing world, infectious diseases remain a major problem.

I. Vaccine-Preventable Diseases

A. Background

Since ancient times, it has been recognized that many diseases induced life-long immunity if one survived the initial encounter. This led to attempts to introduce disease, preferably a mild variety, at opportune times and locations so as to reduce the amount of suffering from the initial encounter. The ancient practice of variolation, which involved intentionally inducing smallpox by applying scabs from smallpox cases with mild disease to those who had not yet suffered from the disease, is an example of this. Once it was realized in 1798 that this could be done safely using a related but less virulent organism, the cowpox virus, the prevention of disease by vaccination with live but altered or weakened (attenuated) organisms was born. In the nineteenth century, the concept of a "killed" vaccine containing nonliving organisms was developed. At about the same time, the concept of immunity was introduced to explain the observations surrounding vaccination.

Since that time, the success of vaccination has been phenomenal. No other medical procedure has saved as many lives. In addition to eliminating the scourge of smallpox altogether, vaccination has controlled eight major infectious diseases in all regions of the world where vaccination is widely practiced: diphtheria, tetanus, yellow fever, pertussis, poliomyelitis, measles, mumps, and rubella. Vaccines are also available to protect against other infectious diseases; many are listed in the following paragraphs.

B. Live Virus Vaccines

Live virus vaccines are manufactured by attenuating the wild virus that causes the natural disease so that the vaccine recipient will develop immunity but not have symptoms of the disease. This attenuation

is accomplished in the laboratory by growing the virus through many generations. Vaccines of this type have many advantages. The immunity they induce, like that of the wild disease, is long-lasting and highly effective in preventing natural disease. Booster doses of the vaccine are seldom necessary. Because those who are immune may not harbor the organism, the organism tends to die out in highly immune populations, protecting even those who are not immune. This phenomenon is known as herd immunity. The disadvantages of these vaccines include the remote chance that the attenuated virus will cause serious disease and the difficulty of transporting and storing the vaccines to keep the altered virus alive. The latter often requires keeping the vaccine constantly frozen or very cold. This is particularly difficult in tropical countries with poorly developed primary health care facilities— the very place where many of the vaccines are needed the most.

Today live virus vaccines are available against measles, mumps, rubella, yellow fever, polio, and smallpox. Under development are vaccines against chicken pox, some respiratory viruses, cytomegalovirus (which causes mild disease in humans but induces congenital defects if a woman is pregnant during her first infection), rotavirus (which is the leading cause of viral gastroenteritis of infants in the world), and hepatitis A (the less severe form of hepatitis that is transmitted via excreta).

Smallpox vaccine and the disease had very special characteristics that contributed to the success of the smallpox eradication campaign: The disease was not extremely contagious (only about 50% of susceptible contacts became ill) and all patients with disease developed recognizable skin lesions. Those who are immune either from wild disease or vaccination are left with life-long immunity that is detectable without medical records by the characteristic scars. The only reservoir for disease in nature is humans. The vaccine is manufactured cheaply, is stable even in very hot conditions, and can be easily administered. No other disease has all these characteristics, and no other disease is close to being eradicated.

Poliomyelitis, however, shows potential for being the next disease to fall to modern technology. A concerted effort on the part of the Pan American Health Organization and its member countries has almost eliminated the disease from the Western Hemisphere. In 1989 only a few cases were detected, where a few years before there were thousands. This is despite the fact that poliovirus vaccine is relatively expensive and requires continuous storage at cold temperatures, and the disease, paralytic poliomyelitis, is only apparent in a fraction of the cases infected with the virus, allowing undetected cases to pass on disease unchecked.

Measles is a much more difficult disease to consider for eradication, because it is so contagious that herd immunity does not appear to protect non-immune individuals unless >95% of the population is immune. The disease still occurs in North America despite a major effort, now in force for nearly 20 years, to immunize all children. This is because occasional vaccine failures and occasional children who miss being immunized keep the population below an immunity rate of 95%. Because 1 in 1,000 cases of the disease in healthy populations will develop complications, many leading to permanent disability, and 1 in 3,000 cases will die from the disease, new strategies to eliminate this disease are worth consideration. In populations suffering from other major health threats, such as those in developing countries, the death toll from measles is much higher.

Rubella (German measles) is a mild disease whose only claim to importance lies in its ability to cause severe deformities and even death in the fetus of a pregnant woman infected with the disease. Without vaccine, >80% of children are infected before reaching adulthood and possess life-long immunity. Unfortunately, this rate is low enough to permit a significant number of infections in pregnant women, especially during an epidemic year. Mass immunization in developed countries has lowered the number of these tragedies by increasing the proportion of pregnant women who are immune and decreasing their exposure by largely eliminating epidemics. Unfortunately, severe disease in unborn children still occurs occasionally, even in well-immunized populations, because of the extreme infectiousness of the disease and the occurrence of susceptible people due to vaccine failure or missed immunization. [See MEASLES.]

C. Toxoid Vaccines

Some microorganisms produce toxins that are mainly or totally responsible for the production of disease symptoms in those infected. Immunity to the toxin, while not protecting against infection with the organism, will protect against the severe effects of the disease. Immunity can be induced by exposing a person to an altered form of the toxin called a toxoid, which has lost all the toxic proper-

ties but is similar enough to the toxin to induce immunity. The organisms responsible for tetanus and diphtheria fall into this category. Tetanus toxoid and diphtheria toxoid vaccines have been remarkably successful in preventing disease due to these organisms. They do, however, have two disadvantages: because they do not prevent infection, the organism remains prevalent even in highly immunized societies so that inadequately vaccinated persons are at constant risk of developing the disease and booster doses of toxoid vaccines are necessary to maintain immunity.

Tetanus is a disease that occurs when the organism *Clostridium tetani,* a common contaminant in the environment, gains access to the body and proliferates, producing a toxin that paralyzes the nerves. The outcome was invariably fatal before modern intensive care treatment became available. The organism grows best in the absence of oxygen, so it flourishes in deep wounds. Following mass immunization in the developed world, the disease has become rare, affecting mainly the elderly who were missed during the immunization campaigns. In underdeveloped countries, the disease is still common, especially in infants who become infected at the time of delivery when unsterile instruments are used to cut the umbilical cord.

Diphtheria is a respiratory disease caused by the organism *Corynebacterium diphtheriae.* It is transmitted by tiny droplets expelled by infected people. Infection is usually in the throat but occasionally in the nose or ear. The toxin produced affects the heart as well as local tissues. The disease is often fatal without antibiotic treatment. Mass immunization has been associated with the organisms becoming much less common possibly because those who are immune to the toxin do not tend to cough as violently when infected and, hence, do not produce as many infectious droplets.

D. Killed Vaccines

As described above, the technique for producing killed vaccines is very old, and most of the vaccines in this category have been around for a number of years. They have the advantage of being sterile, so they are incapable of accidently producing a virulent form of the disease. Unlike toxoid vaccines, these vaccines produce true immunity to the organism, giving them the potential to produce herd immunity. Inactivated polio vaccine is an excellent example of this, having virtually eliminated the disease in countries where it is widely used. These vaccines do, however, have the disadvantage of being relatively unselective, in that the entire organism is injected, both the parts that induce immunity and many other parts that do not. These other parts may induce unpleasant local reactions, as anyone who has had a typhoid or cholera vaccination will confirm. This drawback has led scientists to search for ways to identify and purify the part of the organism that induces immunity and use it alone in the vaccine.

Influenza is unique among vaccine-preventable diseases in that the virus that causes the disease changes constantly, so that immunity to the organism circulating any one year does not guarantee immunity to the organism that will circulate the next year. The killed vaccine must be prepared anew every year, and everyone who wishes to be protected against this respiratory infection must have a yearly vaccination. Because this is logistically difficult on a population basis, only those who are at high risk of complications from the infection are encouraged to undergo the yearly ritual. Complications involve mainly secondary bacterial infections. Those at high risk of severe disease and death include the elderly and those with heart and lung diseases. [*See* INFLUENZA VIRUS INFECTION.]

Pertussis is a bacterial respiratory infection that affects mainly children and produces a characteristic whooping cough (hence, its popular name) that lasts for up to 6 wk. Very young children are prone to severe complications such as brain damage and may die from the disease. Vaccination may not be entirely effective in preventing disease, particularly in those who have not had a complete course of four injections over 1–1.5 yr. However, the vaccine is effective in preventing severe disease and death.

E. Antigen Vaccines

These vaccines induce immunity in the recipient by exposing them to only the part of the organism that incites an immune response (known in immunology as the antigen). This approach has been particularly successful in dealing with bacteria that have an outer polysaccharide coat. This coat can be purified in the laboratory and a vaccine made from it, which will give good immunity to the organism itself. Examples of this type of vaccine are those that give protection against *Haemophilus influenzae* type b, meningococcus, and pneumococcus. The only virus for which this technique has been used successfully is the hepatitis B virus.

Haemophilus influenzae type b and meningococcus cause meningitis, mainly in young children, but meningococcus also causes epidemic meningitis that affects older children and adults as well. Parts of Africa south of the Sahara Desert have frequent epidemics of this type, and travelers there are advised to receive the vaccine.

Hepatitis B is a virus that attacks the liver. The acute attack is rarely fatal, and the main burden of this disease lies in its long-term consequences for those who cannot totally shake off the initial infection but instead become chronic carriers of the organism. These individuals have a high risk of developing chronic liver disease and liver cancer. Many die in early middle age. The organism is spread from person to person through close personal contact, especially sexual contact. Needle-sharing among intravenous drug abusers is another mode of transmission. Carriers who have no idea of their infectious status are frequent sources of infection. The newly developed vaccine has the potential to greatly reduce suffering due to this disease if it can be delivered to those at risk before they become infected. Some feel that the most efficient way to accomplish this is by immunizing all children.

Pneumococcal infections are an important cause of morbidity and mortality worldwide. In the United States, pneumococcus is isolated from 26 to 53% of children with acute ear infections and from 17 to 39% of adults requiring hospitalization for pneumonia. It is also thought to cause over half of fatal pneumonias in children in developing countries, where acute lower respiratory infections are one of the leading causes of death in children. Currently available interventions for the prevention and treatment of pneumococcal disease include antimicrobial agents, immunization, and the use of immune globin preparations. It is recommended that individuals >2 yr old who are at increased risk of developing pneumococcal infections be vaccinated with the currently available 23-valent (containing 23 different antigen types) pneumococcal vaccine. Individuals at increased risk include persons >65 yr of age, persons with chronic heart or lung disease, diabetes, alcoholism, no spleen, or Hodgkin's disease–lymphoma, and persons infected with the human immunodeficiency virus.

F. Live Bacterial Vaccines

Because bacteria do not lend themselves to attenuation nearly as well as viruses, few of these vaccines are available. The notable exception is Bacille Calmette-Guérin (BCG), an attenuated form of the bacteria *Mycobacterium tuberculosis,* the organism that causes tuberculosis. This vaccine has been shown to be very successful in preventing tuberculosis in some trials but has been notably unsuccessful in others. As a result, a considerable amount of controversy surrounds its use. At the present time, it is seldom used in North America except among indigenous people. It is much more popular in some European and developing countries.

Tuberculosis is a bacterial infection that usually attacks the lungs but may infect just about any part of the body. Only the respiratory form of the disease is infectious. Infections last for many years, often silently, and are hard to eliminate, even with modern antibiotics. The disease is spread more easily in crowded conditions. Improved housing and general living conditions may have had more to do with the decline of the disease in developed countries than modern therapy.

G. Immune Globulin

The proteins in an individual's blood responsible for fighting infections are known as immune globulins. These can be easily removed from one person and purified, then injected into another person. This produces what is known as passively acquired immunity, which lasts only a few months. Nevertheless, it is quite effective while it lasts, and it is the only method of preventing some diseases for which no vaccine has yet been developed. It is also effective in preventing disease during the period between immunization and the development of immunity. The most common use of this method of immunization is the prevention of hepatitis A during a discrete exposure such as travel in a country where the disease is common or cohabitation with an acute case of the disease. Specific immune globulins have been purified (using blood taken from many people) for the following diseases: rabies, hepatitis B, chicken pox, tetanus, pertussis, and smallpox.

II. Diseases of Developing Countries

A. Background

As one may have gathered from the above discussion of the vaccine-preventable diseases, they are much more common in the developing world where vaccination is not as widely available. The World Health Organization (WHO) has implemented a campaign to immunize the world's children against diphtheria, pertussis, polio, tetanus, measles, and tuberculosis by the end of the 1990's, and to date great strides have been made. Whereas only 5% of children in the developing world were fully immunized against these diseases in 1974, it is estimated that >70% are now fully protected. Thus, many diseases that were common in these countries are rapidly disappearing. Despite improvements in levels of immunization, diarrhea and acute respiratory disease remain two of the leading causes of death in children and adults in these countries. The majority of enteric and acute respiratory diseases, which have an identified etiologic agent, are due to bacterial and/or viral infection. In the tropics, parasitic diseases also contribute greatly to human illness.

While it is easy to focus on the infectious agents associated with illness and death in the developing countries, one needs to keep in mind the very complex interaction between susceptibility to infectious diseases and proper nutrition, availability of safe water and basic sanitation, health knowledge, and infant-feeding practices. Malnutrition may predispose one to infection (e.g., low levels of vitamin A). Transmission of certain infectious agents (e.g., *Neisseria meningitis*) may be enhanced in crowded living conditions.

B. WHO-Targeted Diseases

Eight diseases have been targeted by WHO as diseases of special concern because they cause an enormous amount of suffering in the developing world; however, because they do not affect the developed world, there is no economic incentive for research into possible preventive measures. No effective vaccine is available to control any of them at the present time. An effort is being made by WHO to reduce the burden caused by these diseases by encouraging research and control programs. Each disease will be discussed individually.

1. Schistosomiasis
Schistosomiasis is caused by a trematode and is associated with contaminated water supplies. Several forms exist in different parts of the world, some of which affect the urinary bladder and kidneys and others of which affect the bowel and liver. The parasite has a complicated life cycle involving fresh-

water snails, which become infected after human waste containing the eggs of the organism contaminates the water in which they live. Once the parasite has passed through the snail, it becomes a free-swimming organism that enters the skin of any person who happens to be bathing, playing, or working in the water. Approximately 200 million people are infected with this parasite, most of them residents of tropical rural and agricultural districts. Drug treatment is quickly effective and safe but expensive. Without treatment, long-term sequelae can be very severe, particularly if a large number of organisms are harbored. The control of this disease depends on a multifaceted approach involving treatment of human waste, health education, treatment of cases, reduction of snail populations, and provision of clean water supplies.

2. Filariasis

A nematode called *Wuchereria bancrofti* causes filariasis, also known as elephantiasis because the symptoms—huge swollen limbs—make the patient look like an elephant. Adult worms live in the human lymphatic system, blocking lymph drainage and causing swelling, while producing many larvae known as microfilariae. These larvae infect mosquitoes when the patient is bitten and undergo part of the life cycle there. The infection is eventually passed on to another person when the mosquito bites again. Adult worms develop in this person in about 1 yr and may produce microfilariae over the next 5–40 yr. It is estimated that 80 million people are infected with this organism throughout the tropical world, but only those with a large number of parasites and a strong reaction to the parasites develop severe symptoms. No host exists other than humans. Treatment is usually aimed at eliminating only the microfilariae, not the adult worms. (The drug that kills the adult worms is quite toxic.) Prevention of this disease depends primarily on control of the mosquito vector.

3. Onchocerciasis

Onchocerciasis is also caused by a filarial nematode, *Onchocerca volvulus*. Known as river blindness, its life cycle resembles that of filariasis except that the vector is a black fly found near freshwater rivers in Africa and South and Central America. The adults reside mainly in the tissues just under the skin, producing disfiguring lesions. The larvae wander throughout the body and cause the major symptom of this disease—eye irritation and eventu-

ally blindness when they invade the eye. Again, many years of exposure are necessary to produce severe symptoms. It is estimated that 17.5 million people are infected with the organism and over 300,000 are blind as a result of their infection. Treatment with a new drug is very effective. The main preventive approach involves vector control and mass drug treatment.

4. African Trypanosomiasis

Better known as sleeping sickness, African trypanosomiasis is caused by the protozoan parasite *Trypanosoma brucei* and occurs only in tropical Africa. The organism enters the body following the bite of its insect vector, the tsetse fly. After forming a sore at the site of the bite, it travels throughout the body, eventually entering the central nervous system, causing a progressive mental deterioration, which leads to death in the absence of treatment. More than 10,000 cases of the disease occur annually in the affected geographic area. Control programs emphasize surveillance and treatment of cases to prevent further transmission of the disease as well as local control of contact between humans and the vector fly.

5. American Trypanosomiasis

Another trypanosome, *Trypanosoma cruzi,* causes a very different disease—Chagas' disease in Central and South America. The life cycle is similar except that the vector is a bloodsucking triatomine bug known as the "kissing bug." The parasite can invade the heart, esophagus, or intestine, causing a variety of symptoms depending on the organ involved. Domestic and wild animals serve as a reservoir for the disease. Approximately 17 million people are infected but only one-third will develop symptoms. In any one year, 200,000 symptomatic cases develop. In some areas of Brazil, Chagas' disease is responsible for 10% of all adult deaths. Treatment is only partially effective. Prevention efforts are based on vector control and improvement in housing, because the insects and humans come in contact mainly in substandard housing. [*See* TRYPANOSOMIASIS.]

6. Leishmaniasis

Several diseases fall under the leishmaniasis category, each caused by a different species of the protozoan parasite *Leishmania*. A number of forms affect only the skin (cutaneous leishmaniasis) or the skin and mucous membranes (mucocutaneous leishmaniasis). These can be very disfiguring and resemble leprosy. Wild animals often serve as disease reservoirs. A more severe form of the disease affects many body organs and is known as visceral leishmaniasis or kala azar. The reservoir for this form is the domestic dog. The insect vector for all forms is the sandfly. All continents are affected, although there is a predilection for warmer climates. Worldwide, about 12 million people are infected, with about 400,000 new cases occurring each year. Only visceral leishmaniasis is fatal. A recent epidemic in India resulted in 20,000 deaths. Treatment can eliminate the disease, but drugs have the potential to cause severe side effects. Disease control involves vector control (which is difficult) and reduction of reservoir animals. Research is underway to develop a killed vaccine. Meanwhile, deliberate infection with a mild cutaneous form at an aesthetically acceptable site is sometimes practiced. [*See* LEISHMANIASIS.]

7. Malaria
Malaria is a mosquito-borne protozoan disease and is well known as one of the world's worst scourges. About 56% of the world's population live in malarious areas and most are subject to frequent infections, which become less severe with increasing age because of previously developed immunity. It is estimated that 100 million symptomatic infections occur annually. Many infants in these areas succumb to their first encounter with the parasite. In addition, many individuals have death hastened because malaria exacerbates other medical conditions. The life cycle involves human to human transmission through a mosquito vector. Treatment in the past was simple and very effective. Recently, strains of malaria that are resistant to the usual treatment drug have become quite common. This has led to the search for new therapeutic agents, but the parasite seems able to develop resistance to many of the new agents as soon as they are put into widespread use. Thus, treatment may present a challenge. Individual prevention for travelers to malarious areas is provided by regularly taking medication to which local parasites are known to be sensitive; this is not practical for residents of such areas. This approach has become much less effective with the emergence of constantly changing parasite resistance. Mosquito control has the potential to protect whole populations but has run into the problem of vector insecticide resistance. Current emphasis is on personal mosquito protection measures, including the use of insecticides and mosquito bednets. The world malaria situation has not improved over the past 15 years, and, in many parts of the world, it has deteriorated. Work is progressing on the development of a malaria vaccine that may offer some hope for future control of this disease. [*See* MALARIA.]

8. Leprosy
Leprosy is the only disease of the eight that is not caused by a parasite. The responsible organism is *Mycobacterium leprae,* a bacterium that is closely related to the one that causes tuberculosis. Depending on the host's immune response, a wide spectrum of disease severity is possible ranging from asymptomatic infection to severe disfiguring illness. The most common symptom is skin lesions, which vary from a reddish or less pigmented patch to multiple nodules with thickening and folding of the skin. Loss of normal sensitivity of the skin is also very common. Severe cases can also have paralysis and bone destruction. The disease can be treated with some difficulty. The main problems are the prolonged time required to complete an effective treatment course and toxic reactions to therapy. Newer drug regimens are showing improved effectiveness in shorter periods of time. It is estimated that 10–12 million people are infected worldwide, mainly in Asia, Africa, and Latin America. No preventive measures are available at present other than early treatment of cases to prevent further spread. Vaccines based on killed *M. leprae* are currently undergoing trials, but results are not expected in the near future.

C. Enteritis

Infectious diarrhea caused by bacteria, viruses, and parasites is estimated to result in the death of 5 million children a year. Rotavirus may cause 30–40% of diarrhea in children <2 yr old in the developing countries; no etiology can be established in 20–40% of children with diarrhea in this age group. Watery diarrhea caused by toxin-producing *Vibrio cholera* may be the best known of the bacterial causes of diarrhea in children and adults; however, diarrhea caused by other toxin-producing bacteria is more common. *Escherichia coli* that produce a heat-labile toxin frequently cause diarrhea in local residents and tourists. Disease caused by *Shigella* species is also toxin-mediated, although some *Shigella* strains invade intestinal mucosal cells. Other bacteria that cause diarrhea include *Salmonella* species, *Campylobacter jejuni, Yersinia enterocolitica,* and noncholera *Vibrio* species. The most important parasites causing diarrhea include the protozoa *Giardia lamblia, Entamoeba histolytica,* and *Cryptosporidium.*

Salmonella typhi and *Salmonella paratyphi* cause enteric fever, a systemic illness with symptoms of headache, malaise, fever, somnolence, abdominal pain, and occasionally diarrhea. The reported incidence of typhoid fever (enteric fever caused by *S. typhi*) in some areas of the developing world approaches 40/100,000 population. The mortality rate among treated patients with typhoid fever in endemic areas is 1–2%.

D. Respiratory Disease

In the developing world, >4 million children <5 yr old die of pneumonia each year, 30% of the deaths that occur in this age group. Like diarrheal disease, pneumonia in children can be caused by a number of infectious agents. Viral causes of pneumonia include measles virus, respiratory syncytial virus, parainfluenza virus, adenovirus, influenza virus, and cytomegalovirus. Bacterial causes include *Streptococcus pneumoniae (Preumococcus), Haemophilus influenzae, Staphylococcus aureus,* group A streptococci, and *Klebsiella.* Other causes include *Mycoplasma pneumoniae, Pneumocystis carinii,* and *Chlamydia.* Although measles pneumonia may be associated with 30% of the respiratory-related deaths in unvaccinated children, pneumonia remains a major cause of death even in populations well vaccinated against measles. *Streptococcus pneumoniae* is thought to be the leading cause of severe and/or fatal pneumonia in children in developing countries. Tuberculosis is also an important cause of respiratory disease in these countries.

III. Diseases of the Developed World

A. Background

In Western Europe and North America, infectious diseases continue to cause illness and death in children and adults, although improved sanitation, availability of clean potable water, immunization, improved access to health care, and early use of effective antimicrobials for treatment have lessened their impact. The only infectious disease category among the leading causes of death in these coun-

tries is influenza–pneumonia, which falls well behind the categories of heart disease, cancer, and other chronic diseases.

B. Pneumonia

Influenza and pneumonia deaths occur primarily in the elderly and persons with underlying chronic heart and pulmonary disease. Influenza virus and the bacteria *S. pneumoniae* are most frequently associated with these deaths. In children <2 yr old, the most frequent cause of pneumonia is respiratory syncytial virus. In older children and young adults, *M. pneumoniae* is the leading cause. However, tuberculosis continues to be a problem in developed countries, particularly in groups such as Native Americans, the homeless, and persons with AIDS.

C. Sexually Transmitted Diseases

Sexually transmitted diseases (STDs), including AIDS, are a major cause of illness in developing countries, and AIDS has become a prominent cause of death. Over 70,000 deaths have occurred in the United States due to AIDS. Chlamydial cervicitis in women (*Chlamydia trachomatis*) is now recognized as the most common STD, which, if left untreated, can lead to infertility. Gonorrhea, caused by the bacteria *Neisseria gonorrhoeae,* and syphilis, caused by the bacteria *Treponema pallidum,* continue to be of concern. Herpes virus and human papilloma virus are also transmitted through sexual contact. Genital herpes infections cause pain and discomfort and can lead to severe infections in infants born through infected birth canals. Infections with certain types of human papilloma virus may be associated with cervical cancer. Other less common STDs in developed countries are lymphogranuloma venereum and chancroid. [*See* AIDS EPIDEMIC (PUBLIC HEALTH); SEXUALLY TRANSMITTED DISEASES (PUBLIC HEALTH).]

D. Role of Infectious Agents in Chronic Disease

The etiology (cause) of most chronic diseases is not well understood. Many factors probably combine to cause them, and infectious processes may play a role in the origin of many chronic diseases.

1. Cancer

Infectious agents, mainly viruses, are well known to be associated with certain tumors, particularly in animals. A few examples of human cancers are firmly linked to viruses: Kaposi's sarcoma and the AIDS virus; hairy cell leukemia and the virus HTLV 1; and liver cancer and hepatitis B virus. Several other tumor–virus links are less direct (e.g., human papilloma virus, cervical cancer). Exactly how these viruses are linked to the tumors is not certain, but it is probably not a simple one-to-one causal relationship. Many other factors must play a role. However, when it is realized that the cause of most cancers is unknown, the possibility that viruses may play a significant role is exciting because viral diseases are sometimes preventable by vaccination.

2. Arthritis

It is well known that infectious agents can cause arthritis. For example, many bacteria can infect a joint causing acute pain, swelling, and destruction. Some viruses such as rubella virus are known to occasionally cause a less acute arthritis. Rheumatoid arthritis is a very common form of arthritis for which no cause has yet been found. Viruses may play at least a partial role in the multifactorial etiology of this disease.

3. Dementia

Although uncommon, infectious dementias have been well described: measles virus occasionally causes a disease called subacute sclerosing panencephalitis and the dementias of Kuru and Jakob–Creutzfeldt disease are due to infectious agents that are not well characterized. Because most other dementias are of unknown etiology, an infectious origin is a distinct possibility. These include such common diseases as Alzheimer's disease and senile dementia. Parkinson's disease, although not a dementia, is a neurodegenerative disease, which is thought to be associated in some way with previous influenza infection. [*See* ALZHEIMER'S DISEASE; DEMENTIA IN THE ELDERLY; PARKINSON'S DISEASE.]

4. Diabetes

It has been postulated that a pancreatic islet cell infection may precede the onset of the juvenile form of diabetes, a disease that causes a large amount of suffering and premature death in the developed world.

5. Cardiovascular Disease

Cardiovascular diseases are, for the most part, of unknown cause. One of the few diseases with at least partially delineated etiology, rheumatic heart disease, is associated with streptococcal infection. It seems unlikely that infectious processes contribute greatly to the remainder of these diseases but only time will tell.

6. Chronic Fatigue Syndrome

Chronic fatigue syndrome is ill-defined and causes virtually no deaths and few hospitalizations in the developed world; however, it is responsible for a considerable amount of long-term suffering. The cause is unknown but a viral etiology has been postulated, and viral illness is probably involved in some way.

Bibliography

Benenson, A. S. (ed.) (1990). "Control of Communicable Diseases in Man." The American Public Health Association, Washington, D.C.

Evans, A. S. (ed.) (1989). "Viral Infections of Humans: Epidemiology and Control," 3rd ed. Plenum Medical Book Company, New York.

Evans, A. S., and Feldman, H. A. (eds.) (1982). "Bacterial Infections of Humans, Epidemiology and Control." Plenum Medical Book Company, New York.

Katz, M., Despommier, D. D., and Gwadz, R. W. (1989). "Parasitic Diseases," 2nd ed. Springer-Verlag, New York.

Maurice, J., and Pearce, A. M. (eds.) (1987). "Tropical Disease Research: A Global Partnership, Eighth Programme Report." World Health Organization, Geneva.

Plotkin, S. A., and Mortimer, E. A. (1988). "Vaccines." W. B. Saunders Company, Philadelphia.

Walsh, J. A., and Warren, K. S. (ed.) (1986). "Strategies for Primary Health Care: Technologies Appropriate for Control of Disease in the Developing World." University of Chicago Press, Chicago.

Infectious Diseases, Pediatric

ANNE M. HAYWOOD, *University of Rochester*

Glossary

Child   Person between 2 and 13 yr of age
Infant   Person between 1 mo and 2 yr of age
Neonate or newborn   Person between birth and 1 mo of age
Preschool-aged child   Person between 2 yr of age and the age of entry into kindergarten or first grade, which is usually 5–6 yr
School-aged child   Person between 5–6 yr and 13 yr of age

UNTIL THIS CENTURY, infections were responsible for a high death rate in children. Sanitation, vaccines, and antibiotics have reduced this death rate in developed countries to the extent that the average child is expected to live to adulthood. Children, however, still have a lot of illness due to infectious diseases and still occasionally die. Different age ranges are vulnerable to different infectious diseases. The major infectious threats to society and to the pediatric population vary over time. This is only partly because of changes resulting from medical advances. The microbes and their virulence and antibiotic resistance change with time. As societies change, the epidemiology of infections changes. Diseases and syndromes continue to be newly recognized as infectious, and microorganisms are constantly being newly recognized as pathogens (i.e.,

as producing disease). The pediatric population will continue to be particularly vulnerable to certain kinds of infections because of their immune system, the nature of their exposures, and the details of their anatomy. However, many of the main concerns of parents and pediatricians will change with time as diseases such as diphtheria and poliomyelitis cease to be a major concern and new concerns such as Reye syndrome and acquired immunodeficiency syndrome (AIDS) arise. Some diseases that have become infrequent may again reappear, therefore, although there have been great advances in infectious diseases, new problems continuously arise.

I. Introduction

Mankind coexists with a wide range of microorganisms, some of which are helpful for humans. Thus, colonization of various parts of the body (e.g., the intestine) with nonpathogenic microorganisms leads to protection against pathogens, to induction of antibodies, and to production of some factors that are required by humans, such as vitamin K. Because it is not in the best interests of a microorganism to kill its host, most harmful microorganisms have evolved to become less virulent for their natural hosts. Infectious diseases result from invasion of a very few of the total number of microorganisms. Some organisms have properties that routinely make them pathogens, whereas other microorganisms create disease only under specific circumstances. [*See* INFECTIOUS DISEASES.]

The microorganisms—viruses, bacteria, fungi, and parasites—produce disease by a wide range of mechanisms. Viruses are too small to be seen by the light microscope. They cannot reproduce independently but must replicate inside cells. They usurp many of the host cell components for their own replication. Viruses usually have a mechanism to attach and enter cells in specific organs. This can result in an acute infection in which the infected cells die and either the host defenses destroy the virus or the virus ultimately kills the host. Viruses can also produce persistent infections in which the virus does not immediately kill the host cell or may kill only a few cells and in which the host is not able to completely eliminate the virus. A persistent viral infection may result in disease at a later time due to cumulative damage and/or may produce chronic disease.

Bacteria (prokaryotes) are much larger and more complex than viruses, and their virulence depends on many mechanisms. For medical purposes, bacteria are divided according to how they react with Gram stain, which in turn depends on properties of the bacterial cell wall. Gram-positive bacteria retain a crystal violet-iodine complex after alcohol treatment and appear blue. Gram-negative bacteria do not retain the crystal violet-iodine complex and therefore appear red after addition of the counterstain safranin. *Staphylococcus* and *Streptococcus* are gram-positive bacteria, whereas *Escherichia* and *Pseudomonas* are Gram-negative bacteria. Bacteria that are able to grow only in the presence of air are aerobic, and those that require the absence of air are anaerobic. Bacteria also can adhere to specific host tissues. Some bacteria secrete toxins, and some are able to invade tissues. Many bacteria have mechanisms that help evade host defenses and, therefore, are factors in determining bacterial virulence. These include IgA-specific protease (an enzyme that digests the immunoglobulin IgA, which is one kind of antibody) and the ability to bind a specific domain of host immunoglobulins to reduce their efficacy. Some bacteria have components on their surfaces such as polysaccharides (long chains of sugars) that make their ingestion by white cells difficult. Bacteria have many enzymes, some of which, such as neuraminidase or collagenase, allow the bacteria to spread through tissues.

The fungi that can cause human fungal infections (mycoses) can be divided into those that cause disease of the skin, nails, and hair and those that can cause systemic infections. Those that cause the superficial infections can be transmitted from child to child or from pet to child. These include tinea corporis, ringworm of the body, tinea capitis, ringworm of the scalp and tinea pedis, the usual cause of athlete's foot. With the exception of *Candida*, most of the fungi that cause systemic mycoses are organisms that live independently in the soil or decaying matter, and the individual fungi are found in specific geographic regions. For instance, *Histoplasma capsulatum* is found in the central and southeastern United States, especially in areas such as chicken houses or caves, where the soil is enriched with bird or bat feces. *Coccidioides immitus,* which causes coccidioidomycosis (valley fever), is found in arid regions of the United States and of Central and South America. Outbreaks of these soil fungi may occur after dust storms.

Parasites include the helminths and the protozoa, which are one-cell eukaryotes. Examples of protozoa that are geographically widespread are *Giardia lamblia, Toxoplasma gondii,* and *Cryptosporidium* species; geographically more limited protozoa are *Entamoeba histolytica* (which cause amoebic dysentery), *Leishmania, Trypanosoma,* and *Plasmodium* (which cause malaria).

The helminths are worms and are particularly common in children. Among the fairly common helminths are *Enterobius vermicularis* (pinworms) and *Toxocara canis,* a dog roundworm that can be transmitted to children. Other helminths have a more limited geographical distribution such as *Schistosoma,* which require snail intermediates.

Microorganisms, especially bacteria, have genetic mechanisms that allow them to change according to differences in the environment, differences in the host, and exposure to antimicrobial drugs. Very often the same disease can be caused by a number of organisms, and the organisms that commonly cause disease vary over time. For instance, the common causes for neonatal septicemia and meningitis have varied over the decades. Group A streptococcus was a common cause in the 1930s and 1940s, *Staphylococcus aureus* became a problem in the 1950s, and then *Escherichia coli* appeared followed by group B streptococcus. In the 1980s, *Staphylococcus epidermidis* and nontypable (unencapsulated) *Haemophilus influenzae* also emerged as causes of neonatal septicemia and meningitis. Microorganisms that were thought to be nonpathogens may change their pattern and become pathogens. Genetic material is transferred between microorganisms and can result in transfer of properties such as toxin production or antibiotic resistance.

Not only do known microorganisms change but microorganisms are newly recognized, and connections between different microorganisms and specific disease entities continue to be discovered. For instance, human parvoviruses were first discovered in the mid-1970s. In the early 1980s, they were recognized to infect red cell precursors and cause profound anemia (aplastic crisis) in children with hemolytic anemias such as sickle-cell disease; they were recognized to cause erythema infectiosum (fifth disease) and to infect fetuses. Rarely a new microorganism appears by a mutation of an old one that gives it completely new characteristics. A virus may spread from other species to humans. For instance, human immunodeficiency virus (HIV), or a variant of it, is thought to have originally existed in monkeys. A fair number of animal viruses are known to have the potential to move into the human population. Thus, despite the advances resulting from antimicrobials and vaccines, there are so many organisms with so much capacity for adaptation and change that infectious diseases will always be a problem. The diseases that pose the greatest problems change with time, and the scourges vary over the decades and centuries.

Whether or not microbial invasion occurs and whether or not it results in disease depends not only on the organism but also on the host. Thus, infectious diseases must be thought of in terms of the host–virus interactions.

The human immune response is comprised of humoral immunity and cellular immunity. Humoral immunity results from antibodies, which are secreted by immune cells. Cellular immunity (cell-mediated immunity) involves participation of cells in recognition and destruction of microbes or of infected cells. The lymphocytes, which are one type of leukocyte (white cell), are divided into B cells and T cells, which are responsible for different aspects of immune defense. The different immune cells interact with each other to develop and control the immune response. The B cells are mainly concerned with making antibodies. Antibodies are proteins called immunoglobulins and include IgM, IgG, IgA, IgD, and IgE. IgM is made early in infection and decays fairly rapidly, so its presence usually indicates active or recent infection. IgG (γ-globulin) is important for longer-term protection and is the main immunoglobulin in serum. If a microbe causes infection, this is usually reflected by the appearance of antibodies (seroconversion) to the specific microbe or by an increase in the antibodies to the specific microbe over the course of an illness. IgA protects the mucosal surfaces and is present in human breast milk and colostrum. The T cells are responsible for much of the cell-mediated immune response and both kill cells carrying foreign antigens and regulate the activities of the B cells and other cells. An antigen is any structure that can be recognized by the immune system and stimulate an immune response. The immune system also includes the polymorphonuclear leukocytes, which phagocytize and destroy organisms, and the monocytes and macrophages, which phagocytize organisms, process antigens, and interact with the rest of the immune system. Complement is the name for a group of serum proteins that undergo a cascade of reactions to make products that assist in the immune response and, in some cases, cause lysis of microorganisms or infected cells. [*See* IMMUNE SYSTEM; LYMPHOCYTES; MACROPHAGES.]

Some aspects of infectious diseases result from the immune response. The immune system recognizes cells that are infected by viruses and in destroying such cells contributes to cell death and to disease. The immune system causes much of the inflammation associated with infections. Some viruses and bacteria contain molecules that are at least partially similar to human molecules, and the immune system recognizes these similar parts; this is referred to as molecular mimicry. Molecular mimicry can confuse the immune system, so that the antibodies a person makes to an invading microorganism may cross-react with his/her own tissues. It is also postulated that there should be the correct balance of humoral immunity and cellular immunity. An imbalance may be relevant to the pneumonias caused by respiratory syncytial virus and *Mycoplasma pneumonia*. This is of concern when designing new vaccines.

Age is a significant factor in terms of the characteristics of the host. Hence, many of the infectious diseases commonly seen in the pediatric age group are different from those seen in adults. The most outstanding difference between the pediatric and adult population is that the pediatric population has yet to undergo the exposures that cause development of immune defenses. Thus, a part of childhood is inevitably spent with the infections that are necessary to gain these immune defenses. In addition to all the immunologic differences, other differences—physiologic and anatomic—have an influence on infectious diseases in children. While some infectious diseases are more severe in children than in adults, other diseases such as poliomyelitis and Epstein–Barr virus are milder if contracted at a very young age. In countries without modern sanitation, poliomyelitis is usually contracted in the first few years of life, when it rarely produces paralytic disease. As sanitation improves, it is contracted later in childhood, when serious disease is more frequent. Therefore, it began to be recorded as a dread disease only in the nineteenth century.

Some diseases have been colloquially referred to as childhood diseases. These diseases are infectious enough that exposure usually occurs during childhood and pathogenic enough that clinically evident disease almost always results from infection and induce life-long immunity. Thus, diseases such as varicella (chicken pox) and measles have been assumed to be an inevitable part of childhood. However, many other organisms and diseases also occur mainly in children, and some of these will be discussed. Fetuses, newborns, infants, and children have very different exposures, and the status of the immune system changes with age. Therefore, pediatric infections must be considered not only as different from adult infections but also different within each age group.

II. Fetal Infections

The fetus is separated from much of the external environment by the fetal membranes. The most common source of infection for the fetus is the mother's bloodstream. The infected blood may leak through the placenta to the fetus or cause a local area of placentitis with subsequent transmission from this area to the fetus.

A. Examples of Fetal Infections

Viruses are frequently a cause of fetal infection. Rubella (German measles) is an important fetal infection because infection of the mother during the first trimester of pregnancy can cause fetal defects. It also can cause intrauterine growth retardation or active general disease at birth. Cytomegalovirus (CMV) is a fairly common viral infection and may be acquired *in utero* or after birth. The result of infection *in utero* varies broadly from no significant disease to very severe general disease with hepatosplenomegaly (enlarged liver and spleen), petechiae (spots on the skin), and major neurologic problems, which include microcephaly (small head), cerebral calcifications visible on X-ray, chorioretinitis, and retardation. Severe disease is most likely to occur if the infection is during the first half of pregnancy and

if it is the mother's primary CMV infection. Human parvovirus B19 has been associated with spontaneous abortions, and several cases of aborted fetuses with evidence of widespread parvovirus infection have been reported.

Most children who are infected with HIV, the virus that causes AIDS, acquired the infection from their mothers, although before 1985 some acquired the virus through blood products. The details of fetal and infant infection are not fully understood, but there does seem to be transplacental transmission and the virus is also excreted in human milk. Probably less than half of the infants of infected mothers become infected. There is an interval before the onset of symptoms, but it is usually shorter than that in adults, so AIDS is often diagnosed before 1 yr of age and the majority of cases have been diagnosed by 3 yr of age. Among the problems that occur in children with HIV infection are serious or recurrent bacterial infections, lymphocytic interstitial pneumonitis, which is a chronic pneumonia, pneumonia due to *Pneumocystis carinii,* an encephalopathy that results in delay in development or deterioration of motor and intellectual abilities, *Candida* esophagitis, recurrent diarrhea, cardiomyopathy, and renal disease. [*See* AIDS EPIDEMIC (PUBLIC HEALTH).]

Bacteria and protozoa can also be transmitted from mother to fetus. Syphilis is transmitted to the fetus in the latter half of pregnancy. It is not always symptomatic. Symptoms, if present, include a rash, which involves palms and soles, rhinitis (snuffles), a saddle nose, nephrotic syndrome, and an osteochondritis or periostitis of the shins or arms so the infant does not want to move the affected limb. *Toxoplasma gondii* is a protozoon that infects many species. Human maternal infection usually occurs either through eating insufficiently cooked meat or by the handling of excrement of infected animals such as kittens. In addition to a general illness, it can cause neurologic damage to the fetus with microcephaly and cranial calcifications.

B. Possible Outcomes of *in utero* Infection

(1) Infection can result in the death of the fetus. This can happen before a woman realizes she is pregnant and go unnoticed, or it can occur later and produce spontaneous abortion or stillbirth. (2) Infected infants often have a low birth weight as a result of intrauterine growth retardation. The gestational age of an infant can be estimated by the development of the infant. Thus, it is possible to classify infants as small, large, or normal for their gesta-

tional age. Infected infants tend to be "small for gestational age." (3) Infection of the fetus can be associated with premature birth. (4) Fetal infection can cause developmental anomalies. Deafness, congenital heart disease, retardation, cataracts, and other developmental anomalies can result from rubella contracted during the first trimester. (5) Infants who have been infected *in utero* can have active disease at birth. The severity of illness varies but can be quite marked. Severely infected infants develop signs of disseminated infection, which include jaundice, hepatosplenomegaly (enlargement of the liver and spleen), and pneumonitis. Depending on the organisms, there may be lesions of the skin or mucous membranes, meningoencephalitis, heart lesions, and particularly eye involvement such as chorioretinitis. (6) Some infections persist after birth. Rubella, herpes simplex, CMV, toxoplasmosis, and syphilis can persist as active infections after birth. Rubella and CMV infections may persist for several years after birth during which time the child is producing virus and is infectious. (7) Many infants infected *in utero* show no deleterious effects of the infection at birth. For those infections that can be persistent, however, problems such as hearing loss may develop later from the continuing subliminal infection.

III. Infections of the Neonate

As a newborn, an infant can become ill with infections acquired prior to, during, or after delivery. Because the exposures during gestation and delivery are different from those in the general environment, many microbes that infect neonates are different from those that infect the older infant and child. The earlier an infection occurs, the earlier the symptoms are likely to become evident and the more severe the illness is likely to be. The neonate does not respond to serious infection as does the older infant or child, rather the neonate responds nonspecifically with many of the same signs that occur when he or she has noninfectious problems. The neonate may have a low temperature or a high temperature, signs of respiratory distress, apnea (periods of cessation of breathing), increased or decreased heart rate, or poor feeding.

A. Immune Defenses

By 8 wk gestation, the liver is hematopoietic (makes red and white blood cells). At 5 mo gestation, the bone marrow takes over this function. At birth, the neonate has small amounts of lymphoid tissue and a fairly large thymus. Thus, even the most premature infant has all the components of the immune system. Nevertheless, the immune system of the neonate is underdeveloped and immature, so the neonate is very susceptible to infection. In many ways, the infected neonate should be considered an immunocompromised patient. IgG is transported across the placenta, but IgM and IgA are not transported. The infant can make IgG as early as 12 wk gestation, but almost all of the IgG in the umbilical cord blood at birth is maternal. IgG has a half-life of roughly 25 days, so after birth the maternal IgG slowly disappears and must be replaced by the IgG synthesized by the infant. This means the levels drop during infancy and for the term infant reach their lowest level about 3–4 mo of age; thereafter, they rise again. The premature infant starts with lower levels; therefore, the IgG may drop to very low levels before the infant's own synthesis raises these levels again. Because IgM does not cross the placenta, any IgM in the cord blood has been synthesized by the infant, and high levels may be a reflection of intrauterine infection. IgA is mostly synthesized after birth. [*See* IMMUNE SYSTEM.]

Some protection is afforded by breast-feeding. Both human milk and colostrum contain not only lymphocytes, phagocytic cells, and immunoglobulins, particularly secretory IgA, but also a variety of other factors that inhibit microbes. The secretory IgA in milk and colostrum is particularly useful in protecting the infant against enteric organisms. The negative aspect of the immune cells in milk is that some viruses such as CMV and HIV are found in white cells, so the infant can be infected with these viruses as a result of breast-feeding.

B. Infections Acquired Around the Time of Delivery

During delivery, the infant is exposed to organisms in the maternal genital tract. If the infant is born soon after rupture of the membranes, he or she is exposed to the organisms present in the cervix and vaginal canal. If a significant delay between rupture of the membranes and delivery occurs, the vaginal organisms have a chance to migrate to the uterus and perhaps produce an infection of the amnion, which is a placental membrane. In this case, the infant is likely to be infected also. The vaginal canal contains many kinds of bacteria both aerobic and anaerobic. Historically, group A streptococcus was the cause of puerperal fever and often of infant infection. Group B streptococcus and *E. coli* have emerged as major pathogens of neonates in recent decades. Among other organisms that women may harbor and that the infant may acquire during birth are herpes simplex, CMV, hepatitis B, *Candida albicans,* nontypable (unencapsulated) *H. influenzae, Neisseria gonorrhoeae,* and *Chlamydia trachomatis.*

Ophthalmia neonatorum (neonatal conjunctivitis) is a suppurative infection of the conjunctiva occurring in an infant <30 days of age. It has long been recognized as a disease that can be acquired by exposure to *Neisseria gonorrhoeae* (gonococcus) during delivery. Gonococcal conjunctivitis usually occurs in the first week after birth, is often severe, and can produce corneal ulcerations. In the era before antibiotics, this commonly resulted in blindness, and the practice of putting silver nitrate solution into a neonate's eyes at birth as prophylaxis for gonococcus was introduced in 1884. Exposure to *C. trachomatis* during parturition may be followed by conjunctivitis 5–14 days after birth or by pneumonitis in the first few months of life. *Chlamydia* conjunctivitis is less severe than gonococcal conjunctivitis but is now recognized to be fairly common.

Herpes simplex virus type 2 is another infection that can be transmitted from the mother to the infant during delivery. It can give a disseminated infection, which may be fatal or result in severe neurologic sequelae. Milder forms of this infection also occur. [*See* HERPESVIRUSES.]

C. Postnatal Infections

Once delivered, the infant becomes exposed to the microorganisms common to his or her community. The infant's closest contact will be his or her mother, and the infant will be lacking antibodies to any infections she acquires after delivery. For infants who must remain hospitalized, the epidemiology of the neonatal intensive care nursery is always a cause for concern, because many other infants may remain for extended periods. Many of these infants require invasive procedures such as insertion of arterial and venous catheters and endotracheal intubation, and these increase the opportunities for infection. If a virulent organism is introduced by one infant, it may colonize the staff or other infants and be a source of difficulty in very vulnerable prematures. For these reasons, there is usually close epidemiologic surveillance of hospital nurseries.

Many infections can be acquired prenatally, perinatally, or postnatally. Hepatitis B can be acquired from the mother perinatally but often can be prevented by γ-globulin and immunization at birth if the infection in the mother is recognized. Infection at this age frequently results the infant becoming a permanent carrier, and some of these individuals develop cirrhosis or hepatocellular carcinoma as adults. Hepatitis B can also be acquired from other children in the child's household after birth. While CMV acquired *in utero* can result in serious disease, CMV acquired perinatally or postnatally is usually asymptomatic. CMV can also be transmitted by blood transfusions, and this can cause serious illness to the premature infant. If the mother is infected with enteroviruses, especially coxsackie B viruses or echoviruses, the infant can be infected *in utero,* during delivery or after delivery. Enterovirus infections acquired during the perinatal period can be quite severe. In the neonatal period, enteroviruses are often acquired from the community and account for many of the hospitalizations in this age group.

D. Sepsis Neonatorum

Sepsis neonatorum, or neonatal sepsis, is defined as disease of infants <30 days of age that involves bacteremia (bacteria in the blood). Bacteria carried by the blood may settle and grow in other organs, and in neonates the bacteria often seed the meninges (the membranes that cover the brain and spinal cord) to cause meningitis. Because the signs of infection in the neonate are general symptoms that can also occur with noninfectious problems, diagnosis of sepsis with just the clinical picture is often difficult. Meningitis also often gives a similar nonspecific clinical picture in neonates; therefore, the two are usually considered simultaneously. Because neonatal sepsis and meningitis are serious, progress rapidly, and give a clinical picture that is not distinctive, infants with signs and symptoms that could result from sepsis are given what is known as a septic work-up, which includes cultures of blood, spinal fluid, and urine. They are then treated with antibiotics until the cultures indicate whether or not infection is present. The most common causes of neonatal sepsis in the 1980s were group B streptococcus and *E. coli. Listeria monocytogenes* can be a cause and is an example of an organism that infects neonates but rarely causes disease in older children unless they are immunocompromised. [*See* MENINGES.]

E. Localized Neonatal Infections

Localized infections can result from the seeding of an organ by bacteria in the blood or may result from entry of bacteria at the local site. In the latter case, the local infection may cause bacteria to enter the blood, so bacteremia may also occur. Neonatal pneumonia may result from fetal infection, from inhalation of infected fluid from the mother's reproductive tract during parturition, or from the environment after birth. Most osteomyelitis involves growing bone and starts in the metaphysis (the part of the shaft of the bone next to the epiphysis, which is at the end of the bone). The most common organism is *S. aureus*. In infants under 1 yr old, the blood vessels go from the metaphysis through the epiphysis; therefore, the bone "growth plate" is at risk and a bone infection can spread to the joint to cause arthritis. Some skin and soft tissue infections are specific to the neonate. Funisitis is an infection of the umbilical cord, and omphalitis is an infection of the umbilicus or surrounding tissues. They are most often due to *S. aureus* or group A streptococcus. Breast abscesses (mastitis) occur in term infants most frequently in the second or third week of life and are most commonly due to *S. aureus*.

As can be seen from the above, infectious diseases of the fetus and neonate are very different from those of the older infant and child. Many diseases depend on the maternal infectious diseases and, thus, the health of the mother and prenatal care are important factors in infectious diseases in this age range.

IV. Infections of Infants and Toddlers

Infants and toddlers are exposed to the microbes in society at large. Many of the problems in this age group are due to immunologic inexperience and anatomic limitations. Otitis media, respiratory infections, and gastrointestinal infections are very common among infants and toddlers. The infections that create emergencies in infants and children are those that compromise the airway, those that involve the central nervous system (meningitis and encephalitis), and those that create shock such as infections due to *Neisseria meningitidis*.

Children in this age group can respond immunologically to most organisms. One exception appears to be the encapsulated bacteria. Several bacteria such as *H. influenzae*, *N. meningitidis* (meningococcus), and *S. pneumoniae* (pneumococcus) may have capsules of polysaccharides, which are repeating sugar units that are often negatively charged. The different capsules determine the bacterial types. Nontypable *H. influenzae* have no capsule. For *H. influenzae*, type b is the type usually associated with serious disease. Unless antibody to the polysaccharides is present, these polysaccharide capsules inhibit phagocytosis (ingestion by white cells) of the bacteria and allow the bacteria to become invasive. For reasons that are not understood, children <2 yr old have difficulty producing antibodies to polysaccharides; therefore, once the maternal antibodies are lost, they are specially vulnerable to the encapsulated bacteria. These bacteria can colonize the upper respiratory tract without producing disease; however, if they penetrate the respiratory mucosa and gain access to the bloodstream, they multiply and seed different organs including the meninges. As children get older, they begin to produce antibodies to the bacterial capsules. *Haemophilus influenzae* infections are a major problem up to school age and especially <2 yr of age, but they are rarely seen in older children and young adults. Susceptibility and ability to produce antibodies, at least for *H. influenzae*, also depends on a child's genetic makeup.

Despite the fact that infants and toddlers appear to respond immunologically to most microorganisms, they are still immunologically inexperienced, and this leaves them susceptible to many common infections.

A. Otitis Media

Otitis media is an inflammation of the lining of the middle ear. It is one of the most common pediatric problems, especially in the first 2 yr of life, although it continues to be a problem until about school age. Some children seem particularly prone to otitis media and have repeated infections up to school age. This becomes a considerable problem to the parents and is of concern because of the possibility of hearing deficit for considerable periods and associated speech and learning problems. Otitis media is related to the eustachian tube dysfunction. The eustachian tube, which connects the middle ear and nasopharynx, should equilibrate middle ear and atmospheric pressure, protect against reflux of nasopharyngeal secretions, and drain the middle ear into the nasopharynx. The eustachian tube of the young child is not only smaller but also shorter and more horizontal than that of the older child. If the eustachian tube either becomes obstructed or is too wide open, an opportunity is presented for bacteria from the nasopharynx to replicate in the middle ear. The most common pathogenic organisms identified in otitis media are *S. pneumoniae,* nontypable *H. influenzae,* and *Branhamella catarrhalis.* Prior to the advent of antibiotics, it was not unusual for the eardrum to rupture during acute otitis media. Because the lining of the middle ear is continuous with that of the mastoid, mastoiditis commonly accompanies otitis media. Again, since the advent of antibiotics, serious infection of the mastoids and possible complications such as meningitis, brain abscess, labyrinthitis, and facial nerve paralysis are rare.

B. Respiratory Infections

Several kinds of respiratory infections are major problems usually only in infants and toddlers. The airways are small in this age group, so inflammation can compromise the air flow. When an infant or toddler has difficulty getting enough air, he or she starts working harder at breathing. This is evidenced by tachypnea (increased rate of breathing) and retractions (pulling in of the regions between and around the ribs). The increased work of breathing creates a need for more oxygen, causing a vicious circle. While respiratory infections are often described in terms of the anatomic location of the infection, many organisms can infect several locations. For instance, the viruses causing croup can spread to cause pneumonia. In children, aspiration of a foreign body always must be considered along with infections as a possible cause for respiratory difficulties. [*See* PATHOPHYSIOLOGY OF THE UPPER RESPIRATORY TRACT.]

1. Upper Airway Infections

Upper airway obstruction, which occurs during infections, is a pediatric emergency. A common upper airway infection is viral croup (laryngotracheitis). Croup is usually caused by one of three types of parainfluenza virus or by respiratory syncytial virus (RSV). Outbreaks of these viruses occur between late fall and early spring. Parainfluenza virus type I is the most common cause of croup, and its peak incidence is in the second year of life. The infection starts in the nasopharynx and then spreads down the respiratory tract. When it reaches the subglottic region, the resultant inflammation may take up a good part of the small airway in this age group. The voice is hoarse and the child has a characteristic cough that sounds like a seal barking. Stridor, which is a crowing sound usually heard on inspira-

tion, suggests some respiratory difficulty. Many cases are mild and can be treated with mist at home, but the severe cases are life-threatening. While older children and adults can be reinfected with parainfluenza viruses and respiratory syncytial virus, partial immunity and a larger airway make it a much less severe infection. Thus, what seems to be just a cold in an older child or adult, when transmitted to an infant or toddler can cause more serious disease.

To be distinguished from croup is epiglottitis, which is an inflammation of the epiglottis and associated structures. It occurs in the infant and toddler age group but is more common in older preschool children. It is usually due to *H. influenzae* type b and is associated with a high fever and the appearance of being quite sick. Because epiglottis makes swallowing difficult, the child often drools and refuses to talk. In contrast to croup, which usually does not progress to serious compromise of the airway and can be treated at home, epiglottitis can progress quite rapidly and constitutes an emergency.

A now rare cause of airway obstruction is diphtheria, which is caused by *Corynebacterium diphtheriae.* It is seen mainly in regions such as Seattle, where there is cutaneous diphtheria among the indigent population on ''skid row'' with occasional transmission to children. Diphtheria toxin can cause a membrane to be formed in the upper respiratory tract, which with inflammation can cause asphyxiation. Diphtheria toxin also can cause myocardial damage.

2. Bronchitis

In children, bronchitis usually follows upper respiratory symptoms and is associated with cough and is most commonly due to viruses including the parainfluenza viruses, influenza viruses, adenoviruses, RSV, and measles. In older children, it can be caused by *M. pneumoniae.*

Pertussis (whooping cough), caused by *Bordetella pertussis,* is one of the few bacterial causes of pediatric bronchitis. At the beginning of the twentieth century, pertussis was an important cause of death. With the use of pertussis vaccination after World War II, its incidence greatly decreased, but it is still a serious illness in children <1 yr of age; it is very contagious. Adults who were vaccinated as children are susceptible and often do not recognize the illness, which may be mild or atypical, so they are likely to unknowingly transmit it to infants. It starts with what seems to be a cold (catarrhal stage) and then, after 1–2 wk, progresses to the paroxysmal stage, during which there are paroxysms of coughing. These paroxysms may be followed by a "whoop," which is the forceful inhalation of air through a partially closed glottis. The whoop is not common in infants; complications in infants include apnea, pneumonia, and atelectasis (incomplete lung expansion). Some central nervous system complications are seizures or an encephalopathy, which can leave permanent neurological damage. The increased pressure that occurs during the paroxysms of coughing can result in hernias or in rupture of blood vessels, including cerebral vessels.

3. Bronchiolitis

Bronchiolitis, as the name implies, is an infection involving the bronchioles. Bronchiolitis is a problem mainly in infants because of their small airways. It is associated with wheezing and tachypnea; apnea also occurs in the small infant. It is most often caused by RSV but also can be caused by other respiratory viruses such as the parainfluenza viruses.

RSV is a major cause of hospitalization for infants. When infants are infected for the first time, it is not unusual for them to develop lower respiratory tract disease—bronchitis, bronchiolitis, or pneumonia. Infants <6 mo old are particularly affected. Maternal antibody does not seem to prevent the disease and possibly makes it worse. In the 1960s, a vaccine was made from inactivated virus. It caused production of antibodies, but the children not only got RSV but were considerably sicker. Efforts are being made to develop a live attenuated vaccine.

4. Pneumonia

When pneumonia occurs at <20 wk of age and is not accompanied by fever, it is likely to be due to *C. trachomatis* acquired during parturition. RSV and the parainfluenza viruses also commonly cause pneumonia in infants; less commonly, adenoviruses or influenza viruses can cause pneumonia. While viruses cause the majority of the pneumonias observed in this age group, bacterial pneumonia also occurs. The most common causes after the neonatal period and <5 yr of age are the encapsulated bacteria *H. influenzae*, type b, and *S. pneumoniae*.

C. Meningitis and Encephalitis

Meningitis is an inflammation of the meninges. Meningitis can be caused by bacteria, viruses, fungi, or parasites. Important for the diagnosis of meningitis is examination of the cerebrospinal fluid, which is obtained by lumbar puncture.

The most common causes of bacterial meningitis in infants and toddlers are the encapsulated bacteria *H. influenzae, S. pneumoniae,* and *N. meningitidis.* Meningitis caused by bacteria is life-threatening. Most bacterial meningitis occurs in children <5 yr of age and is especially common in the child 6–12 mo of age. While *H. influenzae* meningitis rarely occurs in anyone >5 yr, *N. meningitidis* is also seen in teenagers and can cause outbreaks in military camps. Adults also get *S. pneumoniae* meningitis.

Infection in the central nervous system produces cerebral edema. Since the brain cannot expand beyond the skull, the intracerebral pressure increases. In infants, this is reflected by a bulging fontanelle (soft spot). The increased pressure compromises the neurological tissue, and, if the pressure approaches blood pressure, it slows cerebral blood flow. In addition, components of the bacteria are toxic. The cerebrospinal fluid in bacterial meningitis usually contains bacteria and many white blood cells with a predominance of polymorphonuclear leukocytes. Even with antibiotics, some children die and some have long-term neurologic deficits. One of the most common complications is sensorineural hearing loss.

Aseptic meningitis is an inflammation of the meninges in which no bacteria can be isolated and in which the white cells in the cerebrospinal fluid are predominantly lymphocytes and generally fewer in number than in bacterial meningitis. Aseptic meningitis occurs throughout childhood and also in adults. Aseptic meningitis is usually due to viruses, especially enteroviruses, but can result from organisms such as spirochetes (e.g., Lyme disease) or brucella. Before the mumps vaccine, mumps was a common cause of aseptic meningitis. Generally, when viruses produce aseptic meningitis, the disease is self-limited and permanent sequelae are rare.

Encephalitis is an inflammation of the brain. About 1 in 1,000 cases of rubeola (measles) is followed by encephalitis, which was one of the concerns about measles that led to the development of the vaccine. In many cases of encephalitis, a causative agent is never identified. Most encephalitides in which an agent is identified are caused by viruses including the enteroviruses, herpes simplex virus, and the arboviruses, which are arthropod-borne viruses that have an animal host but can be spread to humans by arthropods such as mosquitoes. The age range infected and the likelihood of death or long-term neurologic sequelae depend on the organism causing the encephalitis.

D. Gastroenteritis and Diarrhea

Before modern sanitation in the developed nations and currently in the Third World, a leading cause of infant mortality is diarrheal disease and consequent dehydration. When sanitation is poor, breast-feeding is important because it is a supply of uncontaminated milk and also offers some immunologic protection against intestinal organisms.

The word diarrhea derives from the Greek word meaning to flow. Normally, in addition to the fluids that are taken in orally, the intestine also receives salivary, gastric, pancreatic, and biliary secretions. It absorbs most of these liters of fluid and electrolytes (the body salts such as sodium, potassium, bicarbonate, and chloride), so that very little is lost in the stool. If the process of adsorption is disturbed, considerable amounts of fluid and electrolytes can be lost. In addition, the oral intake of a child with a gastrointestinal infection is usually poor. Infants, of course, have smaller absolute amounts of water in their bodies than do older children or adults. In addition, a greater proportion of their body weight is fluid, and they have a higher surface-to-volume ratio. Therefore, an infant can become critically dehydrated much more quickly than an older child or adult. A great enough fluid loss can lead to shock and death.

Fluid loss can result from damage to the intestinal mucosa so the intestine cannot absorb fluids. This occurs in diarrhea due to viruses, which are the most common cause of pediatric diarrhea in developed nations. Viral gastroenteritis usually involves self-limited vomiting and diarrhea. In the 1970s, when electron microscopy became easily available, different viruses (e.g., rotaviruses, Norwalk virus, caliciviruses) were observed in the stools of individuals during different outbreaks of viral gastroenteritis. Rotaviruses are a major cause worldwide of gastroenteritis in children between the ages of 6 mo and 2 yr. In temperate climates, rotaviruses cause winter outbreaks of gastroenteritis. This infection is so common that studies on potential vaccines are being carried out.

Some bacteria excrete toxins that cause secretion of fluids and salts. The toxins cause a change in a cellular protein. This change leads to a series of biochemical reactions that keep the electrolyte secretory mechanisms of the cell active when they should not be. The resulting intestinal electrolyte secretion, which is accompanied by water loss, results in a copious diarrhea. The diarrheas caused by organisms that produce toxins but do not invade, such as *Vibrio cholerae* and some strains of *E. coli*, are called secretory diarrheas.

Other bacteria that produce diarrhea are invasive (i.e., they are able to enter and move through tissues). Among the organisms that commonly invade the bowel are *Shigella*, enteroinvasive *E. coli*, *Salmonella*, *Campylobacter*, and *Entamoeba histolytica* (which causes amebic dysentery). The invasive diarrheas are seen in infants in developing nations and also occur in children in the developed nations, particularly in day-care centers. When these bacteria invade the bowel mucosa, there is often crampy abdominal pain, tenesmus (straining that is often ineffectual or painful), fever, and general malaise. The stool may have blood and mucus. The disease caused by invasive bacteria is often referred to as dysentery.

E. Examples of Syndromes with Recently Discovered and Unknown Infectious Agents

Roseola infantum (exanthem subitum) is a disease usually seen in children between the ages of 6 mo and 2 yr that was long thought to be infectious. It is associated with several days of very high fever, and then a rash occurs when the temperature drops. This age distribution of the infection suggests that there is protection from maternal antibody during the infants' first months and that the infectious agent is common. Human herpes virus, type 6 (HHV-6), was first discovered in 1986 when it was found in AIDS patients, where it is probably an opportunist. HHV-6 is probably the cause of roseola infantum, because in 1988 it was isolated from children with roseola and seroconversion for HHV-6 was shown to occur during roseola infection.

Kawasaki syndrome has not had a causative agent identified but has an epidemiology that is consistent with its being an infectious disease. It was recognized in Japan in 1961 and in the United States in 1974. The median age of the patient is around 2 yr, and most cases occur at <5 yr of age. It is a systemic vasculitis syndrome, which is important because of the cardiac complications.

V. Infections of the Preschool Child

Because children are developing their immunity to the microbes present in their community and frequently have an infectious disease, it is not surprising that child-to-child spread of infections is common. When children leave their family units to be part of a group of children, the number of infections they acquire increases.

A. Diseases Common in Day-Care Centers

The number of children going to day-care centers is rapidly increasing; thus, studies have been made on the transmission of infectious diseases in day-care centers and more needs to be known. Day-care center size seems to be a factor in transmission of infectious diseases with transmission increasing with increasing size. Transmission is highest among infants and toddlers, because they are still in diapers. Diarrheal disease is easily transmitted when infants are not bowel-trained. Shigella, rotavirus, and *G. lamblia,* a protozoon, are commonly found in day-care centers probably because they take a small inoculum to spread. *Cryptosporodium,* a protozoon first recognized as a human pathogen in 1976, also can cause diarrheal outbreaks in day-care centers. Hepatitis A can be transmitted in day-care centers. Infants who are exposed to other children often have seven or eight or more respiratory infections per year. These respiratory infections are passed around in a day-care center. One result is that children who attend day-care appear to be at increased risk of otitis media. In a day-care center, cases of meningitis caused by the encapsulated bacteria are a cause for concern because secondary cases may occur.

Children in day-care centers often transmit infectious diseases to their families. With hepatitis A, children may be relatively unaffected, whereas adults become sick, so the transmission in the day-care center may be identified by illness in the parents. CMV can spread quite widely in day-care centers without producing acute illness in the children, but the children may infect their mothers, which could damage the fetus if the mothers are pregnant and not immune.

B. Exanthems

Exanthems are rashes associated with general illness. The word enanthem refers to a rash on the oral mucosa. Exanthems occur at all ages and do not particularly belong to this age group. The age at which exanthems are likely to occur depends on the infecting microorganism, but the prevalence of several common exanthems such as measles and chicken pox in the preschool and school age groups tends to make people associate exanthems with this age range. Rashes can be produced during infections by several mechanisms. Infectious agents can be brought to the skin by the blood so that a layer of the skin is infected, as is the case with chicken pox and meningococcemia. Bacteria may make a toxin that is liberated in the blood and affects the skin. This occurs with the rash of scarlet fever and the rash that may appear in toxic shock syndrome. Lastly, it is thought that some rashes are due to immune responses to infecting organisms.

Rashes vary in their appearance. They include erythematous, petechial, and vesicular rashes. Erythematous rashes blanch when pressed and can be macules, which are spots that are not elevated above the skin, or papules, which are elevated above the skin. Erythematous maculopapular (with both macules and papules) rashes occur in many diseases. In measles, an enanthem called Koplik spots appears before the exanthem. The maculopapular rash of measles often coalesces. Measles used to be one of the most common maculopapular rashes; thus, this kind of rash is often called morbilliform (morbilli means measles). Since the advent of measles vaccine, a more common cause of maculopapular rashes is enteroviruses. Petechiae are flat, round red or purple small discrete spots caused by hemorrhage from capillaries or small blood vessels. While there are many reasons for petechiae, in the presence of fever the possibility of bacteremia and especially meningococcemia (which is a serious and rapidly progressing disease) must always be considered. Purpura means hemorrhage into larger areas of the skin than with petechiae and looks somewhat like a bruise. Purpura can be due to the same microbes that cause petechiae. Vesicular rashes begin as a papule and progress to a vesicle (which is a blister that contains fluid) and then to pustules or rupture to produce an ulcer with a crust or scab. The rash of varicella (chicken pox) is vesicular. Urticaria (hives) and erythema multiforme (redness of the skin with many forms) occur during infections and are thought to represent an allergic reaction to the microbes. Rashes can start as erythematous and then progress to petechial or vesicular.

Rashes become distributed on the body in different patterns. The reason for this is not known, but it has diagnostic utility. In the case of measles, the rash appears first on the forehead and works its way down to the feet. The rash of Rocky Mountain spotted fever, on the other hand, initially involves the wrists and ankles and then spreads toward the center (centripetally) to involve the trunk. Only a few rashes appear on the palms and soles. Even in the context of the whole clinical picture, it is often not possible to distinguish diseases that cause rashes from each other without the help of cultures or serology. For instance, it is not possible to diagnose rubella on the appearance of the rash, which can be similar to that caused by enteroviruses.

VI. Infections of the School-Age Child

Whether many infections occur at preschool age or later depends on whether the child goes to day-care or nursery school and whether or not the child has older siblings to expose him or her to infections. Some diseases, however, have a peak incidence at school age, even among children who have had many exposures at early ages.

A. Mycoplasma Pneumoniae

An important cause of pneumonia in the school-age child is *M. pneumoniae*. The first mycoplasma species was isolated from cattle in 1898, but it was not until 1962 that what was known as primary atypical pneumonia was shown to be due to *M. pneumoniae*. Not all infections with *M. pneumoniae* are symptomatic. *Mycoplasma pneumoniae* can reinfect a child who has been previously infected, so the immunity produced by the first exposures does not give complete protection. It has been postulated that infection of the school-age child tends to result in pneumonia more frequently than in the young child because the immune response of the school-age child, while not adequate to prevent infection, does contribute to the pathogenesis.

B. Streptococcal Disease

Streptococci were identified about a century ago, and much is known about the individual bacterial components and the secreted toxins and enzymes; however, why the streptococci vary in virulence and how they cause rheumatic fever and glomerulonephritis, inflammatory disease of the kidney, still is not clarified. Streptococci can be divided into groups according to serologic differences in the cell wall carbohydrates. The streptococcal groups can be subdivided into types according to serologic differences in the M protein, which is a major virulence factor and inhibits phagocytosis. Some strep-

tococci secrete substances that cause β-hemolysis (a clear zone around the colony) on culture plates containing blood.

Pharyngitis is a general term for inflammation of the pharynx (the region between the mouth and nose and the esophagus) and encompasses tonsillitis. Pharyngitis can be due to many microorganisms. When there is associated rhinitis, pharyngitis is more likely to be due to a virus. Pharyngitis without rhinitis can be due to many microbes including adenoviruses and Epstein–Barr virus, *M. pneumoniae*, *Corynebacterium haemolyticum*, and *C. diphtheriae*; however, *Streptococcus pyogenes*, which is a group A β-hemolytic streptococcus, is a frequent cause of pharyngitis. Pharyngitis caused by *S. pyogenes* (strep throat) is sometimes followed by rheumatic fever or glomerulonephritis. Treatment of streptococcal pharyngitis with antibiotics can prevent rheumatic fever. Therefore, identifying streptococcal pharyngitis with cultures or rapid diagnostic methods is important. Rheumatic fever often involves a carditis, which leaves the child with damaged cardiac valves. How streptococcus triggers rheumatic fever is not known. Rheumatic fever and glomerulonephritis appear to be related to the immune response to *S. pyogenes* because they occur after time for development of antibodies. Streptococcal M proteins have antigenic cross-reactivity with the myocardium (i.e., antibodies to M protein react to myocardium). Thus, molecular mimicry may perhaps be a factor in the pathogenesis of rheumatic fever. The genetics of the child also has an influence on whether or not rheumatic fever develops after a streptococcal infection.

Streptococcal infections are good examples of how the diseases caused by a microbe can vary over the decades. Different streptococci produce different toxins. Some produce an erythrogenic toxin, and one of the effects of this toxin is the production of the rash of scarlet fever (scarlatina). At the turn of the century, scarlet fever was a very serious illness, but, even before the advent of antibiotics, it and other streptococcal infections became much less severe, so that scarlet fever is not now considered a major illness. With the use of penicillin for streptococcal pharyngitis, the incidence of rheumatic fever dramatically dropped in developed nations. During the late 1980s, however, reports appeared of outbreaks of severe streptococcal infections and independently of outbreaks of rheumatic fever. This raises the question of whether or not the virulence of streptococci is now again increasing.

C. Reye Syndrome

Reye syndrome is a rare disease first described in 1963. It usually follows a viral illness and occurs most commonly in school-age children, less commonly in preschoolers, and very rarely in adults. Although it can follow a variety of viral infections, it most commonly follows varicella (chicken pox) and influenza, especially influenza B. It starts with intense vomiting; this is followed by neurologic symptoms, which can be mild, such as lethargy, or can progress to various levels of coma or death. Reye syndrome is an encephalopathy (illness of the brain). No evidence indicates immune cells or microorganisms in the brain, so it does not appear to be an encephalitis. The liver is also affected, and the blood ammonia rises. What causes Reye syndrome is not known. In many ways, it is similar to illnesses due to toxins and to aspirin (salicylate) toxicity. One possibility is that as the viruses are broken down in the body, one of their components is toxic. Epidemiologic studies indicated that Reye syndrome is more frequent in children who have received aspirin when they have influenza; therefore, although it is not certain that aspirin plays a role, it is suggested that children sick with influenza or varicella should not take aspirin.

VII. Immunizations

Prevention is a main goal of pediatrics, thus immunization is an important part of pediatric infectious diseases. Immunization can be passive or active.

Passive immunization consists of giving an individual antibodies to a specific organism. The transfer of antibodies from a mother to her infant *in utero* is passive immunization. Passive immunization includes giving antibodies for an anticipated or known exposure to a disease. An example is the use of γ-globulin for hepatitis A exposure. At the beginning of the twentieth century, animals, especially horses, were injected with antigens and the resultant animal antisera were used. Animal serum can cause allergic reactions including immediate acute reactions and a delayed reaction known as serum sickness. Horse antiserum is still used in the management of some of the uncommon diseases such as diphtheria and botulism, although human immune serum is now available for some other diseases such as tetanus. Prior to the advent of antibiotics, serum therapy was used for serious illness such as severe

pneumonia. Research is being carried on to determine if monoclonal antibodies, which are produced by technology developed in the 1970s, could be used as therapy for specific diseases.

Active vaccination consists of giving the person a microbe or some microbial components in a form that is not harmful but that stimulates production of antibodies that protect against the microbe. The basic concept of active vaccination has existed for centuries. The earliest vaccines focused on smallpox. Prior to the nineteenth century, infection with mild strains of smallpox was used as vaccination, which, of course, was very hazardous. In 1798, vaccination with vaccinia virus, which is cowpox virus, was introduced. Smallpox has no animal reservoir and does not cause persistent infection in humans (i.e., has no human reservoir); therefore, the World Health Organization was able to eradicate it as a disease in nature in the 1970s by intensive worldwide surveillance, containment, and vaccination (although it still exists in several laboratories). Most of the currently used vaccines were first used between the 1940s and 1990s. Considerable research is ongoing to develop new vaccines. Thus, for many diseases, experience with the vaccines is limited, and both the nature of the vaccines and the scheduling of vaccinations are still in flux.

Bacterial vaccines consist of a suspension of the inactivated bacterium (e.g., *B. pertussis*), a component of the bacterium (e.g., the capsular polysaccharide of *H. influenzae*), or a toxoid, which is a bacterial toxin that has been denatured by formaldehyde so it is no longer active.

Viral vaccines contain either inactivated (killed) viruses or attenuated viruses. Inactivated viruses cannot replicate but stimulate the formation of antibodies to the components of the virus. Some of the early experimental inactivated viral vaccines resulted in sensitization of the recipient so that the disease was more severe upon subsequent exposure. Attenuated virus vaccines contain live viruses that have been modified (attenuated) so that they do not give disease. The concern with attenuated viruses is that they could revert to their virulent form and that in immunocompromised patients they may cause disease. However, because attenuated viruses replicate in people as virulent viruses do, they are thought to stimulate immunity that is similar to that produced by natural infection.

The immunizations included in routine well-child care in developed countries have caused some childhood diseases to become quite rare. A vaccine combining diphtheria, tetanus, and pertussis (DTP) has been used since the early 1950s and is given to infants. The diphtheria and tetanus component are toxoids. The pertussis vaccine was developed in the 1940s and 1950s and is prepared by treating the whole bacterium with heat and chemicals. It appears to be responsible for most of the reactions to the DTP vaccine. The less serious reactions include local pain and swelling at the injection site, fever, and irritability. Some infants cry persistently or have very high fevers. More serious complications are quite rare but do occur; these include convulsions, collapse or shocklike states, and encephalopathy, which in rare cases results in permanent neurologic damage. The rate of persistent neurologic damage in previously normal children was estimated to be around 1 in 310,000. Because of concern over these reactions, popular acceptance of the vaccine greatly decreased in Britain during the late 1970s. This decrease in vaccine use was followed by an outbreak of pertussis, and the mortality and morbidity of pertussis infection far exceeded that of the immunization. During the late 1980s, work on an acellular pertussis vaccine (one that uses specific pertussis antigens) that causes fewer reactions got underway, and in the 1990s such a vaccine may replace the whole cell vaccine.

Poliovirus enters the alimentary tract, multiplies locally in the adjacent lymph nodes with excretion from the intestine, and then spreads to the bloodstream and the nervous system. The two forms of polio vaccine are the inactivated polio vaccine (IPV) and the trivalent oral polio vaccine (OPV or TOPV). IPV was developed first and was used during the late 1950s. As the name indicates, it uses inactivated virus and is given by injection. OPV contains attenuated virus that does not affect the nervous system. The advantage of OPV is that it initiates infection in the alimentary tract, where it stimulates local immunity, and, because it is also excreted, it tends to immunize the population. The disadvantage is that in rare instances it may revert to the virulent form to cause neurologic disease, usually in an adult. Therefore, IPV rather than OPV should be used in an adult or in a child who is immunocompromised or has a family member that is immunocompromised. Some countries routinely use IPV and some use OPV.

γ-Globulin has been used for at least 50 years to modify or prevent measles infection of susceptible people after exposure. Since introduction of the first measles vaccine in the 1960s, the incidence of measles has greatly decreased. Measles vaccination has been through many changes, and as of 1990 the vaccination procedure is still in flux. From 1963 to 1967, a killed measles vaccine was used, which caused some recipients, when later exposed to measles, to develop an atypical and severe form of the disease. Early live measles vaccine needed further attenuation, which was undertaken about 1970. Thereafter, there were difficulties with inactivation upon storage, and a stabilizer was added in 1979. The age at which the vaccine is given has undergone changes. Natural measles infection is thought to give life-long immunity, but a small proportion of the recipients of measles vaccine are susceptible to measles. Thus, the optimal ages for vaccination and for boosters continue to be evaluated. Measles vaccine is usually given with mumps and rubella vaccines (MMR). Rubella is a mild disease in children, but vaccination is undertaken to prevent spread of rubella in women of childbearing age because of the serious results of rubella in pregnancy described previously. In some countries, rubella vaccine is given to all children so that these children will not spread rubella, and in some countries rubella is given to pubescent girls.

The most recently developed vaccine to be used routinely contains the capsular polysaccharide of *H. influenzae*, type b. The first vaccine released in 1985 was just polysaccharide and was given to children >24 mo of age because younger children do not respond as well to the polysaccharide. The polysaccharide was conjugated to a protein-diphtheria toxoid, a mutant nontoxic diphtheria toxin or meningococcal protein. These conjugated vaccines stimulate antibodies more efficiently than the polysaccharide alone and have been used in children >18 mo of age since the late 1980s. Because these vaccines are still new, much has yet to be learned.

Other vaccines are available for children who have diseases that put them at high risk for specific infections or who are going to travel. Research is ongoing to improve present vaccines and develop new vaccines.

Bibliography

Christie, A. B. (1987). "Infectious Diseases," 4th ed. Churchill Livingstone, Edinburgh.
Committee on Infectious Diseases, American Academy of Pediatrics (1988). "Report of the Committee on Infectious Diseases," 21st ed. American Academy of Pediatrics, Elk Grove Village, Illinois.

Feigin, R. D., and Cherry, J. D. (eds.) (1987). "Textbook of Pediatric Infectious Diseases," 2nd ed. W. B. Saunders Co., Philadelphia.

Mandell, G. L., Douglas, R. G. J., and Bennett, J. E. (eds.) (1990). "Principles and Practice of Infectious Diseases," 3rd ed. Churchill Livingstone, New York.

Mims, C. A. (1987). "The Pathogenesis of Infectious Disease," 3rd ed. Academic Press, London.

Moffet, H. L. (1989). "Pediatric Infectious Diseases: A Problem-Oriented Approach," 3rd ed. J. B. Lippincott, Co., Philadelphia.

Remington, J. S., and Klein, J. O. (eds.) (1990). "Infectious Diseases of the Fetus and Newborn Infant," 3rd ed. W. B. Saunders Co., Philadelphia.

Inflammation

NABIL HANNA, *IDEC Pharmaceuticals, Inc.*

GEORGE POSTE, *Smith-Kline Beecham Pharmaceuticals, Philadelphia*

Glossary

Adhesion molecules Polypeptides expressed on cell membranes and that play a significant role in cellular recognition and activation

Complement Multicomponent system that upon activation exerts biological activities important for host defense mechanisms including lysis of bacteria and foreign cells, chemotaxis, phagocytosis, and immune cell activation; complement proteins are sequentially activated by limited proteolysis; the activation cascade may be triggered by immune complexes, bacterial lipopolysaccharides, and proteolytic enzymes

Cytokines Polypeptides produced by lymphocytes (lymphokines) or mononuclear phagocytes (monokines) that regulate the function of other cells; they play a key role in immune regulation

Hageman factor Plasma protein that upon activation initiates the cascade of the coagulation system

Kinins Vasoactive peptides derived from inactive protein precursors (kininogens); they dilate blood vessels, increase vascular permeability, induce leakage of fluids into the tissue, and contract smooth muscle

Lectin Protein that recognizes specific carbohydrate residues; upon binding to glycoproteins on cell membranes, lectins may activate the cell to secrete cytokines or stimulate cell division

Mast cell Bone marrow-derived cell found in connective tissues that is characterized by the presence of cytoplasmic granules containing chemotactic factors, enzymes, and the vasoactive amines histamine and/or serotonin

Mononuclear phagocyte Widely distributed bone marrow-derived phagocytic cell that participates in killing of microorganisms, production of cytokines, presentation of antigens to lymphocytes, and resolution of inflammation

Neutrophil Phagocytic cell characterized by its polymorphic nucleus and secondary granules containing hydrolytic enzymes; they constitute 60% of the white cells in the blood and play a significant role in host resistance against infections

INFLAMMATION IS A critical body defense reaction that involves complex coordination of a variety of cell types with a diverse array of molecular signals. These events are triggered by either exogenous noxious agents or endogenous immune responses that result in tissue damage. The delicate and well-balanced interplay between inflammatory cells and the molecular mediators (cytokines and lymphokines) that they release is designed to eliminate harmful agents and initiate the repair of the damaged tissue. The initial inflammatory response characterized by redness, warmth, swelling, and pus formation has two major components: fluid exudation and neutrophil infiltration, which are differentially regulated. In most instances, however, the acute inflammatory reaction is not efficient and proceeds to chronic inflammation, which is driven by immunologic processes and is characterized by the dominance of mononuclear phagocytes associated with influx of lymphocytes and fibroblasts. The products of these cellular components contribute to demolition and healing followed by tissue repair and regeneration. Although inherently essential for the body's protection, inflammation when aberrant in magnitude or persistent in duration may lead to considerable normal tissue damage and organ pathologies that are more harmful to the host than the original insult that initiated the reaction. In these cases of uncontrolled chronic inflammatory responses, pharmacological intervention is needed to prevent further tissue damage and organ disfunction.

I. Initiation and Progression of the Inflammatory Response

The significance of the inflammatory response to host survival is represented by an efficient process of immediate deployment of immune and nonimmune mechanisms aimed at preventing the invasion by infecting organisms, limiting the tissue damage inflicted by the initial insult, clearing the debris from the injured areas, and initiating the healing process. The apparent redundancy in mobilizing multiple humoral and cellular elements to the site of inflammation attests to the life-threatening situations that may arise as a result of an ineffective inflammatory response. Therefore, processes of recruitment, amplification, activation, and synergy constitute key events that render the inflammatory response efficient and effective. Alternatively, when aberrant, the same response may lead to tissue damage and disease.

Mediators that are involved in the initiation and progression of the inflammation process include kinins, activated complement and coagulation components, systems derived from the plasma, vasoactive amines derived from mast cells and platelets, prostaglandins and leukotrienes derived from monocytes, neutrophils, platelets, and lysosomal enzymes, and oxygen radicals produced by neutrophils and macrophages.

The initial phase of acute inflammation is exemplified by changes in the microvascular bed leading to fluid exudation. An initial transient vasoconstriction is followed by arteriolar vasodilation and increase in blood flow. The outflow of fluid is caused primarily by escape of intravascular fluid through gaps between endothelial cells in structurally intact capillaries. At the molecular level, several mediators were demonstrated to work in concert in influencing fluid exudation during the inflammatory process. The earliest response of fluid escape takes place across the postcapillary venules and is mediated by the release of the vasoactive amines histamine and serotonin from tissue mast cells. Thereafter, the major mediators that influence fluid extravasation throughout the acute inflammatory response are prostaglandins, leukotrienes, and complement components. Further delineation of the mechanisms that influence fluid escape into the extravascular tissues is warranted before progressing to the cellular component, which constitutes the effector phase of acute and chronic inflammation. Histamine is released from mast cells in response to stimulation by neuropeptides such as substance P, vasoactive intestinal polypeptide, and somatostatin as well as by antigen-induced cross-linking of surface-bound immunoglobulin E (IgE) antibodies. All three neuropeptides will induce a wheal-and-flare reaction when injected into human skin, substance P being 100 times more potent than histamine. The neuropeptide-induced histamine release is unique to skin mast cells and is not detected in mast cells isolated from the lungs or intestine, whose primary role is believed to be IgE-mediated defense. Histamine causes vasodilation and increased vascular permeability by its action on contractile elements of small venules. Upon IgE-dependent activation, mast cells synthesize and release proinflammatory lipid mediators including prostaglandin E_2 (PGE_2), leukotriene C_4 (LTC_4), and platelet-activating factor (PAF). In addition, several preformed enzymes, which are associated with the mast cell granules, are released following cell activation and are believed to be involved in the local formation of vasoactive kinins such as bradykinin. Bradykinin is a nine-amino acid peptide that causes vasodilation, increased vascular permeability of small venules, and pain in nanogram doses. It stimulates phospholipase A_2, which is responsible for the release of arachidonic acid from cellular phospholipids and thereby provides the substrate for the generation of prostaglandins and leukotrienes. However, the significance of bradykinin as a mediator of inflammation has been difficult to substantiate, particularly because of the peptide instability being rapidly degraded by angiotensin II converting enzyme and carboxypeptidases present in blood and most tissues. The kinin system is also activated by the clotting cascade because activated Hageman factor cleaves the high-molecular weight kininogen to bradykinin. Indeed, the link between blood coagulation and inflammation extends beyond the kinin system and involves both platelet factors and complement components known for their effects on vascular permeability and cellular recruitment. Thus, at the site of tissue damage, the subendothelial collagen and basement membrane initiate platelet adhesion and Hageman factor activation, which in turn initiates the clotting pathway. Further platelet aggregation and release reaction are enhanced by thrombin generated during the clotting pathway and by thromboxane A_2 and adenosine diphosphate released from activated platelets. In addition, activated platelets release serotonin and a range of cyclooxygenase and lipoxygenase products, which exhibit vasodilatory and chemotactic properties. This reaction is further amplified by the inflammatory process as stimulated mast cells release PAF and inflammatory cells release thromboxane A_2, both of which are potent platelet activators and cause increased vascular permeability. [*See* KININS: CHEMISTRY, BIOLOGY, AND FUNCTIONS.]

The initiation of the clotting cascade and the activation of the kinin and plasmin systems lead to the activation of the complement cascade, which plays a significant role in host defenses and inflammation. The complement system consists of circulating plasma proteins that upon activation exhibit enzymatic and other proinflammatory activities. Indeed, bradykinin, activated Hageman factor, and plasmin trigger the complement cascade; however, the classic pathway of complement activation occurs following interaction with antibody–antigen complexes. This immune-mediated mode of activation is most significant in the initiation and progression of inflammatory responses against foreign antigens and exogenous microorganisms as well as in the chronicity of autoimmune responses resulting in tissue damage and disease. The complement components most significant in inflammation are $C3_a$ and $C5_a$. They induce the release of PAF and histamine from mast cells causing increased vascular permeability and fluid exudation. $C5_a$ is chemotactic for neutrophils and macrophages recruiting them into the injured tissue. It also causes the release of lysosomal enzymes, leukotrienes, and oxygen radicals, which contribute to the intensity of the local inflammatory response. The recruitment and activation of phagocytic cells is central to the inflammatory response, the elimination of the noxious agent, and eventually to tissue repair. The dynamics of this process is complex and involves multiple mediators and cellular components. Thus, the enhanced vascular permeability mediated by PAF, histamine, thromboxane A_2, $C3_a$, and $C5_a$ facilitate the ex-

travasation of inflammatory cells in response to chemotactic stimuli released in the area of tissue injury. During the early stages of acute inflammation, neutrophils respond to extravascular chemotactic factors such as $C5_a$ and leukotrienes and migrate across the walls of postcapillary venules. First, neutrophils adhere to endothelial cells via specific recognition receptors, a process that most likely results in cell activation and secretion of mediators that cause the separation of the gap junctions between adjacent endothelial cells followed by migration into the extravascular tissue. Further recruitment of inflammatory cells is mediated by chemotactic mediators such as leukotrienes and cytokines released by the activated neutrophils and macrophages already present in the inflamed tissues. In most acute inflammatory responses, the early cellular response is dominated by neutrophils with monocytes and macrophages constituting a minority population. In the absence of a continuous stimulation, the neutrophil response declines rapidly and the lesion becomes dominated by mononuclear phagocytes. The mononuclear response is prevalent also in immune-mediated chronic inflammation and in cases where the neutrophil is incapable of eliminating the invading organisms. The delayed and persistent mononuclear cell response is consistent with the cell's role in immune responses and the synthesis of lipid mediators and cytokines. The later include growth factors involved in cell-mediated immunity and tissue repair and regeneration. [*See* COMPLEMENT SYSTEM; CYTOKINES IN THE IMMUNE RESPONSE; NEUTROPHILS.]

The primary function, however, of the neutrophil and monocyte at the inflammatory site is the phagocytosis and intracellular destruction of the agents that initiated tissue injury. The process of phagocytosis is enhanced by the expression on the membrane of phagocystic cells of specific receptors that recognize antibody- or complement-coated agents or microorganisms. During the phagocytic process and the delivery of the engulfed material to the lysosomes, a burst of oxidative metabolism ensues, which generates oxygen-free radicals believed to be responsible together with lysosomal enzymes for the intracellular killing of microorganisms. At the same time, this respiratory burst provides oxygen radicals needed for lipid oxidation and the generation of proinflammatory leukotrienes and prostaglandins. In addition, the release of connective tissue-degrading enzymes (collagenase, proteoglycanase), lysosomal enzymes and oxygen metabolites to the extracellular environment cause further damage to the connective tissue and basement membrane underlying the endothelial cells. [*See* Respiratory Burst.]

This process of amplification of the inflammatory response, if persistent, may lead to further tissue damage and a transition from acute to chronic inflammatory reaction. The process of chronic inflammation involves a complex interplay between cells and their mediators, all of which contribute to the vascular changes and recruitment of mononuclear phagocytes followed by chronic immune-mediated inflammatory response, which is associated with reparative and proliferative processes. These processes translate into cleaning up of the damaged tissue, healing, repair, and regeneration. The onset of the chronic stage of inflammation is characterized by the prevalence of mononuclear cells, followed by an influx of fibroblasts and lymphocytes. The immune-mediated activation of the mononuclear phagocytes results in the synthesis and secretion of cytokines and growth factors essential for the growth and migration of endothelial cells and fibroblasts. These two cellular elements are required for the development of new blood vessels at the inflammatory site and the synthesis of the connective tissue components, collagens, and glycosaminoglycans. Thus, the formation of granulation tissue consisting of new vascular network, macrophages, lymphocytes, fibroblasts, and newly synthesized connective tissue components is the hallmark of chronic inflammation and the repair process associated with it. [*See* Connective Tissue; Lymphocytes; Macrophages.]

The fundamental role that inflammation plays in host defenses as well as in the progression of immune-mediated diseases justifies a more detailed look into the molecular mechanisms that regulate the complex cellular interactions contributing to the initiation, progression, and outcome of the inflammatory process.

II. Lipid Mediators of Inflammation

Upon activation by chemotactic or immune mediators via specific surface receptors, the neutrophil undergoes several biochemical changes that regulate cellular functions including aggregation, degranulation, release of reactive oxygen species, and the production of PAF, leukotriene β_4, and lipoxyn

A. The biochemical changes include activation of phospholipase A_2(PLA$_2$) and phospholipase C (PLC), turn over of membrane phospholipids, alterations in ion flux and membrane potential, increases in cytosolic calcium, and phosphorylation of cellular proteins and receptors.

The first step in lipid mediators' generation is the mobilization of arachidonic acid from membrane phospholipids mediated by the activation of PLA$_2$ and PLC. Once released, arachidonic acid is oxidized by two major enzymatic pathways—the cyclooxygenase and lipoxygenase pathways—to produce biologically active mediators including eicosanoids, prostaglandins, thromboxanes, and leukotrienes. Although these mediators have a wide range of biologic activities, the focus here will be on their roles in inflammation. Of the blood cells, monocytes and platelets have active cyclooxygenase enzymes, whereas neutrophils and monocytes produce lipoxygenase products. The significance of these products in inflammatory diseases was confirmed by the observations that, when injected *in vivo*, they induce inflammatory responses, they are present in elevated concentrations in inflamed tissues and inhibitors of eicosanoid production or their receptor antagonists exhibit therapeutic benefits in acute and chronic inflammatory diseases.

By adding oxygen molecules to the 9-, 11-, and 15-positions of arachidonic acid, the cyclooxygenase enzyme forms the prostaglandin endoperoxide PGG$_2$. A subsequent reduction of the 15-hydroperoxy group of PGG$_2$ results in the 15-hydroxy compound PGH$_2$, which is converted to prostaglandins and thromboxanes through the action of isomerase enzymes. The nature of the end product produced by different cells varies according to their specific isomerase activity such as blood platelets producing primarily thromboxane A$_2$, endothelial cells produce PGI$_2$, and monocytes producing PGE$_2$. Prostaglandins and thromboxanes are rapidly synthesized and secreted in response to cell activation by stimuli that activate phospholipases and cause the liberation of free arachidonic acid. Once released, however, these potent mediators of inflammation are rapidly degraded to inactive products.

The second pathway of arachidonic acid metabolism to biologically active products is the lypoxygense pathway mediated primarily by the 5-lipoxygenase enzyme. This is a cytoplasmic enzyme that requires phospholipid, Ca^{2+}, and adenosine triphosphate for optimal activity. The enzyme exists in an inactive form and, upon cell stimulation, it translocates to the membrane where the arachidonic acid substrate is released from membrane phospholipids and is accessible for catalysis. Furthermore, the 5-lipoxygenase enzyme is inactivated during catalysis, most likely by the autocatalytic hydroperoxides generated by the enzyme itself. The 5-lipoxygenase reaction is initiated by the addition of oxygen to arachidonic acid to generate 5-hydroperoxyeicosatetraenoic acid (5-HPETE) which generates LTA_4 by the action of LTA_4 synthase. This epoxide is either hydrolized to LTB_4 or, alternatively, glutathione is added to the 6-position by glutathione-s-transferase to produce LTC_4. The cleavage of glutamic acid from the tripeptide moiety of LTC_4 generates LTD_4, which then yields LTE_4 by the loss of glycine. All 5-lipoxygenase products exhibit potent biological activities that influence the nature and intensity of the inflammatory response. They exert their effects locally at the site of synthesis as they are rapidly inactivated in the circulation. A large array of biological effects is mediated by the different eicosanoids.

Most of the cyclooxygenase products contribute to the vascular phase of inflammation. For example, PGE_2, PGD_2, and PGI_2 cause vasodilation but do not increase vascular permeability by themselves; however, they synergize with the proinflammatory mediators leukotrienes, bradykinin, and serotonin to increase vascular permeability. Thromboxane A_2 induces platelet aggregation and release of vasoactive platelet mediators. The contribution of lipoxygenase products to inflammation is exemplified by the ability of the peptidoleukotrienes to increase vascular permeability and cause arteriolar and bronchial constriction. In contrast, LTB_4 is a potent chemotactic agent for neutrophils, eosinophils, and monocytes and promotes the secretion of reactive oxygen species and hydrolytic enzymes from neutrophils. Therefore, LTB_4 is believed to play a critical role in the recruitment and activation of inflammatory cells at the site of tissue injury. A lipid mediator of inflammation that is different from eicosanoids is PAF. It is derived from phosphatidylcholine by the action of 1-O-alkyl-PLA_2. PAF is produced by platelets, neutrophils, and monocytes and induces platelet aggregation, increased vascular permeability, and chemotaxis for neutrophils. Like peptidoleukotrienes, PAF causes bronchoconstriction and is involved in immediate allergic responses and anaphylaxis.

III. Role of Cytokines in Immune Inflammation

Cell activation leading to the release of preformed mediators such as histamine or newly synthesized products including biologically active proteins and eicosanoids constitute key evens in the amplification of the inflammatory response. Unlike neutrophils, lymphocytes and macrophages respond to immune and nonimmune stimuli by producing a wide range of polypeptide mediators called cytokines.

Although most, if not all, cytokines play an important role in homeostasis by regulating the activation, growth, and differentiation of target cells, they can contribute to disease pathogenesis if their production continues uncontrolled. All cytokines exert their action via specific receptors expressed on the surface of the target cells. As such, their functions could be regulated by factors that influence the expression of these receptors. Therefore, it is not surprising to find synergy among cytokines produced by the same or different cell lineages if receptor upregulation results from cytokine-induced activation of the target cells. Because of the complex interplay among cytokines, it is difficult to address their contribution to inflammation and disease pathology in isolation. However, the focus of this discussion will be limited to two cytokines produced primarily by activated monocytes and macrophages: interleukin-1 (IL-1) and tumor necrosis factor (TNF). Both IL-1 and TNF have profound effects on inflammation, tissue remodeling, and repair by influencing the activation, growth, and differentiation of other cells including lymphocytes, neutrophils, endothelial cells, fibroblasts, chondrocytes, and osteoclasts.

IL-1 is a 17-kDa protein produced by monocytes in response to immune and other proinflammatory stimuli. Two different IL-1 proteins, which are the products of two separate genes, IL-1α and IL-1β, have been identified. Each of the IL-1 proteins is expressed as a 31-kDa precursor that is cleaved to its mature form of cell 17-kDa protein before it is released from the cell. The two mature IL-1 molecules bind to the same receptor and share overlapping biological activities, which include the induction of prostaglandin and collagenase synthesis and secretion by synovial cells and chondrocytes, activation of phospholipase A_2, stimulation of fibroblast proliferation, activation of vascular endothelial cells leading to upregulation of adhesion receptors and promotion of lymphocyte activation. IL-1 can mobilize neutrophils from the bone marrow, provides a chemotactic signal that promotes the adherence of neutrophils to vascular endothelium, and augments neutrophil degranulation.

Based on the wide spectrum of target cells responsive to IL-1 and the diverse activities induced by this cytokine, release at the site of inflammation may orchestrate tissue repair, whereas an exaggerated response may result in tissue destruction and organ failure. For example, the effects on synovial cells, chrondrocytes, and osteoclasts promote panus formation, cartilage degradation, and bone resorption, representing key pathologies in the chronic disease rheumatoid arthritis (RA). This is further exemplified by the observation that IL-1 is present at high levels in synovial fluids and is spontaneously released from mononuclear cells isolated from synovial tissues of RA patients. Also, IL-1 induces the release of eicosanoids, collagenase, and proteoglycanase from synovial cells and fibroblasts. These functions promote the degradation of proteoglycan and cause cartilage resorption.

Osteoclast activation and inhibition of bone collagen synthesis by IL-1 enhance bone resorption associated with rheumatic diseases. The IL-1 induction of fibroblast proliferation, which normally constitutes an important component in tissue repair, when persistent may result in panus formation, granulomatous lesions, fibrosis, and joint destruction.

In addition to the specific effects described for RA, IL-1 promotes acute and chronic inflammation by acting on different cells in this complex process. IL-1 acts on endothelial cells and induces the synthesis of procoagulant activity and plasminogen activator inhibitor as well as the expression of surface adhesion molecules that promote the adherence of neutrophils, monocytes, and lymphocytes to endothelial cells. This step is pivotal for the targeted migration and recruitment of inflammatory and immune cells to the site of tissue injury. Representatives of the different classes of adhesion receptors include leukocyte function antigen (LFA) from the integrin superfamily, intercellular adhesion molecule 1 (ICAM-1) from the immunoglobulin supergene family, and endothelial–leukocyte adhesion molecule 1 (ELAM-1) from the L-CAM gene family. In addition to inducing adhesion receptors on endothelial cells, which promotes leukocyte accumulation, IL-1 stimulates fibroblasts, stromal cells, and endothelial cells to produce colony-stimulating

factors required for the production of monocytes and neutrophils by the bone marrow. This ensures a sustained production of inflammatory cells for maintenance of the inflammatory response as long as inducers of IL-1 production persist. The inducers of IL-1 in macrophages are many and include immune complexes, the complement components $C3_a$ and $C5_a$, substance P, TNF, interferon gamma (IFN-γ) (a product of activated T lymphocytes), and endotoxin of Gram-negative bacteria.

TNF is a 17-kDa protein synthesized as a precursor, which subsequently cleaved during processing to yield the mature molecule. The cellular effects of TNF are mediated by binding to specific high-affinity receptors expressed on many cell types and the consequences of receptor binding appears to be tissue-specific. The number of TNF receptors are upregulated by other cytokines such as IFN-γ, which may contribute to the synergy observed between these two cytokines. Although TNF has been demonstrated to play a major role in the pathogenesis of endotoxic shock and to contribute to the cachexia associated with many chronic diseases, its role in inflammation and immunoregulation should not be underestimated. TNF shares many of the bioactivities of IL-1 on endothelial cells, fibroblasts, chondrocytes, lymphocytes, and macrophages. For example, TNF induces procoagulant activity on endothelial cell surface and stimulates IL-1 production, which enhances tissue factorlike procoagulant activity. The intravascular coagulation with increased capillary permeability and escape of macromolecules to the tissues observed in endotoxemia may be the result of the interaction of TNF with the vascular endothelium. Also, TNF stimulates the expression of adhesion molecules as ICAM-1 and promotes neutrophil adherence to endothelial cells, which constitutes a critical factor in the development of inflammation. Also, the TNF-induced production of collagenase and PGE_2 by human synovial cells disrupts the collagen matrix of connective tissues and stimulates bone resorption. Similarly, TNF activates chondrocytes to degrade proteoglycan and to inhibit the synthesis of new proteoglycans, both of which contribute to cartilage destruction and severe impairment in joint function in RA. These activities are further exacerbated by the ability of TNF to induce IL-1 production in fibroblasts and macrophages. The effects of TNF on neutrophils and macrophages are diverse and include chemotaxis, release of reactive oxygen radicals, increase in phagocytic and cytotoxic capacities, and production of PGE$_2$. Thus, the recruitment and activation of neutrophils and macrophages by TNF play a pivotal role in mounting an effective and well-orchestrated response to infection and injury. Furthermore, local and limited production of IL-1 and TNF at the site of injury can reduce tissue damage and promote healing and tissue repair. However, uncontrollable and prolonged exposures to these otherwise beneficial cytokines might cause severe tissue destruction and organ failure and, in the case of TNF, may lead to shock, cardiovascular collapse, acute multiorgan failure, and death. In those cases of aberrant and persistent production of potent proinflammatory mediators, pharmacologic intervention will be needed to control the adverse effects of the body's own defense system.

IV. Adhesion Molecules Involved in Endothelial–Leukocyte Cell Interactions

The focal adhesion of leukocytes to the vascular endothelial cells is pivotal for mounting an effective inflammatory response. The specificity of this reaction resides primarily in the upregulation of adhesion molecules or receptors expressed on endothelial cells adjacent to the inflammatory site. This generates high-avidity binding for normal as well as activated leukocytes that express the respective ligands on their surface. The kinetics and regulation of the expression of adhesion receptors on leukocytes and endothelial cells indicate that these highly specialized recognition structures provide an effective mechanism for the specific mobilization of immune and inflammatory cells to the injured tissue sites. Moreover, many members of these families of molecules do not serve as points of attachment only but, in addition, provide activation signals to the cells resulting in cytokine production and further expression of surface receptors. More direct evidence for the role of adhesion receptors in inflammation was provided by the demonstration that antibodies to these surface molecules inhibited leukocyte–endothelial cell adherence *in vitro* and cellular infiltration into inflamed sites *in vivo*. Also, the genetically inherited disease of leukocyte adhesion deficiency (LAD), in which the surface expression of a family of structurally and functionally related adhesion receptors is deficient, is characterized by recurrent infections, impaired cell accumulation at inflamed sites (i.e., pus formation), and poor wound healing.

The most-studied members of this Leu-CAM family of adhesion molecules are (LFA-1), and complement receptor 3 (CR3). They belong to the integrin superfamily characterized by proteins that span the membrane lipid bilayer to provide a link between the extracellular matrix and the cellular cytoskeleton. The leukocyte integrins are highly suited to adhesion to endothelial cells because they mediate low-affinity interactions that can be transiently stabilized by the clustering of the receptors in the membrane and by cytoskeletal interactions. Both LFA-1 and CR3 are expressed on neutrophils, monocytes, and large granular lymphocytes. In addition, LFA-1, but not CR3, is expressed on T lymphocytes and is believed to participate in antigen-specific T-cell responses.

LFA-1 and CR3 molecules are heterodimers composed of α- and β-chains. Both share a common β-chain coded by a single gene while having specific α-chains coded by separate genes. LAD patients have genetic defects in the common β-chain leading to failure to synthesize this chain; therefore, they lack the ability to express both LFA-1 and CR3. Both receptors bind ICAM-1 on the target cells and, in addition, CR3 binds the cleavage fragment of the third complement component, C3bi. Cell activation and degranulation results in a marked increase in the number of CR3 molecules expressed on the cell surface, which enhances cell adhesion and migration. Studies using antibodies to adhesion molecules demonstrated that the adhesion of neutrophils to human endothelial cells in culture is inhibited by antibodies to the common β-chain (CD18) and the CR3 α-chain (CD11b). Human monocyte adhesion to endothelial cells was inhibited also by β-chain antibodies. Under certain conditions, monocyte adhesion depends on LFA-1 and, to a lesser extent, on CR3, and the opposite is true for neutrophils whose adhesion is primarily CR3-dependent. However, upon IL-1 stimulation of the endothelial cells, both members of the family play significant roles. To add further complexity to this apparently simple adhesion phenomenon, antibody to CD18 complex inhibited the adhesion to endothelial cells of neutrophils stimulated by LTB$_4$, PAF, and TNF but had no effect on cells stimulated by LTC$_4$ or thrombin. Thus, the involvement of specific adhesion molecules in leukocyte–endothelial cell interaction depends on the nature of the activating signal interacting with both the leukocyte and the endothelial cell.

ICAM-1 binds specifically to LFA-1 and CR3 and is strongly expressed on vascular endothelium, tissue macrophages, and dendritic cells. It is weakly expressed on blood mononuclear cells including lymphocytes and monocytes, and enhanced expression is observed following cell stimulation. Similarly, the proinflammatory cytokines IL-1, TNF, and IFN-α selectively induce the expression of ICAM-1 on endothelial cells, a process that requires *de novo* protein synthesis. ICAM-1 is a surface membrane glycoprotein that belongs to the immunoglobulin supergene family with five extracellular immunoglobulin domains and short transmembrane and cytoplasmic domain. ICAM-1 mediates the adhesion among B cells, T cells, and monocytes, and antibodies to ICAM-1 block several lymphocyte immune functions. Similarly, treatment of endothelial cells with anti-CAM-1 antibodies blocks the adhesion of stimulated neutrophils.

Based on studies with specific antibodies against ICAM-1 and LFA-1, it became obvious that inhibition of mononuclear cells or neutrophil adhesion to endothelial cells is not universal, indicating that under certain conditions other adhesion molecules are involved in this process. A recently characterized molecule, which is expressed only on cytokine-stimulated endothelial cells and is involved in neutrophil adhesion, is the endothelial–leukocyte adhesion molecule 1 (ELAM-1). The expression of ELAM-1 is rapid and transient after IL-1 or TNF treatment of endothelial cells, suggesting that it is involved in the early and acute inflammatory response. In contrast, ICAM-1 expression is delayed and persistent, which may contribute to the maintenance of the response. The primary sequence of ELAM-1 predicts an amino-terminal lectinlike domain, which is shared by another cell-surface antigen, MEL-14, which is involved in lymphocyte homing. ELAM-1 is not related to molecules of either the immunoglobulin supergene family or the integrin supergene family. The identification of other cell-surface molecules of similar domain structure suggests the emergence of a new gene family. Two members of this family, ELAM-1 and MEL-14, appear to play a key role in leukocyte–endothelial cell adhesion. Expressed on cytokine-stimulated endothelium, ELAM-1 mediates the adhesion of neutrophils, whereas MEL-14 molecule expressed by lymphocytes, monocytes, and granulocytes mediates lymphocyte adhesion to highly specialized lymph node endothelium. The adhesive properties of MEL-14 appear to be mediated by a lectinlike function recognizing mannose-6-phosphate and certain forms of sialic acid. However, the exact nature of the cell-surface ligands that bind to either ELAM-1 or MEL-14 has not been elucidated.

The pattern of leukocyte accumulation in inflammatory processes may be mediated by the selectivity and kinetics of expression of leukocyte–endothelial cell-adhesion receptors. A focal expression of ELAM-1 at sites of cytokine-mediated endothelial cell stimulation promotes neutrophil adhesion and extravasation. This is further confirmed by the blocking of neutrophil adhesion by monoclonal antibodies to ELAM-1. Also, the rapid induction and decay of ELAM-1 from the cell surface are consistent with its role in acute inflammatory responses. Unlike ELAM-1, ICAM-1 is expressed on multiple myeloid and nonmyeloid cell types and is involved in other cellular interactions involving lymphocytes, monocytes, fibroblasts, and endothelial cells. The expression following cytokine activation is maintained for a longer time than that observed with ELAM-1, suggesting a role in the progression of chronic inflammation. Tissue-specific vascular adhesion molecules are involved in lymphocyte homing and, in certain conditions, may contribute to leukocyte adhesion and migration. The availability of antagonists or blocking antibodies to specific adhesion molecules should assist in defining their contribution to physiologic as well as pathologic disease processes and should provide insight into new therapeutic approaches to acute and chronic inflammatory diseases.

V. Implications for Drug Therapy

Inflammatory responses are essential for host defenses against exogenous noxious agents with the objective of limiting tissue damage, restricting systemic spread to normal tissues, inactivating and clearing toxic products, and promoting tissue repair. However, when aberrant or exaggerated in intensity or duration, the inflammatory process may result in extensive tissue damage and organ malfunction that require pharmacologic intervention. The multiplicity of effector mechanisms that participate in the inflammatory process renders the effective inhibition of this response by a single approach or therapeutic modality difficult to achieve. However, because several of the mediators act synergistically to amplify cell recruitment and activation, interruption of this cascade by antagonizing the effect of key mediators or inhibiting their production will lead to therapeutic benefit to the host. Fluid exudation has been demonstrated to be sensitive to the action of cyclooxygenase inhibitors known in clinical medicine as nonsteroidal anti-inflammatory drugs (NSAIDS). Such agents proved to be effective in reducing the edematous response in chronic diseases including RA. Similarly, histamine receptor antagonists were demonstrated to be effective in inhibiting allergic responses in which histamine release from mast cells is the key vasoactive mediator. Two additional key mediators that are being targeted for therapy of diseases such a asthma are PAF and LTD_4. Receptor antagonists to both mediators are effective in inhibiting fluid exudation in experimental models of acute inflammation and antigen-induced bronchoconstriction, activities that justified their clinical development. However, most tissue damage in acute and chronic inflammatory states is mediated by neutrophils and mononuclear cells and their products. Therefore, inhibition of cellular accumulation in extravascular tissues and antagonizing their products received special attention. Today, steroids are the most effective drugs in clinical use and are capable of inhibiting neutrophil and monocyte infiltration to inflamed sites as well as the production of the cytokines TNF and IL-1. However, the toxic side effects of steroids are prohibitive of widespread clinical use. One of the mechanisms by which steroids prevent cell infiltration is related to the inhibition of phospholipase A_2 and the consequent mobilization of free arachidonic acid resulting in inhibition of cyclooxygenase and lipoxygenase products. The discovery of specific lipoxygenase inhibitors would complement the activities of existing NSAIDS, and by inhibiting both pathways an effective inhibition of the cellular and edematous phases of the inflammatory response will be expected. These strategies together with LTB_4 receptor antagonists, IL-1 and TNF production inhibitors or receptor antagonists, and inhibitors of adhesion molecule interactions should provide significant therapeutic advances for treatment of such diseases as RA, psoriasis, asthma, acute respiratory distress syndrome, inflammatory bowel disease, and myocardial reperfusion injury, among others. Thus, our better understanding of the basic mechanisms involved in the inflammation process resulted in the identification of new molecular targets for therapy of acute and chronic inflammatory diseases.

Bibliography

Barnes, P. J., Chung, K. F., and Page, C. P. (1988). Inflammatory mediators and asthma. *Pharmacol. Rev.* **40,** 49.

Kaplan, A. P., and Silverberg, M. (1988). Mediators of inflammation: An overview. *Methods Enzymol.* **163,** 3.

Proud, D., and Kaplan, A. P. (1988). Kinin formation: Mechanisms and role in inflammatory disorders. *Annu. Rev. Immunol.* **6,** 49.

Spaethe, S. M., and Needleman, P. (1988). Biosynthesis and release of lipid mediators in inflammation. *In* "Cellular and Molecular Aspects of Inflammation" (G. Poste and S. T. Crooke, eds.). Plenum Press, New York.

Stooleman, L. M. (1989). Adhesion molecules controlling lymphocyte migration. *Cell* **56,** 907.

Vernon-Roberts, B. (1988). Inflammation: An overview. *Anti-Rheumatic Drugs, AAS* **24,** 1.

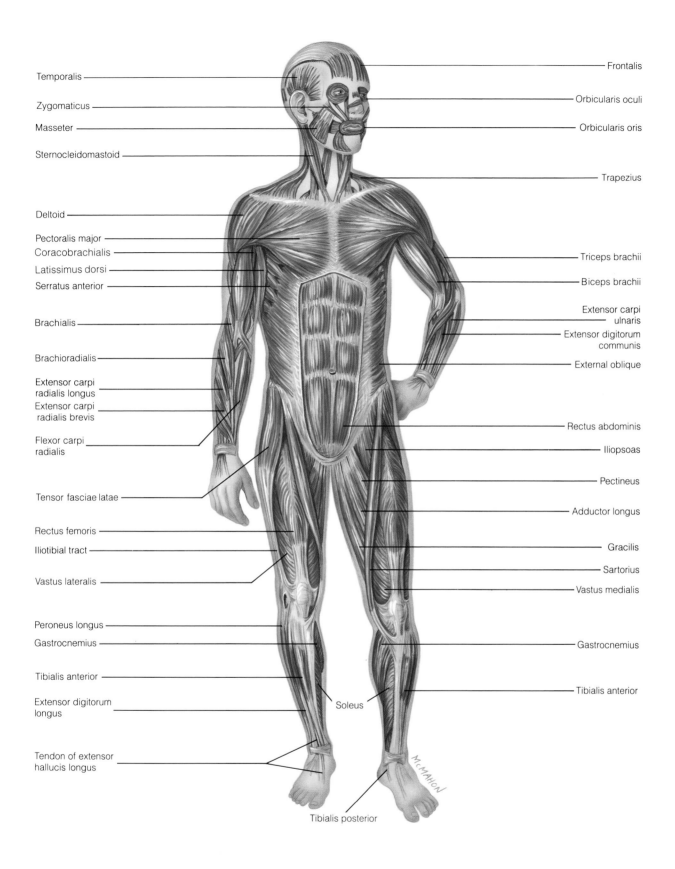

Temporalis

Zygomaticus

Masseter

Sternocleidomastoid

Deltoid

Pectoralis major
Coracobrachialis
Latissimus dorsi
Serratus anterior

Brachialis

Brachioradialis

Extensor carpi
radialis longus
Extensor carpi
radialis brevis

Flexor carpi
radialis

Tensor fasciae latae

Rectus femoris

Iliotibial tract

Vastus lateralis

Peroneus longus
Gastrocnemius

Tibialis anterior

Extensor digitorum
longus

Tendon of extensor
hallucis longus

Frontalis

Orbicularis oculi

Orbicularis oris

Trapezius

Triceps brachii

Biceps brachii

Extensor carpi
ulnaris
Extensor digitorum
communis

External oblique

Rectus abdominis

Iliopsoas

Pectineus

Adductor longus

Gracilis

Sartorius

Vastus medialis

Gastrocnemius

Tibialis anterior

Soleus

Tibialis posterior

COLOR PLATE 1   Anterior view of the major muscles of the human body. [Source: Gaudin, A. J., and Jones, K. C. (1989). "Human Anatomy and Physiology." Harcourt Brace Jovanovich, San Diego, p. 212. Reproduced with permission.]

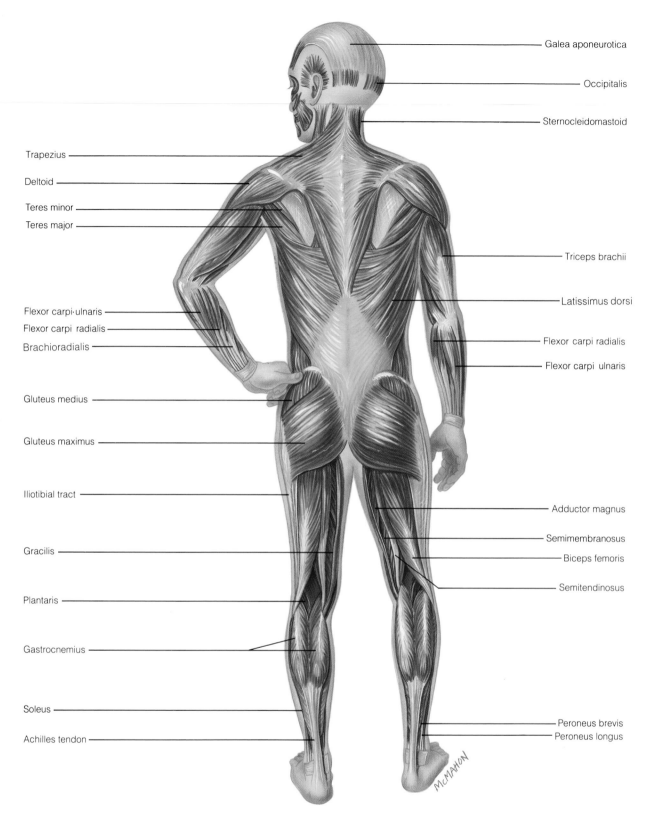

Galea aponeurotica

Occipitalis

Sternocleidomastoid

Trapezius

Deltoid

Teres minor

Teres major

Triceps brachii

Latissimus dorsi

Flexor carpi·ulnaris

Flexor carpi radialis

Brachioradialis

Flexor carpi radialis

Flexor carpi ulnaris

Gluteus medius

Gluteus maximus

Iliotibial tract

Adductor magnus

Semimembranosus

Gracilis

Biceps femoris

Semitendinosus

Plantaris

Gastrocnemius

Soleus

Peroneus brevis

Achilles tendon

Peroneus longus

COLOR PLATE 2 Posterior view of the major muscles. [Source: Gaudin, A. J., and Jones, K. C. (1989). "Human Anatomy and Physiology." Harcourt Brace Jovanovich, San Diego, p. 213. Reproduced with permission.]

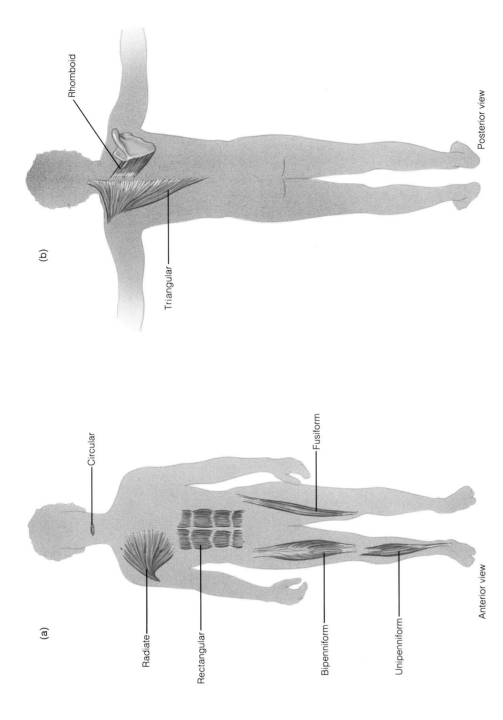

COLOR PLATE 3 Shapes of the muscles. (a) Anterior view. (b) Posterior view. [Source: Gaudin, A. J., and Jones, K. C. (1989). "Human Anatomy and Physiology." Harcourt Brace Jovanovich, San Diego, p. 208. Reproduced with permission.]

(a)

Circular

Radiate

Rectangular

Bipenniform

Unipenniform

Fusiform

Anterior view

(b)

Rhomboid

Triangular

Posterior view

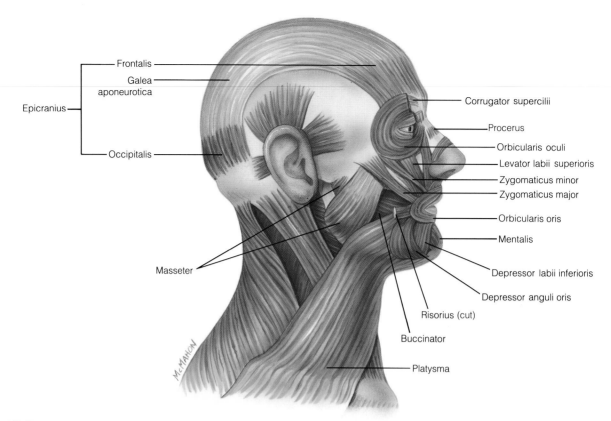

COLOR PLATE 4 Muscles of the head and the neck. [Source: Gaudin, A. J., and Jones, K. C. (1989). "Human Anatomy and Physiology." Harcourt Brace Jovanovich, San Diego, p. 215. Reproduced with permission.]

(a)

(b)

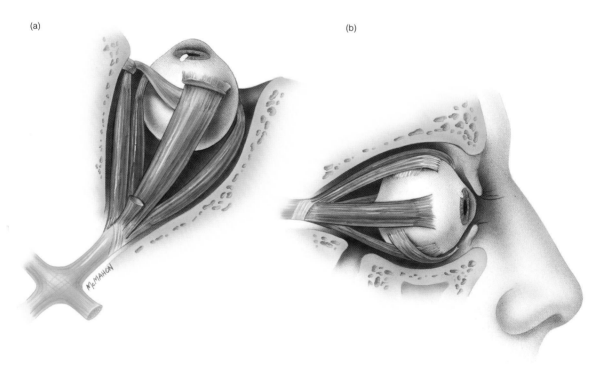

COLOR PLATE 5 Extrinsic muscles of the eyeball. (a) Medial view. (b) Lateral view. [Source: Gaudin, A. J., and Jones, K. C. (1989). "Human Anatomy and Physiology." Harcourt Brace Jovanovich, San Diego, p. 217. Reproduced with permission.]

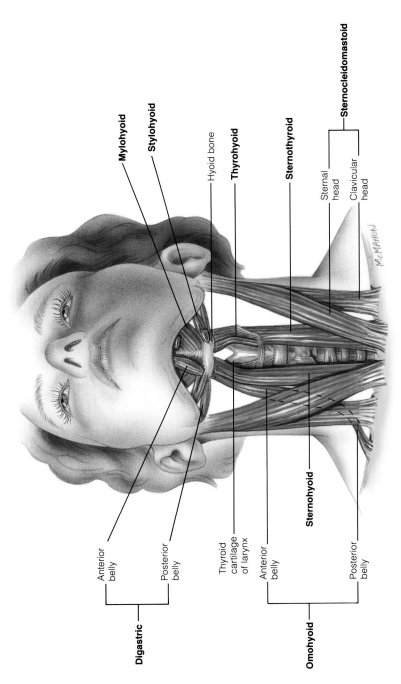

COLOR PLATE 6 Muscles of the neck. [Source: Gaudin, A. J., and Jones, K. C. (1989). "Human Anatomy and Physiology." Harcourt Brace Jovanovich, San Diego, p. 221. Reproduced with permission.]

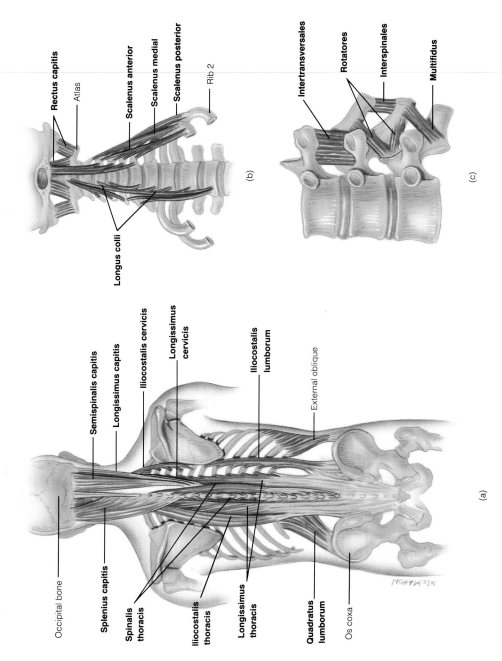

COLOR PLATE 7   Muscles that move the vertebral column. (a) Posterior view. (b) Anterior view. (c) Posterior view of the intervertebral muscles. [Source: Gaudin, A. J., and Jones, K. C. (1989). "Human Anatomy and Physiology." Harcourt Brace Jovanovich, San Diego, p. 223. Reproduced with permission.]

(a)

Occipital bone

Splenius capitis

Spinalis thoracis

Iliocostalis thoracis

Longissimus thoracis

Quadratus lumborum

Os coxa

Semispinalis capitis

Longissimus capitis

Iliocostalis cervicis

Longissimus cervicis

Iliocostalis lumborum

External oblique

McMAHON

(b)

Rectus capitis

Atlas

Scalenus anterior

Scalenus medial

Scalenus posterior

Rib 2

Longus colli

(c)

Intertransversales

Rotatores

Interspinales

Multifidus

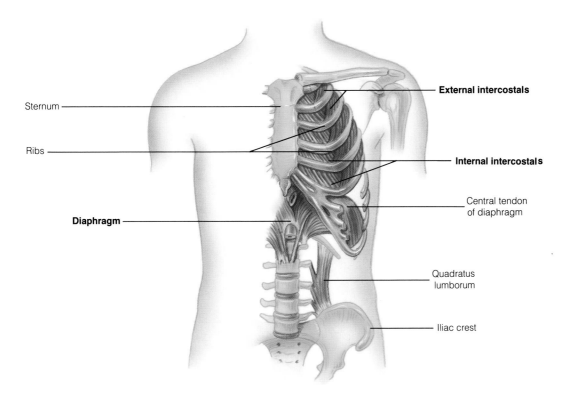

Sternum

Ribs

Diaphragm

External intercostals

Internal intercostals

Central tendon
of diaphragm

Quadratus
lumborum

Iliac crest

COLOR PLATE 8 Anterior view of muscles of the thoracic wall. [Source: Gaudin, A. J., and Jones, K. C. (1989). "Human Anatomy and Physiology." Harcourt Brace Jovanovich, San Diego, p. 225. Reproduced with permission.]

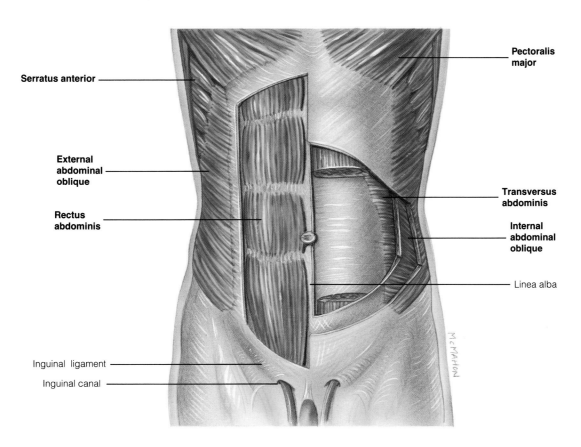

Serratus anterior

External
abdominal
oblique

Rectus
abdominis

Inguinal ligament

Inguinal canal

Pectoralis
major

Transversus
abdominis

Internal
abdominal
oblique

Linea alba

COLOR PLATE 9 Muscles of the abdominal wall. [Source: Gaudin, A. J., and Jones, K. C. (1989). "Human Anatomy and Physiology." Harcourt Brace Jovanovich, San Diego, p. 227. Reproduced with permission.]

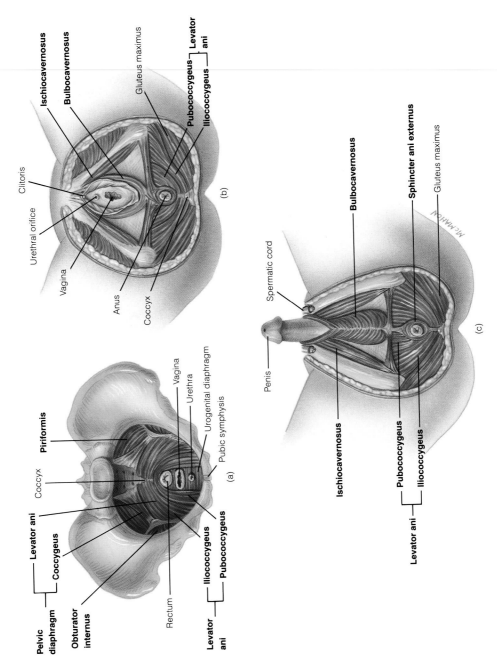

COLOR PLATE 10 Muscles of the pelvic floor. (a) Superior view. (b and c) Inferior view. [Source: Gaudin, A. J., and Jones, K. C. (1989). "Human Anatomy and Physiology." Harcourt Brace Jovanovich, San Diego, p. 229. Reproduced with permission.]

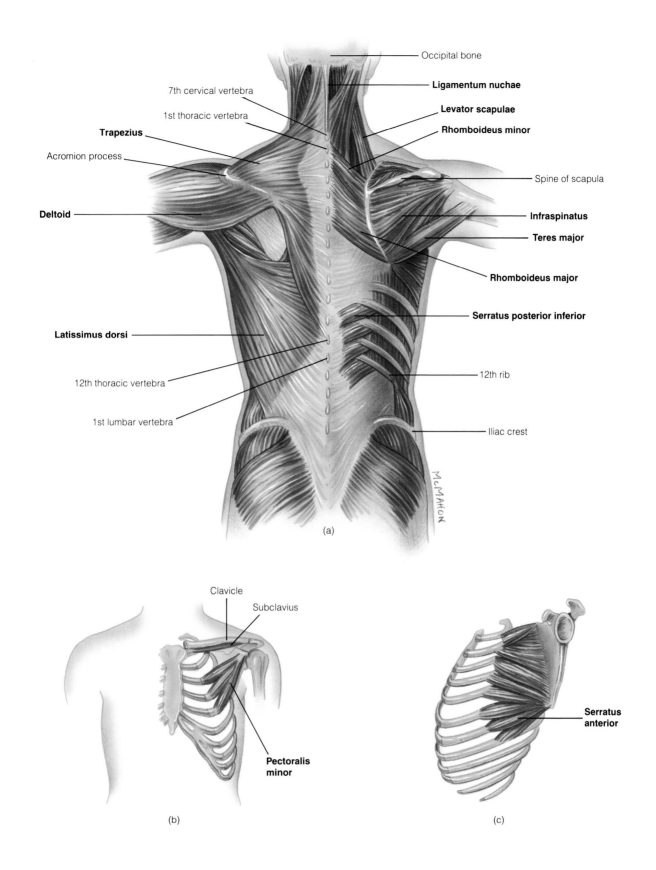

Occipital bone

Ligamentum nuchae

7th cervical vertebra

Levator scapulae

1st thoracic vertebra

Rhomboideus minor

Trapezius

Acromion process

Spine of scapula

Deltoid

Infraspinatus

Teres major

Rhomboideus major

Serratus posterior inferior

Latissimus dorsi

12th rib

12th thoracic vertebra

1st lumbar vertebra

Iliac crest

McMAHON

(a)

Clavicle

Subclavius

Serratus
anterior

**Pectoralis
minor**

(b)

(c)

COLOR PLATE 11 Muscles of the trunk. (a) Posterior view. (b) Anterior view. (c) Lateral view. [Source: Gaudin, A. J., and Jones, K. C. (1989). "Human Anatomy and Physiology." Harcourt Brace Jovanovich, San Diego, p. 229. Reproduced with permission.]

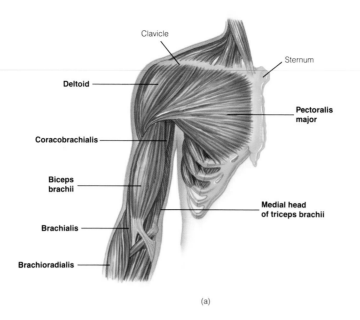

Clavicle

Sternum

Deltoid

Pectoralis
major

Coracobrachialis

Biceps
brachii

Medial head
of triceps brachii

Brachialis

Brachioradialis

(a)

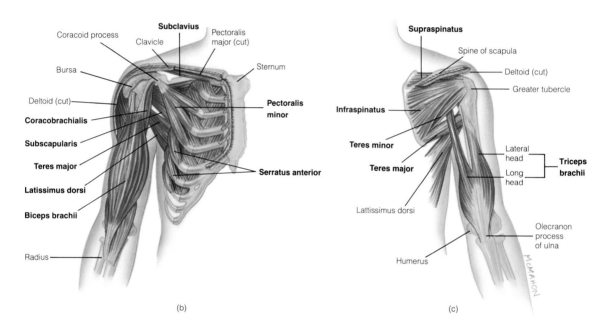

Coracoid process

Subclavius

Clavicle

Pectoralis
major (cut)

Bursa

Sternum

Deltoid (cut)

**Pectoralis
minor**

Coracobrachialis

Subscapularis

Teres major

Latissimus dorsi

Serratus anterior

Biceps brachii

Radius

(b)

Supraspinatus

Spine of scapula

Deltoid (cut)

Greater tubercle

Infraspinatus

Teres minor

Lateral
head

**Triceps
brachii**

Teres major

Long
head

Lattissimus dorsi

Olecranon
process
of ulna

Humerus

McMAHON

(c)

COLOR PLATE 12 Muscles that move the humerus. (a) Superficial posterior view. (b) Deep anterior view. (c) Deep posterior view. [Source: Gaudin, A. J., and Jones, K. C. (1989). "Human Anatomy and Physiology." Harcourt Brace Jovanovich, San Diego, p. 233. Reproduced with permission.]

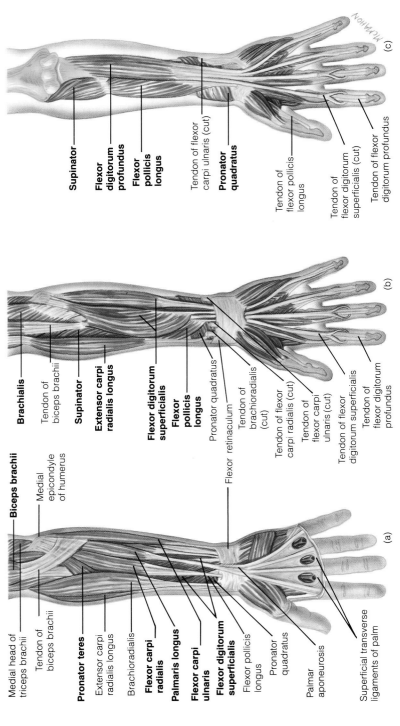

COLOR PLATE 13 (a) Superficial, (b) middle, and (c) deep layers of the anterior muscles of the forearm. [Source: Gaudin, A. J., and Jones, K. C. (1989). "Human Anatomy and Physiology." Harcourt Brace Jovanovich, San Diego, p. 237. Reproduced with permission.]

Medial head of
triceps brachii

Tendon of
biceps brachii

Pronator teres

Extensor carpi
radialis longus

Brachioradialis

**Flexor carpi
radialis**

Palmaris longus

**Flexor carpi
ulnaris**

**Flexor digitorum
superficialis**

Flexor pollicis
longus

Pronator
quadratus

Palmar
aponeurosis

Superficial transverse
ligaments of palm

(a)

Biceps brachii

Medial
epicondyle
of humerus

Brachialis

Tendon of
biceps brachii

Supinator

**Extensor carpi
radialis longus**

**Flexor digitorum
superficialis**

**Flexor
pollicis
longus**

Pronator quadratus

Flexor retinaculum

Tendon of
brachioradialis
(cut)

Tendon of flexor
carpi radialis (cut)

Tendon of
flexor carpi
ulnaris (cut)

Tendon of flexor
digitorum superficialis

Tendon of
flexor digitorum
profundus

(b)

Supinator

**Flexor
digitorum
profundus**

**Flexor
pollicis
longus**

Tendon of flexor
carpi ulnaris (cut)

**Pronator
quadratus**

Tendon of
flexor pollicis
longus

Tendon of
flexor digitorum
superficialis (cut)

Tendon of flexor
digitorum profundus

(c)

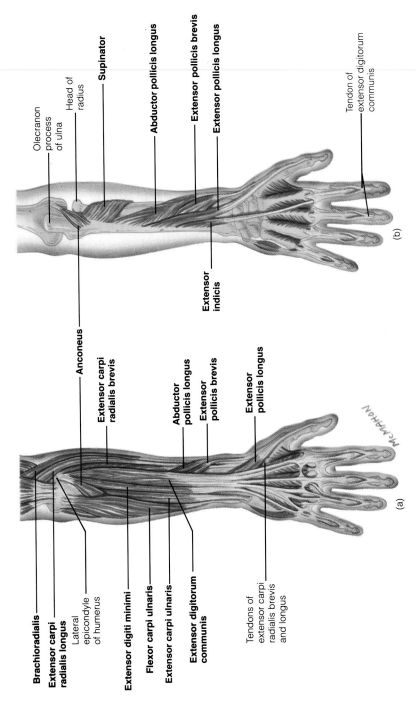

COLOR PLATE 14 (a) Superficial and (b) deep layers of the posterior muscles of the forearm. [Source: Gaudin, A. J., and Jones, K. C. (1989). "Human Anatomy and Physiology." Harcourt Brace Jovanovich, San Diego, p. 239. Reproduced with permission.]

Brachioradialis

Extensor carpi radialis longus

Lateral epicondyle of humerus

Extensor carpi radialis brevis

Extensor digiti minimi

Flexor carpi ulnaris

Extensor carpi ulnaris

Extensor digitorum communis

Abductor pollicis longus

Extensor pollicis brevis

Extensor pollicis longus

Tendons of extensor carpi radialis brevis and longus

(a)

Olecranon process of ulna

Head of radius

Anconeus

Supinator

Abductor pollicis longus

Extensor pollicis brevis

Extensor pollicis longus

Extensor indicis

Tendon of extensor digitorum communis

(b)

McMAHON

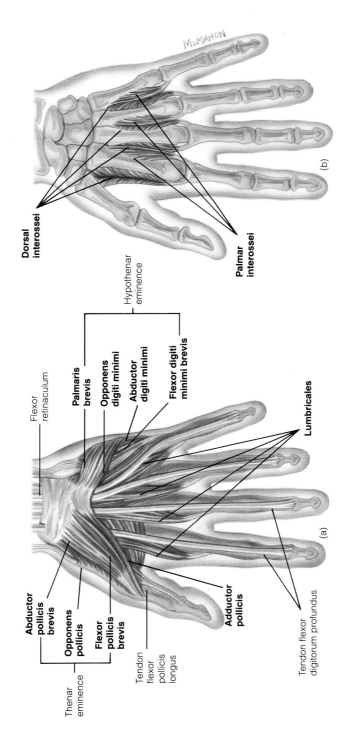

Dorsal interossei

Palmar interossei

(b)

Flexor retinaculum

Hypothenar eminence

Palmaris brevis

Opponens digiti minimi

Abductor digiti minimi

Flexor digiti minimi brevis

Lumbricales

Abductor pollicis brevis

Opponens pollicis

Flexor pollicis brevis

Thenar eminence

Tendon flexor pollicis longus

Adductor pollicis

Tendon flexor digitorum profundus

(a)

McMAHON

COLOR PLATE 15 (a) Superficial and (b) deep intrinsic muscles of the hand. [Source: Gaudin, A. J., and Jones, K. C. (1989). "Human Anatomy and Physiology." Harcourt Brace Jovanovich, San Diego, p. 241. Reproduced with permission.]

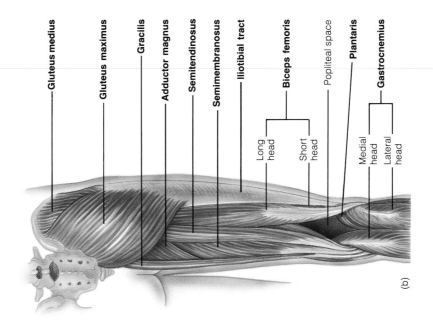

Gluteus medius

Gluteus maximus

Gracilis

Adductor magnus

Semitendinosus

Semimembranosus

Iliotibial tract

Biceps femoris
Long head
Short head

Popliteal space

Plantaris

Gastrocnemius
Medial head
Lateral head

(b)

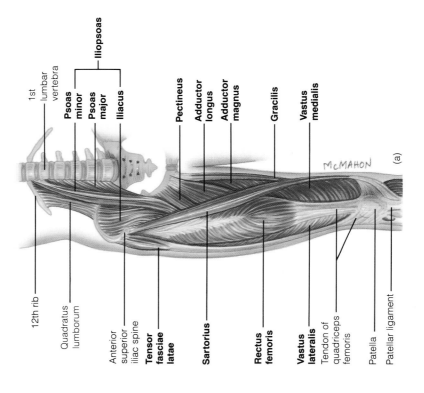

12th rib

Quadratus lumborum

1st lumbar vertebra

Psoas minor

Psoas major

Iliacus

Iliopsoas

Anterior superior iliac spine

Tensor fasciae latae

Pectineus

Adductor longus

Adductor magnus

Sartorius

Gracilis

Rectus femoris

Vastus lateralis

Vastus medialis

Tendon of quadriceps femoris

Patella

Patellar ligament

McMAHON

(a)

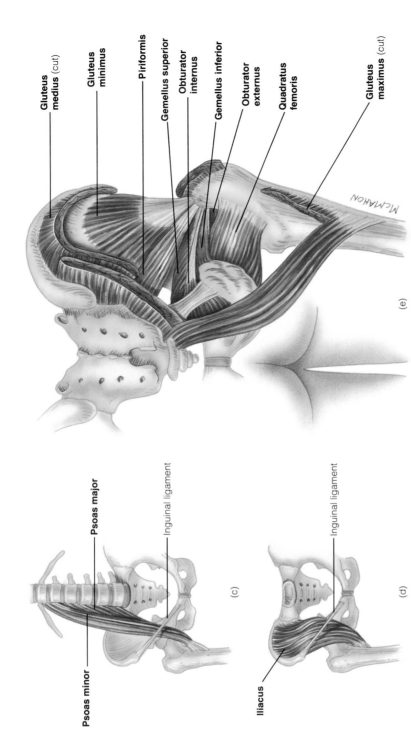

COLOR PLATE 16 Muscles of the hip and thigh. (a) Anterior view. (b) Posterior view. (c and d) Deep anterior view. (e) Deep posterior view. [Source: Gaudin, A. J., and Jones, K. C. (1989). "Human Anatomy and Physiology." Harcourt Brace Jovanovich, San Diego, p. 244–245. Reproduced with permission.]

Gluteus medius (cut)

Gluteus minimus

Piriformis

Gemellus superior

Obturator internus

Gemellus inferior

Obturator externus

Quadratus femoris

Gluteus maximus (cut)

(e)

Psoas minor

Psoas major

Inguinal ligament

(c)

Iliacus

Inguinal ligament

(d)

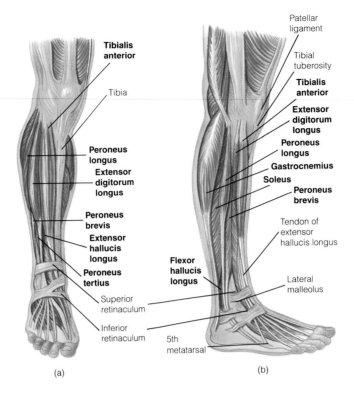

COLOR PLATE 17 (a) Anterior and (b) lateral views of the muscles of the leg. [Source: Gaudin, A. J., and Jones, K. C. (1989). "Human Anatomy and Physiology." Harcourt Brace Jovanovich, San Diego, p. 248. Reproduced with permission.]

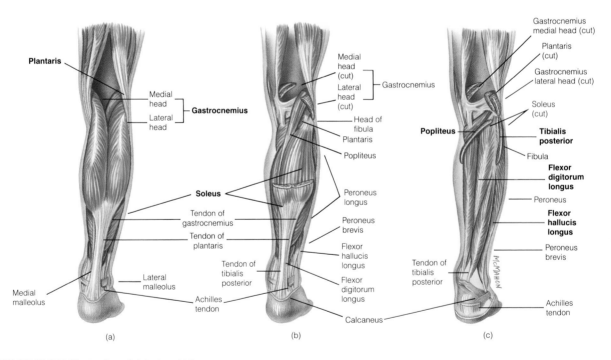

COLOR PLATE 18 (a) Superficial, (b) middle, and (c) deep layers of the posterior muscles of the leg. [Source: Gaudin, A. J., and Jones, K. C. (1989). "Human Anatomy and Physiology." Harcourt Brace Jovanovich, San Diego, p. 249. Reproduced with permission.]

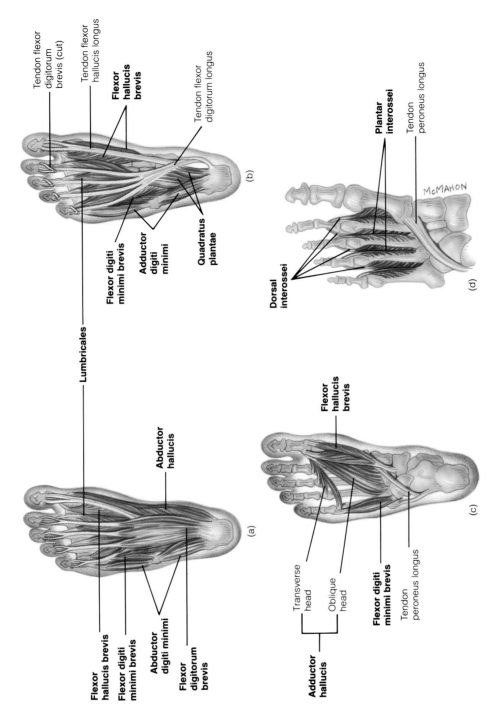

COLOR PLATE 19 (a) Superficial, (b) second, (c) third, and (d) fourth layers of the muscles of the foot, as seen in plantar view. [Source: Gaudin, A. J., and Jones, K. C. (1989). "Human Anatomy and Physiology." Harcourt Brace Jovanovich, San Diego, p. 251. Reproduced with permission.]

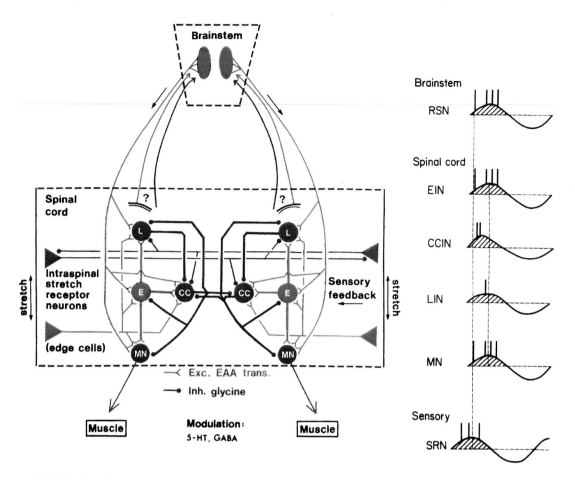

COLOR PLATE 20 The circuitry for initiation, burst generation, and sensory modulation in the lamprey nervous system. The segmental network produces bilateral alternating burst activity, which results in alternating left and right side contractions. The motoneurons (MN) are pure output devices and are indicated in green. In each half cycle, they receive excitation followed by active inhibition in the next half cycle. The excitation is produced by the E interneurons (red; utilizes excitatory amino acids as transmitter) and the inhibition by contralateral inhibitory interneurons (CC), which have glycine (blue) as the transmitter. The third interneuron (L) aids in the burst termination and inhibits the CC interneurons. The brainstem initiates locomotion by exciting cell segmental interneurons through reticulospinal neurons (RSN). As the locomotor activity starts, efference copy signals are fed back from the spinal cord to the reticulospinal neurons, which causes them to become active in a burst fashion. The activity of the spinal cord neurons (excitatory [EIN], lateral [LIN], contralateral [CC] and motoneuron [MN]) on one side of the network is indicated during one cycle as well as the reticulospinal (RSN) neuron. In addition, stretch receptor neurons (edge cells) sense the extension of one side as the contralateral side is shortening due to its muscle contraction. These stretch receptors are of two kinds: one blue, which inhibits contralateral interneuron activity and thereby acts toward a termination of the burst activity on this side, and another red, which excites the interneurons on the ipsilateral side, which is about to start its activity. This sensory circuity can explain the important sensory control of the spinal network, which is a very essential part of the control system. Exc. EAA, synapses with excitatory amino acid transmitters; Inh glycine, inhibitory glycine synapses; 5-HT and GABA, serotonin and γ amino-butyric;SRN, stretch receptor neurons.

Influenza Virus Infection

D. P. NAYAK, *Johnson Comprehensive Cancer Center, UCLA School of Medicine*

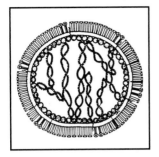

Glossary

+ strand Same polarity as the messenger RNA
− strand Opposite polarity of the messenger RNA
Antigenic drift Minor change in antigenic epitopes due to mutations
Antigenic shift Major change in antigenic properties usually due to gene exchange (reassortment)
Budding Assembly and maturation of virus particles from cell membrane
Core Internal component including ribonucleoprotein and polymerase complex
Epidemics Local outbreaks usually restricted to a limited region
Glycosylation Attachment of carbohydrates to protein backbone
Hemadsorbtion Adsorbtion of red blood cells to virus-infected cells
Hemagglutination inhibition Assay used for measuring antibody titer against flu virus
Influenza Acute respiratory disease caused by a group of viruses belonging to the orthomyxovirus group
Pandemics Global outbreak of a disease; for influenza virus, pandemics are due to antigenic shift

Reassortment Exchange of viral genes often leads to antigenic shift
Replication Synthesis of full-length viral ribonucleic acid
Transcription Synthesis of messenger RNA
Tropism Affinity of a virus for a specific organ or tissue
Virulence Ability of a virus to cause disease.

INFLUENZA IS A contagious, febrile viral disease (commonly known as flu) caused by a virus. It is often accompanied by inflammation of the upper respiratory tract, myalgia, gastrointestinal disorders, and neurological disorders such as headache, prostration, and insomnia.

I. Virus

Influenza viruses belong to the orthomyxovirus group (family orthomyxoviridae [*ortho*, true; *myxo*, affinity for mucoproteins]). The schematic presentation of a virus particle (Fig. 1) as well as the electron micrograph of intact (Fig. 2) and disrupted (Fig. 3) virus particles are shown. These are enveloped, segmented, negative-stranded RNA viruses containing helical nucleocapsids (also known as ribonucleoprotein [RNP]). An influenza virus particle is usually spherical and approximately 110 nm in diameter (Fig. 1 and 2). The envelope consists of a lipid bilayer containing spikes on the outside and a membrane (or matrix) protein (M1) on the inside. The spikes are of two types: the hemagglutinin (HA) and neuraminidase (NA), both of which are anchored in the lipid bilayer. Because of the presence of an envelope, the virus particle is very susceptible to disruption by lipid solvents and detergents. The genetic material (or genome) of the

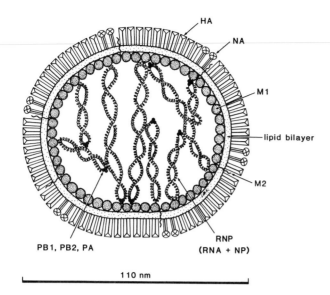

PB1, PB2, PA

110 nm

FIGURE 1 Schematic presentation of influenza virus structure. HA, hemaglutinin; M1, M2, membrane (or matrix) proteins; NA, neuraminidase; NP, nucleoprotein; PA, polymerase acidic protein; PB1, PB2, polymerase basic proteins 1 and 2; RNP, ribonucleoprotein.

virus, enclosed in the envelope, consists of RNA, which is complexed with a nucleoprotein (called NP) to form RNP. The RNA genome contains eight distinct RNA molecules (seven RNA molecules in case of influenza C virus), each of which encodes at least one (or sometimes two) protein. The viral RNA is of negative polarity (or minus-stranded), i.e., of the opposite polarity of the messenger RNA (mRNA) used for the translation of proteins. The viral mRNAs (positive polarity) are made inside the cell after infection and serve as a template for protein synthesis. Each infectious virus particle must carry an RNA-dependent RNA polymerase (also called transcriptase or replicase), which consists of three virally encoded polymerase proteins known as polymerase basic protein 1 (PB1), polymerase basic protein 2 (PB2), and polymerase acidic protein (PA). The viral transcriptase (and not the host RNA polymerase) synthesizes viral mRNAs using the viral RNA as a template. Because the viral mRNA synthesis is the first event that must occur after infection, the pure viral RNA lacking the viral transcriptase is noninfectious. The replication and transcription of viral RNA occur inside the nucleus of the infected cells, whereas the assembly process leading to virion (infectious virus particle) formation takes place on the plasma membrane [*See* DNA AND GENE TRANSCRIPTION.]

A. Nomenclature and Classification

Because of the epidemiological nature of the disease and possible emergence of new pandemic viruses, the World Health Organization in collaboration with national organizations such as the Communicable Disease Center, the National Institutes Health, and others throughout the world have organized a worldwide influenza surveillance program. Consequently, a vast number of influenza virus strains have been isolated from human as well as from a variety of animal species. A uniform system of nomenclature is followed in identifying these strains. This includes type specificity of the virus, place of isolation, isolate number, and year of isolation and is accompanied with subtype specificity when applicable. Some examples are A/Puerto Rico/8/34 (H1N1), B/Great Lakes/1/54, and C/Paris/1/67. In some older prototype strains, a person's name may be used instead of the place of isolation, and sometimes the isolate number may be missing (e.g., A/WS/34, B/Lee/40, C/Cal/78). However, the present nomenclature fails to identify the host from which the virus was isolated (e.g., human, swine, equine, chicken, duck).

On the basis of the antibody response they elicit, influenza viruses have been grouped into *types* A, B, and C. Type specificity is based on the presence of two common major internal antigens: NP and M1, the serological specificity of which is measured by complement fixation or enzyme-linked immunosorbant assay. Therefore, all type A viruses will have similar M1 and NP proteins that are different from those present in type B or type C viruses. However, nucleic acid sequence analyses have predicted that subtle but specific differences are present in the M1 and NP proteins of the type A viruses isolated from mammalian and avian hosts. Influenza type C viruses differ from A and B viruses in some detail but are sufficiently similar to be included within the Orthomyxoviridae family. In addition to M1 and NP proteins, three polymerase proteins, a minor component of the virus particle, are also type-specific by serological analysis.

Subtype specificity is determined serologically by the nature of the major envelope antigens (i.e., HA and NA). Different subtypes are found only among the influenza type A viruses isolated from humans and many other warm-blooded animals. Although influenza viruses are species-specific (i.e., they produce disease and spread within the same host species but do not cause disease in other animal spe-

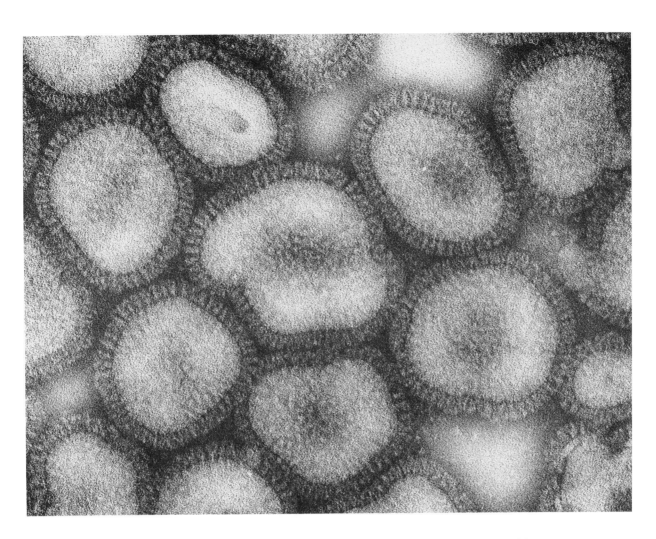

FIGURE 2 Electron micrograph of A/WSN/33 influenza virus by negative staining. ×600,000.

cies) in causing epidemics, they can infect hosts of other species in a limited way. For example, farmers working with animals have been found to possess antibodies against chicken, swine, or equine viruses. According to the present nomenclature, 13 HA subtypes (H1–H13) and 9 NA subtypes (N1–N9) have been found among the influenza type A virus isolates of humans and animals. Individual virus isolates are identified by subtype specificity (e.g., H1N1, H2N2, etc. where H1N1 denotes that the virus isolate possesses HA of subtype 1 and NA of subtype 1). Only three subtypes—H1N1, H2N2, and H3N2—have been isolated from human epidemics and pandemics. Influenza type B and type C viruses do not exhibit any subtype variation. These viruses are also known to cause epidemics in humans only and not in other animals.

B. Morphology and Composition

Influenza viruses are enveloped particles of variable shape (pleomorphic) and are approximately 110 nm in diameter with surface projections or spikes (Figs. 1 and 2). The majority of the particles are approximately spherical, but some are filamentous up to 1 μm long. Usually, fresh isolates are more pleomorphic and filamentous, but upon repeated passages and adaptation in cell culture, they exhibit a rather uniform spheroidal morphology (Fig. 2). An influenza virion (mature infectious virus particle) consists of a core of helical, segmented RNP (nucleocapsid; Fig. 3) enclosed by an envelope.

1. Viral Envelope

The viral envelope consists of a lipid bilayer and proteins. The lipid bilayer is derived from the membrane of the host cell; therefore, its composition varies with that of the host cell in which the virus is

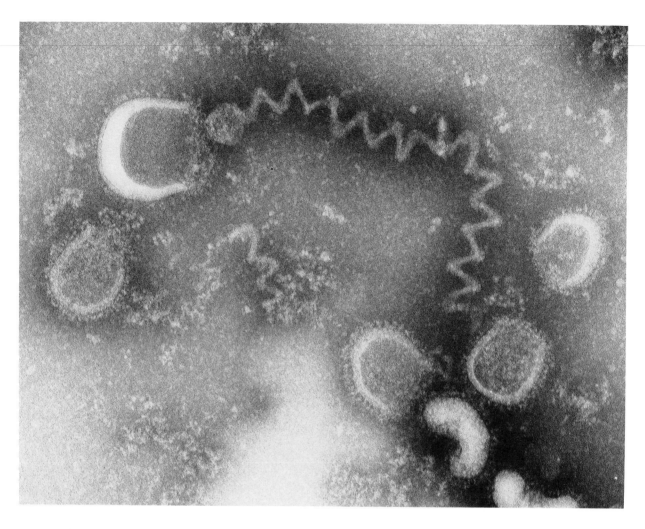

FIGURE 3 Electron micrograph of a detergent (Triton X)-treated A/WSN/33 influenza virus showing helical ribonucleoprotein (RNP). ×291,200.

grown. On the outer surface of virus, three transmembrane proteins are anchored in the lipid bilayer (Figs. 1 and 2). Of these, HA and NA form two types of surface spikes (or projections) (Figs. 1 and 2), and the third protein, M2, is only a minor component of the viral envelope (14–68 molecules per virion) (Fig. 1) with no known specific function. HA spikes are made up of three identical subunits. They are rod-shaped and approximately 16 nm long. NA spikes are made up of four subunits and are mushroom-shaped. HA spikes are five times more abundant than NA.

a. HA HA is the major surface antigen that elicits neutralizing antibodies. It also possesses the receptor-binding site by which the virus binds to the host cell receptor, the first step in the infectious process. HA is also responsible for the fusion of viral envelope with endosomal membrane at low pH (pH ≈ 5) and the entry of the viral RNP into the cell

to initiate the infectious cycle. HA is responsible for the ability of viral particles to agglutinate erythrocytes, by binding two or more of them at the same time, a phenomenon known as hemagglutination, which is commonly used for detection and quantitation of the viral antigen. HA, present on the surface of the virus-infected cells, causes them to bind red cells, a phenomenon known as hemadsorbtion, which can be used for detection and quantitation of virus-infected cells in cell cultures.

HA is a type *I* transmembrane glycoprotein in which the NH$_2$-terminal signal sequence is removed during translocation into the endoplasmic reticulum (ER) and, therefore, is absent in the mature HA. The COOH-terminal hydrophobic domain functions as a transmembrane domain anchoring the protein into the lipid bilayer of the membrane. HA is further cleaved at an internal hydrophobic domain to generate HA1 and HA2, which remain connected by a disulfide bond. The hydrophobic domain then becomes the NH$_2$ terminus of HA2. This cleavage is host cell protease-dependent and occurs during the transport of HA through the membrane-trafficking pathway. The cleavage of HA is not required for intracellular transport, for trimer (spike) formation, for assembly and budding of virus particles, for binding of virus particles to the cell-surface receptor, or for hemagglutination. The cleavage and the free NH$_2$ terminus of HA2 are absolutely essential for the fusion of the virus with the endosomal membrane at low pH and the entry of the viral RNP into the cytoplasm of the infected cell. Therefore, virus particles containing the uncleaved HA are noninfectious, although they can bind to the cell receptor and be internalized into the endosomes. In vitro treatment of the noninfectious virus particles containing the uncleaved HA with a protease, such as trypsin, generates HA1 and HA2 and renders the noninfectious particles infectious.

A typical HA polypeptide of influenza types A and B consists of 566 and 584 amino acids, respectively. For influenza C virus, a single transmembrane protein of 654 amino acids possesses both the receptor-binding (hemagglutination) and receptor-destroying (esterase) activity and is referred to as HE protein (see Section II.B.1.b). The sequence specificity of the cleavage peptide at the junction of HA1 and HA2 varies for different viruses, but it is invariably present in all HAs. Many of the structural features are conserved among the HAs of different virus isolates. HA contains a number of N-linked oligosaccharides, but their number, location, and composition vary with different subtypes and even among different strains of the same subtype. Oligosaccharides, although not essential for intracellular transport and function of the protein, provide the structural stability, protect against host proteases, and may play an important role in immunological modulation of viruses in epidemics.

The three-dimensional structure of HA has been determined by crystallography and X-ray diffraction analysis (Fig. 4). The trimeric HA possesses two distinct structural domains. (1) A triple-stranded coil of α-helix extends from the viral or the cellular membrane. In the monomeric HA, the fibrous stalklike region consists of a five-stranded β-sheet of the NH$_2$ termius of HA1 and the entire HA2, which forms a cylindrical α-helical stem (Fig.

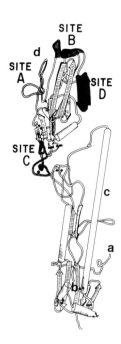

FIGURE 4 Three-dimensional schematic representation of the structure of an HA monomer. a, NH$_2$ terminus of HA2; b, COOH terminus of HA1; c, the fibrous stalk, which forms the stem of the HA trimer; d, globular head possessing four major antigenic epitopes—sites A–D, respectively. •• represents disulfide linkage. [Reproduced, with permission, from P. Palese and D. W. Kingsbury (eds.), 1983, "Genetics of Influenza Virus," Springer-Verlag, Wein and New York.]

4,c). This region of the monomeric HA interacts with other subunits of the trimeric HA to stabilize the stem of the trimeric HA spike. (2) The globular head at the distal end of the stalk consists of the major portion of HA1, forming an eight-stranded β-sheet structure. The globular head possesses the receptor-binding site (Fig. 4,d) and the major epitopes eliciting neutralizing antibodies, whereas the fibrous stem possesses the NH$_2$ terminus of HA2 (Fig. 4,a), which is required for the fusion and the infectivity of virus particles.

b. NA NA (also known as sialidase, EC 3.2.18) accounts for 10–20% of the spikes on the viral envelope. It is a protein with enzymatic activity, an exoglycosidase that breaks the linkage between sialic acid (*N*-acetylneuraminic acid) and an adjacent sugar residue. Major studies on the structure and function of NA have been done with influenza type A viruses.

NA appears to be critical for the release of virus from infected cells and to prevent aggregation among the virus particles by removing sialic acid residues from the surface of both virus and virus-infected cells. NA removes the same sialic acid residues that bind to the receptor-binding site of HA. Therefore, NA plays an important role in spreading the virus infection and possibly in determining the host specificity and virus virulence. Although they do not directly neutralize free virus particles, antibodies to NA may play an important role in reducing virus pathogenicity by restraining virus spread within the same host and its spread within the population. NA may also aid virus particles to move through the sialic acid containing mucous present in the upper respiratory tract, which is the natural route of infection, thus allowing them to find susceptible cells for initiating infection in a host.

NA is a type II membrane protein in which a single extended hydrophobic domain functions in both translocating the protein across the membrane of ER and anchoring it into the lipid bilayer of the membrane. NA spike is made up of four equal subunits and possesses a mushroom-shaped oblong structural head attached to a fibrous stalk. The tetrameric head can be removed by protease treatment without affecting the enzymatic activity or antigenic determinants of NA.

The three-dimensional structure of the globular tetrameric head of N2 NA has been determined by X-ray crystallography. The monomeric head (Fig. 5,a) consists of six propellerlike, four-stranded, antiparallel β-sheets in counterclockwise order; loop structures connecting the β-strands are variable and implicated in antigenic variation. Each monomeric head contains a catalytic site for the enzymatic activity, in a large depression on the top of the head (Fig. 5,a). Therefore, an NA spike will possess four catalytic sites 40° A apart. Like HA, NA also contains four to six N-linked oligosaccharides. The number and nature of oligosaccharides as well as the sites of glycosylation vary for different NA subtypes. Glycosylation is not obligatory for transport of NA to the cell surface but may aid in stabilizing the structure and the enzymatic activity.

Type B virus-infected cells contain an additional protein NB of 100 amino acids, which is synthesized from the same mRNA that also encodes NA. NB is absent in the virus particles and its function is unknown.

c. M2 protein   M2, a small protein of 97 amino acids, is a minor component of the viral envelope. M2 is encoded by a spliced mRNA of segment 7. Like NA, it is a transmembrane type II protein with

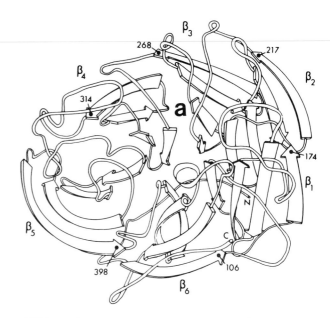

FIGURE 5   Schematic representation of the three-dimensional structure of the monomeric NA head. a, catalytic site for enzymatic activity. [Reproduced with permission, from R. M. Krug (ed.), 1989, "The Influenza Viruses," Plenum Press, New York.]

a signal–anchor domain, but unlike NA, its NH$_2$ terminus is out and COOH terminus in. The definite role of M2 in the infectious process remains undetermined. Changes of M2 affect host range (the cell types that can be infected by a virus strain) and sensitivity to the antiviral agent, amantidine or rimantidine.

d. M1 protein   The M1 protein is a major component of the viral envelope and resides underneath the lipid bilayer (Fig. 1). It is not a transmembrane protein and does not possess signals for translocation into the ER membrane or a transmembrane anchor domain, but it does possess two hydrophobic domains, which appear to interact with the inner-face of the lipid bilayer. It may interact with the cytoplasmic tail of HA, NA, or M2 on one side and RNP on the other side. It forms as an electron-dense layer beneath the lipid bilayer of the plasma membrane (Figs. 1 and 2), and it is postulated to function in the assembly of the viral particles. A defect of M1 protein synthesis has been implicated in the abortive replication of the virus in a number of cell lines.

2. Internal Components

The internal component (core) of an influenza virus particle is called the nucleocapsid, or *RNP*

(Figs. 1 and 3). RNPs consist of minus-strand RNA, NP, and three polymerase molecules (PA, PB1, and PB2). The viral genome consists of eight separate RNA segments, each essential for infectivity (influenza C virus has only seven RNA segments). Isolated RNP segments from virus particles are transcriptionally active and give rise to the synthesis of complementary mRNAs both *in vitro* and *in vivo*. Isolated RNPs introduced into cells can generate infectious particles.

a. NP RNPs possess helical structures (Fig. 3). The size varies from 50 to 150 nm depending on the size of the RNA segments. The ratio of NP–RNA is constant and independent of size. Approximately 1,000 NP molecules per virion are distributed on each of the eight RNP segments, like beads on a string. The RNP is probably coiled, each loop representing 25 nucleotides, arranged as a double-helical ribbon. The 5′ and 3′ termini of RNA form a double-stranded panhandle stem, which appears to be important in initiating transcription of viral mRNA. Three polyerases (PB1, PB2, and PA) are present in the RNP in equal numbers and form complexes located at one end of the RNP (Fig. 1). NP is a basic protein that is used for determining the type specificity of the virus. During transcription, NP probably provides scaffolding function for presenting RNA as a template. Active synthesis of NP is required for the replication of the viral RNA but not for the synthesis of mRNA.

b. Polymerase proteins The transcriptase (also called replicase or RNA-dependent RNA polymerase) consists of three proteins: PB1, PB2, and PA. They are the three largest proteins (approximately 750 amino acids each) and are encoded by the three largest viral RNA segments (segments 1, 2, and 3). These three proteins function as a complex (3P complex) in both the transcription of the viral mRNA and the replication of the viral RNA. Only about 15–20 molecules of 3P complex are present in an infectious virus particle. Both the transcription and replication of influenza virus RNA occur within the nucleus of the infected cell. Each polyermase protein possesses the nuclear localization signal and is actively transported from the cytoplasm into the nucleus. Influenza virus transcription is a unique process that uses as primer a short stretch (10–13 nucleotides) of the newly synthesized capped host mRNA and continues on the viral mRNA. Therefore, the completed viral mRNA will have the first 10–13 nucleotides at the 5′ end derived from the host mRNAs (Fig. 6). The PB2 protein appears to carry on the initiation process, whereas PB1 may be responsible for chain elongation. The function of PA is unknown at present.

c. Viral RNA Influenza A and B viruses contain eight RNA segments, whereas influenza C virus contains seven. Viral RNA segments range from approximately 2.3 to 0.8 kb, the largest ones encoding the polymerase proteins (PB1, PB2, and PA), and the smallest one encoding the nonstructural proteins (NS1 and NS2; see below). Each RNA segment codes for a single mRNA and a single protein except the two smallest RNA segments, each of which encodes two proteins. In these cases, one protein is encoded by the complete mRNA, and the other through a spliced mRNA. Because the viral RNAs present in the viral particles are minus-stranded, they cannot function as mRNAs. The RNAs complementary to the viral RNAs, made in the cells after infection, are plus-stranded and function as mRNAs for the synthesis of viral proteins in infected cells. All viral RNA segments possess similar sequences at their 5′ as well as their 3′ termini, which are partially complementary to each other (Fig. 6) and appear to form a double-stranded panhandle stem, perhaps involved in regulating transcription and replication.

C. Nonstructural Proteins

Two nonstructural proteins (NS1 and NS2), encoded by the smallest RNA segment (segment 8), are found in the virus-infected cells but not in virus particles. Their function is unknown. NS1 is transported into the nucleus and has been implicated in inhibition of the host mRNA synthesis.

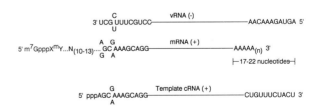

FIGURE 6 Schematic representation of three types of transcript found in influenza virus-infected cells. vRNA (−) is present in virus particles. mRNA (+) is used for translation of viral proteins in infected cells. Template cRNA (+) is used for the synthesis of vRNA (−). N, nucleotides; X, purine; Y, pyrimidine nucleotides.

II. Infectious Process (replication cycle)

The infectious process of influenza viruses has been studied extensively in suitable cell cultures and in experimental hosts such as embryonated chicken eggs. The replication process involves attachment (or adsorption), penetration, and uncoating; transcription and replication of viral RNAs; translation of viral proteins; assembly of viral proteins and viral RNAs into RNPs; and budding of virus particles.

Viral attachment is mediated by the interaction of the receptor-binding site of the HA spike to the host cell receptor, which is sialic acid (or N-acetylneuraminic acid) present ubiquitously in cellular glycolipids and glycoproteins. Single amino acid substitutions in the receptor-binding site of HA can affect the receptor-binding specificity. Immediately following attachment, if the virus is not internalized, NA removes the receptor sialic acids, and the virus moves on to the next receptor. Eventually, NA may remove all the receptors found on the cell surface. This phenomenon is observed when the virus attaches to erythrocytes (because erythrocytes cannot internalize virus particles). Initially, virus causes agglutionation of erythrocytes, but if the virus–erythrocyte complex is left a 37°C, NA will eventually remove all of the receptor sialic acids present on the surface of erythrocytes. These erythrocytes cannot be agglutinated again by the same influenza virus but may be agglutinated by another influenza virus that has a different sialic acid receptor. Thus, viral strains can be arranged in a series in which a strain can agglutinate cells that are not agglutinable by the preceding viruses. This phenomenon is known as receptor gradient. In live cells, virus particles attached to receptor are engulfed into the cell in a vacuole. The vacuole then fuses with an endosome, thereby lowering the pH of the vacuolar content. Around pH 5, the HA spikes undergo destabilization, and the free hydrophobic NH_2 terminus of HA2 interacts with the membrane of the endosome. This process leads to the fusion of the lipid bilayer of the viral envelope with the lipid bilayer of the endosomal membrane, thereby releasing the internal content of the virus (the viral RNP complex containing three polymerase [3P] proteins) into cytoplasm of the infected cell.

After the RNP–polymerase complex is released in the cytoplasm, it is transported into the nucleus of the infected cells, where transcription and replication of the viral RNA occur. The first event after uncoating is the synthesis of plus-stranded viral mRNA (transcription) by the polymerase (3P) complex using the incoming minus-stranded RNP as the template (primary transcription). Transcription requires host cell mRNA synthesis to provide the capped oligonucleotide primer needed for the initiation of viral mRNA synthesis. At the 5' end, the mRNAs made will have a capped oligonucleotide of 10–13 bases derived form the host mRNA, followed by the body of the RNA, which is the complementary form the second nucleotide of viral RNA but lacks the last 17–22 nucleotides and possesses a poly(A) track at its 3' end (Fig. 6).

Replication of viral RNA requires two steps, which are different from the mRNA synthesis. The first step involves the synthesis of the complete (full-length) complementary (+ strand template) RNA (cRNA) without either the capped host mRNA primer at its 5' end or the poly(A) track at its 3' end. This cRNA (+ strand) is used as a template for viral RNA (− strand) synthesis (Fig. 6). The switch from synthesis of mRNA to that of full-length cRNA requires continued synthesis of viral proteins, particularly NP, which binds the newly synthesized RNA to form RNP. Therefore, NP has been implicated in both providing the scaffolding function for presenting the RNA template and the antitermination function for full-length RNA chain elongation. RNA synthesis is coupled with the assembly of viral RNA into RNP. The ratio of viral RNA–cRNA is >10 in infected cells. [see RNA REPLICATION.]

When viruses are passaged at a high multiplicity of infection, they generate defective interfering (DI) virus particles. These defective particles contain shorter viral RNA (DI RNA) segments, which are internally deleted. These DI RNAs are primarily generated from the three largest RNA segments encoding 3P proteins. These DI particles are not infectious and need the helper function of infectious (or standard) influenza virus particles for replication. Therefore, they will replicate only in cells coinfected by infectious virus particles. In such coinfected cells, DI particles replicate, interfere with the replication of infectious virus particles, and become predominant particles at the expense of infectious particles.

Synthesis of viral proteins continues at a high rate through the infectious cycle, although mRNA synthesis is greatly diminished in the late phase. Therefore, viral mRNAs made in the early part of the cycle continue to function in translation in the late phase. 3P proteins, M1, NS1, and NS2 are synthesized on the free cytoplasmic polysomes, whereas HA, NA, M2, and NB are synthesized on polysomes bound to the membranes of the ER. These transmembrane proteins follow the same trafficking pathway as the host transmembrane proteins and are glycosylated during transport. HA and NA appear to form complexes (trimer for HA or tetramer for NA) in the ER membrane soon after their synthesis.

Assembly of the virus particles occurs at the plasma membrane. The site of virus assembly is modified by the accumulation of HA and NA spikes, which exclude the majority of the host cell membrane proteins (patching); subsequently, M1 proteins are attracted on the inner layer of the membrane. Viral RNPs, after their synthesis in the nucleus, are transported into the cytoplasm and eventually accumulate underneath the M1 protein at the assembly site; eventually, virus particles will bud off from the plasma membrane. How each of the eight RNP segments are selected and packaged within a virus particle is unknown. Absence of either NA or HA alone does not appear to prevent budding. In polarized epithelial cells (e.g., epithelial cells of respiratory tract), HA and NA are directed to the apical side of the plasma membrane and viral particles bud only from that side. The external domain of HA appears to possess the signal for sorting and insertion into the apical side of the cells.

Influenza virus kills the infected cells. The mechanism of cell-killing may involve (1) cleavage and degradation of newly synthesized cellular mRNA, which is used as a primer for viral mRNAs; (2) suppressing synthesis of the cellular proteins; and (3) modification of the plasma membrane, which may contribute toward *in vivo* killing of the virus-infected cells by host immune response (cytotoxic T cells).

III. Pandemics and Epidemics

Flu (or influenza) outbreaks appear almost predictably as epidemics in fall and end in spring. Although the severity of outbreaks may vary from year to year, they are responsible for an increase in morbidity and mortality during that period. Because the virus cannot survive outside the host for a long time and the virus does not produce latent or chronic infections in patients, it must persist by spreading from one person to another, evading the immune defenses of the infected persons. Two unique prop-

erties of the virus appear to be at the root of its epidemic behavior. (1) The highly mutable nature of its RNA permits the virus, in the presence of the host immune response, to progressively change its antigenic specificity. Such antigenic drift due to mutation is responsible for the annual epidemics. The frequency of mutation is one per 10^5–10^6 nucleotides per infectious cycle, allowing for a high frequency of antigenic mutations. (2) The segmented nature of the viral genome and the availability of a large gene pool of nonhuman influenza A viruses provide an unusual opportunity for exchanging viral genes by reassortment, giving rise suddenly to entirely new antigenic subtypes. Such antigenic shift is believed responsible for worldwide pandemics.

Epidemics are defined as geographically limited outbreaks of the disease in the human population. Analysis of virus isolates from epidemics have shown that new mutant viruses continue to emerge almost year after year. Although mutations can occur in any of the eight RNA segments, the selection pressure of the host antibodies leads to the emergence of antigenic variants. There are four major antigenic sites (epitopes) on the globular head of HA1 (Fig. 4). For antigenic variants to survive and produce disease, all four of these epitopes must be sufficiently altered in the new variant virus to escape the host immune response elicited by a prior infection. These changes may accumulate over a number of years. These new variants produced by antigenic drifts are responsible for annual epidemics. Because some of the minor antigenic determinants may not be altered and a majority of the human population is likely to have antibodies against the parent virus, the disease produced by the antigenic variants tends to be milder and epidemics appear to be less widespread when compared with the worldwide pandemics.

Unlike epidemics, *pandemics* are global in nature, they rapidly spread from one continent to another, and the entire human population (>80%) is affected within a short period of 1 yr or so. The jet age mobility aids in the rapid spread of the virus. In this century, three major influenza pandemics have been documented. The first, the swine flu pandemic of 1918, was one of the deadliest pandemics in human history. It killed over 20 million people worldwide and 500,000 in the United States alone. Subsequently, two other major pandemics occurred in 1957 (Asian flue, or Japan flu) and in 1968 (Hong Kong flu). The nature of the 1918 virus was later determined indirectly from the antibody analyses of the sera of the people afflicted that year. Viruses have been isolated from the other two pandemics and have been characterized in detail. These analyses show that pandemics are caused exclusively by type A viruses. An H1N? (NA subtype unknown) swine influenzalike virus was responsible for the 1918 pandemic, an H2N2 virus for the 1957 pandemic, and an H3N2 virus for the 1968 pandemics. In addition, in 1946–1947, a major epidemic was caused by an H1N1 virus.

The characteristics of pandemic viruses are different from those of the epidemic viruses. Pandemic viruses possess either an HA or both HA and NA, which are totally different from the pre-existing human viruses in circulation at that time. Because HA elicits the neutralizing or protective antibodies, the entire human population is susceptible to the new virus. In addition to having either a new HA or both HA and NA genes, the new pandemic virus must be virulent (i.e., it must be able to replicate efficiently in humans, spread in the population, and cause disease).

How and why a pandemic virus emerges is not definitely known, but the following scenario is the most likely: Influenza A viruses are present in humans as well as in many other warm-blooded animals and birds. They are species-specific; however, humans are known to become occasionally infected with animal influenza viruses (e.g., farmers have been found to have antibodies against swine, equine, or avian viruses). Animal viruses provide a divergent pool of viral genes (13 HA and 9 NA subtypes); therefore, a person (or an animal) can be infected at the same time by a human and an animal virus, both viruses replicating in the same cell. Because the viral genome is segmented, genetic reassortment will take place during the assembly and packaging of viral RNPs and the budding of the virus particles. Some of these particles will contain some RNA segments from one virus, other segments from the other virus. Some of the new virus particles will retain the virulence properties for humans while replacing their old HA or both HA and NA with those of the animal viruses. This model of gene reassortment between two or more viruses may be responsible for the emergence of pandemic strains and is supported by various observations. Genetic reassortment between two or more viruses is seen very frequently in the laboratory when the same cell (or the same animal) is experimentally infected with two different viruses. In fact, this procedure routinely is applied for making current influenza vaccine in chicken embryos. Each year, the newly isolated candidate virus, which is predicted to cause epidemics next year, is grown together with a high-yielding A/PR/8 virus to generate the high-yielding vaccine virus, which will contain the HA and the NA of the new strain. Furthermore, the evidence for reassortment was found in the H3N2 virus responsible for the 1968 pandemic: it possessed seven genes derived from the H2N2 virus present at that time in the human population, except for the HA gene. This HA gene (H3) has the closest resemblance with the HA of an equine and a duck virus, suggesting that a similar HA gene was acquired by the new pandemic H3N2 virus.

However, as yet no definitive evidence indicates that genetic reassortment between a human and an animal virus actually was responsible for the emergence of either the 1918, 1957, or 1968 pandemic virus. Also, numerous questions remain unanswered. For instance, although many subtypes are present in influenza viruses of animals, why only three HA subtypes (H1, H2, and H3) have been found among human viruses is not clear. Furthermore, conditions leading to the advent of pandemic are not completely understood, making it impossible to determine when the next pandemic will occur.

IV. Disease

Clinical manifestations of influenza virus infection vary greatly from person to person. Many host factors such as age, immunological status (i.e., prior exposure but to antigenically related viruses), physiological state, and many other factors may influence the disease syndrome. Infection may be subclinical (i.e., asymptomatic) or may produce mild symptoms such as fever or cold or may cause an acute prostrating febrile illness accompanied by general and respiratory symptoms. These may include pharyngitis, tracheitis, bronchitis leading to shortness of breath, cough, and myalgia. In severe cases, it may lead to primary viral pneumonia accompanied with subbronchial hemorrhage and even fatal outcome. The incubation period is short (1–2 days); onset is sudden with fever (temperature up to 39°C or even higher), chill, and myalgia. In young children, common symptoms are pharyngotracheobronchitis, croup, and even gastrointestinal disorder. In a typical flu in adult, fever may subside in 3 days, but respiratory symptoms and general weakness may persist for 2–3 wk. The disease may be severe in very young or elderly patients or in patients with other pre-existing underlying conditions. These may include pregnancy, cardiovascular diseases, respiratory conditions such as chronic bronchitis, bronchopulmonary neoplasia, and asthma as well as emphysema in patients with immunocompromised conditions.

The virus is spread by aerosol droplets or by fingertips. During the winter period, indoor living and closeness of people in an enclosed environment may aid in spreading the infection from person to person. No clear correlation exists between the severity of the winter with either the frequency and severity of the epidemics and the clinical disease.

V. Pathogenesis

The virus replicates and destroys the respiratory epithelium, and much of the respiratory symptoms including the severity of the disease appear to be related to the extent of desquamation of respiratory epithelial cells and the site of destruction (i.e., upper, middle, or lower respiratory tract). Experimental studies in animals suggest that influenza is a descending type of infection (i.e., virus infects either the upper respiratory or the tracheal epithelium and later the infection spreads to the bronchial epithelium or even to alveolar cells causing primary viral pneumonia). Although the site of primary infection may vary, depending on the virulence of the virus, dose, and method of inoculation, the ciliated epithelium of the middle respiratory tract apparently is the preferred primary site of initial infection. Usually, the virus is shed for about 1 wk, and for longer times in rare cases with persistent pulmonary disease.

The respiratory symptoms can be explained on the basis of the site and extent of viral pathology, but the mechanism of general symptoms such as myalgia, anorexia, malaise, or gastrointestinal disorder remains unexplained, especially because influenza infection usually does not lead to viremia (presence of virus in blood). A number of unusual conditions such as myopathy, myocarditis, and encephalopathy have been occasionally associated with influenza infection, again without a rational basis for an explanation.

VI. Virulence

Although in an epidemic the variation in the disease syndrome from one person to another is primarily due to host factors (because the same virus strain infects all persons), factors associated with viral virulence are critically responsible for the overall nature and severity of the epidemics. Basically, the properties that regulate replication efficiency and cell-killing ability also determine the virulence of a virus strain. Influenza virus virulence is determined by several viral genes. For example, the cleavage of HA by a host protease appears to be critical in producing infectious virus; the length and the nature of the amino acids of the connecting peptide between HA1 and HA2 have been implicated in virulence. NA is also an important determinant for spreading of virus infection and, therefore, is an important factor in virus virulence. Each of the three polymerase proteins as well as NP, M, and NS1 may also play an important role in virulence and in determining which cells are infected.

All pandemics and most epidemics are caused by influenza A viruses, but influenza B epidemics are not uncommon. Both influenza A and B viruses produce similar symptoms, although influenza B may be more common in children than in young adults. The majority of influenza B virus infections appear to be less severe than influenza A infections, although influenza B virus has the potential to produce severe and even fatal illness. Infections due to influenza C virus are less common and produce milder symptoms than those due to either A or B viruses. They are also less frequently associated with any major epidemics.

VII. Host Response

The progression and outcome of a disease in humans is a complicated process and is based not only on the viral virulence but also on the host response. Following virus infection, initial host response results in the production of interferon, which is likely to slow the process of virus replication. Subsequently, the induction of three major classes of antibodies (IgM, IgG, and IgA) as well as cellular defense mechanisms (cytotoxic T-cell response) aid in limiting the disease and in promoting recovery. IgM is produced after primary infection, and subsequent response is limited to IgG and IgA. Although antibodies are generated against all viral components, the antibodies against HA and NA provide protection and limit the virus spreading within the host. Antibodies against M2 may also aid in preventing viral spread. After recovery, the person becomes solidly immune against reinfection by the same influenza virus. Secretory IgA in the nasal and respiratory mucosa is effective in preventing reinfection. Humoral IgG against HA provides the neutralizing antibodies. Protection against reinfection by the same virus is easily assayed by hemagglutination inhibition. Antibodies against NA do not neutralize virus but limit virus spreading and infection of new cells. Cytotoxic T cells are generated against both the surface (HA) and internal viral components (particularly NP).

VIII. Complications

The most frequent cause of the complication is secondary bacterial infections resulting in bacterial pneumonia. Pneumococci, *Haemophilus influenza*, and staphylococci have been isolated most frequently from fatal cases of influenza. The onset of bacterial infection usually follows after the symptoms of viral infections have subsided and is accompanied by the return of sudden chill, fever, productive cough, chest pain, and shortness of breath. These bacterial complications are more common in patients with pre-existing lung conditions or in immunocompromised patients. The pneumonia due to secondary bacterial complication is essentially the same as the disease caused by the bacteria alone in absence of a virus infection and follows a similar course.

Other uncommon complications are encephalitis, myocarditis, and postinfection neuritis, which is indistinguishable from the Guillain–Barré syndrome. Another serious and often fatal complication called Reye's syndrome has been implicated with influenza B virus (rarely with influenza A virus) and varicella virus along with aspirin (salicylates) medication. It is an acute noninflammatory encephalopathy (cerebral edema accompanied with fatty degeneration of liver). The onset of the disease occurs within 4–7 days of infection with repeated vomiting, disorientation, delirium, and coma.

IX. Diagnosis

A definitive diagnosis based on the clinical symptoms alone cannot differentiate influenza virus infection from other similar flulike conditions. Many other viruses such as respiratory syncytial virus, adenoviruses, coronaviruses, rhinoviruses, and parainfluenzaviruses may cause flulike symptoms. Definitive virological diagnosis can be done only in the laboratory. For rapid laboratory diagnosis, specimens may be collected by swabbing the posterior pharynx; viral antigen can be detected in the nasopharyngeal cells by staining with antibodies conjugated to an immunofluorescence dye. Specimens from swabs can also be used for growing virus in appropriate cell culture madin darby canine kidney (MDCK cells) or chick embryos, in which virus can then be detected by immunofluorescence, hemadsorption, or hemagglutination. Specific antibodies against internal antigens (NP and M1) are used for determining the viral type and antibodies against HA and NA for subtype. Usually, such diagnoses are carried out by public health laboratories and are important in determining the nature of the epidemic and its cause.

X. Treatment

Influenza usually is a short illness with uneventful recovery; therefore, general supportive treatment is recommended in most cases. This includes rest, light diet, and avoiding dehydration. Other symptomatic treatment to reduce fever or suppress cough may be given as required. In secondary bacterial pneumonia, appropriate antibiotics should be used. In addition, in viral or bacterial pneumonia, oxygen therapy may be needed to reduce hypoxemia.

Amantidine (or rimantidine) hydrochloride (2×200 mg/day) has been shown to be beneficial in reducing the intensity of the disease. It is effective against influenza A but not against B and can be used as a prophylactic treatment during an ongoing epidemic in high-risk groups. Amantidine appears to affect the entry of the virus by preventing dissociation of the M protein from RNP during uncoating and may also affect the budding and assembly process. Some of the side effects produced by amantadine are headache, confusion, insomnia, and slurred speech. Amantidine treatment may lead to the emergence of amantidine-resistant virus strains.

XI. Prophylaxis

Because influenza infection leads to a solid immunity against the same virus, many attempts have been made to produce a safe and effective vaccine against flu epidemics. Various kinds of vaccine have been made. The current vaccine available in the United States is either inactivated whole virus vaccine or a subunit vaccine containing components of the viral envelope (HA and NA). It is administered twice 1 mo apart subcutaneously or intramuscularly. Vaccine should be given annually before the flu season (i.e., early fall). Protection is approximately 85%. To be most effective, the vaccine virus must be of the same subtype as the current epidemic; this poses a logistical problem because the vaccine virus strains to be used in the next fall are usually selected by the health authorities earlier in the spring.

Epidemiological criteria, existence of new virus variants currently in circulation (i.e., severity, number of cases, and rapidity of the spread to multiple geographical locations of the disease) are used to predict the next year's epidemic strain. The vaccine strain is then converted to a high-yielding virus for growing in chicken embryos by reassortment with A/PR/8 virus, which grows best in embryonated chicken eggs. Although epidemic virus may differ from the vaccine strain, vaccination, in addition to eliciting the homologous (variant-specific) antibodies, also boosts the level of antibodies against the previous influenza virus and induces significant heterovariant (homosubtypic) as well as homotypic (heterosubtypic) antibodies. This phenomenon, called "original antigenic sin," is due to cross-reactive amnestic response and results from the stimulation of the cross-reactive memory B cells. Hemagglutination inhibition titer correlates with virus neutralization and provides an indicator of protection against a specific subtype virus. To be more effective, the vaccine must produce antibodies against both HA and NA as obtained with the complete unmodified virus. Vaccines containing only the soluble form of HA are much less effective than the intact HA in producing neutralizing antibodies. Cytotoxic T-cell response is produced only inefficiently by the current vaccine.

Vaccination with whole virus sometimes produces side effects including pain at the site of inoculation and febrile condition. Subunit vaccine reduces these toxic reactions greatly. During the 1977 swine influenza campaign, Guillain–Barré syndrome was an unconfirmed complication of vaccination.

In summary, the current vaccine elicits good immune response and is effective in reducing the morbidity and mortality in the target groups. The recommended target groups are

1. adults and children with chronic cardiovascular and pulmonary disorders,
2. residents of nursing homes and other chronic care facilities,
3. medical personnel treating high-risk groups,
4. immunocompromised patients,
5. otherwise healthy individuals >65 yr old, and
6. adults and children with chronic metabolic diseases.

Acknowledgment

This research was supported by grants from the National Institutes of Health. The electron micrographs were kindly provided by Dr. K. G. Murti.

Bibliography

Kilbourne, E.D. (1987). "Influenza." Plenum Medical Book Co., New York.

Krug, R. M. (ed.) (1989). The Influenza Viruses." Plenum Press, New York.

Nayak, D. P. (1978). The biology of myxoviruses. In "Molecular Biology of Animal Viruses" (D. P. Nayak, ed.). Marcel Dekker Inc., New York.

Nayak, D. P. (ed.) (1981). "Genetic Variation among Influenza Viruses." Academic Press, New York.

Nayak, D. P., Chambers, T. M., and Akkina, R. K. (1985). Defective interfering RNAs of influenza viruses: Origin, structure, expression and interference. *Curr. Top. Microbiol. Immunol.* **114,** 104–151.

Nayak, D. P., and Jabbar, M. A. (1989). Structural domains and organizational conformation involved in the sorting and transport of influenza virus membrane proteins. *Annu. Rev. Microbiol.* **43,** 465–501.

Palese, P., and Kingsbury, D. (eds.) (1983). "Genetics of Influenza Viruses." Springer Verlag, Wien and New York.

Selby, P. (ed.) (1976). "Influenza: Virus, Vaccines and Strategy. Academic Press, New York.

Stuart-Harris, C. H., and Porter, C. W. (eds.) (1983). "The Molecular Virology and Epidemiology of Influenza." Academic Press, New York.

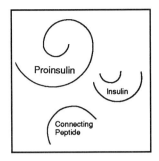

Proinsulin
Insulin
Connecting Peptide

Insulin and Glucagon

J. LARNER, *Department of Pharmacology, University of Virginia School of Medicine*

Glossary

Amino acids Molecules that are the building blocks of proteins. They are characterized by having an acidic carboxyl group next to a carbon with an attached basic amino group. They exist as families. One family has only the basic structure (i.e., glycine and alanine); a second family has an extra acidic carboxyl group (i.e., aspartic acid and glutamic acid); a third has an extra basic group (i.e., arginine, histidine, and lysine); a fourth has a sulfur atom present (i.e., cysteine and methionine); a fifth has a hydroxyl present (i.e., serine, threonine, and tyrosine); a sixth has a ring structure present (i.e., tryptophan, tyrosine, phenylalanine, and proline); a seventh has a fatty backbone (i.e., valine, leucine, and isoleucine). Of importance in this article are cysteine, which has a sulfur (S) atom involved in the formation of S–S bridges, and serine, threonine, and tyrosine, which contain a hydroxyl group capable of being phosphorylated.

Cyclic nucleotides Nucleotides in which the phosphate unit present is joined back on the nucleotide to form a cyclic structure

Dalton Unit of mass, equivalent to one hydrogen atom; abbreviated Da

Dehydrogenase Enzyme in cells that catalyzes (i.e., speeds up) the process of removal of hydrogen atoms from chemical species (e.g., fatty acids)

Dephosphorylation Removal of a phosphate unit. In the present context we refer to the specific phosphorylation of the amino acids serine, threonine, and tyrosine. A phosphate unit can be joined to or removed from the hydroxyl group of the amino acid.

Diacylglycerol Compound composed of glycerol joined to two fatty acids; it remains when inositol and phosphate are removed from an inositol phospholipid

Inositol Cyclic sugar alcohol

Inositol phospholipid Phospholipid that contains inositol

Ketone bodies Completely oxidized, fatty acids, the products of which are CO_2 and H_2O, which are generally not harmful. When fatty acids are not completely oxidized, ketone bodies accumulate, which are remaining sections of fatty acids and include acetone, acetoacetic acid, and hydroxybutyric acid. These products are acidic in nature and disturb the body's metabolic pH and ionic balance.

Nucleosides Nucleotides without the phosphate unit present

Nucleotides Building blocks of nucleic acids

Phospholipid Fatty molecule containing glycerol joined to two fatty acids; it contains a phosphate and an additional unit (e.g., inositol, ethanolamine, or choline)

Phosphoprotein phosphatase Enzyme that catalyzes the removal of phosphate groups from hydroxyl groups on serine, threonine, or tyrosine in proteins

Phosphorylation Addition of a phosphate unit, which is

$$\begin{array}{c} \overset{\displaystyle O}{\underset{\displaystyle OH}{HO-\overset{\|}{\underset{|}{P}}-OH}} \end{array}$$

Polypeptides Small proteins; they, along with proteins, are biopolymers made by assembling one or more of the 20 amino acids in a specific order, or sequence

Processing Proteolytic cleavage of polypeptides at specific sites to convert polypeptide precursors into active products (e.g., hormones)

Protein kinase Enzyme that catalyzes the phosphorylation of hydroxyl groups of serine, threonine, and tyrosine in proteins using ATP as the phosphate donor. Tyrosine protein kinases are specific for tyrosine.

Proteolysis Process of breaking down polypeptides or proteins by breaking the bonds that join the amino acids together. Complete breakdown can occur to amino acids, or partial breakdown to peptides can occur.

S–S bonds Bonds between two sulfur atoms, joining two cysteines together in polypeptides

INSULIN AND GLUCAGON are hormones made in the islets of Langerhans, organized nests of cells interspersed throughout the pancreas, an organ which has a major separate function in food digestion. Hormones are classically defined as chemical messengers, made in an organ of origin termed endocrine, secreted into the bloodstream on demand, then acting at other specific organs, called target organs. Hormones are a class of chemical communicators. Insulin is synthesized in specific cells termed β cells in the central part of the islets of Langerhans, and glucagon is synthesized in specific cells called α cells in the outer part of the islets.

I. Structure

Insulin and glucagon are polypeptides, or small proteins. Insulin is composed of two polypeptide chains, A and B, joined by two S–S bonds. Insulin has a molecular size of about 6000 Da. A precursor, proinsulin, is first synthesized in the β cells of the islets of Langerhans and is then processed proteolytically to yield the mature two-chain molecule. An intermediate connecting peptide is also formed in the processing (Fig. 1). [See POLYPEPTIDE HORMONES.]

Glucagon is a single-chain polypeptide. It has a molecular size of about 3500 Da. It is also processed in the α cells of the islets by proteolytic cleavage of a large precursor, proglucagon (with a molecular size of 18,000 Da) (Fig. 2). A related molecule, glycentin (with a molecular size of about 7000 Da),

somewhat larger than glucagon, but still containing glucagon within it, is formed from proglucagon in the stomach and the gastrointestinal tract.

The three-dimensional structures in space of both hormones have been determined by X-ray crystallography. Insulin exists in the crystalline state as a hexamer, in which six molecules are joined to two atoms of zinc (Zn^{2+}). Glucagon exists as a trimer, with three associated molecules.

II. General Actions *in Vivo*

In the body insulin and glucagon generally have opposing actions. For example, insulin lowers blood glucose concentration, whereas glucagon raises it. Insulin acts to promote anabolic (i.e., building up) processes in target organs by increasing the uptake of glucose, amino acids, fatty acids, and nucleosides into cells and then promoting their conversion to the larger macromolecules: glycogen, protein, lipids, and nucleic acids. As part of its overall effect to increase anabolism, insulin also acts to decrease catabolic (i.e., breaking down) processes as well.

In contrast, glucagon facilitates the breakdown of macromolecules (i.e., catabolism). As part of its control of catabolism, glucagon acts to decrease anabolic processes in target organs.

III. Insulin-Specific Actions *in Vivo*

The major target organs for insulin in the body are liver; striated, or voluntary, muscles (including the heart); and adipose tissue, or fat. Insulin is also known to act on other organs as well, including specific areas of the brain, the kidney, and the eye, but its effects at these sites may be of lesser importance, at least in terms of whole-body metabolism. In terms of carbohydrate and lipid metabolism, insulin stimulates the increased transport of glucose into the cell, where it is built up into glycogen (i.e., animal starch). This occurs to a major extent in muscle, which contains about half of the total body glycogen. A minor portion of the glucose is also broken down to lactate, in a process called glycolysis, and then oxidized to CO_2 and H_2O to provide energy, in the form of ATP, needed to synthesize the glycogen and other anabolic macromolecules (Fig. 3-A). [See GLYCOLYSIS.]

In liver (Fig. 3A) insulin does not increase the transport of glucose into the cells, but it does stimu-

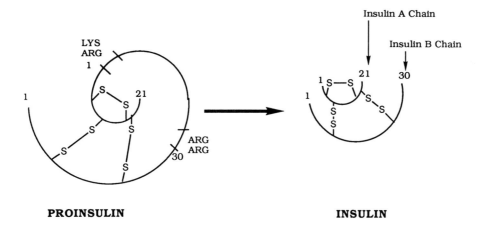

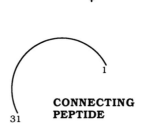

FIGURE 1 The conversion of proinsulin to insulin, together with the connecting peptide. The insulin A chain contains 21 amino acids. The insulin B chain contains 30 amino acids. The connecting peptide contains 31 amino acids. Two basic amino acids, arginine (ARG) and lysine (LYS), are the junction points at which the processing occurs.

late the synthesis of glycogen by activating the enzyme that produces it, glycogen synthase. A principal action of insulin is to decrease the output of glucose, the major source of blood sugar.

In adipose tissue insulin stimulates both the transport of glucose into the cell as well as its conversion to glycogen and fat, the major anabolic end product (Fig. 3A). This pattern is similar to the action in muscle, but the distribution of end products favors glycogen in muscle and fat in adipose tissue. In all three tissues glycogen synthase becomes activated.

In terms of protein metabolism, in all major insulin-sensitive cells and organs insulin stimulates the transport of amino acids into cells and their biosynthesis into proteins (Fig. 3B). Insulin also stimulates the biosynthesis of nucleic acids, both DNA and RNA, from their building blocks, the nucleosides and the nucleotides, leading to cell growth and cell division in sensitive cells (Fig. 3C).

The anticatabolic actions of insulin include decreasing fat breakdown (i.e., lipolysis) in fat (Fig. 3A), decreasing protein breakdown (i.e., proteolysis) in the liver and in muscle (Fig. 3B), and de-

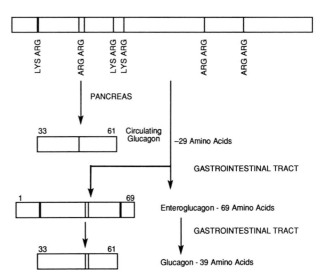

FIGURE 2 The 39-amino acid glucagon and the 69-amino acid enteroglucagon (i.e., glicentin) are processed from the 18,000-Da proglucagon precursor. Two basic amino acids, arginine (ARG) and lysine (LYS) are the junction points at which the processing occurs.

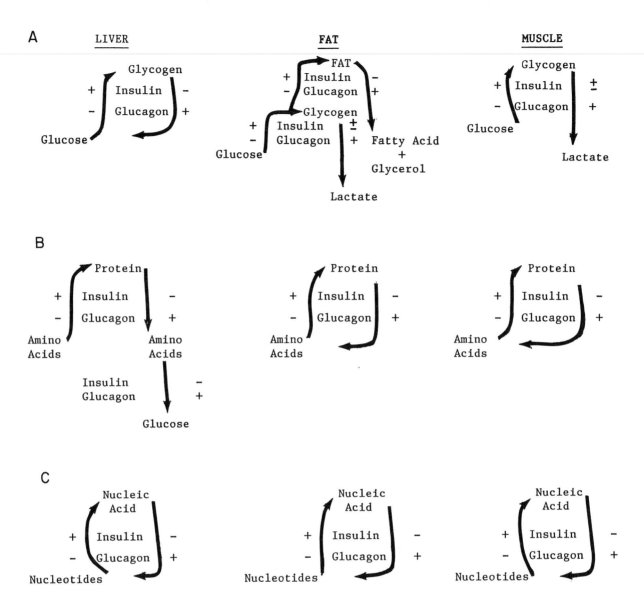

FIGURE 3 Regulation of anabolic and catabolic pathways by insulin and glucagon. (A) Carbohydrate metabolism. (B) Protein metabolism. (C) Nucleic acid metabolism. +, Increase; −, decrease; ±, small variable effect.

creasing glucose formation, chiefly from amino acids, but also from lactate and glycerol (i.e., gluconeogenesis) in the liver (Fig. 3B). To summarize, insulin acts in all tissues to promote an anabolic state. This is accomplished by stimulating the uptake of metabolites into the cells, as well as anabolic reactions, and decreasing catabolic reactions.

IV. Glucagon-Specific Actions *in Vivo*

The major organ in which glucagon acts in the body is the liver, but it also acts in fat and in the heart. In general, catabolic processes are accelerated. In terms of carbohydrate metabolism, these include the breakdown of glycogen to glucose in the liver

(i.e., glycogenolysis), a major effect of the hormone (Fig. 3A), as well as a breakdown to lactate in the heart and in fat. Fats are broken down to fatty acids and glycerol (i.e., lipolysis) in all sensitive tissues, but principally in adipose tissue. Proteins are broken down to amino acids (i.e., proteolysis) (Fig. 3B); nucleic acids (DNA and RNA) are broken down into component nucleotides (Fig. 3C). Glucose formation, chiefly from amino acids, but also from lactate and glycerol (i.e., gluconeogenesis), is stimulated in the liver by glucagon (Fig. 3B). When

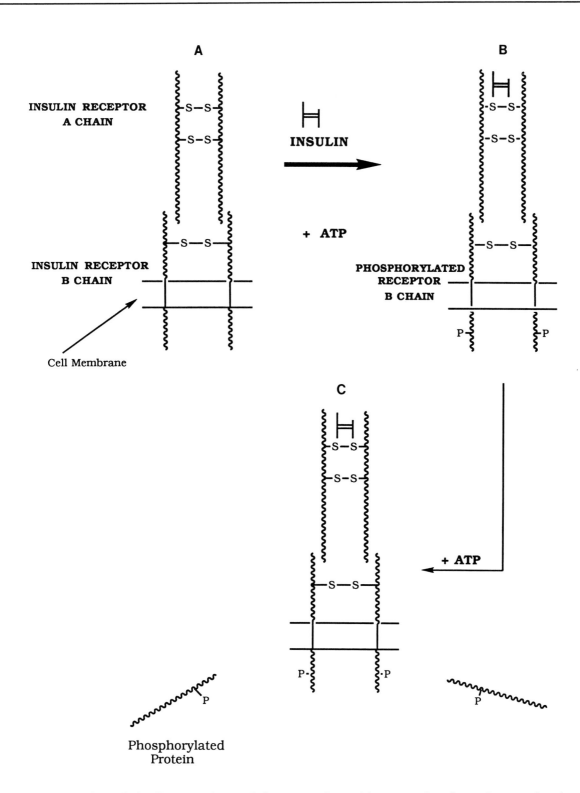

INSULIN RECEPTOR
A CHAIN

INSULIN RECEPTOR
B CHAIN

Cell Membrane

INSULIN

+ ATP

PHOSPHORYLATED
RECEPTOR
B CHAIN

+ ATP

Phosphorylated
Protein

FIGURE 4 Insulin action at the insulin receptor-increased phosphorylation. The insulin receptor and insulin combine (A). As a result there is autophosphorylation of the insulin receptor B chain, altering receptor structure and activating the tyrosine protein kinase (B). This activation leads to increased protein phosphorylation (C).

amino acids are used to form glucose, the nitrogen is released and is converted in the liver to urea. Urea synthesis is thus increased with glucagon action. In general, the anabolic reactions, including glycogen, fat (Fig. 3A), protein (Fig. 3B), and nu-

A   B

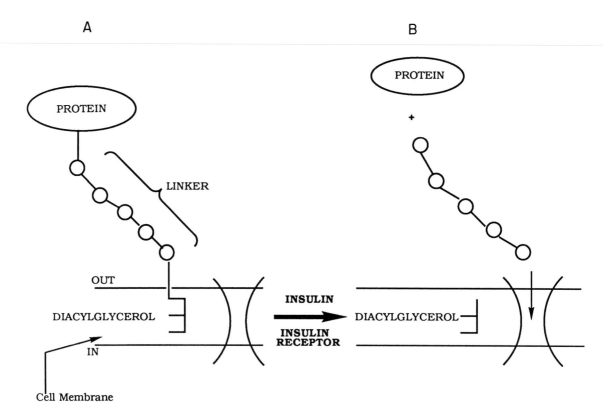

cleic acid synthesis (Fig. 3C), are all inhibited by glucagon. With increased fat breakdown and inhibited fat synthesis, fatty acids are incompletely oxidized to CO_2 and H_2O, and as a result ketone bodies are formed under the influence of glucagon.

V. Diabetic State

The absence of insulin causes diabetes; if the disease is uncontrolled, catabolic processes are set free and run in an unimpeded fashion, leading to cellular and organ breakdown. Catabolic products accumulate in the blood and are excreted in the urine. Insulin is required for life. In the diabetic state an excess of glucagon is present and, together with an absolute or relative deficit of insulin, doubly compounds the accelerated catabolic breakdown of cells and organs and the accumulation and urinary excretion of catabolic products.

VI. Mechanism of Action of Insulin

The initial interaction of insulin with a sensitive cell occurs at a specific receptor on the outside surface of the cell. The structure of the insulin receptor is

FIGURE 5 The action of insulin at the insulin receptor leads to mediator (shown as linker) formation, together with increased diacylglycerol and the release of protein. Mediator (shown as linker) and protein are released outside of the cell. Diacylglycerol is formed in the membrane. Mediator (shown as linker) is then taken up by the cell of origin or by neighboring cells. (A) before insulin action; (B) after insulin action.

shown in Fig. 4A. It is a molecule with four polypeptide chains: two A chains each 130,000 Da in size, located entirely on the outside of the cell, and two B chains, each 95,000 Da in size, positioned both outside and inside the cell, with a section spanning the membrane. The chains are joined by S–S bonds. The binding of insulin occurs essentially to the A chains (Fig. 4B). This interaction leads to a release of latent biochemical, signaling activity in the B chain, together with a structural change. Tyrosine protein kinase in the B chain becomes activated, leading to the autocatalytic phosphorylation of tyrosine residues in the B chain itself by ATP (Fig. 4B). This sets into motion a cascade of phosphorylation reactions on a group of additional polypeptides in the cell (Fig. 4C). This set of increased phosphorylation reactions occurs initially on tyrosine residues of polypeptides and then also on serine and threonine residues. Although the de-

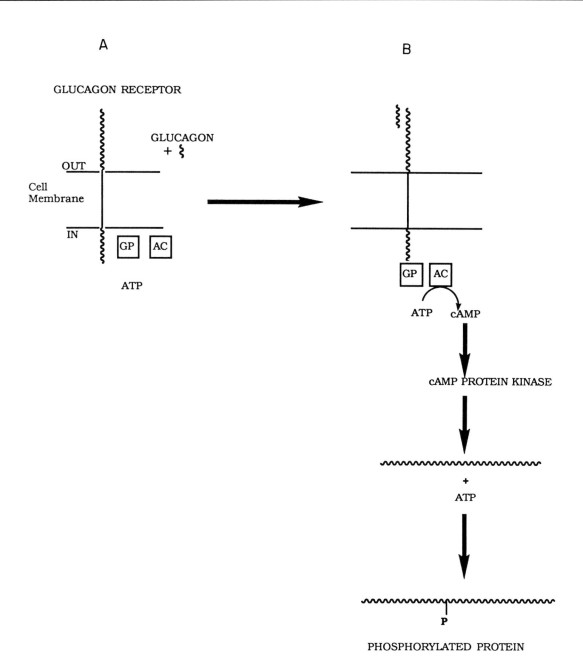

A

GLUCAGON RECEPTOR

GLUCAGON
+

OUT

Cell
Membrane

IN

GP   AC

ATP

B

GP   AC

ATP   cAMP

cAMP PROTEIN KINASE

+

ATP

P

PHOSPHORYLATED PROTEIN

FIGURE 6 Glucagon action at the glucagon receptor-increased protein phosphorylation. The glucagon receptor and glucagon combine, causing the formation of cAMP from ATP by adenylate cyclase (AC). Adenylate cyclase is activated by an intermediate acting G protein (GP). cAMP activates the cAMP-dependent protein kinase, which phosphorylates proteins. (A) Before glucagon action; (B) after glucagon action.

tails are not well understood, it is believed that these phosphorylations particularly on serine and threonine residues play a major role in regulating metabolic activities in the cell, converting them into an anabolic mode.

In addition, a second signaling mechanism comes into play in the cells, forming mediators from a novel set of linker molecules, including phospholipids, which anchor proteins to the outside of cell membranes. The structure of such an external linked protein is shown in Fig. 5A. By an incompletely understood set of reactions connected to insulin receptor activation, a mediator corresponding to the linker region is thought to be formed by processing, on the outside of the cell, and then to be taken into the cell of origin or a neighboring cell by a transporter present in the cell membrane (Fig. 5B).

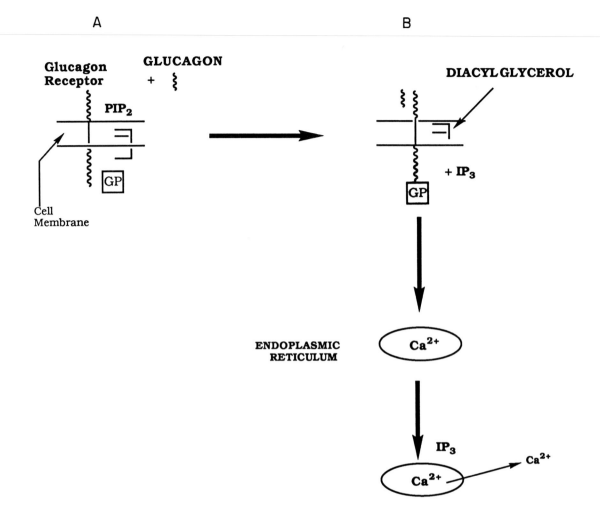

A

B

FIGURE 7 Glucagon action at the glucagon receptor–Ca^{2+} mobilization. The glucagon receptor and glucagon combine, leading to cleavage of the phospholipid phosphoinositol bisphosphate (PIP_2) into diacylglycerol and inositol trisphosphate (IP_3). An intermediate G protein (GP) is involved in this activation step. The IP_3 then releases Ca^{2+} stored in the endoplasmic reticulum into the cytosol, where it is available to activate various enzymes that require it for activity. (A) Before glucagon action; (B) after glucagon action.

Two other products formed are the protein itself, which is released to the outside of the cell, and diacylglycerol in the cell membrane. The diacylglycerol is thought to be a messenger involved in the activation of glucose transport in muscle and fat.

The mediators formed from the linker region are thought to control enzyme activity by decreasing polypeptide phosphorylations, increasing phosphoprotein phosphatase, and decreasing protein kinase reactions. The decreased phosphorylation would lead to increased anabolic reactions in the cell, because, in general, decreased phosporylation increases the activity of biosynthetic enzymes, but decreases the activity of degradative enzymes. Other chemical messengers (e.g., Ca^{2+}) might also play a role in insulin action, but the precise role is not clear. Cyclic nucleotides cAMP and cGMP appear not to be directly involved in insulin action. However, one of the insulin mediators has strong anti-cAMP actions and is thought to be responsible for the known ability of insulin to antagonize cAMP effects in cells.

VII. Mechanism of Action of Glucagon

Glucagon, like insulin, also interacts with a specific receptor on the outside surface of a sensitive cell (Fig. 6A). The receptor is not as well characterized as the insulin receptor, but it is thought to be a single polypeptide chain 60,000 Da in size. It is coupled to at least two intracellular secondary messenger systems within the cell. The first is the enzyme adenylate cyclase, which catalyzes the formation of cAMP from ATP (Fig. 6B). A group of membrane-associated proteins, called G proteins, has been identified as intermediaries in the activation of adenylate cyclase. cAMP is an activator of a protein kinase known as the cAMP-dependent protein kinase. This activation leads to a set of increased phosphorylations of polypeptides on serine or threonine, again controlling metabolism, but in a catalytic direction. For example, several of the degradative enzymes are activated, and several biosynthetic enzymes are inactivated by phosphorylation.

The second signal has also been identified. Inositol trisphosphate is formed by the hydrolysis of diphosphoinositol phospholipid, along with the second product, diacylglycerol (Fig. 7B). Inositol trisphosphate acts, possibly via a specific receptor, on a cellular system, endoplasmic reticulum, that stores Ca^{2+}, causing a release of the stored Ca^{2+} into the cell sap (i.e., cytosol). Cytosolic Ca^{2+} then activates one or more protein kinases, again causing increased phosphorylation of a set of key polypeptides and directing the cell metabolism catabolically. Increased cytosolic Ca^{2+} also leads to an increased concentration of Ca^{2+} in the mitochondria. This activates three dehydrogenase enzymes, thereby increasing the ATP production needed to support the accelerated gluconeogenesis.

VIII. Summary

In summary, both insulin and glucagon act by combining with specific cellular receptors, coupled to several signaling systems, using secondary messengers or mediators to control the state of phosphorylation of key polypeptides within the cell. The state of phosphorylation or dephosphorylation of the polypeptides results in directing the cell toward an anabolic or catabolic activity state. In the liver, muscles (including the heart), and adipose tissue macromolecules are laid down under the influence of insulin as cells grow and divide. In the presence of excess glucagon or in the absence of insulin macromolecular constituents in the organs are broken down, organ functions are compromised, and the cells are damaged. In the body as a whole, storage takes place in the presence of insulin, whereas breakdown occurs in its absence or in the presence of excess glucagon.

Bibliography

Larner, J. (1985). Insulin and oral hypoglycemic agents: glucagon. *In* "Goodman and Gilman's The Pharmacological Basis of Therapeutics" (A. G. Gilman, L. S. Goodman, T. W. Rall, and F. Murad, eds.), 7th ed. pp. 1490–1516. Macmillan, New York.

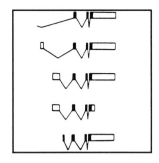

Insulinlike Growth Factors and Fetal Growth

VICKI R. SARA, *Karolinska Hospital, Stockhom*

Glossary

Autocrine Hormone acting on the same cell that produces it

Endocrine Hormone secreted into the circulation and transported to a target site distal to its site of production

Insulinlike growth factor 1 Single-chain molecule with three intrachain disulfide bridges consisting of 70 amino acids (MW 7,646)

Insulinlike growth factor 2 Single-chain molecule with three intrachain disulfide bridges consisting of 67 amino acids (MW 7,471)

Insulinlike growth factor binding proteins Family of polypeptides that specifically bind insulinlike growth factors and function as storage depots and transport proteins for the insulinlike growth factors

Paracrine Hormone having a local site of action on neighboring cells

Somatomedins Generic term earlier used to refer to the family of insulinlike growth factors

Truncated insulinlike growth factor 1 Single-chain molecule with three intrachain disulfide bridges consisting of 67 amino acids (MW 7,367), being identical to insulinlike growth factor 1 but lacking the amino-terminal tripeptide Gly-Pro-Glu

FETAL DEVELOPMENT is the most rapid growth found throughout life. Such intense cell proliferation is only seen again during the uncontrolled growth of tumor cells. In the human fetal brain, for example, in order to achieve the number of neurons present at birth, cells must be formed at a rate of >250,000/min. During this period of organogenesis and rapid development, the fetus is especially vulnerable to any interference in its growth, and permanent and irreversible changes in structure and function may be induced. Thus, fetal life is a critical period for the growth of organs. Although our knowledge has increased enormously in the last years, the mechanisms that regulate fetal growth remain to be completely established. Fetal growth clearly depends both on nutrition for substrates, which provide the building blocks for cellular growth, as well as on growth-promoting hormones, which regulate the rate of cellular growth. Several such growth-promoting hormones have been identified in recent years. The insulinlike growth factors (IGFs) are a family of growth-promoting peptide hormones that are believed to play a major role in the regulation of fetal growth. Other growth factors, such as platelet-derived growth factor (PDGF), nerve growth factor (NGF), epidermal growth factor (EGF), and fibroblast growth factor (FGF) as well as hormones such as steroids and pituitary hormones also regulate development. Growth and differentiation are the result of the interaction of these factors. [*See* TISSUE REPAIR AND GROWTH FACTORS.]

I. Introduction

A. Cellular Growth

To understand the regulation of growth, growth in cellular terms must be considered first. Cellular growth occurs in successive stages, and different growth-regulatory factors are most likely operative at unique stages in the growth process. In simple terms, cellular growth occurs first as rapid cell proliferation with a concomitant increase in cell number. This proliferative phase is followed by an increase in cell size, and finally the cells begin to differentiate and take on the unique characteristics of mature cells. The timing of each of these growth phases appears to be unique for individual organs and cellular types. Each cell type is also specific for other growth characteristics such as life span and renewal ability. Blood cells such as erythrocytes have a short life span and are constantly renewed from their stem cells, whereas neurons do not normally divide during their long life span. Fibroblasts, for example, have maintained the ability to proliferate, which may be evoked as a repair response to injury. Thus, although growth is most rapid during fetal development and statural growth may be complete after puberty in humans, it must be remembered that cellular growth is a continual process throughout life. Different growth-regulatory factors most likely predominate at different stages of the growth process. For example, IGFs stimulate the proliferation of neuronal precursors, whereas NGF induces their differentiation into the cholinergic phenotype.

B. IGFs

IGFs appear to have a general anabolic action and to be involved in several stages of development, especially proliferation as well as differentiation. The IGF family of growth-promoting hormones consists of two homologous peptides—insulinlike growth factor 1 (IGF-1) and insulinlike growth factor 2 (IGF-2)—as well as variant forms of these two peptides. Two forms of IGF-1 have been identified, namely IGF-1 and truncated IGF-1, which lacks the amino-terminal tripeptide. These are both products from the single IGF-1 gene and arise from different cleavage of the IGF-1 prohormone. IGF-1 and truncated IGF-1 may fulfill different biological functions. Since IGF-1 has been isolated from the circulation and is bound by the IGF-binding proteins (BPs), which act as transporters through the circulation, and truncated IGF-1 has been isolated from tissues and is poorly bound by the BPs, the peptides have been hypothesized to represent the endocrine and autocrine–paracrine forms of IGF-1, respectively. Variants of IGF-2 have been identified at both the protein and complementary DNA (cDNA) level but their biological significance is unknown.

II. Biosynthesis of IGFs in Fetal Tissues

IGFs have been found in extracts of fetal tissues as well as the medium of fetal cells and explants *in vitro*. IGF genes are expressed in fetal tissues, where the corresponding proteins are also synthesized. It is well recognized that IGFs are produced in a wide variety of fetal tissues as autocrine or paracrine hormones and have a local growth-promoting action throughout the fetus. Additionally, IGFs circulate as endocrine hormones associated to a binding protein. However, during fetal development, their primary role appears to be as autocrine or paracrine regulators of growth and development.

A. Gene—Protein

The genes for IGF-1 and IGF-2 have been localized to the long arm of human chromosome 12 and the short arm of chromosome 11, respectively. Both genes span a considerable length of genomic DNA and have a discontinuous structure containing several exons (coding segments). The human IGF-1 gene, which consists of at least five exons, is schematically represented in Figure 1. The primary transcript from the IGF-1 gene can be alternatively spliced to result in either IGF-1a or IGF-1b messenger RNA (mRNA), both of which encode IGF-1 precursor proteins differing in their carboxyl-terminal E domains. Following translation, the signal peptide is removed and the IGF-1 prohormone is further cleaved to release either intact or truncated IGF-1. Under certain circumstances, the prohormone form itself may be released from the cell, although its biological relevance remains unknown. The human IGF-2 gene consists of at least nine exons and is schematically shown in Figure 2. Initiation of transcription at different promoter sites in the IGF-2 gene generates alternative mRNAs, which differ in their 5'-untranslated regions but contain the same exons encoding the IGF-2 precursor protein. Following translation, the IGF-2 prohormone is cleaved to release IGF-2 as well as, in some instances, a higher-molecular weight form of IGF-2 with a carboxyl-terminal extension peptide representing the E domain of the prohormone.

The expression of the IGF genes and the biosynthesis of the peptides is both development- and tissue-specific. With the exception of the liver, which provides the primary source for circulating IGF, the greatest abundance of IGF mRNA is observed dur-

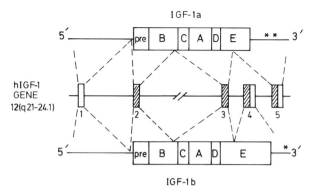

FIGURE 1 Schematic representation of the human IGF-1 (hIGF-1) gene located on chromosome 12 and its alternative IGF-1a and IGF-1b mRNAs, giving positions of the precursor proteins. The gene consists of five numbered exons (□) with coding regions hatched. There is an indeterminate gap (//) between exons 2 and 3. Alternative mRNA splicing results in two IGF-1 transcripts encoding precursor proteins, which both contain intact IGF-1 (B, C, A, and D domains) but differ in the length and structure of their E domains. Asterisk represents polyadenylation sites in the 3'-untranslated regions. [Reproduced, with permission, from V. R. Sara and K. Hall (1990). The insulin-like growth factors and their binding proteins. *Physiol. Rev.* **70,** 591–614.]

ing fetal development. With maturation and slowing of the growth rate, a corresponding decline in IGF gene expression has been observed. Multiple IGF-1 and IGF-2 mRNA species resulting from specific gene processing vary in abundance during development. These reflect the use of alternative 5'-untranslated regions as well as variable polyadenylation sites at the 3'-untranslated region, which could influence mRNA stability, turnover, and translatability. In both humans and rat, IGF-1a mRNA is the predominant IGF-1 gene transcript in fetal tissues. Indeed, the IGF-1a transcript predominates in tissues throughout life, raising the possibility that the functional significance of the IGF-1a prohormone encoded by this transcript relates to the autocrine and paracrine targeting of IGF-1. In contrast, the IGF-1b mRNA encoding the IGF-1b prohormone may relate especially to liver endocrine IGF-1 production. Developmentally specific expression of alternative 5'-untranslated regions of IGF-1 mRNA has been observed in the rat, suggesting specific regulation via different promoter sites during development. Thus, IGF-1 gene expression, involving alternative RNA splicing encoding different prohormones, as well as the use of alternative 5'-untranslated regions, is tissue- and development-spe-

cific. The protein product from expression of the IGF-1 gene in human fetal brain has been isolated and characterized as the truncated form of IGF-1, which lacks the amino-terminal tripeptide Gly-Pro-Glu (Fig. 3). Truncated IGF-1 arises from posttranslational processing of the IGF-1 prohormone. The truncated IGF-1 has been proposed to represent the autocrine–paracrine form of IGF-1, whereas the intact IGF-1 has been proposed to represent the endocrine form. Only low levels of IGF-1 are present in the fetal circulation. The peptide has been isolated from fetal calf and fetal sheep serum and identified as intact IGF-1. Truncated IGF-1 is found in bovine colostrum.

The expression of the IGF-2 gene has been well characterized during development in both humans and rat. IGF-2 gene expression is far more abundant

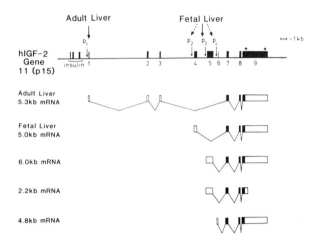

FIGURE 2 Schematic representation of the human IGF-2 (hIGF-2) gene and its mRNA transcripts. The hIGF-2 gene consists of nine numbered exons and is located in close proximity to the insulin gene on chromosome 11. The position of the four promoter sites (P_1, P_2, P_3, P_4) and polyadenylation sites (*) is indicated. Coding regions are shown by black boxes in the mRNAs. The developmental switch in the use of different promoter sites to initiate IGF-2 gene transcription in the human liver is shown. In the adult liver, transcription is only initiated at P_1, which generates an IGF-2 mRNA of 5.3 kb. In the fetal liver, transcription is initiated mainly by P_3 and to a lesser extent at P_2 and P_4, which generates multiple IGF-2 mRNAs of various sizes as shown. The two IGF-2 mRNAs initiated at P_3 (6.0 kb, 2.2 kb) arise from differential polyadenylation. The 5.0-kb and 4.8-kb mRNA are initiated at P_2 and P_4, respectively. [Adapted, with permission, from M. Jansen, P. Holthuizen, M. A. van Dijk, F. M. A. van Schaik, J. L. Van den Brande, and J. S. Sussenbach, 1990, Structure and expression of the insulin-like growth factor II (IGF-II) gene, *In* "Growth Factors: From Genes to Clinical Applications" (V. R. Sara, K. Hall, and H. Löw, eds.), pp. 25–40. Raven Press, New York.]

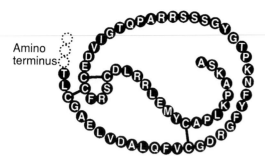

FIGURE 3 The primary structure of truncated IGF-1. The position of the amino-terminal tripeptide of IGF-1, namely GPE, which is lacking in truncated IGF-1, is shown by the broken rings at the amino terminus. Amino acid residues are shown by their single-letter abbreviations and the position of the three disulfide bridges is given.

in the fetus as compared with the adult, and its expression in tissues is repressed with maturation. As illustrated in Figure 2, the use of different promoter sites to initiate gene transcription is developmentally regulated. For example, in humans, promoter sites P2, P3, and P4 are active in the fetus as well as, to a lesser extent, in adult nonhepatic tissues, whereas P1 is only active in the adult liver (Fig. 2). Thus, IGF-2 gene transcripts, which although encoding the same proIGF-2 differ in their 5'-untranslated regions, are produced in a developmentally specific manner. The functional significance is unclear. Similarly in the rat multiple IGF-2 mRNA species arise from the use of alternative promoter regions as well as polyadenylation sites. Abundant mRNA is present in the rat fetal tissues, whereas, with the exception of the choroid plexus and leptomeninges in the brain, IGF-2 mRNA is undetectable in the adult. The protein product from expression of the IGF-2 gene in the human fetal brain is identical to IGF-2 isolated first from human serum in 1978. IGF-2 has also been purified from fetal calf and fetal sheep serum. Apart from species differences in only a few amino acid residues, which occur mainly toward the end of the B domain and within the C domain in IGF-2, no developmental variant of IGF-2 has been identified.

B. Localization

While the timing for the appearance of IGF-1 mRNA is as yet undetermined, the IGF-2 gene begins to be expressed in the human fetus at around 5–18 days postfertilization. By approximately the end of the first trimester, most human fetal tissues are expressing both the IGF-1 and IGF-2 genes. Both IGF-1 and IGF-2 mRNAs have been localized to connective tissues or cells of mesenchymal origin throughout the human fetus. Because these cells are widely distributed, they provide an ideal production site for the widespread paracrine actions of the IGFs. IGF immunoreactivity is not present in identical cells in the human fetus supporting the paracrine concept that the IGFs are synthesized and secreted from specific cells to be taken up and act on neighboring cells in the fetal organs. The strong association between IGF and IGF BP immunoreactivity in human fetal tissues supports the concept that the IGF BPs act as transport proteins to deliver the IGFs to their target cells. The cellular expression of the IGFs temporally depends on the growth and differentiation stage of the cell. Consequently, the IGF genes appear to be expressed in cell types additional to those of mesenchymal origin. In these cells, IGF gene expression occurs in a strict developmentally regulated manner. In rats, IGF-1 and IGF-2 mRNAs are present in the embryo and yolk sac. IGF-2 mRNA predominates in fetal rat tissues and high levels are present in a variety of differentiating tissues especially of mesodermal origin. In tissues derived from the endoderm and ectoderm, IGF-2 mRNA apparently is present in restricted sites such as the gastrointestinal and bronchial epithelium as well as the choroid plexus and pituitary rudiment. The presence of IGF-2 mRNA in these cells occurs only at specific stages in development. Apparently, no striking correlation exists between the expression of the IGF-2 gene and rate of proliferative activity, implying that IGF-2 may be involved in other cellular processes possibly involved with cellular differentiation and trophic outgrowth. Evidence in the rat, for example, indicates that the genes for IGF-2 and its receptor are expressed in muscle not during the proliferative phase of myoblast formation but rather during differentiation. Additionally, muscle IGF-2 gene expression correlates to synaptogenesis, suggesting that IGF-2 acts as a trophic factor to induce innervation. IGF-1 mRNA is less abundant than IGF-2 mRNA in fetal tissues and has consequently received far less attention. However, the pattern of IGF-1 gene expression during rat embryogenesis shows a site-specific expression, which is under strict developmental control. Although the expression of IGF-1 is low, it occurs in highly selective cell groups at specific developmental stages and appears to reflect the rate of mitotic activity in these areas.

III. Circulating Forms of IGFs in the Fetus

Following their secretion from fetal cells, the IGFs are found in the surrounding extracellular fluid or in the fetal circulation as higher-molecular weight complexes of around 40 kDa consisting of IGF associated with the IGF BP. The IGF BP (IGF BP1), which is synthesized by decidual cells, was first isolated from human amniotic fluid where its concentration peaks around midterm. An additional fetal-binding protein (IGF BP2) has now been characterized in both rat and humans. The fetal IGF BPs, which are growth hormone-independent, predominate in the fetal circulation until the higher-molecular weight growth hormone-dependent binding complex (IGF BP3) appears at the end of gestation in humans and around weaning in the rat. This coincides with growth hormone regulation of both growth and IGF-1 production and is presumably coupled to growth hormone receptor maturation. The IGF BPs function as storage forms and transporters of IGFs in the circulation and thereby regulate their availability to the target cells.

The truncated IGF-1 binds only extremely weakly to all species of IGF BP and, therefore, is readily available to cells within the immediate environment where it is secreted. Thus, this IGF-1 variant has been proposed to represent the autocrine–paracrine form whose local biological action is uninhibited by the presence of the IGF BPs. Both IGF-1 and IGF-2 are present during fetal life; however, there are species differences in the predominant fetal form found in the circulation. The rat is the only species where levels of IGF-2 are high during fetal life and fall with maturation in parallel to repression of hepatic IGF-2 gene expression. In contrast, serum IGF-1 levels are low in the rat fetus and increase with advancing age. In the rat, the switch over from IGF-2 to IGF-1 production occurs at the time of onset of growth hormone regulation. In humans and other species examined, the levels of both IGF-1 and IGF-2 are low in the fetal circulation and increase with development. However, the circulating levels are unlikely to reflect local tissue production but rather the hepatic source of endocrine IGF. Consequently, in early development, the physiological significance of the circulating levels of IGF is dubious because these peptides primarily fulfill an autocrine or paracrine role during development.

IV. Regulation of IGF Biosynthesis in the Fetus

Alternative pathways for the biosynthesis of IGF-1 and IGF-2 during development provide a means for regulation at different stages in their biosynthesis. Autocrine and paracrine IGF production on the one hand and endocrine IGF production on the other are probably independently regulated. These regulatory mechanisms change during development. For example, growth hormone is the primary regulator of endocrine IGF-1 production in the adult. Growth hormone rapidly stimulates hepatic IGF-1 gene expression at the level of transcription. In other mature extrahepatic tissues, growth hormone can additionally stimulate local paracrine IGF-1 production. However, the response to growth hormone is developmentally dependent. In contrast to the adult, fetal IGF-1 production as well as growth, is independent of growth hormone regulation. Growth hormone begins to regulate IGF-1 production by the first months of postnatal life in humans and around 20 postnatal days in the rat, which corresponds to the time of maturation of growth hormone receptors. Thus, growth hormone fails to stimulate fetal IGF-1 production and does not elicit a growth response during early development. Growth hormone-related hormones such as placental lactogen may substitute as regulators of IGF production during development. Many trophic hormones such as adrenocorticotrophin, thyroid-stimulating hormone, luteinizing hormone, and follicle-stimulating hormone can, in addition to their classic action on secondary hormone production, stimulate the paracrine biosynthesis of IGF-1 from their target organs. Insulin is also associated with IGF production during fetal development. Fetal hyperinsulinemia is accompanied by elevated IGF levels, whereas fetal pancreatectomy reduces IGF levels. Whether insulin has a direct action on IGF production or this is mediated by insulin action on substrates remains to be clarified.

Glucocorticoids are well established as inhibiting growth during development and have been considered as inducers of maturation. The mechanism for this action has been suggested by the demonstration that glucocorticoids inhibit the expression of both IGF-1 and IGF-2 genes. Growth factors such as PDGF and EGF can stimulate local IGF-1 synthesis in their target cells. The interrelationship between these locally produced growth factors provides a means for intercellular communication to coordinate growth. Nutrition has been suggested to play a primary role in the regulation of IGF production throughout life. The influence of nutrition on growth is well established, and several studies have now demonstrated that both IGF-1 and IGF-2 syntheses during early development are regulated by nutrition. In the neonatal rat, for example, undernutrition impairs growth and reduces IGF production, which in the case of IGF-1 is especially influenced by dietary protein intake and in the case of IGF-2 by caloric intake. Nutrition also regulates the IGF BPs, which increase during starvation, suggesting a protective mechanism to reduce IGF bioavailability.

V. Receptors for IGFs in Fetal Tissues

The biological activity of any hormone depends on the ability of the target cell to respond to the signal in its extracellular milieu. This is a function of the cell receptors as well as postreceptor mechanisms. Two distinct IGF receptors, which are transmembranously located in the plasma membrane, have been characterized in fetal tissues. The IGF-1 receptor consists of two extracellular α-subunits and two transmembranal β-subunits. The hormone-binding site is located in the extracellular α-subunit and the extracellular domain of the β-subunit. The binding of the ligand to this extracellular region of the receptor results in intracellular signal transmission by autophosphorylation of tyrosine and serine residues within the intracellular β-subunit. Receptor aggregation or cross-linking may be involved in transmembranal signaling. This receptor preferentially binds IGF-1, but IGF-2 also cross-reacts to a lesser degree and, at high concentrations, insulin cross-reacts weakly. Several subtypes of the IGF-1 receptor have been suggested, and their appearance may be developmentally regulated. The IGF-2 receptor is structurally unrelated to the IGF-1 receptor, being a single-chain polypeptide consisting of a large extracellular domain. The IGF-2 receptor is identical to the mannose-6-phosphate receptor and contains distinct binding sites for these ligands, which exhibit cooperativity. The physiological significance of the relationship between IGF-2 and mannose-6-phosphate as well as the mechanism of intracellular signal transmission is at present unknown. The IGF-2 receptor has a high affinity for IGF-2 and does not recognize insulin. The existence of subtypes of the IGF-2 receptor may account for

tissue differences in the ability of IGF-1 to cross-react in this receptor.

Both the IGF-1 and IGF-2 receptors are widely distributed throughout fetal tissues. The receptors are expressed from early in development, having been demonstrated on pre- and peri-implantation mouse embryos. Both receptors are usually present in each fetal organ; however, their predominance is development- and tissue-specific. The ontogenesis of the IGF receptors has been followed in several tissues. In the human fetal brain, for example, both receptors are present from early in development, but they alter their characteristics during maturation. The affinity and specificity of the IGF-1 receptor changes toward the end of the first trimester. Because this coincides with the cessation of the rapid proliferative phase of neuronal formation in the cortex and the beginnings of cortical neuron differentiation, the alteration in receptor characteristics may represent maturation of the neuron-specific subtype of the IGF-1 receptor. In contrast to the brain, the IGF-2 receptor predominates in the human fetal liver. There are however species differences in the pattern of receptor expression during development. In the placenta, for example, IGF-1 receptors are present in humans, but IGF-2 receptors predominate in the rat. The IGF receptors appear to be regulated by different mechanisms. The IGF-1 receptor is regulated according to receptor occupancy and is modulated by nutrition. The expression of this receptor parallels the growth rate. The IGF-2 receptor, which is rapidly internalized and recycled, is regulated by insulin and growth hormone. [*See* Cell Receptors.]

VI. Biological Actions of IGFs on Fetal Growth

The IGFs are ubiquitous anabolic hormones, whose biological action depends on the responsiveness and programming of the target cell. In general terms, the biological actions of the IGFs may be summarized as stimulating growth and metabolism and inducing differentiation. These actions are established in a wide variety of fetal cells and explants using purified, generally recombinant, peptides (Table I). Both IGF-1 and IGF-2 stimulate DNA synthesis and cell proliferation in all fetal cell types examined. Meiotic division in oocytes is also in-

TABLE I Biological Actions of IGFs on Fetal Cells and Tissues *in vitro*

Action	Target	Species
Growth	Nervous system	Human
Cell proliferation	Neuroblasts–neurons	Rat
DNA synthesis	Glioblasts–glia	Mouse
RNA synthesis	Muscle	Rabbit
Protein synthesis	Myoblasts	Cow
Metabolism	Liver	Chick
Glucose uptake	Hepatocytes	Quail
Glycogen synthesis	Pancreas	Frog
Differentiation	Islet cells	
	Bone–cartilage	
	Calvaria	
	Osteoblasts	
	Chondrocytes	
	Fat	
	Adipocytes	
	Limb buds	
	Mesenchymal cells	
	Fibroblasts	
	Oocytes	

duced by the IGFs. It has been suggested that the IGFs act as progression factors to stimulate cells through the DNA synthesis phase of the cell cycle once the cell cycle has been initiated by progression factors such as PDGF and FGF. The IGFs additionally induce insulinlike metabolic responses such as stimulating glucose transport and metabolism especially in target organs for insulin, such as adipose tissue and muscle. With the use of receptor antibodies in the majority of fetal cells the growth-promoting actions of the IGFs, as well as those of insulin, are apparently mediated via the IGF-1 receptor. Similarly, in many cells the metabolic actions of the IGFs are mediated via the insulin receptor; however, there are exceptions to this. For example, IGF-2 and insulin have growth-promoting activity through their own receptors in human erythroleukemia cells and teratocarcinoma cells, respectively, and glycogen synthesis in hepatoma cells is mediated via the IGF-2 receptor. The response of cells to the IGFs can be modulated by the presence of the IGF BPs. Truncated IGF-1 as well as mutant IGF-1 variants, which have a marked reduction in their affinity of binding to the IGF BPs, display enhanced biological activity. While the addition of IGF BPs inhibits the receptor binding and biological activity of IGF-1, it has little influence on truncated IGF-1. Thus, the biological actions of the IGFs are modu-

lated by the IGF BPs, which regulate the availability of IGFs for receptor binding of the cell surface and cell responsiveness to the IGFs.

In recent years, it has become evident that the IGFs also act as inducers of differentiation in several cell types (Table I). Within the developing nervous system, for example, the formation of neurites, which is an early marker of neuronal differentiation, is induced by IGF-2 and is also accompanied by an increased expression of IGF-1 receptors. Neuronal progenitor cells are pluripotent in their ability to express different neurotransmitters. During migration and target contact, factors from the surrounding milieu induce the expression of specific phenotypes. IGF-1, for example, induces the catecholaminergic phenotype in avian sympathetic nerve cell precursors and induces oligodendrocyte commitment by bipotential oligodendrocyte-type 2 astrocyte progenitor cells. In addition, IGF-1 appears to be a differentiating factor for ovarian and testicular function. IGF-1 as well as IGF-2 induces erythropoiesis, and IGF-1 stimulates granulopoiesis as well as chemotaxis in endothelial and melanoma cells.

In vivo studies of the biological actions of the IGFs have confirmed the *in vitro* findings. Chronic administration of IGF-1 stimulates the growth of hypophysectomized as well as rapidly growing neonatal and preweaning rats and Snell dwarf mice. IGF-2 has only weak or no growth-promoting action when administered to hypophysectomized rats or in nude mice transplanted with IGF-2-secreting tumors. However, the actions of IGF-2 may prove to be developmentally dependent as IGF-2 induces growth in transplanted rat embryos. During fetal and early postnatal growth, when growth occurs independently of growth hormone, IGF-1 and not growth hormone administration induces growth, involving a general increase in the weight, DNA, and protein content of most organs such as brain, liver, kidney, and spleen. The action of IGF-1 is nutritionally dependent. Consequently, IGF-1 administration fails to stimulate the growth of undernourished developing animals, most likely due to receptor inhibition and elevated IGF BP concentrations, which may bind IGF-1, rendering it unavailable to target cells. In contrast to intact IGF-1, the truncated IGF-1, which only weakly associated with the IGF BPs, has a potent biological action when locally applied to target cells. As illustrated in Figure 4, using an *in vivo* model of intraocular transplantation of

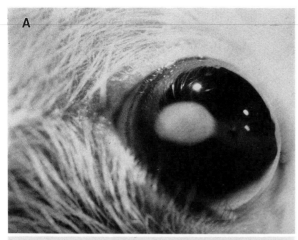

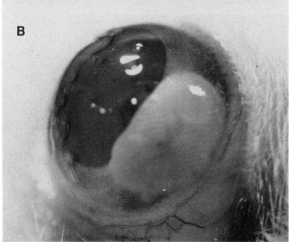

FIGURE 4 The growth-promoting action of truncated IGF-1 on fetal brain tissue grafts. Pieces of embryonic rat parietal cortex (1–2 mm²) were grafted into the anterior chamber of the eye of adult rat hosts. Grafts were treated with vehicle alone (A) or 0.1 μg/μl truncated IGF-1 (B) for 10 min incubation prior to grafting and 5 μl injections on the 5th, 10th, and 15th day after grafting. A and B show the typical appearance of the parietal cortex grafts after intraocular maturation as observed through the cornea of the hosts. Truncated IGF-1 is shown to be a potent stimulator of the growth of grafted fetal cortex cerebri. [Photographs provided by MaiBritt Giacobini and Lars Olson, Department of Histology and Neurobiology, Karolinska Institute, Stockholm, Sweden.]

fetal brain tissue, truncated IGF-1 exerts potent growth-promoting effects on grafted embryonic cerebral cortex. Thus, the biological role of the IGFs during fetal development is established as promoting growth as well as inducing differentiation.

VII. Conclusion

It is now evident that the IGFs play an important role in the regulation of fetal growth and differentiation. From the earliest stages of development, the IGFs are synthesized within fetal tissues to have a local autocrine and paracrine action on cellular growth and differentiation. IGF-1 and IGF-2 are each synthesized as products from a single gene whose expression can be regulated at several stages such as alternative RNA splicing and processing of the precursor proteins. By these mechanisms, diversity in the final protein product and, hence, biological activity can be achieved from expression of the single gene. The use of these alternative biosynthetic pathways appears to be developmentally regulated, with autocrine and paracrine IGF production predominating during fetal development. With maturation, there may be repression of this pathway and switching to a predominance of endocrine production. However, the IGF biosynthetic pathway can also be regulated according to physiological state. Throughout life, paracrine and autocrine production can be elicited as a repair response to local tissue injury. Intensive autocrine IGF production is also found during tumor growth. Neoplastic cells display characteristics typical of a reversal to the fetal state, and IGFs may function as autocrine growth factors stimulating tumor cell proliferation. Fetal IGF biosynthesis is regulated by several factors in a developmentally determined manner. Early in development, substrate availability to the cells regulates autocrine and paracrine production. Higher neural control, such as mediated via growth hormone, develops only with maturation as a secondary regulating system. The production of the IGFs in several fetal cell types, especially of mesenchymal origin, provides for their widespread availability to target cells distributed throughout the fetus. Thus, the IGFs have a very generalized action in the fetus to regulate growth and differentiation.

Bibliography

D'Ercole, A. J. (1987). Somatomedins/insulin-like growth factors and fetal growth. *J. Dev. Physiol.* **9**, 481.

Hill, D. J. (1987). Growth factors in embryogenesis. *Oxf. Rev. Reprod. Biol.* **9**, 389.

Hill, D. J. (1989). Peptide growth factors in fetal development. *In* "The Physiology of Human Growth" (J. M. Tanner and M. A. Preece, eds.). Cambridge University Press, Cambridge.

Sara, V. R., and Hall, K. (1990). The insulin-like growth factors and their binding proteins. *Physiol. Rev.* **70,** 591–614.

Sara, V. R., Hall, K., and Low, H. (eds.) (1990). "Growth Factors: From Genes to Clinical Applications." Raven Press, New York.

Schofield, P., and Tate, V. (eds.) (1990). "The Insulin-Like Growth Factors: Structure and Biological Functions." Oxford University Press, Oxford (in press).

Underwood, L. E., and D'Ercole, A. J. (1984). Insulin and insulinlike-growth factors/somatomedins in fetal and neonatal development. *Clin. Endocrinol. Metabol.* **13,** 69.

Interferons

JOSEPH H. GAINER, *Division of Veterinary Medicine, Food and Drug Administration*

Glossary

Double-stranded RNA (dsRNA) RNA form with two complementary strands of RNA

Heterokaryons Cells containing two or more nuclei of different genetic constitution

Histocompatibility antigen Genetically determined cell-surface antigen

Human leukocyte antigen histocompatibility antigens Cell-surface histocompatibility antigens on human cells that are important in tissue transplantation and controlled by a single gene complex

Pleiotropic Pertaining to the quality of having affinity for manifesting itself in a multiplicity of ways in the phenotype

Pinocytosis Imbibition of liquids by cells especially via minute incuppings in cell surfaces which close to form fluid-filled vacuoles

Reverse transcriptase Enzyme coded by certain RNA viruses that is able to make complementary single-stranded DNA chains from RNA templates and then to convert these DNA chains to double-helical form

Transcription Process involving base pairing, whereby the genetic information contained in DNA is used to form a complementary sequence of bases in an RNA chain

Translation Process whereby the genetic information present in a messenger RNA molecule directs the order of the specific amino acids during protein synthesis

INTERFERONS (IFNs) may be considered to be polypeptide hormones. They were first recognized in 1957 by Isaacs and Lindenman as proteins that had antiviral activity; i.e., they would protect animal cells against infection by viruses when the cells had been treated with the IFN preparation before being infected with virus. The phenomenon of viral interference—that infection of a cell by one virus would prevent infection by another virus—had been known a number of years before the recognition that the viral interference could be mediated by a soluble protein named IFN. IFNs are now known to be a multigene family of regulatory proteins. They act through specific cell receptors that modulate the expression of at least a dozen different cellular genes, whose encoded products profoundly affect a number of biological functions including viral replication, cell growth, and the immune response.

The biological activity of IFNs is very high. Specific activities of about 10^8 international units of antiviral activity per milligram of protein make them among the most potent biologically active compounds known. They show antiviral activity *in vitro* at about 10^{-14} M. Prostaglandins are active at about 10^{-9} M and peptide hormones such as insulin at 10^{-9}–10^{-12} M.

Two systems of nomenclature are used to identify IFNs. One system defines three major types of IFNs: IFN-α, IFN-β, and IFN-γ, classified according to their structure (i.e., their amino acid sequence and their antigenicity). The terms leukocyte and lymphoblastoid IFNs (primarily IFN-α) and fibroblast IFN (primarily IFN-β) refer to preparations named for the cells from which they are derived. They are all type I IFNs binding to a common receptor. Type II, or immune, IFN designates IFN-γ produced by T lymphocytes. IFN-γ binds to a different receptor from that of type I IFNs. There are many subtypes (>15) of human IFN-α, differing by as much as 30% in their amino acid sequences. There is only one type of human (Hu) IFN-β and one type of Hu IFN-γ.

IFNs are assayed by their antiviral activity in tissue cultures. Recombinant DNA technology for IFN synthesis has permitted the production and purification of large quantities of homogeneous IFNs, advancing the knowledge of the chemical and biological properties of these molecules (Table I).

I. Genes of IFNs and Genes Regulated by IFNs

A. IFN Genes

Initial cDNA recombinant plasmids were used as probes to identify the genes and other cDNA clones. With Hu IFN-α, at least 23 separate loci have been identified corresponding to the major human IFN-α family. At least 14 of these contain coding sequences for apparently functional proteins; these have substantial homology and comprise six linkage groups. Apparently, IFN-αA and IFN-α2, differing in one amino acid only, are both represented in each of 14 Hu DNA samples. In addition, two IFN-α loci have identical coding sequences but different flanking sequences. None of these chromosomal genes contains an intron within the coding sequence. Several loci represent pseudogenes. All mammalian species have a large number of IFN-α gene families. None of the Hu IFN-α species except IFN-αH (IFN-α14)contains an *N*-glycosylation site.

A related family of IFN-α genes has been described. Whereas the major Hu IFN-α class genes encode proteins of 165 or 166 amino acids, this second class of Hu IFN-α genes encodes 172 amino acid proteins. Both classes exhibit potent antiviral activity and are coordinately expressed in response to viral induction. Apparently, both IFN-α classes are present in all mammalian species. The major IFN-α class of human genomic DNA appears to be the minor class of bovine genomic DNA. Both classes of Hu IFN-α genes are located on chromosome 9 and do not contain any introns.

Hu IFN-β gene is a single gene also located on chromosome 9; it codes for a protein of 166 amino acids with about 29% homology to Hu IFN-αD or IFN-α1. It also does not contain any introns. Bovine, equine, and porcine IFN-β gene families contain many genes, whereas humans and most other mammalian species probably contain only one. The lion and the rabbit genomes contain two IFN-β genes.

Hu IFN-γ is specified by a unique gene containing three introns located on chromosome 12. It contains little or no homology to IFN-α or IFN-β, although some structural similarity has been reported between Hu IFN-γ and Hu IFN-α or -β. Whereas IFN-α and IFN-β genes exist in most vertebrates examined, the existence of IFN-γ genes in non-mammalian vertebrates has not been seen.

B. Evolutionary Relationships to IFN Genes

The homology between human IFN-α and IFN-β genes indicates that they originated from a common ancestral gene about 300 million years ago, about the time of the divergence of mammals, reptiles, and birds from fish and amphibians. This suggests that amphibia and fish contain only an IFN-α or IFN-β gene homologous to the mammalian counter-

TABLE I Physical and Chemical Properties of IFNs

Type of IFN[a]	Size (kDa)	No. of IFN species/cell	Amino acid composition	Other comments
Alpha (I)	16,000–27,000	10–16; spec. act. of $2-4 \times 10^8$ units/mg protein	Rich in leucine, glutamic acid, glutamine; 165–166 or 172 residues	Conservation of cysteines at positions 1, 29, 98, 99 and 138, 139
Beta (I)	20,000	1; spec. act. of $2-5 \times 10^8$ units/mg protein	Similar to alpha; 165 residues	Functional unit appears to be a dimer
Gamma (II)	15,500–25,000	1; spec. act. of $1-25 \times 10^7$ units/mg protein	NH_2-terminus pyroglutamic acid residue	Functional unit appears to be a dimer or a tetramer

[a] Alternate nomenclature; these are the dedicated names for the IFN receptors.

parts, but not both, and is in accord with weak hybridization of Hu IFN-β, but not IFN-α to amphibia and fish DNA. The major IFN-α species can be then subdivided into two major groups based on divergence of silent sites. Substantial polymorphic variants probably exist among the IFN-α and IFN-β genes; in some individuals, a duplicated Hu IFN-β gene is found. Apparently, little or no clear homology exists between the IFN-α and IFN-β nucleotide sequences and that of IFN-γ.

All three classes of IFN genes (α, β, and γ) appear to be evolving rapidly at a rate comparable with that of immunoglobulin genes. In addition, the 3' noncoding sequence of the IFN-α genes shows much greater divergence than the rest of these genes, a seemingly puzzling observation. The occurrence of apparent natural hybrids among the Hu IFN-α genes is also puzzling.

C. Genes Activated by IFNs

IFNs, like many polypeptide hormones, exert their effects by binding to plasma membrane receptors affecting the transcriptional activity of a specific set of genes. A cell-surface signal is transduced into the nucleus, causing an alteration of gene expression. The IFN-activatable genes are found scattered on various chromosomes. IFNs behave as negative regulators of cell growth as shown by their decrease of the expression of protooncogene c-*myc* mRNA.

IFNs are pleiotropic modifiers of biological responses. A recent concept of endogenous IFNs is emerging. IFNs and IFN antagonists have been discovered in human placental blood in the absence of any infectious disease. Porcine and ovine conceptuses have been reported to secrete an IFN during the preattachment period of early pregnancy; this IFN resembles IFN-α, particularly the 172-residue long IFN-α11 (or IFN-ω).

Biochemical characterization of the antiviral activity of IFNs revealed a new class of 2'–5' A-linked oligonucleotides known collectively as 2–5 A. They are produced from adenosine triphosphate (ATP) by (**ds**)RNA-activated 2–5 A synthetase. Their expression is controlled mainly at the transcriptional level by IFNs, the steady-state intracellular level of active 2–5 A oligonucleotides being controlled by 2'-phosphodiesterases and possibly by phosphatase(s). The following observations are mainly consistent with a role of the 2–5 A pathway in IFN in antiviral activity: (1) 2–5 A oligomers accumulate in IFN-treated cells infected with dsRNA

viruses; (2) viral mRNAs are preferentially degraded in IFN-treated viral-infected cells, and (3) exogenous 2–5 A promotes a transient antiviral activity when introduced into intact cells in the absence of IFN treatment, while 2–5 A antagonists reverse IFN-induced antiviral activity. In c-*myc* mRNA, the target structure could consist of a large intramolecular hairpin structure between regulatory exon 1 and more distal regions of c-*myc* mRNA; the target sequence could be the AU-rich sequence common to many mRNAs coding for competence factors.

II. IFN Receptors

IFN proteins must interact with specific cell-surface receptors to induce biological activity. IFN-α and IFN-β bind to a common receptor, type I, and IFN-γ binds to a separate receptor, type II. Both types of high-affinity receptors have been demonstrated on a variety of cells.

A. IFN Receptor Genes

Some properties of IFN receptors are given in Table II. Genes for the IFN receptor proteins are not linked to the structural genes for hu IFN-α and IFN-β, which is on the short arm of chromosome 9, or to the gene for IFN-γ, which is on the long arm of chromosome 12. Type II receptor for IFN-γ probably requires the contribution of a product from human chromosome 21 locus for the expression of its activity. The degree of IFN sensitivity appears to relate to the number of duplications of chromosome 21 in human cells, the same chromosome that causes Down's syndrome.

B. Properties of IFN Receptors

A variety of cell types have a relatively low number of receptors, 10^2–10^4 per cell, having high affinity, (kDa of 1×10^{-9}–10^{-11} M). The degree of IFN effect, especially in transcription of IFN-inducible genes, appears to relate to the number of receptor sites occupied. Incubation of cells at 37°C with low concentrations of IFN-α and IFN-β leads to a downregulation, i.e., decrease in the number of receptors.

The hu α/β receptor gene has been transfected into mouse cells, which then responded to hu IFN-β by induction of 2',5'-oligoadenylate synthe-

TABLE II Properties of IFN Receptors

Receptor	Interacts with:	Gene	Size	IFN receptor (complex, size)
I	IFN-α and IFN-β	Located on chromosome 21	100-kDa protein	130, 220, 230, and 320 kDa from lymphoblastoid cells
II	IFN-γ	Located on chromosome 6q	68, 80, and 95 kDa protein; from placenta there is a 65-kDa polypeptide core and a 23-kDa carbohydrate, an extracellular domain of 55 kDa and an intracellular domain of 135 kDa	105–125 kDa from foreskin fibroblasts

tase. Antibodies, selected for their ability to block IFN-induced antiviral effects and bind IFN-α, inhibited HLA antigen increase after Hu IFN-α/β treatment of these transformant mouse cells. The size of the unglycosylated receptor is a 100-kDa protein. The IFN-α receptor from human Raji lymphoblastoid cells was expressed in human B cells and detected by binding of IFN-γ to the transfected cells but without biological activity.

C. Receptors and the Mechanism of IFN Action

The conceptual connection between the binding of IFN at the cell surface and the events associated with derepression and repression of host genes remains elusive. Controversy continues as to whether IFNs and/or their receptors achieve their actions directly or indirectly. Some evidence suggests that IFNs need not enter cells to act. Anti-idiotypic antibodies (i.e., antibodies produced against anti-IFN immunogloblin), which mimic the binding site of IFN, not only bind to IFN receptors but elicit a biological response, activating type I and II receptors on human cells. This observation suggests that IFN acts extracellularly, as the antibody could not enter cells by itself.

1. Membrane Effects and Second Messengers

Approaches have been taken to demonstrate transmembrane signaling following IFN treatment of cells. It is difficult to conclude that IFN activity results from rapid alterations in (1) phosphotidylinositol, (2) cytoplasmic alkalinization, (3) cytoplasmic concentrations of free calcium, (4) cyclic adenosine monophosphate (AMP) or cyclic guanosine monophosphate levels, (5) diacylglycerol, or (6) ATP-dependent protein kinase activity.

2. Internalization and Cellular Processing of IFNs

After binding to cell receptors, IFN molecules are internalized into the nucleus within a few minutes by receptor-mediated endocytosis, the receptor alone, or in combination with IFN ligand acting as their own messengers. IFNs bind first to clathrin-coated areas of the plasma membrane or coated pits, specialized cell membrane formations involved in a highly selective and efficient mechanism of the receptor-mediated endocytosis of ligands such as hormones, serum lipoproteins, antibodies, toxins, and viruses. Electron microscopy has demonstrated that unequivocally natural and recombinant IFN molecules entered cells, were processed within the cytoplasm, and moved toward the nucleus, which they entered through nuclear pores and passed into the dense chromatin.

IFN can have a direct physiological effect on the cell nucleus, presumably mediated by nuclear membrane receptors as seen in isolated L 929 cell nuclei treated with murine IFN-β, reducing the energy-dependent efflux of ^{3}H-RNA from these nuclei in a dose-dependent manner, an effect opposite to that noted with insulin treatment. Relatively large amounts of IFN-α and IFN-γ can directly alter the activity *in vitro* of nuclear DNA polymerases α and β of heterologous as well as homologous species, through binding of IFN to the enzyme-substrate complex.

An association between IFN molecules and nuclear proteins or receptors has been shown intracellularly. Two complexes were observed in the nuclear fraction—one of 80 kDa and the other of 235 kDa—whereas the cytoplasmic fraction was a single moiety of 110 kDa. The possible role of IFN-binding nuclear proteins in IFN action is evocative of, but clearly different from, the circumstances with steroid and thyroid hormone gene activation.

Other observations support the possible physiological activity of intracellular IFN molecules, bypassing surface receptors. In heterokaryons treated with IFN, a functioning nucleus of the same animal species as that of the IFN is required for the induction of antiviral activity, indicating that the nucleus plays a vital role in the selective response to IFN. Microinjection of IFN fragments and anti-IFN antibodies into cytoplasm and nucleus prior to IFN treatment inhibit the induction of antiviral resistance. Antiviral resistance has been induced in cells exposed to IFNs encapsulated in liposomes, which presumably would bypass interaction with cell-surface receptors, although it may be difficult to exclude the presence of IFN on the exterior of the liposomes.

It then can be postulated that IFNs can enter cells by two mechanisms: specific receptor-mediated endocytosis and/or bulk pinocytosis.

III. Role of Enzymes in IFN Actions

Two IFN-induced, double-stranded RNA-dependent enzymes that may play important roles in the regulation of viral and cellular macromolecular synthesis and degradation have been identified. They are the $(2'-5'$ A)-oligoadenylate $(2-5$ $A_n)$ synthetase and the P1/eIF-2_α protein kinase. Both the IFN-induced 2,5-A_n synthetase and the IFN-induced P1/eIF-2_α protein kinase must be activated following induction by IFN. Synthetic and natural dsRNAs can fulfill the activation requirement *in vitro*, but IFN treatment alone does not lead to increases in these enzymes. However, infection of IFN-treated cells in culture with encephalomyocarditis virus (EMCV), reovirus, influenza virus, vaccinia virus, or simian virus$_{40}$ (SV$_{40}$) causes a significant increase in the intracellular concentration of both enzymes.

Enzymes, or cellular gene products, have been implicated in the antiviral actions of IFNs directly against either a specific virus family (e.g., Orthomyxoviridae) or against multiple virus families. The expression of a number of other cellular genes, in addition to the enzymes associated with the kinase and the synthetase systems, is regulated by IFN.

The steady-state concentrations of many different cellular gene products are either increased or decreased by IFN treatment (Table III). Some of

TABLE III Cellular mRNAs and Proteins Regulated by IFNs

Cellular gene
A. Increased expression in IFN-treated cells
1. P1/eIF-2_α protein kinase
2. $2',5'$-Oligo(A) synthetase
3. $2',5'$-Oligo(A)-dependent RNase
4. $2',5'$-Phosphodiesterase
5. Protein Mx
6. Class I MHC antigens
7. Class II MHC antigens
8. Class III MHC antigens
9. β_2-microglobulin
10. Guanylate-binding proteins
11. Xanthine oxidase
12. Indoleamine 2,3-dioxygenase
13. Glutathione transferase
14. Metallothionein-II
15. Thymosin B4
16. Tumor necrosis factor receptor
B. Decreased expression in IFN-treated cells
1. c-*myc*; 6 hr, response time of mRNA in Daudi cells
2. c-*fos*; 20 min, response time of mRNA in 3T3 fibroblasts
3. Collagen
4. Epidermal growth factor receptor
5. Transferrin receptor

Abridged from Samuel.

these products (e.g., the major histocompatibility complex [MHC] antigens) appear to play an important role in the antiviral actions of IFNs in the intact animal.

IV. Biological Effects of IFNs

IFNs exhibit numerous biological effects. The antiviral activity of them led to their name; it serves to define the unit of activity of IFN. Purified natural Hu IFN-α was found to exhibit antigrowth activity along with antiviral action. The ratio of antiviral to antiproliferative activity was not constant from one purified fraction to another or from one IFN to another; this observation was confirmed with purified recombinant IFNs and extended to other activities: stimulation of cytotoxic activities of lymphocytes and macrophages, natural killer (NK) cell activity, and increase in the expression of some tumor-associated antigens. With NK cell stimulation, IFN-αJ was found deficient in this activity while exhibiting other actions, and Hu IFN-αD and -αJ lack the ability to stimulate the surface expression of breast and colon tumor-associated antigens, whereas antiviral and antiproliferative activity remain intact. Also, IFN effects on lytic viruses such as vesicular stomatitis virus (VSV) can be dissociated from those on the Moloney leukemia virus. Thus, many IFN effects can be dissociated, due to different molecular mechanisms.

The antiproliferative activity of IFNs was first reported in 1962. No general rules will predict which cells will be inhibited and which will not. Empirical methods have been employed to apply IFN as an antitumor agent; IFNs also modulate cellular differentiation.

A major effect of IFNs is their modulation of antigens of the MHC. All IFNs (α, β, and γ) induce and increase in surface expression of class I MHC antigens including β_2-microgloblin (which is a protein associated with the antigen), whereas class II antigens are stimulated predominantly by IFN-γ with little or no effect by IFN-α or IFN-β. Fc receptor expression is also stimulated.

Often IFN-α/β is synergistic with IFN-γ for antiviral or antiproliferative activities. In some cases, IFNs may be antagonistic. Hu IFN-α/β blocks the IFN-γ-induced stimulation of H_2O_2 by human mononuclear cells; and murine (Mu) IFN-α/β antagonizes downregulation of mannosyl-fucosyl receptors by Mu IFN-γ. One IFN-α species can block the activity of another IFN-α species.

V. Mechanisms of Antiviral Actions of IFNs

The antiviral activity of IFNs, the property that led to their discovery, remains one of the most widely studied aspects of IFN research. IFNs induce activities that inhibit virus multiplication in cell cultures, in animals, and in humans. A wide range of different DNA and RNA viruses is sensitive to the antiviral actions of IFNs. The antiviral properties of natural and molecularly cloned IFNs can differ for different viruses in different animal cell systems, suggesting that multiple mechanisms of IFN action exist. Three lines of evidence initially indicated that the mechanisms of action of type I and type II IFNs may be distinct from each other for a number of different animal viruses: (1) antiviral activities of type I (α or β) and type II (γ) IFNs are synergistic for the inhibition of multiplication of several different viruses; (2) type I and type II IFNs bind to different cell-surface receptors; and (3) type I IFN does not induce an antiviral state when introduced

directly into cells, whereas type II can induce an antiviral state intracellularly. Detailed biochemical studies of the inhibition of VSV by purified cloned IFNs have demonstrated differing molecular mechanisms of antiviral actions of type I IFN compared with type II IFN.

For most animal virus–cell combinations examined, the stage of virus multiplication cycle inhibited with type I IFNs is the synthesis of viral macromolecules. Viruses in this category are the Picornaviridae, Rhabdoviridae, Orthomyxoviridae, Reoviridae, Poxviridae, Adenoviridae, Herpesviridae, and Retroviridae. Inhibition of macromolecular synthesis is often exerted at the step of translation of viral mRNAs into viral polypeptides.

A. Effects of Early Stages of Viral Multiplication Cycles

1. Papovaviridae

SV_{40}, a papovavirus, is inhibited by IFN at the stage of uncoating of the parental SV_{40} virions, a block preventing the subsequent formation of active early transcriptional complexes. This action of IFN seems exceptional.

2. Viruses Other Than Papoviridae

Although most studies revealed that neither type I nor type II IFN significantly affects the early stage of viral multiplication cycles, an effect of IFN on virus penetration by endocytosis has been described. IFN treatment inhibits pinocytosis as measured by uptake of either a virus, VSV, or of horseradish peroxidase. Pinocytosis is inhibited with homologous but not with heterologous IFN preparations.

B. IFN Effects on Viral Macromolecular Synthesis

The synthetic phase of viral multiplication cycle, during which macromolecular synthesis occurs, commences following virion penetration and uncoating. During this stage, viral mRNA synthesis, viral polypeptide synthesis, and progeny genome synthesis occur. With most animal virus–host cell systems studied, IFN treatment of the host cell prior to infection causes a significant reduction in virus-specific macromolecular synthesis, often without adversely affecting overall host macromolecular synthesis. Synthesis of viral polypeptides is inhibited in IFN-treated cells; however, identification of a single principal biochemical mechanism accounting for the selective inhibition in synthesis of viral polypeptides in IFN-treated cells has not been achieved. IFN treatment causes a direct inhibition of viral mRNA translation or, alternatively, affects a step in either the synthesis, modification, or degradation of viral mRNAs, leading to a reduction in viral polypeptide synthesis.

Depending on the genetic system of the virus, the uncoating of the input parental virions, which must occur for viral macromolecular synthesis to begin, is either partial or complete. Partial uncoating yields a subviral particle or nucleocapsid structure containing an activated virion-associated viral transcriptase; complete uncoating yields the free virion genome in a form suitable for either direct translation or transcription by cellular enzymes. VSV, influenza virus, reovirus, and vaccinia virus, all possessing virion-associated RNA polymerase, and EMCV, which does not possess virion-associated polymerase, are among the specific animal viruses studied in great detail with regard to mechanisms by which IFN acts to inhibit viral macromolecular synthesis.

1. Picornaviridae

Picornaviruses (such as poliovirus) have a positive-stranded RNA genome that, following uncoating, is directly translated to produce a large precursor polyprotein that undergoes a complex pattern of posttranslational cleavages; these cleavages yield several mature polypeptides including capsid polypeptides and polypeptides required for virus-specific RNA polymerase activity. The inhibitory action of IFN against a picornavirus—mengovirus and poliovirus—was established at a step in the multiplication cycle following complete uncoating of the virion genome. The synthesis of mengovirus polypeptides was inhibited by type I IFN. The reduction in synthesis of the mengovirus-infected cells is caused by degradation of viral RNA. Activation of the IFN-induced $2',5'$-oligoadenylate synthetase system occurs; this was shown with the EMCV, a variant of mengovirus. $P1/eIF-2\alpha$ protein kinase is activated in some cells treated with IFN and infected with EMC virus.

Natural type I and type II mouse IFNs are synergistic for inhibition of EMCV replication.

2. Rhabdoviridae

VSV, a rhabdovirus, has often been used as a challenge virus in studies of IFN production and action because of its relatively short multiplication time, broad host range, and acute sensitivity to the antiviral effects of IFN in many cell lines. VSV virions possess a virus-coded RNA-dependent RNA polymerase that transcribes the negative-stranded VSV genome into a 47-nucleotide leader RNA and five mRNA species. The monocistronic VSV in RNAs are capped, methylated, and polyadenylated. Specific anti-VSV site of action of IFNs is unclear and difficult to determine. It is independent of cyclic AMP, but type I IFN inhibits VSV progeny virions, and VSV transcription. This type of cell is crucial to effects seen; and the $P1/eIF-2\alpha$ protein kinase and the $2',5'$-oligo(A) synthetase are essential. IFN-αA and IFN-γ are synergistic for VSV inhibition.

3. Orthomyxoviridae

Influenza virus, an orthomyxovirus, possesses a negative-stranded RNA genome; an RNA-dependent RNA polymerase is associated with enveloped virion. Influenza virus occupies a special place in the historical development of IFNs, because IFNs were discovered during studies performed with influenza virus on the mechanism of viral interference. When heat-inactivated influenza virus was incubated with fragments of chick membrane, an interfering substance, later named IFN, appeared in the culture medium. Influenza virus multiplication was then inhibited when fresh membranes were treated with the medium and infected with virus.

The antiviral state in the mouse against influenza virus induced by type I IFN is controlled by a host gene, designated Mx, that encodes a 75-kDa polypeptide. The mouse Mx gene is induced by type I IFN-α and IFN-β but not by type II IFN-γ; the IFN-induced Mx protein accumulates in the nucleus of Mx+ mouse cells. The Mx gene appears to be a hereditary allele originating in inbred mouse strain A2G. Multiplication of a mouse-adapted strain of influenza virus is blocked by IFN at a step following virion attachment and penetration but prior to or including protein synthesis. Another study indicated that the site of IFN action against influenza virus was exerted at an intermediate step between primary and secondary transcription. The exact molecular mechanism by which influenza virus is inhibited by IFN is not yet resolved. [See INFLUENZA VIRUS INFECTION.]

4. Reoviridae

The human reovirus genome consists of 10 distinct segments of dsRNA, each of which is transcribed into single-stranded RNA. The step in the multiplication cycle of reovirus at which the type I IFN-induced inhibition exerts its principal antiviral effect in IFN-sensitive cells is the translation of primary mRNA transcripts into viral polypeptides. Elevated levels of 2',5'-oligo(A)-activated endoribonuclease have been reported in tissue cultures treated with IFN and infected with reoviruses. The enzyme eIF-2α is also involved; both the kinetics of induction and of decay of the IFN-induced P1/eIF-2α kinase correlate with the induction and decay of the antiviral state against reovirus infection. Most evidence is consistent with the notion that phosphorylation of eIF-2α leading to an inhibition of reovirus mRNA translation at the step of polypeptide synthesis may play an important role in the antiviral action of type I IFNs.

Replication of human reovirus is not appreciably inhibited by human type I IFN-α in most human cells so far examined; however, replication is significantly inhibited by both natural and molecularly cloned human type II IFN-γ in human amnion and fibroblast cells. The mechanism of the antireovirus effects of IFN-γ has not yet been fully elucidated.

5. Poxviridae

Vaccinia, a poxvirus, has a large double-stranded DNA genome. Enveloped virion has a DNA-dependent RNA polymerase that catalyzes the synthesis of a family of early mRNAs that encode a number of enzyme activities involved in DNA synthesis. Vaccinia virus mRNAs are capped, methylated, and polyadenylated.

An early study concluded that the inhibitory action of natural mouse IFN is expressed principally at the step of synthesis of viral polypeptides. Vaccinia virus replication and protein synthesis are extremely sensitive to IFN. Activation of 2',5'-oligo(A)-dependent RNase and the cleavage of rRNA and viral RNA correlate with the inhibition of protein synthesis in mouse fibroblasts. In other cells, 2–5 A increases, but viral replication is not inhibited. Therefore, it seems that 2–5 A per se is not obligatory in the inhibition of vaccinia virus by IFN. Reports suggest that resistance of vaccinia virus replication to antiviral effects of IFN is associated with a virus-infected inhibitor of the P1/eIF-2α protein kinase.

6. Adenoviridae

Adenoviruses are "naked" icosahedral viruses that possess a linear double-stranded DNA genome of about 36 kilobases encoding about 50 different polypeptide products. Two temporal stages of viral gene expression, early and late, have been described for productive infection of human cells by adenoviruses.

Adenoviruses are generally weak IFN inducers, and they are more resistant to the antiviral action of type I IFNs in cell culture than those of other animal viruses. This resistance to IFN appears to be caused by the production of virus-associated RNAs, which prevent the activation of the IFN-induced P1/e IF-2α protein kinase.

The IFN-induced expression of the MHC antigens conceivably may contribute to the antiviral and antiproliferative actions of IFN within intact animals, perhaps by the enhancement of the antigen-specific lytic effect of cytotoxic T cells directed against adenovirus-infected or -transformed cells. The cytotoxic T-cell response is virus-specific and histocompatibility antigen-restricted. Adenoviruses, however, appear to possess a mechanism that permits them to evade immune surveillance as well as to antagonize the antiviral action of IFN.

7. Herpesviridae

Herpesviruses have a double-stranded linear DNA in the core of the icosahedral virion, which possess an envelope derived from the nuclear membrane. Herpesviruses are ubiquitous and can produce many diseases of varying severity in human beings.

The yield of infectious herpes simplex virus (HSV) virions in human peripheral blood macrophages is reduced by treatment with IFN, both IFN-α and IFN-β, but not IFN-γ. The type I IFNs appear to act early to inhibit the synthesis of HSV-α and HSV-β proteins, DNA polymerase and thymidine kinase, and γ proteins. Inhibition of HSV translation has also been observed in cells treated with the IFN inducer poly(rI)·poly(rC). Translation of HSV mRNAs encoding gene products also is inhibited.

IFN-α and IFN-β appear to inhibit replication of HSV in human fibroblasts by a mechanism probably different from that observed in macrophages. HSV replication is blocked at a late stage in the viral infection cycle. Both release of total extracellular HSV virions and cell-to-cell fusion are inhibited.

HSV latency and reactivation have been studied with IFNs; they play a crucial role in maintaining persistent HSV infection of macrophages. [*See* HERPESVIRUSES.]

8. Retroviridae

The genome of a retrovirus is a 60S-70S dimer complex of two identical subunits of positive-stranded RNA that resemble mRNA in that they are capped and methylated at the 5'-termini and polyadenylated at the 3'-termini. Most replication-competent retroviruses contain three viral genes arranged (5')-gag-pol-env in the RNA genome. Replication and transformation by retroviruses requires the synthesis of a double-stranded DNA copy of the RNA genome and its integration into the host cell DNA, via a retrovirus-coded RNA-dependent DNA polymerase (reverse transcriptase) that catalyzes the first step in the replication cycle, synthesis of retroviral DNA from the template genomic RNA. Viral gene expression encoded by the integrated provirus is by cellular machinery.

IFN greatly reduces the production of infectious retrovirus particles, both in acutely infected cells and in chronically infected cells. Chronically infected cells are inhibited by IFN in a late stage of virus morphogenesis, either particle assembly or release and maturation. In acute exogenous infection, IFN acts at an early stage in the replication cycle, preventing either the synthesis or integration of proviral DNA.

IFN in cell transformation by oncogenic murine retroviruses prevents the transformation of murine fibroblasts by the Kirsten strain of murine sarcoma virus. IFN appears to block a stage after uncoating but before integration of the provirus.

IFN treatment can affect the expression of v-*onc* and c-*onc* genes. IFN treatment does not appear to affect the c-*myc* transcription rate but, rather, reduces the half-life of c-*myc* mRNA. IFN also inhibits the oncogenic transformation and stable expression of genes of both viral and cellular origin, genes including v-Ha-*ras*, c-Ha-*ras*, and c-Ki-*ras*.

C. Effects of Late Stages of Viral Multiplication Cycles

Virion morphogenesis and release from cells are the last points in the multiplication cycle of a virus where IFN could conceivably act.

1. Retroviridae

Retroviral multiplication is inhibited by IFN at a comparatively late stage in the viral multiplication cycle; inhibition of retrovirus multiplication in IFN-treated, chronically infected cells is observed in the absence of a significant reduction in viral macromo-

lecular synthesis. The mechanism by which the number and the specific infectivity of released retrovirus particles is decreased by IFN treatment is not yet clear.

Early studies with the Moloney MuLV and the Rauscher MuLV indicated that IFN treatment may affect the processing of viral precursor polyproteins with an inhibition of viral protein glycosylation. Defective virions of low specific infectivity deficient in virion envelope glycoprotein Gp 70 have been observed after IFN treatment.

Human type I IFN-αA has a suppressive effect on HIV replication (HIV/HTLV-III) *in vitro*.

2. Viruses Other Than Retroviridae

IFN treatment causes production of defective progeny virions in the case of murine and human cells infected by three different enveloped viruses: VSV, vaccinia, and HSV.

D. Cellular mRNAs and Proteins Regulated by IFN

Enzymes of the P1/eIF-2α protein kinase, the 2–5 A synthetase, and protein Mx are all induced by IFN. These gene products have been implicated in the antiviral actions of IFNs directed against either a specific virus family (e.g., orthomyxoviridae) or multiple virus families.

Twenty-one different cellular gene products, whose biochemical functions have been identified, are either increased or decreased by IFN treatment (Table III). Some of these products, MHC antigens, play an important role in the antiviral actions of IFNs in intact animals.

VI. Treatment of Viral Infections and Other Illnesses with IFNs

A. Effects of Clinical Viral Infections

The discovery that IFNs are natural cell products that have antiviral activity offered the promise that IFNs would indeed be useful in the treatment of clinical viral infections. Antiviral chemotherapeutic research has been slow in the development of useful synthetic antiviral products. IFNs have held promise as antiviral agents in the clinical setting. With the advent of recombinant DNA technology, IFN synthesis became a much more feasible process, although preparation of IFN in cell cultures such as human peripheral blood monocytes obtained from blood banks has also served as a useful IFN source. Toxicity of IFNs, (Table IV), has plagued their usefulness; it has been a more serious problem in cancer patients, where IFN use is at higher levels for longer periods of time. [*See* CHEMOTHERAPY, ANTIVIRAL AGENTS.]

The problem in using IFN as an antiviral substance is how to use it and under what circumstances. Experiments with corneal infection and warts suggest that it can be used effectively when administered locally; however, a protein will unlikely penetrate the skin. Which diseases are best treated with IFN? Experience with hepatitis shows that IFN may be an effective therapeutic agent, not limited to prophylaxis. A recent report has indicated successful treatment of hepatitis C infections with IFN-α.

Just how effective is IFN? Clearly, IFN must be used in doses exceeding millions of units; however, whether or not efficacy is improved with high doses is not yet known. Higher doses are not necessarily more antiviral, as shown by certain studies. IFN is clearly effective for certain infections as shown in studies on the common cold, HSV, and cytomegaloviral infections. The side effects of IFN, its inferiority to other antiviral compounds, and its lack of effect on latency may restrict its use. There is little experience with IFNs in severe acute infections such as encephalitis, hemorrhagic fevers, rabies, and influenza. IFN-α in combination with AZT has been found beneficial in the treatment of HIV infections.

IFN has clear benefits against herpes zoster infections in cancer patients; however, it has been established that the delay in crusting of zoster vesicles is associated with a delay in the synthesis of local IFN and antibody. In the majority of acute viral infections, it seems that endogenous IFN production will have begun and reached its peak by the time symptoms appear; therefore, to be effective, antiviral therapy with IFNs must be given very early to affect the course of a viral infection.

Immunosuppressed patients such as cancer patients and those with chronic lymphatic leukemia, Hodgkin's disease, and multiple sclerosis have relative deficiencies in type I and type II IFN responses. Endogenous IFN responses perhaps can be improved by the priming effect of exogenous IFN.

Well-designed, placebo-controlled, double-blinded trials of IFN in nonlife-threatening infections such as HSV including postneurosurgical reactivation of HSV-1, first and recurring episodes of genital herpes due to HSV-1 or HSV-2, primary varicella (chicken pox) in children with cancer, and zoster in immunosuppressed adults have all documented significant beneficial clinical responses from IFN use.

B. Central Nervous System Illnesses with Possible Uncertain Viral Etiology

1. Subacute Sclerosing Panencephalitis

Subacute sclerosing panencephalitis (SSPE) is a progressive and fatal central nervous system disease of children caused by persistent infection with the measles (rubeola) virus. Attempts at therapy with IFN as well as with other therapeutic agents have been only partially successful at best. Recombinant IFN-α2 was licensed for treatment of certain types of leukemia in 1986; it was the most logical candidate for initial studies. Measles is rather rare in the United States because of extensive vaccination; therefore, SSPE is indeed rare. Thus, further studies with SSPE are advised in the Middle East and Mexico.

2. Multiple Sclerosis

Multiple sclerosis (MS) has had some success in treatment with IFN. Mean exacerbation rates per year were significantly lower for an IFN treatment group than for a control group. In an experiment using IFN-β, the rate of deterioration was greater in controls (0.80) compared with the IFN treated (0.32), although not statistically different. Fevers were significantly more common in treated than in control groups (75% vs. 31%). On the other hand, IFN-γ may play a role in active lesion growth in MS, whereas IFN-α and IFN-β may exert local immunosuppressive activity.

TABLE IV IFN-α Toxicities[a]

Initial injections	Prolonged administration
Chills	Fatigue
Fever	Anorexia
Malaise	Mild neutropenia
Myalgias	Transaminase elevation
Mild neutropenia	Diarrhea

[a] Toxicities of IFN-β and IFN-γ are similar to those of IFN-α.

3. Amyotrophic Lateral Sclerosis

IFN-α, the only IFN tested to date, does not cure amyotrophic lateral sclerosis (ALS). It is possible that another IFN, alone or in combination with IFN-α, may prove to be an effective therapy for ALS.

4. IFNs, Neurologic Disease, and Neuroimmunology

The role of IFNs in the treatment of neurologic disorders takes advantage of their roles as antiviral substances, immune modulators, and an antiproliferative agent. In terms of neuroimmunology, stimulated lymphocytes have been shown to produce neuropeptides such as adrenocorticotropin and γ-endorphin by activation of the propiomelanocortin gene concurrently with the production of IFN. Mice injected with Hu IFN-α showed analgesia and catatonia similar to that induced by β-endorphin and morphine, reversible by naloxone. Competition *in vitro* with ^{3}H-dihydromorphine for binding to the opiate receptor with IFN-α but not with IFN-β or IFN-γ was seen. Some of this information suggests the source of neurotoxicity, particularly of IFN-α. Concerning IFN-γ and the nervous system, decreases in cerebrospinal fluid 5-hydroxyindoleacetic acid in response to treatment have been shown, possibly due to IFN-γ showing increases in tryptophan metabolism. Exacerbation with IFN-γ of such diseases as MS may be related to an alteration in tryptophan metabolism. Increases in corticosteroids after IFN-γ, -β, and -α suggest another mechanism by which these central nervous system effects could occur. These interactions among the IFN, the central nervous system, and the endocrine systems underscore the potential links between the immune system and nervous systems via neuropeptides, IFN, and other cytokines. [*See* PSYCHONEUROIMMUNOLOGY.]

C. Anticellular Actions of IFNs

The anticellular activity of IFNs was first reported in 1962, relatively early in the history of the IFNs, just 5 years after their antiviral effects were reported. This anticellular action of the IFNs has led to extensive research on the potential use of IFNs as anticancer drugs.

Different cell types, either normal or transformed, exhibit a wide range of sensitivity to the growth inhibitory actions of the IFNs. Apparently, the interactions of the receptor-mediated intracellular signal (s) with the biochemical pathways by which cellular proliferation is regulated vary between different cell types in as yet unknown ways. The genes c-*myc* and c-*fos* are cellular protooncogenes believed to be important in the establishment or maintenance of the transformed phenotype; decreases in mRNA levels may indicate decreased transcription or pre-mRNA processing or reduced stability of existing mRNAs (Table III) and a mechanism of the anticellular action of IFNs. Inhibited c-*myc* and c-*fos* contrast with increased expression of many enzymes, MHC antigens, etc. (Table III). Daudi cells are highly sensitive to IFN-α and IFN-β; they are derived from the Epstein–Barr virus. Daudi cells treated with IFNs reveal inhibition of radioactive thymidine into DNA, largely due to decreased thymidine kinase rather than through the inhibition of DNA synthesis. Other changes in the Daudi cell DNA are observed: (1) incompletely ligated DNA fragments, (2) increased susceptibility to endonuclease, (3) impaired histone synthesis, and (4) reduced overall protein synthesis. Gene expression also changes in the Daudi cells following IFN treatment. Genes coding for 2',5'-oligoadenylate synthetase are induced, whereas c-*myc* is inhibited. The mRNA for Epstein–Barr virus nuclear protein also decreases along with that of the c-*fgr* oncogene. Daudi cells can also be inhibited by a tumor promoter 12-*O*-tetradecanoylphorbol 13-acetate instead of IFN; Daudi cells provide some answers as to the mechanism of anticellular actions of IFNs.

1. Treatment of Neoplastic Diseases with IFNs

Enthusiasm for investigations of IFNs in the treatment of malignant disease stem from three factors. First, IFNs have antiproliferative activity *in vitro*. Second, IFNs are potent biologic response modifiers. Third, IFNs as natural proteins produced by the body might have less toxicity than current cytotoxic approaches to systemic cancer treatment.

IFNs have induced partial or complete disease regression in a broad range of human neoplasms. The best response to IFN therapy to date has been for patients with hairy cell leukemia treated with IFN-α. Optimal doses, schedules, and routes of administration of IFNs remain, however, to be defined for most neoplasms. The role of IFNs in combination with other modalities of cancer treatment such as chemotherapy, uses of biological response modifiers, and hyperthermia are now just being defined. The antiproliferative effect of IFNs in human cells *in vitro* and in mice has been enhanced with the addition of tumor necrosis factor. There seems little doubt that IFNs will remain the prototypic biological therapeutic treatment of neoplastic diseases. IFNs have a level and spectrum of therapeutic activity in neoplastic diseases comparable with those of cytotoxic drugs currently in use in oncologic clinical practice. Although side effects of fatigue and anorexia are troublesome with IFN use, when compared with cytotoxic agents, no residual toxicities for vital organs occur. Ultimate therapeutic application will come in combination with the established modalities of surgery, radiation, and chemotherapy; the IFNs will likely play a large role among chemotherapeutic substances.

VII. Poisoning of IFNs by Biological and Chemical Substances

Certain viruses are now shown to inhibit IFN activity, and xenobiotic chemicals as well as deficiencies of protein intake and certain vitamins and minerals aggravate viral infections, in part, through impaired IFN synthesis or action or both. Physical entities such as gamma rays that enhance infectious diseases would also likely inhibit IFN synthesis and/or action.

A. Viral Inhibition of the Antiviral Action of IFN

1. Evasion of the 2',5'A System

DNA viruses are generally less sensitive to IFN antiviral actions than RNA viruses, perhaps in part because little if any dsRNA is synthesized during infection. Also, some DNA viruses have specific strategies for overcoming the antiviral action of IFN. Inhibition of HSV replication was found to require 10 times the amount of IFN needed to inhibit EMCV replication. IFN induces synthesis of 2–5 A-linked oligomers to the same extent, whether the cells are infected with HSV or with EMCV; little RNA cleavage by endonuclease occurs in HSV-infected cells compared with EMCV-infected cells. HSV infection subverts the 2–5 A synthetic system to make nonfunctional 2–5 A oligomer analogues, which do not efficiently activate the endonuclease.

Similarly, in SV_{40}-infected, IFN-treated cells, synthesis of 2–5 A oligomers occurs, but little 2–5 linked oligonucleotides were in the active trimeric or tetrameric form. Instead, analogues similar to

2–5 A but lacking terminal phosphoryl groups were found. Little cleavage of RNase L occurred. RNase L (F) activity is present in EMCV-infected cell extracts, but it is inhibited by both HSV and SV_{40}.

In vaccinia virus-infected HeLa cells, containing high levels of endogenous 2–5 A synthetase, IFN increases the 2–5 A oligomers. Without presenting all details, it is suggested that viral ATPase and phosphatase activities are involved in the vaccinia-mediated inhibition of the 2–5 A system.

One RNA virus, EMCV, possesses a specific mechanism that circumvents the 2–5 A directed inhibition of viral translation. EMCV can replicate in HeLa cells, although they possess high levels of endogenous 2–5 A synthetase. IFN treatment increases the amount of enzyme 3- to 10-fold, but the enzyme is synthesized in an inactive form; therefore, no 2–5 A oligomers are detected. EMCV infection activates the synthetase, and then endonuclease is activated. Without IFN, the endonuclease is rapidly inactivated and EMCV replicates well. IFN prolongs the nuclease activity. How EMCV overrides the endogenous endonuclease activity is unknown, but for this RNA virus EMCV the mechanism of avoidance of 2–5 A mediated block appears to differ from that found in the DNA virus infections described earlier.

2. Evasion of the P1 Protein Kinase System

Some viral-infected cells possess mechanisms to avoid the IFN-induced protein kinase system. The DNA virus adenovirus 5 has genes coding for small RNA molecules called VA 1 RNAs, which are required for efficient translation of viral and cellular genes late in viral infection. This efficiency of translation correlates with inhibition of the kinase that phosphorylates e1F-2α. VA 1 RNA competes with dsRNA and prevents kinase activation in IFN-treated cells, thereby blocking the antiviral activity mediated by this pathway.

Adenovirus reversion of the IFN-induced antiviral state can extend to other IFN-sensitive viruses such as poliovirus and VSV. If adenovirus DNA synthesis is blocked, the reversal of the antiviral state is lost, indicating that the adenovirus reversal effect is a late viral function. Expression of EA 1, an early adenovirus gene, is essential for the reversal of IFN antiviral action.

Vaccinia virus appears to be able to inhibit both IFN-induced protein kinase and the 2',5'-A synthetase pathways. The protein kinase-directed phosphorylation of the e1F-2α was specifically inhibited in poliovirus-infected HeLa cells, but autophosphorylation of the P1 kinase was not inhibited; the mechanism by which inhibition occurs is not known, but it does not appear to rely on phosphatase activity.

B. Chemicals Aggravate Viral-Induced Disease Through Inhibition of IFN Synthesis and Action

The polybrominated biphenyls have been associated with depressed immunologic functions in persons exposed in Michigan to this xenobiotic. A variety of chemicals including xenobiotics such as the dioxins and the polychlorinated biphenyls, diethylstilbestrol, and the salts of several heavy metals have experimentally in laboratory animals been shown to enhance viral-induced disease mortality or leukemogenic viral-induced splenomegalies, in part through the inhibition of IFN synthesis or action or both. Metallic salts shown to enhance experimental virus infections are arsenic, cadmium, cobalt, lead, mercury, and nickel.

Procedures to illustrate viral disease enhancement via chemical exposure have been through feeding the metals to mice, and then challenging the mice later with a virus such as the EMCV or the Rauscher leukemogenic virus. Elevated mortality above the nontreated viral-infected controls signals an aggravation of the viral infection. With the Rauscher virus, splenomegaly is a pathognomonic feature. Indications that IFN is suppressed may be made directly by assaying for IFN or indirectly by testing and comparing the percent survivability after giving polyriboinosinic acid–polyribocytidylic acid (poly I/poly C), which is a synthetic inducer of IFN-β. Arsenicals, cobalt, and lead, but not cadmium, mercury, and nickel, reverse the protective effect of poly I/poly C against EMCV mortality, although all six metallic salts alone enhance EMCV mortality. Arsenicals then are likely inhibiting the action of IFN-β, but the other three salts are not. Sodium arsenite was further shown to directly inhibit IFN-β action *in vitro*, in a dose–response relationship at concentrations from 10^{-4}–10^{-6} M, but at a $10^{-6.5}$ M concentration the sodium arsenite was "pro-IFN-β," it benefited the IFN in reducing VSV plaque formation.

Organic mercurials are generally more biologically active than the inorganic mercurials, and they experimentally have been shown to enhance infections in laboratory animals through depressed IFN and antibody. In the disastrous methyl mercury poisoning in Iraq in the early 1970s, no reports of increased viral or bacterial infections were noted although >500 people died and >5,000 had mercurial poisoning.

C. Chronic Alcoholism, Heavy Cigarette Smoking, and Blood IFN Levels

The IFN system was evaluated in three groups of individuals: (1) chronic alcoholics, (2) heavy cigarette smokers, and (3) control individuals with essentially no smoking or drinking habits. The healthy controls had virtually no IFN or IFN inactivators in their plasma, and their lymphocytes produced normal levels of IFN-α and IFN-γ. In contrast, chronic alcoholics contained significantly higher levels of IFN and IFN inactivators in their plasma, and their lymphocytes produced significantly lower levels of both IFN-α and IFN-γ than those of controls. The presence of IFN and its inactivators is generally indicative of a weaker immune system. Similarly, the smokers had significantly lower IFN-α and IFN-γ levels than those of controls. The IFN-α levels for the alcoholics and smokers were 21 and 30%, respectively, of controls; IFN-γ levels for both test groups were 26% of controls.

D. Mineral and Dietary Deficiencies and IFN

Infectious and parasitic diseases are rampant in malnourished and undernourished people, especially growing children. Protein undernutrition and deficiencies in essential minerals such as copper, iron, selenium, and zinc have been associated with increased incidence and severity of viral infections. The rapidly growing piglet is exquisitely sensitive to the development of iron-deficiency anemia. Iron-deficient baby pigs are anemic and neutropenic, and they produce less IFN-β than iron-supplemented controls; thus, iron deficiency in the baby pig results in less serum IFN-β, explaining why the iron-deficient baby pig would likely be more susceptible to viral infections than would the iron-supplemented pig. The mechanism for the depressed IFN levels in the serum of the iron-deficient baby pig may be, in part, through the depressed number of circulating neutrophils.

VIII. Summary

A discussion is given of the IFNs, their high reactivity, their genes and the genes they regulate, and how they produce the antiviral and anticellular effects that they do produce. Much has been learned about IFNs since they were first described in 1958. Although they are formed and observed in nature as a first response of an animal cell to a virus, they are also observed in other occurrences such as in ovine and porcine conceptuses, or as IFN-γ in its wide effects as an immune modulator.

Discussion continues on the viruses and viral-induced diseases against which IFNs are active, including hepatitis C infection of humans. Certain viruses are capable, however, of avoiding the antiviral action of the IFNs. Chemical toxicosis may reduce the antiviral effectiveness of IFNs as do deficiencies in protein and essential trace minerals. The hope for IFNs as anticancer agents has been realized in that hairy cell leukemia is significantly reduced and treated with IFN-α. The future appears promising for additional important uses of IFNs as well as for a more full understanding of human biology through the understanding of the IFNs.

Bibliography

Borden, E. C. (1988). Effects of interferons in neoplastic diseases of man. *Pharmacol. Therapeut.* **37,** 213.

Chada, K. C., Whitney, R. B., Cummings, M. K., Norman, M., Windle, M., and Istvan, S. (1990). Evaluation of the interferon system among chronic alcoholics. *In* "Alcohol, Immune Modulation, and AIDS" (A. Pawlowski, D. Seminara, and R. Watson, eds.). Alan R. Liss Publishers, New York.

Engel, D. A., Snoddy, J., Toniato, E., and Lengyel, P. (1988). Interferons as gene activators: Close linkage of two interferon-activatable murine genes. *Virology* **166,** 24.

Grossberg, S. E., Taylor, J. L., and Kushnaryov, V. M. (1989). Interferon receptors and their role in interferon action. *Experientia* **45,** 509.

Kende, M., Gainer, J. H., and Chirigos, M. (1984). "Chemical Regulation of Immunity in Veterinary Medicine." Alan R. Liss Publications, Inc., New York.

Mechti, N., Bisbal, C., Leonette, J., Salehzada, T., Affabris, E., Bayard, B., Piechaczyk, M., Blanchard, J., Jeanteur, P., and Lebleu, B. (1988). Interferons and oncogenes in the control of cell growth and differentiation: Working hypothesis and experimental facts. *Biochemie* **70,** 869.

Pestka, S., Langer, J. A., Zoon, K. C., and Samuel, C. E. (1987). Interferons and their actions. *Annu. Rev. Biochem.* **56,** 727.

Samuel, C. E. (1988). Mechanisms of the antiviral actions of interferons. *Prog. Nucleic Acid Res. Mol. Biol.* **35,** 27.

Sikora, K. (1987). Interferon therapy—Molecular mechanisms and clinical applications. *Br. J. Clin. Prac.* **42,** 1.

Smith, R. A. (1988). "Interferon Treatment of Neurologic Disorders." Marcel Dekker, Inc., New York.

Stringfellow, D. A. (1986). "Clinical Application of Interferons and Their Inducers." Marcel Dekker, Inc., New York.

Taylor, J. L., and Grossberg, S. E. (1990). Recent progress in interferon research: Molecular mechanisms of regulation, action, and virus circumvention. *Virus Res.* (in press).

Traugott, M. D., and Leben, P. (1988). Multiple sclerosis: Involvement of interferons in lesion pathogenesis. *Ann. Neurol.* **24,** 243.

Interleukin-2 and the IL-2 Receptor

TADATSUGU TANIGUCHI, *Osaka University*

Glossary

CD3 Cluster of differentiation antigen specifically expressed on T cells; consists of multiple components and is associated with T cell antigen receptor

CD4 Cluster of differentiation antigen expressed on a subset of T cells; majority of CD4$^+$ T cells have helper function; recognizes MHC class II antigens

CD8 Cluster of differentiation antigen expressed on a subset of T cells (usually those that do not express CD4); majority of CD8$^+$ T cells have suppresser–cytotoxic (effector) functions; recognizes MHC class I antigens

Complementary DNA Synthesized by using a messenger RNA as the template

Major histocompatibility complex (MHC) Cluster of genes encoding major histocompatibility antigens (MHC antigens); antigens can be recognized by T cell antigen receptor in the context of MHC molecules.

T cell antigen receptor T cell-specific receptor (a heterodimer) that recognizes antigen in the context of MHC molecules; it is associated with CD3, responsible for transduction of antigen-induced, cellular activation signals

THE IMMUNE SYSTEM involves two classes of lymphocytes that react specifically with antigen: thymus-derived T lymphocytes (T cells) and B lymphocytes (B cells). For the efficient operation of this system, mechanisms that ensure the selective, clonal expansion of antigen-stimulated lymphocytes are required. In such a process, a class of soluble factors, referred to as cytokines, appears to play an essential role. Cytokines include interleukins, interferons, colony-stimulating factors, and others. Through their homologous receptors, they deliver signals for cell growth promotion–inhibition, differentiation, and activation. Among these cytokines, interleukin-2 (IL-2) is known to be a potent growth factor, playing a major role in the proliferation of antigen-stimulated T cells. Expression of both IL-2 and its receptor (IL-2R) is induced in antigen-stimulated T cells. IL-2 also exerts its activity on various cell types including nonlymphoid cells, indicating that it is a multifunctional cytokine. The IL-2 signal(s) is delivered via specific cell-surface receptor (IL-2R) complex composed of two receptor components, IL-2Rα and IL-2Rβ chains. The genes encoding IL-2, IL-2Rα, and IL-2Rβ have been cloned and the complete primary structures of the ligand and receptors deduced. The recombinant DNA technology has made it possible to produce IL-2 in pure and homogenous form, thus opening new possibilities for the use of IL-2 in basic and clinical immunology. [*See* Cytokines in the Immune Response; Lymphocytes; Immune System.]

I. Introduction

Observations made in the 1960s found plant lectins such as phytohemagglutinin to be mitogenic for lymphocytes. Subsequently, the presence of soluble mitogenic factors has been reported in conditioned medium derived from cultures containing a mixture of leukocytes from unrelated individuals (allogeneic mixed leukocyte cultures). These mitogenic factors were initially considered to function merely in amplifying the cellular responses, driven entirely by antigen, and the nature of these factors remained totally elusive. In 1976, it was reported that conditioned media from lectin-stimulated mononuclear cells contained a mitogenic factor that would support the viability and proliferation of primary T cells in long-term culture. The factor was then designated as T cell growth factor (TCGF). The conditioned medium containing TCGF activity made it possible to establish and grow antigen-specific T cell clones in long-term cultures, leading to the realization that antigen was primarily essential to render T cells responsive to mitogenic (antigen-nonspecific) factors such as TCGF, which then provides the proliferation signal(s). Subsequently, a system of nomenclature for a series of soluble factors was developed, based on their ability to act as transmitters of cell-to-cell communication among different populations of leukocytes, and TCGF was put in the group of interleukins ("between leukocytes") and renamed as interleukin-2 (IL-2).

The IL-2 system has been studied in great detail in the context of T cell growth control. In fact, clonal proliferation of T cells is initiated via the process of signal transduction, wherein the specific interaction of the antigen–major histocompatibility complex molecules on an antigen-presenting cell with the CD3 complex–T cell receptor complex (TCR) on a T cell triggers the expression of IL-2 and its homologous receptor IL-2R. Subsequent to IL-2–IL-2R interaction, the activated T cells undergo proliferation (Fig. 1). In effect, the IL-2 system plays an essential role in antigen-specific immune responses; the antigen–TCR interaction confers specificity for a given immune response by selecting the proper T cell, whereas IL-2–IL-2R interaction determines its magnitude and duration by inducing the selected T cell to proliferate extensively. Expression of the IL-2R has been detected on other hemopoietic cells such as B lymphocytes (B cells), natural killer cells (NK cells), thymocytes, and macrophages as well as neural cells such as oligodendrocytes. Thus, IL-2 is also a multifunctional cytokine, exerting various biological activities (i.e., cell growth promotion–inhibition, activation, and differentiation activities) to these cells. [*See* MACROPHAGES; NATURAL KILLER AND OTHER EFFECTOR CELLS.]

II. Structure of IL-2

IL-2 is a glycoprotein produced and secreted primarily by CD4$^+$ (i.e., helper) T cells activated by antigen or mitogen. The complete primary structure of IL-2 was first deduced by cloning and nucleotide sequence analysis of the complementary DNA (cDNA) encoding the biologically active human IL-2. The structure was also determined and confirmed by analyzing purified human IL-2. The human IL-2 precursor consists of 153 amino acids and, upon secretion, its signal sequence (20 amino acids long), which is needed for the protein to cross the membrane through which secretion occurs, is cleaved off to generate the mature form (Fig. 2). IL-2 shows no obvious structural homology to other growth factors and cytokines. The two cystein residues at positions 58 and 105 form a disulfide bond essential for the biological activity of the molecule. The three-dimensional structure of the human IL-2 has been elucidated by X-ray crystallography, revealing that the core structure is composed of antiparallel α-helices without any segments of beta secondary structure. As described below, the functional, high-affinity IL-2R is composed of two receptor components: IL-2Rα and IL-Rβ chains. Both receptor proteins have been shown to interact with different regions of the IL-2 molecule; the N-terminal α-helix (amino acids 11–19) and the neighboring amino acid 20 (Asp), and the second

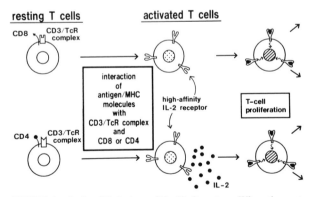

FIGURE 1 A simplified view of the IL-2 system. When the resting T cells are activated by the interactions of antigen–major histocompatibility complex molecules with CD3–T cell antigen receptor (TCR) complex and with CD4 or CD8 molecules, the cells acquire growth competence. The cellular activation signal leads to the expression of IL-2 and IL-2R, and IL-2 transduces the signal for progression of T cell proliferation. The CD3–TCR complex is downregulated following the T cell activation.

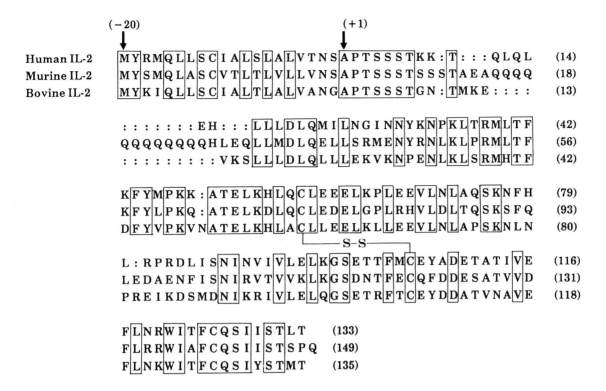

	(−20)	(+1)	
Human IL-2	MYRMQLLSCIALSLALVTNS	APTSSSTKK:T:::QLQL	(14)
Murine IL-2	MYSMQLASCVTLTLVLLVNS	APTSSSTSSSTAEAQQQQ	(18)
Bovine IL-2	MYKIQLLSCIALTLALVANG	APTSSSTGN:TMKE::::	(13)

:::::::EH:::LLLDLQMILNGINNYKNPKLTRMLTF	(42)
QQQQQQQQQHLEQLLMDLQELLSRMENYRNLKLPRMLTF	(56)
:::::::::VKSLLLDLQLLLLEKVKNPENLKLSRMHTF	(42)

KFYMPKK:ATELKHLQCLEEELKPLEEVLNLAQSKNFH	(79)
KFYLPKQ:ATELKDLQCLEDELGPLRHVLDLTQSKSFQ	(93)
DFYVPKVNATELKHLACLLEELKLLEEVLNLAPSKNLN	(80)

S-S

L:RPRDLISNINVIVLELKGSETTFMCEYADETATIVE	(116)
LEDAENFISNIRVTVVKLKGSDNTFECQFDDESATVVD	(131)
PREIKDSMDNIKRIVLELQGSETRFTCEYDDATVNAVE	(118)

FLNRWITFCQSIISTLT	(133)
FLRRWIAFCQSIISTSPQ	(149)
FLNKWITFCQSIYSTMT	(135)

FIGURE 2 Comparison of the amino acid sequences for human, murine, and bovine IL-2 molecules. The three IL-2 molecules each contains 20 amino acids (−1 to −20) at the N-terminal region, which would constitute the signal peptide required for secretion. Coincident amino acids are framed. Presence of a disulfide bond by two conserved cystein residues is indicated as -S-S-. The sequences are presented with one-letter amino acid codes as follows: A, alanine; C, cystein; D, aspartic acid; E, glutamic acid; F, phenylalanine; G, glycine; H, histidine, I, isoleucine; K, lysine; L, leucine; M, methionine; N, asparagine; P, proline; Q, glutamine; R, arginine; S, serine; T, threonine; V, valine; W, tryptophan; Y, throsine.

helix involving residues 33–56, seem to be involved with the binding to the IL-2Rβ and IL-2Rα, respectively.

In addition to the characterization of human IL-2, the murine and bovine IL-2 cDNAs have been cloned and their structure extensively analyzed. The murine IL-2 shows 63% amino acid homology to the human IL-2, and it contains a unique stretch of 12 consecutive glutamine repeats. The repeats are, however, not essential, because the deletion of the repeats within the IL-2 molecule by genetic engineering does not largely affect the biological activity. The amino acid sequence of bovine IL-2 is 65 and 50% homologous to the human and murine IL-2, respectively (Fig. 2).

III. Structure of IL-2R

IL-2 delivers its signal by virtue of its interaction with a specific cell-surface receptor (IL-2R), expression of which is induced in activated T cells (Fig. 1). The IL-2R is also expressed on various cell types as described above. The IL-2R exists in three different forms defined on the basis of its affinity to IL-2 (i.e., low-, intermediate-, and high-affinity forms with respective dissociation constants [K_d] of 10^{-8} M, 10^{-9} M, and 10^{-11} M. The high-affinity IL-2R, which is primarily responsible for the IL-2 signal transduction, is composed of two distinct IL-2-binding membrane components, the IL-2Rα and IL-2Rβ chains, each of which respectively manifests low and intermediate affinities when expressed singly. Evidence also suggests that IL-2Rβ, but not IL-2Rα, transduces the IL-2 signal(s) in the absence of other receptor components at high IL-2 concentrations. High- and intermediate-affinity IL-2Rs can internalize bound IL-2, whereas low-affinity IL-2R (i.e., IL-2Rα) cannot.

Human IL-2Rα is a 55-kDa glycoprotein with a backbone polypeptide consisting of 251 amino acids containing a single membrane-spanning region (Fig. 3). The primary structure of the IL-2Rα shows no significant homology with other known receptor

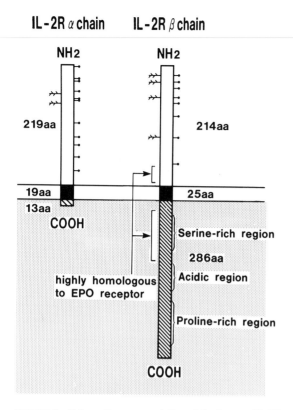

FIGURE 3 Schematic representation of the human IL-2Rα and IL-2Rβ chains. Symbols on the right and left sides of each column represent the positions of cystein residues and N-glycosylation sites, respectively. The IL-2Rβ chain shows structural homology with the erythropoietin (EPO) receptor, particularly at restricted regions. The cytoplasmic region of the IL-2Rβ contains regions characterized by abundance of particular amino acids.

molecules. As depicted in Fig. 3, it contains a very short cytoplasmic region consisting of only 13 amino acids. Initial expression studies of the cloned cDNA for the human IL-2Rα in various host cells revealed that the IL-2Rα constitutes a low-affinity IL-2R with a K_d of $\sim 10^{-8}$ M in fibroblast cells, whereas it formed high-affinity IL-2R when expressed in lymphoid cells ($K_d \sim 10^{-10}$ M). These observations suggested the presence of an additional membrane component(s) limited to lymphoid cells, thereby forming the functional high-affinity IL-2R complex. The conjectured membrane component has been identified as a novel IL-2R chain, designated as IL-2Rβ. The IL-2Rβ is a 70–75-kDa glycoprotein structurally unrelated to IL-2Rα. The predicted mature form of the human IL-2Rβ consists of 525 amino acids, of which the N-terminal 214 residues represent the cystein-rich extracellular region, followed by a single transmembrane region consist-

ing of 25 amino acids. The cytoplasmic region of the IL-2Rβ is 286 amino acids long and it is far larger than that of the IL-2Rα, suggesting the importance of this region is signal transduction (Fig. 3). However, this region contains no characteristic sequences (motifs) such as a tyrosine kinase domain essential for many growth factor receptors. The primary structure of the cytoplasmic region is well conserved between human and murine IL-2Rβ's.

The IL-2Rβ shows structural homologies with other hemopoietic cytokine receptors such as those for erythropoietin (EPO), interleukin-3 (IL-3), interleukin-4 (IL-4), interleukin-6 (IL-6), interleukin-7 (IL-7), and granulocyte–macrophage colony-stimulating factor. For example, the amino acid homology is 19% between the murine IL-2Rβ and EPO receptor. Because many of the cytokine receptors are not fully characterized as yet, it may turn out that the IL-2Rβ belongs to a new large family of receptors that have evolved from a common ancestral gene. Some, if not all, of those receptors possibly utilize common or similar intracellular pathways of signal transduction.

At present, as to whether or not another specific membrane component is involved in the IL-2R complex is not clear. A number of studies suggest that the IL-2R may be a multichain complex, but further work is required to clarify this issue.

IV. IL-2–IL-2R Interaction and Signal Transduction

The functional importance of the IL-2Rα and IL-2Rβ components have been extensively studied by expressing the cloned cDNAs in various host cells. The IL-2Rα manifests the low-affinity from of IL-2R ($K_d \sim 10^{-8}$ M), both in lymphoid and fibroblast cells. In contrast, the IL-2Rβ manifests intermediate affinity to IL-2 when expressed in lymphocytic cells such as the human leukemic T cell line, the Jurkat line, and the mouse thymoma line, EL-4. However, IL-2Rβ shows no IL-2-binding activity when expressed in fibroblast cells (mouse L929, NIH3T3 cells, monkey COS cells, etc.). This observation is consistent with the idea that the IL-2Rβ requires another lymphoid-specific molecule(s) to become a functional receptor. When both IL-2Rα and IL-2Rβ are expressed simultaneously, "high-affinity" receptor is generated even in fibroblast cells ($K_d \sim 10^{-10}$ M), indicating that association of the two molecules can occur to generate a high-affinity IL-2 binding site in the absence of such a conjectured associate molecule(s). However, such a receptor neither transduces the IL-2 signal nor internalizes IL-2, suggesting the absence of an IL-2Rβ-coupling molecule(s) in fibroblasts.

To elucidate further the role of the IL-2Rβ in the IL-2 signal transduction, a cDNA expression system was established in which the human IL-2Rβ transduces a growth signal upon IL-2 stimulation in mouse cell lines. In fact, the cDNA linked to an expression vector was introduced in mouse mast cell progenitor cell line (IC-2) and pro-B cell line (BAF-B03). Both cell lines are IL-3-dependent for cell growth. Although these cell lines expressed a large number of endogenous IL-2Rα, they could not respond to IL-2, indicating that the IL-2 signal is not transduced by the IL-2Rα alone. When the human IL-2Rβ was expressed in these cells by introducing the cDNA, high-affinity IL-2R was generated in conjunction with the endogenous IL-2Rα and, consequently, the cells became responsive to IL-2 as well as to IL-3. Some of the other IL-3 dependent cell lines can switch to IL-2 dependence. Collectively, there may be a common growth signal transduction pathway(s) for these cytokines in a wide range of hemopoietic cells. To identify the critical region(s) of IL-2Rβ for the growth signal transduction, mutant IL-2Rβ cDNAs containing deletions within the region encoding the cytoplasmic region were generated. They were each expressed in BAF-B03 cells and the response of the cells expressing various IL-2Rβ mutants to IL-2 was examined. This study revealed that a restricted cytoplasmic region of the IL-2Rβ (i.e., the region encompassing the "serine-rich region" [Fig. 3]) is essential for IL-2 signal transduction. However, the mutants deficient in signal transduction still formed high-affinity IL-2Rβ and internalized IL-2 in BAF-B03 cells, indicating that the ligand internalization process may not directly couple with the intracellular signal transduction pathway. The critical region is well conserved between the human and mouse IL-2Rβ. Interestingly, this region also shows a high degree of homology with the EPO receptor (Fig. 3). Presumably, this region couples with a yet unknown protein(s), which further drives the downstream signaling pathway(s).

Reportedly, IL-2 induces rapid phosphorylations at tyrosine residues of several proteins in the IL-2 responsive cells. Growth factor-induced tyrosine phosphorylations have been known to be essential in signal transduction by receptors containing intrinsic tyrosine kinase. The IL-2Rβ, however, contains no obvious kinase motif within the identified cytoplasmic region essential for signal transduction. It has also been reported that protein kinase C, calcium mobilization, and phosphatidylinositol turnover are not directly involved in the IL-2 signal transduction. Hence, the nature of the molecule(s) coupled with the IL-2Rβ in the intracellular cascade of signal transduction is unknown at present. Recently, evidence has been obtained for the physical association of IL-2Rβ with cytoplasmic tyrosine kinase, p56^{lck}. Thus, such tyrosine kinase molecule might regulate the IL-2 system.

V. Structure and Expression of IL-2 and IL-2R Genes

Following the activation T cells, expression of IL-2 and IL-2R genes is induced. The controlled, coordinated expression of these genes plays a crucial role in the regulation of T cell clonal proliferation. The genes encoding IL-2, IL-2Rα, and IL-2Rβ have been cloned and analyzed; in humans, they are mapped on chromosomes 4q, 10p, and 22q, respectively. The genomic organization of these genes is depicted in Fig. 4. In normal T cells, expression of IL-2 and IL-2Rα genes is tightly regulated; both genes are not expressed in resting T cells but can be induced by antigens or mitogens. The induced gene expression is primarily controlled at the transcriptional level, and both genes contain, within their immediate 5' flanking region, regulatory DNA sequences, which are activated to promote transcription in induced T cells. The activation appears to be mediated by a number of transcription factors, some of which become activated or *de novo* synthesized upon induction. One such factor, termed NFκB, or NFκB-like factor (the structure of NFκB is still unknown but may include a number of related factors), is activated in mitogen-induced T cells and regulates both IL-2 and IL-2Rα genes. The NFκB factor has been known to activate the immunoglobulin κ chain enhancer and either NFκB or its homologue seems to activate a wide range of genes operating in cytokine systems. In contrast to IL-2 gene, expression of the IL-2Rα gene is not restricted to T cells, and little is known about the regulation mechanism of this gene in non-T cells. Expression of IL-2Rβ gene is somewhat unique in T

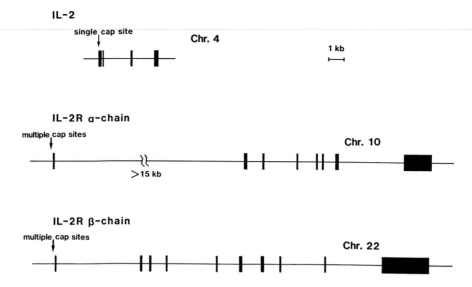

IL-2

single cap site

Chr. 4

1 kb

IL-2R α-chain

multiple cap sites

Chr. 10

>15 kb

IL-2R β-chain

multiple cap sites

Chr. 22

cells in that the gene (and the product) is constitutively expressed in a certain population of resting CD8+ (i.e., cytotoxic) T cells and is undetectable in CD4+ (i.e., helper) T cells. The gene is expressed at high levels in both T cell populations upon mitogenic induction. The IL-2Rβ gene is also expressed constitutively in a variety of normal cells such as NK cells and macrophages. The promoter region of the human IL-2Rβ gene shows no similarity to the human IL-2 and IL-2Rα gene promoters, suggesting the operation of distinct gene regulation mechanism.

Given the critical role of the IL-2 system in T cell proliferation, dysregulation of the IL-2 and IL-2R genes may cause immunological disorders or malignancies under certain circumstances. Aberrant activation of IL-2 and IL-2R genes has been reported in the case of adult T cell leukemia (ATL), a fatal hematological disorder that is endemic in southwestern Japan and the Caribbean islands. ATL is believed to be caused by the retrovirus human T cell leukemia virus type I (HTLV-1). The leukemic T cells usually display a large number of the IL-2Rα and some cells produce and respond to IL-2. The HTLV-1 encodes no obvious oncogenes, but it does encode a transcriptional activator protein referred to as *tax*-1 (or p40[tax]). By an as yet unknown mechanism, *tax*-1 activates its own long terminal repeat sequence as well as cellular genes such as IL-2 and IL-2Rα genes. Thus, viral infection of T cells possibly results in aberrant activations of the IL-2 system, which may be a crucial event in the development of ATL. In this context, IL-2-dependent T cell lines become independent of exogenous IL-2 and acquire

FIGURE 4 Organization of the chromosomal genes for the human IL-2, IL-2Rα, and IL-2Rβ. The rectangles represent exons; regions corresponding to exons are expressed in the mature messenger RNA. The cap sites correspond to the messenger RNA transcription initiation sites.

tumorigenic properties when IL-2 gene is transfected and constitutively expressed. In addition, some leukemic T cell lines appear to grow by producing IL-2 and responding to it.

VI. Therapeutic Implications

The availability of basically unlimited amounts of recombinant IL-2 has made it possible to explore a practical approach for adoptive cancer immunotherapy. The recombinant IL-2 can be used to generate lymphokine-activated killer cells (lymphocytes that eliminate a broad range of tumor cells) or to grow tumor-infiltrating lymphocytes (cytotoxic T lymphocytes that infiltrate in tumor lesions, therefore restricting their target to a given tumor) *in vitro,* and these cells are then reintroduced into cancer patients. The potentiation of the host immune response thus induced by IL-2 is under clinical investigation and promising results have been obtained in some types of cancer. Approaches to immunotherapy may also include the use of combinations of cytokines, such as the use of interferons or tumor necrosis factors in conjunction with IL-2.

Antibodies against IL-2 or IL-2R's might be potential agents in preventing allograft rejection as well as in suppressing autoimmune diseases. Bacterial toxin-IL-2 conjugates, which are produced by recombinant DNA techniques; may provide another approach to immunosuppression. In this context, IL-2 linked to diphtheria toxin reportedly is quite effective in immunosuppression in experimental systems, suggesting the possibility of clinical applications.

Finally, IL-2 has been used as an adjuvant for stimulating the immune response to vaccines. Two reports have appeared in which the IL-2 gene was inserted with antigen-coding DNA sequences into vaccinia virus. The produced IL-2 appears to allow an effective immune response.

Bibliography

Green, W. C., and Leonard, W. J. (1986). The human interleukin-2 receptor. *Annu. Rev. Immunol.* **4,** 68.

Hatakeyama, M., Tsudo, M., Minamoto, S., Kono, T., Doi, T., Miyata, T., Miyasaka, M., and Taniguchi, T. (1989). Interleukin-2 receptor β chain gene: Generation of three receptor forms by cloned human α and β chain cDNA's. *Science* **244,** 551.

Rosenberg, S. A. (1988). The development of new immunotherapies for the treatment of cancer using interleukin-2. *Ann. Surg.* **208,** 121.

Smith, K. A. (1988). Interleukin-2: Inception, impact and implication. *Science* **240,** 1169.

Taniguchi, T. (1988). Regulation of cytokine gene expression. *Annu. Rev. Immunol.* **6,** 439.

Taniguchi, T., Matsui, H., Fujita, T., Takaoka, C., Kashima, N., Yoshimoto, R., and Hamuro, J. (1983) Structure and expression of a cloned cDNA for human interleukin-2. *Nature* **302,** 305.

Waldmann, T. A. (1989). The multi-subunit interleukin-2 receptor. *Annu. Rev. Immunol.* **58,** 875.

Interpersonal Attraction

ELLEN BERSCHEID, *University of Minnesota*

Glossary

Affiliation Fact of interaction and association with another, or the expressed desire to interact with another; may or may not be associated with attraction

Attitude Predisposition to respond in favorable or unfavorable ways toward the object of the attitude

Close relationship Relationship in which the partners are highly interdependent on each other, where interdependence is evidenced by the fact that the activities of each partner are strongly influenced by the activities of the other; usually but not always associated with attraction

Closed-field interaction settings Interaction settings in which interaction with a specific other is virtually mandated by the environmental aspects of the setting or by the social norms governing it

Hypothetical construct Internal state of the individual assumed to exist for explanatory and predictive purposes; inferred on the basis of various kinds of observable evidence

Liking Mild form of interpersonal attraction, usually evidenced by a positive attitude toward the other, often with the belief that the other possesses positive qualities

INTERPERSONAL ATTRACTION is a psychological hypothetical construct that refers to the positivity of an individual's cognitions about, feelings precipitated by, and overt behaviors directed toward another person. An individual's attraction to another is thus regarded as a relatively stable internal state of the individual that is inferred on the basis of a variety of evidence. Within social psychology, the discipline that examines the ways in which an individual's behavior both influences and is influenced by the behavior of other people and within which most theory and research on interpersonal attraction has been conducted, attraction traditionally has been closely identified with the construct of attitude, or the disposition to respond in a favorable or unfavorable manner toward another person or object. Research on interpersonal attraction has not been closely identified with any one theory or with a single means of measurement.

I. The Biological Roots of Interpersonal Attraction

Positive and negative sentiments toward others have their source in the human's most basic and fundamental biological needs because the satisfaction of these needs from birth to death critically depends on the actions of other people. The importance of others for the individual's survival and well-being is reflected in the facts that humans are among the most social creatures in the animal kingdom and that they possess a finely honed propensity for making discriminative evaluative (good or bad) judgments of other people. These evaluative judgments have their source in the individual's belief that the other will act to further the individual's well-being or, conversely, might be a source of harm. Thus, matters of interpersonal attraction are of the highest importance to the welfare and survival of the species. For a species to survive, its members need to find food, avoid injury, reproduce, and rear the young. All of these adaptive behaviors engage issues of interpersonal attraction for the individual and present concerns for human society.

Most human behavior, then, takes place in a social context permeated with the causes and consequences of interpersonal attraction. Although attempts to identify the laws governing attraction began at least with Aristotle and continued through the early philosopher psychologists, the use of systematic observation to uncover these laws is relatively recent. J. L. Moreno's development of sociometry, a self-report questionnaire technique designed to assess an individual's preferences for interacting with others within a specific group, was one of the first empirical investigations of attraction. With the publication of Moreno's book, "Who Shall Survive?," in 1934, sociometric measures became an integral part of the emerging discipline of social psychology. Interest in the antecedents, consequences, and correlates of interpersonal attraction surged in the 1960s, partially because social psychologists recognized that attraction is central to virtually all social psychological phenomena, including group dynamics, social perception, and socialization processes.

Until recently, most work on attraction was (1) theoretically derived from general theories of social behavior, rather than specific theories of attraction; (2) directed toward understanding the antecedents and consequences of attraction toward strangers in closed-field settings where the individual has little alternative but to interact with another, as opposed to open-field settings where the exercise of choice is possible; and (3) thus conducted outside the context of ongoing relationships in naturalistic settings, usually within laboratory settings with college students. These efforts resulted in a robust literature of theory and investigation, which is described in many reviews and usually occupies a standard chapter in contemporary social psychology texts. In the late 1970s, however, attraction theorists and researchers shifted their attention to an examination of the role attraction plays in the formation, maintenance, and dissolution of ongoing relationships. This shift came about partially because it was the natural next step to take from examining initial encounters between people but also because of public and governmental demand for information about ongoing relationships, especially close relationships.

II. The Conceptualization and Measurement of Attraction

A. Attraction as an Attitude

Attitude has been the most central construct within social psychology, and so it was perhaps inevitable that attraction initially would be defined as attitude. The widespread conceptualization of attraction as an attitude toward another resulted in attraction often being measured just as attitudes are typically measured, through verbal self-report. Responses to such questions as "How much do you like X?" are usually facilitated by the presentation of simple bipolar scales, whose anchors may range from "like very much" to "dislike very much." Another common means of measuring attraction asks the respondent to evaluate the person's properties (e.g., kind, honest, warm), and then the known positivity of the affective loadings of each property ascribed to the other are summed or combined in some way to arrive at an attraction assessment.

The conceptualization and measurement of attraction as an attitude carried with it a number of assumptions that retarded the investigation of several important attraction issues. First, the attitude construct carries the assumption of stability; thus, the question of how attraction toward another waxes or wanes over substantial time frames is only now receiving attention. Second, the assumption has been that the constellation of cognitions and behaviors that comprise an attitude toward another are affectively homogeneous in tone, or that the properties the other is believed to possess, the emotions and feelings precipitated by the other, and the favorability of the actions taken toward the other are all of the same level of positivity. In fact, of course, ambivalence toward others is common. Ambivalence manifests itself in affectively mixed views of another's qualities, in experiencing a range of both positive and negative feelings in association with him or her, and in behaving in both favorable and unfavorable ways toward the other. The necessity to take a more complex view of the affect generated by another has become clear as researchers have turned to questions of attraction in long-term relationships where ambivalence is frequently apparent.

Another consequence of the identification of attraction with the attitude construct, along with the investigation of attraction in brief encounters between strangers, has been a focus on the mild forms of attraction, principally liking or disliking, and a neglect of such strong forms as love or hate. The recent tendency to investigate attraction as it occurs within the context of different types of relationships has also fueled interest in the many forms attraction may take, and particularly in the strong forms. [See ATTITUDE AND ATTITUDE CHANGE.]

B. Attraction as Emotion and Feeling

The strong forms of attraction are often associated with the experience of such strong emotions as passionate love, hatred, jealousy, contempt, and joy. The renaissance of theory and research in emotion within psychology in the 1970s, along with an interest in attraction as it is manifested in close relationships, has led many attraction theorists and researchers to focus on the antecedents and consequences of the emotions as they are experienced in an ongoing relationship with another. Although emotion is also a hypothetical construct that refers to an internal state inferred on the basis of various kinds of observable evidence, the construct of emotion, unlike the construct of attitude, does not carry the assumption of stability, and, in fact, the experience of an emotion is relatively short-lived. Moreover, emotion is not regarded as a state always reportable by the individual, and thus self-reports of emotion are not taken at face value. As a consequence, other kinds of evidence, including discharge of the sympathetic portion of the autonomic nervous system, changes in facial musculature and temperature, and other nonvoluntary and nonverbal behaviors, as well as more easily observable and volitional actions, are often considered necessary to the measurement of emotion. In addition, because virtually all contemporary theories of emotion emphasize the adaptive function of emotion in fulfilling the individual's biological needs, special attention is given to the individual's motivations (e.g., his or her needs, plans, and goals and their satisfaction or frustration by others) as they underlie affective phenomena, as well as to the social context in which the emotion is experienced. Furthermore, emotion researchers have documented that affectional space is better defined by two relatively independent dimensions, one positive and one negative, rather than by one bipolar dimension ranging from positive to negative. Thus, ambivalence, or the simultaneous experience of both positive and negative feelings toward another, is easily recognized. Finally, emotion theorists preserve distinction among the different varieties of emotion, while recognizing that all need to be explained under one theory; in searching for the common denominators of the different emotions, however, emotion researchers do not gloss over their differences.

Some attraction theorists recently have attempted to extend emotion theory and research to account for the experience of positive and negative emotions in close relationships. Others have sought to determine how such long-lasting feeling states as moods, precipitated by events extraneous to the relationship with another, influence attraction for that other. Still others are attempting to outline the patterning of emotion in close relationships.

C. Other Views of Attraction: Affiliation

Attraction has been assessed in a number of other ways over the years. Galton, for example, thought attraction toward one's dinner partner could be measured by the lean toward the partner, or the degree of deviation from a 90° plumb line. Pupil dilation, duration of eye gaze, and relaxation of skeletal musculature are among the many means by which investigators have hoped to measure attraction. Each is influenced by factors other than attraction, however, thus proving to be an unreliable indicator when taken by itself.

Attraction was initially assumed to be both necessary and sufficient for affiliation, or for an individual to attempt to interact with another. Thus, Moreno's sociometric technique, where each member of a group indicates with which other members he or she would like to engage in a specific activity and with which other members he or she would not like to engage in that activity, was originally used as a measure of attraction. For many years, in fact, the sociometric literature and the interpersonal attraction literature were simply two different labels for the same body of research; however, researchers gradually distinguished two separate issues: (1) the nature of the role attraction plays in attempts to interact with another and (2) the identification of determinants of affiliation other than attraction.

Attraction is not a necessary condition for affiliation; many factors other than liking another prompt affiliation attempts and actual interaction. Environmental conditions, for example, often facilitate and sometimes force interaction with others. In closed-field settings, interaction with specific others is virtually mandated; work settings and classrooms are examples. In an open-field setting, such as a cocktail party, people have more choice with whom they affiliate and, thus, there is likely to be closer correspondence between attraction and affiliation.

The development of theory and research on ingratiation, or attempts to interact with another to augment or maintain power in a relationship by inducing the other to like oneself, describes another set of conditions under which affiliation may occur without attraction. A wide variety of other social motivations may lead to affiliation without attraction, and investigators have explored some of these. One important factor has been identified as the need for social comparison, or the desire to evaluate the correctness of one's opinions and beliefs through comparison with those held by others. It has been demonstrated, for example, that when an individual is uncertain about the validity of his or her opinions and there is no nonsocial means of evaluation, others will be sought out for social validation. With whom one will attempt to affiliate under these circumstances encompasses a very large body of theory and research. Emotional arousal, particularly anxiety, also has been shown to lead to affiliation under some circumstances, namely when interaction under these conditions is not likely to lead to embarrassment and when the other is in a similar situation and state to oneself.

Much current theory and research on affiliation focuses on the effects of loneliness, a state that is associated with poor physical and psychological health. The development of scales to measure loneliness and of therapies designed to ameliorate it are relatively recent by-products of loneliness research. An interest in social networks, or the number and nature of persons with whom an individual actually interacts and the frequency of those interactions, has been spurred by research demonstrating the efficacy of social support in reducing the harmful effects of physical and psychological stress. Indeed, current interest in the dynamics of close relationships has been prompted in part by evidence indicating that the existence of such relationships is associated with physical and mental health. The mechanisms by which they do so (e.g., the role a close relationship plays in the functioning of the immunological system) are currently under examination.

Attraction is also not a sufficient condition for affiliation. Even in open-field settings, people do not always attempt to interact with those to whom they are attracted. One of the most important factors governing the attraction–affiliation relationships is the anticipation of acceptance or rejection by the other should an interaction attempt be made. For example, anticipation of social rejection by persons possessing higher social desirability than one's own often leads to interaction with persons whose social desirabilities are approximately equal. This "matching hypothesis of social choice" has been confirmed in a number of experiments, many of them conducted in opposite-sex dating situations.

III. Theoretical Approaches to Understanding Attraction

A. Cognitive Consistency Theories

The cognitive consistency theories (general social psychological theories that dominated research in social psychology through the 1960s and early 1970s when experimental research on interpersonal attraction burgeoned) assume that people try to keep their thoughts about themselves and other people and objects in a psychologically consistent relationship and, thus, will strive to obtain consistency and to reduce inconsistency when it occurs. Heider's balance theory, for example, proposed that sentiment (e.g., liking or disliking) toward another and feelings of belongingness with that other tend toward a harmonious, balanced state such that people tend to like those with whom they perceive they are joined in some way and, if they do not, liking will either be induced or they will attempt to break the associative bond. A number of studies have confirmed this hypothesis. It has been demonstrated, for example, that the mere prospect of spending substantial time interacting with another induces liking for that other prior to the interaction and prior to the receipt of specific information about the attributes of the other.

Heider's theory was subsequently modified by others on the basis of their own experimental findings. It was demonstrated, for example, that certain situations that are pleasant and balanced in Heider's terms are regarded by most people as unpleasant. For example, people prefer positive sentiment relationships with others, even if they are not in an associative bond with them, suggesting that people prefer to like others, rather than to dislike them, even when disliking them may present a more cognitively balanced situation.

Newcomb's strain-toward-symmetry theory adopted Heider's general position but focused on attraction as it occurred within a group. His classic longitudinal study of patterns of attraction in a group of male college students residing in a dormitory examined the hypothesis that cognitive balance in a group will increase with members' length of acquaintance. As reported in *The Acquaintance Process* (1961), Newcomb initially found little relationship between the men's actual attitude similarity and their liking for one another; however, over time, the pattern of attraction among the men became cognitively balanced in that they tended to like those who shared their own attitudes and beliefs and to dislike or become indifferent toward those men who did not. This was also the first study to demonstrate the importance of attitude similarity in generating attraction between two people, now one of the most corroborated findings in the attraction area.

Cognitive dissonance theory also generated some attraction research. Perhaps the most important finding was that although an individual's liking for another is popularly thought to be caused by the other person's objectively positive characteristics, it is also significantly influenced by the individual's behavior toward that other, behavior which may be under the control of forces extraneous to the individual's relationship with the other. Thus, it has been demonstrated, for example, that if an individual is led to harm another through external forces, the dissonance generated by cognitions such as "I am a kind person" and "I have hurt another" will often be reduced by coming to believe that the other deserved to be harmed and by derogating the victim; the insult follows, not precedes, the injury as a consequence of dissonance reduction processes.

B. Social Exchange Theories

Reinforcement theories, which have dominated all of psychology, have been most influential in attraction research, the general hypothesis being that people like those who reward them and that the more an individual is rewarded by another, the greater the attraction generated. The reinforcement theories were developed largely with infrahuman animals. When they were translated for application to human social behavior, they became known as social exchange theories because they assume that to understand social phenomena in general and attraction in particular, one must understand the principles that underlie the rewards and punishments people exchange when they interact.

Adapting B. F. Skinner's theory of operant behavior and incorporating some concepts from elementary economics, George Homans developed one of the first social exchange theories. In this theory, people are viewed as reward-seeking and punishment-avoiding creatures who try to maximize their rewards and minimize their costs in social interaction to obtain the most "profit" they can. Expression of esteem for another is regarded as a "generalized reinforcer," one that an individual gives to another who rewards him or her so that the other will continue to give rewards.

Of particular interest to researchers was the hypothesis of "distributive justice," which posited that people expect that the rewards received by each person in a social interaction should be proportional to his or her costs, and that each person's profits should be proportional to his or her investments (e.g., time, money, and other personal resources, including social status and talent) such that the greater the investments, the greater the profit. When the rule is violated, anger and a variety of attempts to obtain social justice are predicted.

The equity theories further elaborated when individuals would regard their exchanges with another as inequitable and detailed how attempts to achieve equity would be made. Equity theories propose that inequity exists for an individual when his or her ratio of outcomes to inputs in a relationship with another and the other's ratio of outcomes to inputs are unequal. Such situations often lead an individual to increase or decrease inputs, depending on whether the ratio is personally advantageous or disadvantageous. Research generated by the equity theories has demonstrated that while people are motivated to obtain high profits in their social interactions, they are also sensitive to social justice, often modifying their own behavior when injustice exists.

Early equity research largely focused on social exchange in work settings or other situations in which money was the principal reward exchanged.

Recent work has extended equity predictions to a wide array of settings including close relationships, where the rewards may be diverse. Many of these studies pit equity against the sheer magnitude of rewards received from another to predict attraction and relationship growth. Although the matter is not settled, and although it is apparent that equity is a potent consideration, the amount of reward received from another often predicts relationship stability and satisfaction better than equity. Moreover, with the new focus on ongoing relationships, two other social justice norms are receiving attention: the distribution of rewards according to need, not desert (a "communal" exchange norm), and equal distribution of rewards (the "equality" exchange norm). These norms appear to be adhered to in different types of relationships and sometimes at different stages of the same relationship.

Perhaps the most influential of the exchange theories has been Thibaut and Kelley's theory of social interdependence, which focuses on the profitability or "outcomes" of each behavior an individual may perform in a social relationship and predicts that the behavior an individual actually performs is a function of the configuration of *both* persons' outcome matrices rather than a function of the individual's outcome matrix alone. This theory formally separated the issue of attraction to another from the issue of the stability of the relationship, hypothesizing that attraction is a function of the degree to which the individual's profits from the relationship are above his or her comparison level, or the level of profits the individual believes he or she deserves, while the stability of a relationship is a function of the comparison level for alternatives, or the level of profits available in the individual's best alternative relationship. Thus, this theory predicts that people may dissolve a satisfying relationship with a person they like to enter an even more profitable relationship, and they may also stay in relationships with people they don't like and in which they receive poor outcomes simply because they have no better alternative. This theory has recently been extended and elaborated, in part to account for the fact that people will often "transform" the raw outcome matrix of rewards and costs by the adoption of "exchange rules," such as a communal rule or an equality rule, which benefit the less powerful person in the relationship.

In experimentation associated with the exchange theories, a good deal has been learned about the kinds of exchanges that are likely to generate attraction in specific circumstances; however, there has been dissatisfaction with all of the exchange theories because, although useful in understanding attraction and social exchange phenomena in a broad sense, their theoretical precision is illusory in practice. The problem lies in determining what is a reward for whom under what circumstances and then in quantifying the degree of reward represented in each exchange. People exchange many different kinds of rewards and punishments in social interaction (e.g., love, information, status, services) and reducing these to a common standard for quantification and predictive purposes is difficult. In response to dissatisfaction with the exchange theories of attraction, numerous other theoretical approaches have been developed. These are typically addressed to a particular kind of attraction (e.g., love), to a particular phenomenon associated with attraction (e.g., relationship stability; the development of closeness in a relationship), or to attraction as it occurs in a particular kind of relationship (e.g., courtship; marital).

IV. Varieties of Attraction

At the present time, there is no generally accepted taxonomy of the specific forms attraction to another may take; however, theory and research within social psychology have focused largely on four types of attraction. Although conceptually separable, and apparently differing somewhat in their causal antecedents and consequences, these types are no doubt often blended in most attraction experiences in naturalistic situations.

A. Liking

Despite the fact that the general theories guiding most attraction research were assumed to be applicable to all varieties of attraction across all social relationships and contexts, they have most often been used to predict a mild form of attraction—liking—and, until recently, usually liking between two strangers in their initial encounter with one another. Within early attraction theory and research, liking was almost synonymous with the word attraction, especially given the definition of attraction as a positive attitude toward another. Several factors associated with liking have been extensively investigated.

1. Physical Proximity

Many studies have found an association between the sheer physical distance between one person and another and the extent to which they find each other attractive. In classrooms and apartment houses, for example, physical proximity has been found to be predictive of friendship development. Distance is not only a causal factor in attraction but it is also, of course, a consequence, as studies of affiliation have documented. The physical distance between two people usually approximates the degree to which they are accessible to each other for interaction, and it is this accessibility, which is usually but not always associated with physical distance, that is facilitative of attraction, primarily because it facilitates the exchange of rewards. Moreover, physical proximity lowers the costs of the exchange; physical distance in a relationship usually adds costs, thereby reducing the profitability of the relationship. Moreover, and as previously discussed, cognitive consistency processes likely will produce attraction for those in close physical proximity, especially if they are compelled by extraneous forces to interact. And, finally, evidence indicates that familiarity itself, which increases with proximity, leads to attraction, as will be discussed later.

Although accessibility for interaction may be a necessary condition for the exchange of rewards, the interaction also may lead to the exchange of punishments. Thus, the contact hypothesis (i.e., interaction between hostile groups may ameliorate their bad feelings for one another) has not always been confirmed, and greater attention is now paid to the content of the interaction and to providing a context that is conducive to the exchange of rewards rather than punishments.

2. Similarity

Few variables have been so thoroughly investigated in the attraction literature as similarity, especially attitude similarity. Correlational studies of friends and spouses have shown that they tend to be more similar to each other than chance would dictate on virtually every dimension, including background, religion, education, height, political affiliation, and so forth. At least part of this association can be accounted for by the fact that people are more frequently thrown together in time and space with those similar to themselves and that similar people are thus more accessible for interaction than dissimilar persons. Similarity also has been shown, however, to be a causal antecedent of attraction.

For example, a large body of experimental work has demonstrated that attraction, at least to a stranger, is associated with the proportion (rather than the number) of similar, as opposed to dissimilar, attitudes shared with another.

A number of factors help account for the similarity—attraction relationship: (1) similarity often signals that rewards will be forthcoming in the relationship, particularly similarities that signal similar preferences for activities and people; (2) similarity of attitude provides social validation for one's views, which itself has been shown to be rewarding; and (3) people assume that similar people will like them (and that dissimilar people will not) and it is known that the anticipation of being liked by another generates attraction in return, following the "reciprocity-of-liking" principle where attraction breeds attraction.

3. Physical Attractiveness

Another factor whose role in generating attraction has been extensively examined is the physical attractiveness of the other. There appears to be a substantial amount of agreement in our culture about who is attractive and who is not, agreement that ranges across age, sex, educational, and social groups. A very large body of experimental work has shown that the physically attractive, from infancy to late adulthood, receive preferential treatment from others and are, other things being equal, more likely to be liked. One important mediating factor of this preferential treatment is a physical attractiveness stereotype such that people infer a number of desirable, but less visible, qualities from a physically attractive appearance. A recent meta-analysis of a large number of these studies indicates that inferences from physical attractiveness to social competence variables are most likely but almost nil to variables associated with integrity and concern for others. Evidence indicates that physically attractive people are indeed more competent in social interaction, perhaps as a result of preferential treatment. Finally, association with physically attractive others has been shown to increase an individual's status in the eyes of others under some circumstances.

Physical attractiveness has been shown to be an especially potent factor in date and mate selection, again with the physically attractive preferred but with assortative mating taking place along this dimension (e.g., significant positive correlations between the physical attractiveness levels of spouses). The impact of physical attractiveness beyond the early stages of a relationship has not been investigated as much as warranted by the mounting evidence that indicates that the receipt of individuating information about another attenuates the influence of social stereotypes, nor has the influence of other morphological factors been extensively investigated, the exceptions being male height and obesity, with young children sensitive to the latter and imputing a negative value to it.

4. Familiarity

Two different lines of investigation suggest that, other things being equal, familiarity leads to attraction. Much research has supported the hypothesis that simple repeated exposure to a person (or other stimulus, such as a work of art or a piece of music, for example) is a sufficient condition for increased attraction. The notion underlying this work is that familiar people are "safe" people, unlikely to be sources of harm. Although exposure to a stimulus does lead to a long-term increase in attraction under a wide range of conditions, there may be short-term decline immediately following many sustained exposures. Moreover, although repeated exposure may be associated with increases in liking, it may be negatively associated with another form of attraction in humans, sexual attraction, just as familiarity has been shown to be negatively associated with sexual attraction in infrahuman animals (i.e., the Coolidge effect).

The other line of research that suggests that familiarity may lead to attraction stems from Bowlby's attachment theory. Attachment, in fact, is now viewed by some theorists as a type of attraction, different from liking in a number of respects, including the fact that one may be "attached" to punishing persons one does not especially like.

B. Attachment

Attachment theory maintains that the primary purpose of affectional bonds is to promote the individual's physical proximity to other members of the species who can and will act to protect the individual from survival threats. That most animals are born with an instinctive proximity-promoting mechanism that leads them to stay physically close to those who can provide food, shelter, and protection was first demonstrated by Lorenz in his classic imprinting experiments. Later, the importance of this mechanism was demonstrated with monkeys,

showing that failure of the affectional system to develop normally in the infant–mother relationship impaired later socioemotional adjustment.

Human infants also quickly develop a strong preference for a person who meets certain minimum visual, tactile, and auditory requirements and who is also the first person they have become familiar with, and they exhibit attachment behavior toward that person. Attachment behavior has two distinctive features: (1) it results in the restoration or maintenance of proximity to a specific other, and (2) it is usually exhibited toward an older, stronger, or more dominant individual by a younger or weaker one. Although attachment behavior is especially evident between a child's first and third years, it may characterize people of all ages and is likely to be exhibited when the individual is threatened, ill, or otherwise distressed. Only recently, however, have investigators begun to examine this hypothesis as well as to explore how the patterns of childhood attachment may influence adult relationships, especially the marital relationship.

Attachment, in any case, is conceptually distinct from liking, at least when liking is defined as a favorable attitude toward another. It is also believed to be somewhat different from liking in its causal antecedents, with familiarity playing a stronger role in attachment and the actual receipt of rewards in interaction playing a stronger role in generating liking.

C. Altruistic Love

Altruistic love focuses on promoting the welfare of another in contrast to satisfying one's own needs and receiving rewards from another. The circumstances under which people will act to facilitate another's welfare have been extensively investigated and encompass many more determinants than attraction to another (with social norms and sanctions being, for example, important determinants). Thus, attraction is not a necessary condition for a person to exhibit altruistic love toward another. It also is not a sufficient condition; although the conferring of benefits upon people one likes and finds attractive has been found to be one dimension differentiating acquaintances from loved persons, persons who are strongly attracted to one another, as evidenced by other behaviors, do not always promote each other's welfare and sometimes are destructive of it. Altruistic love, then, is conceptually separable from other forms of attraction although often associated with them.

D. Romantic Love

As previously noted, theorists and investigators initially assumed that the causal antecedents of the strong forms of attraction were the same in kind but differing in magnitude from the mild forms. Anecdotal observations, however, that in heterosexual dating relationships greater and greater liking usually leads only to a great deal of liking—not love, and especially not to romantic love—forced recognition that the antecedents of romantic love may be very different in kind. Rubin was the first to demonstrate that the two are at least measurably different in heterosexual relationships with his successful development and validation of Loving and Liking Scales. Interest in the romantic variety of attraction has increased dramatically in the past decade for a number of reasons, including evidence that romantic love remains a prerequisite for marriage in Western culture, and it actually has increased in importance over the past few decades. The view that romantic love is a necessary condition for marriage has also now extended to Japan, China, and other Eastern cultures.

Many theories of romantic love have been offered. They often differ from general theories of attraction in their special emphasis on the role that deprivation plays in making a social stimulus event a reward, as well as in their emphasis on the role of fantasy, idealization of the other, and anticipation of reward (as opposed to the actual receipt of rewards) in generating this variety of attraction. The element of sexual desire, while recognized, particularly in its association with physical attractiveness, has not been investigated as much as warranted. Unlike liking, romantic love is also associated with the experience of strong emotions, both positive and negative. Moreover, unlike liking, which tends to increase with familiarity, romantic love appears to decrease in intensity over time, with its symptoms virtually disappearing, according to one study that examined couples in Japan and the United States, after 6 yr of frequent interaction. The general theories of emotion are being used to understand emotion, including romantic love, as it occurs in relationships. Interest has been shown in assessing different romantic "love styles." Finally, several sociobiological theories of romantic love have recently been offered. The general thesis of these theories is that the behaviors generated by romantic love cause people to attract mates, reproduce, and invest in the survival of offspring; thus, reproductive success is believed to be associated with the experience of romantic love.

V. Attraction and Other Relationship Phenomena

Theory and research on interpersonal attraction shifted in the 1980s from a focus on identifying the causal determinants of attraction between people in their first encounter with each other to examining the role attraction plays in further relationship development and its association with a number of other relationship phenomena characteristic of ongoing relationships in naturalistic situations.

A. Theories of Adult Relationship Development

Social psychologists have become increasingly interested in devising conceptual frameworks within which the development of a relationship and the reasons for its growth, maintenance, or dissolution may be explored. Although people develop relationships with only a small portion of people to whom they feel initial attraction, attraction is almost universally regarded as an important condition facilitating the growth of a relationship, especially under open-field conditions where people have some choices about interacting with specific others. Moreover, the same conditions that promote an individual's attraction to another, particularly the individual's evaluations of the profits, or rewards relative to costs, currently received in a relationship, as well as forecasts of future profits, are the same conditions currently theorized to promote relationship growth and development. Thus, mutually rewarding relationships tend to further develop, whereas unrewarding relationships tend to be terminated. All theorists recognize, however, that attraction need not be characteristic of a developing relationship nor even a characteristic of an established relationship.

Most theorists of relationship development view relationships as proceeding from a superficial stage, such as that typical of the initial encounters, to, in some cases, a very close relationship. The specific stages through which a relationship passes, and the kinds of processes and events presumed to be critical to each stage, differ among theorists. Appearing in several theories, however, are (1) an explorative stage where people try to identify the rewards that may be available to them in the relationship, often through the process of self-disclosure; (2) a conflict, bargaining, and negotiation stage as they try to mesh their goals and objectives and set the terms of the relationship; and (3) a commitment stage where each expresses, directly or indirectly, an intention to continue the relationship and cease exploration of alternative relationships. Many of the processes believed to be associated with these stages have been the subject of a good deal of theory and research in themselves, with self-disclosure, conflict resolution, and commitment processes each representing robust literatures.

Self-disclosure (the revelation of information about oneself), for example, which was initially believed to be important to relationship growth and an important component of the first relationship development theory (Altman and Taylor's social penetration theory) has now been revealed to be important only in the very early period of a relationship, with some evidence suggesting that disclosure reaches asymptote as early as 6 wk of acquaintance. Evaluative self-disclosure (revealing personal feelings), as opposed to descriptive self-disclosure (revealing facts about oneself), has been shown to occur more often between spouses than between strangers and may be important in relationship maintenance. Attraction and self-disclosure may be mutually facilitative under certain conditions (depending on the content of the disclosure); for example, people tend to like others who reciprocate disclosures.

Intimacy has recently been differentiated from self-disclosure and conceived to be a process in which the individual expresses, either verbally or nonverbally (e.g., eye contact, touch, facial expression), important self-relevant feelings and information to another and, as a result of the other's response, comes to feel known, validated (in terms of world view and self-worth), and cared for. Intimacy, at least as defined in this way, is strongly associated with attraction to the other.

B. Other Recent Relationship Theory and Research

Effort has also been directed toward clarifying the ambiguous terminology that was long associated with relationship phenomenon. Kelley and his colleagues, for example, published *Close Relation-*

ships in 1983, which presented a conceptual and methodological framework intended to facilitate the study of close relationships and which explicated the relationship property of closeness. Close relationships are viewed as ones in which the participants are highly interdependent on each other, where that interdependence is revealed in their interaction sequences with each other by evidence that the causal impact of each person's activities on the other is frequent, is to diverse kinds of activities, and is strong and that this pattern has characterized the relationship for a relatively long period of time. Again, a close relationship need not be a happy or healthy relationship in which the principals like each other, experience positive emotions in association with each other, or behave in positive ways toward each other. Berscheid, however, has presented an auxiliary model based on this conception of closeness and on current emotion theory and research to make predictions about attraction and other emotional phenomena within close relationships; some confirmatory empirical evidence has now been obtained. Further research examining the processes and correlates of closeness will be facilitated by the recent development of an inventory that allows investigators to assess the degree of interdependence within a relationship.

Other work taking place within the relationship context will help illuminate attraction processes in ongoing relationships. For example, the examination of the emotional "physiological linkage" between distressed and nondistressed marital couples has reliably demonstrated that during discussions of problems, distressed couples are more linked emotionally, at least as evidenced by physiological indicants of emotion. Moreover, in unhappy marriages there is more reciprocity of negative affect than of positive affect, revealing that disaffection tends to become stable.

Some of the most valuable recent work is directed toward understanding emotional phenomena as they occur in relationships. Turning from regarding attraction as an undifferentiated "favorable" orientation toward another, then to examining different varieties of attraction, investigators are now more closely examining the specific emotional experiences that lead to attraction or that are associated with different types of attraction. Much of this work analyzes people's implicit understanding of and beliefs about the elicitation and expression of emotion in relationships.

Another line of work, initiated by Tesser's self-evaluation maintenance theory, is directly relevant to attraction in that it posits that persons in close relationships may be threatened by the partner's superior performance in a domain highly relevant to the individual's self-esteem, although positive emotions are experienced when close associates perform well on tasks irrelevant to self-esteem and identity. Some confirmatory evidence is available, suggesting that a division of talents and tasks in a close relationship, preventing unfavorable self-comparisons on dimensions important to self-identity and esteem, may be important in maintaining attraction to another.

Finally, a good deal of current work is directed toward childhood relationships. In fact, an improved understanding of the developmental significance of a child's social relationships, particularly of the importance of peer relationships, has been characterized as one of the most significant advances of psychology in the past two decades. Developmental psychologists are tracing the effects of a child's early social relationships on later social and emotional functioning. They have shown that the failure of a child to become securely attached (in Bowlby's sense) to its caretaker has devastating consequences for the child, including difficulty in later developing satisfying peer relationships. The effect of attachment patterns in these early relationships on adult relationships is now gathering much attention, and evidence suggests that the effects of early social relationships are long-lasting and difficult to change. Friendship development in children is also receiving special scrutiny because now a good deal of evidence indicates a positive correlation between social rejection in childhood and problems in later life. A formal integration of the adult and childhood literatures on attraction has not yet been accomplished but rapprochement, to the mutual benefit of each of these areas of inquiry, is beginning.

Bibliography

Berscheid, E. (1983). Emotion. *In* "Close Relationships" (H. H. Kelley, E. Berscheid, A. Christensen, J. H. Harvey, T. L. Huston, G. McClintock, A. Peplau, and D. Peterson, eds.), pp. 110–168. W. H. Freeman, New York.

Berscheid, E. (1985). Interpersonal attraction. *In* "Handbook of Social Psychology," 3rd ed. (G. Lindzey and E. Aronson, eds.), pp. 413–484. Random House, New York.

Clark, M. S., and Fiske, S. T. (eds.) (1982). "Affect and Cognition: The Seventeenth Annual Carnegie Symposium on Cognition." Erlbaum, Hillsdale, New Jersey.

Clark, M. S., and Reis, H. T. (1988). Interpersonal processes. *Annu. Rev. Psychol.* **39,** 609–672.

Gergen, K. J., Greenberg, M. S., and Willis, R. H. (eds.) (1980). "Social Exchange: Advances in Theory and Research." Plenum, New York.

Hartup, W. W. (1989). Social relationships and their developmental significance. *Am. Psychol.* **44(2),** 120–126.

Huston, T. L., and Levinger, G. (1978). Interpersonal attraction and relationships. *Annu. Rev. Psychol.* **29,** 115–156.

Kelley, H. H., Berscheid, E., Christensen, A., Harvey, J. H., Huston, T. L., Levinger, G., McClintock, E., Peplau, L. A., and Peterson, D. (1983). "Close Relationships." W. H. Freeman, New York.

Sternberg, R. J., and Barnes, M. L. (eds.) (1988). "The Anatomy of Love." Yale University Press, New Haven.

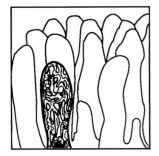

Intestinal Blood Flow Regulation

CHING-CHUNG CHOU, *Michigan State University*

Glossary

Arterioles Last small branches of the arterial system, which act as control valves of the circulation and control the volume of the blood flowing to the capillaries

Capillaries Function is to exchange fluid, oxygen, nutrients, etc., between blood and the interstitial spaces

Cardiac output Volume of blood (L/min) being pumped into the aorta by the heart

Ischemia Partial or complete deficiency of blood supply

Oxygen consumption Volume of oxygen consumed by the body or an organ

THE SMALL INTESTINE and colon receive approximately 10–15% and 3% of the cardiac output, respectively, and consume about equal shares of the total oxygen consumed by the body at rest. In contrast, the heart receives only 4.5% of the cardiac output, whereas it consumes 11% of the total body oxygen consumption. The relatively high blood-flow-to-oxygen-consumption ratio in the small in-

testine may reflect its primary function (i.e., absorption of nutrients and fluids). The small intestine absorbs about two times the plasma volume of fluid each day (i.e., 7–8 L). The microcirculation of the intestinal mucosa has several characteristic features that optimize the ability of its tissue to move fluid, electrolytes, and nutrients between the blood and epithelial cells. On the basis of weight, the intestinal blood flow in resting conditions ranges from 30 to 100 ml/min/100 g tissue weight. This resting flow, which is 10 times that of flow to the skeletal muscle, increases 30–130% during nutrient absorption. As in any organ, blood flow to the small intestine is determined by the pressure drop across the intestine and the resistance to blood flow offered by the intestinal vessels. Because arterial and venous pressures remain relatively constant under physiological conditions, blood flow is primarily determined by vascular resistance, which in turn is primarily determined by the diameter of arterioles. Thus, vasodilation increases blood flow (hyperemia), whereas vasoconstriction decreases blood flow. Vascular resistance is regulated by extrinsic (also called central or neurohumoral regulation of blood flow) and intrinsic (also called local regulation of blood flow) mechanisms. Resistance changes caused by extrinsic factors usually subserve homeostasis of the whole body (e.g., maintenance of mean arterial pressure). Resistance changes caused by local factors serve to maintain local homeostasis and functions (i.e., absorption, secretion, and motor activity of the small intestine). This chapter primarily describes the local factors unique to the regulation of blood flow in the small intestine. Most of the information is derived from animal studies because studies on human subjects are difficult and scarce. Owing to limitation in space, description in some sections cannot be detailed. The bibliography provides sources of additional information.

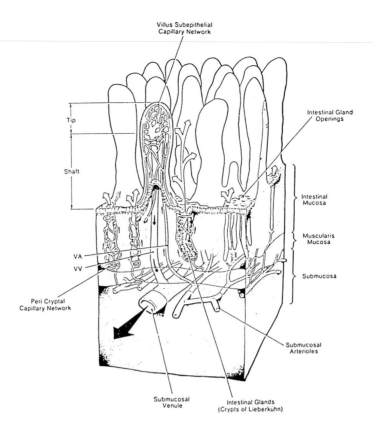

FIGURE 1 Model of mucosal microcirculatory patterns typical of human small intestine. VA, villus arteriole; VV, villus venule. *Solid arrows* indicate direction of blood flow, and *open arrows* indicate directions of intestinal secretion. Note the "fountain" arrangement of villus subepithelial capillary network. (From J. R. Casley-Smith, J. R. and Gannon, B. J. (1984). Intestinal microcirculation: Spatial organization and fine structure. *In* (A. P. Shepherd and D. N. Granger, eds. "Physiology of the Intestinal Circulation Raven" Raven Press, New York.)

I. Anatomy and Function of Intestinal Circulation and Microcirculation

The small and large intestines receive their blood supplies primarily from the superior and inferior mesenteric arteries. The blood perfused through the intestine is drained by the portal vein, which perfuses the microcirculation of the liver. A significant feature of the intestinal circulation is the anatomic presence of anastomotic connections at several levels of vessel branching (vascular arcades), from the main mesenteric arteries to the extensive submucosal vascular plexus (Fig. 1). This arrangement serves to protect intestinal tissues from ischemia in case of constriction or mechanical obstruction at any level of the arterial network.

As shown in Fig. 2, the small arteries at the mesenteric margin of the intestine penetrate the muscle layer of the intestine (1A), without branching in this layer, to join the submucosal arteriolar plexuses (submucosal arterioles, 2A–4A). The capillary network of the mucosal layer (villi and crypts) (Fig. 1) receives its blood supply from this plexus. The microvascular architecture of villi widely varies among different species. In human villi, a single eccentrically located arteriole (VA in Fig. 1 and 3A in

Fig. 2) passes to the villus tip, where it breaks up in a fountain-like pattern, from which subepithelial capillaries are formed and cascade down the exterior surface of the villus. The capillary blood drains into an eccentrically located venule (VV in Fig. 1 and CV in Fig. 2).

The capillaries form an extensive network just beneath the mucosal epithelium, in both the villus and crypt, where fluid is absorbed and secreted, respectively. The intestinal mucosa has a high capillary density and consequently a large capillary surface area. Furthermore, these capillaries are fenestrated (openings in the capillary endothelium), and these fenestrations usually face the basal aspect of the epithelial cells. All these characteristic features of the mucosal capillaries facilitate fluid transport between the epithelial cells and blood and thus pro-

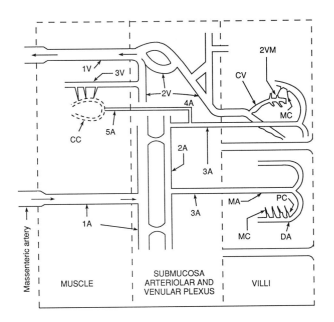

FIGURE 2 Schematic microcirculatory anatomy of the small intestine. 1A, 2A, 3A, 4A, 5A, the first, second, third, fourth, and fifth order arterioles; 1V–3V, the first to third order venules; MA, the main arteriole of the villus; DA, distributing arterioles; PC, precapillary sphincters; MC, mucosal capillaries; VM, mucosal venules; CV, collecting venules; CC, capillaries of the smooth muscle. (Vascular classification based on Gore, R. W., and Bohlen, H. G. (1977). *Am. J. Physiol.* **233**, H685–H693.)

ies (CC) run parallel to the surrounding smooth muscle fibers. The small veins (CV and 3V) draining the capillary networks of the mucosal and muscle layers empty into the submucosal venular plexus (2V), from which the blood is drained out of the gut via 1V.

Blood flows to different segments and different tissue layers of the small intestine vary. There is a gradient of blood flow from duodenum to ileum, with the proximal intestine (duodenum) having the highest blood flow on a tissue weight basis. Under resting conditions, about 75–80% of the blood perfusing the gut wall is distributed to the mucosa, 5% to the submucosa, and 20–25% to the muscle layer. The villi receive 60% (114 ml/min/100 g tissue weight) of the mucosal flow, with the crypts receiving the remainder. The relative distribution of blood flow to different sections and layers of the intestinal wall reflects the difference in metabolic activities of the cells in that tissue. When calculated as flow per unit weight, the flow to intestinal muscle is only $\frac{1}{2}$ to $\frac{1}{10}$ of mucusal flow, and oxygen consumption by the muscle is $\frac{1}{4}$ to $\frac{1}{3}$ of mucosal oxygen consumption. The dependence of blood flow on tissue functions and metabolism becomes more apparent when the activity of a given tissue is increased after a meal.

mote intestinal absorption and secretion. Under resting conditions, only 20–30% of the intestinal capillaries is perfused by the blood. The number of perfused capillaries is controlled by the microvascular element, microfilaments, or a single spiralling smooth muscle fiber (precapillary sphincter) encircling the origin of the villous capillary (PC in Fig. 2). Thus, whereas relaxation of the arterioles increases blood flow, relaxation of the precapillary sphincters increases the number of capillaries being perfused.

The fountain-like arrangement of the villus microcirculation provides anatomical basis for countercurrent exchange and multiplication in the villus. Thus, diffusible chemicals such as water can diffuse between parallel vessels in which blood is flowing in opposite directions (e.g., between arterioles and venules or capillaries) according to osmotic or concentration gradient. The countercurrent exchanger and multiplier create a hyperosomatic environment in the villous tip during active absorption of NaCl; the hypertonicity enhances absorption of water from the gut lumen. The muscle layer receives an independent blood supply that branches from the submucosal arteriolar plexus (4A); and the capillar-

II. Techniques and Importance of Tissue Blood Flow

Intestinal blood flow can be determined by measuring the arterial inflow of a main artery perfusing the intestine with an electromagnetic flow probe and flowmeter or by measuring venous outflow after cannulating a vein that drains a segment of intestine. These techniques measure blood flow to the entire intestine, and the values obtained are usually defined as total intestinal blood flow or total blood flow to the gut wall (in ml/min or ml/min/100 g tissue weight). The intestine is composed of three tissue layers (i.e., mucosa, submucosa, and muscularis layers), each having different functions. Measurement of total intestinal blood flow alone is inadequate when one attempts to determine the relations between blood flow and absorption or between blood flow and motor activites. One must measure blood flow to the mucosal or muscularis layer separately in the study. The microsphere technique is the one most commonly used to measure compartmental blood flow within the gut wall. In general, microspheres having a diameter of 15 μm and labeled with isotopes of various energy levels (e.g., [85]Sr, [141]Ce) are injected into the left ventricle. The basic assumptions are that the spheres behave like blood cells and are delivered from the heart to, and are trapped in, the mucosal and mucularis tissues in proportion to the blood flow to these tissues. Thus, one type of radioactively labeled sphere is injected under resting condition, and spheres labeled with another type of radioisotope are injected under an experimental condition. After the experiments, the intestine is removed and the gut wall separated into three layers. The radioactivity of both isotopes in these tissue samples is then determined, and from these data blood flows to these tissues under the two conditions are calculated. Other techniques available for determination of compartmental blood flow within the gut wall are microscopical technique, Laser-Doppler velocimetry, hydrogen clearance, and the clearance of various substances such as inert gas and lypophobic solutes.

III. Overview of Regulation of Intestinal Blood Flow

As in any organ of the body, intestinal blood flow is regulated by extrinsic and intrinsic factors. The extrinsic mechanisms are those originating outside of the intestine, including the autonomic nervous system and circulating chemicals (hormones). The intrinsic mechanisms are those originating in the intestine and can modify the extrinsic mechanisms. A good example is that physical exercise stimulates the sympathetic nervous system, thereby decreasing intestinal blood flow. However, exercise performed 60–90 minutes following a meal does not significantly influence the postprandial increase in intestinal blood flow. The intrinsic mechanisms maintaining the hyperemia are strong enough to oppose the sympathetic influence. In this section, only some of the unique characteristics of the extrinsic and intrinsic regulation will be described.

A. Extrinsic Mechanisms (Central or Neurohumoral Regulation)

1. Sympathetic Neural Regulation

The intestinal blood vessels are richly innervated by sympathetic fibers, stimulation of which produces pronounced constriction of the arterioles, precapillary sphincters, and veins. The decreased blood flow, however, is short-lived; within 2 to 4 min intestinal blood flow returns toward prestimulation levels despite continued sympathetic stimulation. This phenomenon is called "autoregulatory escape" and has been demonstrated in many species, including humans. The mechanisms involved in the escape are local intrinsic factors. In contrast to the response of the arterioles, the constriction of the precapillary sphincters and veins persists and does not exhibit this escape. Sixty to seventy-five percent of the circulating blood volume is localized in the veins, constriction of which (venoconstriction) displaces blood to the heart, thereby increasing the cardiac output.

The sympathetic vasoconstriction and venoconstriction in the splanchnic organs are important features in the maintenance of overall cardiovascular homeostasis. The splanchnic circulation (which includes blood flow to the entire gastrointestinal tract, pancreas, spleen, and liver) receives 25% of the cardiac output and contains 20–25% of the total blood volume. Under stress, both splanchnic blood flow and volume can be diverted to more vital areas (e.g., to the brain and heart during hemorrhagic shock and to the skeletal muscles during physical exercise). In humans, splanchnic blood flow reduces in proportion to the increase in total body oxygen consumption, heart rate, and plasma norepinephrine concentration during exercise, as a result of increases in sympathetic neural activity. Maximal splanchnic vasoconstriction may redistribute about 1,150 ml/min of blood flow to working skeletal muscle.

2. Parasympathetic Neural Regulation

The vagus nerves innervating the small intestine do not appear to carry any specific vasodilator fibers to the blood vessels, and stimulation of the vagus nerves below the level of the heart does not alter intestinal blood flow. The parasympathetic nervous system, however, plays a significant role in the regulation of gastrointestinal functions. The action on these functions can indirectly influence intestinal blood flow (see Section IV). For example, electrical stimulation of the vagal fibers innervating the colon does not elicit any blood flow change, apart from those caused by an increase in colonic motility.

3. Circulating Chemicals

The response of intestinal blood flow to circulating chemical is similar to that in other organs. Thus, vasoconstrictors such as norepinephrine, vasopressin, and angiotensin decrease intestinal blood flow, whereas vasodilators such as glucagon increase intestinal blood flow.

B. Intrinsic Mechanisms

The intestine can regulate its blood flow when extrinsic nervous and circulating humoral influences are eliminated. The local factors regulating the intestinal blood flow are as follows:

Tissue metabolism: Intestinal absorption, secretion, and motility are oxygen-consuming processes. According to the metabolic theory of blood flow regulation, an increase in the tissue oxygen demand will result in a decrease in tissue oxygen tension (Po_2) and an increase in the concentration of vasodilator metabolites in the interstitial fluid. This would lead to relaxation of the arterioles and precapillary sphincters, thereby increasing blood flow and the number of the capillaries being perfused. These two events tend to increase oxygen delivery to tissues and serve to maintain tissue Po_2 at a level that does not limit the rate of aerobic metabolism.

Autoregulation: Autoregulation is the capacity of circulation to maintain relatively constant blood flow in the face of widely varying arterial pressures, and it is present in the intestinal circulation.

Reactive hyperemia: When intestinal blood flow is arrested by arterial occlusion for a few seconds to few minutes, the flow on release exceeds the control flow for a short period before returning to preocclusion levels. This occurs after ischemia or strong tonic contraction of the intestinal muscle.

Autoregulatory escape: This has been described.

Venous-arteriolar response: An increase in intestinal venous pressure produces vasoconstriction in the arterioles and precapillary sphincters, thereby decreasing blood flow and capillary density. This phenomenon is best explained by the myogenic theory, which hypothesizes that the vascular smooth muscles respond to an increase or decrease in vascular transmural pressure with contraction or relaxation, respectively. This intrinsic mechanism may be important in the intestine, because the intestine is coupled in series with the liver. An increase in hepatic vascular resistance will cause increases in portal (intestinal) venous pressure. The venous-arteriolar response tends to maintain mean capillary pressure and thus keep the transcapillary exchange of fluid relatively constant.

Local nerves and chemicals: These will be discussed in Section V, D.

IV. Relations Between Intestinal Functions and Blood Flow

A. Absorption

1. Influence of Blood Flow on Absorption

Absorption of highly diffusible (and lipid-soluble) substances (e.g., tritiated water) is significantly dependent on blood flow whereas the absorption of less diffusible substances (e.g., ethanol) is only partially dependent on blood flow. An increase in blood flow does not increase the absorption of water-soluble and actively absorbed nutrients such as glucose. A decrease in blood flow also does not reduce the absorption unless the flow is reduced by more than 50% of resting values, when both the absorption rate of glucose and intestinal oxygen consumption decrease in a linear fashion as blood flow is reduced. This indicates that availability of oxygen is a limiting factor in glucose absorption during low blood flows. The absorption of amino acids and electrolytes also decreases under low blood flow conditions and after total ischemia of 10-min to 1-hour duration.

2. Influence of Absorption on Intestinal Hemodynamics

During nutrient absorption, the following phenomena occur in the intestinal mucosa. The arterioles and precapillary sphincters relax, thereby increasing blood flow and the number of capillaries perfusing the tissues. Entrance of the absorbed fluid into the interstitial tissue increases the tissue fluid volume, thereby increasing tissue hydrostatic pressure and decreasing tissue oncotic pressure. All these capillary and interstitial forces favor movement of fluid from the interstitial tissue to the capillary, thus promoting fluid absorption into the blood circulation. In addition, lymphatic flow is increased as a result of an increase in tissue fluid pressure.

B. Secretion

Active intestinal secretion, as induced by cholera, is accompanied by an increase in intestinal blood flow; the vasodilation occurs exclusively in the mucosa, suggesting again the close relationship between tissue function and blood flow. Secretion of fluid from the interstitial tissue to the gut lumen decreases interstitial fluid volume. As a result, tissue hydrostatic pressure decreases and oncotic pressure increases. Intestinal secretion is accompanied by increases in blood flow and capillary density. All these changes tend to increase capillary filtration of fluid into the interstitial tissue. As a result of decrease in tissue pressure, lymph flow almost ceases, and all the capillary filtrate is then available for secretion into the gut lumen.

C. Motility

1. Influence of Motility on Blood Flow

Muscle contractions can affect blood flow by at least two opposing mechanisms. An increase in contractions increases intestinal luminal pressure and exerts extravascular compression on the blood vessels encased within the gut wall, thereby interfering mechanically with its blood flow. An increase in contractions also increases metabolic requirements of the contracting muscle, which in turn leads to an increase in blood flow to the muscle.

a. Rhythmic Contractions Rhythmic segmental contractions are the most common type of intestinal motor activity after a meal. These contractions may increase, decrease, or have no effect on mean blood flow (ml/min), depending on the pattern, strength, frequency, and duration of the contractions. Mild contractions generally do not alter mean blood flow, whereas high-amplitude contractions may either increase or decrease mean blood flow, depending on the balance of the two opposing mechanisms described above. A decrease in mean blood flow is usually accompanied by an increase in gut wall tension, as indicated by an increase in basal lumen pressure. Mild distension of the gut lumen by intestinal contents often produces rhythmic contractions and increases local blood flow. The increased blood flow is confined only to the muscle layer of the gut wall and is most likely due to an increase in metabolic requirements of the contracting muscle. Although rhythmic contractions produce diverse effects on mean blood flow, they usually produce cyclic changes in instantaneous blood flow that parallel the rhythmicity of the contractions. During contraction, arterial inflow decreases while venous outflow increases; the reverse occurs during relaxation. The contraction squeezes blood out of the gut wall, thereby increasing venous outflow, while preventing arterial blood from flowing into the gut wall.

b. Tonic Contractions Tonic contractions is not a common type of intestinal motor activity under physiological conditions, but when it occurs, the tonic concentrations always decrease mean blood flow as a result of their compressing effect on the blood vessels within the gut wall. The decrease in the total gut blood flow is primarily due to a decrease in mucosal flow; flow to the muscle layer is relatively unchanged. After tonic contraction, the local flow markedly increases above precontraction levels. This post–tonic hyperemia appears to represent a reactive hyperemia comparable to that seen in skeletal muscle after tetanic contraction.

2. Influence of Blood Flow on Motor Function

Increasing intestinal blood flow does not alter intestinal motility, but severe ischemia can alter electrical and contractile activities of the intestine. The effects depend on the duration and severity of the ischemia. Ischemia produces an immediate transient increase in contractile and electrical activities of intestinal muscles lasting for few minutes. This increase in activity appears to be due to stimulation of local intestinal nerves. After this transient increase, contractile activity disappears for as long as the insult lasts. Recovery of motility to its normal state after re-establishment of normal circulation depends on the duration of ischemia. If circulation is restored within 1 to 3 hours, the electrical and spontaneous motor activities return to normal within 5 min. Four hours is the critical duration of ischemia for irreversible depression of the contractile and electrical activities.

V. Postprandial Regulation of Intestinal Blood Flow

A. Overall Cardiovascular Responses to Meals

It has long been thought that the drowsiness one feels after a meal may be due to a redistribution of blood away from the brain to the intestine. In 1786, Heberden described for the first time the symptom of angina pectoris as follows: "Those who are afflicted with it are seized, while they are walking, and more particularly when they walk soon after eating, with a painful and most disagreeable sensation in the breast." Thus, the cardiovascular response to eating and digestion is a topic of interest not only to the general population but also to clinicians. The cardiovascular response to meals varies with species and the type of ingested food and can be divided into two phases anticipation/ingestion and digestion/absorption phases.

During anticipation and ingestion of a meal, the activity of the entire cardiovascular system changes in response to a generalized activation of the sympathetic nervous system. Thus, cardiac output, heart rate, arterial blood pressure, and blood flow to the heart increase, whereas flow to the kidneys decreases. Intestinal blood flow either decreases or does not change. These changes, however, subside 30 min after a meal, with the exception of blood flows to the digestive organs, limbs, and kidneys. Brain blood flow does not change during the digestion/absorption phase, and therefore, drowsiness after a meal is not due to a decrease in brain blood flow.

Blood flow to the stomach increases on entrance of food into the stomach; the hyperemia, however, lasts only 5 minutes. The transient gastric hyperemia is not due to gastric distension but appears to result from stimulation of local mucosal nerves by the nutrients. As the gastric hyperemia subsides, blood flow to the superior mesenteric artery starts to increase, reaching its maximum 30–90 min after feeding and lasting for several hours. The increase in the superior mesenteric arterial flow is primarily due to increases in intestinal and pancreatic blood flows. This postprandial intestinal hyperemia during digestion/absorption phase has been observed in humans and many animal species. Because cardiac output is postprandially unchanged, the intestinal hyperemia is compensated by a decrease in blood flows to the skin and skeletal muscle under resting conditions. However, in ambulatory conditions the increased intestinal flow should come from an increased cardiac output. Kidney blood flow increases only after ingestion of a protein-rich meal, which may last for several hours.

B. Localization of Intestinal Hyperemia

In dogs, the hyperemia occurs only in the intestinal segment exposed to nutrients. Blood flows to the duodenum and jejunum increase 30 min after a meal; flow to the terminal ileum does not increase until 45–90 min after a meal; and colon blood flow does not change during these periods. Furthermore, the hyperemia during nutrient absorption is confined to the mucosal layer; flow to the muscularis layer is unchanged unless there are significant changes in local motor activities (see Section IV,C). In cats and rats, however, introduction of nutrients to the duodenum may elicit diffuse intestinal hyperemia involving segments of the intestine that are not exposed to nutrients. Furthermore, the hyperemia may also involve both the mucosal and muscularis layers.

C. Luminal Stimuli Responsible for Intestinal Hyperemia

Intestinal hyperemia after a meal is produced by the presence of food that has been predigested with pancreatic enzymes; undigested food does not produce hyperemia. Of the breakdown products of carbohydrates, proteins, and fats, oleic acid (a major long-chain fatty acid in milk) is the strongest vasodilator, whereas amino acids (except alanine and phenylalanine) and caproic acid (a short-chain fatty acid) do not increase intestinal blood flow. Glucose is a weak vasodilator. A protein-rich meal, however, produces intestinal hyperemia greater than, or similar to, that produced by a carbohydrate-rich meal. It is most likely that the postprandial intestinal hyperemia results from a synergistic effect of all three dietary components. Bile not only plays an important role in digestion and absorption of fats, it also significantly enhances the vasodilation produced by glucose and renders oleic acid, caproic acid, and amino acids vasoactive. Although bile itself in the jejunal lumen does not alter jejunal blood flow, bile in the ileal lumen produces local hyperemia, suggesting that absorption of bile salts in the ileum is related to the local hyperemia. The carbohydrates not absorbed by the small intestine are converted to volatile fatty acids in the colon. These volatile fatty acids increase blood flow in the colon.

D. Mechanisms of Postprandial Intestinal Hyperemia

Postprandial intestinal hyperemia is not influenced by elimination of intestinal nerves nor by anticholinergic or antiadrenergic drugs, indicating that the mechanisms involved are local and intrinsic to the intestine. Figure 3 outlines the possible steps and mechanisms leading to the intestinal hyperemia after entrance of breakdown products of food into the gut lumen. Presence of food in the gut lumen initiates and stimulates intestinal absorption, secretion, and motility via local nerves and chemicals. The influence of these functions on blood flow (Section IV) and the local intrinsic mechanisms (Section III,B) are part of the events leading to postprandial

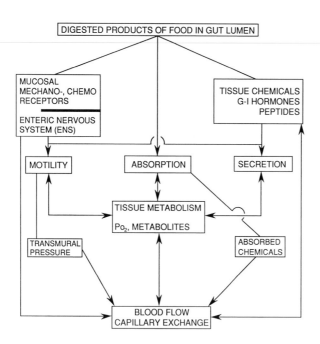

FIGURE 3 Possible pathways and mechanisms involved in regulation of intestinal blood flow during nutrient absorption.

intestinal hyperemia. It must be noticed here that the mechanisms involved are complex with many factors playing a role and that there is controversy on some factors. The controversy is likely due to the difference in the type of nutrients and animal species used in the experiments.

1. Mucosal Neural Receptors and the Enteric Nervous System

There are about 1 million neurons in the enteric nervous system (ENS), which is primarily composed of the myenteric and submucous nerve plexuses within the gut wall. The plexuses are interconnected, and some of their neurons have processes that originate in receptors on the mucosal surface. These receptors sense the composition of the intestinal contents (chemoreceptors, osmoreceptors) and mechanical stimuli (mechanoreceptors). Mechanical stimulation mimicking movement of chyme over the mucosal surface or presence of hypertonic glucose solutions stimulate these mucosal neural receptors, activating a reflex confined to the ENS to produce local hyperemia. Recent studies further show that the hyperemia produced by oleic acid plus bile is mediated by an axon reflex that involves capsaicin-sensitive mucosal afferent nerves. The nerve fibers involved are noncholinergic and nonadrenergic, and the neurotransmit-

ter acting on the vascular smooth muscle appears to be vasoactive intestinal peptide. The vasodilation evoked by oleic acid, however, involves more than just this neural factor.

2. Absorption

Absorptive processes somehow play a role in the initiation and maintenance of the hyperemia. For example, absorbable glucose or its analogue 3-0-methylglucose in the gut lumen increases local blood flow, but another analogue, 2-deoxyglucose (which is nonabsorbable) does not. As described in Section I, the countercurrent exchanger and multiplier in the villi make the interstitial fluid hypertonic during active absorption of NaCl, and intestinal blood vessels respond to hypertonicity with vasodilation. The vasodilator nutrients may act directly on the villous or submucosal vasculature to increase blood flow, once absorbed into the interstitium. Exposure of intestinal vessels by intra-arterial infusions of the common intestinal contents such as bile salts, oleic acid, hydrogen ions, CO_2, and caproic acid increase blood flow, whereas glucose or amino acids do not.

3. Secretion and Motility

These are discussed in sections IV,B and C.

4. Gastrointestinal Hormones and Peptides

Almost all hormones and peptides released from the intestine are vasodilators, the exception being somatostatin, the only peptide known to produce intestinal vasoconstriction. Postprandial blood concentration of these peptides, however, is not high enough to affect intestinal blood flow. Therefore, rather than act as circulating hormones, they may act as the paracrine or neurotransmitter to play a role in regulating postprandial intestinal blood flow. The paracrine are chemicals that act on target cells (e.g., vascular smooth muscle) adjacent to the site of their release by diffusion from the site to the target cells. The following peptides have been proposed to play this role: cholecystokinin, secretin, substance P, and vasoactive intestinal peptide. [*See* PEPTIDE HORMONES OF THE GUT.]

5. Local Tissue Chemicals

The small intestine releases a variety of local tissue chemicals that play significant roles in regulation of intestinal functions. Serotonin (also known as 5-hydroxytryptamine), bradykinin, and histamine have been proposed to play a role in postprandial intestinal hyperemia. Although both histamine H_1 and H_2 receptors are present in the intestinal vasculature, only H_1 receptors modulate the postprandial intestinal hyperemia.

Prostaglandins (PGs) are a family of unsaturated fatty acids derived from arachidonic acid. Principal PGs synthesized by the intestine are PGI_2, PGE_2, $PGF_{2\alpha}$, TXA_2, and PGD_2. The first two PGs are vasodilators, the next two vasoconstrictors, whereas PGD_2 has variable vascular effects on intestinal blood flow. Presence of digested food in the jejunum increases local production and release of these PGs. Food-induced intestinal hyperemia is enhanced when mefenamic acid is given to inhibit synthesis of these PGs but is attenuated by administration of their precursor, arachidonic acid. This appears to indicate that the endogenous PG system acts to limit the hyperemia. The primary PG involved in the inhibition appears to be TXA_2, which is a potent vasoconstrictor. Arachidonic acid, the precursor of PGs, is a regular constituent of food. The amount of arachidonic acid contained in the ingested food could influence postprandial intestinal blood flow.

6. Tissue Metabolism

As in any organ, intestinal blood flow regulation is closely linked to the metabolic status of the intestine. During nutrient absorption, intestinal blood flow increases in parallel with the increased local oxygen consumption. Furthermore, almost all stimuli described above that enhance and inhibit postprandial intestinal hyperemia also enhance and inhibit oxygen consumption, respectively, in parallel fashion. Therefore, there is a strong possibility that these enhancers or inhibitors act indirectly via intestinal functions and oxidative metabolism to affect blood flow. For example, mefenamic acid increases intestinal glucose absorption and oxidative metabolism. The mechanism by which mefenamic acid enhances the food-induced hyperemia may be due to its enhancing action on glucose absorption and oxidative metabolism. As shown in Fig. 3, most factors (e.g., the ENS, gastrointestinal peptides, and histamine) involved in the regulation of postprandial intestinal hyperemia also alter intestinal functions. In addition to their direct vasodilatory action on the vascular smooth muscle, they could affect intestinal functions and metabolism to alter blood flow.

Of all the possible mechanisms described in this section, tissue metabolism is the most important

factor regulating the intestinal hyperemia occurring after a meal. It has been estimated that two-thirds of the hyperemia can be attributed to tissue metabolism. A good example to support this thesis is the following finding. Glucose (which is actively absorbed and metabolized) produces a greater intestinal hyperemia than does 3-0-methylglucose (which is actively absorbed but not metabolized). The former increases oxygen consumption, whereas the latter does not.

The mediators between metabolism and blood flow are tissue oxygen concentration and vasodilator metabolites. The mucosal tissue Po_2 decreases during glucose absorption, and there is an inverse correlation between changes in mucosal tissue Po_2 and intestinal blood flow in rats. As much as one-fourth of the hyperemia induced by glucose may be attributed to a reduction in Po_2. The possible vasodilator metabolites involved in regulation of intestinal blood flow are K^+, H^+, adenosine, and the level of osmolality. The food-induced increases in intestinal blood flow and oxygen consumption are accompanied by an increase in adenosine production in the intestine. Furthermore, adenosine antagonists have been shown to inhibit the food-induced hyperemia.

The primary function of the circulation is to provide nutrients, particularly oxygen, to functioning tissues. The increased oxygen demand in response to an increase in tissue functions can be met by either an increase in blood flow or by an increase in tissue oxygen extraction, or both. In the intestine, the contribution of each in the maintenance of adequate oxygenation depends on the type of nutrients and the duration of the presence of food in the lumen. In the absence of bile, the increased oxygen consumption during glucose absorption is accompanied by an increase in the number of perfused capillaries with minimal or no increase in blood flow; the increased oxygen consumption, therefore, is met by an increase in tissue oxygen extraction. The increased oxygen consumption during absorption of oleic acid, however, is accompanied by a decrease in tissue oxygen extraction and no change in the perfused-capillary density, with a marked increase in blood flow. The oleic acid–induced increase in oxygen consumption, therefore, is primarily met by an increase in blood flow. Placement of predigested food containing equal amounts of carbohydrates, proteins, and fats produces an increase in oxygen consumption, which is maintained at a certain level for the entire 30-min placement period. The increased oxygen consumption during the initial 15 min, however, is met by an increase in blood flow, but that during the last 10 min is met by an increase in tissue oxygen extraction. The blood flow returns to resting levels during the last 10 min. The above phenomena clearly show that the intestinal arterioles and the precapillary sphincters are harmoniously regulated to maintain adequate oxygenation of intestinal tissues when the tissue activity is increased during absorption of nutrients.

Bibliography

Chou, C. C. (1983). Splanchnic and overall cardiovascular hemodynamics during eating and digestion. *Fed. Proc.* **42,** 1658–1661.

Chou, C. C. (1984). Distribution of blood flow in the bowel wall: Measurement and physiological characteristics. *Fed. Proc.* **43,** 8–10.

Chou, C. C. (1989). Gastrointestinal circulation and motor function. *In* "Handbook of Physiology, section 6: The Gastrointestinal System, vol. 1. Motility and Circulation" (J. D. Wood, ed.), pp. 1475–1518. American Physiological Society, Bethesda, Maryland.

Chou, C. C., and Kvietys, P. R. (1981). Physiological and pharmacological alterations in gastrointestinal blood flow. *In* "The Measurement of Splanchnic Blood Flow" (G. B. Bulkley and D. N. Granger, eds.), pp. 475–509. Williams and Williams, Baltimore, MD.

Chou, C. C., Mangino, M. J., and Sawmiller, D. R. (1984). Gastrointestinal hormones and intestinal blood flow. *In* "Physiology of the Intestinal Circulation" (A. P. Shepherd and D. N. Granger, eds.), pp. 121–130. Raven Press, New York.

Gallavan, R. H., Jr., and Chou, C. C. (1985). Possible mechanisms for the initiation and maintenance of the postprandial intestinal hyperemia. *Am. J. Physiol.* **249** (Gastrointest. Liver Physiol. 12), G301–G308.

Granger, D. N., Kvietys, P. R., Kortuis, R. J., and Premen, A. J. (1989). Microcirculation of the intestinal mucosa. *In* "Handbook of Physiology, section 6: The Gastrointestinal System, vol. I. Motility and Circulation" (J. D. Wood, ed.), pp. 1405–1474. American Physiological Society, Bethesda, Maryland.

Lundgren, O. (1984). Microcirculation of the gastrointestinal tract and pancreas. *In* "Handbook of Physiology, section 2: The Cardiovascular System, vol. IV, Microcirculation" (E. M. Renkin and C. C. Michel, eds.), pp 799–863. American Physiological Society, Bethesda, Maryland.

Shepherd, A. P., and Granger, D. N., eds. (1984). "Physiology of the Intestinal Circulation" Raven Press, New York.

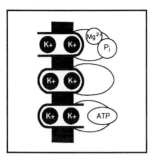

Ion Pumps

R. W. ALBERS, *National Institute of Neurological Diseases and Stroke*

Glossary

Calciosome Form of endoplasmic reticulum specialized for the uptake, storage, and release of calcium ions

Domain As applied to protein structure, refers to a compactly folded region of a protein, usually composed of a single polypeptide chain; may also refer to the portion of the overall protein structure that performs a discrete function (e.g., the substrate binding domain)

Endoplasmic reticulum Intracellular organelle consisting of a highly convoluted membrane which, depending on cell type and state, could encompass several percent of the intracellular volume and a major fraction of the cell membranes. The portion specialized for synthesis of secretory and membrane proteins is called rough, because its outer surface is studded with ribosomes. A portion (smooth) is specialized for the synthesis of phospholipids and other membrane components; another portion is specialized for the storage and release of calcium ions.

Ionophoric Related to the transport of ions. "Ionophore" refers to a class of molecules, either synthetic or naturally occurring, that facilitate the movements of ions across cell membranes, by forming either lipid-soluble complexes or transmembrane channels. In the case of the relatively large transport proteins, the ionophoric function probably involves only part of the molecule: the ionophoric domain.

Membrane potential Voltage difference between the two sides of a cell membrane

Plasma membrane Outer limiting membrane of cells

Sarcoplasmic reticulum In skeletal and cardiac muscle, a form of endoplasmic reticulum specialized for the regulation of muscle contraction by the uptake, storage, and release of calcium ions (cf. calciosome)

Siemen Standard unit of electrical conductance; the reciprocal of ohmic resistance

Site-directed mutation Change in the structure of a gene at a specific site, produced by a laboratory procedure. Systematic changes in the protein coding part of the gene are often used to examine the relationship between amino acid sequence and protein function

Turnover rate Measure of the catalytic activity of an enzyme molecule in terms of the number of molecules of substrate transformed per unit of time

ALL CELLS ARE bounded by membranes and contain several types of intracellular organelles that are also membrane bounded. Nearly all of these membranes can move ions selectively and unidirectionally to produce higher effective ion concentrations on one side relative to the other. These capabilities define ion transport and are mediated by proteins that span cell membranes. These ion pumps can selectively bind ions from dilute solutions, transport these ions across cell membranes, and release them into local environments that contain relatively high concentrations of the same ions.

I. Overview of the Role of Ion Transport in Physiology

Primary transport processes are those that are driven by energy derived from biochemical reactions. The presently known primary transport processes in animal cells act on ions. There are also many secondary transport processes that couple the transmembrane movements of one or more ions to the transport of another ion or molecule. Proteins that mediate the movements of two or more transport fluxes in the same direction are called symporters, or cotransporters. Proteins that couple the movement of two transport fluxes in opposing directions are called antiporters, or exchangers. Secondary transport processes can achieve the "uphill", or concentrative, transport of one ion or molecule by using energy derived from the movement of another from higher to lower concentration. The term active transport is often used to describe any uphill transmembrane movement of ions or molecules, whether mediated by primary or secondary transport mechanisms.

Pumps have evolved to move ions through membranes in specific directions and, in so doing, to produce large concentration differences across membranes. During the transport process ions bind to transport proteins at specific sites, stoichiometrically, and with high affinity. In this respect the ion-binding sites of transport proteins have properties similar to those of receptors or enzyme substrate sites.

Membrane channels are another class of proteins that mediate flows of ions across membranes. They are quite distinct from pumps in both function and structure. Channels selectively regulate membrane permeability and thereby influence the rates of diffusive ion movements. They do not control the direction of ion flows nor do they create ion concentration differences, but they do regulate the rates of ion flow and therefore the rates at which the concentration differences formed by pumps are dissipated.

The selective absorption and excretion of ions and molecules across cellular membranes are fundamental to nutrition, biosynthesis, and homeostasis. Theories of the origin of life usually postulate the existence of adequate concentrations of all of the necessary elements for the evolution of increasing complexity in a "primeval soup" and thus obviate the necessity for transport processes. Present conditions on the planet are quite different: the ability to acquire and retain essential molecules is practically a definition of cellular life. With the exception of certain lipids and macromolecules that can form aggregate structures spontaneously, all of the molecules and ions of cells are dependent on membrane transport processes for either their acquisition or retention. As discussed in the next section, the transport of most organic molecules is driven by the simultaneous transport of one or more inorganic ions.

Apart from nutritive functions, ion transport is essential at several levels of energy metabolism. In animal cells proton pumping couples oxidative metabolism to ATP synthesis, to produce the oxidative phosphorylation processes of mitochondria. Ion transport plays an equally important role in linking ATP phosphorylation energy to many different transport processes that occur via sodium ion concentration differences across the plasma membrane and via proton concentration differences across various intracellular organelles. [*See* ATP SYNTHESIS BY OXIDATIVE PHOSPHORYLATION IN MITOCHONDRIA.]

Cellular information processing is also fundamentally dependent on the ion concentration differences created by active transport. Intracellular Ca^{2+} and proton concentrations are links in the regulation of other processes (e.g., protein synthesis, muscle contraction, secretion, and endocytosis). The plasma membrane electrical potential, which is the basis of the nerve and muscle action potentials, is maintained by the Na^+, K^+-pump. Another essential aspect of membrane transport physiology is volume regulation. Because of the ability of cells to concentrate ions and molecules, osmotic forces develop within membrane compartments and water movements occur. These are controlled by the highly integrated activities of ion pumps and channels, some of which are discussed in Section IV.

II. Principles of Ion Transport

The environment of mammalian cells is generally characterized by a total solute concentration that is about 300 mOsm. Na^+ is the predominant external cation, and K^+ is the predominant internal cation. These are dynamic steady-state conditions that are only maintained by a continuous supply of metabolic energy. Although lipid bilayers that form the matrix of cell membranes are nearly impermeable to ions, functional cell membranes vary greatly in their ion permeability, primarily because of different membrane proteins which form ion channels. Channel densities have been measured in many different kinds of cells and have ranged from less than 10 to more than 300,000 per square micron. In electrical terms the ion conductances of single channels also vary from approximately 2 to more than 100 pS. A channel with 10 pS conductance will conduct about 3 million ions per second if the voltage difference across the membrane is about 50 mV.

The maximum rate of the ATP-dependent Na^+, K^+ pump is less than 1000 ions per second. In contrast, the observed conductances of some single-membrane channels are about 10^7 per second. Thus, to counterbalance the current flow through a single continuously open Na^+ channel could require the operation of 10,000 pump molecules.

A. Why Are Pumps So Much Slower Than Channels?

Movements of ions through channels are nearly as fast as the rates calculated from their free diffusion in water. This suggests that interactions between the ions and the protein surfaces of channels are minimal. In contrast, transport proteins and their substrate cations must interact strongly on at least one side of the membrane. The bound cations must then undergo transport and be discharged from the pump on the other side of the membrane. To be effective, uptake must occur from relatively dilute solutions and discharge must occur into relatively concentrated solutions. Finally, the transport system must be recycled to its original state. In general terms, ion transport can be represented as occurring in four stages, as indicated in Fig. 1.

To make this an effective transport system, two conditions are necessary:

1. At least one of these steps must do work. In thermodynamic terms there must be a negative free-energy change during the transformation of at least one of the states of the pump molecule into the next. If this process is to continue through many cycles, there must be a correspondingly continuous input of free energy. In the case of primary transport, this will come from a chemical reaction of, for example, ATP, with one of the pump states, followed by regenerative reactions during the other state transformations. In the case of secondary transport, this will come from binding or uptake of a second ion at the side of the membrane where it

1) $E_U + M^+_{(1)} = E_U M^+$ ⟵ uptake from dilute solution

2) $E_U M^+ \quad = E_D M^+$ ⟵ transport

3) $E_D M^+ \quad = E_D + M^+_{(2)}$ ⟵ discharge

4) $E_D \quad = E_U$ ⟵ recycle to original state

FIGURE 1 Generalized reaction mechanism for type 2 transport ATPases. Uptake (U) occurs on side 1 and discharge (D) occurs on side 2 of the membrane. E and M^+, transport protein and the transported cations, respectively.

exists at a high concentration, followed by its discharge on the other side, where the concentration is lower.

2. The transport cycle must proceed fast enough to achieve the physiological requirements. In the steady state, in which all stages are proceeding at the same rate, one or more stages may be rate controlling. For example, as written in Fig. 1, stages 2 and 4 are conformational transitions of the transport protein that occur at rates that are independent of the concentrations of any other reactants. If the intrinsic rates of either of these transitions are too slow in the forward direction (i.e., from left to right), or too rapid in the reverse (i.e., from right to left), most of the transport protein will accumulate as $E_U M^+$ or E_D, respectively (see Fig. 1), and the regeneration of cation-transporting sites will be impaired. Conversely, if the forward rate of the stage 4 transition is much faster than the reverse, at low concentrations of M^+, the ready availability of cation-transporting sites may permit the overall transport rate to become more responsive to the concentration of M^+ on side 1.

B. How Are Pumps Matched to Their Work Loads?

Theoretically, the limits of the work available from a primary transport system, and hence the maximum achievable concentration differences, can be calculated if both the number of ions transported in each direction and the chemical reaction (i.e., energy source) are known. Thus, if the overall process is electroneutral, the minimum energy requirement to transport 1 mol equivalent of a cation into a 10-fold higher concentration is about 1400 calories. Since about 12,000 calories are available from the hydrolysis of 1 mol of ATP, an ATP-dependent pump that uses one molecule of ATP for each cation transported might maintain transmembrane concentration differentials of about 10^8 under these conditions. If the work of transport includes a 58-mV membrane potential and each cation has one charge, the maximum sustainable concentration difference would be reduced by 10-fold. If two such cations were transported per ATP hydrolyzed, it would be reduced to 10^4. [*See* ADENOSINE TRIPHOSPHATE (ATP).]

Different functional requirements cause transport systems to operate under different constraints. For example, the transport requirements for Ca^{2+} are quite different among fibroblasts, secretory cells, smooth muscle, heart muscle, and skeletal muscle. In each case Ca^{2+} participates in the motility, secretion, and contraction processes. In the case of heart and skeletal muscles, large changes in the intracellular Ca^{2+} concentration must occur in fractions of a second throughout the cell, whereas in many types of secretion, the changes are relatively slow and localized near the plasma membrane. Thus transport requirements often differ in different parts of the same cell. In neurons Na^+ enters and must be removed along the length of the axons to maintain excitability, and also it may enter via neurotransmitter symporters and the Na^+–Ca^{2+} exchanger in pre- and postsynaptic processes of the same neurons. Perhaps as a consequence of these varying requirements, different cell types often express more than one type of pump for a given ion.

C. How Do Ions Interact with Transport Proteins?

This is one of the least well-understood aspects of ion transport biology. A major part of channel proteins is considered to be involved in the formation of hydrophilic pores bounded by transmembrane peptide segments. Channel proteins respond to regulating factors (e.g., neurotransmitter molecules, polypeptide hormones or changes in transmembrane voltage) by transitions between conformations with open and closed states and, sometimes, by assuming inactivated states. However, as mentioned in Section I, channel interactions with ions are weak relative to those of transport proteins, and most of channel selectivity can be accounted for simply on the basis of the sieving effects of pore dimensions and electrical charge.

The structures of transport proteins are entirely different from those of channel proteins and their functions are more complex. Sequence data, now available from several dozen transport proteins, uniformly indicate the presence of multiple (i.e., 6–12) transmembrane segments. However, with the pos-

sible exception of the type 1 ATPases (see Section III, A), the sequences do not reveal the repetitive structure that would suggest the presence of massive pore-forming domains or subunits. Moreover, considering the relative complexity of their mechanisms, in most cases the subunit structures of ion pumps are surprisingly simple. For example, a single peptide of 100 kDa constitutes the whole structure of many of the type 2 ATP-dependent primary transporters. Some transporters, including the type 1 and probably also the type 3 ATPases, are composed of multiple dissimilar subunits which associate to form three distinct parts: a globular cytoplasmic domain that possesses the ATPase activity, a transmembrane channel, and a neck region connecting the two. A major question for each transport protein is whether the events that are crucial for transport occur within a globular cytoplasmic domain or within a transmembrane domain.

The binding of ions to transport proteins seems similar in almost every way to the binding of substrates to enzymes. These interactions are saturable and reversible, and can induce catalytic activity and/or conformational transitions in the transport proteins. Usually, if more than one ion is transported per cycle, the binding is positively cooperative (i.e., binding one ion facilitates the binding of the next).

There are circumstances in which ion binding to a transport protein is not freely reversible: the binding site becomes occluded so that both access and egress from the site is slow. This is best documented for the case of K^+ binding to the Na^+, K^+ pump, as depicted in Fig. 2. Under certain conditions the rate of exchange of the bound K^+ for K^+ in solutions is less than 1 per second. This rate is increased dramatically in the presence of either (1) ATP or (2) Mg^{2+} and inorganic phosphate. These observations are interpreted to mean that K^+ is occluded within the protein structure during the actual transport event. Release at the cytoplasmic side is promoted by ATP binding to the pump protein; this mechanism of release probably corresponds to the K^+ discharge events that occur during the normal direction of pump operation. Release of K^+ to the outside is promoted by Mg^{2+} and phosphate binding to cytoplasmic pump sites; this corresponds to a reversal of the normal K^+ uptake events. The kinetics of K^+ release from the occluded state strongly support a model in which a pathway through the protein is filled by two K^+ in tandem (Fig. 2). The path is accessible from either side, but not simultaneously. Because of experiments that show that the

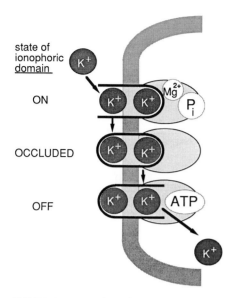

FIGURE 2 Formation of the "occluded" complex of K^+ with the ATP-dependent Na^+,K^+-ATPase. The rate of release of ions from occluded binding sites is much slower than their reversible dissociation from electrostatic or chelating sites, on the order of seconds to hundreds of seconds. In the case of the Na^+,K^+ pump, the rate of release of occluded K^+ is greatly accelerated by the addition of either Mg^{2+} plus orthophosphate or of ATP. In the first instance the rate of release of one-half of the total occluded radioactive K^+ is suppressed by the presence of nonradioactive K^+ in the solution. This is evidence for the tandem nature of the two sites in each ionophoric domain and represents the conditions which, in normal pump function, permit the binding of extracellular K^+. P_i, inorganic phosphate.

binding order of the two K^+ is "first in, last out," they appear to migrate in tandem on the same path. Occlusion probably occurs subsequent to a reversible binding step, and concurrently with a catalytic event. The reversible binding sites and the occlusion sites are thus identical except for their conformations, according to this model.

The amino acid sequence of an ion transport domain has not been identified with certainty for any transport protein. Because of its high binding affinity, the ATP-dependent calcium pump has provided a good model for investigating this problem. Some possibilities have been eliminated. Several other types of calcium-binding proteins contain a special sequence known as the "EF hand," but this does not occur in the known sequences of calcium pump proteins. Some other prospective Ca^{2+}-binding sites have been ruled out by studies in which alteration of the relevant amino acids has not altered the Ca^{2+}-dependent capabilities of the pump. Mutations which have been found to be effective in inactivating the calcium pump include changing several negatively charged residues that may be part of the

transmembrane segments. This has reinforced the hypothesis that the ionophoric structures of some transport proteins involve ion binding within spaces bounded by their transmembrane segments.

D. What Constitutes a Transport Event?

Of all the mammalian transport proteins, most is known about the mechanism of the Na^+,K^+ pump (Fig. 3). The principle that the energy supply and the ionophoric events are coupled together through a cycle of changes in protein conformation is well

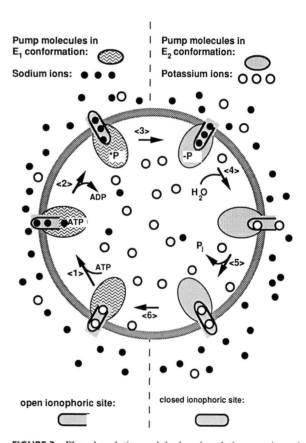

FIGURE 3 Phosphorylation and dephosphorylation reactions alternate with conformational changes of the ATP-dependent Na^+,K^+ transport protein. The reaction mechanism is arbitrarily divided into six stages as numbered. K^+ displacement from the pump to the cytoplasm is promoted by ATP binding in stage 1. Na^+ binding leads to reversible phosphorylation of an aspartyl residue at the catalytic ATPase site (stage 2). Subsequent to phosphorylation, the Na^+ become occluded in stage 3. However, this can only be demonstrated if stage 3 is blocked by an inhibitor such as oligomycin. Otherwise stage 3 leads to a conformational transition of the pump protein to E_2, which permits the dissociation of Na^+ extracellularly. In stage 4 K^+ binding catalyzes hydrolysis of the aspartyl phosphate and subsequent K^+ occlusion (stage 5). This is normally followed by the reversion of the E_2 conformation to E_1 (stage 6) which permits the dissociation of K^+ cytoplasmically.

documented in this case. Probably all of the type 2 ATP-dependent transporters operate by analogous mechanisms. Although it is reasonable to expect that these general principles apply to other classes of transporters, at present little evidence is available.

III. Biochemical Characteristics of Ion Pumps

This survey of the biochemistry of major transport proteins provides an introduction that will permit discussion of how these systems are used within cells to meet various physiological requirements.

A. Primary Active Transport Systems

1. ATP-dependent Na$^+$,K$^+$ pump

The ATP-dependent Na$^+$,K$^+$ pump is the principal primary active transport system in the plasma membrane of nearly all animal cells. The operation of this pump creates the high K$^+$ and low Na$^+$ concentrations that are characteristic of cytoplasm. ATP-dependent Na$^+$,K$^+$ pumps located, for example, in the basolateral plasma membranes of kidney epithelia also maintain the characteristically high Na$^+$ of the extracellular environment. An adequate Na$^+$ concentration difference across cell membranes comprises an important store of potential energy that is directed by other membrane processes to diverse ends (see Section IV).

The overall transport process is

$$3\ Na^+_i + 2\ K^+_e + ATP_i$$
$$= 3\ Na^+_e + 2\ K^+_i + ADP_i + P_i$$

where "i" and "e" refer to intra- and extracytoplasmic compartments, respectively. This system can be measured in plasma membrane fragments as an enzyme reaction (ATPase) that hydrolyzes ATP and requires Na$^+$, K$^+$, and Mg^{2+} for full activity. The purified protein consists of equimolar amounts of two subunits: an α subunit of 100 kDa, which accounts for the known catalytic functions, and a β subunit of unknown function. The minimal functional unit is probably α_2–β_2. A unique characteristic of Na$^+$,K$^+$-ATPases is inhibition by cardiac glycosides (see Section IV).

The Na$^+$,K$^+$ pump is representative of the "P" or type 2 ATPases that operate by a mechanism involving cyclic phosphorylation and dephosphory-

lation of an aspartyl residue at the catalytic site (Fig. 3). A general characteristic of type 2 ATPases is inhibition by low concentrations of vanadate.

It was possible to show that Na$^+$ and K$^+$ activate, respectively, phosphorylation and dephosphorylation of the ATPase protein. The existence of an intervening protein conformational change was deduced from observations that, although the phosphorylation step is essentially irreversible in the native enzyme, certain inhibitors of the ATP hydrolysis make this step readily reversible. Because the same aspartyl residue on the protein is phosphorylated in both cases, a mechanism involving covalent bond migration was ruled out. Subsequently more direct evidence for the conformational nature of the step was obtained, for example, from a correlation of reversible changes in the protein tryptophan fluorescence with the occurrence of the transition.

2. ATP-Dependent H$^+$,K$^+$ Pumps

Parietal cells of the stomach can secrete an essentially isotonic solution of hydrochloric acid. The acid is supplied by an H$^+$,K$^+$-ATPase present in the membrane of an intracellular canalicular system of these cells. Equal numbers of H$^+$ and K$^+$ are transported. The transport cycle appears to be similar to that of the Na$^+$,K$^+$ pump, with protons substituting for Na$^+$. The catalytic subunits of H$^+$,K$^+$-ATPase and Na$^+$,K$^+$-ATPase are 60% similar. The existence of a subunit homologous to the β-subunit of the Na$^+$,K$^+$ pump is uncertain.

3. ATP-Dependent Ca^{2+} Pumps

Because Ca^{2+} functions primarily as an intracellular messenger, responding to a variety of receptor-coupled regulatory mechanisms, the important parameter to be regulated in Ca^{2+} metabolism is not the transmembrane concentration difference, but, rather, the actual concentration of free Ca^{2+} in the cytoplasm. The messenger functions of calcium are diverse, depending on cell type and developmental stage. They may even be different in different parts of a given cell. Moreover, some of these functions involve rapid changes of cytoplasmic Ca^{2+} within localized intracellular cell regions, whereas others may modulate relatively long-term Ca^{2+} homeostasis. To respond to this diversity, cells express several different calcium pumps. There are ATP-dependent Ca^{2+} pumps in both plasma membranes and endoplasmic reticulum (ER). These are different, but structurally related, and belong to the type 2 class of ATPases.

a. ER Ca^{2+} Pump The ER Ca^{2+} pump was first isolated from skeletal muscle and is usually referred to as the sarcoplasmic reticulum (SR) calcium pump. SR is a form of ER specialized for the rapid release of Ca^{2+} into, and removal of Ca^{2+} from, skeletal muscle cytoplasm. ATP-dependent Ca^{2+} pumps constitute more than 70% of the protein of these membranes, consisting of a single-polypeptide subunit of about 110 kDa. The existence of a Ca^{2+} pump subunit homologous to the β subunit of the Na,K-ATPase is uncertain.

The overall transport process is

$$2\,Ca^{2+}_c + ATP_c = 2\,Ca^{2+}_1 + ADP_c + P_c$$

where "c" and "l" refer to cytoplasmic and SR luminal compartments, respectively. This pump can rapidly reduce cytoplasmic Ca^{2+} to less than 1 μM, which is required to bring about muscle relaxation.

Two variants of the SR calcium pump have been identified with amino acid sequences that are about 85% similar. One variant is found in the fast-twitch form of skeletal muscle and has not been shown to interact with any regulatory system. However, the variant that occurs in slow skeletal and cardiac muscle fibers is associated with another integral membrane protein: phospholamban. When SR membranes containing both proteins are phosphorylated by cAMP-dependent protein kinase, the catalytic rate of the calcium pump is increased. In intact cardiac muscle preparations increased phosphorylation of phospholamban has been shown to correlate with increased rates of relaxation and contraction. One proposed mechanism is that phopholamban acts to suppress the calcium pump and that the cAMP-dependent phosphorylation can release this suppression. This mechanism may account for the epinephrine-induced acceleration of cardiac relaxation.

Ca^{2+} is also sequestered in endoplasmic compartments of many other types of cells from which, in some cases, it can be released by inositol triphosphate to initiate a variety of regulatory events. There is evidence that the ER of other tissues contain's calcium pumps that are similar or identical to those of the SR. For example, ER of rat liver contains a protein that is immunologically related to skeletal SR Ca^{2+}-ATPase. Moreover, cDNAs encoding the slow-twitch form of the SR calcium pump have been identified in libraries derived from brain, kidney, and stomach mRNAs.

b. Plasma Membrane Ca^{2+} Pump A calmodulin-activated ATP-dependent calcium pump occurs in the plasma membrane of erythrocytes, neurons, cardiac muscle, renal epithelia, and many other cell types. Two forms have been detected in a cDNA library from rat brain. Both appear to consist of single-peptide chains of about 130 kDa and contain, near their C-termini, sequence similarities to the calmodulin binding sites of other proteins.

The plasma membrane Ca^{2+} pump protein exists in an inactive state in isolated cell membranes and can be transformed to an active state by interaction with calmodulin. It also can be activated in the absence of calmodulin by the addition of certain fatty acids and acidic phospholipids, or by limited proteolysis. In some cells (e.g., vascular smooth muscle) there is evidence for a different calcium pump regulation involving protein kinase C. The overall transport process of the plasma membrane calcium pump is reported to be

$$Ca^{2+}_c + H^+_e + ATP_c$$
$$= Ca^{2+}_e + H^+_c + ADP_c + P_c$$

4. ATP-Dependent H$^+$ Pumps

Plants, fungi, and bacteria have ATP-dependent proton pumps localized in their plasma membranes which function in cellular electrolyte metabolism in a manner comparable to the role of the Na$^+$,K$^+$ pump in animal cells. In the several cases that have been studied, the structures and amino acid sequences of these proton pumps are related to the those of the type 2 ATPases (see Section III, A). Other than the gastric H$^+$,K$^+$-ATPase, examples of this form of proton pump have not been described for animal cells. However, several other ATP-dependent proton pumps are found as described below.

a. Type 1 Proton Pumps that Participate in ATP Synthesis Most of the energy used by animal cells is extracted from carbohydrates and fats by oxidative reactions (i.e., by the transfer of electrons to oxygen). The process by which the energy of oxidative metabolism is captured and conserved as high-energy phosphate bonds involves the transfer of water molecules rather than electrons. Efficient energy transfer between these two fundamentally different types of chemical reaction arises from a combination of two different membrane-pumping mechanisms: (1) oxidative reactions, which are electron-powered pumps that can drive protons out of the mitochondrial matrix and (2) an ATP-dependent proton pump operating in reverse (type 1 ATPase), which condenses ADP and orthophosphate to form ATP and is driven by the electrochemical force of protons flowing back out of the mitochondrial membranes.

b. Type 3 Proton Pumps of Vacuolar Organelles A different type of ATP-dependent proton pump operates to inject protons into various intracellular organelles. Most of these are part of, or derived from, Golgi membranes, and are known collectively as the vacuolar apparatus. They interact with plasma membranes and participate in endo- and exocytosis. In many of these cases, the resultant proton concentration differences are used to accumulate other molecules within the organelles via antiporters.

Internal acidification functions to activate various degradative enzymes in lysosomes. In the chromaffin granules of the adrenal medulla, a proton concentration difference is coupled through a symporter to concentrate the catecholamines which they store. In endosomes, which mediate the cellular uptake of various regulatory peptides, the low pH promotes dissociation of the peptides from their receptors.

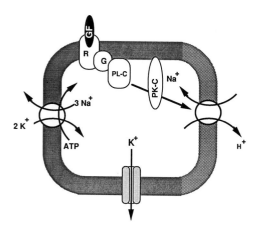

FIGURE 4 The sodium/proton antiporter is subject to regulation by a variety of growth factors (GF) which bring about the phosphorylation of a regulatory site on the antiporter. This, in turn, increases the affinity of the ionophoric site for protons, and thus initiates an increase in cytoplasmic pH. The energy for this process derives from the influx of Na^+. This remains effective only if the ATP-dependent Na^+, K^+ pump is competent to maintain the required low level of intracellular Na^+. R, plasma membrane receptor; G, GTP-binding transducer protein; PL-C, phospholipase C; PK-C, protein kinase C.

These pumps form a distinct class of vacuolar, or type 3, ATPases that characteristically maintain an intraorganelle pH of 5–6. Structurally, they appear to consist of a complex of several different subunits, ranging in size from 15 kDa to more than 100 kDa and an overall molecular weight of at least 400,000. There is no evidence for phosphorylation of the catalytic site in the mechanism of these pumps. Some of the smaller subunits probably form a transmembrane proton channel. The similarity of some of their properties to those of the type 1 mitochondrial ATPases has complicated the purification and identification of the subunits. However, distinctive and highly conserved subunits of the vacuolar ATPases have been identified in species ranging from yeast to mammals.

B. Secondary Active Transport Systems

1. Antiporters

a. Na⁺/Ca²⁺ Antiporters

$$3\ Na^+_e + Ca^{2+}_c \leftrightarrow 3\ Na^+_c + Ca^{2+}_e$$

This type of system has been studied extensively in heart muscle plasma membranes (i.e., the sarcolemma), where it is readily measured in isolated vesicles. It appears to provide a mechanism for rapidly reducing cytoplasmic Ca^{2+} levels, although conditions under which it may contribute significantly to Ca^{2+} influx have also been proposed. Similar or identical systems have been demonstrated in membrane vesicles of other tissues including brain and kidney. Perhaps the highest capacity for Ca^{2+} uptake by this mechanism has been demonstrated in ram sperm flagellae.

Although these antiporters are usually assayed by measuring the rate of uptake of Ca^{2+} into membrane vesicles that have previously been loaded with Na^+, they can operate in either direction. The stoichiometry appears to be three Na^+ per Ca^{2+}. A high turnover rate of about 1000 per second has been estimated for the bovine cardiac Na^+/Ca^{2+} antiporter. Assuming a 100-kDa protein, these data permitted an estimate that the exchanger constitutes only 0.1–0.2% of the sarcolemmal protein.

b. Na⁺/H⁺ Antiporters The process mediated by this protein is a tightly coupled neutral exchange of sodium ions for protons (Fig. 4):

$$Na^+_e + H^+_c \leftrightarrow Na^+_c + H^+_e$$

Antiporters of this type have been found in plasma

membranes of all mammalian cells examined, but not in intracellular membranes. Amiloride is a relatively selective and potent inhibitor of these antiporters and is often used to define the system. In several cases the ratio of the two cations exchanged was found to be 1:1. Because the exchange of H^+ and Na^+ is driven by their combined electrochemical forces, experimental manipulation of these concentration differences can produce an exchange in either direction. The physiological concentration differences normal for most cells produce Na^+ influx and H^+ efflux. When maximally activated, the turnover number is estimated to be 2000 per second. Analysis of the kinetics of interaction of Na^+ with the extracellular aspect of the antiporter indicates a single saturable binding site. In contrast, the interaction of protons with cytoplasmic binding sites is highly cooperative: The antiporter switches from virtually "off" to completely "on" as the H^+ concentration increases 10-fold (the pH decreases from 7 to 6). This suggests the existence of a distinct modifier site that is only accessible to cytoplasmic protons.

If quiescent cells (in culture) are stimulated by various growth factors, a sustained decrease in cytoplasmic proton concentration occurs. The affinity of the antiporter for internal H^+ is apparently increased at this modifier site. A remarkable range of growth factors and peptide hormones produce this activation in different cell types. Although these antiporters are not driven by ATP, regulation of the affinity of the modifier site for protons by growth factor-activated phosphorylation sometimes occurs: cytoplasmic ATP depletion has been shown, in some cases, to reduce this affinity.

An amiloride-sensitive Na^+/H^+ antiporter has been cloned from human genomic DNA. Transfection of cDNA derived from this gene into antiporter-deficient cells restored the ability of these cells to display the amiloride-sensitive exchange and to regulate their internal pH. The coding portion of this cDNA specifies a single peptide of 894 residues (molecular weight, 99,354).

c. Proton-Dependent Antiporters The Golgi-related intracellular organelles contain H^+-dependent antiporters that are ultimately driven by type 3 H^+-ATPases. One of the best characterized is the amine transporter from chromaffin granules. In the presence of external ATP, chromaffin granules from adrenal medullary cells can concentrate biogenic amines such as norepinephrine and serotonin to produce transmembrane concentration differences of 4000:1. The uptake depends on the H^+ concentration difference generated by ATPase. However, even in absence of ATP, spontaneous amine efflux from the granules is low. This is attributed to a kinetic barrier induced by the maintenance of a low intravesicular pH. A rapid net efflux of amines from these granules can be induced by adding agents which deplete the proton concentration difference (e.g., nigericin or NH_3).

d. Cl^-/HCO_3^- Antiporter The process mediated by this protein is a tightly coupled neutral exchange of chloride for bicarbonate:

$$Cl^-_e + HCO^-_{3c} \leftrightarrow Cl^-_c + HCO^-_{3e}$$

Erythrocyte membranes permit rapid exchange of bicarbonate for chloride during passage through the lungs. Other tissues, (e.g., placental epithelia) have similar capabilities which result from the presence of a Cl^-/HCO_3^- antiporter (Fig. 5). In erythrocytes this is the major membrane protein, with about 10^6 copies per cell. It is frequently termed the band 3 protein because of its relative position in gel electrophoretic patterns. This protein mediates a 1:1 exchange of monovalent anions. The rates are highest with Cl^- and HCO_3^-, but Br^- HSO_4^- and $H_2PO_4^-$ are also exchanged. With Cl^- and HCO_3^-, the rate is about 10^5 per second at 37°C. The rate of net Cl^- conductance is estimated to be only 0.01% of the exchange rate.

The cDNA-derived sequence for the mouse band

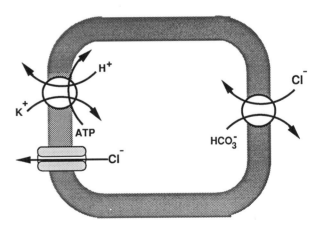

FIGURE 5 Hydrochloric acid is secreted into the stomach from specialized epithelial cells. The ATP-driven exchange of extracellular K^+ for intracellular H^+ can create a ratio of proton concentrations of 10^5. The protons originate from intracellular buffers such as the H_2CO_3/HCO_3^- couple. As the intracellular protons are removed, more carbonic acid dissociates and the accumulating bicarbonate probably exchanges for extracellular chloride as shown.

3 protein consists of 929 amino acids. A proposed membrane topology includes eight to 10 transmembrane segments, and a cytoplasmic N-terminal domain of more than 373 residues.

In addition to this electroneutral exchanger, a Na^+-dependent component of Cl^-/HCO_3^- exchange has been identified in some cells. This system has not been characterized biochemically. Several additional types of anion exchangers have been characterized physiologically in kidney proximal tubules. Among these are systems that will exchange chloride for various mono- and dicarboxylic acids and for urate.

2. Symporters

Plasma membranes contain an array of Na^+-linked symporters which are ultimately driven by the ATP-dependent Na^+,K^+ pump.

a. $Na^+,K^+,2\,Cl^-$ Symporters

The process mediated by this protein is a coupled neutral cotransport of Na^+, K^+, and Cl^-:

$$Na^+_e + K^+_e + 2\,Cl^-_e \leftrightarrow Na^+_c + K^+_c + 2\,Cl^-_c$$

These systems mediate the entry of salt into cells. In conjunction with the ATP-dependent Na^+,K^+-pump, the net effect is the accumulation of KCl. Most of the available information is derived from studies on intact cells. These symporters are characteristically inhibited by "loop diuretics" such as bumetanide and furosemide. Studies of the binding of radioactive bumetanide have led to an estimated turnover rate of 4000 per second in duck red blood cells. This type of symporter mechanism is activated by different stimuli in different cell types (see Section IV, B). Recognition of this has unified many previously disparate observations of these ion fluxes.

Although ATP is not utilized in this symport mechanism, ATP depletion of cells has been shown to reduce its activity. This is taken as evidence that the activity of the symporter may be increased when it is phosphorylated by ATP through a phosphokinase which, in turn, may be sensitive to regulatory factors. Examples of evidence for regulation through cytoplasmic second messenger systems include stimulation by phorbol esters, elevated cytoplasmic Ca^{2+}, cAMP, and cGMP. Exogenous factors that have been shown to activate in various cells include insulin, vasopressin, atrial natiuretic factor, and β-adrenergic agonists.

b. Na^+, Glucose Symporters

These proteins mediate the voltage-dependent cotransport of Na^+ and glucose:

$$2Na^+_e + glucose_e \leftrightarrow 2\,Na^+_c + glucose_c$$

They occur in cells that can concentrate glucose (e.g., kidney and intestinal epithelia). Another transport protein—the facilitated glucose carrier—occurs in cells that do not have this concentrative ability, (e.g., erythrocytes and neurons). The latter system displays selectivity toward sugars, but is not coupled to ion transport.

The intestinal brush border Na^+,glucose symporter is currently the only symporter from mammalian sources that has been cloned and for which structural information is available. The cDNA-derived amino acid sequence of the rabbit symporter consists of a single peptide of 662 residues (molecular weight, 73,080). It may contain as many as 11 transmembrane segments, of which five are amphipathic. Other functionally-related transport proteins that have been sequenced include the mammalian facilitated glucose carrier and several bacterial sugar transport proteins. No sequence similarities occur between the Na^+, glucose symporter and any of these.

c. Other Na^+-Dependent Symporters

Amino acid transport into mammalian cells is mediated by several systems that have been extensively characterized in whole-cell studies. These systems differ in their amino acid selectivities and cation dependencies. System A is a Na^+-driven symport for neutral α-amino acids. In intact cells or reconstituted vesicles this system can be experimentally defined as the rate of Na^+-dependent uptake of α-aminoisobutyrate. System A copurified and immunoprecipitated with a protein of approximately 120 kDa from Ehrlich ascites cell membranes. Similar symporters have been characterized in hepatocytes, renal and intestinal epithelia, and astrocytes.

Nucleosides are taken up by several different transport systems, including one or more Na^+ symporters.

The kidney proximal tubule brush border is the site of reabsorption of many different metabolites recovered from the glomerular filtrate. In addition to glucose, this is the site of Na^+-dependent uptake of amino acids, mono- and dicarboxylic acids, phosphate, and sulfate.

d. Na$^+$,Cl$^-$, GABA Symporter

$$2\,Na^+_e + GABA_e + Cl^-_e \leftrightarrow$$
$$2\,Na^+_c + GABA_c + Cl^-_c$$

The principal inhibitory neurotransmitter of mammalian brain is considered to be γ-aminobutyrate (GABA). Both neurons and glial cells appear to have the ability to concentrate GABA. The glial and neuronal uptake systems may differ, because the two cell types display different sensitivities to GABA uptake inhibitors. Most of the information about the operation of GABA uptake has been derived from experiments with plasma membrane vesicles derived from partially purified preparations of nerve endings (i.e., synaptosomes). The GABA symporter, purified to apparent homogeneity from a detergent extract of rat brain membranes, appears to be an 80-kDa glycoprotein. This protein, when reconstituted into liposomes, catalyzes the accumulation of GABA into these vesicles only when both Na$^+$ and Cl$^-$ are present in the external medium. The turnover number of this system has been estimated to be about 2.5 per second.

e. Other Cl$^-$-coupled Symporters

Several Na$^+$ symporters apparently also cotransport Cl$^-$. These include systems that mediate recovery of catecholamines and glycine after their release at nerve endings. Some of these may prove to belong to the more complex category discussed in the following section.

3. Complex Transporters

Transporters that combine antiport and symport functions constitute a growing category of recent discoveries. In several cases these are plasma membrane systems that are driven by a sodium concentration difference. In some cases a physiological advantage of coupling to an additional ion may be that a higher concentration difference of the driven species can be established than would otherwise be possible.

a. Na$^+$,HCO$_3^-$/H$^+$,Cl$^-$ Transporter

This system was first characterized in invertebrates, but has more recently been shown to occur in mammalian tissues, including human fibroblasts. Although it can operate in either direction, it may provide an important mechanism to remove protons from cells and, simultaneously, to accumulate bicarbonate.

b. Na$^+$,Glutamate/K$^+$ Transporter

Glutamate is the principal excitatory neurotransmitter released by neurons in mammalian brain. There is now evidence that the glial cells that surround neurons are important for the rapid removal of extracellular glutamate by uptake through a transporter that mediates the following events:

$$3\,Na^+_e + glutamate^-_e + K^+_c \leftrightarrow$$
$$3\,Na^+_c + glutamate_c + K^+_e$$

A glycoprotein that mediates this reaction when reconstituted into liposomes has been partially purified from rat brain.

c. Na$^+$,Serotonin, Cl$^-$/K$^+$ Transporter

This transporter occurs in platelets, which store and release serotonin, and probably also in some neurons in which serotonin functions as a neurotransmitter. Serotonin uptake is driven by an inward flux of Na$^+$ and Cl$^-$ which requires a counter flow of either K$^+$ or H$^+$. Equivalent serotonin transport in the reverse direction requires reversal of both the Cl$^-$ and the Na$^+$ concentration differences. A turnover number of 8 per second has been estimated.

IV. Physiological Integration of Ion Pump Functions

A. Na$^+$ Metabolism

Multicellular animals, for the most part, maintain extracellular environments (e.g., serum and extracellular fluids) that are high in Na$^+$ and low in K$^+$, whereas cytoplasm is maintained relatively low in Na$^+$ and high in K$^+$. ATP-dependent Na$^+$,K$^+$ pumps create both of these environments. Pumps located, for example, in the basolateral plasma membranes of kidney and intestinal epithelia maintain the extracellular conditions by energizing the secretion and absorption of Na$^+$ (Fig. 6). The density of sodium pumps in the more distal segments of kidney tubules is regulated by the level of secretion of the adrenal steroid hormone, aldosterone.

Na$^+$,K$^+$ pumps in the plasma membranes of most cells maintain low intracellular Na$^+$ and high intracellular K$^+$. These concentration differences comprise an important store of potential energy that is ultimately transformed to physiological uses by other membrane proteins, principally via channels and secondary transport systems (Fig. 6).

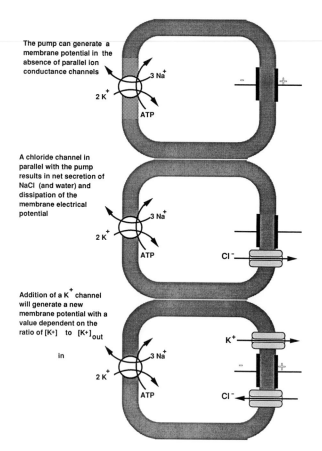

The pump can generate a membrane potential in the absence of parallel ion conductance channels

$3 Na^+$
$2 K^+$
ATP

A chloride channel in parallel with the pump results in net secretion of NaCl (and water) and dissipation of the membrane electrical potential

$3 Na^+$
$2 K^+$
ATP
Cl^-

Addition of a K^+ channel will generate a new membrane potential with a value dependent on the ratio of $[K+]_{in}$ to $[K+]_{out}$

$3 Na^+$
$2 K^+$
ATP
K^+
Cl^-

FIGURE 6 (A) In certain cells (e.g., neutrophils) the parallel ion conductances are so small that the membrane potential is produced mostly by the electrogenic effect of the Na^+,K^+ pump. (B) Because the Na^+,K^+ pump transports 50% more Na^+ than K^+, the functioning of anion channels in parallel with the pump can produce a net efflux of NaCl and water. (C) However, in most excitable cells (e.g., nerve and muscle) the functioning conductance channels of the resting state of the membrane are predominantly selective for K^+, and the resting cell membrane potential is determined by the ratio of intracellular K^+ to extracellular K^+.

B. K^+ Metabolism

K^+ is the predominant intracellular cation and is also acquired through the operation of the ATP-dependent Na^+,K^+ pump. Many cells also have a plasma membrane ($Na^+,K^+,2 Cl^-$) symporter. This system can function to change the total cytoplasmic electrolyte and so permit cells to regulate their volume in response to alterations in external osmolarity. In concert with the ATP-dependent Na^+,K^+ pump, the operation of this symporter results in the net uptake of KCl and water. Kidney and intestinal epithelia regulate the whole body balance of K^+ absorption and secretion, primarily by regulating the amount retained.

In brain tissues the extracellular accumulation of K^+ that results from nerve impulses can lead to membrane depolarization and spontaneous triggering of nerve impulses—a dysfunction that is often associated with convulsive activity. Certain types of astrocytes are thought to ameliorate this phenomenon through a "spatial buffering" process that permits the rapid diffusion of K^+ into their surrounding cytoplasm.

C. Ca^{2+} Metabolism

Ca^{2+} participates in a multitude of functions. Intracellularly, Ca^{2+} participates in second messenger systems that link membrane receptors to the initiation of cell responses. These responses range from muscle contraction to the regulation of secretion and, in neurons, possibly to some of the processes underlying memory. Extracellularly, Ca^{2+} is a factor in such processes as cell adhesion and blood clotting and as a structural element of bone. Reflecting this diversity is the wide range of Ca^{2+} transport systems already discussed. Figure 7 summarizes some of the ways in which these systems can interact. In general, cells in the resting state maintain internal free Ca^{2+} levels of less than 0.1 μM. Cells contain a number of different Ca^{2+}-binding proteins that act both as Ca^{2+} buffers and as regulators of different systems. For an example, the plasma membrane ATP-dependent Ca^{2+} pump is activated by the Ca^{2+}-binding protein calmodulin and is further activated by regulatory phosphorylation.

D. Anion Metabolism

Mammalian metabolism generates CO_2, which, in most cells, is rapidly equilibrated to H^+ and HCO_3^- by carbonic anhydrase. Although Cl^- is the major extracellular anion, a Cl^-/HCO_3^- antiporter does not function at a high level in most nonerythroid cells. It is found in the basolateral membranes of kidney proximal tubule epithelia and in the brush border membranes of intestinal epithelia. In red cells the function of this antiporter appears to be to facilitate CO_2 transport from tissues to lungs.

There are reports of both Na^+,Cl^- and K^+,Cl^- cotransport in various cells, but it is not clear to what extent these are distinct from the $Na^+,K^+,2 Cl^-$ cotransport (see Section III, B, 2, a).

There are several other anion transporters with differing specificities that function in the transport of metabolites. For example, kidney epithelial cells

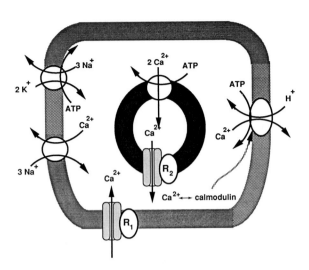

FIGURE 7 A generalized scheme of cellular Ca^{2+} metabolism. In erythrocytes the major mechanism for Ca^{2+} extrusion is a calmodulin-activated ATP-dependent pump. Skeletal muscle contraction and relaxation are functions of the release and reuptake of Ca^{2+} from the sarcoplasmic reticulum, a specialized form of endoplasmic reticulum, the major protein of which is an ATP-dependent Ca^{2+} pump. In heart muscle an important pathway for Ca^{2+} removal is the combination of Na^+/Ca^{2+} exchange driven by the Na^+,K^+ pump. The latter mechanism is evidently the basis for the therapeutic action of cardiac glycosides (digitalis) in increasing the heart rate: partial inhibition of the sodium pump increases cytoplasmic Na^+. This slows the Na^+/Ca^{2+} exchange mechanism that increases the steady state cytoplasmic Ca^{2+} which, in turn, increases the force of contraction of cardiac muscle. R_1 and R_2, generic receptors that regulate Ca^{2+} channels in plasma membranes and endplasmic reticulum.

contain a Na^+, lactate symporter at their brush border and they also contain a lactate/OH^- antiporter, which perhaps mediates lactate reabsorption. Another distinct Na^+ symporter in kidney proximal tubules is effective for di- and tricarboxylic acids such as succinate and citrate. Cells must be able to retain inorganic phosphate, and this has been demonstrated in some mammalian cells to be mediated by a Na^+ cotransport system.

E. Proton Metabolism

1. Regulation of Cell pH

Most cells contain carbonic anhydrase, which rapidly equilibrates metabolic CO_2 with water to form bicarbonate. Given that most cells have a plasma membrane potential that is -60 mV or more, a passive distribution of H^+ would theoretically produce an internal pH greater than 1 unit more acidic than the extracellular pH of 7.4. For mammalian muscle cells internal pH (at rest) is 7.0–

7.1. Thus active proton secretion is necessary to maintain the resting state. The Na^+/H^+ antiporter is a major pathway for extruding protons in most cells. However, several other mechanisms have also been shown to have important roles. Thus HCl is effectively secreted in exchange for $NaHCO_3$ by the $Na^+,HCO_3^-/H^+,Cl^-$ transporter. There is also evidence for an Na^+,HCO_3^- cotransporter. Given a sufficiently maintained Cl^- concentration difference, the Cl^-/HCO_3^- antiport can serve to remove alkali.

As noted in Section III, B, 1, b, Na^+/H^+ antiporters are activated by several different growth factors and mitogens. It has recently been shown that, at least in certain cells, the $Na^+,HCO_3^-/H^+,Cl^-$ transporters and the Cl^-/HCO_3^- antiporters are activated by the same factors. In a case in which the three systems coexist in the same cell, the pH-regulating activity of these transporters can be dissociated from the growth stimulating response and thus is not causally related. The growth factor stimulation of transporters is more likely an anticipatory preparation of the cell for increased metabolic activity.

2. Epithelial Proton Transport

In the proton transport mechanisms discussed in Section IV, E, 1, an effective proton concentration difference of 10-fold (equal to 1 pH unit or approximately 58 mV) is maintained by various secondary transporters. In gastric cells that secrete HCl at a pH near 1 (a proton concentration difference of nearly 10^6), the operative mechanism is the ATP-dependent H^+,K^+ pump. In other proton-transporting epithelia, the mechanism is not entirely clear. Thus, in the kidney, urinary acidification can occur in collecting ducts to a pH of 4.5, equivalent to a concentration difference of nearly 10^3. The turtle urinary bladder epithelium has been studied as a functional analog for mammalian urinary collecting tubules. These cells contain ATP-dependent proton pumps in endocytic vesicles. CO_2 stimulates acid secretion by these cells, an event that is accompanied by fusion of the pump-containing vesicles with their apical membranes. Considering the inhibitor sensitivity of this pump, it may belong to the class of vacuolar proton ATPases discussed in Section III, A, 6b.

F. Regulation of Cellular Volume

Mammalian cells are bounded by nearly inelastic plasma membranes of relatively fragile structure.

Cell membranes are usually crenulated to varying extents, only expanding to the elastic bounds of their membranes in pathological situations. The lower limit to cell volume is evidently set by the total content of osmotically active impermeant molecules and ions and their counter-ions. Changes in volume can occur as a result of changes in the content of osmotically active molecules. For example, intestinal epithelia increase in volume during Na^+-dependent amino acid uptake. Because the extracellular milieu is closely regulated, most cells do not ordinarily experience large changes in external osmolarity. However, this parameter is readily manipulated experimentally. Most cells respond to moderate external changes in osmotic pressure first by swelling or shrinking as expected for osmometers. This response is followed by a return toward their original volumes. These secondary responses are termed regulatory volume increase and regulatory volume decrease. These must be accompanied by a proportionate change in the amount of total osmolyte, which is predominantly KCl. Several mechanisms have been implicated for different cell types.

Regulatory volume decrease seems to occur in most cells by an activation of K^+ and Cl^- conductance channels. The response can be induced by volume increases as small as 5%. The activation mechanisms are not known although there is evidence that the release of Ca^{2+} from internal stores may be involved. Possible triggers include receptors coupled either to stretch-sensitive conductance channels or to cytoskeletal proteins that transmit changes in mechanical tension. Regulatory volume increase in some cells results from an activation of the Na^+/H^+ antiporter, leading to transient alkalinization. In the presence of external CO_2, which can freely cross the cell membrane, this can lead to an increase in total cytoplasmic bicarbonate with some ultimate readjustments by the Na^+,K^+ pump and the Cl^-/HCO_3^- antiporter. A more direct pathway for increasing the total electrolyte is by way of the $Na^+,K^+,2 Cl^-$ transporter, which couples to the Na^+,K^+ pump to produce a net increase in cytoplasmic KCl and water (Fig. 8). In some cells there is evidence for Na^+/Cl^- cotransport unlinked to K^+ and conversely K^+/Cl^- cotransport unlinked to Na^+. It is not certain whether these are distinct systems or different modes of operation of the same transporter.

G. Regulation of Total Body Fluid

Similar mechanisms operating in epithelial cells may regulate the total body electrolyte and fluid volume (Fig. 9). In mammalian kidneys an ultrafiltrate of blood is formed in the glomeruli. Thus, rather than selectively removing impurities from the blood, the kidney essentially removes all molecules smaller than about 5 kDa and selectively reabsorbs selected constituents, of which the major components are water, glucose, and salt. Once again the primary driving force is the ATP-dependent Na^+,K^+ pump. Water transport is always secondary to the transport of solutes. In the inner zones of the kidney, an arrangement of parallel loops of tubules and capillaries constitutes a countercurrent mechanism for removing water from the glomerular filtrate: NaCl is secreted from the tubules into the interstitial tissue; because of a relatively slow blood flow through this tissue, the extracellular salt concentration can accumulate to several times the normal 300 mosm. Cells in this part of the kidney synthesize impermeant intracellular molecules such as sorbitol, apparently to partially compensate for the high external osmolarity. However extracellular hyperosmolarity is maintained, and water flows rapidly through the tubular cells and perhaps also through paracellular channels down the osmotic gradient. More than 90% of the water and Na^+ is reabsorbed in the first two segments of the kidney tubule: the proximal and thin loop regions. The distal segments of each kidney tubule exercise the final control on the volume and composition of the urine. Their basolateral membranes contain a Na^+,K^+

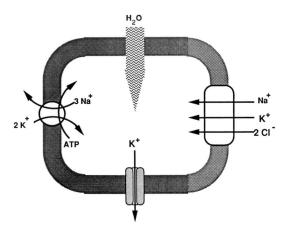

FIGURE 8 Exposure of cells to a hypertonic environment causes cell shrinkage. Subsequently, $Na^+,K^+, 2 Cl^-$ cotransporters participate with the Na^+,K^+ pump to produce a net influx of salt and water and a return toward the original volume. In many cells, (e.g., brain astrocytes, which are highly permeable to K^+) the net effect is to increase the total cellular KCl.

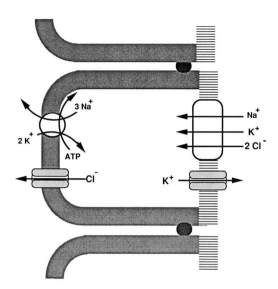

FIGURE 9 NaCl is efficiently absorbed by kidney and intestinal epithelial cells. These cells contain $Na^+, K^+, 2Cl^-$ cotransporters, among others, in their brush border membranes which face the lumen (i.e., kidney tubule or intestine). These cells are joined by specialized junctions slightly below the brush border membranes. The remainder of the cell surface, called the basolateral membrane, contains a different array of proteins, including the Na^+, K^+ pump. The coordinate action of these two transporters can have different effects, depending on the parallel conductive channels. With adequate parallel K^+ or Cl^- conductances (see Fig. 7) the net effect is to reabsorb NaCl and water from the lumen into the extracellular tissue spaces.

pump that is subject to regulation by the adrenal hormone aldosterone, and their apical membranes contain channels for which the water permeability is regulated by the posterior pituitary hormone vasopressin.

H. How are Transport Systems Integrated into the Functional Requirements of Cells, Tissues, and Organisms?

As information is rapidly accumulating about the genes that specify transport systems, some insights into regulation at the genetic level should be forthcoming. We know, for example, that genes for multiple forms of both Na^+, K^+ and Ca^{2+} pumps exist and are differentially expressed in different cell types. Some evidence is at hand to indicate that different forms are subject to different second messenger regulation. Many outstanding problems remain. For example, in kidneys only the Na^+, K^+ pumps of distal tubules and perhaps collecting ducts appear to be regulated by aldosterone. Nevertheless, rat kidney tubules apparently express only one form of the pump. Much of the focus of current research in transport biology is on problems of this nature.

Bibliography

Albers, R. W., Siegel, G. J., and Stahl, W. L. (1989). Membrane transport. *In* "Basic Neurochemistry," 4th ed. Raven, New York.

Hoffmann, E. K., and Simonsen, L. O. (1989). Membrane mechanisms in volume and pH regulation in vertebrate cells. *Physiol. Rev.* **69,** 315–382.

Semenza, G., and Kinne, R. (Eds.) (1985). Membrane Transport Driven by Ion Gradients. *Ann. N.Y. Acad. Sci.* **456.**

Skou, J. C., Norby, J. G., Maunsbach, A. B., and Esmann, M. (eds.) (1988). The Na^+, K^+-Pump. *Prog. Clin. Biol. Res.* **268.**

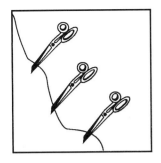

Isoenzymes

ADRIAN O. VLADUTIU, GEORGIRENE D. VLADUTIU, *State University of New York at Buffalo*

Glossary

Alleles One or more alternative forms of a gene at the same site of a chromosome, which determine alternative characteristics in inheritance

Antigenicity Property of some substances (mainly proteins) to elicit antibodies after introduction into a foreign (other than self) organism

Chromatography Separation of a mixture of substances based on differences in their relative affinities for two different media: the mobile phase (solvent) and the stationary phase (sorbent)

Electrophoresis Technique used to separate charged particles in solution by the differences in their rates of migration in an applied electric field

Epigenetic Modification of gene products by events or factors that occur after transcription and translation of the gene

Isotachophoresis Electrophoresis in which ionized particles move with equal velocity in an electrical field and are separated only by their charge

Locus Place on a chromosome occupied by a particular gene or its alleles

Michaelis constant Experimentally determined substrate concentration at which the enzymatic reaction proceeds at one-half its maximum velocity

Polymorphism Occurrence in the same population of two or more alleles at a locus, with at least one allele having a frequency >1%

Translation Formation of a peptide chain on the mRNA template from individual amino acids

ISOENZYMES (Gk. *izos*, equal; *zymos*, leaven), also known as isozymes, are enzymes that catalyze the same reaction (in a single species) but are different in structure, hence in some of their properties, because they are encoded by distinct genes. These genes can be at different loci or can represent different alleles at the same locus. This definition is in accordance with the recommendations of the International Union of Biochemistry, Commission on Biochemical Nomenclature, to restrict the term isoenzyme only to those forms of the same enzyme originating at the level of genes that encode the structure of the enzyme-protein in question.

Most enzymes have multiple molecular forms, and the differences between the various forms of the same enzyme can be genetically determined or due to epigenetic changes of the enzyme. Although the latter forms are not isoenzymes as just defined, from an operational point of view they have been often called isoenzymes. Newly discovered variants of enzymes cannot be considered isoenzymes until their genetic control is determined; however, this term is sometimes used in a general sense to signify different forms of an enzyme regardless of the mechanisms generating the enzyme diversity.

In 1948, Warburg suggested that enzymes that catalyzed the same reaction in different tissues might be organ-specific. The discovery of the heterogeneity of human lactate dehydrogenase (LD) in 1957 as well as the introduction of histochemical techniques for the detection of esterases led to the proposal of the term "isozyme" by Markert and Møller in 1959 to describe different proteins with similar enzymatic activity.

I. Derivation of Isoenzymes

More than one-third of human enzymes are coded for by multilocus genes. The genes that code for specific isoenzymes can be on the same chromosome (e.g., chromosome 1 for human salivary and pancreatic amylase and for the soluble and mitochondrial forms of guanylate kinase, and chromosome 4 for the four loci of alcohol dehydrogenase) or on different chromosomes (e.g., nicotinamide adenine dinucleotide (NAD)-dependent malate dehydrogenase, mitochondrial form on chromosome 7 and cytoplasmic form on chromosome 2; hexosaminidase alpha subunit on chromosome 15 and beta subunit on chromosome 5). Isoenzymes controlled by genes at multiple loci and the chromosome locations of these loci are listed in Table I.

TABLE I Human Isoenzymes with Known Chromosomal Locations

Isoenzyme	Chromosome location
Acid phosphatase 1, soluble	2p25 or 2p23
Acid phosphatase 2, lysosomal	11p11
Adenylate kinase 1	9q34
Adenylate kinase 2	1p34
Adenylate kinase 3	9p24–p13
Alcohol dehydrogenase 1, alpha polypeptide	4q21–q25
Alcohol dehydrogenase 2, beta polypeptide	4q21–q25
Alcohol dehydrogenase 3, gamma polypeptide	4q21–q23
Alcohol dehydrogenase 5, chi polypeptide	4q21–q25
Aldolase A	16q22–q24
Aldolase B	9q21.–3–q22.2
Aldolase C	17
Alkaline phosphatase, intestinal	2q34–q37
Alkaline phosphatase, liver, bone, kidney	1p36.1–p34
Alkaline phosphatase, placental	2q37
Amylase, pancreatic, alpha 2A	1p21
Amylase, pancreatic, alpha 2B	1p21
Amylase, salivary	1p21
Arylsulfatase A	22q13.31–qter
Arylsulfatase B	5p11–q13
Arylsulfatase C	X
Creatine kinase BB	14q32.3
Creatine kinase MM	19q13
Enolase 1	1pter–p36.13
Enolase 2	12p11–qter
Fucosidase, alpha 1, tissue	1p34
Fucosidase, alpha 2, plasma	6
Glucose-6-phosphate dehydrogenase	Xq28
Glucosidase, alpha AB	11q13–qter
Glucosidase, alpha C	15
Hexosaminidase A, alpha polypeptide	15q23–q24
Hexosaminidase B, beta polypeptide	5q13
Isocytric dehydrogenase, mitochondrial	15

(table continues)

TABLE I *(continued)*

Isoenzyme	Chromosome location
Isocytric dehydrogenase, soluble	2
Lactate dehydrogenase A (H)	11p15.1–p14
Lactate dehydrogenase B (M)	12p12.2–p12.1
Lactate dehydrogenase C(X)	11
Malate dehydrogenase 1, soluble	2p23
Malate dehydrogenase 2, mitochondrial	7p13–q22
Mannosidase A	15q11–qter
Mannosidase B	19
Monoamine oxidase A	X
Monoamine oxidase B	12
Peptidase A	18q23
Peptidase B	12q21
Peptidase C	1q25 or 1q42
Peptidase D	19q12–q13.2
Peptidase E	17
Peptidase S	4p11–q12
Phosphofructokinase L	21q22.3
Phosphofructokinase M	1 cen–q32
Phosphofructokinase P	10pter–p11.1
Phosphofructokinase X	12
Phosphoglucomutase	1;4;6
Phosphorylase, brain	20
Phosphorylase, brainlike	10
Phosphorylase, liver	14
Phosphorylase, muscle	11q12–q13
Protein kinase C, alpha polypeptide	17q22–q24
Protein kinase C, beta polypeptide	16p12–q11.1
Protein kinase C, gamma polypeptide	19q13.4
Pyruvate kinase L	1q21
Pyruvate kinase M	15q22–qter
Superoxide dismutase	21q22.1
Thymidine kinase	16

Isoenzymes can also originate from mutations of alternative forms of genes (alleles) at the same locus ("allelozymes") (e.g., allelic variants of placental alkaline phosphatase), or they can arise from the modification of the structure or expression of genes in somatic cells—such an event can result from malignant transformation.

Oligomeric isoenzymes are another type of isoenzyme made up of subunits that are associated in various combinations producing hybrid molecules when the subunits are different. These molecules are formed by combinations of units derived from different loci or from multiple alleles (e.g., creatine kinase [CK], LD, and hexosaminidase). Lactate dehydrogenase in serum and organs is a tetramer of two different polypeptide subunits, H (from heart) and M (from muscle), that combine randomly producing homopolymeric and heteropolymeric molecules. Thus, five different LD isoenzymes are formed (Fig. 1), from LD-1 (four H subunits) to the most cathodal LD-5 (four M subunits). Lactate dehydrogenase-2, LD-3, and LD-4 are hybrid molecules, with the subunit composition H_3M, H_2M_2, and HM_3, respectively. An additional LD isoenzyme, LD-C (LD-X), composed of four identical subunits (C), which are different from the other two subunits of LD, is characteristic for postpubertal testis. Creatine kinase exists as three different dimers composed of two subunits, B (from brain) and M (from muscle), namely CK-1 (CK-BB), CK-2 (CK-MB), and CK-3 (CK-MM). Hexosaminidase exists as two main forms: B, with a structure of four beta subunits, and A, with a structure of one alpha and two beta subunits. Because of their different genetic control, the isoenzymes differ in their amino acid structure; however, the amino acid sequence of the active site is usually highly conserved and is the same for different isoenzymes of a single enzyme.

Isoenzymes that are produced as a result of allelic variations are distributed in the population as dictated by the Mendelian laws of inheritance. However, in contrast to classical genetics, allelozymes are inherited codominantly, i.e., the products of both alleles are expressed. The hereditary patterns of these isoenzymes can thus be used to trace their genetic inheritance. On the other hand, isoenzymes that have arisen from genes at multiple loci have been distributed throughout the species population over the course of evolution, resulting in all individuals possessing essentially the same complement of isoenzymes. The existence of multiple gene loci for specific isoenzymes has been hypothesized to have arisen by two different evolutionary processes. In one process, catalytic activity may have arisen from two or more unrelated events leading to the establishment of two different genes, each with the capacity to code for a different enzyme-protein that catalyzes the same reaction. The two genes thus formed can undergo mutations and will then code for different enzyme-proteins. In the other process, a single ancestral gene coding for a single enzyme may be duplicated. The two resultant genes would then be modified favorably for the evolving organism, cell, or organ function. This latter process is responsible for the occurrence of isoenzymes controlled by genes at different loci. Although great change in the gene products over time could be predicted to occur, the similarities in the two gene products should be great enough to ascertain their common ancestry. Such ancestral interspecies homologies exist, for example, in the isoenzymes of LD. For some isoenzymes (e.g., alpha-amylase), the two loci occurring after gene duplication remain closely linked.

Genetically determined changes in isoenzymes between individuals can explain abnormalities in metabolism manifested as hereditary metabolic diseases and can account for individual sensitivity to various drugs. Genetic deficiency of certain isoenzymes can also occur (e.g., deficiency of either the H or the M subunits of LD).

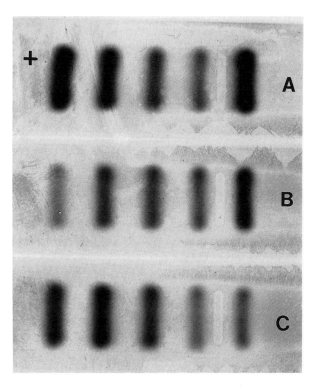

FIGURE 1 Lactate dehydrogenase isoenzyme pattern in serum as seen after electrophoresis in agarose gel (barbital buffer pH 8.6, nitroblue diformazan dye). The most anodal band (left) represents LD-1. Lane A, patient with myocardial infarction (increased LD-1); Lane B, patient with liver damage (increased LD-5); Lane C, normal individual (LD-2, probably originating from dead red cells, has the highest activity).

II. Structure, Function, and Developmental Aspects

Isoenzymes by definition must differ in primary structure to some degree, however small. Full elucidation of these differences requires the application of physical and chemical methods of protein analysis to purified isoenzyme preparations. The complete amino acid sequencing of isoenzymes has been achieved for several isoenzymes, such as human carbonic anhydrases I and II (differing in about 90 of 260 amino acid residues). Some information about the primary structure of an isoenzyme can be gained by identifying the carboxy- or amino-terminal residues and the short amino acid sequences adjacent to these residues. One of the most sensitive and specific methods for analyzing subtle differences in primary structure among isoenzymes is the "fingerprinting" technique, which compares two-dimensional maps of the peptides in a partial hydrolysis of the enzyme-protein. This method is especially useful for identifying single amino acid substitutions in products of allelic genes. The elucidation of the secondary and tertiary structures of isoenzymes can be achieved by X-ray crystallography. Differences in special arrangements also may exist and can provide important information about the structure and function of isoenzymes.

Because isoenzymes are coded for by different genes, they are expected to have different properties, such as substrate concentration optimum, electrophoretic mobility, resistance to inactivation, sensitivity to inhibitors, Michaelis constant for substrate(s), relative rate of activity with substrate analogs, and sometimes a different extent of substrate specificity (e.g., LD-C [LD-X] isoenzyme found in the primary spermatocyte acts on alpha-hydroxy butyrate and alpha-hydroxy valerate). Isoenzymes of LD and carbonic anhydrase, for example, can have different functions because of different properties. The difference in charge distribution over the surface of the molecule affects the location of the molecule within the cell.

Some isoenzymes show tissue or cell specificity and can be predominantly represented in cells at each stage of their differentiation, from embryo to adult. For example, changes in LD and CK isoenzyme distribution occur during development. Lactate dehydrogenase-C (LD-X) appears in testis and sperm only after puberty. The relative proportions of cytoplasmic and mitochondrial forms of certain isoenzymes change during differentiation.

Alkaline phosphatase exists as several isoenzymes in various organs: the liver–bone–kidney-unspecified form, the intestinal form, and the placental form. Aldolase is present in three isoenzyme forms: A, predominant in muscle; B, in liver; and C, in nervous tissue. The presence of different isoenzymes in certain tissues, cells (e.g., enolase, LD,

phosphofructokinase, CK, hexokinase, pyruvate kinase, and amylase) or subcellular structures such as lysosomes and mitochondria (e.g., the NAD-dependent malate dehydrogenase, isocitrate dehydrogenase and adenylate kinase isoenzyme systems, with one isoenzyme in the mitochondria and another in the cytosol), implies a specialized role of each isoenzyme in cell metabolism. Indeed, isoenzymes allow a fine adaptation of metabolic patterns to ontogenic development or to changes in the environment. One physiological advantage of the isoenzymes is to create a compartmentalization of the same catalytic reaction to occur as part of different metabolic pathways. Compartmentalization allows precision in the maintenance of metabolic patterns unachievable by a single enzyme. The early embryonal liver has three aldolase isoenzymes and hybrid tetramers (i.e., types A, B, and C), whereas at birth only one type of aldolase (type B) predominates. Isoenzymes of LD, for example, are represented in different percentages in various organs (Fig. 2), and this differential distribution has functional significance. A correlation exists between the properties of isoenzymes predominant in certain tissues and the metabolic patterns in these tissues. Creatine kinase-1 (CK-BB) is found mainly in the brain and is the major CK isoenzyme in prostate, thyroid, bladder, and stomach, whereas CK-3 (CK-MM) is the

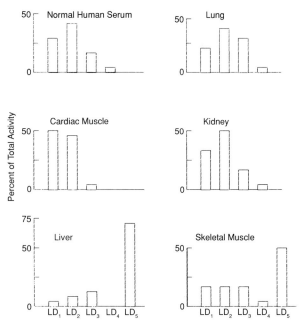

FIGURE 2 Lactate dehydrogenase isoenzyme content in different human organs.

predominant CK isoenzyme in the heart and striated muscle. There are also age-related changes of isoenzymes in certain cells. The proportions of the isoenzymes in a tissue or organ depend on their half-lives, i.e., rates of synthesis and degradation.

Many isoenzymes are named according to their tissue or subcellular structure of origin (e.g., placental and liver alkaline phosphatase) or, for those isoenzymes that can be separated by electrophoresis, are numbered according to their electrophoretic mobility. Because most enzymes are negatively charged, the most anodic isoenzyme is numbered as the first (e.g., CK-1 [CK-BB] and LD-1 migrate most anodally in electrophoresis).

III. Methods of Analysis

Separation of isoenzymes is performed by various techniques; some of the most commonly used are zone electrophoresis (e.g., in agarose, polyacrylamide, or starch gels), elution-convection electrophoresis, isoelectric focusing, and discontinuous electrofocusing. Isotachophoresis separates isoenzymes on the basis of their differences in net mobility. These methods as well as various types of ion-exchange chromatography (e.g., diethylaminoethyl cellulose or hydroxylapatite) take advantage mainly of differences in the net electric charge of isoenzyme molecules at a given pH and, to some extent, of differences in molecular size and shape. However, many amino acid substitutions do not alter the net charge of the enzyme; hence, not all isoenzymes can be separated by electrophoresis. Therefore, other methods such as high-performance liquid chromatography (HPLC) or affinity chromatography (based on differences in ligand binding) must be used to separate certain isoenzymes. Once separated, the isoenzymes can be identified on the basis of their properties. For colorimetric detection in serum or tissue extracts, histochemical techniques are used to show enzymatic activity by producing a colored precipitate (e.g., reduction of a tetrazolium salt such as triphenyltetrazolium) or a fluorescent substance when fluorogenic substrates (e.g., alcohols) are used or when the product of the enzymatic reaction is fluorescent (e.g., reduced NAD). The intensity of the color or fluorescence can be measured, and a semiquantitative determination of the relative proportions of the different isoenzymes obtained. This technique, known as the zymogram technique, or enzymoelectrophoresis, is commonly

performed in clinical laboratories for diagnostic purposes, with the use of scanning densitometers and fluorimeters. Other methods make use of inhibition or denaturation of enzymes by heat, concentrated urea, or other organic compounds; reactivity with different substrates, coenzyme analogues, or measurements of pH optima; Michaelis constants; etc. [See HIGH PERFORMANCE LIQUID CHROMATOGRAPHY.]

Immunological techniques are also used to separate and to quantitate isoenzymes because most of the isoenzymes differ with respect to their antigenicity. Monoclonal antibodies raised against pure isoenzymes are particularly useful to study isoenzyme subunit structure and can detect subtle differences between various forms of enzymes (e.g., the isoenzymes of alkaline phosphatase, the hybrid isoenzymes of CK, or the two subunits of LD). Immunoinhibition techniques (i.e., measurement of the residual activity after treatment with specific antibodies) can be used for isoenzyme quantitation (e.g., aldolases, CK, hexokinases, LD, and placental alkaline phosphatase). Immunological precipitation followed by measurement of the enzyme catalytic activity or measurement of the amount of radioactive label bound to the enzyme allows quantitation of isoenzymes and expression of their amount in units of activity or in mass units. Some individuals, especially those with autoimmune manifestations, have autoantibodies against certain isoenzymes (e.g., alkaline phosphatase, CK, and LD), and these antibodies can be used to reveal differences in the structure of isoenzymes. [See MONOCLONAL ANTIBODY TECHNOLOGY.]

IV. Practical Applications of Isoenzyme Analysis

In addition to its theoretical importance for understanding various processes (e.g., the transformation of one gene into another), the regulation of gene expression, and the significance of metabolic pathways in different tissues, the analysis of isoenzymes has a wide array of practical applications. These applications range from their use as markers in human–mouse cell hybrids, made for human chromosome mapping, to their use as diagnostic indicators in both hereditary and nonhereditary diseases. Because isoenzymes are valuable markers of gene function, they can be used to study population genetics (for instance, to identify new phenotypes and genetic polymorphisms). With isoenzyme analysis, the measurement of allele frequencies in a population facilitates the study of evolutionary selective pressure. [See POPULATION GENETICS.]

The organ or tissue specificity of isoenzymes allows them to be used to pinpoint organ damage caused by various diseases. For example, the CK-2 (CK-MB) isoenzyme, which is found primarily in the myocardium, is increased in the serum of patients with myocardial damage. Indeed, measurement of this enzyme (and of its isoforms) in serum is one of the most valuable tests for the diagnosis of myocardial infarction, which occurs after the sudden obstruction of an artery that supplies blood to the heart. Measurement of LD isoenzymes is also useful for diagnosis of myocardial infarction and other diseases and conditions (e.g., red cell, kidney, muscle, or liver damage; some leukemias; etc.). The diagnostic specificity is further increased by simultaneously measuring several different isoenzymes in serum and other body fluids.

Isoenzyme analysis has also come to play an increasingly important role in forensic science. It is possible, for example, to use isoenzymes to distinguish species and tissues. In addition, blood sources (menstrual, childbirth, or vascular) can be distinguished based on differences in LD isoenzyme patterns. The identification of semen from sexual assault victims is facilitated by the use of electrophoretic techniques in the assessment of several isoenzyme systems, including acid phosphatase and LD. Isoenzyme markers are also valuable tools in paternity testing based on allelic variations as an index for testing.

Isoenzymes can be used for studying cell differentiation and alterations of this process. Because new patterns of isoenzymes commonly occur when fully differentiated cells undergo malignant transformation, isoenzymes are valuable for the diagnosis of malignancy as well as for understanding some mechanisms of malignant transformation. Enzyme complements of tumors tend to be similar so that in the enzymatic sense tumors resemble each other more than they do normal tissues, or more than normal tissues resemble each other. Some ubiquitous isoenzymes (e.g., aldolase, LD, and beta-hexosaminidase) are altered in tumors. A shift in the LD isoenzyme pattern toward the cathode with a preponderance of the muscle type (LD-5) is observed in patients with cancer. This shift is believed to be more toward a fetal pattern, as fetal LD contains more M subunits. Aldolase A, normally predominant in muscle, is elevated in cancer tissue, especially in carcinoma of the gut, pancreas, and liver. This increase is probably due to the resurgence of fetal aldolases and the expression of the enzyme in a tissue different from that in which it is normally expressed. Creatine kinase-1 (CK-BB) isoenzyme is increased in malignant tumors such as anaplastic small-cell carcinomas and adenocarcinomas of the prostate, whereas mitochondrial CK ($CK-M_1$ and $CK-M_2$) are increased in the serum of individuals with metastases to the liver (especially of melanoma, a tumor of the skin). In cancer cells, a recrudescence of the embryonic pattern of isoenzymes characterizes the cells during their early stages of differentiation. Deletion of a repressor protein allows the reappearance of fetal isoenzymes in tumor cells, a phenomenon common to malignancies of many tissues. It is conceivable that a common carcinofetal or carcinoplacental gene (re)expression, similar to that occurring for carcinoembryonic proteins, occurs for the isoenzymes as well. Modifications of the genes by mutations and alterations of the gene product by posttranslational events can also occur. Examples of carcinoembryonic "isoenzymes," which occur only in cancer tissue but not in normal adult organs (excluding pregnancy), include placental protransglutaminase, "isoenzyme" III of branched-chain amino acid transaminase, "isoenzyme" V of 5'-nucleotide phosphodiesterase, "isoenzyme" P of diaphorase, cystinaminopeptidase, 17-beta-hydroxysteroid dehydrogenase, and oncofetal "isoenzyme" of gamma-glutamyltransferase (Table II). [See MALIGNANT MELANOMA.]

A significant number of relatively rare inborn errors of metabolism result from a mutation leading to the deficiency or modification of a specific isoenzyme. The functional consequences of these mutations vary depending on the nature of the alteration. When allelic variation of isoenzymes results in disease, usually the catalytic activity of the mutant enzyme is very low or absent. The recognition of a specific isoenzyme deficiency not only provides a firm biochemical basis for a clinical diagnosis but also provides the means to distinguish variants of some metabolic disorders. Ultimately, the biochemical classification of subtypes leads to an understanding of the molecular pathology of individual forms of a disorder. The most common enzyme deficiency in man that leads to disease is that of glucose-6-phosphate dehydrogenase (G-6-PD). This enzyme has a role in glucose metabolism, specifi-

TABLE II Isoenzymes Found in Serum and Organs of Individuals with Malignant Tumors

Isoenzyme	Organ normally found	Organ of malignancy
Aldolase isoenzymes A and C	liver, muscle, intestine	liver, colon, stomach, pancreas, rectum
Alkaline phosphatase		
Alpha-L		liver metastases
Placental	placenta	testis
Regan isoenzyme	placenta	pancreas, uterus, ovary, breast
Creatine kinase isoenzyme I (BB)	brain, lung, thyroid	kidney, testis, ovary, stomach, hematopoietic system
Galactosyl transferase isoenzyme 2		colon, rectum, breast, lung, pancreas
Hexosaminidase isoenzyme B	fetal liver	colon, rectum
Glutamyltransferase variants I, II, III	fetal liver	liver
Phosphoglycerate mutase, M-type	muscle	brain
Branched-chain amino acid transferase type III	placenta	liver

cally in the hexose-monophosphate shunt pathway in red cells. Allelozymes of G-6-PD have been detected based on altered catalytic activity, altered Michaelis constants for and reactivity with substrates, altered electrical charge, and stability. Most allelozymes have likely arisen from point mutations in the enzyme's structural gene. Varying degrees of severity of hemolytic anemia (lysis of red cells occurring especially after taking certain drugs) may develop in individuals with G-6-PD deficiency, depending on the nature of the aberration. [*See* METABOLIC REGULATION.]

A number of disorders arise from alterations of isoenzymes coded for by multiple gene loci due to mutations at these loci. The metabolic effects of these alterations consequently are manifested only in the tissues or organs in which the affected isoenzyme is present. In some muscle disorders, a tissue-specific isoenzyme is deficient in either glycogen or fatty acid metabolism leading to symptoms of skeletal muscle injury. Other tissues in patients may be involved if the affected isoenzyme is normally

found in significant amounts in those tissues as well. Examples of these disorders include glycogenosis (deposition of glycogen in certain tissues) type V (muscle phosphorylase deficiency), glycogenosis type VII (muscle phosphofructokinase deficiency), and glycogenosis type VI (liver phosphofructokinase deficiency).

More than 30 disorders of lysosomal enzymes alone have been described; most of these follow a recessive mode of inheritance with the exception of two X-linked disorders (Hunter's syndrome and Fabry's disease). Some lysosomal isoenzyme deficiencies can lead to progressive neurodegenerative disease usually (but not always) manifested in childhood. The most known example is Tay-Sachs disease (deficiency of alpha subunit of beta-hexosaminidase A). Examples of other lysosomal storage disorders in which a single genetically distinct isoenzyme is affected include mannosidosis (mannosidase A deficiency), metachromatic leukodystrophy (arylsulfatase A deficiency), and Maroteaux-Lamy disease (arylsulfatase B deficiency).

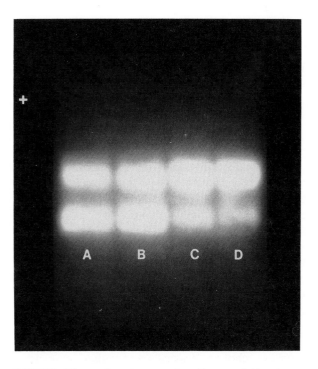

FIGURE 3 Electrophoretic separation of hexosaminidase isoenzymes in peripheral blood leukocytes performed in cellulose acetate gel (sodium citrate buffer pH 5.5, methylumbelliferone fluorogenic indicator). The most anodal band (top) represents hexosaminidase A, and the cathodal band hexosaminidase B. Lanes A and B, a Tay-Sachs disease carrier with reduced hexosaminidase A activity; lanes C and D, a noncarrier.

Allelic mutations of lysosomal enzymes can lead to quite different clinical phenotypes such as in the case of beta-glucocerebrosidase deficiency in Gaucher disease, for which three different clinical phenotypes exist depending on the location of the intra-allelic mutation.

Although no effective treatment exists for most inborn errors of metabolism, a positive diagnosis can provide information that facilitates assessing the prognosis and developing the medical management plan. Furthermore, a predictive value for the risk to future offspring can be obtained. Carrier detection is possible in many disorders in which the affected isoenzyme is normally detectable in peripheral blood leukocytes. Tay-Sachs disease carrier screening is possible by evaluating the relative concentration of hexosaminidase A as a percentage of total hexosaminidase A + B. Figure 3 depicts the carrier test when performed by the electrophoretic method. Prenatal diagnosis is also possible for many metabolic disorders, based on the absence of the affected isoenzyme in cultured amniotic fluid cells obtained by amniocentesis (aspiration of amniotic fluid).

V. Nonisoenzymic Molecular Forms of Enzymes

Multiple molecular forms of enzymes can also arise due to posttranslational modification of the enzyme-protein. Although these forms are not isoenzymes (i.e., they are not coded for by separate genes), historically and from an operational point of view they have often been referred to as isoenzymes because they have many characteristics similar to isoenzymes (i.e., they can exist in different cell organelles and in different metabolic compartments within a single cell [e.g., the cytoplasmic and mitochondrial forms of aspartate aminotransferase]). Posttranslational mechanisms of producing enzyme variants include the formation of aggregates of a single unit (homopolymers), or of an enzyme with nonenzymatic proteins (e.g, glutamate dehydrogenases, cholinesterases), and other epigenetic modifications of the initial protein such as combination of the protein with nonprotein molecules (e.g., sialic acid residues for phosphatases and fucosidases, phosphate groups for proteinases). Different forms of enzymes arise when they bind to immunoglobulins; the larger forms thus produced ("macro" enzymes) have been found in the serum of some individuals for CK and LD isoenzymes as well as for enzymes such as amylase and aspartate aminotransferase. Alteration of carbohydrate side chain, acylation, deamidation, sulfhydryl oxidation, and partial cleavage of the polypeptide chain are other mechanisms of producing nonisoenzymic molecular forms of enzymes. The removal of some residues (e.g., amide groups for amylase and carbonic anhydrase) can explain differences in function because, for example, deamidation and phosphorylation may influence secretion and uptake of enzymes. Partial proteolysis of the initial enzyme polypeptide can produce so-called isoforms of enzymes when only one amino acid is removed (e.g., isoforms of CK-2 [CK-MB], CK-MB$_1$, and CK-MB$_2$, which differ by a terminal lysine and can be separated electrophoretically). It is also conceivable that conformational changes of the polypeptide chains can lead to "conformational enzymes," or "conformers," with the same molecular weight and amino acid sequences but different properties, such as electrophoretic mobility. The different forms of enzymes that are generated through posttranslational modifications of the protein-enzymes in general do not differ in their catalytic characteristics but have biological significance and practical applications similar to those of isoenzymes. Once their genetic control has been elucidated, some of the different molecular forms of enzymes may prove to be, in fact, isoenzymes.

Bibliography

Blaten, V., and Van Stirteghem, A., eds. (1986). Plasma isoenzymes: The current status. *In* "Advances in Clinical Enzymology," Vol. 3. S. Karger, Basel.

IUPAC-IUB Commission on Biochemical Nomenclature. (1977). Nomenclature of multiple forms of enzymes. *J. Biol. Chem.* **252,** 5939.

Lehman, F. G. (1979). Enzyme and isoenzyme diagnosis of cancer. *In* "Advances in Clinical Enzymology" (E. Schmidt, R. W. Schmidt, I. Trautschold, and R. Friedel, eds.), pp. 171–195. S. Karger, Basel.

Moss, D. W. (1982). "Isoenzymes." Chapman & Hall, London.

Scandalios, J. G., Whitt, G. S., and Rattazzi, M. C., eds. (1977–1987). "Isozymes." Current Topics in Biological and Medical Research, Vols. 1–16. Alan R. Liss, New York.

Taylor, C. B. (1980). "Isoenzymes." Chapman & Hall, London.

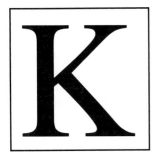

Keratinocyte Transformation, Human

MICHAEL REISS, *Yale University School of Medicine*

I. Experimental Systems for the Study of
   Human Keratinocytes
II. Malignant Cell Transformation
III. Malignant Transformation of Human
   Keratinocytes

Glossary

Carcinogenesis Process of development of malignant tumors in epithelia; causes include physical agents, such as ultraviolet light, chemical carcinogens, and viruses

Carcinoma Malignant tumor originating in any epithelial tissue

Epidermal cell Type of epithelial cell that is the predominant constituent of the epidermis; the epidermis is the keratinizing epithelium that forms the outer layer of skin

Fibroblast Type of cell that is the predominant constituent of connective tissue

Keratinocyte Type of cell that is the constituent of any keratinizing, stratified squamous epithelium such as, for example, the epidermis, esophageal epithelium, or cervical epithelium

Malignant transformation Process whereby a normal cell (i.e., one that is subject to all the normal spatial constraints of growth and fully sensitive to humoral factors that control cell growth and differentiation) converts to one that has sufficiently escaped normal controls to be able to proliferate more rapidly and differentiate at a lower rate than its neighboring cells, invade adjacent structures, and metastasize to other parts of the organism

Proto-oncogenes Normal cellular genes that play a role in cell growth or differentiation; these genes can undergo structural or functional changes, which lead to cell transformation, thus becoming mutant genes, or oncogenes

THE VAST MAJORITY OF HUMAN CANCERS originate in epithelial tissues. Therefore, an understanding of the process of malignant transformation of epithelial cells is of paramount importance for cancer control. In general, malignant transformation refers to the multistep process whereby a normal cell acquires the properties of a neoplastic cell. These steps include 1) loss of the normal controls of proliferation and differentiation; 2) invasion of surrounding tissues; and 3) metastasis. In a more restricted sense, the term transformation is often used to denote those properties of cells in culture that correlate with their ability to form tumors in animals.

The keratinocyte is the cell type that constitutes stratified squamous epithelium. This is the type of nonsecreting, multilayered epithelium that lines the skin, oropharynx, esophagus, and vagina (including the vaginal portion of the cervix uteri). Its primary function is to protect underlying tissues. Occasionally, other types of epithelium undergo conversion to a stratified squamous variety under the influence of toxic agents; this process is referred to as squamous metaplasia. Malignant tumors that are derived from keratinocytes are called squamous cell carcinomas (SqCCs) and can originate in the skin, oropharynx, esophagus, and uterine cervix. Such tumors can also arise in metaplastic epithelium. Therefore, SqCCs can occasionally be found in the bronchi, gallbladder, renal pelvis, bladder, or uterus.

I. Experimental Systems for the Study of Human Keratinocytes

A. General

Because carcinogenesis in humans cannot be studied *in vivo*, *in vitro* systems for the cultivation of epithelial cells had to be developed. Methods have been established for the cultivation of keratinocytes (from skin, oral mucosa, and uterine cervix), bronchial epithelial cells, urothelial cells, and mucosal cells from the gastrointestinal tract. Epidermal keratinocytes have been used most extensively because they are readily available. Numerous techniques for the cultivation of primary human keratinocytes have been developed and have undergone steady improvements during the past 15 years. As a result, our knowledge of the biology of this type of cell, as well as the changes it undergoes in pathological conditions, has increased considerably. A variety of factors have been identified that control the growth as well as the differentiation of keratinocytes. Furthermore, the pathway of differentiation has been characterized in detail as a consequence of the development of numerous antisera and monoclonal antibodies to specific components of the cell membrane and cytoskeleton. In addition, the genetics of the keratin gene family and their transcriptional regulation are being studied in depth. Changes in each of these aspects of keratinocyte biology have been associated with transformation of keratinocytes.

B. Keratinocyte Culture

Two fundamentally distinct methods have been developed to establish primary cultures of human primary keratinocytes.

1. Explant Cultures

In this method, small fragments of tissue comprising both epidermis and underlying dermis are plated onto a culture dish, dermis side down. Keratinocytes grow out from the edges of these fragments and eventually cover the culture surface. This technique is simple and reproducible; however, cells can only be subcultured a few times. Moreover, because of the presence of dermal cells, one does not obtain a pure population of keratinocytes. For both of these reasons, this culture method is inadequate for most biochemical and molecular biological experiments.

2. Disaggregated Cultures

This technique involves separation of the dermis from the epidermis by trypsin-digestion, followed by disaggregation of the epidermal layer into single cells. The culture is initiated by plating the keratinocytes onto a special substrate, such as growth-ar-

rested fibroblasts, which is necessary for the cells to attach. Although the method is technically more difficult and less reproducible than the explant technique, it provides a pure population of epidermal cells that can be expanded by serial subculturing; this allows investigators to perform biological and biochemical experiments. Optimal growth of human keratinocytes in culture requires a concentration of <1.0 mM Ca^{2+} and the presence of at least two of the following three components: epidermal growth factor (EGF) (or its functional homologue, transforming growth factor alpha [TGFα]), insulin (or its homologue, insulin-like growth factor-1 [IGF-1]), and bovine pituitary extract. Normally, the monolayer of cells that is attached to the culture dish is covered by the medium. Under these conditions, keratinocytes undergo only partial keratinization. In order to obtain cultures of epidermal cells that mimic full-thickness epidermis *in vivo* more closely, two technical innovations have been introduced in recent years. First, cells are provided with a substitute for the normal basement membrane, such as a collagen gel in which fibroblasts are mixed or a de-epidermized pig dermis. Secondly, cultures are maintained at the interface between the medium and air, so that the cells at the bottom of the culture are bathed in medium but the upper layers are exposed to air. As a result, cultures stratify and form a keratinizing epidermal equivalent. These organotypic epidermal equivalents provide a way to detect changes in tissue architecture associated with transformation (see Fig. 1).

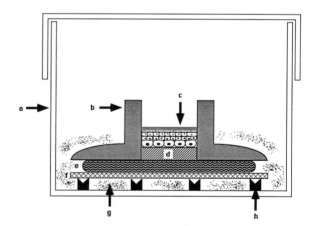

FIGURE 1 Example of organotypic culture system for human keratinocytes (see Fusenig, 1986). a, tissue culture dish; b, silicon chamber; c, organotypical cell culture after recombination with dermis; d, collagen gel; e, de-epidermized dermis; f, membrane filter; g, culture medium; h, projections in Petri dish.

Keratinocyte cultures contain three populations of cells that can be distinguished on the basis of their proliferative potential: 1) Holoclones—these form progressively growing colonies and probably represent true tissue stem cells; 2) Paraclones—these undergo a maximum of 15 cell doublings, after which all of the colonies cease to grow; 3) Meroclones—these have a growth potential that is intermediate between holoclones and paraclones. Between 5 and 100% of clones form progressively growing colonies.

C. Epidermal Cell Grafts

Sheets of cultured human epidermal cells can be removed from the culture dish and transplanted onto immunodeficient animals or man. These grafts are able to reconstitute normal epidermis *in vivo*. This technique has been exploited primarily for the treatment of massive burns; however, the same approach can be used to study the effects of carcinogenic events on the patterns of growth and differentiation of keratinocytes *in vivo*. Primary cultures are established *in vitro*. The cells are then exposed to the transforming agent and transplanted onto mice that are immunodeficient, so that they do not reject the grafts. In this way, tumor formation can be assessed *in vivo*, in a setting that closely resembles the natural conditions of skin (see Fig. 2).

II. Malignant Cell Transformation

A. General

Transformation of cells requires multiple steps. Each step consists of a genetic change that confers

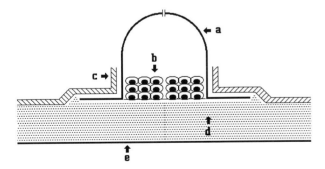

FIGURE 2 Transplantation chamber for grafting of human keratinocytes onto mice (see Fusenig, 1986). a, silicon transplantation chamber; b, keratinocytes engrafted on back of mouse; c, back skin of the recipient mouse; d, graft bed of granulation tissue; e, muscle fascia.

a growth advantage. The agents that cause these mutations are called carcinogens. The proliferative advantage of the mutated cells becomes apparent under selection pressure by the appropriate environmental factor; these are called tumor promoters. Successive steps do not have to occur in a specific order; rather, the accumulation of a sufficient number of genetic changes results in the phenotype that is recognizable as malignant. The number of genetic events required for the transformation of normal epithelial cells is estimated as variable but is probably more than two.

B. Transformation of Rodent Fibroblasts

In cultures of rodent fibroblasts, many changes have been observed that correlate to a greater or lesser extent with the acquisition of the neoplastic phenotype as manifested by the ability of "transformed cells" to form tumors when injected into animals. The major *in vitro* characteristics of transformation of rodent fibroblasts can be summarized as follows:

1. In contrast to normal cells that have a finite life span in culture, transformed rodent cells usually can be passaged indefinitely *in vitro*. This property is called immortalization.
2. Transformed cells generally require lower concentrations of serum or specific growth factors for optimal proliferation in culture. Sometimes, malignant cells are entirely independent of specific growth factors because they produce the factor themselves (autostimulation).
3. Normal fibroblasts grow in culture until they cover the entire surface of the dish and then undergo growth arrest. In contrast, transformed fibroblasts continue to proliferate under these conditions and form small mounds of cells, piled up on top of each other, that are macroscopically recognizable as foci. This property of transformed cells is referred to as loss of density-dependent growth arrest.
4. Normal mesenchymal cells need to attach to a surface prior to initiating cell proliferation. When they undergo transformation, cells are often able to proliferate even in the absence of a suitable substrate. In assays for such anchorage-independent growth, cells are artificially kept in suspension by using a semisolid growth medium (such as soft agar). Spherical colonies can be detected macroscopically after several weeks in culture.
5. Transformed rodent cells often acquire chromosomal abnormalities such as deletion, duplication, and translocation of chromosomes.

C. Properties of Human Epithelial Tumor (Carcinoma) Cells

In general, an *in vitro* parameter of transformation can be accepted as valid only if it is present in human tumors or tumor cell lines. The properties of human carcinoma cells and transformed rodent fibroblasts clearly do not always overlap.

1. Immortality

Although transformation of rodent fibroblasts usually results in an indefinite life span in culture, this is an extremely rare phenomenon among transformed human cells. In addition, only a small proportion of explants from human tumors give rise to immortal cell lines in culture. Whether this represents a deficiency in our cultivation techniques or signifies that immortality is not always a property of human tumor stem cells remains to be determined.

2. Reduced Growth Factor Requirements

As transformed rodent fibroblasts, many cell lines derived from human carcinomas have been shown to possess less stringent requirements for growth factors than primary cells. In fact, tumor cells themselves often constitutively produce one or more growth factors, which stimulate their own growth (autocrine growth stimulation).

3. Anchorage-Independence

Anchorage-independent growth is a consistent marker of transformation in rodent cells. However, not all cell lines derived from human tumors are anchorage-independent, and some lines are tumorigenic in animals but still require a substratum for growth *in vitro*. For example, most human SqCC cell lines are not capable of sustained growth in semisolid medium, although they were derived from invasive cancers.

4. Chromosomal Changes

Rodent cells often acquire chromosomal abnormalities during prolonged cultivation *in vitro*. In general, the karyotype of human cells is more stable in culture than that of rodent cells. However, most human tumor cell lines display chromosomal abnormalities.

5. Alterations of Morphology

The morphology of human epithelial tumor cell lines *in vitro* is not recognizably different from that of normal cells, at least in submerged cultures.

However, preliminary studies of human SqCCs in organotypic cultures have shown that the architecture of the tumor tissue is grossly distorted compared to normal epidermis and resembles that of carcinoma in situ. Organotypic cultures may represent an important advance in assessing transformation of human epithelial cells.

III. Malignant Transformation of Human Keratinocytes

A. General

The information on transformation of human keratinocytes has lagged behind that on rodent fibroblasts for a number of reasons, including the following:

1. Technical difficulties in cultivating primary epithelial cells *in vitro*
2. Resistance of human cells to spontaneous malignant transformation
3. The fact that *in vitro* assays for transformation of rodent fibroblasts cannot be applied to human keratinocytes
4. Technical difficulties of testing for transformation *in vivo* because it requires grafting of the keratinocytes onto animals
5. Most evidence suggesting that human epithelial cells require several steps to become fully transformed

Most of our current knowledge of transformation of human keratinocytes is derived from comparisons of the biology of SqCC lines with that of normal primary human keratinocytes in culture. Because some of these biological differences have been defined, these newly developed parameters are being used to test the "transforming" activity of carcinogens, physical agents, or viral genes. These new parameters must be validated by correlation with *in vivo* tests. The assays that have been developed to test specifically for transformation of keratinocytes include the following:

1. Extended life span in culture
2. Increased growth rate
3. Decreased ability to undergo terminal differentiation in semisolid medium, in response to calcium and after treatment with phorbol esters.
4. Increased ability for clonal growth in semisolid medium
5. Tumorigenicity after injection into immunodeficient animals

6. Tumorigenicity after grafting onto immunodeficient animals
7. Disrupted architecture in organotypic culture

The transformation of human keratinocytes has been studied in a number of different model systems and is briefly discussed in the following sections.

B. Simian Virus 40 (SV40)

Infection of human keratinocytes with the SV40 virus leads to cultures in which the SV40-infected cells gradually dominates, suggesting that they possess a slightly faster growth rate. In addition, the cultures loose their requirement for a feeder layer and serum. The cells have an extended culture life span, but after 100–150 doublings, they undergo a "crisis." During this period, most cells are lost, but a small fraction of cells survive and give rise to permanent cell lines. Postcrisis cells are anchorage-independent and are resistant to differentiation induced by high cell density or phorbol esters. Surviving cells are assumed to acquire additional genetic changes during the crisis period that allows them to survive. The SVK14 cell line arose from an SV40-infected primary culture without going through a crisis period. This cell line also has a reduced requirement for feeder cells, growth factors, and anchorage but is not tumorigenic when injected into mice.

Transfection of human keratinocytes with subgenomic fragments of SV40, which contain the transforming regions but are incapable of replicating, also resulted in immortal lines with a slightly decreased ability to form cornified envelopes but remained anchorage-dependent and not tumorigenic. SV6-1 *Bam*/HFK is such a fetal keratinocyte line transfected with SV40 early region. SV6-1 *Bam*/HFK began to form SqCCs in animals at passage 45, suggesting that secondary genetic changes had been acquired during cultivation *in vitro*.

C. Ad12-SV40 Hybrid Virus

Rhim and colleagues infected primary human keratinocytes with hybrid Ad12-SV40 to generate clonal cell lines. Apart from being immortal, these cell lines were aneuploid but not anchorage-independent nor tumorigenic. As no transcripts from the Ad12 genes were detectable in these lines, the SV40 genes alone were probably responsible for the phenotype of these cells. Within a week, superinfection of these cells with Kirsten murine sarcoma virus (KiSV) resulted in the appearance of foci of morphologically altered cells that were anchorage-independent and induced invasive squamous carcinomas in nude mice. KiSV infection alone did not lead to any detectable phenotype. Apparently, the concurrent expression of SV40 early genes and the transforming genes from KiSV were required to achieve the fully malignant phenotype.

D. Adenovirus E1A

Green and coworkers infected paraclones from primary human keratinocyte cultures with a retroviral vector containing the adenovirus E1A gene. They found that the oncogene transformed the cells very efficiently. Approximately 50% of the paraclones became immortalized, whereas their normal culture life span is limited to 15 doublings. When grafted onto nude mice, these cells formed a disorganized epidermis but showed no evidence of invasion.

E. Human Papillomaviruses (HPVs)

SV40 and adenoviruses are not associated with the development of human tumors *in vivo*. In contrast, mounting epidemiological and molecular evidence suggests that HPVs play a role in the genesis of human cancers, especially squamous carcinoma of the anogenital region. HPV types 6, 11, 16, 18, or 33 have been clearly associated with the development of anogenital tumors. In general, HPV types 6 and 11 can be detected in benign hyperplastic lesions of the anogenital region. In these lesions, the cells still undergo complete keratinization, and the virus undergoes its complete vegetative reproductive cycle. More recent studies have shown that HPV DNA (usually types 16, 18, or 33) is integrated and transcribed in the vast majority of invasive cervical and penile SqCCs. This circumstantial evidence suggests that HPV may be involved in the multistep process that leads to the malignant transformation of keratinizing epithelium in some cases of human SqCC. Several recent studies have been conducted in order to define the biological effects of different types of HPV on human keratinocytes. [*See* PAPILLOMAVIRUSES AND NEOPLASTIC TRANSFORMATION.]

1. HPV-11

This type of HPV is often found in benign hyperproliferative lesions of the anogenital region, condylomata acuminata (CA). Various types of human skin were exposed to extracts of CA and grafted beneath the renal capsule of athymic mice. In all cases, the CA-treated grafts were larger than the controls, suggesting a higher growth rate. In approximately half the cases of foreskin or vulvar skin grafts, the grafts were papillomatous and expressed HPV-11 capsid antigen; however, in no case did invasive tumors form.

2. HPV-16

This type of HPV is often present in invasive cervical carcinomas. A number of investigators have shown that transfection of primary human keratinocytes with HPV-16 resulted in a number of immortalized cell lines. Chromosomal aberrations, believed to have occurred as secondary events, were seen in the immortal lines after prolonged cultivation. Transfection of primary keratinocytes with HPV-6, -11, -16, and -18 DNA gave rise to 30–60 times more colonies than uninfected control cells. In addition, the transfected cells gave rise to numerous large colonies, compared to none for the controls. This suggested that the transfected cells had acquired a prolonged life span in culture. Furthermore, the cells transfected with HPV-16 or -18 gave rise to colonies that failed to undergo terminal differentiation in the presence of calcium and serum, in contrast with surrounding untransfected cells. Permanent cell lines have been derived from these transfected cells and were not tumorigenic by injection in nude mice. Thus, the nontumorigenic types HPV-6 and -11 increased the plating efficiency of keratinocytes, and the tumorigenic types HPV-16 and -18 conferred a prolonged life span and resistance to terminal differentiation onto primary human epidermal cells. The E6 and/or E7 gene of HPV-16 is responsible for immortalization of keratinocytes. Immortalization of primary human keratinocytes can be achieved by transfection with subgenomic fragments of cervical carcinoma DNA that contained the HPV-16 E6 and E7 genes. Primary human keratinocytes transfected with HPV-16 form dysplastic epidermis if the cells are grown on collagen rafts at the air–liquid interface.

3. HPV-18

Transfection of primary human keratinocytes with HPV-18 also resulted in an extended life span, changes in morphology, and resistance to induction of differentiation by calcium and phorbol esters.

F. Chemical Carcinogens/Physical Agents

The information on the effects of chemical or physical agents on transformation of human keratinocytes is limited. Primary human keratinocytes treated with chemical carcinogens, such as aflatoxin B, N-methyl-N'-nitro-N-nitrosoguanidine (MNNG), propane sulfone, β-propiolactone, or ultraviolet light, acquired an extended life span in culture, became anchorage-independent *in vitro,* and were invasive in an *in vivo* assay. Furthermore, primary human keratinocytes that had been immortalized by infection with a hybrid Ad12-SV40 virus became anchorage-independent and tumorigenic after subsequent treatment with MNNG. [*See* CARCINOGENIC CHEMICALS.]

G. Summary

DNA viruses such as SV40 and HPV-6, -11, -16, and -18, or chemical carcinogens such as, for example, MNNG increase the growth rate and extend the life span of primary human keratinocytes *in vitro* but are not sufficient for the cells to become tumorigenic. In addition, HPV-16 and -18, which are associated with cervical carcinomas, cause keratinocytes to become resistant to the induction of terminal differentiation.

The combination of SV40 genes and either a retroviral mutated *ras* gene or a chemical carcinogen such as MNNG appear to be sufficient to induce tumorigenicity in cultured keratinocytes. Thus, a minimum of two mutations appear to be required for transformation of human keratinocytes.

Bibliography

Chang, S. E. (1986). *In vitro* transformation of human epithelial cells. *Biochim. Biophys. Acta* **823,** 161–194.

Fusenig, N. E. (1983). Malignant transformation of epithelial cells in culture. *Prog. Clin. Biol. Res.* **132B,** 91–104.

Fusenig, N. E. (1986). Mammalian epidermal cells in culture. In "Biology of the Integument—Vol. 2 Vertebrates" (J. Bereiter-Hahn, A. G. Matoltsy, K. S. Richards, eds.), pp. 409–441. Springer-Verlag, Berlin, Heidelberg, New York, Tokyo.

Howley, P. (1987). The role of papillomaviruses in human cancer. In "Important Advances in Oncology 1987" (V. T. DeVita, Jr., S. Hellman, and S. A. Rosenberg, eds.), pp. 55–73. J. B. Lippincott Company, Philadelphia.

Rheinwald, J. G., and Green, H. (1975). Serial cultivation of strains of human epidermal keratinocytes: The formation of colonies from single cells. *Cell* **6,** 331–344.

Ruddon, R. W. (1987). "Cancer Biology." Oxford University Press, New York, Oxford.

Temin, H. M. (1984). Do we understand the genetic mechanisms of oncogenesis? *J. Cell. Physiol.,* Suppl. **3,** 1–11.

Zur Hausen, H., and Schneider, A. (1987). The role of papillomaviruses in human anogenital cancer. *In* "The Papovaviridae" (P. M. Howley and N. P. Salzman, eds.), pp. 245–263. Plenum, New York.

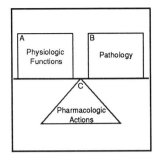

Kinins: Chemistry, Biology, and Functions

NATHAN BACK AND GURRINDER S. BEDI, *State University of New York*

I. Kinin-Forming Protease Cascade System
II. Assay of Kinins
III. Physiological Functions of Kinins
IV. Kinin Receptors
V. Kinins in Pathology

Glossary

Agonists Chemical compounds that mimic some of the biological effects of endogenous compounds (as hormones) by interaction with the appropriate receptor

Antagonists Compounds that compete for the agonist binding sites on receptors, thereby inhibiting the characteristic action of a specific agonist. Ideally the antagonist would not have any pharmacological activity and would possess receptor specificity

Hageman factor Primary plasma protein factor. A primary plasma coagulation protein, converted into a protease by surface contact with negatively charged agents, that activates other vascular protease systems

Polypeptide Linear polymer consisting of L-α-amino acids linked together by n-1 peptide bonds

Prostaglandins Group of biologically active substances formed from arachidonic acid (fatty acid) that are potent chemical transmitters of cellular signals mediating diverse physiological and pathological functions. Leukotrienes also are formed from arachidonic acid via an alternate pathway

Receptors Macromolecular components of cells with which most biologically active compounds interact with varying degrees of specificity, evoking a response characteristic for the compound

Second messenger Designation of biochemical mechanism, activated by agonist complexation with target cell receptor, by which characteristic action of the agonist is mediated

KININS ARE low-molecular-weight (LMW) polypeptides formed by the action of specific and general proteases on specialized glycoprotein substrates containing the specific kinin structural moiety. Kinins have a high affinity for specific kinin receptors on membrane surfaces of various target cells. The kinin–receptor interaction produces an array of biological responses affecting vascular and smooth muscle, ion transport, regional tissue blood flow, select biochemical pathways, and the cardiovascular system. A specific kinin-forming system is present in the vascular system and a variety of tissues. Components of this system include kinin-forming enzymes (kallikreins), kinin-donor substrates (kininogens), vasoactive polypeptides (kinins), and kinin-destroying enzymes (kininases). At times this system is referred to as the kallikrein-kininogen-kinin (K-K-K) system. Endogenously formed kinins function physiologically to mediate glandular and reactive hyperemia, regulate tissue blood flow, and effect transepithelial ion transport. Although their role in physiology awaits full confirmation with the aid of specific kinin antagonists currently under development, kinins are presumed involved in the pathogenesis of diverse pathologic conditions including allergies, inflammation, malignancy, and a variety of shock states.

I. Kinin-Forming Protease Cascade System

A. Mammalian Kinins

Kinins refer to an entire family of naturally occurring, smooth muscle stimulating peptides isolated from tissues, blood, and secretions of various mammalian species. The term derived from the first kinin of this peptide family to be discovered, bradykinin, so named to describe the biologically active

Encyclopedia of Human Biology, Volume 4. Copyright © 1991 by Academic Press, Inc. All rights of reproduction in any form reserved.

589

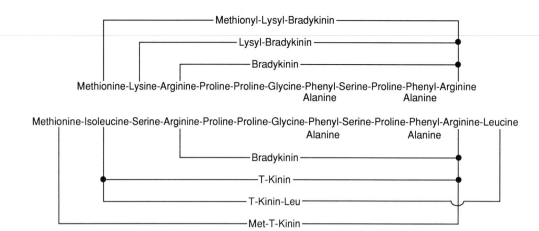

FIGURE 1 Primary structures of bradykinin and related mammalian kinins.

compound released from plasma by the action of proteolytic enzymes that caused a slow onset (brady) contraction (kinin) of isolated perfused smooth muscle. Bradykinin ultimately was found to be a nine-amino-acid polypeptide (nonapeptide) with the following sequence: Arg-Pro-Pro-Gly-Phe-Ser-Pro-Phe-Arg. The kinins thus far isolated represent a group of straight-chain LMW polypeptides similar to bradykinin in both structure and pharmacologic activity. The amino acid sequence of some known mammalian kinins is shown in Fig. 1. [*See* Peptides; Smooth Muscle.]

B. Pathways of Kinin Formation

Kinins are formed by the action of general and specific kinin-forming proteases on appropriate kinin-donor glycoprotein substrates known as *kininogens*. The most specific kinin-forming proteases are the tissue and plasma kallikreins that selectively cleave kininogen substrates to release kinin moieties. Because kinins are formed and released at the cellular site of action without the involvement of special glands, they often are referred to as "local hormones." The characteristic biologic actions indeed are localized and ordinarily are short-lived, terminated by kinin-destroying enzymes (kininases) present in plasma, tissue, and secretions. Thus, the plasma and tissue pathways of kinin formation are embodied in a cascading sequence of limited proteolysis in which the product of one reaction acts as a catalyst (protease) for the next reaction (Fig. 2).

C. Pharmacologic Action, Physiology, and Pathology

In addition to contracting and relaxing smooth muscles, kinins are potent vasodilators, causing a pro-

found but transient hypotension. Kinins also increase (capillary) vascular permeability, effect ion transport across intestinal and renal tubular epithelium, stimulate prostaglandin synthesis, and evoke pain (Fig. 3). Proposed physiologic role(s) of the kinin-forming system generally, and of kinins in particular, include mediation of glandular functional and reactive hyperemia and regulation of tissue local blood flow. Ordinarily, the concentration of kinins in tissues and circulating blood is low. Under certain physiologic and pathologic circumstances, kinin-forming protease systems are activated, resulting in increased kallikrein and subsequently increased kinin levels. Specific and more general kallikrein inhibitors and kinin-destroying enzymes, present in plasma, control or otherwise regulate kallikrein activity and kinin levels, respectively. Changes in component levels of kinin-forming protease systems have been noted in such diverse pathologic conditions as acute pancreatitis, allergies, diabetes, hypertension, inflammation, malignancy, and shock, *inter alia* (Table I).

D. Kallikreins

Two distinct types of kallikreins have been identified: plasma and tissue (glandular) kallikrein. These types differ with respect to physical-chemical properties, distribution through various body compartments, and biological functions. Plasma kallikrein is present as an inactive proenzyme termed *prekallikrein*. It is the major circulating protease that, in the presence of high-molecular-weight (HMW) kininogen, activates Hageman factor, an activator protease common to these vascular protease cascades

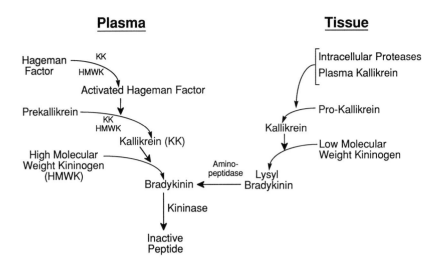

FIGURE 2 Tissue and plasma pathways in the formation, bioconversion, and degradation of kinins.

(i.e., the blood coagulation, fibrinolysin, and kinin-forming systems). Plasma kallikrein cleaves HMW kininogen to form the nonpeptide bradykinin (Fig. 4).

Tissue kallikrein, although expressed primarily in the pancreas, kidney, and submaxillary salivary gland, is present also in the brain, stomach, intestine, adrenal gland, and prostate. Molecular cloning has revealed the existence of multiple members of the tissue kallikrein gene family. The physiological substrate of tissue kallikrein is cleaved at a C-terminal Arg-Ser bond and a Met-Lys bond to yield lysyl bradykinin (kallidin). This lysyl bradykinin is rapidly biotransformed into bradykinin by amino-peptidase that splits off the terminal lysyl amino acid (see Fig. 2).

E. Kininogens

Kininogens are the natural substrates of the kallikrein family of proteases. Kininogen is the collective term for the glycoproteins containing the bradykinin peptide sequence. Three types of mammalian kininogens have been isolated thus far (Fig. 5). They differ in their molecular sizes and by their susceptibility to various proteases, in particular, the kallikreins. Initial studies identified a HMW and LMW kininogen with apparent molecular masses of 120 kDa and 68 kDa, respectively. HMW kininogen is cleaved by plasma kallikrein to yield bradykinin, whereas glandular kallikrein cleaves LMW kininogen to form lysyl bradykinin. Recently, a third kininogen species of LMW was discovered in rat plasma with an apparent mass of 68 kDa. Although this kininogen was found to resist cleavage by all kallikreins, high concentrations of trypsin or catalytic concentrations of tumor-derived cathepsin D–like acid proteases released from this substrate a kinin peptide structure with the novel Ile-Ser-bradykinin sequence.

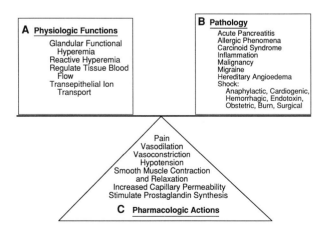

FIGURE 3 Pharmacological actions and proposed involvement of kinins in physiology and pathology.

TABLE I Pathologic Conditions in Which the Kinin-Forming System and Kinins are Presumed to Play a Role in the Disease Process

Acute Pancreatitis	Infection
Allergic Phenomena	Inflammation
Arthritis	Malignancy
Asthma	Migraine
Burns	Shock States
Carcinoid Syndrome	Anaphylaxis, Burn, Cardiogenic,
Cirrhosis of the Liver	Endotoxin, Envenomation,
Diabetes	Hemorrhage Hyperfibrinolysis,
Dumping Syndrome	Obstetric, Peptone, Surgical
Hypertension	Pain

FIGURE 4 Sites on high molecular weight kininogen cleaved by plasma kallikrein to release bradykinin. The Arg-Ser bond first is cleaved thereby freeing the light chain (LC) that contains a histidine-rich portion (HRP). The Lys-Arg bond then is cleaved, liberating bradykinin from the heavy chain (HC) that contains the cysteine proteinase inhibitor (CPI) domains.

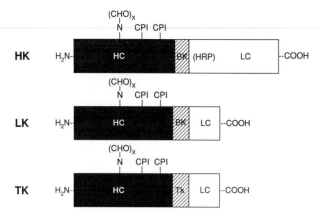

FIGURE 5 Representation of high molecular weight kininogen (HK), low molecular weight kininogen (LK) and T-kininogen (TK). Bradykinin (BK) is contained within both the HK and LK whereas TK contains the T-kinin (Tk) peptide with the Ile-Ser-bradykinin sequence.

Structurally, all three kininogens share in common a heavy-chain (HC) portion of equal size followed by a kinin moiety and then a light-chain (LC) portion of varying sizes. The HCs have two tandemly positioned functionally active cysteine proteinase inhibitory (CPI) domains that render all kininogens strong inhibitors of such proteinases as cathepsin L and papain. The HCs (and LCs of HMW kininogen) contain carbohydrate (CHO)$_x$ units whose functions as yet are not determined. The LC of HMW kininogen also has a histidine-rich portion (HRP) responsible for the blood coagulation contact activation functions of that molecule. The preferred designation of HMW and LMW kininogens and the rodent kallikrein-resistant kininogen is H-kininogen (HK), L-kininogen (LK), and T-kininogen (TK), respectively. The term *thiostatin* for TK, however, is preferred by some investigators because a physiological kininogenase has not been isolated as yet, thereby raising the question of the active kinin-donor function of this kininogen form.

The complete amino acid sequences of HK, LK, and TK have been deduced by molecular cloning and sequence analysis of their cloned cDNAs. In the rat, the major difference between the kallikrein-sensitive and kallikrein-resistant kininogens resides in the sequence immediately flanking the bradykinin moiety. Both HMW and LMW kininogens from the rat contain the Leu-Met-Lys-bradykinin-Ser sequence. Lack of the Lys-Arg linkage in TK apparently renders the molecule resistant to kallikrein cleavage.

F. Mechanism of Kinin Formation

The mechanism of kinin formation by kallikrein has been elucidated. HMW kininogen circulates in plasma complexed with prekallikrein and coagulation factor XI [plasma thromboplastin antecedent (PTA)] in a 1 : 1 molar stoichiometry. HMW kininogen augments the attachment of prekallikrein and factor XI to clot-initiating surfaces. This binding presumably positions the substrates of Hageman factor (factor XII) in a conformation that facilitates their cleavage. Once a small amount of activated Hageman factor (HFa) has been generated, it acts as a protease to convert prekallikrein to kallikrein (see Fig. 2). The kallikrein thus formed then cleaves HMW kininogen at two sites, Lys-Arg and Arg-Ser, to liberate the nonapeptide bradykinin (Fig. 4). Human LMW kininogen is identical to human HMW kininogen from the amino terminus to 12 residues beyond the bradykinin moiety. The LMW and TK both have short LCs (Fig. 5). The mechanism of cleavage of LMW kininogen remains to be clarified. TK is the major kininogen of rats. It lacks the Lys-Arg bond in sequence around the bradykinin moiety, as indicated above.

G. Kininases

Kininases is a collective term that describes an important group of kinin-destroying enzymes. Kininases are distributed ubiquitously in biological fluids and tissues and play a vital role in a number of physiologic functions. Of the two best-known kininases, kininase I is responsible for more than 90% of the kinin-destroying activity in plasma. Kininase II cleaves both kinin and angiotensin-I, thereby forming the potent vasopressor angiotensin-II peptide of the renin-angiotensin-aldosterone system. Kininases are present in plasma and in such tissues as brain, lung, liver, kidney, and dental pulp. Pathological conditions have been attributed to alterations in both tissue and plasma kininase levels.

II. Assay of Kinins

Kinin levels in biological material (blood, tissue, secretions) ordinarily are extremely low. Moreover, the biological half-life for most kinins is short, as low as 15 sec for bradykinin, because of rapid destruction by kininases. Thus kinin assay methods should be highly sensitive, specific, rapid, and reproducible. Although it is also desirable for the method to be relatively convenient to carry out, this feature often is sacrificed in favor of sensitivity and specificity. Traditionally, kinins (and kininogen in terms of bradykinin equivalents) have been determined by bioassay technique. The bioassay method has been supplemented with or replaced by radioimmunoassay (RIA). High-performance liquid chromatographic (HPLC) developments have permitted both the separation and quantitation of individual kinins. Preparation of immuno-reagents to bradykinin by both polyclonal and monoclonal antibody production techniques has enabled the further development of an enzyme-linked immunosorbent assay (ELISA) for qualitative and quantitative estimation of kinins.

A. Bioassay Method

Although *in vivo* bioassay methods involving blood pressure responses are in practice, one of the earliest and most popularly used methods is based on dose-dependent isolated smooth muscle responses (contraction or relaxation) of test kinin samples appropriately processed for assay. The muscle (rat uterus in estrus, rat duodenum, guinea pig ileum, etc.) is suspended in a bath containing a physiological medium (Tyrodes) with the upper end of the muscle attached by a thread to a force transducer for polygraph recording of the biological response. Potency of the test samples is compared with a reference standard, usually bradykinin. Although the method may suffer from variability, sensitivity, and specificity problems, various techniques involving manipulations of the muscle preparation and perfusing medium tend to overcome these problems. The method also allows for the study of kinin antagonists. A modification of the method has been developed whereby blood kinin levels are estimated by passing the blood directly over (superfusion) the muscle. More than one muscle type can be used in a cascade form of arrangement.

B. Radioimmunoassay

The development of immunologic techniques has permitted the production of antibodies against bradykinin complexed with a carrier such as bovine serum albumin. RIA technique involves the incubation of bradykinin antisera with radiolabeled antigen (^{125}I-bradykinin) together with the test kinin samples. The separation of the free radiolabeled bradykinin from the labeled bradykinin–antibody complex is accomplished by precipitation of the complex with sheep immunoglobulin (IgG). The radioactive pellet then is counted on a gamma-isotope counter. The RIA method is sensitive and rapid but does require more specialized technique and equipment. [*See* RADIOIMMUNOASSAYS.]

C. High-Performance Liquid Chromatography

High-Performance Liquid Chromatography procedures have been recently reported for the quantitation and identification of various natural kinins. These methods involve rapid, high-resolution separation of kinins, using both ion-exchange and reverse-phase columns. Kinins can be run native or derivatized for greater sensitivity using various derivatizing reagents [e.g., phenylisothiocyanate (PITC)]. Samples in physiological buffers can be run directly on an ion-exchange column with an aqueous buffer solvent system or, after prior sample treatment, on a reverse-phase column with a simple organic solvent and water mobile phase. The elution profile of the peaks is manipulated by adjusting the solvent percents, pH, and flow rate, until maximum resolution of the desired kinin peaks are obtained. Kinins are quantitated by comparing the height or areas of the peaks to the respective reference standard peaks.

The HPLC method is fast, sensitive, and suitable for routine quantitative and qualitative analysis, particularly when mixtures of kinin are involved. One drawback is interference of kinin peaks by overlapping impurities and peptides. This can be remedied by initial treatment of the sample using various simple, quick, and inexpensive cleanup procedures. [*See* HIGH PERFORMANCE LIQUID CHROMATOGRAPHY.]

D. Enzyme-Linked Immunosorbent Assay

Rapid and highly sensitive immunologic quantitation of kinins by ELISA has been achieved using both polyclonal and monoclonal antibodies directed against kinins. Perhaps the most frequently used ELISA for kinin estimation is a competitive ELISA. This technique involves binding purified kinin to a solid carrier surface (e.g., the well of a plastic ELISA plate). Then the sample containing kinin is added to the well, followed by antibodies produced against kinin. The kinin in the sample competes with the kinin immobilized on the well surface for the available antibody binding sites. Any antibody that binds to the free kinin in the sample washes off, whereas the antibody that binds to the immobilized kinin remains in the well. Next, a secondary antibody (antibody produced against the first antibody), which is complexed with an enzyme, is added. This antibody–enzyme complex binds to the anti-kinin antibody already bound to the immobilized kinin. Finally, a substrate is added and the color intensity, which is inversely proportional to the initial kinin concentration in the sample, is photometrically measured. Thus, ELISA permits rapid, specific, and sensitive estimation of kinins in biological fluids and tissues. ELISA can also be applied to the estimation of kininogen by using antibodies produced against kininogen or, using the kinin ELISA system, by determining the kinin released from kininogen by tryptic digest.

III. Physiological Functions of Kinins

Kinins share some pharmacological activities in common with other substances such as histamine, serotonin, prostaglandins, leukotrienes, and angiotensin. Despite this, it has been possible to ascribe certain physiological functions to these small kinin peptides (see Fig. 3).

A. Regulation of Blood Flow

Kinins, particularly bradykinin, are considered local regulators of blood flow in various glands in response to glandular secretory activity. Kinins also are the presumed mediators of functional vasodilation in nonendocrine glands. Ample evidence supports the role of bradykinin as the chemical mediator of nerve-induced vasodilation in the submaxillary salivary gland. During glandular activity, kallikrein in the glandular secretion acts on plasma kininogen substrate to form kinins. This takes place within the interstitial spaces of the gland, and the formed kinins dilate surrounding blood vessels. The vasodilator response is rapidly terminated by kininases. The tissue distribution pattern of kallikrein provides further support that kinins are important in the local regulation of blood flow, not only in exocrine organs, but also in vascular beds of many tissues and nonexocrine organs such as the kidneys.

B. Ion and Water Transport

Kinins also are presumed involved in the transepithelial transport of electrolytes and water in the kidney. Urinary kallikrein, synthesized in the kidney, is localized in the distal nephron, a segment associated with the regulation of sodium and water excretion. Immunohistochemical studies have identified the distal nephron as the site of origin of urinary kinins. Although kinins may increase sodium ion excretion and diuresis by inhibiting the luminal reabsorption of sodium, these effects also may be caused by, in part, a kinin-mediated increase in renal blood flow. Kinins also release renin, stimulate prostaglandin synthesis, and antagonize the hydroismotic action of antidiuretic hormone (ADH). Thus a combination of both direct and indirect actions of kinins may be involved in regulating kidney function.

C. Gastrointestinal Function

Kinins affect gastrointestinal motility and function. Kinins increase mucosal blood flow (hyperemia) and enhance the transepithelial transport of sodium and chloride. These ion changes also may influence glucose and amino acid transport across the epithelium. Because kinins have been shown to release prostaglandins from intestinal smooth muscle, quite possibly some of these intestinal actions of kinins may be due to prostaglandins that have similar effects on ion transport and motility. It also has been suggested that prostanglandins, by positive feedback mechanisms, may release kinins at gastrointestinal sites.

D. Effect on Vascular Tissue and Myocardium

Kinins lower the systemic arterial blood pressure in all animal species studied. This hypotensive action results from a decrease in total peripheral resistance, most probably at the arteriolar level. Infusions of bradykinin increase the heart rate and elevate cardiac output by an increased stroke volume and a diminished systemic vascular resistance. Vascular responses to kinin in various organs and tissue vascular beds are not uniform but rather display a pattern of selective vasoconstriction and vasodilation. Thus, bradykinin dilates veins of the skin, thereby increasing cutaneous blood flow, but does not affect blood flow to the skeletal muscle. Blood flow through the kidney is affected variably by bradykinin. Although vasodilation appears to be the dominant vascular response to kinin, vasoconstriction has been reported in both arterial and venous blood vessel beds of various animal species. Additionally, under specific blood perfusion and local hormone regulatory conditions, the customary vasodilator action of kinin may reverse itself, a phenomenon possibly involving release of specific vasoconstrictor prostaglandins (thromboxane A_2) and/or the presence of more than one type of kinin receptors (see Section IV).

E. Regulation of Blood Pressure

The potent vasodilator action of kinins, the renal site of kallikrein production, and the intimate relation between the kinin-forming and angiotensin-renin systems have led investigators to consider the role of kinins in the regulation of blood pressure. However, at the present time, there is evidence suggesting that kinins may be only involved indirectly in the maintenance of the normotensive state. Thus, antagonists that block kinin receptors that mediate vascular responses (B_2 receptors; see Section IV), given in doses that inhibit the hypotensive action of bradykinin, were reported to increase the hypertensive effects of such pressor agents as vasopressin. Bradykinin receptor antagonists also increase blood pressure in normotensive rats, but this effect may relate to the release of epinephrine. The receptor specificity of bradykinin antagonists thus is in question. Also, reductions in urinary excretion of kallikrein have been reported in experimental hypertension models and in some patients with hypertension. It is appealing to advocate a significant role for kinins in the regulation of systemic blood pressure, particularly in light of their effects on hemodynamics and sodium and water transport. However, further studies are required to clarify and establish the precise position of kinin within the array of the many hormonal and nervous factors controlling and modulating blood pressure.

F. Extravascular Smooth Muscle

The effect of kinin on extravascular smooth muscle also depends on the muscle and animal species studied. Most characteristically, kinins contract smooth muscle of the intestine, bronchioles, and uterus. Indeed, strips of the isolated perfused guinea pig colon or rat uterus are used to bioassay the relative potencies of kinin preparations. However, such tissues as the rat duodenum and hen rectal caecum are relaxed by kinins. The sensitivity of respective smooth muscles also varies depending on the animal species. Thus, the isolated rat uterus in estrus is most sensitive to kinins as are the bronchial smooth muscles, particularly of the guinea pig. Other substances such as histamine, prostaglandins, and leukotrienes have similar bronchiolar constrictor action that increases resistance to air flow into and out of the lungs.

G. Central and Peripheral Nervous Systems

The presence of bradykinin-related peptides in brain tissue has suggested a possible central nervous system function for kinins. Injection of bradykinin directly into the ventricles of the brain elicits both central nervous system stimulation and depression in the cat and vocalization in the dog. Intravenous administration of bradykinin causes drowsiness in both the cat and the rabbit and roosting in baby chicks. With regard to the peripheral nervous system, low concentrations of bradykinin were reported to depress peripheral nerve excitability. Pre- and postsynaptic bradykinin receptor sites have been identified in the rat vas deferens. Bradykinin also stimulates sympathetic ganglia, resulting in the release of catecholamines. This particular effect complicates our understanding of the possible role(s) of kinin within the peripheral nervous system.

Thus, there is considerable evidence implicating kinins in the physiology of the renal, gastrointestinal, cardiovascular, pulmonary, central nervous system, and autonomic nervous system. Kinins also are reported to stimulate DNA and protein synthesis, promote glucose uptake by peripheral tissue, and induce cell proliferation. Kinins also may help initiate the development of new blood vessels (angiogenesis). Kinin action at the cellular level is expressed via specific receptors, as discussed below. Ultimately, the function of kinins will be understood within the context of their interaction with chemical mediators and neurotransmitter substances of other biochemical systems.

IV. Kinin Receptors

Considerable attention has been directed toward clarification of the mechanism of action of kinins at the cellular and molecular level. It is now generally accepted that there are macromolecular components termed *receptors* present on the membrane surface of target cells, which usually initiate responses when occupied by biologically active agents (agonists). The magnitude of these responses is proportional to the concentration of the agonists, the relative affinity of the agonists for the receptor, and the number of receptors occupied. [*See* CELL RECEPTORS.]

Plasma membrane receptors for bradykinin were recognized more than a decade ago. Kinin–receptor interactions are accepted as responsible for the biochemical and physiological changes characteristic for such kinin actions as vasodilation, hypotension, increased vascular permeability, smooth muscle contraction and relaxation, and pain. The existence of specific kinin receptors was demonstrated by both *in vivo* and *in vitro* studies. Thus, low doses of bradykinin injected into the marginal ear vein of rabbits caused vasoconstriction. This vasoconstrictor response occurred in the denervated ear and also could not be blocked by antihistamines and agents that block receptors of the sympathetic nervous system. The vasoconstrictor response thus was attributed to the stimulation of specific kinin receptors situated along the vascular tissue. Subsequent *in vitro* studies with isolated perfused rabbit aorta strips showed that the bradykinin-induced contraction could not be either blocked or reversed by pretreatment with specific inhibitors of serotonin (5-hydroxytryptamine), angiotensin-II, and the sympathomimetic norepinephrine.

From these and additional studies with any other isolated smooth muscle preparations including cat ileum, rat uterus, guinea pig ileum, and rat duodenum, the concept of specific kinin receptors was established. Additionally, the presence of more than one type of kinin receptor was proposed based on differences in tissue responses to bradykinin as noted with aorta, uterus, and duodenum tissue of

the rat. Studies with bradykinin analogues and antagonists confirmed the presence of at least two types of receptors, termed B_1 and B_2, in rabbit aorta and other tissues. The B_1 receptor was identified by its sensitivity to the octapeptide agonist, des-Arg^9-Leu^8-Bk [bradykinin without the terminal arginine and with leucine in place of phenylalanine (see bradykinin structure, Fig. 1)]. However, B_2 receptors appear to have a higher affinity for lysyl bradykinin (kallidin) and bradykinin. The relative potencies of these and other agonists and antagonists have helped classify the type and tissue/cell distribution of kinin receptors. In particular, responses to bradykinin and des-Arg^9-BK have permitted the classification of a kinin-induced response as a B_1 or B_2 response. These kinin-agonist sensitivity studies have shown that the B_2 receptors are distributed widely throughout the cardiovascular system in various mammalian species. The B_2 receptor is assumed to be the principle mediator of kinin-stimulated vasodilation. Recent evidence indicate the existence of a more heterogeneous population of B_2 receptors mediating kinin responses. This evidence was provided by structure-activity studies with bradykinin analogues using the isolated rat vas deferens smooth muscle preparation. Many of these analogues were found to antagonize such kinin-induced actions as smooth muscle contraction, increased vascular permeability, venous constriction, and prostaglandin release. B_1 receptor responses, contrarily, appear to be more difficult to evoke. B_1 receptors apparently can be induced *in vivo* by noxious stimuli such as the injection of Triton X-100 in the uteri of rats or by the intravenous injection of *Escherichia coli* lipopolysaccharide into rabbits. These observations have raised questions regarding the existence of two distinct groups of kinin receptors.

Bradykinin receptors characteristically are of high affinity and are present in small concentrations 1.6×10^{-13} moles/mg protein). Only little progress has been made toward a detailed description of the structural components of the receptor, primarily because of their low concentrations in the target tissues.

The transfer of information from such biologically important molecules as kinins from the external cell surface to the cell interior is mediated, in many instances, by "second messengers" formed through the binding of the ligand to the specific plasma membrane receptors. Little is known or understood about the cellular events mediated by the bradykinin receptors. Responses to bradykinin have been linked to prostaglandins, cyclic nucleotides and nucleotide cyclases, G proteins, and calcium. Bradykinin has been shown to release prostaglandins from such diverse tissues as the kidney, heart, lung, blood vessels, spleen, duodenum, and uterus. Bradykinin also influences prostaglandin synthesis by stationary cell cultures of mouse and human synovial fibroblasts, pig aorta endothelium, rabbit reno-medullary interstitium, and human umbilical vein smooth muscle. Bradykinin-stimulated synthesis of prostaglandins in isolated cells was blocked by mepacrine, an inhibitor of phospholipase A_2, the enzyme that makes available arachidonic acid for conversion to prostaglandins. Thus it was concluded that bradykinin promotes prostaglandin synthesis by stimulating phospholipase A_2 that acts, in turn, on membrane phospholipids to release the arachidonic acid, the immediate precursor for prostaglandin synthesis. Indeed, it has been demonstrated that bradykinin releases arachidonic acid from phospholipids from many of the cell cultures listed above. In point of fact, bradykinin is one of the few peptide hormones shown to stimulate phospholipase A_2 activity. Bradykinin also stimulates acyl hydrolase activity in the perfused rabbit kidney and cultures of dog kidney epithelial cells. The question remains whether bradykinin activates phospholipase A_2 by interacting directly with the enzyme or whether the enzyme is a second messenger activated by the interaction of bradykinin with its receptor, similar to adenyl cyclase.

Activators of adenyl cyclase as well as inhibitors of the cyclic AMP (cAMP) phosphodiesterase potentiate bradykinin-induced relaxation of the rat duodenum muscle. This observation suggests that the effect of bradykinin on the rat duodenum is possibly linked to the adenyl cyclase system. Bradykinin treatment of guinea pig lung slices, human synovial fibroblasts and umbilical artery, and cell cultures of rat mesentery resulted in an accumulation of intracellular cAMP. The anti-inflammatory agent indomethacin, a potent inhibitor of the cyclooxygenase pathway to prostaglandin synthesis, was shown to block the bradykinin-stimulated accumulation of cAMP in isolated tissue and cell culture preparations. Although this suggested that bradykinin stimulated the accumulation of cAMP via an indirect prostaglandin release mechanism, the increase in cGMP level of tissues incubated with bradykinin was not affected by indomethacin. Additionally, maximal bradykinin-mediated increases in cGMP in

rabbit renal medullary slices preceded increases in prostaglandin synthesis. The synthesis of cGMP and prostaglandin was not interrelated. It has been suggested that perhaps some of the physiological effects of bradykinin on renal function are mediated by cGMP and other effect by prostaglandins.

V. Kinins in Pathology

Changes in tissue and plasma levels of components of the vasopeptide kinin-forming system have been reported in many pathological conditions and disease states (Table 1). Biochemical defects in all these conditions have been linked to activation and/or release of proteases of vascular cascades or intracellular proteases that digest appropriate substrates leading to generalized proteolysis, hemorrhage, formation of potent chemical mediators (as kinins), and in some instances, shock (Fig. 6).

A. Shock

There is direct and compelling evidence that associates kinins with the pathogenesis of shock, irregardless of the initiating stimulus. The essential feature of shock is a profound disturbance of the microvasculature (Fig. 7). Hypotension, one of the cardinal symptoms of shock, leads to decreased blood flow through the tissues (tissue underperfusion), resulting in low levels of tissue oxygen (hypoxia), a subsequent increase in acid metabolites (thereby de-

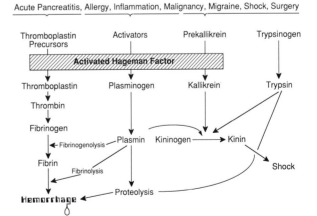

FIGURE 6 Biochemical defects encountered in various pathological conditions that may lead to general proteolysis, hemorrhage, and shock, resulting from activation of Hageman factor–dependent protease cascades.

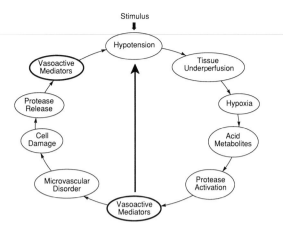

FIGURE 7 Cycle of biological events associated with the initiation development, and attenuation of shock hypotension.

creasing the pH of the cell environment), and protease activation with the formation of vasoactive mediators that affect the microvasculature and cell competence. Experimental studies have demonstrated that the pH change is a sufficient stimulus to activate, via the Hageman factor, three major vascular protease cascades: the blood coagulation, fibrinolysis, and kinin-forming systems (Fig. 8). Although activation of these protease cascades may occur simultaneously, activation also may be sequential because the central protease of each of the three systems (thrombin, plasmin, kallikrein) is capable of activating the other two systems at several sites. These three systems indeed were found to be activated in a variety of experimental shock models (anaphylactic, burn, hemorrhagic, cardiogenic, traumatic) as reflected by dynamic changes in these system plasma component levels during shock (commonly referred to in the early literature as Back's triade). Kinins were shown clearly to play a pivotal role in the shock pathology. Prophylactic and therapeutic treatment with protease inhibitors also prevented or reduced the shock-induced biochemical changes and also significantly decreased the morbidity and mortality in all shock models studied, thereby providing further evidence of the initiating role of proteases.

B. Acute Pancreatitis

Human pancreatitis, in its most acute form, is a most serious and potentially life-threatening condition. Although it is brought on most commonly by chronic intake of alcohol and the presence of stones in the gallbladder and common bile duct, it can also be caused by gastrointestinal ulcers and trauma or

cancer of the pancreas. The fulminating acute attack is characterized by marked tachycardia and severe abdominal pain and distension and may also be associated with hypotension, shock, and respiratory distress syndrome (shock lung). Autodigestion of the pancreas occurs because of pancreatic proteolytic enzymes activated within the gland rather than in the intestine after secretion from the gland. The pain, fluid and electrolyte loss, and ensuing shock, as well as damage to the capillary endothelium in the lungs, are probably caused by kinin mediators released into the blood by the action of pancreatic kallikrein and trypsin on kininogen substrates. Edema seen in acute pancreatitis may be due to pancreatic duct obstruction coupled with the kinin-induced increased vascular permeability. The damming effect on pancreatic secretion results in further activation and release of kinin-forming enzymes, liberating kinins locally into the gland as well as systemically. Although acute pancreatitis is a rare, poorly defined, and complex condition that also may occur without apparent cause, the role of proteases in its developing pathology appears established. Thus, the use of protease inhibitors as aprotinin (Trasylol) has been found of great therapeutic benefit in reducing both the severity and mortality of the disease.

C. Inflammation

Kinins have been found directly involved in clinical symptoms associated with a number of inflammatory disorders. Kinins were present in nasal secretions obtained by lavage from sensitive subjects

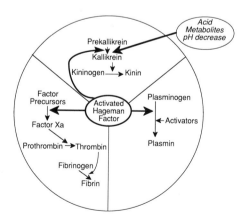

FIGURE 8 Interrelations among the kinin-forming, blood coagulation, and fibrinolysin protease cascade systems and the site of action of cell acid metabolites.

challenged with the allergen. These kinins occurred both immediately after the challenge and during the more delayed phase of the allergic response. There also was a direct correlation between the onset of symptoms and extent of kinin production. Nasal instillation of kinin also causes a rhinorrhea similar to that seen in rhinitis. Although other mediators as histamine, leukotrienes, and prostaglandins also were present in the allergic rhinitis secretions, only kinins were found during clinical or experimental rhinovirus colds. During symptomatic rhinovirus infections, both bradykinin and lysyl bradykinin (kallidin) were generated, and vascular permeability of nasal vessels increased as evidenced by the presence of albumin in the nasal secretions. Kinins also were identified in bronchial secretions obtained from subjects with asthma, bronchitis, and pneumonitis. Bradykinin administered by inhalation to human asthmatic subjects caused bronchial spasm. These results together with finding of kininogen and kallikrein in the bronchial secretions provide clear evidence of kinin involvement in inflammatory airway diseases. The advent of the new competitive antagonists of bradykinin should help contribute to our understanding of kinin participation in such allergic and nonallergic inflammatory diseases. [*See* Inflammation.]

D. Hereditary Angioedema

Hereditary angioedema is an inflammatory condition in which Hageman factor–dependent protease cascades are defective. The disease is characterized by sudden episodes of edema of the larynx with the threat of asphyxiation and edema of the bowel wall resulting in considerable abdominal pain. Swelling of the extremities and face also occurs. These attacks of edema are attributed to bradykinin formed by the plasma kallikrein cleavage of kininogen. The bradykinin apparently increases vascular capillary permeability, causing edema in the extravascular spaces.

E. Migraine

It has been proposed that bradykinin may be responsible for the pain, vasodilation, edema, and depression of the migraine syndrome. Intraspinal injection of bradykinin into normal subjects evokes migraine-like headache symptoms. Furthermore, cerebral administration of bradykinin led to an increase of the permeability of the "blood-brain bar-

rier," suggesting that kinins may have a mediator function in vasogenic brain edema. Kinin-forming proteases and kinin have been reported present in cerebrospinal fluid.

F. Hypertension

Although bradykinin probably does not contribute directly toward the regulation of blood pressure in the normotensive state, studies continue to explore kinin involvement in hypertension. Many experimental and clinical studies of hypertensive subjects have reported reduced urinary excretion of kallikrein and kinin, higher than normal activity of kininase II, decreased synthesis of kininogens, and increased kallikrein inhibitor activity. Unfortunately, there have been conflicting biochemical data on changes in kinin-forming component levels in hypertension. Results apparently depend on the experimental hypertension model and other experimental conditions. Thus differences in urinary kallikrein excretion in hypertensive rats were found to be genetic-based, whereas in some strains of spontaneously hypertensive rats decreased urinary kallikrein excretion and sodium/water retention were clearly correlated with blood pressure increases. Treatment of hypertensive patients with kallikrein significantly reduced blood pressure. Also, the intimate biochemical interrelation between the kinin-forming and renin-angiotensin systems has rendered difficult a clear and unequivocal determination of the etiologic role of kinin in hypertension. Results with inhibitors of angiotensin-converting enzyme (kininase II) in the clinical treatment of hypertension have suggested that part of the therapeutic response may have been related to the potentiated action of kinin. The use of specific kinin receptor antagonists may shed light on the kinin–hypertension relation. Kinin antagonists have been shown to increase blood pressure both in normotensive and some hypertensive rat models. Considerably more study is needed to establish the extent to which kinins participate in regulating and moderating hypertension. [*See* Hypertension.]

G. Neoplasia

Kinin-forming and more general protease cascade systems have been assigned a variety of roles in the neoplastic processes. Proteases appear to be essential both in generating and controlling complex biological events such as morphogenesis, fertilization, cell replication, and migration. With regard to neoplasia specifically, proteases have been shown to be involved in tumor promotion (tumorigenesis), neoplastic transformation, tumor nutrition and growth, replication of cells, distribution pattern and frequency of metastases, and formation of new vessels in tumors (angiogenesis, neovascularization). Through the process of limited proteolysis, proteases are presumed to provide chemical messengers (mediators) capable of initiating, regulating, and terminating cell processes that regulate many of the above functions.

Many of the proteases and protease inhibitors isolated from neoplastic tissues and cells have been shown to enhance or inhibit tumor viability and growth and also alter tumor metastatic potential. Tumor fragments also stimulated angiogenesis in experimental vascular models, an effect also reported earlier with kallikrein. Components of both the fibrinolysin and kallikrein–kinin systems were reported in various experimental models of ascites (fluid) tumors. Recently, the kinin peptide Ile-Ser-bradykinin was isolated from the ascites fluid of a patient with ovarian carcinoma. Both [hydroxypropyl^3]-bradykinin [(Hyp^3)-bradykinin] and bradykinin were identified and purified in human gastric cancer ascites fluid. Changes in kinin system component levels correlated with the kinetics of experimental tumor growth. Protease inhibitors such as ε-amino caproic acid (EACA), pepstatin, and also the anticoagulant heparin suppressed *in vivo* tumor growth. A kinin-forming system and two new kinins also were purified and characterized from a stationary cell line of fibroblasts, a cell type often comprising part of the heterogeneous cell population of tumors.

All components of both a specific alkaline and acid protease kinin-forming system (kallikreins, kininogens, kinins, kininases) also have been isolated from both experimental and human clinical solid tumors. During solid tumor growth, levels of prekallikrein and kininogen decreased, whereas kininase activity increased. The serine protease inhibitor aprotinin (Trasylol) and heparin significantly inhibited solid experimental tumor growth and reduced the magnitude of the component level changes. The novel kinin peptide, Ile-Ser-bradykinin, has been reported isolated from human ovarian carcinoma ascites fluid. A transplantable rodent liver tumor (Morris hepatoma) was found to synthesize enormous amounts of LMW kallikrein-resistant kininogen (TK), at levels almost 20 times that of the normal liver. The purpose of such high kininogen production by the tumor, similar to that observed with inflamed livers, remains to be elucidated.

The involvement of kinins in the neoplastic process may be viewed within the context of their pharmacologic actions (i.e., increased permeability, vasodilation, pain, and angiogenesis). Tumor tissue often is hypervascular, with blood vessels that are hyperpermeable, resulting in formation of edema and/or ascites. The lack of a lymphatic system in solid tumors also provides a mechanism for accumulation of proteins in the extravascular spaces. Formed kinins may be responsible for the increased vascular permeability observed in both solid and ascites tumors and may cause the formation of ascites. Kinins also may contribute to the neovascularization vital in maintaining the viability of the tumor by providing blood, nutrition, and a variety of growth-stimulating factors. Further study is required to determine the role(s) kinins may play in neoplasia.

H. Carcinoid Syndrome

The carcinoid syndrome describes a cluster of clinical manifestations of metastatic carcinoid of the liver arising from tumors of the gastrointestinal tract, pancreas, bile duct, and bronchi. The clinical features of the syndrome include vasomotor symptoms such as cutaneous flushing, rapid pulse, hypotension, headache, intermittent diarrhea, asthma, abdominal pain, and collagen deposits on the surface of heart valves. Kinins have been implicated in the etiology of the vasomotor and some of the other symptoms. Typical flushes in patients with carcinoid syndrome have followed bradykinin injection. During a carcinoid flush, arterial kallikrein levels were increased 10-fold, and bradykinin was found in the venous blood of patients. The bradykinin was presumed to arise either from activation of the vascular kinin-forming system or by the action of tumor kallikrein on kininogen substrate. Both serotonin and histamine have been considered as mediators of the vasomotor component of the carcinoid syndrome. Quite possibly several other chemical mediators (autocoids) may interact together with kinin on target organ receptors to produce symptoms of the syndrome.

I. Postgastrectomy Dumping Syndrome

Clinical symptoms of the postgastrectomy dumping syndrome are similar to those encountered in the carcinoid syndrome, notably the vasomotor features of increased peripheral blood flow and cutaneous flushing. The gastrointestinal hypermotility and diarrhea also occur. The syndrome occurs when carbohydrate-rich food quickly enters the intestine of individuals who have had subtotal or total surgical removal of their stomachs. Although serotonin is considered responsible for many of these symptoms, bradykinin formed by intestinal mucosal protease action is another candidate mediator. Plasma bradykinin levels were found elevated at the height of the dumping syndrome response.

J. Diabetes Mellitus

Diabetic patients often present with atherosclerosis and other vascular disorders. These vascular lesions are presumed to be caused by the activation of the blood coagulation and fibrinolysin cascade systems, resulting in diabetes-associated hypercoagulability and thrombosis. Recent studies of the kinin-forming cascade system in diabetic patients reported increases in levels of plasma prekallikrein, kallikrein, and kallikrein inhibitor. However, in diabetic patients with microvascular lesions, levels of plasma prekallikrein, HMW kininogen, and kallikrein inhibitor were decreased. The data suggest that the diabetic state activates the kinin-forming system, particularly in response to insulin deficiency, resulting in the formation of kinins that may cause some of the vascular defects noted. Plasma kinin levels were higher in diabetic patients with orthostatic hypotension compared with the control group of patients. Pancreatic glandular kallikrein, however, is decreased in the diabetic patient, resulting in lowered kinin production. Because kinins have been shown to promote the uptake of glucose by peripheral tissues (an insulin-like activity), the decreased kinin production in the diabetic further exacerbates the glucose metabolism defect characteristic of insulin-dependent diabetics. The ultimate clinical significance of the kinins in diabetes mellitus awaits further study.

Acknowledgment

This work was support in part from U.S. National Institutes of Health grant CA 382700.

Bibliography

Anumula, K. R., Schulz, R., and Back, N. (1989). Quantitative determination of kinins released by trypsin using enzyme-linked immunosorbent assay (ELISA) and identification by high performance liquid chromatography (HPLC). *Biochem. Pharmacol.* **38,** 2421–2427.

Bedi, G. S., and Back, N. (1989). Further characterization of monoclonal antibodies against rat plasma kallikrein, rat low molecular weight kininogen and synthetic bradykinin. *In* ''Kinins V.'' Advances in Experimental Medicine and Biology Series, vol. 247B. (K. Abe, H. Moriya, and S. Fujii, eds.), p. 223–230. Plenum Press, New York.

Bedi, G. S., Balwierczak, J., and Back, N. (1983). A new vasopeptide formed by the action of a Murphy-Sturm lymphosarcoma acid protease on rat plasma kininogen. *Biochem. Biophys. Res. Commun.* **112,** 621–628.

Colman, R. W., and Muller-Esterl, W. (1988). Nomenclature of kininogens. *Thromb. Haemost.* **60,** 340–341.

Di Sabato, G., ed. (1988). Biochemistry of inflammation. *In* ''Methods in Enzymology,'' vol. 163, p. 210–230. Academic Press, New York.

Gauthier, F., Gutman, N., Moreau, T., and Moujahed, A. (1988). Possible relationship between the restricted biological function of rat T-kininogen (Thiostatin) and its behaviour as an acute phase reactant. *Biol. Chem. Hoppe Seylers* **369,** 251–255.

Kaplan, A. P., and Silverberg, M. (1987). The coagulation-kinin pathway of human plasma. *Blood* **70,** 1–15.

Miller, D. H., and Margolius, H. S. (1989). Kallikrein-kininogen-kinin systems. *In* ''Endocrinology'' (L. J. Degroot and G. F. Cahill, Jr., eds.), p. 2491–2503. W. B. Saunders, Philadelphia.

Muller-Esterl, W. (1989). Kininogens, kinins and kinships. *Thromb. Haemost.* **61,** 2–6.

Steranka, L. R., Farmer, S. G., and Burch, R. M. (1989). Antagonists of B_2 bradykinin receptors. *FASEB J.* **3,** 2019–2025.

Stewart, J. M., and Vavrek, R. J. (1989). Development of competitive antagonists of bradykinin. *In* ''Kinins V.'' Advances in Experimental Medicine and Biology Series, vol. 247A. (K. Abe, H. Moriya, and S. Fujii, eds.), p. 81–86. Plenum Press, New York.

Lactation

JAMES L. VOOGT, *University of Kansas*

I. Development of the Human Breast
II. Milk Synthesis and Its Hormonal Control
III. The Milk-Ejection Reflex
IV. Lactation and Fertility
V. Incidence and Advantages of Breast-Feeding

Glossary

Alveolus Functional unit of the lactating mammary, composed of a sphere of epithelial cells surrounding a lumen

Bromocriptine Drug originally extracted from plants that imitates the action of dopamine

Colostrum Thick yellowish fluid secreted by the breast before and for a few days after parturition

Ducts Hollow tubelike structures made up of epithelial cells through which milk passes on its way to the nipple

α-Lactalbumin Protein necessary for the synthesis of lactose (the primary sugar found in milk), stimulated by prolactin

Lactogenesis Final differentiation of the lobuloalveolar tissue around the time of parturition and onset of first milk synthesis

Lactotrophe Cell of the anterior pituitary that secretes prolactin

Mammogenesis Development of the mammary gland

Milk sinuses Areas in the mammary gland where milk accumulates due to the convergence of numerous ducts

Myoepithelial cell Elongated cell containing contractile elements surrounding an alveolus

Nulliparous Having never given birth

Oxytocin Peptide hormone from the posterior pituitary gland that causes milk ejection

Parturition Act of giving birth

"LACTATION IS THE FINAL PHASE of the complete reproductive cycle of mammals. In almost all species the newborn are dependent on maternal milk during the neonatal period; in most the young are dependent for a considerable time. Adequate lactation is therefore essential for reproduction and the survival of the species and, biologically, failure to lactate can be just as much a cause of failure to reproduce as is failure to mate or to ovulate. In view of this necessity for lactation, it is not surprising that the lactating mother will, if necessary, produce milk at the expense of her own body tissues and that suckling young, like foetuses, can metabolically cannibalize the maternal organism. Correspondingly, the reproductive cycle is usually in abeyance during at least the early stages of lactation."

The above paragraph, from a World Health Organization Technical Report (1965), points out that survival of mammals clearly depends on lactation. Although it is true that many individual human babies have survived without ever receiving human milk, human lactation is fundamental to the well-being of the newborn for many reasons discussed later.

This article will describe the physiology of human lactation, including morphology of the lactating breast, hormonal control of milk synthesis and milk ejection, and how lactation inhibits fertility. The article will close with a description of the incidence and advantages of breast-feeding.

I. Development of the Human Breast

There are two periods during a woman's lifetime when mammary tissue shows its greatest developmental changes. The first of these occurs at the time of puberty in conjunction with the onset of secretion of the ovarian hormones estrogen and progesterone. The second occurs during pregnancy, when the breast becomes fully developed in preparation for the synthesis of milk that occurs following the birth of the infant. [*See* PUBERTY.]

The mature human breast is made up of both functional (potentially milk producing) and nonfunctional tissue. The functional tissue is characterized by ducts and lobular *alveolar* structures. Fat and connective tissue contribute to the dome-shaped appearance of the breast.

At the beginning of adolescence, ducts in the rudimentary mammal begin to grow and bifurcate. End buds form at the terminus of a duct, and these end buds become alveolar buds as adolescence proceeds. A cluster of alveolar buds around a terminal duct makes up a lobule. During the menstrual cycle, new alveolar budding occurs while older alveolar tissue atrophies and disappears. The peak in mitotic activity of the glandular tissue of the breast occurs during the luteal phase, which is the period of time between ovulation and menses. In general nulliparous women have virginal lobular development, whereas parous women have larger and more extensively developed lobules.

During early pregnancy there is increased ductal proliferation. Alveolar budding and lobule formation increase rapidly, and the epithelial cells that make up each alveolus increase in number and in size due to cytoplasmic enlargement. By midpregnancy the lobules are so large and numerous that they completely envelop the terminal portion of the ducts. During the second half of pregnancy proliferation of new lobules is slowed greatly. Instead, differentiation of the newly formed structures occurs, and the cells become secretory. The amount of secretion that occurs during the last half of pregnancy is minimal compared to lactation. However, the cells of the alveolus secrete into the lumen a yellowish fluid that contains numerous proteins and often can be expelled through the nipple. Mature milk is not secreted during pregnancy even though all the cellular machinery is present. Sex steroids, which are present in large amounts during the last half of pregnancy, account for this inhibition. During lactation milk secretion occurs in parallel with continued growth and differentiation of glandular tissue. However, major morphological changes do not occur after *parturition*.

Mammary gland growth and development in preparation for lactation is the result of complex interaction among hormones originating in the pituitary, ovary, and placenta. *Mammogenesis* requires sex steroids, lactogenic hormones (of which prolactin is the major one), and growth factors. Although the classical experiments have been done in animals, in which various combinations of endocrine gland ablation and hormone replacement were performed, it is generally accepted that ovarian estrogens are primarily responsible for epithelial proliferation, particularly ductal growth in both women and other mammals. Progesterone also plays a role in mammogenesis, but it is primarily involved in tissue differentiation. Neither sex steroid is effective in vivo in the absence of the pituitary gland. Corticosteroids from the adrenal cortex and insulin from the pancreas also are necessary for mammary growth.

Both *prolactin* and placental lactogen, which has structural and functional homology with prolactin, have profound effects on breast tissue proliferation and differentiation. Maternal serum prolactin levels gradually increase throughout pregnancy, as do placental lactogen levels. The physiological importance of placental lactogen is unknown since there have been cases of pregnant women who did not express immunologically recognizable placental lactogen during pregnancy, but experienced normal lactation. However, they may have produced an immunologically different but biologically active placental lactogen.

More recent advances in cell and molecular biology suggest that *growth factors* may play a major role in mammary gland development, and estrogens may be acting via these factors. Epidermal growth factor appears to be a leading candidate as a promoter of mammary tissue growth, perhaps acting in synergism with prolactin. [*See* EPIDERMAL PROLIFERATION.]

II. Milk Synthesis and Its Hormonal Control

Final differentiation of the lobulo-alveolar tissue of the mammary gland occurs towards the end of pregnancy and for a few days after parturition, then synthesis of milk begins. This process is known as *lactogenesis*. Before the beginning of milk synthesis and secretion, which occurs several days after parturition, the mammary gland secretes *colostrum*. This thick, yellowish fluid contains a high concentration of antibodies, proteins, fat-soluble vitamins, and some minerals, and is the first nutrient that the offspring receives. Colostrum also contains a zinc-binding factor that facilitates the absorption of zinc in the intestine by the newborn. In humans, the off-spring receives immunoglobulin of the IgG class in utero and is born with high titer of antibodies in the blood. However, colostrum does contain high levels of a modified IgA molecule, which is resistant to the changes in pH and proteolytic enzymes found in the gut of the newborn, and affords protection against pathogenic organisms.

A. Composition of Milk

Proteins, sugars, and fats in the milk are the sources of nutrition for the newborn. Milk proteins are synthesized from amino acids, which are taken up from the blood into the epithelial cells of the alveolus. *Casein* is the major milk protein synthesized. Although casein messenger RNA (mRNA) can be detected early in pregnancy, appreciable amounts of casein are not synthesized until after parturition. A second protein, *α-lactalbumin,* is not present until lactation begins, and its mRNA is present only after parturition. The total time required for protein synthesis, starting with amino acid uptake until the protein is secreted into the alveolar lumen is 30 minutes.

The major milk sugar is *lactose,* which is uniquely found in milk. Two proteins are important for its synthesis: α-lactalbumin and galactosyl transferase. The complex formed by these proteins, called *lactose synthetase,* is found in the Golgi vesicles and catalyzes the reaction of UDP-galactose and glucose to form lactose. The vesicles move to the luminal border of the alveolar cell and release lactose into the lumen by exocytosis.

The lipid content of milk varies greatly from one woman to another, between 2% and 5.3%. By far *triglycerides* comprise the majority of the lipids in milk. The fat content of human and cow's milk differs greatly. Human milk has a much greater concentration of linoleic acid, an essential fatty acid, and a larger amount of oleic acid, whereas cow's milk has more short-chain fatty acids. Human milk also has much higher concentrations of *cholesterol,* which may be important for myelin synthesis in the developing central nervous system. Women who have low-fat diets produce milk that is lower in fat than women on high-fat diets. [*See* CHOLESTEROL.]

The mechanism for secretion of milk fat into the lumen of the alveolus is still uncertain. Lipid droplets, in a manner not well understood, become enveloped in the apical plasma membrane, and are extruded from the cell. [*See* LIPIDS.]

B. Diet during Lactation

In general lactating women need more of each nutrient than do nonlactating women. How much more depends on the degree of lactation. Completely breast fed babies take in 300–1000 ml of milk per day. Part of the energy needed for milk synthesis can come from mobilizing the body fat the woman stored during pregnancy. Once these body stores of fat are used up, if the maternal weight falls below the ideal weight for height, increased daily intake of energy will be required. However, in general the extra energy needs of lactation can be met and are compatible with continual gradual weight loss. Moderate to severe dietary restrictions for weight loss purposes are not compatible with full lactation, especially during the first few weeks postpartum when full lactation is still being established.

In addition to the increased energy demands, an increase in protein and calcium intake is of major importance during lactation. The easiest and most efficient way to meet this demand is for the lactating woman to drink an extra 3–4 cups of whole milk per day. This will provide the extra energy, protein, and calcium needed. In order to receive the additional amounts of vitamins, such as ascorbic acid, vitamin E, and folic acid, a small increased intake of citrus fruits and leafy vegetables is important.

C. Hormones and Initiation of Lactation

The two hormones most important for the initiation of lactation are *prolactin* from the anterior pituitary gland and *glucocorticoids* from the adrenal cortex. Prolactin increases the expression of the casein gene. Transcription of casein mRNA occurs more quickly and to a greater degree than does translation to casein itself in response to prolactin. By themselves glucocorticoids do not affect casein mRNA, but synergize with prolactin. *Progesterone* is inhibitory to accumulation of casein mRNA, accounting for the lack of milk synthesis during pregnancy, when progesterone levels are very high. Once delivery of the placenta occurs, the main source of progesterone is lost, and increased casein synthesis occurs. Within 24 hours after delivery, progesterone levels in the maternal blood and mammary secretions fall precipitously. This fall always preceeds the increase in lactose, indicating that the secretory activity of the breast changes from colostrum to true milk within 48 hours of delivery. The timing of these events suggests that the major factors involved in the initiation of lactation are falling progesterone levels and presence of adequate amounts of prolactin.

D. Hormones and Maintenance of Lactation

Several hormones with widely varying biological functions in nonlactating organisms are important for maintenance of full lactation. These include prolactin, growth hormone, thyroid hormones, insulin, and glucocorticoids, although the relative importance of each varies greatly in different species.

1. Prolactin

Inhibition of prolactin secretion by ergot alkaloids, particularly *bromocriptine,* leads to a decrease or cessation of lactation in all species except ruminants. In women, the effect of prolactin inhibition on lactation due to bromocriptine treatment is very rapid, especially in early lactation. Stimulation of the nipple is necessary for prolactin release in lactating women. The neural input to the hypothalamus during suckling evokes production of chemical signals that reach the *lactotroph* cells in the anterior pituitary via the hypophysial portal blood. These signals include a decrease in the secretion of *dopamine,* a prolactin inhibitor, into the portal blood and an increase in secretion of factors or hormones that stimulate the lactotrophs to release prolactin. Basal- and suckling-induced prolactin levels vary during the course of lactation. For the first several weeks, basal prolactin levels are 2–3-fold higher than in nonlactating women, but by 8–10 weeks, levels have returned to baseline. Each nursing period causes a rapid and very substantial increase (10–20-fold) in prolactin secretion, which returns to basal levels after the suckling stimulus is removed. As lactation progresses, the prolactin response to suckling becomes less, but some prolactin still is necessary to maintain lactation. Correlations have been established between the amount of milk produced and the amount of prolactin released during suckling at given stages of lactation. It is quite possible that the prolactin response depends on the suckling intensity of the baby.

Prolactin has many lactogenic actions in the breast. Prolactin stimulates the synthesis of several fatty acids that are specific to milk. It also stimulates the protein casein by increasing the expression of casein mRNA. Prolactin regulates the formation of α-lactalbumin. In general, progesterone inhibits the action of prolactin on protein synthesis.

The epithelial cells of the breast alveoli have high affinity binding sites (receptors) for prolactin. Prolactin receptors also are found in the kidney, liver, adrenals, prostate, ovaries, and lymphocytes. The mechanism for signal transduction used by these receptors is largely unknown. Prolactin does not act by stimulating adenylate cyclase activity or guanylate cyclase activity in mammary tissue. Cyclic GMP can induce some increase in casein mRNA. Prostaglandins have been ruled out as the second messenger. The polyamine spermidine may play a minor role, but is not the main factor in mediating the effect of prolactin. The calcium-calmodulin system may have a function in mammary tissue differentiation in response to prolactin, but by itself cannot fully account for the action of prolactin. It is unclear how prolactin stimulates milk synthesis; several signal transduction mechanisms may be involved.

2. Growth Hormone (GH)

There is no conclusive evidence that GH release from the anterior pituitary increases during lactation or during the suckling period. Human growth hormone is capable of binding to prolactin receptors in a variety of species, but a definitive role for GH via its own receptors in human lactation is lacking. It clearly does have an important role in milk production in ruminants.

3. Adrenal Steroids

Removal of the adrenals reduces milk synthesis, primarily due to a reduction in mammary gland enzymatic activity and casein mRNA. Injection of glucocorticoids, but not mineralocorticoids, restores lactation. Each suckling period increases the secretion of the anterior pituitary hormone adrenocorticotropic hormone (ACTH), which in turn increases glucocorticoid release from the adrenals.

4. Thyroid Hormones

Removal of the thyroid significantly lowers milk production in a variety of animal species. In contrast, the concentrations of the two thyroid hormones, thyroxine and triiodothyronine, in serum are correlated negatively with milk production during early lactation in ruminants. Suckling in a variety of species, including humans, does not increase the release of thyroid stimulating hormone (TSH) from the anterior pituitary. [*See* THYROID GLAND AND ITS HORMONES.]

5. Insulin

Very little is known about the effect of insulin on milk synthesis in women. In rodents, insulin is stimulatory to milk production, and plasma insulin levels are higher during lactation than during a nonlactation period. However, it is not clear whether insulin has a direct action on the alveolar cells of the breast, or is merely permissive, allowing the various other hormones to exert their milk-producing EFFECTS. [*See* INSULIN AND GLUCAGON.]

III. The Milk-Ejection Reflex

A distinction must be made between milk secretion and milk ejection or milk letdown. Secretion is the process in which epithelial cells synthesize the various components of milk and release these components into the lumen of the alveolus. Milk ejection is the process of expelling the milk from the thousands of alveoli and terminal ducts into the larger ducts and expulsion from the breasts.

In simplest terms the milk-ejection reflex consists of the following sequential events: stimulation of nerve endings in the nipple; nerve impulses ascending via the spinal cord to the hypothalamus; nerve impulses on the *oxytocin* neuron originating in the hypothalamus and terminating in the posterior pituitary; release of oxytocin from these neurons into the blood; transport of the hormones to the breast where they cause the *myoepithelial cells* to contract and expel the milk from the aveoli and ducts. This is truly a neuroendocrine reflex because both neurons and hormones play a necessary role in bringing about the appropriate response.

The myoepithelial cells are the primary contractile tissue in the breast. These elongated cells form networks around each alveolus, and although present in nulliparous women, they greatly increase in number during pregnancy due to estrogen stimulation. At the onset of lactation, numerous myofilaments form in the cells, and this is the first time contractile activity of the cells is evident. Oxytocin receptors are present on myoepithelial cells only for a short period of time before parturition and these persist during the entire length of lactation.

The nipple in humans contains 10–12 openings of the *galactophores,* which are ducts connected to the *milk sinuses* in the breast. Sebaceous glands at the tip of the nipple are associated with the openings. Connective tissue and smooth muscle of the nipple are important for nipple erection, aiding at-

tachment by the baby. At the base of the nipple, in the pigmented areolar area, are additional openings from galactophores. These raised areas are called *Montgomery's glands,* and are connected to secretory alveoli.

The frequency and duration of nursing varies greatly from one species to another. In some, such as the agile wallaby, the young are continuously attached to the nipple for many weeks, whereas rabbits only nurse once a day. Rats nurse their pups 12–18 hours/day. Women vary greatly, but on an average nurse every 4–6 hours. However, in some cultures, such as the !Kung tribe of Africa, the baby may suckle several times per hour, 24 hours/day.

In humans, the *mechanism of sucking* consists of the baby forming a "teat" by drawing in both the nipple and surrounding areolar area into the buccal cavity. The base of the "teat" is compressed by the upper gum and the tip of the tongue, and the tongue moves from the base, stripping the milk out of the sinuses and the galactophores. Forming a suction or vacuum normally contributes little to milk flow. The baby alternately sucks for a minute or two, pauses for 20–30 seconds, and sucks again. As the nursing period progresses, the pauses get longer and the sucking activity decreases. Usually 10–14 days are required postpartum before full and sustained milk production occurs, at which time 120–180 ml of milk is obtained per feeding, or about 500–1000 ml per day.

One of the questions is the importance of active milk ejection in transfer of milk to the baby. Can babies obtain milk without milk ejection? The answer is yes, but to a much lesser degree. Using a breast pump will remove some milk without active milk ejection, but the presence of maternal oxytocin greatly increases the amount. However, during a normal suckling episode, there may be no detectable increases in blood oxytocin levels associated with the release of milk. This may be explained by the finding that oxytocin release often occurs during the presuckling period, in response to the sight or sound or touch of the baby. This oxytocin may be enough to initiate milk letdown. In many women, however, oxytocin levels are elevated during suckling, with the presence of seemingly random pulses of oxytocin. There is general agreement that oxytocin is an essential component of milk removal in women, and in its absence, normal lactation cannot continue.

The oxytocin, in response to stimulation of nerve endings in the nipple, comes from two *hypotha-*

lamic nuclei (clusters of densely packed cell bodies of neurons) containing *magnocellular neurons* that terminate primarily in the posterior pituitary. The cell bodies of the magnocellular neurons are the sites of oxytocin and antidiuretic hormone (ADH) synthesis. In primates, about 80% of the oxytocin neurons are in the paraventricular, and 20% in the supraoptic nuclei. These *neurosecretory* neurons send unmyelinated fibers to the *posterior pituitary,* passing through the external zone of the median eminence and the infundibulum or pituitary stalk. In the pituitary, the neurons often have swellings or dilatations called *Herring bodies,* where oxytocin is stored or degraded. The nerve terminals are in direct contact with fenestrated (leaky) capillaries; as oxytocin is released, it diffuses into the capillaries and is carried throughout the body.

Elegant studies have been done in the rat relating suckling behavior to the firing of oxytocin neurons and subsequent milk ejection. Initially when pups attach to the nipple, background firing of the neurons occurs at a steady rate. Then a burst of firing of the neurons occurs for 2–4 seconds, followed by an increase in intramammary pressure 12–15 seconds later, which signals milk ejection. This delay is the time required for the released oxytocin to reach the myoepithelial cells and initiate contraction. This neuronal behavior is repeated many times during one suckling period, with many separate milk ejections. How applicable this pattern is to that found in women is unknown.

The failure to lactate successfully in women is due to complex and as yet poorly understood reasons. In addition to physical factors such as nutrition or drugs, emotional elements play a very critical role in carrying out a normal lactation. Even mild emotional stresses inhibit the milk-ejection reflex, creating a vicious cycle such that the longer it takes for the suckling baby to get milk, the more agitated the baby becomes. This increases the anxiety level in the mother, resulting in total inhibition of milk ejection. Without a doubt, successful nursing periods require a peaceful and relaxed state of mind.

IV. Lactation and Fertility

The primary function of lactation is to provide nourishment for the offspring, without which most mammalian species would soon become extinct. A sec-ond important function of lactation in many species is to alter or regulate reproductive activity. In general, suckling results in prolonged periods of ovulatory failure and infertility. The duration of this delay in reproductive activity varies greatly between individual women and between cultures. In general, the more extensive is the breast-feeding, the longer is the period in which there is no ovulation. Over 90% of women who do not breast-feed at all have their first menstruation within 4 months postpartum, whereas this same percentage for women who continue to breast-feed is not reached until 18–20 months postpartum. There is a linear relationship between the average months of breast-feeding and the average months of amenorrhea (absence of menstruation).

Ovarian follicular development during lactational amenorrhea is inhibited. Plasma estradiol is lower than seen during the preovulatory stage just preceeding ovulation, and formation of luteal structures in the ovary never occurs. More than half of the non-breast-feeding women ovulate within 8 weeks of postpartum; however, often their luteal phases are not sufficient to maintain a pregnancy for several months. Similarly, some breast-feeding women who do not ovulate, even before there is a resumption of menstrual periods, have shortened luteal phases.

The primary cause for lactational amenorrhea is the reduction in secretion of *luteinizing hormone* (LH), which is the anterior pituitary hormone partially responsible for follicular development and entirely responsible for ovulation. Occasionally during this anovulatory period there is an increase in basal and pulsatile LH secretion, followed by a rise in estradiol from the ovary. If the frequency of breast-feeding continues to be the same, this elevated LH and estradiol will again decline. However, if frequency is reduced at this time, LH may remain elevated, estradiol levels may increase further, indicating increased follicular development, and ovulation may result.

There has been a continuing question in many species whether the elevated prolactin seen during lactation directly suppresses LH secretion. In women, there is no evidence that prolactin acutely inhibits LH secretion. The inverse relationship between prolactin and resumption of menstrual cycles may reflect the amount of suckling activity. The greater the suckling activity, the higher the level of prolactin, and vice versa. A decline in prolactin may reflect a decline in suckling, and it is probably the decline in suckling that allows an increase in LH secretion, not the decline in prolactin.

An interesting finding shows that continuation of breast-feeding at night prolongs the length of lactational amenorrhea. The preovulatory LH surge usually begins at night in women with normal menstrual cycles. In lactating women, failure to breast-feed at night may allow the LH surge to begin and the result is ovulation. If suckling does occur, the preovulatory LH surge may be inadequate for ovulation or for formation of a normal corpus luteum.

It is not clear how suckling inhibits LH secretion, but the location of its effect is most likely the hypothalamus. Because there are many neurochemicals in the hypothalamus that are involved in regulating LH secretion, alteration in any one of these by suckling would alter the normal pattern of LH secretion.

V. Incidence and Advantages of Breast-Feeding

In the two decades before 1970, over 80% of all babies in the United States were artificially fed. This trend has reversed such that by the late 1970s over 50% of all babies 2 months of age were being breast-fed. In addition, more women are breast-feeding for a longer period of time.

Historically, abandonment of breast-feeding began in higher socioeconomic levels first, followed by decreases in lower socioeconomic levels. As the trend was reversed, women in upper socioeconomic levels were the first to show an increased incidence of breast-feeding, followed by progression down class lines. A significant boost to breast-feeding came from the American Academy of Pediatrics, which in 1978 recommended that all physicians encourage mothers to breast-feed their babies.

Many advantages of breast-feeding have been proposed. It is often difficult to document whether these advantages are real because a comparable control group of babies is often difficult to find. The following are some of the proposed advantages.

1. Breast milk is nutritionally superior to any alternative.
2. Breast milk is bacteriologically safe and always fresh.
3. Breast milk contains a variety of antiinfectious factors and immune cells.
4. Breast milk is the least allergenic of any infant feed.
5. Breast-fed babies are less likely to be overfed.
6. Breast-feeding promotes good jaw and tooth development.
7. Breast-feeding *generally* costs less than the commercial infant formulas currently available.
8. Breast-feeding automatically promotes close mother–child contact.
9. Breast-feeding is generally more convenient once the process is established.

As more research is done on human lactation, more advantages of breast-feeding may become established. Hopefully, this will help lead to a worldwide sustained increase in the number of mothers choosing to lactate. The positive impact on the well-being of the baby and the negative impact on population growth will be substantial.

Bibliography

Cowie, A. T., Forsyth, I. A., and Hart, I. C. (1980). "Hormonal Control of Lactation." Springer-Verlag, Berlin, Heidelberg, and New York.

Knobil, E., and Neill, J. D. (1988). "The Physiology of Reproduction," vol. 2. Raven Press, New York.

Larson, B. L. (1978). "Lactation," vol. 4. Academic Press, New York and London.

Neville, M. C., and Daniel, C. W. (1987). "The Mammary Gland." Plenum Press, New York and London.

Vorherr, H. (1974). "The Breast." Academic Press, New York and London.

Worthington-Roberts, B. S., Vermeersch, J., and Williams, S. R. (1985). "Nutrition in Pregnancy and Lactation." Times Mirror/Bosby, St. Louis.

Waletzky, L. R. (1976). "Symposium on Human Lactation." U.S. Department of Health, Education, and Welfare. DHEW Publication No. 79-5107.

WHO Technical Report (1965). "Physiology of Lactation." WHO Tech. Rep. Ser. 305.

Tyson, J. E., ed. (1978). "Neuroendocrinology of Reproduction." W. B. Saunders Company, London, Philadelphia, Toronto.

Lactose Malabsorption and Intolerance

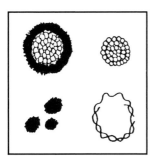

JOSE M. SAAVEDRA, JAY A. PERMAN,
Johns Hopkins University

I. Lactose and Milk
II. Physiology of Lactose Digestion and Absorption
III. Lactose Malabsorption
IV. Lactase
V. Lactase Deficiency
VI. Lactose Intolerance
VII. Diagnosis of Lactose Malabsorption
VIII. Management of Lactose Intolerance

Glossary

Congenital lactase deficiency Extremely rare autosomal recessive condition in which lactase activity is extremely low at birth and throughout life

Lactase deficiency Lower than expected level of lactase activity in the brush-border membrane; the activity of lactase can only be measured directly in mucosal tissue obtained by small bowel biopsy; lactose tolerance tests are indirect measurements of total intestinal lactase activity; lactase deficiency is most often transient and secondary to gut injury

Lactase nonpersistence Synonyms include adult or primary lactase deficiency, ontogenetic adult lactase deficiency, primary adult hypolactasia, and racial–ethnic lactase deficiency; this term characterizes the normal postweaning decline in lactase activity occurring in most mammals, including many humans; because this condition is the norm, it should not be considered a deficiency

Lactase persistence Less common condition describing the minority of humans who have maintained their lactase activity at levels found in early infancy

Lactose intolerance Clinical manifestations of lactose malabsorption, including diarrhea, bloating, flatulence, and abdominal pain; this term is not interchangeable with lactose malabsorption, because

a significant number of individuals with low lactase activity and resultant malabsorption may not necessarily be symptomatic in response to a specific amount of lactose or lactose-containing products

Lactose malabsorption Impairment in hydrolysis of lactose digestion to its constituent monosaccharides, glucose and galactose; a more appropriate term is lactose maldigestion

LACTATION is the essential characteristic of all mammals. Thus lactose, the principal carbohydrate in milk, constitutes the single most important source of carbohydrate of lactating newborns, including humans. Following the development of dairying, milk from other mammals has served humans as an important source of high-quality protein and fat and as a versatile ingredient of more complex foods developed by humans for their consumption.

All normal infants have the capacity to digest lactose regardless of their diet or ethnic origin. As a rule, this capacity disappears at the time of weaning, with only a minority of human populations maintaining the ability to digest lactose. A number of acquired conditions may also remarkably decrease the capacity to digest lactose, the most common of these being acute infectious diarrhea. Despite this, milk remains a high-quality, food staple in industrialized societies with strong dairy industries, in part due to its high palatability, protein quality, and calcium content. Milk and milk products may also constitute an important part of the diets of developing countries and, sometimes, the basis of food-relief programs worldwide.

The amount and extent of consumption of dairy products in different populations responds to a number of cultural, economic, agricultural, and political factors and may have far-reaching implications. Thus, the study of lactose and its digestion is

Encyclopedia of Human Biology, Volume 4. Copyright © 1991 by Academic Press, Inc. All rights of reproduction in any form reserved. **611**

pertinent not only to the sciences of food and nutrition but to the fields of medicine, biology, history, anthropology, and other social, political, and economic sciences as well.

I. Lactose and Milk

Lactose is a disaccharide composed of one molecule of glucose and one of galactose. It is present in milks of nearly all species of mammals, and it is found in very few other natural sources. Of all mammalian orders yet investigated, only the milks of some pinnipeds, such as the California sea lion, fur seals, and walruses of the superfamily *Otarioidea,* lack lactose.

The environment likely plays an important role in influencing the composition of milk. Thus, aquatic and arctic animals will produce milk of very high caloric density, which is supplied by their relatively high fat content (Table I). Human milk, on the other hand, has the highest concentration of lactose as well as the lowest concentrations of protein, calcium, and phosphate of known mammalian milks. Although speculative, it is provocative to relate these differences to the comparatively slow postuterine growth and high glucose requirement of the relatively large human brain.

Humans are the only mammals that will continue consuming milk as part of their diet beyond the age of weaning. Evidence indicates that domesticated animals were milked dating back to 4,000–3,000 B.C. in Southwest Asia and northern Africa; from these

TABLE I Composition of Milks of Selected Mammals[a]

	Lactose	Fat	Protein
Sea lion	0.0	36.5	13.8
Fur seal	0.1	53.3	8.9
Polar bear	0.3	33.1	10.9
Blue whale	1.3	42.3	10.9
Mouse	3.0	13.1	9.0
Dog	3.1	12.9	7.9
Brown bear	4.0	3.2	3.6
Goat	4.6	4.3	3.2
Indian elephant	4.7	11.6	4.9
Cat	4.8	4.8	11.0
Llama	6.0	2.4	7.3
Horse	6.2	1.9	4.5
Chimpanzee	7.0	3.7	1.2
Human	7.2	3.9	1.1

[a] Grams per 100 ml or grams per 100 g. No distinction or rounding off is made due to multiple sources and methodologies used in compiling these values.

areas, dairying spread to Eurasia and sub-Saharan Africa. Though milk and particularly dairy products are a staple in every continent of the world, consumption of these products, particularly of milk, grew significantly with the advent of modern dairy farming, which is relatively recent in origin. It is only within the past 100 years that specific advances in farm processing, conservation and preservation in conditions of sterility, refrigeration, and transportation have caused milk and milk products to become an important article of commerce, dramatically increasing its consumption in developed countries. The wide adoption of pasteurization and laws requiring specific food values in dairy products have also benefited the entire dairy industry.

II. Physiology of Lactose Digestion and Absorption

Dietary carbohydrate is usually ingested in the older child and adult as a mixture of starch, the disaccharides sucrose and lactose, small amounts of glucose and fructose, and nonabsorbable forms including stachyose, raffinose, some components of dietary fiber, and very small amounts of trehalose. Except for the monosaccharides, all other carbohydrates require digestion (hydrolysis) to monosaccharides prior to absorption, as shown in Fig. 1. Starch, a mixture of amylose and amylopectin, is hydrolyzed in the intestinal lumen by salivary or pancreatic alpha-amylases. The final products of amylose hydrolysis are maltose and maltotriose, and those of amylopectin are maltose, maltotriose, branched alpha-dextrins of six to ten glucose units, and small amounts of glucose. The further hydrolysis of these oligosaccharides and dietary disaccharides is carried out by the intestinal brush-border glycosidases. These glycosidases are proteins anchored to the microvillar surface of the intestinal cell, with most of their mass and catalytic sites protruding toward the small intestinal lumen. [*See* CARBOHYDRATES (NUTRITION).]

The major intestinal glycosidases are as follows. 1) The sucrase–isomaltase complex consisting of two subunits, sucrase, which hydrolyzes sucrose into glucose and fructose, and isomaltase (also called alpha-dextrinase), which hydrolyzes branched dextrins. Both subunits will also hydrolyze maltose and maltotriose. 2) The maltase–glucoamylase complex, consisting of two subunits with maltase activity. These subunits are very similar, but probably not identical, and hydrolyze the

INTESTINAL LUMEN · · · · · · · · · · · · · · · BRUSH BORDER · · · ENTEROCYTE

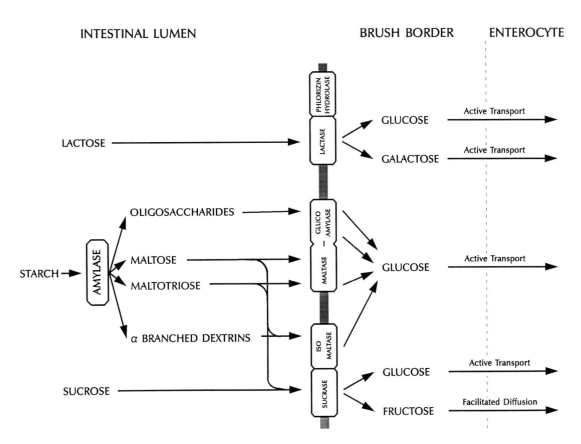

FIGURE 1 Major pathways of digestion and absorption of dietary carbohydrates. Trehalase is not shown. See text for explanation.

oligosaccharides maltose and maltotriose to glucose. 3) Trehalase, which hydrolyzes trehalose, a sugar of little dietary or clinical significance found in some mushrooms. 4) The lactase–phlorizin hydrolase complex, consisting of one unit of phlorizin hydrolase (also called glycosylceramidase) and one unit of lactase. The first unit can hydrolyze ceramides, phlorizin, and other beta-glycosides, and its nutritional and clinical importance has not yet been determined. Lactase hydrolyzes lactose into glucose and galactose as well as other beta-glycosides and is the focus of this paper. The first three complexes are composed of alpha-glycosidases, the last one of beta-glycosidases.

The glycosidases, except for trehalase, are synthesized as large polypeptide chains with molecular weights ranging from 200,000 to 250,000. These precursor forms are later cleaved into final forms either extracellularly, as for alpha-glycosidase complexes, or intracellularly, as in the processing of the beta-glycosidase complex.

The rate of hydrolysis of most oligosaccharides usually exceeds the rate of absorption of the resultant monosaccharides, whose transport becomes the rate-limiting step for final absorption. However, hydrolysis of lactose occurs at approximately one-half the rate of sucrose hydrolysis, even in apparently normal individuals, thus becoming the rate-limiting step for final absorption of glucose and galactose. This relatively low activity is an important factor in explaining the apparently high vulnerability of lactase activity to any mucosal intestinal injury.

The activity of lactase–phlorizin hydrolase and sucrase–isomaltase complexes is highest in the jejunum and decreases toward the distal small bowel. Glucoamylase activity, on the other hand, increases along the small intestine, achieving maximal activity in the distal ileum.

III. Lactose Malabsorption

A. Pathophysiology

The pathophysiologic events, which follow impaired hydrolysis and absorption of lactose, are shown in Fig. 2. If no hydrolysis occurs in the

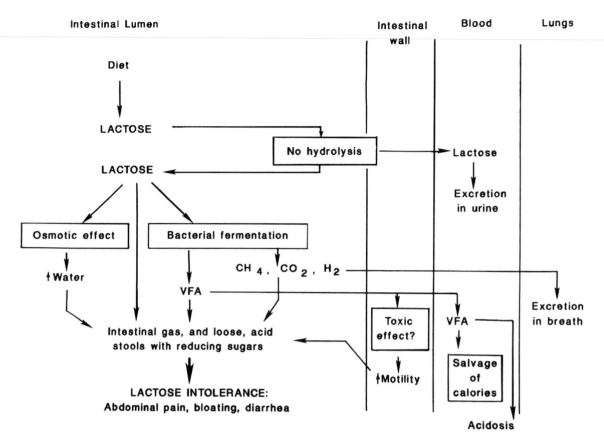

FIGURE 2 Pathophysiologic events following lactose malabsorption. VFA indicates volatile fatty acids, see text for explanation. (Reproduced, with permission, from the *Annual Reviews of Nutrition*, Vol. 9 © 1989 by Annual Reviews Inc.)

proximal intestine, the concentration of lactose in the distal lumen will increase. A portion of unhydrolyzed lactose may permeate the mucosa and enter the circulation, particularly under conditions resulting from intestinal mucosal damage. The urinary excretion of small amounts of lactose or galactose have occasionally been used as an indicator of lactose malabsorption.

The unabsorbed lactose will also exert an osmotic effect, increasing the amount of water and sodium in the lumen. The increase in volume of luminal contents may in turn accelerate intestinal transit and potentially cause a decrease in absorption of water and nutrients. Larger than normal amounts of protein, calcium, magnesium, and phosphate have been found in the ileum of lactase-deficient subjects studied by infusion-aspiration techniques, but the clinical significance of this has not been determined.

Upon reaching the colon, the unabsorbed carbohydrate will be subject to fermentation by bacteria. The products of fermentation include gases such as methane, carbon dioxide, hydrogen, and volatile fatty acids, particularly acetate, butyrate, and propionate. The passage of gas through the intestinal wall into the bloodstream, followed by excretion in

the lungs, is the basis for the breath-hydrogen test, which detects lactose malabsorption, as discussed below. It is estimated that 65–85% of carbohydrate reaching the colon as mono- or disaccharides can be metabolized to gas and short-chain organic acids. Absorption of organic acids by the colonic mucosa may constitute a significant pathway for salvage of energy, which has escaped small bowel absorption. This mechanism may be particularly important in neonates, who are known to malabsorb significant amounts of lactose while continuing to grow normally with no evidence of intolerance. However, should large amounts of organic acids be absorbed into the bloodstream at a rapid rate, they may overwhelm the capacity to metabolize these products and result in metabolic acidosis.

These processes correlate clinically with the presence of watery acid stools, which also contain reducing sugars representing the unabsorbed mono- and disaccharides. The formation of large amounts of gas, leading to abdominal distension, borborygmi, flatulence, and abdominal pain, accompanies the diarrhea. These symptoms constitute lactose intolerance.

B. Nutritional Implications

In studies with infusion-aspiration techniques, abnormally high quantities of protein, calcium, magnesium, and phosphate have been identified in the ileum of lactase-deficient subjects. On the other hand, several studies have shown that the absorption of protein, fat, and vitamins is not significantly affected when similar diets containing or lacking lactose are fed.

Although not a uniform finding, some investigators have found an association between osteoporosis and lactose malabsorption, particularly in postmenopausal women. A lower intake of milk and milk products has been postulated as the reason for this association. Some evidence indicates that lactose might enhance the absorption of calcium. On the other hand, blacks, who have an increased prevalence of lactose malabsorption, usually have a decreased prevalence of osteoporosis. Regardless of the mechanisms, the association between malabsorption of lactose and osteoporosis does not appear to be one of cause and effect.

Finally, some reports have suggested that individuals who absorb lactose may have a higher risk of developing ocular cataracts. Senile cataracts may be a consequence of persistence of lactase activity, high milk consumption, and a decreased capacity to utilize the galactose deriving from digested lactose. These findings await definitive proof.

C. Malabsorption and Milk Consumption

As discussed below, milk and milk products not tolerated in large quantities can be well tolerated by many individuals at lower doses. In this way, many malabsorbing populations can continue to consume nutritionally significant amounts of milk. Lactose malabsorbers are frequently unaware of their condition because they consume small amounts of milk, and children who develop symptoms such as abdominal pain after consuming milk continue to drink it because they are unaware of its relationship to their discomfort. Thus, milk consumption may remain surprisingly common, despite the high prevalence of lactose malabsorption in many populations.

Consumption may be related to a multiplicity of factors, including the availability of lactose and lactose-containing products, the way the milk products are processed, other components of the diet with which milk and dairy products are consumed, and cultural acceptability of these products. Other economic factors, such as the aggressiveness with which the benefits of milk and milk products are advertised, may play a role in societies with a strong dairying industry.

IV. Lactase

A. Ontogenetic Development and Decline of Lactase Activity

Lactase is the last of the disaccharidases to appear in the brush border of the fetal intestine. At 23 wk gestation, lactase is at approximately 10% of full-term levels, increasing to 30% between 26 and 34 wk gestation, and to 70% by 34–35 wk gestation.

It has been estimated, based on intestinal length and lactase activity, that a premature newborn born at 6 mo gestation could absorb only 3% of the lactose supplied with the amount of milk necessary to provide the calorie requirement for growth. Lactase activity peaks in the immediate postnatal period.

Premature infants frequently demonstrate large amounts of reducing sugars and short-chain fatty acids in their stool together with hydrogen in their breath when ingesting breast milk or lactose-containing formulas, indicating significant lactose malabsorption. Nevertheless, they are unique in that they do not show any clinical sign of intolerance and will gain weight adequately despite carbohydrate malabsorption. It is postulated that these infants have a particular capacity for "colonic adaptation." By absorbing the volatile fatty acids produced by fermentation, the infant can salvage a significant amount of calories, which would otherwise have been lost in the stool. Thus, to withdraw lactose from the diet of these infants, particularly from those who are being breast-fed is not justifiable.

Some evidence indicates that premature infants may quickly increase their lactase levels beyond those levels expected for a fetus of similar gestational age. Certain postnatal events, particularly the early introduction of enteral feedings to premature infants, may probably influence lactase activity independently of postconceptional age.

Lactase activity is high throughout lactation, and age-related prevalence data in humans suggest a decline in the postweaning period. Many studies show that the onset and rate of decline in lactase activity are influenced by ethnic background. Lactose malabsorption may occur as early as 2–4 yr in Thais and as late as 15 yr in Finns. The mechanisms responsible for this decline are discussed below.

B. Genetics of Lactase

Most humans and mammals are characterized by the phenotypic expression of low lactase levels in the postweaning period, or lactase nonpersistence (Table II). Thus, philogenetically, lactase nonpersistence in adult humans is the norm. However, lactase activity persists at the same levels found in early infancy in a minority of normal individuals. Several family studies indicate that the phenotypic expression of lactase nonpersistence in adult life is genetically determined. These observations are explained by a Mendelian type of inheritance consisting of two autosomal alleles: a recessive allele for lactase nonpersistence and a dominant one for lactase persistence. In other words, normal adult humans with low lactase activity are homozygous for the recessive autosomal allele, which determines the normal postweaning decline of lactase activity, and subjects with persistence of high lactase activity in adult life are either heterozygous or homozygous for a dominant allele. Thus, the ontogenetic postweaning decline in lactase activity can be described as a genetically determined, "hard-wired" developmental program. The actual synthetic and post-translational mechanisms responsible for this decline are only recently being elucidated.

High prevalence of persistence of lactase activity is observed in certain populations in Africa and central and northern Europe, whose ancestors were pastoralist tribes producing and consuming substantial amounts of dairy products. This phenomenon may be explained by a "cultural historical hypothesis" that suggests certain environmental conditions in these areas conferred significant selective advantages to early dairying societies for whom milk provided a major source of nutrition not readily available from other foods. This in turn conferred selective advantages on individuals with lactase persistence.

Assuming that rickets and osteomalacia may have been significantly strong selective factors in areas of northwestern Europe, where solar irradiation is low, and based on the observation that lactose may increase the absorption of calcium in subjects with high levels of lactase activity, the "calcium absorption hypothesis" has been suggested. This advantage in individuals with lactase persistence may have led to another form of selective pressure, which resulted in a high prevalence of lactase persistence in northern Europe.

The significant geographic distance between the northern European and African populations with high prevalence of lactase persistence may agree with the suggestion that different selective mechanisms operated in each of them. The genetic determinants of the capacity to absorb lactose, together with migratory phenomena, explain the geographic distribution of populations with differing prevalences of lactose malabsorption.

C. Biosynthesis and Processing of Lactase

In recent years, significant advances have occurred in understanding the mechanisms of synthesis and subsequent processing of lactase–phlorizin hydrolase in animals and humans. Initially, three separate beta-galactosidase activities were found in intestinal mucosa of both rats and humans. Only one of these, the mature form of lactase–phlorizin hydrolase, appears to be present in the brush-border membrane. Lactase is synthesized as a glycosylated

TABLE II Prevalence of Lactose Malabsorption in Selected Healthy Adult Populations

	Individuals with malabsorption (%)
North America	
Indian (Arizona)	95
Black	70–75
White	5–10
Eskimo (Canada)	73–100
Mexico	74
Peru	80
Brazil	45–95
Denmark	3
Finland	8–17
Italy	39–100
Africa	
Bantu	100
Yoruba	83
Fulani	60
Israeli Jews	45–85
India	50–100
China	76–100
Thailand	100
Japan	75–100

precursor of molecular weight 220,000, which undergoes intracellular cleavage (probably in a two-step fashion), to yield the mature form of lactase–phlorizin hydrolase of molecular weight 130,000.

The enzymatic, chromatographic, and electrophoretic characteristics of mature lactase are identical in infants and adults with lactase nonpersistence, implying no difference in the type of mature lactase synthesized and processed in adults as an explanation of the lower activity. Recent evidence in rats suggests that the postweaning decline of the specific activity of lactase is related to a decrease in the amount of mature lactase, associated with an increase in the amount of the precursor form of the enzyme. Lactase mRNA is present in both suckling and adult intestine, suggesting that the age-related changes in specific activity of lactase may result from alterations in translation or posttranslational processing of the enzyme precursor.

Other factors associated with a decrease of specific lactase activity are related to cytokinetic[1] mechanisms. Lactase activity has been shown to increase as the enterocytes mature and travel along the villus length. Activity reaches a maximum at the tip of the villus. Hence, the acceleration of enterocyte migration rate as rats reach the age of weaning may be responsible for a decrease in lactase activity due to the relative immaturity of enterocytes reaching the villus tip. A reduced life span of the enterocyte in the postweaning stage may also be an important factor. Some investigators believe that both changes in translational and posttranslational processing and cytokinetic factors explain the maturational decline in lactase activity.

Hormonal influences also affect lactase activity. Corticosteroids may have a regulatory effect, as there appears to be correlation between increasing levels of steroids and the gestational rise in lactase. In contrast, thyroid hormones appear to have a negative regulatory effect; they have been shown to cause an early decline in lactase activity in nursing rats and further decrease lactase levels in adult rats. On the other hand, the postweaning decline of lactase activity can be prevented by thyroidectomy, and the increase in lactase activity, seen during starvation, has been suggested to result from a decline in the concentration of circulating thyroid hormones. The mechanisms for these effects and their clinical significance are still unclear. [*See* THYROID GLAND AND ITS HORMONES.]

1. Related to intestinal cell turnover or renewal.

D. Lactose Intake and Lactase Activity

Lactase is not an inducible enzyme. Lactase activity cannot be maintained or increased in humans or experimental mammals by consuming lactose or dairy foods. Early studies of the prevalence of lactose malabsorption in different racial and ethnic groups showed that malabsorption was more prevalent in populations with lower consumption of dairy products. The "adaptive hypothesis" was advanced, which suggested that the persistent dietary consumption of lactose would maintain or increase the levels of enzyme activity. Experimental studies in rats and rabbits given very high doses of lactase show statistically significant increases in lactase activity. However, this apparent induction of lactase may not be substrate-specific and is far smaller than the 10-fold increase in lactase activity seen in suckling versus postweaning lactase levels.

Overwhelming evidence supports the fact that the maturational decline of lactase activity is independent of diet. When isografts of fetal mouse intestine were implanted into adult mice and never exposed to dietary substrate, they followed the same changes in disaccharidase activities as those observed in normally developing animals. The prolonged feeding of milk to rat pups delays but does not prevent the maturational decline of lactase activity. Many studies in humans have documented that withdrawal of lactose from the diet in human adults will not induce a decrease in the enzyme activity and, conversely, that the continued consumption of dairy products does not increase lactase activity. Lack of induction of lactase activity with consumption of milk or dairy products must not be confused with the apparent improved tolerance or "adaptation" to dairy products, which may occur with continued consumption. This phenomenon is discussed below.

V. Lactase Deficiency

A. Congenital Lactase Deficiency

This deficiency is an extremely rare autosomal recessive condition, which is characterized by isolated absence of lactase activity in the presence of normal levels of other disaccharidases and normal small bowel morphology. It is characterized by severe diarrhea, dehydration, malnutrition, and large amounts of lactose present in the stool. The deficiency is present at birth and persists throughout life, and treatment requires removal of lactose from the diet.

B. Secondary Lactase Deficiency

Any condition that injures the intestinal mucosa can lead to a transient or secondary decrease in lactase activity. The most common cause of secondary lactase deficiency is acute diarrheal disease. The prevalence of lactose malabsorption in infants and children with acute gastroenteritis has been found to be 50% or more. Of the infectious etiologies, rotaviral diarrhea appears to be one of the most commonly associated with malabsorption. Giardiasis and ascariasis have also been associated with lactose malabsorption. Immune-mediated injury of the small bowel and inflammatory bowel disease may, likewise, be associated with secondary lactase deficiency.

Development of lactose malabsorption during infantile diarrhea may significantly complicate the dietary management of this illness, particularly in developing countries where malnutrition coexists. Nevertheless, infants who are breast-fed exclusively often appear to continue to tolerate milk even in the presence of significant lactose malabsorption, as will children with mild gastroenteritis. On the other hand, significant lactose loads may cause large increases in stool output in infants with severe gastroenteritis, which may increase the risk for dehydration and acidosis. In these situations, lactose-free formulas may be recommended.

VI. Lactose Intolerance

A. General Considerations

Symptomatology resulting from suboptimal digestion and absorption of lactose includes watery acid diarrhea, abdominal distension, flatulence, and pain. The presence of these symptoms, most of them subjective, indicates intolerance, which may occur only in a subset of individuals who are lactose malabsorbers. In a group of individuals with low lactase levels, the threshold dose of lactose found to elicit symptoms of intolerance ranged from 3 to 90 g of lactose. Symptoms of intolerance to a 12-g dose of lactose have been found to occur in 0–75% of subjects with low lactase activity. Numerous studies indicate that a history of symptoms of intolerance to milk products does not accurately predict

the capacity of an individual to digest a finite quantity of milk, and, conversely, a history of "tolerating milk products" by an individual is not predictive of their capacity to absorb lactose. Thus, the diagnosis of lactose malabsorption utilizing direct or indirect methods is a standardized way of classifying an individual as an "absorber" or "malabsorber" of lactose but may not necessarily be reflective of the actual degree to which an individual can consume or tolerate lactose as part of their daily diet.

B. Relation of Malabsorption to Intolerance

The discrepancy between malabsorptive and intolerant states is attributable to a number of factors aside from the degree of lactase activity, as illustrated in Fig. 3. First, malabsorption is often diagnosed on the basis of a lactose tolerance test with "unphysiologic" loads of lactose, e.g., 24–50 g (equivalent to the amount found in 2 glasses (16 oz.) to 1 liter of milk). The majority of individuals diagnosed as being lactose malabsorbers will not develop symptoms of intolerance following the consumption of a single 8-oz. glass of milk, which contains approximately 12 g of lactose. Only one-fifth to one-third of malabsorbers will develop symptoms following this "physiologic" dose of lactose. Subsequently, even individuals with lactose malabsorption by objective testing may be able to consume significant amounts of dairy products without symptoms.

Second, the amount of unabsorbed carbohydrate reaching the colon is dependent on the amount and rate at which the substrate is delivered to it. Hence, any mechanism that slows down the rate of substrate delivery to the small bowel will decrease the chances of developing symptoms of intolerance. This may explain the apparent improvement in ab-

Low	Lactase Activity	High
High	Lactose Load	Low
Rapid	Gastric Emptying	Slow
Rapid	Small Bowel Transit	Slow
Absent	Colonic Adaptation	Present

SYMPTOMS   NO SYMPTOMS

FACTORS WHICH DETERMINE LACTOSE INTOLERANCE

FIGURE 3

sorption when enteral feeding formulas are delivered to the stomach by continuous drip instead of boluses, and why intolerance improves when lactose is given as part of a meal or with whole milk, which delays gastric emptying. In contrast, aqueous solutions of lactose or skim milk result in more rapid gastric emptying and more symptoms. Fat content, osmolarity, and the presence of other sugars conceivably may delay gastric emptying. Thus, the vehicle of delivery of lactose to the small bowel may be of significant importance to the development of symptoms of intolerance. Rapid gastric emptying of lactose will increase the likelihood of malabsorption, as seen in some patients who have undergone surgical gastrectomy. Lactase-deficient subjects will also empty lactose from the stomach faster than equivalent doses of lactose, presumably due to a lesser osmotic regulatory effect of lactose as compared with glucose on the rate of gastric emptying. It is also possible that individual and/or diet-related variations in the rate of small bowel transit and "colonic filling" may lead to different degrees of tolerance.

Third, because intolerance is primarily based on subjective symptomatology, we can expect significant differences in sensation and perception, which may be influenced by psychological factors. Because the factors mentioned above can significantly affect the symptoms related to malabsorption, intolerance to lactose can only be convincingly demonstrated with double-blind challenges.

C. Adaptation to Lactose Consumption

Several investigators have noted that prolonged exposure to lactose by malabsorbing humans and animals may lead to improved tolerance. For example, malabsorbing Nigerian blacks introduced to ice cream had resolution of symptoms of intolerance after several months of consumption. Diarrhea in lactase-deficient rats ceases over time after continued ingestion of lactose. Prolonged exposure of lactose to colonic flora possibly may induce quantitative or qualitative changes in fermentation, gas production, and diarrhea, which may lead to improved tolerance. Potential mechanisms include the fact that prolonged fermentation will decrease fecal pH, inhibit bacterial metabolism, and limit the rate of production of gas and acid, or that an acid colonic pH may result in conversion of enteric flora to other acidophilic types, which may use alternate fermentative pathways. It has also been postulated

that bacterial lactase could be inducible, thereby increasing digestion of lactose in the colon.

VII. Diagnosis of Lactose Malabsorption

A. Screening Techniques

Two simple methods used to screen for malabsorption of lactose or other carbohydrates are the presence of reducing sugars in stool and fecal pH. The first method is performed with a Clinitest tablet, which is a modification of the Benedict test in which a color change will denote the presence of reducing sugars. Stool pH can be measured with either a pH meter or indicator paper. A pH of <5.5 is indicative of significant fermentation.

B. Enzyme Assay

The assay for brush-border lactase activity is the only test that directly measures activity of the enzyme. It requires intestinal tissue obtained by peroral biopsy of the small bowel from which mucosal homogenates are obtained and activity of lactase–phlorizin hydrolase is measured in units per gram of wet tissue or units per gram of tissue protein. Disadvantages of this method include the need for an invasive procedure and sophisticated laboratory techniques. Furthermore, the measurement establishes the degree of activity only at the level of bowel from where the sample is taken. Thus, this test may not reflect the entire capacity of the small bowel to hydrolyze a lactose load.

C. Radiographic Studies

In these tests, lactose is administered together with a fine barium-sulfate suspension. Changes in transit and dilution of the barium as it progresses through the bowel (as seen on X-rays) are used as indices of absorption. These studies are limited by the subjective nature of the interpretation and the exposure to radiation. In addition, small bowel pathology precludes accurate interpretation.

D. Blood Tests

Measurement of blood glucose or galactose after ingestion of lactose was, for some time, the most common indirect test of lactase activity. Blood samples are taken at various time intervals, >2 hr after ingestion of the test dose. A rise in glucose of >20 or 25 mg/dl at any time has been used as a criteria for normal lactose digestion. The glucose test is obviously subject to many other absorptive or hormonal factors, which may affect glucose metabolism. Plasma galactose has also been measured in blood. Taking advantage of the fact that ethanol inhibits hepatic conversion of galactose to glucose, an oral test dose of lactose together with alcohol is given, and plasma galactose is measured serially. A rise of <5 mg/dl is considered consistent with lactose malabsorption. Glucose and galactose measurements require serial blood drawing, and relatively large test doses of lactose are necessary for adequate interpretation.

E. Breath-Hydrogen Tests

At present, the most widely used, sensitive, inexpensive, and noninvasive test available for the diagnosis of lactose maldigestion is the breath-hydrogen test. It has become the ''gold standard'' for this purpose. The test is based on the fact that hydrogen in breath results exclusively from the fermentation of carbohydrate by colonic bacteria. A stoichiometric relationship exists between the amount of hydrogen produced in the colon and the amount of unabsorbed carbohydrate, which serves as a substrate. It has been estimated that between 14 and 21% of hydrogen produced in the colon passes into the portal circulation and is excreted in breath. Samples of breath are taken at varying intervals, following the administration of the test carbohydrate, and analyzed by gas chromatography, using relatively inexpensive, commercially available, instruments.

The original, relatively cumbersome techniques of continuous collection of breath have been replaced by sampling of end-expiratory[2] air using masks, cannulas, or nasal prongs. The samples can be collected in plastic syringes or stored in specialized containers for ≤ 1 mo. An increase in breath-hydrogen concentration of $>10–20$ ppm above the baseline value are considered abnormal and indicative of malabsorption. Correlation with mucosal lactase activity is well established.

A number of factors can interfere with interpretation of the breath-hydrogen test and need to be taken into consideration. Between 2 and 27% of in-

---

2. End-expiratory refers to the last portion of air exhaled in a single expiratory movement.

dividuals in some populations may not have hydrogen-producing flora, and antibiotic therapy may cause quantitative or qualitative changes in the flora, which may also invalidate the procedure. The acidification of colonic contents by continued ingestion of carbohydrate may inhibit bacterial hydrogen production. Cigarette smoking during the test has been shown to abruptly increase the hydrogen concentration, although it returns to baseline shortly after smoking ceases. Finally, because hydrogen is absorbed and transported in the bloodstream, any impairment in circulation will decrease its concentration in breath.

The breath-hydrogen test is generally performed using 2 g/kg body weight of lactose with a maximum of 50 g in an aqueous solution. This amount of lactose is the equivalent of that in a quart of milk and is unphysiological. Due to the relatively high sensitivity of the breath-hydrogen test, a more ''physiological'' amount of lactose can be given. A 12-g dose of lactose, the amount present in an 8-oz. glass of milk, is used by some investigators to determine if an individual has malabsorption. Whether an individual is truly intolerant to lactose or milk can be determined by a double-blind food challenge.

VIII. Management of Lactose Intolerance

Because many individuals who are classified as having lactase nonpersistence can still tolerate significant amounts of milk, total elimination of lactose or dairy products is, therefore, generally unnecessary. The amount of lactose should be limited to the threshold above which symptoms occur. A number of other strategies are available to facilitate the consumption of dairy products and limit the associated symptoms. Consuming milk as part of a meal may improve tolerance. Whole milk or chocolate milk are associated with less symptoms than skim milk, presumably attributable to decreased rates of gastric emptying.

Significant recent advances in enzyme technology have permitted the development of milk in which the lactose has been prehydrolyzed and is better tolerated. Wider acceptance may be hindered by its increased cost and sweetness as compared to regular milk. Lactase of yeast and fungal origin is also commercially available in tablet and liquid form. These can be used either to pretreat milk or by ingesting the enzyme just prior to intake of milk or milk products.

Improved tolerance of dairy products such as yogurt and other fermented foods has been demonstrated. Two reasons are advanced to explain this phenomenon: 1) Decreased concentration of lactose is found in these products due to the fermentation by bacteria used in yogurt cultures (*Streptococcus thermophilus* and *Lactobacillus bulgaricus*). 2) These organisms contain beta-galactosidase activity, which appears to survive gastric digestion and enhance lactose digestion *in vivo*. Using breath-hydrogen production as a measurement of malabsorption, the digestion of lactose in yogurt and subsequent tolerance is improved compared with milk. Unflavored yogurt appears to have an improved effect on tolerance compared with flavored yogurt. "Frozen yogurt," as manufactured in the United States by current commercial methods, contains very little lactase activity. The means of processing and preserving yogurt products may significantly alter the degree of absorption and tolerance. Sweet milk acidophilus, containing *Lactobacillus acidophilus,* has been associated with some modest improvement in absorption and tolerance, but results of studies evaluating such products have not been very consistent.

Total elimination of lactose or dairy products from the diet may be particularly unwarranted in individuals where such restriction may actually compromise nutritional status. Dairy products, as consumed in North America, represent a significant proportion of high-quality protein and may contribute to up to 10% of total energy derived from food, 75% of calcium, 37% of riboflavin, 34% of phosphorus, 20% of magnesium, and 20% of vitamin B_{12} available in the typical diet.

Because of the high prevalence of lactose malabsorption in developing countries, the appropriateness of using milk or milk supplements in food-relief or supplemental feeding programs has been questioned. Nevertheless, no major adverse experiences have been identified despite widespread use of milk in many populations, including hospital patients and preschool and school-age children. The benefits gained by malnourished children from small but nu-

tritionally significant amounts of milk have been well-documented, and children with kwashiorkor can be successfully rehabilitated with milk and milk-containing products. Therefore, to discourage the use of milk or milk products as a food staple, or as the basis of a supplemental or nutrition-relief program, solely on the basis of the high prevalence of lactose malabsorption in certain populations seems inappropriate.

Similarly, milk consumption should not be discouraged as part of the treatment of childhood malnutrition as long as it is available and inexpensive. Efforts should be directed toward the development of adequate alternatives to milk for infant-weaning diets in areas where milk is costly or not available. Also, alternatives may need to be developed for the dietary management of severe acute infantile diarrhea, where continuation of milk as a sole source of nutrition may worsen the clinical and nutritional outcome.

Bibliography

Flatz, G. (1989). The genetic polymorphism of intestinal lactase activity in adult humans. *In* "The Metabolic Basis of Inherited Disease," 6th ed. (C. R. Scriver, J. B. Stanbury, J. B. Wyngaarden, and D. S. Fredrickson, eds.), pp. 2999–3006. McGraw-Hill, New York.

Larson, B. L., and Smith, V. R., eds. (1974). "Lactation. A Comprehensive Treatise. Vol. III. Nutrition and Biochemistry of Milk Maintenance." Academic Press, New York.

Paige, D. M., and Bayless, T. M., eds. (1981). "Lactose Digestion. Clinical and Nutritional Implications." The Johns Hopkins University Press, Baltimore.

Saavedra, J. M., and Perman, J. A. (1989). Current concepts in lactose malabsorption and intolerance. *Annu. Rev. Nutr.* **9,** 475–502.

Scrimshaw, N. S., and Murray, E. B. (1988). The acceptability of milk and milk products in populations with a high prevalence of lactose intolerance. *Am. J. Clin. Nutr.,* Suppl. **48,** 1083–1159.

Semenza, G., and Auricchio, S. (1989). Small-intestinal disaccharidases. *In* "The Metabolic Basis of Inherited Disease," 6th ed. (C. R. Scriver, J. B. Stanbury, J. B. Wyngaarden, and D. S. Fredrickson, eds.), pp. 2975–2997. McGraw-Hill, New York.

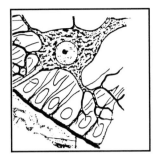

Laminin in Neuronal Development

HYNDA K. KLEINMAN, BENJAMIN S. WEEKS, *National Institutes of Health and National Institute of Dental Research*

I. Basement Membranes
II. Laminin
III. Cellular Receptors
IV. Summary

Glossary

Basement membrane Thin sheets of extracellular matrix, which underlies epithelial and endothelial cells and surrounds nerve, fat, and muscle cells

Extracellular matrix Acellular material composed of collagens, proteoglycans, and glycoproteins that is located outside of cells and serves both structural and biological roles in tissues

Integrins Family of cell-surface molecules that function as receptors for several adhesive molecules such as laminin, fibronectin, collagen, fibrinogen, vitronectin, and thrombospondin

Laminin Large (M_r = 800,000) basement membrane glycoprotein

NEURONAL CELL INTERACTIONS with the extracellular matrix and, particularly, with the basement membrane glycoprotein laminin are important for axon guidance and growth during nerve development and regeneration. *In vitro*, laminin has been shown to promote both peripheral and central neuronal cell adhesion, process outgrowth, and survival. *In vivo*, laminin has been demonstrated in regions of developing nervous tissue, where both peripheral and central axons are formed and supplements of laminin as nerve guides can promote regeneration of nerves. Recent molecular biological studies have defined the structure of laminin and allowed for the identification of some of its biologically active sites. This chapter will review the recent progress in understanding the interactions of

neuronal cells with laminin including the cellular receptors and active sites on laminin. Progress is also beginning to be made on the intracellular signaling involved in the strong cellular response to laminin and on the use of laminin *in vivo* to facilitate repair.

I. Basement Membranes

A. Tissue Localization

Basement membranes are the thin sheets of extracellular matrices that underlie epithelial and endothelial cells and surround nerve, fat, and muscle cells, as demonstrated by numerous immunocytochemical studies (Fig. 1). These structures are found in all tissues. In the developing embryo, basement membranes are the first extracellular matrix to be synthesized and laminin is the first compound of the basement membrane to be detected. In both the embryonic and adult peripheral nervous system, an abundant basement membrane is produced by the Schwann cells, which surround the axons (nerve fibers) (Fig. 2), and is present at the neuromuscular junction, where nerve signals are transferred to the muscle. In the central nervous system, basement membranes are only present in contact with nerves during embryonic stages and not in the adult tissues. In developing tissue, basement membranes likely form barriers that help to define tissue boundaries and act as tracts along which cells migrate and differentiate. In the adult, basement membranes function as molecular filters in capillaries and in the kidney to restrict protein loss and also serve as barriers to separate different cell types. They also are very biologically active and likely hold epithelial cells in tissues and maintain them in a differentiated polarized state. Recent studies using basement membranes, either formed by cultured cells or re-

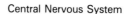

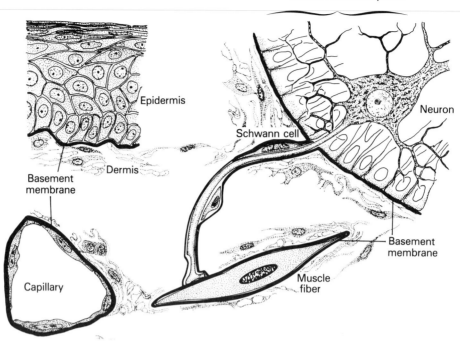

constituted *in vitro,* have demonstrated the ability of this extracellular matrix to promote cell differentiation *in vitro* and *in vivo.* [*See* EXTRACELLULAR MATRIX.]

FIGURE 1 Schematic model of tissue distribution of basement membrane. [Reprinted, with permission, from Leblond and Inoue, 1989, *Am. J. Anat.* **185,** 367–390.]

B. Structure and Composition

Basement membranes are thin (40 nm) sheets of extracellular matrix in which the details can only be seen with the electron microscope (Fig. 2). They generally contain a central region that is more dense (lamina densa) than the region adjacent to the cells or to the underlying stroma (lamina lucida). The reason for the different zones within the basement membrane is unclear but could be due to variations in the amounts and/or types of the components in the matrix. Basement membranes are composed of several matrix-specific components, including the structural protein collagen IV, the glycoproteins laminin and entactin, and a large heparin sulfate proteoglycan (Table I). These components are present in all basement membranes in varying proportions. Other components are present in tissue-specific distributions such as bullous pemphigoid antigen, which is present in skin. Few of these tissue-specific components have been described due to the problems of limited tissue availability, but many are likely to exist. The major components of

the basement membrane have all been shown to interact with each other at specific sites on the molecules. Laminin, collagen IV, and the heparin sulfate proteoglycan can also self-aggregate *in vitro.*

C. Biological Activity

As demonstrated from various *in vitro* studies, basement membranes are biologically active when used as a substratum for cell lines or primary cells. The basement membranes used for *in vitro* studies are either reconstituted extracts from basement

TABLE I Basement Membrane Components

Component	Size	Function
Laminin	800 kDa	adhesion, migration, neurite outgrowth, differentiation, growth
Collagen IV	~350 kDa	structure, cell adhesion
Entactin–nidogen	158 kDa	adhesion
Heparin sulfate proteoglycan	~800 kDa	adhesion, filtration

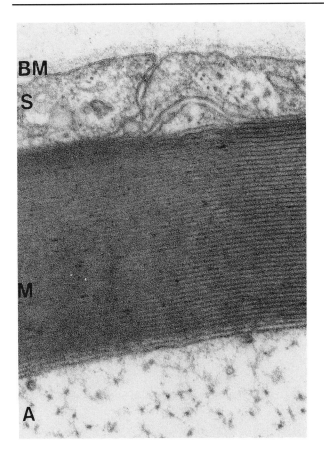

FIGURE 2 Electron micrograph of a basement membrane from spinal nerve root. A, axon; BM, basement membrane; M, myelin; S, Schwann cell. [From J. Wrathall, Georgetown University.]

II. Laminin

A. Structure

Laminin is the major and ubiquitous glycoprotein found in basement membranes. It apparently has several different isoforms. The form initially isolated from neutral extracts of the Englebreth-Holm-Swarm tumor (EHS) and from cultured teratocarcinoma cells is composed of three large polypeptide chains designated A (M_r = 400,000), B1 (M_r = 210,000), and B2 (M_r = 200,000). These chains are held together by disulfide bonds and by coiling of a portion of the chains to form a unique cross-shaped structure (Fig. 3). Both electron microscopic observations and computer analyses of the nucleotide sequences of the three chains have been used to define the structure of the molecule. It is composed of several rodlike and globular structures. The A chain contains 3,060 acids. The amino terminus of the A chain has three globular domains separated by three cysteine-rich segments of approximately 50 amino acids, which are repeated multiple times. When

membrane-producing tumors or intact basement membranes from tissues such as amnion. When in contact with such basement membranes, epithelial cells quickly become highly differentiated. For example, mammary cells establish polarity, form ductules, and demonstrate an 80-fold increase in casein synthesis over that observed with cells grown with normal plastic culture dishes. When neural-derived cells are placed on a basement membrane substratum in culture, they also remain highly differentiated. Cultured dorsal root ganglia cells extend long processes on a basement membrane matrix with accompanying Schwann cells, which elongate, wrap around the neural processes, and produce myelin (the sheath surrounding the neural processes) within 4 days in culture. *In vivo*, this matrix has been found to promote the survival of explants in the central nervous system and to promote repair when placed in nerve guides. The major biologically active species for neural cells in basement membranes is laminin.

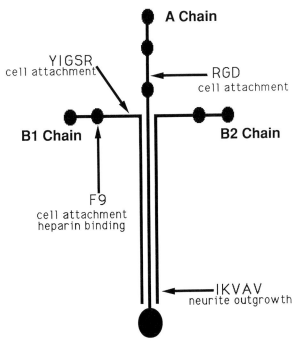

FIGURE 3 Schematic model of the laminin model. Laminin is composed of three chains, A, B1, and B2, which are held together by disulfide bonds. The location of various active synthetic peptides are shown. Y1GSR (tyr-ile-gly-ser-arg), RGD (arg-gly-asp), and IKVAV (tyr-ile-gly-ser-arg) are designated by the single letter amino acid code.

aligned, the eight cysteines of each repeat are in the same position and have considerable homology to similar repeats in epidermal growth factor (EGF) and in transforming growth factor (TGFα). The long arm of the A chain has two domains. The sequence predicts an α-helical structure with heptad repeats that can participate in a coiled-coil and a large (M_r = 100,000) carboxyl-terminal globule. The B1 and B2 chains are homologous to the A chain and to each other except as listed below. At the amino terminus (short arm), both chains contain only two globules and two EGF-like, cysteine-rich repeats. Both chains lack a large carboxyl-terminal globule as well. In addition, the B1 chain contains a 30 residue interruption in the coiled-coil domain, which includes six cysteines. The function of this interrupting sequence is not known, but in the electron microscope a bend in this approximate location on the long arm is observed.

Not all laminin molecules are composed of three chains. Several cultured cells, including Schwann cells, produce only the B chains, which form a Y-shaped molecule. Some tissues also produce laminin with a different structure. Analyses of messenger RNA for all three laminin chains in various adult murine tissues indicated very low levels of the A chain and varying proportions of the B1 and B2 chains, suggesting that isoforms of laminin likely exist.

Homologues of laminin also have been described. In placenta, a species designated M has an intermediate size between the A and B chains. Glioma cells have been found to synthesize a variant form of laminin, which has a very different A chain and/or lacks the A chain. At synaptic sites, a laminin chain, designated S-laminin, with 40% homology to B1 has been identified. Whether this S-laminin chain assembles with homologues of B1 and A or with the already described B1 and A chains is not yet known. At least three other homologues of laminin have been described: merosin, from muscle, has homology to the A chain. An additional alternate form of the A chain has been observed in adipocytes. A portion of the heparin sulfate proteoglycan has sequence with strong homology to the amino-terminal portions of the laminin chains. Thus, laminin has multiple forms and may be composed of different gene products in various tissues. Such modifications of laminin may serve to regulate cellular behavior at different times during development and in repair.

B. Localization of Laminin in Nervous Tissue

In vivo laminin is present in the peripheral nervous system where it is produced by the Schwann cells and to a lesser extent in the brain where it is only present at certain times. Schwann cells produce laminin in culture but do not assemble a basement membrane unless axons are in contact with them. *In vivo,* peripheral nerves are ensheathed in a laminin-rich matrix produced by the Schwann cells. When peripheral nerves are damaged, regeneration rapidly occurs along the tracts of basement membrane left behind. These data are consistent with *in vitro* studies where neurons follow the "path" of laminin. It has been postulated that the peripheral nervous system can regenerate after damage due to the presence of a basement membrane, unlike the central nervous system, which lacks a basement membrane and cannot regenerate. In the central nervous system, laminin is present in the capillary basement membrane, in the external limiting membrane (pia mater), and at the ependymal layer during development. Laminin is also present in the migratory pathways of neural crest cells. Patches of laminin are observed in several regions of the brain parenchyma during neural development, but with maturation laminin levels are reduced in adult tissue. Upon injury to adult brain, a transient appearance of laminin has been reported. Astrocytes also secrete laminin transiently after hemisection of the spinal cord. Thus, laminin is present in the peripheral nervous system and in embryonic brain, where it presumably functions to promote neural cell adhesion, survival, and growth. Although reduced in adult brain parenchymal tissue, the mechanisms to respond to laminin are in place. Exogenous laminin can promote rapid growth and regeneration and increase transplant survival when added exogenously (see below).

C. Biological Activities of Laminin

Laminin exhibits numerous biological activities on various cell types, including stimulating cell adhesion, migration, growth, differentiation, neurite outgrowth, collagenase IV and tyrosine hydroxylase production, and the promotion of the metastatic phenotype (Table II). When coated on a substratum, such as a culture dish, cell attachment occurs rapidly (30–60 min), requires low concentrations of laminin (1–10 μg/cm^2), and results in cell

TABLE II Biological Activities of Laminin

Adhesion[a]
Morphology
Migration[a]
Growth
Survival[a]
Differentiation
Neurite outgrowth[a]
Collagenase IV
Tyrosine hydroxylase[a]
Tumor metastases[a]

[a] Demonstrated with neural or neural-derived cells.

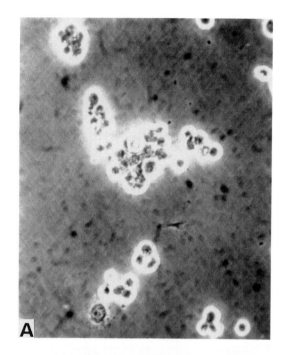

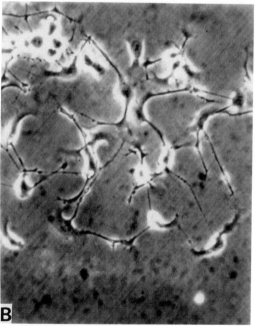

FIGURE 4 Effect of laminin on NG108-15 neuroblastoma × glioma process formation. Laminin (50 ng/ml) was added to a 16-mm culture dish and the cells were photographed 1 hr later. A, no additions; B, laminin.

spreading or other morphological changes depending on the cell type (Fig. 4). For example, Sertoli cells become more columnar, whereas Schwann cells become very elongated.

Progress has been made in identifying the active sites on laminin for these biological activities using proteolytic cleavage, fragment- and site-specific antibodies, and synthetic peptides. Protease digestion of laminin followed by microscopic, immunochemical, and biological assays demonstrated two major active domains. A pepsin-generated fragment (P1) containing the three short arms (amino portions) was found to promote cell adhesion and growth and inhibit tumor metastases. Antibodies to this fragment also inhibit cell adhesion. A heparin-binding fragment produced by elastase digestion (E8) and containing the end of the long arm was found to promote cell adhesion, bind heparin, and stimulate neurite outgrowth. Both polyclonal and monoclonal antibodies reactive with this region blocked cell adhesion and neurite outgrowth. A monoclonal to a domain just above the terminal, heparin-binding globule in this domain and an antibody to a peptide located just above the globule blocked laminin-mediated neurite outgrowth. These data suggest that the active site for neurite outgrowth is located at the end of the long arm just above the globule. Additional sites may exist in the cross region.

Synthetic peptides have been used to further define the biologically active sites on laminin. To date, five active peptides have been described. These peptides have been localized and identified from fragment and antibody studies. On the B1 chain, two active sites have been identified. The YIGSR-containing peptide (tyr-ile-gly-ser-arg) is localized within the inner EGF-like repeat and has been

found to promote cell adhesion and migration and block B16F10 melanoma colonization of lungs and endothelial cell differentiation *in vitro*. It can also block both cardiac and neural crest cell explant outgrowth into basement membrane gels. This peptide has some ability to promote neural cell adhesion. It is partially active for adhesion with NG108-15 neuroblastoma and PC-12 cells and is fully active with type I astrocytes. Cerebellar and septal neurons do not bind to this peptide at all. Another active peptide, designated F9, is localized to the inner globule on the short arm of the B1 chain. This peptide promotes cell adhesion and heparin binding. It is not active with neural cells. Located in the EGF-like repeat region between YIGSR and F9 is the active peptide PDSGR, which is similar in activity to YIGSR but somewhat less active. Within the central EGF-like repeat of the A chain is an RGD-containing sequence. This sequence has been found to be present and active for cell adhesion in many adhesion glycoproteins including fibronectin, vitronectin, thrombospondin, and fibrinogen. This peptide promotes cell spreading and can block B16F10 melanoma cells from colonizing the lungs. This sequence in laminin is also biologically active, as demonstrated by competition studies. This sequence can promote the adhesion of NG-108-15 neuroblastoma cells and some processes are observed but most neural cells are unresponsive. It should be noted that all three sites described above are contained within the P1 fragment of laminin, which is active for cell adhesion and for blocking tumor cell colonization of lungs. [*See* PEPTIDES.]

Two of the active synthetic peptides derived from laminin sequences have been shown to promote neural cell responses. Both peptides are localized to the end of the long arm of laminin (E8 domain), which is known to promote cell adhesion and neurite outgrowth. One peptide, RNIAEIIKDI, is from the B2 chain and has been shown to promote cerebellar neuron differentiation. Another synthetic peptide from the A chain, designated PA22-2 (RARKSIKVAVSADR), has been found to be active with several types of primary and established neuronal cells (Fig. 5). This peptide promotes neurite outgrowth with PC12 cells. The long processes observed with this peptide are indistinguishable from those observed on laminin and nearly the same number per cell are observed. Primary cells such as type I astrocytes, septal cells, and cerebellar neurons are all strongly affected by this peptide. Two studies suggest that this site on laminin is biologically important. This peptide can effectively block PC12 cell response to laminin in a dose-dependent manner, whereas the YIGSR peptide is in-

active. In addition, antibody raised against this peptide blocks peptide-mediated process formation by >90% and laminin-mediated process formation by approximately 45%. These data suggest that this site on the A chain is a major neurite-promoting site on laminin, but other sites may exist. This peptide also has other activities with non-neuronal cells. It promotes cell adhesion, spreading, and migration and promotes murine melanoma colonization of lungs presumably due to increased collagenase IV synthesis. Thus, PA22 has several activities in common with the B1 chain peptides F9, PDSGR, and YIGSR and the A chain peptide ALRGDQ, but it is distinct in its ability to promote neurite outgrowth. Because laminin is a highly immunogenic molecule, even in the same species, it is possible that active synthetic peptides may be used therapeutically in nerve regeneration studies.

D. Effects of Laminin on Neuronal Cells

Of the many extracellular matrix components that promote neurite outgrowth, laminin is the most potent and has the most rapid effects. In addition, neurite outgrowth is observed in the presence of picomolar amounts of laminin, a concentration well below that of other active matrix components. Most neuronal cells respond to laminin, and pluripotent teratocarcinoma cells are able to form neurons in the presence of laminin (Table III). Both peripheral

TABLE III Neuronal Cell Response to Laminin

In vitro	*In vivo*
Central nervous system	
Neural crest	Optic nerve
Septal cells	Spinal cord
Hippocampal neurons	Septal cells
Retinal ganglia	Dopamine neurons
Optic lobe	Serotonin neurons
Corpus striatum	Norepinephrine neurons
Hypothalamic neurons	Retinal neurons
Mesencephalic neurons	
Peripheral nervous system	
Dorsal root ganglia	Sciatic nerve
Sympathetic cervical ganglia	
Ciliary ganglia	
Spinal sensory ganglia	
Cell lines	
NG108 neuroblastoma × glioma	
Neuroblastomas: LAN-5, BE2, KCNR, AS, C1300, NGP, P38[2]	
PC12	
Rugli	

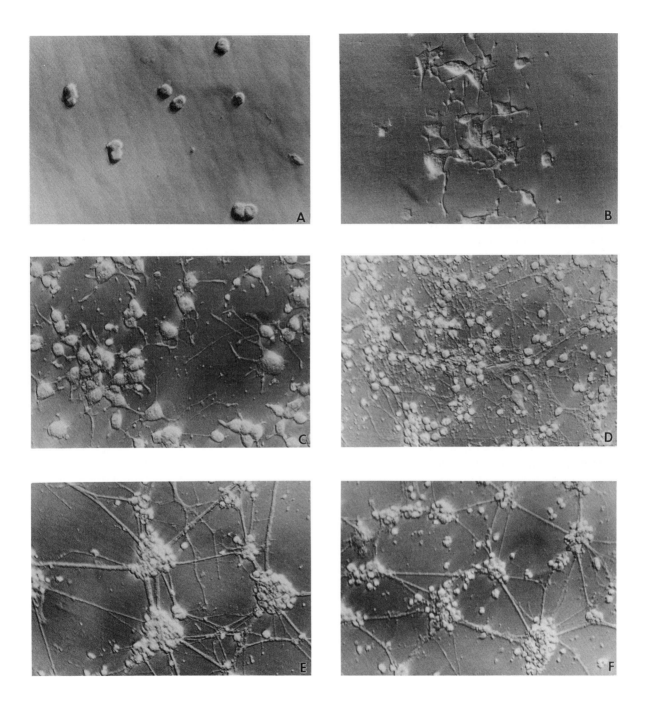

FIGURE 5 Effect of laminin and peptide PA 22-2 on PC12 and cerebellar neuronal cell differentiation. PC12 cells were primed with NGF for 24 hr, then replated on either plastic (A), laminin (1 μg/16 mm culture dish) (B), or laminin A chain peptide PA 22-2 (20 μg/16 mm culture dish) (C). Freshly isolated primary cerebellar neurons were plated on either polylysine (D) or laminin (E) and peptide PA 22-2 (F) as described above. [Reprinted, with permission, from Sephel *et al.*, 1989, *Biochem. Biophys. Res. Commun.* **162,** 821.]

and central neuronal cells, as well as established cell lines, respond to laminin with rapid adhesion and extension of neural processes. The effect of nerve growth factor (NGF) is further augmented by the presence of laminin. The response to laminin is seen with all neuronal cells tested to date, and the effects of polylysine substrates are also further augmented by laminin.

In vivo laminin and basement membrane have been used to promote repair. For example, rolled amnion membranes (with the epithelial cells removed), when implanted into the lesioned brain promoted cholinergic axons to extend long processes. The reconstituted basement membrane gel (Matrigel) has also been found to greatly increase the survival and growth of fetal neurons transplanted into the striatum. Matrigel has also been fashioned into hollow nerve guides and found to promote regeneration of several different types of severed nerves. When placed inside a biodegradable nerve guide, laminin accelerated the regeneration of peripheral nerve and promoted regeneration of optic nerve. Laminin also guides and facilitates fiber outgrowth from fetal serotonin and norepinephrine neurons transplanted into adult brain. These studies demonstrate that basement membrane and laminin are active *in vivo* and support the idea that these materials may be used in repair. It should be noted also that a major problem in nerve repair is the formation of scar tissue. As laminin greatly inhibits fibroblast growth during repair, laminin would have two powerful effects: promoting neuron growth and blocking scar formation.

Not much is known about the events occurring in the cell after interaction with laminin. It is known that protein synthesis is not required because the cells can form processes in the presence of cycloheximide. Cytoskeleton-disrupting agents disrupt the organization of the cells and lead to reduced or aberrant processes, suggesting that an intact cytoskeletal system is required. Several studies have suggested a role for calcium and for phosphorylation of cytoskeletal proteins in the cellular response to NGF. While the intracellular signaling has not yet been studied in detail, large changes in calcium are not observed in response to laminin. Some changes in the phosphorylation state of proteins occur with laminin, but the exact role of kinases and phosphatases is not known. An understanding of the intracellular events in the neural response to laminin would be useful in regeneration studies and in degeneration diseases.

III. Cellular Receptors

Because laminin contains multiple active sites and various forms of laminin have been described, it is not unexpected that several specific cell-surface re-

TABLE IV   Cellular Receptors for Laminin

Receptor	Source
Neural cells	
Gangliosides	Neuronal
CSAT-140 kDa[a]	Peripheral nerve
110 + 180 kDa[a]	NG108-15
120 + 200 kDa[a]	B50
120 + 140 + 180 kDa[a]	PC-12 cells
67 + 120 − 140 kDa[a]	Brain
Non-neural cells	
32/67 kDa	Macrophages
Sulfatides	Red blood cells
120 + 150 kDa[a]	Platelets
118 kDa	Trichomonads

[a] Denotes integrins or possible integrins.

ceptors that mediate the cell response to laminin exist (Table IV). Several laminin-binding, cell-surface components have been identified, including a 32/67-kDa protein, a 110-kDa protein, gangliosides, sulfatides, and integrins. Antibody to the 32/67-kDa protein does not affect laminin-mediated neurite outgrowth, but antibody to the 110 kDa and to the integrins and exogenously added gangliosides have been found to block the neural cellular response to laminin.

The integrins are a family of heterodimeric molecules, which function as receptors for several adhesive molecules including laminin, fibronectin, collagen, thrombospondin, tenascin, fibrinogen, and vitronectin. Integrin receptors are composed of one of several homologous α-subunits (α_1, α_2, α_3, etc.), which associate noncovalently with a β-subunit to form receptor complex. Many α- and four β-subunits have been described. To date, only the β_1- and α_1-, α_2, α_3-, and α_6-subunits have been found to function as laminin receptors. The integrins thus represent a large family of matrix receptors (Table IV).

Several lines of evidence suggest that integrins are involved in the neural cell response to laminin. Both polyclonal and monoclonal antibodies to the β_1-component block laminin-mediated neurite outgrowth. In addition, a monoclonal antibody to the α_1-subunit blocks PC-12 cell neurite outgrowth on laminin. Various different integrinlike molecules have been isolated from cultured neuronal cells. Because several different integrins have been isolated from neural cells, it is possible that each cell has specific receptors. It is also likely that more than one site on laminin is involved in the cellular response. It should be noted that integrins have not yet been isolated from neural tissues.

IV. Summary

Laminin, a major basement membrane glycoprotein, is shown to have potent effects on both peripheral and central neurons *in vitro* and *in vivo*. The effects are rapid, require low concentrations of laminin, and include promoting neural cell adhesion, migration, survival, and axonal process formation. More than one site on laminin is involved in the cellular response as are multiple cellular receptors. It is possible that different cells use different active sites on laminin at various times during development, in aging, and in repair. The identification of active sites on laminin and demonstration of biological activity *in vitro* should facilitate its use on patients with nerve damage.

Bibliography

Akiyama, S. K., and Yamada, K. M. (1987). Fibronectin. *In* "Advances in Enzymology and Related Areas of Molecular Biology" (A. Meister, ed.) John Wiley & Sons, Inc., New York, **57,** 1.

Kleinman, H. K., Cannon, F. B., Laurie, G., Hassell, J. R., Aumailley, M., Martin, G. R., and Dubois-Dalcq, M. (1985). Biological activities of laminin. *J. Cell. Biochem.* **27,** 317.

Letourneau, P., Rogers, S., Peck, I., Palm, S., McCarthy, J., and Furcht, L. (1988). Cell biology of neuronal interactions with fibronectin and laminin. *In* "Current Issues in Neural Regeneration Research," p. 137. Alan R. Liss Inc., New York.

Martin, G. R., and Timpl, R. (1987). Laminin and other basement membrane components. *Annu. Rev. Cell Biol.* **3,** 57.

Sanes, J. R. (1989). Extracellular matrix molecules that influence neuronal development. *Annu. Rev. Neurosci.* **12,** 521.

Sephel, G. C., Burrous, B. A., and Kleinman, H. K. (1989). Laminin neural activity and binding proteins. *Dev. Neurosci.* **11,** 313.

Language

MARY ANN ROMSKI, ROSE A. SEVCIK, AND DUANE M.
RUMBAUGH, *Georgia State University–Emory University
Language Research Center*

Glossary

Grammar   Composed of morphology (the study of
the smallest units of meaning) and syntax (the rules
governing the ordering of elements into phrases,
clauses, and sentences)

Phonology   Study of speech sounds, their production, and how they function in the language
system

Pragmatics   Rules governing the use of language in
context

Semantics   Study of meaning (i.e., the relationship
between a speaker's language and his/her perception of reality)

LANGUAGE IS A socially shared code or conventional system for representing concepts through the
use of arbitrary symbols and rule-governed combinations of symbols. It is typically divided into four
components: phonology, grammar (comprised of
morphology and syntax), semantics, and pragmatics. For decades behavioral and biological disciplines alike have focused investigative energies on
the study of language, its origins, and its development. In this synopsis the varied aspects of language that have been addressed during the last two
decades are highlighted.

I. Theories of Language Acquisition

The modern study of child language acquisition has
focused on exploring the processes by which children develop language. The proposed theories have
generated a significant amount of lively debate
about these processes.

A. Traditional Perspectives

The theories of language acquisition proposed by
Skinner, a behaviorist, and his contemporary,
Chomsky, a linguist, in the late 1950s provided a
foundation for debates about how language is acquired. Skinner's original theory argued vigorously
that adult language skills were shaped almost solely
by the environment. Chomsky's theory, on the
other hand, emphasized the importance of linguistic
structure and asserted that language (i.e., linguistic
structure) was an innate capacity unique to humans.
A child, he argued, was born with a biological capacity for learning rules for combining words. For
Chomsky the environment played a relatively minor
role in the way children learned to speak in sentences. A third account of language development
was offered by Piaget, a Swiss genetic epistemolo-
gist. His theory placed language development
within a more general framework of cognitive development. From a Piagetian perspective children
map language onto what they already know about
their world conceptually. Many alternative explanations have since been advanced to prove, disprove,
or modify these theories to some degree.

B. Contemporary Accounts

In the early 1970s many scholars objected to the
intense emphasis on formal linguistic structure, par-

ticularly in relation to the way it accounted for initial language development. The primacy of syntax gave way to a broadening interest in the early acquisition of words themselves and the communicative context in which children learned to talk. It was argued that the meanings of words and the social conventions for using language played roles as important as learning the rules for combining words. The 1970s saw a proliferation of empirical studies that sought to resolve some of these issues.

More recently, there has been a renewed interest in linguistic structure and its learnability. Chomsky's earlier notions have been refined in the form of the learnability theory. This theory asserts that if language is acquired by learning syntactic rules, then those rules must, in some sense, be grasped from the linguistic data provided by the child's environment. A trend toward an examination of children's lexicons, with an emphasis on verbs and their role in learnability, is also apparent in the current literature.

A contemporary competing viewpoint to language learnability theory, which accounts for the social component of language, is derived from the parallel distributed processing informational model. This model emphasizes both structural and social functions in language learning. The proponents argue that linguistic structure emerges from the communicative function served by the structure. In this model the only innate structure necessary is the powerful parallel distributed processing learning mechanism.

Theories that account for the emergence of a competent adult language user are plentiful. However, the explorations of language learning are far from complete. For every argument there is a counter-argument with contradictory evidence. Theoretical as well as empirical investigations of the language acquisition process will most definitely continue to generate significant debate as scientists wrestle with the explanations of our complex system of communication.

II. Biological Bases of Language Acquisition

Implicit within the theories that include an innate component is the notion of the biological base for language development. Two broad issues emerge concerning the biological foundations of language development: the relationship between language and its correlates in the brain, as observed during human discourse, and the potential biological uniqueness of language to our species.

A. Neurological Correlates

While language itself is an observable phenomenon, a large quantity of research has focused on specifying the relationships between the observed behavior and its correlates in the brain. It is estimated that in about 85% of the population language resides in the left cerebral hemisphere of the brain. The most common method of investigating language laterality has been dichotic listening. In this experimental paradigm different auditory material is simultaneously presented to each ear, and differences in response or error rate are noted. Research findings suggest a right ear, or left hemisphere, advantage for the processing of linguistic information. Specific structures in the human cerebral cortex (e.g., Broca's area) are also known to be associated with particular aspects of adult language. Recent evidence suggests that the right hemisphere of the brain also plays a role in controlling language, most likely the emotional dimensions.

In the late 1960s it was proposed that there is a critical period during which language development could occur. Lateralization for language would take place by puberty, and after that language could no longer be acquired. Recent advances in technology have permitted more sophisticated investigations of brain–language behavior relationships during the developmental process. Using electrophysiological procedures (i.e., auditory-evoked potentials), it was shown that children as young as 16 months can discriminate known from unknown spoken words and that these effects are lateralized in the brain.

The use of such refined methodologies suggests that language lateralization under typical conditions takes place well before puberty, or, in fact, is probably complete at birth. In the late 1970s the language development of an adolescent named Genie, who had spent her childhood isolated in a small room in her parents' home, was examined. Even given intense instruction, Genie, an adolescent, did not master language structures typical of young children. These findings provided additional support for the early lateralization of language. Even though brain–behavior research has advanced considerably over the last two decades, understanding of the relationship between the brain and observable language behavior is still in its infancy.

B. Origins of Language

Questions about the biological bases of language development have generated two lines of inquiry about its origins. Some scientists have examined the extant natural communication systems of other creatures to provide a comparative perspective into the evolution of our auditory language system. Another area of investigation that has received intense scrutiny in the last two decades is language training research with marine animals and great apes.

Research has explored the natural communication systems of a wide range of animals. While earlier work focused on been communication and bird song, more recently scientists have studied the vocalization systems of lower primates. It was found that the alarm calls of the vervet monkey (*Cercopithecus aethiops*) in the natural environment convey specific information about either the type of predator that is nearby or the type of evasive action the monkey should take to avoid the specific predator. Such findings suggest that animals have elaborate systems of natural communication that facilitate survival in their respective habitats.

A second route for studying animal communication has been to teach components of human language and communication to marine animals and great apes. It has been possible to teach bottlenose dolphins (*Tursiops truncatus*) to carry out complex commands presented via either computer-generated acoustic signals or a gestural language. These dolphins were able to carry out novel auditory or visual commands. Two California sea lions (*Zalophus californianus*) were taught a gestural language similar to the one used with bottlenose dolphins. Here, too, carrying out commands was the desired response. At the end of 2 years, the sea lions had learned approximately 20 gestures and were responding to them by carrying out three-item commands. While these studies, in concert, suggest that marine animals are capable of responding to gestural or acoustic patterns, the animals themselves did not generate spontaneous linguistic communication as we know it.

Young common chimpanzees (*Pan troglodytes*) do not have the specialized physical mechanisms necessary to produce speech. However, they have the capacity to learn alternative communication systems, such as complex manual signs or computer-based visual–graphic symbols.

In the late 1970s new research sparked debate concerning the earlier findings on the language abilities of the great ape and whether or not other explanations could account for these findings. While questions were raised as to the apes' grasp of syntax, convincing evidence was obtained that two common chimpanzees could acquire word meanings and use computer-based visual–graphic symbols to convey information to humans as well as to each other.

Recently, research efforts have focused on the language learning abilities of the rare bonobo or pygmy chimpanzee (*Pan paniscus*). Findings with a young male bonobo have revealed that he understands spoken English words and learned symbols via observation, rather than through intense training efforts, as used in earlier studies. These findings have been replicated with additional infant bonobos and suggest that the capacities for auditory language comprehension might not be completely unique to humans. Ongoing investigations are examining the abilities of the young male bonobo to understand more linguistically complex spoken utterances. The preliminary data indicate that he can comprehend novel combinatorial information and suggest that the bonobo's speech comprehension skills could far exceed their symbol production abilities.

The compelling findings of the last decade, then, suggest that great apes share at least some of the components that permit the emergence of language competence in humans. The future should reveal more detailed explanations about the origins of language, as well as the differential capacities of various mammalian species to learn language.

III. Current Perspectives about the Language Acquisition Process

Anyone observing a child beginning to talk will agree that it is a most remarkable process. While one might receive the impression that the child becomes a competent communicator overnight, a careful review of the evidence suggests that this perception is misleading.

A. Language in Infancy

Infants actually begin communicating very early. Research over the last decade suggests that the stage for language development emerges from birth. Advances in methodologies have permitted scientists to measure changes in the physiological state (e.g., heart rate) and correlate them with environmental changes (e.g., the mother's voice versus a stranger's voice). Mothers seem to interpret what their infants want from early in the infant's life. Thus, even though infants are not talking, their behaviors are interpreted as communication by their mothers. As infants proceed through their first year, they begin to perceive speech sounds and later to comprehend speech in the context of intonational, gestural, and facial cues. Concurrently, infants proceed through a series of stages, during which their vocal abilities become more complex. Before youngsters are 9 months of age, they are using their vocalizations and gestures to communicate a variety of intents to their caregivers.

B. Development of Meaning

Infant communicative interactions lead to the use of first words and to examinations of what these words mean. Of all of the language subsystems, semantics is most closely tied to cognitive development and what a child knows about his or her world. The conceptual basis of a child's early word meanings has generated considerable debate in the language acquisition literature. A semantic feature theory suggested that early word aquisition is based on perceptual attributes. For example, the word "doggie" might be used to refer to all four-legged animals, based on the four legs characteristic of dogs, cats, cows, and the like. A relational theory argued that the child's meaning of words is initially based on functional relationships, followed by a perceptual analysis. According to a prototype theory, semantic categories are not organized by a single defining feature, but by categories occurring at both a basic object level and a superordinate level.

More contemporary debate has focused on the findings of systematic experimental studies, as well as the methodological issues related to the way children use words, especially concrete words, such as nouns. Recently, some attention has been given to the role of relational words and verbs in the language acquisition process.

Children's vocabularies expand at an amazingly rapid rate. By the time they reach school age, children have acquired vocabularies of about 8000 words. One recent theoretical explanation for this remarkable process is described as fast mapping. Researchers have hypothesized that children quickly recognize a word as a word, know some-

thing about its grammatical role, and have a partial sense of the word's meaning after only brief exposure to it. While learning individual words is a continuing process throughout the life span, children quickly learn to combine words to create more complex communications.

For some time television was not thought to have a significant role in language learning. With the impact of television on modern society, empirical studies are suggesting that some types of television viewing could facilitate the language acquisition process. For example, novel words can be learned via viewing some specially designed children's television sequences. Investigations in this area are continuing.

C. Syntactic Development

Because syntax has been central to linguistic theories of language acquisition, intense emphasis has been placed on studying the unfolding of syntactic skills. Empirical data have advanced significantly since the pioneering research of Roger Brown, who studied the early language development of three young children. Brown provided a detailed analysis of their early development of grammatical morphemes (e.g., present progressive tense marker "-ing"), which has served as a basis for much of the later empirical work in this area of investigation. Brown devised a measure, termed mean length of utterance in morphemes, to serve as an index of early syntactic growth. When children produce longer sentences, the type and the complexity of the sentences expand as well. As syntactic development continues into the school years, the production of more complex grammatical constructions (e.g., passive constructions, coordinations, and relative clauses) emerge.

Traditionally, the empirical study of syntax has focused on spontaneous speech production skills, with relatively minimal attention on comprehension. In a few studies the production of sentences actually preceded comprehension of them. Using a creative behavioral testing methodology, it has been shown that infants as young as 17 months of age might comprehend word order before they even begin to produce two-word utterances. Such findings suggest that the young child is absorbing a large amount of sophisticated linguistic information at a very young age, and perhaps such comprehension is setting the stage for later productive syntactic abilities.

D. Later Development of Oral and Written Language

As the child progresses through school, important syntactic and vocabulary skills continue to develop. Not until adolescence are some figurative language concepts (e.g., metaphors and idioms), completely mastered.

Oral language skills are the basis from which literacy skills emerge. The development of literacy skills is now thought to be a gradual process that begins in infancy, when most infants and toddlers enjoy book reading. Long before some children begin school, they are immersed in activities that permit the development of an awareness of print uses and conventions.

Literacy has multiple interwoven dimensions, from reading comprehension to the mechanisms of written language production (e.g., spelling and punctuation), which must be in place to produce a literate individual. The development of reading skills is thought to result from a synthesis of a complex network of perceptual and cognitive acts along a continuum from word recognition and decoding skills to comprehension and integration. Writing skills include development of the mechanics of written productions as well as expository writing skills. [*See* READING PROCESSES AND COMPREHENSION.]

Literacy is a dominant social policy issue today. Efforts are under way to increase literacy awareness in general. Because a large number of adults are functionally illiterate, special attention is being given to adult literacy. While some individuals have not had adequate instruction or opportunity to learn to read, others show evidence of developmental dyslexia. This term refers to the continuum of impairments in the acquisition and development of reading and writing skills despite instruction and opportunity. The scientific study of these impairments offer valuable insight into the developmental reading process. [*See* DYSLEXIA–DYSGRAPHIA.]

E. Language through the Life Span

While it was once thought that language learning was complete by the end of childhood, current views suggest that learning language is an evolving process that continues and changes, albeit subtly, across the life span. Adolescents and adults can acquire a variety of language skills, such as a foreign language, effective speaking and listening skills, coherent writing skills, and socially polite language conventions. These skills are typically not formally learned, but, instead, are acquired in context.

The multicultural nature of society highlights the consideration of second language acquisition across the life span. Children exposed to two languages during early childhood appear to learn both languages with a minimum of delay. The key to their bilingual development seems to be the pattern of use within the language community. Even young monolingual children can learn a second language relatively easily, and the sequence of second language development mirrors that found in the acquisition of the initial language. Second language learning is sometimes not as facile for adolescents and adults as it is for young children.

Another multicultural issue that emerges during the school years and continues into adulthood is dialect. Dialects are associated with relatively stable characteristics of the speaker (e.g., race, gender, age, social class, and geographic origin). Adults' language use can be greatly influenced by the social contexts in which they communicate. While some speakers use one dialect exclusively during their lifetime, others might find it necessary to learn more than one dialect. Variations in language use might also take place in business and social situations. For example, a speaker might use different registers in conversation based on the gender of the listener.

With an aging population changes that take place in language as a part of the natural aging process are becoming an increasing focus of investigation. Access to the vocabulary frequently described as "word finding," becomes more difficult as we age. Changes in the ability to understand complex linguistic materials and to use complex syntactic forms also occur. While we currently know little about the effects the typical aging process has on language, the shifting population dynamics will likely result in continued emphasis on this particular area of investigation. [*See* AGING AND LANGUAGE.]

IV. Language in Special Populations

Given the importance placed on language by society, it is of particular consequence when language is in some way impaired, be it at birth or after it has already been acquired. Studying the language patterns of these special populations provides scientists with a unique route for understanding the language learning process and the relationship between the brain and behavior. [*See* SPEECH AND LANGUAGE PATHOLOGY.]

A. Congenital Language Disabilities

Researchers have estimated that 8–10% of school-aged children demonstrate patterns of language development that might be termed "delayed" or "disordered." Any number of congenital disabilities can contribute to an impairment of language, including autism, cerebral palsy, sensory impairments (i.e., deafness), mental retardation, and specific language impairment. Each of these disabilities may have a unique detrimental effect on language acquisition. The type of impairment, its severity, the age of the child at the time of detection, and the child's environment interact to determine the seriousness of the language impairment across the life span.

Intervention research has advanced the treatment of congenital language disabilities considerably. Children who, 20 years ago, were not given an optimistic prognosis for language development are now making impressive gains in learning to communicate. One of the most significant advances in this decade relates to the integration of advancing computer technology into treatment approaches for severe congenital language disabilities. Augmentative communication systems permit children who have not learned to speak (e.g., those with cerebral palsy or mental retardation) to communicate via alternative output modes. Advances in synthetic speech intelligibility have provided an interface between their means of communication and our otherwise auditory world. When technology is paired with experimentally validated treatment approaches, children who were previously unable to communicate develop functional means of language expression.

B. Acquired Language Disabilities

After an individual has developed language, a variety of events, including strokes, progressive neurological diseases, and traumatic brain injuries, can impair language skills. Characterizing the language deficits associated with these various acquired disorders has been a focus of considerable investigation. Research with human adult brains that have suffered cerebral insults clearly indicates that specific areas of the cerebral cortex are associated with specific aspects of adult language. Loss of previously acquired language due to brain damage, known as asphasia, has been divided into several types, based on the site of lesion and the person's residual receptive and productive language skills. For example, Broca's aphasia is primarily an impairment of expressive grammatical abilities associated with frontal lobe damage specifically to Broca's area. Specific language disorders associated with dementing diseases (e.g., Alzheimer's disease) frequently resemble aphasia during the early stages. As the dementia progresses, the language disturbance is much different from aphasia, as it is the result of more widespread damage to the brain. Advancing diagnostic tools permit scientists to gain an increasingly sophisticated understanding of neurological impairment and language breakdown.

V. Conclusions

In conclusion, language acquisition is an amazing and extremely complex process that unfolds not only in infancy, but over the course of human development. Our collective knowledge about the biological bases of language, the process of language acquisition, and the language of special populations has advanced significantly during the last two decades. Probing and creative questions, coupled with more sophisticated methodologies, frequently using advanced computer technology, have permitted scientists to uncover previously unattainable information. Even so, the theoretical and empirical data bases are far from complete. Scientific investigation about language processes should easily continue into the 21st century.

Acknowledgments

The preparation of this article was supported by National Institute of Child Health and Human Development grant 06016, which supports the Language Research Center, cooperatively operated by Georgia State University and the Yerkes Regional Primate Research Center of Emory University. Support, in part, also comes from National Institutes of Health grant RR-00165 to the Yerkes Primate Center and from the College of Arts and Sciences, Georgia State University. Thanks also go to Krista Wilkinson for her helpful comments on the manuscript.

Bibliography

Berko-Gleason, J. (1989). "The Development of Language." Merrill, Columbus, Ohio.

Clark, E., and Clark H. (1977). "Psychology and Language." Harcourt Brace Jovanovich, New York.

Nelson, K. (1985). "Making Sense: The Acquisition of Shared Meaning." Academic Press, Orlando, Florida.

Rice, M., and Schiefelbusch, R. (1989). "The Teachability of Language." Brookes, Baltimore, Maryland.

Romski, M. A., Lloyd, L. L., and Sevcik, R. A. (1988). Augmentative and alternative communication issues. *In* "Language Perspectives: Acquisition, Retardation and Intervention" (R. L. Schiefelbusch and L. L. Lloyd, eds.), 2nd ed., pp. 343–366. Pro-Ed, Austin, Texas.

Savage-Rumbaugh, E. S. (1986). "Ape Language: From Conditioned Response to Symbols." Columbia Univ. Press, New York.

Language, Evolution

PHILIP LIEBERMAN, *Brown University*

Glossary

Formant frequencies Frequencies at which maximum acoustic energy will pass through the supralaryngeal vocal tract. The value of the formant frequencies depends on the shape and length of the vocal tract. The different speech sounds of human language are determined, in part, by their particular formant frequencies

Speech encoding Process by which human speech achieves a rapid transmission rate. The formant frequency patterns that determine individual sounds are melded together and subsequently decoded by the listeners

Supralaryngeal vocal tract Airway above the larynx including the mouth and nose and the pharynx in modern human beings. The vocal tract acts as a variable frequency-sensitive acoustic filter

HUMAN LANGUAGE, as a complete system, is a unique attribute of modern *Homo sapiens*. Many scholars, therefore, claim that the origin of human language is outside the range of Darwinian evolution. However, recent studies show that human linguistic ability consists of a number of biological factors; the evolution of these components can be traced in the fossil record and in the homologous organs and behavior of closely related living species. One component of language, the ability to acquire and use words, is present in a reduced degree in apes. Word ability may derive from the brain mechanisms that underlie associative learning in most animals, the difference between human beings and intelligent animals (e.g., chimpanzees) being a matter of degree. Speech and syntax appear to be the factors that qualitatively differentiate human beings from all other living species. The specialized anatomy and brain mechanisms that underlie speech and syntax are present only in human beings. However, these "unique" aspects of human language evolved from homologous anatomical structures and brain mechanisms and probably existed in the earliest dated specimens of anatomically modern *Homo sapiens* 100,000 years ago.

Human speech achieves a high rate of data transmission by a process of "encoding"; anatomy that was initially adapted for swallowing and respiration was adapted to make sounds that enhance this process. Brain mechanisms that were initially adapted to facilitate precise manual maneuvers were adapted to allow the production of the complex muscular maneuvers that underlie speech. These brain mechanisms also appear to be implicated in syntax. The neural mechanisms implicated in human speech and syntax include phylogenetically older parts of the brain that derive from reptiles as well as newer parts that achieved their present proportions in *Homo sapiens*. Moreover, the brain mechanisms involved in human language also play a part in other aspects of cognition.

I. Human Speech

Although human language serves cognitive ends, it is a medium for communication that uses acoustic signals to transmit information. Visual signals usually play a supplementary role in the form of facial and manual gestures and can be the primary mode for the transfer of information for deaf people using sign language. However, the vocal signals that are the primary linguistic channel have significant advantages, freeing the hands for other tasks and operating at greater distances to more recipients. Moreover, human speech, the system of particular speech sounds that anatomically modern *Homo sapiens* produce and perceive, allows information to be transmitted at rates between 15 and 30 items per second. The rate at which nonspeech sounds can be transmitted usually cannot exceed seven to nine items per second; at the rates at which the sounds of speech are transmitted, nonspeech sounds merge into an undifferentiated buzz because the rate exceeds the fusion frequency of the auditory system. The high data transmission rate of human speech allows us to communicate complex sentences that otherwise would take so long to transmit that we would forget the beginning of the sentence before we heard its end. It also contributes to biological fitness when simpler utterances are rapidly transmitted (e.g., The tree is falling to your left).

A. Formant Frequency Encoding and the Source-Filter Theory

The process by which human speech achieves its high data transmission rate involves formant frequency encoding. Figure 1 shows a schematized normal human supralaryngeal vocal tract. The airway above the larynx acts as a variable acoustic filter. A given airway shape and length allows maximum acoustic energy through at particular "formant frequencies." The formant frequency pattern is a major determinant of particular speech sounds. The vowel [i] of the word *seed,* for example, has formant frequencies that differentiate it from the vowel [a] of *father.*

The puffs of air that exit the larynx provide the acoustic energy that the supralaryngeal vocal tract filters (along with acoustic energy generated at constrictions in sounds such as [s], [t], [k]). The process is in many ways analogous to the way that a pipe organ functions (i.e., the musical quality of a note is determined by the length and shape of a particular pipe). As we talk, we continually change the shape and make small adjustments in the length of the supralaryngeal airway by moving our tongue, jaw, lips, and larynx. The velum of the human vocal tract can also be raised to seal off the nasal cavity; this allows us to produce nonnasal sounds in which the acoustic effects of the nose are not superimposed on the formant frequency patterns generated by the rest of the supralaryngeal vocal tract.

Figure 2 shows formant frequency patterns for

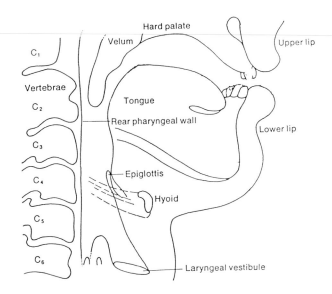

FIGURE 1 Schematized normal human supralaryngeal vocal tract. [From Lieberman, P. (1984). "The Biology and Evolution of Language." Harvard University Press, Cambridge, Massachusetts, with permission.]

the syllables [ba] and [da]. The arc-like formant frequency patterns for the first, second, and third formant frequencies (F1, F2, and F3) move from ones determined by the shape of the vocal tract at the moment of the "burst" (schematized by the squares) that occurs as the "stop" consonants [b] and [d] are released to the vowel [a]. Human listeners "hear" the entire syllable [ba] or [da] even when they can listen to the first 20 msec or so of the acoustic signals that correspond to these formant patterns. Humans appear to be equipped with neural devices that are "tuned" to respond to the encoded signals that the human vocal tract can gener-

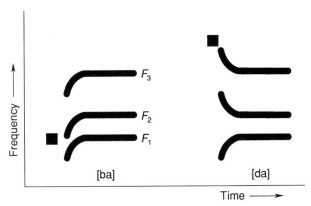

FIGURE 2 Formant frequency and bursts for the syllables [ba] and [da].

ate. The acoustic speech signal is, in effect, encoded into syllable-sized units that are transmitted at a slower rate and then "decoded" into the individual sounds of speech. The perceptual process involves our being able to perceive and interpret the formant frequency and burst patterns in a special "speech mode."

B. Nasalization, Formant Frequency Perception, and Normalization

The ability to produce sounds that are *not* nasalized enhances the perceptibility of the sounds of human speech. The formant patterns of the changing supralaryngeal airway are not obscured by the nasal airway's acoustics; nasalized vowels, for example, are confused 30–50% more often than nonnasal vowels. The vocal anatomy of nonhuman primates and archaic hominids inherently yields signals that are always nasalized. Moreover, their vocal anatomy also imposes another limitation relative to that of anatomically modern *Homo sapiens*—they cannot produce the vowel [i]. Formant frequency patterns are indirectly related to the sounds of human speech—they must be "normalized" by means of another perceptual process in which listeners take account of the presumed length of a speaker's vocal tract. The length of the supralaryngeal airway differs from different adult human speakers; it also doubles in length from infancy to maturity. Consequently, the "same" speech sound may have different absolute formant frequencies when it is produced by different speakers. Human listeners when they interpret formant frequency patterns unconsciously adjust for the length of the vocal tract that produced the speech signal to which they are listening. A number of acoustic cues and procedures appear to play a part in this perceptual process. However, one of the optimal cues is a token of the vowel [i]. The formant frequency pattern of an [i] inherently specifies vocal tract length. Tests of the perception of human speech show that the vowel [i] has the lowest error rate for all vowels and serves to adjust the listeners' frame of reference to a particular vocal tract length.

II. Evolution of Vocal Tract Anatomy

Figure 3 shows a schematized view of a nonhuman supralaryngeal airway. The tongue is long and thin in profile and is positioned almost entirely within

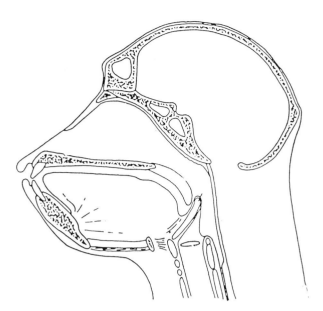

FIGURE 3 Schematized view of a nonhuman supralaryngeal airway. [From Lieberman, P. (1984). "The Biology and Evolution of Language." Harvard University Press, Cambridge, Massachusetts, with permission.]

the mouth. The larynx is positioned closed to the entrance to the nose and can lock into the nose forming a sealed airway that leads from the nose to the animal's lungs, allowing it to simultaneously drink and breathe. Modern human beings are the only mammals in which the larynx cannot be locked into the nose. Food and drink must be carried over the entrance to the larynx, which has retreated down into the neck away from the entrance to the nose. Therefore, human beings can choke to death when food becomes lodged in the larynx. The base of the human skull has been restructured in the course of evolution consistent with this change; it is more flexed compared with those of living nonhuman primates or fossils like australopithecine, *Homo erectus,* or classic Neanderthal hominids. The distance between the anterior edge of the hard palate (which marks the forward margin of the mouth) and the anterior edge of the foramen magnum (where the spinal column enters the skull base) is long in these archaic fossils to accommodate a nonhuman supralaryngeal airway in which the larynx is positioned near the base of the skull in proximity to opening to the nose. Their basicraniums are likewise unflexed and resemble those of living species of nonhuman primates. Quantitative studies show that the basicraniums of these fossils group with those of living apes and human infants who have supralaryngeal airways optimally adapted for swallowing rather than speech. [*See* COMPARATIVE ANATOMY.]

Comparative studies of the relation between the soft tissue of the supralaryngeal airway and the morphology of the skull base and mandible indicate that functionally modern supralaryngeal vocal tracts appeared in the earliest reliably dated specimens of anatomically modern *Homo sapiens,* Skuhl V and Jebel Kafzeh, who lived about 100,000 years ago in Israel. Earlier African fossils like Broken Hill (which cannot be accurately dated) also had modern-like supralaryngeal airways. The fossil evidence for a modern human supralaryngeal vocal tract is consistent with other evidence that points to the African origin of anatomically modern *Homo sapiens*. In contrast, the classic Neanderthal population of western Europe that persisted until 35,000 years ago lacked a modern human vocal tract. They were, in all likelihood, displaced by anatomically modern hominids of African origin who possessed the anatomy necessary for the production of modern human speech.

III. Neural Mechanisms for Speech and Syntax

The modern supralaryngeal airway has a number of negative qualities compared with the nonhuman configuration [i.e., death by choking on food lodged in the larynx, impeding airflow, less effective chewing because of reduced jaw length, the risk of infection and death from impacted teeth (the longer nonhuman jaws do not crowd teeth)]. The only function of the modern human supralaryngeal vocal tract that enhances human biological fitness is fully articulate speech. However, this entails having the neural mechanisms that allow modern human beings to produce the complex articulatory maneuvers that underlie speech. Nonhuman primates are not able to produce these complex, voluntary, muscular maneuvers. Therefore, the presence of a modern supralaryngeal airway in the early examples of anatomically modern *Homo sapiens* is an *index* for the concurrent presence of these specialized neural mechanisms; their modern airways otherwise would have reduced biological fitness.

The neural mechanisms that regulate the articulation of speech are "lateralized"—they are associated with the motor control of dominant hand ma-

neuvers. Comparative studies of other mammals indicate that the preadaptive basis for the neural mechanisms that regulate human speech production may be the brain mechanisms that facilitate the precise manual maneuvers associated with tool use and tool making. Studies of the deficits associated with various types of brain damage in human beings are consistent with these data; these studies furthermore indicate a linkage between the neural structures that underlie human speech production and the ability to use and comprehend syntax. Nonhuman primates who lack the ability to control speech production likewise are unable to use complex syntax productively, although they can communicate with words specified by manual sign language or other nonvocal systems. The preadaptive basis for the evolution of the brain mechanisms that underlie human syntactic ability may be speech production. Therefore, the enhancement of speech production abilities in anatomically modern *Homo sapiens* may be the key to the evolution of human language.

Although speech and syntax are unique attributes of human language, their evolution appears to follow from Darwinian mechanisms. Older structures were modified and adapted to facilitate new activities that contributed to biological fitness. The evolution of the human supralaryngeal vocal tract involved modifying the mouth, tongue, and larynx, which were initially adapted for respiration and ingesting food and water. The resulting speech-producing system is compromised for these basic vegetative functions. The brain mechanisms that regulate the production of speech appear to have evolved from ones that were first adapted for facilitating manual maneuvers. Although neocortical regions of the brain like Broca's area are involved in speech and syntax, older subcortical pathways that derive from reptiles also take part in regulating speech production and the comprehension of sentences that have complex syntax.

IV. Words

The ability to use words appears to derive from different brain mechanisms than speech and syntax. Lexical abilities, for example, deteriorate in conditions like Alzheimer's dementia, whereas speech and syntax are preserved. The reverse occurs in conditions like Parkinson's disease. The neural bases for words may be distributed networks that have a primary cognitive function in associative learning. Significantly, nonhuman primates who can neither talk nor use complex syntax can productively acquire and use and even transmit lexical skills to other nonhuman primates. There are important quantitative differences—their vocabularies are limited compared with even 5-year-old human children. However, they clearly are able to manipulate words, although their words obviously are not marked for syntactic functions. Therefore, the neural bases for words probably have a long evolutionary history that antedates the appearance of hominids. [*See* ALZHEIMER'S DISEASE; PARKINSON'S DISEASE.]

Bibliography

Holloway, R. L. (1985). The poor brain of *Homo sapiens neanderthalensis:* See what you please . . . *In* "Ancestors: The Hard Evidence" (E. Delson, ed.), pp. 319–324. Arthur D. Liss, New York.

Kuhl, P. K. (1988). Auditory perception and the evolution of speech. *Hum. Evolution* **3,** 21–45.

Laitman, J. T., and Reidenberg, J. S. (1988). Advances in understanding the relationship between the skull base and larynx with comments on the origins of speech. *Hum. Evolution* **3,** 101–111.

Lieberman, P. (1984). "The Biology and Evolution of Language." Harvard University Press, Cambridge, Massachusetts.

Lieberman, P. (1991). "Uniquely Human: The Evolution of Speech, Thought and Selfless Behavior." Harvard University Press, Cambridge, Massachusetts.

MacNeilage, P. F., Studdert-Kennedy, M. G., and Lindblom, B. (1987). Primate handedness reconsidered. *Behav. Brain Sci.* **10,** 247–303.

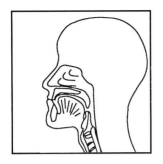

Laryngeal Reflexes

GAYLE E. WOODSON, *University of California, San Diego; VA Medical Center, San Diego*

Glossary

Apnea Cessation of airflow due to lack of inspiratory effort (central apnea) or obstruction of the upper airway (obstructive apnea)

Deglutition Act of swallowing

Glottis Portion of the larynx that opens and closes to regulate airflow and vibrates during the production of sound; the lateral edges of the glottis are formed by the true vocal folds

Hypercapnea Higher than normal concentration of carbon dioxide in the blood

Hypoxia Lower than normal concentration of oxygen in the blood

Phonatory Pertaining to production of the voice by the larynx

THE LARYNX IS richly supplied with sensory receptors. In fact, there are several times more sensory fibers from this small organ than from the lungs, organs with internal surface areas of several square meters. The role of sensory information from the larynx in the function of the organism is currently only partially understood; however, it is well known that reflexes from laryngeal stimulation clearly produce responses in heart rate and blood pressure. Laryngeal sensation is also important in regulating speech, swallowing, and breathing. In the sleep apnea syndrome, breathing is frequently interrupted during sleep, due to collapse of the throat (obstructive apnea) or cessation of breathing efforts (central apnea). Failure to respond to laryngeal sensation of airflow and pressure may play a role in this phenomenon.

The larynx is also the efferent limb of many reflexes, and closure of the larynx or changes in the size of its opening can result from respiratory stimuli, such as obstruction or asphyxia, as well as sounds and tactile or chemical stimuli. Some cases of asthma may not consist purely of bronchoconstriction but may also involve partial closure of the larynx.

Laryngeal reflexes are not fully developed at birth but undergo, in the first few months of life, profound changes in latency, threshold, and central inhibition. These changes occur during the period that corresponds to the peak incidence of sudden infant death syndrome (SIDS), also known as crib death. SIDS is the sudden unexplained death of an infant, usually during sleep.

I. Respiratory Reflexes

A. Airflow Control

The larynx is situated between the upper and lower respiratory tracts. This location, as well as the ability of this organ to effect profound changes in airflow in a very brief interval of time, suggest that the larynx may be important in controlling airflow and significantly influencing the rate of breathing. The larynx does move predictably with the breathing cycle. Some data regarding respiratory reflexes that lend support to the concept of the larynx as a regulator of flow are reviewed below.

The larynx is normally open. Its lumen, the glottis, becomes wider as the vocal folds abduct (move apart) during inspiration and gradually narrows, but does not close, as the folds adduct (move toward each other) during expiration. Inspiratory abduction of the larynx appears to be a primary action of breathing, as the muscle that opens the larynx (posterior cricoarytenoid [PCA]) consistently begins to contract before the diaphragm with each inspiration. The degree of laryngeal movement depends on respiratory demand. Upper airway occlusion leads to stronger contraction of the diaphragm with inspiration. There is also a reflexive increase in the strength of PCA contraction, so that the larynx opens more widely. Abduction of the larynx facilitates breathing by decreasing resistance to airflow. With very strong respiratory demand, the PCA also contracts during expiration, after the diaphragm has ceased contracting. The resultant decrease in resistance permits faster outflow of air, shortening the duration of expiration and increasing the rate of breathing. In animals, upper airway obstruction can increase inspiratory PCA activity by as much as twofold, due to the stimulation of negative pressure receptors in the upper airway. Sensation of upper airway pressure is primarily mediated by the superior laryngeal nerve. Positive pressure does not affect PCA activity.

A second laryngeal muscle, the cricothyroid (CT), appears to be an accessory muscle of respiration. The action of this muscle is to increase the anterior–posterior diameter of the larynx, lengthening and tightening the vocal folds. When this muscle contracts alone, it pulls the vocal folds closer together, resulting in increased airway resistance. When it works in concert with the PCA, however, it can result in maximal glottic dilation and minimal resistance to airflow. The CT does not usually contract during quiet breathing but becomes active with increasing breathing effort. Upper airway obstruction recruits CT contraction in inspiration, whereas voluntary deep breathing results in CT activity in both phases of respiration. The strongest stimulus for CT appears to be upper airway pressure, as mediated by the superior laryngeal nerve.

The effect of changes in blood concentrations of oxygen and carbon dioxide on the PCA and CT are complex. Excess carbon dioxide in the blood (hypercarbia) results in increased activity of both muscles in inspiration and recruits activity during expiration as well. The result of this reflex is a maximal decrease in resistance to airflow in both phases of respiration. Even while breathing through a tracheotomy, bypassing the sensory receptors in the upper airway, both muscles increase activity in proportion to the respiratory drive to inspiratory muscles such as the diaphragm. Maximal inspiratory laryngeal muscle activity occurs during upper airway occlusion in the presence of hypercarbia.

Animal experiments indicate that the effect of hypoxia on laryngeal resistance depends on the integrity of the tenth cranial nerve, the vagus. The vagus is a very complex nerve that carries sensory input from and motor output to important systems in the body: cardiovascular, respiratory, and digestive. When this nerve is intact, hypoxia recruits PCA activity in both inspiration and expiration, decreasing laryngeal resistance to airflow. If the vagus is severed, below the takeoff of the recurrent laryngeal nerve (RLN), PCA response to hypoxia in inspiration does not change, but the response in expiration is abolished. This implies that the control of laryngeal activity in expiration is under different control.

The preponderance of experimental evidence suggests that the larynx does play a role in regulating breathing, as reflex responses of the larynx to negative upper airway pressure or changes in blood gas concentrations usually result in resistance changes that would have a beneficial effect on ventilation. [*See* RESPIRATORY SYSTEM, ANATOMY; RESPIRATORY SYSTEM, PHYSIOLOGY AND BIOCHEMISTRY.]

B. Cough

Cough is an important mechanism for ejecting mucus and foreign matter from the lungs and possibly for maintaining patency of the small air sacs (alveoli) in the lungs. Cough may be voluntary or in response to stimulation of receptors in the larynx or lower respiratory tract. Cough consists of three phases: inspiratory, compressive, and expulsive. During the first phase, the larynx opens very widely to permit rapid and deep inspiration. In a voluntary cough, the degree of laryngeal opening and inspiratory effort appear related to the amount of force intended to be generated. During the compressive phase, the vocal folds are tightly closed and then expiratory muscles are strongly activated. The expulsive phase begins when the larynx suddenly opens, permitting a sudden rapid outflow of air, in the range of 6–10 liters/sec.

A maximally effective cough requires tight laryngeal closure, hence cough is impaired by laryngeal paralysis. Certain stimuli result in a less than effective cough. For example, stimulation of the larynx sometimes results in only an expiratory effort, without a compressive phase.

The threshold of stimulation required for a cough varies with sleep states. In deep sleep, stimuli must result in arousal before cough can be produced.

C. Apnea

The obvious advantage of apnea, in response to laryngeal stimulation, is that it prevents the aspiration of the stimulating material into the lower airway. This occurs in response to diverse chemical agents such as ammonia, phenyl diguanide, and cigarette smoke. Water in the larynx can also inhibit inspiration under certain conditions including general anesthesia and during an upper respiratory infection. In infants, water in the larynx can produce prolonged apnea. The usual response to water in the larynx of a normal conscious adult is vigorous coughing.

D. Laryngospasm

Laryngospasm is the forceful and prolonged closure of the larynx in response to direct mechanical stimulation. It is most often seen in the operating room, in response to endotracheal intubation for mechanical ventilation, and is most likely to occur under very light anesthesia. It occurs in the conscious state under certain pathological conditions. Sometimes, an upper respiratory infection can markedly decrease the threshold for laryngospasm. This lower threshold can persist for many months, resulting in frequent episodes of frightening airway obstruction. Patients who have sustained damage to laryngeal nerves may also be very susceptible to spasm, presumably due to aberrant regeneration of nerve fibers. In rare cases, laryngospasm can represent a form of epilepsy.

Laryngospasm can be a very frightening experience, because it results in total upper airway obstruction, making breathing impossible. The threshold for laryngospasm varies with the concentration of oxygen in the blood. Spasm is most likely and most severe when blood is very well oxygenated. This is fortunate, because it means that as blood oxygen falls, due to the airway obstruction, the severity of the spasm decreases, and the larynx eventually opens to permit breathing. In normal adults, the hypoxia stimulates breathing efforts as well, hence laryngospasm is not often fatal; however, in infants, because ventilatory response to hypoxia is not sustained, laryngospasm could be fatal.

E. Bronchoconstriction

Mechanical or chemical stimulation of the larynx produces, after a significant delay, bronchoconstriction, which persists after cessation of the stimulus. This constriction leads to increased airway resistance. Teleologically, it is difficult to explain the reason for this reflex. It is possible that increased rigidity stabilizes the bronchial walls and prevents collapse during forceful exhalation. And because the smaller cross-sectional diameter does result in increased velocity of airflow, the force and expulsive properties of exhalation are increased. Bronchoconstriction slightly reduces the amount of dead space, and the response may also be protective, to prevent inspiration of extraneous particles into the alveoli.

II. Deglutition Reflexes

Deglutition, or swallowing, is a complex act that may be divided into three phases: oral (mouth), pharyngeal (throat), and esophageal. The oral phase, consisting of chewing and otherwise preparing food for swallowing, is largely under conscious control. The last phase, the esophageal phase, is the least complicated, as ingested material is simply propelled along by peristaltic waves of contractions. Between these two phases is the pharyngeal phase, totally under involuntary control, and by far the most complicated and treacherous part of a swallow. The larynx plays a vital role in this middle phase.

The pharyngeal phase of the swallowing reflex is initiated in response to stimulation of mechanoreceptors or chemoreceptors in the pharyngeal and laryngeal mucosa by the presence of food, saliva, or other material. Taste buds in the larynx may also be sensitive to the very small amounts of chemical stimuli that leak out of the mouth during chewing and enter the pharynx. Quantities that are not sufficient to trigger the swallow reflex may actually affect the oral phase, influencing the time at which food is actually ejected from the mouth into the larynx.

As in the esophageal phase, constrictor muscles in the pharynx propel food through the throat. But for food to reach the esophagus, it must cross the pharynx without spilling into the lower airway or squirting back out the nose or mouth. Prevention of ejection of food is accomplished by contraction of muscles in the soft palate and around the tonsils. Respiration is halted during swallowing to lessen the chance of aspiration of material into the trachea, but protection of the lower airway from leaking or spilling of food is primarily accomplished by the larynx.

A. Laryngeal Elevation

The first line of defense is laryngeal elevation. Contractions of extrinsic laryngeal muscles pull the larynx upward and forward under the base of the tongue. This shortens the path from the back of the mouth to the opening of the esophagus and removes the larynx from the center of the food trajectory. Laryngeal elevation also has the effect of opening the pyriform sinuses, mucosal lined sacs on either side of the larynx. Swallowed food is propelled through these sinuses, around, rather than over, the top of the larynx, further lessening the chance of a spill into the airway.

B. Laryngeal Closure

Another action of the larynx during swallowing is closure of the laryngeal opening, on two levels. The vocal folds are adducted to close the glottis. The supraglottis is also constricted, bringing the false vocal folds together. Thus, any matter that spills into the laryngeal inlet does not get through into the trachea.

Effective laryngeal closure during a swallow requires normal strength and innervation of the laryngeal muscles. The rest of the reflex arc, however, must also be intact, because it is very difficult, if not impossible, to coordinate voluntary closure of the larynx at just the right point in the swallow. Impaired sensation of the larynx or pharynx blocks the afferent limb of the reflex and can lead to severe aspiration during or after a swallow. Central connections are also vital, and central lesions can impair the swallow even when sensation and motor innervation are intact.

III. Cardiovascular Reflexes

Stimulation of the larynx can produce changes in heart rate and blood pressure. This effect is most noticeable in response to endotracheal intubation during induction of general anesthesia; however, it may be more significant in naturally occurring cir-

cumstances. For example, obstructive sleep apnea is a disorder in which upper airway patency is not maintained during sleep, resulting in negative airway pressure. This can stimulate negative pressure receptors in the larynx so strongly that cardiac arrhythmias occur. If the resultant slowing of the heart rate or skipping of beats is so severe that it reduces cardiac output, it can result in lowering of blood pressure. However, the direct result of laryngeal stimulation on blood pressure is elevation.

The pathways responsible for mediating the cardiovascular responses are not fully understood. The afferent limb is the superior laryngeal nerve, because transection of this nerve abolishes the cardiovascular response to laryngeal stimulation. Also, electrical stimulation of the superior laryngeal nerve affects heart rate and blood pressure. The efferent limb for bradycardia (slowing of the heart rate) is the vagus nerve. Intervening connections as well as the efferent limb for blood pressure elevation are not known. However, evidence indicates that a sympathetic response to laryngeal stimulation may be mediated through respiratory control mechanisms. Recordings from sympathetic roots in the neck demonstrate phasic activity with breathing. Electrical stimulation of the superior laryngeal nerve suppresses inspiratory sympathetic activity. This suppression depends on blood levels of CO_2, as hypocapnea blocks the effect. Thus, the effects of laryngeal stimulation on sympathetic output are seemingly mediated through central inspiratory control mechanisms.

IV. Phonatory Reflexes

Reflexes are important in the control of laryngeal activity during phonation. The primary mechanism for controlling phonatory output during speech is auditory feedback. Other factors are also involved, as evidenced by the fact that deaf patients who lost their hearing after they learned to speak can maintain fairly normal speech patterns for years after losing hearing. This indicates that other feedback mechanisms are being used for control. It is also true that trained singers can produce a specific pitch and loudness, even when they cannot hear their own voices.

Sensory receptors in the larynx have been demonstrated to respond to contact of mucosal surfaces, air pressure, stretch of joint capsules, and tension in muscles. Receptors are present in mu-

LARYNGEAL REFLEXES

651

cosa, submucosa, tendons, perichondrium, muscles, and joint capsules. All of these supply information that could be used to control phonatory output. Study of the precise mechanisms involved is quite difficult, because control of the larynx during speech is largely unconscious. Experimental studies on the effects of various nerve lesions on the voice cannot be performed in humans and would be irrelevant in animals, because speech is a uniquely human function.

One reflex that can noninvasively be demonstrated in both animals and humans is the phonatory stapedial reflex. The stapedius is a tiny muscle that connects one of the tiny bones in the ear to the skull. Contraction of this muscle tightens the chain of three small middle ear bones responsible for transforming sound waves into the air to fluid waves in the inner ear. This tightening reduces the amount of sound that is transmitted. The stapedial reflex is the contraction of this muscle in response to a loud sound, presumably to protect the inner ear from excessive stimulation. The phonatory stapedial reflex is the contraction of this muscle just prior to the onset of phonation. Because it occurs before sound is transmitted, it is not an acoustic reflex but is something the organism does to protect the inner ear from its own voice, or possibly to reduce competing sound input from other sources.

Laryngeal muscles are activated prior to the onset of phonation. This is called prephonatory tuning and is the reason why we are able to determine the pitch we will make before we actually utter a sound. Trained singers and actors have a great deal of fine control over this, but the activity is used everyday, unconsciously, in normal speech.

Much remains to be learned about the control of the larynx during phonation. Increased knowledge of phonatory reflexes likely will allow their exploitation to improve vocal function.

V. Developmental Changes in Reflexes

Respiratory reflexes undergo profound changes during development. Research in the last decade indicates that fetal breathing movements occur *in utero*, long before birth. These movements are not constant and do not result in ventilation, but may be important in the development of the lungs and respiratory muscles in preparation for breathing after birth. At birth, control of breathing movements undergoes a fundamental change, but the control of

breathing and the response to laryngeal stimulation are not yet mature. Developmental changes occur over the first few months of life. The time course of these changes seems to parallel the incidence of SIDS, which is the unexplained death of healthy infants, usually during sleep. Because SIDS commonly occurs during this phase of maturation, aberrations in development of respiratory reflexes are suspected as the most likely cause.

In utero, breathing movements do not occur continuously but only during specific states of brain wave activity. Levels of blood oxygen are much lower in the womb than after birth, because the fetus receives all of its oxygen via the placenta, and hypoxia does not stimulate breathing activity. In fact, respiratory muscle activity in the fetus is abolished by decreasing levels of oxygen in the blood. This makes sense in the womb, because the muscle activity does not result in ventilation and only serves to consume precious oxygen.

Immediately after birth, breathing becomes a continuous activity, in wakefulness as well as sleep. The infant is exposed to much higher levels of oxygen in the blood, because oxygenation is accomplished much more efficiently by the lungs than by the placenta. Decreases in oxygen produce an immediate increase in efforts to ventilate the lungs. The mechanism of this profound change in control is unknown. It does not appear to be the result of a developmental change in neural circuits, because the transition always occurs at birth, even when premature or postmature.

Newborns do respond to hypoxia with increased ventilation, but the response is not the same as that in mature animals. Mature animals have a sustained ventilatory increase in response to hypoxia, which lasts until the onset of hypoxic coma. In newborns, the increased ventilation in response to hypoxia is only a transient phenomenon. Within a few minutes, hypoxia not only fails to stimulate ventilation, but actually suppresses it. In the first few months of life, the breathing response to hypoxia gradually changes to the mature pattern. Any event that even transiently impairs ventilation could block breathing long enough to reach the phase of depression of respiration by hypoxia. Thus, the infant would not have sufficient respiratory drive to resume breathing, resulting in asphyxia. Therefore, central or obstructive apnea in response to laryngeal stimulation could have a fatal outcome.

Newborn animals and humans have significant periods of apnea in response to the application of water to the larynx, known as the laryngeal chemoreflex. The stimulus appears to be the lack of chloride ions in water rather than low osmolarity, because isotonic glucose has the same effect as water. Resumption of breathing is almost always preceded by swallowing, indicating the importance of the swallow in clearing the stimulating liquid from the area. The role of blood oxygen levels in modulating this reflex are not clear. Patterns of apnea in response to water seem similar in animals before and after denervation of arterial chemoreceptors, indicating that falling levels of oxygen are not the stimulus for resumption of breathing at the end of the apnea interval. Changes in blood oxygen levels in the intact anesthetized animal, however, do significantly influence the laryngeal chemoreflex, because the duration of resultant apnea is shorter in the presence of early hypoxia. If hypoxia is severe and prolonged enough to suppress respiration, however, ventilation may be suppressed indefinitely; hence, the apneic response has a higher fatality rate in the presence of hypoxia. The laryngeal chemoreflex becomes weaker with increasing age and is not seen in mature animals, except in certain pathologic states. The central mechanisms of maturation are not known.

Another laryngeal reflex that changes considerably with development is the laryngeal adductor reflex, which consists of closure of the glottis in response to mechanical stimulation. Excessive and prolonged reflex laryngeal closure is known as laryngospasm, and it results in total airway obstruction. The laryngeal adductor reflex can also be elicited by electrical stimulation of the superior laryngeal nerve. Experiments in puppies have demonstrated a period of transient hyperexcitability of this reflex between 50 and 75 days of postnatal life. This peak of excitability apparently is caused by the interaction of different maturational changes in the central processing of this reflex. The ability to temporally summate stimuli develops during this interval, such that successive stimuli have an additive effect. Low-level spatial summation also develops. Both of these factors, which increase the likelihood of an adductor response, persist into adulthood. Thus, they can explain the increase in excitability during this period but not its subsequent decrease. The decrease is probably due to the development of central inhibitory pathways, because the increasing influence of central inhibition with age inversely correlates with excitability of the reflex. In the experiments on puppies, central inhibition decreases from birth to 50 days, only to increase again thereafter. There is also an abrupt increase in central latency of the crossed adductor reflex after 50 days of age, probably due to the addition of additional neurons into the reflex arc. This means that reflex motor output from a laryngeal stimulus will reach the contralateral laryngeal muscles later than the ipsilateral muscles. This could possibly decrease the resultant obstruction from a given stimulus.

Maturation of respiratory control in infancy, including the role of the larynx, is highly complex, and much more research will be required to fully elucidate this process. A better understanding of this process of maturation will not only elucidate the pathogenesis of SIDS but will also contribute to our understanding of respiration in general.

Bibliography

Kirchner, J. (1987). Laryngeal reflex systems. *In* "Laryngeal Function in Phonation and Respiration" (T. Baer, C. Sasaki, and K. Harris, eds.), pp. 65–70. Laryngeal Function in Phonation and Respiration, College Hill Press, Boston.

Mathew, O. P., and Sant'Ambrogio, F. (1988). Laryngeal reflexes. *In* "Respiratory Function of the Upper Airway" (O. P. Mathew and G. Sant'Ambrogio, eds.), pp. 259–302. Marcel Dekker, Inc., New York.

Mortola, J., and Fisher, J. (1988). Upper airway reflexes in newborns. *In* "Respiratory Function of the Upper Airway" (O. P. Mathew and G. Sant-Ambrogio, eds.), pp. 303–358. Marcel Dekker, Inc., New York.

Larynx

GAYLE E. WOODSON, *University of California, San Diego*

Glossary

Articulation Enunciation of words and sentences; phonated sound, produced by laryngeal vibration, is shaped into spoken language by the lips, tongue, palate, and throat

Aspiration Inhalation of food, secretions, or foreign matter into the lungs

Glottis Vocal apparatus of the larynx, consisting of the true vocal folds (commonly known as the vocal cords) and the space between them

Phonation Production of vocal sound; in humans, this is accomplished by vibration of the vocal folds of the larynx

Resonance Prolongation and intensification of a sound by the transmission of its vibrations to a resonant structure; vocal resonance is the amplification and modification of the voice by the induction of sympathetic vibrations in the chest, head, and throat

Stridor Noise generated by breathing through a constricted airway

THE LARYNX, also known as the voice box, is an organ located at the intersection of the upper respiratory and digestive tracts, at the caudal end of the pharynx. It is just anterior to the upper end of the esophagus and serves as a valve in the opening between the pharynx and trachea.

The larynx is best known for its role as the sound source in speech production. Its most vital function, however, is that of protecting the lower airway from the aspiration of secretions or ingested food or water, particularly during the act of swallowing. Coughing, an important mechanism for cleaning the lungs, requires tight closure of the larynx. The larynx also likely plays a role in regulating airflow during breathing; however, the significance of this function has not yet been established. Observed movements of the larynx with respiration in humans could be important in regulating airflow or may only be useless motion, a phylogenetic leftover from the function of the larynx in lower animals.

The structure of the upper aerodigestive tract (Fig. 1) seems far from ideal, as ingested food and water must traverse the upper airway to reach the alimentary tract, without spilling into the trachea. In fact, any slight derangement of the complex act of deglutition can result in significant and dangerous aspiration. Breathing must be interrupted during a swallow. Speech is quite difficult during chewing and impossible during swallowing. The precarious configuration of the mouth, nose, and throat makes more sense when it is seen as a product of its evolution and embryology. The following discussion will describe the anatomy and development of the larynx and outline its function in breathing, swallowing, and speech.

I. Anatomy

The larynx has a skeleton, made up of several cartilages and one bone. These elements are strung together in series by ligaments, joints, and sheets of connective tissue and suspended from the mandible and the base of the skull. Motion of the skeleton is effected by "intrinsic" muscles, which both arise and insert on laryngeal cartilages, as well as "extrinsic" muscles, which connect the larynx to other structures. Also, because of the way the laryngeal skeleton is interconnected, traction along the ros-

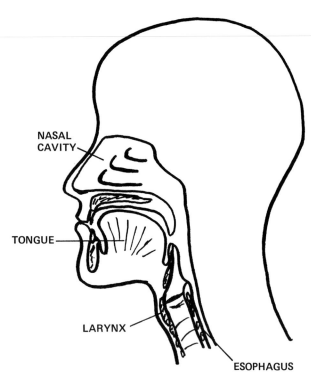

FIGURE 1 Upper aerodigestive tract: sagittal section.

tral–caudal axis, which occurs during inspiration, will result in motion of the cartilages with respect to one another, producing changes in glottic diameter. The musculoskeletal structure is draped in mucosa. Over the free edge of the vocal fold, the histology of this covering layer is highly specialized for vibration. This section will detail the skeleton, muscles, nerve supply, and major subdivisions of the larynx, as well as the histologic architecture of the vocal fold. [*See* CARTILAGE; CONNECTIVE TISSUE.]

A. Subdivisions of the Larynx

1. Supraglottis

The supraglottis is the upper part of the larynx. The most cephalad portion of this section is the epiglottis, a cartilaginous structure that projects into the lumen of the pharynx. The aryepiglottic folds are bands of mucosal-covered tissue that extend from the epiglottis to the posterior portion of the glottis. The pyriform sinuses (Fig. 2), recesses that extend downward and are open posteriorly to the level of the esophageal inlet, are lateral to the aryepiglottic folds on either side of the larynx. The false vocal cords are interior and inferior to the aryepiglottic folds.

2. Laryngeal Ventricle

This pouch is located between the true and false vocal folds on each side (Fig. 2). The pouch ascends lateral to the false folds for a variable distance. In lower animals, it may function as a resonator. Its function in humans is not known, but it may serve as a reservoir for mucus to lubricate the glottis.

3. Glottis

The glottis (Fig. 3) is made up of the true vocal folds, those parts of the larynx that vibrate to produce sound during phonation. These structures are comprised of muscle, stretching across the laryngeal opening from anterior to posterior, and draped in mucosa. The anterior ends of the vocal folds are attached to the interior surface of the thyroid cartilage in the midline, with virtually no space between them. The posterior ends are mobile, rotating in an axial plane. The glottis is closed by approximation of the folds in the midline. Laryngeal opening is affected by abduction of the posterior ends of the vocal folds, creating a roughly triangular opening.

4. Subglottis

That portion of the larynx caudal to the glottis, above the trachea, is known as the subglottis. The skeletal support of the subglottis, the cricoid cartilage, is the only complete ring in the human respiratory tract, hence it is the only point where the lumen is fixed, changing only with swelling of the lining mucosa. It is also the level of the respiratory tract with the smallest cross-sectional area, being smaller than the trachea. Certainly individual bronchioles and alveoli are smaller, but due to branching, they are much more numerous, so that total area is greater. Because of this, a single foreign

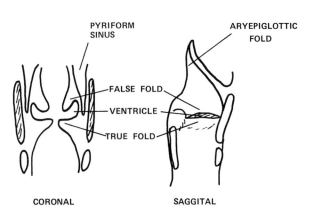

FIGURE 2 Vertical sections through the larynx.

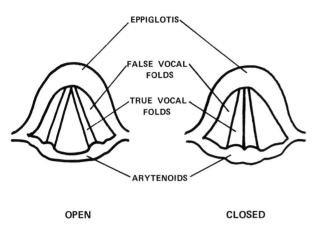

FIGURE 3 Glottis: seen from above.

body, which is small enough to pass through the subglottis, will not cause total airway obstruction.

B. Skeleton

1. Hyoid Bone

The most cephalad component of the laryngeal skeleton is the hyoid bone (Fig. 4), which is roughly U-shaped and lies in an axial plane, with the opening posterior. The two free ends project posterolaterally as the greater cornua, while two small bumps on the superior anterior surface are the lesser cornua. This bone not only supports the larynx but also maintains patency of the hypopharynx. The hyoid bone is connected inferiorly throughout its width to the thyroid cartilage by the broad thyrohyoid membrane.

2. Thyroid Cartilage

Below the hyoid bone is the thyroid cartilage (Fig. 4), which is narrower and taller than the hyoid

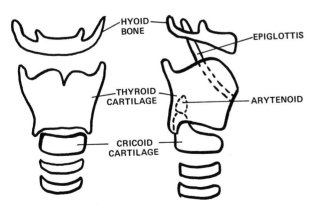

FIGURE 4 Laryngeal skeleton, anterior and lateral views.

and also opens posteriorly. The two halves of the thyroid cartilages fuse anteriorly to form a sharp angle (90 degrees in males; 120 degrees in females). This prominence is commonly referred to as the Adam's apple. Posteriorly, the edges of the cartilage project as the superior and inferior cornua. The thyrohyoid ligament connects the superior cornu of the thyroid cartilage to the hyoid bone, while the inferior cornu articulates with the cricoid cartilage.

3. Epiglottis

The epiglottis is a leaf-shaped piece of fibroelastic cartilage (Fig. 4). The "stem" of this "leaf" points inferiorly and is attached to the inner surface of the thyroid cartilage in the superior midline. The hyoepiglottic ligament runs from the inner anterior surface of the hyoid bone to the anterior upper epiglottis. The superior free end of the epiglottis projects into the hypopharynx, just behind the base of the tongue, and forms the upper boundary of the larynx.

4. Cricoid

The cricoid cartilage (Fig. 4), caudal to the thyroid, is a complete ring, wider in back than in the front. Anteriorly, it is a few millimeters high and has a smooth, curved surface. It is 2–3 cm high posteriorly, where the superior surface is flattened centrally to provide an area of articulation for the arytenoid cartilages. There are also superior facets, posterolaterally, for articulation with the inferior cornua of the thyroid cartilage. The cricothyroid joints allow the cricoid and thyroid to rotate with respect to each other in a sagittal plane, opening or closing the anterior cricothyroid space. Tension in the vocal folds is markedly influenced by the angle of the cricothyroid joint. When the cricoid and thyroid cartilages are pulled together anteriorly, this increases the distance between the anterior thyroid cartilage and the posterior cricoid cartilage, stretching the vocal fold.

5. Arytenoids

The two arytenoid cartilages are the most mobile portions of the laryngeal framework (Fig. 4). Each is a roughly pyramidal or pear-shaped mass, with a wide base for articulation with the posterior–superior surface of the cricoid. The cricoarytenoid interface is a complex synovial joint, permitting both rotation of the arytenoid in a roughly axial plane and gliding medially or laterally along the facet of the cricoid. A firm posterior cricoarytenoid ligament prevents anterior displacement. The base of the arytenoid projects anterior and medial as the vocal process, which attaches to the posterior end of the vocal fold. The vocal folds are fixed to the thyroid cartilage anteriorly. Movement of the vocal folds to open or close the glottis occurs by abduction or adduction of the arytenoids.

6. Sesamoid Cartilages

There are also two other small sesamoid cartilages in each side of the larynx, the corniculate and cuneiform, located superior to the arytenoid. These may serve to support the aryepiglottic folds.

C. Major Connective Tissue Components

1. Thyrohyoid Membrane

The thyrohyoid membrane is a sheet of fibroelastic tissue. Its vertical dimension extends from the posterior surface of the hyoid bone to the superior edge of the thyroid cartilage. It does not attach directly to the hyoid bone, as a bursa is interposed, which enhances vertical mobility of the larynx. The lateral edges of this membrane are thicker, forming the thyrohyoid ligaments.

2. Conus Elasticus

The conus elasticus is another fibroelastic structure in the larynx. Anteriorly, it attaches to the midline of the lower border of the thyroid cartilage. Posteriorly, it connects to the vocal process of the arytenoid. Laterally, it is attached along the superior surface of the cricoid. Medially, its free edge forms the vocal ligament of the true vocal fold.

3. Quadrangular Membrane

The quadrangular membrane, another sheet of fibroelastic tissue, provides support for the superior portion of the larynx. Anteriorly, it attaches to the epiglottis, while posteriorly it is connected to the arytenoid and the corniculate cartilage. Its superior and inferior free edges are draped in mucosa to form the aryepiglottic fold and the false vocal fold, respectively.

D. Intrinsic Laryngeal Muscles

The intrinsic muscles of the larynx consist of five paired muscles (Fig. 5) and one single muscle, which straddles the midline. They function to adduct, abduct, or tense the vocal folds or to constrict the supraglottis.

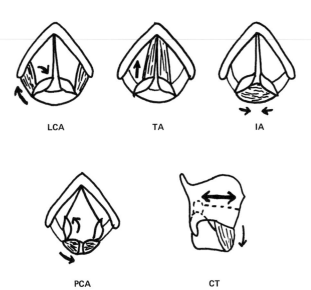

FIGURE 5 Laryngeal muscles. CT, cricothyroid; IA, interarytenoid; LCA, lateral cricoarytenoid; PCA, posterior cricoarytenoid; TA, thyroarytenoid.

1. Posterior Cricoarytenoid Muscle

The posterior cricoarytenoid muscle is the only muscle that abducts the vocal folds. It originates from the posterior surface of the cricoid and inserts onto the muscular process, on the posterior surface of the arytenoid. Contraction of this muscle rotates the vocal process laterally and also produces lateral and posterior gliding displacement of the arytenoid, thereby abducting the vocal fold.

2. Lateral Cricoarytenoid Muscle

The lateral cricoarytenoid muscle origin is on the lateral aspect of the cricoid, with insertion on the posterior muscular process of the arytenoid. When it contracts, the muscular process is pulled forward, rotating the vocal process medially and adducting the vocal folds.

3. Thyroarytenoid Muscle

The thyroarytenoid muscle arises from a point on the inner surface of the most anterior portion of the thyroid cartilage. In females, this point is about midway along the vertical dimension of the cartilage. In males, this point is about one-third up from the inferior edge. Insertion is on the vocal process. Contraction of the thyroarytenoid muscle pulls the vocal process anteriorly, adducting the glottis and tightening the vocal fold. Changes in vocal fold tension are important in determining the frequency of vocal fold vibration during phonation. The thyroarytenoid is considered by some authors to consist of two separate muscles: the medial thyroarytenoid, also known as the vocalis, and the lateral thyroarytenoid.

4. Cricothyroid Muscle

The cricothyroid muscle runs upward from the anterior surface of the cricoid muscle to the thyroid cartilage. Contraction of this muscle pulls these two cartilages together anteriorly, resulting in lengthening and tightening of the vocal fold. The effect of cricothyroid contraction on abduction or adduction of the vocal folds is controversial. Experiments have shown that direct electrical stimulation of the cricothyroid muscle during stimulated posterior cricoarytenoid muscle contraction results in greater dilation of the glottis than that produced by posterior cricoarytenoid action alone. Similar contraction, induced in the absence or any other intrinsic muscle activity, is reported to result in vocal fold adduction. Naturally occurring contraction of the cricothyroid, however, has not been shown to significantly affect vocal fold position.

5. Interarytenoid Muscle

The interarytenoid muscle is the only unpaired intrinsic laryngeal muscle. It stretches between the two arytenoids, connecting their posterior and lateral borders. Contraction of this muscle pulls the arytenoids together, adducting the vocal folds.

6. Aryepiglottic Muscle

The aryepiglottic muscle is a very small band of muscle fibers originating from the lateral epiglottis and attaching to the contralateral arytenoid. Contraction of this muscle narrows the laryngeal inlet above the glottis.

E. Extrinsic Laryngeal Muscles

Other muscles, which connect the larynx to other structures, can cause movement of the larynx. In particular, several muscles connect the hyoid bone to the mandible and base of the skull, including the mylohyoid, the digastric muscle, and the stylohyoid muscle. Contraction of these muscles in various combinations can produce elevation or depression of the larynx or movement anteriorly or posteriorly. Because of the way the laryngeal cartilages are interconnected, extrinsic muscle activity can cause changes in the diameter of the laryngeal lumen.

F. Laryngeal Nerves

Nerve supply to the larynx is provided by branches of the vagus nerve: the superior laryngeal nerve and the recurrent laryngeal nerve.

1. Superior Laryngeal Nerve

The superior laryngeal nerve leaves the vagus nerve high in the neck, just below the nodose ganglion, and soon divides into two branches, the internal and external. The internal branch is sensory and carries afferent fibers from receptors in the superior larynx, from the vocal folds and upwards. It enters the larynx through the lateral portion of the thyrohyoid membrane. The external branch is primarily motor, supplying innervation to the cricothyroid muscle, but has been shown in cats to contain sensory fibers as well, from receptors in the cricoarytenoid joint. The presence of sensory fibers in the external branch of the superior laryngeal nerve in humans is suspected but has not been confirmed.

2. Recurrent Laryngeal Nerve

The recurrent laryngeal nerve carries motor fibers to all of the intrinsic muscles of the larynx, except for the cricothyroid muscle. It also carries sensory fibers from the larynx below the vocal folds, and from the trachea. Due to peculiarities in the embryologic development of the larynx, the nerve follows a long and circuitous route, which differs in the right and left sides. On both sides, recurrent laryngeal nerve fibers remain within the vagus nerve throughout its cervical course. On the right side, the recurrent laryngeal nerve separates from the vagus in the upper mediastinum, then curves around the subclavian artery, and ascends in the tracheoesophageal groove to reach the larynx, entering behind the cricothyroid joint. The left recurrent laryngeal nerve leaves the vagus at a lower point in the mediastinum and then curves around the ligamentum arteriosum, which is a fibrous band connecting the aortic arch to the pulmonary artery. The nerve then ascends in the neck, as on the right side, to reach the larynx.

G. Mucosal Cover

The larynx is covered mostly by respiratory epithelium, with numerous mucous glands. An exception is the free edge of the vocal fold, which bears squamous epithelium, without mucous glands.

The mucosal cover of the vocal fold is highly specialized for periodic vibration, as it is very loosely connected to the underlying muscle. This permits a very wide range of movement of the mucosa with respect to the muscle. The tissue between the epithelium and the muscle is often referred to as "Reinke's space," although it is not really a potential space. It is best termed the lamina propria and contains three layers: superficial, intermediate, and deep. The deep layer is continuous with the conus elasticus and is also known as the vocal ligament. The three layers have different mechanical properties, due to the varying densities of elastic and collagenous fibers. Collagenous fibers, which are responsible for most of the stiffness of the tissue, are most numerous in the deep lamina propria, with the density decreasing toward the free edge of the vocal fold. The density of elastic fibers is greatest in the intermediate layer and gradually decreases toward the epithelium as well as the muscle. The concentration of both elastic and collagenous fibers is lowest in the superficial lamina propria, so that this layer has the lowest impedance to vibration. The boundaries between layers is not sharp. There is a gradual transition, which allows for impedance matching between the epithelium and muscle. Thus, vibration of the low-density, flexible epithelium is not inhibited by contact with the stiffer and more massive muscle tissue.

II. Embryology

A. Origin of the Lower Respiratory System

The primordium of the respiratory system, the respiratory diverticulum, first appears in the embryo at an age of 3 wk. The diverticulum begins as an outpouching of entoderm from the primitive foregut. This elongates to form the trachea and lung buds. The opening between this diverticulum and the foregut is the primordial larynx. The larynx develops from this entoderm, as well as mesodermal elements of branchial arch origin. The embryology of the larynx is best understood as a product of pharyngeal development.

B. Formation of Branchial Arches

In the fourth or fifth week, the pharynx forms several outpouchings. At the same time, four grooves, known as pharyngeal clefts, appear on the ectodermal surface. As the grooves deepen and the pouches expand laterally, endoderm and ectoderm approximate very closely, dividing the interposed mesoderm into paired bars known as branchial arches. In humans, there are five bars, while lower vertebrates may have six or more. These arches differentiate into the musculoskeletal and connective tissue elements of the upper aerodigestive tract.

C. Segmentation

In the embryo, each arch is associated with its own segmental nerve. This association is preserved throughout development such that muscles derived from a given arch are always supplied by the nerve of that arch. The embryonic arterial system also develops as five aortic arches. Arteries, which develop from these, are also associated with mesodermal products of corresponding branchial arches. In humans, branchial arches are numbered one through four, and six; the fifth arch is omitted for better comparison with development in lower species. The first arch gives rise to the mandible and most of the middle ear ossicles. The second, third, fourth, and sixth arches all contribute to development of the larynx.

D. Laryngeal Development

The hyoid bone is formed by the second and third branchial arches. The second arch contributes to the upper part of the bone, and the third arch to the lower portion and greater cornua. The motor nerve of the second arch is the facial nerve, cranial nerve VII. This innervates, in addition to facial muscles, the stylohyoid muscle and the posterior portion of the digastric muscle, both of which attach to the hyoid bone. The thyroid cartilage is derived from the fourth arch, and its segmental nerve is the superior laryngeal nerve. The sixth arch develops into the cricoid cartilage, and its nerve is the recurrent laryngeal.

The long and circuitous route of the recurrent laryngeal nerve is a product of its relation to its segmental artery. During development, the embryonic aortic arches move caudally into the thorax while the branchial arches remain more cephalad. On the left side the sixth aortic arch forms the fetal ductus arteriosus (Fig. 6), connecting the pulmonary artery to the descending aorta. After birth, this vessel degenerates, but persists as a fibrous band between the two great vessels. The recurrent laryngeal nerve, the nerve of this arch, is pulled down

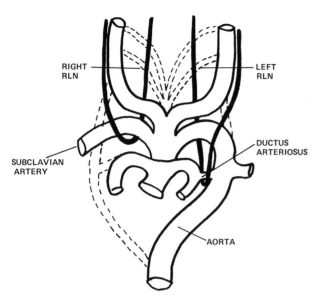

FIGURE 6 Embryology of the recurrent laryngeal nerves. Segmental arteries that disappear in development appear as dashed lines. The ductus arteriosus fibroses after birth to form the ligamentum arteriosum. RLN, recurrent laryngeal nerve.

into the thorax as its artery descends during development. On the right, the embryonic sixth arch totally disappears, so that the sixth segmental nerve is subject only to the pull of the fourth arch artery. Hence the course of the right recurrent laryngeal nerve is shorter, looping around the subclavian artery.

III. Phylogeny

The protective, sphincteric function of the larynx is not only the most vital role of the larynx, it is also the function that developed first in evolution. Embryologically, lungs develop as an offshoot of the digestive tract. This is because in primitive animals, such as the lungfish, the mouth is used to force air into the lungs with positive pressure. The larynx first developed as a sphincter to seal the entrance into the airway while the fish is eating or breathing.

Muscles to actively open the larynx are not present in the most primitive known larynx, that of the bichur lungfish, but are found in the African lungfish and higher animals. Cartilages, which support the larynx and prevent collapse during inspiration, first appear in amphibians, who, like their predecessors the lungfish, inspire by forcing air from the mouth into the lungs. The larynx actively

opens just prior to air injection and quickly snaps shut at the termination of inspiration, trapping air in the lungs. Thus, the breathing cycle contains a breath-holding phase between inspiration and expiration. This is useful in preventing the entry of water or food into the lungs. Tight laryngeal closure occurs, however, even when the animal is not eating or submerged. Therefore, breath-holding may either represent only an evolutionary artifact, or serve to facilitate gas exchange by increasing intrapulmonary pressure or prolonging transit time of air in the lungs.

In reptiles, injection of air into the lungs by positive intraoral pressure is replaced by the use of negative intrathoracic pressure to draw air in. As in amphibians, the larynx opens just prior to inspiration and closes to maintain lung inflation during a breath-holding phase before expiration. The breath-holding phase can be quite lengthy in duration, permitting continuing gas exchange during prolonged underwater dives; however, breath-holding is seen in all reptiles, even those that are entirely terrestrial. Again, the functional significance of laryngeal adduction during expiration is not known.

In mammals, end-inspiratory breath-holding is only an occasional phenomenon. The larynx still opens actively during inspiration but, in general, does not completely close during or after expiration. A notable exception to this is found in newborn lambs. During the first few hours of life, tight laryngeal closure is involved in producing intermittent short periods of extremely high intrathoracic pressures. These pressure changes appear to play a role in clearing interstitial fluid from the lungs. If a tracheotomy is opened, which precludes the production of high intrapulmonary pressures, fluid clearance is delayed. Whether or not a similar phenomenon occurs in humans or other mammals is not known.

Active opening of the larynx during inspiration increases with respiratory demand. Animals adapted for running, such as the gazelle, have larynges with very long segments of arytenoid cartilage in the vocal folds, permitting extremely wide dilation of the laryngeal opening. In most mammals, however, the larynx is also adapted for phonation. The adaptation involves an increase in the membranous, vibratory portion of the vocal fold. This of necessity decreases the cartilage portion and, hence, decreases maximum possible airflow. Speech is a sophisticated adaptation of phonation. Although birds and some primates can be taught to use simple phrases, speech and language occur spontaneously only in humans.

IV. Functions of the Human Larynx

A. In Breathing

1. Protection

By far the most important role of the larynx is to prevent the aspiration of food, secretions, or foreign matter into the lungs. Aspiration can lead to asphyxia, but more commonly it leads to inflammation or infection of the lungs. Repeated bouts of aspiration can lead to permanent lung damage. The aspiration of refluxed gastric material, which is very acidic, is particularly dangerous. The sphincteric function of the larynx is activated when pharyngeal or supraglottic mucosa is stimulated by matter aspirated through the mouth or nose, or refluxed from the stomach. Other noxious stimuli can produce laryngeal constriction. [*See* LARYNGEAL REFLEXES.]

2. Cough

Coughing is a very important mechanism for cleansing and preventing collapse of the lungs. Although a weak cough is possible even in the presence of a tracheotomy, a strong effective cough requires tight laryngeal closure. Whether occurring reflexively or as a voluntary act, a cough consists of three phases: inspiration, pressure building, and explosive exhalation. In the first phase, the glottis opens very widely to permit rapid inhalation and expansion of the lungs. The true and false vocal folds then close very tightly to prevent the egress of air, while expiratory muscles contract, producing extremely high intrathoracic pressure. The larynx then rapidly opens widely, resulting in a sudden high velocity outflow of air. The pressure-building phase serves to distend alveoli, while the outflow of air carries mucus and any foreign material into the pharynx, where it is swallowed or expectorated.

3. Regulation of Airflow

The larynx is better adapted than any other portion of the respiratory tract for producing sudden alterations in resistance to airflow. It can open widely or close completely in a very brief interval of time. That the larynx is an important valve for occluding airflow is well established. Its role in regulating the rate of airflow and breathing is suspected but not as yet proven. Evidence for this function of the larynx largely consists of observations of laryngeal movement with breathing.

The vocal folds begin to abduct just prior to inspiration. During exhalation, the vocal folds gradually adduct, but not completely. The degree of laryngeal movement varies with respiratory demand. That is, movements are very slight or absent with quiet breathing but increase with increased respiratory effort, whether voluntary or in response to blood gas changes or exercise. Some authors contend that inspiratory opening of the larynx is secondary to passive tension, induced by descent of the larynx during inspiration, and that expiratory adduction is a passive phenomenon. Phasic inspiratory contraction of the posterior cricoarytenoid muscle, the only laryngeal abductor, has been conclusively demonstrated in humans; therefore inspiratory laryngeal abduction is an active phenomenon. Expiratory contraction of laryngeal adductors during expiration has also been observed in conscious humans, and variations in the intensity of contraction have been shown to correlate with both expiratory resistance to airflow, as well as the duration of expiration. Adductor activity is, however, not seen during sleep. Therefore, expiratory closure of the larynx appears to be a mechanism for regulating the duration of expiration during wakefulness, but not during sleep.

The resistance of the larynx to airflow varies with the direction of airflow, being greater in inspiration. The majority of resistance in inspiration is due to the vocal folds. When the glottic inlet is compromised, such as by laryngeal edema or vocal fold paralysis, inspiratory resistance can become great enough to result in audible stridor, before expiratory resistance is affected.

B. In Swallowing

Swallowing is one of the most primitive human functions. Although we usually give it little thought, a swallow is an extremely complex act requiring a high degree of coordination. There are three phases of swallowing. The first phase is the oral stage, wherein food is chewed, mixed with saliva, formed into a bolus, and propelled into the pharynx. This part of swallowing is subject to conscious control. The second phase, the pharyngeal phase, is entirely involuntary and lasts less than a second. During this interval food is propelled down the pharynx and into the esophagus. In the third phase, food travels down the esophagus and into the stomach.

If all goes well, ingested material will completely pass into the esophagus and down to the stomach, without spilling into the lungs or refluxing back out the nose. Significant dysfunction, however, can result from even minor defects in the strength or timing of component events. For example, if the upper esophageal sphincter does not relax at the very end of the pharyngeal phase, the swallowed material cannot enter the esophagus and remains in the pharynx. The pharyngeal phase is undoubtedly the most treacherous segment of the swallow, as malfunction at this level can lead to serious or even fatal impairment of respiration. The larynx plays a crucial role, as it is responsible for preventing aspiration.

The larynx offers several levels of protection: diversion of material around the larynx, closure of the supraglottis, and closure of the glottis. The epiglottis is responsible for diversion. This structure is often regarded as a "lid" for the larynx, but in actual fact, it does not close snugly over the airway. Its major function is as a baffle, which diverts material flowing from the mouth around the larynx, through the pyriform fossae, and into the esophagus. At the initiation of a swallow, the larynx is pulled upward and forward, under the base of the tongue. This motion both pulls the larynx out of the mainstream of the swallow and deflects the epiglottis posteriorly. Supraglottic closure is accomplished by adduction of the false vocal folds. The final barrier to aspiration is adduction of the true vocal folds.

C. In Speech and Singing

By far the best known function of the larynx is that of a sound generator in the production of speech. It is less well recognized by most people that the larynx does not perform this task alone. The human voice is the result of the combined and highly coordinated interaction of the larynx with the lungs, diaphragm, abdominal muscles, throat, neck muscles, and mouth. Singing and drama teachers, have been keenly aware of this for centuries.

Speech as well as singing can be divided into three components: phonation, resonance, and articulation. Phonation is the production of sound by vibration of the vocal folds. Resonance is the modulation of that sound, with selective amplification of certain component frequencies. If the larynx vi-

brated in isolation from the rest of the body, the resulting sound would in no way resemble the human voice. Vocal training, for singing as well as acting or public speaking, concentrates heavily on refining and maximizing resonance, so that the loudest and most pleasing sound can be produced, with the least amount of strain or pressure on the larynx. Articulation is the shaping of sound by the lips, tongue, teeth, palate, and throat to produce words. Articulation, and not phonation, is the adaptation that characterizes human speech.

1. Phonation

The glottis produces sound in much the same manner as a reed wind instrument. Expired air generates vibration of the free edges of the vocal folds, resulting in alternating compression and rarefaction of air. The amplitude, or loudness of the sound produced is determined by the difference between the peak and trough air pressure values generated. For normal phonation to occur, four conditions must be met: adequate breath support, optimal vocal fold position, flexible vocal folds, and control of vocal fold length and tension.

To initiate phonation, first the vocal folds are positioned and tuned. Then exhalation causes pressure beneath the vocal folds to rise. When this pressure becomes high enough, the vocal folds are pushed apart, and pressure then decreases rapidly. The vocal folds then return to the midline. Forces that restore vocal fold position include the sudden decreases in pressure just after opening, tension and elastic forces in the vocal fold, and the Bernoulli effect. This last phenomenon refers to the tendency for the walls of a narrow channel to collapse when air rushes through at high speed. Once the folds are again approximated, pressure in the trachea begins to build again, and the cycle is repeated.

a. Breath Support
The air pressure that provides the driving power for phonation is generated by expiration. In quiet breathing and speech, this is largely a passive force, as energy stored in the rib cage and diaphragm is released by relaxation of inspiratory muscles. The deeper the inspiration, the louder and longer one can speak. For louder sounds (higher pressure), or for longer duration of phonation, active contraction of abdominal and intercostal muscles is required.

In general, respiratory effort is sustained, and the

periodic pressure fluctuations of the sound produced are due to the vibrations of the vocal fold. In singing, however, expiratory effort is also cyclically varied, to superimpose slower frequency pressure changes in the vocal output. This technique is known as vibrato, and it provides some perceived amplification.

b. Vocal Fold Position Proper positioning of the vocal folds is crucial: They must be lightly touching or very close, but not tightly compressed together. The vocal folds can then close to periodically stop airflow, resulting in sudden very brief drops in air pressure. Loose approximation allows the glottis to easily swing open to produce wide fluctuations in pressure and, hence, a loud sound, even at low air flows. If the vocal folds are too far apart, vocal efficiency falls, and a higher flow of air is required to produce excitation of the vocal folds. The voice becomes breathy sounding. With wide glottic openings, the airflow required for periodic sound exceeds the capacity of the lungs and abdominal muscles, and the voice may dwindle to a coarse, weak whisper.

c. Vocal Fold Mobility Because the key process in phonation is vibration of the vocal folds, much depends on the ability of the vocal folds to vibrate. Optimal vibration is achieved when the epithelium is thin and flexible and separated from the underlying muscle by a normal lamina propria (see Section I.G). Edema can increase the impedance of the mucosa and severely limit vibration, as is commonly seen in laryngitis. Scarring, from trauma, infection, or surgery, can tether the mucosa, and this also limits voice production. Many other defects can impair the vibratory capacity of the vocal fold and produce significant vocal impairment.

d. Vocal Fold Tension The pitch of the voice is primarily determined by how fast the vocal folds vibrate. When the vocal folds vibrate very rapidly, the pitch is high. Slower variations result in lower pitch. The physical properties of a given larynx determine the range of pitch it can produce. In infancy, the larynx is very small and produces a high pitched cry. As the child grows, the voice gradually deepens. At puberty, gender differences in the voice appear, as there is a rapid increase in the size of the larynx in boys. This rapid change often outstrips the capacity for adaptation, and there are fre-

quent sudden pitch changes, or "cracking of the voice." Adult males, with longer and heavier vocal folds, and a deeper supraglottis, produce the lowest human pitches. Females have smaller larynges and a higher pitch. Within each sex, of course, there is great variation in ranges, including sopranos (high) and altos (low) in women, and tenors (high) and basses (low) in men.

Within the physical constraints of each larynx, there is normally a wide dynamic range of pitch production, making singing and dynamic inflection of the voice possible. Changes in the length and tension of the vocal fold alter its resonant frequency, producing an increase or decrease in pitch, just as tightening and shortening a guitar or piano string produces a higher frequency sound. The larynx, however, is a more complex musical instrument. Frequency can also be controlled by changing the thickness of the fold. Furthermore, pitch changes can be effected by limiting how much of the vocal fold participates in vibration.

Control of pitch is obviously essential in singing. It is also very important in conversational speech, as patterns of voice inflection add another dimension to spoken language. A flat, monotone voice is expressionless, while the introduction of pitch changes can totally change the meaning of an utterance. In Chinese dialects, differences in intonation are used to form totally different words from the same phonetic sounds.

2. Resonance

As mentioned before, resonance is necessary to produce a human voice from the raw sound product of phonation. The chest, neck, and head all play a role in this process. The thyroid cartilage itself also participates. Resonance not only gives the voice its characteristic acoustic pattern but can also be used to amplify the voice. A great deal of classical vocal training involves modulation of the upper aerodigestive tract to achieve optimal resonance.

3. Articulation

Much of articulation is controlled by the lips, tongue, teeth, and pharynx to form the various consonants and vowels. The larynx too, however, also participates in articulation, by coordinating the beginnings and endings of phonation to coordinate with upper articulators, producing voiced and unvoiced sounds.

V. Summary

Lack of a larynx does not preclude life, as one can breathe through an opening in the trachea. The quality of life, however, is significantly affected by this organ. Although small, it plays significant roles in breathing, eating, speaking, and singing. Normal swallowing requires precise laryngeal movement, highly coordinated with those of the pharynx; thus, even a mild neurologic disorder can result in significant impairment. Human speech involves high speed vibration of the laryngeal true vocal folds, which requires an intact larynx with healthy, flexible tissues. The larynx may also serve as a regulator of respiration.

Bibliography

Bartlett, D. B. (1989). Respiratory functions of the larynx. *Physiolog. Rev.* **69,** 33–57.

Hirano, M., Kakita, Y., Ohmary, K., and Kurita, S. (1982). Structure and mechanical properties of the vocal fold. *In* "Speech and Language: Advances in Basic Research and Practice," Vol. 7, pp. 271–297. Academic Press, New York.

Proctor, D. F. (1980). "Breathing, Speech and Song." Springer-Verlag, Vienna.

Sasaki, C. (1985). Laryngeal physiology. *In* "Surgery of the Larynx" (B. Bailey and H. Biller, eds.). W. B. Saunders, Philadelphia.

Learned Helplessness

JANE PENAZ EISNER AND MARTIN E. P. SELIGMAN,
University of Pennsylvania

Glossary

Cognitive deficit Organism's difficulty learning that its actions control an outcome; this difficulty follows experience with uncontrollable events

DSM-III-R Revised third edition of the *Diagnostic and Statistical Manual of Mental Disorders* of the American Psychiatric Association

Escape/avoidance training Experimental procedure in which an organism learns to escape and avoid shock, often by moving from one side of a shuttlebox to the other

Experimentally naive Term used to describe an organism that has received no prior treatment in the laboratory

Explanatory style Habitual manner in which a person explains the causes of events. A person with a pessimistic explanatory style explains the causes of negative events as permanent (stable), present in all situations (global), and self-originated (internal). He or she explains the causes of positive events as transient (unstable), present in few situations (specific), and externally originated (external). Conversely, a person with an optimistic explanatory style explains the causes of negative events as unstable, specific, and external; the causes of positive events are explained as stable, global, and internal

Motivational deficit Organism's failure to initiate responses aimed at controlling an outcome; this failure follows experience with uncontrollable events

Shuttlebox Chamber containing two sides separated by a barrier. An animal turns off, or escapes, shock by jumping across the barrier from one side to the other. It can avoid shock altogether by jumping across the barrier before shock begins

Yoked group Group of subjects that receives the same physical events as does the experimental group. Unlike the experimental group, however, this group is unable to influence these physical events with its actions

LEARNED HELPLESSNESS, described in humans and infrahumans, occurs when prior experience with uncontrollable events leads to the expectation that events will be uncontrollable in the future. The main behavioral manifestations of helplessness—motivational and cognitive deficits—result from this expectation. The motivational deficit in helplessness consists of lowered response initiation: for example, the organism fails to attempt escape from the aversive event. The cognitive deficit of learned helplessness involves the organism's failure to learn that its actions can control the events. Other helplessness symptoms include loss of self-esteem; appetite and sleep disturbance; loss of aggression; anxiety, sadness, and hostility; norepinephrine and serotonin depletion; and immunological deficits.

I. Discovery of Learned Helplessness in Dogs

A. Escape/Avoidance Training with Experimentally Naive Dogs

Experimentally naive dogs show a typical pattern of behavior when they receive escape/avoidance training in a shuttlebox. The first shock sends the animal scrambling frantically. Eventually, on this first trial, the dog accidentally crosses the barrier and thus escapes the shock. During the second trial, the dog again scrambles but crosses the barrier more quickly than it did the first time. This pattern continues until the dog finally learns to avoid the shock altogether.

B. Escape/Avoidance Training in Dogs Pretreated with Inescapable Shock

In 1967, Bruce Overmier and Martin Seligman found that dogs given inescapable shock before escape/avoidance training behave differently from experimentally naive dogs. Initially, a dog given prior inescapable shock reacts as does an experimentally naive dog to shock in the shuttlebox. However, in sharp contrast to the experimentally naive dog, the pretreated dog soon ceases running and remains silent until the shock ends. The dog fails to cross the barrier and, consequently, fails to escape the shock. Unlike the struggling naive dog, the pretreated dog becomes passive and seems to accept the shock. On subsequent trials of shock, the pretreated dog again fails to attempt escape; indeed, the animal will tolerate as much shock as the experimenter gives it. In its failure to seek escape from the shock, the pretreated dog shows a clear motivational deficit.

Dogs pretreated with inescapable shock also show a notable cognitive deficit. Occasionally, a pretreated dog accidentally will cross the barrier and thus escape shock. However, the animal will then frequently revert to accepting the shock on succeeding trials. Unlike the naive dog, the pretreated dog does not learn that crossing the barrier will lead to shock termination.

C. The Triadic Design

Does the shock itself or the uncontrollability of the shock cause the motivational and cognitive deficits observed in pretreated dogs? The triadic design directly tests the idea that the uncontrollability of an outcome, rather than the outcome *per se,* causes these deficits. Three groups are compared in the triadic design: an escapable group, an inescapable group, and a naive group. The escapable group is pretreated with an outcome that it can control through a specific response. The inescapable group consists of subjects yoked to those in the escapable group. Each subject in the inescapable group re-

ceives exactly the same outcome as does its counterpart subject in the escapable group. However, unlike those in the escapable group, subjects in the inescapable group cannot control the outcome with any of their responses. The naive group receives no pretreatment. Later, all three groups are tested for deficits in a new task.

Martin Seligman and Steven Maier in 1967 used the triadic design to compare the escape/avoidance behavior of three groups of eight dogs: a group receiving escapable shock, a yoked group exposed to inescapable shock, and a group that received no shock. Placed in a hammock, each dog in the escapable group was trained to turn off shock by pressing a panel with its nose. Dogs in the yoked group, each also placed in a hammock, received shocks of the same amount, duration, and pattern as those received by the escapable group. Unlike dogs in the escapable group, however, dogs in the yoked group could not stop the shocks by pressing a panel. A naive group of dogs received no shock in the hammock. Twenty-four hours later, all dogs received escape/avoidance training in a shuttlebox. Dogs in the escapable and naive groups readily learned to jump the barrier quickly to escape shock. In contrast, dogs in the inescapable shock group responded much more slowly; six of them completely failed to escape the shock. These results indicate that the inability to control shock, not the shock itself, produces failure to respond.

D. The Learned Helplessness Theory

The term *learned helplessness* denotes the deficits that follow uncontrollable events and also implies the process by which prior exposure to uncontrollable events interferes with escape and avoidance learning in dogs. In 1967, Seligman and Maier proposed a theory to account for the motivational and cognitive deficits observed in learned helplessness. A simple, yet provocative notion lays the foundation for the learned helplessness theory: When an organism encounters an outcome that is independent of its responses, the organism learns that its responses and the outcome are independent. Having learned that responses are futile because the outcome is uncontrollable, the organism comes to expect that future responding also will not affect the outcome.

This expectation of noncontingency between response and outcome produces behavior characterized by certain motivational and cognitive deficits.

Motivationally, the organism will fail to initiate responses designed to control the outcome. If its haphazard actions affect the outcome, the organism will manifest a cognitive deficit: difficulty learning that its response controls the outcome. Figure 1 summarizes the four steps that comprise the learned helplessness theory.

II. Learned Helplessness in Other Nonhuman Species

A. Cats

When cats first receive inescapable shock in a hammock, they later fail to escape shock in a shuttlebox. In fact, the cats passively accept the shock. This behavior mirrors the motivational deficits of learned helplessness in dogs.

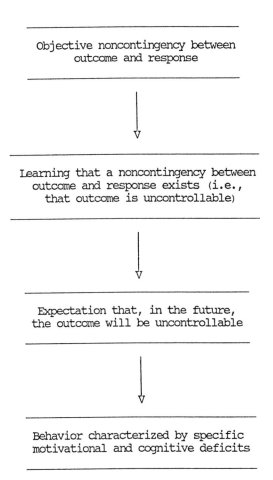

FIGURE 1 Flow of events leading to helplessness deficits.

B. Fish

Goldfish pretreated with inescapable shock are slower to avoid shock in an aquatic shuttlebox than are experimentally naive goldfish. Although their behavior is not identical to that of pretreated dogs, the fish clearly show impaired escape and avoidance response.

C. Mice

Mice exposed to inescapable shock subsequently escape flooding water less effectively than do mice that received either escapable shock or no shock. The inefficient escape behavior of pretreated mice indicates a motivational deficit similar to that manifested by dogs receiving prior inescapable shock.

D. Rats

Learned helplessness effects emerge in rats only when a difficult response is used to test for deficits. For example, when rats first receive inescapable shock, they will show no deficits if a single bar press or crossing to the other side of a shuttlebox allows them to escape shock. Apparently, these are simple responses readily performed by the rat under any circumstances. However, when the response required to terminate shock is more difficult, learned helplessness deficits arise in the rat. So, when the rat must press the bar three times or cross the shuttlebox barrier and then return, the pretreated rat fails to escape shock. By comparison, both experimentally naive rats and rats that have first received escapable shock easily learn these more challenging responses to escape shock.

III. Learned Helplessness in Humans

A. Motivational Deficit

Donald Hiroto documented a learned helplessness effect in humans in 1974. Employing the triadic design, he assigned subjects to one of three groups: an escape group, an inescapable group, and a control group. In the escape group, subjects received loud noise that they could turn off by pressing a button. Subjects in the inescapable group were exposed to the same loud noise, but they could not control its termination. In the control group, subjects received no noise. All three groups subsequently underwent escape/avoidance training in a new situation using a hand shuttle. An analogue of the shuttlebox, the hand shuttle is a rectangular box containing a handle that subjects can move from one side of the box to the other to terminate noise. Both the escape and control groups easily learned to terminate noise by moving the handle. In contrast, subjects in the inescapable group failed to escape and avoid the aversive noise. Rather, they passively accepted it. This motivational deficit is identical to that found in animals pretreated with inescapable shock.

B. Cognitive Deficit

Further work replicates the cognitive deficit of helpless animals in humans. Subjects first received escapable, inescapable, or no noise and then solved a set of 20 anagrams. Each anagram had letters arranged in the pattern 53124. For example, the word *NOISE* was represented as anagram ISOEN. To solve this anagram successfully, subjects had to arrange the letters ISOEN in the pattern 53124, in which N was the fifth letter, O the third letter, I the first letter, S the second letter, and E the fourth letter. Compared with subjects exposed to escapable noise or no noise, subjects pretreated with inescapable noise solved fewer anagrams correctly. In addition, subjects exposed to inescapable noise discerned the pattern in the anagrams after an average of approximately seven successes. In sharp contrast, subjects who received escapable noise or no noise noticed the pattern after only three successes on average. Thus, inescapable noise retarded subjects' ability to learn a new task.

IV. Learned Helplessness as a Model of Depression

Depression is a mental and behavioral condition characterized by sadness and loss of interest in pleasurable activities, hopelessness about oneself and the world, passivity and suicidal thoughts, and appetite and sleep disruption. A common disorder, clinical depression affects up to 26% of females and 12% of males at some point in their lives. Tragically, depression is the most lethal mental disturbance, not uncommonly resulting in death by suicide. [*See* DEPRESSION; SUICIDE.]

Learned helplessness offers a model for a subset of depressions. At the root of a "helplessness depression" is the generalized expectation that negative life events are uncontrollable. According to the

learned helplessness model of depression, this expectancy arises from experience with uncontrollable aversive events and then produces the characteristic symptoms of depression.

Experimental work supports the idea that experience with uncontrollable aversive events leads to symptoms of depression. To meet the DSM-III-R criteria for depression, a person must have at least five of the following nine symptoms: (1) diminished interest in usual activities, (2) appetite disturbance and weight loss or gain, (3) sleep disturbance, (4) psychomotor retardation or agitation, (5) loss of energy, (6) difficulty thinking or attending, (7) depressed mood, (8) feelings of worthlessness, and (9) suicidal thoughts. Animals exposed to uncontrollable events show the first six of these symptoms, thus meeting the diagnostic criteria for depression. The last three symptoms—depressed mood, feelings of worthlessness, and suicidal thoughts—cannot be assessed in animals. However, humans exposed to inescapable noise display depressed mood and feelings of worthlessness as well as the cognitive and motivational deficits. In short, experience with uncontrollable negative events produces eight of the nine symptoms of depression.

In addition to the similarities between helplessness and depressive symptomology, the neurochemical changes associated with depression parallel those in learned helplessness. Norepinephrine and serotonine depletion is a correlate of depression in humans. In the rat, learned helplessness occurs simultaneously with serotonin and norepinephrine depletion. This finding suggests that learned helplessness and depression are accompanied by similar neurochemical changes. [*See* DEPRESSION: NEUROTRANSMITTERS AND RECEPTORS.]

V. Physiology of Learned Helplessness

A. The Neurochemistry of Learned Helplessness

The injection of the antidepressant desipramine into the rat frontal neocortex reverses previously established learned helplessness. Experimental work suggests the following model to explain this finding. Desipramine injection increases the amount of norepinephrine available at synapses in the frontal neocortex. This surplus norepinephrine stimulates frontal neocortex neurons with norepinephrine receptors, causing them to release serotonin onto cells in another area of the neocortex. The serotonin-stimulated cells then release GABA onto neurons in the hippocampus. These hippocampal neurons, in turn, activate a serotonergic synapse in the septum. Activation of this septal serotonergic synapse results in the reversal of helplessness deficits. [See HIPPOCAMPAL FORMATION; NEOCORTEX.]

In addition to desipramine, several other therapies for depression beneficially affect helpless rats. Tricyclic compounds, monoamine oxidase inhibitors, and several other compounds used to treat depression in humans all successfully reverse helplessness in rats. In addition, electroconvulsive shock and REM sleep deprivation, two nonpharmacological treatments for depression, also break up helplessness in rats. These findings indicate that depression and helplessness can be reversed by the same types of therapy, thus lending further support to the learned helplessness model of depression.

Further experiments with rats also suggest a neurochemical model for the prevention of helplessness by desipramine. When injected into the frontal neocortex, lateral geniculate body, and hippocampus, desipramine increases the amount of norepinephrine available in these areas. The pathway divides into two parts. Along one path, hippocampal cells, now stimulated by increased norepinephrine, release norepinephrine into the septum. In the second path, frontal neocortex and lateral geniculate body neurons, also excited by the increased norepinephrine, release GABA into the hippocampus. Within the hippocampus, GABA stimulates a second, distinct population of cells, which then release norepinephrine into the septum. The release of norepinephrine into the septum, occurring along two pathways, underlies the prevention of learned helplessness.

B. Learned Helplessness and Physical Illness

Animal studies indicate that learned helplessness impairs the body's defenses against physical illness. In one study, sarcoma tumor cells were injected into adult rats. Twenty-four hours later, these rats received inescapable, escapable, or no shock. During a 30-day period, only 27% of rats exposed to inescapable shock rejected the tumor implant. In contrast, 63% of rats given escapable shock and 54% of rats given no shock rejected the tumor. Thus, exposure to inescapable shock, an uncontrollable aversive event, markedly compromised the rats' ability to combat the establishment and consequent spread of a lethal, malignant tumor.

Further work demonstrates that early helplessness experiences impact tumor rejection in adult rats. Weanling, 21-day-old rats received either helplessness or mastery training. Helplessness training consisted of exposure to uncontrollable shock, whereas mastery training involved experience with controllable shock. Two months later, the same rats (now adults) were injected with tumor cells. No differences in tumor rejection emerged between the helplessness and mastery groups. In a second experiment, adult rats who had received early helplessness or mastery training were first injected with tumor cells and then given inescapable shock. Of the rats trained in mastery, 68% successfully rejected the tumor, whereas only 30% of those trained in helplessness rejected the tumor. Thus, when rats exposed to early helplessness training are later confronted with an uncontrollable stressor, they revert to helplessness behavior and demonstrate diminished defenses against cancer.

How might learned helplessness contribute, physiologically, to physical illness? One answer to this question involves immunological functioning. Normally, the immune system recognizes foreign cells and destroys them. Experimental evidence indicates that learned helplessness suppresses the activity of T lymphocytes—white blood cells with many functions, including recognition and lysis of some tumor cells. One study examined the *in vitro* proliferation of lymphocytes taken from rats. Twenty-four hours after exposure to escapable shock, inescapable shock, or no shock, all rats received brief inescapable shock. Then, blood samples were taken from each rat and treated to separate lymphocytes from the other constituents of blood. T-cell mitogens were added to the lymphocytes to stimulate lymphocyte proliferation. Compared with lymphocytes of the escapable shock or no shock groups, lymphocytes derived from the inescapable shock group showed suppressed proliferation in response to the mitogens. This result indicates that learned helplessness suppresses the proliferation of T lymphocytes, an early aspect of the immunological response to foreign invaders. [See LYMPHOCYTES.]

VI. The Reformulated Theory of Learned Helplessness

Studies of learned helplessness in humans, although replicating many of the animal findings, reveal several important inadequacies of the original learned helplessness theory. First, the theory does not predict when helplessness deficits will endure and when they will be transient. Second, it does not explain why helplessness deficits sometimes generalize to many different kinds of outcomes and at other times occur only in relation to a specific outcome. Third, the original theory does not explain why people, on perceiving that they are helpless, experience a loss of self-esteem. Finally, the theory does not explain why some individuals are more susceptible to helplessness than are others.

In 1978, Lyn Abramson, Martin Seligman, and John Teasdale reformulated the original learned helplessness theory to resolve these inadequacies. According to the reformulated theory, when a person finds that his or her actions cannot control an outcome, that person tries to explain the cause of this noncontingency. The causal explanation, or attribution, that a person makes for the failure falls along three dimensions: the stable-unstable, global-specific, and internal-external dimensions. First, causes can be described as permanent (stable) or transient (specific). A person who explains the failure with a stable cause will expect future failures and will experience chronic helplessness deficits. Second, causes can be construed as affecting many aspects of a person's life (global) or only few areas (specific). When a person believes that the cause of the failure globally affects his or her life, that person will expect failure in many areas of life. This expectation will result in helplessness deficits across many areas of life. Third, causes can be viewed as internal or external to the individual. A person who explains the failure with an internal cause will lose self-esteem.

The reformulated theory argues that a person's vulnerability to helplessness stems directly from his or her explanatory style, the habitual manner in which that person explains the causes of bad events. Explanatory style ranges from pessimistic to optimistic in quality. A person who tends to explain the causes of bad events as stable, global, and internal in nature has a pessimistic explanatory style; conversely, a person who habitually explains the bad events by unstable, specific, and external causes has an optimistic explanatory style. The reformulated theory asserts that compared with individuals with an optimistic explanatory style, those with a pessimistic explanatory style are more likely to experience pervasive and chronic symptoms of helplessness when faced with uncontrollable negative events.

Figure 2 summarizes the flow of events that eventually leads to the deficits in learned helplessness according to the reformulated theory. The individual first encounters an event that is uncontrollable and then perceives that his or her actions do not influence the event. The individual posits an explanation for the cause of this failure. When the person explains the failure with a stable cause, he or she will expect that outcomes will be uncontrollable in the future. This expectancy produces the deficits that characterize learned helplessness. The stability of the cause determines the chronicity of these deficits, the globality of the cause determines their pervasiveness, and the internality of the cause determines the extent of self-esteem loss.

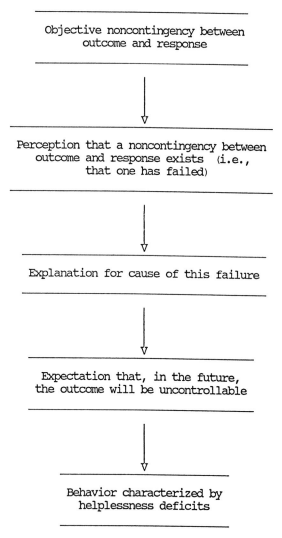

Objective noncontingency between outcome and response

Perception that a noncontingency between outcome and response exists (i.e., that one has failed)

Explanation for cause of this failure

Expectation that, in the future, the outcome will be uncontrollable

Behavior characterized by helplessness deficits

FIGURE 2 Flow of events leading to helplessness deficits: The reformulated theory of learned helplessness.

VII. The Reformulated Learned Helplessness Model of Depression

According to the original learned helplessness model of depression, the cause of a "helplessness" depression is a person's expectancy that future negative events are uncontrollable. The reformulated learned helplessness model of depression posits this same expectation as the immediate cause of depression; however, the reformulated model also describes explanatory style as a more distal risk factor for vulnerability to depression. Specifically, the reformulated model proposes that whether or not a person chronically expects future failure depends on his or her explanatory style. A person with a pessimistic explanatory style chronically expects failure, whereas a person with an optimistic explanatory style does not. Accordingly, the reformulated theory predicts that persons with a pessimistic explanatory style are more vulnerable to depression than are persons who have an optimistic explanatory style.

Evidence confirms this prediction. A review of more than 100 studies involving more than 15,000 subjects shows that depressed individuals, students, patients, and children tend to have a more negative explanatory style than do their nondepressed counterparts. In addition, depressed patients, when contrasted with other psychiatric patients, believe they have little control over important life goals. Most convincingly, compared with nondepressed individuals with an optimistic explanatory style, nondepressed individuals with a pessimistic explanatory style more often become depressed when they experience aversive events. This work shows that a negative explanatory style is a risk factor for depression.

Acknowledgements

This work was supported by a National Science Foundation Graduate Fellowship to Jane Penaz Eisner and by NIMH Grant MH19604, NIMH Grant MH40142, NIA Grant AG05590, and a MacArthur Foundation Research Network on Determinants and Consequences of Health-Promoting and Health-Damaging Behavior Grant to Martin E. P. Seligman.

Bibliography

Abramson, L. Y., Seligman, M. E. P., and Teasdale, J. (1978). Learned helplessness in humans: Critique and reformulation. *J. Abnorm. Psychol.* **87,** 32–48.

Hiroto, D. S., and Seligman, M. E. P. (1975). Generality of learned helplessness in man. *J. Pers. Soc. Psychol.* **31,** 311–327.

Maier, S. F., and Seligman, M. E. P. (1976). Learned helplessness: Theory and evidence. *J. Exp. Psychol.* [*Gen.*] **105,** 3–46.

Overmier, J. B., and Seligman, M. E. P. (1967). Effects of inescapable shock upon subsequent escape and avoidance learning. *J. Comp. Physiol. Psychol.* **63,** 23–33.

Peterson, C., and Seligman, M. E. P. (1984). Causal explanations as a risk factor for depression: Theory and evidence. *Psychol. Rev.* **91,** 347–374.

Porsolt, R. D. (1979). Animal model of depression. *Biomedicine* **30,** 139–140.

Seligman, M. E. P. (1975). "Helplessness: On depression, development, and death." Freeman, San Francisco.

Seligman, M. E. P., and Maier, S. F. (1967). Failure to escape traumatic shock. *J. Exp. Psychol.* **74,** 1–9.

Sherman, A. D., and Petty, F. (1980). Neurochemical basis of the action of antidepressants on learned helplessness. *Behav. Neural Biol.* **30**(2), 119–134.

Visintainer, M., Volpicelli, J. R., and Seligman, M. E. P. (1982). Tumor rejection in rats after inescapable or escapable shock. *Science* **216,** 437–439.

Weiss, J. M., Glazer, H. I., and Pohoresky, L. A. (1976). Coping behavior and neurochemical change in rats: An alternative explanation for the original "learned helplessness" experiments. *In* "Animal Models of Human Psychobiology" (G. Serban and A. King, eds.). Plenum, New York.

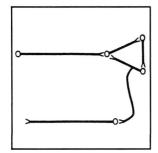

Learning and Memory

JOE L. MARTINEZ, JR., *University of California, Berkeley*

Glossary

Amnesia Inability to store new memories or to recall old memories

Dementia Impairment of intellectual capacity

Diencephalic Referring to a part of the forebrain that includes the thalamus and hypothalamus

Dual trace mechanism Idea that reverberating assemblies of neurons form the neural basis of the memory trace; a short-term trace set up by a new experience will decay unless it is exercised through repetition to become a long-term trace; hence, one trace can have a dual function being the substrate for both short- and long-term memory

Episodic memory Memory of a particular event

Flashbulb memories Extraordinary events that happen once are vividly remembered for a lifetime (e.g., assassination of John F. Kennedy, President of the United States)

Habit Learned combination of movements that do not require much attention (e.g., knitting)

Iconic memory Memory of very short duration (e.g., mental image produced by a nighttime lightning flash)

Intermediate-term memory Memory that is longer than short-term memory but shorter than long-term memory (e.g., remembering the location of your car in a parking lot after a day of work)

Memory Knowledge of an event or simply knowing a fact; memory is a cognitive representation

Neurotransmitter Chemical that serves as the basis of neuron to neuron communication and travels across the synaptic cleft between neurons

Procedural memory Collection of different abilities (e.g., skill-learning such as is required to play tennis)

Semantic memory Generalized memory without knowledge of where the item was learned (e.g., we all know that a red light means stop but do not know where we learned this concept)

Short-term memory Memory that decays rapidly with time and may be completely forgotten (e.g., a phone number given to you by the information operator—you must dial the number rapidly before short-term memory decays)

ANIMALS LEARN and remember. This is true for simple animals such as sea slugs (*Aplysia*), which are found in the coastal waters of California, and it is true for more complex animals such as humans. Of course the capacity and nature of learning and remembering found in a sea slug is not comparable with that found in humans if one considers a complex behavior such as language. Nevertheless, both sea slugs and humans are capable of learning simple reflexes, and we have learned a great deal about the neural mechanisms of learning and memory from studying animals that are simpler than humans. The search for an understanding of the biological basis of learning and memory through the use of animals and humans is motivated by a desire to alleviate crippling diseases of memory that result from brain injury or are associated with degenerative processes such as Alzheimer's disease.

I. What Is Learning and Memory?

To understand learning and memory, we must explore the phenomenon. First, learning and memory is unique for each individual: One's earliest memories of family are not shared by others. Second, everyone knows things in common; thus, we can all agree that a red light means stop. Third, memories last for a long time; for example, everyone can remember their first-grade teacher. Fourth, humans in particular can remember a great deal of information; the average person can recognize tens of thousands of words. Fifth, any person has almost instant access to this vast knowledge base. Consider for example the word "zenitram." Most people have never encountered this word (Martinez spelled backward) and can tell you so after only a moment's thought. In the parlance of computer language, the instant knowledge that you have never seen "zenitram" before means that you scan your memory banks in parallel fashion. If you had to compare "zenitram" serially to every word you know beginning with "abalone" and ending with "zygote," then the time required to scan the thousands and thousands of words you know would be considerable. These are some of the facts of learning and memory.

II. How Are Memories Stored?

Learning and memory has some logical constraints that need to be defined. The process of learning new information is called acquisition. Once information is acquired it must be stored somewhere for later use. Calling up information for use at a later time is called retrieval. Learning and memory must include at least these three basic processes: acquisition, storage, and retrieval. Each of these processes may be studied independently. For example, consider trying to memorize a poem in two different situations. First, drink six cans of beer and study the poem, or study the poem first and then drink six cans of beer. The result may be the same the next morning, one may not remember the poem, but the mechanism for the amnesia seems different. In the first case, one may not clearly see the page or even particularly care if the poem is memorized. The material is not remembered because of a deficit in acquisition of the new knowledge. In the second case, the poem is learned in an alcohol-free state, but the alcohol acted on some neural process involved in

the storage of the information. The poem is not remembered the next day because of a storage deficit. We have all experienced retrieval failure commonly known as knowledge on the "tip-of-the-tongue"; one cannot retrieve the information. Also, students frequently have retrieval failure while taking tests. Students are frequently observed approaching professors after an exam and lamenting that they really knew much more than they put on the test, but they could not access all of their information. This common experience exemplifies the learning performance distinction. Memory exists in someone's head stored as a physical entity inside neurons and between neuronal networks; however, memory is not directly observable. Memory can only be inferred from performance and, as we all know, performance is affected by things other than memory.

III. Functional Localization

Rene Descartes said in 1650, "When the mind wills to recall something, this volition causes the little gland, by inclining successively to different sides, to impel the animal spirits toward different parts of the brain, until they come upon that part where the traces are left of the thing which it wishes to remember. . . ." The little gland Descartes was referring to is the pineal gland, where he thought the soul resided. The interesting part of this quote however is "that part where the traces are left." As Descartes knew, memory is a thing, in a place, in a brain. The search for memories in the brain began early in this century and did not meet with initial success. Lashley, a famous pioneer in this field, said in 1950, "I sometimes feel, in reviewing the evidence on the localization of the memory trace, that the necessary conclusion is that learning just is not possible . . . Nevertheless, in spite of such evidence against it, learning does sometimes occur." Today memories have been localized to actual cells in invertebrates and to brain regions in vertebrates. The early failures were caused by the distributed nature of the networks that underlie memory traces.

IV. Neuroanatomy

To proceed further with this analysis of human memory, we need to sketch the basic outlines of the organization of the brain. If this sketch were a map

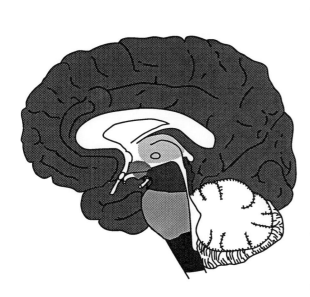

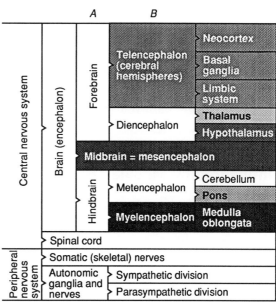

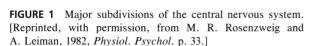

FIGURE 1 Major subdivisions of the central nervous system. [Reprinted, with permission, from M. R. Rosenzweig and A. Leiman, 1982, *Physiol. Psychol.* p. 33.]

of the United States, we would only be identifying major features such as the Great Lakes, the Mississippi River, the Rocky Mountains, and the Grand Canyon.

The brain is made up of cells, as are all organs of the body. The cell type that underlies behavior is a neuron. There are billions of neurons, and each neuron can connect with other neurons thousands of times. The connections between neurons are specialized contacts known as synapses. Interestingly, neurons do not touch each other at these contacts but communicate with each other by releasing special chemicals called neurotransmitters. As described later, synapses are very important for memory.

Figure 1 shows a diagram of a human brain that has been cut in half. The highest level of the central nervous system is called the neocortex, and the lowest level is the spinal cord. The concepts highest and lowest come from evolution. The neocortex is the most recently evolved part of the human brain and, hence, is thought to be responsible for uniquely human behaviors such as language and complex memory storage. By contrast, a lizard does not have much of a neocortex and is not capable of conversation, but it is capable of some learning and it does have a thalamus, hypothalamus, cerebellum, and medulla oblongata. Also outlined in

Fig. 1, but not shown as a diagram, is the peripheral nervous system. Surprisingly, the peripheral nervous system is important for memory and will be discussed later. [*See* BRAIN; NEOCORTEX; SPINAL CORD.]

Cortical areas are important for human memory and include the temporal lobe, frontal lobe, parietal lobe, occipital lobe, and cerebellum. Damage to these areas produce profound deficits in human memory, as discussed later.

V. How Do Neural Networks Form Memories?

As noted earlier, memory traces are composed of circuits or networks of neurons. The idea was clearly elaborated in the 1949 postulation of a dual trace mechanism. With this concept, memory is first encoded as a reverberating cellular assembly of neurons. The activity of this assembly decreases over time unless exercised by repetition; this assembly forms the basis of short-term memory. Through exercise, this assembly becomes a functionally intact system through changes in the synaptic contacts between neurons to become long-term memory. Thus, associative strength is conveyed by the activity of the memory trace; however, reverberatory activity is not an intrinsic property of the memory trace, and it must be regulated by external processes. Normally, the importance of the to-be-remembered event is conveyed by the strength unconditioned stimulus. For example, if one burns a hand by touching a hot stove, then some time will pass before this happens again, because the strength of the unconditioned stimulus—the searing heat—strongly reinforces the memory. However, as any cook knows, you will burn your hand again.

VI. The Strength of Memory Is Influenced by Modulatory Systems

On the other hand, sometimes events that happen only once in our lifetimes are remembered with extraordinary fidelity for a lifetime. These events may be uniquely individual, such as witnessing the birth of a child, or they may occur for an entire population, such as occurred when President John Kennedy was assassinated. These are called flashbulb memories. In this circumstance, exercise is not a central concept, because the event only occurred once. A modulatory system conveys the salience of the event for the organism. As autonomic arousal is a concomitant of these unique events that are learned in one trial, it is reasonable to suggest that the modulatory substances might be hormones. Thus, the response of the peripheral nervous system to stimulus events that are learned influences the strength of memory in the central nervous system. This idea is depicted in Fig. 2, where a modulatory input affects the memory trace and the resultant behavioral output.

If a modulatory input can make memory traces stronger, then this input may make memory traces weaker as well. It is easy to conceive of situations where it would be better not to remember. For example, the pain of childbirth does not deter many women from having a second child, and some re-

searchers think that certain hormones such as oxytocin that are present in high concentrations during childbirth actually produce a kind of amnesia for the event. It is well known that hormones both enhance and impair learning and memory in animals, and this notion of making memory traces both stronger and weaker is easily understood in neurobiological terms as either increased or decreased reverberatory activity of the memory trace. Therefore, memory storage involves two distinct processes. One process is the generation of the memory trace itself. Neurons within the memory trace change as a result of reverberatory activity to maintain the integrity of the trace. The second process involved in memory storage concerns the associative strength that is conveyed by the modulatory input.

VII. Different Types of Memory

One of the major dichotomies of memory was alluded to in the preceding section: short-term memory and long-term memory. Short-term memory is best exemplified by trying to remember a telephone number long enough to successfully complete a call once the number is given by the information operator. On the other hand, practice can produce long-term memory and it is possible to recall one's own home telephone number. Iconic memory is experienced in a lightning storm in the evening. Just after lightning strikes, we have a picture in our mind of the surrounding scene, but it decays rapidly. Some think that there is even intermediate-term memory. For example, at the end of a work day you can remember where you parked your car. Different parts of the brain and even different molecular mechanisms are probably involved in these different kinds of memory. In addition, other memory dichotomies are important to know.

One early distinction was between habit and memory. Habit is stimulus-response learning, whereas memory is a cognitive representation. A more recent dichotomy based on habit versus memory is procedural knowledge and declarative knowledge. Procedural knowledge is "knowing how" to play tennis, whereas declarative knowledge is exemplified by stating a fact, or "knowing that." A conceptually different dichotomy is semantic memory and episodic memory. Semantic memory is generalized memory, as discussed earlier, where we can all agree that red means stop; however, no one knows exactly where they learned this concept. Ep-

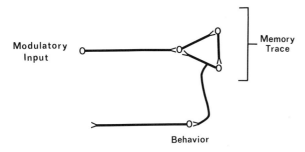

FIGURE 2 Simple conceptualization of a memory circuit containing a modulatory input that regulates the strength of the reverberatory activity.

isodic memory, on the other hand, is recalling a particular event. These memory dichotomies are useful in studying memory pathology, which surprisingly tells us much about the cognitive structure of memory and brain localization of memory.

VIII. Memory Pathology

There are several celebrated case histories in which a patient received localized brain damage and then exhibited particular memory impairment. The most important of these will be briefly reviewed.

A. Medial Temporal Amnesia

The patient H. M. was an epileptic whose disease could not be controlled with medication. Finally, to reduce the occurrence of seizure activity, surgeons removed portions of his temporal lobes on both sides of his brain. The brain regions removed included most of his hippocampus and amygdala. Following recovery, H. M. could not learn new information, even though his IQ was 118, which is above average. He could not remember the names of new people he met, and he kept returning to the house where he used to live; however, H. M. could learn backward mirror drawing even though he did not remember that he had experience on the mirror-drawing task. Thus, the damage to his temporal region produced a deficit in his ability to learn declarative but not procedural knowledge.

This case implicates the hippocampus and the amygdala as being important for memory storage but not as sites of memory, because the remote memories of H. M., such as where he used to live, were intact. An alternative view is that memories are stored in other regions of the brain but damage to the hippocampus and amygdala impairs the ability of the brain to retrieve the information. [*See* HIPPOCAMPAL FORMATION.]

B. The Hippocampus

The patient R. B. suffered ischemia (insufficient blood supply) during cardiac surgery and woke up amnesic. As with H. M., this patient had an above average IQ of 111 and had trouble learning new information. Following his death, a detailed analysis of his brain was conducted and, despite widespread brain damage, there was a striking loss of some specialized neurons called pyramidal cells in the hippo-campus. Thus, damage to a restricted portion of the hippocampus is associated with medial temporal amnesia.

C. Diencephalic Amnesia

The patient N. A. was a popular, intelligent young man in the Air Force who had done quite well in school and had a bright future. While he was assembling a model airplane, a roommate was playing with a miniature fencing foil near N. A. The roommate tapped him on the shoulder and thrust forward as N. A. turned around. The foil passed through N. A.'s nose into his brain. N. A., like H. M., has severe amnesia and can only remember events before his injury; he cannot store new information. He is popular among the staff of the outpatient treatment center he uses, but he has no close personal relationships. Sexually active before his injury, he no longer has girlfriends. He made a date with a girl once but he forgot and never pursued the matter further. Computer tomography revealed damage in the left mediodorsal thalamic nucleus. This case indicates that damage to a diencephalic structure is associated with severe amnesia.

D. Korsakoff's Syndrome

Korsakoff's syndrome results from a lack of thiamine, mainly as a result of poor nutrition due to alcoholism. As there are some 10 million alcoholics in the United States today, there are always new Korsakoff patients. These patients fail to recall past events or learn new information very well. They often do not think anything is wrong with them, and they confabulate to hide their memory deficiency. Korsakoff's syndrome is perhaps due to damage to the medial thalamic and mammillary nuclei; thus, Korsakoff's syndrome may be considered diencephalic amnesia.

E. Alzheimer's Disease

Another name for Alzheimer's disease is senile dementia. It is characterized by a progressive intellectual decline. One of the earliest behavioral signs of Alzheimer's disease is memory loss of recent events. Interestingly, another early sign is loss of peripheral sympathetic responsiveness, suggesting a modulatory deficit in memory function. Memory function is continuously lost as the disease progresses so that eventually the patient does not know where they are or the current year. Alzheimer's disease will become a major problem in the United States: in 1890, 2% of the population was >65 yr of age; by the year 2020, some 20% will be >65 yr old, and probably 10% of these people will suffer from Alzheimer's disease. In general, the brains of Alzheimer's patients show cortical atrophy. Glucose utilization is reduced in posterior parietal cortex and some portions of the temporal lobe. Studies at the cellular level show accumulations of neurofibrillary tangles, which disrupt the orderly flow of cell metabolism, and senile plaques, which contain a protein called amyloid. The function of amyloid is unknown at this time, but most researchers think it is the remnants of the neuron's fight for survival with the Alzheimer's disease process. Alzheimer's brains also are deficient in the important neurotransmitter acetylcholine. Many hope that some cognitive function can be restored to Alzheimer's patients by development of drugs that act to stimulate or replace lost cholinergic function. Another treatment of the future is brain grafts, in which fetal brain material is injected into the affected area to grow and reconnect with existing neurons and hopefully to restore behavioral function. The cause of Alzheimer's disease is unknown and may be due to a viral infection, the accumulation of toxic metals such as aluminum, or an autoimmune process, where the body attacks its own brain with antibodies. [*See* ALZHEIMER'S DISEASE.]

IX. Conclusion

Human memory is not a single thing. Different types of memory exist, and different parts of the brain are involved in different memories. Memories are physical entities that exist in brains and result from changes in individual neurons that influence the activity of many neurons or neural networks. Studies of memory pathology revealed that two areas of the brain—the medial temporal cortex, particularly the hippocampus, and the diencephalon—are important for memory, because damage to these areas results in severe amnesia. The incidence of Alzheimer's disease will dramatically increase in the next 30 years, and increased understanding of the molecular basis of memory through research with animals and humans will hopefully lead to the development of therapies to treat this devastating illness.

Bibliography

Alkon, D. L. (1987). "Memory Traces in the Brain." Cambridge University Press, Cambridge.

Gormezano, I., Prokasy, W. F., and Thompson, R. F. (eds.) (1987). "Classical Conditioning," 3rd ed. Laurence Erlbaum Associates, Hillsdale, New Jersey.

Martinez, Jr., J. L., and Kesner, R. P. (eds.) (1986). "Learning and Memory: A Biological View." Academic Press, Orlando, Florida.

Neisser, U. (1982). "Memory Observed." W. H. Freeman and Co., San Francisco.

Olton, D. S., Gamzu, E., and Corkin, S. (eds.) (1985). "Memory Dysfunctions: An Integration of Human Research from Preclinical and Clinical Perspectives," Vol. 444. Annals of the New York Academy of Sciences, New York.

Squire, L. R. (1987). "Memory and the Brain." Oxford University Press, New York.

Weingartner, H., and Parker, E. S. (eds.) (1984). "Memory Consolidation." Laurence Erlbaum Associates, Hillsdale, New Jersey.

Woody, C. D. (1982). "Memory, Learning, and Higher Function." Springer-Verlag, New York.

Leishmaniasis

KWANG-POO CHANG, *University of Health Sciences, Chicago Medical School*

I. The Diseases
II. Causative Agents
III. Molecular Mechanism of Parasitism

Glossary

Amastigote   Mammalian stage of *Leishmania*
Espundia   Mucocutaneous leishmaniasis
Kala azar   Visceral leishmaniasis
Leishmania   Genus in the Trypanosomatidae family of protozoa
Promastigote   Insect stage of *Leishmania*
Sandfly   Insect vector of leishmaniasis

LEISHMANIASES ARE vector-borne and largely zoonotic diseases caused by parasitic protozoa in the genus *Leishmania*. Endemic areas for the diseases include Central and South America, the Mediterranean regions, North and Central Africa, The Middle East, Central and western Asia, and the Indian subcontinent. The parasites are transmitted by the blood-sucking sandflies. In the vector is the promastigote stage of the parasites, which multiply and develop mostly as motile flagellated cells in the alimentary canal. Upon delivery into the mammalian hosts, promastigotes infect macrophages wherein the parasites differentiate into the amastigote stage and multiply as such in intracellular vesicles, the phagolysosome. This intralysosomal parasitism results in immunopathology manifested as cutaneous or visceral lesions, depending largely on the species of the parasites involved. All forms of the diseases follow a chronic course. The clinical symptoms range from self-healing, inocuous skin infection to facial disfiguring mucocutaneous lesion to fatal consequence of the visceral form. Current methods of diagnosis and treatment are outdated and often unreliable. Active research is underway to provide new methods through better understanding of these parasites and their interactions with the host. This is made possible by advances in parasitology for the maintenance and cultivation of *Leishmania* and their vectors under laboratory conditions. The studies of these materials have provided not only new leads for practical applications but also novel discoveries of general importance in biochemistry, immunology, and cell and molecular biology. Understanding the unique mechanism of parasitism through such basic research is expected to foster the rational development of effective chemo- and immunoprophylaxis and -therapy of leishmaniasis.

I. The Diseases

A. Epidemiology

Leishmaniasis is a collective term for human diseases caused by a group of parasitic protozoa in the genus *Leishmania*. While the diseases are manifested clinically in different forms (see below), they are all vector-borne and mostly zoonotic. The parasites are naturally transmitted by different species of blood-sucking sandflies among reservoir animals (e.g., rodents, edentates, canines). Leishmaniasis is thus widespread in tropical, subtropical, and temperate regions of all continents (except Australia) where different combinations of vector species and reservoir animals exist. Humans are accidental hosts, except in India where reservoir animals have not been found. Sandflies transmit leishmaniasis from animal to human, from human to human, and possibly from human to animal. Transmissions via congenital route, blood transfusion, coitus, and other means of contacts are rare. [*See* ZOONOSES.]

In different endemic areas, the natural life cycle of all *Leishmania* species is identical. They are parasites with two morphological forms (called digenetic): promastigotes and amastigotes in the insect and mammalian hosts, respectively. In the vector, *Leishmania* multiply and develop extracellularly as promastigotes in the digestive tract. They are delivered into the skin of the mammalian host when the female fly takes blood meals. These promastigotes apparently infect macrophages (white blood cells that penetrate tissues and engulf parasites), wherein they differentiate into amastigotes and multiply intracellularly as such. Heavily parasitized cells eventually die and degenerate, releasing amastigotes to infect other macrophages. The number of infected macrophages thus increases locally in the skin or in the visceral organs in cutaneous and visceral leishmaniasis, respectively. The infected cells in circulation or in the subcutaneous tissues are taken up by the sandfly vector along with the blood meals. In the sandfly gut, amastigotes differentiate into promastigotes, thereby completing the cycle of transmission. This cyclic transmission of *Leishmania* between the sandfly vector and the mammalian host perpetuates leishmaniasis in all endemic regions. [*See* MACROPHAGES.]

B. Pathology

Infection of mammalian hosts by different *Leishmania* species culminates in various forms of leishmaniasis. Simple cutaneous leishmaniasis is marked by the development of skin lesions, which are self-limiting and self-healing in the course of several months. Mucocutaneous leishmaniasis is complicated by the spread of the skin infection to the nasopharyngeal cavity, often resulting in facial disfigurement. Visceral leishmaniasis, or kala azar, is potentially a fatal disease, manifested as chronic disorders of hematologic and hepatosplenic functions. The clinical manifestations of visceral leishmaniasis also include fever, anemia, leucopenia, and cachexia. Clinical symptoms and pathology are overlapping or intermediate in different cases. In India, post-kala azar dermal leishmaniasis may develop months or years after cure of the visceral disease. Diseases in the reservoir animals are either inapparent or severe but may differ completely from human leishmaniasis. In the Mediterranean region, overgrowth of toe nails is a peculiar sign found only in canine leishmaniasis.

In all cases of leishmaniasis, macrophages are the exclusive host cells. Cutaneous forms of the diseases have infected cells in nodular or craterlike lesions. In visceral leishmaniasis, the major sites of infection are spleen, liver, and bone marrow, but infected cells must be present also in skin or in circulation as a source of parasites for transmission by the vector. Where symptoms are manifested in humans or in other animals, the pathology is due not only to the destruction of macrophages by the infection but also to the immunopathological responses of the host to parasite antigens. These antigens are present in circulation and on the surface of infected macrophages and other cells. Antibodies produced by the host are ineffective against intracellular amastigotes but, instead, damage cells and tissues, to which these antigens adhere. The interactions of these antigens with other immune cells possibly also result in the production of interleukines and cachectin or tumor necrotic factor, which exacerbate the diseases. The severity of the immunopathology in leishmaniasis is manifested as a spectrum of individual immune responses, ranging from anergy to hyperactivity. The outcome of host immune response varies with the diseases of concern. It is often effective in resolving skin lesion in simple cutaneous leishmaniasis but not in visceral and mucocutaneous leishmaniasis. Infections in the latter cases become chronic, and immunopathology is invariably accompanied by marked immunosuppression. Secondary infections are common and attributable partly or wholly to the fatal consequence seen in untreated cases of visceral leishmaniasis.

C. Control

Means of diagnosis, chemotherapy, and vector control for leishmaniasis are available but are largely outdated and, thus, ineffective. These difficulties are further complicated by the sylvatic cycle of these diseases, namely their transmission by the sandfly vector among wild animals serving as the reservoir. Thus, epidemics often flare up sporadically in endemic regions as well as in forestlands explored in the developing countries for agricultural or industrial uses.

Diagnostic procedure for leishmaniasis is cumbersome and sometimes dangerous and often lacks sensitivity. In most endemic regions, this is based largely on the direct examination of biopsied samples or after cultivation of these samples for the presence of parasites. Biopsies of patients are performed by spleen or bone marrow or skin punctures with the inherent risks of secondary infections and injuries. Less invasive means of diagnosis on the basis of detecting either antigens or antibodies are being developed for wide applications and better specificity.

Chemotherapy of leishmaniasis relies mainly on antiquated pentavalent antimonials (Pentostam or sodium stibogluconate, Glucantime) or amidine compounds (pentamidine), which are not always effective and are difficult to administer and toxic. In addition, no effective vaccine is available for these diseases. [*See* CHEMOTHERAPY, ANTIPARASITIC AGENTS.]

Despite the antiquity of the available tools, visceral leishmaniasis has been virtually eradicated in densely populated regions of China by an integrated control program, consisting of massive treatment of patients, eradication of sandflies, and elimination of dogs. However, control of leishmaniasis with silvatic cycle is difficult.

II. Causative Agents

A. Biological and Taxonomical Status

Leishmania belong to a group of unicellular protozoa known as hemoflagellates or trypanosomatid flagellates (Kinetoplastidae). Many digenetic species of these protozoa live alternately in two hosts of different species and cause diseases in animals and plants; however, a large number lives only in insects as monogenetic species. Some species of the digenetic *Leishmania* are parasites of lizards and are nonpathogenic to humans. The pathogenic species include *L. major, L. braziliensis,* and *L. donovani,* which cause simple cutaneous, mucocutaneous, and visceral leishmaniasis, respectively. Many other species causing similar diseases have been described and often are named after the countries of their origin. Differentiation of *Leishmania* species from various geographic regions proves difficult on the basis of biological or clinical criteria alone. Modern immunological, biochemical, and genetic methods have been applied in an attempt to characterize these geographic isolates. *Leishmania* species and subspecies have been thus differentiated by zymodeme and schizodeme analyses, which are based on the electrophoretic mobility of their isoenzymes and the endonuclease restriction fragments of their mitochondrial or kinetoplast DNA (see below), respectively. Species-specific monoclonal antibodies have also been produced as diagnostic tools. Available data indicate that *Leish-* *mania* speciation is more complicated than previously thought.

B. Cultivation of *Leishmania in vitro* and *in vivo*

With very few exceptions, all species of *Leishmania* can be isolated and maintained in pure culture as promastigotes and in laboratory animals as amastigotes. In 1908, Nicolle began to use a simple blood agar slant of Novy and McNeil known as NNN medium for *Leishmania*. This medium or its modified versions has since been widely used until now for the isolation and cultivation of promastigotes in the endemic regions. Commercially available media, originally designed for growing mammalian cells or insect cells, have been increasingly used to grow promastigotes for laboratory research and for diagnosis. With the advent of tissue culture techniques, methods have been developed to grow amastigotes in macrophages and cell lines of these phagocytes. It is also possible to grow cutaneous *Leishmania* of South American origin as "axenic amastigotes" under macrophage-free conditions. Hamsters and mice are the animal models for experimental leishmaniasis, and several species of sandfly vectors have been successfully raised in insectarium. Thus, the entire life cycle of *Leishmania* can be duplicated under laboratory conditions for detailed investigations.

C. Cell Biology

All members of *Leishmania* are microscopic. Promastigotes are elongated or spindle-shaped motile cells of variable length, each equipped with a flagellum at the anterior end. The flagellum, originating from a flagellar reservoir, is as long as or longer than the body length. The rotatory motion of the flagellum propels the cell forward. In the vector, the flagellum sometimes becomes broadened at the distal end for attachment to the gut wall of the fly. In the mammalian host, promastigotes differentiate into amastigotes, which are nonmotile oval or round bodies found in the macrophages. The flagellum is still present in the amastigote but is nonfunctional and concealed within its flagellar reservoir.

The cellular ultrastructure of *Leishmania* is similar to that of other trypanosomatid protozoa. Like typical eukaryotic cells, they are enclosed by a plasma membrane. Present in the cytoplasm of each cell are a nucleus, mitochondria, endoplasmic reticulum, Golgi apparati, lysosomes, ribosomes, and other cell organelles. Microtubules and mitochondrial DNA are especially abundant in these and other related protozoa. In addition to the flagellar and mitotic microtubules, the microtubular cytoskeleton exists just beneath the plasma membrane. These microtubules run parallel in a spiral course along the longitudinal axis of the cell body. These subpellicular microtubules are responsible for shaping the cells. The mitochondrial DNA in all trypanosomatid protozoa constitutes up to 20% of the total cellular DNA and is condensed into regularly arranged fibrous arrays in a special section of the mitochondrion called the kinetoplast. Each cell has a single kinetoplast visible by light microscopy. It is often situated above or beside the nucleus and beneath the basal body of the flagellum. This unusual mitochondrial DNA is a distinctive cytological feature for this group of protozoa.

D. Biochemistry

Leishmania promastigotes are aerobic cells using glucose or proline as the energy sources. Adenosine triphosphate (ATP) is produced via glycolysis and the tricarboxylic acid cycle, although these pathways are somewhat modified. *Leishmania* synthesize macromolecules such as proteins, lipids, carbohydrates, and nucleic acids from small substrates. Similar to other trypanosomatid protozoa, *Leishmania* have a number of biochemical peculiarities.

Most striking is their inability to synthesize purine and heme *de novo*. These deficiencies are manifested as nutritional requirements of these protozoa for hemin or hematin and a purine nucleobase or nucleoside in addition to amino acids, vitamins, and an energy source when grown as promastigotes in chemically defined media. The biochemical basis of these nutritional requirements is lack of the enzymic machinery necessary for these pathways. When other enzymes or molecules of *Leishmania* are analyzed biochemically, they are often found to differ from their counterpart of the mammalian host in one aspect or another.

Much attention has been devoted to *Leishmania* enzymes and molecules thought to be important in their survival in the host. Instead of glutathione, *Leishmania* contain glutathionyl-spermidine conjugates known as trypanothione. Oxidation and reduction of these molecules are important for the maintenance of the cellular redox potential. Super-oxide dismutase of the bacteria type is present in *Leishmania*. This enzyme in conjunction with trypanothione peroxidase probably forms the main line of defense in *Leishmania* against the microbicidal oxidative metabolites of the host, because the catalase activity is very low or absent in these parasites. Most unusual are the findings of an array of *Leishmania* ectoenzymes (cell surface enzyme) for example, Zn-proteinase, acid phosphatase, protein kinases, 5'-nucleotidase, 3'-nucleotidase–nuclease, ATPases (cation and glucose transporters). The Zn-proteinase is a glycoprotein of 63 kDa and is ubiquitously present in all pathogenic species and constitutes up to 1% of the total cellular protein. This and other ectoenzymes are thought to play important roles in the procurement of nutrients by these parasites and/or their defense against the host antimicrobial activities.

Most unique immunologically and biochemically in *Leishmania* are their carbohydrates and lipid-carbohydrate conjugates (i.e., lipophosphoglycans [LPG], phosphatidylinositol glycans [GPI], and asparagine-linked oligosaccharides). The LPG is the most abundant surface molecules of *Leishmania*, constituting about 10^6 copies per cell. The basic structure of LPG consists of the following components: terminal repeating units of di-, tri-, and/or tetrasaccharides, a heptasaccharide core, and a phosphatidylinositol-containing lipid anchor linking to the plasma membrane. The presence of phosphorylated hexoses, nonacetylated glucosamine, and lysoalkylglycerol moieties is the outstanding features of LPG. Different *Leishmania* species are marked by the heterogeneity of this and other related surface molecules. Multifarious functions ascribed to LPG include metal chelation, free radical scavenging, inhibition of protein phosphorylation, and immunogenicity of its unique epitopes. The *Leishmania* GPI is similar antigenically and biochemically to this structure initially found in the variant surface antigen (VSG) of the African trypanosomes (now also known to be a lipid anchor for a variety of mammalian membrane glycoproteins). The basic structure of GPI includes diacylglycerol, phosphatidylinositol, ethanolamine, nonacetylated glucosamine, and glycans consisting of mannose and galactose residues. This structure with myristic acids in the diacylglyceride moiety anchors the Zn-proteinase of *Leishmania* to the plasma membrane and modulates its enzymatic activity as well as its immunogenicity. The asparagine-linked oligosaccharides are synthesized via the tunicamycin-sensitive dolichol pathway and are of the high mannose type in Zn-proteinase of *Leishmania*. Heterogeneity of the oligosaccharide length, unusual glycosidic bonds, and the presence of a terminal glucose are the unique properties of the glycans in this glycoprotein. Elucidation of the pathways and their enzymes for the biosynthesis of these unusual macromolecules is likely to uncover additional biochemical oddities.

The unusual biochemistry of *Leishmania* is undoubtedly associated with their evolution as a digenetic parasite to live alternately in the insect gut and in the phagolysosomes of the mammalian macrophages during their life cycle. Loss of certain biosynthetic pathways and amplifications or modifications of others are expected during the evolution of such parasitism. Knowledge gained here is expected to explain the molecular pathogenicity of *Leishmania,* thereby helping the formulation of rational approaches to effective immuno- and chemoprophylaxis and -therapy of leishmaniasis.

E. Molecular Biology

The haploid genome size of the trypanosomatid protozoa including *Leishmania* is in the order of 10^5 kilobase (kb) pairs. Nuclear genes are arranged in "chromosomes," which are resolvable by pulsed field gel electrophoresis into some 20 chromosomal bands within the range of 50–2,000 kb. The mitochondrial DNA exists as maxicircles and minicircles. These circular DNAs are condensed into a large network within the kinetoplast. There are 10–20 copies of the mitochondrial genome-equivalent maxicircles (ca. 20 kb each) and up to 10^4 copies of minicircles (ca. 1 kb each) per cell. Additional genetic elements of *Leishmania* are extrachromosomal DNA of chromosomal origin, most of which exist as supercoils of 30 to >100 kb. These extrachromosomal circles become amplified up to 100-fold upon selection for resistance to toxic compounds such as heavy metals (arsenite), drugs (methotrexate, difluoromethylornithine), and antibiotics (tunicamycin). Linear DNA thought to be of viral origin has been reported in some *Leishmania* species.

Molecular cloning and sequencing of complete maxicircle genes (e.g., those for the subunits of ATPase and cytochrome b) and some nuclear genes (e.g., those for tubulins, heat-shock proteins, dihydrofolate reductase-thymidylate synthase, glucose transporter, cation transporter) reveal several try-

panosomatid-specific features of *Leishmania*. The codons (triplets of bases specifying amino acids) are biased for guanine and cytosine in the third position and no introns (noncoding sequence) within the genes. The regulatory sequences, such as promoters and polyadenylation sites, appear to differ from other eukaryotes. There are multiple copies of a miniexon (coding sequence) gene. This gene produces a transcript of 107 nucleotides, from which a 35-nucleotide sequence is derived and transpliced onto all messenger RNAs (mRNAs) and are essential for their translation. This discontinuous transcription of two different sequences to create functional mRNAs and their mode of splicing in *trans* (i.e., not in the same molecule) represent a regulatory mechanism unique to all trypanosomatid protozoa. Most novel is the discovery of RNA editing, which coverts mitochondrial nonsense transcripts of cryptogenes into functional mRNAs by polyuridination. Missing uridine residues are enzymatically inserted precisely into the pre-edited RNAs. The insertion is guided by template transcripts derived from other genes of maxicircles and minicircles. Thus, the transcriptional product of one gene regulates the formation of mRNA from another gene. This extraordinary phenomenon is unprecedented, and its discovery introduces a new concept of transcriptional regulations in molecular biology.

Leishmania along with related trypanosomatid protozoa emerge as a useful model for studying special aspects of molecular biology, especially in the areas of RNA splicing, RNA editing, and DNA amplification. Experimental approaches developed in molecular genetics provide fresh avenues to address the practical questions of *Leishmania* virulence, drug-resistance, and efficient productions of *Leishmania* molecules for vaccination and other purposes. The recent success of introducing foreign genes into *Leishmania* by electroporation represents a step forward in the direction of *Leishmania* genetic research. [*See* PARASITOLOGY, MOLECULAR.]

F. Immunology

It has long been known that patients recovered from leishmaniasis generally develop life-long immunity to the infection by the same species of parasites. In certain endemic areas of the Middle East, inoculation of children with lesion-derived parasites had been (and is probably still) in practice to avoid possible facial scars due to simple cutaneous leishmaniasis. In Africa, Israel, and the Soviet Union, human vaccinations with killed promastigotes have been tried on a limited scale with unsuccessful or inconclusive results. Very little is known about human immune responses to *Leishmania* antigens.

Immune responses are demonstrable in patients of leishmaniasis, especially the visceral and diffused cutaneous forms. Anti-*Leishmania* antibodies are detectable with the progression of the disease; the immunoglobulin M (IgM) appears before IgG. However, most of the antibodies are nonspecific and produced possibly by polyclonal B cell activation. The amount of IgG produced is large, such that they are precipitable from serum by formalin, providing a crude diagnostic means known as the formal gel test in certain endemic regions of visceral leishmaniasis. While *Leishmania*-specific antibodies are present and useful for serodiagnosis of leishmaniasis, they are not correlated with the recovery of patients from the infection. Instead, this is accompanied by positive reactions to leishmanial antigens in the skin test for delayed-type hypersensitivity and in the T-cell proliferation assay *in vitro*. Cure of leishmaniasis spontaneously or after chemotherapy is thus thought to be T cell-mediated, as is true for most infectious diseases caused by intracellular parasites.

Animal immunity to leishmaniasis has been studied mainly in mouse models. Susceptible mice can be immunoprophylactically protected against leishmaniasis with purified *Leishmania* antigens, such as LPG, its combination with gp63 (Zn-proteinase) (see Section II,D), and other antigens of unknown biochemical properties. The mechanism of this immunity can only be speculated based on the current hypothesis. Different *Leishmania* antigens are thought to stimulate T-helper lymphocytes (CD4 marker positive T cells) to differentiate into either Th1 or Th2 cells. Th1 cells releasing interleukin-2 (IL-2) and γ-interferon (IFN) are protective, whereas Th2 cells releasing IL-3 and IL-4 are disease-exacerbating. The lymphokines from Th1 cells affect other immune cells. Such interactions are detrimental to the parasites through several possible mechanisms. Activation of macrophages by IFN, for example, is known to stimulate their respiratory burst and other antimicrobicidal activities, thereby killing the amastigotes therein. This and other mechanisms of protective immunity seen with *Leishmania* antigens are under investigation for rational design of a safe vaccine for leishmaniasis. [*See* IMMUNOLOGY OF PARASITISM.]

III. Molecular Mechanism of Parasitism

It is extraordinary for *Leishmania* to live alternately in the sandfly gut and in the phagolysosome of the mammalian macrophages—a disparate ecological niche extremely hostile to invading microorganisms. *Leishmania* must have evolved mechanisms not only to neutralize the arrays of microbicidal factors but also to thrive in such extreme environments. Some of these molecules and their possible functions have already been alluded to in Section II,D. A repertoire of genes, which are evolved for switching on and off by signals provided during their transits from one environment to another, are predictable. Some products of these genes may predispose *Leishmania* to survive the changing environments. Others may be involved in the evasion of host immunity or invasion of host cells. Additional products may be further needed for *Leishmania* to multiply under harsh conditions. What are the signals that turn on the genes for these products? How are they regulated and coordinated? These questions of considerable biological and practical interests are currently under investigation.

Leishmania, however, do not act alone but must interact with both of their insect and mammalian hosts to set the stage for the chronic course of leishmaniasis. The drama of such interactions reaches the climax when the molecules from all these players meet and interact at the site of the sandfly bites. Attempts to re-enact the play *in vitro* so far only tell us a small portion of the story, summarized as follows: (1) predisposition of promastigotes to act in the mammalian host after their multiplication and development in different regions of the sandfly gut; (2) further orientation of the invasive parasites in response to the new environment after delivery into the mammalian host; (3) interactions of these parasites with the mammalian humoral and cellular components; and (4) modulation of the host immune system by the sandfly factors delivered together with the promastigotes. Exactly how these multifarious factors from different players interact remains unclear, but promastigotes are ultimately ingested by macrophages via receptor-mediated endocytosis. While all types of receptors known to exist on the surface of macrophages have been implicated in their binding of promastigotes, those for complement component 3 appear to play a predominant role. Leishmanial ligands for this binding have been ascribed to their surface glycoconjugates (i.e., the Zn-metalloproteinase and LPG). The possible contributions of these and other ectoenzymes–surface molecules to the intralysosomal survival and multiplication of *Leishmania* have been discussed in Section II,D.

The molecular mechanism of *Leishmania* differentiation and the host–parasite molecular interactions in leishmaniasis are mysteries in parasitology. Their elucidation is crucial for understanding the evasive strategies of these parasites for their invasion of extreme environments. These subjects bear on the evolution of intracellular parasitism as well as practical application for controlling the diseases.

Bibliography

Chang, K.-P., and Bray, R. S. (ed.) (1985). "Leishmaniasis." Elsevier, Amsterdam.

Chang, K.-P., and Snary, D. (ed.) (1987). "Host–Parasite Cellular and Molecular Interactions in Protozoal Infections." NATO ASI Series H: Cell Biology, Vol. 11. Springer-Verlag, Berlin.

Chang, K.-P., Chaudhuri, G., and Fong, D. (1990). Molecular determinants of *Leishmania* virulence. *Annu. Rev. Microbiol.* **44,** 499–529.

Fairlamb, A. H. (1989). Novel biochemical pathways in parasitic protozoa. *Parasitology* **99,** S93–S112.

Hart, D. T. (ed.) (1989). "Leishmaniasis. The Current Status and New Strategies for Control." NATO ASI Series A: Vol. 163. Plenum, New York.

Louis, J., and Milon, G. (1987). Immunobiology of experimental leishmaniasis. *Ann. Inst. Pasteur/Immunol.* **138,** 737–795.

Molyneux, D. H., and Ashford, R. W. (1983). "The Biology of *Trypanosoma* and *Leishmania,* Parasites of Man and Domestic Animals." Taylor and Francis, London.

Peters, W., and Killick-Kendrick, R. (ed.) (1987). "Leishmaniasis." Academic Press, New York.

Simpson, L., and Shaw, J. (1989). RNA editing and the mitochondrial cryptogenes of kinetoplastid protozoa. *Cell* **57,** 355–366.

World Health Organization Expert Committee (1984). "The Leishmaniases." Technical Report Series 701. World Health Organization, Geneva.

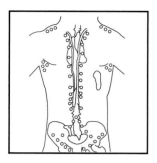

Leukemia

HOWARD R. BIERMAN, *Institute for Cancer and Blood Research*

Glossary

Alleles Alternative forms of genes occupying an identical site(locus) on sister homologous chromosomes

Base pair One of the three components of nucleotides. Four bases have been identified: adenine, which is always matched with the base thymine, and guanine, which is always paired with cytosine

BCR Refers to the entire gene on chromosome 22. BCR refers to the 5.8-kilobase segment in the region of the breakpoint in CML. The function of this gene is unknown

CLL Chronic lymphocytic leukemia

CML Chronic myeloid leukemia (myelogenous, neutrophilic, granulocytic)

Heterozygous Having nonidentical alleles at a single locus

Homozygous Having two identical alleles at a single locus

Isochromosome Symmetrical chromosome composed of duplicated long or short arms formed after misdivision of the centromere in a transverse plane

Megakaryoblast Precursor marrow cell, 15–50 μm in diameter, containing a large oval nucleus with several nucleoli; cytoplasm is nongranular and intensely basophilic. It matures into megakaryocytes from which many platelets arise and enter the blood

Nucleotide Basic building block of DNA and RNA; compound composed of a base, a sugar, and a phosphate group

Oncogene Gene, essential for the development of cancer in conjunction with other factors

PMN Polymorphonuclear leukocyte—a mature leukocyte; nucleus is formed into multiple lobes joined by thin filaments of a coarse condensed chromatin. The cytoplasm contains fine, specific granules that vary in size

Protooncogene Normal gene that through point mutation or some other alteration becomes an oncogene capable of causing malignancy

Ras genes Family of genes that code for proteins located at the inner surface of the cell membrane. These proteins are thought to play a part in intracellular growth-signal transduction.

THE HEMATOLOGIC MALIGNANCIES consist of the leukemias and the lymphomas and, less commonly, multiple myeloma. The myelodysplastic syndromes terminating in leukemias are also included, as well as polycythemia vera and selected monoclonal gammopathies. Despite single clone dominance, each leukemia and lymphoma is a spectrum of clinical expression and variation in symptoms, severity, and survival. Considerable similarity and interchange of the biology and clinical presentations occur within these disorders.

Leukemia is the abnormal proliferation in the bone marrow of a clone of cells of hematopoietic origin. The marrow consists of myeloid, lymphoid, megakaryocytic, and erythroid cells.

There are 24 types of leukemia each named after the dominant cell (Table I). Myeloproliferative dis-

TABLE I Classification of the Leukemias (Modified from the French-American-British (FAB) Classification)

FAB class	Dominant cell type	Chromosome rearrangements[a]
	Acute Nonlymphocytic leukemia (ANLL)	
	Poorly differentiated/(immature)/(acute)	
M1	Myeloblastic without maturation	Additional 8 deletion −7;less −5
	30–80% myeloblasts	Abnormal promyelocytes
M2	Myeloblastic with maturation	t(8;21)(q22;q22)
	90% myeloblasts	
	30% or more granulocytes	
	Myeloblasts (basophilic)	t(6;9)(p23;q34)
	(eosinophilic)	
M3	Promyelocytes	t(15;17)(q22;q11-12)
M4	Myelomonocytes	inv(16)(p13;q22)
M5	Monocytes (histiocytic)	t(9;11)(p22;q23)/t(11q)del(11q)
	80% or more monoblasts, promonocytes, or	
	monocytes	
M6	Erythroleukemia	
	50% or more erythroid	
M7	Megakaryoblastic	
	30% or more megakaryocytoblasts	
	Myeloid	
	Chronic myeloid leukemia (CML)	t(9;22)(q34;q11)
	(Chronic neutrophilic leukemia)	
	Well differentiated (mature). Primarily polymorphonuclear	Philadelphia chromosome, 95%
	neutrophils, less metamyelocytes, and few myelocytes	simple and complex translocations of
		chromosome #22 and #2 to #21
	Myeloblastic conversion	5(9;22), i(17q)
	"Blast" crisis	
	Acute Lymphocytic Leukemia (ALL)	
	Poorly differentiated lymphoblasts	
	B cell	
L1	Common lymphoblastic (common ALL) stem cell—early	t(1;19)(q23;p13)
	B cell	t(4;11)q21;q23)
	Childhood; less in adults	
L2	Large lymphoblasts with irregular nuclei and shrunken	t(9;22); del(6q)
	cytoplasm	(q15-q21)—near haploid
L3(rare)	Large B cell (Burkitt cell)	t(8;14)(q24;q32)
	Fine nuclear chromatin, prominent nucleoli, vacuolated	t(2;8)(p12;q24)
	cytoplasm	t(8;22)(q24;q11)
	Lymphoblastic crisis of chronic myelogenous leukemia	+8, +Ph' rare −7
		t(9;22), i(17q)
	Plasma cell	
	B cell	
	Chronic lymphocytic leukemia (CLL)	(q13;q32)
	Mature lymphocytes	(q32;q13); t(11;14)
		14q + (q32); t(14;19)
	Leukosarcoma/lymphosarcoma lymphocyte small,	
	cleaved cell	
	Prolymphocyte	
	Waldenström macroglobulemia	
	Plasmacytoid lymphocyte	
	Hairy cell leukemia/reticuloendotheliosis	
	Lymphocytes with cytoplasmic projections	
	T cell	t(11;14)(p13;q11)
		t(8;14)(q24;q11)
		inv(14)(q11;q32)
	Adult T cell (HIV-AIDS)	
	Cutaneous T-cell lymphoma with circulating T lympho-	
	cytes (Sezary syndrome)—mycosis fungoides	
	T-cell lymphocytosis	

[a] + sign, additional; − sign, missing whole chromosomes; t, translocation; inv, inversion; p, short arm of chromosome; q, long arm of chromosome; ( ), used to surround structurally altered chromosome(s); (;), separates structural rearrangements involving more than one chromosome; del, deletion; i, isochromosome; M1–M5, degree of marrow myeloid–monocyte involvement; L, degree of lymphoid involvement.

New Cases Each Year: 40,000
                     31,000 in 1984

8% of all cancers

Race: White
Age: 3–14, 20–30s and 70s
Sex: Males more than females
Family History: Leukemia and
         Lymphoma
Past Medical History:
    Impaired immune response:
        T-Cell function
    Infections: Epstein Barr Virus (EBV)
                Mononucleosis
                Burkitt's Lymphoma
                HIV (AIDS Virus)
Chemical: Benzene, anticancer agents
Radiation Exposure:
    Radiation therapy
    Radiologists
    Radiation research scientists

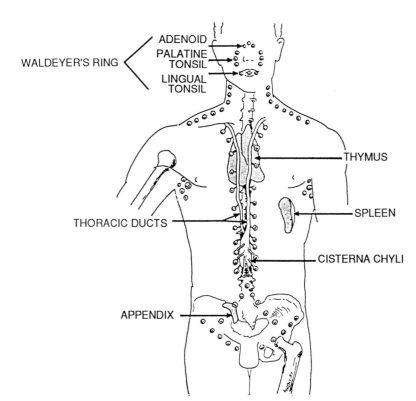

FIGURE 1 Predictive Profile of Leukemias and Lymphomas.

orders are derived from immature myeloid stem cells, giving rise to usually neutrophilic or, rarely, eosinophilic, basophilic, and megakaryocytic leukemias. The lymphoproliferative leukemias are similarly derived from progenitor lymphoid cell line. Abnormal proliferation in the erythrocyte lineage gives rise to polycythemia and erythroleukemia.

I. Incidence

The incidence of the leukemias is about 8–10 per 100,000 U.S. inhabitants per year and represents 3% of all cancers (Fig. 1). Acute leukemias in the United States and western European countries occur in 3.5 per 100,000 persons annually, increasing with age except between ages 2 through 9, because of high frequency of the common type lymphoblastic leukemia (ALL) at that age. Ratio of ALL to AML (acute myeloid leukemia) is 4 : 1 under the age of 15; AML is more common in adults. Acute leukemia occurs less frequently in non-Caucasian children, but it is similar in adults of different races. It occurs more frequently in industrialized countries and in urban areas. However, racial, ethnic, and economic differences may reflect sophistication of medical care and accuracy of diagnosis.

II. Biology of Normal Hematopoiesis

The normal leukocyte count in the peripheral blood is 5,000–11,000/cmm (per cubic millimeter). An average adult of 70 kg has approximately 6 liters of blood. Consequently, in the circulation, at any given time in the normal individual, there are 40–70 billion leukocytes.

Leukocytes in the normal human have different life spans in the blood. Polymorphonuclear leukocytes (PMN) live from 8 to 12 days, lymphocytes from 80 to 500 days, erythrocytes to 120 days, and platelets about 9 days. There is constant renewal to maintain a constant concentration of leukocytes; for instance, the production rates of PMNs averaging 1.6×10^9/kg/day. Lymphocytes, monocytes, eosinophils, and basophils have less rapid renewal rates (Fig. 2).

In normal cell division, the two chromatids of chromosomes (DNA) separate and are segregated equally to the daughter cells, each receiving a full set of genes. Genes are transcribed into RNA, which is then processed in various steps to messen-

MARROW AND/OR LYMPH NODES   PERIPHERAL BLOOD

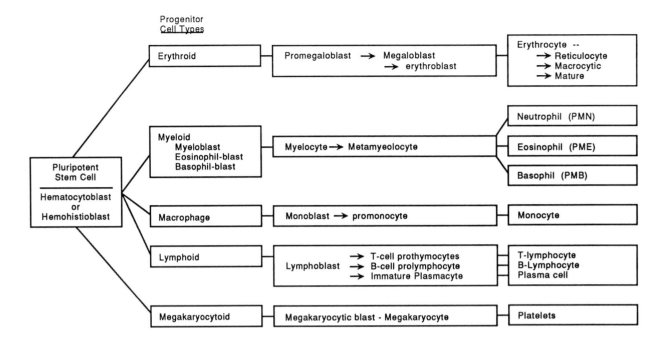

FIGURE 2 Schematic progressive development of hematopoietic elements from the pluripotent stem cell to the mature circulating blood cells. Multiple growth factors are involved at each phase either singly or in concert (colony stimulating factors), interleukins, erythropoietin, hormones, and platelet-derived growth factors.

ger RNA (mRNA). mRNA is transported to the cytoplasm where it combines with ribosomes to be translated into specific proteins.

Normal granulocytopoiesis is initiated at the primitive hematopoietic stem cell and progresses from the myeloblast to the myelocyte and to the metamyelocyte. The metamyelocyte does not undergo further cell division and matures to the PMN. Normal human leukocytes may undergo a maximum of 50 stem cell divisions. [*See* HEMOPOIETIC SYSTEM.]

III. Hematopoiesis in the Leukemias

The untreated clinical course of the leukemias may be extremely short, giving rise to the "acute" appellation. However, effective treatment has largely displaced the acute and chronic designations by describing the morphology of the dominant leukocyte. Both designations are used interchangeably here.

Leukemias and lymphomas follow the general pattern of progression of cancers. Leukemias usually progress from the mature well-differentiated leukocytes to the more aggressive immature cell type, which has a proliferative advantage over its corresponding normal cell. The single clone of leukemic cells, therefore, soon dominates the hemato-

poietic system, interfering with the proper normal function. [*See* LYMPHOMA.]

In leukemic granulocytopoiesis, the metamyelocytes ignore the maturation signals and continue to divide. They keep ignoring the maturation signal and divide incessantly, thereby increasing the population of leukemic cells. Leukemic lymphopoiesis behaves similarly. Leukemic and lymphoma cells divide at rates equal to or slower than normal, but by avoiding differentiation, which terminated their lifespan, continue to increase in cell number. In the less frequent, subleukemic (aleukemic) leukemia, the leukocyte count is normal or less than normal, but the dominant cells express greater immaturity.

Leukemia previously has been considered to be derived from a single cell type, but this dogma is no longer tenable. In fact, in chronic myelogenous leukemia during a blastic conversion (blast crisis), the leukemic cells may have characteristics of lymphoblasts such as the presence of the enzyme terminal dioxynucleotidyl transferase (TdT), which is characteristic of lymphoblastic leukemia in children.

The primitive/stem cell leukemias are undifferentiated; the cell type often cannot be identified. Leukemias of the poorly differentiated cell (myeloblast or lymphoblast) have a highly aggressive course, often drug resistant and with short untreated survival. In the chronic leukemias with full maturation, CML and CLL, a smaller population of immature cells grows and differentiates. The course is often placid, with a longer survival.

IV. Pathobiology of the Leukemias

Cancer is essentially a genetic disorder, either inherited, acquired, or both. The leukemias and lymphomas are examples of disorders of gene regulation, primarily characterized by DNA alterations, often as small as a single nucleotide.

A. Etiology

1. Genetics

An inherited genetic predisposition to an increased risk for developing cancer in general or, specifically, leukemia or lymphomas depends on the inheritance of predisposing genes, of low penetrance detectable from the occurrence of cancer in the family history, combined with DNA damage caused by exogenous agents. Cancer susceptibility genes may enhance the sensitivity to chemicals, radiation, and viruses. The exact influence of each factor is unknown. Identified predisposing genetic conditions in acute leukemias are immune deficiency disorders (ataxia-telangiectasia, Wiskott-Aldrich syndrome), Fanconi's anemia, and Down's syndrome.

The acquired genetic factors are well documented. Many exposures causing genetic damage occur in the work place, at home, in edible foods, cosmetics, and other drugs that we use freely.

2. Chemicals

Chemicals, viruses, and radiations, either manmade or of extraterrestrial sources such as solar ultraviolet rays and cosmic rays, may cause DNA damage. This insult may result in gene alteration (mutation) with transformation of normal cells into leukemic cells. Heavy metals, organic chemicals, solvents, and toxins are often associated with leukemogenesis. These toxins can produce neuropathies, gastrointestinal disturbances, and metabolic and endocrine abnormalities that obscure and complicate the clinical picture of the leukemias. Benzene is an extremely potent leukemogenic agent; it also causes severe depression of erythrocyte and platelet precursors in the marrow, resulting in aplastic anemia, often leading to leukemia. [See CARCINOGENIC CHEMICALS.]

3. Radiation

Ionizing radiation can produce DNA damage. More than a million Americans have been exposed to greater than normal amounts of radiation; 82% of the exposure coming from natural sources. Cosmic rays are emitted as a steady drizzle of gamma rays and heavy particles. They are intensified by episodes of solar flares approximately every 11 years. The ozone of the earth's atmosphere shields us from this ultraviolet radiation coming from the sun. Thinning this ozone shield increases the risk of cancer. Other radiation comes from rocks, soil, and ground water. Radium, as well as its decay product radon, if sufficiently concentrated, is the greatest single source of radiation to the population. Eighteen percent of the exposures is manmade from (medical) radiation, radioactive sensors, and smoke detectors.

Therapeutic radiation, particularly in association with antineoplastic chemical therapy, is associated with an increased risk of leukemia. Radiation of large body surface areas, such as in Hiroshima, will cause leukemias and lymphomas. Radioactivity, however, does not always cause harmful effects. Controlled radiation for diagnostic medical use has proved lifesaving. [See RADIATION INTERACTION PROPERTIES OF BODY TISSUES.]

4. Viruses

One virus type, the human T-cell leukemia virus (HTLV), is implicated in the origin of a T-cell leukemia in humans. A gene of this virus codes for a protein that promotes activity of other genes (transactivation) and may be involved in the pathogenesis of the adult T-cell leukemia, perhaps by activating a cellular oncogene. [See ONCOGENE AMPLIFICATION IN HUMAN CANCER.]

The human HTLV belongs to the family of retroviruses, which contains many members capable of inducing leukemias and other tumors in animals.

Many retroviruses are oncogenic because they contain one or more oncogenes in their own genome.

Viral oncogenes are derived from protooncogenes, which regulate normal cell growth and differentiation. The c-myc, fos, kim, ras, and erb-B are protooncogenes producing proteins to control cell multiplication. Alteration of these regulatory genes is intimately involved in cancer development. The beta chain of platelet-derived growth factor is the c-sis protooncogene; epidermal growth factor receptor is the c-erb-b gene and the colony-stimulating factor receptor is the c-fms gene. When protooncogenes are activated and become oncogenes, they keep the cell in continuous state of growth and block differentiation.

More than 50 viral oncogenes have been identified; they are produced from protooncogenes mainly as a result of the action of damaging agents such as ionizing radiation or chromosomal rearrangements.

5. Chromosomes

Human cells contain 23 pairs of chromosomes, which are highly coiled DNA ribbons associated with proteins. DNA is a nucleic acid polymer. A gene is a segment of DNA. Genetic instability may involve multiple genes on the same or different chromosomes that influence enzyme performance, chemical/drug sensitivity or resistance, or system function. Each chromosome contains a large number of genes. All leukemic and lymphoma cells carry rearranged genes, often a result of translocation (i.e., an exchange between two chromosomes of different pairs, Table I). [*See* CHROMOSOMES.]

Deviations from the normal diploid number of chromosomes, deletions, translocations, and abnormal fragility of chromosomes are common features of all malignancies, including leukemias and lymphomas. [*See* CHROMOSOME PATTERNS IN HUMAN CANCER AND LEUKEMIA.]

Karyotypical abnormalities play a critical role in the pathological biology of neoplasia. Homogeneous regions and double minute chromosomes are most frequently found in tumor cells. These may play a significant role, in the amplification (i.e., the increase in the number of copies) of some genes and protooncogenes important in cancer development.

Chromosomal translocations often place protooncogenes nearby genes coding for immunoglobulins or other molecules produced in large quantity in hematopoietic cells, causing the activation of the protooncogene to oncogene. The first translocation cloned was t(8;14) in Burkitt lymphoma, specifically between chromosome 8 and chromosome 14, bringing together the c-myc protooncogene locus (chromosome 8) and an immunoglobulin locus (chromosome 14). This finding was soon followed by the translocation t(9;22) in chronic myelogenous leukemia (CML), in this case the Abelson (ABL) protooncogene on the long arm of chromosome 9 is translocated to the proximity of a gene called BCR on chromosome 22 (Table I). The exact role of these various genes in the genesis of the leukemias and lymphomas are not fully understood but suspected to be causal.

B. Kinetics

The leukemias have become examples of cancer growth.

The major portion of the growth process of the leukemias and lymphomas is silent until clinical detection. Leukemogenesis is a multistep complex process, beginning with the activation of protooncogenes to active oncogenes, followed by steps of unknown nature. Once the leukemic cell line is established in humans, its doubling time requires many months to years until a sufficient cell mass is produced to become clinically evident. This requires approximately 1 billion cells (1 ccm if all packed together). A 1-ccm leukemic node containing only neoplastic cells would permit an estimation of growth rate as it subsequently enlarges, if there were no cell loss. In reality, there is significant leukemic cell loss so that a retrospective extrapolation of the time of the onset of that leukemia is extended backward.

The level of differentiation dominates the leukemic state because the maturation level determines the clinical manifestations. The more mature the leukemic cells, the longer doubling times, the better the prognosis. It requires at least 30 doubling times to reach 1 billion cells. If the doubling time is 50 days, it takes 4 years from the onset of that leukemia and/or lymphoma. However, if the doubling time of the immature aggressive leukemic cell is shorter (e.g., 20 days), the time of onset would be more recent.

The attrition of leukemic cells may be due to diminished blood supply, normal immunologic defense, or genetic instability. As the nutritional demands exceed the supply through circulation, it decreases the cytokinetic growth rate, causing starvation of the leukemic cell burden. This converts the anticipated exponential rate toward a sigmoid-shaped curve, best described by the Gompertzian equation.

When the division rate of the cell population decreases, the clinical appearance of that lymphoma or leukemia is delayed. Then many cells are resting in the G_0 phase of the growth cycle, but unpredictably they may suddenly start dividing.

C. Morphology

Stages of differentiation are reflected in the morphologic characteristics of the hematopoietic cells. Immunochemistry further delineates various antigens and cell surface markers. Cytochemical stains are used to identify the leukemic cells expressing specific enzymes such as myeloperoxidase or an esterase (chloroacetate esterase) or nonspecific cytoplasmic esterase (alpha-naphthylacetate esterase). The leukemic cells can be arrested at various levels of maturation as seen in the blood and marrow. Multiple copies of certain genes detected by the amount of their protein products by recognizing DNA amplification may be present at various stages. The large nuclei reflect polyploidy—an increased number of chromosomes. The cells of myeloblastic leukemia can be distinguished by a translocation between chromosome 8 and 21 and by prominent azurophilic granules and Auer rods (phospholipids).

1. Cell Surface Markers

Significant cell surface molecules on lymphoid and myeloid cells assist in evaluating the biologic and clinical behavior of the leukemias and lymphomas. Monoclonal antibodies define these unique cell surface antigens and categorize them as cluster designations (CD) groups. Many useful markers for identifying stages of differentiation are CD 10, 19, 20, 24, and immunoglobulins anti-sig, anti-kappa, anti-lambda, and anti-mu. Markers enable the definition of stages of T and B lymphocytes and myeloid differentiation of leukemias and the lymphomas and relate them to the clinical presentations of the disease. About 95% of chronic lymphocytic leukemias (CLL) are of B-cell origin. A minority of undifferentiated and differentiated leukemias and non-Hodgkin lymphomas are of T-cell lineage. T-cell ALLs comprise 15–20% of these cases. Forty percent of lymphomas express the common ALL antigen (CALLA). Less than 10% of the T-ALLs are CALLA-negative.

2. Immunologic Phenotypes

Lymphocytic leukemias and lymphomas being neoplasms of the immune system are often characterized by immune deficiency with generalized hypoglobulinemia. Patients with CLL and other lymphoid malignancies often express autoimmunity—self-production of antibodies against their own tissues. Overproduction of specific monoclonal immunoglobulins may occur frequently. Waldenström described a specific macroglobulin caused by a lymphocyte-plasma cell neoplasm named after him. On occasion the neoplasm has multiple clones each expressing a different globulin and giving rise to polyclonal gammopathy.

3. Heterogeneity

Gene rearrangements and irregularities of gene expressions produce a mixed population of neoplastic cells. As a result, a specific leukemia contains cells that grow rapidly and other cells not at all—resting; another group of cells secrete proteins—enzymes, hormones, antigens, and immunoglobulins that can be used as markers of specific cell types. Some cells show a propensity to metastasize and others do not. Patients with slow-growing leukemia, as in CLL, with minimal involvement of vital organs, may have a prolonged survival. However, survival may be exceedingly short if the leukemia is rapidly proliferating, as in myeloblastic leukemia, and readily involving the liver, brain, or lungs. This spectrum of biologic behavior in the leukemias also includes immunocompetence, drug sensitivity, resistance, and impact on other vital functions.

4. Metastases

The leukemias have access to all tissues through the circulation. Invasion of the lungs, liver, brain, and gastrointestinal and genitourinary tracts is characteristic. The severity of the involvement is variable. Multiple factors determine the extent of the cellular proliferation outside the bone marrow and the resulting impairment of function. [*See* METASTASIS.]

V. Natural History of the Untreated Leukemias

A. Diagnosis

It must be realized that at the time of clinical diagnoses, the leukemia is advanced. A detailed history is required, including genetic background, environmental exposures (occupational and home), drug/medications, and dietary preferences. Careful physical examination is essential in searching for nodal and organ enlargement, hematologic phenomena, muscle tenderness, or wasting.

The diagnoses of leukemias and lymphomas depend on biopsies of the tissues, be it bone marrow, blood, or lymph nodes. The site of biopsy is determined after physical examination of the patient, often by radiological imaging, isotope scanning, X-ray, computerized tomography, magnetic resonance imaging, and liver and spleen scans.

Bone marrow aspiration and biopsies are easily obtained from children or adults. The adult marrow is approximately 1,400 g, the size of the normal adult liver. Cellular abnormalities are revealed by chromosome analysis to detect locations of chromosome breaks in peripheral blood, bone marrow aspiration and biopsy by studies of biochemical expression of genes, identification of T and B cells, analysis of markers using monoclonal antibodies, and electron microscopy.

B. Clinical Description

Preclinical leukemia is often silent. Initial clinical presentation is easy fatiguability, weakness, and minimal but progressively severe anemia. Platelet deficiencies with defect of blood coagulation occurs early in the aggressive leukemias, later in the well-differentiated ones.

The identification of the "suspect" cell in transition to malignancy is the target of diagnosis. A careful review of the leukocytes from the peripheral blood, bone marrow, lymph nodes, and involved organs will establish the diagnoses in most patients.

There are atypical cases. An unexplained lymphocytosis may present and remain essentially unchanged for many years. A slow gradual involvement of lymph nodes, liver, spleen, and marrow occurs to some patients justifying the term of "smoldering leukemia." A more-determined evolution of the dominant leukocytes may occur subsequently, leading to a firm diagnosis of lymphocytic leukemia. A marked decrease in circulating leukocytes, erythrocytes, or platelets (pancytopenia) may be an early finding in diagnoses of subleukemic (aleukemic) leukemia. More commonly, hyperleukocytosis soon appears. Composite lymphomas are represented by combinations of leukemia and lymphoma, often observed within a single or group of lymph nodes. Generally, this indicates a more ominous course.

In mature granulocytic leukemia, the PMN lifespan in the blood increases to 26 days, often causing an increase in cell number by 10-fold. The circulation is slowed because of the increased viscosity of the blood. Extremely high leukocyte counts (greater than 100,000/cmm) may result in pulmonary or cerebral hypoxia and hemorrhage. Prompt systemic therapy often followed by radiation can relieve this unfortunate complication, which may become disabling and frequently fatal. The tissues become infiltrated with leukocytes; central nervous system involvement of the brain, meninges, and the spinal cord produces meningeal leukemia/lymphoma. Infiltration of the lung may result in shortness of breath and encourages bacterial growth. The gastrointestinal tract, liver, spleen, and kidneys may also be infiltrated. Alterations of the immune system may impair immunity, predisposing the leukemic patient to infections. Decreases in platelets result in skin hemorrhages. Hemorrhages of the mouth, lungs, and gastrointestinal and urinary tract further reduce the number of erythrocytes, causing anemia, which impairs the cardiovascular function, exquisitely sensitive to hemoglobin and erythrocyte deficiency.

Quite often immune deficiencies, alcoholism, diabetes mellitus, and cardiopulmonary disorders will interact unfavorably with the leukemias, particularly in patients older than 40. Dysfunction of the autonomic nervous system, which controls many organs of the body, can be detected by alteration of pupillary response to light, sweat glands, and bladder function. The lack of sexual prowess in the adult is often an initial complaint of autonomic dysfunction.

Because nerve function declines with age, it may be difficult to dissociate the effect of the leukemic process on the nervous system from the age factor. In slowly progressive leukemias, polyneuropathy may occur usually with gradual onset and varying symptomatology. This may be expressed as a somatosensory or diffuse motor neuropathy developing into muscular atrophy (amyotrophy) and autonomic dysfunction. The diffuse neuropathy presents primarily as burning pain. Weakness with painless arthropathy may occur. There may be lack of sweating, and bladder and bowel control may be delayed. If the process is aggressively abrupt, sudden mononeuropathy or radiculopathy may occur. Weight loss is common in the leukemias.

The myelodysplastic syndromes, which are characterized by bone marrow failure, often become preleukemic and may persist for many years as ill-defined myeloproliferation or lymphoproliferation. Suddenly the course changes, evolving into an aggressive leukemia. Myelodysplastic syndrome may be borderline malignant (e.g., polycythemia), in which the number of red cells markedly increases. This syndrome has a slow progression at first but may lead inexorably to granulocytic leukemia.

The increased destruction of leukocytes during the leukemic process with and without therapy results in high uric acid levels in the blood and tissues (hyperuricemia). Uric acid crystals may precipitate in the renal tubules and impair kidney function. Increased levels of phosphate and potassium in the blood and urine are elevated and variable. High blood levels of calcium demand prompt therapeutic measures to control the effects of skeletal and leukocyte destruction.

VI. Treatment

The extent of the disease (i.e., the sensitivity of the normal and leukemic tissues to therapeutic attack) determines its curability. The aim of cytotoxic chemotherapy administered within specific guidelines is to achieve maximum benefit with minimum distress. To reduce the leukemic cell population and to avoid damage to normal tissues is difficult in all patients but is often successful. The rate of proliferation is often slower than the normal counterpart cell. Stopping cell proliferation is only one target; another equally attractive target is to overcome the inability for maturation.

Treatment modalities greatly vary and must be evaluated by carefully monitoring all biomedical systems. Options widely vary. Surgery, radiation, chemical therapy, biotherapy, and combinations of these are employed. In childhood leukemias, the drugs of choice are methotrexate, mercaptopurine, thioguanine, cyclophosphamide, and cytosine arabinoside. Radiation therapy combined with intrathecal methotrexate is employed to eradicate central nervous system leukemia. In adults, the same agents are used in the aggressive leukemias with the addition of vincristine, daunorubicin, and L-asparaginase. Corticoids (prednisone or dexamethasone) are employed liberally.

In adults with CLL, the chemotherapeutic agents are chlorambucil, cyclophosphamide, and other lympholytic agents. Irradiation of the blood outside the body destroys lymphocytes without toxicity to the rest of the body. Radiation may be employed to localized masses of lymphocytes. In CML, busulfan and hydroxyurea are primary agents assisted on occasion by other oral alkylating agents. In blast crisis, the treatment is as for an aggressive leukemia. In hairy cell leukemia (leukemic reticuloendotheliosis), interferon is the agent of choice; surgical removal of the spleen (splenectomy) is often dramatically beneficial. Infection, uncontrolled hemorrhage, immunosuppression, and impairment of vital systems are common fatal complications with and without therapy.

Proper selection of radiation, chemicals, dosage, and scheduling is essential for optimal therapy. Untoward sensitivity to chemotherapy and failure to repair the normal cell population will require blood and platelet transfusions for rapid replenishment. Analysis by flow cytometry to detect abnormal chromosome numbers can indicate the aggressiveness of the lymphomas or leukemias or cancer in general. So is the detection of gene amplification. Drug-resistant tumor cells generally display increased copies of the dTMP-synthase (TS) genes. The early detection of resistant cells lead to improved treatment protocols.

Granulocyte transfusions to counter infections are no longer frequently employed. Patients with bacterial, fungus, parasitic, and viral infections with granulocytes less than 500/cmm require appropriate cultures from blood and urine to identify the infectious agent before starting patients on antibiotic therapy. Intervention must be prompt to avoid serious infections and death.

Immunoglobulin protection is critical in treatment of the leukemias to avoid increased susceptibility to infections. Intravenous administration of immune globulin can restore normal humoral antibody function. IgG_4 blocks IgE-mediated histamine release and may offer symptomatic relief in the granulocytic leukemias.

Adequate nutrition must be maintained with supplemental vitamins and minerals in addition to normal protein, carbohydrate, and fats. The hormonal and metabolic systems must be adequately controlled, so that the patient remains as close to a normal state as possible.

Autologous marrow transplantation may achieve 30–50% cures of ALL in childhood in selected patients during second remission on chemotherapy or at the first sign of hematologic relapse. In adults, the acute leukemias are also responsive to marrow transplantation, although the cure rate has not been conclusively established. [*See* BONE MARROW TRANSPLANTATION.]

All therapies of leukemia are judged by benefit-to-risk ratio. Side effects of antileukemic agents differ widely and may require an acceptable metabolic expense to achieve a cure. Functional impairment of hematopoiesis, renal, liver, and nervous systems may occur. Furthermore, the risk of subsequent neoplasm later in life is increased. Life today for a threat tomorrow is a reasonable risk.

It is important for the therapy of the leukemias to consider strongly the emotional and psychological impact on the patient, family, and friends. It is devastating when the patient cannot continue normal occupation and activity and particularly educational pursuits, so that all effort may be made to maintain normal procedures. Truthful information of the leukemias is an important aspect of treatment and requires a thorough comprehension of the risks, both fact and fancy, to understand the reality of increased susceptibility to infection and to dispel the myth of confusion and false remedies.

Bibliography

Bierman, H. R. (1990). "Protect Yourself from Cancer." (*in press*).

Darnell, J., Lodish, H., and Baltimore, D. (1986). "Molecular Cell Biology." Scientific American Books, New York.

DeVita, V. T., Jr., Hellman, S., and Rosenberg, S. A. (1982). "Cancer: Principles and Practice of Oncology." Lippincott & Co., Philadelphia.

Greaves, M. F., Chan, L. C., Furley, A. J. W., and Molgaard, H. V. (1986). Lineage promiscuity in hematopoietic differentiation and leukemia. **Blood, 67,** 1–11.

Heisterkamp, N., Jenster, G., ten Hoeve, J., Zovich, D., Pattengal, P. K., and Groffen, J. Acute leukaemia in bcr/abl transgenic mice, *Nature* (*London*). **344,** 251–253.

Quinn, T. C., Zacarias, F. R. K., and St. John, R. K. (1989). HIV and HTLV-1 infections in the Americas: A regional perspective. *Medicine* **68,** 189–209.

Rapaport, S. I. (1987). "Introduction to Hematology," 2nd ed. Lippincott & Co., Philadelphia.

Sandberg, A. (1983). "The Chromosomes in Human Cancer and Leukemia." Elsevier, New York.

Stomatoyannopoulos, G., Nienhuis, A. W., Leder, P., and Majerus, P. W. (1989). "The Molecular Basis of Blood Disease." Saunders Co., Philadelphia.

Tannock, I., and Hill, R. (1987). "Basic Science of Oncology." Pergamon Press, New York.

Williams, W. J., Beutler, E., Erslev, A. S., and Rundles, R. W. (1983). "Hematology," 2nd ed. McGraw Hill, New York.

Life-Cycle Transitions

MARGARET LOCK, *McGill University*

Glossary

Analogy Partial similarity between like features of two things on which a comparison may be based
Cosmos World or universe regarded as an orderly, harmonious system
Matriliny Tracing of descent through the mother's line of a family (cf. patriliny)
Ontology Branch of metaphysics that studies the nature of existence or being
Relativism Theory holding that criteria of judgement are relative, varying with individuals and their environments
Shitsuke Japanese word used to describe the way in which children are trained to embody correct behavior

AGING IS RECOGNIZED in all cultures, but comparative analysis shows that understanding of the process varies according to cultural conceptions about time, the body, personhood, the relationship of individuals to society, and to the cosmos.

Examples are presented from nonliterate societies to show how passage of individuals through the life cycle is predominantly a social process, which is reinforced through ritual activity. The cultural construction of the individual is examined next with special reference to Japan, where, in common with the majority of the world's cultures, the social order and not individual people is thought of as primary. These examples are then contrasted with the situation in the contemporary West, where the notion of clearly bounded, autonomous individuals is recognized as primary. Understanding of the life cycle is fragmented in the West today so that the process of obtaining individual psychological maturity is visualized relatively independently of biological aging. Both of these processes are in turn isolated from the larger social sphere. This situation is illustrated with particular reference to middle age and menopause.

I. Introduction

In all cultures, the process of aging is recognized and socially marked. Although biological aging is normally a slow, continuous progression, it is usually recognized socially as a series of relatively discontinuous stages. Passage from one stage of the life cycle to another has traditionally been achieved through participation in rituals and ceremonies designed to facilitate the transition and to have it socially recognized. The interrelationship of biological aging with the procession of human beings through the socially marked life cycle provides the subject matter of this chapter. The creation of mature persons in any one culture and the narratives that are generated to explain the process of embodiment and becoming a mature person have a profound effect on aging itself. Equally important, aging has an influence on the concept of self and the meanings that are attributed to growing older.

In these scientifically oriented times, we often visualize aging simply as biological change, a process of growth and maturation followed by a long decline ending in death. When we limit our thinking to biological parameters, the experience of aging is usually more or less the same for the entire human species, but the moment we try to situate this knowledge in the larger domain of the conscious world of human activity, some major difficulties arise. More than 30 years ago Jean-Paul Sartre posed the problem this way:

> . . . if after grasping ''my'' consciousness in its absolute interiority and by a series of reflective acts, I then seek to unite it with a certain living object composed of a nervous system, a brain, glands, digestive, respiratory, and circulatory organs whose very matter is capable of being analysed chemically into atoms of hydrogen, carbon, nitrogen, phosphorus, etc., then I am going to encounter insurmountable difficulties. But these difficulties all stem from the fact that I try to unite my consciousness not with *my* body but with the body *of others*. In fact the body which I have just described is not *my* body such as it is *for me*. I have never seen and never shall see my brain nor my endocrine glands. But because I . . . have seen cadavers of men dissected, because I have read articles on physiology, I conclude that my body is constituted exactly like all those which have been shown to me on the dissection table or of which I have seen colored drawings in books (''Being and Nothingness,'' 1956, p. 401).

Sartre highlights an important point, which, in our eagerness for scientific accuracy and replication, we often minimize or ignore to our peril. The development of biological knowledge as we know it today is largely due to the application of an approach in which the body is objectified and systematically reduced to smaller and smaller units for the purposes of analysis. This method is extremely powerful analytically and accounts largely for the rapid advances that have recently been made in

connection with human biology. But if we are to talk about human beings, and not simply about biology, then we must resort to other forms of description and analysis. As Sartre states, the subjective, unique ''me'' cannot be substituted for the universal ''body'' based on anatomical and physiological terminology.

The human body is both a product of nature *and* culturally produced; i.e., we all learn to experience and understand our bodies as a result of being socialized into a culture. This process of embodiment does not simply produce a veneer, glued onto the ''real'' biological base; on the contrary, culture and biology stand in a dialectic relationship with each other such that biology is modified by culture and culture constrained by biology. This dynamic relationship is in constant engagement, day by day and throughout the life cycle.

The body, therefore, is the locus of the active self throughout the life span, but the self is not just an extension of the physical characteristics of the organism that is its ''carrier.'' The very notion of self is culturally constructed and is in turn linked to ideas about gender, age, social roles, and class differences in any given society. Furthermore, the physical body itself, and the way in which it is covered, decorated, and used in daily life, is a symbolic construct, loaded with meaning, and not merely a flesh and blood entity. Gender, age, social roles, and class differences account in part for the way in which symbolic meanings are attributed to the body, and fundamental to socialization, to becoming a social being, is the process of taking on and absorbing culturally appropriate meanings and behaviors in connection with one's physical body and self. Our attention in this chapter will on the whole be limited to aging and movement through the life cycle, and less emphasis will be given to gender, class, and social roles and how these variables contribute to the meanings that are associated with the physical body.

II. The Cultural Construction of Time

A focus on the life cycle means inevitably that the constitution of human experience in relation to the concept of the passage of time takes center stage. We learn to think of time as a concrete thing, as a duration that can be measured, but in fact we *create* time by marking out intervals between objects and events. An awareness of time is held in common by all human beings; it represents an attempt to establish some kind of basis for ordering and accounting for sequences of events in the phenomenal world. Our experience of time, however, is not by means of our senses: we cannot touch, see, hear, smell, or taste it. According to the anthropologist Edmund Leach, we recognize time in three ways.

First, we are aware of repetition, such as drops of water falling off a roof. To recognize these drops as different from one another, we must distinguish and, hence, define time intervals between them. Such time intervals are durations that always begin and end with the same thing: a pulse beat, a clock strike, New Year's Day.

Second, we recognize aging or entropy. All living things are born, grow old, and die. Leach stresses that aging and interval are two different kinds of experience but that human beings tend to lump them together and to conflate birth and death in some mystical way, so that these events are often thought of as the same kind of experience, in other words, as a return to an original state: ashes to ashes and dust to dust.

Our third experience of time is the rate at which time passes. There is no intrinsic reason why we should think of time as a constant flow. Most people appear to have the subjective impression, for example, that time passed slowly during childhood, that childhood days were long intervals of time, but why should time not slow down occasionally, stop, or even go in reverse? Many different kinds of pictorial metaphors have been produced throughout history to represent time. Some are images of a linear progression, others of cycles, others of spirals—a combination of cyclical and linear movements in space—and still others of discontinuous events, a going back and forth like the movement of a pendulum. There may well be other metaphorical representations for time.

Reliable ethnographic work on conceptions of time is still rather sparse, but apparently more than one representation of time is usually used in all cultures, resulting in a complex belief system, which is inevitably visualized as ''natural'' and ''real.'' In any society, repetition, entropy, and rate must be accounted for in connection not only with physical events but with the social life of human beings. First of all, there is the repetitive character of daily life— the routines that are part of day-to-day living and that are linked with the cycles of light and dark and with the changing seasons. This part of human life is often visualized as cyclical, as something that re-

turns, a continuous flow that does not lead anywhere. Alternatively, it can be conceptualized as oscillations back and forth between light and dark or the wet and dry seasons of, for example, many tropical and semitropical environments. This type of image encourages a representation of time as a discontinuity of repeated contrasts.

Second, there is the life of individuals, which in technological societies is usually regarded as irreversible, a one-way passage from infancy to old person. In the majority of the cultures of the world, where emphasis is less on individuals and more on the continuity of groups such as the family, clan, or lineage, there is a strong sense of continuity over succeeding generations of individuals. There is also an awareness of repetition in that as one cycle ends, another cycle begins. In societies such as these, strong links are usually made between grandparents and grandchildren. Each individual can, therefore, be visualized both as the latest link in a long line of ancestors and often, simultaneously, as part of the repetitious, cyclical return of, or alternatively oscillation between, young and old and back to young.

A third way in which time is conceptualized is in terms of the continuity of society itself and the institutions of which it is composed. Awareness of social continuity must, of course, be cultivated by the individuals who participate in the society. It is they who create and recreate the myths and historical narratives, who build the monuments, preserve the artifacts, and keep the genealogies. Although a linear representation of historical time seems natural and correct to most people today, cyclical and discontinuous notions of history have apparently been the most usual ones in the past. There is an increasing awareness of how the representation of history is not actually a ''true'' narrative about the past, but a recreation of what is supposed to have happened in the past, a projection backward that is highly colored, try as we might to strive for accuracy, by our present beliefs and condition. This awareness has disrupted our sense of security about historical time as linear and has opened the doors to some radical questioning of our contemporary assumptions, including the notion that clear and direct links should be made with events in the past.

Shared representations of time in any given society have a profound influence on the way in which, first, the life span and the biography of individuals and, second, cohesion and continuity of social groups are conceived of with respect to nature and the cosmos. In contemporary Western society, the dominant image of time, whether in terms of evolution, history, or the individual life cycle, and despite the presence of the competing discourse alluded to above, is linear. Moreover, each of these linear paths is visualized by most people as very different orders of time in terms of duration and also relatively independent of one another. Only at a very general level do we see connections between the life of any one individual and either evolution or history as when we say, for example, that phylogeny recapitulates ontogeny, or history is made by humans.

This preponderance of linear thinking apparently is of rather recent origin. When it is coupled with the reductionistic thinking characteristic of much modern science, an emphasis on individualism common to contemporary Western cultures and, further, with an unexamined belief that the linear paths of both evolution and history are those of progress from simple (primitive) to complex (superior), then we arrive at a very special understanding of the human life cycle, which is remarkably different from that of other parts of the world. To try and understand just how different, we will briefly examine the notion of time, the life cycle, life-cycle transitions, and the idea of individual and social self in comparative perspective.

III. Ecological Time and Structural Time: The Nuer

The Nuer are pastoralists, herders of cattle, who reside in the region of the upper Nile in what is now Sudan, about 500 km south of Khartoum. The anthropologist Evans-Pritchard who worked among the Nuer was able to discern two major ways in which time is conceptualized in this culture: a kind of ecological time, in which relatively short periods are framed within the annual solar cycle, and structural time, a system that connects people one with another and that appears to focus more on society than on nature.

In the ecological cycle the basic markers are the events that control the movements of people and cattle within the environment, among which water or the alternation between the wet and dry seasons is the most important. In tune with these two seasons, life oscillates between residence in villages and dry season camps, and between scarcity and abundance of food. The Nuer further divide

up their life by marking the relationship of full moons to the time at which, for example, the planting of millet or making the first fish dams is started, or moving from the dry season camp back to the village. However, Evans-Pritchard pointed out that emphasis is given to the activity rather than to celestial time, so that people say: "Since we are breaking up camp it must be *dwat* [May–June]." Moreover, duration is imprecise, days and weeks are not marked, and time cannot be "lost" or "saved."

Intertwined with the ecocycle is a second system of time reckoning designed to organize the movement of Nuer males through the life cycle. The Nuer have little interest in chronological age but instead come to understand themselves as individuals in terms of their fixed relationship to other members of Nuer society. This is accomplished by means of age-sets, which are fixed, stratified, and segmented groups through which every male member of Nuer society must move (age-set systems have been used extensively in the past in both Europe and Asia in addition to East Africa). Young men between the ages of about 14 and 16 yr are initiated together into an age-set, at the end of the rainy season, at which time six long and deep parallel incisions are made across the forehead from ear to ear with a knife. This traumatic procedure is followed by a period of seclusion and various ritual sacrifices, which culminates in a communal celebration. A 3- or 4-yr gap then ensues, followed by the initiation of another age-set. Age-mates created through a shared initiation experience have life-long bonds and obligations to each other. The age-set is internally stratified into juniors and seniors, and together they pass through the life-cycle stage of Nuer male society—boy, warrior, and elder. Age-sets form one key principle of cohesion in Nuer social life and are associated with a set of rules about behavior whose complexity is staggering, most particularly because they must be coordinated with other kinds of social ties involving obligations to affines, lineage, and clan. In this type of society, biological age itself is not very important; the more social phenomenon of age *differences* concerns people, by which someone judges who they are and what position they occupy in Nuer society. Furthermore, a long duration of time, historical time, is calculated in terms of its relationship to the initiation of a particular age-set. Hence, people estimate how many age-sets ago something happened, not how many years.

IV. Dual Aspects of Life-Cycle Transitions

Evans-Pritchard's study is now recognized as an anthropological classic, and although it has some obvious shortcomings (almost no mention of women, for example), it remains as one of the few studies where the social significance of life-cycle transitions are examined in detail. A more recent, detailed study has been carried out by Hugh-Jones among the Pira-parana Indians of Northwest Amazonia, and her findings corroborate, in general, those of Evans-Pritchard. She interprets the Indians' understanding of the life cycle as a dual process concerned, on the one hand, with the physiological and spiritual reproduction of the individual and, on the other, with the reproduction of the social structure.

This dual aspect of the process of aging appears to be shared in common by many other societies of the world. A girl's puberty ritual among the Ndembu of Zambia is used as an illustration of the way in which the individual and the social aspects of this particular life-cycle transition are negotiated. A pubescent girl is wrapped in a blanket and laid at the foot of a *mudyi* tree. This tree is conspicuous for the white latex that it exudes in milky beads if the thin bark is scratched. For this reason, it is called the milk tree. Most Ndembu women can attribute several meanings to this particular tree: they say that the milk tree is "senior" for this particular ritual. Other rituals have other seniors taken from the natural world, which anthropologists call dominant symbols. Secondly, the milk tree stands for human breast milk and for the breasts themselves. The ritual is performed when the girl's breasts begin to "ripen" and not after her first menstruation (at which time another less elaborate ritual is carried out). The main theme of this puberty ritual is that of nurturance between mother and child. The women also describe the tree as "the tree of a mother and her child," which highlights the most important social tie in the Ndembu community. One ritual specialist described the tree in the following way:

> The milk tree is the place of all mothers of the lineage. It represents the ancestress of women and men. The milk tree is where our ancestress slept when she was initiated. . . . One ancestress after another slept there down to our grandmother and our mother and ourselves the children. That is the place of our tribal custom. . . . , where we began, even men just the same, for men are circumcised under a milk tree ("The Forest of Symbols," 1967, p. 21).

Therefore, the milk tree stands for social organization, particularly for the matriliny, the principle on which the continuity of Ndembu society depends. The inheritance of property, residence rights, and incumbents of official positions are all decided on the basis of the matriliny, and hence it confers social order to Ndembu life. The milk tree also represents tribal custom itself, moral order. In this sense, the tree is thought to produce harmony and cohesion among people. Furthermore, just as the child depends on its mother for nourishment, the tribesman is said to drink from the breasts of tribal custom. Nourishment and learning are associated through reference to the tree and its symbolic meanings.

During the actual ritual itself, the women dance around the tree, while the initiate lies at the hub of their whirling circle for many hours. In performing their dance, the women set themselves up specifically in opposition to the men, they taunt the men with obscene songs and will not let them join the dance. Although harmony is emphasized, in this particular ritual, it is especially harmony and solidarity of women in opposition to men. Moreover, for long periods during the ritual the girl herself is the focus of events and picked out as different, for that day, from all other women. A young milk tree is chosen for the ceremony—a sapling, which symbolizes the new social personality of the girl who is transformed by the ritual into a woman. The girl is initiated alone, and most Ndembu believe that it is the most thrilling and gratifying day of her life, the culmination of rigorous and stressful training that she has successfully completed in the previous months. The milk tree is also known as the "place of death," because both boys and girls are thought to die there and be reborn as adults. During some sections of the ceremony, the girl is placed in opposition to all other women and, most particularly, to her own mother, who may not dance in the ring. The mother must give up her child, but toward the end of the ritual she re-establishes her relationship with her daughter, but as a grown woman. The oppositions between the girl and her mother and between the girl and women are finally resolved in a harmonious solidarity of women.

Throughout the ritual, the ambivalences and contradictions are evident. By the way the Ndembu talk about it, the tree represents for them linking and unifying aspects of the entire society. But also during the course of the ritual action around the tree, incipient and unresolvable oppositions in connection with gender and age, which are inevitable in any society, are acted out, although the Ndembu do not appear to recognize this explicitly.

Study of the milk tree enables elaboration on how symbols associated with rituals used in connection with life-cycle transitions appear to function. Ritual symbols serve to link together in people's minds a series of ideas, events, or objects through the process of analogy. Thus, the milk tree stands for women's breasts, motherhood, the principle of matriliny, the specific matriliny of the girl going through the ritual, the girl herself, learning, and the unity and persistence of Ndembu society. Moreover, themes of nourishment and dependence run through all these diverse phenomena.

A polarization of meaning is also apparent in connection with symbols. All dominant Ndembu symbols possess two clearly distinguishable poles of meaning. At one pole, a cluster of symbols refers to the moral and social orders of Ndembu society, to the principles of social organization, and to the values and norms that are inherent in structural relationships. At the other pole, the symbols are associated with natural and physiological phenomena and processes. These poles are described as the ideological and sensory poles, respectively.

At the sensory pole, meaning is closely related to the outward form of the symbol. Thus, one meaning of the milk tree—breast milk—is closely related to the milky latex that exudes from the tree. Similarly, another tree, used in other rituals, is selected because the red gum it produces is associated with blood. The sensory pole arouses emotions, desires, and feelings and is successful because it is "flagrantly" physiological. The objects and meanings associated with this pole represent things that are common to all men or women as individuals: blood, male and female genitalia, semen, urine, breast milk. Through ritual, they are explicitly linked with the social values associated with the ideological pole, with the continuity of social groups: domestic, village, and tribal. Based on many years of research among the Ndembu, one researcher criticized the psychoanalytic notion that conflict can be reduced to what is essentially an intrapsychic phenomenon, the product of an essential tension between society and the biophysical drives and early conditioning in the primary family. Ndembu rituals are concerned with the basic needs of social existence (hunting,

agriculture, female fertility, favorable climatic conditions, and so on) and with shared values on which communal life depends (generosity, comradeship, respect for the elders, the importance of kinship, hospitality, etc.). These social needs should be thought of as having an importance equal to those of individual needs, and they should not be relegated to a secondary position, as may be the case when using a psychoanalytic framework.

The Nuer and the Ndembu examples alert us to the importance of social bonds and to the creation of a "socially correct" person in these societies and others like them. One researcher, who did some research into the local conception of the life cycle among the Ilongot in the northern Phillipines, came to the conclusion that in that society newborn children are thought of as a homonculuslike figure, a pale and miniature sketch of the adult he or she is to become. Although the life cycle is divided clearly into childhood, youth, and adulthood, the notion of the creation of an individual and autonomous adult does not exist; on the contrary, people are first and foremost recognized as social beings, clearly situated in a nexus of kin and other human relationships. Emphasis in all these societies is on reproduction, continuity, and replication of the social system, and the rituals of life-cycle transitions serve above all to immerse the individual deeper into the social system of which he or she is a part.

The above examples are, of course, taken from nonliterate cultures. Clearly, literacy and its associated technological changes has had a major impact on the way in which societies are organized and reproduce, but in many literate societies too, emphasis is given to social continuity, such that the individual explicitly takes second place to the social order.

V. Cultural Construction of the Person

In contrast to the societies described above, where it is considered natural for individuals to be immersed in society, sociologists and psychologists, following the Western philosophical heritage, usually conceptualize the relationship of the individual to society as one of an opposition. The demands of the social and moral order are visualized as being in an inevitable tension with the egocentric drives, impulses, needs, and desires of individuals. This opposition, while fundamental to Western epistemology, is also apparently highly unusual. An anthropologist has argued that the Western conception of the person as a bounded, unique, integrated, motivational, and cognitive universe, a dynamic center of awareness, emotion, judgement, and action, is a rather peculiar idea within the context of the world's cultures.

Japan illustrates the way in which, even in a technologically complex society, the concept of the individual is understood in a dramatically different way from that in the West. In Japan, although the concept of individualism has been debated vigorously for well over 100 years, the family is still considered the most natural and fundamental unit of society, not the individual. Traditionally, the source of tension was not between individuals and society but between the family and the state. The Confucian system of ethics, which came originally to Japan from China, had a profound effect on Japanese history and especially on the structure of social relationships. Japanese children are raised to be sensitive to group relationships and believe that the needs of the groups should take place over individual aspirations. Japan has been described as a culture of "social relativism" in which the person is visualized as acting within the context of social relationships and never simply as an autonomous person. An individual's sense of personhood is understood in terms of how he or she best fits into various groups; even one's physical body is thought to belong to the family.

To understand a little of how this culture of social relativism is acquired by individual Japanese, a comparative research project was carried out into early child-rearing practices in Japan and the United States. By 3 or 4 mo of age, infants were responding to their mothers in characteristic ways, which depended on the styles of caretaking employed by their mothers. While American mothers in this study generally encouraged their babies to be vocal and physically active, Japanese mothers aimed to soothe and quiet their babies. The maintenance of what is thought of as the natural state of a Japanese infant—its gentleness and passivity—was thought to be best for optimal health and development of the child. Therefore, Japanese mothers rocked and lulled their infants much of the time. Unlike American mothers in the study, they did not place their babies facing them to talk at them very often. For very small Japanese babies, continual body contact with another human body virtually all the time is considered ideal, thus the baby is usually either on someone's back or in their arms. Although

some behavior patterns are changing, adult Japanese will have almost all been through a socialization experience in which they rarely, if ever, slept or bathed alone as young children. The majority will have experienced a very close relationship with their mother in which she regarded her baby primarily as an extension of herself.

In America, the mother views her baby, at least potentially, as a separate and autonomous being who possesses from birth a distinct personality and who should learn to do and think for himself or herself as soon as possible. In Japan, in contrast, the mother views her baby much more as an extension of herself, and psychologically the boundaries between the two of them are blurred. An ancient and much quoted adage in Japan states that the "soul of a three year old must last for one hundred years," implying that early socialization is of the greatest importance and something that cannot be undone if handled incorrectly. Japanese parents employ what is known as *shitsuke,* a concept that contains the idea of putting into the body of a child the patterns of living, ways of conduct of daily life, and a mastery of manners and correct behavior, to socialize their children. The raising of children is explicitly recognized as a vital investment in the future and the only effective means of ensuring continuity of the Japanese culture. Therefore, the molding of people in Japan is seen as a skill to be cultivated with a good deal of time and careful attention. This early molding of behavior is designed to lead ultimately to the internalization of a basic moral style in which one is made highly sensitive to both the verbal and nonverbal cues of other people and in which individuals must bend (like a bamboo) to accommodate themselves to the needs of others, especially those designated as more mature and experienced. This process is explicitly carried out throughout the entire compulsory education system in Japan; the school has never been a place simply to learn academic skills, but is an institution tuned to continue the moral training begun during early socialization at home and, hence, correct social and moral behavior is part of the core curriculum.

The Japanese traditionally used an age-grade system as a means of relating the individual life cycle to a process of social maturing. Today, only vestiges of this system remain, but a keen sensitivity to age remains evident. People who are approximately at the same stage of the life cycle have what is regarded as a natural affinity for one another, including similar philosophical and moral outlooks, and a socially mature person in Japan is someone who above all else is able to use their experience to participate fully in social life. Individual values are also prized, so long as these do not lead to nihilistic self-centeredness or selfishness; pushing for individual rights over those of the family or striving for autonomy are thought of as selfish. The relationship between individuals and society is usually thought of today as complementary, and the dominant view in Japan is that only from a corporate group can one be assured of getting sensitive responses to one's individual needs. An awareness of self is endlessly reconfirmed through active engagement with others.

Accounts about the relationships of the individual to society in Japan and among the Ndembu and the Nuer are the results of statements given by and observations of members of these societies. These accounts are necessarily oversimplified, narratives of what is supposed to happen. In reality, of course, there is not such a close fit nor a completely harmonious blending of the dual aspects of the life cycle, the social, and the individual. Nevertheless, what is striven for as the ideal stands in strong contrast to the situation in both Europe and North America, where the life cycle is understood in a very different way. To better appreciate the difference, a brief historical summary is given.

VI. Culture of Oppositions: Mind versus Body, Individual versus Society

The sources of the dominant Western notion of a self were set in Plato's "Republic," where the idea of self-mastery through reason was first clearly expressed. To be master of oneself was, according to Plato, to have the higher part of the soul rule over the lower; i.e., reason should rule desire. When this was the case, then unity with oneself, a sense of calm, and a collected self-possession would follow naturally. There have always been dissenting opinions to the belief of reason in the West, ranging from early Christians to the Romantics, Nietzsche, and, at present, several strands of thought in the contemporary social sciences and philosophy. Nevertheless, the notion of a natural opposition within the individual between reason and desire has remained a dominant belief throughout Western history.

It was in the seventeenth century, during what has come to be known in retrospect as the Englightenment, that major transformations took place that further solidified the opposition set out by Plato. The mathematician–philosopher René Descartes most clearly reformulated the opposition, which would later become the foundation for contemporary biological thought. Descartes was determined to hold nothing as true until he had established the grounds of evidence for accepting it as such. The only category that he accepted on faith was the idea of the consciousness of his own being as stated in his most famous dictum *Cogito, ergo sum* (I think, therefore I am). From this position, Descartes proceeded to argue for the existence of two classes of substance that together constitute the human organism: the palpable body—a product of nature—and the intangible mind, which he claimed God had given to men in order for them to examine and understand nature. Implicit in this concept was the notion of an opposition between the body, associated with desire and unruliness, and the mind, associated with rationality and order. Such an opposition has never been so clearly formulated in any other major philosophical tradition, and it gradually opened the door to the reductionistic approach characteristic of modern Western science. Central to this approach is the notion of an objective observer, one who makes use of his mind to systematically uncover the laws of nature.

The position of Descartes and other like-minded thinkers represented a clear ontological break with the earlier philosophical tradition. Whereas it had previously been assumed that reality was ordained by a larger cosmic order visualized as essentially good, in the new vision the source of moral order was now found within the individual; an important power had been internalized. From now on, to know things, to grasp reality, one must have a correct representation within one's mind of the world "out there." Matter of all kinds, including the human body, must be demystified, denuded of obfuscating spiritual qualities. Descartes insisted that reason should at all times control the passions, and in taking this position he firmly located the moral order inside the heads of individuals, in the "self."

John Locke was one of the first philosophers to develop his ideas around the notion of a thoroughly disengaged, rational, internalized, and disciplined self. From the late seventeenth century, the notion of a clearly bounded "I," a self with a unique identity that remains essentially stable throughout the life span, took root for the first time. Following the Cartesian split, this self was conceptualized as having an internalized monitoring system, independent of both the physical body and the social order.

From the middle of the eighteenth century onward, when these ideas about the relationship of the individual to society were fused with the emerging theories of natural and social evolution, the stage was fully set for the appearance of the divided modern self—one in which each clearly bounded individual is visualized as part of a new kind of secular cosmic order, an unfolding from primitive to sophisticated, in terms of both body and mind.

The late eighteenth century brought one more change that is of great significance for our present understanding of the human body, namely the development of the pathological anatomy associated with Bichat, which forms the basis for modern medicine as it is practiced today. Because the rational physician observer was now able to peer directly into the interior of the body and read the signs revealed there to make a diagnosis, he was able to ignore the patient. Hence, the patient as a person was no longer held as an authority on his or her own body. This approach gave rise to what is now known as biomedicine, and it exerts a profound influence on how we view the life cycle today.

Although these historical changes did not directly cause in any simple way the situation in which we find ourselves today, they are part of a highly complex, somewhat random set of events that can help us understand our present attitude toward the process of aging and the life cycle.

VII. The Fragmented Life Cycle

At present, no current theory in the West accounts for the progression of a social being through the life cycle. We use several different languages and, hence, representations of the human being that do not overlap or intersect to any significant degree. The language of biology, of physical aging, is one construction, but this is often reduced to information about cells and organs. Because of its close association with medicine, there is a tendency to highlight pathology, and only rarely does the whole person make an appearance in this discourse. A second language is that of psychology, almost completely divorced from biology, and grounded in the twin assumptions of autonomous individuals whose needs are placed in opposition to those of society,

but who are, at the same time, further fragmented within themselves in terms of reason–mind versus passion–body. An effort to harmonize the individual with the needs of society, so characteristic of life-cycle transitions of the other societies examined above, is nowhere evident. Movement through the life cycle is a personal or, at most, a family event, a series of crises in which the individual must ideally achieve more and more autonomy. Associated physiological changes are either largely ignored or managed through technology and medical care.

We also have a language that we use today in connection with the development of societies, but this is usually reduced to economics and is largely divorced from ideas about morals and human relationships. Thus, the person is fragmented into various parts: medical–mechanical, psychological–rational, and consumer bodies, wherein our needs and desires are housed. Most medical, psychological, and economic discourse is without context of time and space, and each of these languages is usually considered to be of universal applicability. Thus, they do not relate well to individuals in a specific context nor to a specific stage in their life course. Furthermore, to return to the quote by Sartre at the beginning of this chapter, they do not take account of the subjective experience of any one individual. These languages are each thought of as metanarratives, at a greater level of abstraction and sophistication than individual understanding and experience. Because they are assumed to be grounded in rational deduction and the scientific method, they are often thought of as factual, as truths, or at least on the path to truth.

VIII. Identity and Aging: The Psychology of the Life Cycle

One of the only comprehensive attempts in the West to deal with the whole life cycle in psychological terms is the work of Erik Erikson. A follower of Freud, Erikson saw the life cycle not as a continuum but as a series of predetermined steps or stages by which the individual seeks contact in an ever-widening radius of social relationships. Each phase of the developmental process is thought to have a turning point, a crisis that poses the solution to a specific task. The solution is prepared for in the preceding stages and worked out further in succeeding ones.

The crisis associated with the first part of the life

cycle, infancy, is the question of basic trust versus mistrust, and its successful resolution depends on the relationship that is established between infant and mother. Early childhood, the second stage, is marked by increasing muscular and locomotor coordination as well as maturing cognitive discrimination. At this stage, the child's will comes up against the parents' will, and successful resolution means that the child develops a sense of autonomy as opposed to shame and doubt. Therefore, successful development of autonomy is associated with will power.

The third stage is associated with play, the attainment of a certain independence, the development of the beginnings of a conscience, and a dawning understanding of the fact that the child must eventually take his or her place in the adult world. The child becomes aware at this time that his genitals are "inferior" compared with those of his father. The task at this stage of the life cycle is the resolution of initiative versus guilt, which is associated with an emerging purposefulness. The fourth stage is "birth" into a community, i.e., school. Here the young person must learn to be competent, and this is achieved by dealing with the crisis of industry versus inferiority. The next stage is adolescence, when energy is focused on the question of identity formation versus identity confusion. The question of a career must be dealt with, and successful resolution of this stage means that the youth learns fidelity, including a sense of being true to himself, to significant others, and to a belief system.

The sixth stage is that of young adulthood, in which the conflict of intimacy versus isolation should be resolved in terms of a mature love, including not only sexual satisfaction but also shared procreation, work, and recreation. During the seventh stage of adulthood, the primary concern is with "generativity" versus "stagnation," which, if dealt with successfully, results in the ability to give care. The last stage is old age. Here the crisis is one of integrity versus despair, and the resulting strength is wisdom. Integrity implies an emotional and cognitive integrity, which results from traversing the life cycle, this in spite of the inevitable decline of bodily and mental functions.

Erikson's scheme is clearly linear and emphasis is given from as early as the second stage to the importance of developing a sense of autonomy. The beginning and end is demarcated by the span of one individual life time, and hence there is no feeling for a continuity with some larger social or cosmic order

by placing the individual life cycle into a larger framework. If the crises are successfully dealt with, then an individual achieves a sense of continuous development and mastery, which can theoretically be maintained even in the face of impending biological demise.

Erikson's model, therefore, is overwhelmingly concerned with individual development, and "resolution of the conflictual tasks" lie not in going through any socially meaningful ritual but in learning to internalize certain attitudes. Although this scheme is loosely allied to biological aging, it in no way deals specifically with, for example, the onset of menstruation or the birth of a baby (this may be in part due to the fact that Erikson's scheme is gender blind). Symbolic associations are not made between the biology of the individual and the moral order but tend, in contrast, to be limited to the achievement of an internal psychological order in which will power, purposefulness, initiative, and industry dominate.

In an effort to avoid the rather clinical and pathological focus embedded in the Eriksonian scheme, more emphasis has been given recently to normal development. However, the most influential of these models have been derived from cognitive psychology, and make little or no effort to reinsert either the embodied person or the social order into the analysis.

IX. Medicalization of the Life Cycle

In the introduction to one of the most used textbooks on human behavior written for medical students, Simons states:

> [The] cycle of life and death, this imperceptible yet sometimes all too sudden unfolding of the generations, is the great mystery of our existence. It is the privilege—and responsibility—of the physician to be able to participate in the significant events of this mystery from the beginning to the end. Our role was not always so central. In centuries past we were preceded first by the lawgiver, then by the priest, and most recently by the family of relatives and friends . . . But in our age of mobility and modern science, it is the physician to whom we look increasingly to guide us through the awesome experiences of birth and parenthood, of growing up and aging, of separation and loss, and of illness and death ("Understanding Human Behavior," 1985, p. 2).

Medicalization of the life cycle is one logical outcome of the drive to master nature associated with post-Enlightenment thinking. Only at the turn of the century, for example, the death of infants came to be seen as a problem about which something could be done. Public health and educational measures were first institutionalized at that time to preserve the life of infants, and doctors added their medical authority, largely through trying to establish a regulated approach to breast-feeding. Prior to this time, a high infant mortality rate was considered inevitable, and hence it was assumed that the life span did not really commence in earnest until the child was well over 1 yr old.

Very recently, the life cycle has been extended backward even further, earlier than birth, and this can be attributed largely to developments in technology and medicine: the recent discovery of the fetus as a patient is due largely to the creation of the fetal monitor and other technologies for viewing the fetus. Until very recently, mother and fetus have been regarded as essentially one, but now they are often set up in opposition to one another in both medical and moral battles. Therefore, the life span now starts *in utero*. At the other end of the life span, the lengthening of life by means of respirators and organ transplants has led to arguments about what is death and by what criteria it should be defined. These arguments are often elaborated around definitions based on neurological criteria in isolation from any larger social context, and hence attempts are made to reduce what should be social and moral judgements to biological measurements and medical knowledge.

Menopause, a life-cycle transition of relatively recent discovery, will be discussed in closing as an example of the complex position in which we now stand with respect to the process of aging—biological and social.

Although the end of menstruation must inevitably be marked subjectively by women, this event has not traditionally been ritually or socially marked. Two reasons probably account for this. First, throughout most of human history, women have spent the greater part of their reproductive lives either pregnant, lactating, or in a state of amenorrhea. Apparently, in hunting and gathering or pastoral societies where people are required to be physically very active, individual body weight may quite frequently drop below that which is necessary to continue menstruation. Moreover, because the production of many children has more often than not been the ideal, most women have experienced a

considerable number of pregnancies followed by lactation periods of several years or more. This situation means that, in contrast to women living in middle class, urban families in industrial societies, most other women experience rather few menstrual cycles during their reproductive life. Middle-aged Samburu women living in northern Kenya recall, for example, that they have had about 10 or 12 menstrual cycles during their reproductive life. Most women, therefore, do not know when they cease to menstruate, they can only be aware of this in retrospect, after it is evident that they have not become pregnant for some considerable time. In the majority of societies where biological age is not marked, women simply have a general idea that they are about at the stage when they will cease to have children.

There have been very few anthropological studies on the experience of menopause in non-Western societies. The assumption has tended to be that women in "primitive" societies do not suffer at menopause because they experience more freedom and power as old women than they ever did previously. Several issues become conflated when this assumption is made. First, as pointed out above, a woman probably does not know when she is menopausal. Second, if we confine suffering to the experience of uncomfortable symptoms such as the hot flash, there is insufficient data to draw any conclusions. Furthermore, when working in societies where women have passed most of their reproductive life without menstrual cycles, we cannot assume that they have similar estrogen profiles as do women who have one or two children. Several recent studies indicate that the vasomotor symptoms regarded as characteristic of menopause in the West are absent or reported rather infrequently in other cultures. Clearly, further research is in order, but the possibility of biological variation must be considered. In addition to gene pools, other variables including diet, exercise, parity, lactation, and disease history are all possible contributing factors to the menopausal experience. Culture, therefore, influences not only the meanings attached to life-cycle experiences but also indirectly influences what happens at the biological level.

At the social level, although the dominant stereotype that is held in connection with menopause in the West is that it is a time of loss, a crisis, we now have data from many sources to indicate that this is a very oversimplified view and by no means always the case. Recent work carried out by social scientists working with large nonclinical populations have found that the vast majority of middle-aged women do not express regret at reaching menopause, do not report a poorer health status, and do not evidence increased use of medical services. Women who do experience distress are most usually those who have had hysterectomies and/or oopherectomies and women who have experienced distress at previous transitions in the life cycle. The picture in nonindustrial societies has been similarly oversimplified. If a woman has produced several children, she can usually (but not always) expect support in her old age, and hence she can look forward to the future as a time when she need no longer work or bear children. On the other hand, if she has produced no children, or only girls, then most likely she can only look forward to a lonely and poverty-stricken old age, and hence menopause is an unwelcome sign. The social status of older women must be established for each society and in each individual case. Moreover, it cannot be assumed in any society that the biological changes of menopause coincide very closely with major social transitions such as children leaving home, parent deaths, and so on. This too must be empirically established. In the majority of societies where social transitions take precedence over biological changes, rituals are focused neither on middle age nor the end of female reproduction but on entry into old age and what can be expected for the last part of the life cycle, and this usually takes place some time after menopause.

It is only over the course of the past 100 years in the West, with the rise of the gynecological profession and of leisured women of the middle class, that the end of menstruation has gradually become a life-cycle transition on which considerable attention has been focused. We will turn briefly to the contemporary competing narratives that account for distress at menopause, because they are illustrative of the extent to which we focus on pathology and psychological losses and, secondly, of how the subjective experience of any one individual is not linked in a positive way to social events, history, or society at large.

X. Menopause as a Deficiency Disease

Considerable disagreement exists among clinicians as to how to account for the fact that some women and not others experience distress at menopause, and these accounts depend in turn on how menopause is defined. When using the biomedical approach, menopause is equated with a somatic process in which dropping estrogen levels are said to lead to a state of "estrogen deficiency." The external marker for this event is, of course, the end of menstruation. Although the relationship between falling estrogen levels and physical or psychological distress is not at all well understood, a causal relationship is often assumed and described as a "menopausal syndrome." The narrow definition of this syndrome limits it to vasomotor complaints and vaginal dryness, while a broad-spectrum estrogen hypothesis, in contrast, assumes that experiential, behavioral, vasomotor, and also a range of other somatic complaints can be attributed to dropping estrogen levels. A third explanation suggests that if psychological symptoms occur at menopause then these are due to the indirect result of estrogen withdrawal leading to a "domino effect" such that hot flashes lead to sleep loss, which in turn can cause depression.

All of these approaches have in common a focus on a correlation between endogenous estrogen levels and unwanted physical and/or affective states. Recently, the estrogen-deficiency model has been elaborated in yet another way. Emphasis in this latest approach is less on symptomatology associated with the end of menstruation, and more on possible long-term risks associated with what is described as a "chronic" state of estrogen deficiency. Estrogen replacement therapy in this latest approach has been described as a "type of immunization" against coronary heart disease and osteoporosis. This approach strongly reinforces the notion that an estrogen-deficiency state is a "normal" part of female aging and that virtually all women, therefore, should expect to be medicated throughout the second half of their life cycle. Statements to the effect that life expectancy rates have lengthened considerably over the past 100 years are often used by supporters of this immunization approach to reinforce the erroneous notion that postmenopausal life is a relatively new phenomenon and that while estrogen deficiency is "normal" today, it is, nevertheless, evolutionarily speaking, rather "abnormal." The focus in all of these narratives is on chemistry and cellular changes and on professional control of the "problem." The only links to society are in terms of the cost that untreated women may be in terms of health care expenditures for chronic disease that should have been avoided. All of the narratives are justified in terms of either basic science (often poorly executed) or statistics, and subjective accounts are often assumed to be irrelevant. The individual woman is rendered passive in the extreme. [*See* ESTROGENS AND OSTEOPOROSIS.]

XI. Accounting for Distress at Menopause: Psychosocial Explanations

Psychiatric and social science explanations for menopausal symptomatology can be more or less characterized in three ways: Psychodynamically oriented clinicians assume that menopause can act as a stressor leading to a reoccurrence of unresolved psychosocial conflicts from earlier stages in the life cycle. In contrast, psychologists often focus on social changes more directly associated with mid-life such as role loss due to the end of fertility. Other researchers emphasize that coincidental psychosocial stressors predispose certain women to psychological symptomatology at this stage of the life cycle. Biological changes are regarded as relatively unimportant in this model, because it is generally assumed that for most women somatic symptomatology will be absent or minimal. In each of these narratives, emphasis is on intrapsychic processes and on appropriate adaptation of the individual to their current situation.

In contrast to the biomedical approaches, the accounts given by psychiatrists and social scientists tend to pay more attention to the importance of the subjective experience of menopause, of thinking in terms of multicausality, and of being sensitive to psychosocial variables. Whereas in the hormone-deficiency model there is a tendency to assume a rather direct relationship between chemistry and behavior, in the social science approaches, despite an emphasis on multicausality, there is a tendency to assume a simple relationship between "stressors" on the one hand and experience and behavior on the other. In theory, in the biomedical model everything beyond individual chemistry remains unexamined, whereas in the social science model emphasis is on the psyche and its potential for adjustment and adaptation, while the physical body is often ignored completely. In other words, Cartesian dualism haunts the usual explanations for distress at menopause.

Menopause research to date, therefore, fragments a complex biosocial and biopsychological phenomenon, focusing on one or two aspects as though they were representative of the whole or uninfluenced by other factors. Obviously no simple relationship exists among physical symptomatology, the subjective experience of menopause, and social and environmental variables. Symptom reports cannot, for example, be considered straightforward reflections of biological reality. Many biological processes are not subjectively recognized or labeled and, therefore, are in essence "unspeakable." Alternatively, a number of different physiological processes may be associated with the same physical symptom. Socialization into a particular culture exerts a strong influence on the recognition and labeling of somatic sensations; moreover, it cannot be assumed that middle age is necessarily a crisis, either biological or social.

XII. Conclusions

Basic research into the biological processes associated with aging are slowly accumulating, although there is an urgent need to carry out more studies cross-culturally. If the findings of this kind of research are to be applied, then they must be contextualized in specific social contexts and with reference to the subjective life experiences of individuals. This can only be done successfully if the importance of the cultural construction of concepts about time, the body, personhood, the relationship of individuals to society, and the cosmos are recognized as central in this exercise.

Bibliography

Erikson, E. (1963). "Childhood and Society." W. W. Norton and Co., New York.
Evans-Pritchard, E. E. (1940). "The Nuer." Oxford University Press, Oxford.
Hugh-Jones, C. (1979). "From the Milk River: Spatial and Temporal Processes in Northwest Amazonia." Cambridge University Press, Cambridge.
Leach, E. R. (1966). "Rethinking Anthropology." The Athlone Press, London.
Levins, R., and Lewontin, R. (1985). "The Dialectical Biologist." Harvard University Press.
Lock, M. (1986). Ambiguities of aging: Japanese experi-

ence and perceptions of menopause. *Cult. Med. Psych.* **10(1),** 23–46.

Lock, M., Kaufert, P., and P. Gilbert. (1988). "Cultural Construction of the Menopausal Syndrome: The Japanese Case." *Maturitas* **10,** 317–332.

Simons, R. C. (1985). "Understanding Human Behavior in Health and Illness," pp. 2–5. Williams and Wilkins, London.

Taylor, C. L. (1989). "Sources of the Self: The Making of Modern Identity." Harvard University Press, Cambridge.

Turner, V. (1967). "The Forest of Symbols: Aspects of Ndembu Ritual." Cornell University Press, Ithaca, New York.

Limbic Motor System

GERT HOLSTEGE, *Rijksuniversiteit Groningen, The Netherlands*

Glossary

Hemiplegia Paralysis of one side of the body
Palsy Paralysis
Vocalization Nonverbal production of sound
Periaqueductal gray (PAG) Gray matter around the aquaduct of Sylvius in the mesencephalon
Nucleus raphe magnus (NRM) Group of midline cells in the ventral tegmentum of the most rostral medulla oblongata
Nucleus raphe pallidus (NRP) Group of midline cells in the ventral tegmentum just caudal to the nucleus raphe magnus
Nucleus raphe obscurus (NRO) Group of midline cells in the dorsal tegmentum, dorsal and caudal to the nucleus raphe pallidus
Rhythm generators Group of cells that fire in a rhythmical way causing rhythmical movements

THE LIMBIC SYSTEM is closely involved in the elaboration of emotional experience and expression and is associated with a wide variety of autonomic, visceral, and endocrine functions. The limbic system consists of several cortical and subcortical structures, although there is no agreement on exactly which structures belong to it. Some authors argue that the use of the term limbic system should be abandoned; nevertheless, many scientists still use this term. Subcortical regions usually included in the limbic system are the hypothalamus and the preoptic region, the amygdala, the bed nucleus of the stria terminalis, the septal nuclei, and the anterior and mediodorsal thalamic nuclei. The limbic system has extremely strong reciprocal connections with mesencephalic structures such as the periaqueductal gray (PAG) and the laterally and ventrally adjoining tegmentum (Nauta's limbic system–midbrain circuit). More recent findings strongly support this concept and has led to considering the mesencephalic PAG and large parts of the lateral and ventral mesencephalic tegmentum to belong to the limbic system.

I. Introduction

Hemiplegic patients with damage to corticobulbar fibers, resulting in a complete central paresis of the lower face on one side, are able to smile spontaneously (e.g., when they enjoy a joke). On the other hand, in cases with postencephalitic parkinsonism, patients are able to show their teeth, whistle, and frown (i.e., they have no facial palsy, but their emotions are not reflected in their countenance and they have a stiff, masklike facial expression [poker face]). Patients with irritative pontine lesions sometimes suffer from nonemotional laughter and crying, and patients with pseudobulbar palsy (e.g., with lesions in the mesencephalon) often suffer from uncontrollable fits of crying or laughter. Such fits are usually devoid of grief, joy, or amusement; they may even be accompanied by entirely incompatible emotions. Fits of crying and laughter may occur in the same patients; other patients may show only one of them. Crying and laughter belong to an expressive behavior, which in animals is called vocalization. In many different species, stimulation in the caudal part of the PAG produces vocalization. Recently, vocalization was based on a specific final common pathway, originating from a distinct group of neurons in the PAG that project to the nucleus retroambiguus (NRA), which in turn has direct access to all vocalization motoneurons. In all likelihood, in humans this projection forms the final common pathway for laughing and crying. The vocalization neurons in the PAG receive their afferents from structures belonging to the limbic system but not from the voluntary system. All this clinical and experimental evidence shows that a complete dissociation exists between the voluntary and emotional or limbic innervation of motoneurons.

Although the limbic system is well known to exert a strong influence on somatic and autonomic motoneurons, lesion–degeneration fiber-tracing studies did not reveal strong limbic projections to levels caudal to the mesencephalon. This led to the idea that the limbic pathways to caudal brainstem and spinal cord were multisynaptic (i.e., they involve several neurons in series). Since 1975, this view changed dramatically, mainly because new tracing techniques became available. The most interesting projections discovered with these new tracing techniques are the limbic system projections to the nucleus raphe magnus (NRM) and pallidus (NRP) as well as to the adjacent ventral part of the caudal pontine and medullary reticular formation. These findings are important because NRM, NRP, and adjoining reticular formation in turn project diffusely but very strongly to all parts of the gray matter throughout the length of the spinal cord. Therefore, the diffuse brainstem–spinal projections will be discussed in the framework of the descending limbic motor control systems.

II. Diffuse Pathways to the Spinal Cord

A. Projections from NRM, NRP, and Nuclei Raphe Obscurus and Ventral Part of the Caudal Pontine and Medullary Medial Reticular Formation

Retrograde and anterograde tracing studies indicate that the NRM, NRP, and nuclei raphe obscurus (NRO) with their adjoining reticular formation send fibers throughout the length of the spinal cord, giving off collaterals to all spinal levels. These descending systems are extremely diffuse and are not topographically organized. Furthermore, a strong heterogeneity exists in these projections, in which the NRM and adjoining reticular formation project to the dorsal horn (Fig. 1, left), and the NRP, NRO, and ventromedial medulla project to the intermediate zone and the ventral horn, with the motoneuronal cell groups (Fig. 1, right).

Physiological studies are consistent with the anatomy of the descending pathways outlined above. Electrical stimulation in the NRM and the adjacent ventral part of the upper medullary medial reticular formation inhibits neurons in the caudal spinal trigeminal nucleus and spinal dorsal horn, producing inhibition of nociception. The diffuse organization of the raphe and ventromedial medulla projections to the motoneuronal cell groups suggests that they do not steer specific motor activities such as movements of distal (arm, hand, or leg) or axial parts of the body but that they have a more global effect on the level of activity of the motoneurons.

Many different neurotransmitter substances exist in this area, of which serotonin (5HT) is the best known. Serotonin plays a role in the facilitation of motoneurons, probably directly by acting on the Ca^{2+} conductance or indirectly by reduction of K^+ conductance of the motoneuron membrane. Thus, serotonin enhances the excitability of the motoneurons for inputs from other sources such as red nucleus or motor cortex. Several other peptides are also present in the spinally projecting neurons in the ventromedial medulla and NRP and NRO. Many neurons contain substance P, thyrotropin-releasing hormone, somatostatin, methionine, and leucine-enkephalin, whereas relatively few neurons contain vasoactive intestinal peptide and cholecystokinin. Most of these peptides coexist to a variable extent with serotonin in the same neuron. It must be emphasized that a major portion of the diffuse descending pathways to the dorsal horn and the motoneuronal cell groups is not derived from serotonergic neurons. Possible neurotransmitter candidates for these pathways are acetylcholine and somatostatin, but γ-aminobutyric acid (GABA) may also play an important role. Colocalization of serotonin and GABA in spinal cord-projecting neurons in the ventral medulla in the rat has been demonstrated. This means that some terminals taking part in this diffuse descending system may have inhibitory as well as facilitatory effects on the postsynaptic element (i.e., the motoneuron), although the majority is probably either facilitatory or inhibitory. Spinal motoneurons display a bistable behavior; i.e., they can switch back and forth to a higher excitable level. Bistable behavior disappears after spinal transection but reappears after subsequent intra-

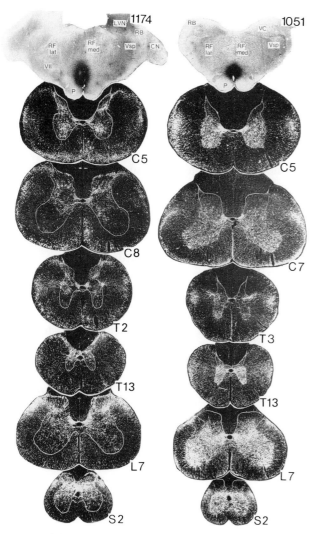

FIGURE 1 Brightfield photomicrographs of autoradiographs showing tritiated leucine injection sites in the raphe nuclei and darkfield photomicrographs showing the distributions of the labeled fibers in the spinal cord. On the left, an injection is shown in the caudal NRM and adjoining reticular formation. Note that labeled fibers are distributed mainly to the dorsal horn the intermediate zone and the autonomic motoneuronal cell groups. On the right, the injection is placed in the NRP and immediately adjoining tegmentum. Note that labeled fibers are not distributed to the dorsal horn but very strongly to the intermediate zone and autonomic and somatic motoneuronal cell groups. [Reproduced, with permission, from Holstege and Kuypers, 1982.]

The following abbreviations pertain to those used throughout the figures: AA, anterior amygdaloid nucleus; AC, anterior com-

missure; ACN, nucleus of the anterior commissure; AD, anterodorsal nucleus of the thalamus; AH, anterior hypothalamic area; AL, lateral amygdaloid nucleus; AM, anteromedial nucleus of the thalamus; AP, area postrema; AV, anteroventral nucleus of the thalamus; BC, brachium conjunctivum; BIC, brachium of the inferior colliculus; BL, basolateral amygdaloid nucleus; BM, basomedial amygdaloid nucleus; BNSTL, lateral part of the bed nucleus of stria terminalis; BNSTM, medial part of the bed nucleus of the stria terminalis; BP, brachium pontis; CA, central amygdaloid nucleus; CC, corpus callosum; Cd, caudate nucleus; CGL, lateral geniculate body; CGM, medial geniculate body; CI, capsula interna; CI, inferior colliculus (Fig. 9); CL, claustrum; CM, centromedian thalamic nucleus; CN, cochlear nuclei; CO, cortical amygdaloid nucleus; CR, corpus restiforme; CS, superior colliculus; CSN, nucleus raphe centralis superior; CU, nucleus cuneatus; CUN, cuneiform nucleus; DBV, nucleus of the diagonal band of Broca; DH, dorsal hypothalamic area; DMH, dorsomedial hypothalamic nucleus; EC, external cuneate nucleus; ECU, external cuneate nucleus; En, entopeduncular nucleus; F, fornix; G, nucleus gracilis; GP, globus pallidus; HC, hippocampus; IC, inferior colliculus; IN, interpeduncular nucleus; IO, inferior olive; IVN, inferior vestibular nucleus; KF, nucleus Kölliker-Fuse; LL, lateral lemniscus; LOTR, lateral olfactory tract; LP, lateral posterior nucleus of the thalamus; LRN, lateral reticular nucleus; LTF, lateral tegmental field; LVN, lateral vestibular nucleus; MB, mammillary body; MC, nucleus medialis centralis of the thalamus; MD, nucleus medialis dorsalis of the thalamus; MesV, mesencephalic trigeminal tract; ML, medial lemniscus; MLF, medial longitudinal fasciculus; mot V, motor trigeminal nucleus; MTF, medial tegmental field; MVN, medial vestibular nucleus; NCL, nucleus centralis lateralis; NLL, nucleus of the lateral lemniscus; NRM, nucleus raphe magnus; NRP, nucleus raphe pallidus; NRTP, nucleus reticularis tegmenti pontis; NTB, nucleus of the trapezoid body; NTS, nucleus tractus solitarius; nVII, facial nerve; OC, optic chiasm; OT, optic tract; P, pyramidal tract; PAG, periaqueductal gray; PBL, lateral parabrachial nucleus; PBM, medial parabrachial nucleus; PC, pedunculus cerebri; PEA, anterior part of periventricular hypothalamic nucleus; PON, pontine nuclei; PP, posterior pretectal nucleus; Pt, parataenial nucleus of the thalamus; Pu, putamen; PV, posterior paraventricular nucleus of the thalamus; PVA, paraventricular nucleus of the thalamus (anterior part); PVN, paraventricular hypothalamic nucleus; R, reticular nucleus of the thalamus; RE, nucleus reuniens of the thalamus; RF, reticular formation; RFlat, lateral reticular formation; RFmed, medial reticular formation; RM, nucleus raphe magnus; RN, red nucleus; RP, nucleus raphe pallidus; RST, rubrospinal tract; S, solitary complex; SC, nucleus subcoeruleus (Fig. 3F); SC, superior colliculus (Fig. 3A); SC, suprachiasmatic nucleus (Fig. 5); SI, substantia innominata; SM, stria medullaris; SN, substantia nigra; SO, superior olivary complex; SON, supraoptic nucleus; ST, subthalamic nucleus; STT, stria terminalis; TMT, mammillothalamic tract; VA, ventroanterior nucleus of the thalamus; VB, ventrobasal complex of the thalamus; VC, vestibular complex; VL, ventrolateral nucleus of the thalamus; VM, ventromedial nucleus of the thalamus; VMH, ventromedial nucleus of the hypothalamus; VTN, ventral tegmental nucleus; ZI, zona incerta; III, oculomotor nucleus; IV, trochlear nucleus; Vm, motor trigeminal nucleus; V princ, principal trigeminal nucleus; Vsp, spinal trigeminal complex; V sp.cd., spinal trigeminal complex pars caudalis; VI, abducens nucleus; VII, facial nucleus; Xd, dorsal vagal nucleus; XII, hypoglossal nucleus venous injection of the serotonin precursor 5-hydroxytryptophan. Thus, intact descending pathways are essential for this bistable behavior of motoneurons, and serotonin is one of the neurotransmitters involved in switching to a higher level of excitation. GABA may possibly be involved in switching to a lower level of excitation.

In summary, the diffuse descending pathways originating in the ventromedial medulla, including the NRP and NRO, have very general and diffuse facilitatory or inhibitory effects on motoneurons and probably also on interneurons in the intermediate zone. Although most of the terminals have either a facilitatory or an inhibitory function, recent evidence suggests that terminals also exist with both facilitatory and inhibitory functions.

B. Projections from the Dorsolateral Pontine Tegmental Field

Many neurons in the locus coeruleus and/or nucleus subcoeruleus and ventral part of the parabrachial nuclei project diffusely to all parts of the spinal gray matter. Many neurons in the locus coeruleus, subcoeruleus, and parabrachial nuclei contain noradrenaline or acetylcholine, but not both. Electrical stimulation in the area of the locus coeruleus and subcoeruleus produces a decrease in input resistance and a concurrent nonselective enhancement in motoneuron excitability, indicative of an overall facilitation of motoneurons. Furthermore, evidence indicates that in the rat noradrenergic fibers derived from the locus coeruleus and descending via the ventrolateral funiculus have an inhibitory effect on nociception. Thus, neurons in the area of the locus coeruleus and subcoeruleus have an effect on the spinal cord, which is strikingly similar to that obtained after stimulation in NRM and NRP and their adjacent reticular formation.

C. Projections from the Rostral Mesencephalon and Caudal Hypothalamus

Dopamine-containing neurons located in the border region of the rostral mesencephalon and dorsal and posterior hypothalamus, extending dorsally along the paraventricular gray of the caudal thalamus, project throughout the whole gray matter at any level of the spinal cord. The distribution of the dopaminergic fibers (using the neurotransmitter dopamine) in the spinal gray strongly resembles that of noradrenergic fibers (using norepinephrine) in the

spinal cord. Functionally, there is also a resemblance between noradrenergic and dopaminergic fiber projections to the spinal cord. Infusion of dopamine in the spinal cord increases (sympathetic) motoneuron activity and has an inhibitory effect on noxious input to the spinal cord.

III. Projections from the Mesencephalon to Caudal Brainstem and Spinal Cord

The PAG and adjoining tegmental field project to a large number of structures in the caudal brainstem and spinal cord. In this section, a short description of some of these projections (summarized in Fig. 2) is given.

A. Descending Projections to the NRM, NRP and Ventral Part of the Caudal Pontine and Medullary Medial Tegmentum

Anterograde and retrograde tracing studies indicate that an enormous number of HRP-labeled neurons in the PAG and laterally and ventrolaterally adjoining areas project in the same basic pattern to the NRM, NRP, and ventral part of the caudal pontine and medullary medial tegmentum (Fig. 3). On their way to the medulla, they give off fibers to the area of the locus coeruleus and nucleus subcoeruleus. Neurons in the ventrolateral portion of the caudal PAG and the ventrally adjoining mesencephalic teg-

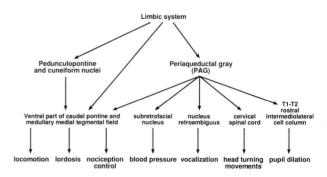

FIGURE 2 Schematic overview of the descending projections from the PAG and pedunculopontine and cuneiform nuclei to different regions of the caudal brainstem and spinal cord. The functions in which each of the projections might be involved are also indicated. It should be emphasized that these functional interpretations are only tentative.

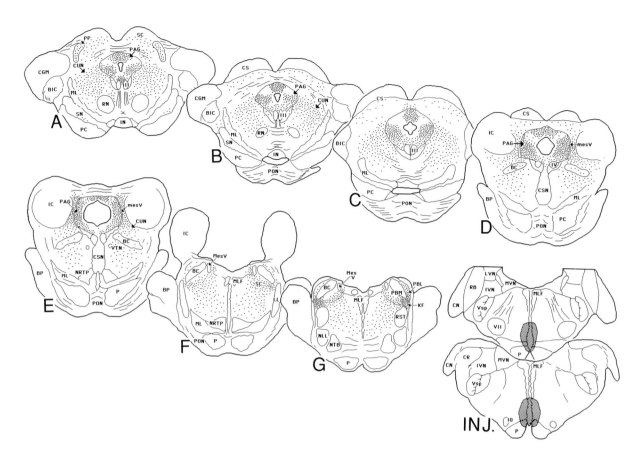

FIGURE 3 Schematic drawings of horseradish peroxidase (HRP)-labeled neurons in mesencephalon and pons after injection of HRP in the NRM–NRP region. Note the dense distribution of labeled neurons in the PAG (except its dorsolateral part) and the tegmentum ventrolateral to it. Note also the distribution of labeled neurons in the area of the ventral parabrachial nuclei and the nucleus Kölliker–Fuse [Reproduced, with permission, from Holstege, 1988].

mentum send fibers to the NRP. A mediolateral organization exists within the descending mesencephalic pathways. The main projection of the medially located neurons (i.e., neurons in the medial part of the dorsal PAG) is to the medially located NRM and immediately adjacent tegmentum. On the other hand, neurons in the lateral PAG, the laterally adjacent tegmentum, and the intermediate and deep layers of the superior colliculus project mainly laterally to the ventral part of the caudal pontine and upper medullary medial tegmentum, with virtually no projections to the NRM. Figure 4 schematically shows this mediolateral organization in the ipsilaterally descending pathways from the dorsal mesencephalon.

1. Involvement of the Descending Mesencephalic Projections in Control of Nociception

In animals as well as in humans the PAG is well known for its involvement in the supraspinal control of nociception. The strong impact on nociception is partly mediated via its projections to the NRM and adjacent reticular formation, because in cases with reversible blocks of the NRM and adjacent tegmentum, PAG stimulation results in reduced analgesic effects. However, the analgesic effects do not completely disappear after blocking the NRM and adjacent tegmentum, which suggests that other brainstem regions also play a role.

2. Involvement of the Descending Mesencephalic Projections in the Lordosis Reflex

Stimulation in the PAG also facilitates the lordosis reflex. Lordosis, a curvature of the vertebral column with ventral convexity, is an essential element of female copulatory behavior in rodents. The lordosis reflex is facilitated by stimulation of the ventromedial hypothalamic nucleus and the PAG.

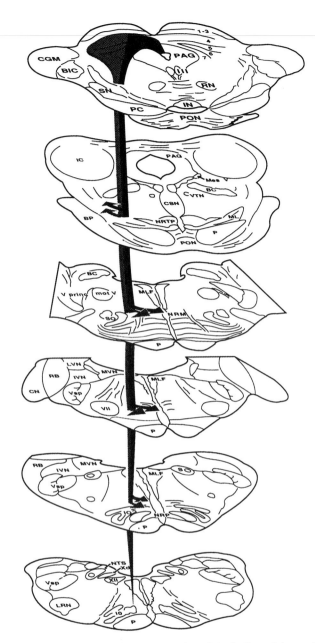

FIGURE 4 Schematic representation of the ipsilateral descending pathway, originating from the intermediate and deep layers of the superior colliculus and dorsal PAG. The mediolateral organization of this descending system is illustrated. The lateral (gray) component projects to the lateral aspects of the ventral part of the medial tegmentum of caudal pons and medulla oblongata. The medial (black) component projects to the medial aspects of the medial tegmentum, including the NRM. A similar mediolateral organization exists for the descending pathways originating in the more ventral part of the mesencephalic tegmentum. [Reproduced, with permission, from Cowie and Holstege, 1991 (submitted).]

Stimulation of the L1–S1 dermatomes can elicit the lordosis reflex, but until recently it was thought that the reflex could only take place when there was a facilitatory forebrain influence. However, it was recently demonstrated that the lordosis reflex can also be elicited in decerebrate rats. Also, descending fibers in the ventrolateral funiculus play a role in the facilitation of the reflex, but these fibers do not originate from neurons in the ventromedial hypothalamic nucleus or the PAG, because none of the two structures project directly to the lumbosacral spinal cord.

Perhaps the lordosis reflex should be considered as a spinal reflex, in which the L1–S1 cutaneous input from flank, rump, tailbase, and perineum serves as the afferent loop, and the fibers of the back and axial muscle motoneurons form the efferent loop. Both loops are interconnected by spinal interneurons and short and long propriospinal pathways. However, the cutaneous afferents produce lordosis behavior only when the membrane excitability of the motoneurons is high. This level of excitability is determined by descending pathways, which originate in the NRP and adjoining ventromedial medullary tegmental field, travel via the ventrolateral funiculus, and project diffusely to all inter- and motoneuronal cell groups in the ventral horn throughout the length of the spinal cord (see Section II). The NRP and its adjoining ventral part of the medial tegmental field receive their afferents from various structures in the medial limbic system, such as the PAG and anteromedial hypothalamus, but not from the ventromedial hypothalamic nucleus (see Fig. 5). Thus, a concept is put forward in which the ventromedial hypothalamic nucleus controls the lordosis reflex by means of its projections to the ventral part of the medullary medial tegmentum, using the anteromedial hypothalamus and/or the PAG as relay structures. The medullary ventromedial tegmentum increases the excitability of the motoneurons to such a level that cutaneous L1–S1 afferent stimulation, which is otherwise ineffective, results in lordosis.

3. Involvement of the Descending Mesencephalic Projections in Locomotion
Just lateral to the brachium conjunctivum, just ventral to the cuneiform nucleus, and just rostral to the parabrachial nuclei the so-called pedunculopontine nucleus is located, which contains cholinergic neurons. Stimulation in the pedunculopontine nu-

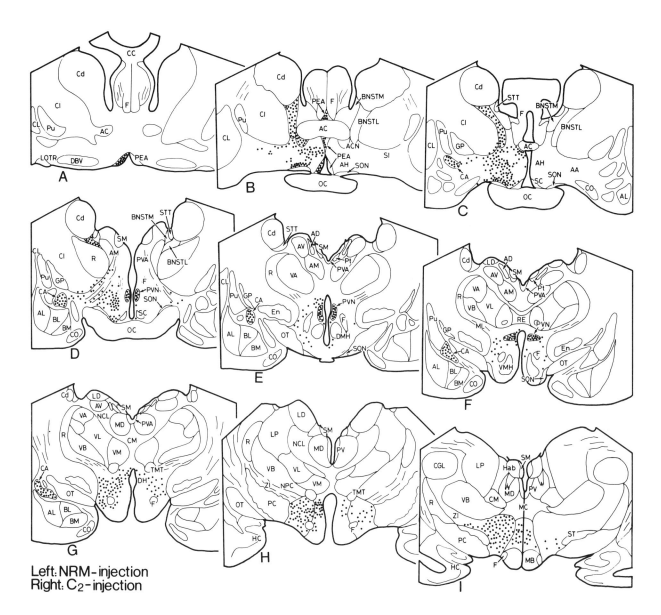

Left: NRM-injection
Right: C₂-injection

FIGURE 5 Schematic drawing of horseradish peroxidase (HRP)-labeled neurons in the hypothalamus, amygdala, and bed nucleus of the stria terminalis. On the left in each drawing the pattern of distribution of labeled neurons after a large injection of HRP in the NRM, rostral NRP, and adjoining tegmentum is indicated. On the right, the pattern of distribution of HRP-labeled neurons after hemi-infiltration of HRP in the C₂ spinal segment is shown. [Reproduced, with permission, from Holstege, 1987].

cleus induces locomotion in cats, which is the reason that this area is also termed the mesencephalic locomotor region (MLR). The MLR not only comprises the pedunculopontine nucleus but extends into the cuneiform nucleus, which is located just dorsal to the pedunculopontine nucleus.

The descending projections from this area are organized similar to those from the PAG and adjacent tegmentum. The mainly ipsilateral fiberstream first descends laterally in the mesencephalon and upper pons and then gradually shifts medially to terminate bilaterally, but mainly ipsilaterally in the ventral part of the caudal pontine and medullary medial tegmental field. Electrical stimulation or injection of cholinergic agonists (with the effect of the neurotransmitter acetylcholine) in this same caudal brainstem area resulted in locomotion, which could control or override the stepping frequency induced by the MLR. Thus, locomotion, elicited in the MLR, is based on the MLR projections to the medial part of the ventral medullary medial tegmentum and on the diffuse projections from the latter area to the spinal cord.

The afferent connections of the MLR are derived from lateral parts of the limbic system, such as the bed nucleus of the stria terminalis, central nucleus of the amygdala, and lateral hypothalamus. Strong projections are also derived from the entopeduncular nucleus, the subthalamic nucleus, and the substantia nigra pars reticulata, but motor cortex projections to the MLR are very scarce. These findings indicate that the MLR is influenced by extrapyramidal and lateral limbic structures and virtually not by somatic motor structures.

B. PAG Projections to the Ventrolateral Medulla; Involvement in Blood Pressure Control

Neurons in the rostral part of the ventrolateral tegmental field of the medulla (subretrofacial nucleus) are essential for the maintenance of the vasomotor tone and reflex regulation of the systemic arterial blood pressure. Neurons in the rostral part of the subretrofacial nucleus project specifically to the neurons in the intermediolateral cell column (IML), innervating the kidney and adrenal medulla, whereas neurons in the caudal part of it innervate more caudal parts of the IML, with neurons innervating the hindlimb. Recently, Bandler et al (1991) has shown that neurons in the dorsal portions of the caudal half of the PAG have an excitatory effect on the neurons in the subretrofacial nucleus (increase in blood pressure), whereas neurons in the ventral part of the PAG have an inhibitory effect (decrease of blood pressure). The same authors have also shown that neurons in the subtentorial portion of the PAG project to the rostral part of the subretrofacial nucleus and send fibers to the IML motoneurons that innervate the kidney and adrenal medulla. On the other hand, neurons in the caudal part of the pretentorial PAG project to the caudal subretrofacial nucleus, which in turn project to IML motoneurons innervating the hindlimb. In conclusion, a precise organization exists in the mesencephalic control of blood pressure in different parts of the body. All these projections take part in a descending system involved in the elaboration of emotional motor activities.

C. PAG Projections to the NRA; Involvement in Vocalization

In many different species, from leopard frog to chimpanzee, stimulation in the caudal PAG results in vocalization (i.e., the nonverbal production of sound). In humans, laughing and crying are examples of vocalization. A specific group of neurons in the lateral and to a limited extent in the dorsal part of the caudal PAG send fibers to the NRA in the caudal medulla. The cell group in the PAG differs from the smaller cells projecting to the raphe nuclei and adjacent tegmentum or the larger cells projecting to the spinal cord. The NRA in turn projects to the somatic motoneurons innervating the pharynx, soft palate, intercostal and abdominal muscles, and probably the larynx. Direct PAG projections to these somatic motoneurons do not exist. In all likelihood, the projection from the PAG to the NRA forms the final common pathway for vocalization (Fig. 6). This finding is important, because it shows that a specific expressive motor activity (fixed action pattern) such as vocalization is based on a distinct descending pathway, suggesting that all the other specific motor activities displayed during expressive behavior are based on separate descending pathways.

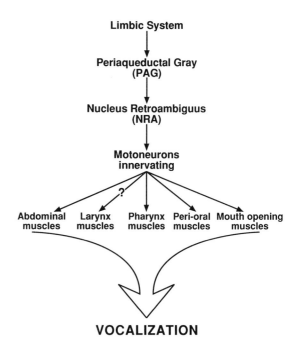

FIGURE 6 Schematic representation of the pathways for vocalization from the limbic system to the vocalization muscles. [Reproduced, with permission, from Holstege, 1989].

D. PAG Projections to the Spinal Cord

Only limited PAG projections to the spinal cord exist. Some neurons in the lateral PAG and laterally adjacent tegmentum send fibers through the ipsilateral ventral funiculus of the cervical spinal cord to terminate in laminae VIII and the adjoining part of VII. Few fibers descend ipsilaterally in the lateral funiculus to terminate in the T1-T2 IML. The projections to the spinal cord may play a role in the defensive behavior observed while stimulating the PAG. For example, the projection to the medial part of the intermediate zone of the cervical cord may be involved in the contralateral head turning movements as part of defensive behavior, whereas the projection to the T1-T2 IML may produce pupil dilation.

IV. Projections from the Hypothalamus to Caudal Brainstem and Spinal Cord

The descending hypothalamic projection systems differ greatly, depending on which part of the hypothalamus is considered. In this section, the hypothalamus will be subdivided into the anterior paraventricular and lateral hypothalamus.

A. Projections from the Anterior Hypothalamus–Preoptic Area

Neurons in the anterior hypothalamus–preoptic area project strongly to the caudal brainstem but not to the spinal cord (Fig. 5). Neurons in the medial part of the anterior hypothalamus–preoptic area project (via a medial fiber stream) to the PAG, the dorsal and superior central raphe nuclei in the pontine tegmentum, and the ventromedial tegmentum of caudal pons and medulla, including the NRM and NRP. The anterior hypothalamus–preoptic area receives afferent fiber from many sources, including caudal brainstem structures such as the lateral parabrachial nucleus, the solitary nucleus, and neurons in the ventrolateral medulla, which suggests that it is involved in cardiovascular regulation. Moreover, application of cholinergic drugs in the anterior hypothalamus results in an emotional aversive response, which includes defense posture and autonomic (e.g., cardiovascular) manifestations.

B. Projections from the Paraventricular Hypothalamic Nucleus

The paraventricular hypothalamic nucleus (PVN) is best known for its projections to the hypophysis. More recently it has been shown that PVN neurons other than those projecting to the hypophysis project to the caudal brainstem and spinal cord. These fibers descend via the medial forebrain bundle and more caudally via a well-defined pathway through the ventrolateral brainstem into the lateral and dorsolateral funiculus of the spinal cord throughout its total length. Via this pathway, the PVN sends fibers to the NRM, rostral NRP and adjoining reticular formation, and specific parts of the medullary lateral tegmental field, such as the noradrenergic brainstem nuclei A1, A2, and A5 areas, the parasympathetic motoneurons in the salivatory nuclei, the rostral half of the solitary nucleus, in all parts of the dorsal vagal nucleus, and in the area postrema (Fig. 7). In the spinal cord (Fig. 8), the PVN projects bilaterally, but mainly ipsilaterally, to the marginal layer of the caudal spinal trigeminal nucleus and to laminae I and X throughout the length of the spinal cord, the thoracolumbar (T1–L4) intermediolateral (sympathetic) motoneuronal cell group, and to the sacral intermediomedial and intermediolateral (parasympathetic) motoneuronal cell groups. Finally, the PVN projects to the nucleus of Onuf, which contains motoneurons innervating the pelvic floor muscles, including the urethral and anal sphincters. The function of this pathway is unknown, but one could speculate that, in the light of the diffuseness of the projection (the PVN projects to all autonomic neurons in the brainstem and spinal cord), it is a general function, similar, for example, to the hormone adrenocorticotropin. The PVN contains a large number of transmitter substances, and evidence indicates that the PVN pathway to the caudal brainstem and spinal cord contains vasopressin and oxytocin as neurotransmitters. However, other as yet unidentified transmitter substances also play a role in this pathway. The PVN is believed to play an important role in cardiovascular regulation as well as in the feeding mechanism. The neural circuitry for this feeding system is believed to start in the noradrenergic neurons of the locus coeruleus that project to the PVN. The PVN neurons in turn innervate neurons in the dorsal vagal nucleus, which play a crucial role in the noradrenaline-elicited eating response.

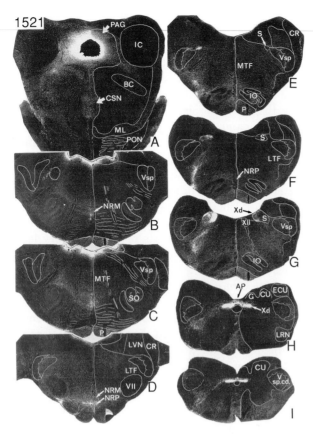

FIGURE 7 Darkfield photomicrographs of nine brainstem sections of a cat with an injection of ³H-leucine in the area of the PVN of the hypothalamus. Note the distinct descending pathway in the area next to the pyramidal tract and its fiber distribution to the NRM–NRP, dorsal vagal nucleus, area postrema, and rostral solitary complex. [Reproduced, with permission, from Holstege, 1987].

C. Projections from the Medial Part of the Posterior Hypothalamic Area

The posterior hypothalamic area sends fibers via a medial pathway to the caudal raphe nuclei and adjoining reticular formation and to the spinal cord. The caudal hypothalamic projections to the NRM are weaker and those to the NRP are much stronger than other hypothalamic projections to this area. In the spinal cord, caudal hypothalamic fibers terminate in the upper thoracic intermediolateral cell column and in lamina X.

The dorsomedial region of the caudal hypothalamus plays an important role in temperature regulation, and it contains the primary motor center for the production of shivering. Shivering is an involuntary response of skeletal muscles, which are usually under voluntary control, and all skeletal muscle

groups can participate. The strong caudal hypothalamic projections to the NRP and adjacent tegmentum possibly play an important role in this "shivering pathway," similar to its role in the descending pathways involved in locomotion, which is also a rhythmical activity.

D. Projections from the Lateral Hypothalamic Area

Functional and anatomical studies on the lateral hypothalamus have always been difficult, because the fibers of the medial forebrain bundle pass through it. This important fiber bundle not only contains fibers originating in the lateral hypothalamus but also in

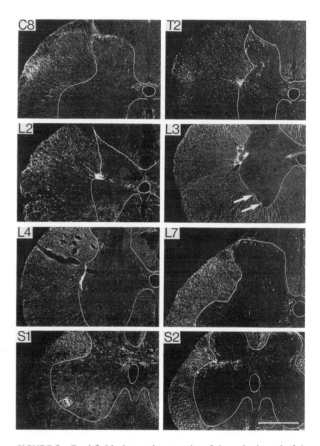

FIGURE 8 Darkfield photomicrographs of the spinal cord of the same cat as illustrated in Figure 5, with a ³H-leucine injection in the area of the PVN of the hypothalamus. Note the projection to lamina I (C8, T2, and L7), the sympathetic intermediolateral cell group (T2, L2, L3, and L4), the nucleus of Onuf (S1), and the parasympathetic intermediomedial and intermediolateral cell group (S2). The arrows in L3 probably indicate projections to distal dendrites of the motoneurons located in the sympathetic intermediolateral cell group. [Reproduced, with permission, from Holstege, 1987].

many other areas, and stimulation or lesions in this area not only affect lateral hypothalamic neurons but also fibers derived from many other limbic structures. The lateral hypothalamus sends fibers to the PAG, the cuneiform nucleus, the parabrachial nuclei and nucleus Kölliker-Fuse, the nucleus subcoeruleus, the caudal pontine and medullary lateral tegmental field, and the ventral part of the caudal pontine and medullary medial reticular formation. Few fibers terminate in the area of the NRM and none in the NRP. Some fibers terminate in the periphery of the dorsal vagal nucleus and in the rostral half of the solitary nucleus. The rostral portion of the lateral hypothalamus also projects strongly to Barrington's nucleus or the M-region, and stimulation of this part of the hypothalamus produces micturitionlike bladder contractions. Only the caudal portion of the lateral hypothalamus sends fibers throughout the length of the spinal cord via the lateral and dorsolateral funiculi to the intermediate zone, lamina X, and the thoracolumbar sympathetic IML. In summary, the lateral hypothalamus has direct access to autonomic motoneurons in brainstem and spinal cord, and indirect access, via premotor interneurons, to the somatic motoneurons of brainstem and spinal cord.

Many of the brainstem motoneurons are involved in activities such as swallowing, chewing, and licking. Interestingly, the lateral hypothalamus is involved in feeding and drinking behavior and salivation. It is probably also involved in cardiovascular control and defense behavior (see Section V).

V. Projections from Amygdala and Bed Nucleus of the Stria Terminalis to Caudal Brainstem and Spinal Cord

The central (CA) and medial amygdaloid nuclei and the lateral (BNSTL) and medial parts of the bed nucleus of the stria terminalis are considered to form a single anatomic entity. The projections from CA and BNSTL to the caudal brainstem are virtually identical. Both structures send many fibers to the lateral hypothalamic area, and via the medial forebrain bundle, into the lateral part of the mesencephalon, pons, and medulla oblongata (Fig. 9). At mesencephalic levels fibers terminate in the PAG (except its dorsolateral part), the ventrolaterally adjoining nucleus cuneiformis and pedunculopontine nucleus, and the mesencephalic lateral tegmental

field. In the pons, fibers terminate laterally in the tegmentum (i.e., the medial and lateral parabrachial nuclei, the nucleus Kölliker-Fuse, the nucleus subcoeruleus and possibly the locus coeruleus). At caudal pontine and medullary levels the main termination area is the lateral tegmental field, including the rostral and caudal parts of the solitary nucleus and the peripheral parts of the dorsal vagal nucleus. An exception is the fibers that branch off from the lateral descending fiber bundle to terminate in the ventral part of the caudal pontine and upper medullary medial tegmentum. A few fibers terminate in the NRM, but none in the NRP (Fig. 9). Both CA and BNSTL send a few fibers to the intermediate zone of the C1 spinal cord but not beyond that level.

Neurons in CA and BNSTL receive many afferent fibers from other (basolateral and basomedial) amygdaloid nuclei, whose connections are not re-

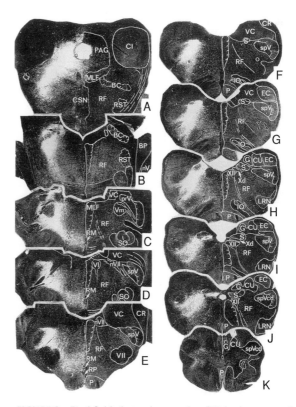

FIGURE 9 Darkfield photomicrographs of 11 brainstem sections of a cat with an injection of ^{3}H-leucine in the bed nucleus of the stria terminalis, which is almost identical to the projections from the central nucleus of the amygdala and the lateral hypothalamus. Note the strong projection to the PAG, with the exception of its dorsolateral part. Note also the strong projection to the bulbar lateral tegmental field and the projection to the ventral part of the caudal pontine and upper medullary medial tegmentum. [Reproduced, with permission, from Holstege (1991).

Descending Pathways from Limbic System to Caudal Brain Stem and Spinal Cord

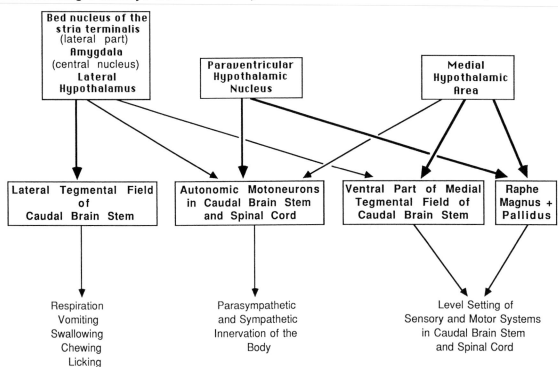

FIGURE 10 Descending pathways from limbic system to caudal brain stem and spinal cord. Schematic overview of the mediolateral organization of the limbic system pathways to brainstem and spinal cord and its possible functional implications. The strongest projections are indicated by thick arrows. [Reproduced, with permission, from Holstege, 1988].

ciprocal. Apparently, both CA and BNSTL serve as "output nuclei" for other parts of the amygdala–bed nucleus of the stria terminalis complex to reach the caudal brainstem. A great similarity exists between the caudal brainstem projections originating in the CA and BNSTL on the one hand and the lateral hypothalamic area on the other. In addition, all three areas have very strong mutual connections.

The direct projections from CA, BNSTL, and the lateral hypothalamus to the caudal brainstem lateral tegmental field may form the anatomic framework of the final output of the defense response of the animal. Electrical stimulation in the amygdala, bed nucleus of the stria terminalis, lateral hypothalamus, and PAG elicits defensive behavior. In fact, a column of electrical stimulation sites exists from CA, BNST, lateral hypothalamus, and PAG through the lateral mesencephalic tegmentum into the lateral tegmentum of the caudal brainstem, which elicits defensive behavior. The initial phase of the defense response in cats is arrest of all spontaneous ongoing activities, and the whole attitude of the animal changes to one of attention. The arousal is followed by orienting or searching movements toward the contralateral side, frequently accompanied by sniffing, swallowing, chewing, and twitching of the ipsilateral facial musculature. Later in the

defense reaction, the cat retracts its head and crouches with the ears flattened to a posterior position. The cat growls or hisses, the pupils are dilated and there is piloerection, elevation of blood pressure with bradycardia, increased rate of breathing, alteration of gastric motility, and secretion. On stronger stimulation, an "affective" attack may take place, in which the cat strikes with its paw with claws unsheathed, in a series of swift, accurate blows. If the stimulus continues, the cat will bite savagely. Many of the activities in the beginning of the defense response are coordinated in the caudal brainstem lateral tegmental field. The observation that part of this behavior appears to be ipsilateral corresponds with the predominantly ipsilateral projection of CA, BNSTL, and lateral hypothalamus to the caudal brainstem lateral tegmentum.

Figure 10 gives a schematic overview of the descending projections to the caudal brainstem from hypothalamus, amygdala, and BNST. Similar to the descending projections from the mesencephalon, a mediolateral organization exists within this descending system in which the medial hypothalamus forms the medial, and the lateral hypothalamus, amygdala, and BNST the lateral component. The PVN, with its direct projections to all preganglionic (sympathetic and parasympathetic) motoneurons in brainstem and spinal cord, occupies a separate position within this framework.

VI. Projections from Prefrontal Cortex to Caudal Brainstem and Spinal Cord

The prefrontal cortex projects directly to the caudal brainstem. In the rat, the medial frontal cortex sends fibers to the solitary nuclei (NTS), the dorsal parts of the parabrachial nuclei, the PAG, and the superior colliculus, while the insular cortex projects also to the NTS and PAG. Electrical stimulation of the rat's insular cortex leads to elevation of arterial pressure and cardioacceleration. In the cat, the orbital gyrus, anterior insular cortex, and infralimbic cortex project to the nucleus tractus solitarius. In the monkey, the dorsolateral and dorsomedial prefrontal cortex projects to the locus coeruleus and nucleus raphe centralis superior.

VII. Conclusions

Holstege has brought up the concept of a third motor system, which represents the descending pathways from the limbic system to caudal brainstem and spinal cord. In this concept (Fig. 11), the second motor system is formed by the pathways that originate in the motor cortex, red nucleus, and brainstem nuclei such as vestibular nuclei, dorsal part of the medial tegmentum, interstitial nucleus of Cajal, etc. The first motor system in this concept is formed by the interneurons in brainstem and spinal cord, projecting to the motoneurons (Fig. 11).

The third motor system, which consists of thin fibers, is discovered only recently. The development of the immunohistochemical techniques has revealed a large number of neurotransmitters or neuromodulators within the central nervous system. Interestingly, with the exception of acetylcholine, glutamate, and aspertate, all these new monoamines and peptides were found in the third motor system.

The third motor system does not overlap in its

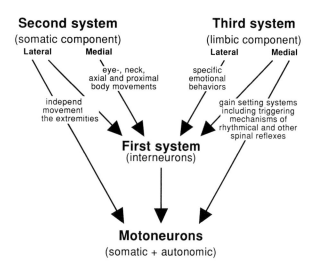

FIGURE 11 Schematic overview of the three subdivisions of the motor system [Reproduced, with permission, from Holstege, 1990].

projections with the second motor system. An exception to this rule is the monoaminergic projections that originate in the raphe nuclei and locus coeruleus–subcoeruleus complex and send fibers to many structures in the central nervous system, including some belonging to the second system.

A mediolateral organization is present within the third motor system. The medial component originates in the medial portions of hypothalamus and in the mesencephalon and terminates in the area of locus coeruleus–subcoeruleus and in the ventral part of the caudal pontine and medullary medial tegmental field. The latter structures determine the final output of this system. The lateral component originates laterally in the limbic system (i.e., in the lateral hypothalamus, central nucleus of the amygdala, and bed nucleus of the stria terminalis). These structures project to the lateral tegmental field of caudal pons and medulla but not to the somatic motoneurons in this area. How far the prefrontal cortex plays a role within these systems remains to be determined. This general subdivision into medial and lateral third motor systems has some exceptions. (1) Within the PAG and lateral adjacent tegmentum, some specific groups of neurons are probably related to specific functions such as vocalization, head movements involves in emotional behavior, or blood pressure control. They may serve as final common pathways for especially the lateral component of the third motor system. (2)

Some of the fibers of the lateral component of the third descending system terminate in the ventromedial tegmentum at levels around the facial nucleus. Neurons in this area in turn project diffusely to the dorsal horn of the spinal cord. Via these fibers, the lateral component structures may have some control over nociception.

The functional implications of the third system motor pathways differ, depending on whether they belong to the medial or lateral system. The medial system, via its projections to the locus coeruleus–nucleus subcoeruleus and NRM and NRP–NRO and the diffuse coeruleo- and raphe-spinal pathways, has a global effect on the level of activity of the somatosensory and motoneurons in general by changing their membrane excitability. In other words, the emotional brain has a great impact on the sensory as well as on the motor system. In both systems, it sets the gain or level of functioning of the neurons. The emotional state of the individual determines this level. For example, many forms of stress such as aggression, fear, and sexual arousal are well known to induce analgesia, while the motor system is set at a "high" level and motoneurons can easily be excited by the second motor system. In this concept, the brainstem structures, which project diffusely to all parts of the spinal cord, can be considered as tools for the limbic system controlling spinal cord activity. The diffuse descending system is also used to trigger rhythmical (locomotion, shivering) or other (lordosis), in essence, spinal reflexes. Whether functions such as locomotion, shivering, and lordosis use different or the same diffuse pathways from the caudal brainstem to the spinal cord is not yet clear. If they use the same pathways, the differences lie in the function of the spinal generators for each of these functions. The lateral component of the third motor system projects to the caudal brainstem lateral tegmental field, which contains first motor system interneurons involved in specific functions such as respiration, vomiting, swallowing, chewing, and licking. These activities are displayed in the beginning of flight or defense response and can be easily elicited by stimulation of the lateral parts of the limbic system. Therefore, the lateral component of the third motor systems seemingly is involved in more specific activities related to emotional behavior.

Bibliography

Bandler, R., Carrive, P., and Zhang, S. P. (1991). Integration of somatic and autonomic reactions within the midbrain periaqueaductal grey: Viscerotopic, somatotopic and functional organization. *In* "Role of the Forebrain in Sensation and Behavior" (G. Holstege, ed.). *Progr. Brain Res.*, Elsevier, Amsterdam.

Besson, J.-M., and Chaouch, A. (1987). Peripheral and spinal mechanisms of nociception. *Physiol. Rev.* **67**, 67–186.

Heimer, L., de Olmos, J., Alheid, G. F., and Zaborsky, L. (1991). "Peristroika" in the basal forebrain; Opening the border between neurology and psychiatry. *In* "Role of the Forebrain in Sensation and Behavior" (G. Holstege, ed.). *Progr. Brain Res.*, Elsevier, Amsterdam (in press).

Holstege, G. (1987). Some anatomical observations on the projections from the hypothalamus to brainstem and spinal cord: An HRP and autoradiographic tracing study in the cat. *J. Comp. Neurol.* **260**, 90–126.

Holstege, G. (1988). Direct and indirect pathways to lamina I in the medulla oblongata and spinal cord of the cat. *In* "Descending Brainstem Controls of Nociceptive Transmission," Vol. 77 (H. L. Fields and J. M. Besson, eds.), pp. 47–94. *Progr. Brain Res.* Elsevier, Amsterdam.

Holstege, G. (1989). Anatomical study of the final common pathway for vocalization in the cat. *J. Comp. Neurol.* **284**, 242–252.

Holstege, G. (1991). An anatomical review of the descending motor pathways and the spinal motor system. Limbic and non-limbic components. *In* "Role of the Forebrain in Sensation and Behavior" (G. Holstege, ed.). *Progr. Brain Res.*, Elsevier, Amsterdam.

Holstege, G., Griffiths, D., De Wall, H., and Dalm, E. (1986). Anatomical and physiological observations on supraspinal control of bladder and urethral sphincter muscles in the cat. *J. Comp. Neurol.* **250**, 449–461

Kuypers, H. G. J. M. (1981). Anatomy of the descending pathways. *In* "Handbook of Physiology, Section I, The Nervous System, Vol. II, Motor Systems (R. E. Burke, ed.), pp. 597–666. Washington American Physiological Society, Washington, D.C.

MacLean, P. D. (1952). Some psychiatric implications of physiological studies on frontotemporal portion of limbic system. *EEG Clin. Neurophysiol.* **4**, 407–418.

Nauta, W. J. H. (1958). Hippocampal projections and related neural pathways to the mid-brain in the cat. *Brain* **80**, 319–341.

Nieuwenhuys, R., Voogd, J., and Van Huijzen, C. (1988). "The Human Central Nervous System," 3rd. rev. ed. Springer, Berlin, Heidelberg, New York, and Tokyo, 437 pp.

Price, J. L., Russchen, F. T., and Amaral, D. G. (1987). The limbic region. II: The amygdaloid complex. *In* Handbook of Chemical Neuroanatomy. Vol. 5. Integrated Systems of the CNS. Part I, Hypothalamus, Hippocampus, Amygdala" (A. Björklund, T. Hökfelt, and L. W. Swanson, eds.), pp. 279–388. Retina Elsevier Science Publishers, Amsterdam.

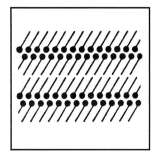

Lipids

DONALD M. SMALL AND R. A. ZOELLER, *Boston University Medical Center*

Glossary

Amphiphile Molecule having a hydrocarbon (lipophilic) part and a water soluble (hydrophilic) part; these molecules have physical properties which attract them to both lipid and water and thus tend to form interfaces between hydrocarbon and water

Hydrocarbon chain Elongated chain of methylene groups—CH_2—CH_2—CH_2—CH_2—that are common to most lipid molecules

Lipoproteins Small (80–2000 Å diameter) physical aggregates of lipids and proteins that circulate in the plasma and lymph of animals. Lipoproteins target important lipids and lipid soluble vitamins to tissues, which require them for cell repair, maintenance, and growth

Liquid crystals States of matter with physical properties between crystals and liquids. Crystals have long range order in all three spatial dimensions and liquids have no long range order. Liquid crystals can have long range order in the one or two dimensions, and thus are intermediate states of matter between crystals and liquids

Micellar solution Thermodynamically stable, isotropic liquid composed of small aggregates of lipids in equilibrium with monomers

Pheromones Organic molecules, often volatile, that are secreted by some animals and sensed by others. Pheromones provide a mechanism for transmitting chemical signals through air

Subcell A term used to define the geometry of hydrocarbon chain packing in lipids. Since long hydrocarbon chains consist of many repeats of—CH_2—CH_2—it was recognized that a repeating unit two carbons long could be related geometrically to its nearest neighbors to define a 3-dimensional repeating unit of hydrocarbon chain packing within the whole unit cell. Common subcells include triclinic parallel, orthorhombic perpendicular, and hexagonal

Unit cells Smallest structural unit that repeats itself within a crystalline structure. Geometrically the unit cells can be divided into seven general categories depending upon the dimensions and symmetry; these include cubic, tetragonal, orthorhombic, monoclinic, triclinic, hexagonal, and rhombohedral

X-ray diffraction Technique used to define crystal and liquid crystalline structures

A LIPID can be defined as any molecule of intermediate molecular weight between 100 and 5,000 that contains a substantial portion of aliphatic or aromatic hydrocarbon. Included are hydrocarbons, steroids, soaps, detergents, and more complex molecules such as waxes, triacylglycerols (fats and oils), phospholipids, sphingolipids, fat soluble vitamins and lipopolysaccharides.

I. Introduction

Lipids are a major class of biologically essential organic molecules found in all living organisms, functioning as barriers, receptors, antigens, sensors, electrical insulators, biological detergents, and membrane anchors for proteins. The major lipid species to be discussed in this chapter are listed in Table I. Phospholipids play a critical role in maintaining the integrity of all living things, plant and

I. Hydrocarbons (normal, branched, saturated, unsaturated, cyclic, aromatic)
II. Substituted hydrocarbons
    A. Alcohols
    B. Aldehydes
    C. Fatty acids, soaps, acid soaps
    D. Amines
III. Waxes and other simple esters of fatty acids
IV. Fats and most oils (esters of fatty acid with glycerol)
    A. Triacylglycerols
    B. Diacylglycerols
    C. Monoacylglycerols
V. Glycerophospholipids (diacyl; 0-alkyl, acyl; di-0-alkyl)
    A. Phosphatidic acid
    B. Choline glycerophospholipids
    C. Ethanolamine glycerophospholipids
    D. Serine glycerophospholipids
    E. Inositol glycerophospholipids
    F. Phosphatidylglycerols
    G. Lysoglycerophospholipids
VI. Glycoglycerolipids (including sulfates)
VII. Sphingolipids
    A. Sphingosine
    B. Ceramide
    C. Sphingomyelin
    D. Glycosphingolipids [ceramide monohexosides (cerebrosides), ceramide monohexosides sulfates, ceramide polyhexosides]
    E. Sialoglycosphingolipids (gangliosides)
VIII. Steroids [sterols, bile acids, cardiac glycosides, sex and adrenal hormones]
IX. Other lipids [vitamins (A, D, E, K), eicosanoids, acyl CoA, acyl carnitine, glycosyl phosphatidylinositols, lipopolysaccharides, dolichols]

[a] The chemical classification of lipids given in the table is necessarily incomplete and somewhat arbitrary. It progresses from hydrocarbons to more complex chemical structures. Simple esters and glycerol esters yield, on hydrolysis, the alcohol and/or glycerol and fatty acid. The major membrane lipids, glycerophospholipids, yield fatty acid (or aldehyde), glycerol, phosphate, and the appropriate base (choline, ethanolamine, etc.). Sphingolipids yield the base sphingosine and a fatty acid on hydrolysis.

animal, because they form the barrier separating the living cell from the extracellular environment. Fats, oils, and waxes are stored in cytoplasmic droplets and represent a major source of potential cellular energy. Components of fats (i.e., fatty acids, fatty alcohols, and glycerol) are used as building blocks for membranes during growth and repair. Some fish and copepods can accumulate massive amounts of oils or waxes to change their buoyant density, enabling them to float in the sea at a given depth. Waxes from the prene gland of aquatic birds coat the feathers to increase the feather–water surface tension and thus prevent immersion. Many hormones are lipids [e.g., steroid hormones (cortisol, estrogens, progesterones, androgens), prostaglandins, and leukotrienes]. Some lipids are given off in a gaseous state as pheromones that attract or distract other organisms. In higher animals, lipids are transported to and from cells in the form of small aggregates called *lipoproteins*. Lipids in the brain, spinal cord, and nerves are ordered in a way that permits the transmission of electrical impulses without short-circuiting. Lipids also play a role in many diseases that afflict humans (e.g., atherosclerosis, obesity, gallstone disease, Reyes syndrome, and the familial lipidoses and lipoproteinemias).

Lipids also function importantly in industry and modern technology. Petroleum, for example, is still the world's major source of energy. Many plastics, insecticides, herbicides, cosmetics, and drugs are derived from petroleum. However, oil spills and some of the synthetic petroleum byproducts are major sources of environmental pollution. The food industry uses lipids derived from plant or animal fat not only for their nutritional value, but also as emulsifiers, stabilizers, and moisturizers. Lipids are used as lubricants and as constituents of cosmetics and many pharmaceutical preparations. Lipids can be spread on water to produce wave damping in harbors, to prevent excessive water evaporation from reservoirs, or to control insect breeding in ponds and lakes. Many lipids form liquid crystals, intermediate states of matter between crystals and liquids. Specific characteristics of certain liquid crystal substances, such as reflected color or optical transmittance, have permitted their use as temperature sensors and display systems [e.g., the liquid crystal display (LCD)].

In this chapter we will discuss the overall physical properties of lipids, their extraction from tissues, and their chemical analysis and finally describe the chemical structure, metabolism, physical properties, and known biologic importance of the major individual chemical classes of lipids.

II. Physical Properties of Lipids

The physical properties of lipids coupled with the reactivity of their substitutent groups determine the biological function and biochemistry of lipids. The aliphatic hydrocarbon chain is an important constitutent of many lipid molecules, from substituted alkanes through complex glycosphingolipids. A general knowledge of the properties of the hydrocarbon chain in the solid, liquid crystal, and liquid state is fundamental to understanding the behavior of any given lipid class, for there are striking similarities between chain packing in normal alkanes and chain packing in aliphatic lipids.

Most lipids, because of their hydrocarbon chains, are elongated molecules that align along their long axis and pack in layers more or less normal to their molecular axis. X-ray diffraction has been extensively used to study the structure of the hydrocarbon chain, the packing between chains, and the crystalline and liquid crystalline packing of lipid molecules. The hydrocarbon chain, in its rigid, most stable crystalline state, forms a zigzag, all-*trans* arrangement from carbon to carbon (Fig. 1). The distance between carbons is about 1.533 Å, similar to the carbon–carbon distance in the diamond. The carbon–carbon bond angle slightly varies but approximates 112°. Thus, the distance between every other carbon atom is 2.54 Å, and the increment along the chain for each carbon atom is 1.27 Å. In crystals, these values vary slightly from atom to atom, particularly at the ends of the chain, but the general structure holds.

A. Packing of Lipids in the Solid State

Like crystalline hydrocarbons, hydrocarbon chains of more complex lipids pack into two distinct classes of subcells. The first class is characterized by dense, tightly packed chains in which there is specific chain–chain interaction. In the second class, the chains are more loosely packed, and specific chain–chain interaction is partially lost because of partial rotation of the chain and/or of its —CH$_2$— groups. The first class packs in

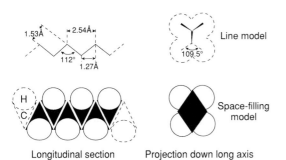

FIGURE 1 Line and space-filling models of the hydrocarbon chain in its most stable zigzag, all-*trans* conformation. Carbon–carbon distances and bond angles are given. (From: Small, D. M. (1986). Handbook of Lipid Research, vol. 4, "The Physical Chemistry of Lipids from Alkanes to Phospholipids." Plenum, New York, N.Y.)

orthorhombic, triclinic, or monoclinic subcells, whereas the more loosely packed chains align in hexagonal or nearly hexagonal subcells (Fig. 2). The volumes, chain surface areas, and nomenclature of these crystalline states are compared with liquid chains in Table II.

As a general rule, crystalline aliphatic long chained, molecules lie parallel to form extended lamellae or sheets of molecules. The repeating unit may consist of one, two, or more. These sheets are arranged in layers. The chains may be perpendicular or tilted to the plane of the layer.

B. The Liquid Crystalline State of Lipids

In nature, lipid aggregates are not typically crystalline but rather are found as liquid crystal–like arrays. The liquid crystal state is the fourth state of matter (the others being gas, liquid, and solid) and has properties of both crystals and liquids. This intermediate state combines both order and fluidity and thus is particularly suited to cellular function, especially the formation of membranes and other ordered structures. At least four factors can induce liquid crystalline formation in a particular lipid: temperature, pressure, magnetic field, and electrostatic field. Furthermore, certain molecules may be induced to form liquid crystals by the addition of water.

Many if not most biologic lipids form liquid crystals. For instance, when heated cholesterol esters, phospholipids, soaps, and glycosphingolipids transform from crystals to liquid crystals. Hydration causes soaps, acid soaps, phospholipids, monoacylglycerols, detergents, and glycosphingolipids to form hydrated liquid crystals. Liquid crystals formed by lipids have been classified depending on their degree of long-range order in Table III. Long-range order is determined by X-ray diffraction. Liquid crystals that have three-dimensional (3D) long-range order pack in a cubic lattice and could be considered crystalline. However, because the chain packing is liquid, they are classified as liquid crystals (see Fig. 3). Lipids that form cubic crystals (the viscous isotropic state) include hydrated soaps, hydrated monoacylglycerols, and some hydrated phospholipids. Two-dimensional (2D) liquid crystals are ordered in rectangular or hexagonal 2D lattices. The rectangular lattice is also called the "ribbon-like structure," in which finite bundles of polar head groups pack in ribbons of infinite length. Ribbon-like structures are formed by certain dry soaps and slightly hydrated phospholipids. Hexagonal 2D

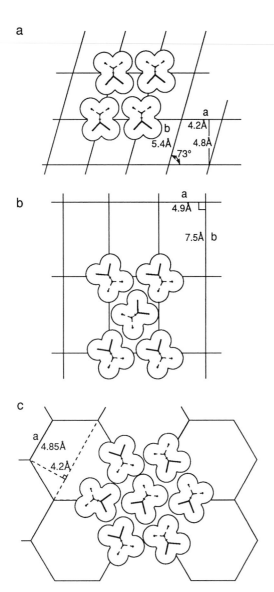

FIGURE 2 Hydrocarbon chain packing in the solid state. (a) Typical triclinic subcell. Note that the two-dimensional lattice is oblique. This represents a tightly packed chain having a specific chain–chain interaction. The mean volume per —CH$_2$— group is about 24 Å^3, and the area perpendicular to the chain is about 18.9 Å^2 per —CH$_2$— group. (b) Orthorhombic perpendicular subcell. Here the two-dimensional lattice is rectangular but also represents a tightly packed lattice with specific chain–chain interaction. The volume and area per —CH$_2$— group are similar to triclinic subcell. (c) Hexagonal subcell. The two-dimensional lattice is hexagonal, hence its name, giving rise to a 4.2-Å reflection between the plane of chains. (Chain packing is loose and specific chain–chain interaction is lost, because of the ability of the carbon atoms to rotate several degrees. Assuming that the C–C bond distance along the chain is 1.27 Å and the center-to-center distance of the chain is 4.85 Å, the volume of the —CH$_2$— group is about 25.9 Å and the area perpendicular to the chain about 20.4 Å. (From: Small, D. M. (1986). Handbook of Lipid Research, vol. 4, "The Physical Chemistry of Lipids from Alkanes to Phospholipids." Plenum, New York, N.Y.)

liquid crystalline states are comprised of long rods or cylinders packed in a hexagonal array. When no or little water is present, the polar entities of the molecules align with water to form the cores of the rods, and the liquid hydrocarbon chains are the continuous matrix. This is called the "Hex II phase." Lipids forming Hex II phases include some anhydrous soaps and dry and hydrated phospholipids and acid soaps. In higher water concentration, particularly when the polar group tends to interact strongly with water, a hexagonal phase may form in which the hydrocarbon chains form the center of the rod and the polar groups are on the exterior of the rod with water as the continuous matrix (Hex I phase). Lipids forming Hex I include hydrated soaps, detergents, and some lysophosphatides.

Lamellar liquid crystals are 1D (i.e., the only order is the stacking between the lamellae). In 1D liquid crystals (smectic, layered, lamellar), the acyl chains may be either crystalline (gel state) or liquid. In the gel state, the chains are packed in a hexagonal or a near-hexagonal chain lattice (see crystalline state above). When the chains are melted (liquid), the state is called L$_\alpha$ or lamellar liquid crystalline. Molecules that form the gel state include hydrated potassium, ribidium and cesium soaps, and many phospholipids. The lamellar crystalline phase is formed by hydrated soaps, acid soaps, phospholipids, monoacylglycerols, and glycosphingolipids.

Liquid crystals that have no long-range periodic order are truly ordered liquids but have been called liquid crystals because of their optical properties. They have no true X-ray diffraction and show only X-ray scattering with maxima corresponding to a molecular length. The particular lipids that exhibit this kind of liquid crystalline behavior are generally sterol esters and, in particular, cholesterol esters.

C. The Liquid State—Melts, Solutions, and Suspensions

Liquid states of melted lipids show the general properties of liquids including fluidity, cohesiveness, relative incompressability, and rapid molecular motions. Because only a 10–20% change in volume occurs when a lipid melts from a solid, some short-range order must remain in the liquid. It has been shown that many molten lipids form nonideal liquids consisting of clusters of a few hundred aligned molecules separated by more disordered molecules. These clusters may act as nucleating centers for the formation of more ordered liquid

TABLE II Comparison of Nomenclature for Hydrocarbon Chain Packing[a]

Aliphatic chain interaction	Subcell lattice	Mean vol—CH_2— (Å^3)	Mean surface area/—CH_2— (Å^2)	Motion	Common names[b]
Specific	Orthorhombic perpendicular	~24	~18.8	Very restricted	Alkanes: orthorhombic perpendicular, triclinic, monoclinic, β, γ
	Orthorhombic parallel				Acids and amides: A, B, and C forms
	Monoclinic parallel				Mono-, di-, and triacylglycerols: β and β' forms
	Triclinic parallel				Phospholipids: subphases, crystalline phases, L.C. (lamellar crystalline)
	Hybrid cells				
Nonspecific	Hexagonal or near-hexagonal	~25.5	~20	Restricted	Alkanes: rotator phase, α
					Acids, alcohols, glycerides: α phase
					Phospholipids: gel phase, ordered phase, hexagonal phase, L_β
					Soaps: gel phase
					Smectic B of some liquid crystals
Liquid	No lattice, but domains of roughly ∥ chains	29–30	~23 or greater	Fluid	Alkanes, acids, alcohols, di- and triacylglycerols: melt, neat liquid, isotropic liquid phase
					Monoacylglycerols and phospholipids: liquid crystal phase (e.g., lamellar or L_α cubic, hexagonal I, and hexagonal II)
					Soap: neat, viscous isotropic or middle phase, liquid phases
					Cholesterol esters: liquid crystal, fluid crystal, or mesophase (smectic A and C, nematic, cholesteric), ordered (as opposed to isotropic liquid)

[a] From: Small, D. M. (1986). Handbook of Lipid Research, vol. 4, "The Physical Chemistry of Lipids from Alkanes to Phospholipids."

[b] Nomenclature for different states of lipids is complicated and not consistent between lipid classes. This table should be helpful in orientation, since many aliphatic molecules undergo transitions from various crystalline states to more liquidlike states.

crystals or crystals. In mixtures of different lipids, some domain separation of individual lipids may occur in the liquid state. Fluidity (1/viscosity) and diffusivity in liquid and liquid crystal systems are directly proportional to the free volume of the molecule (i.e., that volume above the minimum volume at zero fluidity). The free volume is a function of the temperature, and thus fluidity increases with temperature.

The apparent solubility of lipids in aqueous systems is variable, extending from virtually insoluble (large hydrocarbons) to very soluble (soaps and detergents). In fact, the true solubility of most lipids in water is very low. Many lipids [e.g., alkanes, fatty alcohols, and fatty acids (see below)] have some measurable solubility in true solution. The solubility is less below the melting point than above it, and at any temperature it decreases as the length of the hydrocarbon chain or the size of the hydrophobic part increases (i.e., the longer the chain or the

larger the hydrophobic moiety, the lower the solubility. Some of the more polar lipids (soaps, detergents, and bile salts) form optically clear aqueous solutions in which the apparent solubility may be as high as 60 g/100 g solution. These lipids, in fact, have a low monomer solubility but spontaneously form small aggregates of molecules called *micelles* when their true (monomer) solubility is exceeded. These solutions are called *micellar solutions*.

A micellar solution is a thermodynamically stable system formed spontaneously in water above a critical concentration and temperature (Fig. 4). The solution contains small aggregates (micelles) whose molecules are in rapid equilibrium with a low concentration of monomers. This low concentration of monomers is, in fact, the true solubility of the lipid and is called the critical micellar concentration (CMC). Above the CMC the excess lipid forms micelles. Micellar solutions can solubilize other less soluble lipids in mixed micelles. Micelles can be

TABLE III Properties of Liquid Crystals and Ordered Fluid Structure of Lipids[a]

| | Liquid crystals | | | | | Isotropic liquids | | | |
| | 3D | 2D | | 1D | | ordered liquids | | None (liquids) | Nonideal liquids |
Degree of periodic long-range order	Cubic	Rectangular	Hexagonal	Lamellar "Gel," L_β, $L_{\beta'}$ Smectic B	Smectic A and C L_α	Nematic	Cholesteric	Micellar solution	Melts
Structure	Cubic	Rectangular	Hexagonal	"Gel," L_β, $L_{\beta'}$ Smectic B	Smectic A and C L_α	Nematic	Cholesteric	Micellar solution	Melts
Freedom of movement	Extremely restricted in all directions	1 dimension	1 dimension		2 dimensions	3-dimensional but anisotropic		3-dimensional	3-dimensional
Fluidity (gross)	Nearly rigid	Very low	Very low	Low	Moderate	Moderate	Moderate	High	High
Chain packing	Liquid	"Liquid"	Liquid	Hexagonal or pseudo-hexagonal	Liquid	Liquid	Liquid	Liquid	Liquid
Gross aspect	Very viscous Clear	Very viscous Opaque	Very viscous Nearly clear	Viscous Nearly clear	Moderately viscous Clear	Viscous, fluid turbid	Viscous, fluid, brilliant colors	Fluid	Fluid
Polarizing microscopic aspect	Viscous isotropic with angular bubbles	Anisotropic	Anisotropic middle soap texture	Anisotropic neat texture	Anisotropic neat on smectic texture	Anisotropic nematic texture	Anisotropic smectic texture, focal conics with $(-)$ sign	Isotropic round bubbles	Isotropic round bubbles
X-ray long-spacings index into:	3D cubic	2D-rectangular	2D-hexagonal	1D-lamellar	1D-lamellar	Scattering with maximum approximating molecular length			
Short A	~4.6 fringe	~4.6 fringe (?) 4.8	~4.6 fringe	4.2 sharp Bragg reflections	~4.6 fringe	4.6 fringe	4.6 fringe[b]	4.6 fringe	4.6 fringe

[a] From: Small, D. M. (1986). Handbook of Lipid Research, vol. 4, "The Physical Chemistry of Lipids from Alkanes to Phospholipids."
[b] For cholesterol or cholesterol esters the fringe is 5Å.

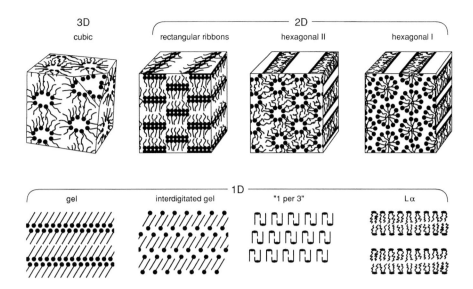

FIGURE 3 Some liquid crystalline states with various degrees of dimensional order. Polar groups are represented by *dots,* the hydrocarbon chains by *wriggles or lines,* and water by *blank areas.*

The three-dimensional cubic phase unit cell formed by egg lecithin at low hydration is centered cubic. The structure consists of rigid rods of finite length, all identical and crystallographically equivalent, joined three by three to form two interwoven three-dimensional networks. The hydrocarbon chains are fluid.

Liquid crystals with two-dimensional order, rectangular or hexagonal. The two-dimensional centered rectangular phase of the anhydrous sodium soaps has the loci of the polar groups arranged in infinitely long ribbons of finite width. The two hexagonal phases have the loci of the polar groups arranged in infinitely long rods packed hexagonally. Hexagonal II is the water-in-oil type observed in low water–phospholipid systems and in the anhydrous soaps of divalent cations. Hexagonal I is the oil-in-water type observed in lipid-water systems containing monovalent soaps, lysolecithin, or aliphatic detergents.

Lamellar systems are generally one-dimensional. However, if the chains are packed in a 2D hexagonal array, two-dimensional order is perpendicular to the chain axis, and these lamellar systems called "gels" (Table 3) are ordered. Lamellae may be fully or partially interdigitated layers or bilayers. Transitions between interdigitated layers and bilayers would change the lamellar thickness greatly, and such transitions may have a biological role.

spherical structures in pure water, but when salt is added they often enlarge and assume cylindrical or discoid shapes.

Lipids may also be suspended in aqueous systems. Such lipids are very insoluble as monomers, but by adding energy to the system relatively stable suspensions of lipid in water can be achieved. For instance, emulsions containing a core of very insoluble lipid and a surface of an emulsifier can be formed, or a course suspension of lamellar liquid crystals (see above) can be ultrasonically irradiated to produce a suspension of small (25-nm) unilamellar vesicles. Such emulsions and vesicles have been widely used as models of lipoproteins and membranes.

D. Transitions Between States

Solid–state (polymorphic), solid–liquid (chain melting), liquid crystal–liquid crystal, and liquid crystal–liquid transitions are all first-order phase changes and are accompanied by a sharp change in volume and excess specific heat.

In Fig. 5 polymorphic and melting transitions of the common phospholipid, dipalmitoyl phosphatidylcholine, a predominant phospholipid in the alveolar linings of the lung are shown. In the hydrated state, this phospholipid undergoes two polymorphic transitions and a melting transition. The first transition at 25°C is the change of a tightly packed crystal to a more disordered lamellar crystalline structure ($L_{\beta'}$). The second transition (at 35°C) of smaller heat and volume change is a polymorphic transition involving both the structure of the bilayer and the packing of the chains. The third transition (41.5°C) is the chain melting transition, which converts the $P_{\beta'}$ phase into an L_α liquid crystalline phase, the well-known gel-to-liquid crystal transition. Note that, at each transition, there is a fairly sharp change in volume and an excess specific heat. At a much higher temperature (not shown), there is a transition from lamellar liquid crystal to hexagonal II liquid crystal. Such transitions occur in almost all

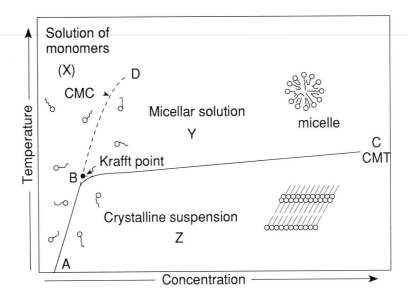

FIGURE 4 Behavior of soluble lipids at high water concentration as a function of temperature and concentration. The temperature is plotted on the ordinate and increasing concentrations of soluble lipid (amphiphile) on the abscissa. Micelles occur only in region Y. In this region the amphiphile is above a concentration indicated by the curve *BD* and above a temperature indicated by *BC*. *BD* is the critical micellar concentration (CMC). Thus, the CMC varies along *BD* with the temperature. The solution must also be above a certain temperature (*BC*), which indicates the transition temperature from the crystalline state to a micellar solution. This temperature is called the critical micellar temperature (CMT). The point *B* at which the CMC and CMT curves meet is termed the Krafft point. It can be considered a triple point and indicates the CMT at the CMC. (From: Small, D. M. (1986). Handbook of Lipid Research, vol. 4, "The Physical Chemistry of Lipids from Alkanes to Phospholipids." Plenum, New York, N.Y.)

lipids with long aliphatic chains. In general, about 0.18 kcal/mol CH_2 is needed to change the volume by 1 $Å^3$.

The effect of chain length, polar substitution, or double bonds on the transition temperatures is illustrated in Fig. 6, which plots the chain-melting transition (crystal to liquid chain) for increasing chain length and for a variety of different molecules having different substituents. Note that as the chain length increases, the melting transition increases. Double bonds, triple bonds, and halide substitutions at the end of the chain decrease the melting temperature, but polar substituents, particularly those that can form hydrogen bonds or form ionic bonds, increase the melting transition. The order of increasing melting temperatures for a given chain length is as follows: alkanes < ethyl esters < aldehydes < methyl esters < normal alcohols < carboxyls < sodium carboxylates. Interaction of water with the polar groups tends to decrease the melting transition.

E. Surface Behavior of Lipids

A major property of lipids is their potential for accumulating at air or oil/water interfaces. The interaction of a specific lipid with an aqueous interface depends on the hydrophilic–liphophilic balance (HLB) of the lipid (i.e., the relative strengths of the hydrocarbon and water-seeking parts). Thus, when a neat drop of lipid contacts a water surface of limited area, one of three events will occur: (1) a very lipophilic lipid, like cholesteryl oleate or linoleoyl palmitate (a wax ester), will simply sit on the water as an intact droplet or lens; (2) a more hydrophilic lipid, like oleoyl alcohol, will spread to form a continuous insoluble monolayer of molecules in equilibrium with the remainder of the droplet; or (3) a highly hydrophilic lipid, like potassium oleate, will spread to form an unstable film from which molecules desorb into the water.

F. Lipid Classification Based on Physical Interaction with Water

Because biological systems are aqueous systems, a classification based on the behavior of lipid in water and at aqueous interfaces (Table IV) is given to help one predict how certain classes of lipids will assemble and distribute in cells. This classification gener-

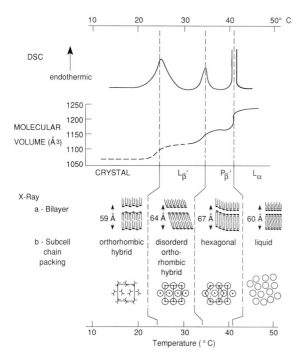

FIGURE 5 Composite diagram of differential scanning calorimetric (DSC), volumetric, and X-ray data of hydrated dipalmitoyl phosphatidyl-choline, DPPC, showing phases and transitions. DPPC forms a series of lamellar structures that have decreasing order as the temperature is increased. The most stable state below 25°C is the 3-dimensional crystal state, having orthorhombic hybrid tight chain packing. At about 25°C this structure undergoes a polymorphic transition to a more hydrated and tilted structure ($L_{B'}$) with loosened chain packing. This chain packing is a disordered 2-dimensional orthorhombic, which is also called pseudo- or near-hexagonal. At ~35°C, a transition ($L_{B'} \rightarrow P_{B'}$) of lower enthalpy and volume change occurs. In $P_{B'}$ the chains are packed in a true hexagonal lattice, but the bilayer undulates periodically to form what is called the "rippled" bilayer phase. At 41.5°C this phase undergoes chain melting to the L_{α} liquid crystal phase. One degree of order, between layers, is preserved, but the chains are in a disordered liquid state. (From: Small, D. M. (1986). Handbook of Lipid Research, vol. 4, "The Physical Chemistry of Lipids from Alkanes to Phospholipids." Plenum, New York, N.Y.)

form micelles. They also partition into membranes and can be membrane disruptive. Class II and III lipids do not partition into intracellular oil droplets.

III. Extraction and Analysis of Lipids from Cells and Tissues

Lipids are contained in a variety of special compartments within cells and tissues (e.g., intracellular droplets, membranes, bound to protein). Extraction requires their solubilization and removal into a liquid phase. Mixtures of organic solvents, e.g., chloroform : methanol (2 : 1 v/v), are used to remove virtually all the lipids from homogenized tissue.

The general procedure (see Fig. 7) requires mechanical disruption of the tissue; adequate chloroform : methanol (2 : 1 v/v) is then added to the tissue to incorporate all the tissue water into a single chloroform : methanol : water phase. Lipids are ex-

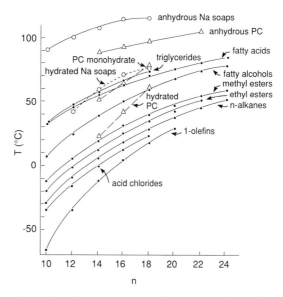

FIGURE 6 Effects of polar substituents on melting of the hydrocarbon chain for a variety of lipids. The major chain-melting transition (i.e., to liquid chain) temperatures for a variety of lipids are plotted against the number of carbons in the aliphatic chain. The stronger the interaction of the polar group, the higher the melting point. Note that the melting transitions increase in temperature with increasing hydrocarbon chain length, even in water. The presence of water, however, lowers the chain transition when compared with the dry lipid. Anhydrous lipids (●); saturated phosphatidylcholines (PC) (△); sodium (Na) soaps (○). (From: Small, D. M. (1986). Handbook of Lipid Research, vol. 4, "The Physical Chemistry of Lipids from Alkanes to Phospholipids." Plenum, New York, N.Y.)

ally applies to lipids with melted chains. Nonpolar lipids have extremely limited solubility in membranes and will usually be found in intracellular droplets. All other lipids are amphiphiles (i.e., they have affinity for both oil and water). The interaction with water (or hydrocarbons) is determined by the nature of the polar group (water-liking) and the mass of the hydrocarbon part. Class I polar lipids have limited solubility in membranes and partition between membranes and intracellular droplets. Class II lipids are membrane formers. Class III lipids have measurable monomer solubility and can

TABLE IV Classification of Biologically Active Lipids[a]

Class	Surface Properties[b]	Bulk Properties	Examples
Nonpolar	Will not spread to form monolayer	Insoluble	Long-chain, saturated or unsaturated, branched or un-branched, aliphatic hydrocarbons with or without aromatic groups (e.g., dodecane, octadecane, hexadecane, paraffin oil, phytane, pristane, carotene, lycopene, gadusene, squalene) Large aromatic hydrocarbons (e.g., cholestane, benzopyrenes, coprastane, benzophenanthrocenes) Esters and ethers in which both components are large hydrophobic lipids (e.g., sterol esters of long-chain fatty acids, waxes of long-chain fatty acids, and long-chain normal monoalcohols, ethers of long-chain alcohols, sterol ethers, long-chain triethers of glycerol)
Polar Class I: Insoluble, non-swelling am-phiphiles	Spreads to form stable monolayer	Insoluble or solubility very low	Triglycerides, diglycerides, long-chain protonated fatty acids, long-chain normal alcohols, long-chain normal amines, long-chain aldehydes, phytols, retinols, vitamin A, vitamin K, vitamin E, cholesterol, desmosterol, sitosterol, vitamin D, un-ionized phosphatidic acid, sterol esters of very short-chain acids, waxes in which either acid or alcohol moiety is less than four carbon atoms long (e.g., methyl oleate)
Class II: Insoluble swelling amiphiphilic lipids	Spreads to form stable monolayer	Insoluble but swells in water to form lyotropic liquid crystals	Phosphatidylcholine, phosphatidylethanolamine, phosphatidyl inositol, sphingomyelin, cardiolipid, plasmalogens, ionized phosphatidic acid, cerebrosides, phosphatidylserine, monoglycerides, acid soaps, α-hydroxy fatty acids, monoethers of glycerol, mixtures of phospholipids and glycolipids extracted from cell membranes or cellular organelles (glycolipids and plant sulfolipids)
Class IIIA: Soluble amphiphiles with lyotropic mesomorphism[b]	Spreads but forms unstable mono-layer because of solubility in aqueous sub-strate	Soluble, forms micelles above a critical micellar concentration: at low water concentrations forms liquid crystals	Sodium and potassium salts of long-chain fatty acids, many of the ordinary anionic, cationic, and nonionic detergents, lysolecithin, palmitoyl and oleyl coenzyme A and other long-chain thioesters of coenzyme A, gangliosides, sulfocerebrosides
Class IIIB: Soluble amphiphiles, no lyotropic meso-morphism	Spreads but forms unstable mono-layer due to solubility in aqueous sub-strate	Forms micelles but not liquid crystals	Bile salts, sulfated bile alcohols, sodium salts of fusidic acid, rosin soaps, phenanthrene sulfonic acid, penicillins, phenothiazines

[a] From: Small, D. M. (1986). Handbook of Lipid Research, vol. 4, "The Physical Chemistry of Lipids from Alkanes to Phospholipids."
[b] Lyotropic mesomorphism mean the formation of liquid crystals on interaction with water.

tracted into this phase, and the insoluble precipitate (proteins, nucleic acids, polysaccharides, etc.) is removed by filtration or centrifugation. Excess water containing some acid and salts is then added to form a two-phase system, the top being methanol : water–rich and the bottom being chloroform-rich. Virtually all the lipids partition into the chloroform phase, which may then be removed and concentrated by evaporation. At this point an aliquot may be removed for dry weight to give the total lipids per weight of tissue. The chloroform extract is then analyzed by thin-layer chromatography (TLC) or high-performance liquid chromatography (HPLC) to separate the different lipid classes (e.g., hydrocarbons, waxes, acyl glycerols, and the other lipid classes noted in Table 1). Each class is then collected and further separated into subclasses. For instance, phospholipids may be separated by TLC or HPLC into a number of specific phospholipid classes [e.g., phosphatidylcholines (PC), sphingomyelins (SM), phosphatidylethanolamines (PE), phosphatidylserines (PS), phosphatidylinositols

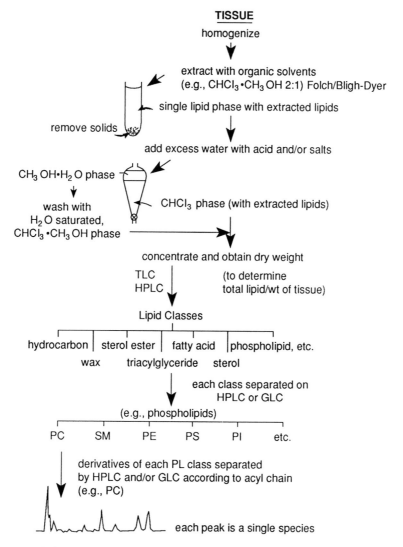

TISSUE
homogenize

extract with organic solvents
(e.g., CHCl$_3$•CH$_3$OH 2:1) Folch/Bligh-Dyer

single lipid phase with extracted lipids

remove solids

add excess water with acid and/or salts

CH$_3$OH•H$_2$O phase

wash with
H$_2$O saturated,
CHCl$_3$•CH$_3$OH phase

CHCl$_3$ phase (with extracted lipids)

concentrate and obtain dry weight

TLC
HPLC

(to determine
total lipid/wt of tissue)

Lipid Classes

| hydrocarbon | sterol ester | fatty acid | phospholipid, etc. |

wax   triacylglyceride   sterol

each class separated on
HPLC or GLC

(e.g., phospholipids)

PC   SM   PE   PS   PI   etc.

derivatives of each PL class separated
by HPLC and/or GLC according to acyl chain
(e.g., PC)

each peak is a single species

GLC/Mass Spectrometry or NMR to define chemical structure.
X-ray to define crystal structure and conformation.

FIGURE 7 Procedure for isolation and analysis of lipids from animal tissues.

(PI), phosphatidyl glycerols (PG)]. Also each individual class of phospholipid may be modified to remove interferences from the headgroups and further separated according to fatty acid makeup on different gas-liquid or HPLC columns. Each peak may be a unique chemical species or a mixture of similar species. Individual peaks are harvested and analyzed by capillary gas–liquid chromatography/mass spectrometry or by nuclear magnetic resonance to define the chemical structure. If adequate quantities are obtained, X-ray crystallography may be carried out to define the crystal structure and conformation of the lipid molecule in the crystal.

Certain classes of lipids, such as triacylglycerols, may be directly analyzed by gas–liquid chromatography according to total number of carbons and double bonds. Each individual peak may then be isolated, hydrolyzed enzymatically, to release the fatty acids in the α and β positions. These fatty acids may be methylated and analyzed by gas–liquid chromatography and mass spectrometry to give the individual racemic triacylglycerols. Stereoisomers of acylglycerols may be isolated using chiral column chromatography and mass spectrometry. Thus, from a tissue, all the lipids may be extracted by appropriate solvents, isolated, reseparated, and analyzed so that the individual molecular species of the lipids can be determined. [*See* HIGH PERFORMANCE LIQUID CHROMATOGRAPHY.]

The complexity of individual lipids is astonishing. A sample of adipose tissue fat or even butter fat may contain a dozen major triglycerides and several hundred minor ones. A plasma membrane will contain most of the phospholipid classes, and within each class, dozens of individual species abound. The biological importance of this enormous complexity is not understood.

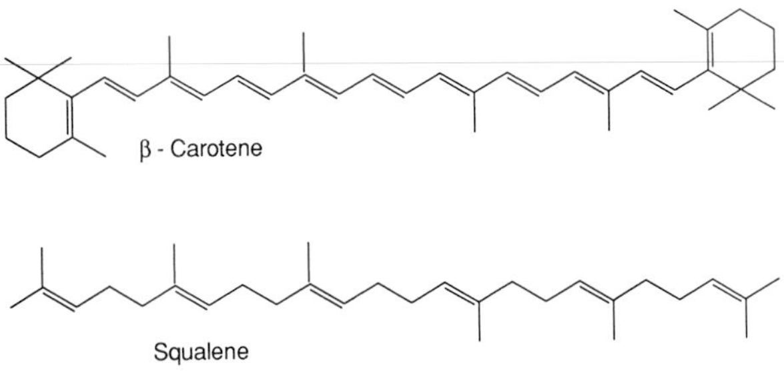

IV. Occurrence, Metabolism, Physical Properties, and Biological Importance of Major Lipid Classes

A. Hydrocarbons

Hydrocarbons contain only hydrogen and carbon. They may be saturated or unsaturated, branched or unbranched, cyclic or aliphatic, or may exhibit a combination of these characteristics. The source of most hydrocarbon is petroleum, but a particular group of 40-carbon polyenoic compounds composed of isoprene units

$$
\begin{array}{c}
CH_3 \\
| \\
(-CH_2-C=CH-CH_2-)
\end{array}
$$

are synthesized from mevalonic acid in plant cells (phytoene, phytofluenes, lycopene, etc.) to produce the carotenes (Fig. 8). These complex hydrocarbons are ingested by animals, absorbed into the gut cell where they are oxidatively cleaved in the center to produce vitamin A, a 20-carbon polyisoprenoid alcohol. In animals, a similar pathway from mevalonic acid leads to the formation of the 30-carbon polyisoprenoid squalene (Fig. 8), which is then oxidized and cyclized to form steroids.

1. Alkane Metabolism

Although mammalian cells typically see little alkane (from the diet or generated metabolically except for squalene), they are capable of incorporating simple, straight-chain alkanes and assimilating them into complex lipids after the appropriate modifications (Fig. 9). The initial step in alkane assimilation is the hydroxylation of a terminal methyl group to form the fatty alcohol (reaction 1), with the primary sites of hydroxylation being the intestine and liver. The activity has been partially characterized in microsomes from rabbit mucosa and requires a unique cytochrome P450 as well as NADPH-cytochrome c reductase. The enzyme system responsi-

FIGURE 8 Line models of β-carotene and squalene, showing the similarity of their structures.

ble for the hydroxylation of alkanes is different from that used to hydroxylate and detoxify aromatic hydrocarbons such as benzo-pyrene and xylenes because these substrates do not compete with alkanes in hydroxylation assays.

B. Substituted Hydrocarbons

Alcohols ($R-CH_2OH$), aldehydes ($R-CHO$), acids ($R-COOH$), and amines ($R-CH_2NH_2$) are examples of substituted hydrocarbons. Such molecules are usually found in low concentrations in cells, as they are rapidly metabolized. Fatty acids are perhaps more abundant than the other substituted hydrocarbons, and they have the potential to reach high local concentrations during fat catabolism.

The major saturated fatty acids in higher animals are palmitic (16 carbons) and stearic (18 carbons) followed by smaller amounts of 12-, 14-, and 20-carbon fatty acids (Table V). The major monounsaturated fatty acids are oleic acid, which has 18 carbon atoms and a *cis* double bond at carbon 9 (ω9), and vaccenic acid, which has the double bond at carbon 11 (ω7). The major polyunsaturated fatty acids are linoleic (18:2 ω6), arachidonic (C20:4ω6), eicosapentenoic (C20:5ω3), and docosahexenoic (C22:6ω3). In polyunsaturated fatty acids, the double bonds are usually three carbons apart ($-CH_2-CH=CH-CH_2-CH=CH-CH_2-$). Arachidonic acid is shown in Fig. 10. Abbreviated nomenclature for fatty acids, numbering carbons from either the carboxyl (Δ system) or from the terminal methyl (ω system), is used in Table V.

Fatty acids are essential building blocks for membrane lipids. Certain fatty acids are not synthesized by animals and must be obtained in the diet. Some of these dietary fatty acids, especially linoleic acid (18:2 ω6), are essential fatty acids necessary for

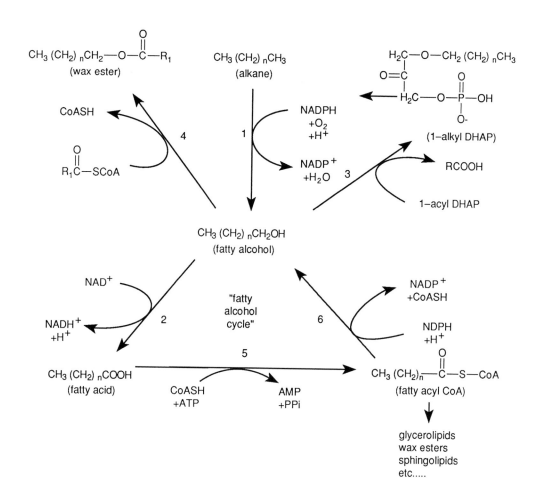

FIGURE 9 Fatty alcohol metabolism in animal cells. 1, alkane hydroxylation system; 2, fatty alcohol: NAD^+ oxidoreductase; 3, alkyl DHAP synthase; 4, acyl CoA: fatty alcohol acyltransferase; 5, acyl CoA synthase; 6, acyl CoA reductase; DHAP, dihydroxyacetonephosphate. $n = 14–18$.

proper growth and membrane function. Arachidonic acid (20:4 ω6), derived from linoleic acid by chain elongation and desaturation, is a major precursor of eicosanoids such as prostaglandins, prostacyclin, thromboxanes, and leukotrienes.

1. Metabolism of Fatty Alcohol and Fatty Acid

The major source of long-chain fatty alcohol in mammalian tissues (excluding the possibility of heavy dietary intake of foods containing high levels of fatty alcohols or wax esters) is the reduction of long-chain fatty acyl-CoA (Fig. 9, reaction 6). This reaction is catalyzed by a microsomal enzyme system, fatty acyl-CoA reductase, which uses acyl-CoA as substrate and is specific for NADPH.

As they are generated, free fatty alcohols are metabolized by one of three pathways. The two major metabolic fates for fatty alcohols are ether lipid biosynthesis and fatty acid formation. Fatty alcohols are incorporated into ether lipids during the first committed step in ether lipid biosynthesis catalyzed by the peroxisomal enzyme, alkyldihydroxyacetonephosphate synthase. This enzyme catalyzes a unique replacement reaction in which the fatty acid moiety of 1-acyldihydroxyacetonephosphate is replaced with a fatty alcohol, forming an ether bond (Fig. 9, reaction 3). The second metabolic fate, oxidation of long-chain fatty alcohol to the fatty acid, is performed by fatty alcohol: NAD^+ oxidoreductase (Fig. 9, reaction 2). This is a membrane-bound activity that appears to be composed of two oxidation activities: a fatty alcohol dehydrogenase, which converts the fatty alcohol to the intermediate aldehyde (RCHO), and an aldehyde dehydrogenase, which forms the final product. This oxidation reaction is not the reverse of the acyl-CoA reductase reaction described previously (reaction 6). Both hu-

TABLE V Names, Formulas, and Selected Properties of Some Common Fatty Acids

Fatty acid	Chemical name	Δ Formula[a]	ω Formula[b]	MW	mp (°C)	Solubility μM (25°C)
Saturated						
Lauric	Dodecanoic acid	12.0	—	200.31	44.2	11.5
Myristic	Tetradecanoic acid	14.0	—	228.36	54.4	0.79
Palmitic	Hexadecanoic acid	16.0	—	256.42	62.9	0.12
Stearic	Octadecanoic acid	18.0	—	284.47	69.6	(18×10^{-3})c
Arachidic	Eicosanoic acid	20.0	—	312.52	75.4	(27×10^{-3})c
Behenic	Docosanoic acid	22.0	—	340.57	80.0	
Lignoceric	Tetracosanoic	24.0	—	368.62	84.2	
Monounsaturated						
Myristoleic	cis-9-Tetradecenoic acid	14.1 9C	c14:1ω5	226.34		(2.00)c
Palmitoleic	cis-9-Hexadecenoic acid	16.1 9C	c16:1ω7	254.40		(0.30)c
Oleic	cis-9-Octadecenoic acid	18.1 9C	c18:1ω9	282.45	13.4α; 16.3β	(45×10^{3})c
Elaidic	trans-9-Octadecenoic acid	18.1 9T	t18:1ω9	282.45	46.5	
cis-Vaccenic	cis-11-Octadecenoic acid	18.1 11C	c18:1ω7	282.45	14.5	
Petroselinic	cis-6-Octadecenoic acid	18.1 6C	c18:1ω12	282.45	30	
Erucic	cis-13-Docosenoic acid	22.1 13C	c22:1ω9	348.55	34.7	(7×10^{-3})c
Polyunsaturated						
Linoleic	cis-9,12-Octadecadienoic acid	18.2 9C12C	c18:2ω6	280.44	−5	
γ-Linolenic	cis-6,9,12-Octadecatrienoic acid	18.3 6C9C12C	c18:3ω6	278.44		
Linolenic	cis-9,12,15-Octadecatrienoic acid	18.3 9C12C15C	c18:3ω3	278.44	−10 (−11.3)	
ETA[d]	cis-5,8,11-Eicosatrienoic acid	20.3 5C8C11C	c20:3ω9			
Arachidonic	cis-5,8,11,14-Eicosatetraenoic acid	20.4 5C8C11C14C	c20:4ω6	304.5	−49.5	
EPA[e]	cis-5,8,11,14,17-Eicosapentaenoic acid	20.5 5C8C11C14C17C	c20:5ω3	302.5		
DHA[e]	cis-4,7,10,13,16,19-Docosahexaenoic acid	22.6 4C7C10C13C16C19C	c22:6ω3	328.5		

[a] Formulas are shown as the number of carbon atoms followed by the number of double bonds. The position of each double bond is indicated by the number of the two doubly bonded carbon atoms, counting from the carboxyl carbon, and specified as to cis (C) or trans (T) configuration. Thus, oleic (cis-9-octadecenoic) acid is 18.1 9C.

[b] The number of the carbon atoms starts from the methyl end, and the location of the first (or only) double bond is indicated by a single number as a suffix preceded by "ω." Oleic acid is therefore designated as c18:1ω9, which informs the reader that the most distal double bond is nine carbons from the methyl terminus. It is assumed that unless otherwise specified, all multiple double bonds in polyunsaturated fatty acids have the 1,4-relation to each other.

[c] Extrapolated from solubility of shorter chain acids.

[d] ETA, eicosatrienoic acid, an acid that accumulates in essential fatty acid deficiency.

[e] Found in marine animals in high concentration. EPA, eicosapentenoic acid; DHA, docosahexaenoic acid

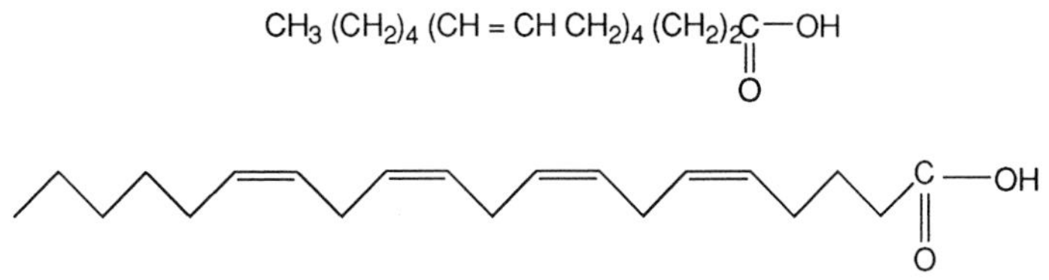

$$CH_3\,(CH_2)_4\,(CH = CH\,CH_2)_4\,(CH_2)_2\!\!\underset{\underset{O}{\|}}{C}\!\!-OH$$

All *cis* - 5, 8, 11, 14 - Eicosatetraenoic Acid
(Arachidonic Acid)

FIGURE 10 Line model of a polyunsaturated fatty acid.

man and rodent enzyme systems are specific for NAD^+ (not $NADP^+$) as cosubstrate, and free fatty acid (not fatty acyl-CoA) is generated. Studies using both human and rodent mutant cell lines that are defective in this oxidase activity have shown that only one activity is involved in the oxidation of the long-chain fatty alcohols such as palmitate. Other activities are responsible in the oxidation of shorter-chain substrates such as decanol and hexanol.

The oxidation of long-chain fatty alcohol and the reduction of fatty acyl-CoA form a metabolic "cycle." The importance of this cycling between fatty alcohol and the fatty acid is unknown; however, a disruption in this cycle, in which the oxidation of the fatty alcohol is greatly reduced, has been recently described in patients suffering from a genetic disorder, Sjögren-Larsson syndrome. These patients have increased amounts of long-chain fatty alcohols in their tissues and display neuromuscular dysfunctions as well as severe congenital icthyosis (scaling of the skin), suggesting that fatty alcohol oxidation, or cycling, is important in the proper functioning or development of the neuromuscular tissues and skin.

The third, and quantitatively least important, pathway for fatty alcohol metabolism in mammalian cells is wax ester formation. The synthesis of wax ester has not been well characterized, but most studies describe a membrane-bound enzyme that uses free fatty alcohol and fatty acyl-CoA as co-substrates (Fig. 9, reaction 4).

The primary sources of fatty acids in animal cells are dietary intake of fat and oil (triacylglycerols) and/or *de novo* biosynthesis from the combined actions of acetyl-CoA carboxylase and fatty acid synthase (Fig. 11). Malonyl-CoA, formed by the action of the cytosolic enzyme, acetyl-CoA carboxylase.

$$CH_3\!\!-\!\!\underset{\underset{O}{\|}}{C}\!\!-\!\!S\!\!-\!\!CoA + HCO_3^- + ATP \xrightarrow[\text{Carboxylase}]{\text{Acetyl-CoA}}$$

$$^-OOC\!\!-\!\!CH_2\!\!-\!\!\underset{\underset{O}{\|}}{C}\!\!-\!\!S\!\!-\!\!CoA + ADP + P_i + H^+$$

is used by the multifunctional polypeptide, fatty acid synthase, to build a fatty acid in two-carbon units by the repetitive cycling of a reaction series described in Fig. 11. Fatty acid synthesis is initiated by covalent attachment of malonate and acetate to the fatty acid synthase polypeptide. Malonate and acetate are condensed with the loss of CO_2, forming a β-keto acyl intermediate (reaction 1). The β-keto oxygen is then removed in reactions 2 through 4 to form the saturated fatty acyl intermediate. Reactions 1 through 4 are repeated six times, with malonate being used to add two carbons to the carboxyl end of the growing fatty acyl chain during each new cycle. The acyl chain remains covalently attached to the enzyme throughout the process and is finally released as the free fatty acid, palmitate, by a thiolase activity associated with the fatty acid synthase enzyme. Palmitate can then be activated to the acyl-CoA by the enzyme acyl-CoA synthase (see Fig. 9, reaction 5) and further metabolized. The smallest functional unit of fatty acid synthase is a homodimer. Both acetyl-CoA carboxylase and fatty acid synthase are cytosolic enzymes, and their activity levels greatly vary, depending on dietary and hormonal state of the animal.

Animal cells do not synthesize only palmitate. This fatty acid can be modified (in the form of acyl-CoA) by a fatty acyl-CoA elongation system (which elongates the fatty acid by two carbons) and/or the stearoyl-CoA desaturase system (which places a *cis*-double bond between C9 and C10) (Fig. 12). The

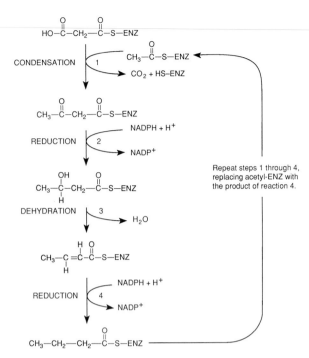

FIGURE 11 Biosynthesis of fatty acid in animal cells. Reaction 1, β-ketoacyl synthase; reaction 2, β-ketoacyl reductase; reaction 3, β-hydroxyacyl dehydrase; reaction 4, enoyl reductase. ENZ represents the fatty acid synthase polypeptide. Acetate and malonate must be covalently attached to the fatty acid synthase polypeptide (through reactions which are not described in this text) prior to use in reaction 1. One molecule of malonate is used during each cycle, but the acetyl-S-ENZ intermediate is utilized only once, in the first cycle of fatty acid chain formation. The butyryl intermediate (the product of reaction 4) is used in its place during the next round of chain elongation. This is repeated for each round, using intermediates associated with the growing fatty acid chain (e.g., hexanoyl-S-ENZ, octanoyl-S-ENZ, etc.) as substrate for reaction 1. To form one molecule of palmitic acid, seven malonate molecules and one acetate molecule are utilized. The growing fatty acid chain and the intermediates remain covalently attached to the polypeptide throughout the process.

major products of endogenous fatty acid synthesis in mammalian cells are, depending on the modification(s) and the order in which they are performed, palmitate (16 : 0), palmitoleate (16 : 1ω7), stearate (18 : 0), oleate (18 : 1ω9), and vaccenate (18 : 1ω7).

Mammalian cells break down saturated fatty acids in two-carbon units to yield acetyl-CoA in an energy-yielding process called β-oxidation. Chemically, the process is the reverse of fatty acid synthesis with the generation of acetyl-CoA and a fatty acyl-CoA, which is two carbons shorter, but the mechanism and the enzymes involved are different (Fig. 13). The reactions are carried out by a series of

enzymes and not one multifunctional protein. Also, the acyl intermediates are not covalently attached to the enzymes, but remain in the CoA form throughout the reaction series. [*See* FATTY ACID UPTAKE BY CELLS.]

Until recently, β-oxidation was thought to be the responsibility of the mitochondrion, but the peroxisome has its own unique β-oxidation system. β-oxidation in the two organelles serves different functions. The peroxisomal system is responsible for the oxidation of long-chain (C14–C18) and very long-chain (C20–C26) fatty acids, but it is unable to oxidize fatty acids shorter than eight carbons and must pass these on to the mitochondria for further oxidation. It has been estimated that peroxisomes are responsible for 30–60% of the β-oxidation occurring in some cells. Mitochondria are capable of completely breaking down fatty acids of 18 carbons or less, but are unable to handle the longer-chain fatty acids. Mutant cells, which lack peroxisomes, accumulate very long-chain fatty acids to 10 times normal values, demonstrating the role this organelle plays in their metabolism.

2. Behavior and Distribution

Primary aliphatic alcohols, aldehydes, and amines are Class I polar amphiphiles, and these distribute between membranes and oil droplets in cells. Strictly speaking, protonated fatty acid (at pH < 3) is also a Class I amphiphile. However, the behavior

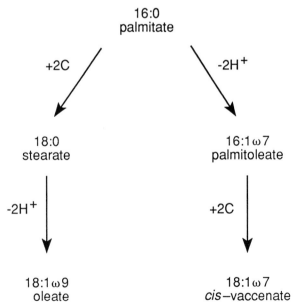

FIGURE 12 Fatty acid modification pathways.

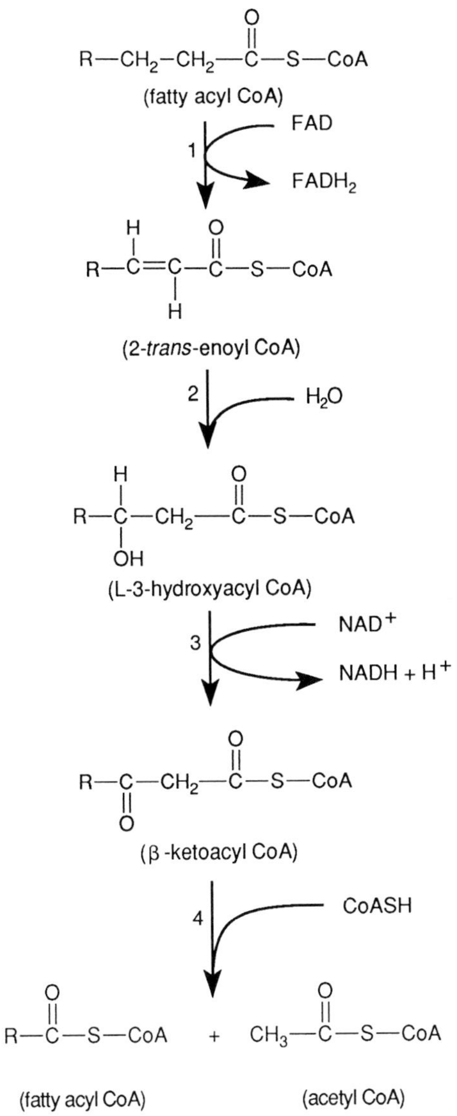

$$R-CH_2-CH_2-\overset{\displaystyle O}{\overset{\|}{C}}-S-CoA$$

(fatty acyl CoA)

1, FAD → FADH_2

$$R-\overset{\displaystyle H}{\underset{}{C}}=\overset{}{C}-\overset{\displaystyle O}{\overset{\|}{C}}-S-CoA$$
$$\quad\quad\underset{H}{}$$

(2-*trans*-enoyl CoA)

2, H_2O

$$R-\overset{\displaystyle H}{\underset{OH}{C}}-CH_2-\overset{\displaystyle O}{\overset{\|}{C}}-S-CoA$$

(L-3-hydroxyacyl CoA)

3, NAD⁺ → NADH + H⁺

$$R-\overset{}{C}-CH_2-\overset{\displaystyle O}{\overset{\|}{C}}-S-CoA$$
$$\quad\underset{O}{\overset{\|}{}}$$

(β-ketoacyl CoA)

4, CoASH

$$R-\overset{\displaystyle O}{\overset{\|}{C}}-S-CoA \quad + \quad CH_3-\overset{\displaystyle O}{\overset{\|}{C}}-S-CoA$$

(fatty acyl CoA)　　　　(acetyl CoA)

FIGURE 13 β-Oxidation in animal cells. 1, acyl CoA dehydrogenase; 2, enoyl CoA hydratase; 3, L-3-hydroxyacyl-CoA dehydrogenase; 4, thiolase.

of fatty acids is more complex because the carboxyl can ionize (RCOOH ⇌ RCOO⁻ + H⁺).

The pKa of monomeric fatty acid in solution is about 4.5, like acetic acid. Practically, however, the solubility of long-chained fatty acids is very low (e.g., palmitic acid ~0.1 μM; see Table V), and they strongly partition into membranes. When fatty acids are titrated in membranes or in lipoproteins, their apparent pKa is ~7.4, and thus at this pH they are half-ionized. Only above pH 8.5–9 are fatty acids fully ionized. Fatty acids at high alkaline pH are soaps and behave as Class III polar amphiphiles

forming micelles in water and disrupting membranes. However, at normal extra- and intracellular pH (7.4–6.8), fatty acids are about half-ionized and behave as Class II amphiphiles (i.e., as membrane formers). In cells and in plasma of animals, there are proteins that compete with membranes and lipoproteins for binding fatty acids. A class of small intracellular proteins called fatty acid binding proteins (FABP) is present in many cells, especially those in which active fatty acid flux occurs (liver, gut, heart, etc.). These proteins bind one or two molecules of fatty acid. The function of the FABPs is not clearly known, but they may act to bind excess fatty acid produced during active intracellular fatty acid flux. Plasma albumin also binds up to 8–10 fatty acids in several distinct binding sites. These sites compete with membranes, lower pH, and higher fatty acid/albumin ratios favoring distribution into membranes. Although fatty acid concentrations in quiescent cells are low, during certain metabolic events local fatty acid concentration may become quite high. For instance, during exercise and fasting, intracellular triglyceride deposits are rapidly hydrolyzed, generating potentially high intracellular concentrations. In the human intestinal lumen, as much as 50 g of fatty acid may be released after a fatty meal and enter the intestinal cells. There it is resynthesized to triglyceride and secreted in chylomicrons into the lymph. Chylomicrons circulate to capillaries where lipoprotein lipase rapidly hydrolyzes triglyceride to fatty acid and partial glycerides, creating a high local concentration of fatty acid at the lipoprotein surface and endothelial cell membrane, far exceeding the capacity of albumin to remove it. If fatty acid was fully ionized and acted like a Class III lipid, membranes of the capillary would be readily dissolved by the detergency of the soaps. Fortunately, at pH 7.4, fatty acid is a membrane former (Class II lipid) and, as such, relatively nondisruptive to native cellular membranes.

C. Waxes, Esters, and Ethers

Waxes are esters of long-chain alcohols and long-chain acids (e.g., R—COOCH_2—R′). Their formation in animal cells has been described above.

Many plankton and higher members of the aquatic food chain, including coral, mollusks, fish, sharks, and even whales, store large quantities of waxes. Plankton appear to synthesize waxes that are used for both energy and buoyancy. When large amounts of liquid waxes are synthesized by plankton in the ocean depths, the density of the organism is lowered, permitting it to float to warmer regions near the ocean surface. Waxes are also present in skin lipids, beans, seeds, and leaves, serving to make the outer surface nonwettable. Ethers (R—CH_2OCH_2—R') are also present in skin lipids, forming a barrier against water loss.

Waxes are almost nonpolar molecules. They tend to accumulate in intracellular droplets and have almost no solubility in membranes.

Acylglycerols and Fats

Many complex lipids have a backbone of glycerol, a three-carbon polyalcohol. When a substitution is made at one end of the glycerol molecule, the 2-carbon (C-2) becomes optically active, as, for instance, in glycerophosphate. The standard nomenclature used is the *sn* terminology. A single substitution of octadecanoic acid to form an ester bond, e.g., at the 1-position, produces 1-octadecanoyl-*sn*-glycerol (Fig. 14). When two fatty acids are reacted with glycerol to form ester bonds, a diglyceride is formed (e.g., 1,3-diacyl-*sn*-glycerol or 1,2-diacyl-*sn*-glycerol). Diglycerides are biochemical intermediates in many lipolytic reactions, are critical building blocks used in the synthesis of more complex phospholipids and triacylglycerols, and are second messengers for some membrane-triggered reactions.

A triacylglycerol is formed when all three hydroxyls form ester bonds with fatty acids. If all three acids are the same, the triacylglycerol is called a simple triglyceride (e.g., triolein). If one of the fatty acids is different, it becomes a complex triglyceride. The properties of glycerides (e.g., melting point) depend greatly on the fatty acid chains involved. Triacylglycerols are the major storage lipids of plants and higher animals. Both plant oils (olive, corn, safflower) and animal fats (lard, suet, tallow) are nearly pure mixtures of complex triglycerides. In animals, adipose tissue is the main source of fat, but skeletal muscle, heart, liver, and bone marrow often contain appreciable triacylglycerols in intracellular oil droplets.

1. Glyceride Metabolism

The early stages of *de novo* triacylglycerol biosynthesis, the attachments of the first two acyl groups, are shared with the synthesis of phospholipids. Phosphatidic acid, a key intermediate in glycerolipid biosynthesis, is converted to 1,2-diacyl-

1 - Octadecanoyl - *sn* - Glycerol
(D - α - Monostearin)

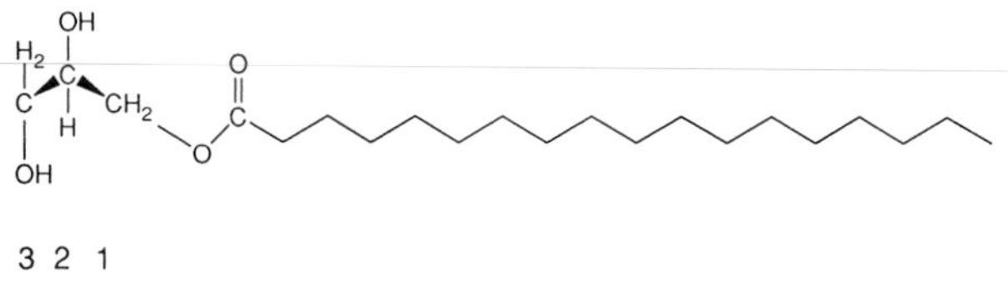

FIGURE 14 Line model of a monoacylglyceride, showing the stereospecificity of the glycerol conformation. With the glycerol carbon in the 2 (middle) position in the plane of the page and the carbons 1 and 3 behind the plane of the page, if the OH points up then the carbon on the right is designated carbon 1 and the one on the left carbon 3. Glycerol carbon numbers are shown below the glycerol structure.

sn-glycerol by removal of the phosphate moiety by phosphatidate phosphohydrolase. It is this step that seems to control the rate of triacylglycerol synthesis. The cellular levels of this enzyme are increased by glucagon and increased concentrations of cAMP, whereas insulin counteracts these effects. Short-term activation is observed with vasopressin, agents that affect the mobilization of calcium, and high levels of free fatty acids. Translocation of the phosphohydrolase from the cytosol to the endoplasmic reticulum, where it is in better contact with its substrate, is thought to be the mechanism behind activation by these latter agents. Triglycerides are formed by acylation of the diacylglycerol by acyl-CoA : diacylglycerol acyltransferase, which must compete for diacylglycerol with enzymes involved in phosphatidylcholine and phosphatidylethanolamine biosynthesis.

A separate pathway for triacylglycerol biosynthesis is found in the intestinal mucosa, which is involved in the absorption of dietary triacylglycerols. In the intestinal lumen, triacylglycerols are hydrolyzed by pancreatic lipase and colipase at the 1 and 3 positions to form 2 moles of fatty acid and 1 mole of 2-acyl-*sn*-glycerol. These products are solubilized by bile salts in mixed micelles and transported to the enterocytes that line the gut. They pass through the brush border where the fatty acid and monoacylglycerol are absorbed into the cell. The 2-acyl-*sn*-glycerols are then acylated in sequential steps, first by monoacylglycerol acyltransferase, then by diacylglycerol acyltransferase to reform triacylglycerol. The fatty acids used during the acylations are primarily those that have also been absorbed from the intestinal lumen.

Diacylglycerol is formed as an intermediate during *de novo* synthesis of certain phospholipids and triacylglycerols (see above). It can also be formed during the degradation of either triacylglycerol, by triacylglycerol lipase, or phospholipids, by phospholipase C. Once thought to be unimportant intermediates in lipid metabolism, diacylglycerols have attracted a great deal of interest because of their role as a second messenger in the activation of protein kinase C.

2. Behavior and Distribution

Monoacylglycerols are generally observed in extremely small levels in mammalian tissues. As previously mentioned, monoacylglycerols are generated during the breakdown of triacylglycerols by pancreatic lipase in the intestinal lumen before transport into the mucosal cell. They are generated during lipolysis of triacylglycerols in plasma by lipoprotein lipase and hepatic triglyceride lipase. Monoacylglycerols can also be generated in cells during the stimulated breakdown of phospholipids, although the exact mechanism involved in their formation and their effect on cellular processes (if any) are not known. Some are also transiently formed during lipolysis of adipose fat. Unlike diacylglycerols, monoacylglycerols have not been shown to be second messengers.

Triglycerides are weak Class I amphiphiles. They partition poorly into membranes (~3 mol% with respect to phospholipids) and are largely found in intracellular droplets of many cells, especially adipose tissue and muscle. In plasma they form the cores of chylomicrons and large VLDL. The more saturated fatty acids in the glyceride, the higher the melting point.

Diglycerides are strong Class I amphiphiles and partition strongly into membranes. Their presence in membranes of phospholipids lowers the $L_\alpha \Rightarrow$ Hex II transition temperature, a property that might favor membrane fusion.

Monoacylglycerols are Class II amphiphiles and form membranes readily. They also form other liquid crystal phases (cubic, Hex II) under appropriate conditions, Biologically, they would be found almost exclusively in membranes or the surfaces of lipoproteins. Albumin may also bind some monoglycerides in plasma.

E. Glycerophospholipids

These complex lipids have a phosphate at the 3-position of glycerol and acyl or alkyl groups on at least one (usually both) of the other glycerol carbons. The phosphate esterified to the *sn*-3-glycerol may be free (phosphatidic acid) or esterified to other small molecules (Fig. 15) to form phosphatidylcholine (lecithin), phosphatidylethanolamine, phosphatidylserine, phosphatidylglycerol, phosphatidylinositol. All these phospholipids have two acyl chains but different head groups, some charged (PS, PI, PG) and some zwitterionic (PC, PE) at pH 7. In some cases ether bonds replace ester bonds in their association with glycerol, as exemplified by plasmalogens. One ether-linked phospholipid, platelet activating factor, is a powerful hormone and causes marked vasoactivity. When one of the acyl groups is removed, a phospholipid with a single hydrocarbon chain is formed, called a *lysocompound*. Lysophospholipids act as Class III detergents because of their large water-soluble head group. A rare genetic deficiency of the enzyme lecithin, cholesterol acyltransferase, which transfers the fatty acid on the *sn*-2-position of lecithin to cholesterol to form a cholesterol ester and lysolecithin, has been described. Deficiency of this enzyme results in accumulation of lecithin and free cholesterol in plasma lipoproteins.

Although all phospholipids are Class II lipids (membrane formers), they differ in some important ways. PE has the smallest head group and tends not to be charged at neutral pH. The head group also does not hydrate as extensively as PC or the charged PLs. Thus, if the volume of the acyl chains is increased—by increasing temperature or by increasing the length and unsaturation of the chains—these lipids readily undergo a $L_\alpha \Rightarrow$ Hex II transition. They tend to be found on the inner surface of the plasma membrane. PC is zwitterionic and has very stable properties. It hydrates to about 9–11 molecules of H_2O and forms stable bilayers that do not readily form Hex II at normal biological temperatures. It is concentrated on the outside monolayer of the plasma membrane.

The charged phospholipids (PS, PI, PG) form stable bilayers but swell in water much more than the zwitterionic PC. Because they carry a (−) charge, they interact with divalent cations like calcium. At low Ca^{2+} concentrations, some of the negative charge is shielded and hydration is attenuated. At Ca^{2+} concentration of 2 mM–10 mM, saturated and unsaturated PS is converted from L_α to a dehydrated crystalline $Ca(PS)_2$ bilayer. Saturated PS also interacts with Li^+ to form crystalline bilayers. Fusion between membranes in cells (e.g., between a secretory vesicle and the plasma membrane to affect exocytosis) has been related to both local Hex

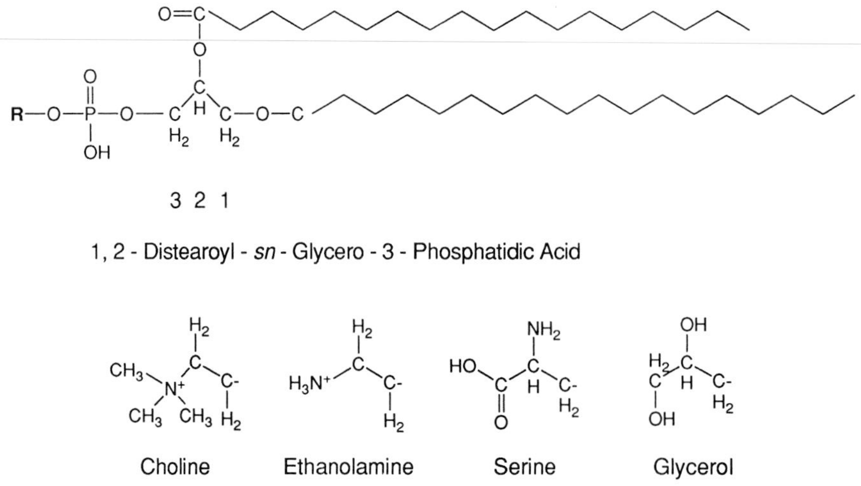

1, 2 - Distearoyl - *sn* - Glycero - 3 - Phosphatidic Acid

Choline   Ethanolamine   Serine   Glycerol

II formation or calcium-charged PL interaction at the site of fusion.

F. Sphingolipids

Sphingolipids are formed by the addition of fatty acids to the base sphingosine. When a fatty acid is linked to sphingosine through an amide bond at the 2-position, a molecule called a *ceramide* is formed. Substituted ceramides are important substituents of skin lipids. Elevated ceramides have been found in a disorder characterized by the inability of lysosomes to catabolize ceramide, known as Farber's disease. When ceramide is esterified with phosphocholine, sphingomyelin is formed. When sphingomyelin catabolic enzymes are absent, sphingomyelins accumulate in tissue, giving rise to Niemann-Pick disease.

Sphingosine can react at the 1-position with sugars to form specifically linked sugar moieties. The simplest moiety is psychosine, in which a monosaccharide is linked at the 1-position to sphingosine. When a monosaccharide is linked to the 1-position of a ceramide, a cerebroside is formed, such as galactocerebroside (Fig. 16). Different sugars may be added to form a variety of neutral glycosphingolipids. Complex charged glycocerebrosides, called *gangliosides,* are formed when sialic acid (negatively charged) is added to the sugars. These molecules in low concentration are firmly anchored to the outer surface of many plasma membranes and appear to act as both antigens and receptors for certain toxins, antibodies, and lectins. Specific enzymes are necessary for the catabolism of each of these glycolipids; when an enzyme is either absent

FIGURE 15 Line models of some phospholipids. If R = hydrogen, the compound is the phosphatidic acid as indicated. If the polar moieties shown below are substituted at R, the compound is the corresponding phosphatidyl-R (i.e., if R is choline, the compound is phosphatidylcholine).

or defective, the nonmetabolized glycolipid molecule accumulates, resulting in a lipid storage disease.

1. Sphingolipid Metabolism

Sphingolipids have recently attracted a great deal of attention because of their possible role in the attenuation of signal transduction through the regulation of protein kinase C activity as well as their function as the lipophilic portion of glycosphingolipids. Sphingolipid formation is initiated by the condensation of palmitoyl-CoA and serine to form dihydrosphingosine (3-oxosphinganine), followed by reduction of the ketone moiety by a NADPH-requiring enzyme, forming sphinganine (Fig. 17, reactions 1 and 2). In most sphingolipids, the sphingosines contain a *trans*-double bond produced by a poorly characterized flavoprotein (reaction 3). The primary amide group, originally associated with serine, is then acylated to form ceramide (reaction 4). The acylation reaction can involve either fatty acyl-CoA or free fatty acid as the cosubstrate, depending on the enzyme involved. The final step in sphingomyelin biosynthesis involves the transfer of choline, from either CDP-choline or phosphatidylcholine to ceramide. Although there is still uncertainty with respect to the true choline donor, evidence tends to support phosphatidylcholine as the physiologically relevant donor.

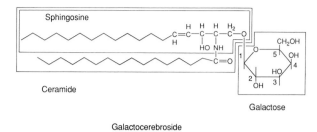

FIGURE 16 Line model of a cerebroside. Sphingosine is D-*ery-thro*-2-amino-4-octadecene-1,3-diol, the ceramide shown is 2-stearoyl sphingosine, and the galactocerebroside is 1-galactosyl ceramide.

Sphingomyelin is broken down in animal cells by sphingomyelinases, resulting in the formation of ceramide and phosphocholine. There are two forms of this enzyme, lysosomal and non-lysosomal. Although the relative importance of these two forms is not clear, mutations in the lysosomal sphingomyelinase observed in humans with Niemann-Pick disease result in the accumulation of sphingomyelin in reticuloendothelial cells. Ceramide, one of the products of sphingomyelinase, can be either reincorporated into sphingolipids or deacylated by ceramidase to form free fatty acid and sphinganine. Again, there are two forms of ceramidase, one active at acid pH and another active under neutral conditions, and both are capable of catalyzing the reverse of the catabolic reaction to form ceramide. A deficiency in humans in the acidic form of the enzyme (Farber's lipogranulomatosis or Farber's disease) results in accumulation of ceramide in the tissues and mental dysfunction in the patient. [*See* SPHINGOLIPID METABOLISM AND BIOLOGY.]

2. Glycosphingolipid Metabolism

Glycosphingolipids represent an extremely complex array of molecules that are characterized by carbohydrate chains attached covalently to a lipophilic anchor, ceramide. These compounds are classified according to the makeup of the carbohydrate chain, which can consist of more than 15 sugar residues and is usually branched. They are typically found on the outer aspect of the plasma membrane, with the carbohydrate chains extending into the external environment. Considering that more than 100 varieties of glycosphingolipids have been described, it is not surprising to find that these compounds have been implicated in many cellular functions including cell–cell communication, regulation of cell growth and receptor function, and antigenic specificity.

Ceramide is a pivotal intermediate, standing at the branchpoint between sphingomyelin and glycosphingolipid synthesis. The first step in the biosynthesis of glycosphingolipids is the glycosylation of ceramide, using either UDP-glucose or UDP-galactose, to form glucosylceramide and galactosylceramide, respectively (Fig. 17). It is glucosylceramide from which most (although not all) of the more complex glycosphingolipids are derived. A thorough description of the synthesis of all the glycosphingolipids is impossible, but we can generalize the process by stating that the more complex glycosphingolipids are formed from glucosylceramide by the sequential addition of various sugars, from sugar nucleotides, to the nonreducing end of the carbohydrate chain. Each step involves the addition of a single sugar to the growing chain by a glycosyltransferase.

Sulfatides (sulfated glycosphingolipids) are found in significant levels only in nervous tissue, kidney, testes, and the gastrointestinal tract, in which it is hypothesized they play a role in metal ion transport. In the formation of sulfatide (3′-sulfogalactosylceramide) galactosylceramide is sulfated by the microsomal enzyme galactosylceramide sulfotransferase, which uses 3′-phosphoadenosine-5′-phosphosulfate as the sulfate donor. This enzyme is membrane-bound and can also sulfate lactosylceramide (Gal(β1-4)GlcCer).

Probably as well studied as the synthesis of glycosphingolipids is their catabolism. This is due to the description of a variety of human disorders in which these compounds are not broken down. These disorders are typically caused by a deficiency in one of the glycosidases that break down the carbohydrate chain. The result is a buildup of glycosphingolipids within the tissues of the patients and the manifestation of a variety of neuromuscular dysfunctions. The complexity of the subject matter does not allow for a detailed description of this area, and the reader must be referred to other reviews covering the subject.

3. Behavior and Occurrences

Sphingolipids are found in most cells and especially in brain and nerve tissue. The ceramide molecule is probably in low concentration and, as such, would dissolve into membranes. In Farber's disease the high melting point of the noncatabolized ceramides (60–80°C) appears to allow crystals of

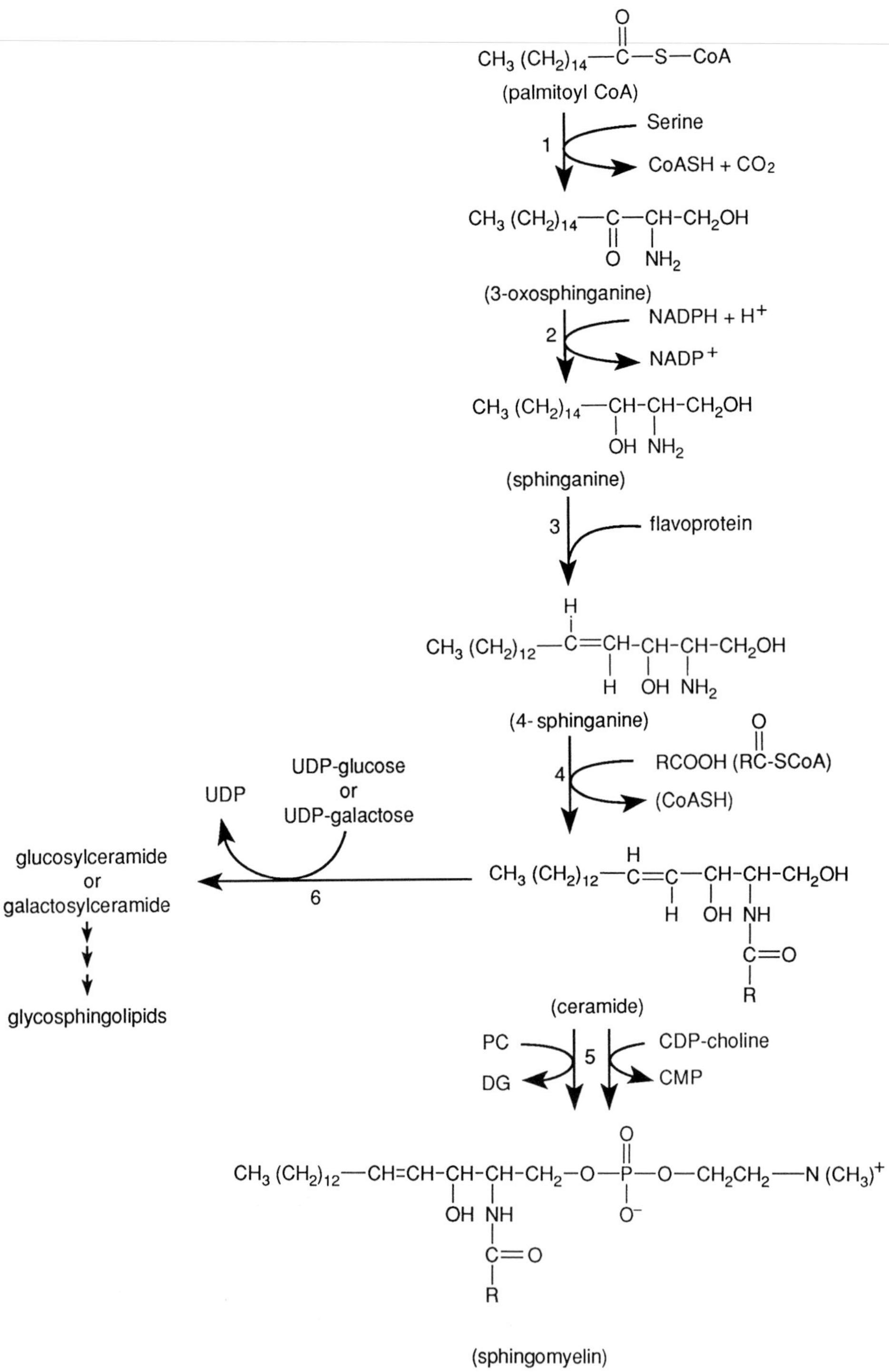

FIGURE 17 Sphingomyelin biosynthesis in animal cells.

this substance to deposit in cells. Sphingomyelins are widely distributed in plasma membranes, particularly in the outer leaflet. The ceramide hydrocarbon part tends to give sphingomyelins higher chain melting than the similar phosphatidylcholines.

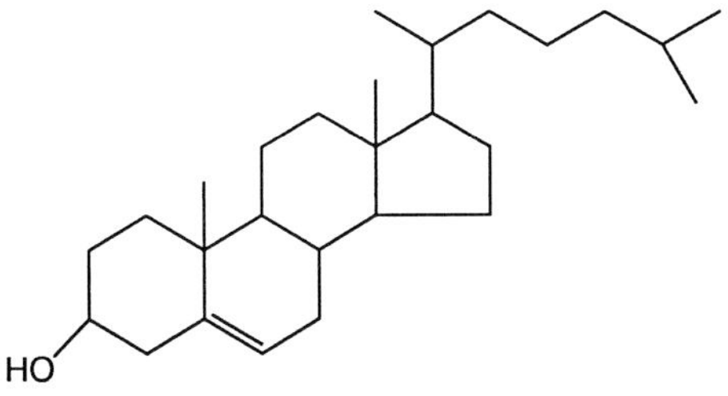

Cholest - 5 - en - 3β - ol
(Cholesterol)

FIGURE 18 Line model of cholesterol.

Thus, the gel $\Rightarrow L_\alpha$ transitions for many sphingomyelins are above body temperature. If in high enough concentration, these molecules may form patches of gel phase in membranes. However, because plasma membranes are usually rich in free cholesterol, the molar cholesterol/phospholipid ratio being about 0.5:1, cholesterol interacts with sphingomyelin to produce a very viscous, but not quite solid, membrane.

Glycosphingolipid behavior depends on the number of sugars and the presence of charge (sialic acid). Cerebrosides are found in fairly high concentration in nervous tissue. They have high solid $\Rightarrow L_\alpha$ transitions and may separate in phospholipid membranes to form local solid patches. In low concentration, they are dissolved in membranes. In cerebrosidoses, these lipids form massive solid and liquid crystal deposits. Di- and tri-glucosyl ceramides are less abundant and are generally partitioned in membranes. Gangliosides are present in small amounts in cell surfaces as receptors and cell markers, anchored by ceramide. They appear to be exclusively on the surface and do not transfer between cells. In the gangliosidoses, such as Tay Sachs disease, these molecules are not broken down and accumulate in high concentrations. They appear to act as Class III molecules, and they disrupt and dissolve the complex membranes of the brain, leading to severe early neurologic dysfunction and death.

G. Steroids

Steroids are molecules of ubiquitous biologic origin and great interest. They have been defined as "all those substances that are structurally related to the sterols and bile acids to the extent of possessing the characteristic perhydro-1,2-cyclopentano-phenan-

throne ring system." This polycyclic structure, consisting of four linked rings, is illustrated in the structure of cholesterol (Fig. 18), a molecule that may modulate membrane fluidity, permeability, fusibility, and thickness. Although cholesterol appears in low amounts in some primitive animals such as sponges, it becomes the predominant sterol in higher animals. [*See* CHOLESTEROL.]

Steroid hormones (i.e., testosterone, androgens, estrogens, progesterones, cortisol, cortisone, aldosterone) are formed from cholesterol. These molecules exert major effects in regulating metabolism in higher animals. [*See* STEROIDS.]

Plant sterols are present in vegetable oils. Normally, these molecules are not absorbed by the intestine of humans. However, a rare genetic condition that permits their absorption leads to β-sitosterolemia, a disease in which large amounts of plant sterols deposit in the body tissues. More complex molecules with a steroid nucleus, such as digitalis, are found in plants. Digitalis is a strong stimulant for heart contractions and has been used for centuries to combat heart failure. This agent also complexes with cholesterol in a 1:1 compound and has been used in the analysis of cholesterol.

Fatty acids form esters with cholesterol. These interesting molecules are often stored in organs, such as the adrenal and the corpus luteum of the ovary, where they serve as precursors for steroidal hormones. They also accumulate in certain disorders (e.g., cholesteryl ester storage disease, atherosclerosis, familial hypercholesterolemia, and Tangier disease). Cholesterol esters form both cholesteric and smectic liquid crystals, which can be identified by polarizing microscopy in living tissues.

Bile acids are formed by partly oxidizing the terminal side chain from 27 carbon atoms to 24 carbon atoms and by adding -OH groups to various positions in the ring. Thus, bile acids such as cholic and chenodeoxycholic acids are formed. The alkali metal salts of hydroxylated bile acids are natural detergents (Class III amphiphiles) synthesized in the liver and secreted into bile. They solubilize phospholipid and cholesterol in the bile of higher animals, thus permitting secretion of cholesterol into the gut. They aid in the absorption of fat and fat-soluble vitamins in the intestine. Bile acid deficiency may lead to cholesterol gallstone formation, and abnormal bile acid metabolism is associated with cholestanol storage in the disease cerebrotendinous xanthomatosis. In biochemical studies, bile acids have been used to solubilize membranes and membrane proteins. [*See* BILE ACIDS.]

H. Other Lipids

Eicosanoids refers to a large number of oxygenated fatty acids principally derived from the 20-carbon fatty acids, arachidonic acid, eicosatrienoic acid, and eicosapentenoic acid (Table V). These oxygenated derivatives are present in low concentration, are generally unstable and have a very short lifetime, and act as local hormones to influence contractility, membrane permeability, and many other cellular functions. *Acyl-coenzyme A* and *Acyl-carnitine* are key intermediates in fatty acid metabolism, usually present in low concentrations but under certain conditions may accumulate and act as Class III amphiphiles, disrupting some cellular functions. CDP-diacylglycerol is an intermediate in phospholipid synthesis and most probably partitions into membranes. *Glucosyl phosphatidylinositols* are a recently described class of lipid moieties that are attached to certain proteins and act as membrane anchors for the protein (e.g., alkaline phosphatase). It is interesting to note that the two fatty acid chains of PI appear to be adequate to anchor large proteins in membranes. Such proteins may be released from the membranes by phospholipase C, which hydrolyzes the PI, leaving diglyceride in the membrane. *Lipopolysaccharides* are a large class of bacterial lipids that are complex in nature and may act in higher animals as antigens, transforming factors, and mutagens. *Dolichols* are derived through the isoprenoid synthetic pathway and consist of 15–19 isoprene units (75–90 carbon atoms) to which phosphate is esterified to the terminal alcohol group. They are found in the endoplasmic reticulum and Golgi membranes and function as anchors for glycosylation in the lumen of these organelles. Their organization within these membranes is not known. These 75–95 Å long molecules are either buried in the center of the bilayer, or fold back and forth within the membrane three or four times, or their hydrophobic parts must be shielded by hydrophobic domains of proteins on either the cytoplasmic or luminal surface. The hydroxyl group of the dolichol, which is usually esterified to phosphate, appears to be on the luminal side of the endoplasmic reticulum or Golgi, and to this a large number of sugars are attached by specific transferases (up to 14), which are then translocated *en block* to protein during the post-translational glycosylation process. Little is known about the physical properties of the dolichol phosphates and the dolichol glycophosphates. *Vitamins A, D, E, and K* are also derived through modifications in the isoprenoid pathway. They are generally Class I lipids and are either bound to membranes or to specific carrier proteins.

Bibliography

Abrahamsson, S., Dahlen, B., Lofgren, H., and Pascher, I. (1978). Lateral packing of hydrocarbon chains. *Prog. Chem. Fats Other Lipids* **16,** 125.

Bligh, E. G., and Dyer, W. J. (1959). A rapid method of total lipid extraction and purification. *Can. J. Biochem. Physiol.* **37,** 911.

Feiser, L. F., and Feiser, M. (1959). "Steroids." Reinhold, New York.

Folch, J., Lees, M., and Sloane-Stanley, G. H. (1957). A simple method for the isolation and purification of total lipids from animal tissues. *J. Biol. Chem.* **226,** 497.

Hanahan, D., ed. (1978, 1983). "Handbook of Lipid Research Series," vols. 1–3. Plenum Press, New York.

Hilditch, T. P., and Williams, P. N. (1964). "The Chemical Constitution of Natural Fats," 4th ed. John Wiley & Sons, New York.

Mead, J. F., Alfin-Slater, R. B., Howton, D. R., and Popják, G. (1986). "Lipids. Chemistry, Biochemistry and Nutrition." Plenum Press, New York.

Myant, N. B. (1981). "The Biology of Cholesterol and Related Steroids." Wm. Heinemann Medical Books, London.

Ness, W. R., and McKean, M. L. (1977). "Biochemistry of Steroids and Other Isopentenoids." University Park Press, Baltimore.

Paoletti, R., and Kritcehvsky, D., eds. (1963–1989). "Advances in Lipids Research Series," vols. 1–23. Academic Press, San Diego.

Scriver, C. R., Stanbury, J. B., Wyngaarden, J. B., and Frederickson, D. S., eds. (1989). "The Metabolic Basis of Inherited Disease," 6th ed. McGraw-Hill Information Service Co., New York.

Small, D. M. (1986). The physical chemistry of lipids from alkanes to phospholipids. *In* "Handbook of Lipid Research Series," vol. 4. (D. Hanahan, ed.). Plenum Press, New York.

Vance, D. E., and Vance, J. E. (1985). "Biochemistry of Lipids and Membranes." Benjamin/Cummings Publishing Co., New York.

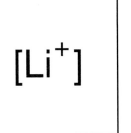

Lithium Effects on Blood Cell Production and on the Immune System

ARTHUR H. ROSSOF, *MacNeal Cancer Center; Rush University*

I. Initial Observations Made by Psychiatrists
II. Hematologic Research to Explain the Observed Granulocytosis
III. Exploitation of the Granulopoietic Effect of Lithium in Selected Hematologic Diseases and Cancer
IV. Immunologic Effects of Lithium
V. Concluding Remarks

Glossary

Bipolar affective disorder Psychiatric disorder (manic depressive illness) in which there are very wide mood swings, varying from severe depression with suicidal thoughts to agitated flamboyant periods in which unsound judgments are frequently made.

Colony stimulating factors General term, adopted from *in vitro* experiments and referring to a large group of materials that are growth factors for various hematopoietic cell populations

Cytotoxic chemotherapy Drug treatment which kills cells, generally used in reference to the drugs used to treat cancer by killing the malignant cells

Neutrophil (neutrophilic polymorphonuclear leukocyte, PMN) One of the types of white blood cells produced in the bone marrow. It commonly accounts for nearly half the circulating white blood cell population. It is characterized by a multilobed nucleus and a cytoplasm filled with conspicuous fine granules. These cells have a very brief half-life in the circulation and they provide a major part of the first line of defense against invading microorganisms, particularly bacteria. They are capable of phagocytosis and a number of chemical processes which are lethal to invading microorganisms

IN 1949, THE VALUE of lithium salts in the management of "psychotic excitement" was first reported. The Food and Drug Administration of the United States granted approval in 1970 for the use of lithium carbonate in the management of manic-depressive illness, now known as bipolar affective disorder. This simple salt is still marketed and continues to be the major drug in the psychiatric armamentarium for the management of patients with this bipolar illness. The marked efficacy of this agent in controlling the manic phase of this disorder and in generally smoothing the course of this potentially destructive illness has excited psychiatrists to evaluate the efficacy of lithium carbonate in other psychiatric disturbances, primarily other depressive disorders, alcoholism, and obsessive–compulsive disorders. None of these other disorders are managed with lithium carbonate as reliably as the bipolar illness, and its use in them, therefore, remains investigational.

Because the free lithium ion (Li^+) is a small monovalent alkali ion, similar to H^+, Na^+, and K^+, which are widespread in the intra- and extracellular fluids of the body, effects of various kinds can be anticipated in diverse organs. Renal, neurologic, and endocrinologic side effects have been reported, and some are anticipated with lithium carbonate use; the drug turns out to be remarkably safe as long as its serum level is kept within the therapeutic range (0.6–1.0 meq/liter). The rapid and accurate determination of serum [Li^+] has become a standard procedure in well-equipped clinical chemistry laboratories around the world.

I. Initial Observations Made by Psychiatrists

In the late 1960s, a number of reports appeared in the psychiatric literature indicating that patients treated with lithium carbonate were likely to experience an abnormal elevation of white blood cells (WBC), owing to an increased number of the neutrophilic polymorphonuclear (neutrophil [PMN]) cells, without changes of other cell types. Normally, neutrophils provide one of the first lines of defense against infection, particularly by bacteria. [*See* NEUTROPHILS.]

These observations were made by psychiatrists during the formal clinical evaluation of lithium carbonate as an investigational new drug. Thus, it can be assumed that these blood studies were being done more frequently than would have been dictated by the clinical circumstances alone, and any deviations from normal required explanation and would be factored into the overall regulatory concerns regarding safety versus efficacy. Accordingly, the patients were referred to infectious disease experts to determine whether or not they had occult infections and to hematologists to explain the changes in WBC composition.

II. Hematologic Research to Explain the Observed Granulocytosis

The search for an occult infection proved negative. Hematologists, however, faced an unusual new challenge, because neither the clinical appearance of the patients, the peripheral blood characteristics, nor the bone marrow examinations suggested leukemia or any other fundamental disorder of PMN production. Furthermore, many kinds of drugs can cause impaired production of PMNs with a diminution of the total WBC concentration in the bloodstream. At the time these observations were made, a great deal was already known about the proliferation and distribution of PMNs in the body.

Stress or hormones and drugs of the corticosteroid class cause release from the bone marrow of the pool of mature PMNs not yet released to the circulation, the storage pool. Exercise or epinephrine cause release off the vessel walls of the marginal granulocyte pool, the postmarrow granulocyte pool that is adherent to blood vessel walls and not directly measured when analyzing one's blood counts. Pharmacologic manipulation of these PMN pools is neither practical nor safe over long periods of time, is not known to be associated with clinical benefit, and merely redistributes these cells among the known pools of mature PMNs.

Hematologists provided evidence indicating that simple redistribution was not the explanation for the observations and that the total body mass of PMNs was expanded under Li$^+$ stimulation. The mechanism(s) by which Li$^+$ stimulated the production of PMNs was, however, unknown. Considerable research has been devoted to this topic over the past decade, but a single unifying mechanism of action cannot be described. Evidence indicates a direct effect of Li$^+$ on various bone marrow stem cell populations (the precursor cells), on bone marrow stromal cells, which produce the microenvironment for the blood cells, on the production and elaboration of normal physiologic regulators such as the colony-stimulating factors, and on lymphocytes, which may have a regulatory influence on PMN proliferation. It became clear that the normal background low-level concentration of Li$^+$ in the body contributes to the normal regulation of granulocytic, eosinophilic, and erythroid precursors. It was also found that the effect of lithium can also be produced by two other monovalent cations: rubidium and cesium.

Occasionally, some pathology was observed in lithium-treated persons. There were some cases of anemia, probably unrelated to the lithium therapy or representative of rare idiosyncratic reactions.

Several cases of leukemia have been reported in persons taking lithium, and some data suggested that Li$^+$ can stimulate leukemic clones *in vitro*. The association with leukemia, however, has been assumed to be fortuitous as the multiple forms of leukemia are not uncommon and lithium use is widespread. Still, one cannot exclude the possibility that lithium could rarely stimulate a known or unknown leukemic clone to more rapid proliferation and clinical deterioration. [*See* LEUKEMIA.]

III. Exploitation of the Granulopoietic Effect of Lithium in Selected Hematologic Diseases and Cancer

One can readily envision several potential applications of lithium therapy for existing and anticipated neutropenic conditions. This section is devoted to commentary on each of these applications.

A. Management of Neutropenic Disorders

The circumstances under which the various neutropenic conditions arise is imprecise. Some are acquired, others congenital; some are familial, others sporadic. Some are associated with known causes, some are idiopathic (i.e., no cause known). The prognosis for many forms is indefinite, whereas the course for some is clear. Some are associated with other clinical problems, thus lending themselves to a discrete clinical definition. Some cases are obscured by antiquated terminology emphasizing other clinical features such as aplastic anemia.

This imprecision obscures the study and analysis of these diseases and the reporting of responses to hematopoietic stimulants such as lithium. Nonetheless, an impressive number of cases with granulocytopenia respond to lithium therapy. The most widely recognized responses are seen in the disorder known as Felty's syndrome, in which granulocytopenia is associated with rheumatoid arthritis. Although not all persons with Felty's syndrome respond to lithium, the number is sufficiently high enough to urge clinical investigators to attempt lithium therapy in this condition. In one case, a patient with Felty's syndrome was pretreated with lithium so that he could be treated with very myelosuppressive cytotoxic cancer chemotherapy. The observed blood count response was typical of hematologically normal persons receiving the same chemotherapy.

Granulocyte responses to lithium also have been seen in the congenital familial disorder known as the Shwachman syndrome, which is characterized by pancreatic insufficiency and granulocytopenia.

Lithium effects were studied in another disorder known as cyclic hematopoiesis (formerly known as cyclic neutropenia), in which the reproduction of all bone marrow-derived hematopoietic elements and the number of the end cells in circulation vary cyclically. PMN cycles are most evident because their normal half-life in the circulation is briefer than that of the other blood cells. An animal model of this disease exists in the grey collie dog. This disease is a fundamental disorder of marrow function, which can be corrected by bone marrow transplantation. Oral lithium therapy in the grey collie dog alters the cycling pattern and, in most cases, protects them from severe life-threatening granulocytopenia. In humans, however, this syndrome is generally unresponsive to lithium, although in one patient it increased the PMN counts at their lowest point, without stopping the cycling.

Some effects of lithium were observed in hairy cell leukemia, a rare form of chronic leukemia associated with a low number of other types of blood cells. The low PMN count causes these patients to be susceptible to infectious complications. Several patients have responded to oral lithium carbonate

with an increase of their PMNs and, occasionally, of other blood cell types and a decrease of circulating hairy cells.

Other patients with a variety of classifiable and nonclassifiable granulocytopenic disorders have been treated with lithium carbonate with variable results.

Although in most cases accurately classifying neutropenic patients is not possible, in virtually all cases of chronic or cyclic granulocytopenia, it is reasonable to offer a trial of oral lithium carbonate, because reports of beneficial responses can be found in the literature, the treatment is rarely associated with severe toxicity, the cost is low, and response, if it is to occur at all, is likely to be seen in 1–2 wk.

B. Adjuvant Management of Patients Receiving Myelosuppressive Anticancer Therapy

Undoubtedly, the most common form of granulocytopenia is that caused by cytotoxic chemotherapy or radiotherapy to kill cancer cells. These forms of treatment injure many cells because with today's technology they cannot be specifically targeted to the malignant cells. The bone marrow cells that proliferate rapidly are especially susceptible to enhanced injury and lethality by these modalities of cancer care. A number of cases, most anecdotal and uncontrolled, have been reported in which patients receiving anticancer therapy were also treated adjunctively with oral lithium carbonate. Animal models have been developed, especially in mice and dogs, in which the behavior of blood PMNs and marrow stem cells have been determined following myelosuppressive therapy and lithium.

These animal experiments are generally positive, and many encouraging reports have come from various trials in humans.

The treatment of acute leukemia in adults consists of an effort to ablate the bone marrow with intensive cytotoxic chemotherapy. Normal myeloid elements are sacrificed along with the neoplastic cells. These special circumstances provide an exceptional opportunity to evaluate drugs that may protect the normal hematopoietic tissue. A successful experiment could have any of the following end points: less reduction of the blood counts; less duration of blood count suppression; more rapid and/or higher recovery of normal blood counts; fewer days with fever, infection, or on antibiotics; or fewer days in the hospital. Several studies, including well-controlled randomized studies, have failed to indicate a clear role for lithium in achieving any of these desirable end points in the human affected by adult acute leukemia.

In "solid tumor" chemotherapy, however, tumor mass is generally situated away from the marrow space and normal bone marrow function is generally preserved. In most solid tumor studies reported, in adults or children, some beneficial result has been seen in the lithium-treated patients. The overall significance remains unclear, however, as treatment results are not necessarily improved, competing causes of mortality are not erased, and survival differences are not regularly identified or reported. Detailed quality-of-life studies have not yet been performed, but conceivably they would yield data suggesting an improved status in lithium-treated patients, especially if infectious morbid complications were reduced. Some transplant surgeons have used oral lithium in organ-transplant patients in whom immunosuppressive drugs have caused granulocytopenia, but no uniform results were reported.

These various phenomena can produce contrasting systemic effects. Confirmed or putative serious immunologic reactions, either suppressing or enhancing the immune system, have been reported only rarely in the vast number of patients receiving lithium.

B. Special Considerations Regarding Lithium and the Acquired Immunodeficiency Syndrome

As most of the immunologic effects of lithium are immunostimulatory, there may be a potential role of this agent in the repair of the immunodeficiency that characterizes acquired immunodeficiency syndrome (AIDS). Although this concept has not yet been tested, lithium has been given to AIDS patients with granulocytopenia due to zidovudine (AZT) therapy with good results. If lithium is found also to reverse some of the immunodeficiency that characterizes AIDS, it may have a unique double benefit in granulocytopenic AIDS patients receiving AZT.

IV. Immunologic Effects of Lithium

Entirely separate from the issues pertaining to lithium effects on bone marrow, various investigators have explored the effects of lithium on the cells of the immune system: lymphocytes and monocyte–macrophages. [*See* LYMPHOCYTES; MACROPHAGES.]

A. General Immunologic Observations

Many studies have evaluated either immunocompetent cell function in patients receiving lithium or studies in which normal immunocompetent cells were exposed under finite conditions to graded concentrations of Li^+. The reported effects include augmentation of thymidine incorporation in lymphocytes by mitogen or antigen, increased immunoglobulin production, increased E-rosette formation, enhanced proliferation in mixed lymphocyte cultures, enhanced suppressor cell activity, enhanced interferon production, increased natural killer cell activity, and increased interleukin-2 production following mitogen stimulation. Monocytes produce more interleukin-1 and tumor necrosis factor and have enhanced phagocytosis.

V. Concluding Remarks

The monovalent cation of the lightest metal has wide biological properties when provided in pharmacologic quantity. Its effects on blood cells, particularly on granulocyte production and on lymphocyte function, are important and frequently unique side effects that can be exploited easily and inexpensively in the proper clinical circumstances.

Bibliography

Barr, D. B., Koekebakker, M., Brown, E. A., and Falbo, M. C. (1987). Putative role for lithium in human hematopoiesis. *J. Lab. Clin. Med.* **109,** 159–163.

Boggs, D. R., and Joyce, R. A. (1983). The hematopoietic effects of lithium. *Semin. Hematol.* **20,** 129–138.

Gallicchio, V. S. (1988). Lithium and granulopoiesis: Mechanisms of lithium action. *In* "Lithium: Inorganic Pharmacology and Psychiatric Use" (N. J. Birch, ed.). Proceedings of the Second British Lithium Congress, September 6–9, 1987, pp. 93–95. IRL Press Limited, Oxford, England.

Gupta, R. C., Robinson, W. A., and Kurnick, J. E. (1976). Felty's syndrome—Effect of lithium on granulopoiesis. *Am. J. Med.* **61,** 29–32.

Irvine, A. E., Crockard, A. D., Desai, Z. R., Ennis, K. T., Fay, A. C., Morris, T. C. M., and Bridges, J. M. (1986). Lymphocytes from patients receiving lithium do not inhibit CFU-C growth. *Br. J. Haematol.* **62,** 467–477.

Pazdur, R., and Rossof, A. H. (1981). Cytotoxic chemo-therapy for cancer in Felty's syndrome: Role of lithium carbonate. *Blood* **58,** 440–443.

Rossof, A. H., and Robinson, W. A. (eds.) (1980). "Lithium Effects on Granulopoiesis and Immune Function," Vol. 127. *Advances in Experimental Medicine and Biology,* Plenum Press, New York.

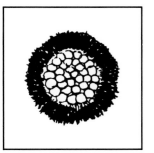

Liver Cancer and the Role of Hepatitis B Virus

WILLIAM S. ROBINSON, *Stanford University School of Medicine*

Glossary

Cirrhosis Pathologic state of the liver in which reaction to hepatocyte injury or necrosis results over months or years in features including loss of microscopic lobular architecture, liver regeneration in the form of regenerative nodules of hepatocytes, and scarring or fibrosis

Hepadnaviruses Family of small DNA viruses that share ultrastructural, molecular, antigenic, biological, epidemiologic, and pathogenic features with the prototype hepatitis B virus; that employ a reverse transcriptase step in replication of their circular DNA genome; and chronic infection with many (and perhaps all) hepadnaviruses is associated with an increased risk of developing primary liver cancer

Hepatitis B vaccine Hepatitis B surface (envelope) antigen (HBsAg) particles, immunization which can induce protective immunity

Hepatitis B virus Small DNA virus that infects human liver cells, sometimes causing acute viral hepatitis (hepatitis B) and sometimes causing chronic or persistent infection that can be associated with chronic hepatitis B, cirrhosis, and primary liver cancer

Insertional mutagenesis Viral integration that results in altered function or expression of a cellular gene (proto-oncogene or tumor suppressor gene), thus imparting neoplastic properties to the cell

Oncogenes Viral or altered cellular genes, the expression of which in a cell has the potential for oncogenic transformation of the cell

Primary liver cancer (PLC), hepatocellular carcinoma (HCC), or hepatoma Malignant tumor arising clonally in the liver from a hepatocyte or the primary parenchymal cell of the liver

Viral hepatitis Pathologic state of the liver in which liver cell injury or necrosis resulting from acute or chronic viral infection leads to reactive inflammation and repair

Viral integration Insertion of viral DNA into chromosomal DNA of an infected cell by interruption of the chromosomal DNA and covalent joining of the ends of viral DNA to chromosomal DNA ends

PRIMARY LIVER CANCER (PLC) or hepatocellular carcinoma (HCC) is one of the most common cancers of man. Although this cancer occurs in all parts of the world, the highest prevalence is in regions in which hepatitis B virus (HBV) infections are most common (eastern Asia and sub-Saharan Africa). There is strong epidemiologic evidence that persistent HBV infection is the most important risk factor for the development of HCC. Recent research has provided much information about the molecular and genetic structure of HBV, its molecular mechanism of replication, its phylogenetic relationship to other viruses, and its molecular state in cells of HCC. This information suggests possible mechanisms by which HBV infections can lead to HCC. Because of the important role of HBV infections in development of HCC, control of this virus through HBV vaccination of high-prevalence populations offers the best approach for prevention of HCC.

I. Introduction

Hepatitis B virus (HBV) infections commonly occur at early ages and become persistent in populations in which this virus is highly endemic, such as in eastern Asia and sub-Saharan Africa. Macronodular cirrhosis develops in a significant fraction of such persistently infected individuals, usually over the course of many years. Long-standing persistent HBV infection is associated with a greatly increased risk of hepatocellular carcinoma (HCC) compared with the risk in uninfected populations, and the presence of macronodular cirrhosis appears to further increase the risk of HCC by at least 10-fold. The strong association of HCC with persistent HBV infection in man and with other natural hepadnavirus infections in woodchucks, ground squirrels, and ducks suggests an important role for infection with members of this virus family in the development of HCC. HCCs in man often arise from adenomatous foci that form within regenerative nodules of cirrhotic liver, and the finding of a clonal pattern of hepadnavirus DNA integration in cellular DNA of many HCCs in man and in other hosts naturally infected with hepadnaviruses indicates that these tumors arise from a single cell with integrated virus.

The most important risk factor for HCC in man appears to be chronic or persistent hepatitis B virus infection, however, several other specific factors are known to increase the risk of developing liver cancer and almost all HCC cases appear to be associated with HBV or one of the other recognized risk factors. It has been estimated that 80% of all HCCs in the world occur in HBV-infected individuals and careful prospective studies have indicated that 40% or more of middle-aged Chinese males with chronic HBV infection can be expected to die of HCC. Since the prevalence of this infection can be 10% or more in high-prevalence populations, the number of liver cancers in large populations is very great and accounts for HCC being among the most common cancers of man. Important risk factors associated with the remaining 20% of HCC cases include chronic alcohol-induced liver disease, chronic non A, non B hepatitis (caused by persistent infection with a newly discovered virus with properties of a *Togavirus* or *Flavivirus* that for several years has been designated posttransfusion non A, non B hepatitis virus and is now called hepatitis C virus), hemochromatosis (liver injury caused by excessive dietary iron or genetically determined excessive intestinal absorption of iron, which is deposited in the liver over many years), dietary aflatoxins (carcinogenic toxins produced by fungal organisms that infest improperly stored grain, peanuts, and other food), and possibly other factors associated with a very small fraction of HCC cases. The important risk factors for HCC all lead to the common pathologic effect of chronic liver cell injury or necrosis and the common reactive processes of inflammation, liver regeneration, and, in humans, frequently cirrhosis. A large majority of HCCs in different settings and associated with the important risk factors occur in individuals with cirrhosis. [*See* TOGAVIRUSES AND FLAVIVIRUSES.]

HCC has a worldwide distribution and is among the most common cancers of man. However, there is an uneven geographic distribution of HCC in the world because the most important risk factors for HCC are unevenly distributed. The highest prevalence of HCC is in eastern Asia (China and adjacent regions) and sub-Saharan Africa, and these are areas of the world with the highest prevalence of chronic HBV infection. The prevalence of HCC is much lower in the United States and western Europe, where HBV infections are much less common and more important risk factors for HCC are alcoholic liver disease and probably chronic non A, non B hepatitis.

This article will consider the role of hepatitis B virus in the development of HCC by reviewing important features of the virus, describing the evidence that HBV infection plays an important role in HCC, and considering possible carcinogenic mechanisms for the virus.

II. Hepatitis B and Related Viruses (Hepadnaviruses)

HBV infection was recognized as a risk factor for HCC soon after the serendipitous discovery of hepatitis B surface antigen (HBsAg) in human serum in the mid 1960s, and its later association with acute serum hepatitis (now known as hepatitis B) and eventually its identification as a viral antigen. This led to rapid advance in understanding the epidemiology of the virus, the course of infection, its associated diseases, and the physical nature of the virus. Many primary HBV infections were then shown to be asymptomatic, infections could persist for many years, and the virus was found to be hepatotropic (infects primarily liver cells), although less frequent and less permissive infection sometimes occurs in bone marrow, circulating leukocytes, B-cells and T-cells, and pancreas. Virtually all active infections were accompanied by high concentrations of HBsAg and often infectious virus in the blood. Persistent infections were often associated with minimal liver disease or with chronic hepatitis that sometimes led to cirrhosis, and the risk of hepatocellular carcinoma was shown to be more than 200 times higher in some such "HBsAg carrier" populations than in similar uninfected populations. Studies of geographic distribution revealed HBV in all regions of the world but unevenly distributed, with the highest prevalences of serum HBsAg being in eastern Asia and sub-Saharan Africa (e.g., approximately 10% in China compared with 0.1% in the United States), where most primary infections occur at very early ages, are silent or subclinical, and almost all become persistent. Between 10 and 50% of HBV infections that become persistent in different high-prevalence geographic regions are acquired perinatally from infected mothers and very few are intrauterine infections. In Western countries most primary HBV infections are in adults (acquired by sexual contacts or percutaneous exposure), they are more often associated with clinically apparent acute hepatitis B, and only 5 to 10% become persistent.

There are estimated to be more than 200 million HBsAg carriers in the world, and these are the only reservoir of HBV for new human infections. The associated chronic liver disease and HCC are among the most important human health problems in high-prevalence regions. The more recent discovery of several closely related viruses with similar epidemiologic features and associated with similar disease syndromes, including HCC occurring naturally in certain rodent and avian species, indicates that there may be many members of this virus family, which has been named Hepadnaviridae for hepatotropic DNA viruses. The hepadnavirus family includes hepatitis B virus of man, woodchuck hepatitis virus (WHV) of *Marmota monax*, ground squirrel hepatitis virus (GSHV) of *Spermophilus beecheyi*, duck hepatitis B virus (DHBV) found in several varieties of domestic ducks, and similar viruses in tree squirrels and herons. Less well documented findings in other rodents, marsupials, and cats suggest that other hepadnaviruses may have been detected.

A characteristic feature of all hepadnaviruses is a moderately narrow host range or spectrum of ani-

mals that can be infected with a particular virus. For example, HBV infects only man, chimpanzees, and possibly other great apes, but not monkeys or other species. All hepadnaviruses tend to selectively infect liver cells, which then produce large amounts of noninfectious viral envelope particles (as well as infectious virus) that can be detected in high concentrations (up to 500 μg/ml) in the blood and that commonly cause persistent infections with viral forms in liver and blood continuously for years and often for the lifetime of the host. Infections with hepadnaviruses may be associated with acute and chronic hepatitis characterized by liver cell necrosis, inflammatory reaction, lymphocytic infiltration and liver regeneration, immune complex (viral surface antigen–antibody) mediated disease, and hepatocellular carcinoma.

Hepadnaviruses share unique features of virion (virus particle) size and ultrastructure with an envelope surrounding an electron-dense spherical nucleocapsid or core (see Figs. 1 and 2); characteristic polypeptide and antigenic composition; common virion DNA size, structure, and genetic organization; and an unusual mechanism of viral DNA replication that includes reverse transcription (synthesis of viral DNA on an RNA template) of a greater than genome length viral RNA transcript. The virion DNA is among the smallest of all known animal viruses, consisting of a 3200-base-pair circular molecule that contains a single-stranded region of different length in different molecules (see Fig. 3), reflecting the fact that DNA molecules are packaged into virions before their replication is complete. Neither viral DNA strand is a covalently closed circle and single breaks are present in the two strands with the 5' ends at unique positions in the nucleotide sequence approximately 200 nucleotide pairs apart (see Fig. 3). The virion contains DNA polymerase activity that catalyzes the repair of the single-stranded region to make fully double-stranded relaxed circular DNA molecules, and this enzyme appears to be the reverse transcriptase involved in viral DNA replication in infected liver cells. The viral genome contains four genes that encode (1) the hepatitis B surface antigen proteins of the virion outer layer or envelope (the S-gene), (2) the major structural polypeptide of the viral core or nucleocapsid with hepatitis B core antigen (HBcAg) and hepatitis B e antigen (HBeAg) specificities (the C-gene), (3) the reverse transcriptase and a protein primer for viral DNA synthesis (the P-gene), and (4) a protein called the X-protein that can transactivate

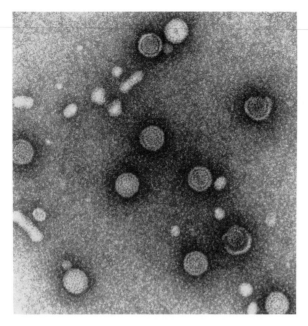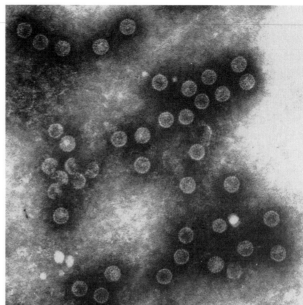

transcription controlled by kB-like enhancer elements that are DNA sequences that specifically bind the transcription factor NFkB (the X-gene) (see Fig. 3). The hepadnavirus genome is unusually compact and efficiently organized in that much of the genome is utilized for multiple functions, including overlapping genes, so that the same nucleotide sequence encodes more than one protein and all cis-acting regulatory sequences (e.g., transcriptional enhancer and promoter elements) are contained in genomic sequences also encoding protein (see Fig. 3).

Viral genome replication involves conversion of infecting virion DNA molecules to covalently closed circular molecules in liver cell nuclei; formation of a greater than genome length RNA transcription with a terminally repeated sequence and shorter transcripts that function as messenger RNAs; packaging of the long transcript, newly made viral reverse transcriptase, and protein primer in viral core particles found in hepatocyte cytoplasm; and synthesis of new viral DNA molecules by reverse transcription of the long RNA transcript, utilizing a viral-encoded protein primer in forming the first synthesized viral DNA-strand and a capped oligoribonucleotide derived from the 5' end of the long RNA template as primer for synthesis of the second (+) DNA strand exclusively within cytoplasmic viral core particles (see Fig. 4).

Hepadnaviridae appear to be phylogenetically related to members of two other virus families: cauli-

FIGURE 1 Electron micrograph of hepatitis B viral forms in the blood of an affected patient. Electron micrographs of virions (left) and virion cores after detergent (NP 40) treatment of virions (right) are shown (experiment by June Ameida).

flower mosaic virus, and Retroviridae and related transposable elements. Several factors indicate that these virus families are related: similarity in gene number, function, and order; shared genome nucleotide sequence homology in genomic regions of similar function; and utilization by all of a reverse transcriptase step in genome replication. However, the form of the genome packaged in virions may be different (linear RNA in the case of retroviruses and nicked or gapped double-stranded circular DNA in hepadnaviruses and cauliflower mosaic virus) and details of the replication mechanisms are different. Reverse transcription is an unusual mechanism of genome replication for DNA viruses (known at this time only for hepadnaviruses and cauliflower mosaic virus) and undoubtedly it is used because of their evolution from a common ancestor shared with retroviruses and different from the evolutionary pathway of other DNA viruses. The relationship of hepadnaviruses with retroviruses is of particular interest because many retrovirus infections in animals are associated with neoplasia or malignant tumor formation, and intense investigation of these viruses has led to understanding certain viral oncogenic mechanisms.

A notable difference in hepadnavirus and retrovirus replication is in the requirement for DNA

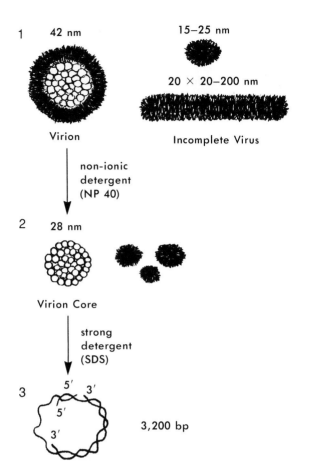

FIGURE 2 Schematic representation of hepatitis B viral forms found in the blood of infected patients.

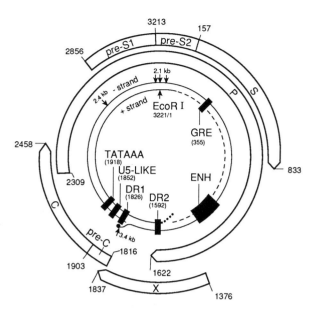

FIGURE 3 Circular map of the HBV (adw2) genome. The inner circles represent the virion DNA stands and the broken (dashed) line in the short (+) DNA strand represents the region within which the 3′ end of the + strand may occur in different molecules, and the corresponding region of the long strand is that which may be single-stranded in different molecules. A line of dots represents the oligoribonucleotide primer covalently attached to the 5′ end of the + DNA strand and a single dot represents the protein primer covalently attached to the 5′ end of the − DNA strand. The large arrows represent the recognized functional open reading frames (ORF) with the direction of transcription from the minus DNA strand indicated. The small arrows indicate the 5′ ends of the three major size classes of transcripts of 3.4, 2.4, and 2.1 kb, all of which are identically terminated near the poly (A) addition signal (TATAAA). The nucleotide sequence locations of the initiation and termination codons of each ORF are given with reference to map position 1 at the single *Eco*RI cleavage site in this DNA. The map position of the first nucleotide of the glucocorticoid responsive element (GRE), DR2, DR1, the U5-like sequence, and the poly (A) addition sequence (TATAAA) are indicated, as is the general region (1182–1216) exhibiting enhancer (ENH) activity.

genome integration in cellular DNA. The orderly integration of retroviral DNA "provirus" with a preserved genome sequence that includes a terminally repeated sequence (the long terminal repeat or LTR) containing cis-acting transcriptional and other regulatory elements and the expression of viral genes and synthesis of new RNA genomes from the integrated provirus are integral features of retrovirus replication that occur in all infected cells. This feature is not shared by hepadnaviruses, whose replication appears to involve only episomal (unintegrated) viral nucleic acid forms. Hepadnavirus DNA has been found to be integrated in cellular DNA of some chronically infected livers (as well as many HCC) but viral integration appears to be an uncontrolled event involving a mechanism called illegitimate recombination and integration is not required for virus replication. Most of the evidence of the fine structure of hepadnavirus integrations comes from investigation of those in human and woodchuck HCCs and there has been much less investigation of viral integrations in nontumorous liver. However, there is no apparent difference in integrations of HCC and nontumorous liver. The viral genome sequence is rarely preserved in integrations, but sequence rearrangements, including deletions and duplications of viral sequences (as well as flanking cellular DNA rearrangements), are the rule and would preclude functioning of such integrations in virus replication as is the case with retroviruses. Whether viral integration occurs in every hepadnavirus-infected cell *in vivo* has not been clearly established because it is difficult to de-

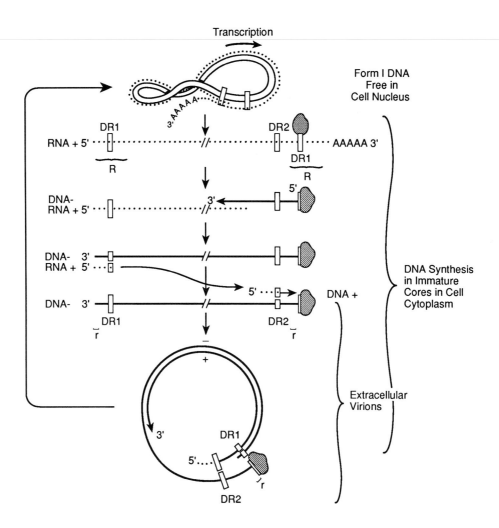

tect viral integration in cells in which virus is replicating and appropriate cell culture systems to investigate this question have not been available for the mammalian hepadnaviruses.

Viral DNA appears to be integrated in many different (possibly random) sites in cellular DNA. In cirrhotic liver, HCCs arise in adenomatous foci within regenerative nodules and examination of individual cirrhotic nodules of HBV-infected livers has revealed that the DNA of many nodules contains no integrated virus, other nodules contain a clonal pattern of viral DNA integration (i.e., viral integration in the same cellular DNA site in many cells of the nodule, implying that the nodule was formed by proliferation and clonal expansion of an original hepatocyte that contained a viral integration), and still others contain a nonclonal pattern of viral integration (many integrations at different cellular DNA sites). Thus the state of HBV DNA is not

FIGURE 4 Scheme of proposed mechanism of hepadnavirus DNA replication. DR1 and DR2 represent 12-nucleotide-pair direct repeat sequences. R represents the approximately 200-nucleotide terminal redundancy in the long RNA transcript and r represents the short terminal redundancy of the − DNA strand. Solid lines represent DNA strands, dotted lines RNA, and the stippled area the protein primer for DNA strand synthesis.

the same in all regenerative nodules of HBV-infected cirrhotic liver. While hepadnavirus integration does not appear to favor identifiable cellular DNA sites, there are preferred sites for integration within the viral genome. In 50% or more of viral integrations, the site in the viral genome that joins cellular DNA is very near the nucleotide positions of the 5′ ends of the two virion DNA strands (see Fig. 3), suggesting that the ends of the DNA strands may be involved in the integration mechanism.

III. Hepadnavirus Infection and HCC

The best evidence that chronic hepadnavirus infection plays an important role in the development of HCC is the epidemiologic association in all the well-studied hepadnavirus–host systems. This association was first suggested for chronic HBV infection and HCC in man, and the recognition of HCC in woodchucks and ducks prompted the search for and discovery of hepadnaviruses in these hosts.

A. Geographic Correspondence of HBV Infections and HCC

Epidemiologic evidence indicates that there is a geographic correspondence of relative rates of HBV and HCC. HCC in man has a worldwide distribution and, numerically, is one of the major cancers in the world today. Although HCC is rare in most parts of the world, it occurs commonly in sub-Saharan Africa, Southeast and eastern Asia, Oceania, Greece, and Italy, and in Alaskan Native Americans. Geographic areas with the highest incidence of HCC are also areas where hepatitis B virus infection is common and persistent HBV infections occur at the highest known frequencies. Within the limits of the data available, there appears to be a good correlation between the worldwide geographic distribution of HCC and active HBV infection (serum HBsAg positive), with the highest frequency of both occurring in sub-Saharan Africa and eastern Asia. Apparent unusual local exceptions to this close correspondence in high-prevalence regions suggest the possibility of other important factors such as dietary aflatoxin in these populations. In certain areas of Asia and Africa, where the prevalence of chronic HBV infection is the highest, HCC is the most common cancer in males.

B. Increased Incidence of HCC in Hepadnavirus-Infected Populations

The incidence of HCC has been shown to be much higher in hepadnavirus-infected compared to uninfected humans, woodchucks, ground squirrels, and some duck populations. This appears to be true for HBV in man in both high- and low-incidence HCC populations. A prospective study of more than 22,000 male government workers in Taiwan has shown the incidence of HCC to be more than 200-fold higher in HBsAg positive than in HBsAg negative individuals, and a prospective study in Japan

yielded similar results. The few cases of HCC in HBsAg negative patients in the large prospective study had serum anti-HBc, indicating past HBV infection. These studies document that HBV infection precedes the development of HCC and they quantitate the risk. HCC incidence in HBsAg carriers in these populations rises steeply with age after 40 years, and in HBsAg carriers with cirrhosis (characterized by regenerative nodules and fibrosis) the risk of HCC appears to be more than 10-fold higher than in carriers without cirrhosis. The risk of HCC is significantly higher in males than in female carriers, and this sex difference is greatest in the presence of cirrhosis.

Some evidence suggests that most HBV infections in HCC patients occurred early in life and had continued for many years before HCC developed. The high incidence of persistent HBV infection in mothers of HCC patients, in contrast to that in fathers, suggests that transmission from mothers to newborn or infant children is a frequent mode and time of HBV infection in HCC patients. If HBV infection does occur frequently at very early ages in HCC patients in high-HCC areas, the age distribution of patients with clinically recognized HCC would suggest that these tumors usually appear after continuous HBV infection of 30 or more years. Few cases of HCC occur in children. Up to 90% of HCC patients have coexisting cirrhosis. Projections made from the large prospective study in Taiwan have suggested that over 40% of middle-aged male HBsAg carriers in Taiwan will die of HCC and less than 60% from all other causes.

Although the preceding findings make a persuasive case for a role of HBV in HCC, the association with two other mammalian hepadnaviruses is even stronger. More than 90% of both wild-caught WHV-infected and colony-born experimentally infected woodchucks died of HCC within 2 to 3 years. Analysis of food given to such captive woodchucks has revealed no detectable aflatoxin or other recognizable carcinogen. These animals have active hepatitis with significant components of inflammation and regeneration, but cirrhosis has not been observed. In a colony of captive ground squirrels infected with GSHV in the wild, no HCCs were seen before 4.5 years but by 7 years more than two-thirds of the infected animals had died of HCC. As with infected woodchucks, infected ground squirrels have significant hepatitis but not cirrhosis. In both the woodchuck and ground squirrel colonies, HCC developed in a few HBsAg negative, anti-HBc/anti-HBs positive animals (evidence of past infection) but in no animals without any serum viral marker. These studies suggest that chronic hepadnavirus infection alone, without a recognizable cofactor, can result in HCC and this process appears to be the most efficient in the case of WHV-infected woodchucks. In individuals (humans, woodchucks, or ground squirrels) with evidence of past hepadnavirus infection (serum HBsAg negative, and anti-HBs and/or anti-HBc positive), the risk of developing HCC is apparently much lower than in those with chronic infection (serum HBsAg positive), but significantly higher than in individuals with no marker of either ongoing or past infection.

The association of DHBV infection and HCC in ducks is less clear. DHBV was first found in small brown domestic ducks in a region of China where HCC is common in these animals and in man. In addition, there appears to be a high content of aflatoxin in many human and animal foods in that region, a factor that complicates the assessment of the role of hepadnaviruses in HCC. Although the DHBV was first discovered in these ducks, no correlation of virus infection with HCC in this location has been reported. Infected brown ducks from this region of China (and not uninfected animals) followed prospectively in Japan were noted to develop HCC. On the other hand, white Pekin ducks infected with DHBV in the United States have not been observed to develop HCC. Whether this apparent difference is related to a critical difference in virulence of Chinese and U.S. DHBV strains, susceptibility of different hosts (e.g., Chinese brown versus Pekin ducks) or to other nonviral factors is not clear. The apparent differences in incidence and time of onset of HCC in the different hepadnavirus–host systems provide an opportunity to investigate viral virulence and host susceptibility factors in the development of HCC. The ability to infect woodchucks with GSHV, and different duck varieties with different DHBV strains, should permit direct comparison of different viruses in the same host.

IV. The Presence of Virus in Some (but not all) HCC

Hepadnaviruses have been found in the cells of many, although not all, HCCs of hepadnavirus-infected hosts. Hepadnavirus DNA has been found in approximately 85% of HCCs but not in the remaining 15% of surface antigen carrier humans, wood-

chucks, and ground squirrels. Thus persistence of viral DNA in tumor cells detectable by Southern blot analysis does not appear to be essential for the development of HCC in hepadnavirus-infected liver. Episomal viral DNA forms have been observed in some HCC and integrated viral DNA can be detected in most (75 to 85%) although not all HCC. As described earlier, HCCs often arise within adenomatous foci in regenerative nodules of cirrhotic liver and the state of HBV DNA is not the same in all such nodules. It is not known whether the state of HBV DNA in such cirrhotic nodules influences the risk of neoplastic change. However, the occurrence of HBV DNA integrated in most HCC of HBV-infected cirrhotic liver suggests that HCCs arise most often within regenerative nodules with HBV integrations, and the smaller number of HCC without integrations may arise from cirrhotic nodules without viral integrations. Viral DNA integration occurs in a few (usually 1 to 4) specific cellular DNA sites in many cells of individual tumors, indicating that tumors are clonal, but integrations always occur at different cellular DNA sites and on any of several different chromosomes (found so far on 12 different chromosomes in 26 different HCCs in published reports) in different tumors. When multiple tumors are present in a liver, the same clonal integration pattern is usually found in different tumor nodules, indicating that they are metastases arising from the same original tumor, but occasionally integration sites can be different, indicating a multicentric origin. Hepadnavirus integrations in HCC can lead to deletions and rearrangements of cellular DNA at sites of integration. Among the most dramatic examples are chromosomal translocations, all different, reported for three different HCC (involving chromosomes 17 and 18, 5 and 9, and X and 17), which suggest postintegration rearrangement of integrated viral and flanking cellular sequences. Whether any of these translocations had an oncogenic effect or were incidental to the HCC is unknown.

Staining of HCC tissue with immunofluorescent- and immunoperoxidase-labeled antibodies to viral antigens has demonstrated that in patients with HBsAg in the blood and in whom nontumorous liver cells are positive for HBsAg and/or HBcAg, tumor cells appear most often to be negative for these antigens, although some studies have reported small numbers of HBsAg positive cells in tumors. HBcAg has been detected even more rarely. Thus few tumor cells appear to express either viral gene product in amounts that can be detected by this method and cells within the same tumor do not express these antigens in a uniform way.

V. Viral Oncogenic Mechanisms

The hepadnavirus association with HCC raises the question of whether these viruses act by any oncogenic mechanism known for other tumor viruses. The cells of monoclonal tumors are likely to contain one or more common genetic alterations that cause the neoplastic phenotype. The first identified genes recognized to impart neoplastic properties to cells, that is, to cause rapid neoplastic transformation in cell culture and tumors *in vivo*, were certain retroviral genes called viral oncogenes that were carried into the cell by virus infection. The neoplastic phenotype of such cells requires persistence and continuous expression of the viral oncogene in the cells. Retroviral oncogenes were soon found to have nucleotide sequence homology with specific cellular genes (now designated proto-oncogenes), leading to the concept that retroviral oncogenes were derived from these cellular genes by recombination between viral and cellular genomes during virus infection. Although viral oncogenes share a high degree of sequence homology, they are not identical to proto-oncogenes but contain point mutations, deletions, or genetic substitutions that in some cases have been shown to result in a gene product that has an altered control mechanism so that it manifests increased biological activity. Proto-oncogenes commonly have regulatory functions in normal cells and the expression or function of proto-oncogenes can sometimes be altered by mutations in a way that imparts neoplastic properties to the cells, and such mutated or "activated" proto-oncogenes are called cellular oncogenes. Such mutations may result in altered protein structure and activity or in altered regulation and overproduction of the gene product or oncoprotein. The effects of oncogenes often appear to involve disturbance of a signal transduction pathway that transmits signals from extracellular factors that in turn affect cell growth or function in the cell nucleus, where the gene expression is regulated. [See ONCOGENE AMPLIFICATION IN HUMAN CANCER.]

This pathway includes cell surface membrane receptors that are proteins with cell surface, transmembrane, and cytoplasmic domains, and interaction of the cell surface domain with extracellular factors such as growth factors often triggers protein (tyrosine) phosphorylation catalyzed by the cytoplasmic domain of the receptor. Membrane-associated guanine nucleotide binding proteins with GTPase activity, called G-proteins, may be involved in transmitting the signal to the cytoplasm. Cytoplasmic proteins such as serine–threonine protein kinases in the family of protein kinase C proteins appear to be involved in relaying the signal to the nucleus. Nuclear factors involved in the signal transduction pathway are proteins that regulate transcription and possibly DNA replication by directly binding to specific DNA sequences with cis-acting regulatory functions (e.g., transcriptional enhancer or promoter elements) or by complexing with or activating other proteins to form active regulatory factors. These proteins are frequently phosphoproteins and their function may be regulated by phosphorylation. Some 23 retroviral oncogenes have been identified and the functions of the cellular homologues of some of them include growth factors (sis), tyrosine kinase–growth factor receptors (erb B), hormone receptors (erb A), G-proteins (ras), nonreceptor tyrosine kinases (src), serine–threonine kinases (mos and raf), nuclear proteins (myc, myb, fos, and jun), including recognized transcription factors (jun and fos), and others with as yet unknown functions. Activation of the cellular counterparts (proto-oncogenes) of more than 10 viral oncogenes has been implicated in oncogenesis.

Many retroviruses do not contain oncogenes and some cause cancer by another mechanism known as insertional mutagenesis. Insertion of retroviral DNA near or within a proto-oncogene can activate transcription of the proto-oncogene by action of a viral promoter or a viral transcriptional enhancer. Insertion of viral DNA can also result in proto-oncogene mutations. New proto-oncogenes have been discovered by their proximity to retroviral integrations in tumors arising through insertional mutagenesis. Another oncogenic mechanism is inactivation of cellular genes that normally appear to suppress neoplastic behavior of cells (tumor suppressor genes) by insertional mutagenesis. There is usually a prolonged time lag between virus infection and development of tumors that arise through the mechanism of viral insertional mutagenesis. The reason is that retroviruses integrate in many cellular genome sites at random, but only integrations that can activate proto-oncogenes or inactivate tumor suppressor genes are effective and these are rare.

Members of several different DNA virus families

in addition to hepadnaviruses are associated with a variety of tumors. For example, Epstein–Barr virus (EBV) infection is associated with naturally occurring nasopharyngeal carcinomas and B-cell lymphomas (e.g., Burkitt's lymphoma) in man, papillomavirus infections are associated with naturally occurring anogenital and cutaneous tumors in man and other animals, and polyoma viruses (e.g., SV40 and mouse polyomavirus) and adenoviruses transform cells in culture and produce tumors in experimentally infected rodents but are not associated with naturally occurring tumors. The DNAs of polyoma viruses and adenoviruses become integrated in the chromosomal DNA of infected cells by illegitimate recombination (a phenomenon unrelated to virus replication), often leading to rearranged viral sequences unlike the orderly integration of retroviruses. Certain of these viruses (polyomaviruses, adenoviruses, and papillomaviruses) are known to contain viral oncogenes (usually more than one gene of these viruses plays a role in cell transformation) and unlike retroviral oncogenes that were derived from cellular genes, those of DNA tumor viruses have no cellular counterpart. The produts of some of these viral genes are membrane proteins that regulate protein phosphorylation reactions and others are nuclear proteins. Polyoma virus middle T-antigen interacts with and stimulates the protein kinase activity of the product of the cellular proto-oncogene src. SV40 large T-antigen protein and adenovirus E1b (55K) protein complex with the p53 protein product of a tumor suppressor gene. Similarly, SV40 large T-antigen, adenovirus E1A transforming protein, and human (HPV-16) papillomarvirus E7 protein each complex with the Rb protein product of the retinoblastoma (Rb) suppressor gene. By forming these complexes the viral gene products are thought to impair the antiproliferative action of the tumor suppressor gene products, resulting in neoplastic behavior of the cell. A characteristic feature of retroviruses and DNA viruses with oncogenes is that they transform cells in culture and/or induce tumors *in vivo* rapidly after virus infection. [*See* EPSTEIN–BARR VIRUS; PAPILLOMAVIRUSES AND NEOPLASTIC TRANSFORMATION.]

An oncogene has not been identified in EBV and its role in the neoplasms associated with EBV infection has not been determined. However, in almost all B-cell lymphomas associated with EBV infection, translocations involving the c-myc oncogene on chromosome 8 and one of the chromosome segments bearing immunoglobulin genes on chromosomes 2, 14, or 22 are present. The mechanisms that triggers the critical translocation and the role of EBV in this event are uncertain.

A. Do Hepadnaviruses Contain an Oncogene?

There is no evidence that hepadnaviruses carry an oncogene capable of transforming cells and causing a tumor. The very long time (10 to 100% of the life expectancy of the host) between infection and appearance of HCC and the failure of hepadnaviruses to transform or induce proliferation of liver cells in culture are against the presence of an oncogene. Three viral genes (S or surface gene, C or core/e gene, and P or polymerase gene) encoding virion structural proteins are unlikely oncogene candidates. The role of the X-gene is not known although it is expressed in the liver of at least some HBV-infected patients. The X-protein of HBV expressed in cells in culture can activate transcription controlled by kB enhancer, which is a common enhancer sequence known to regulate the immunoglobulin kB chain gene, the β-interferon gene, HIV-1, SV40, and several other viral and cellular genes. Whether the HBV X-protein directly interacts with the kB sequence or activates an NFkB-like cellular factor is unknown. Although there is no kB sequence in the HBV genome, the isolated X-gene does appear to activate HBV transcription to a small extent and this may or may not represent a mechanism that regulates HBV replication. However, the possibility that the X-gene activates a cellular gene that plays a role in HBV replication or a cellular gene that could have oncogenic effects must also be considered.

Expression of the isolated X-gene in NIH 3T3 cells has been shown to make them tumorigenic in nude mice. However, this gene is often interrupted or missing in HCC with integrated viral DNA. This is because the X-gene spans the DNA region of virion DNA where viral DNA joins cellular DNA in viral integrations, suggesting that expression of the complete X-gene product in naturally occurring HCC could not be a common event, although X-protein antigen appears to have been detected in some HCC. The viral enhancer, the X-promoter, and a limited part of the X coding sequence, however, are more often retained in viral integrations and may sometimes be positioned to influence cellular gene expression or sometimes result in X-cell gene fusion proteins. Some such fusion proteins appear to have retained the transcriptional transactivator function of the X-protein. Duck hepatitis B virus contains no X-gene and thus the X-gene could play no role in DHBV-associated HCC in ducks, as it could not in HCC in mammalian hosts without viral integrations.

B. Hepadnavirus Integrations Altering Expression of Cellular Genes

Among all hepadnavirus integrations in HCC only a very few have been found to be within or near known cellular proto-oncogenes. These include a single integration in a human HCC that is fused in frame with the proto-oncogene erb A, the product of which is related to steroid and retinoic acid receptor protein. The tumor tissue was not available for studies to assess the level of expression of this cellular gene and it is not known whether the integration and the alteration of the erb A gene in this case played any role in the neoplastic change. This integration, although intriguing, appears to be an exception and of unknown biological significance, and viral integrations within or near erb A have not been found in other HCC.

Viral integrations have been found in the genomic domain of the proto-oncogene c-myc in 2 of 31 HCC of WHV-infected woodchucks examined. In one, viral integrations within the untranslated region of the proto-oncogene c-myc exon 3 resulted in overexpression of a long c-myc viral cotranscript. In the second, a single insertion of highly rearranged viral sequences 600 base pairs upstream of c-myc exon 1 was associated with increased levels of normal c-myc mRNA. In both cases, viral enhancer insertion and disruption of normal c-myc transcriptional or posttranscriptional control appeared to be involved in c-myc activation that could contribute to the genesis of HCC. The number of HCCs with c-myc activation by WHV integrations, although small, is much greater than expected by chance if viral integrations occur at random cellular DNA sites, suggesting that this could be a significant mechanism in a fraction of woodchuck HCC. c-Myc rearrangements have not been found in human HCC.

An HBV integration has been found associated with a deletion of cellular DNA at chromosome 11p12 in the region of a tumor suppressor gene, the loss of which is associated with Wilm's tumor and hepatoblastoma. The functional significance of this HBV integration has not been determined and it is

premature to conclude that it played a role in the HCC in which it was found. With these exceptions, hepadnaviruses have not been found to integrate in the proximity of known proto-oncogenes or tumor suppressor genes or to alter their expression.

C. Pathogenesis of Hepadnavirus-Associated HCC without Implicated Viral Integrations

We have seen that neither persisting viral genetic material nor an effect of viral integrations on a common cellular sequence appears to be responsible for a common genetic alteration important for the genesis of HCC. In addition, no common mutation or chromosomal alteration, such as translocation not associated with a viral integration and activating a proto-oncogene, has been found in HCC as described for certain other tumors, including some that are known to be associated with virus infection (e.g., EBV and B-cell lymphomas).

Since all monoclonal neoplasms would appear to contain genetic alterations that critically alter cell functions (e.g., activation of proto-oncogenes or inactivation of tumor suppressor genes), it is important to investigate HCC for such mutations and determine how chronic hepadnavirus infection may play some role without a direct effect of retained integrated viral sequences, since viral integrations cannot be implicated in most hepadnavirus-associated HCC. Carcinogenesis in most systems appears to be a multistep process and more than one mutagenic event thus may commonly be required for cells to manifest a fully malignant phenotype. Liver regeneration, which is a feature of chronic hepatitis and cirrhosis associated with chronic HBV infection, involves proliferation of many cells with HBV integrations, and such integrations have been shown to be unstable and can lead to new mutational events in cellular genes or regulatory elements through postintegration rearrangements of cellular sequences at sites of viral integrations. Viral sequences appear to be lost or deleted at some such sites of rearranged cell DNA. If oncogenic mutations arose by such a mechanism, the resulting HCC might have no detectable viral integration (as is the case for approximately 15% of hepadnavirus-associated HCC) or have one or more viral integrations away from the site of the critical mutation(s), as current evidence suggests may be the case for the majority of hepadnavirus-associated HCC.

Another observation of uncertain biological significance is that HBV DNA appears to transform

NIH 3T3 cells in culture with a significant frequency. The transformed cells grow suspended in soft agar and form tumors in nude mice. A subgenomic fragment of HBV DNA containing the transforming activity has several potential stem and loop configurations similar to "insertion sequence-like" elements found in fragments of herpes simplex virus DNA, which transform cells in culture. It is unclear, however, whether HBV DNA transforms 3T3 cells by such a mechanism, and whether transformation of 3T3 cells by HBV DNA has biological significance in relation to HCC formation.

On the other hand, it is not clear that the virus need be involved in such a direct way. There may be mutagenic mechanisms in chronic hepadnavirus-infected liver not directly related to virus integration that could lead to the accumulation of mutations over a long period with a chance that oncogenic mutations eventually occur. A common feature of the important risk factors for HCC, including chronic hepatitis B, chronic non A, non B hepatitis, alcoholic liver disease, hemochromatosis, and cryptogenic cirrhosis, is that they are associated with chronic hepatocellular injury and necrosis, resulting in inflammation and liver regeneration that continues over many years. This process commonly leads to cirrhosis in man, a pathologic lesion of the liver characterized by fibrosis or scarring and regeneration in the form of regenerative nodules or foci of liver cells arising through proliferation (and at least sometimes clonal proliferation) of liver cells. Sixty to 90% of HCC in different settings and associated with different risk factors occur in individuals with cirrhosis. HCCs appear to arise in adenomatous foci within the regenerative nodules of cirrhotic liver. Since several diverse initiating factors resulting in a common pathologic process all appear to lead to the development of HCC, it seems possible that this common pathologic process is in some way carcinogenic and the carcinogenic mechanism thus may not depend specifically on which of the diverse factors initiates the process of liver injury. Therefore it is important to identify mutagenic mechanisms that may be present during hepatocellular injury and the reaction to injury initiated by any of the foregoing risk factors for HCC.

For example, a mutagenic factor that is undoubtedly present in chronic hepatitis and developing cirrhosis is oxidants (active oxygen, radicals, and organic peroxides) generated by phagocytic and other cells during chronic inflammation. Such oxidants can cause DNA single-strand breaks, can activate expression of proto-oncogenes c-fos and c-myc, and can be strong tumor "promoters" in some models. Antioxidant compounds can block the tumor "promoter" activity of such oxidants. Consistent with this model are recent interesting findings with transgenic mice carrying HBV sequences encoding pre-S + S polypeptide (see Fig. 3) and no other viral gene. This polypeptide was overexpressed in the liver of these transgenic mice, and HBsAg accumulated in hepatocytes resulting in hepatocyte injury, necrosis, inflammatory response, and liver regeneration. At several months of age, these animals developed cirrhosis and HCC. This would appear to be an example of chronic hepatocellular injury leading to reactive inflammation, liver regeneration, cirrhosis, and HCC, and the only clear role of the virus would appear to be in providing a mechanism for liver injury that initiated other events, including inflammatory response and hepatocyte proliferation. Mutation ("initiation") and stimulation of cell proliferation ("promotion") also appear to be important for successful chemical carcinogenesis in several tissues including liver.

Although there is no evidence for a common chromosomal alteration (e.g., translocation, deletion, or point mutation) involving a known oncogene in most HCC (exception being the c-myc and erb A rearrangements described earlier), DNA studies have shown loss of alleles due to deletions on chromosomes 11p and 13q at a high frequency in HBV-associated HCC. That finding, together with the observation that HBV integrations have been found more often in chromosome 11 (although none has been reported in chromosome 13) than in any other chromosome in HCCs, suggests that deletions in 11p or in 13q could play a role in many HCC.

Other recent studies have shown in isolated cases the possible activation of apparently previously unrecognized proto-oncogenes detected by cell transformation assays. NIH 3T3 cells transformation by the DNA of a few HCC suggests that cellular genes with transforming activity not present in normal liver may have been detected. The genes appear not to be related to previously recognized oncogenes and are not related to HBV integrations. What role such cellular genes may play in HCC is unclear because it is not certain that factors that "transform" cells like NIH 3T3 (which already have properties of autonomous growth in culture) will necessarily be tumorigenic for normal hepatocytes. It is also unclear whether hepadnavirus infection plays a role in the mutations in such genes. However, these and other experimental approaches need to be extended to attempt to identify one (or more) common tumorigenic mutation(s) in HCC and determine how hepadnaviruses may play a role.

VI. Control of HCC

Although tumorigenic mechanisms leading to HCC and the precise role of HBV in development of most HCC are unknown, the overwhelming epidemiologic evidence described earlier indicates that persistent HBV infection is the most important risk factor for HCC in man. This suggests that prevention of new HBV infections may eventually prevent a majority of HCC. Although measures such as identification of infected individuals and avoidance of the kind of contacts that result in HBV transmission can reduce infection rates to some extent, the development of effective vaccines for HBV represents the most important tool for control of this virus. Vaccines consisting of HBsAg particles purified from the plasma of human HBsAg carriers and HBsAg produced by recombinant DNA methods in yeast or vertebrate cells in culture have proven to be highly immunogenic and provide excellent protection against HBV infection. Remarkably these vaccines are also highly immunogenic in newborn infants (many vaccines are poorly immunogenic in the first year of life when infants are immunologically immature) and can reduce perinatal infections by 70 to 80% when newborn infants of serum HBsAg positive mothers are vaccinated within 7 days of birth and subsequent vaccine doses are given 1 and 6 months later. When vaccination is combined with immunoglobulin preparations with high titers of anti-HBs (hepatitis B immune globulin or HBIG) given intramuscularly to infants within a few hours after birth, perinatal infection rates can be reduced by 95% or more.

The goals of HBV vaccination in geographic regions of high HBV prevalence are to interrupt the cycles of infection that lead to viral persistence in 5 to 20% of these populations and the sequelae, including HCC, by immunizing all newborn infants. Unfortunately only HBV vaccine alone can be used in many such areas because of the great expense and limited supply of HBIG, and thus HBIG is not available for optimum protection of newborn infants. In low HBV prevalence countries, newborn infants of all HBsAg positive mothers should be given HBIG and HBV vaccine just after birth to

prevent HBV infection, and all other susceptible (serum HBsAg, anti-HBS, and anti-HBc negative) children and adults with significant risk of HBV exposure (e.g., health care workers and children in households with an HBV carrier) should be given HBV vaccine. This approach promises to be the most effective way to attempt to control and eventually eliminate hepatitis B virus and thus prevent HCC. In high HBV prevalence populations it is expected to take several generations to achieve universal vaccination of all newborns and thus eliminate most new HBV infections because of difficulties in manufacturing sufficient quantities of vaccine at a feasible cost and in delivering vaccine to all who require it. Because of the very large number of existing HBV carriers of all ages in these populations, most of whom will remain infected for life, HBV-associated HCCs, most of which occur after age 35, can be expected to continue to occur for several years after new infections in children have been stopped. A reduction in HCC incidence following reduction in HBV carrier rates in populations through vaccination programs will further confirm the importance of HBV infections in the development of HCC. HCC is the only cancer in man today for which the most important approach for control is vaccination.

Bibliography

Beasley, R. P., Lin, C. C., Hwang, L. Y., *et al.* (1981). Hepatocellular carcinoma and hepatitis B virus: A prospective study of 22,707 men in Taiwan. *Lancet* **2,** 1129–1133.

Bishop, J. M. (1987). The molecular genetics of cancer. *Science* **235,** 305–311.

Dejean, A., Bougueleret, L., Grzeschik, K. H., and Tiollais, P. (1986). Hepatitis B virus DNA integration in a sequence homologous to v-erb-A and steroid receptor genes in a hepatocellular carcinoma. *Nature* **322,** 70–72.

Edmondon, H. A., and Craig J. R. (1987). Neoplasms of the liver. *In* "Diseases of the Liver" (L. Schiff and E. R. Schiff, eds.). 6th ed., pp. 1109–1158. Lippincott, Philadelphia.

Hsu, T., Moroy, T., Etiemble, J., Louise, A., Triepo, C., Tiollais, P., and Buendia, M. A. (1988). Activation of c-myc by woodchuck hepatitis virus insertion in hepatocellular carcinoma. *Cell* **55,** 627–635.

Marion, P. L., Van Davelaar, M. J., Knight, S. S., Salazar, F. H., Garcia, G., Popper, H., and Robinson, W. S. (1983). Hepatocellular carcinoma in ground squirrels persistently infected with ground squirrel hepatitis virus. *Proc. Nat. Acad. Sci. U.S.A.* **83,** 4543–4546.

Maynard, J. E., Kane, M. A., and Hadler, S. C. (1989). Global control of hepatitis B thru vaccination: Role of hepatitis B vaccine in the expanded program on immunization. *Rev. Infectious Diseases* (Suppl. 3), S574–S578.

Nagaya, T., Nakamura, T., Tokino, T., Tsurimoto, T., Imai, M., Mayumi, T., Kamino, K., Yamamura, K., and Matsubara, K. (1987). The mode of hepatitis B virus DNA integration in chromosomes of human hepatocellular carcinoma. *Genes Dev* **1,** 773–782.

Popper, H., Roth, L., Purcell, R. H., Tennant, B. C., and Gerin, J. L. (1987). Hepatocarcinogenicity of the woodchuck hepatitis virus. *Proc. Nat. Acad. Sci. U.S.A.* **84,** 866–870.

Robinson, W. S. (1989). Hepadnaviruses and their replication. *In* "Virology" (B. N. Fields and D. N. Snipe, eds.) 2nd ed. Raven Press, New York.

Rogler, C. E., Sherman, M., Su, C. Y., Shafritz, D. A., Summers, J., Shows, T. B., Henderson, A., and Kew, M. (1985). Deletion in chromosome 11p associated with a hepatitis B integration site in hepatocellular carcinoma. *Science* **230,** 319–322.

Seeger, C., Ganem, D., and Varmus, H. E. (1986). Biochemical and genetic evidence for the hepatitis B virus replication strategy. *Science* **232,** 477.

Summers, J., and Mason, W. S. (1982). Replication of the genome of a hepatitis B-like virus by reverse transcription of an RNA intermediate. *Cell* **29,** 403–415.

Szmuness, W. (1978). Hepatocellular carcinoma and the hepatitis B virus: Evidence for a causal association. *Progr. Med. Virol.* **24,** 40–69.

Yokosuka, O., Omata, M., Zhou, Y. Z., Imazeki, F., and Okuda, K. (1985). Duck hepatitis B virus DNA in liver and serum of Chinese ducks: Integration of viral DNA in a hepatocellular carcinoma. *Proc. Natl. Acad. Sci. U.S.A.* **82,** 5180–5184.

Locomotion, Neural Networks

STEN GRILLNER, GRIGORI ORLOVSKI, *Nobel Institute for Neurophysiology, Karolinska Institutet, Stockholm*

Glossary

Basal ganglia Large groups of nerve cells in the forebrain that are important for motor coordination

Midbrain locomotor region Midbrain area that elicits locomotion when stimulated repetitively; it controls the spinal cord networks via relay neurons

Pyramidal tract Pathway from the cerebral cortex to the spinal cord, also referred to as the corticospinal tract; it is important for the visual guidance of the limb movements during locomotion

Spinal pattern generator Network of interneurons in the spinal cord that when activated can produce the coordinated muscle activity underlying locomotion

THE GENERATION OF locomotor movements, which are adapted to the environment and the needs of an animal or human, is not a trivial task for the central nervous system. Hundreds of different muscles must be coordinated in each step cycle to provide the propulsive thrust while maintaining good control of the body position so that it will not fall in one direction or another. Several control systems are linked to achieve a good and well-adapted behavior. The basic system for propulsion consists of a spinal cord network of nerve cells, which causes an activation of different muscle groups in a sequence appropriate for stepping movements. This network can be turned on by simple signals from the brainstem locomotor region, and the level of activity can likewise be set by brainstem neurons. Sensory signals informing about the progress of the movement play an important part in the control system and interact directly with the spinal network. The principles underlying the central nervous control of this complex behavior is reviewed in this chapter.

I. Motor Pattern in Walking and Running

Walking and running are the two main forms of locomotion in humans. Walking is normally used between 0 and 7 km/hr and running from around 5 km/hr to extreme values of near 45 km/hr toward the end of a 100-yard race. The basic features of these two forms are the same (Fig. 1): the legs perform alternating stepping movements. The step cycle of each leg consists of a support and a swing phase. At the same time, many differences exist between the motor patterns of walking and running, both in relation to timing of various events within the step cycle and in relation to the force trajectories and the flexion–extension movements at the different joints. These small but numerous distinctions indeed make walking and running two different forms of locomotion, and a gradual transition from one pattern to another is not observed.

In walking, one or two legs are in contact with ground throughout the step cycle (single support phases or double support phases). In running, some phases in the step cycle are without any support. Both in walking and running, the knee and hip antigravity muscles (extensors) are activated during

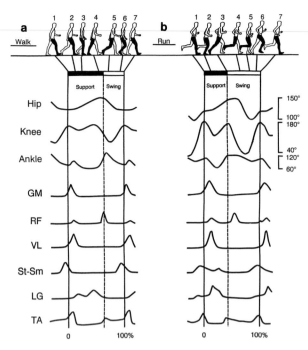

FIGURE 1 Walking and running in humans—coordination in the step cycle. The upper row shows the body and leg positions at various phases of the step cycle of the right leg. The main phases of the step cycle of this leg are indicated (i.e., the support phase and swing phase). The graph shows the joint angle excursions at the main joints of the right leg (hip, knee, and ankle) as well as the activity (rectified and filtered electromyograms) of some muscles of this leg. GM, gluteus maximus (hip extensor); RF, rectus femoris (hip flexor); VL, vastus lateralis (knee extensor); St-Sm, semitendinosus and semimembranosus (knee flexor); LG, lateral gastrocnemius (ankle extensor); TA, tibialis anterior (ankle flexor). [Adapted, with permission, from J. Nilsson, A. Thorstensson, and J. Halbertsma, 1985, *Acta Physiol. Scand.* **123**, 457–475.]

and slightly before the support phase, while the main activity in the ankle extensors is somewhat delayed in relation to other extensors. The hip and knee extensors do not allow these joints to flex markedly under the action of the body weight. By the end of the support phase, these muscles provide the propulsive force moving the body forward. An important source of this force development is a group of ankle extensors strongly contracting at the end of the support phase, which results in an ankle extension that pushes the body forward.

In the swing phase, the leg is transferred forward in relation to the body, and this work is done mainly by a group of hip flexors. In walking, the lower part of the leg is moderately elevated above ground during the transfer, whereas in fast running it is lifted much higher due to a stronger knee flexion. This relates to different biomechanical advantages at high and low speeds of progression. At the end of the swing phase, the foot touches ground. In walking, this occurs with a "heel strike," and at the moment of contact the foot points upward. Soon after this, the whole sole of the foot gets in contact with the ground because of an ankle extension. This extension is delayed to some extent by a group of ankle flexors, which become active shortly before the heel strike and act like a spring softening the heel strike. In running, on the contrary, the front part or the entire sole of the foot touches ground initially and all extensor muscles are coactivated. This contributes to a different force trajectory in walking and running (see Fig. 2).

The main features that make walking and running motor patterns in humans different from those of other mammals are the following. (1) Due to the vertical body position, the hip excursions during the step cycle in relation to the body axis proceed within the limits of 90°–180° in contrast to, for instance, tetrapods such as the cat, the hindlimbs of which are much more flexed during locomotion, and the hip angle varies within the limits of 70°–110°. (2) A co-activation of all extensors at the moment of foot contact is typical for quadrupeds, but in humans a mixed muscle synergy becomes active at the moment of heel strike (hip and knee extensors plus ankle flexors), while the ankle extensors come into action later, by the end of the support phase. (3) Because of the upright, vertical body position during locomotion in humans, there are specific requirements for the maintenance of equilibrium which are different in the case of quadrupedal locomotion.

The ability to locomote on two legs appeared in humans (hominids) about 1 million years ago, as demonstrated in fossil traces of walking *Australopithecus*. Many advantages are related to the upright position of the body and the bipedal gait. Although one might be impressed with our ability to locomote on two legs, it should be recalled that birds, reptiles, and even dinosaurs have been bipedal, sometimes moving very fast and effectively (e.g., the ostrich at 60 km/hr).

Much data have been accumulated demonstrating that the neural systems controlling locomotion have great similarities among all vertebrates, although they are adapted to the particular anatomical features of each species. Unfortunately, data about the nervous control of locomotion in humans are more scarce than, for instance, in animals such as the cat or lamprey, but the latter systems provide basic principles also relevant to the human system.

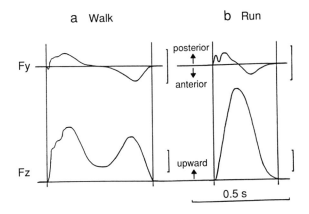

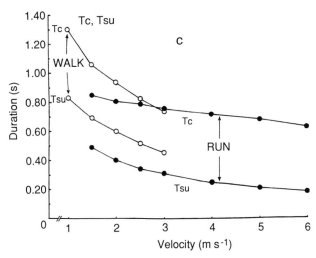

FIGURE 2 Some dynamic and kinematic characteristics of the step cycle are different in walking and running. a, b. Ground reaction forces of the right foot developed in the support phase during walking and running at the same speed (2 m · sec⁻¹). The vertical bars represent 500 Newtons. The upper records show the horizontal force developed with an initial braking of the forward movement in the first part of the support phase and a push-off in the latter part. In the curve of vertical force (Fz) there are characteristically two peaks in walking against one main peak during running. c. Stepcycle (Tc) and support phase (Tsu) duration is plotted versus velocity of locomotion for walking and running. Note the difference in these characteristics of the step cycle for walking and running, even for equal speeds of locomotion. [Adapted, with permission, from J. Nilsson and A. Thorstensson, 1989, *Acta Physiol. Scand.* **136**, 217–227.]

II. Different Components of the Locomotor Control System

During locomotion several different tasks must be achieved:

1. *Propulsion*. The legs must be moved in a fashion such that the body moves forward. This is achieved by

neural networks in the brainstem and spinal cord, activating the relevant muscle groups via motoneurons.

2. *Visuomotor coordination*. With each step, the foot must be positioned on an appropriate place on the ground—not in a hole but on a suitable surface. This adaptation is mediated via the cerebral cortex and the pyramidal tract.

3. *Steering control*. As a rule, we locomote to reach a particular place; therefore, we must be able to turn and follow different paths, walk uphill or downhill, etc. This control most likely involves both the cerebral cortex and the midbrain.

4. *Equilibrium control*. In bipedal locomotion, there are phases in the step cycle in which we would immediately fall if we stopped the ongoing leg movements; this means that we maintain equilibrium due to the motion. A refined control system is therefore required to prevent us from falling. One sensor here is the vestibular apparatus, but visual information and mechanoreceptors also play an important role.

5. *Predicted and unexpected perturbations*. During practically each step, unexpected events will occur. We must be able to compensate for such perturbations. In most cases, we predict what will occur and then compensate accordingly before the fact, as when we walk on a slippery surface or see an obstacle. Some perturbations are unpredicted, like when we stumble on an object. In that case, a variety of reflexes help us make a first stereotyped compensation (like a first aid), which is followed by a volitional adaptation to the event.

6. *Combination with other movement patterns*. In many cases, the locomotor movements are combined with other simultaneous movements, like walking and playing an instrument, or running fast and jumping. In these cases, the motor centers controlling the different movements must be coordinated in an optimal fashion. In this category, we should also consider the locomotor modifications that we use when we walk in a stooping fashion, wear shoes with high heels, or imitate the walking of Charlie Chaplin.

In a sense, one can say that the locomotor behavior is generated by a family of control systems taking care of different aspects like propulsion, equilibrium, and visuomotor coordination (1–6 above). We shall now consider some of these systems.

III. Generation of the Basic Propulsive Synergy

Mammals without a cerebral cortex (e.g., decorticate rat) behave close to normal on a cursory examination. They move around, "sleep," search for food, eat, etc. This motor behavior is, as a first approximation, adapted to their biological needs.

Although their visuomotor coordination is significantly impaired, they can to some extent process visual information, presumably in the midbrain, to select different targets in the environment. Such decorticate animals move in a seemingly normal way, but they do not respond to the environmental situation in a normal fashion. For instance, they become aggressive in "inappropriate situations" and may even attack their companions. Thus, they initiate walking spontaneously and adapt the movements to their "assumed needs."

The situation is different in mammals, which lack the entire forebrain (telencephalon and diencephalon) but have intact mesencephalon and lower brainstem. On one hand they can generate the entire propulsive synergy, they can walk and run, and they maintain their equilibrium to some degree, but they behave in a robotlike fashion. They do not avoid obstacles, and they would try to walk through a wall if they happened to encounter one.

The stepping movements themselves are produced by a number of leg muscles rhythmically activated in different parts of the step cycle (Fig. 1). The pattern of activity of various muscles is determined by commands coming from networks of neurons in the spinal cord, which together form a "pattern generator" for each leg. These pattern generators are autonomous in the sense that, when activated, they can generate the basic locomotor pattern without assistance of higher brain centers. Locomotor centers in the brainstem initiate locomotion by activating the spinal pattern generators and by controlling their level of activity. Conversely, a blockade of brainstem locomotor centers prevents normal initiation of locomotion. The brainstem–spinal cord circuitry can thus elicit well-coordinated locomotion, which moves the animal forward.

Below we will consider the organization of the systems initiating locomotion and controlling the stepping movements.

A. Initiation of Locomotion

In all cases of vertebrates, locomotion can be initiated by stimulation in the upper part of the brainstem in an area referred to as the midbrain (mesencephalic) locomotor region (MLR) (Fig. 3a). A local electric or biochemical stimulation will give rise to locomotion. A weak stimulation elicits slow walking, whereas stronger stimulation can elicit trot and gallop in different types of mammals. Thus, the basic locomotor synergy with hundreds of different muscles coordinated in the specific way characteristic of locomotion comes in operation under the effect of a very simple command. Most likely locomotion is normally initiated by a simple stimulus of this type. To regulate the speed (intensity) of locomotion the brain merely has to modify the level of activity in the MLR. The problem of coordinating the different muscle groups is taken care of by the preformed pattern generating circuitry in the spinal cord. [*See* SPINAL CORD.]

The MLR includes the pedunculopontine nucleus, which utilizes acetylcholine as a transmitter, and the cuneiforme nucleus. They send their axons to a cell group in the lower brainstem (N. reticularis magnus and gigantocellularis), which in turn sends axons down in the ventral part of the spinal cord to the spinal locomotor circuitry. This projection is most likely the main pathway for activation of locomotion; however, other cell groups in the midbrain may also contribute, such as the direct noradrenergic (i.e., utilizing noradrenaline as transmitter) projection to the spinal cord from locus coreuleus, which may be activated by stimulation of MLR.

The MLR is itself subject to a tonic γ-aminobutyric acid (GABA) inhibition from forebrain structures (Fig. 3a), and injection of drugs that block this inhibition will also increase nerve cell activity in the MLR, thereby eliciting locomotion. The forebrain neurons responsible for this effect are located in three different output nuclei of the basal ganglia (ventral pallidum, N. entopeduncularis and substantia nigra, pars reticularis). The initiation of locomotion probably depends on a cessation of activity in these neurons, removing the inhibition from the neurons in the midbrain locomotor area. Several lines of evidence indicate this: we know that these neurons are active at resting conditions, that they utilize GABA as an inhibitory transmitter, and that injection of GABA antagonists into the midbrain locomotor area gives rise to locomotion in an inactive preparation. This mode of initiation of motor activity by a removal of inhibition from the midbrain locomotor center is similar to that suggested for saccadic eye movements. In addition to an inhibition of the midbrain locomotor region, which is well documented, a direct excitation from certain forebrain areas appears probable.

The different basal ganglia output nuclei are in their turn controlled by N. accumbens or neostriatum. Dopamine stimulation of the former nu-

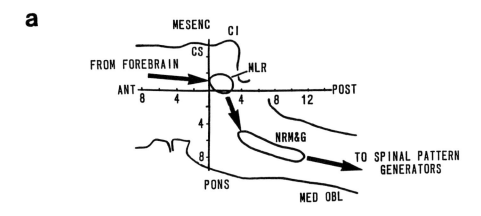

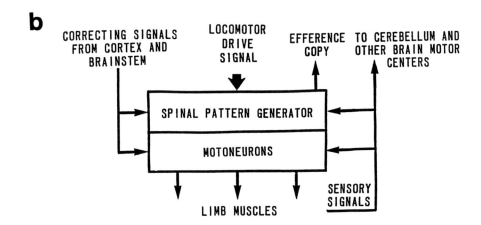

FIGURE 3 Neural organization of the locomotor control system. a. Longitudinal section of the brainstem with the midbrain (mesenc) and the lower brainstem (pons and med. obl.). The midbrain locomotor region (MLR) is indicated and its connection to the lower brainstem locomotor region in nuclei reticularis magnus (NRM) and gigantocellularis (G). This area projects through the ventral quadrant of the spinal cord to the spinal pattern generators for locomotion (I, colliculi inferior; CS, colliculi superior). b. Schematic representation of the spinal network organization. The spinal pattern generator circuitry is activated by the locomotor drive signal from the lower brainstem. This pattern generator activates the different motoneurons and thereby muscles in sequence. During the ongoing movement, sensory signals arising from the movement can act at several different levels: the motoneuron, the spinal pattern generator, and, by relay pathways, on cerebellum and other brain motor centers. Correcting signals from a variety of descending pathways can modify the activity of individual groups of motoneurons or even the central pattern generator.

cleus can also initiate locomotor activity. It belongs to those parts of the forebrain that relate to the older primitive part of the cerebral cortex such as the hippocampus, gyrus cinguli, and also N. amygdala. Initiation of locomotion during normal behav-

ioral conditions likely originates in these and in subcortical structures like in the hypothalamic nuclei, responding to hunger, thirst, and threat. [*See* BRAIN; HIPPOCAMPAL FORMATION.]

B. Spinal Locomotor Network

1. Central Pattern Generator

The isolated spinal cord from all classes of vertebrates can be made to produce the output pattern that should have resulted in movements more or less close to locomotor movements—had the muscles been connected to the nervous system. This pattern is referred to as "fictive locomotion" in contrast to real locomotor movements. Detailed studies have been performed to corroborate this finding, and the efferent pattern has been sampled by recordings from the nerves projecting to the different muscles or by microelectrode recordings from the cell bodies of the motoneurons. Such data are sufficient to conclude that interneurons residing in the spinal cord are connected to a neural network that can activate the different motoneurons in the appropriate order. Such a network is sometimes referred to as a central pattern generator (CPG) for locomotion, simply testifying to the fact that, deprived of phasic feedback from sensory afferents, the spinal cord can generate a well-organized output pattern resembling that of normal locomotion.

The spinal CPG for locomotion comes into operation under the excitatory drive of signals originating in the locomotor centers of the lower brainstem (Fig. 3b). These commands produce tonic excitation of some groups of spinal interneurons and inhibition of others, thus transferring the network into an unstable (oscillatory) state. The locomotory rhythm is produced by the specific interaction of various groups of interneurons as well as by the specific properties of the membrane of these cells (see below). In the isolated spinal cord, the CPG can also be activated by glutamatergic or noradrenergic agonists simulating the action of the fibers descending from the brain. The utilization of such preparations have allowed a rapid progress of an understanding of the organization and operation of spinal locomotor generators, in both lower and higher vertebrates.

During the locomotor cycle, each motoneuronal group is activated in a specific fashion with one or sometimes two bursts. They receive phasic input from excitatory interneurons of the CPG, which thus are responsible for generating the burst of activity in the motoneurons, and input from inhibitory interneurons, which assures that motoneurons are not being activated in the inappropriate phase. The motoneurons are pure output devices and are not themselves interconnected in a generator network. When it comes to understanding the exact neuronal mechanisms responsible for generation of rhythmic patterns in the spinal cord, little information is available on mammals. On the contrary, the segmental network responsible for generating the rhythmic pattern underlying swimming in a lower vertebrate—the lamprey—has now been mapped, as well as the simpler network controlling swimming in the frog embryo. Although the spinal neural mechanisms controlling stepping in mammals and humans will differ in complexity, the basic organization will most likely resemble that of the simpler network of lower vertebrates. The advantage of these preparations is that they have fewer nerve cells and can be maintained *in vitro* for a long period of time, while their isolated nervous system can produce the motor pattern normally resulting in lo-

comotion. Color Plate 20 shows the general organization of the lamprey CPG known to produce alternating burst activity in two symmetrical (left and right) groups of segmental motoneurons. Two groups of interneurons (left and right) mutually inhibit one another. Each group of interneurons contains both excitatory (E) and inhibitory (CC, L) cells.

The network is turned on by a tonic excitatory inflow (from brainstem locomotor centers) to all network interneurons, so that they discharge action potentials when not actively inhibited. The pattern itself is due to the synaptic interaction between the interneurons in the network and the membrane properties of the individual cells (see Color Plate 20 and Fig. 4). Essentially, when one side is active, the other becomes inhibited and vice versa. The activity on the first side is terminated by several factors: (1) an adaptation of the response to continuous excitation; (2) the high threshold for activation of L cells, which when activated will inhibit the CCs; (3) depolarization plateaus, which are reset with a time delay. These mechanisms add to terminate the activity of the CCs, and thereby the group of interneurons on the other side will cease to be inhibited and therefore become active and take over. In this way, an alternating burst activity will arise.

2. Sensory Control of the Different Phases of the Step Cycle

Although the motor pattern and the actual leg movements can be performed without any sensory information, these movements are functional to only a limited extent. In almost each step, minor adjustments are normally made to optimize the movement and compensate for unexpected perturbations; thus, the sensory information from each limb is indispensable. The information regarding the movements in the most proximal joint, the hip, is particularly important because it signals how far the overall limb movements have proceeded during the support or the swing phase, respectively. For instance, when the leg is stretched backward in the end of the support phase, the information related to hip position will be transmitted by the afferents to the spinal cord and influence the central circuitry so that it facilitates the onset of the flexor muscle activity, which initiates the swing phase (Fig. 5). Due to this feedback, the frequency of stepping and the speed of locomotion are automatically regulated in accordance with the counteraction of external forces to the forward motion of an animal. As a

a

NMDA induced TTX-resistant oscillations

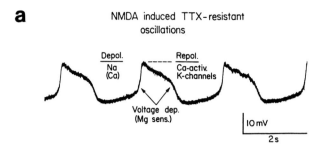

b

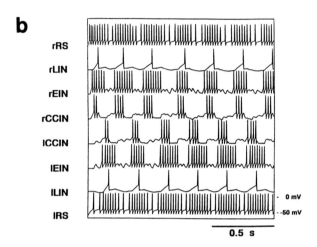

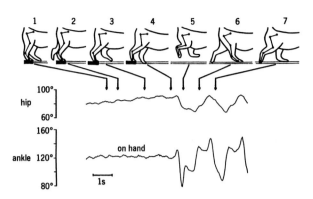

FIGURE 5 Sensory control of the spinal pattern generators. The transfer of the limb forward (i.e., the swing phase of the step) starts when the limb is sufficiently deflected backward as sensed by receptors influenced by hip movements. The spinal cat (complete transection of the spinal cord at lower thoracic level) was walking on the moving treadmill belt. During the period of time corresponding to frames 1–4, the right limb was held by the experimenter, and its movement backward was slowed while the other limb continued to walk. The transfer movement in a "handheld" limb started ones when the hip extension reached 90°, as in normal (subsequent) step cycles. Because the spinal cord was transected, all sensory interactions take place in the spinal cord. [Reproduced, with permission, from S. Grillner and S. Rossignol, 1978, *Brain Res.* **146,** 269–277.]

FIGURE 4 Vertebrate spinal cord networks depend on NMDA receptor activation, which can induce long-lasting plateaulike depolarizations or pacemakerlike potentials. The oscillations arise through an interaction between NMDA channels and other ion channels in the membrane. (a) Shows such oscillations; (b) shows a computer simulation of the neural network shown in Color Plate 20. All simulated neurons have been given their specific membrane properties, and the synaptic interaction is mediated by conventional inhibitory and excitatory ion channels. L, left; R, right; TTX, tetrodotoxin. Abbreviations as in Color Plate 20.

result, the locomotion decelerates when running uphill and accelerates when running downhill. A whole set of afferent signals from the limb contributes in the same way to the adaptation of the step cycle to the environment. In the support phase, receptors sensing the overall force exerted by the limb extensors will assure that the limb is not flexed while it is carrying the main load of the body. In the swing phase, due to a system of special reflexes, the spinal network is capable of a considerable modification of the limb trajectory. In Fig. 6 it is shown how the spinal cat (i.e., with the spinal cord transected at thoracic level) oversteps an obstacle.

Exactly how these different sensory afferents are

connected to the spinal cord locomotor network in mammals is unknown, how this type of control can be solved in a simple vertebrate is known. Let us again look at the lamprey segmental network including both the sensory neurons and CPG, which produces alternating contractions on the left and right side of the body (see Color Plate 20). When one side of the body is contracting, the other side (including its stretch receptors located in the spinal cord itself) is extended. The increased level of activity in the spinal stretch receptors will synaptically add excitation to all neurons located on the same side, and at the same time inhibit all neurons on the contralateral side, i.e., the side that is actively contracting. The net result of the sensory input is thus to inhibit the active side and excite the side that is to become active. In this way, the sensory input will always adapt the central activity of the network. For instance, if the movements are delayed by an increased external resistance, the ongoing contraction will be prolonged so as to compensate for the perturbation. As we can see in this example from the lamprey, a fairly simple neuronal arrangement can account for the sensory regulation. In the control of the mammalian limb, it would be plausible that the afferents activated when the

a

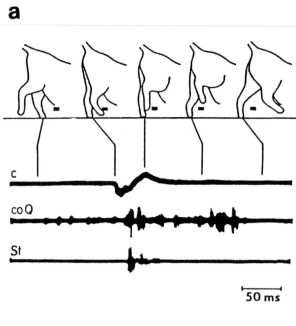

b

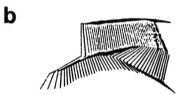

c

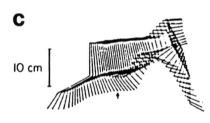

10 cm

50 ms

FIGURE 6 The spinal cat is capable of overstepping an obstacle while walking on a moving treadmill belt. (a) Touching an obstacle at the dorsum of the paw (moment of touching is shown in trace C), reflexively activates both the knee extensor quadriceps in the contralateral limb (coQ) and the knee flexor semitendinosus (St) in the ipsilateral limb. As a result, the foot trajectory during the swing phase is changed considerably. Compare stick diagrams for normal swing (b) and for reflex overstepping (c). In the latter case, the instant of touching the obstacle is indicated by an arrow. These results show how the spinal cord circuitry is programmed to elicit compensatory adjustment of the limb movement at a minimal delay.

hip is in a posterior position will excite the interneurons that contribute to limb flexion and, at the same time, inhibit the interneurons controlling the extensor activity.

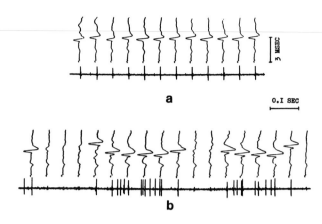

FIGURE 7 The central pattern generator produces the step-linked changes of the efficiency of transmission in reflex pathways. This record illustrates a single spinal interneuron mediating inhibitory influences of muscle afferents from the knee extensor (quadriceps) on motoneurons of the antagonistic muscles, the knee flexors. A brainstem–spinal cord preparation producing fictive locomotion was used. A continuous repetitive electrical stimulation of muscle afferents (group 1a) in the nerve of quadriceps was performed. At rest, the neuron responded to each stimulus applied to the nerve (a), but during fictive locomotion (b) (evoked by MLR stimulation) it responded to stimuli applied during one part of the cycle, while it did not respond to those applied in another part. Neuron activity was recorded continuously (horizontal trace), while the vertical traces were triggered by stimuli applied to the nerve. [Reproduced, with permission, from A. G. Feldman and G. N. Orlovsky, 1975, *Brain Res.* **84,** 181–194.]

3. Gating of Sensory Information in the Step Cycle

In the course of the step cycle, the spinal CPG not only controls the motoneurons but also produces some additional actions necessary for stepping coordination. Among these actions is the rhythmical reorganization of the system of connections between various groups of spinal neurons. In other words, the spinal network modifies itself in the course of the step cycle. An example of such a reorganization is the "gating mechanism," which allows afferent and other signals to reach their target cells in one phase of the cycle, while in the other phase the pathway for these signals becomes inefficient (Fig. 7).

IV. Interaction between Spinal and Supraspinal Motor Centers—Cerebellum

The spinal locomotor CPGs are sending signals about various aspects of their activity to higher

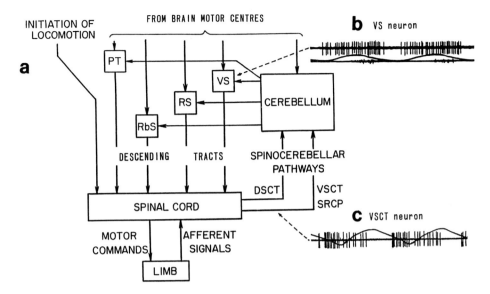

FIGURE 8 Cerebellum plays an important role in the coordination of spinal locomotor mechanisms and supraspinal motor centers. a. Scheme of the cerebellar inputs and outputs. When the spinal pattern generator is activated by the command coming from the MLR (initiation of locomotion), it sends information about its operation back to cerebellum, through the ventral spinocerebellar tract (VSCT) and spinoreticulocerebellar pathway (SRCP). At the same time, through the dorsal spinocerebellar tract (DSCT), the cerebellum receives sensory information monitoring the stepping movements of the limb. On the bases of this information, the cerebellum performs rhythmical action on the "output" neurons of various brain motor centers, i.e., on the neurons giving rise to descending tracts (vestibulospinal [VS], reticulospinal [RS], rubrospinal [RbS], and pyramidal [PT]). As a result, all commands addressed to the spinal cord become linked to the ongoing stepping movements. b. Activity of a descending VS neuron. This modulation is due to the cerebellum output and is similar but with different phase relations in all fast descending pathways. c. The efference copy information conveyed from the spinal cord to cerebellum is illustrated by a recording from VSCT neurons in a walking preparation in which the afferent activity from the hindlimb had been abolished by a transection of all dorsal roots.

brain centers, in particular to cerebellum (Fig. 8c). At the same time, the signals from receptors located in the stepping limbs convey information about the ongoing movement. They are used by the brain to organize and time its activity in a strict relation to the activity of the spinal CPGs and the actual ongoing limb movements. A leading role in the coordination of various spinal and supraspinal motor centers during locomotion is played by cerebellum (Fig. 8a). Input signals enable the cerebellum to send a rhythmic output to the brainstem and cortical motor centers (Fig. 8b), which participate in the control of locomotion. As a result, all commands addressed

from these centers to the spinal cord may become linked to a definite phase of the step cycle. This system is used for achieving a perfection of the locomotor movement generated by the spinal cord and for organizing the visuomotor coordination in locomotion (see below).

V. Maintenance of Equilibrium during Locomotion

In bipedal locomotion, the main aim of the stepping movements is to move the body forward; however, these movements can also be considered as a means of preventing the body from falling in one direction or the other because the equilibrium during bipedal locomotion is of a dynamic nature. The direction and amplitude of the leg movement in the swing phase are strictly linked to the motion of the body. The locomotor program contains features that seem to maximize the stability during movement. For instance, in each step the body is moved toward the side of the supporting limb to assure maximal lateral stability. The trunk movements and those of the arms also serve to provide stability.

Also, various compensatory movements are aimed at maintaining the equilibrium during locomotion despite different perturbations. The type of compensation depends on both the nature of the disturbance and the current state of the locomoting human (i.e., on the phase of the step cycle), as shown in Figure 9. The changes of the response are produced by gating mechanisms most likely controlled by the CPG.

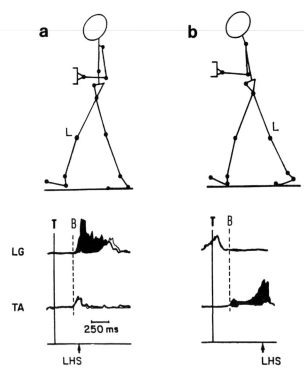

FIGURE 9 Compensation of an equilibrium disturbance occurring in different phases of the step cycle, evoked in walking humans by voluntary hand movements. A man walking on a treadmill belt was instructed to pull a handle in response to a sound stimulus. In (a) and (b) two different positions of the left leg (L) are shown, as well as the activity of the ankle extensor, lateral gastrocnemius (LG) and the flexor, tibialis anterior (TA) (electromyograms were rectified and filtered). In each graph, the curves of muscle activity in normal cycles and in cycles including postural disturbances are superimposed, and the area in between them is shaded to show the compensatory muscle activity. The moments of the sound signal (T) and of the beginning of voluntary activity in the elbow flexor, m. biceps (B) are indicated. LHS is the moment of left heel strike (beginning of the support phase). One can see that the step cycle modification necessary to compensate for the equilibrium disturbance starts simultaneously with voluntary hand movement. These modifications occur differently in different phases of the step cycles. [Adapted, with permission, from H. Forssberg and H. Hirschfeld, 1988, Phasic modulation of postural activation patterns during human walking. *In* "Progress in Brain Research," Vol. 76.]

VI. Visuomotor Coordination in Locomotion

So far we have dealt with the movements in relation to a body-centered coordinate system. As we move in the real world, however, we must continuously position each foot in each step so as not to step in a hole or on a slippery surface. This requires a per-

ception of the outside world by vision and an adaptation to an extrinsic coordinate system. In mammals, this adaptation needs the integrity of the cerebral cortex and particularly the direct projections from the cerebral cortex to the spinal cord (pyramidal tract). It has long been known that after a lesion in the pyramidal tract mammals will be unable to achieve a good visuomotor coordination during a task such as walking up a ladder or on a stony beach, but they are still able to walk perfectly well on a floor.

During walking on a flat surface, the pyramidal tract neurons exhibit rather low activity related to the step cycle. As the positioning of the foot gets progressively more difficult, however, this activity increases considerably (Fig. 10). Different pyramidal tract neurons will make "corrections" or slightly modify different aspects of the step cycle: to position the foot somewhat more to the left or right or a little bit forward or backward, etc. Thus, the visuomotor control signals coming via the pyramidal tract could be considered as just superimposing extra excitation or inhibition on a group of motoneurons in addition to the signals coming from the spinal locomotor pattern generator. For the accurate positioning of the foot, the central nervous

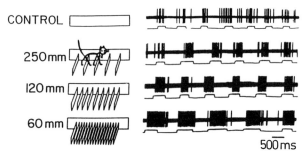

FIGURE 10 Rhythmical activity in a corticospinal (pyramidal tract) neuron considerably increases when the cat has to accurately position the foot on the ground. An intact cat with a microelectrode implanted into the motor cortex either walked on a smooth surface (control) or was forced to walk over barriers located at different regular intervals (the intervals indicated on the left). With smaller intervals, positioning the foot became more difficult; evidence indicates that the more difficult the positioning task, the higher level of activity in the pyramidal tract neuron. The upper curve shows activity of a corticospinal neuron projecting to the spinal forelimb center, the lower one the swing (up) and support (down) phases of the step cycle of the corresponding limb. [Adapted, with permission, from J. N. Beloozerova and M. N. Sirota, 1989, Role of motosensory cortex in control of locomotion. *In* "Stance and Motion," (V. S. Gurfinkel and M. E. Joffe, eds.), Plenum, NY.]

system must estimate if the desired position is different from the point that would be reached in normal walking on a flat floor. This latter point can be regarded as a zero-point in a coordinate system around which the brain can decide to shift the position of the limb. Thus, the task for the central nervous system during positioning is indeed rather complex. As we move forward, we identify a point where to position the leg at a later moment in time and how different this point is in relation to the estimated unperturbed landing point. During everyday walking, we make those adjustments without much notice, and we rarely look down to check where we put our feet—the actual positioning relies on predicted motor commands and without visual guidance. The complex positioning processing comprises visual perception of desired target points during ongoing movements and calculation of the correct motor commands to modify the trajectory of stepping appropriately. Visual information is transferred from the primary visual areas in the occipital lobes to the specialized motion analyzing areas in the parietal lobe, which in their turn connect to the motor areas in the frontal cortex. Exactly how this processing comes about is unknown, and projections from the parietal lobe to the deeper forebrain structures and the brainstem may also play an important role. In addition to the cerebral processing, the visual area underneath superior colliculus in mesencephalon may also play a role as it does in lower vertebrates.

So far, we have considered the strategy used for corrections of the limb movement on a flat surface, but in reality the third dimension is as important. Figure 11 shows how a cat oversteps an obstacle due to an increased activity in pyramidal tract neurons responsible for excitation of elbow flexors.

Yet another aspect of the control of locomotion is the modification of the direction of walking. This relates to the overall problem of how our nervous system decides which path to take and controls the movements; here we must admit that our ignorance is great and it is beyond the aim of this chapter to forward the current ideas of how this can come about.

The motor system responsible for the positioning of the limb during locomotion can presumably be used in isolation. This would result in movements of the limb toward different points in space from a stationary position, i.e., what is normally referred to as reaching. Our voluntary reaching movements are probably derived from the complex positioning

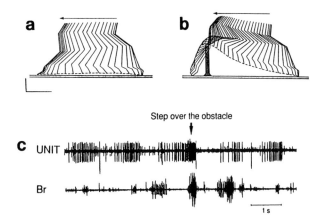

FIGURE 11 The burst of activity in a pyramidal tract neuron, controlling elbow flexion, considerably increases when the limb oversteps an obstacle. An intact cat with a microelectrode implanted into the motor cortex was walking on a treadmill. (a) and (b) show stick diagrams of the forelimb motion during stepping in a normal cycle (a) and when overstepping the obstacle (b). In the step over the obstacle (c), the activity of elbow flexor, m. brachi (Br) increased considerably as well as the burst of activity in the corticospinal neurone (unit). [Adapted, with permission, from T. Drew (1988) Motor cortical discharge during voluntary gait modification. *Brain Res.* **457**, 151–182.]

movements used during locomotion and possibly also the grasping movements.

VII. Development of Locomotion

The human fetus performs movements in the womb from around week 12, and they gradually become more organized. At birth, the baby is still helpless and unable to walk, stand, or even sit; however, with support it can perform step-like movements with the legs. This is interpreted to mean that the spinal machinery for locomotion is present but that it cannot be controlled volitionally by the baby nor can the baby maintain the balance of the body. During the first year, the baby acquires a gradually improving equilibrium control, which enables it to sit, then stand, and finally take a few steps without support. At around 1 yr old with great variability, the infant starts to walk and the walking movements are gradually perfected throughout adolescence (see Fig. 12). The nervous system is subject to a gradual maturation and the development of the basic locomotor ability mainly reflects a maturation process rather than a learning experience. Many vertebrates and even mammals such as deer and horses walk and even run within the first few hours after deliv-

ery. In such species, an evolutionary pressure may have enabled the animals to run away from predators to survive. The maturation process of the central nervous circuitry responsible for the movement repertoire has thus proceeded further at birth in these animals than in humans. Humans have a maturation process of the nervous system prolonged over more than a decade, which is exceptional in the animal kingdom. Such a protracted development may, however, be beneficial in terms of a long period of learning of different skills before adulthood is reached.

Bibliography

Armstrong, D. M. (1988). The supraspinal control of mammalian locomotion. *J. Physiol.* **405,** 1–37.

Arshavsky, Yu. I., Gelfand, I. M., and Orlovsky, G. N. (1986). Cerebellum and rhythmical movements. *In* "Studies of Brain Function," Vol. 13. Springer Verlag, Berlin and Heidelberg.

Cohen, A. H., Rossignol, S., and Grillner, S. (eds.) (1988). "Neural Control of Rhythmic Movements in Vertebrates." John Wiley and Sons, New York.

Grillner, S. (1985). Neurobiological bases of rhythmic motor acts in vertebrates. *Science* **228,** 143–149.

Grillner, S., and Dubuc, R. (1988). Control of locomotion in vertebrates: Spinal and supraspinal mechanisms. *In* "Advances in Neurology," Vol. 47, "Functional Recovery in Neurological Disease" (S. G. Waxman, ed.). Raven Press, New York.

Grillner, S., Wallén, A., Brodin, P., and Lansner, L. (1991). Neuronal network generating locomotor behaviour in lamprey: circuitry, transmitters, membrane properties and simulation. *Annu. Rev. Neurosci.* **14,** 169–199.

Nilsson, J., Thorstensson, A., and Halbertsma, J. (1985). Changes in leg movements and muscle activity with speed of locomotion and mode of progression in humans. *Acta Physiol. Scand.* **123,** 457–475.

Noga, B. R., Kettler, J., and Jordan, L. M. (1988). Locomotion produced in mesencephalic cats by injections of putative transmitter substances and antagonists into the medial reticular formation and the pontomedullary locomotor strip. *J. Neurosci.* **8,** 2074–2086.

Roberts, A., Soffe, S. R., and Dale, N. (1986). Spinal interneurons and swimming in frog embryos. *In* "Neurobiology of Vertebrate Locomotion" (S. Grillner, P. S. G. Stein, D. G. Stuart, H. Forssberg, and R. M. Herman, eds.), pp. 279–306. Macmillan, London.

Shik, M. L., and Orlovsky, G. N. (1976). Neurophysiology of locomotor automatism. *Physiol. Rev.* **56,** 465–501.

Stein, P. S. G. (1983). The vertebrate scratch reflex. *Symp. Soc. Exp. Biol.* **37,** 383–403.

FIGURE 12 Schemes for positions of the body during phases of the step. n_A, downward push in the thigh of the rear leg; C, thrust to the rear by the rear leg; m, limit of raising of the knee to the rear; D, thrust to the rear; E, the last dynamic element of the support period. [Reproduced, with permission, from N. Bernstein, 1967, "The Coordination and Regulation of Movements," Pergamon Press, New York.]

Longevity

ROBERT J. MORIN, *Harbor–UCLA Medical Center*

I. Predictors of Longevity
II. Determinants of Longevity
III. Means to Enhance Longevity

Glossary

Acetylcholine Chemical that enables nerve signals to be transmitted from one neuron to another

Atherosclerosis Diseased arteries containing fatty-fibrous obstructions that lead to heart attacks and strokes

Autoimmunity Reaction of the body's immune cells and proteins against its own tissues

Catecholamines Chemicals that function as transmitters of nerve signals

Cellular immunity Immunity that is mediated by T-lymphocytes in the blood

Cirrhosis of the liver Fatty degeneration and scarring of the liver

Dendrites One of the two types of branching protoplasmic processes that extend from neurons and connect them to each other

DNA Deoxyribonucleic acid, the chemical component of the cell nucleus that contains the genetic information

Elastic recoil Capacity of the lungs to expand after air is expelled

Glomerulus Portion within the nephron that filters the blood

Glycosylation Formation of bonds between sugar and proteins or other molecules

Humoral immunity Immunity that is mediated by antibodies in the blood that are produced by B-lymphocytes

Hypothalamus Portion of the midbrain that secretes hormones controlling the pituitary gland

Immune complexes Large protein substances found in the blood when antibodies bind to antigens

Mitogen Chemical that stimulates cells to divide

Nephrons Microscopic units inside the kidneys through which blood is filtered and urine is carried away

Systolic volume Volume of blood pumped during each heartbeat

Vital capacity Maximal amount of air that can be expired after a deep breath

LONGEVITY CAN BE DEFINED as either life expectancy or the maximum life span that can be obtained. *Life expectancy* at birth, or average life span, meaning the age to which 50% of the population survives, is determined by the totality of all causes of death. Average life span has increased from about 20 years in ancient Rome to about 75 years today. This is primarily due to decreased infant and maternal mortality, conquest of infectious diseases, and reduced environmental hazards. *Maximum life span,* which is the oldest age that any humans survive to, is a function of the aging processes, and may be more closely related to genetic than to environmental factors. This is about 110 years, and has not changed much since the origin of our species (Fig. 1).

I. Predictors of Longevity

Studies of population groups in the United States have revealed a number of factors that are statistical predictors of the life expectancies of individuals within our society. Some of the most significant predictors are as follows:

1. Heredity: There is a general correlation between the life spans of parents and their offspring. Studies of twins indicate that identical twins show much closer life spans than do fraternal twins. The genetic influence on longevity may be determined by the inheritance of protective factors against the major killer diseases e.g., more efficient receptors to remove low-density lipopro-

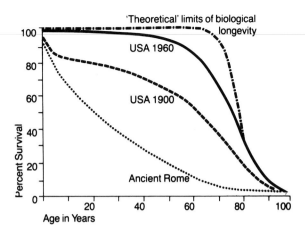

FIGURE 1 Survival curves in human populations: ancient Rome, U.S.A. 1900, and U.S.A. 1960, and the curve resulting if all diseases were cured but the basic aging process was still continuing ("theoretical" limits). [Reproduced from "Maximum Life Span," by Roy L. Walford, M.D., by permission of W. W. Norton & Company, Inc. Copyright © 1983 by Roy L. Walford.]

teins that cause atherosclerosis, or enhanced resistance to cancer due to the presence of genes expressing a strong immune response.

2. Sex: At birth the life expectancy of females is presently 8 years more than males. The ratio of males/ females at birth is 105/100. By age 30, the numbers of males and females have equalized, and at age 65 only 70% of males are still alive, as compared to 84% of females. During the first half of the life span, the mortality of males is greater due to more accidents, homicides, and suicide. In the second half of life, men succumb at greater rates to coronary heart disease, lung cancer, and cirrhosis of the liver. Atherosclerotic coronary heart disease accounts for 40% of the sex difference in longevity. Up to half this difference may be due to men's smoking habits—more men smoke, they smoke more cigarettes, tar content is higher, and they inhale more. The increased incidence of lung cancer is also due to smoking and to increased exposure to industrial carcinogens. Cirrhosis of the liver is caused in large part by alcohol abuse, which is greater in men than in women. The sex gap in longevity is greater among the blue-collar population, most likely because of the poorer health habits of the men.

The above behavioral components account for only part of the sex difference in longevity. Women have two sets of X-linked genes, but men have only one set of these genes, and therefore have an increased mortality from a number of X-linked defects. The enzyme DNA polymerase alpha, which is essential for DNA replication and repair, is located on the X chromosome, and therefore its activity is less susceptible to mutational alteration in women. Another sex-linked enzyme,

dehydroepiandrosterone (DHEA) sulfatase, which converts DHEA to its active form, may enable this hormone to exert several longevity-enhancing effects. Unlike most other X-linked genes, the cells of females have two functional copies of this sulfatase gene.

Another possibility is that the Y chromosome of males may have some deleterious effects. A group of Amish men who have a deletion of part of their Y chromosome have been found to live longer than the women of this population. It seems probable that both genetic and behavioral factors are responsible for the present greater life expectancy of females versus males.

3. Body weight: It seems clear from numerous epidemiologic studies that people who are overweight, particularly at a younger age, tend to die sooner. This is true even after exclusion of obesity-associated disorders such as hypertension and hyperglycemia. A few studies had reported an excessive mortality in underweight persons, but this association probably is explainable by underlying illness causing weight loss in a number of these persons. Thinness in smokers may be associated with increased mortality. Except for this, the minimum mortality rate seems to occur at weights 10% below the U.S. average weight.

4. Exercise: Although there is as yet no conclusive evidence of a relationship between exercise and longevity, older persons who are more active do tend to live longer. This may be a reverse cause and effect relationship, in that those whose health is better are able to exercise more. It also may be related to the lower incidence of obesity in the exercising population.

5. Socioeconomic factors: Manual laborers have a 37% higher mortality rate than average, whereas professionals have a mortality rate 20% lower than average. This is most likely due to lifestyle and behavioral factors such as less smoking, better diet, and better medical care in the latter groups. The observed higher mortality in blacks versus whites may be more a function of these lifestyle factors than of genetics. Individuals of higher intelligence have greater longevity, probably again secondary to higher socioeconomic status and healthier lifestyles.

6. Retirement: The degree of work satisfaction shows a positive association with longevity. Those who continue to work show a lower mortality rate than those who retire. Cause and effect have not been established, and this association may be explained by the fact that more people whose health is good continue to work, and that some people retire because of ill health.

7. Geography: The states with the highest longevities are Nebraska, Colorado, South Dakota and Minnesota, and those with the lowest are the southeastern states in Georgia, North and South Carolina, and Alabama. The reasons for the disparities are most likely due to environmental or lifestyle differences between these areas.

II. Determinants of Longevity

Cardiovascular diseases, including heart disease, stroke, and other blood vessel disorders, are responsible for almost 50% of deaths. Coronary heart disease, which is the major contributor to this category, has fortunately shown a decline in incidence since its peak in the mid 1960s. Cancer accounts for about 23% of all deaths, equivalent to about 500,000 cancer deaths in the United States in 1988. Although cancer death rates have been increasing, the rate for those under age 55 has been decreasing since 1950. If lung cancer, caused by smoking, is excluded, cancer death rates show a 13% decrease from 1950 to the present. Accidents are the next leading cause of death, claiming 140,000 lives/year in the United States. Prior to age 40, accidents are the number one cause of death. Motor vehicle fatalities, the major contributor to this category, have decreased by 31% since 1970, due to safer motor vehicles and improved driving habits. Other leading causes of death in order of frequency are chronic obstructive pulmonary diseases, pneumonia and influenza, diabetes mellitus, suicide, and chronic liver disease.

If all cardiovascular disease were conquered, 12 years would be added to the average life expectancy. If cancer was eliminated, 2 additional years would be added. If all disease were eliminated, life expectancy would remain at about 100 years. The reason for this limitation is the aging process, which after sexual maturity begins to produce a progressive decline in physiological functions. Accompanying aging is an increased susceptibility to a number of degenerative disorders, resulting in an exponential increase in mortality rate with age. After age 30 there is an 0.8–0.9%/year loss of physiological function, and the rate of mortality doubles every 7 years. The remaining discussion will deal primarily with the causes, effects, and possible amelioration of the aging process.

A. Theoretical Causes of Aging

The cause of aging is presently unknown, and there are therefore many theories to attempt to explain it. One leading theory holds that inside each of our cells there is a built-in genetic program, sort of a biological clock, that limits the life span of our somatic cells and of the whole organism. Data providing some support for this theory has been provided by Hayflick, who observed that fibroblasts in culture could double a maximum of 50 times, after which they died. Cells from older individuals show a lesser ability to replicate than those from younger individuals. Cells from diabetics, who show many features of accelerated aging, have a lesser replicative ability, and in patients with the rare disease Werner's syndrome who show markedly accelerated aging, replicative ability is also markedly diminished. Although there is a definite association between life span and replicability of cells, there is no evidence that aging of organs or the whole organism is due to limited cell replicability, and in fact cell populations tend to have life spans well in excess of the life span of the organism. Other postulated genetic mechanisms involve changes in the loci that code for proteins that correct the subtle damages that occur during the life span of the cell (particularly those enzymes that repair DNA). There is a correlation between DNA repair capacities of different organisms and their life spans. Shorter-lived species, such as shrews and mice, have a very low repair capacity, whereas long-lived species, such as elephants and humans, have the greatest DNA repair capacity. Along similar lines is the error theory, which postulates that the accumulated errors that occur during cellular metabolism, such as losses of essential DNA sequences and inaccuracies in protein synthesis, are responsible for cellular aging.

Another theory postulates that there is sort of a master control biological clock located in the hypothalamus, and that the age-related changes in cell function are secondary to hormonal changes mediated by the central nervous system. The decreased levels of neurotransmitters produced in the brain as we age may cause decrements in hypothalamic function. This could result in altered pituitary hormone secretion, which in turn would result in altered production of adrenal, thyroid, and gonadal hormones, leading to dysfunction of the various target organs. There is also some evidence that the pituitary may secrete a hormone that results in decreased oxygen consumption, with a resulting gradual loss of cell functions. Since many organisms that age do not have complex neuroendocrine systems, this theory lacks universal applicability. It is possible that the observed decrements in the neuroendocrine system may be the same result of the same process of genome-mediated changes in all aging cells. On the other hand, genomic expression

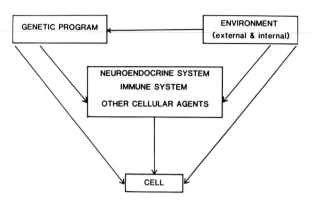

FIGURE 2 Pathways by which the genome and environment can regulate aging processes. [Reprinted with permission from "Experimental Gerontology," vol. 22: Mietes, J., Goya, R., and Takahashi, S. (1987), Why the neuroendocrine system is important in aging processes, Pergamon Press, New York.]

in the "command cells" of the neuroendocrine system may be more critical to aging than programmed decrements in other cells (Fig. 2). [*See* HYPOTHALAMUS.]

Other lines of evidence suggest that decline in the immune system may be a primary cause of aging. With age there is a marked age-associated decline in both humoral and cellular immunity and an increase in autoimmunity. These changes lead to degeneration in other organ systems and a predisposition to a number of age-related diseases. The immune system is genetically regulated by a group of genes called the major histocompatibility complex. It has been found that animals in which this complex is functioning at the highest level tend to have longer life spans. The immune system, however, is regulated by both hormonal and neural influences, and the decline of immune function could be secondary to these. All organisms that age do not have complex mammalian types of immune systems.

Another currently popular theory postulates that free radicals are a major cause of the aging process. Free radicals are molecules with an unpaired electron that are produced from molecular oxygen during the course of normal oxidative metabolism and by exposure to ionizing radiation. They are highly reactive, and produce damage to cell membranes, DNA, proteins, and other cell components. Fortunately the body has some defense mechanisms to get rid of these free radicals. The superoxide free radicals formed from oxygen are converted by the enzyme superoxide dismutase into hydrogen peroxide. This itself can give rise to free radicals, but is degraded further by the enzymes catalase and glu-

tathione peroxidase into harmless water. These defense mechanisms are not 100% efficient, and so some superoxide and hydrogen peroxide remain, and some are converted to the particularly harmful hyroxyl free radical. Antioxidants can boost the efficiency of the free radical removal mechanisms and/or limit the damage they produce. Although antioxidant administration has extended the average life expectancies of mice and rats, these treatments have not extended their maximum life spans, indicating minimal effect on the aging process itself. It is possible that free radicals accelerate the age-associated degenerative disorders such as cancer, atherosclerosis, renal disease, etc., rather than influencing aging alone.

According to the cross-linkage theory of aging, chemical bonds progressively form between adjacent strands of collagen, other protein molecules, and DNA, which reduces the functioning level of these essential molecules. Recent evidence suggests that nonenzymatic glycosylation of proteins and DNA may be responsible for cross-linkage. Consequences of cross-linkage are impaired intracellular transport and loss of elasticity of tissues. It has not yet been demonstrated that the degree of cross-linkage occurring is sufficient to result in all the age-associated functional deficits. Cross-linkage cannot explain the great diversity of life spans between species, since it has not been demonstrated to occur at greatly accelerated rates in short-lived species.

Another theory holds that accumulation of waste products by cells over time impedes their function and results in aging. For example, large accumulations of the lipid peroxidation end product lipofuscin can occur in tissues with age, particularly in the heart and nervous system. This accumulation is not quantitatively consistent, however, and does not occur universally. It has not yet been demonstrated that accumulations of lipofuscin or other waste products interferes with cell function to the degree necessary to cause aging.

B. Effects of Aging

Loss of function with age occurs at different rates in different organ systems (Fig. 3). Functions that are relatively well preserved are the velocity of nerve signal transmission and the basal metabolic rate. Functions that decline more rapidly are the pumping efficiency of the heart, maximum breathing ca-

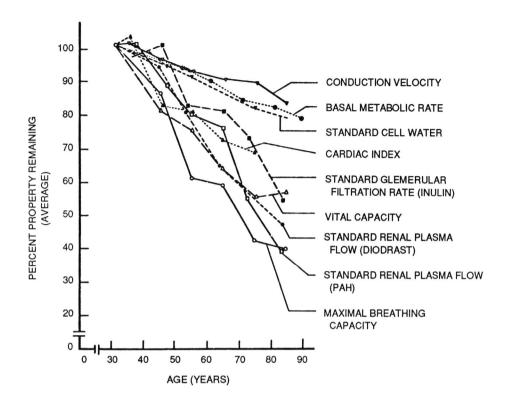

FIGURE 3 Average age decrements in a number of physiologic characteristics in normal healthy men. These data are calculated from cross-sectional observations, and may differ from measurements made longitudinally. [Reprinted with permission from Shock, N. L. (1985), The physiological basis of aging, in "Frontiers in Medicine" (R. J. Morin and R. J. Bing eds.). Human Sciences Press, N.Y., 1985.]

pacity and kidney filtration rate. This is an overall average loss of about 0.5–0.75%/year.

With regard to the cardiovascular system, the influences of lifestyle and the occurrence of heart disease make it difficult to distinguish those changes in the heart due to the aging process alone. Some changes definitely related to age are a decreased maximal heart rate, an increased end systolic volume, and a decreased ejection fraction (percentage of blood pumped with each beat). Hearts of older persons show losses of cardiac muscle cells, increases in fibrous tissue, and lipofuscin and amyloid accumulation. The loss of muscle cells is associated with a resulting compensatory enlargement of the remaining cells. From age 30 to 80 the left ventricular wall becomes 25% thicker with age, and the incidence of all types of atherosclerosis increases. The increased vascular rigidity causes increases in systolic and diastolic blood pressure with age. The ef-

fect is somewhat greater in women. [See ATHEROSCLEROSIS.]

The pulmonary system undergoes considerable decline with age. The lungs become more rigid due to increasing fibrosis. The small air spaces become dilated and there is an increase in the size and number of alveolar pores. The elastic fibers in the pulmonary blood vessels are replaced by fibrous tissue, resulting in increased pulmonary vascular resistance. The vital capacity, which is a good general measure of pulmonary function, shows a considerable decline with advancing age. Due to decreased elasticity and fibrosis, elastic recoil is decreased, flow rates of air exchange are decreased, and residual volume is increased. Due to decrease of the alveolar capillaries we also see a reduced diffusion capacity for both oxygen and carbon dioxide. [See PULMONARY PATHOPHYSIOLOGY.]

In the kidney there is a progressive loss of functioning nephron units with age, reaching 30–40% loss by age 85. The kidney shows fibrosis, vascular degeneration with loss of elasticity, and degeneration of the glomeruli. Membranes become thickened and therefore less permeable. Other age-related changes are deposition of amyloid, immunoglobins, and lipofuscin in the glomeruli. These pathologic changes lead to considerable reduction in glomerular filtration rate and tubular function. The reduction may be more than 40%, and renal blood flow is also correspondingly decreased. This causes decreases in both excretory and resorptive capacities of the kidney.

The brain shows a number of age-related degenerative changes. There is loss of brain mass and significant loss of neurons in some areas of the brain. The most striking change in neural cells is a loss of the numbers of dendrites and their branches. This may account for more loss of function than the actual loss of neurons. [*See* Brain.]

Neurotransmitter metabolism is altered, with decreased activity of the enzyme that produces acetylcholine, as well as localized decreases in content and turnover of catecholamines. The end results of these morphological and chemical changes are an increased reaction time to stimuli, a mild memory loss, and an increased difficulty in learning new concepts. The speed of processing new information by the brain is somewhat slowed, but in the absence of disease most intellectual functions remain intact.

Most endocrine functions show a decline with age. The body's ability to metabolize glucose decreases with age, resulting in progressively higher postprandial blood glucose levels. This impairment in glucose tolerance in elderly persons is similar to what is seen in younger diabetic persons. Much of this decreased glucose tolerance may not be due to aging alone, but to associated factors such as obesity, physical inactivity, and drugs that impair glucose metabolism. Ovarian function decreases, estrogen levels fall, and the gonadotropic hormones FSH and LH, which control the ovary, increase in the blood after menopausal age. The sex hormone testosterone levels in males begin to decrease after age 50. This decrease may be gradual or rapid, and does not occur in all men. The levels of the male hormone produced by the adrenal gland, dehydroepianodrosterone (DHEA), decrease dramatically with aging in both sexes. This may be a key hormone in the aging process. Giving DHEA to experimental animals has prolonged life span and decreased the incidence of diabetes, obesity, and cancer. Higher blood levels of DHEA in humans are associated with a lower incidence of breast cancer in women and less cardiovascular disease in men.

Decline in immunologic function, as mentioned previously, may be central to the aging process. After sexual maturity the thymus gland begins to involute. T-lymphocytes, which mature in the thy-mus, and are the cellular mediators of immunity, are correspondingly decreased. The ability of T-lymphocytes to proliferate in culture in response to mitogens (such as plant lectins) decreases with age. This is probably related to the loss of thymic hormones, and the resulting impaired T-lymphocytic maturation. The ability of a subset of lymphocytes called T helper cells to produce interleukin-2 (a T-cell growth factor), also declines with age. An increased incidence of autoimmune disorders is associated with aging, due to a diminished ability of the immune system to distinguish the body's own proteins from foreign substances. There are also increases in autoantibodies and immune complexes and decreases in normal antibodies. The above changes in the immune system cause an increased suceptibility to infection and possibly contribute to the increased incidence of cancer and other disorders. [*See* Thymus.]

III. Means to Enhance Longevity

Epidemiologic studies of human populations indicate the unhealthy lifestyles can account for as much as 50% of premature mortality. For maximal prevention of the major killer diseases it is important that each individual obtain an accurate professional assessment of his or her risk factors for these diseases, and then adapt those changes specifically designed to reduce these risks. The following preventive measures, however, may prevent or delay the major disorders and thereby extend healthful longevity in the general population.

1. Cardiovascular disease: elimination of smoking; reduction of saturated fat, cholesterol, sugar, salt, alcohol, and caffeine; increase in fiber, fish, and vegetables; maintenance of ideal weight; regular exercise; avoidance of excessive stress; and control of high blood pressure and diabetes by medications if necessary.

2. Cancer: elimination of smoking and of smokeless tobacco; avoidance of excessive alcohol consumption; avoidance of excessive radiation exposure such as sunlight, X-rays, and radon; avoidance of exposure to industrial chemicals; reduction in fat intake; maintenance of ideal body weight; increased intake of fruits and vegetables high in fiber, beta carotene, and vitamins A, C, and E; avoidance of salt-cured and smoked foods and charcoal broiled meat; periodic physical exams; knowledge of cancer warning signs.

3. Accidents: reduction in excessive alcohol intake; increased use of safety belts for motorists and helmets

for cyclists; increased use of smoke detectors and flame-resistant materials in homes.

4. Chronic obstructive lung disease: avoidance of smoking and industrial pollutants.

5. Diabetes mellitus: maintenance of ideal body weight.

6. Chronic liver disease: avoidance of excessive alcohol intake; hepatitis immunization.

7. Infectious diseases: adequate immunization; avoidance of exposure.

To achieve significant extension of life span past the age of 100 it will be necessary to find means to slow the aging process. At present there are no established means of slowing the aging process in humans. There are, however, some interesting theoretical possibilities derived from animal experiments and other data. One means that has worked consistently in lower species, including protozoa, nematodes, fish, and rodents, is caloric restriction without essential nutrient restriction. Inhabitants of the Japanese island of Okinawa have the world's highest longevity. Okinawans eat less calories than the rest of the Japanese people, and have 30–40% less disease. Caloric restriction in mice and rats, whether started very early in life or after sexual maturation, has been found to decrease mortality rates, increase maximal survival times, and decrease the incidence of late-life diseases. The mechanism of these effects is not yet established, but a leading possibility is the preservation of immune function that results from caloric restriction in rodents. There is both a delayed maturation of the immune system and a slower age-related decline in immune capacity. Proliferative rate of T-lymphocytes in response to mitogens is higher in caloric-restricted animals. Other possible mechanisms are an enhanced capacity to repair DNA and decreased rate of free radical generation in calorie restricted animals.

Lowering of body temperature in cold-blooded animals also has a significant prolonging effect on life span. Fish maintained at 15°C, for example, live much longer than those at 20°C. Although this means would have limited direct applicability to humans, discovery of the mechanisms for this effect might lead to other more practical methods.

Enhancement of immune function might have significant antiaging effects. The Snell-Bag dwarf mice have a genetic defect that results in some signs of premature aging. If these mice are treated while they are young with early lymphocytes from young normal mice, the premature aging can be prevented.

If newborn thymus glands are transplanted into old mice, their immune function can be restored. Thymic hormone administration to old mice can enhance T-cell function and numbers, restore antibody producing capacity of other cells, and extend maximal life spans.

Exercise has been proposed as a possible means for slowing aging in humans. Many of the age-associated physiological decrements are similar to those resulting from prolonged bedrest in younger persons, and these can be reversed by exercise. The maximum oxygen consumption, a function of cardiovascular fitness and lung function, declines about 1%/year, but this can be largely prevented by an aerobic conditioning program. The age-associated insulin resistance of tissues and accompanying glucose intolerance can be significantly improved by exercise. Other age-related decrements partially reversible by exercise are muscle strength, immune function, and visual reaction time. A lifetime progam of moderate exercise in rats has been shown to prolong their life spans. It seems probable that exercise in humans improves health and well-being, delays or prevents several serious disorders, and delays a number of the manifestations of the aging process, but its effect on extending the number of years of maximal life span remain uncertain.

In summary, we have identified all the known factors that limit both our average life expectancy and maximal life span. The factors predisposing to the major life-threatening disorders are well characterized, and a number of measures are known to reduce the risks of developing these and to thereby extend longevity. Although we have extensive knowledge concerning the effects of the aging processes, their exact cause is as yet unknown. Research is producing some encouraging possibilities, and also some potential means for slowing these processes.

Bibliography

Bortz, W. M. (1982). Disuse and aging, *JAMA* **248**, 1203–8.

Hayflick, L. (1985). Theories of biological aging, *Experimental Gerontolology* **20**, 145–59.

Haynes, S. G., and Feinleib, M. (1980). "Epidemiology of Aging." National Institutes of Health, Bethesda, Md.

Manson, J. E., Stanpfer, M. J., Hennekens, C. H., and Willet, W. C. (1987). Body weight and longevity, *JAMA* **257**, 353–58.

Masoro, E. J. (1987). Biology of aging, *Archives of Internal Medicine* **147,** 166–69.

Masoro, W. J. (1988). Food restriction in rodents: An evaluation of its role in the study of aging, *Journal of Gerontology* **43,** 59–64.

Morin, R. J. (1985). ''Frontiers in Longevity Research.'' Charles C. Thomas, Springfield, Ill.

Morin, R. J., and Bing, R. J., eds. (1985). ''Frontiers in Medicine.'' Human Sciences Press, New York.

Morrison, S. D. (1983). Nutrition and longevity, *Nutrition Reviews* **41,** 133–42.

Wilson, D. L. (1988). Biomarkers of aging, *Experimental Gerontolology* **23,** 4–5.

Low Temperature Effects on Humans

W. J. MILLS, *University of Alaska, Anchorage* and

R. S. POZOS, *University of Washington School of Medicine*

Glossary

Acclimatization Adaptive changes that occur in the lifetime of an organism in response to change in the natural climate, as a physiological change occurring within the lifetime of an organism that reduces the strain caused by stressful changes in the natural climate

Adaptation Change that reduces the physiological strain produced by a stressful component of the total environment; it may occur within the lifetime of an organism (phenotypic) or be the result of genetic selection in a species or subspecies (genotypic)

Cold (1) Privation of, or a relatively low, degree of heat; (2) slowing of molecular motion; (3) absolute cold (absolute zero) is the moment that all motion of molecules cease (equal to −273.16°C (−459.69°F)

Frostbite Condition where heat is lost and the area cooled to the point of ice formation in the affected tissues

Hypothermia Condition of generalized whole body heat loss; described is the condition of a temperature-regulating animal, when the core tempera-

ture is >1 standard deviation below the mean core temperature of the species in resting conditions in a thermal, neutral environment; in the human, this generally is considered to be in the range of 35°C (96°F); some variation may be expected in the elderly

I. Introduction

Over 1 million years ago, *Homo erectus*, the first human, appeared in the lush tropical regions. This earliest man shared with Modern man, *Homo sapiens*, many anatomical and cultural characteristics. He walked upright, unlike his apelike predecessors, hence his name. He used tools, developed social interaction, and dispersed knowledge. Most importantly, he used fire. Primordial man has always sought some form of thermal comfort and has modified his environment in many ways to meet this demand. Some of man's primitive drives such as appetite for food and water are intimately tied in with his quest for thermal comfort.

Homo erectus was primarily a tropical being in that he inhabited warm or tropical regions of the world, but unlike those before him, he was able to inhabit a large area of the wold. Eventually, he successfully ventured into the cold northern regions during the Ice Age, a period of major geological upheaval and climactic change. His successes were due in part to his ability to maintain his tropical climate by way of fire, clothing, and other adaptations, which both culturally and socially are taken for granted today.

First man's battle with cold, his advances and retreats to and from the far northern and frigid areas, are well documented from excavated fossil remains. From this early period in human history through the present time, man has persisted in his battle with cold. Assisted with technological adaptations and with applied additions to his own microclimate, the human species has survived the ice ages when so many other species perished.

Neanderthal, Cro-Magnon and Modern man followed the northern migration patterns of *H. erectus*. Excursion over the ages into the cold zones caused constant exposure to cooling and cold and exerted a long-term effect on the emerging human relative to his ability to physiologically adapt to harsh cold climates. The studies of the Australian Aborigines, the Alakalug peoples of Tierra del Fuego in Southern Chile, the Bushmen of the Kalahari Desert, and the Eskimos and Indians of the high Arctic have demonstrated man's ability to adapt or acclimatize to his environment and to live in low-temperature areas without harm.

The human animal is a warm-blooded (tachymetabolic) organism, existing in a state of homeothermy, capable of maintaining a relatively constant internal core temperature between 36 and 39°C. This state of warmth, to remain constant, requires that in any environment to which a human is exposed, no matter how extreme, the amount of heat produced by metabolism and behavioral strategies must be equal to the heat transferred from human to the environment. This critical balance is kept and constantly adjusted by the temperature-regulating controls discussed elsewhere in this chapter. Any continuous downward heat loss from this accepted "normal" range, if not checked, will cause a drop in body (core) heat called hypothermia. Persistent or overwhelming loss of heat will result in harm to segmental body areas or the body as a whole.

The loss of heat is not the only major consequence following low-temperature exposure. Because the chemical reactions of the body depend on a certain temperature, once the temperature of the body cools the cells are no longer able to function as effectively. This effect, although it can be deadly, also allows humans who have been sufficiently cooled to have a certain amount of time in which the process is reversible. This is because the cold-induced decrease in cellular function also decreases the demand for oxygen in the cells; therefore, cold cells can withstand a period of low oxygen (hypoxia) for extended variable times, whereas in a comparably warm situation the same amount of hypoxia would be lethal.

When the interior temperature of the body cools, it causes multiple body systems to falter, and function deteriorates or fails. This failure includes deterioration of cerebral activity and function, and disturbance of thermal regulation. Other organ systems fall victim to hypoxia, eventually anoxia, and include cardiovascular collapse, respiratory failure, renal shutdown, and eventual loss of neuroregulation and failure of the metabolic system. Extreme heat loss, freezing injury, or severe hypothermia may cause irrevocable loss of function, anatomical destruction, or death.

Hypothermia has been a major problem for centuries and is still a problem today. This is particularly so in military campaigns in winter or at high altitude, even in temperate zones. Xenophon, a Greek statesman and general, described the travail of his army in the retreat from Persia in a harsh winter march in 400 B.C. Charles XII, King of Sweden, led his armies in numerous winter campaigns in Poland, Russia, and Norway from 1697 to 1718, resulting in great loss of men to cold. The overwhelming effect of severe cold, with decimation of an army, is related in the surgical memoirs of Baron D. J. Larrey, Napoleon's Chief Surgeon of the Grand Army, in his description of the retreat of the French from Russia in 1812–1813. This description of the advances, the retreats, and the clinical pattern of cold injury, especially hypothermia and freezing injury, is a classic of required reading for the student of the cold effects on humans. Throughout history to the present time, including the American Revolutionary and Civil wars, the Crimean War, World War I, World War II, the Korean War, the Falkland War, and, more recently, the war in Afghanistan, cold has immobilized, destroyed, and obliterated major armies and often the civilian population. The problem of armies in retreat, pursued by overwhelming cold winter weather and an unrelenting enemy on all sides, is a formula of disaster not yet solved.

The hypothermic threat is not the province of the military alone. The onset of hypothermia in the field may be sudden, as immersion in cold, icy, or near-frozen waters, or insidious, as when hunters, fishermen, skiers, or outdoors enthusiasts are exposed to wind, windstorms, fire or flood, rain, snow, or falling temperatures. Natural disasters (especially in winter) such as earthquakes cause humans to face their old nemesis—cold stress. The loss of heat from cold exposure may be enhanced by a mental or physical deterioration from alcohol or drug abuse, illness, cardiorespiratory failure, or severe trauma associated with fluid or blood loss.

Any or all of the above conditions may leave the victim unable to maintain homeothermic control and may permit loss of core temperature. Eventually, general body cooling, with heat loss and exhaustion of the caloric reserve occurs. Soon thereafter, if heat loss is not reversed, general levels of incompetence, semicoma, coma, and eventually death will follow.

II. Physiological Basis of Thermoregulation in the Cold

To prevent the dire consequences of cold, the body controls any heat loss by way of physiological and psychological reactions. For these reactions to be triggered, the brain initiates various responses depending on how cold the person is becoming. This response depends on the fact that the brain monitors two different temperatures: the core and the peripheral. The core temperature has been defined as the temperature of the brain, heart, and lungs. The peripheral temperature is the temperature of the skin and the muscles. The temperature from both areas, the core or the periphery, are sensed by specialized nerve endings that are able to respond to changes in temperature. Thus, with the thermal receptors on the skin the body is designed with an early warning system relative to cold. The brain monitors the difference between the peripheral and core temperatures and then, based on these differences, will initiate certain reactions. The reference point that the hypothalamus uses to define a fall or rise in the core temperature is called the set point. According to this theory, the brain has a fixed temperature around which it attempts to maintain the temperature of the core.

The major driving force for physiological changes to occur is usually the change in peripheral temperatures. If the peripheral temperatures begin to fall, then the brain will monitor this drop relative to core temperature. Depending on the degree of the drop of the peripheral temperature, certain reactions will occur. These changes, which include changes in muscle tone, metabolism, and behavior, will attempt to increase internal core temperature until it reaches the set-point temperature. In the event that those responses are not effective in terms of minimizing thermal loss, then the core temperature will fall. Peripheral temperatures can fall to very low levels, even to freezing levels, and yet tissue damage can be recovered, even reversed, in many situations. However, the delicate core organs such as the brain, heart, and kidneys do not have that kind of resistance to the cold and cannot tolerate such wide temperature swings as can the skin.

If the core temperature drops, the body will continue to respond to the cold stress. Although hypothermia is defined as a 2°C drop in core temperature, smaller changes in core temperature have been associated with subtle changes in physiological systems, enough to cause a decrement in performance. Not all systems respond to the same degree with a drop in core temperature. The brain and nervous tissue seem to be the most sensitive, followed by other organ systems.

III. Measurement of Core Temperature

The core temperature is important, and there are various ways to measure it. Although the core temperature is the temperature of the brain and the heart, usually rectal temperatures are taken as the indication of core temperature. This temperature tends to be the easiest to monitor in subjects but may give readings more reflective of the temperature of the blood and metabolism from the legs and the hips than of the entire body. Core temperature should ideally be measured in the heart. Because this is not easy to accomplish, the temperature of the esophagus is used. Because the thermometer is placed close to the heart even though it is in the esophagus, it gives the most accurate nonblood measurement of the core temperature. Actual core temperature is measured in the heart by using a small wire thermometer called a thermistor, which monitors the temperature in the heart. Another temperature site that is used to measure core temperature is that of the tympanic region. Because the tympanic region is close to the brain, this is arguably a good measurement of brain temperature. Although this might be the case, it is difficult to place and maintain the thermistor close to the tympanic membrane. Clinically, monitoring the temperature of the urinary bladder is thought to be as accurate as that of the esophagus. For all purposes, the temperature of the rectum is used as standard for human experiments, even though it does have limitations. Attempting to measure core temperature by placing a thermometer in the axilla gives a false reading because it is an indication of peripheral and not of core temperature.

All the changes that occur to ward off the effects of the cold are based on physical processes that determine heat transfer. Understanding these mechanisms will allow for a thorough understanding of the human's reaction to the cold and evaluating the effectiveness of various rewarming strategies. Overall, the net transfer of heat is determined by the metabolism of the body as well as by the heat generated or lost by convection, conduction, radiation, and evaporation. For normal thermal comfort as well as survival most of the heat that we generate by way of metabolism must be lost to maintain a core temperature of 37°C. In a cold stress situation, these processes must be understood to minimize heat loss.

A. Metabolism

Metabolism is usually defined as the total amount of heat produced by the chemical processing of the food that we eat as well as by the work being performed. Overall, the body does not generate a uniform temperature due to regional differences of metabolism. Because food is processed primarily by the liver and associated digestive glands, heat is generally highest in the region of the liver. Those areas with more muscle mass will be warmer than those with less muscle mass because muscles generate heat when they contract. An example of how the body generates heat by way of metabolism during a cold stress is shivering. Because any kind of exercise will increase a person's basic metabolism, shivering, which is an involuntary muscular contraction, also increases metabolism. Shivering is a very potent generator of body heat because it eventually can involve the entire body with waves of muscular contraction. In addition to shivering, any kind of hormonal control of metabolism will play a role in generating heat. Hormones such as thyroid, epinephrine, insulin, glucagon, and sex hormones will all influence the rate at which various cellular reactions occur and, consequently, the amount of heat produced.

As a consequence of metabolism, the body constantly generates a certain amount of heat. This heat is lost primarily through the respiratory system and the skin. The heat is transferred to the skin through circulating blood, which has a high heat capacity. Thus, the circulation becomes very important in any heat-transfer situation. As the heat is transferred to the skin through the blood, it is further transferred to the environment through radiation, convection, conduction, and evaporative heat loss.

B. Radiation

Radiation influences heat transfer between the skin and the environment. The amount of heat transferred depends on the temperature gradient between the radiating surfaces and their emissivity as well as the exposed surface area. Any mechanism that can be used to minimize the effective radiating surface of the human body is important in minimizing cold stress. To minimize the body area, one can curl up in a ball, thus decreasing heat transfer. In addition, as the legs are brought in close proximity to the other parts of the body, they will radiate heat to each other. Because air is a poor heat absorber, the human body radiates heat to solid objects such as walls and vice versa. In all situations, the body is interacting with the environment so that the heat transfer between the body surface and the environment is an ongoing two-way process. An example of the use of radiation to warm the human body is the use of radiant heat lamps to warm infants in a hospital.

C. Convection

Convection refers to the heat-transfer process in which one of the heat-transfer objects is in motion relative to the other. As an example, the movement of air or water over the body will promote a large heat transfer from the body. This heat transfer depends on the temperature difference between the skin and the medium it is in and the movement of the other medium. Humans have used this concept for thousands of years. Any shelter that minimizes wind will therefore decrease the heat transfer from the body. Also, any way in which the temperature difference between the skin and the air is minimized will decrease heat transfer from the body. In animals, fur and feathers have also been used to minimize heat transfer. Except for a vestigial reaction in humans in which the hair stands up in response to cold exposure, we have no physiological way to trap air. Clothing is our way of trapping air and represents a simple but effective way of making up for the loss of fur. Wind chill factor refers to the additional effect of the wind increasing the amount of heat transfer. This concept is accurate only when the wind whips across the naked skin. Any kind of clothing that prevents the winds from increasing heat transfer will blunt the effects of the cold wind.

D. Conduction

Humans usually lose heat through convection and radiation. Conduction is a process that usually does not have much of an effect on humans unless a large area of the body is in contact with another thermal object (such as lying on the cold ground) for a significant period of time. This concept is used in a number of ways. In physiological systems, fat is used as a poor conductor. Because of this fact, it is a good insulator; therefore, fat protects (insulates) the body from the cold. In addition, clothing, which can also be a good insulator, is also used to ward off the effects of the cold. Many layers of clothing or other strategies used to trap air and, therefore, have them act as an insulator are very effective in minimizing heat loss.

E. Evaporative Heat Loss

Evaporative heat loss is a process in which the body evaporates the water that is secreted. This process of evaporation requires heat from the body, thus decreasing the temperature of the body. Although this process is thought to occur primarily on the skin, the lungs are another important source of heat transfer by this method. Any way in which the cold air is warmed before it interacts with the warm humid environment of the lungs will minimize heat loss.

F. Summary

All of the above processes are working simultaneously in humans. In considering any kind of overall response to the cold, all of them must be considered. In certain situations, some are more important than others. For example, in cold water, heat loss by convection and conduction is more important than heat loss by radiation. In a person who is running, evaporative heat loss from the lungs and skin is more important than heat loss by radiation. In addition to these considerations, human attempts to find some reasonable compromise between thermal comfort, survival, and appearance. In life and death situations, all that counts is thermal survival, but in less stressful environments, the appearance of clothing becomes another concern for humans as they work or recreate in the cold.

In summary, most of the processes that are involved in heat transfer in our body are used to dissipate heat. Thus, in a certain sense, humans are considered a tropical animal in that they try to dissipate heat that they produce by various mechanisms. The importance of these factors is that these processes must either be blunted or minimized to increase our ability to withstand the cold.

In addition to these physical processes, a number of physiological reactions play a role in human ability to handle the cold. Because food represents the major source of energy that we must have to survive and stay warm, the physiological explanation of what triggers food appetite is important. Many explanations have attempted to describe the physiological basis of appetite. Some include an internal monitoring of glucose in the body, increased cold stress, or genetic factors, but there is still no consensus as to what drives our food appetite. Food appetite as well as fat deposition are examples of our need to gather and store the energy required to maintain constant thermal balance. This drive might run counter to Modern man's desire to maintain a trim profile. [*See* APPETITE.]

IV. Peripheral Response to Acute Low Temperature (Nonfreezing)

Human response to the cold involves peripheral and core-initiated responses. A number of strong peripheral responses are mediated by way of neural reflexes. In most cold situations, the skin, fingers, and toes are the first to respond to the cold stress. Cold causes the skin thermal receptors to send electrical signals to the brain, which causes the blood flow to the skin to decrease. This effect is mediated by a part of the nervous system called the sympathetic nervous system. This allows for the increase of insulation of the human body by having the limbs become cold, therefore minimizing heat transfer away from the body. Some psychological and physiological responses are associated with this decrease in skin temperature. Overall, as a consequence of cold stress, metabolism increases to generate more heat. This might not be a general rule because old people and certain aboriginal groups do not seem to respond to a cold stress with an increase in metabolism.

The general cooling effect is well demonstrated on the extremities, especially in hands and hand contents. Manual performance and dexterity is decreased as heat is lost. This hand activity is lessened primarily from impairment of function of nerves and vessels. Individuals whose hands are unaccustomed to moderate or severe cold stress also report decrease or loss of cutaneous sensation. Skin sensitivity and proprioceptive sense is diminished. In the hand, slow gradual cooling may cause more discomfort than rapid cooling, especially on deep structures.

The cold effect on peripheral and digital nerve function is to slow nerve conduction. If the local tissue temperature remains 4°C, all cold changes may be reversed. Continued and lowered temperatures may result in sensory and motor nerve damage.

Peripheral cold certainly affects muscle. "Cold" muscle is liable to tears, especially if during exercise, work, or activity the muscles are not subject to warm-up and stretching. Cold causes a decrease in muscle contraction and power. Because this activity is not accompanied by a significant decrease in blood flow, this condition is considered a direct affect of cold on muscle fibers. In addition, cold causes failure of nerve conduction and decreases neural musculature transmission.

This diminished or altered muscle function in the cold often results in injury. On the ski slopes, it is not unusual to find a rate of injury inversely proportional to the fall in ambient temperature. This pattern is associated with an increase in joint stiffness and resistance to motion, apparently due to (1) lowered tissue temperatures in structures around the involved joint (e.g., knee joint), (2) lowered joint temperatures, (3) lowered intraarticular joint temperature, and (4) increased viscosity of synovial fluid.

Predictably, one of the most immediate symptoms of cold insult or heat loss is that of cutaneous response. The perceived sensation of "cold" is related to a drop in the average temperature of the skin. The subsequent sensation of discomfort may be due to small vessel vasoconstriction, with discomfort increased by shivering. This kind of pain is an effective warning system that alerts the person that the environment is posing a significant cold stress.

This immediate response to cold, with small vessel vasoconstriction, is labeled cold-induced vasoconstriction (CIVC). This initial response is followed by a vasodilation. This pattern may repeat itself a number of times and has been called the "hunting reaction," because it seems as if the tem-

perature of the finger is searching for some optimal kind of temperature. The mechanism of this is not clear. One hypothesis is that the nerves that control the blood vessels initially cause the blood vessels to vasoconstrict and that when there is less flow due to cold, the nerves themselves eventually stop functioning, which causes the blood vessels to open. Another explanation deals with the control of special small blood vessels that short circuit the superficial arteries and veins of the extremities. These are called arterial venous anastomoses (AVAs). When AVAs are open, warm arterial blood is directly conveyed to the superficial veins of fingers and forearm. These veins are warmer than the adjacent skin. They form a "heating glove," which ensures that the underlying nerves, muscles, and other subcutaneous tissues are kept at an optimal high temperature for best function. When AVAs close, due to cutaneous cold response, functioning of the deeper tissues will diminish.

Some individuals demonstrate an abnormal response of digital arterial constriction in response to cold stress. The response to cold is not of the usual CIVC variety. Here, there is loss of the usual hunting reaction and the blood vessels stay vasoconstricted. This response is often aggravated by vibration, as by operating a snow machine, or even the repetitive action of chipping ice on a sidewalk. This vascular response may have a normal or genetic etiology. This interesting finding is widespread in northern latitudes and occurs commonly in young healthy females. It affects the small arteries and arterioles of the distal areas of the body, especially fingers and hands. The digital response is often a waxy pallor, followed by redness and then cyanosis. These changes in color are due to changes in blood flow to the digit. If the skin is blue (cyanotic), it means that there is some form of vascular stasis. The insult need not be triggered by extremely low temperatures and can be induced by cool temperatures near 20°C, with the blanching following a drop of only a few degrees or by an increase in wind, with even a mild chill factor. The digital blanching may be seen in a peripheral nerve or digital nerve distribution pattern. The pattern of blanching is variable because the entire hand may blanch or just the index fingers. This condition can predispose to cold injury. Whether this is a related Raynaud's phenomena or a basic labile vasomotor response to cold is an area of current investigation. The prevention and treatment of these responses include biofeedback training or efforts at self-regulation of the temperature-regulating mechanisms, suggesting that the brain plays a major role in controlling these kinds of responses.

Associated with the cold stress to the periphery is pain. This acts as a warning to humans because it indicates that the environment is potentially thermally threatening. Besides the hands, feet, the face, and the head, other major sites of cold injury exist because heat can be easily lost from these areas.

Cold stress on the skin also induces changes in metabolism and respiratory drive. A cold stress commonly increases the metabolism of a cold-stress subject. The cold might cause some form of increase in muscle firing at a subtle level to increase body heat, followed secondarily by changes in hormonal control of metabolism. In addition, cold stress on the skin will increase the heart rate and blood pressure. A significant cold stress will cause a person to stop breathing in inspiration for a short period of time.

The respiratory system is also susceptible to environmental cold. Environmental cold is responsible for respiratory tract problems worldwide. Generally, the upper respiratory tract is able to warm and humidify inspired air, avoiding cold damage to lung tissue. However, in the asthmatic patient, airway resistance is often increased during or after exercise. This condition appears to be a result of the removal of heat and moisture from the respiratory tract. Most data indicate that the resulting bronchospasm is secondary to heat loss and not to humidity. This condition of cold-induced asthma seems separate, as an entity, from exercise-induced asthmatic change.

The response of the respiratory system to cold stress is important in a number of environments to which man normally does not belong. Divers need to take their gas mixtures with them for breathing. In most cases, those gases are dry and cold. This condition results in a state called diver's bronchorrhea, occurring secondary to loss of heat from the respiratory tract, which then results in secretion of fluid in the airways and pulmonary edema.

Another interesting peripheral effect of the cold deals with psychological changes associated with cold. In the far northern regions of our world, it has been suggested that cold, often associated with high latitude and the period of winter darkness, may produce neurochemical changes in the brain, resulting in episodes of depression and acts of suicide. Expo-

sure to a cold environment in areas of stark isolation and sparsely populated regions often precipitates episodes of psychic stress. This effect is seen primarily in those not accustomed to cold and barren regions. Small groups, or individuals alone, who lack winter experience, on hiking, hunting, or skiing outings, fall victim to cold stress. The phenomenon is found in untrained Arctic and Subarctic military groups and may include mental and physical disorientation, fear, panic, hallucinations, and stress ulcers. Visual, auditory hallucinations are commonly reported by individuals who have been cold-stressed. The etiology of cold-associated cerebral, physical, and psychic distress may be (1) an underlying neurochemical basis, (2) an underlying genetic predisposition, or (3) a failure of situational adaptation to a cold climate.

In addition to the effect of cold on the skin and the peripheral circulation, it also affects certain proteins in the blood. These groups of serum proteins, which can reversibly gel or precipitate due to cold, are called cryoglublins. These responses to cold have been termed cold hypersensitivities and include cold urticaria, paroxysmal cold hemoglobinuria, cold erythema, and cryoglobulinemia.

Cold urticarial responses are fairly common, especially in children, they are seen often in cold-climate areas. Local signs of redness, itching, wheals, and edema appear on cold-exposed skin. Symptoms of headache, malaise, tachycardia, and even collapse or anaphyllactic shock are occasionally seen. The condition occurs in two forms: primary acquired cold urticaria, the most common form, and familial cold urticaria, the less common type not associated with histamine release as in the primary acquired form. The cold urticarias are associated with the presence of cryoglobulins, cryofibrinogen, and cold agglutinins.

Paroxysmal hemoglobinuria is a form of autoimmune hemolytic anemia. In this event, IgC antibodies bind to red cells, at low temperature, and fix the red cell complement. This causes hemolysis when the temperature is raised to 30°C. Following the exposure of the individual to cold, autoantibodies in the intravascular compartment attach to red cells. When the temperature is elevated, hemolytic action is mediated by the complement sequence.

Cold erythema has been listed as a form of hypersensitivity reaction demonstrating purpura, severe pain, muscular spasm, and sweating. Serum globulins precipitate when the blood is chilled. Treatment, as with other hypersensitivities, includes removal from the cold environment from the start, although most of the signs and symptoms appear exacerbated following warming.

Cryoglobulinemia is a hypersensitivity response, with cryoglobulins found in the serum. Other cryoproteins in the serum are cryofibrinogens and cryogamma-globulins, the latter a gamma-globulin readily precipitated by reduced temperature. These cryoproteins are often associated with cold urticaria. The proteins appear to transfer cold sensitivity and active compliment cascade. Hive formation may be due to cold-dependent anaphylatoxin release. Cryoglobulinemia is also associated with cutaneous vasculitis and cold urticaria.

V. Chronic Effects of Peripheral Low Temperature (Nonfreezing)

In addition to the acute effects of cold on the periphery, a set of responses occur in persons who have been exposed to cold–wet for a period of time. If not corrected, these conditions will cause changes in the nerves, blood vessels, and eventually the muscles.

Chilblains or pernio is a condition characterized by localized itching and painful erythema on the fingers and toes and sometimes on ears or the nose. The problem is produced by exposure to cold, damp conditions. There is often fissuring of the skin and sometimes the appearance of red painful discrete nodules or plaques, with cyanosis, especially in the mottled lower legs. Children and women with a history of outdoor exposure, in the wet environ, are readily affected, although it is a common lesion in sealers, cold-climate wintertime fishermen, or persons with habitual occupations in cold, damp, or wet conditions. This condition is found in the more temperate climates of northeast Europe in chronic form but is worldwide in temperate and northern zones. An abnormal vascular response or prolonged vasospasm is also considered in the etiology. A predisposition probably exists in persons with increased peripheral vascular tone. The cold exposure causes increasing vasoconstriction, resulting in skin lesions from tissue anoxia. This increase in the vascular tone is due to the nervous system causing the blood vessels to vasoconstrict.

The condition occurs in an acute and chronic form. The chronic form is a disease of small blood vessels of the skin and may result in ulceration or necrosis. Although multiple causes are listed for chilblains, cold is the primary agent, associated with an increase in humidity. The initial phase of finger redness, tenderness, and swelling, with fingers warm to the touch, is followed by vasodilation and edema. This progresses to digital tenseness, blisters, and ulcers.

This condition involves primarily the lower leg and toes, hands, and fingers. The lesions may be bilateral, symmetrical, single, or multiple. Erythema and cyanosis precedes the edematous patches hours after the insult. The patchy areas are pruritic, and victims describe a localized burning sensation after rewarming. In advanced stages, vesicles, hemorrhages, or ulcers are evident.

Severity is increased by previous skin disease or injury, poor state of nutrition, or sepsis. The lesions may recur in winter and disappear in warm weather. The pathological microsections demonstrate an inflammatory response that involves arterioles in the superficial and deep dermis. The blood vessels are occluded, resulting in tissue death in the superficial and deep tissues of the affected part.

Treatment may include drugs that inhibit the effects of the sympathetic nervous system that are causing the blood vessels to vasoconstrict as well as traditional warming methods such as whirlpool heating. Removal from the cold or the wet occupation and antibiotics, if indicated, are also part of the treatment protocol. Patients remain sensitive to the cold for a considerable time after the treatment.

Trenchfoot is a nonfreezing cold injury (also called immersion injury) resulting from salt or freshwater exposure or long-standing damp tropical immersion of the foot. It has been the bane of any person who has to spend a considerable amount of time in damp environments. The condition results from exposure to temperatures near, but not at, the freezing level. It follows exposure to wetness and cold. The level of low temperatures causing the injury is not fixed, except that the range is above freezing. The occurrence may result from exposure to any wet condition, where the temperature of the extremities and skin is greater than that of the environment, and allows transfer of heat to the environment. The problem is worsened by constrictive footgear, wet socks, liners, or boots in a constant wet or damp environment. It is aggravated by a fixed or sedentary position such as that in a trench or boat, with moderately or nearly immobile extremities and impaired joint motion and circulation.

Trenchfoot is a nonfreezing injury that is due to

the effects of cold on the neurovascular components of the foot. The cardinal causes are immobility, heat loss, standing for prolonged periods of time, peripheral edema, and clothing or shoe constriction—all in a wet, cold environment. Early signs and symptoms include swelling, boot tightness, numbness, and paresthesias. Numbness in any form of cold injury is an early sign of nerve impairment and should never be overlooked or ignored as an early progressive sign of cold injury, either at, above, or below freezing temperatures. It is one of the earliest signs of impending tissue injury from cold.

Trenchfoot is occasionally labeled in stages. In Stage I, the beginning or inflammatory stage, the involved part is hot and dry with complaints of burning and early edema. Cyanosis may be present. Upon warming, tingling, numbness, and deep pain occur. Stage II has been termed as postinflammatory, the extremity being cool or cold, often with pallor and hyperhydrosis. The moisture is marked, and the cyanosis is most evidence when the person is standing.

Evidence of sympathetic nerve irritation is demonstrated by irritability, excessive sweating, and pallor. Proximal hyperesthesia and distal anesthesia are common. The most important finding in this condition is edema and swelling. The obvious skin tenseness and tissue expansion of extremity contents represent deep compartment pressure increases. This is often due to vessel contracture or occlusion and intercellular, or intercompartmental, expansion of muscle. The deep tissue pressure further causes anoxemia (decreased oxygen in the blood) of tissue and occlusion of the vascular tree with collapse of involved small vessels and capillaries. This often results in severe loss of capillary perfusion with the compartment pressures rising to high levels (60–90 mm of mercury). If not corrected, the tissues will die due to lack of adequate perfusion. Microscopic and tissue examinations find most of the changes in the blood vessels. The vascular tree is often engorged and the blood vessel lumen may be filled with incomplete plugs of fibrin, or hyaline material, or even total plugging. Dilatation of the vessels and marked vasconstriction are both present. Organized thrombi are seen in the area of involvement as well as in the tissues above the level of tissue damage. The blood vessel cells are swollen, and there is often cellular destruction of various layers of the blood vessels. The changes appear to be inflammatory in origin, due to the action of cold on the tissues and, even more likely, to local hypoxia, vascular deprivation, and tissue destruction. Nerves in the area reveal swelling, edema, and degeneration of the axis cylinders with demyelinazation of the nerves, with later lipoid phagocytosis. As the signs and symptoms indicate, the sympathetic nerve components are not significantly altered so that they can influence the vascular tone compounding the damage.

Restoration of vessel patency and circulatory return is the physiological basis of treatment. Treatment must include (1) release of tissue compartment pressure, usually by fasciotomy; (2) drugs that promote vessel dilatation and loss of sympathetic affect, permitting the tissues to dry, and aid in relieving edema and pain; and (3) drugs that relieve severe pain.

Human response to the cold is in response to input from the periphery in nonfreezing injury. However, the most life-threatening situation occurs when the nonfreezing injury causes a decrease in the core temperature.

VI. Effects of Low Temperature (Nonfreezing) on the Core (Hypothermia)

Hypothermia is a condition of abnormally low core temperature. It is considered to be a condition of a temperature-regulating animal, when the core temperature is >1 standard deviation below the mean core temperature of the species in resting condition in a thermal neutral environment. This has been computed to be 35°C (96°F). These definitions do not take into account that there is a diurnal variation in body temperature so that there is a fluctuation of at least 1°C during a 24-hr period of time. The problem of the terminology is more complicated because hypothermia may be more often an effect of the rate of core temperature falling than of the actual temperature.

Keeping all of the above caveats in mind, hypothermia is defined as a 2°C drop in rectal or core temperature. It is clinically defined as beginning at 35°C, and it may go as low as 18°C. Persons may survive severe hypothermia depending on the kind of care and rewarming they receive. Cases of revival from hypothermia as low as 18°C have been reported. Generally, hypothermia is a slow process that allows the rescuers a window of time to successfully rewarm the victim.

Hypothermia has also been divided into primary hypothermia, where the body with normal thermal regulation is exposed to overwhelming cold, and secondary hypothermia, where mild or moderate cold exposure causes hypothermia as a result of abnormal thermogenesis. In this situation, the normal physiological responses to cold are not operating effectively.

Examples of primary hypothermia are classified on the basis of the environment in which they occur because the various thermal–atmospheric characteristics in those environments are unique and present their own set of challenges. These are defined as air exposure, immersion, submersion, divers, and mountain hypothermia.

Examples of secondary hypothermia refer to the reduction in the body's temperature based on the fact that the underlying physiological thermal regulating systems have been compromised for any number of reasons. For example, secondary hypothermia would be a decrease in core temperature due to drug-induced impairment, muscle degeneration, vascular insufficiency, nutritional deficiency, or neural deterioration due to strokes, spinal cord injury, etc. These examples demonstrate how much the core temperature of the body is determined by a complex set of interacting physiological systems.

Developing signs of heat loss and consequent degeneration of the victim's capabilities occur in sequence. The temperature level of the specific symptom or sign, however, is subject to the state of health, the age, and the physiological variations of each individual. The first sign of heat loss may be either the peripheral CIVC or the peripheral vessels or the closure of AVAs as well as the appearance of "goose pimples." The latter appearance demonstrates the elevation of body hair, by action of the erector pilae muscles, to trap air against the skin as a heat-loss barrier. This reaction is a vestigial reaction in that it is not efficient in terms of trapping air—but it does act as a sign of cold stress. This is followed by increased muscle tension and a sensation of deep cold.

Stumbling and incoordination occur near 35°C. Between 35° and 34°C, a cardinal sign of dangerous heat-loss level appears—dysarthria. The victim may speak as if intoxicated or as with a "mouthful of mush." Continued heat loss is followed by muscle rigidity, disorientation, and diminished visual acuity. In many individuals, at the level of 32°–31°C, the hypothermic victim may be approaching a semicomatose state. Various victims demonstrate

bizarre behavior patterns including undressing. This paradoxical undressing is thought to occur at low levels of hypothermia and might be related to a paradoxical whole body vasodilation giving the person a false sense of warmth.

When the core temperature drops to 29°C, the victim may lose total thermal control. At this level, the victim may slide into a "metabolic icebox," where he or she may remain for hours, thermally incompetent, usually comatose, and with diminished function of all necessary systems. Cerebral and cardiorespiratory functions are particularly depressed, so that aerobic metabolism needs are also diminished, permitting, in this low-temperature metabolic icebox, a trace of life, until ultimately over hours (if on land, and much less in water immersion) cardiorespiratory function fails. Not all victims of hypothermia demonstrate these clinical signs at exactly those temperature levels listed. The response to cold varies with the state of health, physical fitness, age, and possibly exposure (e.g., dry cold vs. humid cold).

Initially during cold stress, metabolism increases primarily induced by shivering. Heart rate, blood pressure, and consequently oxygen consumption also increase. As the body is no longer able to withstand the thermal stress and hypothermia begins, oxygen consumption will decrease.

At 30°C, oxygen consumption is decreased by about 50%. Because the mechanisms that control the acid-base level of the blood decrease, acidosis begins to appear. Eventually, the mechanisms that control the organized contraction of the heart are affected, and electrical signs in the electrocardiogram reflect these changes. Finally, ventricular fibrillation, an uncoordinated, lethal contraction of the ventricles, occurs. This is not due directly to the cold but to changes in the electrolytes and pH (acid level) of the blood.

The nervous system is obviously affected early on in any kind of cold stress. Both peripheral and central components of the nervous system are affected. Of these, the peripheral nervous system is usually affected first. The peripheral nervous system components, once cooled, will in essence anesthetize the area in question; therefore, the hypothermic individuals may be unaware of damage to their periphery. In addition, such persons might have become disoriented and wandered off completely oblivious to their plight.

Although this is a very sensitive system in terms of oxygen demand, the cold protects the nervous system from ischemic damages to the extent that at 15–20°C, cerebral circulation may be stopped for approximately 1 hr with minimal damage to the nervous tissue.

The respiratory rate of the hypothermic individual will eventually be high and then it will fall. At 30°C, the rate will be 7–15 min. Fluid builds up in the lungs (pulmonary edema) due to a decrease in the function of the pulmonary cells.

Cold affects the gastrointestinal system so that the motility of the gut decreases. In many experiments of cold stress, subjects complain about constipation after the cold stress. This more than likely is due to dehydration associated with the cold but also may include some decrease in the motility of the gastrointestinal system induced by the cold.

As is commonly observed, cold causes a diuresis (urination). If the hypothermia is prolonged, cellular damage can occur in kidneys, thus halting the production of urine.

Cold has an effect on the hormones of the body. Whether these effects are a direct result of the cold or of just the stress of the cold is not clear. Steroids are elevated in cold-stressed subjects. Thyroid hormone is also elevated, especially in subjects who have been in the cold for prolonged periods of time. Insulin, which is needed for the proper regulation of blood glucose levels is also affected by the cold. Whether or not this is due to the fact that the pancreas is affected by the cold, causing a secondary effect on insulin production, has not been determined. Those hormones involved in preparing the organism for stress such as epinephrine and norepinephrine are also elevated.

One of the most important factors associated with hypothermia is dehydration. Diminishing water intake associated with an increase in fluid needed for adequate metabolic function and production to combat cold may result in metabolic acidosis and mental deterioration. The dehydration increases as individual competence fails, and mental alertness falters, resulting in a decrease in total body water intake and storage.

The dehydrated state increases as cold diuresis increases. Associated with dehydration, is the increasing state of electrolyte imbalance, that is preceded by the hypovolemia, or decreased volume of circulating fluids, shunted into the body interior by peripheral vasoconstriction or A-V shunting. The hypovolemia is brought on partially by the fluid shifts from vasoconstriction of peripheral blood vessels, especially in the extremities. The increased

urinary, respiratory, and dermal loss associated with dehydration and hypovolemia also contributes to decreased cardiac output.

Acidosis is common in the hypothermic state and may increase at lower hypothermic levels indicated by low pH values. A change in the electrolytes of the body is associated with changes in the pH of the blood. Of these electrolytes, the most important is potassium. If the level of potassium is too high, then it can trigger cardiac erraticism and heart death. In hypothermic individuals, the serum potassium level is often elevated. The level of potassium occasionally may be in the lethal range (>5–6 meq/liter) when it is associated with extremity freezing.

It is interesting to note that as more water is shifted from the periphery or lost by respiration, or urination, less water is taken in, due to increasing mental deterioration and failure of neuroregulating temperature and fluid control systems. Once the peripheral vasoconstriction occurs in the extremities and shifts fluid to the core, heat loss is temporarily prevented; however, the hypothalmic "computer" may interpret this shift as regional overhydration and may decrease the secretion of the antidiuretic hormone, thus promoting diuresis. Cold-induced failure of renal absorption of fluid causes the cold diuresis discussed above, resulting in volume depletion. Even without a major diuresis, the hypovolemia from peripheral shunting will decrease circulating fluid, and heat loss will continue.

Shivering at onset and until cessation may result in increased lactic acid and metabolite production and will contribute to further pH alterations, change in electrolyte concentration, and severe fluid balance disturbance. Further body cooling, with contained peripheral vessel vasoconstriction results in decreased oxygen supply to the periphery and organs and contributes to metabolic acidosis.

The problems presented by the individual in the hypothermic state are (1) a lowered body (core) temperature; (2) decreasing function of the metabolic system; (3) dehydration and hypovolemia; (4) loss of caloric reserve in long-standing exposure; (5) enzyme system dysfunction; (6) hypoxia of tissues generally and gradual utilization of anaerobic metabolism; (7) metabolic acidosis; (8) renal dysfunction; (9) increasing loss of neural regulation; (10) gradual slowing of activity and function of the cardiorespiratory and cerebral centers; and (11) fluid shifts and electrolyte imbalance and acid-base imbalance, usually in the form of increasing acidosis.

The hypothermic victim is in a state of lowered metabolism with the victim in a metabolic icebox, remaining there as long as the core temperature remains at a reversible level (21°C [70°F] ??), with little downward deviation permitted, at least at this present state of the art. Below 70°F (21°C) resuscitation of the hypothermic individual is most difficult, and cerebral electrical activity, normal cardiac rhythm, and renal function may be irreversibly lost.

A. Various Kinds of Hypothermia

Air hypothermia or exposure is the most common type of hypothermia. Victims include the homeless, those who suffer from the cold Russian or European winter when there is limited food, and, surprisingly, natives in relatively warm environments who suffer from malnutrition. This is because the core temperature is warmer than the temperature of the environment, so that there is a net heat transfer away from the subject, even in tropical regions.

Immersion hypothermia has a rapid onset. This is because water conducts heat from the body 40 times faster than does air. An air temperature of 50°F is not the same as a water temperature of 50°F. In air, the person can survive indefinitely with various layers of clothing. In the water, unless the person is able to get out of it, hypothermia will occur in approximately 60 min. One of the safeguards against hypothermia shivering can work against someone in cold water. Any kind of movement in cold water causes the person to lose heat. Although shivering is effective in increasing body heat, it does this by causing the body to move; therefore, a shivering subject in cold water will cool down faster than one who does not shiver. Another safeguard in cold air might also work against someone in cold water. Cold on the skin will increase metabolism but will also cause an inspiratory arrest. If a person falls into cold water and has an inspiratory arrest, he or she could aspirate cold water and drown.

Submersion hypothermia refers to situations in which the person drowns in cold water. This drowning in cold water allows the person 45–60 min to be revived. In warm water, the time is no more than 4–6 min before death occurs. The explanation for this is that the person who drowns in cold water usually aspirates cold water and internally cools their core, because the heart continues to beat for 5–8 min after the person is completely submerged. This internal cooling drops the temperature of the

heart and the brain. In addition, the cold water cools the periphery and allows for a rapid external cooling. This proposed sequence is effective only in small, thin children, who can withstand the decrease in oxygen to their brain better than adults.

Mountain hypothermia refers to a situation in which the person is losing heat due to the cold environment, but, in addition, a decreased amount of oxygen in the mountain air does not allow the body to generate sufficient heat. The dry cold air also induces a dehydrated state very quickly. Thus, the person suffers from low oxygen levels and dehydration as well as cold, simultaneously.

Divers hypothermia refers to a situation in which divers become hypothermic due to external cooling of their skin as well as breathing cool dry air from their tanks. In some cases, divers maintain that they become cold from the inside out, thus bypassing many of nature's safeguards and ensuring that an insidious development of hypothermia occurs. In another group of divers, who use a different combination of gases such as helium-oxygen, helium, which has a very high rate of heat conduction, will rapidly make the diver hypothermic.

Exercise-induced hypothermia refers to a situation in which a person is exercising and is copiously sweating and becoming dehydrated in a cold environment. This situation is usually seen in marathon athletes or others who have extended physical performances in cold air. They usually tend to have little adipose tissue, which therefore puts them more at risk of hypothermia.

In all situations of primary hypothermia, the compounding effects of fatigue and lack of sleep promote an early onset of hypothermia. The lack of sleep predisposes the victim to hypothermia because the sleep cycle is directly tied to the temperature-regulating cycle of the body. When a person does not sleep, they alter the temperature-regulating cycle of the body so that core temperature might already have significantly decreased, which would predispose him or her to hypothermia.

Elderly hypothermia refers to the reported situation in which the elderly who are not on medication become hypothermic. Some data indicate that normal elderly victims do not vasoconstrict nor shiver as early as do younger individuals when exposed to the cold. This is coupled with a decreased body mass, which can generate less heat, and possibly a decreased sensitivity to cold. Whether this is normal concomitant of aging or is unique to special subgroups of the elderly has not been extensively studied. The concern in this case deals with the fact that this type of hypothermia is insidious with the elderly becoming cold over time.

Pediatric hypothermia refers to situations in which children become hypothermic in the air. Sometimes this is considered a separate category because newborn infants do not shiver when they are exposed to cold. In lieu of shivering, they generate some of their heat through metabolizing brown fat, which is a specialized kind of fat. This fat is strategically located in the thorax region of children and generates heat when activated by certain neurochemicals. Children are susceptible to hypothermia because they have a large surface area to volume ratio, which thermodynamically allows them to cool faster than adults. In addition, they are unable to control their movement out of the cold as well as adults. Because adults sense and respond to cold stress differently than do children, it is not uncommon for children to become hypothermic while their parents think that the temperature of the environment is not cold.

Secondary hypothermia involves the entire spectrum of any kind of pathology that can influence any aspect of the temperature-regulating system, thus predisposing the person to hypothermia. Specific examples include patients who suffer from spinal cord injuries. Because the brain is unable to receive input from the thermal receptors of the body in an organized fashion, a person with spinal cord injury cannot physiologically temperature-regulate below the site of the lesion; therefore, in cold situations, a person with spinal cord injury will not shiver or vasoconstrict below the site of the lesion. The major defense is to regulate the core temperature based on behavioral strategies.

Any medication that influences the function of the brain or spinal cord has the potential of influencing temperature regulation. Of these, the most common is alcohol. Alcohol causes a peripheral vasodilation, which gives the person a false sense of warmth. Although the person is warm on the periphery, heat is escaping the core. A person who is not severely intoxicated probably will respond to the cold stimulus by having an adequate response such as vasoconstricting, etc. The ethanol in low levels will, nevertheless, allow the person to make poor judgment so that they end up in hypothermic situations. Ethanol in high levels will cause the person to lose their ability to temperature-regulate, and they assume the temperature of the environment. In a recent study, most of the individuals who entered

the emergency room for the treatment of hypothermia were significantly intoxicated.

Persons who suffer from thyroid disease, in which there is a deficiency of the hormone or its receptor, are prime candidates for hypothermia. Because the various thyroid hormones regulate body metabolism and, hence, body heat, hypothyroidism will produce hypothermic states. Diabetic patients who do not adequately monitor glucose levels will be prime candidates for hypothermia, because they do not have enough energy to generate body heat.

B. Prevention

To minimize the effects of a cold environment, the best protection is proper hydration, clothing, and nutrition. A simple measure such as wearing a life jacket can increase a person's chance of surviving in cold water. Most hypothermic accidents involve not being aware of the warning signals from the body such as cold feet and hands, shivering, and discomfort and a change in attitude. In addition, most victims overdress and then begin to sweat and take off their clothes and become hypothermic. In many cases, there is an underlying drug-associated problem.

C. Rewarming

Human response to the cold does not stop when they become hypothermic. If a person is hypothermic, they can be rewarmed. A severely hypothermic person might look as if they are dead because their body is cold to the touch and rigid and has no detectable heart beat or respiration. The clinical literature has many examples of individuals who have been severely hypothermic and were mistaken for dead—but who nevertheless were successfully rewarmed!

Regardless of the definition of hypothermia, the problem is heat loss from the core, dehydration, and cardiovascular erraticism. The cold heart is reportedly extremely sensitive to mechanical disturbances. This effect probably is due to electrolyte imbalance. Although whether or not one should perform cardiopulmonary resuscitation (CPR) on a hypothermic victim is controversial, due to the likelihood of inducing ventricular fibrillation, the consensus is that it should be initiated and maintained as if the person was normothermic.

Thus, the primary effort is the restoration of heat

and fluid and electrolytes in the physiological manner and to constantly monitor cardiac function. In individuals who have an underlying cause of pathology that does not allow them to generate sufficient heat, the major cause of the body's inability to respond to cold stress is important to discern in advance of any rewarming strategy. Rewarming is the immediate reaction to helping a hypothermic victim. It is important to recognize that prior to choosing the method of heating (warming) time must be allotted to diligently search for (1) the cause of hypothermia; (2) all underlying factors that might prohibit an adequate result; and (3) the total physical, chemical, and physiological states of the patient's condition.

In addition, there is an interesting thermal artifact called afterdrop, in which the core temperature of the hypothermic victim who is being rewarmed continues to fall. This continued fall in temperature is perhaps caused by a physical effect of the difference in temperature among various parts of the body and is not due to cold blood from the legs returning to the heart. Whether or not the afterdrop plays a major role in the pathogenesis of hypothermia is still controversial. Afterdrop has been used to explain the sudden death of hypothermic victims who are beginning to be rewarmed and then suddenly die. Although this could occur, potassium levels in the blood more likely increase, which could induce death by way of cardiac standstill.

Many tragedies occur during rewarming because rescuers think that the patient must be rewarmed without considering the underlying cold-induced pathologies. There is reason to believe that the method of warming and even the depth of hypothermia to the level of 21.1–23.8°C (70–75°F) is no more important than having total physiological control of the patient. Many current rewarming methods that are considered controversial might produce better results if, prior to warming, effort was made to have complete control of the blood gases, pH, electrolytes, and associated injury and to begin correction of all deficits, particularly dehydration and hypovolemia.

The rewarming strategies all involve using the major physiological systems that promote hypothermia. For mild hypothermia, there is passive external rewarming. Basically, the patient is rewarmed on the basis of their physiology. The patient is covered with a blanket, and the heat loss due to convection, conduction, and radiation is minimized. As the patient shivers, their core temperature increases. In mild hypothermia, having the person walk, which also increases their metabolism, is physiologically sound and very effective. Having the person drink fluids to correct the hydration is essential; the use of alcohol drinks is strictly forbidden. (The often used example of the St. Bernard dog bringing brandy to snowbound victims did not consider that the solution was primarily a glucose solution with a small amount of alcohol!)

Active external rewarming is a situation in which external heat is applied. One such example is hot baths. Such a method will cause a peripheral vasodilation and possibly hypotension and afterdrop. For this reason, active external rewarming is not considered safe unless all physiological parameters are being monitored. Another example of active external rewarming is the use of body-to-body contact. This method does not allow for warming of the core and actually may cause both persons to become hypothermic—it is not recommended. Body-to-body contact is effective only if several (3!) normothermic people are warming the hypothermic victim at the same time.

The many methods of actively rewarming the core are called active internal rewarming. These include warming the body by way of (1) the respiratory system, in which hot humidified air is given to the patient; (2) the cardiovascular system, in which heated intravenous blood is given to the patient or the entire blood of the person is drained, warmed, and then put back into the patient; and (3) the internal cavities of the body such as putting warm fluid into the cavities of the body such as the gastrointestinal system (periotoneal lavage), urinary bladder, or rectum. Other techniques such as diathermy have also been used to rewarm the core.

Although drugs might be used to stabilize the heart or theoretically assist in warming the patient, most drugs used depend on the temperature of the body for their effectiveness. If the body is too cold, the drug will have no effect. Thus, using drugs on a hypothermic victim is extremely risky because the effect is not predictable and could be lethal.

Whether the patient should be warmed fast or slow depends on the environment and the depth of hypothermia. In a clinical situation, the patient can be warmed safely by any of a number of methods because they are physiologically monitored. In the field, no active rewarming should take place until the patient can be monitored safely.

Primitive man was able to handle the cold environment because of various physiological and technological strategies. These technological advances have not stopped. Although humans have used the basic ideas of clothing and habitation to stay in a zone of thermal comfort, they have used new technologies to allow them to pursue investigations in various environments. The advent of a tight-fitting rubber suit that allows in a small amount of water that is subsequently warmed by body metabolism has allowed divers to investigate cold bodies of water. In another example, by designing a waterproof flotable suit that does not allow entry of water, divers are able to survive being thrown into cold water (0°C) for prolonged periods of time (hours) versus perishing in a short period of time. The development of new kinds of synthetic materials that trap air allow people to have thermally comfortable clothes without the bulk of animal derived materials such as wool or down. Whether or not they are thermally as effective as natural fibers is still hotly debated. The design of these garments should be one in which the core is protected. Thus, a design in which the thorax and head is protected follows the strategy of nature. The fingers and toes should not be overly protected because they must act as a sensitive thermal antennae for the body.

When asleep, a person's temperature normal drops and then cycles. During this period, people complain about cold toes and feet. This effect is reportedly more common among women than men. Using electric blankets assists humans in minimizing the thermal discomfort sometimes associated with the physiological drop in core temperature. Using synthetic material in sleeping bags also allows humans to sleep in outdoor cold environment. Again, whether or not these bags are more effective than the old traditional types is not a hotly debated issue.

To date, there has been no major change in human's inherent physiology to withstand cold stress. A person who spends a considerable amount of time in a cold environment commonly has different responses to the cold than one who does not. The mechanisms behind this physiological adaptation are not clear. Although this area is actively studied there is still no consensus as to whether or not humans can physiologically adapt to the cold. It is difficult to do studies on humans who have been exposed to long periods (months) of cold to investigate whether or not they have a genetic ability to react to the cold by dropping their core temperature, etc. With the ever-advancing inroads of civilization, the chances of ever being able to answer this question are quickly becoming slim. Therefore, studies in the field are urgently needed to answer some of these questions.

At present, it is thought that by understanding more about the hormonal control of thermal regulation it would be possible to have humans rapidly adjust to a cold climate. Most strategies in this area are involved with increasing human core temperature by way of hormonal alterations to minimize the effects of cold stress, increasing subtle muscle contractions, or biofeedback. For example, a person can at will increase blood flow to their fingers and can also stop shivering for short periods of time. Understanding the mechanism behind these observations might allow humans to function longer in the cold. Another proposed strategy is to have a person become a hibernating animal at night by somehow dropping their core temperature—by chemical methods—so that they would not lose as many calories when they sleep. Conceivably, there might be subgroups of people who are more effective in storing calories and, hence, heat, which would allow them to survive prolonged cold periods. Identification of their genetic profile might be used eventually in addressing this problem.

The use of drugs to assist humans in fighting the effects of cold is as old as humankind. The use of alcohol gives a false sense of warmth by promoting a vasodilation. Niacin will cause an overall body flushing, which gives a sense of warmth. However, whatever is used to keep a person warm must be something that can be used in any climate—hot or cold. Simplistic answers that do not consider the long-term development of the human thermoregulatory system and their modern lifestyle will not be that effective. For example, a pilot flying over the Arctic must be concerned about ditching in cold water; therefore, a hypothermic suit is worn. The problem is that the suit is too hot in the cockpit and, consequently, the pilot will not wear the suit as he or she flies. Thermal comfort remains one of our basic drives.

Cold will always present humans with a challenge. With the occurrence of natural catastrophes such as earthquakes, hurricanes, or volcanic-induced global cooling, humankind is again reduced to the state of its primitive ancestors. An unseasonably cold winter blast becomes newsworthy as it presents danger to the young, elderly, and homeless and curtails economic development.

In the event a nuclear winter ever overtakes the globe, humankind will have to regress once again toward warding off the effects of cold to maintain its species and culture. In that event, the challenges that faced armies in the past will pale by comparison to what the armed forces of the future would endure to survive, much less engage in battle.

Cold has been used by humans to their advantage. It has been used as an anesthetic and, in that regard, might have been used earlier than alcohol. It has been effective in cooling patients who must undergo cardiac surgery allowing for manipulation of the heart, which would not be possible otherwise. The transport of various internal organs is always done with the organ in a cold state (e.g., kidneys can survive approximately 72 hr at a temperature of 4–10°C. In addition, it might be effective in assisting persons who suffer from cerebral strokes by minimizing the damage caused by strokes. Theoretically, by cooling the brain after such an injury, some of the damage caused by a decrease in oxygen might be minimized. Finally, it might be somehow involved in promoting a decrease in metabolism of humans as they continue to explore cold environments such as the ocean, planets, and space. [*See* Stroke.]

VII. Peripheral Effects of Low Temperature (Freezing)

Frostbite differs from all other cold conditions described in that it results from exposure to temperatures well below freezing and results in true tissue freezing. In many cases, individuals who die from hypothermia are thought to have died from being "frozen to death"—this rarely occurs. What does occur is that many individuals whose core temperature is falling also suffer from frostbite. The two terms hypothermia and frostbite are not synonymous but may occur in the same individuals, they are both manifestations of the same process. As the body cools, it will shunt blood from the periphery to the core. This shunting of blood allows the limbs, especially the digits, to become susceptible to freezing injury.

Frostbite occurs when heat loss is sufficient enough in the exposed area to allow ice crystals to form in the extracellular spaces. Exposure time resulting in tissue freezing can be measured in hours, minutes, or seconds. The two variables that are of importance are (1) duration of exposure and (2) temperature.

Once concept of tissue freezing is that regardless

of the medium at which the freezing takes place, it is a physical phenomenon in which all available water is removed from solution and relocated as inert foreign bodies—ice crystals. At very high rates of freezing, ice crystals form throughout the tissues, usually intracellular. At slow rates of freezing, ice crystals form in the extracellular spaces. Terrestrial freezing rates are most often well within the slow-freezing range; therefore, extracellular ice crystals are usually the major concern for studying cold stress.

During slow freezing as the extracellular ice crystals grow, extracellular solutions are condensed, causing a hypertonic solution that causes increased osmotic pressure, extracting more intracellular water, that in turn freezes. This sequence progresses until all water, not otherwise bound to protein, sugar, and other molecules, has been appropriated. The cell is then left as a sheet of highly concentrated solute, centered between ice crystals. Some of the crystals are large, often larger than the original cell. After thawing, water may be reabsorbed and the cell apparently becomes normal again. The result, however, is determined by the duration of freezing, the depth of freezing, and the use of varied methods of thawing.

Despite the mechanical pressure on cells of the mechanical formation of ice, much more cellular damage probably results from cellular dehydration, secondary to electrolyte concentration, affecting nerve, muscle, and erythrocytes. All biochemical, anatomical, and physiological sequellae of freezing are directly and indirectly the consequence of this physical event of cellular dehydration.

Freezing injuries may be categorized as to how it came about and then whether or not there had been a freezing and a thawing sequence. In addition, due to the osmotic forces involved in the process of freezing, there is edema formation, which causes an increase in pressure. This physiological process is extremely important if the pressure is increasing in the wrist or other areas where there is fascia that allows the pressure to increase dramatically, causing compression of the nerves and blood vessels. This phenomenon is called the compartment compression syndrome.

Thus, frostbite is categorized as (1) true frostbite, superficial or deep; (2) a mixed injury, immersion injury (wet cold injuries) followed by freezing (this injury is disastrous and often quite painful, resulting in considerable tissue loss); (3) freezing, thawing by any means, and refreezing again, generally disastrous in result, with total tissue destruction and early mummification of distal tissues occurring a week or less; (4) high-altitude environment injury with hypoxia and often dehydration of tissues, due to general dehydration and hypovolemia (decrease in total body fluids), with extremity freezing (prognosis is poor, especially if associated with compartment pressure syndrome); (5) extremity fracture dislocation, or severe extremity injury and superimposed freezing (the results are poor if the fracture dislocation is left unreduced, and the best results seem to follow rapid rewarming between 32° and 38°C in water); and (6) extremity compartment pressure from any cause, followed by freezing (very poor results are seen if compartment pressures are not relieved; this pressure is reduced by surgically opening the space—fasciotomy—where the tissue pressure is high).

Present knowledge would indicate that the pathophysiological changes of frostbite occur in two stages. The first stage is the cellular stage, wherein cellular changes occur during the course and duration of freezing. These changes are (1) structural damage by ice crystal growth; (2) protein denaturation from electrolyte concentration, which in affect is any nonproteolytic change in the chemistry, composition, or structure of a native protein, which causes it to lose some or all of its specific characteristics; (3) pH changes (acidosis) intra- and extracellularly; (4) dehydration, intracellular from extracted water; (5) loss of protein-bound water; (6) rupture of cell membranes in deep, extended freezing or refreezing; (7) damage to vital intracellular organs; (8) abnormal persistent cell wall permeability; (9) destruction of essential enzymes; (10) ultrastructural damage to capillaries; (11) injury to the endothelium of the arterial venous walls; and (12) consistent mitochondrial damage to muscle cells.

The second stage of injury after freezing occurs during the thawing and in the postthaw stage (the vascular injury). These are (1) progressive tissue ischemia, arachidonic acid (fatty acid) cascade with increased production of prostaglandin and thromboxane, which permit progressive capillary, arterial, and venous thrombosis (Treatment at this segment consists of nonsteroidal drugs and other drugs inhibiting thromboxane formation); (2) early circulatory stasis; (3) corpuscular aggregation of red cells; (4) venuole and arterial obstruction blocking the lumen; (5) red blood cells "piling up" back to the capillary bed; (6) hyaline plugs form in the vascular tree; (7) marked tissue edema; (8) anoxia;

(9) tissue compartment space pressure; (10) capillary and peripheral vessel collapse; and (11) thrombosis, ischemia, necrosis, and gangrene.

The effect of low temperatures, on exposed hands and feet can be detected often by roentgenographic examination, but only six months or more after cold injury. In the early stages, unless severe infection is present or amputation has occurred, the x-ray plane film may be read as negative, other than the presence of disuse osteoporosis. However, six months post injury, and beyond, apparently related more to depth of injury, than method of thawing, and interesting changes appear on the x-ray of the part. There are often shown fine, irregular lytic, or cystic areas in the region of the articular surfaces of the digital joints. These changes are also seen in the adjacent peri-articular osseous regions of the metacarpal or metatarsal structures, or their corresponding phalanges, at small joint areas.

These small lytic areas extend into the joint cavity and are lined with fibrous tissue. They cause joint swelling, pain and limited joint motion. These lytic areas may enlarge, remain the same, fill completely, or enhance continued joint narrowing and destruction.

This phenomenon described above is shown very often on the x-rays of small children. There is evidence of destruction, even dissolution of the phalangeal epiphyses, and often, in severe extended injury, absence or marked destruction of the cartilagenous carpi in the very young.

These changes of cold induced epiphyseal necrosis correlate with the clinical appearance of shortened phalanges and the appearance of "stubbiness" of hand and digits, with often increase in finger width. Epiphyseal plates disappear, with joint space narrowing, with often a varus or valgus deformity of the distal phalanges.

It is thought that the debilitating change in children's extremities is due to the direct effect of cold on the cartilage cells (chondrocytes) or on the physis (growth plate) and on the cartilagenous articular surface. Further this injury may be complicated by destruction of the vascular supply to the adjacent osseous structures and ischemia of the epiphysis, resulting in osteocyte death and bone necrosis.

Early demonstration of the effect of cold on extremities, may be gained by utilizing radio-isotope examination of the involved part. This method permits evaluation of capillary perfusion of the area of injury, hand or foot. A useful and favored isotope is Technetium 99 M per technetate.

In the adult, this same change causes avascular necrosis of any osseous structures of hand and foot, particularly of phalanges, metatarsal, and tarsi, again due to the direct effect of cold on bone and cartilage, associated with deprivation of capillary circulation.

Many thawing methods are utilized by the victims or helpful rescuers, but not all methods are equal. In fact, some exacerbate the syndrome. Research is continuing as to the most effective method of rewarming. Some of the older methods have been shown to be deleterious and cause tissue damage.

Rapid rewarming in warm water (32.2–41.1°C [90–106°F]) is currently recommended as the treatment of choice. Rapid rewarming by external means appears to provide better results than any other method, but without question does not always give protection from tissue loss, especially in deep or long-duration injury.

Spontaneous thawing permits variable results, and those results are often determined by the depth of injury, the duration of freezing, and the patient's activity during survival and rescue and thawing. This approach is used when no facilities are available for clinical treatment. Examples are the use of room temperature (45–90°F), thawing during foot travel or rescue transportation, or even the use of a sleeping bag. It is important to emphasize that the limb or digit in question should not be thawed if freezing may occur again.

The following techniques have been used but are *not* recommended because they cause tissue damage. Rubbing ice and snow on the frozen extremity increases cooling, promoting more dehydration and ice crystal formation. There is no sound physiological explanation why this method should be used. Thawing by excessive heat such as campfire heat, oven heat, engine exhaust (temperatures >48.8°C [>120°F]) generally results in heat injury (burning) to a part already injured by cold because the person is insensitive to any kind of stimuli. The results are disastrous, often resulting in major amputation.

At present, rapid rewarming is favored, this method seems to demonstrate the greatest tissue preservation and the most adequate early function, especially in deep injury. Results by gradual thawing vary in deep injury but seem satisfactory in the superficial injury patients. Ice and snow thawing gives variable results, most often poor, with marked loss of tissue.

It has been stated that rapid rewarming by internal means (warm intravenous fluids or arterial line fluids) at temperatures of 37.1–41.14°C (100–106°F) is more physiological and may be a method of choice in dissolving ice crystals and restoring cellular hydration. Although this method appears most logical and has been a method of choice for over 10 years at The University of Alaska High Latitude Center in the treatment of combined hypothermia and freezing injury, the results are still no better than by rapid rewarming methods. In addition, the placement of an arterial line, especially in the area of ankle and wrist, may cause local arterial spasm and further decrease digital profusion. The ideal method is obviously not yet at hand (at least for thawing of the frozen part) but the tissue loss is less now than in past decades, regardless of thawing method. As a consequence of physiologically derived rewarming techniques, major above-knee or below-knee amputation or amputation at wrist or forearm level are much fewer in number than they were when rubbing snow on a frostbitten extremity was the treatment of choice.

VIII. Core Effects of Low Temperature (Freezing)

Frostbite represents a selected frozen injury to the digits or the limbs. On the other hand, limited experiments have taken place with whole body freezing. In this case, the person is dead and then frozen. The person is then kept frozen with the hope that eventually some new cure might be found for correcting the disease that was responsible for their death. At present, there is no known study in which frozen primates have survived for a long period of time. Further research into frostbite show various methods that might be used for whole body rethawing. Studies using amphibians, who become frozen during the winter and then rethaw, might also give clues as to how to rethaw the frozen warm-blooded animal.

Bibliography

Burton, A. C., and Edholm, O. G. (1955). "Man in a Cold Environment." Monograph of the Physiological Society. Edward Arnold, Publishers Ltd., London.

Danzl, D., Pozos, R. S., and Hamlet, M. (1989). Accidental hypothermia. In "Management of Wilderness and Environment Emergencies" (P. S. Auerback and E. C. Geehr, eds.). C. V. Mosby Co., St. Louis.

Gallin, J., Goldstein, I., and Syderman, R. (1988). "Inflammation, Basic Principles and Clinical Correlates." Raven Press, New York.

Laursen, G., Pozos, R., and Hempel, F. (eds.) (1982). "Human Performance in the Cold." Undersea Medical Society, Bethesda, Maryland.

LeBlanc, J. (1975). "Man in the Cold." American Lecture Series, Chas. C Thomas, Springfield, Illinois.

Lloyd, E. L. (1986). "Hypothermia and Cold Stress." Croom Helm, London.

MacLean, D., and Emslie-Smith, D. (1977). "Accidental Hypothermia." Blackwell Scientific Publications, Oxford, London.

McCauley, R., Hing, D., Robson, M., and Heggers, T. (1983). Frostbite, a rational approach based on the pathophysiology. *J. Trauma* **23**, 2.

Meryman, H. T. (ed.) (1966). "Cryobiology." Academic Press, London and New York.

Mills, W. (1979). *In* "Back to Basics" (I. J. Cohen, ed.) EM Books, New York.

Mills, W. Accidental hypothermia: management approach. (1980). *Alaska Medicine* **22,** pp. 9–11.

Paton, B. (1983). Accidental hypothermia. *Phramacol. Ther.* (Great Britain) **22**, 331–377.

Pozos, R. S., and Wittmers, L. E. (eds.) (1983). "The Nature and Treatment of Hypothermia." University of Minnesota Press, Minneapolis.

Robertshaw, D. (ed.) (1974). "Environment Physiology." University Park Press, Baltimore.

Sutton, J. R., Houston, C., and Coates, G. (eds.) (1987). "Hypoxia and Cold." Praeger, New York.

Vanggard, L. Surgical Captain, Danish Navy Office, Chief of Defense. Personal Communication, Feb. 1985.

Whayne, T., and DeBakey, M. (1958). "Cold Injury, Ground Type." Medical Dept., U.S. Army, Clinical Series.

Lymphocyte-Mediated Cytotoxicity

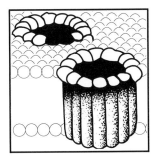

DAVID M. OJCIUS, ZANVIL A. COHN, AND
JOHN DING-E. YOUNG, *Rockefeller University*

Glossary

Antibody Serum proteins secreted by B lymphocytes, which are able to bind a given antigen specifically

Antigen Any substance capable of eliciting an immune response

CD Cluster of differentiation. Cell surface proteins that regulate lymphocyte interactions. All T lymphocytes express CD3, most helper T lymphocytes express CD4, and most cytotoxic T lymphocytes express CD8

Complement Ensemble of enzymatic and pore-forming proteins that is able to lyse bacteria and other foreign invaders. The foreign cell must first be recognized by antibody

Exocytosis Fusion of granules with the cell plasma membrane, which causes the granular contents to be released to the extracellular medium

Interleukin-2 Growth factor that stimulates lymphocyte proliferation and activation

MHC Major histocompatibility complex. A set of genes that encodes for cell surface proteins involved in antigen presentation. Only antigens presented in association with MHC proteins are recognized by T lymphocytes

CYTOTOXIC LYMPHOCYTES kill cancerous and virally infected cells through at least two mechanisms. The mechanism that is currently best understood involves secretion from the killer cells of a pore-forming protein called *perforin*. In its mono-

meric form, perforin binds to the surface of the target cell, where it inserts into the plasma membrane and polymerizes into pores as large as 20 nm in diameter. Salts flow through the pores into the target cell, and water follows because of the difference in osmotic pressure that ensues; the target cell swells and eventually bursts. Likewise, killer cells have been observed to activate endonucleases (that cut DNA molecules) inside of the target cell, with resulting fragmentation of the target cell DNA into multiples of about 200 base pairs. The molecular components of the latter mechanism of cytotoxicity are still in the process of being characterized.

I. Introduction to Cytolytic Lymphocytes

A. Lymphocyte Subsets

The immune response is a highly specific system that has evolved to defend the body against pathogenic microorganisms and tumor cells. The core of this system is comprised of white blood cells called *lymphocytes*, which fall into two main lineages. *B lymphocytes*, which originate in the bone marrow, are the precursors of antibody-secreting cells and are thus responsible for the humoral branch of immunity, with immune responses mediated by antibodies. *T, or thymus-dependent, lymphocytes* carry out the cellular immune responses, with responses, as the name suggests, mediated by cells. This broad category of lymphocytes can in turn be divided into a series of subsets. *Helper T lymphocytes (T_H)* assist B lymphocytes as well as other T lymphocytes in the development of their response to an antigen. They do so by producing growth factors known as *lymphokines* that trigger, for instance, the maturation of B lymphocytes into antibody-producing cells. Conversely, *suppressor T lymphocytes* dampen the response of other lymphocytes once the antigen has been eliminated. *Cytotoxic T lymphocytes* (CTL) represent the last major subset of T lymphocytes and can be thought of as the cellular immune system's equivalent of antibodies. These cells recognize other cells that display foreign antigens on their surface and kill them. Usually, the targets are the body's own cells that have become cancerous or virally infected. The CTL are also directly involved in the rejection of grafts. A closely related class of cytotoxic cells are the *natural killer* (NK) cells, whose lineage is still controversial and which use a quite different recognition system from those of the CTL. [*See* Lymphocytes; Natural Killer and Other Effector Cells.]

B. Stages in the Lytic Process

1. Target Cell Recognition and Binding

When a cell becomes infected by a foreign pathogen, the protein antigens making up the pathogen are processed within the cell and fragments are displayed on the cell surface for presentation to T lymphocytes. These fragments are recognized by antibody-like molecules on the surface of T lymphocytes called the *T-cell receptors* (TCR). The most prevalent TCR is composed of two chains, named α and β, which, like antibodies, have a variable portion on their amino-terminal domain and a constant portion on their carboxyl domain. Following these domains on the carboxyl side are a transmembrane region that anchors the TCR to the T-lymphocyte surface and then a short cytoplasmic tail probably involved in cell activation. The combining site for antigen depends on both variable regions, which form a three-dimensional template complementary in shape and chemical properties to the antigen that the TCR specifically recognizes. The high specificity for antigen is a fundamental characteristic of both antibodies and TCR, which can distinguish between antigens that differ by only one amino acid residue. Unlike antibodies, though, the TCR responds only to antigens on a target cell surface. [*See* T-Cell Receptors.]

Perhaps to avoid being overwhelmed by freely circulating antigens, TCR recognizes foreign antigens only when they are presented as part of a complex with extremely polymorphic (i.e., highly variable in different individuals) proteins coded for by the *major histocompatibility class* (MHC) *genes*. In humans, the MHCs are called *human leukocyte antigen complex* (HLA)-A, HLA-B, and HLA-C, and

they were originally discovered because of their ability to provoke allograft rejection. CTL requires, more precisely, that the antigen be presented with class I MHC proteins, which are found on the surface of all cell types in the body. This is in contrast with the requirements of T_H cells, which, in keeping with their more restrictive, regulatory role, recognize antigens complexed with the class II MHC proteins that are expressed predominantly by a small number of *antigen presenting cells*, such as dendritic cells, macrophages, and B lymphocytes. T_H is a potent producer of interleukin-2 (IL-2), and this growth factor, along with interaction of CTL precursors via their TCR with foreign antigen, causes the precursors to develop into highly effective CTL. [*See* Interleukin-2 and the IL-2 Receptor.]

Preceding the specific antigen recognition step between TCR and foreign antigen, CTL binds to the target cell (TC) nonspecifically through a type of accessory molecule on CTL known as *CD2*, which is a surface marker involved in T-cell activation, and *lymphocyte functional antigen-1* (LFA-1), which on the target cell binds LFA-3 and intercellular adhesion molecule-1 (ICAM-1), respectively. CD2 is present on CTL and most NK cells, whereas LFA-3 and ICAM-1 have a broad tissue distribution. The strength of these cell-to-cell interactions is enhanced by the binding of another surface marker, CD8, on CTL to monomorphic determinants expressed on class I MHC products.

Natural killer cells are mainly involved in the control of tumor progression and metastasis spreading and seem preferentially to lyse tumor cells expressing low levels of MHC. Consistent with this view, loss of MHC has indeed been demonstrated in many tumor cell lines. Besides not requiring expression of MHC molecules on the TC to mediate cytolytic reactions, NK cells do not use TCR to identify their targets. Potential NK recognition receptors are the subjects of current investigation.

2. Granule Exocytosis and Delivery of the Lethal Hit

Electron micrographs of CTL that have been cultured *in vitro* show the presence of granules in the cytoplasm that are darkly stained when the cells are treated with osmium. Simple binding of CTL to TC via the accessory molecules appears not to have any observable morphological effects on the CTL. However, the subsequent antigen-dependent interactions result in a dramatic rearrangement of the CTL cytoskeleton and granules. The granules, the microtubule-organizing center, and other cytoskeletal elements become preferentially oriented toward the TC contact zone. In addition, there is often pronounced interdigitation of the membranes of the CTL and TC, possibly to maximize the contact area between the two cells. All these morphological changes are accompanied by an abrupt increase in the cytoplasmic calcium concentration of CTL, which is the driving force for exocytosis of the granular contents. On delivery of the "lethal hit," the undamaged CTL becomes physically detached from the TC and is able to embark on a new round of TC binding and degranulation. Similar stages leading up to exocytosis have been identified for NK cells.

3. Target Cell Lysis

Research in the 1960s demonstrated that the TC swells and then ruptures after contact with a CTL (Fig. 1). Subsequent investigations revealed that concomitant with the swelling, there is an outflow of inorganic ions from the TC, followed by the release of larger molecules, including proteins and DNA fragments. These and other studies led to the proposal that the CTL secretes *channel-forming*

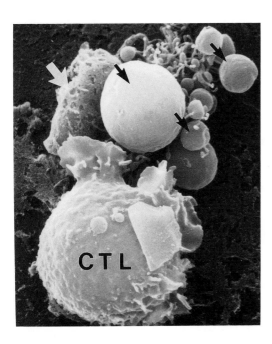

FIGURE 1 Aftermath of an interaction between a cancerous target cell and a cytotoxic lymphocyte. The target cell has burst, leaving behind the target cell membrane (large white arrow) and the nucleus and cellular debris (small arrows). The cytotoxic cell, on the other hand, is unharmed. (Scanning electron microscopy was performed by Dr. Gilla Kaplan, The Rockefeller University.)

proteins into the extracellular space between the CTL and the TC, which produces a limited gateway for small ions through the TC plasma membrane. Thus, small ions are driven down their electrochemical gradient into the TC, while larger molecules like proteins are held back by the membrane. Because of the osmotic pressure difference that develops, water flows passively into the cell; the cell swells and eventually bursts. Further support was given to the "pore-formation model" by electron microscopic observations that *tubular lesions* become deposited on the TC membrane after contact with CTL. These doughnut-shaped protein complexes were reminiscent of antibody-induced complement lesions (see Section B,2), which were previously known to lyse targets through a similar effect.

There is one important morphological difference, however, between the effects of complement and CTL on TC. Delivery of the lethal hit initiates a process of nuclear and plasma membrane blebbing in the TC that precedes release of cytoplasmic protein, and this pattern is not seen for complement. Moreover, CTL causes a breakdown of TC DNA into fragments that are multiples of about 200 base pairs, suggesting that an internal TC endonuclease is activated. These observations have been confirmed for NK cells, but not complement, and have led to an alternative view of lymphocyte-mediated lysis. According to the "internal disintegration" or "*induced suicide model*," lymphocytes trigger an autocatalytic cascade within TC that results in DNA fragmentation and culminates indirectly in membrane damage.

Several lines of evidence therefore indicate that the pore-formation model cannot readily account for all the observed effects of cytotoxic lymphocytes. Results from attempts to compare the kinetics of protein release and DNA release and/or fragmentation have depended on the cell preparations and exact experimental conditions used, so that, on the whole cell level, this issue is far from being resolved.

II. Characterization of Granule Components

A. Granule Lytic Properties

Most of the events after CTL exocytosis are generally thought to be independent of the presence of the effector cell and to be largely due to the secreted

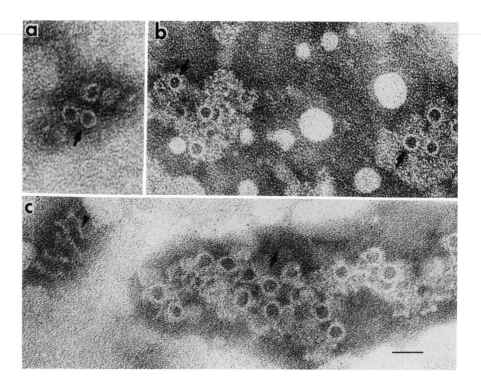

granular components. With the advent of new cell cloning techniques, it has become possible to establish *in vitro* cell lines of CTL and NK cells, from which large quantities of granules can be obtained through simple subcellular fractionation procedures. CTL and NK granules thus derived have potent cytolytic activity, which is strictly dependent on the presence of calcium. As the granules display neither MHC restriction nor antigen specificity, they are effective at lysing a wide variety of targets, including nucleated tumor cells, enucleated cells, erythrocytes, and even artificial single membrane vesicles. The lesions produced have internal diameters ranging from 15–17 nm and resemble those produced by intact lymphocytes.

B. Perforin

1. Pore-Forming Activity

Perforin has been purified from CTL and NK granules by biochemical methods or by binding it to specific antibodies attached to agarose. The purified monomeric perforin has a molecular weight of 66,000 to 75,000 daltons. In the presence of calcium, the perforin monomer inserts into membranes, where it polymerizes to form doughnut-shaped structures similar to those previously described for intact lymphocytes and their granules

FIGURE 2 Cylindrical lesions produced on lipid membranes by purified perforin. The arrows point to cross sections of the lesions, while the arrowheads point to longitudinal views. [From J. D.-E Young *et al*, (1986). *Proc. Nat. Acad. of Sci.* **83**, 5668–5672.]

(Fig. 2). These complexes have molecular weights exceeding 1 million daltons.

Purified perforin has potent hemolytic activity; 1 ng of perforin can completely lyse 10 million red blood cells. Perforin can also lyse lipid bilayers and nucleated tumor cells. However, in agreement with previous observations that nucleated cells can repair membrane damage caused by other pore-forming substances, more perforin is required to lyse nucleated cells than to lyse erythrocytes.

Biophysical measurements indicate that the channels formed by purified perforin are voltage-insensitive, have multiple sizes, ranging up to about 20 nm in diameter, and are not selective for any ion. When the monomers are first added to lipid bilayers, the membrane current increases in discrete steps, implying that discrete ion channels are being incorporated into the membrane. The model that has emerged from these studies has the perforin monomer binding to the membrane, inserting into the membrane, and then aggregating like barrel staves surrounding a central pore that grows in diameter through the progressive addition of monomers. The monomers traverse the membrane and are assumed to polymerize in such a way that the hydrophobic amino acid residues on the outside of the barrel face the oil-like acyl chains of the membrane, and the hydrophilic residues line the water-like interior of the channel.

Blood serum proteins like heparin and high- and low-density lipoproteins inhibit perforin-mediated lysis. These inhibitors prevent only perforin binding to the target cell membrane, because once the perforin is bound it can proceed to polymerization and pore formation even in the presence of the inhibitors. Hence, in the small space comprising the contact zone between a CTL and a TC, perforin is able to bind to the TC membrane, thus lysing the TC, whereas surrounding cells outside the contact zone are protected from the cytotoxic properties of perforin by the blood serum proteins.

2. Homologies with the Membrane Attack Complex of Complement

When a foreign cell enters the body, it is first identified by antibody, but it is then destroyed by other mediators. Prominent in the humoral branch of immunity is an intricately linked system of enzymes and pore-forming proteins called *the complement system*. The membrane attack complex of complement, which is directly responsible for lysing the foreign cells, consists of the five terminal components (C5b, C6, C7, C8, and C9) of the complement cascade, and it lyses TC by forming tubular complexes on the TC membrane. C9 is the protein component mostly responsible for the formation of the lesion, and purified C9 can in fact polymerize under certain conditions into tubular lesions resembling the membrane attack complex. [*See* COMPLEMENT SYSTEM.]

As mentioned above, the lesions produced by cytotoxic lymphocytes are morphologically and functionally similar to those produced by complement. As it turns out, the individual effector molecules are also structurally related. Monoclonal antibodies prepared against target cell membranes damaged by cytotoxic lymphocytes cross-react with antigens expressed on the lesions due to complement. More specifically, antibodies against perforin cross-react with all the components of the membrane attack complex (i.e., C5b-6, C7, C8, and C9).

Unlike the membrane attack complex, however, which is a heteropolymer formed by several complement proteins, the perforin lesion is a homopolymer consisting of only one monomeric species.

Monomeric C9 binds to a preassembled C5b-C6-C7-C8 complex on the target cell membrane, and the circular lesion is formed after 10–20 C9 monomers have polymerized. In contrast, monomeric perforin binds to membrane lipids in the absence of any protein receptor, so that the lesion is formed by polyperforin alone (Fig. 3).

Among the various components of the complement system, the one assumed to be most closely related to perforin is C9, which has a reduced molecular weight of 70,000–75,000 daltons and a nonreduced one of 62,000–66,000 daltons. Studies with polyclonal antibodies have borne out this assumption.

The similarity between C9 and perforin is confirmed by the study of their respective genes, which reveal significant homology between their amino acid sequences. There is also homology, albeit to a lesser extent, between perforin and C6, C7, and C8. Functionally, both C9 and perforin form ion nonselective, voltage-resistant channels that remain permanently open, which favors the role that these stable pores play in lysing cells.

3. Resistance of Cytolytic Lymphocytes to Self-Lysis

A salient feature of lymphocyte-mediated lysis is the ability of effector cells to lyse many TC successively without self-injury. CTL are remarkably resistant not only to other CTL, but also to CTL granules and to purified perforin itself. Moreover, the degree of resistance correlates well with the cytotoxic potential of the CTL (i.e. the more aggressive the CTL, the more resistant it is to lysis by perforin). Although it is clear that CTL and NK cells must be equipped with some mechanism to protect themselves against perforin, little is known about its molecular basis. According to one model currently under investigation, CTL protection is mediated by a surface protein that prevents binding of perforin to the CTL membrane.

4. Physiological Role of Perforin in Lymphocyte-Mediated Killing

Perforin was originally isolated from cytotoxic cell lines grown *in vitro*, and attention only recently shifted to the distribution of perforin in the body. As expected, perforin is completely absent from nonlymphocyte populations and resting T lymphocytes. Resting T lymphocytes can be induced to produce perforin by stimulation with IL-2 or antibodies against the TCR complex, and this production can

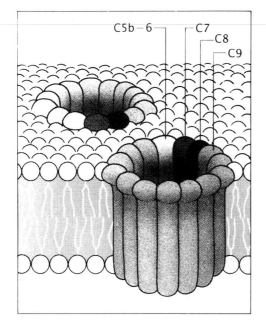
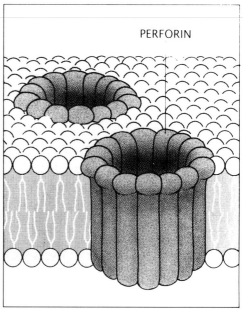

FIGURE 3 Differences on a molecular level between pores formed by complement and those formed by perforin. The complement lesions are formed by a heteropolymer consisting of the five terminal components of the complement cascade. The lesions deposited by lymphocytes are due to perforin alone. [From J. D.-E Young and Z. A. Cohn (1988). *Sci. Am.* **256**, 38–44.]

be blocked somewhat by *cyclosporin A*, a drug often used clinically to prevent graft rejection.

All NK cells apparently produce perforin. Perforin has been shown to be present in CTL under pathological conditions such as autoimmune disorders (diseases resulting from a failure of the immune system to distinguish foreign antigens from self-antigens, causing cytotoxic lymphocytes to react destructively against the body's own antigens) and viral infections, where large amounts of IL-2 might be expected to be present, but the distribution of perforin in primary CTL under less intensive stimulatory conditions is still unclear. [*See* AUTOIMMUNE DISEASE.]

C. Serine Esterases and Proteoglycans

A family of proteolytic enzymes known as serine esterases is also present in CTL and NK cytoplasmic granules, and their release has been widely used as an indicator for effector cell exocytosis. Serine esterases were initially thought to be specific to cytotoxic lymphocytes, but recently they have also been found in other subsets of T lymphocytes such as T_H cells. Similarly, cytotoxic granules are rich in compounds such as proteoglycans. For the most part, however, the possible function of these proteins in lymphocyte-mediated cytotoxicity is still shrouded in mystery. Significantly, purified serine esterases alone display no cytolytic activity, and inhibitors specific for proteolytic enzymes have

no effect on the cytolytic activity of isolated granules. According to the concensus presently emerging, serine esterases may play a more indirect role (e.g., in processing perforin during its biosynthesis and packaging into cytoplasmic granules, or in helping CTL detach from TC after delivery of the lethal hit), and proteoglycans may stabilize perforin in the low pH environment of the granules.

III. Cell Killing in the Absence of Lymphocyte Exocytosis

Reservations concerning the universality of the role of perforin in lymphocyte-mediated cytotoxicity were raised by the observation that depending on the nature of the CTL and the TC, killing can in some cases proceed in the absence of calcium. Under such conditions, killing must take place by a mechanism that is not only independent of perforin, but even of granular components altogether, for lymphocyte exocytosis is obligatorily dependent on the presence of calcium. Subcellular fractionation studies have revealed the existence of yet another class of cytotoxins in CTL, whose activity is calcium-independent. The one best characterized to date, named *leukalexin*, has been detected in granules, in the cytosol, and on the CTL surface. Its mode of action is slower than that of perforin, and it causes DNA fragmentation in a step-ladder fashion, suggestive of the internal disintegration model of killing mentioned earlier. It is still unknown how leukalexin and leukalexin-related cytotoxins exert their action. It is thought that their effect might be mediated through specific receptors on the TC surface, but no direct evidence is yet available to support this hypothesis.

Thus, although NK cells are known to lyse TC only via exocytosis, a more complex picture is emerging for CTL. Until now, perforin is the only CTL toxic agent that has been purified, fully characterized functionally, and whose amino acid and cDNA sequences are available, but it is already clear that CTL can apply several different mechanisms and/or cytotoxins to lyse TC. None of these mechanisms need be mutually exclusive, and they may all contribute simultaneously to damage TC. A better understanding of the alternative mechanisms and the circumstances under which they are employed will have to await a complete molecular characterization of all the components involved.

Bibliography

Henkart, P. A. (1985). Mechanism of lymphocyte-mediated cytotoxicity. *Annu. Rev. Immunol.* **3,** 31–58.

Ojcius, D. M., and Young, J. D.-E (1990). Cell-mediated killing: Effector mechanisms and mediators. *Cancer Cells* **2,** 138–145.

Podack, E. R. (1985). Molecular mechanism of lymphocyte-mediated tumor cell lysis. *Immunol. Today* **6,** 21–27.

Tschopp, J., and Jongeneel, C. V. (1988). Cytotoxic T lymphocyte mediated lysis. *Biochemistry* **27,** 2641–2646.

Young, J. D.-E (1989). Killing of target cells by lymphocytes: A mechanistic view. *Physiol. Rev.* **69,** 250–314.

Lymphocyte Responses to Retinoids

DIRCK L. DILLEHAY, *Emory University School of Medicine*

Glossary

Antibody-dependent cellular cytotoxicity Cytolytic reaction in which a leukocyte effector cell kills an antibody-coated target cell

Antigen Substance that is capable of inducing the formation of antibodies

Chemoprevention Use of a chemical agent to inhibit the occurrence of cancer

Cytokines General term for glycoprotein molecules other than antibodies that are produced by various cell types that mediate the transfer of signals between cells

Delayed-type hypersensitivity Cell-mediated immune reaction that can be elicited by subcutaneous injection of antigen with subsequent cellular infiltrate and edema occurring approximately 24–48 hr after antigen challenge

Hybridoma Cell lines created *in vitro* by fusing two different cell types, usually lymphocytes, one of which is a tumor cell

Interleukins Group of cytokines that control the growth and differentiation of B and T lymphocytes (e.g., IL-1, IL-2)

Mitogen Substance that induces cells (e.g., lymphocytes) to undergo cellular division. Phytohemagglutinin (PHA) and concanavalin (Con A) are T-cell mitogens. Pokeweed mitogen (PWM) is a T-cell-dependent B-cell mitogen. *Staphylococcus aureus* Cowan strain A (SAC) is a B-cell mitogen

Natural cytotoxicity Spontaneous cytolytic reaction between nonimmune lymphoid cells and target cells

RETINOIDS ARE a class of compounds consisting of natural and synthetic analogues of vitamin A (retinol). Retinol and/or its metabolites [e.g., all-*trans*-retinoic acid (RA) and vitamin A esters] play an important role in normal growth, bone formation, vision, reproduction, and regulation of growth and differentiation of epithelial tissues. Therapeutically, retinoids are important in treating various dermatologic diseases and have potential in treating preneoplastic and neoplastic diseases. Currently, retinoids are being investigated for their use in dermatology and oncology, as well as for their possible use as therapeutic agents in inflammatory diseases, rheumatoid arthritis, and immunosuppressive disorders. Retinoids are potent biological response modifiers of the immune system and play an obligatory role in maintaining this system.

I. Retinoids

A. Chemistry

Retinoid is a generic term, referring to the class of compounds consisting of the natural forms of retinol and synthetic analogues with or without vitamin A activity. The most important dietary sources of vitamin A are the carotenoids, which are naturally occurring compounds found in high concentration in green leafy vegetables and colored fruits. Less than 10% of the 600 known carotenoids serve as precursors for vitamin A. Beta-carotene is the carotenoid with the most pro–vitamin A activity. Other natural sources of vitamin A in the diet are the long-chain retinyl esters found in animal tissues. Most retinoids are crystalline in pure form and undergo degradation on exposure to heat, light, or air. Retinoids are lipophilic, unsaturated compounds with the basic structure consisting of a cyclic end

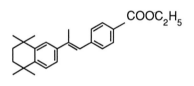

Cyclic   Polyene   Polar
end group   side chain   end group

CH$_2$OH

all-*trans*-Retinol

CONH

OH

all-*trans*-N-(4-Hydroxyphenyl) retinamide

COOH

all-*trans*-Retinoic acid

COOC$_2$H$_5$

etretinate

COOH

13-*cis*-Retinoic acid

COOC$_2$H$_5$

arotinoid ethyl ester

FIGURE 1 Chemical structures of retinoids.

group, a polyene side chain and a polar end group (Fig. 1). [*See* VITAMIN A.]

More than 2,000 synthetic retinoid analogues have been developed in an attempt to increase therapeutic efficacy and decrease toxicity. Modifications of the main units of the retinoid molecule have resulted in new compounds with increased or decreased vitamin A activity. Modifications of the carboxyl end group and/or ring structure have yielded compounds with less toxicity and significantly more biological activity. The first generation of retinoids comprised compounds that resulted from chemical manipulation of the carboxyl end group. Clinically, 13-*cis*-retinoic acid (cRA) is the most useful compound of this group. The second generation of retinoids resulted from modification of the cyclic end group of RA with various substituted and nonsubstituted ring structures. These compounds (e.g., etretinate) are effective in dermatology and hold promise in oncology. The third generation of reti-

noids consist of analogues resulting from cyclization of the polyene side chain. This group includes the arotinoids, which are retinoidal benzoic acid derivatives and which are highly effective at low concentrations in inhibiting the growth of chemically induced skin papillomas in mice.

In the past, clinical investigations generally focused on the therapeutic effects of vitamin A and its esters, RA, cRA, and an aromatic derivative of retinoic acid (etretinate). In cancer research, these compounds have had their greatest effect on cancer of epithelial tissues (e.g., skin, bladder, and mammary gland). Recently, retinamides [e.g., *N*-(4-hydroxyphenyl)retinamide (4-HPR)] and arotinoids have been found to have potent cancer preventive effects in certain animal models. Because of such findings, these compounds have been entered into clinical trials. 4-HPR, one of the most efficacious retinoids for chemoprevention of chemically induced mammary cancer in rats, is currently being evaluated for its cancer chemopreventive properties in women at high risk for mammary cancer. Another chemopreventive trial with 4-HPR is being carried out in bladder cancer patients. Investigations into the clinical effectiveness of arotinoids are just beginning; however, in *in vitro* assays for vitamin A activity (e.g., chick skin keratinization assay), arotinoids are 1,000-fold more active than RA and retinol. This group of retinoids may provide the most clinically effective compounds developed thus far.

The clinical usefulness of retinoids is limited by their toxicity. Toxic side effects in humans include bone hypertrophy, hypercholesterolemia, hypertriglyceridemia, headache, and eye, mouth, and skin irritation. Except for the bone abnormalities, all these clinical signs and symptoms are reversible. Retinoids are also teratogenic and thus should be used with proper birth control in women of childbearing age. Despite the toxic side effects, the newer retinoids (e.g., arotinoids) have improved therapeutic indices (ratio between lowest dose of retinoid producing toxic effects and dose causing 50% regression of papillomas in treated animals). Through continued chemical manipulation, newer compounds are likely to be developed with more favorable therapeutic margins.

B. Physiology and Metabolism

After dietary intake of retinol or beta-carotene, these compounds are hydrolyzed in the intestine, absorbed by enterocytes, and esterified with long-chain fatty acids. Via microscopic fatty particles, the chylomicrons, the retinyl esters are transported in the blood and lymphatics to the liver. The chylomicrons are taken up by liver cells, and the retinyl esters are hydrolyzed to retinol, which is again esterified and stored in hepatocytes or transferred to the hepatic fat-storing cell (Ito or stellate cell). In the hepatocyte, retinol is bound to retinol-binding protein (RBP) and may be secreted into the circulation bound to RBP. Human serum contains approximately 3 μg/ml (10^{-6} M) retinol. Retinol transfers from RBP to the cellular target and binds to a glycoprotein intracellular receptor, cellular retinol binding protein (cRBP). Intracellularly, retinol is oxidized to RA, which is the most active cellular metabolite of retinol.

In contrast to beta-carotene and retinyl esters, which are cleaved before absorption, RA is absorbed directly and is transported in the blood by serum albumin. RA is not stored in the liver but delivered directly to the tissues. RA like retinol also has a cellular binding protein, cellular RA-binding protein (cRABP), which is involved in intracellular transport. These binding proteins appear to mediate the biologic activity of retinoids, because the affinity with which synthetic retinoids bind to these proteins correlates with their biologic activity. There are several cell lines (e.g., HL-60, a human promyelocytic leukemia line), however, that lack detectable levels of cRABP but respond to RA by differentiating into mature granulocytes. They probably contain other binding proteins, which constitute a large family to which the RABP belongs.

II. Lymphocyte Biology

The immune system is one of the most complex biological systems known, and numerous questions concerning the basic mechanisms involved in its normal function remain unanswered. The cells of the immune system are derived from stem cells in the bone marrow and include lymphocytes, macrophages, natural killer (NK) cells, and polymorphonuclear granulocytes. Lymphocytes are the primary cells of the immune system and consist of two groups, T and B lymphocytes. Both cell types are of similar size and appearance but are distinguished by their different immunologic functions and by the presence of different cell surface glycoproteins. [*See* LYMPHOCYTES.]

A. T Lymphocytes

Approximately 20% of the peripheral blood leukocytes are lymphocytes, of which 95% are composed of T and B cells and the remaining 5% are large granular lymphocytes in which is found most of the NK cell activity. T-cell precursors migrate from the bone marrow to the thymus for further proliferation and differentiation, giving rise to mature T cells. Most of the T cells within the thymus die, presumably because of autoreactivity; only 1% leave the thymus as mature T cells. In the thymus, the maturing T cells acquire unique T-cell receptors (TcR) that bind to a specific antigenic determinant, associated with major histocompatibility complex (MHC) molecules (see below). With the advent of monoclonal antibodies having specificities for cell surface molecules, lymphocytes can now be classified according to their repertoire of cell surface antigens, cluster determinants (CD). Most human T lymphocytes express three surface glycoproteins: CD2, CD3, and CD5; CD3 is associated with TcR. Mature peripheral T cells can be divided into two subpopulations based on different functional characteristics and the presence of either CD4 or CD8 cell surface antigens. T lymphocytes involved in helping/inducing other lymphocytes are CD4-positive, whereas lymphocytes involved in cytotoxic/suppression events are CD8-positive. [See CD8 AND CD4: STRUCTURE, FUNCTION, AND MOLECULAR BIOLOGY; T-CELL RECEPTORS.]

The process by which helper T cells are activated to initiate an immune response to antigen involves interactions with antigen-processing cells (APC) such as the macrophage or B cell. To become activated, T lymphocytes must recognize, via their TcR, and bind to antigen that is embedded within a cell surface glycoprotein (MHC class II antigen) on the surface of the macrophage or of the B cells. The MHC genes encode class I and class II cell surface molecules with which T lymphocytes interact during the immune response. Class I antigens are expressed on the surface of all nucleated cells, whereas class II antigens are found on cells of the immune system and some endothelial cells. Helper T lymphocytes are class II restricted and recognize antigen only in association with class II molecules, whereas cytotoxic T lymphocytes recognize antigen mostly in association with class I molecules.

The precise mechanisms by which interactions between antigen and membrane receptors result in lymphocyte proliferation and differentiation are not completely known. Recent evidence, derived from many cell types, including T cells, indicates that intracellular secondary messengers play an important role. After binding of the helper T cell to the antigen–class II complex on the surface of the APC, the lymphocyte TcR/CD3 complex transduces the extracellular signal to the nucleus via changes in secondary messengers, which include activation of the phosphoinositol pathway and protein kinase C and increases in intracellular Ca^{2+}. These biochemical processes result in functional changes in the helper T cell and in its production of several lymphokines. The antigen-activated helper T cell responds by producing IL-1 receptors and secreting interferon (IFN)-γ, which activates the APC. The

APC then produces IL-1, which further activates the helper T cell to secrete IL-2. The IL-2, in turn, causes proliferation and differentiation of antigen-activated T and B cells, which express IL-2 receptors. Thus, the helper T lymphocyte is a crucial cellular component in initiating the early events of the cellular and humoral immune response. [See INTERLEUKIN-2 AND THE IL-2 RECEPTOR.]

Cytotoxic T cells recognize and kill target cells that express foreign antigens in association with class I antigen. The lytic event involves close cellular contact and release of perforins, which alter cellular permeability leading to target cell lysis. Cytotoxic T cells play an important role in killing viral-infected cells and neoplastic cells. The helper T cell, which recognizes the same antigen as the cytotoxic T cell, releases IL-2 during activation. This results in clonal expansion of cytotoxic T cells, thus enhancing the cytotoxic response. Suppressor T cells are less well-characterized. The available evidence, however, indicates that suppressor T cells down-regulate immune responses through the release of soluble suppressive factors.

B. B Lymphocytes

The main function of B lymphocytes is to produce antibodies to antigens on microorganisms and other substances potentially harmful to the body. B cells begin development in the fetal liver, then migrate to bone marrow for maturation before their entry into circulation. The bone marrow is the major production site of B cells, which comprise between 5 to 15% of circulating lymphocytes. B cells are identified by the presence of immunoglobulin (Ig) on their cell surface. Other B-cell surface molecules include class II MHC antigens, CD19 and CD20. CD38 is a molecule expressed on pre–B cells, which are precursors of mature B cells, and plasma cells, which are the antibody-producing cells obtained from B cells. B lymphocytes, like T lymphocytes, acquire specific antigen receptors (surface Ig) during differentiation and become activated after binding and processing antigen in the presence of helper T cells and/or APCs. The release of B-cell growth and differentiation lymphokines by helper T cells results in the generation and clonal expansion of terminally differentiated plasma cells. Some B cells do not differentiate into plasma cells and remain as memory B cells. Subsequent exposure to the same antigen causes the memory B cells to produce a more rapid and highly titered antibody response. The antibody secreting plasma cell is the terminally differentiated effector cell in the B-lymphocyte population, whereas cytotoxic T cells are the effector cells in the T-lymphocyte population. [*See* B-CELL ACTIVATION, IMMUNOLOGY.]

C. Cytokines

Cytokines are a group of important molecules that regulate cellular proliferation and differentiation and are secreted by different cell types. Cytokines function similar to hormones in that they are active at low concentrations (10^{-10}–10^{-15} M) and require cellular receptors to transmit their effects. Lymphokines are a group of cytokines, which are produced predominantly by helper T cells (CD4$^+$) in response to antigenic stimulation, and affect macrophages, lymphocytes, and hematopoietic cells. Many lymphokines have multiple, overlapping functions that affect more than one cell type. The lymphokines that transmit signals between leukocytes are referred to as interleukins (IL). They have been numbered sequentially as they have been isolated and characterized. IL-1 is a cytokine that is produced predominantly by macrophages. To date, eight ILs have been described, and with further research into lymphocyte growth and function, it is likely that other ILs will be discovered. [*See* CYTOKINES IN THE IMMUNE RESPONSE.]

Helper T cells become activated after interactions with APCs and IL-1 and produce several different lymphokines (Fig. 2). IL-2 and IL-4, secreted by helper T cells, cause proliferation of T and B lymphocytes. IL-3 and granulocyte macrophage-colony stimulating factor (GM-CSF) stimulate the growth and differentiation of early hematopoietic progenitor cells in the bone marrow to produce mature hematopoietic cells. B-cell growth and differentiation are also regulated by several different lymphokines produced by helper T cells. IL-2 and IL-4 provide initial signals for B-cell growth. IL-4 also induces MHC class II molecules on B cells and enhances IgG and IgE production. IL-5 enhances IgA and IgM production and affects B-cell growth but at a later stage than IL-4. The terminal differentiation of fully activated B cells to Ig-producing plasma cells is caused by IL-6. IL-7 is a cytokine produced by stromal cells of the bone marrow and causes proliferation of pre–B cells and immature CD4$^-$/

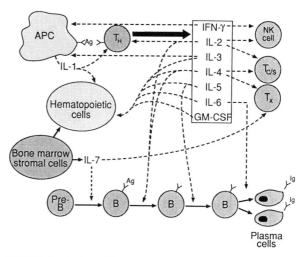

FIGURE 2 Activated helper T cells produce lymphokines that regulate the growth and function of various cell types. (APC, antigen-presenting cell; T$_H$, helper T cell; T$_{C/S}$, cytotoxic/suppressor T cell; NK, natural killer; B, B cell; T$_x$, thymocyte; Ag, antigen; Ig, immunoglobulin; IL, interleukin; IFN, interferon; GM-CSF, granulocyte macrophage-colony stimulating factor.)

CD8$^-$ thymocytes. Thus, through their production of multiple lymphokines, helper T cells play a central role in the growth, differentiation, and function of T and B lymphocytes. [*See* HEMOPOIETIC SYSTEM.]

Interferons (e.g., IFNα, β, and γ) comprise another group of cytokines that are produced primarily by leukocytes and affect cells of the immune system. IFNγ is produced by activated helper T cells and large granular lymphocytes and is a potent inducer of MHC class II antigens and NK cell activity. IFNα is important in mediating the antimitogenic and antiviral effects of interferons and increases the expression of MHC class I molecules. IFNβ is similar to IFNα but is produced primarily by fibroblasts [*See* INTERFERONS.]

The mechanism(s) by which cytokine production is regulated at the molecular level during the immune response is unknown; however, it appears that after helper T-cell activation, expression of lymphokine genes is regulated at the transcriptional level. The maximum production of lymphokine-specific mRNA occurs approximately 5–8 hr after lymphocyte activation. Research is rapidly expanding in the lymphokine field. Information obtained on the function and regulation of these molecules will provide a better understanding of their role in immunologic disease and cancer.

III. Retinoids and Lymphocyte Function

The antineoplastic effects of retinoids have been known for many years. In experimental animals, retinoids can prevent the occurrence and inhibit the growth of epithelial and mesenchymal tumors caused by viruses, chemicals, or radiation. In humans, retinoids are beneficial in the treatment of precancerous lesions (e.g., leukoplakia) and certain neoplastic diseases (e.g., adult cutaneous T-cell lymphoma). The mechanism(s) by which retinoids exert their neoplastic effects is unknown; however, recent evidence suggests two major mechanisms. Retinoids may suppress carcinogenesis by acting directly on the tumor cells or indirectly through potentiation of the antitumor immune response. The main effector cells of the antitumor response are T lymphocytes. It is possible that these two mechanisms are not mutually exclusive because there is evidence to support both. Because retinoids inhibit the growth of syngeneic tumors in mice through potentiation of cytotoxic lymphocyte responses, it is plausible that retinoids may similarly potentiate the antitumor immune response in human patients.

Because retinoids are being considered for use in humans as cancer chemopreventive agents and thus are to be administered during a long period, it is important to better understand their effects on the immune system. Although numerous studies have evaluated the *in vivo* and *in vitro* effects of retinoids on various immune responses in mice, fewer studies have been accomplished with humans. Nevertheless, the data indicate that, in general, the effect of retinoids on immune responses is biphasic (e.g., a high concentration produces one effect and a lower concentration produces the opposite). It is also apparent that different retinoids have different effects on immune responses. Thus, the often conflicting results described for retinoid effect on immune responses may be due to the use of different experimental conditions.

A. Cellular Immunity

The evidence for a possible physiological role of retinoids in human immune responses comes from epidemiologic evidence, which suggests a correlation between vitamin A deficiency and an increased susceptibility to infectious diseases. Hypovitaminosis A in humans is usually associated with protein-energy malnutrition, which can also cause an increase in susceptibility to infectious disease. Thus, it is unclear whether a deficiency of vitamin A or protein/calories is the cause. Animal studies, however, have confirmed the clinical observation that vitamin A deficiency decreases immune responses and increases infectious disease susceptibility. The cellular mechanisms involved in this clinical observation are not entirely known.

Retinoid effects on human lymphocyte function have been evaluated primarily through the use of mitogen proliferation assays. Lymphocyte proliferation induced by mitogens is considered to be an *in vitro* model for the events that take place *in vivo* when lymphocytes are stimulated by antigen. Retinoids affect lymphocyte mitogenesis differentially depending on the phenotype of the lymphocytes being evaluated and their site of origin. Also, the concentration of retinoid has an important effect on lymphocyte function, with high doses ($>10^{-5}$ M) inhibiting proliferation and low doses potentiating proliferation. The mitogenic responses of tonsillar lymphocytes and thymocytes stimulated by phytohemagglutinin (PHA), a T-cell mitogen, are potentiated by 10^{-5}–10^{-7} M RA. RA also augments the proliferative response of peripheral blood lymphocytes (PBL) to PHA if the PBLs are depleted of macrophages. Furthermore, T-cell-enriched populations obtained from PBL are even more responsive to RA in its potentiation of PHA-induced mitogenesis than PBL depleted of macrophages. These data indicate that in lymphocyte mitogen assay systems, T cells are directly responsive to RA.

The retinoid-induced potentiation of lymphocyte mitogenesis also varies depending on the type and dose of mitogen. Superoptimal concentrations of mitogen results in the greatest RA-induced potentiation of PBL and adenoidal lymphocyte proliferation (Fig. 3). Lymphocytes stimulated with either T-cell (PHA, Con A) or T-cell-dependent B-cell mitogens (PWM) are more responsive to retinoid augmentation of proliferation than lymphocytes stimulated by B-cell mitogens (SAC). Retinoids appear to have no effect on B-cell mitogenesis in humans or mice. Thus, it appears that retinoids selectively potentiate T-cell-dependent mitogenesis and that the helper T cell may be an important target cell.

The concentration and type of retinoid also affects lymphocyte mitogenesis differentially. Retinoid-induced potentiation of lymphocyte mitogenesis occurs at concentrations ranging from 10^{-6} to 10^{-13} M, with lymphocytes from tonsils being more responsive to retinoids than lymphocytes isolated from other sites. Retinoid concentrations greater

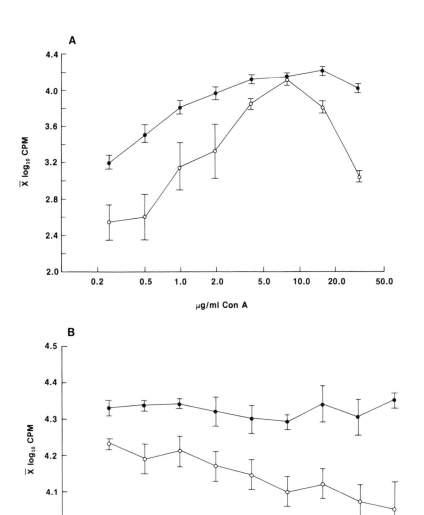

FIGURE 3 RA-induced potentiation of lymphocyte proliferation in the presence of Con A (**A**) or PWM (**B**). PBL and adenoidal lymphocytes were cultured with varying concentrations of Con A or PWM, respectively, in the presence (●) and absence (○) of 10^{-7} M RA. The proliferative response of lymphocytes was measured by the incorporation of [^{3}H] thymidine. Each value represents the $\log_{10}$ mean counts per minute (cpm) of quadruplicate samples. *Vertical bars* are standard error of $\log_{10}$ mean cpm. [Reproduced, with permission, from D. L. Dillehay, et al. *Clin. Immunol Immunopathol* **50**, 100–108, 1989.]

than 10^{-5} M usually inhibit lymphocyte mitogenesis. Most *in vitro* studies have evaluated the effects of only RA on lymphocyte function. A few studies, however, have compared the affects of other retinoids. Retinol at physiologic concentrations (10^{-6} M) has no effect on human lymphocyte proliferation induced by PHA. 4-HPR and cRA potentiate lymphocyte proliferation to a similar degree as RA. However, in-depth evaluations of retinoid structure–function relations using lymphocyte mitogenesis assays have not been done.

The *in vivo* effects of retinoids on human lymphocyte mitogenesis are variable. PBLs from cancer patients treated with cRA have increased proliferative responses to PHA. No change in lymphocyte proliferation induced by PHA or Con A, however, was observed with PBL from dermatologic patients treated with cRA or etretinate or from healthy volunteers administered cRA. This discrepancy may be due to differences in the health status of the patient, retinoid dose, or length of retinoid treatment. Other effects of cRA on human lymphocytes include increases in the percentage of helper T cells in healthy adults and patients with dermatologic diseases.

The cellular mechanisms by which retinoids potentiate lymphocyte proliferation are not entirely clear; however, evidence suggests a direct effect on the cells. For instance, RA increases IL-2 receptor expression on proliferating human thymocytes. Also, mice treated with retinoids have an increased number of IL-2-producing cells and an increased production of IL-2. Thus, retinoid-mediated potentiation of lymphocyte mitogenesis appears to be due to changes in IL-2 production and/or receptor expression by the lymphocytes.

Retinoids also affect other cell-mediated immune responses by human lymphocytes [e.g., antibody-dependent cellular cytotoxicity (ADCC) reactions and natural cytotoxicity]. *In vitro*, a high concentration of RA (5×10^{-4} M) decreases ADCC and natural cytotoxicity. This inhibition is due to a specific effect on effector cells and may be related to the capacity of high concentrations of RA to modulate expression of IgG and complement receptors. Retinoids also affect human NK cell activity. NK cell activity of dermatologic patients treated with etretinate was increased, whereas patients treated with cRA had either no change or a decrease in NK activity. These findings may be due to the effects of retinoid on cell surface receptors and/or lymphokine production. In support of this, the production of IL-2 and IFNγ, lymphokines that potentiate NK activity, is affected by retinoids.

Retinoids also influence another parameter of T-cell-mediated immunity (e.g., skin test reactions). Patients with cancer or dermatologic disease treated with cRA or etretinate for 1 month had an increased delayed-type hypersensitivity reaction to various recall antigens (e.g., mumps, tuberculin, tetanus, and diphtheria). The delayed-type hypersensitivity response to dinitrochlorobenzene sensitization was also increased in patients treated with cRA or etretinate. Taken together, these findings further substantiate the hypothesis that retinoids enhance cell-mediated immunity.

B. Humoral Immunity

The humoral immune response begins with antigen recognition by B cells, which then proliferate and differentiate in response to lymphokines produced by helper T cells and ends with the production of antigen-specific antibodies by plasma cells (see Fig. 2). Assays to measure the humoral immune response include determining serum antibody levels and quantitating the number of lymphocytes capable of producing antibody in response to antigen.

Most research into the effects of retinoids on humoral immunity has been performed with laboratory animals. These studies have shown that vitamin A–deficient animals have decreased humoral immune responses and that retinoid-supplemented animals have increased responses. Retinoids appear primarily to enhance the humoral response to antigens that require the presence of T cells.

Only a few studies have evaluated the effects of retinoids on the humoral immune response in humans. Using human tonsillar lymphocytes, which are composed of highly reactive T and B cells, 10^{-5}–10^{-7} M RA enhances antibody-producing cells in response to sheep erythrocytes, a result caused by a direct effect on B cells. Another *in vitro* study shows that RA increases antibody secretion from a human-human hybridoma produced from B cells from patients with combined variable immunodeficiency (CVI). The B cells from these patients have defects in maturation and secrete low levels of antibody. RA induces differentiation of these B-cell hybridoma cells to a more mature phenotype as evidenced by their increased production of antibody and increase in CD38 (a preplasma cell surface marker). However, antibody secretion from B cells from healthy individuals is not affected by RA. Furthermore, serum antibody levels in dermatologic patients treated with cRA or etretinate for up to 12 weeks are unchanged. Thus, it appears that retinoids affect antibody production only by B cells that are in an immature stage such as are those from the B-cell hybridomas of CVI patients and those B cells isolated from tonsils. These findings are consistent with the well-known differentiating property of certain retinoids.

Taken together, the effects of retinoids on cell-mediated and humoral immune responses of humans appear to involve primarily events concerning the T lymphocyte; however, retinoids can also enhance the responses of less mature B lymphocytes.

IV. Mechanism of Retinoid Action

The major physiological function of retinol and its metabolite RA is in the promotion of differentiation and maintenance of epithelial cells. Evidence has accumulated that retinoids also affect nonepithelial cells (e.g., T and B lymphocytes). The effects of retinoids on lymphocyte function vary and depend on the type of functional assay, on the concentration and route of administration of retinoid, and on the phenotype of the lymphocytes assayed. Although the mechanisms by which retinoids affect the immune system are unclear, it appears that they alter lymphocyte function by changing the stage of differentiation and/or growth through effects on genomic expression.

A. Cellular Level

During a cell-mediated immune response, retinoids act during or before the induction phase. In the lymphocyte mitogenesis assay, for example, if retinoids are added alone or several hours after the mitogen, no effect on lymphocyte proliferation is observed. Potentiation of lymphocyte proliferation by retinoids depends on the near concurrent addition of retinoid and mitogenic stimulus. Sidell and Ramsdell documented that one mechanism involved in RA enhancement of human thymocyte proliferation induced by IL-2 or PHA involves an increase in IL-2-receptor expression by the thymocyte blasts. This retinoid-induced increase in expression of the IL-2 receptor appears to occur only at certain maturational stages of lymphocyte development because it was not observed with PBL blasts. RA also affects other growth factor receptors, such as those for epidermal growth factor, nerve growth factor, GM-CSF, sarcoma growth factor, and IL-1. Expression of the human MHC class II antigen (HLA-Dr) and transferrin receptor was increased on peripheral blood mononuclear cells from healthy volunteers given cRA. Thus, modulation of the number and/or affinity of cell surface molecules and receptors for growth factors may mediate the effects of retinoids on lymphocyte function.

Another mechanism by which retinoids may alter cellular immune responses is through modulation of immunoregulatory lymphokine/cytokine production. Few studies have evaluated the effects of retinoids on lymphokine production, and most of these have involved mice. With isolated cells, RA has no effect on IL-2 production by mouse or human lymphocytes; however, splenocytes from mice dosed with RA have an increase in IL-2 production. The production of IL-1 and IL-3 by murine peritoneal macrophages and splenocytes, respectively, is increased in mice dosed with RA or arotinoids. Arotinoids, which appear to have potent antineoplastic effects, are more effective than RA in enhancing IL

production (IL-1, IL-2, IL-3) in mice. Physiological concentrations of RA also increase the production of IL-1 and IL-3 by cultured human peripheral blood mononuclear cells and by a murine macrophage cell line, P388D$_1$.

Interferons are another group of lymphokines that affect various cell-mediated immune responses (e.g., lymphocyte proliferation and cytotoxicity and NK cell and macrophage activity). RA decreases the production of IFNα and γ by mitogen-activated human PBL. Because IFN suppresses cellular proliferation, inhibition of its production by RA may cause an increase in lymphocyte proliferation. In contrast, lymphocytes from mice dosed with RA had an increase in production of IFNγ. These findings indicate that the mechanism of retinoid activity at the cellular level probably involves their capacity to modulate both the production of lymphokines and/or expression of lymphokine receptors.

Retinoids affect the cellularity of the thymus, spleen, and lymph nodes in mice. At high concentrations ($>10^{-5}$ M), these compounds are toxic, decrease lymphocyte numbers, and inhibit immune responses. In physiologic concentrations, however, retinoids increase the cellularity of lymphoid organs and stimulate immune responses. Although the cells causing the increased lymphoid cellularity have not been adequately studied in mice, some evidence suggests that this increased cellularity may be due to an increase in mature helper T cells. Similar observations have been made in humans. Watson et al. reported that in normal humans administered cRA for 9 months, the CD3$^+$ lymphocyte population remained constant but the number of helper T cells (CD4$^+$) increased and the numbers of cytotoxic T cells (CD8$^+$) decreased. In patients with acne vulgaris that was treated for 8 weeks with cRA, increased numbers of CD4$^+$ cells were also found. Because retinoids increase the production of ILs that are produced predominantly by CD4$^+$ cells and because retinoids increase the number of CD4$^+$ cells in vivo, it seems likely that the CD4$^+$ (helper T cell) lymphocyte may be one of the target cells mediating retinoid action. However, it has clearly been demonstrated that retinoids can also modulate macrophage function.

B. Molecular Level

The molecular mechanisms of action of retinoids on cells of the immune system, as well as other cell types, have not been fully elucidated. Evidence is accumulating, however, to suggest that the end point of retinoid action is modulation of the expression of specific genes (Fig. 4). Regulation of genomic expression, either enhancement or repression, has been demonstrated for approximately 40 gene products. It has been hypothesized that direct regulation by retinoids occurs similar to gene regulation as described for steroid hormones.

Retinol and RA can be found within nuclei bound to physically distinct chromatin acceptor sites. Both cRBP and cRABP appear to be involved in the transfer of retinol and retinoic acid, respectively, from the cytosol to the nuclear binding sites. There are two distinct human RA receptors, RARα and RARβ, within the nucleus that are encoded by two distinct genes on two different chromosomes and share homology with the steroid/thyroid nuclear receptor multigene family. Preliminary evidence indicates that receptor expression varies between tissues. Determining the temporal sequence of RAR expression in lymphocytes may answer questions concerning the varied responsiveness of lymphocytes at different maturational stages to the effects of retinoids. After binding of RA to the nuclear RAR, the RAR appears to act as a transacting en-hancer factor for gene expression. The target genes activated by the RAR are unknown.

Acknowledgment

Work by the author reviewed herein was supported by grants RR07003, CA09467, and CA34968 from the National Institutes of Health.

Bibliography

Brand, N., Petkovich, M., Krust, A. et al. (1988). Identification of a second human retinoic acid receptor. *Nature* **332,** 850–853.

Dennert, G. (1985). Immunostimulation by retinoids. *In* "Retinoids, Differentiation and Disease" (J. Nugent and S. Clark, eds.). Ciba Foundation Symposium, Pittman, London.

Dillehay, D. L., Cornay, W. J., Walia, A. S., and Lamon, E. W. (1989). Effects of retinoids on human thymus-dependent and thymus-independent mitogenesis. *Clin. Immunol. Immunopathol.* **50,** 100–108.

Dillehay, D. L., Shealy, Y. F., and Lamon, E. W. (1989). Inhibition of moloney murine lymphoma and sarcoma growth *in vivo* by dietary retinoids. *Cancer Res.* **49,** 44–50.

Dillehay, D. L., Walia, A. S., and Lamon, E. W. (1988). Effects of retinoids on macrophage function and IL-1 activity. *J. Leukocyte Biol.* **44,** 353–360.

Hill, D. L., and Grubbs, C. J. (1982). Retinoids as chemopreventative and anticancer agents in intact animals (*review*). *Anticancer Res.* **2,** 111–124.

Isakov, N. (1988). Regulation of T-cell-derived protein kinase C activity by vitamin A derivatives. *Cell Immunol.* **115,** 288–298.

Lippman, S. M., and Meyskens, F. L. (1988). Vitamin A derivatives in the prevention and treatment of human cancer. *J. Am. Coll. Nutr.* **7,** 269–284.

Sidell, N., and Ramsdell, F. (1988). Retinoic acid upregulates interleukin-2 receptors on activated human thymocytes. *Cell Immunol.* **115,** 299–309.

Watson, R. R., Jackson, J. C., Alberts, D. S., and Hicks, M. J. (1986). Cellular immune functions of adults with a daily, long-term, low dose of 13-*cis* retinoic acid. *J. Leukocyte Biol.* **39,** 567–577.

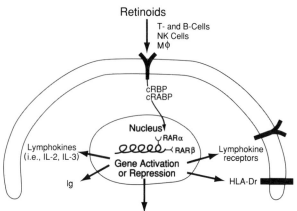

FIGURE 4 Model to explain the effects of retinoids on cells of the immune system. (Mφ, macrophages; NK, natural killer; RAR, retinoic acid receptor; cRPB, cellular retinol binding protein; cRABP, cellular retinoic acid binding protein; Ig, immunoglobulin; HLA-Dr, human MHC class II antigen; IL, interleukin.)

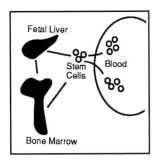

Lymphocytes

J. F. A. P. MILLER, *The Walter and Eliza Hall Institute of Medical Research*

I. Cells Involved in Immune Responses
II. Genes and Molecules Governing Antigen Recognition
III. Major Histocompatibility Complex and Antigen Presentation
IV. Immunological Self-Tolerance
V. Diseases of Immunological Aberrations
VI. Conclusions

Glossary

Antibody Protein produced by B lymphocytes able to combine with an antigen that has stimulated its production

Antigen Substance that can induce a detectable immune response when introduced into the body

Antigen-presenting cells Cells specialized to present antigen or antigen fragments to lymphocytes

B cell (B lymphocyte) Lymphocyte derived from the bone marrow and able to produce antibody after antigenic stimulation

Clone Group of cells that are the progeny of a single cell

Delayed-type hypersensitivity T cell-mediated immune reaction that can be elicited by subcutaneous antigen injection, with a subsequent cellular infiltrate and edema (i.e., fluid extravasation) that are maximal 24–48 hours after antigen challenge

Dendritic cells Cells with elongated processes that can contact numerous lymphocytes and trap antigen–antibody complexes effectively, holding antigen on their surface for prolonged periods

Endocytosis Process by which material external to a cell is internalized within a particular cell

Immunoglobulin Glycoprotein (i.e., a polypeptide to which sugar residues are attached) which functions as an antibody molecule

Lymphokine Soluble product of lymphocytes, not specific for an antigen

Macrophage Phagocytic cells derived from bone marrow

Major histocompatibility complex (MHC) Locus (i.e., group) of genes encoding molecules expressed on cell surfaces and involved in the rejection of transplants. A variety of antigen fragments, or peptides, can associate with the MHC molecules and provoke a T cell response

Phagocytosis Engulfment of microorganisms or other particles by leukocytes (i.e., white blood cells) or macrophages

Plasma cells Antibody-synthesizing cells derived from B lymphocytes

Polypeptide Polymer of amino acids linked together by peptide bonds; proteins are composed of polypeptide chains

T cell (T lymphocyte) Lymphocyte derived from the thymus and responsible for cellular reactions in immune responses

Tolerance State of specific nonreactivity induced by antigen and resulting from the elimination or silencing of the clone of lymphocytes able to respond to that antigen

LYMPHOCYTES ARE the cells that initiate specific immune responses. Two major subsets exist: T cells derived from the thymus and B cells from the bone marrow. On their surface both cells have receptors, that is, molecules capable of binding foreign organisms or antigenic determinants; as a result the cells are activated to produce a response that eliminates the antigen. The receptor molecules have unique binding sites which are predetermined; they are coded by genes that have been inherited and rearranged during development of the lymphocytes. Cells with receptors that can bind to self components (i.e., of the same individual) are eliminated to ensure self-tolerance, that is, a lack of response to molecules of the individual itself. T cells are responsible for cellular immune responses, which include direct killing of infected cells or cells in foreign transplants and the induction of an inflammatory response. B cells are responsible for antibody production.

To survive in a world teeming with infectious organisms, multicellular organisms have had to evolve various defense strategies. These can be classified under two major groups: constitutive and adaptive. The former are innate, as they do not require the entry of something new into the body to activate them. Conversely, the adaptive mechanisms are acquired, becoming established only after the introduction of some foreign material or antigen. This intrusion elicits an immune response comprising a complex series of cellular events designed to eliminate the antigen. The constitutive mechanisms are nonspecific, being unable to distinguish one kind of organism from another. By contrast, the adaptive mechanisms are specific; that is, the reaction provoked by antigen is directed specifically toward that antigen and discriminates among unrelated antigens. The faculty of immunological memory, or anamnesis, is primarily dependent on the property of specific discrimination and is a hallmark of the adaptive immune response. It ensures that a second dose of antigen evokes a more rapid and more pronounced response than the first. Lymphocytes are the cells primarily responsible for the specificity of immunological responses.

I. Cells Involved in Immune Responses

The constitutive nonspecific mechanisms are a function of the cells that scavenge or engulf foreign substances; they are the mononuclear phagocytes, which include tissue macrophages and the various monocytes and polymorphonuclear leukocytes of the blood. The adaptive specific mechanisms include responses in which discrimination between antigenic determinants depends on sets of complementary recognition molecules, either free in serum or attached to the membrane of lymphocytes. These have the appearance of a roughly spherical cell with a darkly staining round nucleus and a generally thin rim of cytoplasm, and may vary in size from small cells (5–8 μm in diameter) to large cells (8–20 μm) (Fig. 1). [*See* LYMPHOCYTES; PHAGOCYTES.]

There are two distinct but closely related sets of

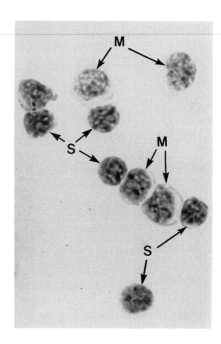

FIGURE 1 Small (S) (diameter 5 μm) and medium (M) (diameter 8 μm) lymphocytes from lymph.

reactions produced by lymphocytes: cellular and humoral. Such a separation of labor has a physiological basis in the two families of lymphocytes, each being derived from the same parental precursor, or stem cell, which matures (i.e., differentiates) under separate influences in two distinct sites (i.e., microenvironments). One is the thymus, a lymphoid organ situated deep to the breastbone or sternum and above the heart (Fig. 2), and in mammalian species the other is the bone marrow. The stem cells themselves originate first in the yolk sac of the embryo, then in the liver in fetal life, and finally in bone marrow. Those which find their way to the thymus produce T cells, which eventually migrate out to populate certain specific parts of the lym-

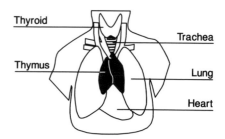

FIGURE 2 The position of the thymus in the human.

phoid tissues, termed T cell-dependent. These include the paracortical areas in the lymph nodes and the periarteriolar lymphocyte sheaths in the spleen (Fig. 3). The mature T cells recirculate from the blood into these tissues and return to the blood either directly, in the case of the spleen, or via lymphatics, in the case of other lymphoid tissues (Fig. 3). Most do not return to the thymus. [*See* THYMUS.]

The recirculation of T cells affords a physiological mechanism enabling the recruitment of cells competent to initiate immune responses into sites (e.g., lymph nodes) (Fig. 4) regionally stimulated by infectious organisms or antigen. After antigenic stimulation the T cells are involved in cellular immune responses, so called because the resulting immune status can be transferred from immune to nonimmune animals by the T cells, but not by serum. Other lymphocytes that differentiate from stem cells not in the thymus, but in the bone marrow in mammals or in the bursa of Fabricius in birds (an organ analogous to the thymus), are termed B cells. These populate the B cell-dependent parts of

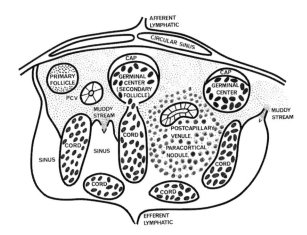

FIGURE 4   Lymph node. The T lymphocytes populate the cortex (stippled area) entering from the blood through the postcapillary venules (PCV) and from other lymph nodes via afferent lymphatics. If unstimulated by antigen, they migrate (via "muddy streams") to the sinuses and out by efferent lymphatics, which eventually end up in the main lymph duct (i.e., the thoracic duct), which empties into the blood. The B lymphocytes are localized in the primary follicles in unstimulated nodes. Following antigenic stimulation T cells enlarge to large pyroninophilic cells (seen around the postcapillary venule on the right), forming a paracortical nodule, and divide to give rise eventually to a progeny of small lymphocytes that recirculate from the lymph to the blood and back to the lymph nodes or the spleen. Antigen that stimulates T cell-dependent antibody formation is associated with T- and B-cell cooperation (presumably in an area where T and B cells meet, as around the follicles) and with the formation of germinal centers (i.e., secondary follicles) composed predominantly of B blasts (large cells) and antigen-capturing dendritic cells. Antibody-secreting cells derived from stimulated B cells are called plasma cells; these accumulate in the cords.

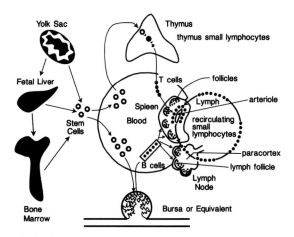

FIGURE 3   Lymphocyte development and migration. Stem cells originating in the embryonic yolk sac, fetal liver, or bone marrow disseminate in the bloodstream; some migrate to the primary lymphoepithelial organs (i.e., the thymus and the bursa, or the bursa equivalent in species other than birds). There, they differentiate along lymphoid pathways and eventually migrate to populate the secondary lymphoid tissues (i.e., the spleen, lymph nodes, and lymphoid aggregates along the alimentary canal). Thymus-derived cells or T cells migrate specifically to the T cell-dependent areas of these tissues (e.g., the area around the arterioles in the spleen and the paracortex in the lymph nodes). They pass through these tissues and recirculate from the blood to the lymph and back to the blood via the main thoracic duct. They are responsible for cellular immune responses. B cells populate the follicles and some recirculate. They are the precursors of antibody-forming cells, which include plasma cells.

the lymphoid tissues, which are distinct from those populated by T cells and include the follicles and the medullary cords, where plasma cells accumulate (Fig. 4). A few B cells recirculate, as do T cells. B cells are the precursors of antibody-forming cells, which include plasma cells. These have in their copious cytoplasm a great deal of RNA, indicating that they are synthesizing antibodies which are protein molecules, known collectively as immunoglobulin (Ig). B cells and plasma cells are involved in humoral immunity, so called because the immune state is transferable to nonimmune recipients by serum, which contains Ig. [*See* B-CELL ACTIVATION, IMMUNOLOGY.]

T and B lymphocytes cannot readily be distinguished on morphological grounds, but can be identified by a distinct pattern of cell surface markers. B cells have Ig molecules on their surface; T cells lack these. A surface glycoprotein known as Thy-1, with a size of 25 kDa, is present on T cells, but not on B cells (Thy-1 is also present on brain cells). In addition, so-called cluster designation (CD) molecules not only distinguish T from B cells, but also differentiate various functional cell subsets within each group (see Section IV).

II. Genes and Molecules Governing Antigen Recognition

How do T and B cells discriminate among the great variety of foreign molecules that confront them? It can be estimated that a human can differentiate among 1 million or more different antigens. It is now clear that each lymphocyte is restricted, prior to antigen encounter, to the extreme position of being able to produce receptors with only one specificity with respect to binding antigen. The appropriate cell is selected by antigen and then divides to give rise to a clone of daughter cells, all with the same specificity. This process is known as clonal selection (Fig. 5). To enable T and B cells to recognize antigen, receptor molecules peculiar for each cell type are displayed at the cell surface: Ig on B cells (the Ig class on unstimulated B cells generally being that designated IgM and IgD) and a so-called T-cell receptor (TCR) on T cells.

Ig molecules have been fully characterized chemically and are generally made up of two heavy polypeptide chains (of about 428 amino acids) and two light ones (of about 214 amino acids) joined together

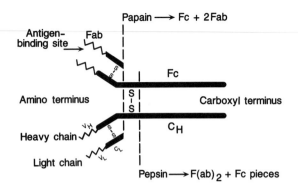

FIGURE 6 The immunoglobulin (antibody) molecule, which is made up of two heavy (H) chains and two light (L) chains, joined together by disulfide bonds (S–S). Each chain of a given class has a constant amino acid sequence in its carboxyl terminus (C_L and C_H) and a variable sequence in its amino terminus (V_L and V_H). The variable portions form the antigen-binding site. The enzyme papain splits the molecule, to create three fragments: an Fc (crystallizable) fragment and two Fab (antibody-binding) fragments. Pepsin splits at another site as shown.

by disulfide bonds. Each chain of a given class is divided into Ig domains with a constant amino acid sequence in the carboxyl terminus (i.e., the end of the chain that has a free α-carboxyl group) and a variable sequence in the amino terminus (i.e., the end with a free α-amino group) (Fig. 6). The portions that vary among Ig molecules of a given class (V regions) form the combining site, which is spe-

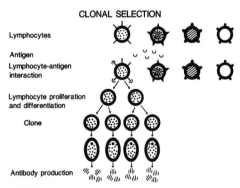

FIGURE 5 Mature lymphocytes are clonally individuated in terms of their antigen receptor specificity. A cell having a receptor that fits a given antigen is selected by the antigen and activated to divide and produce a clone of cells, each with the same specificity. In the case of B cells, shown here, the membrane receptor is identical in its binding site to the antibody eventually secreted by members of the clone.

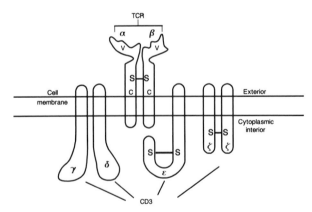

FIGURE 7 The antigen-binding receptor on T cells (TCR) and its associated CD3 molecular complex. The TCR is composed of two disulfide (S–S)-linked polypeptide chains: α and β. As with the immunoglobulin molecule, each chain has a constant amino acid sequence in its carboxyl terminus (C) and a variable sequence in its amino terminus (V). The antigen-binding site is formed by the variable portions. The CD3 complex is intimately associated with the TCR and is composed of four polypeptide chains: γ, δ, ε, and the disulfide-bonded homodimer ζ. All of the chains are transmembrane proteins.

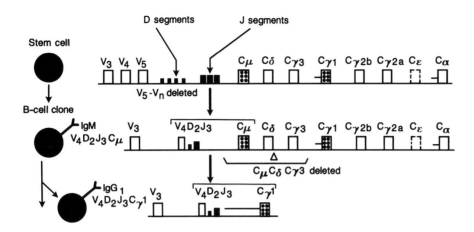

FIGURE 8 The immunoglobulin (Ig) heavy-chain structural genes. The germ-line heavy-chain locus is shown at the top. Below is shown the heavy-chain locus in a differentiated B cell that expresses IgM and has subsequently switched to IgG$_1$. V (variable), J (joining), and D (diversity) segments are joined to form the differentiated V region gene, which is joined to the relevant C (constant) segment, a mosaic of separate V_H, D_H, J_H, and C_γ elements.

cific for a given antigenic determinant. Ig-like domains are found in a large number of other cell surface recognition structures, including the T-cell receptors (TCR), molecules encoded by the major histocompatibility complex (MHC), some CD molecules, Thy-1, and many of the molecules that mediate cell adhesion phenomena. All of these molecules are grouped together as members of the Ig supergene family.

The TCR is made up of two polypeptide chains (generally, α and β chains of 45–50 kDa and, less frequently, γ and δ chains) linked to each other by disulfide bonds. The TCR is closely associated on the T-cell membrane with the molecule CD3 (Fig. 7), a complex of four polypeptides: γ, δ, ε, and the disulfide-bonded homodimer ζ (of 15–25 kDa). CD3 is thought to act as a signal transducer for the TCR, enabling the T cell to sense the occupation of its TCR by antigen.

How can one reconcile the apparent infinite diversity of foreign antigens that must be recognized by the immune system with the clearly finite amount of structural genetic information embodied in the DNA of the fertilized ovum? Such diversity is generated by a variety of mechanisms, which fall into two broad groups. First, individuals inherit from their parents germ-line V genes (V for variable) which code for the combining sites of antigen receptors on both T and B cells. Second, somatic mechanisms (i.e., those occurring in cells not destined to become germ cells) operate to increase diversity during the final assembly of the genes in lymphocyte differentiation. The primary mechanism is recombination; thus, in differentiating T and

TABLE I Potential Amino Acid Sequence Diversity in Immunoglobulin and T Cell Receptor Genes[a]

Segments	Immunoglobulin		T cell receptor	
	Heavy chain	Light chain	α	β
V	250–1000	250	100	25
D	10	0	0	2
N	V–D, D–J	None	V–J	V–D, D–J
J	4	4	50	12
V region combinations	62,500–250,000		2500	
Total combinations	~10^{11}		~10^{15}	

[a] The pairing of random V regions (between heavy and light chains or between α and β chains) generates the V region combination numbers listed. Further diversity is generated by joining V genes to different D and J gene segments, the addition of up to six nucleotides at each junction (N), variability in the 3' joining positions in V- and J-gene segments, and translation of D regions in different reading frames.

B cells a series of DNA rearrangements occurs: Separate germ-line elements—V, D, and J—are rearranged and joined together and to a C (constant region) element, giving rise to the active gene, which is a mosaic of these units (Fig. 8). Hence, an enormous diversity can be generated by such rearrangements in the genes coding for the binding site (Table I). In the Ig molecules further density is generated by somatic mutation in V genes during a B-cell response. This, however, does not appear to be the case for TCR V genes in ongoing T-cell reactions.

There is a marked difference in the form of antigen that can be recognized by T and B cells. By means of their surface Ig molecules, B cells can perceive and bind soluble antigen with varying degrees of strength. T cells, on the other hand, can see antigen only if it is associated with cell surface glycoproteins encoded by the MHC. This phenomenon is known as MHC restriction, and the MHC molecules involved act as restriction elements.

III. Major Histocompatibility Complex and Antigen Presentation

The MHC is a genetic locus (i.e., a group of related genes) encoding groups of highly polymorphic glycoproteins, which were originally discovered because they could elicit the rejection of foreign tissue grafts. The locus consists of several closely linked genes which can be classified as encoding the so-called class I and class II molecules. In humans three genes, termed HLA-A, HLA-B, and HLA-C, encode the class I molecules, and at least six other genes (e.g., HLA-DP, HLA-DQ, and HLA-DR), encode the class II molecules. At each gene one of several alternative forms (i.e., alleles) can be found and are designated by numbers (e.g., HLA-A1). The class I protein molecule is composed of a 45-kDa polypeptide chain which spans the cell membrane with three Ig-like extracellular domains. It is found on the surfaces of a wide variety of cells in association with the small molecule known as β_2-microglobulin (Fig. 9).

The class II molecule also spans the membrane and consists of two noncovalently associated chains, α (28 kDa) and β (34 kDa). Each has two Ig-like extracellular domains, one homologous to the V region and one homologous to the C region (Fig. 9). Class II molecules are found mainly on B cells, macrophages, and dendritic cells. Both classes of

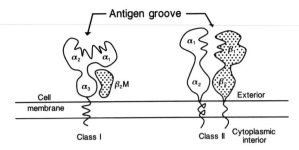

FIGURE 9 The major histocompatibility complex classes I and II molecules. The class I molecule has three external domains—α_1, α_2, and α_3—a transmembrane portion, and a cytoplasmic tail. It is associated on the cell surface with the molecule β_2-microglobulin (β_2M). The polymorphic region (i.e., the site of the molecule in which there are differences in amino acid sequence among unrelated individuals) is situated in the α_1, and α_2 domains, where there is a groove in which may be found peptides derived from the processing of self or foreign components. The class II molecules are made up of two polypeptide chains, α and β, which span the membrane; each has two external domains, α_1 and β_1 being polymorphic and accommodating antigen fragments.

MHC molecules bind peptides, which can be self-derived or foreign. For example, recent X-ray crystallographic analysis of class I molecules has revealed a peptide-binding cleft, usually occupied by electron-dense material, which is likely to be derived from self components.

T cells do not "see" naked antigenic determinants in isolation, but recognize relatively short peptide fragments which become wedged in the groove of the MHC molecules or restriction elements at the cell surface (Fig. 9). In fact, the strength of binding of a particular peptide to a given class II molecule correlates reasonably well with the ability of the complex to elicit a T-cell response. The class II genes have thus been referred to as immune responsiveness (Ir) genes; presumably, the class I genes act similarly. For T cells to respond to antigens, therefore, these must be processed (i.e., broken up into small peptides) and presented as peptides in association with MHC molecules. Such antigen processing is usually a function of the so-called antigen-presenting cells, among which are macrophages, a variety of dendritic cells and B cells themselves. In these cells foreign antigen must compete with self components for binding sites on class I or class II molecules if they are to be successfully displayed for T-cell recognition.

A dichotomy exists in the manner in which such antigens are processed intracellularly: Exogenous antigens taken up by cells are packaged within

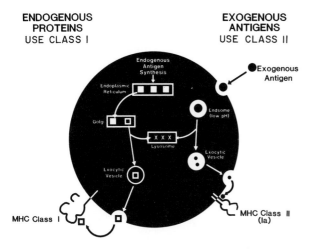

FIGURE 10 Pathways of antigen processing by antigen-presenting cells. Endogenous antigens (e.g., self components or viral peptides) are synthesized in the endoplasmic reticulum and transported to the Golgi apparatus. From there the peptides reach the cell surface together with major histocompatibility complex (MHC) class I molecules. Exogenous antigens enter the cell via endocytic vesicles, in which acidification and proteolysis occur. Resulting peptides become associated with class II molecules, which are either recycled from the surface in the same endosomes or newly synthesized and released from the Golgi apparatus.

acidic endocytic vesicles (endocytosed), degraded by the enzymes contained in lysosomes, and finally associated with class II molecules. On the other hand, peptides derived from endogenous protein synthesis (i.e., from self components or from virus expression within the cells) generally become associated with class I molecules (Fig. 10). This immediately raises a question as to how T cells distinguish peptides derived from self components (to which reaction must not occur) from peptides derived from foreign antigens, which must be eliminated. This issue is addressed in Section IV.

IV. B and T Cell Subsets and Cellular Interactions

B cells may be divided into subsets according to the class of Ig they bear on their membrane or produce after activation (e.g., IgM, IgD, IgG, IgA, and IgE). For example, immature B cells display only IgM, but mature B cells display both IgM and IgD. Each Ig class has distinct biological properties (e.g., IgA is secreted in mucosa, whereas IgE is involved in immediate hypersensitivities and allergy). B-cell subsets can also be defined by monoclonal antibod-

ies (i.e., antibodies in which all Ig molecules are identical) which can be directed specifically to the CD molecules. For example, the CD5$^+$ subset predominates over the CD5$^-$ subset in neonates, whereas it is a minor subset in the adult, in which it is found mostly in the peritoneal cavity.

Two major categories of T lymphocytes have been cloned and identified according to criteria based mainly on function, MHC restriction, and CD molecular markers. Helper T cells help B cells produce antibody molecules of certain classes and help other T cells respond. They may be involved in setting up inflammatory responses, as in the reactions called delayed-type hypersensitivity and in activating macrophages to kill more effectively any organisms they might have ingested. Cytotoxic, or killer, T cells kill foreign target cells after making intimate contact with them and are involved in the rejection of foreign tissue transplants. They may also play a role in eliminating incipient tumor cells. The existence of a third category of T cells, suppressor T cells, has been inferred from experiments using functional criteria, but has yet to be cloned.

T cells bearing the CD8 molecule are restricted by MHC class I molecules and generally function as cytotoxic cells. Those that bear the CD4 marker are class II restricted and usually function as helper cells. The CD8 molecule is a disulfide-bonded glycoprotein of 70 kDa and consists of two chains, α (34–38 kDa) and β (30 kDa). It is a member of the Ig supergene family, with one extracytoplasmic domain closely resembling an Ig light-chain V region. The CD4 molecule is a 52-kDa glycoprotein of 435 amino acids with little homology to CD8. Both CD4 and CD8 are believed to assist T-cell activation by stabilizing low-affinity interactions between the TCR on the T cell membrane and the MHC–peptide complex on the antigen-presenting cell (Fig. 11). [*See* CD8 AND CD4: STRUCTURE, FUNCTION, AND MOLECULAR BIOLOGY.]

Complex interactions involving cells, their surface components, and nonantigen-specific products called cytokines, or lymphokines, occur during immune responses. For example, the cytokine interleukin-1 (IL-1) is released from some antigen-presenting cells and causes T cells to expose on their surface a receptor able to bind a second lymphokine, IL-2. IL-2, which is produced by T cells themselves, can now bind to the IL-2 receptor and stimulate T-cell proliferation. When B cells present antigen in association with MHC class II molecules to T cells, a conjugate is formed and a variety of T-

Antigen-presenting cell

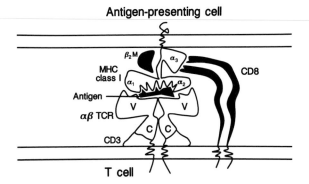

T cell

FIGURE 11 T-cell receptor (TCR)–antigen–major histocompatibility complex (MHC) interaction. Different shading indicate the different molecules. The TCR, consisting of α and β chains and the associated CD3 molecular complex, has a site in its V (variable) domains that binds antigen, occupying the groove formed by the α_1 and α_2 portions of the MHC class I molecule present on the surface of an antigen-presenting cell. The CD8 molecule has an affinity for some structure on the α_3 domain of the MHC molecule and enhances the binding of the T cell to the antigen presented by the antigen-presenting cell. Antigen presentation involving MHC class II molecules is similar to that shown here except that it involves the accessory T-cell molecule, CD4, not CD8. β_2M, β_2-Microglobulin; C, constant region.

cell lymphokines [e.g., γ-interferon (IFN-γ) and IL-4] are involved in activating the B cell to respond. The major effect of these lymphokines is to promote the proliferation of activated B cells and to implement the switch from an IgM-producing B cell to an IgG-secreting one. At least seven interleukins and other factors such (e.g., IFN-γ and tumor necrosis factors α and β) have been cloned by genetic engineering techniques. T lymphocytes produce some 10 distinct lymphokines after activation, including IL-2 through IL-6, IFN-γ, granulocyte–macrophage colony-stimulating factor, and lymphotoxin (Table II).

These glycoproteins play a major role in immunoregulation, having diverse, and at times opposing, effects on the growth, differentiation, and activation of hematopoietic and lymphoid cells. They are difficult to detect in serum, even during immune responses, probably because they are produced locally at the site of cell interactions and are rapidly consumed by their targets. [*See* CYTOKINES IN THE IMMUNE RESPONSE; INTERLEUKIN-2 AND THE IL-2 RECEPTOR.]

TABLE II Lymphokines and Their Activities

Lymphokine	Sources	Target cells	Activities
γ-Interferon	T cells	T cells, B cells, macrophages, monocytes, fibroblasts, neutrophils, endothelial cells, and natural killer cells	Enhancement of function and surface antigen expression
Interleukin-2	T cells	T cells, B cells, monocytes, and natural killer cells	Growth and differentiation
Interleukin-3	T cells	Multiple cells: macrophages, granulocytes, eosinophils, erythroid cells, megakaryocytes, mast cells, and B cells	Growth and differentiation
Granulocyte–macrophage colony-stimulating factor	T cells, macrophages, and endothelial fibroblasts	Granulocytes, monocytes, macrophages, eosinophils, and endothelium	Growth and differentiation
Lymphotoxin	T cells	Tumor cells	Cytotoxicity
Interleukin-4	T cells and bone marrow stromal cells	T cells, B cells, monocytes, and mast cells	Growth and differentiation
Interleukin-5	T cells	T cells, B cells, and eosinophils	Growth and differentiation
Interleukin-6	T cells, macrophages, and nonlymphoid cells	Lymphoid and nonlymphoid cells	Growth and differentiation
Interleukin-7	Bone marrow stromal cells	Pre-B cells and various thymus cells	Growth
Tumor necrosis factor α	T cells, B cells, and macrophages	T cells, B cells, monocytes, neutrophils, and fibroblasts	Growth and activation
Tumor necrosis factor β	T cells and macrophages	T cells, B cells, monocytes, neutrophils, and fibroblasts	Growth and activation

V. Immunological Self-Tolerance

To maintain health, the immune system must not only resist infections, but also ensure that it does not react against self components and destroy the body's own cells. Since individual clones of lymphocytes have predetermined reactivity, they should contain a self-reactive cell, with the potential to cause damage by reacting against its own body. Such cells must therefore be eliminated or functionally silenced. The process that ensures a lack of self-reactivity is referred to as immunological tolerance. It is now clear that this tolerance is mediated by more than one mechanism. In the case of T cells, there is convincing evidence for clonal elimination of self-reactive cells within the thymus. As the stem cell differentiates and expresses its rearranged TCR genes, the newly arising T cell is screened for self-reactivity by what might be a variety of intricate mechanisms. Clones of T cells that recognize self components are deleted (i.e., clonal elimination). Other T cells are allowed to mature and emigrate.

Not all autoantigens, however, are represented within the thymus. Hence, other mechanisms must exist to purge or silence T cells that would react to self components not intrinsic to the thymus. Among these should be mentioned suppression by another class of T cells, the so-called suppressor T cells, which have not yet been cloned or fully characterized. In the case of B cells, the term "clonal anergy" has been coined to designate those cells which, in experimental situations, are physically present, but impaired in their capacity to respond to autoantigens by a mechanism that has not yet been clarified.

VI. Diseases of Immunological Aberrations

Nonreactivity to self is not absolute, particularly in the B cell compartment, since levels of IgM autoantibodies (antibodies directed against self components) are sometimes found in healthy individuals. Under conditions in which tolerance breaks down completely or fails to be implemented, however, a variety of autoimmune disease states result, including rheumatoid arthritis, insulin-dependent diabetes mellitus, and systemic lupus erythematosus. [See Autoimmune Disease.]

Congenital immunodeficiency states could result from genetic defects which interfere with some of the steps leading to normal development of the immune system. Defects may thus occur at the stage of stem cell differentiation or T- or B-cell development. Among the diseases which have been characterized are severe combined immunodeficiency, in which both T- and B-cell functions are impaired; hypogammaglobulinemia, resulting from the failure of B cells to produce Ig molecules; and various forms of T-cell defects. Acquired defects also occur, notably the acquired immunodeficiency syndrome (AIDS), in which the AIDS virus infects and kills $CD4^+$ T cells, thus leaving their host unable to combat a variety of infections.

VII. Conclusions

Immune responses are the result of an extremely complex series of cellular and subcellular events and interactions between cells. These include antigen-presenting cells, T and B lymphocytes, monocytes, macrophages, and other mononuclear and polymorphonuclear cells, as well as their products (e.g., antibodies and lymphokines, or cytokines). The functions of this cellular network are to maintain the integrity of the body and to resist infections. The capacity to respond to foreign intrusion, however, allows the system to reject tissue transplants. Reactions must be regulated at various levels, and responsiveness is dependent on gene constitution. Inadequate regulation or failure of the induction of tolerance to self components results in various forms of immunological aberrations, as seen in autoimmune diseases. Failure of the system to develop at one stage or another is associated with immunodeficiency diseases, as is infection by viruses such as AIDS, which specifically invade and kill an important subset of T lymphocytes.

Bibliography

Benacerraf, B., and Unanue, E. R. (1984). "Textbook of Immunology," 2nd. ed. Williams & Wilkins, Baltimore, Maryland.
Ford, W. L. (1980). The lymphocyte: Its transformation from a frustrating enigma to a model of cellular function. *In* "The Blood, Pure and Eloquent: A Story of Discovery, of People, and of Ideas" (M. M. Wintrobe, ed.). McGraw-Hill, New York. 1980.
Paul, W. E. (ed.) (1988). "Fundamental Immunology," 2nd ed. Raven, New York.
Roitt, I. M. (1984). "Essential Immunology," 6th ed. Blackwell, Oxford, England.

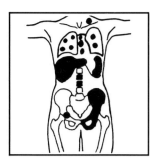

Lymphoma

HOWARD R. BIERMAN, *Institute for Cancer and Blood Research*

Glossary

ALL Acute lymphocytic leukemia
Antigen Any substance capable of initiating an immune response eliciting the formation of antibodies and immune cells
CLL Chronic lymphocytic leukemia
CMV Cytomegalovirus
CT Computerized tomography
Deletion Loss of a part of a chromosome or gene
Differentiation Change in gene expression or depression; appears to be necessary for diversification along a cell line
DLCL Diffuse large cleaved cell lymphoma
DNA Deoxyribonucleic acid
DTIC Dimethyl-triazeno-imidazole-carboxamide
EBV Epstein-Barr virus
Epitope Unique antigenic structure
Exons DNA that specifies the sequence of the encoded mRNA
HD Hodgkin disease
Histiocytes Tissue macrophages derived from monocytes, inhabiting the spleen, lymph nodes, liver, pleural and peritoneal spaces, and the pulmonary alveoli

Introns DNA that interrupts the coding sequence of the gene
Karyotype Chromosome arrangement based on banding by special stains (e.g., fluorescent and enzyme digestive color banding.)
Lymphoblastic Immature large lymphocyte; large nucleus with coarse chromatin and dense nuclear membrane; scanty cytoplasm
Lymphokines Biologically active mediators of cellular immunity capable of eliciting an amplified inflammatory reaction (e.g., macrophage, chemotactic, mitogenic factors, interferon, and transfer factor)
Macrophage Large marrow-derived phagocytic cells that function as accessory cells in induction of immune response and as effector cells in inflammatory responses
Maturation Quantitative change in a cell line
MHC Major histocompatibility complex; set of genes that code for histocompatibility and related markers
MRI Magnetic resonance imaging
NHL Non-Hodgkin lymphoma
Pruritis Itching of skin
TdT Terminal deoxynucleotidyltransferase
Undifferentiation Not yet committed to erythrocyte and leukocyte progenitors
Waldeyer ring Lymphoid tissue of the oropharynx: palatine tonsils, pharyngeal tonsils, adenoids, base of tongue, and posterior palatine glands

LYMPHOMAS are monoclonal lymphoproliferative neoplasms comprised of abnormal proliferation of either B or T lymphocytes and histocytes of the immune system. Lymphomas and the leukemias are closely related (Fig. 1).

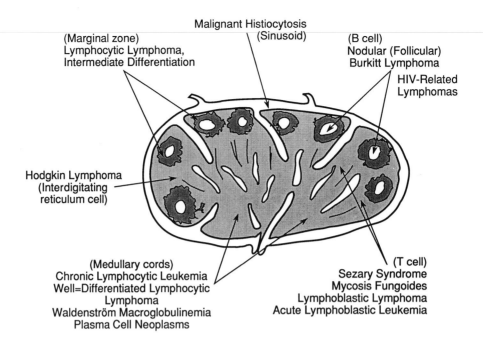

(Marginal zone)
Lymphocytic Lymphoma,
Intermediate Differentiation

Malignant Histiocytosis
(Sinusoid)

(B cell)
Nodular (Follicular)
Burkitt Lymphoma

HIV-Related
Lymphomas

Hodgkin Lymphoma
(Interdigitating
reticulum cell)

(Medullary cords)
Chronic Lymphocytic Leukemia
Well=Differentiated Lymphocytic
Lymphoma
Waldenström Macroglobulinemia
Plasma Cell Neoplasms

(T cell)
Sezary Syndrome
Mycosis Fungoides
Lymphoblastic Lymphoma
Acute Lymphoblastic Leukemia

I. Incidence

The annual incidence of malignant lymphomas in the United States is about 33,000 cases, of which Hodgkin disease accounts for 25%. Hodgkin disease and non-Hodgkin lymphomas (NHL) occur more commonly in white populations than nonwhite. NHLs exhibit an age-specific incidence, which is roughly linear from childhood to age 75. The incidence curve of Hodgkin lymphoma is bimodal, with distinct peaks at the age range of 15 to 21 and 40 to 45 and thereafter parallels other lymphomas.

II. Classification

A. Hodgkin Lymphoma

Hodgkin disease is divided into five histologic types based on the relative proportion of lymphocytes, fibrosis, and Reed-Sternberg (RS) cells to neutrophils, eosinophils, and plasma cells (Fig. 2A). The more mature lymphocytic population in types 1 and 2 provides a better prognosis.

1. Lymphocyte predominance—effacement of normal node architecture; lymphocytic capsular infiltration
2. Nodular sclerosis—fibrosis with or without prominent internodule features
3. Mixed cellularity—small and large poorly differ-

FIGURE 1 Schematic model of normal lymph node. Lymphomas are related to the dominant lymphocyte compartment in the node. [Modified from: Mann, R. B., Jaffe, E. S., and Berard, C. W. (1979). *Am. J. Pathol.* **94,** 103–192 and Berard, C. W. (1981). *Ann. Intern. Med.* **94,** 218–235.]

entiated lymphocytes along with eosinophils, neutrophils, and plasma cells
4. Lymphocyte-depleted—highly pleomorphic, poorly differentiated lymphocytes
5. Unclassified—ill-defined criteria

B. Non-Hodgkin Lymphoma

The cell types of NHL have prognostic significance. The more mature lymphocytes have the better prognosis (see Table I). Clinical prognosis based on proliferative activity can be divided into three groups: low grade (favorable), intermediate, and high grade (unfavorable (Fig. 2B).

High-grade undifferentiated lymphomas include Burkitt and non-Burkitt types and histiocytic or large lymphoid cell tumors. Cutaneous T-cell lymphomas, include mycosis fungoides and Sezary syndrome, predominantly T-helper cell phenotype CD4(T4), and a few are CD8(T8) (suppressor).

All pediatric NHLs are diffuse, grouped histologically into one of three major categories:

1. Lymphoblastic lymphomas—T cell. Localized above the diaphragm with or without involvement outside the chest. Derived from early developing

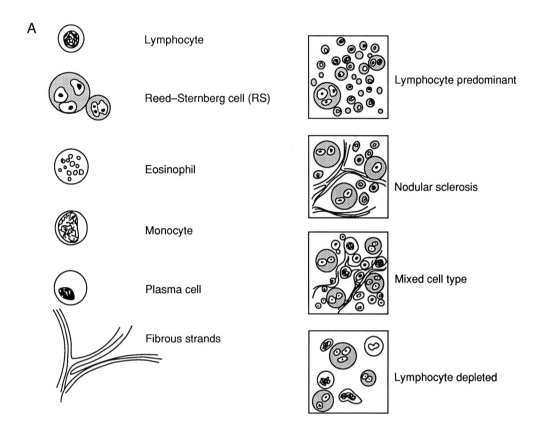

A

Lymphocyte

Reed–Sternberg cell (RS)

Eosinophil

Monocyte

Plasma cell

Fibrous strands

Lymphocyte predominant

Nodular sclerosis

Mixed cell type

Lymphocyte depleted

FIGURE 2A Hodgkin disease is divided into five histologic types based on the relative proportion of lymphocytes, fibrosis and Reed-Sternberg cells in relation to neutrophils, eosinopils and plasma cells. The more mature lymphocytic population in types 1 and 2 provides a better prognosis. (1.) Lymphocyte predominance—Effacement of normal node architecture; lymphocytic capsular infiltration. (2.) Nodular sclerosis—Fibrosis with or without prominent internodule features. (3.) Mixed cellularity—Small and large poorly differentiated lymphocytes along with eosinophils, neutrophils, and plasma cells. (4.) Lymphocyte depleted—Highly pleomorphic, poorly differentiated lymphocytes. (5.) Unclassified—Ill-defined criteria.

thymocytes and contain terminal deoxynucleotidyl transferase (TdT), an enzyme unique to the thymus and representing high proliferative capability
2. Nonlymphoblastic lymphomas including undifferentiated Burkitt and non-Burkitt pleomorphic B cell, arising within the abdominal cavity, gastrointestinal tract, or nasopharynx (Waldeyer ring)
3. Histiocytic or large lymphoid cell—B cell

Histopathologic description of undifferentiated lymphomas in children disclose that the majority are derived from B cells and can be shown to have immunoglobulins (Ig) on the cell surface. Most histiocytic or large lymphoid cell tumors express B-cell characteristics. 15 percent of the diffuse, poorly

differentiated large cell lymphomas are of T-cell origin.

III. Normal Immunopoiesis

The normal immune system primarily consists of humoral and cellular components. Humoral immunity involves soluble molecules, Ig. A molecule of an antibody specifically binds to a limited region of the antigen (epitope). Cellular immunity refers to intact cells capable of recognizing antigens through molecules held tightly to their surface. Although they act like Ig, they never break their association with the cytoplasmic membrane.

Clonal selection is the basis for the functioning of the immune system. Human and mammalian immunopoietic systems produce B lymphocytes studded with Ig molecules projecting from their surface. For any given cell, all these molecules have the same antigen-binding specificity. For an organism not exposed to antigen, any specificity is present in only a small subset of the entire pool of B cells. Binding of the antigen to the antibody on the surface of the B cell activates it and initiates its multiplication. The cell will generate a population of cells

B $\longrightarrow$ Proliferation & Differentiation $\longrightarrow$

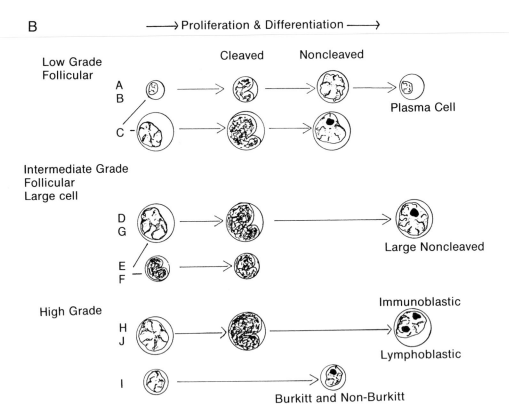

FIGURE 2B Schematic morphology related to histopathologic classification (Table I). A,B and E,F represent small lymphocytes. (Adapted from: Sell, S. (1987). "Immunology, Immunopathology and Immunity."

(clone) all synthesizing that specific antibody. The antigen, therefore, has selected that subset of B lymphocytes to produce the entire pool of production of that antibody, consequently, the term *clonal selection*.

B lymphocytes arising from the marrow, without any exposure to antigen, will have a short lifespan, approximately from 5 to 20 days. If during that time an antigen binds to the surface antibody, the lifespan of some cells emerging from the clone is markedly extended. These long-lived lymphocytes are called memory B cells. The mature progeny of activated B cells is the plasma cell that secretes Ig. [*See* LYMPHOCYTES.]

The large number of different Ig molecules that are present before encountering an antigen is generated at two distinct levels: diversification of DNA sequence in the germ-line DNA, and somatic events. The somatic events are extensive gene rearrangements and somatic mutations. Both multiple germ-line genes and somatic variations are involved in the heterogeneity of the immune response, including neoplasia and in particular the leukemias and the lymphomas.

The initial antibody response to an antigen is IgM. When the organism encounters the same anti-

gen, subsequently, there is a greater response of IgG (secondary response). Circulating memory cells, with surface Ig, recognize the reinvading antigens and engender a more rapid, hugely effective response.

The immune system also contains T lymphocytes that also carry surface antigen-binding molecules. These molecules (receptors) act with ligands on the surface of other cells, a cell-to-cell interaction. Two different types of T lymphocytes exist: CD4 helper, and CD8 suppressor/cytotoxic. Cytotoxic T lymphocytes respond to the foreign antigens on their surfaces and kill cells (killer T cell). The helper T lymphocytes (T_h) are required to allow the B-cell response to the antigen. The third type, the suppressor T cell (T_s) inhibits B-cell activity and dampens the response to antigens. Three subsets of T lymphocytes are distinguished for the surface proteins they contain. [*See* LYMPHOCYTE-MEDIATED CYTOTOXICITY.]

The primary organs of the immune system are the

TABLE I Non-Hodgkin Lymphoma(NHL)—Working Formulation

Immunologic phenotype	Histologic type	Cell markers
B	I. Low Grade	
B	A. Small lymphocytic	CD20(B1), T-helper cells: CD5(T1), CD3(T3), CD4(T4), (T10). Variable CD25(Il2-receptor), T9, Ia, T12
B	B. Follicular, predominantly small, cleaved cell	CD20, kappa/lambda, CD10(CALLA).
B	C. Follicular mixed, small and large cell	
	II. Intermediate Grade	
B	D. Follicular, predominantly large cell	CD20, kappa/lambda, CD10(CALLA).
B or T	E. Diffuse, small cleaved cell	
B or T	F. Diffuse mixed, small and large cell	
B or T	G. Diffuse large cell, cleaved or non-cleaved	T-cell non-Hodgkin lymphoma, diffuse large cell lymphoma. (DLCLs) diffuse poorly differentiated lymphocytic lymphomas 15% T-cell origin of DLCLs. CD7, CD5(T1), CD4(T4), CD3(T3), Ia, and/or T11. CD20(B1), CD21(B2), CD19(B4), CD24(Ba-1), weak sig, CD5(T1).
	III. High Grade	
B or T	H. Diffuse large cell immunoblastic	
B	I. Small noncleaved cell and undifferentiated Burkitt and non-Burkitt	Ia(HLA-DR), CD7(A-1); CD10(CALLA); CD20(B1), CD19(B4), sig.
T	J. Lymphoblastic: convoluted or nonconvoluted	CD4(T4), T6, T8, T10, CD3(T3), CD5(T1), T9 (non-convoluted) and CD7. Overlap with T-ALL. 40% CD10(CALLA) (common acute lymphoblastic leukemia antigen). Commonly in children and young adults: CD5(T1) antibody, T6, T9, T10, CD7, Ia. T-ALL and T-CLL overlap.
	IV. Miscellaneous	
B or T	Composite	
T	Mycosis fungoides—Sezary	Cutaneous T-cell lymphomas (CTCL's) with mycosis fungoides and Sezary syndrome. T-helper cell phenotype: CD5(T1), CD3(T3), CD4(T4) and T11.
B	Plasmacytoma	
B	Multiple Myeloma	
B	Macroglobulinemia—Waldenström	
B	Heavy chain disease	
B or T	Unclassifiable	
Monocyte/ Macrophage	Other Histiocytic	Ia(HLA-DR), CD21(B2), CD19(B4), CD24(Ba-1); lack CD5(T1), sig, AAT, lysozyme Kp1.

bone marrow and the thymus; secondary organs are peripheral lymphoid tissue, lymph nodes, and the spleen. There are approximately 10,000 lymph nodes in humans. The spleen makes up 25% of the peripheral lymphoid tissue. Two fluid systems include the blood and the lymph, which are intimately related. Lymph is a special filtrate from the blood and occupies extracellular spaces within and around tissues. Lymph flows at low pressure through thin-wall lymphatic vessels, eventually re-entering the venous circulation. Accounting for 20 to 30% of the nucleated cells in the blood, lymph contains 99% lymphocytes.

Foreign antigens, microbial, viral, chemical, and other harmful agents in the blood and lymph are filtered out by lymph nodes. The spleen is an additional safeguard immunologic sieve. In the lymph nodes or spleen, macrophages engulf the antigens, whereas the antibody-secreting B lymphocytes or mature T lymphocytes attack the antigen in various forms. [*See* IMMUNE SURVEILLANCE.]

IV. Pathobiology of Lymphomas

Lymphomas are a disorder of immune gene regulation and are associated with karyotypic abnormalities (Fig. 2B).

Protooncogenes regulate and control normal cell growth. Alteration of DNA (mutation) of the protooncogenes with decreased expression causes the formation of an oncogene with malignant potential. Abnormal lymphoid cell growth is a progressive accumulation of gene products, giving growth advantage to lymphocytes and leading to the ultimate production of lymphomas. The aggressiveness of the resultant neoplasm depends on many factors (e.g., chromosomal alteration, amplification of DNA segment, oncogene number and/or activity). DNA amplification is clinically important and is measured by flow cytometry. [*See* ONCOGENE AMPLIFICATION IN HUMAN CANCER.]

The lymphomas share the characteristics of all malignancies. The neoplastic cells arise from a single cell (monoclonality). Proliferation progress through several stages to unrestrained growth rates. In the lymphomas, the most malignant clone will replace all others and progressively reduce normal function.

Lymphomas are derived from the B and T lymphocytes in the lymphatic system—the thymus, lymph nodes, gastrointestinal tract, and other peripheral lymphoid tissues. Lymphoma involves the bone marrow and may become indistinguishable from a lymphocytic leukemia and often is recognized in this combined or composite state (Table II).

A. Etiology

Close relatives of patients with lymphoma have an approximately three-fold higher risk than normal to develop Hodgkin disease. The major histocompatibility (H-2) gene locus in mice increases the susceptibility to the development of lymphomas as well as leukemias. This has not been demonstrated as yet in humans. [*See* LEUKEMIA.]

Radiation exposure to large body surface areas often cause lymphomas (e.g., atomic bomb explosions at Hiroshima, Nagasaki, and Chernobyl). Radiation to the thymic and thyroid region induces lymphomas and leukemia in both susceptible mice and humans. Radiogenic lymphomas may occur with combinations of antineoplastic chemotherapy.

An extensive literature of animal and human investigations in lymphomas has searched for possible etiologic viruses, bacteria, chemical agents, and other substances. Epstein-Barr virus (EBV) infection can coexist with mediastinal Hodgkin lymphoma, but the viral role in the tumor is unknown. The African variety of Burkitt lymphoma is universally associated with EBV; the American type is only sporadically involved. Patients with advanced Hodgkin disease exhibit severe impairment of cell-mediated immunity.

Human immunodeficiency virus (HIV) is cell-specific for the T_4 lymphocyte. 40 percent of patients who develop acquired immune deficiency syndrome (AIDS) will subsequently develop Kaposi sarcoma and/or lymphoma. [*See* ACQUIRED IMMUNODEFICIENCY SYNDROME (VIROLOGY).]

1. Chemical Agents

Many chemicals can induce lymphomas, leukemias, and other neoplasms. Other chemicals can

TABLE II Presenting as Composite Leukemia and Lymphoma

Lymphoma	Leukemia
Small lymphocytic (B cell)	Chronic lymphocytic leukemia
Plasmacytoid lymphocytic (Waldenström macroglobulinemia)	Small lymphocytic, plasmacytoid
Small cleaved cell, follicular	Lymphosarcoma cell leukemia (B-cell); leukosarcoma[a]
Myeloma	Plasma cell leukemia
Small lymphocytic (T-cell)	T-cell leukemia
Small non-cleaved cell diffuse (Burkitt or Burkitt-like) (B-cell)	Acute B-cell leukemia (L3)
Lymphoblastic lymphoma	Acute T-cell lymphoblastic leukemia
Cutaneous T-cell lymphoma	Sezary cell leukemia
Adult T-cell lymphoma	Adult T-cell leukemia

[a] Abnormal lymphoma cells circulating in the blood were previously designated as leukosarcoma.

induce lymphomas in rodents but are inconclusive in humans. Duration and amount of exposure are essential in relating oncogenes to chemicals. Immunosuppressive agents employed therapeutically in humans in transplantation (cyclosporine, azathioprine) and in antineoplastic therapy (alkylating agents and myelosuppressives) can induce lymphomas and leukemias. [*See* CARCINOGENIC CHEMICALS.]

Hydantoin, an anticonvulsive drug, may cause a hypersensitivity reaction with lymphadenopathy. The histopathology is often indistinguishable from Hodgkin lymphoma. The etiology of this pseudolymphoma is uncertain.

2. Chromosomal Abnormalities

Chromosomal translocations alter the quality or quantity of proteins specified by the normal cellular genes involved, in some cases resulting in a lymphoma or other diseases. Genetic rearrangement is intimately involved with abnormal cellular proliferation. In T-cell lymphomas, gene rearrangements involve T-cell receptor (alpha and beta genes), which probably occur at chromosomal breakpoints to produce the abnormal rearrangement of genes. Changes in the structure or copy number of cellular protooncogene products eventuating in karyotypic abnormalities lead to aberrations of cell growth. The molecular anatomy of translocation, deletion, chromosomal fragility, fragmentation, and instability profoundly involve the biology of the lymphomas. [*See* CHROMOSOMAL ANOMALIES.]

DNA damage may cause mutations and/or chromosome alterations. When breakage of the DNA strand involves a substitution of a single nucleotide or deletion of the damaged segment, genetic instability may invite further DNA damage. In some lymphomas, the breakpoints may be highly heterogeneous. These mutations may occur at the premalignant stage and progress to become clinically detected lymphomas. DNA damage may be repaired, preserving an architecturally normal chromosome.

In lymphomas, chromosomal translocations usually occur close to a transcriptionally active Ig gene and a cellular protooncogene. Chromosomes that contain rearrangements do not express Ig genes. About 80% of follicular, small cleaved cell lymphomas is associated with a t(14;18) translocation, involving an Ig heavy chain locus on chromosome 14 and the cellular oncogene bcl-2 located on chromosome 18. This translocation results in deregulation of the bcl-2 oncogene. This plays an important role

in the pathogenesis of follicular cleaved cell lymphoma and perhaps also in Hodgkins disease.

In Burkitt lymphoma, the characteristic t(8;14) translocation results in the juxtaposition of the c-myc oncogene on chromosome 8 and the Ig heavy chain locus on chromosome 14; in some cases it is translocated to the kappa or lambda light chain loci on chromosomes 2 or 22, t(2;8) or t(8;22). The translocation usually is juxtaposed with the 5' part of the c-myc locus with a 5' segment of the heavy chain gene. The c-myc protein is critical for cell proliferation. Expression of this gene prevents cellular differentiation.

C-myc and its relatives are also primarily amplified in AIDS-related lymphomas (Burkitt-like) and when involved in small-cell carcinoma, or neuroblastoma (N-myc) and other solid tumors. Consequently, translocations specific to Burkitt lymphoma are karyotypic markers of chromosome rearrangements leading to the specific oncogene location close to a gene that is both transcriptionally active and rearranged in the normal cell from which the tumor is derived.

Translocation of the long arm of chromosome 14(14q+) is frequently seen in ataxia-telangiectasia. An X-linked recessive syndrome is associated with a variety of sequelae including fatal infection with infectious mononucleosis or lymphoid malignancy, particularly Burkitt lymphoma.

In chronic lymphocytic leukemia and multiple myeloma, there is a t(11;14) translocation to or nearby the Ig heavy chain locus on chromosome 14 or to the bcl-1 locus on chromosome 11.

3. Dysmorphology

Six well-described histopathologic classification systems are available (Table I). The National Cancer Institute established an eclectic "Working Formulation" to guide the clinician in selecting the most reliable and effective treatment program. Selected markers as surface molecules on lymphoma cells assist in evaluating their biologic and clinical behavior, the diagnosis, and the prognosis. These cell markers define differentiation and function of the lymphomas and relate them to the unusual clinical presentations. Monoclonal antibodies can define and classify these unique cell surface antigens. Although there are exceptions, in general, therapy and prognosis are more effective and favorable with the less aggressive lymphomas.

Aggressiveness of the lymphomas are also revealed morphologically (Fig. 2). Normal small lymphocytes become small cleaved and/or large cleaved lymphocytes. These are usually NHLs related to slowly growing leukemias (CLL). The cells, either small or large, may become large noncleaved lymphocytes and develop into immunoblastic lymphoma. All the noncleaved lymphocytes are Burkitt-variant lymphoma. The largest noncleaved lymphoma evolves into a much more aggressive lymphoma.

4. Immunology

Immunoglobulin genes are discontinuous segments of DNA (germ-line configuration) that must be assembled properly to synthesize intact Ig. An Ig molecule is a specific product of lymphoid cell differentiation consisting of two polypeptide (heavy) chains and two polypeptide (light) chains linked by disulfide bridges and hydrogen bonding. The heavy chain gene is located on chromosome 14; light chain genes are on chromosomes 2 (kappa) and 22 (lambda). Chromosome 14 is commonly involved in neoplasms of B-cell origin.

Immature lymphocytes fail to express surface Ig. Consequently, they do not synthesize light chains and cannot adequately respond to antigens. The result is a deficiency or absence of antibodies.

5. Lymphoma-Associated Diseases

Some patients with specific conditions develop a predisposition to develop a lymphoma.

1. Immune deficiency associated with an increased risk of developing NHL, Kaposi sarcoma, and leukemias:
   a. Inherited immunologic deficiencies in children
   b. AIDS, HIV infection
   c. Iatrogenic induced immunosuppression caused by organ transplantation, therapeutic radiation, and antineoplastic agents
   d. Chronic stimulation may cause lymphoid neoplasia, lymphomas of the small intestine or thyroid, angioblastic lymphadenopathy, and immunoblastic sarcoma
2. The following is a list of diseases with a predisposition to develop lymphomas:
   Ataxia-telangiectasia
   Wiskott-Aldrich syndrome
   Swiss-type agammaglobulinemia
   Common variable immunodeficiency disease
   Acquired hypogammaglobulinemia
   Renal transplant recipients
   Sjögren's syndrome
   Rheumatoid arthritis
   Systemic lupus erythematosus
   Klinefelter's syndrome
   Chédiak-Higashi syndrome
   Selective IgM deficiency

V. Clinical Description and Natural History of Lymphomas

Lymphomas and the lymphoproliferative leukemias are closely interrelated (Table II). The lymphomas frequently present with lymphadenopathy, weakness, and weight loss. Lymphomas are generally more extensive than the findings indicate, so physicians must anticipate other preclinical possibilities. Knowledge of contiguous pathways of extension of lymphomas is of great value in selecting the proper therapeutic fields for radiation and for evaluating therapy.

The spread of lymphomas depends largely on the following features: Abnormally increased tumor vascularity is due to enlarged arterial, venous, and lymphatic channels. Abnormal endothelail junctions increase vascular permeability and, with decreased tumor cell cohesiveness, encourage dissemination. Necrosis results from the tumor exceeding its vascularity. The larger the tumor, the greater the risk of intratumor hemorrhage, encouraging the intravascular release of tumor cells.

A. Hodgkin Lymphoma

Hodgkin disease starts in a single group of lymph nodes and spreads predictably and progressively by lymphatic extension to the next contiguous lymph node area (Fig. 3). The majority of cases arise unifocally and remain confined within the lymphatic system for a prolonged period. The primary lesion most often involves cervical, supraclavicular or mediastinal nodes—50–60%, less often abdominal periaortic nodes—25–40% (iliac, celiac-portal, and splenic). Mesenteric nodes in Hodgkin lymphoma are uncommon—about 2%. Extranodal involvement, less than 8%, is primarily in the lungs, liver, and gastrointestinal tract and is evidence of far advanced disease.

Vascular invasion is ominous because it exceeds the contiguity of the lymphatic system. Despite dissemination, usually by hematogenous spread, the primary pattern persists. Nodes in the cervical area involve the mediastinal area and then may involve peripheral node areas, on the same side of the diaphragm (Fig. 3). Subsequently both sides of the dia-

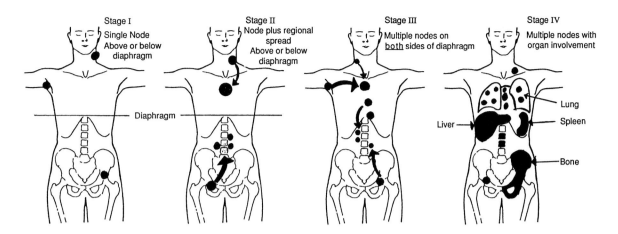

FIGURE 3 Staging of Hodgkin disease by the extent of the lymphoma as it extends from its unifocal node onset to a multinodal and extranodal condition.

phragm become involved. The lymphomatous lymphocytes escape into many extranodal locations (e.g., liver, spleen, stomach, or small intestine, and less frequently the lung, testes, thyroid, brain, and bone marrow).

There is a close interrelation between staging and histopathology (Tables I and III). The histologic diagnosis of Hodgkin disease demands the presence of RS cells. The RS cell is large, with abundant acidophilic cytoplasm. It usually contains two or more nuclei or a lobated nucleus with prominent round nucleoli surrounded by a halo. There are three variants. The major histologic subtypes primarily differ in the number of RS cells and the amount of fibrosis. However, the RS cell is not pathognomonic of the disease; it has also been described in other malignancies and infectious mononucleosis. The diagnosis requires meticulous histopathologic studies of all involved tissues.

Serum copper, haptoglobin, and erythrocyte sedimentation are often elevated. Most patients have defective cell-mediated immunity, resulting in frequent infections with tuberculosis, viruses, and intracellular parasites such as *Pneumocystis carinii*. Splenectomy after staging and/or radiation or chemical therapy will further suppress the serum Ig levels, increasing the risk of pneumococcal and *Haemophilus influenzae* septicemia and pneumonia.

B. Non-Hodgkin Lymphoma

The predictability of spread is uncertain in NHL. Although the disease originates in single nodes, it is usually multifocal and not restricted to contiguous node extension. Hematogenous dissemination to many organs is commonly found. At diagnosis, less than 10% of NHLs is localized as compared with 50% of Hodgkin lymphoma.

The spread of Hodgkin lymphoma is centripetal (from external lymph nodes toward the lymph nodes surrounding the vertebrae). NHL, however, is centrifugal and rarely localized. It more commonly involves the gastrointestinal tract, marrow, liver, and the testes. Abdominal lymph node involvement is frequent.

Major classifications of NHLs are based on morphologic, immunologic, functional, and clinical heterogeneity (Table I). The monoclonality of B cells and their nodal localization in the follicles permits rough prognostic estimates of the clinical course

TABLE III Hodgkin and Non-Hodgkin Lymphoma—The Ann Arbor Classification

Stage I:	Involvement of a single lymph node region or localized involvement of a single extralymphatic organ or site.
Stage II:	Involvement of two or more lymph node regions on the same side of the diaphragm, or localized involvement of a single associated extralymphatic organ or site and its regional lymph nodes with or without involvement of other lymph node regions on the same side of the diaphragm.
Stage III:	Involvement of lymph node regions on both sides of the diaphragm, which may also be associated with an extralymphatic organ or site by involvement of the spleen or both.
Stage IV:	Disseminated involvement of one or more extralymphatic organs, with or without associated lymph node involvement, or isolated extralymphatic organ involvement with distant nodal involvement.

(Fig. 1). Low-, intermediate-, and high-grade types are defined by their clinical behavior.

Lymphomas with a nodular or follicular architecture have a favorable (low-grade) clinical course. The nonaggressive, well-differentiated mature lymphocyte of many types of NHL are characterized by a progressive accumulation of immunologically incompetent lymphocytes that infiltrate the marrow and peripheral lymphoid tissues. Lymphomas with diffuse patterns often have a more rapid, aggressive course (intermediate- and high-grade). Small cell tumors have a more favorable prognosis; large cell tumors are more aggressive. Involvement of the bone marrow and liver is not as ominous as it is in Hodgkin lymphoma. Marrow involvement is often associated with the central nervous system. Patients with follicular small-cell lymphoma have marrow involvement in 40–60% at diagnosis. In diffuse large-cell lymphoma, marrow involvement occurs in 15–20%. Survival in these patients is poor.

Lymphoblastic lymphoma is most commonly seen in children and commonly young men with mediastinal and marrow involvement. The majority are of T-cell lineage and are aggressive.

C. Burkitt Lymphoma

Burkitt lymphoma is a rapidly progressive neoplasm, similar to L3 acute lymphocytic leukemia with the characteristic t(8;14) translocation. This lymphoma was first described by Burkitt in 1958 as a jaw sarcoma of East African children. It also occurs endemically in New Guinea and sporadically in North and South America and Europe. It is strongly suspected to be of viral etiology, EBV, involving an insect vector, the mosquito associated with the c-myc oncogene.

Burkitt lymphoma is high-grade with distinctive clinical-pathologic features. This tumor is composed of intermediate and blastic lymphocytes regardless of the characteristic convoluted nuclei. Chronic immunosuppressive therapy, for instance after kidney transplantation, increases the risk of central nervous system involvement.

The AIDS-related lymphomas are Burkitt-like. They are highly aggressive with extranodal presentation in the central nervous system, marrow, and perivascular and mucocutaneous junctions frequently obscured by infections. Primary central nervous system lymphomas are the most common (10%) noninfectious space-occupying lesions in pa-

tients with AIDS. It is the fourth leading cause of death in these patients.

Spontaneous regression will occur in 20–30% of patients. Transformation from small to a large cell type will occur in approximately 5%, indicating increased aggressiveness. Immunophenotyping is important in assessing prognosis and predicting responses to therapy. A study of DNA amplification may define the potential growth activity of the lymphoma, the detection of chemotherapeutic resistant cells, and a clear biomedical threat of recurrence. These data improve the selection of the most appropriate therapy.

VI. Heterogeneity

Heterogeneity in the behavior of the leukemias and the lymphomas is due to DNA alterations in the primary and metastatic cell population. The variability in clinical lymphoma presentation, cell cycle, growth rate, and survival is created by the genetic chimerism (karyotypic aneuploidy). In slow-growing lymphomas, only 1–2% of the cells are dividing. In the activity growing type, 30% of the cells are dividing.

Genetic instability is expressed as subclones with properties differing from the initial malignant clone, leading progressively to less differentiated forms and aggressive lymphomas.

A. Disorders of Secretory B Cells

1. Myeloma
Myeloma is a clone of lymphoid cells arrested or frozen at the terminal phase of B-cell maturation, which often fails to express adequate humoral immunity. This impaired immunity is due to a decrease of effective functional antibody production either from overproduced Ig or particularly with deficient or inadequate production, hypogammaglobulinemia.

2. Waldenström Macroglobulinemia
Macroglobulinemia was first described by Waldenström in 1944. Waldenström macroglobulinemia is a variant form of myeloma with slightly different morphological and clinical presentation. In this disorder, the neoplastic plasmacytoid lymphocytes infiltrate the marrow, liver, spleen, and lymph nodes, featuring an overproduction of IgM.

IgM macromolecules are normally composed of five identical monomers, each consisting of two μ (heavy) chains and two light chains. Macroglobulins are predominantly restricted to the intravascular pool. The increased production of this large monoclonal Ig results in hyperviscosity of the blood and the symptoms of this disorder.

3. Heavy Chain Disease

Heavy chain disease is a rare condition characterized by excess abnormal free heavy chains of an Ig in association with malignant lymphoma or myeloma. The urine contains gamma chains in malignant lymphoma and alpha chains in patients of abdominal lymphoma with diffuse lymphoplasmacytic infiltration of the small intestine. A deletion in the variable domain of the Ig prohibits light chain production. No myeloma paraprotein is detectable in 1% of patients despite the marrow presence of myeloma-characteristic plasma cells containing monoclonal Ig.

VII. Staging—Defining the Extent of Lymphoma

The clinical course depends on histopathology (Table I) and the extent of the lymphoma when first diagnosed. The prognosis worsens as the stage of lymphoma increases. Staging determines the optimal therapy (Table III).

Patients without symptoms of lymphoma are considered substage A (e.g., IIA). Symptoms of fever, night sweats, and greater than 10% body weight loss are substage B (e.g., IIIB). Lymphokines produce these symptoms. Noncleaved and more aggressive types of lymphocytes secrete lymphokines.

A. Recommended Staging Procedure

Staging must be preceded by complete history and physical examination, including all node-bearing areas and the size of liver and spleen. Duration of the presence or absence of systemic symptoms must be documented.

Chest film and bilateral iliac crest bone marrow biopsies are essential for defining pulmonary and hematopoietic involvement. The collective adult marrow is approximately 1,400 g, equivalent in size to the human liver. The largest available site for bone marrow biopsy and aspiration is the pelvis.

Bipedal lymphangiography and/or abdominal CT scan will reveal periaortic lymphadenopathy. The CT, MRI, and isotope scans are to detect involvement of nodes and organs (i.e., lungs, gastrointestinal tract, liver, spleen) and central nervous system. Adequate surgical biopsy, reviewed by an experienced hematopathologist, is imperative. If disease is present in the oropharyngeal nodes, the patient should have an upper gastrointestinal examination because of frequent association of these two sites.

Laboratory work must include a complete blood count, liver and renal function tests, and quantitative copper and zinc. Evaluation of immune competence requires quantitative IgG, IgA, and IgM and serologic evidence of infection with EBV and cytomegalovirus.

An exploratory laparotomy with splenectomy is indicated whenever a significant change in treatment plan may result. Less aggressive staging is required when the disease has advanced further than suspected. The pathologic stage will change after laparotomy in approximately one-third of the patients.

VIII. Treatment

Radiation therapy alone is indicated in patients with pathologic stages 1 and 2 disease and for asymptomatic patients with disease confined to the upper abdomen. Radiation of the node-bearing areas of the chest (called a mantle) provides a disease-free survival exceeding 85% at 5 years. Radiation therapy alone for stage 1 will show a 72%, 10-year survival; stage 2 affords 31%, 10-year survival. Cure is significantly less with more extensive disease and B symptoms. Consequently, combined radiation and chemical therapy is indicated in Hodgkin lymphoma.

Many chemotherapeutic agents and combinations are available for treatment of Hodgkin lymphoma, NHL, and myeloma. Combined radiation and chemical therapy can be significantly effective and requires careful monitoring. The chemical therapy most commonly employed is nitrogen mustard, vincristine, procarbazine, prednisone (MOPP). Many other agents are frequently used.

The following are some of the antineoplastic agents and combinations. Many variations in dose and schedule are employed:

- ABVD: adriamycin (doxorubicin), bleomycin, vinblastine, and DTIC

- M-BACOD: methotrexate with leucovorin rescue, bleomycin, doxorubicin, cyclophosphamide, vincristine, and prednisone
- MOPP: nitrogen mustard, vincristine, procarbazine, and prednisone
- MOPP/ABV: nitrogen mustard, vincristine, procarbazine, prednisone, adriamycin (doxorubicin), bleomycin, and vinblastine
- ProMACE: prednisone, methotrexate, doxorubicin, cyclophosphamide, and etoposide (VP-16)
- CHOP: cyclophosphamide, doxorubicin, vincristine, and prednisone

Fifty to sixty percent of patients with lymphomas have bone marrow involvement. Most patients with follicular lymphomas have stage III or IV disease, and almost all patients with diffuse large cell lymphomas are initially treated with chemotherapy, regardless of the stage. The route is usually intravenous or oral, particularly in children. Infusion of the spinal canal with antilymphoma (intrathecal) agents is used when indicated.

An increased risk of second malignancies accompanies chemotherapy and radiation; consequently, chemotherapy may be delayed until relapse. Patients with early-stage high-risk disease should be considered for initial combined therapy or intensive chemotherapy alone. Combination chemotherapy is the appropriate treatment for patients with stage IIIA, involving periaortic, iliac and/or mesenteric nodes or IIIB non-Hodgkin disease.

Composite lymphomas are represented by combinations of two lymphomas. The diagnoses are often observed within a single or group of lymph nodes. Generally, this indicates a more ominous course and often requires a wider spectrum of therapy.

Low-grade lymphomas may be treated with oral chlorambucil and corticoids. Myelomas are often treated with phenylalanine mustard and prednisone. Myelomas may also benefit from alpha-interferon therapy as an alternative.

Interferons are immunomodulatory compounds capable of altering expression of cellular antigens. Increasing the expression of tumor cell antigens potentiates the attraction of a monoclonal lymphoma antibody. The more direct delivery of antilymphoma agents by the arterial, venous, and lymphatic route will enhance the therapeutic effectiveness and reduce toxicity. [*See* INTERFERONS.]

Removing of macroglobulin by selective centrifugation (plasmapheresis) can safely reduce the viscosity of the blood, thereby avoiding the potentially fatal circulatory occlusive complications.

After radiation and chemical therapy, 30% of patients have residual masses, 10% of whom have active disease. Chemotherapy followed by radiation to all known areas of disease may result in better disease-free and overall survival. Therefore, it is often wiser to evaluate with caution rather than embark on another intensive program.

Drug resistance requires monitoring to maintain the successive momentum of therapy. Appropriate change in chemotherapy and other modalities ensures benefit. Supportive therapy with antibiotics, isoelectrolyte fluids, blood, and blood products (erythrocytes, platelets, plasma proteins) are valuable and often lifesaving. Knowledge of the lymphomas and its treatment builds confidence and motivation. Patients often become more determined to help themselves.

Overtreatment may result in sterility, organ toxicity, complicated infections with immunoincompetence, and significant risk of subsequent leukemias and other multiple primary malignancies in both children and adults. Risk-to-benefit ratio remains the best guide.

Bibliography

Berard, C. W. (1981). A multidisciplinary approach to non-Hodgkin's lymphomas. *Ann. Intern. Med.* **94,** 218–235.

Bishop, J. M. (1983). Cellular oncogenes and retroviruses. Annu. Rev. Biochem. **52,** 301–354.

Canellos, G. P. (1986). Malignant lymphoma, Oncology 12, XI, medicine. *Sci. Am.* 1–15.

Freedman, A. S., and Nadler, L. M. (1987). Cell surface markers in hematologic malignancies. *Semin. Oncol.* **14,** 193–212.

Holt, J. T., Morton, C. C., Nienhuis, A. W., and Leder, P. (1987) Chap. 10 Molecular mechanics of hematological neoplasms. pp. 347–375. *In* "The Molecular Basis of Blood Disease" (G. Stomatoyannopoulos, A. W. Nienhuis, P. Leder, and P. W. Majerus, eds.), Saunders Co., Philadelphia.

Kaplan, H. S. (1980). Hodgkin's Disease,'' 2nd ed. Harvard University Press, Cambridge, Massachussetts.

Mann, R. B., Jaffe, E. S., and Berard, C. W. (1979). Malignant lymphomas—A conceptual understanding of morphologic diversity. *Annu. J. Pathol.* **94,** 103–192.

MKSAP. (1989). Hematologic Malignancies and Oncology,'' Part B, Books 4 and 5. American College of Physicians. Philadelphia.

National Cancer Institute. (1982). Sponsored study of

classifications of non-Hodgkin's lymphomas. *Cancer* **49,** 2112–2135.

O'Neill, B. P., and Illig, J. J. (1989). Primary central nervous system lymphoma. *Mayo Clin. Proc.* **64,** 1005–1020.

Rose, N., and Israel, M. A. (1987). Genetic abnormalities as biological tumor markers. Semin. Oncol. **14,** 213–231.

Rosenberg, S. A. (1986). Oncology 12, medicine, vol. 2. *Sci. Am.* N. Y.

Biology of cancer, Parts 1 & 2. (June 1989). *Semin. Oncol.* **16.**

Sell, S. (1987). ''Immunology, Immunopathology and Immunity.'' Elsevier Science Publishing Co., New York

Stetler-Stevenson, M., Crush-Stanton, S., and Crossman, J. (1990). Involvement of the bcl-2 gene in Hodgkin's disease. *J. Natl. Cancer Inst.* **82,** 855–858.

Macrophages

DAVID S. NELSON,* *Kolling Institute of Medical Research*

Glossary

Antibody Protein present in blood serum or body fluids as a result of antigenic stimuli or occurring naturally

Endocytosis Process whereby materials are taken into a cell through the plasma membrane

Exocytosis Process whereby materials are released by a cell through its plasma membrane

Golgi apparatus A cellular organelle responsible for the formation of secretory products.

Lymphocyte White blood cell usually formed in lymphoid tissue.

Lymphokine Substance released by lymphocytes that affects cellular immunity by stimulating monocytes and macrophages.

Lysosome A cytoplasmic, membrane-bound particle, containing hydrolyzing enzymes.

Phagocytosis The process of ingestion and digestion of solid substances by cells.

* Deceased—November 2, 1989—Royal North Shore Hospital.

MACROPHAGES, or mononuclear phagocytes, are a family of cells principally characterized by a high capacity for endocytosis (both phagocytosis and pinocytosis) and having a mononuclear morphology. Together with polymorphonuclear leukocytes, which are morphologically different, functionally rather different, and shorter lived, they are often referred to as "professional phagocytes." They are phylogenetically ancient, as specialized phagocytes are formed in simple invertebrates and appear early in ontogeny. Macrophages usually also possess some other characteristics: surface receptors for immunoglobulins (e.g., IgG and IgE) and complement components, the capacity to secrete lysozyme (the enzyme lysozyme), cytochemically detectable nonspecific esterase activity, more or less abundant lysosomes, and a tendency to adhere readily and firmly to glass and a variety of plastic surfaces.

By virtue of their phagocytic capacity, they are important in the defense against infection and as scavenger cells. They are also important in the resistance of cancer and as accessory cells in the induction and the expression of immunity. As a family they can synthesize and secrete a wide range of biologically active and important molecules. They are diverse and widely distributed in the body and are prominent in a variety of pathological conditions.

Much of our knowledge of macrophage function is derived from studies using laboratory animals, especially mice. Except for blood monocytes (young members of the family) and, to a lesser extent, pulmonary alveolar macrophages, human macrophages are obviously less amenable to study. There is no reason to expect that the human macrophage family differs greatly from its murine counterpart.

I. Life History and Distribution

Macrophages are produced in the bone marrow from stem cells, which proliferate and differentiate, giving rise, in turn, to monoblasts, promonocytes, and monocytes. The monocytes enter the circulation, some circulating freely and some remaining as a marginal (noncirculating) pool that can be called on for a rapid supply. The efficient production of monocytes–macrophages requires the operation of growth factors known (because of their effects on bone marrow cultures) as colony-stimulating factors (CSFs). The two principal CSFs are macrophage CSF (M-CSF, or CSF-1), which is macrophage lineage specific, and granulocyte–macrophage CSF, which acts on both macrophage and granulocyte lineages. Both can be produced by a variety of cells, but it is noteworthy that they can be produced by macrophages, providing a positive-feedback loop, and by stimulated T lymphocytes, providing a link between cell-mediated immune reactions and the supply of monocyte–macrophages.

Monocyte production and differentiation are also promoted by interleukin-3 (IL-3, another lymphocyte product), a factor-increasing monocytopoiesis, and a distinct macrophage differentiation-inducer. Monocytopoiesis is inhibited by prostaglandin E_2 (PGE_2), tumor necrosis factor α (TNF-α), and a monocyte production inhibitor. As products of macrophages, PGE_2 and TNF-α can provide negative feedback. Stem cells give rise to monocytes in about 6 days, and the normal daily production of monocytes is about 7×10^6/hour/kg. The half-life of monocytes in the circulation is normally about 3 days, and the normal monocyte count (in adults) is 3–4×10^8/liter, giving a total circulating monocyte population of about 1.7×10^9.

Most tissue macrophages, both normal and inflammatory, are derived from blood monocytes. As mature macrophages can divide, a small proportion can also be produced locally. Mature macrophages—which differ morphologically and functionally from monocytes—are found in all normal tissues, with notable concentrations in the liver (as Kupffer cells), the spleen and the lymph nodes, the bone marrow, the pulmonary alveoli, and the pleural and peritoneal cavities. The microglia of the central nervous system is also part of the mononuclear phagocyte system. Surprisingly, large numbers of macrophages (on the order of 5–10×10^{10}) have been isolated from the placenta, but estimates of the macrophage content of other human organs and tissues are very difficult to make. So, too, are estimates of the life span of macrophages in tissues, but observations on experimental animals suggest that it may be on the order of weeks, rather than days.

Monocytes and macrophages are morphologically and functionally heterogeneous. Some of the heterogeneity reflects differences in maturity; some reflects the effects of stimulating or activating factors (see Section V). Under pathological conditions giant cells are formed by the fusion of macrophages, and epithelioid cells may be derived from macrophages. The cells grouped together as veiled or dendritic cells, including Langerhans cells in the skin, could be derived from, or closely related to, macrophages, but are often not phagocytic and thus lack one of the chief characteristics of macrophages.

The fate of macrophages that have reached the end of their life span is not known. Dead or effete alveolar macrophages could be moved up the respiratory tree by ciliary action, and macrophages in the intestinal wall could migrate through the gut epithelium into the lumen. In other sites they could be disposed of by younger macrophages in a scavenging role.

II. Morphology

As seen on routine stained blood films, monocytes are round cells 15–22 μm in diameter, with a single nucleus that may appear round, oval, reniform, or folded. In the nucleus the chromatin is finely stranded; the nucleus is surrounded by a moderate amount of pale blue-staining cytoplasm, with some granules. Mature macrophages (e.g., from the peritoneal or pleural cavity) are larger, with more granular cytoplasm. In tissue sections macrophages are usually irregular in shape and are sometimes difficult to distinguish without the use of special stains or markers. Cultured macrophages, viewed by phase-contrast microscopy, usually have irregular surfaces, with hyaloplasmic veils or ruffles, and granular cytoplasm. The veils can be seen particularly well by using scanning electron microscopy (Fig. 1). Transmission electron microscopy usually shows the presence of abundant lysosomes, a prominent Golgi apparatus, mitochondria, a varying amount of rough endoplasmic reticulum and (depending on the source and condition of the cells) inclusions derived from phagocytosed material. Giant cells (i.e., heterokaryons) are larger, according

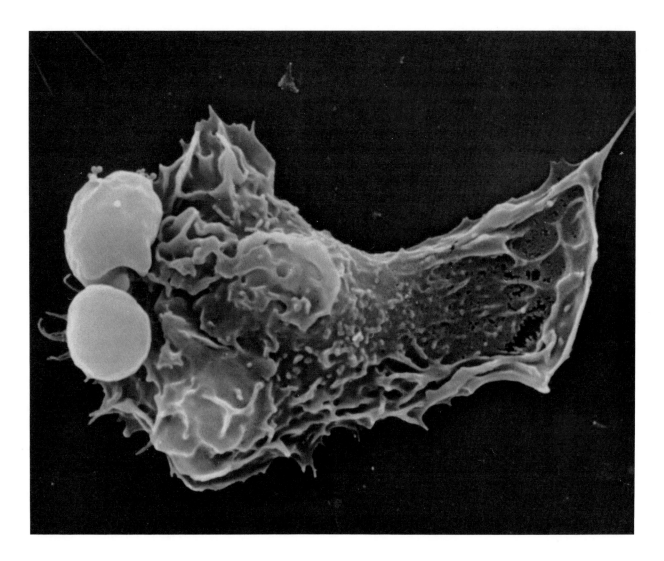

FIGURE 1 Scanning electron micrograph of a cultured macrophage, to which two antibody-treated sheep red blood cells have adhered prior to phagocytosis. (Courtesy of M. Nelson and S. Doyle.)

to the number of cells that have fused, with the nuclei usually arranged peripherally.

Special stains can be used to show the presence of enzymes, such as nonspecific esterase (a useful marker), peroxidase (in monocytes, but not in more mature cells), lysosomal enzymes (e.g., acid phosphatase), or surface markers (see Section III).

III. Surface Receptors and Other Molecules

The principal surface receptors of macrophages are listed in Table I. Monocyte–macrophages can also exhibit a number of other surface markers, although not all are necessarily present on all members of the family, nor are they necessarily unique to mononuclear phagocytes (Table II). Like virtually all nucleated cells, they carry class I antigens of the major histocompatibility complex (MHC). When appropriately stimulated [e.g., by γ-interferon gamma (IFN-γ)], they readily express MHC class II antigens.

IV. Phagocytosis

The process of phagocytosis can be considered to have four stages: (1) movement of the macrophage toward the particle (e.g., the microorganism or the cell) to be ingested, usually in response to chemotactic (i.e., attractive) stimuli; (2) attachment of the

TABLE I Receptors of Macrophages

Immunological receptors for
    Fc portion of immunoglobulin G (three types)
    Fc portion of immunoglobulin E
    Complement components (complement receptor 1 for C3b
       and C4b; complement receptor 3 for C3bi; C5a; and C1q)
Receptors for chemotactic factors
    C5a (above) and factors listed in Section IV,A
Receptors for stimulatory and regulatory cytokines
    γ-Interferon, other interferons
    Granulocyte–macrophage and macrophage colony-stimulat-
       ing factors
    Monokines (e.g., interleukin-1 and tumor necrosis factor α)
    Interleukin-2
    Other macrophage-activating or -stimulating factors (listed in
       Section V)
Hormone receptors
    Glucocorticoids
    Insulin
    1,25-Dihydroxyvitamin D_3
    Neuropeptides (e.g., β-endorphin, dynorphin, Leu-enkepha-
       lin, neurotensin, substance P, and substance K)
Lectinlike receptors for
    Mannosyl and fucosyl residues
    β-Glucan
Receptors for adherence and clotting-related factors
    Fibronectin
    Thrombin
    Fibrinogen/fibrin
Receptors for other proteins
    Lactoferrin
    Transferrin
    Modified lipoproteins (''scavenger receptor'')
    Apolipoproteins B and E
    α_2-Macroglobulin–proteinase complexes

particle to the surface of the phagocyte; (3) ingestion; (4) intracellular disposal.

A. Chemotaxis

Chemotactic stimuli to which macrophages respond
include: the split product of the fifth component of

TABLE II Surface Markers on Monocyte–Macrophages

5'-Nucleotidase (an ectoenzyme; in immature cells only)
Leukocyte function antigen 1 (CD[a]11a plus CD18)
p150, 95 (CD11c plus CD18)
CD4
CD11b plus CD18 (Mac1, Mo1, OKM1, and CR3[b])
CD14 (FMC17 and FMC32)

[a] CD, Cluster designation antigens are detected by monoclonal antibodies; other names are generally derived from the monoclonal antibodies
used to detect them (e.g., FMC17 is the Flinders Medical Centre antibody
17).
[b] CR3, Complement receptor 3 (see Table I).

complement (C5a), chemotactic lymphokines (e.g.,
macrophage chemotactic factor and macrophage
procoagulant-inducing factor), leukotriene B4,
formylated peptides (e.g., formyl–methionine-
leucine–phenylalanine), thrombin, and fibrinogen
cleavage products. In each case the chemotactic
factor binds to a receptor on the macrophage surface, from which signals are transmitted to the cell's
interior (i.e., signal transduction) through changes
in the metabolism of phosphoinositides, the intracellular calcium concentration, and the activity of
protein kinase C. Movement is preceded by polarization of the cell, so that it adopts a wedge shape
with a broad leading edge. Movement toward the
chemotactic stimulus requires, of course, the expenditure of energy.

B. Attachment

Attachment (Fig. 1) is commonly mediated by antibodies directed against surface antigens of the particle (e.g., the cell or the bacterium) being ingested.
The Fab portion of the immunoglobulin binds to the
particle, and the Fc portion binds to an Fc receptor
on the macrophage. Particles may also be bound by
way of complement receptors (especially CR1) and,
in the absence of antibodies or complement, by lectinlike receptors reacting with surface sugars on the
particle. Attachment requires no expenditure of energy by the macrophage.

The term ''opsonization'' is generally used for
the promotion of phagocytosis by antibodies, complement, or other soluble factors. The critical effect
of opsonization is to promote the attachment of the
particle to the phagocyte. [*See* PHAGOCYTES.]

C. Ingestion

The process of ingestion involves the enclosure of
the particle by the macrophage cell membrane,
forming a vesicle—the phagosome—which buds off
into the interior of the cell. The wall of the phagosome is thus cell membrane, inside-out. Ultrastructurally, two basic ways of forming a phagosome can
be seen: the ''crater,'' in which the particle sinks
into a surface depression, which then closes over it,
and the ''zipper,'' in which surface projections of
the cell membrane rise and close over the particle.
Ingestion does not necessarily follow attachment.
In particular, particles attached solely by means of
complement may not be ingested unless the macro-

phage is also stimulated in one of the ways listed in Section V. It is not clear whether particles can be ingested after attachment by way of IgE.

Ingestion requires the expenditure of energy to move the particle/phagosome by means of the cell's microfilaments (i.e., the cell's muscles). Energy is usually obtained from sugars and related molecules through oxidative metabolic pathways. There is usually release of reactive oxygen intermediates/metabolites, such as superoxide anion. The occurrence, extent, and vigor of phagocytosis can thus often be conveniently measured in the laboratory by assays of chemiluminescence triggered by these intermediates. [In chemiluminescence a sensitive chemical (e.g., luminol) emits light when it reacts with the oxygen intermediates. The minute amounts of light are measured with the aid of photomultiplier tubes.] Studies with experimental animals, however, show that some macrophages (e.g., peritoneal macrophages living in an environment low in oxygen) can obtain energy from metabolism that does not require oxygen.

D. Intracellular Fate

The fate of the ingested particle is determined by both its nature and the state of the macrophage. Commonly, there is fusion of the phagosome with a lysosome (i.e., a subcellular membrane-bound collection of acid hydrolases, enzymes that break down various types of macromolecules). This might ensure the digestion and complete destruction of the ingested material, but undigestible material can be excreted from the cell (i.e., exocytosis) or can persist as microscopically or ultrastructurally visible debris (referred to as residual bodies, myelin figures, etc.).

Some microorganisms are readily destroyed after ingestion by normal macrophages. Others are adapted to a life of intracellular parasitism within macrophages, because they prevent lysosome–phagosome fusion, leave the phagosome, or are resistant to the lysosomal environment (see Sections V and IX). Such microorganisms are destroyed only when the macrophages are activated. In some of the latter cases the antimicrobial mechanisms available to macrophages include not only lysosomal enzymes, but, probably more importantly, the production of reactive oxygen intermediates. Oxygen-independent mechanisms that are also available include acidification in the phagosome, cationic antimicrobial peptides, iron-binding proteins, inter-

feron(s), enzymes (e.g., lysozyme, also called muramidase), and perhaps arginase.

V. Activation and Stimulation

The term "activation" was originally applied to the changes in macrophages that allowed them to destroy certain intracellular microorganisms that were otherwise adapted to survive and proliferate within "normal" macrophages. This destruction is immunologically nonspecific, in that macrophages activated to destroy one microorganism can also destroy any of a variety of unrelated microorganisms. There is no specificity, even when the macrophages have been activated as a result of a specific immune response to one microorganism. Activation in this sense is accompanied by the acquired ability to recognize and destroy cells that have been malignantly transformed (i.e., cancer cells) and by a variety of morphological and biochemical changes. It is both convenient and desirable to use the term "activation" for such effects.

Without undergoing these comprehensive changes, however, macrophages can also be stimulated to exhibit increases in one or more of several activities, for example, the secretion of important products (e.g., IL-1), the capacity for phagocytosis,

TABLE III Agents That Stimulate Macrophages

Physiological factors causing full activation
  γ-Interferon gamma
  Granulocyte–macrophage and macrophage colony-stimulating factors
  Other undefined lymphokines
Other physiological factors
  Antigen–antibody complexes (e.g., immunoglobulins G and E)
  Complement components (e.g., C5a and C1q)
  Interleukin-1
  Interleukin-2
  Interleukin-4
  Tumor necrosis factor α (i.e., cachectin)
  Procoagulant-inducing lymphokine
  Thrombin, other proteases
  1,25-Dihydroxyvitamin D_3
  Somatotropin
  Platelet-activating factor
  Adherence
Nonphysiological(?) factors
  Gram-negative bacterial endotoxin (e.g., lipopolysaccharide)
  Muramyl dipeptide
  Polyribonucleotides
  Phorbol esters, calcium ionophore

TABLE IV Accompaniments of Macrophage Activation

Activation of protein kinase C, inositol phospholipid metabolism, and intracellular Ca^{2+} mobilization
Increased intracellular pH
Stimulation of arachidonic acid metabolism
Increased capacity to produce reactive oxygen intermediates
Production of neopterin (2-amino-4-oxo-6-trihydroxypropylpteridine)
Increased production and secretion of neutral proteases (e.g., plasminogen activator, elastase, and collagenase)
Increased lysosomes and lysosomal enzymes
Decreased transferrin receptors
Decreased mannose–fucose receptors
Synthesis of novel proteins (recognized electrophoretically or by monoclonal antibodies)

or the capacity to adhere to glass or plastic. Again, it is both convenient and desirable to use the term "stimulation" in a defined way, specifying both the stimulus and the response, for example, macrophages stimulated by bacterial endotoxin (e.g., lipopolysaccharide) to produce IL-1.

It must also be emphasized that all populations of macrophages are heterogeneous. When physically separated (e.g., on the basis of size or density), different components of a normal, stimulated, or activated population differ in their effector capacities (e.g., antimicrobial, antitumor, and secretory).

Table III lists some of the agents that can activate or stimulate macrophages. Table IV lists some of the changes that accompany activation.

In experimental animals monocytes or young macrophages (i.e., newly arrived in tissues or in inflammatory lesions) are much more readily stimulated than are old or resident cells. In culture macrophages usually require two stimuli to become fully activated: one provided by a lymphokine (e.g., IFN-γ) and another generally provided experimentally by lipopolysaccharide. The physiological counterpart of lipopolysaccharide is not known. Human monocytes appear to respond to IFN-γ alone, although a second signal might be provided by serum components in the culture medium or by the adherence to glass or plastic.

VI. Extracellular Killing

Activated macrophages can recognize cancer cells and kill them or exert a cytostatic effect on them (i.e., stop their growth). The mechanism of recognition is unknown. Cytotoxic/cytostatic effector

mechanisms available to activated macrophages include the production of IL-1, TNF-α, a cytolytic protease, arginase, and undefined products leading to iron loss and secondary metabolic changes in target cells (see Table V). Reactive oxygen intermediates do not appear to be important.

Activated macrophages could also recognize and destroy sperm extracellularly. The mechanisms are unknown, but this might be an important factor in the infertility of pelvic inflammatory disease.

Whether or not they are activated, macrophages can also recognize targets by virtue of antibody (i.e., antibody-dependent cell-mediated cytotoxicity). Targets include not only tumor cells, but also metazoan parasites (e.g., microfilariae and schistosomulae). Reactive oxygen intermediates are important in such killing, although other mechanisms might be involved.

TABLE V Secretory Products of Macrophages[a]

Inflammatory monokines, growth factors, and hormones
  Interleukin-1
  Tumor necrosis factor α (i.e., cachectin)
  Granulocyte–macrophage, macrophage, and granulocyte colony-stimulating factors
  Erythropoietin
  Erythroid colony-potentiating factor/tissue inhibition of metalloproteinases
  Factor-inducing monocytopoiesis
  Platelet-derived growth factor
  Fibroblast growth factor
  Fibroblast-activating factor(s)
  Transforming growth factors α and β
  α-Interferon
  Interleukin-6 (i.e., $β_2$-interferon)
  Somatotropin
  β-Endorphin
  Adrenocorticotropic hormone
  Bombesin
  1,25-Dihydroxyvitamin D_3
Coagulation factors
  Intrinsic pathway (i.e., Factors V, IX, and X and prothrombin)
  Extrinsic pathway (i.e., Factor VII)
  Surface activities (i.e., tissue factor and prothrombinase)
  Prothrombolytic activity (i.e., plasminogen activator)
  Antithrombolytic activity (i.e., plasminogen activator inhibitor and plasmin inhibitor
Components of the complement system
  All components of the classical pathway (i.e., C1–C9)
  Alternative pathway [i.e., factors B and D and factor P (properdin)]
  Inhibitors [i.e., factor I (C3b inactivator) and factor H (β1H)]

TABLE V *(continued)*

Other enzymes
  Proteolytic enzymes (i.e., plasminogen activator, collagenase, elastase, stromelysin, gelatinases, angiotensin convertase, cytolytic proteinase, and cathepsins D, L, and B)
  Lipases (i.e., lipoprotein lipase and phospholipase A_2)
  Lysozyme (i.e., muramidase)
  Lysosomal acid hydrolases [i.e., proteases, (deoxy)ribonucleases, phosphatases, glycosidases, and sulfatases]
  Arginase
Inhibitors of enzymes or cytokines
  Protease inhibitors [i.e., α_2-macroglobulin, α_1-protease (or -trypsin) inhibitor, plasminogen activator inhibitor, plasmin inhibitors, tissue inhibitor of metalloproteinases]
  Phospholipase inhibitor (i.e., lipomodulin/macrocortin)
  Interleukin-1 inhibitor(s)
Proteins of the extracellular matrix and cell adhesion
  Fibronectin
  Gelatin-binding protein
  Thrombospondin
  Heparin sulfate proteoglycans
  Chondroitin sulfate proteoglycans
Other binding proteins
  For metals: transferrin and acidic isoferritins
  For vitamins: transcobalamin II
  For lipids: apolipoprotein E and lipid transfer protein
  For growth factors: α_2-macroglobulin and interleukin-1
  For inhibitors: transforming growth factor β-binding protein
  For biotin: avidin
Bioactive lipids
  Cyclooxygenase products (i.e., prostaglandin E_2, prostaglandin $F_{2\alpha}$, prostacyclin, and thromboxane)
  Lipoxygenase products (i.e., monohydroxyeicosatetraenoic acids, dihydroxyeicosatetraenoic acids, and leukotrienes B_4, C, D, and E)
  Platelet-activating factor
Other bioactive low-molecular-weight substances
  Glutathione
  Purine and pyrimidine products (i.e., thymidine, uracil, uric acid, deoxycytidine, cAMP, and neopterin)
  Reactive oxygen intermediates (i.e., superoxide, hydrogen peroxide, hydroxyl radical, singlet oxygen, and hypohalous acids)
  Reactive nitrogen intermediates (i.e., nitrites and nitrates)

[a] Adapted from C. F. Nathan, *J. Clin. Invest.* **79,** 319–326 (1987) and D. A. Rappolee and Z. Werb, *Curr. Opin. Immunol.* **1,** 47–55 (1988).

VII. Accessory Cell Activity

Macrophages, acting as accessory cells, can play important parts in the induction and expression of specific immune responses. At the simplest level they break down large complex pieces of foreign material (e.g., cells and bacteria) into components (e.g., protein and polysaccharide) that are recognizable by the lymphocytes of the immune system programmed to respond to them. They can serve as intermediaries, in ways that are poorly understood, between helper T lymphocytes and B lymphocytes that will give rise to antibody-producing plasma cells in response to certain thymus-dependent antigens. More particularly, however, they process protein antigens for presentation to committed T lymphocytes in the induction and the expression of cell-mediated immunity. In essence processing involves the breakdown of proteins to peptides that then become associated with MHC antigens on the macrophage surface. In this case the macrophages are said to be antigen-presenting cells.

Cell-mediated immune responses fall into two broad classes: the production of cytotoxic T lymphocytes and the development of delayed-type hypersensitivity (DTH). The latter is an awkward term for immunity expressed by a different class of T cells (T_{DTH}), with the production of lymphokines, whose effects include macrophage activation. The targets of cytotoxic T lymphocytes are commonly virus-infected cells: Much of the antigen processing occurs in those cells, and the MHC antigens are of class I. In the case of DTH, processing is carried out mainly by macrophages, and the MHC antigens are of class II. In the latter case antigens that have been processed can also be presented on the surface of other cells, notably dendritic cells, which could themselves lack a processing capacity. The requirements for antigen processing and presentation appear to be similar for both the induction and the expression of cell-mediated immunity.

Macrophages also contribute to the induction and the expression of cell-mediated immunity by producing molecules that provide important nonspecific signals to the responding T cells. IL-1 is the best-characterized such molecule, but there may be others (e.g., IL-6). This signal might be essential for T cell responses, or it might merely be a powerful amplifying signal.

Macrophages can also exert depressive effects on immune responses. This is most readily seen in culture, when macrophages are present in excessive numbers or have been stimulated or activated. "Suppressor macrophages," whose effects are generally nonspecific, could act in part by producing PGE_2 and in part by producing other, undefined, molecules. They might also be involved in the operation of suppressor T lymphocytes.

VIII. Secretion

As a family macrophages can secrete a surprisingly large range of biologically important products, some of which are listed in Table V. Because of the number of these products, the importance and the diversity of their effects, and the number and the widespread distribution of macrophages, the mononuclear phagocyte system can be viewed as a diffuse and important endocrine organ. Discussion here is restricted to one example of an important macrophage product—IL-1—and some general remarks.

The term "IL-1" is applied to a pair of polypeptides—IL-1α and IL-1β—which share 26% amino acid homology, but have essentially identical effects. IL-1α tends to remain cell associated, whereas IL-1β is secreted. Both are initially produced as large precursor molecules (with a molecular weight of 31,000) which are processed to smaller (i.e., molecular weight 17,500) final products. IL-1 is not uniquely a macrophage product, being produced by, for example, vascular endothelial cells, some epidermal cells, and natural killer cells, but macrophages seem to be the most important source. Stimuli that induce the production of IL-1 by macrophages include the factors listed in Table III, phagocytosis generally, leukotrienes, mast cell heparin, hyaluronic acid, collagen, acute phase proteins (i.e., proteins that appear in serum or plasma after the onset of infection or tissue damage), retinoic acid, epinephrine, and, somewhat strangely, breast milk.

Table VI lists some of the principal biological effects of IL-1. These effects are extremely varied, and IL-1 is therefore a major contributor to both normal homeostasis and pathological reactions (e.g., inflammation). Clearly, regulatory mechanisms must exist to control these activities. Inhibitors of IL-1 have been described, but the regulatory pathways are not well understood.

Some general points can be made about IL-1 and other macrophage products:

1. Some macrophage products have autocrine effects; that is, they can act on the cells that produce them. Thus, IL-1 can induce the production of further IL-1, as well as TNF-α, PGE$_2$, CSFs, etc. Other macrophage products that also act on macrophages include platelet-activating factor, 1-25-dihydroxyvitamin D$_3$, PGE$_2$, and TNF-α.
2. Macrophage products could regulate the production of macrophages (e.g., CSFs, PGE$_2$, and factor-increasing monocytopoiesis).
3. The effects of macrophage products may be amplified by other cells; for example, IL-1 induces the production of CSFs by vascular endothelial cells and fibroblasts, and both IL-1 and TNF induce procoagulant activity (stimulants of the blood clotting system) in endothelial cells.
4. Some products are not unique to macrophages; the importance of macrophages in their production could lie in the body's ability to cause local accumulations of macrophages, allowing the products to achieve high concentrations.

TABLE VI Some Effects of Interleukin-1[a]

Promotion of cell growth
  Lymphocytes, fibroblasts, keratinocytes, mesangial cells, and glial cells
Inhibition of cell growth/cytotoxic effects
  Tumor cells (some), pancreatic islet β cells, and thyroid cells
Other effects on blood cells
  Lymphocyte chemotaxis
  Neutrophil and monocyte thromboxane syntheses
  Basophil histamine release
  Eosinophil degranulation
  Increased bone marrow neutrophil production
  Increased colony-stimulating factor production
  Macrophage interleukin-1 and tumor necrosis factor α production
Vascular effects
  Expression of endothelial cell leukocyte adherence receptors and leukocyte adherence
  Increased production of prostaglandins and I$_2$
  Increased production of platelet-activating factor
  Syntheses of β_1-interferon and β_2-interferon (i.e., interleukin-6)
  Increased endothelial procoagulant activity
  Increased endothelial plasminogen activator inhibitors
  Smooth muscle cell proliferation
  Capillary leak syndrome and hypotension
Effects on the central nervous system
  Fever
  Sleep
  Anorexia
  Increased adrenocorticotropic hormone and neuropeptide production
Proinflammatory and degradative effects
  Increased fibroblast synthesis of collagen, collagenase, and prostaglandin E$_2$
  Increased chondrocyte synthesis of proteoglycan, proteases, plasminogen activator, and prostaglandin E$_2$
  Osteoclast activation and bone resorption
Other metabolic effects
  Hepatic synthesis of acute phase proteins (e.g., α_1-antitrypsin)
  Decreased plasma zinc and iron
  Decreased synthesis of adipocyte lipoprotein lipase
  Increased muscle protein catabolism
  Increased intestinal mucus production

[a] Adapted in part from C. A. Dinarello, *FASEB J.* **2,** 108–115 (1988).

5. Macrophage products may have contrasting, antagonistic, or synergistic effects. Thus, IL-1 can inhibit the proliferation of some cells (e.g., tumor cells) but stimulate others (e.g., fibroblasts and lymphocytes). Macrophages can produce both procoagulant factors and plasminogen activator, which leads to the breakdown of fibrin clots. IL-1 can stimulate fibroblasts to synthesize both collagen and the enzyme that breaks it down, collagenase, and macrophages themselves can produce collagenase. IL-1 and TNF-α not only both have autocrine effects, but they can act synergistically in stimulating PGE$_2$ production.

6. There are important interactions between macrophages and the endocrine system generally. Macrophages are responsive to various hormones (e.g., glucocorticoids, insulin, vitamin D, and somatotropin). They are also a source of hormones (Table V). In addition, they can influence the production of hormones: IL-1 stimulates the release of multiple hormones from the pituitary gland and stimulates steroidogenesis in Leydig's cells of the testes, but inhibits the response of ovarian granulosa cells to other hormones and is toxic to pancreatic islet β cells (which produce the hormone insulin) and thyroid cells. Macrophage CSF could also be important in the development of the placenta.

IX. Physiological Roles

It is apparent from this catalog of activities that the mononuclear phagocyte system as a whole has the potential to play major roles in many bodily functions. Perhaps this is best reflected in the observation that there are no congenital diseases characterized by an absence of monocyte–macrophages; their absence might be incompatible with life. Our knowledge of the secretory activities of macrophages has expanded so much recently that it has not been fully integrated into general physiology. The roles of macrophages in immunity and as scavenger cells have, however, become clearer.

In acquired immunity to infection (i.e., that expressed as a result of a specific immune response), macrophages have three major roles: as accessory cells in the immune response (both induction and expression); as effectors of acquired immunity, operating as phagocytes in conjunction with antibody; and as effectors of cell-mediated immunity, when activated as a result of a DTH reaction. Human infections in which the last activity is of critical importance include mycobacterial infections (e.g., tuberculosis and leprosy), salmonella infections (e.g.,

typhoid), Legionnaire's disease, brucellosis, toxoplasmosis, psittacosis, histoplasmosis, leishmaniasis, and trypanosomiasis.

There is good experimental evidence that macrophages are important in natural immunity to many infections (i.e., in resistance expressed in the absence of an overt immune response). One particularly important factor is the speed with which macrophages can be mobilized and accumulated at the site of infection. Direct evidence in humans, however, is not readily obtainable.

Paradoxically, macrophages, especially when not stimulated, may serve as reservoirs of infection. This has most recently come to light in the case of infection with the human immunodeficiency virus [HIV, also called the acquired immunodeficiency syndrome (AIDS) virus]. Even more paradoxically, antibody can, in the case of some viruses, potentiate virus infection of macrophages. A notable example is dengue, in which a second infection can lead to the severe and potentially fatal dengue hemorrhagic fever. [See ACQUIRED IMMUNODEFICIENCY SYNDROME (VIROLOGY).]

Macrophages clearly also play a role in natural and acquired resistance to some cancers in experimental animals. While this is likely to be true also for some human cancers, direct evidence is clearly hard to obtain. Once again, however, there is evidence that macrophages can, in certain circumstances, paradoxically enhance the survival and the growth of some experimental cancers.

The scavenger activities of macrophages are extremely important in the bodily economy. They phagocytose and digest many aged, effete, or damaged cells. This is most notable in the case of red blood cells, some 2 million of which are destroyed and replaced each second. Here, one of the most important activities of macrophages is the conservation of iron and its return to the pool available for new red blood cell formation. Cultured macrophages can discriminate between young and old red blood cells, phagocytosing the latter, but the recognition mechanism is not known.

The handling of lipids by macrophages is also very important; by virtue of their receptors for lipoproteins and modified lipids, they are involved in both transport and metabolism. [See LIPIDS.]

X. Macrophages in Pathology

Macrophages are so prominent and numerous in all diseases characterized by the pathological changes termed "chronic inflammation" that a listing of diseases would be almost impossible. Some common examples include rheumatoid arthritis and other inflammatory arthritides, in which the production of factors such as IL-1 may well be critically important in destructive damage to the joint; chronic interstitial lung diseases (e.g., sarcoidosis), in which growth factors produced by macrophages may be important in inducing fibrosis; demyelinating diseases (e.g., some forms of peripheral neuritis and perhaps multiple sclerosis), in which macrophages may remove the myelin sheaths from nerve fibers; and atherosclerosis, in the lesions of which lipid-laden macrophages are prominent, and in which growth factors produced by macrophages might contribute to the formation of the lesions.

Even the normal phagocytic activities of macrophages may be important in diseases affecting other cells. In autoimmune blood diseases (e.g., some forms of hemolytic anemia) antibodies to blood cells promote their premature phagocytosis by macrophages, and macrophages can be viewed as the ultimate destroyers and the graveyard of the affected cells. [See AUTOIMMUNE DISEASE.]

Somewhat less obviously, macrophages and their products could be essential effectors of damage incidental to other diseases. A notable example is malaria: It now seems that TNF-α is a critical factor in the pathogenesis of many features of severe malaria, including cerebral damage, abortion, and anemia. TNF-α also appears to be able to induce macrophages to phagocytose blood cells indiscriminately and thus to be involved in the hemophagocytic syndrome that can follow some viral infections and accompany some malignant diseases. Under another name—cachectin—TNF-α has important metabolic effects and may be responsible for the wasting (i.e., cachexia) of malignant disease. [See MALARIA.]

Bibliography

Dinarello, C. A. (1988). Biology of interleukin 1. *FASEB J.* **2,** 108–115.
Friedman, H., Escobar, M., and Reichard, S. M. (eds.) (1980–1988). "The Reticuloendothelial System: A Comprehensive Treatise," 10 vols. Plenum, New York.
Gordon, S., Keshav, S., and Chung, L. P. (1988). Mononuclear phagocytes: Tissue distribution and functional heterogeneity. *Curr. Opin. Immunol.* **1,** 26–35.
Hogg, N. (1987). Human mononuclear phagocyte molecules and the use of monoclonal antibodies in their detection. *Clin. Exp. Immunol.* **69,** 687–694.
Horwitz, M. A. (1988). Intracellular parsitism. *Curr. Opinion Immunol.* **1,** 41–46.
Nathan, C. F. (1987). Secretory products of macrophages. *J. Clin. Invest.* **79,** 319–326.
Nelson, D. S. (ed.) (1989). "Natural Immunity." Academic Press, San Diego.

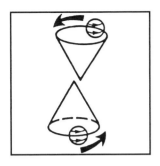

Magnetic Resonance Imaging

GEORGE KROL, *Memorial Sloan–Kettering Cancer Center*

Glossary

B0 External static magnetic field, generated by main magnet

B1 Radiofrequency magnetic field

Blood–brain barrier Semipermeable barrier maintaining chemical balance between blood and tissue within the central nervous system; damage may occur under pathological conditions (e.g., neoplasm or infection), resulting in leakage of intravascular components into surrounding tissue

Echo time Time interval between a 90° pulse (applied) and a spin echo signal (received); abbreviated TE

Flip angle Angle of deflection of magnetization vector M0 from its position of alignment with the main magnetic field; proportional to amplitude and duration of B1

Fourier transform Mathematical process interconverting shape to a multitude of sine waves of appropriate amplitude

Gauss Unit of strength of magnetic field; abbreviated G; 1 T equals 10,000 G

Gradient Increase (or decrease) of the strength of a static field along a given spatial plane; used for localization of the point within the imaging sample; achieved by gradient coils

Gradient echo Echo generated by a change in gradient direction; accentuates fluid spaces and flow

Gyromagnetic ratio Characteristic magnetic property of the nucleus; varies with the type of element

Hertz Unit of frequency, one cycle per second; abbreviated Hz; kHz equals 10^3 Hz; 1 MHz equals 10^6 Hz

M0 Directional vector of the magnetized sample before application of the radiofrequency pulse, aligned with the B0 field

M1, M2, M3 Positional variants of M0, modified (i.e., deflected) by radiofrequency stimulus

Matrix Net of coordinates dividing the imaged area into elements of equal size (e.g., 360 × 360 or 128 × 256)

Pixel Smallest component of the image (matrix)

Precession Circular motion of the vector of magnetization around the vector of the main static field

Pulse sequence Combination of radiofrequency energy pulses with defined timing and direction; used in the acquisition of a specific magnetic resonance image

Relaxation Process of the return of the longitudinal or transverse vector of magnetization (*M*) to its original position after discontinuation of the radiofrequency pulse

Repetition time Time interval needed for completion of the sequence (usually between successive 90° radiofrequency pulses); abbreviated TR

Spatial resolution Small distance between two points, still visually distinguished as two rather than one

Spin–echo Commonly used in clinical practice sequence, providing both T1 and T2 information; abbreviated SE

T1 Time, after discontinuation of the radiofrequency stimulus, needed for longitudinal component of magnetization to return to 67% of its original value

T2 Time interval required for the transverse component of magnetization to decay 67% from its value immediately after a 90° pulse

Tesla Unit of magnetic field strength; abbreviated T. The field of clinical magnetic resonance is usually between 0.35 and 1.5 T; 1 T equals 10,000 G

Vector Characteristic possessing quantitative value and direction; usually expressed by an arrow
Voxel Smallest volume element of the imaged sample (see pixel)

NUCLEAR MAGNETIC RESONANCE refers to the rapid oscillation of atomic nuclei which occurs when certain elements placed in a strong magnetic field are subjected to a radiowave of appropriate frequency. In the process an irradiated sample emits a signal which can be detected by external sensors. The duration, amplitude, and pitch of the signal are analyzed and utilized in clinical imaging and magnetic resonance spectroscopy.

I. Introduction

Nuclear magnetic resonance (NMR) was discovered independently by two researchers, Bloch and Purcell, in 1946. They noticed that a sample of matter placed in a strong magnetic field and treated with a radiofrequency (RF) wave emitted a signal of matching frequency, which could be detected and registered by a receiver. The dependence of amplitude and duration of the signal on the structure of the sample of origin was later used for analysis of the chemical compositions of various compounds and became the basis for NMR spectroscopy. The first images of the human body were produced in the late 1970s by Damadian and associates. A rapid development and clinical utilization of MR followed. Presently, MR imaging constitutes an integral part of diagnostic evaluation of the human body.

II. Theory

A. Structure of Matter and Effects of the Magnetic Field

The basic building blocks of matter—atoms—are composed of nuclei, which carry a positive charge, and electrons, which are negatively charged. The major components of the atomic nuclei—positive protons and neutral neutrons—exhibit rapid rotation around their axes. In the process a minute magnetic field, called magnetic moment, is created. In elements with a paired number of neutrons and protons the magnetic moments cancel each other, re-

sulting in zero value of the moment of the entire nucleus. In others, including isotopes of the principal elements of biological tissues (e.g., hydrogen, carbon, and sodium), an unpaired neutron or proton creates a net spin of the nucleus. The resulting magnetic moment has a quantitative value and direction. Atomic nuclei possessing the magnetic moment respond to external magnetic force in a manner similar to a compass needle responding to the magnetic field of the earth: They align along the north–south longitudinal lines of magnetic force. An element must possess magnetic moment in order to exhibit the MR phenomenon.

In an undisturbed situation the magnetic moments of individual nuclei point randomly, and the total sum is zero. When a strong external magnetic field (B0) is applied, the vectors of the individual nuclei tend to align along the direction of the field, pointing toward north or south poles. This process, called magnetization, proceeds exponentially: It is rapid at the beginning and slows down as equilibrium is approached (when the external magnetic force is discontinued, magnetization also decays exponentially at an equal rate). In the equilibrium phase a fixed fraction of the south- and north-pointing nuclei flip randomly (i.e., resonate) between the two directions. North orientation is more stable and attracts more nuclei. The sum of all unopposed north-pointing vectors constitutes a magnetization vector of the sample M0 (Fig. 1).

The vectors of the individual nuclei also exhibit a rotational motion around the axis of the main magnetic field (B0), called precession. The angular speed of precession is expressed by the Larmor equation and is proportional to the strength of the external magnetic (B0) field and the gyromagnetic ratio, which varies with the type of element (Fig. 2).

B. Effect of Radio Frequency Pulse

We now consider a sample that has been placed in the magnetic field and has developed net magnetization (M0). The individual magnetic moments of the nuclei contributing to M0 precess independently of each other. Since there is no net magnetization in the horizontal plane, the transverse vector equals zero and mean magnetization, M0, does not precess. When an RF pulse of appropriate energy is directed toward such a sample, an increasing number of vectors start precessing in coherence around the axis of the main magnetic field. The effect is additive, resulting in a buildup of transverse magne-

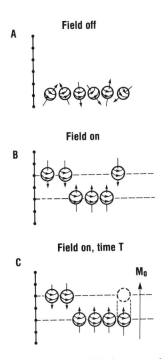

FIGURE 1 (A) With the magnetic field off magnetic moments of the protons point in different directions. (B) They align along north–south orientation when the external magnetic field is applied. (C) After a variable time interval a certain number of protons change orientation to the more stable north-pointing direction. The sum of all unopposed north-pointing magnetic moments constitutes the magnetization (M0) of the sample.

tization. The energy needed to induce coherence (i.e., resonant frequency) is characteristic for a given element and is proportional to the strength of the magnetic field. For hydrogen, the most abundant element within biological tissues, the resonant

frequency is 42.58 MHz/T. The RF pulse, applied in the direction perpendicular to the M0, causes a deviation of the M0 vector away from B0. The angle of deflection is proportional to the duration of the pulse and is expressed in degrees. Commonly, the vector M0 is deflected 90° or 180° (Fig. 3).

Upon discontinuation of the RF stimulus, there is a gradual return of magnetization to the original pre-excitation position. The rate of the return is governed by several factors, the most important being a loss of energy due to interaction with adjacent structures (e.g., spin–lattice and longitudinal relaxation time) and a loss of coherence of individual nuclei secondary to inhomogeneities of the local magnetic properties of the sample and that of the main magnetic field (spin–spin and transverse relaxation time). The rates of longitudinal and transverse relaxation are termed T1 and T2, respectively, and are usually expressed in milliseconds.

C. T1 and T2 Relaxation

T1 relaxation time reflects the rate of dissipation of magnetization from the excited nuclei into the adjacent environment (i.e., the lattice). It is the time interval needed to achieve 63% gain (or loss) of longitudinal magnetization (4 T1 = 98%). T1 depends on the chemical structure of the lattice. Dissipation of energy in an environment containing medium-sized molecules (e.g., cholesterol) is faster, because internal frequencies of those molecules are in the range of precession (i.e., Larmor) frequency. The energy exchange becomes progressively less efficient as the disparity of frequencies increases. Magnetization decays are slower in solids and pure liquids, because of mismatching frequencies of lattice

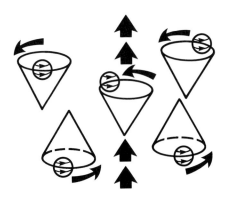

FIGURE 2 When placed in a strong magnetic field, magnetic moments precess around the main magnetic field (B0). The frequency of the precession can be calculated from the Larmor equation, $\omega = \gamma B$, where ω is the frequency, γ is the gyromagnetic ratio, and B is the magnetic field.

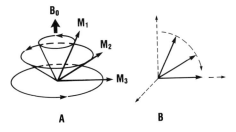

FIGURE 3 As a radiofrequency pulse perpendicular to the direction of the main magnetic field is applied, magnetization M0 is deflected away from B0 to assume new directions (M1, M2, M3...). The angle of deflection, called the flip angle, is directly proportional to the duration of the radiofrequency excitation; that is, doubling of a stimulus, causing 90° deflection of M0, will result in a 180° directional change.

molecules and the precession rate of magnetized protons. T2 relaxation time refers to a rate of loss of transverse component of magnetization (i.e., coherence). This is due to inhomogeneity of the magnetic environment, characteristic to a given lattice in which resonance takes place. The loss of coherence is relatively rapid in solid tissue and impure liquids (short T2), but is maintained longer in pure fluid (long T2). The T2 relaxation rate is significantly faster than T1.

D. Measurements of T1 and T2 Relaxation Times

The rate of dissipation of magnetization is reflected by the loss of amplitude of the signal, after discontinuation of the RF pulse (i.e., free induction decay). A variety of pulse sequences have been designed to measure T1 and T2 relaxation times. Generally, they are composed of a combination of 90° and 180° RF pulses, applied to the imaging sample at various echo time (TE) and repetition time (TR) intervals (Fig. 4). By far the most commonly used in clinical practice is a spin–echo (SE) sequence, which provides both T1 and T2 information. Others include special sequences sensitive to T1 contrast differences (e.g., inversion recovery) or moving protons (e.g., gradient-recalled acquisition at steady state and fast low-angle shot).

E. Image Formation

The imaging process consist of three basic steps: acquisition of data from a sample, computer reconstruction, and image display. During acquisition information in the form of the amplitude and decay rates of the signal is gathered from each individual point (i.e., voxel) of the sample. To achieve spatial

information, additional magnetic fields, created by gradient coils, are consequently superimposed on the main field. The strength of such a modified B0 field increases gradually along the gradient plane (~1 G/cm). The application of the gradient causes nuclei located in the stronger parts of the field to precess faster than those located in the weaker field. When an RF pulse is applied, only protons precessing with corresponding frequency will respond, limiting the imaging volume to a selected plane (slice-selective gradient Z). During application of the second gradient (X), protons in a higher field develop an angle advantage, differentiating the plane into columns of the same phase (i.e., phase encoding). Finally, the third gradient, orthogonal to the previous ones (frequency gradient Y) differentiates the protons along the rows of an imaging volume. Thus, each imaging point of the sample is ascribed a specific combination of phase and frequency in relation to the X, Y, and Z coordinates (Figs. 5 and 6). In practice the gradient fields are switched on temporarily, during or immediately after application of the RF pulse or receipt of the free induction decay signal.

Each imaging point is characterized by a sine wave of a specific amplitude, duration, and frequency. The information pertaining to all points within the sample is analyzed by a computer system, using a mathematical process, two-dimensional Fourier transform (FT). FT decodes the information contained within sine waves and transforms it into a graphic representation of an object (Fig. 6).

F. Image Display

Ideally, a large number of points within an imaging volume should be sampled, thus providing high resolution. In practice the entire imaging volume is divided into a limited number of voxels of equal

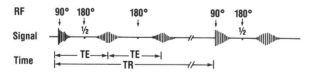

FIGURE 4 Spin–echo sequence. The initial 90° is followed by a 180° refocusing stimulus after a time interval of half-echo time (TE). At TE maximum signal is returned. The 180° pulse may be applied again (i.e., a second echo). After a prescribed repetition time interval (TR), the sequence is repeated. The formula for spin–echo sequence is $I\ N(H)(1 - e^{-TR/T1})(e^{-TE/T2})$, where I is the intensity and $N(H)$ is the proton density. Manipulation of TR and TE intervals results in a shift of the balance of the equation toward either T1 (short TE and TR) or T2 (long TE and TR).

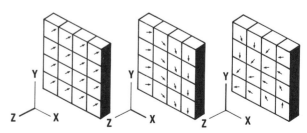

FIGURE 5 Consecutive application of orthogonal gradients along the X, Y, and Z axes ascribes a specific combination of frequency and phase to each imaging point (i.e., voxel).

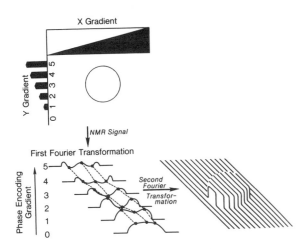

FIGURE 6 Free induction decay signals from each imaging point arrive in the form of sine waves. Information is then decoded, using Fourier transform, and translated into a two-dimensional image of an object. [From Harms *et al.*, Principles of NMR. *RadioGraphics* **4** (Spec. Vol.), 42 (1984).]

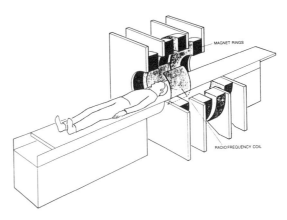

FIGURE 7 Resistive MR unit. The static magnetic field (B0) is created by magnet rings. A radiofrequency (RF) coil generates the RF pulse. Gradient coils (not shown) are superimposed between the magnet rings and the RF coil. [From I. L. Pycket (1982). NMR imaging in medicine. *Sci. Am.* **246**].

shape and volume. Each voxel has a specific position in space, in reference to the *X, Y,* and *Z* coordinates. Each voxel is sampled at least once, emitting a characteristic MR signal. After a multitude of the signals from all voxels is processed via FT, it is displayed as a two-dimensional image. Each voxel has its representative pixel. The amount of brightness displayed by a pixel depends on the intensity of the MR signal. An absent signal results in black and a strong signal results in bright (white) appearance of the pixel. Intermediate intensities are displayed as various shades of gray. The size of the matrix is controlled by an operator. Usually, a multitude of 64 is used (e.g., 128 × 128, 128 × 256, or 256 × 256). The higher number implies that the size of the voxels is smaller and resolution is greater. Of course, this requires more signal probing, extending the acquisition time.

III. Instrumentation

An MR scanner consists of four basic elements: an external magnet (B0), gradient coils, RF coils, and a computer system with image display. The main magnet can be permanent, or an electromagnet (Fig. 7). In the latter case, the coil can be resistive or superconducting. The permanent and resistive magnets generate a magnetic field in the kilogauss range. Superconductive magnets, which operate on the basis of the decreased resistance to circulation

of current at a low temperature, produce higher and more homogeneous magnetic fields (i.e., 0.5–2.0 T). Frequent replenishments of the coolant (e.g., liquid nitrogen or helium) are required to maintain the temperatures close to absolute zero.

Gradient coils modify magnetic fields and create a gradient along the *X, Y,* and *Z* axes. Usually, a coaxial selenoid system is used. A linear gradient of 1 G/cm is considered to be adequate.

RF coils generate RF energy in the plane perpendicular to the direction of the main field. They vary in size and shape, depending on the unit and the manufacturer.

Surface coils are used frequently in clinical practice to examine a selected part or region of the body. They provide better resolution by creating a steeper gradient.

The computer system conducts all automatic functions, including application of the RF pulses and gradients. High capacity enables it to process a large amount of data. In the majority of clinical units, the computer operations are simultaneous with acquisition. Visual display of the image is either instantaneous or is accomplished shortly after the examination has been completed.

IV. Clinical Applications

A. Imaging Parameters

For the purpose of MR investigation, biological tissue can be divided into three basic elements: fat (fatty tissue), impure liquids (all other soft or solid

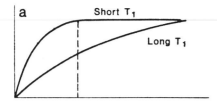

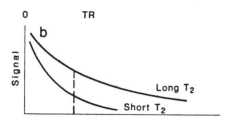

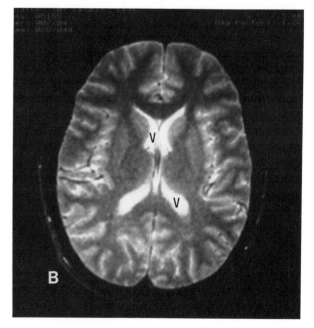

FIGURE 8 Differentiation among various normal and/or pathological tissues depends on differences in the relaxation times. (a) On a spin–echo sequence tissue with a short T1 has a brighter signal intensity than tissue with a long T1. (b) The reverse is true when T2 is measured. The signal information should be obtained in the range of expected maximum differences in T1 and T2 relaxation times between examined tissues (dashed lines). TR, Repetition time; TE, echo time. [Adapted from Budinger, T. F., and Lauterbur, P. C. (1984).]

tissues), and pure liquid (fluid containing spaces; e.g., the cerebrospinal fluid). The MR signal generated from these three elements varies in intensity, allowing for differentiation among various tissues (Fig. 8).

Although many sequences have been designed for MR, an SE sequence with short and long TR/TE intervals is utilized most routinely in clinical imaging. An SE sequence is both T1 and T2 dependent. The balance can be shifted either way by appropriate adjustment of the TR and TE intervals (shorter TR/TE, T1 weighting; longer TR/TE, T2 weighting). T1-weighted images provide good differentiation of fatty muscle tissue and body fluids (high, intermediate, and low intensity, respectively). T2-weighted sequences are more sensitive to changes in the water content of the tissues. Pure fluid (e.g., cerebrospinal fluid within the ventricles of the brain) is homogeneously hyperintense. Brain tissue has intermediate intensity and appears gray.

B. Normal Anatomy

The structures on an MR image are defined as hypo-, iso-, or hyperintense. The relaxation time of

FIGURE 9 Axial spin–echo images of a normal human head. (A) A T1-weighted sequence with a TR–TE ratio of 600 : 20 msec. Cerebrospinal fluid within the ventricular system is hypointense (V), and subcutaneous fat is hyperintense (F). Gray matter (dots) has a longer T1 relaxation time than white matter (O) and appears slightly darker. Blood vessels are hypointense (arrowheads). (B) A T2-weighted image with a TR–TE ratio of 2000 : 80 msec. Cerebrospinal fluid (V) is now hyperintense in reference to the brain tissue.

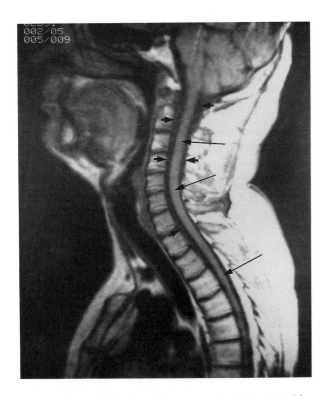

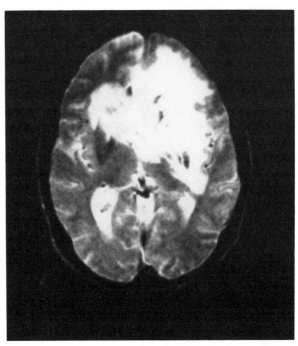

FIGURE 12 A T2-weighted image of the human head. Bifrontal hyperintensity represents a neoplasm (i.e., a lymphoma) and edema of the brain tissue, complicating the acquired immunodeficiency syndrome (AIDS).

FIGURE 10 Sagittal spin–echo sequence of a T1-weighted image of the cervical spine. The TR–TE ratio is 600 : 20 msec. Bone marrow within the vertebral bodies is hyperintense, due to the high content of fat tissue. The spinal cord (long arrows) is of intermediate intensity, surrounded by hypointense cerebrospinal fluid (short arrows).

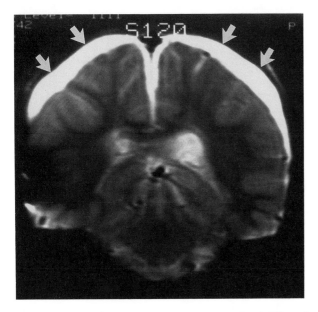

FIGURE 11 A T1-weighted image of the human head. Bilateral collections of abnormal intensity are shown, along the convexities (arrows). The diagnosis was intracranial hemorrhage (i.e., a chronic subdural hematoma.

cerebrospinal fluid (i.e., pure liquid) is long. Therefore, cerebrospinal fluid has a low intensity on T1-weighted images and a high intensity on T2-weighted images. Solid tissues (e.g., compact bone) contain few mobile protons, generating a low-level signal. A signal originating from air spaces (e.g., lungs, paranasal sinuses) is either very low or absent. Fat tissue has short T1 and intermediate T2-weighted sequences. The white matter of the brain consists mostly of conducting fibers, rich in myelin. The gray matter has a significantly higher content on water. Both are of intermediate intensity, but have slightly different relaxation times and can be distinguished on MR image. In general, flowing blood appears to be hypointense. This is due to the influx of nonmagnetized protons into an imaging section between the time of application of consecutive RF pulses (Fig. 9). [*See* ADIPOSE CELL; BRAIN.]

The components of the spine (i.e., the vertebrae, the spinal cord, and the cerebrospinal fluid) are clearly depicted in Fig. 10.

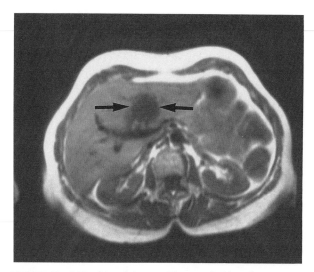

FIGURE 13 MR of the abdomen. The rounded hypointense area (arrows) represents a neoplasm.

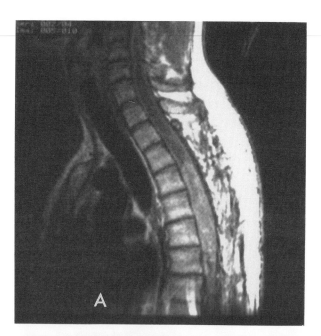

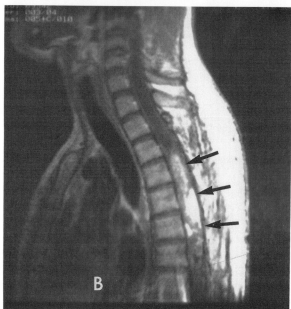

FIGURE 14 T1-weighted images of the thoracic spine (A) without contrast and (B) with contrast. There is enhancement of the spinal cord tumor (arrows in B).

C. Pathological Conditions

Pathological process is identified on MR by either abnormal intensity or configuration. A variety of pathological conditions (e.g., tumors, inflammatory processes, or hemorrhages) might display both. A range of intensities of lesion could be wide and sometimes characteristic for a particular disease entity (Figs. 11–13). The majority of the central nervous system abnormalities appear to be hypointense on T1-weighted images and hyperintense on T2-weighted images.

D. MR Contrast Enhancement

Substances with paramagnetic properties are capable of accelerating the process of energy transfer into the environment from resonating protons. This results in shortening of the relaxation times. Many agents are experimental at the present time. A compound of heavy metal, gadolinium, has been approved for human use (gadolinium–DTPA) (diethylenetriamine pentaacetic acid). When injected intravenously, the contrast leaks out from the circulation into the pathological tissue at the site of the damaged blood–brain barrier in the central nervous system. Accumulation of a contrast agent in the abnormal tissue shortens local relaxation times (predominantly T1), frequently facilitating the diagnosis and outlining the true extension of the pathological process (Fig. 14). [*See* BLOOD–BRAIN BARRIER.]

V. Summary

MR has emerged as a new imaging technique in the early 1980s. It became a method of choice in a radiological assessment in some disease entities and alleviated the need for more invasive procedures in many patients. There is no known direct hazard from exposure to magnetic field or RF waves in diagnostic range. MR imaging has established itself firmly as an integral part of the diagnostic evaluation of human disease.

Bibliography

Bloch, F. (1946). Nuclear induction. *Phys. Rev.* **70,** 460–473.

Bradley, G. W., Waluch, V., Yadley, A. R., and Wyckoff, R. R. (1984). Comparison of CT and MR in 400 patients with suspected disease of the brain and cervical spinal cord. *Radiology* **152,** 695–702.

Bradley, W., Newton, H. T., and Crooks, L. E. (1983). Physical principles of nuclear magnetic resonance. *In* "Advanced Imaging Techniques" (H. T. Newton and D. G. Potts, eds.). Clavadel, San Anselmo, California.

Budinger, F. T., and Lauterbur, C. P. (1984). Nuclear magnetic resonance technology for medical studies. *Science* **226,** 288–298.

Bydder, G. M., Steiner, R. E., Young, I. R., Hall, A. S., Thomas D. J., Marshall, J., Pallis, G. A., and Legg, N. J. (1982). Clinical NMR imaging of the brain: 140 cases. *Am. J. Nucl. Reson.* **3,** 459–480.

Kramer, M. D. (1984). Basic principles of magnetic resonance imaging. *Radiol. Clin. North Am.* **22,** 765–778.

Krol, G., Sze, G., Malkin, M., and Walker, R. (1988). MR of cranial and spinal meningeal carcinomatosis: Comparison with CT and myelography. *Am. J. Neuroradiology* **9,** 709–714.

Purcell, E. M., Torrey, H. C., and Pound, R. V. (1946). Resonance absorption by nuclear magnetic moments in a solid. *Phys. Rev.* **69,** 37–38.

Pykett, L. I. (1982). NMR imaging in medicine. *Sci. Am.* **246,** 78–88.

Sze, G., Krol, G., Zimmerman, R. D., and Deck, M. D. F. (1988). Intramedullary disease of the spine: Diagnosis using gadolinium-DTPA enhanced MR imaging. *Am. J. Neuroradiology* **9,** 847–858 and *Am. J. Roentgenology* **151,** 1193–1204.

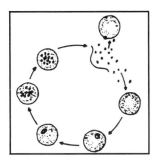

Malaria

WALLACE PETERS, *London School of Hygiene and Tropical Medicine*

Glossary

Anopheles Genus of mosquitos (Insecta), which includes the species that transmit malaria parasites to humans

Gametes Male (microgamete) or female (macrogamete) germ cells

Haplotype Particular combination on a chromosome of alleles of the genes concerned with histocompatability

Merozoites Products of schizogony

Monokines, lymphokines Substances released from monocytes or lymphocytes (white blood cells) when they come into contact with certain antigens; they mediate responses in other cells of the immune system such as macrophages

Parasitophorous vacuole Vacuole in a host cell that houses a parasite

Plasmodium Genus of Protozoa in the phylum Apicomplexa, which includes blood-dwelling parasites of hosts ranging from birds and reptiles to the primates

Poikilothermic Taking on the temperature of the environment

Quartan Associated with fever peaking every 72 hours

Rhoptries Organelles at the apical end of the motile stages of malaria parasites; they are believed to produce proteins that assist the penetration of new host cells

Schizont; schizogony Stage in which asexual division of a malaria parasite occurs; process of such division inside a host cell

Sporozoite Stage of a malaria parasite infective to humans that develops in the mosquito vector

Tertian Associated with fever peaking every 48 hours

Zoonosis Human infections that are acquired from animal reservoirs

Zygote Product of fusion of a male and female gamete

MALARIA is an infection associated with fever and anemia caused by invasion of the red blood cells by parasitic Protozoa of the genus *Plasmodium*. The four species that are pathogenic for humans are all transmitted by female *Anopheles* mosquitoes. At least one-half of the world's human population lives in parts of the tropics and subtropics where malaria is transmitted. Although all four malaria species can cause serious illness, often of a chronic nature, one of them, named *P. falciparum* (the "malignant tertian" parasite), continues to kill more than a million people each year. It is understandable, therefore, that malaria has had a profound influence on the evolution of humans and other aspects of human biology.

I. The Malaria Parasites

A. Parasitological Aspects

1. Life Cycles

The parasites that affect humans belong to two subgenera called *Plasmodium* (*Plasmodium*) and *Plasmodium* (*Laverania*) (Table I). A number of *Plasmodium* species of subhuman primates can also parasitize humans. Although such zoonotic infections are probably rare, they are of interest in that they reflect the evolution both of the parasites and their hosts.

TABLE I Species of *Plasmodium* Found in Humans and the Type of Fever They May Cause

Subgenus	Species	Type of fever
"Human" species		
(*Laverania*)	*falciparum*	"malignant" tertian (or quotidian in early stages)
(*Plasmodium*)	*vivax*	"benign" tertian
	ovale	"benign" tertian
	malariae	quartan
Other species		
(*Plasmodium*)	*knowlesi*	quotidian
	simium	tertian
	cynomolgi bastianellii	tertian
	brasilianum	quartan

Plasmodium undergoes a complex life cycle, partly in poikilothermic *Anopheles* mosquitoes (the females of which require blood as a source of protein to develop their eggs) and partly in warm-blooded primates. The infective stages, sporozoites, are injected together with saliva when the insect feeds and are carried passively to the liver where they enter the cytoplasm of liver cells (hepatocytes), by a route and mechanisms that are not yet entirely clear. Within the hepatocytes, a single schizont develops from each sporozoite during a period of 1–3 weeks, each forming several thousand daughter cells (called *merozoites*) that invade red blood cells within the hepatic capillaries (sinusoids) (Fig. 1).

The parasites undergo further cycles of asexual schizogony within the erythrocytes during a period of 48–72 hours after which they disrupt the host cell, allowing each of the eight to 24 progeny (also termed *merozoites*) to enter a fresh erythrocyte (red blood cell). This process continues until the host either overcomes the infection or succumbs to its pathological effects. The latter situation rarely arises other than in infections with *P. falciparum,* which causes damage to the circulation of the brain and other organs. In two species, *P. vivax* and *P. ovale,* a proportion of the sporozoites that enter hepatocytes round up and remain for months or years as dormant, uninucleate cells ("hypnozoites"), which, triggered by stimuli so far unidentified, subsequently produce further waves of schizonts, which generate new crops of erythrocyte-invading merozoites.

The erythrocytic stages are programmed to undergo, at certain stages of the infection, a genetic change by which they produce not asexual but sex-

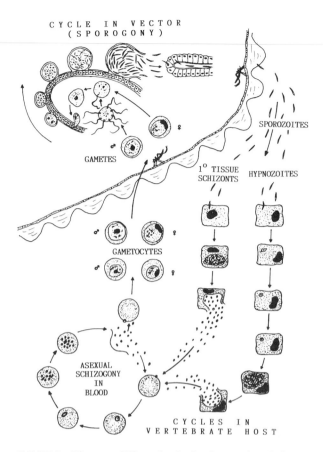

FIGURE 1 Diagram of life cycle of relapsing species of *Plasmodium* (*Plasmodium*) such as *P. vivax.* Note that the hypnozoite stage does not exist in *P. (P.) malariae* or *P. (Laverania) falciparum.* [Reproduced from Peters, W. (1987). Academic Press].

ual stages ("gametocytes"), which only develop further if they are taken up together with blood into the midgut of a susceptible female *Anopheles.* The gametocytes then mature into gametes that fuse, forming motile zygotes ("ookinetes") that penetrate midgut epithelial cells. They pass right through these to exit from the other side and form cysts beneath the membrane that lines the outer surface of the midgut epithelium. Thousands of sporozoites develop within these "oocysts." Once mature they leave the oocysts, migrate through the insect's body cavity, and finally congregate within the salivary glands from which they can subsequently gain access to a new mammalian host when the insect takes a further feed.

2. Ultrastructure and Physiology of the Mammalian Stages

During the long process of evolution, malaria parasites have developed a remarkable series of structural adaptations that enable them to invade and grow within the various types of host cells that they encounter during the different stages of their complex life cycle. The hosts, both vertebrate and invertebrate, in their turn have evolved ways of destroying, or at least tolerating, the invaders. This account will be limited to a review of the host–parasite interfaces in humans.

a. Merozoite–Erythrocyte Interface In the initial process of erythrocyte invasion, the merozoite induces the erythrocyte membrane to form a ring-like junction that progressively moves along the parasite, "squeezing" the merozoite into a parasitophorous vacuole within the host cell. In this process the parasite is preceded by a layer of host cell membrane, so that when newly invaginated, it comes to be surrounded by a complex consisting of its own tissues plus erythrocyte membranes. The nature of these rapidly changes. Some of the parasite membranes degenerate, leaving the vacuole lined by two bilayers, one belonging to the parasite and one to the host. The constituents of the host component become changed, and spectrin, which normally coats the cytoplasmic side of the red cell, is removed. Parasite-derived materials probably become incorporated into the parasitophorous vacuole as the parasite grows. Included in the parasitophorous vacuole membrane there is probably an ATP-driven protein pump for transferring essential nutrients from the erythrocyte cytoplasm into the parasite or for exporting unwanted materials. Increased synthesis of this protein may be responsible for the enhanced export of antimalarial drugs such as chloroquine from the parasites in multiple-drug resistant strains of *P. falciparum*.

b. Parasite "Poaching" of Host Cell Enzymes Once within the erythrocyte, the parasites make the maximum use of several host cell enzymes, "switching off" the genes that are capable of producing the corresponding parasite enzymes at other stages of the life cycle. An example of this is the use by *P. falciparum* (and probably other species) of the host enzyme glucose-6-phosphate dehydrogenase (G6PD), which the parasite requires for a step in energy production via the reduction of glutathione in the pentose phosphate pathway. It also uses the host cell's superoxide dismutase to reduce potentially toxic oxygen derivatives, to the destructive action of which both the parasites and their host cells are highly susceptible. In the presence of G6PD-deficient host cells, the parasites can, under certain circumstances, switch on their own G6PD gene (see below), as they do normally at other stages of the cycle (e.g., intrahepatic schizogony).

A further example of the use of host cell enzymes is the dependence of the avian parasite *P. lophurae* on the host cell enzyme pantothenate kinase to convert pantothenate (a vitamin) to coenzyme A, which is required in many enzymatic reactions. This may also occur in *P. falciparum*.

c. Parasite Proteins in Erythrocyte Envelopes Marked changes are induced in the host erythrocytes by a malaria infection. Thus microscopic examination of suitably stained thin blood films reveals punctate spots, known as Schüffner's dots (in *P. vivax*), James' spots (in *P. ovale*), Maurer's spots, and Maurer's clefts (in *P. falciparum*). At the electron microscope level, small protuberances ("knobs") with a complex structure are seen over the surface of red cells infected with *P. falciparum* (Fig. 2), and even smaller protrusions are found on red cells containing *P. malariae*. They do not correspond to the changes observed with the light microscope. In *P. falciparum*–infected red blood cells, the knobs are sites of adherence of the parasitized erythrocytes with the walls of capillaries, notably in the brain. No such sequestration of infected red cells has been reported in association with *P. malariae* infection, but it may well take place on a smaller scale at sites so far unidentified. The Schüffner's dots of *P. vivax* are formed by a red cell modification that may help to increase the surface area of the infected cells, favoring exchanges with the environment.

The "knobs" on *P. falciparum*–infected red cells are also associated with novel proteins rich in the amino acid histidine synthesized by the parasites, which also have other histidine-rich proteins in certain of their own organelles (the "rhoptries" of the merozoites, which are formed in the intraerythrocytic schizonts). The latter proteins seem to be associated with the invasion of fresh host cells. By sophisticated techniques involving the use of specific antibodies, a large number of parasite-derived proteins have been identified on the surface of the host cells. Some of these are glycoproteins, which can act as antigens that elicit antibody formation. It is not known whether the parasites transfer these proteins to the outside of their host cells' membranes during growth or whether the proteins are released on rupture of the red cell at schizog-

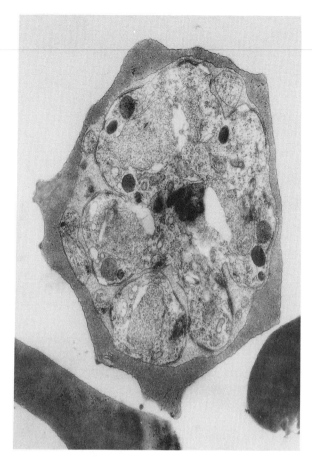

FIGURE 2 Electron micrograph of a human erythrocyte containing a maturing schizont of *Plasmodium (Laverania) falciparum* (×20,000. By courtesy of Dr. D. S. Ellis).

ony, subsequently to be absorbed onto the surface of newly invaded (as well as some uninfected) red cells from the plasma. It has to be assumed that at least some of the parasite proteins in the host cell membranes are of benefit to the parasites, perhaps by facilitating the passive or active import of substrates essential to the parasites. At the same time, the parasites acquire most of the amino acid they require by digesting host cell cytoplasm and hemoglobin within their digestive vesicles.

B. Malaria as a Disease

1. Distribution

Malaria caused by *P. vivax* was formerly to be found in parts of the world where the summer temperature exceeds 16°C, while falciparum malaria required the higher temperature of 21°C for its transmission. Changes of land usage leading to the reduction of surface waters associated with increas-

ing urbanization during the past century and, especially, the application of intensive control measures during the past 50 years have combined to restrict somewhat the boundaries in which malaria remains endemic or to eliminate it entirely. In the bulk of the tropics, however, humans' best efforts to date have been frustrated for many reasons: technical, logistical, financial, and social. The parasites and their vectors have shown a remarkable resilience and genetic adaptability that has ensured and is likely to go on ensuring their survival in the foreseeable future. Indeed the anticipated warming of the earth's surface caused by the "greenhouse effect" may well extend the areas of suitable temperature further outward into the northern and southern hemispheres, so that malaria will once again become commonplace in territories from which it had once disappeared, and perhaps even beyond (Fig. 3).

2. Epidemiological Considerations

Depending on the intensity of its transmission, malaria is classified within a spectrum that ranges from hypoendemic to holoendemic, these terms being defined on the basis of specific factors such as the degree of splenic enlargement at different ages of the indigenous population and the prevalence and intensity of malarial parasitaemia (i.e., presence of parasites in the blood). Hypoendemic malaria is characterized by its instability and tendency to occur in periodic epidemics that may be devastating in their effects. In certain highland areas of the main island of New Guinea, for example, epidemics have been known to wipe out 1% of the population who, being only intermittently infected, have a low level of immunity. On the contrary, holoendemic malaria, which is characteristic of the coastal areas of the same island and of much of tropical Africa, presents a remarkably stable situation in which the level of acquired immunity in the adult population is very substantial, but at the cost of a high mortality in infancy and early childhood. It is in such areas that the interaction of several genetic factors in the human host with the parasite, especially its intraerythrocytic stages, is believed to have led to forms of "balanced polymorphism," which are discussed further below.

3. Human Population Movement

a. Within Countries ("Transmigrance") The existence of malaria as well as other diseases (e.g., sleeping sickness) has undoubtedly played an im-

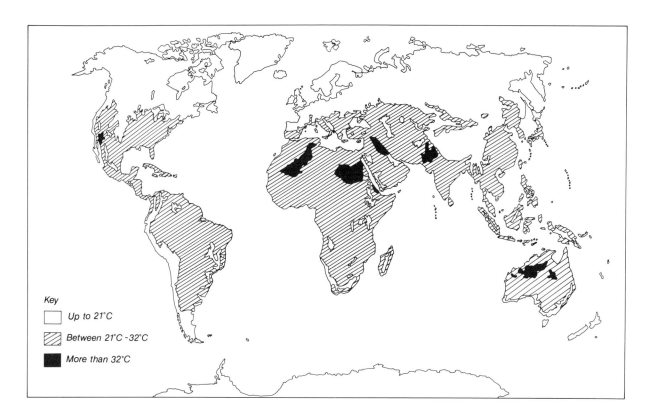

FIGURE 3 Map of the world to show the 21 and 32°C isotherms (July in the northern, January in the southern hemisphere), these representing the approximate limits between which falciparum malaria can be transmitted. (Original by Andrea Darlow).

Key
☐ Up to 21°C
▨ Between 21°C - 32°C
■ More than 32°C

portant role in patterns of human migration and settlement in the past, and it still does so. A current example is the movement of large numbers of Nepalis into the valleys from the hills north of the Terai, which was formerly highly malarious, after the success of the antimalarial campaign that started in Nepal in the early 1950s. From a wild, almost uninhabitable jungle, the Terai has been transformed into spectacularly fertile and productive agricultural land, densely populated by former hill tribes. The other side of the picture, however, is presented by the devastating effect malaria is having on impoverished migrants from the arid, malaria-free northeast states of Brazil to the highly endemic terrain of Amazonas to which they have been attracted by the promise of newly opened territory, fertile for agriculture. For others who have followed the trail in search for gold and other precious minerals, the chance of making a fortune was considerably less than that of dying of malignant tertian malaria.

b. International Movement The impact of malaria on human communities is well exemplified by the plight of refugees obliged to flee for political reasons in their own lands, to seek shelter in neighboring countries where they are exposed to a level of malaria transmission against which they may have little or no immunity. Epidemics of falciparum malaria have struck, for example, Ethiopian refugees sheltering in camps in the Sudan when they became flooded in the rainy season, thus providing exceptionally good breeding grounds for anopheline mosquitoes. In all likelihood there will soon be a significant increase of malaria in Afghanistan, as the refugees return to their homelands carrying with them the malaria parasites, both relapsing *P. vivax* and drug-resistant *P. falciparum,* that they have acquired in refugee camps in neighboring Pakistan. What effects such exposure to an increased level of infection will have during the next decades in the absence of a health infrastructure adequate to cope with the situation remains to be seen. In Thailand the most intense levels of drug resistance ever observed in *P. falciparum* exist in the vast camps sheltering refugees from Kampuchea; these parasites are progressively becoming disseminated from the original foci, affecting the lives of many indigenous Thais.

Today's exponential increase in international air travel has added a new dimension to the possibilities for the further global dissemination of malaria parasites, both through the human host and the vector. At least one focus of drug-resistant falciparum malaria in West Africa is suspected to have originated from nonimmune Europeans who arrived there directly from an endemic area of southeast Asia, carrying a subclinical infection. The supposition is that the highly efficient local anophelines transmitted infection from them to local residents. Vectors too may be transported by sea or air, the latter being the most likely vehicles. The introduction in the 1930s of West African *Anopheles gambiae* to Brazil, where it became established and vastly outperformed the local anophelines as a malaria vector, is a classic of malaria history. In recent years, ''airport malaria'' has acquired a special connotation, namely, of infections initiated in people living near airports in countries otherwise nonendemic, by the bites of anopheline ''stowaways'' that acquired falciparum malaria before boarding the aircraft, usually in West Africa. To what extent such insects carry malaria parasites from one endemic country to another in this manner is unknown, but this phenomenon may well play a role in the rapid dissemination of drug-resistant strains that is a prominent feature of the current scene (see below).

II. Host Responses to Malaria Infection

A. Pathological Effects

The pathological effects of malaria are due partly to changes associated with the invasion and destruction of erythrocytes and partly to the adherence of parasitized erythrocytes within the capillaries of vital organs such as the brain. Infection with *P. vivax, P. ovale,* or *P. malariae* is usually relatively benign, in contrast to that with *P. falciparum.*

Destruction of erythrocytes every 48 or 72 hours by more or less synchronous waves of merozoites leads to anemia with its associated pathophysiological sequelae including anoxia (oxygen deficiency to the tissues) due to loss of oxygen-carrying red cells. Phagocytic macrophages that remove from the circulation both parasites and red cell debris accumulate in the spleen and liver, which subsequently become enlarged. Intraerythrocytic malaria parasites take up and metabolize large amounts of glucose, which they pirate from their hosts who may, in con-

sequence, suffer from a measure of transient hypoglycemia (low blood glucose). Toxic materials also accumulate in severe malaria, leading to a cascade of sometimes lethal pathophysiological changes (see below).

The severity of malaria infection is influenced to some degree by the nutritional status of the host, but it is not, as might be anticipated, necessarily more severe in malnourished individuals. Paradoxically it seems that the parasites as well as the hosts may suffer from deprivation of such substances as, for example, *p*-aminobenzoic acid (an important constituent of milk) or dietary iron. Supplementing these may increase, rather than decrease, the levels of parasitemia. Thus malnutrition or famine have contrasting effects. On the one hand, they act as a significant reducer of human population levels in inhospitable terrain such as drought-stricken states of India in days past or, as has occurred in recent times, in such countries as Ethiopia and Mozambique; on the other hand, they may help to protect the survivors by damping otherwise lethal falciparum malaria infections, one of nature's checks and balances.

B. Host Defense Mechanisms

In general terms, the human organism responds to invading malaria parasites by increasing the production of B lymphocytes, which synthesize a variety of antibodies in response to specific antigens, including those present on the surface of the invaders. These antibodies function either by counteracting products of the pathogens (e.g., exotoxins) or by exposing the invaders to attack by the other major component of the host response, the T cells. Although humoral immunity (the production of antibodies) against malaria infection was the first to be highlighted, it has become abundantly clear in the past few years that protection is afforded mainly by the cellular response, aided and abetted by the humoral component. [*See* LYMPHOCYTES.]

1. Humoral Immunity

It is not surprising, in the light of the complex morphological and physiological changes that the malaria parasite undergoes during the course of its development within both human and invertebrate hosts, that the antibodies produced in response to the different parasite stages are also different and, to a marked degree, stage-specific.

a. Primary Defenses Against Sporozoite Invasion
Although the initial humoral response to any acute infection of the blood is the production of immunoglobulin M (IgM), the major component is immunoglobulin G (IgG). This is apparently also the major component of the antibody response to initial invasion of the circulation by infective sporozoites. IgG produced in response to a primary infection helps to block the penetration of sporozoites into host hepatocytes, but the level of this IgG is not totally sufficient to protect a host on its first infection. It seems that multiple sporozoite challenges are required before a significant protective level of antisporozoite antibodies builds up, and the continuity of this protection probably requires frequent "topping up" by further sporozoite inoculations. The antisporozoite antibodies do not react with young intrahepatic schizonts but do so with more mature ones. Antibodies against intraerythrocytic asexual parasites do not react even with the older liver schizonts.

b. Protection of the Red Blood Cells
First IgM, then IgG, is produced in response to the parasites in erythrocytes, reacting both against parasite antigens expressed at the host cell surface and against antigens present in debris released when red cells rupture. Some of these antibodies act by immobilizing free merozoites or perhaps by preventing their attachment to the surface of new host erythrocytes. Immobilized merozoites thus become exposed to the phagocytic action of circulating macrophages or those fixed in the reticuloendothelial organs (e.g., the spleen or liver).

c. Cutting the Chain of Transmission
Other antibodies are produced in response to the presence of gametes in circulating erythrocytes and to the gametocytes, which are liberated from the red cells after they have been taken up into a mosquito's midgut. These antibodies immobilize the male gametes (microgametes) and probably also block penetration of any free microgametes into the female gametes (macrogametes), thus helping to interrupt the chain of transmission.

2. Cell-Mediated Immunity

a. Role of the Spleen
Of all the reticuloendothelial organs, the spleen plays the primary role in freeing the human host of parasitized erythrocytes. It does this by phagocytosing them as they pass through the capillary vessels (splenic sinusoids) and then destroying the enclosed malaria parasites within the cytoplasm of the splenic macrophages. Paradoxically, however, this function in turn appears to trigger other intraerythrocytic parasites (at least in the case of *P. falciparum*) to transfer proteins to the surface of the host cells, forming the "knobs" by means of which the latter adhere within capillaries of other organs such as the brain.

b. Monokines and Lymphokines
A complex interaction between B and T cells leads to the production of a number of monokines and lymphokines, which are the substances primarily concerned with the uptake and killing, through the production of toxic oxygen radicals, of the intracellular malaria parasites. The substance responsible for attacking the intrahepatic schizonts appears to be gamma-interferon, but interleukins (notably IL-2) are the main effectors in the killing of intraerythrocytic stages. [*See* INTERFERONS; INTERLEUKIN-2 AND THE IL-2 RECEPTOR.]

C. Parasite Defense Mechanisms

To appreciate the subtle interaction of host and parasite biology, consideration has to be given also to the efficient mechanisms that the latter use to ensure their survival in the presence of host immune attack.

1. The Parasite Surface
The induction by intraerythrocytic *P. falciparum* of "knobs" (Fig. 2) described above may well be a mechanism to prevent the sequestration in, and subsequent destruction by, the spleen and its contained macrophages. In fact, "knobless" variants produced when *P. falciparum* is passaged in splenectomized chimpanzees or *Aotus* monkeys are not sequestered *in vivo* and do not attach to tissue culture cells *in vitro* as do "knobbed" variants. Sequestration of knobbed, infected red cells may backfire against the parasites because the disturbance of the cerebral circulation that can follow may kill the host, who then becomes literally a "dead end" also for malaria parasites.

2. Antigenic Variation
Strain diversity in *P. falciparum* has been recognized clinically for many years, but recent studies have demonstrated clearly that the diversity is at the level of the surface of the infected erythrocyte.

Evidence is only now beginning to accumulate to show that antigenic variation occurs also in *P. falciparum* itself. The possible value of this variation in evading host immune responses is self-evident, but much remains to be clarified for this and the other malaria parasites of humans. It is likely that antigenic variation occurs in the relapsing species, *P. vivax* and *P. ovale,* to permit the emergence of consecutive waves of intraerythrocytic parasites from dormant hypnozoites. Either antigenic variation or sequestration in immunologically privileged sites may also account for the persistence of infection with *P. malariae* in some people for decades.

III. Malaria and the Evolution of *Homo sapiens*

A. Evolution of *Plasmodium*

Although the precise origins of protozoan parasites in the genus *Plasmodium* (which includes all those that affect the primates including *Homo sapiens*) must remain obscure, classical taxonomic techniques, together with deductions from our increasing knowledge of the DNA of such organisms, are beginning to transform earlier speculations into likelihoods. An affinity of the subgenus *Laverania* (which includes the malignant tertian *P. falciparum* of humans and the closely related *P. reichenowi* of the chimpanzee and gorilla) with certain species of avian malaria now seems likely, suggesting that there may be an evolutionary connection between these parasites. The family Haemoproteidae (possibly an ancestor of the family Plasmodiidae), which contains blood parasites of lower mammals such as bats, includes also the genus *Hepatocystis*. Members of this genus probably parasitize lemurs and several species of Old World monkeys in Africa and Asia, as well as the orangutan. Because the evolution of parasites closely follows that of their hosts, a study of malaria parasites can throw interesting light on the evolution of the Hominoidea (Fig. 4).

B. The Integrated Evolution of *Plasmodium* and Humans

Whether a particular parasite can associate with a given host depends on many factors in both. In the course of evolution, certain families of parasites such as the Haemoproteidae, for reasons we cannot now determine, have not succeeded in establishing

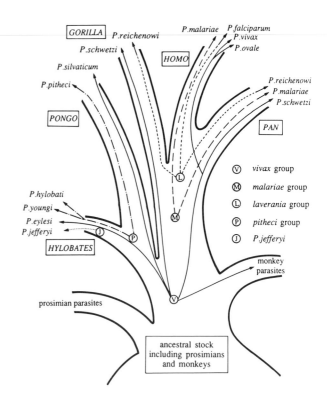

FIGURE 4 The possible evolution of malaria parasites in the subgenera *Plasmodium (Plasmodium)* and *Plasmodium (Laverania)* in apes and humans [Modified from Peters, W., et al. (1976). *Philos. Trans. R. Soc. Lond.* [B] **275,** 439–482.]

themselves in humans or the higher apes. Both haemoproteids and plasmodids feed on hemoglobin during part of their life cycle. It is interesting to note that evidence from studies on hemoglobin suggest that the human–chimpanzee and gorilla lines split from their common Hominidae ancestor between 1 and 2 million years ago, i.e., in the late Miocene or Pliocene periods. DNA data, however, place this split earlier, between 5 and 6 million years ago. This would date the origin of the subgenus *Laverania* at its most recent in the mid-Pliocene period and *P. malariae* well before that. The *P. vivax* group, which gave rise to parasites of monkeys and Pongidae (Fig. 4), must have arisen far earlier, perhaps in the late Eocene or Oligocene periods when the primitive Anthropoidea were first evolving. One of the most recent species of this line, *P. vivax,* has been unable to establish itself in the negroid race of *Homo sapiens,* although it is widely distributed in people of other races. This failure is due to the lack of an erythrocyte surface protein related to the Duffy factor in most blacks. This, however, has not prevented the evolution of the *vivax*-like parasite *P.*

ovale, which is widely distributed in the African population and may have evolved either from *P. vivax* or from the latter's common ancestor with *P. schwetzi.* The prehominid ancestors of *Homo sapiens* must certainly have harbored *vivax*-like parasites, because this line, as stated above, arose at an early stage of plasmodial evolution. This probably took place in Africa (although an Asian site of origin is still proposed by some) to yield several species in Old World monkeys (e.g., *P. cynomolgi*), *P. silvaticum* in *Pongo pygmaeus,* and *P. schwetzi* in the gorilla and chimpanzees.

If Africa was truly the ''cradle'' of *Plasmodium,* the transcontinental migrations of monkeys, apes, and prohomids from at least the Miocene through to the Pleistocene period must have been accompanied by the distribution of the parasites to Europe, across Asia to the Pacific, and probably across the Bering land bridge in the late Pleistocene. Whereas *Plasmodium* thrived and speciated in Asia, it failed to do so in the New World. Although several of the monkey parasites can still give rise to zoonotic infection in humans (e.g., *P. cynomolgi bastianelli* and *P. knowlesi*), it appears that the relatively benign, quartan *P. malariae* (which is believed to have evolved at a later stage than the *P. vivax* group, and which is essentially a parasite of humans and the chimpanzee), was the source, in relatively recent times, of *P. brasilianum* in monkeys of the New World, a ''zoonosis in reverse,'' or anthroponosis. Malaria as a disease does not appear to have been recognized in the New World in pre-Columbian times and may have been taken there by the early human invaders. Whether the *vivax*-like monkey parasite *P. simium* also arose in this manner is less obvious. The malignant tertian *P. falciparum* was certainly introduced by humans who remain its sole host, the transport of slaves from West Africa being a ready source of infection for indigenous anopheline vectors. A possible alternative route for a still earlier introduction of malaria to the New World was with voyagers from the Pacific, a theory that was reinforced by the success of the famous Kon-Tiki expedition.

In its turn, *P. falciparum* has had a marked influence on the direction of human evolution by virtue of its relative inability to thrive in erythrocytes that contain certain enzyme or hemoglobin peculiarities. The fact that *P. falciparum* in most people still produces severe, often fatal infection if untreated is taken to imply that it is the most recent species to have been acquired by *Homo sapiens,* because the

mark of an old-established parasite (at least a successful one) is its tendency toward a more benign relation with its host, the ultimate being a form of commensalism.

IV. Human Genetic Polymorphism and Malaria

A. Genetics of the Immune Response

An attempt has been made to define for one or other of the histocompatibility (HLA) haplotypes a relation with malaria and other diseases. Although certain haplotypes are especially prevalent in parts of Africa that are highly endemic for malaria, or Sardinia, which was formerly malarious, no clear association between the two has been found to date. Similarly, the distribution of two Ig haplotypes unique to Africa, $Gm^{1,5,6,14}$ and $Gm^{1,5,6}$, is closely correlated with that of hyper- and holoendemic malaria, suggesting that malaria may play a role in their natural selection.

B. Erythrocyte Phenotypes

1. Sickle Cell Trait

It is a curious paradox of nature that the existence of a gene that produces a potentially harmful sickle-cell variant of hemoglobin, (HbS), has been able to extend its prevalence in humans by conferring on them a relative protection against the otherwise lethal effects of *P. falciparum.* Sickled erythrocytes containing *P. falciparum* are rapidly sequestered to the detriment of the parasites. Although it is still not clear at the molecular level how this happens, there is abundant evidence that the heterozygous state HbS/HbA has offered a selective advantage for survival, especially in West Africa, although the homozygous condition itself produces severe sickle-cell disease. This state of ''balanced polymorphism'' has permitted the extension of the HbS gene to such a degree that its frequency is now greater than 0.1 in much of equatorial Africa and is greatly elevated in other areas such as the eastern part of the Arabian peninsula and the Indian subcontinent. Two distinct mutant phenotypes are recognized in Africa on the basis of DNA sequences, one focused around Sierra Leone and the other centered on the mouth of the Zaïre River. The latter type occurs in the Middle East and India. Whether this mutant gene diffused there from

West Africa or arose as an independent event is unknown. The migration of Africans to the New World has disseminated the HbS gene widely there also, even in the absence of malaria. In the nonmalarious areas, there is an even greater chance that children will be born with the homozygous HbS/HbS condition. [*See* SICKLE CELL HEMOGLOBIN.]

2. Thalassemia

One type of thalassemia, particularly that due to the absence of one of the genes for the alpha chain (α-thalassemia, another condition that is associated with sickling), may also offer a selective advantage in the presence of falciparum malaria. Thus the α-thalassemia-2 condition occurs at a high rate in Africans and those of African descent, including a third of North American blacks. The other type of thalassemia, β-thalassemia, also confers some protection against malaria.

3. Other Hemoglobinopathies

P. falciparum does not develop as well in erythrocytes that contain fetal hemoglobin (HbF). This declines at a slower rate than normal in infants who are heterozygous for β-thalassemia, thus perhaps conferring some degree of selective advantage to such infants. The parasites also grow less well in the erythrocytes of people with homozygous HbC or HbE genes but not in those of heterozygotes. There is no clear evidence, however, that these hemoglobinopathies have given any real survival advantage in human populations of malaria-endemic areas.

4. Hereditary Ovalocytosis

In this condition, which is common in parts of southeast Asia and Melanesia, the erythrocytes assume an oval form caused by changes in their cytoskeleton. These changes are detrimental to invasion by malaria parasites so that the levels of parasitemia of both *P. falciparum* and *P. vivax* are lower in the blood of individuals with this mutation. It is uncertain whether ovalocytosis itself is detrimental to survival of the parasite, but the fact that the trait has not penetrated the totality of the population would suggest that it is so and that its presence is limited by another form of "balanced polymorphism." In Papua New Guinea, gene frequencies are as high as 0.22–0.50 in the coastal areas where malaria is hyper- to holoendemic, but very low or nil in the mountains where malaria is hypoendemic or absent.

5. Duffy Factor

Two red cell surface antigens, Fy^a and Fy^b, are the main components of the Duffy blood group system, and Fy^a has recently been identified as a glycoprotein. Erythrocytes that lack these factors (FyFy) are not invaded by merozoites of *P. vivax*. This condition is very common in blacks who are, consequently, largely refractory to this parasite. The Fy antigens, which appear to act as merozoite receptors, affect not the attachment of the parasites to the hosts but the stage of penetration of the red cells. It has been noted above that the *vivax*-like parasite *P. ovale* replaces *P. vivax* in areas where the homozygous FyFy condition is prevalent, whereas *P. falciparum* is commonly the dominant species. There is undoubtedly a subtle and fascinatingly complex balance between coexistence or mutual exclusion of the different species of *Plasmodium* and the innate genetic polymorphisms of the hosts' erythrocytes in many parts of the world.

C. Enzyme Variants

Marked variations in the genes coding for erythrocytic glucose-6-phosphate dehydrogenase (G6PD) occur in different geographical regions, and a deficiency of this enzyme is a common trait. As with some of the hemoglobinopathies noted above, the frequency of G6PD deficiency tends to be highest in areas where falciparum malaria is or was endemic. There is good evidence that malaria in those with the heterozygous state Gd A$^-$/B is less severe than in those with normal G6PD or the homozygous A$^-$/A$^-$ condition, giving rise, as with HbS, to a state of balanced polymorphism in highly endemic areas such as West Africa. The mechanism responsible for limiting the growth of *P. falciparum* in G6PD-deficient erythrocytes is probably a reduction in maintenance of reduced glutathione, which protects the cell and its parasites against oxidant stress to which they are very susceptible. In the absence of host G6PD, *P. falciparum* apparently has the ability to switch on genes of its own for the synthesis of this enzyme because, in time, the parasites can adapt to existence in G6PD-deficient host cells if at least a proportion of normal cells is present.

The relative mildness of falciparum malaria in nonimmune, black U.S. soldiers infected in Vietnam compared with their white colleagues led to the observation that American blacks tend to have lower levels of adenosine triphosphate in their erythrocytes. Subsequently it was shown that they also have reduced pyridoxal kinase levels, and it is suggested that the parasites depend normally on their host cells for this enzyme, much as they do for G6PD. Linked to the action of this enzyme is pyridoxine oxidase, a deficiency of which has been identified in some individuals with thalassemia. It is thus possible that a deficiency of these two host enzymes plays a role in protection against *P. falciparum*, contributing to the relative ability of Africans and their descendants to survive this infection.

V. Man versus Parasite

Although the organisms causing malaria were first recognized by Laveran in 1880 and their mode of transmission by mosquitoes was described before the turn of this century by Ross, the disease has remained as one of the most widespread and life-threatening maladies of humankind up to the present day. Two main lines of attack have been evolved and deployed to counter malaria. The first is directed at the vectors and the second at the parasites. In the early 1950s a brave attempt to eradicate malaria completely was sponsored by the World Health Organization, using both approaches, the application of the then newly discovered, synthetic insecticide, dichloro-diphenyl-trichlorethane (DDT) to kill the vectors, and the potent, synthetic antimalarial drug, chloroquine, to kill the parasites. Some significant gains were made, and malaria was eliminated in some of the marginal areas of its distribution, but little lasting impact was made on the disease in its heartlands (i.e., the equatorial belts of Africa and South America, much of southeast Asia, and the islands of the southwest Pacific). A third line of attack is now the focus of attention, vaccination of humans against *Plasmodium*.

A. Attack on the Vectors

Meticulous entomological research on the breeding habits of the *Anopheles* that serve as malaria vectors has shown that some species should be relatively easy to control by restricting the aquatic sites available to them, but others that can adapt to a wide variety of ecological niches pose almost insurmountable obstacles to malaria control by simple water management. In this event a variety of insecticides can be deployed to reduce the population of larval or adult mosquitoes.

Among the first successful adulticides was pyrethrum, a readily produced plant product, with the aid of which the exotic, African *Anopheles gambiae* was eliminated from its new stronghold on the east coast of Brazil in the 1930s. When DDT was synthesized just before World War II and the immense potential of its prolonged insecticidal action was realized, it was adopted for use as a tool for stopping malaria transmission. The principle was to kill the female mosquitoes that entered houses and bit infected people before the parasites could mature in the insects and be passed to new hosts. This scheme was successful for about 20 years from the early 1950s, but sadly, it has now largely been abandoned for two principal reasons. Firstly, a high proportion of the mosquitoes became resistant to DDT. Secondly, it was discovered that DDT was entering major food chains and could pose a threat to the environment (e.g., by damaging the eggs of birds that had fed on contaminated insects). Both these problems were aggravated, if not primarily caused, by the massive deployment of DDT and related insecticides, not so much for malaria control as for the control of agricultural insect pests. DDT and other chlorinated hydrocarbons were partly usurped by new generations of insecticides; the organophosphates (relatives of a family of potent nerve gases) and carbamates proved to be excellent in activity but far more costly than DDT. The pendulum has now swung back. A new generation of insecticides has recently been developed based on the chemical structure of the active constituents of pyrethrum. However, these, too, have many disadvantages, not the least being their high price, beyond the purses of many of the Third World countries that most need them. A recent development in larval control is the application of the metabolic products of a bacterium, *Bacillus thuringiensis,* which is highly toxic to many aquatic insects.

B. Antimalarial Drugs

It is an interesting commentary on the relation of humans with our environment that we frequently have to return to nature to achieve what all our ingenuity has failed to do. The return to pyrethrum (at least as a starting point for a new generation of insecticides) is one example. The return to other plant derivatives to kill malaria parasites is another.

Plants that can be used to decrease fever have been a feature of native folk remedies from time immemorial. Because malaria was the cause of many fevers in the tropics and subtropics, it is not surprising that some plants used in traditional medicine have since been found to possess specific antimalarial properties. The most widely known antimalarial plant in the western world was quinine, which was brought to Europe some 4 centuries ago. The development, during the 20th century, of a number of synthetic antimalarial drugs, the 8-aminoquinolines pamaquine and primaquine, mepacrine (atebrine), and 4-aminoquinolines such as chloroquine, opened the way to the prevention and treatment of malaria on the grand scale, without the side effects commonly produced by quinine. Subsequently, a new family of synthetic antimalarials including proguanil and pyrimethamine were developed. These combined the ability to prevent the establishment of infections in the liver with that of killing any parasites that escaped from that organ to establish infection in the blood.

Chloroquine, pyrimethamine, and primaquine (which prevents relapses of *P. vivax* by killing hypnozoites) came to be used on a massive scale as the second limb of the attack on malaria during the eradication era of the 1950s and 1960s. One reason why the eradication campaign failed was that *P. falciparum,* the malignant tertian parasite, evolved a mechanism for overcoming the action of chloroquine and pyrimethamine. Quinine, until recent times, retained its ability to destroy parasites that resisted the synthetic compounds. An intensive program was established in 1963 by the U. S. Army for other alternative antimalarials to combat the drug-resistant parasites from which their troops had suffered badly in Vietnam. The most interesting compounds that have emerged from their screen of more than 300,000 candidates turned out to be either analogues of quinine or compounds that bear an analogy to another natural antimalarial, febrifugine.

The latest family of antimalarials and one that shows great promise for the treatment of severe, multiple drug-resistant falciparum malaria, is derived from a common plant, *Artemisia annua Linn.,* that was used as a febrifuge in traditional Chinese medicine for some 2 millenia. The active ingredient, a sesquiterpene lactone called artemisinin, has been modified to yield more active, semisynthetic derivatives, which are currently undergoing clinical trials, as well as forming the template for novel series of synthetic antimalarials based on the active center of the artemisinin molecule. [*See* CHEMOTHERAPY, ANTIPARASITIC AGENTS.]

C. Development of Antimalarial Vaccines

Stimulated largely by our failure to control malaria with insecticides or drugs, humans have reverted again to a naturalistic approach, an attempt to exploit the body's own methods of attacking invading *Plasmodium* (i.e., development of antimalarial vaccines). Research on vaccination has been able to blossom through the availability of two models, the first the mouse, which can serve as a host for species of *Plasmodium* such as *P. berghei,* and the second, a system for maintaining *P. falciparum* in continuous culture. The first model enabled investigators to identify the repetitive dodecapeptides on the invasive sporozoites against which the human body forms antibodies and to reproduce them either by peptide synthesis or genetic engineering. Subsequently similar synthetic antigens were developed against the sporozoites of *P. falciparum.* Unfortunately, those concerned underrated the importance of cell-mediated responses in the establishment of protective immunity against malaria parasites. The vaccines based on the synthetic peptides have proved to give little protection, and attention has had to be diverted to antigens that will trigger appropriate, protective T-cell responses, rather than just antibodies.

Other researchers have focused on antigens of the intraerythrocytic asexual stages. Their efforts have revealed a host of molecules associated with the surface of infected red cells, and the race is currently on to determine which, if any of them, will stimulate a combination of cell-mediated and humoral responses that will be sufficiently powerful to protect people against the parasites. Attention has also been directed to the identification of antigens that will block the transmission of the parasites through the mosquito vectors.

In practical terms, the technical and logistic problems facing the development of malaria vaccines are enormous. If and when potent vaccines are produced, their testing in humans in all the stages from safety and efficacy in nonimmune volunteers and field studies under conditions of natural exposure, through to their deployment in large indigenous populations in endemic Third World countries will demand enormous resources in terms of funding, expertise, and organization. Sadly, the main endpoint of the malaria eradication program was the near elimination of malariologists with the practical clinical and field experience to carry out these crucial stages of vaccine development. Between the marvels of molecular biology and genetic engineering and the realities of human life, the gap is ever-widening.

D. Future Impact of Malaria on Human Biology

The main purpose of this encyclopedia is to take a broad look at all aspects of human biology. The aim of this chapter has been to relate the biology of humans to that of *Plasmodium,* to show how important a role the parasites have played in human evolution and how they affect our health and progress in the modern world. We have tried to show how nature does not readily accept human intervention in her schemes, that the precarious balance between malaria parasites and their human hosts cannot readily be manipulated, for all human intelligence and ingenuity, because nature will often find ways of countering our most erudite moves. We have shown how a brave attempt to eliminate malaria as a source of human suffering was frustrated by the emergence of resistance to insecticides by the vectors and to drugs by the parasites. We have indicated also that the parasites can modify their antigens to avoid the response of the host. If malaria vaccines are forthcoming, what will be their fate? How long will they survive nature's counterattacks?

Today half the world's population is threatened by malaria. More than a million die from it every year. It would be a bold writer who promised that we would be freed of this scourge even a century from now—a century, after all, is only a grain in the sands of time where the evolution of humans and *Plasmodium* is concerned.

Bibliography

Bruce-Chwatt, L. J. (1965). Paleogenesis and paleo-epidemiology of primate malaria. *Bull. W. H. O.* **32,** 363–387.

Clark, I. A., Hunt, N. H., and Cowden, W. B. (1986). Oxygen-derived free radicals in the pathogenesis of parasitic disease. *Adv. Parasitol.* **25,** 1–44.

Edelstein, S. J. (1986). "The Sickled Cell. From Myths to Molecules." Harvard University Press, Cambridge, Massachusetts.

Evered, D., and Whelan, J., eds. (1983). "Malaria and the Red Cell," CIBA Foundation Symposium No. 94. Pitman, London.

Garnham, P. C. C. (1966). "Malaria Parasites and Other Haemosporidia." Blackwell, Oxford.

Gillett, J. D. (1981). Increased atmospheric carbon dioxide and the spread of parasitic diseases. *In* "Parasitological Topics; A Presentation Volume to P. C. C. Garnham, F. R. S., on His 80th Birthday" (E. U. Canning, ed.), pp. 106–111. Society of Protozoologists Special Publication No. 1, Allen Press, Lawrence, Kansas.

Godson, G. N., et al. (1984). Structure and organization of genes for sporozoite surface antigens. *Philos. Trans. R. Soc. Lond.* [B]**307,** 129–139.

Luzzatto, L., ed. (1981). "Clinics in Haematology. Vol. 10 No. 3. Haematology in Tropical Areas." W. B. Saunders, Philadelphia.

Mattingly, P. F. (1983). The paleogeography of mosquito-borne disease. *Biol. J. Linnean Soc.* **19,** 185–210.

Peters, W. (1987). "Chemotherapy and Drug Resistance in Malaria," 2nd ed. Academic Press, London.

Prothero, R. M. (1961). Population movements and problems of malaria eradication in Africa. *Bull. W. H. O.* **24,** 405–425.

Weatherall, D. J. (1987). Common genetic disorders of the red cell and the 'malaria hypothesis.' *Ann. Trop. Med. Parasitol.* **81,** 539–548.

Wernsdorfer, W. H., and McGregor, I., eds. (1988). "Malaria. Principles and Practice of Malariology." Churchill Livingstone, Edinburgh.

World Health Organization. (1986). Resistance of vectors and reservoirs of disease to pesticides. *Tech. Rep. Series No. 737,* World Health Organization, Geneva.

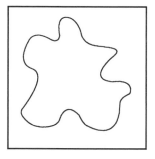

Malignant Melanoma

S. EVA SINGLETARY AND CHARLES M. BALCH, *The University of Texas M. D. Anderson Cancer Center*

Glossary

Acral lentiginous melanoma Type of melanoma characterized by intradermal spread, arising mainly on the nail bed or skin of digits
Epidermis Outermost layer of the skin derived from the embryonic ectoderm
Melanin Dark pigment of skin, hair, choroid coat of the eye, and substantia nigra of the brain; produced by polymerization of oxidation products of tyrosine and dihydroxyphenyl compounds
Melanoma Malignant tumor of skin with melanin-pigmented cells
Metastasis Growth of malignant cells distant from the site of origin
Ultraviolet radiation Radiation beyond the violet end of the light spectrum, with wave lengths of 1800–3900 Å

MELANOMA REPRESENTS only 1–3% of all malignancies, but the rate of increase in its incidence is greater than that of almost any other cancer, doubling every 6–10 years. Melanoma is a potentially curable cancer, if diagnosed and properly treated at an early stage. Thus, identification of the etiological factors and the early recognition of melanoma can change the natural history of this disease.

I. Embryology

Melanoma develops from the malignant transformation of the melanocyte, a cell derived from the primitive neural tube of the embryo which produces the pigment melanin. During early gestation these cells migrate to the skin, uveal tract of the eye, meninges, and ectodermal mucosa, according to a genetically predetermined pattern. When the neural cells reach their destination in the skin at the basal layer of the epidermis, a commitment to differentiation occurs in the form of a melanocyte.

The melanin is synthesized within an intracellular organelle called the melanosome. The amino acid tyrosine is oxidized to dihydroxyphenylalanine (dopa), which is subsequently converted to dopaquinone and then polymerized to melanin. The melanin is then released from the long threadlike processes of the melanocyte and is phagocytized by surrounding epithelial cells, the keratinocytes.

II. Epidemiology

It is estimated that by the year 2000, one in 90 whites in the United States will develop a melanoma during their lives; in comparison, only one in 1500 was at risk in 1935.

Overall, the incidence of melanoma among males and females is approximately equal. Men have a higher incidence on the trunk, but women have a larger number of melanomas, at the lower extremities. The majority of patients are initially diagnosed between the ages of 30 and 60 years. Melanoma can occur in infants and young children, but the annual risk is less than one per million in those younger than 18 years.

Melanoma occurs more frequently in whites than in nonwhites. When melanoma occurs in nonwhites, the lesions are often in unusual sites (e.g., the palmar surface of the hand, the plantar surface of the foot, and the mucous membranes).

An estimated 11% of melanoma cases might be hereditary. The sex distribution of patients with familial melanoma is similar to that for sporadic (i.e., nonhereditary) melanoma, but the age at diagnosis is approximately 10 years younger. Multiple primary melanomas also occur more frequently in persons with familial melanoma (10–15% of these patients) than in those with sporadic melanoma. A familial history of melanoma increases a person's risk for developing melanoma by two to eight times. However, the occurrence of an autosomal-dominant hereditary form of melanoma, the dysplastic nevus syndrome, increases the risk of melanoma by seven to 70 times. Patients with this syndrome typically have numerous moles that are usually large (5 mm or greater diameter), with irregular pigmentation and borders, and located predominantly on the trunk, buttocks, or lower extremities (i.e., dysplastic moles). The relative lifetime risk for developing melanoma in large congenital moles, which are pigmented nevi present at birth, is estimated to be 6.3%. The melanoma risk associated with small congenital moles is controversial.

Males and females with fair complexions (especially those with red hair) and a tendency to sunburn after a relatively brief exposure to bright sunlight have a two- to threefold increased risk for developing melanoma. An association between melanoma and early childhood exposure to excessive sun has been suggested based on patterns of migration to and from high-risk geographic areas.

An increased risk of melanoma is also seen in immunosuppressed patients, such as renal transplant recipients, and in patients with leukemia or lymphoma.

III. Etiology

Exposure to the ultraviolet (UV) radiation in sunlight is considered the major cause of cutaneous melanoma. Supporting this association is the known sensitivity of DNA to damage by UV radiation, the promotion by UV radiation of melanoma development in strains of mice prone to spontaneous melanoma, the sporadic induction of melanomas in animals through the use of exposure to both chemical carcinogens and UV light, and the activation of the suppressor pathway of the immune response by UV radiation, thus blocking the anticancer effect of the immune system.

Evidence in humans suggesting some relationship between UV light and melanoma includes (1) an inverse relationship between the incidence of melanoma and the degree of natural skin pigmentation, (2) the preferential occurrence of melanoma in anatomic sites exposed to sunlight, (3) the increased incidence in light-skinned populations, especially in areas near the equator, (4) the cyclic fluctuation in melanoma incidence, correlating with cyclic increases in UV exposure from sunspot activity, which disrupts the ozone layer of the atmosphere, and (5) the increased rates of melanoma in patients with xeroderma pigmentosum, an autosomal-recessive hereditary defect in the capacity to repair UV radiation-induced DNA damage.

Several aspects of melanoma occurrence however, are inconsistent with the UV radiation hypothesis. (1) The incidence varies both within and among countries with increasing hours of sunlight. In Australia both Queensland and Western Australia have a high incidence of melanoma, but the incidence is reversed with respect to the duration of sunlight. In Europe a higher rate of melanoma is seen in the Nordic countries than in the southern countries. This relationship could be influenced by the populations' different ethnic backgrounds and skin color. (2) The majority of melanomas are diagnosed in the fourth decade of life, in contrast with the more common skin cancers that are most often detected in later years of life, when the cumulative exposure of the patient to sunlight is greater. (3) Melanoma is more common in urban, indoor workers than in individuals with outdoor occupations. (4) Some melanomas occur in relatively unexposed sites of the body (e.g., the palms, soles, mucous membranes, and genitalia).

Other environmental factors have been investigated as potential causes of melanoma—such as female sex hormones, exposure to fluorescent light, occupational exposure to petroleum products, trauma, and radiation therapy—but there is no sufficient evidence that they play a significant role.

IV. Host–Tumor Relationship

The clonal evolution hypothesis of tumor progression assumes that most neoplasms arise from a single altered cell, with sequential selection of subpopulations of heterogeneous clones. Tumor progression in the human melanocyte system has been extensively studied in tissue culture and in immunosuppressed mice. The first step of tumor progression is the formation of a common melanocytic nevus of structurally normal melanocytes with no architectural abnormality. The melanocytic nevus then develops persistent structural atypia, followed by a dysplastic change, with both architectural and cellular atypia. Like normal melanocytes, these nevus cells have a finite life span in culture, show no consistent chromosomal abnormalities, and are not tumorigenic in nude mice. In contrast with normal melanocytes, nevus cells can form colonies in soft agar. [*See* TUMOR CLONALITY.]

As the tumor progresses early primary melanoma cells acquire a prolonged, but not indefinite, life span in culture but are still not tumorigenic in nude mice, nor do they metastasize *in vivo*. The melanoma cells next enter the vertical growth phase, making the tumor thicker. Vertical growth phase melanoma cells grow indefinitely in culture, are tumorigenic in nude mice, and are capable of generating metastases, the final stage of tumor progression. Metastatic cells grow more rapidly in culture and in nude mice. Both the vertical growth phase cells and metastatic cells show nonrandom abnormalities of chromosomes 1, 6, and 7. These genetic alterations indicate possible sites of genes that are important for the malignant transformation of melanoma. Genes coding for proteins that are potentially involved in malignant transformation (i.e., oncogenes) have not been clearly identified.

Tumor progression might also affect the regulation of growth and differentiation of melanocytes and melanoma cells. Metastatic melanoma cells require fewer exogenous growth factors, because they produce their own factors. Growth factors and surface molecules responding to them (i.e., receptors) expressed in melanoma cells include nerve growth factor and its receptor, epidermal growth factor, transforming growth factor α, platelet-derived growth factor, transferrin receptor, and insulin receptor. Markers for the differentiation of melanoma cells have been identified by tyrosinase activity, melanosomes, and melanocyte antigens. The unique features of melanoma cells might provide research tools for novel biological approaches to therapy. [*See* NERVE GROWTH FACTOR; TRANSFORMING GROWTH FACTOR-α; TRANSFORMING GROWTH FACTOR-β.]

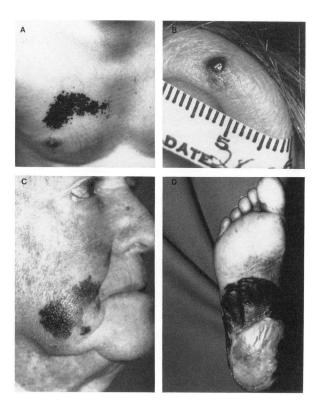

FIGURE 1 The four major growth patterns are (A) superficial spreading melanoma, (B) nodular melanoma, (C) lentigo maligna melanoma, and (D) acral–lentiginous melanoma.

V. Growth Patterns

The four major growth patterns are superficial spreading melanoma, nodular melanoma, lentigo maligna melanoma, and acral–lentiginous melanoma (Fig. 1). Each of these growth patterns has unique clinical features and specific histological appearances.

A. Superficial Spreading Melanoma

Superficial spreading melanomas constitute approximately 70% of all melanomas and generally arise in a preexisting nevus. The common history is a slowly evolving change of the precursor lesion over 1–5 years, followed by a rapid growth phase. The typical lesion is deeply pigmented, with variegated colors ranging from black, brown, or reddish-pink to white patches of pigment regression. Initially, superficial spreading melanomas are predominantly flat, because of a radial (i.e., horizontal) growth pattern. With invasion of the deeper layers of the dermis (i.e., vertical growth), the melanoma's surface can become irregular and more elevated. As the lesion undergoes growth expansion, the periphery of the melanoma is often notched or indented.

B. Nodular Melanoma

Nodular melanomas account for 15–30% of melanoma cases and occur more frequently in males than in females. These tumors are more aggressive than superficial spreading melanomas and usually arise without evidence of a preexisting lesion. They occur most commonly on the trunk or the head and neck. Lesions that are polypoid, with a stalk or cauliflower appearance, are particularly virulent. Nodular melanomas are usually blue–black, but they might lack pigment (i.e., be amelanotic). These melanomas grow vertically into the dermis without horizontal spreading.

C. Lentigo Maligna Melanoma

Lentigo maligna melanomas constitute only 4–10% of melanomas. They are typically located on the face or neck in an older age group (>65 years), although a few might occur on the back of the hands or the lower legs. They are generally large (i.e., greater than 3 cm), flat, tan-colored lesions that have been present for many years. The borders of the lesions are convoluted, with prominent notching that generally represents areas of regression. They have a lesser propensity for metastasis. The histological diagnosis of lentigo maligna melanoma requires the presence of severe changes caused by sun exposure in both the epidermis and the dermis.

D. Acral–Lentiginous Melanoma

Acral–lentiginous melanomas make up only 2–8% of melanomas in whites, but constitute 35–60% of melanomas in blacks, Orientals, and Hispanics. These melanomas characteristically occur on the palms of the hands, on the soles of the feet, or beneath the nail beds. Acral–lentiginous melanomas resemble tan or brown flat stains, with irregular borders on the palms or soles. They are generally large (i.e., average diameter, 3 cm) and develop over a short time (average, 2.5 years). Ulceration is not uncommon in neglected lesions and is often mistaken for a fungal infection. A small number of these lesions have a flesh-colored appearance that can be misdiagnosed as a granuloma. Subungual melanomas appear as brown–black discolorations under the nail bed and are most often diagnosed in older patients (median age, 55–65 years). More than 75% of subungual melanomas involve either the great toe or the thumb.

VI. Staging Systems

Any lesion that undergoes a change in size, contour, configuration, or color should be considered a possible melanoma, and a biopsy should be performed. After the biopsy the pathologist measures the thickness of the melanoma (Breslow's microstaging) and determines the level of invasion (Clark's microstaging).

Breslow's microstaging system quantitatively assesses the maximum tumor thickness (in millimeters) with an ocular micrometer. Clark's system categorizes five levels of invasion that reflect increasing depths of penetration into the underlying dermis or subcutaneous tissue. Clark's level I melanomas are confined to the epidermis above the basement membrane, which separates the epidermis from the dermis; level II tumors invade through the basement membrane into the papillary dermis; level III lesions have tumor cells at the junction of the papillary and reticular dermis; level IV tumors extend into the recticular dermis; level V melanomas penetrate through to the subcutaneous fat.

The staging system (Table I) adopted by the American Joint Committee on Cancer is based on

TABLE I Staging of Melanoma[a]

Stage	Criteria[b]
IA	Localized melanoma ≤0.75 mm thick or level II (T1, N0, M0)[c]
IB	Localized melanoma 0.76–1.50 mm thick or level III (T2, N0, M0)
IIA	Localized melanoma 1.5–4.0 mm thick or level IV (T3, N0, M0)
IIB	Localized melanoma >4 mm thick or level V (T4, N0, M0)
III	Limited nodal metastases involving only one regional lymph node basin, or <5 intransit metastases, but no nodal metastases (any T, N1, M0)
IV	Advanced regional metastases (any T, N2, M0), or any patient with distant metastases (any T, any N, M1 or M2)

[a] Adopted by the American Joint Committee on Cancer.
[b] When the thickness and level of invasion criteria do not coincide within a T classification, the thickness of the lesion should take precedence.
[c] T, Primary tumor; N, regional nodes; M, distant metastases.

the microstaging of the primary melanoma (Breslow's tumor thickness and Clark's levels of invasion) and knowledge of the patterns of metastasis. The regional lymph nodes are the most common first site of metastasis. Melanoma cells can also become entrapped within dermal lymph vessels near the primary tumor (i.e., satellites) and between the primary tumor and the regional lymph nodes (i.e., in-transit metastases).

Stages I and II consist of clinically localized disease, subdivided into A and B according to microstaging (see Table I). Patients with stage III disease have nodal metastases in a single lymph node basin, whereas those with stage IV melanoma have one of the following: (1) distant metastases, (2) nodal metastases involving two or more lymph node basins, (3) five or more in-transit metastases, or (4) nodal metastases either larger than 5 cm in diameter or fixed to the surrounding tissues.

VII. Prognostic Factors

A. Localized Melanoma (Stages I–II)

The thickness of the tumor is the single most important prognostic factor in localized melanoma (Table II). Although the histological level of invasion (Clark's system) is also significant, tumor thickness is a more accurate predictor of the risk for metastasis. Other, secondary, factors include ulceration of the lesion, sex of the patient (females have a better survival rate), and anatomic location of the lesion

TABLE II Predictors of Clinical Course for Patients with Melanoma

Clinical stage	Unfavorable predictors
Localized melanoma (stages I–II)	
Tumor thickness	>1.5 mm
Ulceration	Present
Sex	Male
Location	Trunk
Regional metastasis (stage III)	
Number of nodal metastases	>1
Ulceration	Present
Extension outside node	Present
Distant metastasis (stage IV)	
Number of distant sites of metastasis	>1
Location	Visceral
Remission duration	<12 months

(the extremities are more favorable sites than the trunk).

B. Regional Lymph Node Metastasis (Stage III)

The number of lymph node metastases has a direct correlation with survival. Patients with only one positive node have a reasonable prospect of cure (i.e., 40% alive after 10 years), whereas only 13% of the patients with two or more positive nodes are alive after 10 years. Other prognostic factors include extension of the tumor outside the lymph node capsule, tumor thickness, and ulceration of the primary lesion.

C. Distant Metastasis (Stage IV)

Dominant prognostic factors in patients with distant metastasis are (1) the number of metastatic sites (one, two, or three or more), (2) the remission duration (shorter than or longer than 1 year), and (3) the sites of metastasis (visceral versus nonvisceral). The median survival is 7 months for patients with one site of metastasis, 4 months for those with two sites, and 2 months for those with three sites. There are no features of the primary melanoma that predict the clinical course of the disease once distant metastases develop.

Bibliography

Albino, A. P. (1987). The role of oncogenes in the pathogenesis of malignant melanoma. *In* "Basic and Clinical Aspects of Malignant Melanoma" (L. Nathanson, ed.). Nijhoff, Boston, Massachusetts.

Balch, C. M., and Milton, G. W. (eds.) (1985). "Cutaneous Melanoma: Clinical Management and Treatment Results Worldwide." Lippincott, Philadelphia, Pennsylvania.

Balch, C. M., Houghton, A., and Peters, L. (1989). Cutaneous melanoma. *In* "Cancer: Principles and Practice of Oncology" (V. T. DeVita, ed.), 3rd ed. Lippincott, Philadelphia, Pennsylvania.

Herlyn, M., Clark, W. H., Rodeck, U., Mancianti M. L., Jambrosic, J., and Koprowski, H. (1987). Biology of tumor progression in human melanocytes. *Invest. Cell Pathol.* **57,** 461.

Veronesi, U., Cascinelli, N., and Santinami, M. (eds.) (1987). "Cutaneous Melanoma: Status of Knowledge and Future Perspective." Academic Press, London.

Malnutrition

BUFORD L. NICHOLS, JR., *Baylor College of Medicine*

I. Marasmus
II. Kwashiorkor
III. Pathophysiology of Macronutrient
   Deficiencies
IV. Immune Function
V. Role of Infection

Glossary

Amino acids Amphoteric organic acids containing the amino group NH_2. There are 20 amino acids that are the building blocks of protein

Antibody Any of the body immunoglobulins that are produced in response to specific antigens and that counteract their effects especially by neutralizing toxins, agglutinating bacteria or cells, and precipitating soluble antigens

Complement Complex series of enzymatic proteins occurring in normal serum that in combination with antigen-antibody complex produces the destruction of the bacteria acting as antigens

Edema Abnormal excess accumulation of serous fluid in skin or in a serous cavity

Fat Ester of glycerol with fatty acids, usually oleic, palmitic, or stearic

Glucose Monosaccharide occurring in certain foodstuffs, especially fruits, and in normal blood. It is the usual form in which carbohydrate is assimilated and used by the body

Kwashiorkor Severe protein-energy malnutrition in infants and children that is characterized by the presence of edema and is associated with a diet high in carbohydrate and low in protein

Marasmus Condition of protein-energy malnutrition occurring especially in children and usually caused by a diet deficient in calories and proteins but sometimes by disease or parasitic infection

Phagocytic Capable of functioning as a cell that engulfs bacteria and consumes debris

Protein-energy malnutrition Complex nutritional disorder that may result from an inadequate diet or from disease, or from a combination of both

T cells Lymphocytes that either pass through the thymus or are influenced by it on their way to the tissues

Turnover Simultaneous synthesis and degradation of a metabolic end-product (e.g., protein) that occurs without any change in the quantity (mass) of that product

PROTEIN-ENERGY MALNUTRITION (PEM) is a leading cause of death worldwide among children younger than 5 years of age. PEM results from deficiencies in dietary protein and calories. Primary PEM is due to social or economic factors that result in a lack of food. Secondary PEM occurs in infants with various diseases associated with increased food requirements (infection, diarrhea, cancer), increased protein and caloric loss (malabsorption, cystic fibrosis), reduced food intake (anorexia, cancer, fasting), or a combination of these three variables.

I. Marasmus

Marasmus is the most frequent form of primary PEM and is caused by severe dietary energy deficiency. Synonyms for this condition are growth failure, atrophy, athrepsia, and failure-to-thrive, all of which describe the wasting of the body. Many secondary forms of marasmic PEM are associated with diseases such as cystic fibrosis, tuberculosis, or celiac disease. The principal sign of marasmus in a child is weight loss, with the most severe cases weighing below 60% of that normally expected for age. Linear growth is also slowed, but weight for height is lower than normal, showing that body mass loss exceeds coexisting dwarfing. There is a visible loss of subcutaneous fat and a wasting of muscle mass that can be confirmed by measurements of skinfolds and limb circumferences. The head looks large but is proportional to body length. Skin lesions are scarce, and edema is absent. Atrophy of the surface of the tongue is common and cracks at the corners of the mouth are frequent. The complaint leading to a health examination is usually diarrhea. Marasmus rarely occurs in young infants that are exclusively breast-fed. Recent weaning of the child followed by feeding of dilute formula or milk in an unsanitary environment is a common cause of marasmus in developing countries. Marasmus may develop in older children fed insufficient food.

II. Kwashiorkor

Kwashiorkor, the edematous form of PEM, is due to severe depletion of dietary protein and energy. It is the most serious of the prevalent forms of malnutrition in the developing world and often accompanies an underlying disease. Critically ill hospitalized patients with cancer, burns, acute and chronic infections, anorexia, inflammatory bowel disease, and postoperative surgical conditions are predisposed to this form of PEM, because they are frequently fed inadequate quantities of protein and nonprotein energy.

The major manifestation of kwashiorkor is edema. Body weight ranges from 80 to 60% of the expected weight for age; when the weight is less than 60% of that expected, the term *marasmic kwashiorkor* is used. There is a relative maintenance of subcutaneous fat but a marked atrophy of muscle mass. Edema varies from a mild minor puffiness of the skin of the foot to severe edema involving the eyelids and is associated with low serum albumin concentrations. The hair is sparse and falls out easily. Skin changes are common and range from dark scaly patches to a raised red rash. Peeling of the skin surface occurs in the most severe form of kwashiorkor. Cracks at the corners of the mouth and atrophy of the surface of the tongue are universal. There is a large, soft liver, and lymph nodes and tonsils are atrophic. A mild bronchopneumonitis is

usually present. The stomach is distended, bowel motility decreased, and diarrhea is always present.

III. Pathophysiology of Macronutrient Deficiencies

A. Background

Protein-energy malnutrition is produced by reduced dietary protein and energy intake, caused by early weaning or diarrhea and other infections. The metabolic adaptation to reduced dietary protein and calorie intake is typical of starvation. The immune dysfunction and continued exposure to infectious diseases increase the morbidity and mortality of PEM.

B. Endocrine Adaptation

Plasma growth hormone levels are elevated in PEM. Somatomedin C, the mediator for growth hormone, is abnormally low. With a diet adequate in protein, growth hormone levels decline and somatomedin levels rise to normal within a week. Blood insulin levels are low and unresponsive to food intake, resulting in a glucose intolerant or "diabetic" state. Insulin response to dietary carbohydrate returns to normal when protein intake is restored to normal. Adrenal steroid levels are increased and increase further in response to the stress of infection. Thyroid hormone levels are within the normal range, but there are increased concentrations of inactive forms. These adrenal and thyroid adaptations promptly respond to adequate dietary intake. The net effect of all these endocrine adaptations is a restriction in glucose use by tissues requiring glucose. Simultaneously, stored lipids and amino acids from muscle are mobilized to provide energy leading to the wasting of fat and muscle tissue. [*See* ENDOCRINE SYSTEM.]

C. Metabolic Substrates

There are several stages of metabolic adaptation in PEM. The first is a reduction in voluntary energy expenditure, the second is a reduction in the gain of body mass and length, and the last is a reduction in resting energy expenditure. In marasmus there is a more efficient reduction in resting energy expenditure than in kwashiorkor.

Inadequate protein and energy intake results in

growth failure, especially when infection increases the rate of energy loss. Body protein stores may be sustained if intake has not been exceeded by excessive losses. In both healthy and PEM children, the efficiency of dietary protein storage is greater than 95% of intake. Protein turnover is a measure of total body protein synthesis and degradation. In normal adults, the rate of turnover is only modestly influenced by starvation, but the rate of turnover in kwashiorkor is strikingly reduced. In contrast, turnover rates in normal children are increased in response to the stress of infection.

The turnover of individual blood proteins is reduced under modest dietary restrictions. In both well-nourished and malnourished children, a reduced protein intake for a few days results in reduced serum albumin synthesis by the liver and, if the reduction continues, decreased liver synthesis of proteins that transport iron, vitamin A, and thyroid hormone. The synthesis of serum proteins associated with infection (e.g., complement proteins) is preserved under these conditions. In addition to the liver adaptations, there is increased breakdown of muscle proteins; in the marasmic child, these mechanisms maintain the serum levels of free amino acids and transport proteins secreted by the liver. In the child with kwashiorkor, adaptation fails to sustain serum amino acids and liver protein synthesis, and the serum levels of muscle-derived amino acids decrease. In marasmus, the synthesis of urea from amino acids and its serum concentration are normal, but in kwashiorkor, serum urea levels are low, reflecting the unavailability of amino acids. Fatty liver occurs in kwashiorkor as a result of failure of the liver to synthesize and secrete fat-binding and transporting apolipoproteins. Infections such as measles or viral diarrhea decrease the synthesis of pancreatic and intestinal digestive enzymes and can impair subsequent dietary rehabilitation.

Body water is increased in PEM syndromes because of body fat loss and fluid retention. Body concentrations of potassium are reduced owing to a loss of muscle and other lean tissues. In the child with kwashiorkor, there are also defects in the maintenance of normal potassium concentrations in soft tissues. Serum zinc and copper concentrations are reduced in kwashiorkor but are normal in marasmus. If adequate dietary supplements are not provided, serum zinc and copper concentrations fall rapidly during nutritional rehabilitation.

IV. Immune Function

White blood cell function is abnormal in PEM. Severely malnourished children fail to respond to tuberculin or fungus skin tests, which reveals an inability of their system to demonstrate prior sensitization. These children do not react to new immunizations, indicating that new lymphocytes are unresponsive to new antigens. Thymus, lymph node, tonsil, and spleen are atrophic in malnourished children. Total circulating white blood cells are reduced, with a reduction of the number of thymus-derived cells, which have a poor response to antigenic activation. This immunodeficient status improves with dietary rehabilitation within 2 weeks, and the white blood cell count recovers within 4 weeks. [See NUTRITION AND IMMUNITY.]

The concentration of antibody-producing white blood cells in lymphoid tissues or in peripheral blood is normal. Antibody synthesis is increased as reflected in elevated serum IgA, IgM, IgG, IgD, and IgE levels, especially in children with kwashiorkor. The serum antibody response to vaccines for yellow fever, influenza, and typhoid is reduced, but the response to cholera vaccine is not. The recovery of normal antibody response is dependent on adequate dietary protein intake. Malnourished children have reduced secretory immunity, shown by low SIgA (secreted IgA) levels in nasal washings, duodenal fluids, and tears, despite elevated serum IgA. The reduction is reflected in the diminished response of SIgA to oral measles and polio vaccines. The loss of secretory immunity probably contributes to the increase in respiratory and intestinal infections.

The response of circulating phagocytic white blood cells to bacteria is normal, but that of tissue phagocytic macrophages is reduced in PEM. The circulating white blood cells are able to phagocytize bacteria but have reduced ability to kill the pathogens, *Staphylococcus aureus, Escherichia coli,* and *Candida albicans.*

All PEM subjects have depressed concentrations of specific complement factors, but the levels of several important factors are more profoundly depressed in kwashiorkor. The complement factors are essential for phagocytoses of microorganisms and the binding of white blood cells to infected tissues. Most respond to dietary therapy within 4 weeks, but the levels of some remain low for several months. Recovery of complement levels occurs more quickly in children who receive diets with adequate protein.

V. Role of Infection

A triangle of interactions exists among host immunity, infection, and malnutrition. PEM may be initiated by primary dietary deficiencies or illnesses that lead to low food intake and increase energy expenditure. The impaired host defenses associated with PEM are responsible for increased frequency and severity of infection. Gastrointestinal infections also cause poor digestion and malabsorption.

Acknowledgment

This work is a publication of the USDA/ARS Children's Nutrition Research Center, Department of Pediatrics, Baylor College of Medicine and Texas Children's Hospital, Houston, Texas. This project has been funded in part with federal funds from the U.S. Department of Agriculture, Agricultural Research Service under Cooperative Agreement no. 58-7MN1-6-100. The contents of this publication do not necessarily reflect the views or policies of the U.S. Department of Agriculture, nor does mention of trade names, commercial products, or organizations imply endorsement of the U.S. government.

Bibliography

Behrman, R. E., and Vaughan, V. C. (eds.) (1987). "Nelson Textbook of Pediatrics," 13th ed. (Sect. 3.15–3.18). pp. 138–141. W. B. Saunders, Philadelphia.

Brasel, J. A. (1980). Endocrine adaptation to malnutrition. *Pediatr. Res.* **14,** 1299–1303.

Frenk, S. (1986). Metabolic adaptation in protein-energy malnutrition. *J. Am. Coll. Nutr.* **5,** 371–381.

Keusch, G. T., and Farthin, M. J. G. (1986). Nutrition and infection. *Annu. Rev. Nutr.* **6,** 131–154.

Nichols, V. N., Fraley, J. K., Evans, K., and Nichols, B. L. (1989). Acquired monosaccharide intolerance in infants. *J. Pediatr. Gastroenterol. Nutr.* **8,** 51–57.

Suskind, R. M. (1981). Malnutrition and the immune response. *In* "Textbook of Pediatric Nutrition," pp. 241–262. Raven Press, New York.

Waterlow, J. C. (1986). Metabolic adaptation to low intakes of energy and protein. *Annu. Rev. Nutr.* **6,** 495–526.

Malnutrition of the Elderly

LOUISE DAVIES, *Royal Free Hospital School of Medicine, London*

Glossary

Elderly people A much-disputed term (i.e., should it be reckoned biologically or chronologically?); in the United Kingdom, it is arbitrarily applied to "persons of pensionable age" (women 60+ yr, men 65+ yr); sometimes it is divided into "young old" and "old old," but this categorization may be influenced by the life expectancy in different societies

Malnutrition Disturbance of form or function due to lack of (or excess of) calories or of one or more nutrients; in many countries, undernutrition is more common in old age than overnutrition, but the term covers both

Overt Obvious; evident; not concealed

Risk factor Major, identifiable biological or environmental circumstances or event that increases the risk of malnutrition and, therefore, suggests the need for prevention and special care and attention

TO DATE, few large-scale studies have investigated the nutritional status of elderly men and women. In the absence of famine or other disasters, very little overt malnutrition has been found among the noninstitutionalized, who form the vast majority of most elderly populations (e.g., 96% in the United Kingdom; 95% in the United States). Diagnosed malnutrition is more prevalent in advanced old age, frequently linked with disease states and with those persons in need of institutional care. At present, old

people are disproportionately large consumers of costly health and welfare resources. Population forecasts predict a worldwide increase in numbers surviving into old age as life expectancy increases; therefore, more demands on the already overstretched social and medical services are probable. This chapter illustrates the importance of prevention of malnutrition through early recognition and treatment; the identification or risk factors, followed by prompt action by elderly people themselves, helped by professional and lay carers.

I. Types of Malnutrition

The majority of elderly men and women maintain an adequate nutritional status even into extreme old age; however, in times of national food shortages, deprivation can especially affect the oldest people. In countries with plentiful food supplies, nutritional deficiencies can occur in old age due to changes in economic circumstances and way of life or because of age-associated diseases, confusion, or disabilities, which may lead to alterations in dietary intake, absorption, and metabolism of nutrients.

In some countries, overnutrition is as much of a problem as undernutrition. Loosing weight in old age is difficult: calorie requirements may be low due to an age-associated drop in metabolic rate and a decline in physical activity; therefore, weight loss is likely to be slow in proportion to the effort made. There is a danger that in cutting down further on calories, the intake of essential nutrients will also be lowered. Menus for old people need to be expertly designed and monitored, but slimming diets should only be recommended if medically essential (e.g., if obesity is seriously hampering surgery or mobility, or exacerbating illness exists). It is important to encourage increase in physical activity, both for the overweight and for those with a diminished appetite.

In placing an emphasis on preventative action, following recognition of risk factors, it should be realized that not just one type of malnutrition exists; at least four types can manifest themselves at any stage. Although they are distinct, they are often interrelated. Each type calls for a different response in terms of practical prevention.

A. Specific Malnutrition

Specific malnutrition includes diet-related disease such as osteomalacia, pellagra, scurvy, or anemia. It is frequently associated with—but may be masked by—a non-nutritional disease or disability, which may produce similar signs and symptoms. [See MALNUTRITION.]

Practical prevention of specific malnutrition involves skilled diagnosis at an early stage, followed by medical treatment and dietary intervention. After reassessing dietary practices, unambiguous and easy-to-follow advice needs to be given to the patient.

B. Sudden Malnutrition

Sudden malnutrition may follow a physical or mental trauma such as surgery, acute illness, bereavement, or an unwelcome retirement. Its practical prevention requires prompt dietary and social action including, where possible, help from neighbors, friends, and family to re-establish healthy eating patterns.

C. Recurrent Malnutrition

Recurrent malnutrition is a severe malnutrition that follows a worsening cycle of illness and poor nutrition. Practical prevention includes community support services arranged before the patient's discharge from a hospital or other medical care facility. Long-term planning is essential to ensure the continued provision of simple basic food.

D. Long-Standing Malnutrition

Long-standing malnutrition involves a long, latent period between nutritional deficiencies and their clinical appearance. A frequent cause is an inadequate diet during a long-standing dementia of the Alzheimer type. In other cases, a seemingly well person may just not be choosing a balanced diet over a period of months or even years. Therefore, there are plenty of opportunities for successful prevention. Practical prevention includes recognition of the risk factors of potential malnutrition by all those in contact with the elderly person. Social and nutritional intervention to prevent the crisis is also warranted.

When dealing with any type of malnutrition, some important questions must be posed; for instance, under what conditions is it best *not* to prescribe vitamins, supplementary feeds, puree or soft diets, and nasogastric feeds?

II. Difficulties in Assessing Nutritional Status

Although very little overt malnutrition has been found in some studies on noninstitutionalized elderly men and women, more longitudinal studies are needed.

In 1988–1989, a concerted action project on nutrition in the elderly in the European community started to collect information on diet intake, nutritional status (including anthropometrical and biochemical parameters), physical activity, life-style, health, and performance among about 2,500 randomly selected elderly, born in 1913–1918, in 17 regions in Europe; this is planned as a longitudinal study. Cross-cultural studies on Food Habits and Health in Later Life have also been started by the subcommittee on Nutrition and Aging of the International Union of Nutritional Sciences; these studies will include information about past food habits, and the influence of food beliefs on food choice.

Meanwhile, diagnosis of malnutrition in the elderly is not easy. This leads to the query: What might be the extent of unrecognized or subclinical malnutrition in elderly populations?

Some 20 years ago, a large-scale U.K. survey on elderly men and women living in their own homes or with families—on which much of our current information is based—concluded that, apart from the small numbers demonstrably malnourished, there were others for whom more could have been done and others still whose margin of safety must have been narrow. Why is it so difficult to detect malnutrition, especially in its early stages? The survey report gave some of the answers.

A. Clinical

On their own, clinical signs and laboratory findings were often unreliable indicators of malnutrition. In most instances, it is not possible to attribute particular clinical signs to specific nutrition deficiencies: they may be related to disorders other than malnutrition. For instance, many causes—including partial gastrectomy—contribute to the proliferation of pathogenic bacteria in the gut, leading to decreased absorption of nutrients; they may present as diarrhea.

B. Biochemical

Caution was required in the interpretation of biochemical findings, because normal ranges for the elderly are not well established. In many instances, the reference ranges are merely extrapolations from studies on younger adults. Moreover, other factors apart from nutritional status could influence the results.

C. Anthropometric

Normal anthropometric measurements may need to be reinterpreted for the elderly. Age brings with it alterations in the proportion of fat and lean body mass, without necessarily affecting weight; it also may result in redistribution of fat at different sites of the body. Essential measurements such as height may be difficult to ascertain on, for instance, an old person crippled with arthritis. Other anthropometric measurements suggested for the elderly include waist, hip and mid-arm circumferences, hip length, waist–hip ratio, and mid-arm muscle area.

D. Dietary

Similarly, dietary intakes are difficult to establish: relying on fading memory to recall foods and amounts eaten is notoriously inaccurate. Searching questions about the frequency of consumption of specific foods generally must be included.

III. Nutritional Requirements and Recommended Daily Allowances

Recommended Daily Allowances (RDAs) for the elderly have been based on the findings of studies on younger adults. While maintaining intakes for vitamins, minerals, and other nutrients, they take into account the average rate at which activities decline in old age. However, many older people—particularly in rural communities—maintain an active lifestyle; consequently, their calorie needs and intake are much greater than the recommendations. Similarly, in the United Kingdom, all women >55 yr of age are classified as sedentary, but the self-reported activities of some would challenge this premise.

RDAs are being reassessed, and the appropriate committees are being asked to consider the need for additional age bands, and also for distinction to be made between requirements for the ''healthy'' and the ''unhealthy.'' For example, protein requirements are increased by fever, infections, surgical procedures, and pressure sores; malnutrition itself—including obesity—can contribute to the development of pressure sores. Studies have shown that elderly men and women with ''health worse than the average'' consume fewer calories than the rest, but their intakes may need to increase if they increase their activities with wheelchair mobility or rehabilitation. Similarly, intakes may fall with dementia—due to memory loss, disorientation, or indifference to food—but energy needs of the confused person may actually rise with agitation and consequent wanderings around the room. More research is needed on the stress effects of disease. Do they cause alterations? And if so, to what extent—in both dietary intake and biochemical values?

For the individual, nutritional requirements change with factors that are likely to vary more widely with advancing age: activity, weight, metabolism, absorption, physical–mental state, drug therapy, and adaptation. No wonder it is difficult to interpret the true significance of measured dietary intakes. It has been found that elderly people with a diet better than average have better than average health. But the question has been asked: Is it possibly the better health and greater physical activity that has led to the good appetite, rather than vice versa?

From the foregoing discussion, it is not surprising that most studies recommend a combination of investigations in the search for malnutrition. When malnutrition is suspected, a thorough assessment should be made, calling for a nutritional, medical, and clinical history as well as a physical examination with, when necessary, anthropometric measurements, specific biologic and hematologic tests, and sometimes specialized hospital investigations. It is also recommended that a clinical history should include investigations into social, family, and emotional problems and a history of food aversions or fad diets. In many instances, aggressive dietary treatment based on these combined findings have yielded excellent results.

Such assessments call for team expertise, whether it be in residential homes, in hospitals and nursing homes, with a high proportion of frail elderly with increased risks of malnutrition, or in a day hospital, which can monitor more active elderly people brought in from their homes in the surrounding community. They emphasize the need for collaboration among physician, dietitian, nurse, laboratory staff, physiotherapist, occupational therapist, speech pathologist, and social workers. But even after all this activity, we are left with the problem of identification of marginal deficiencies and subclinical malnutrition. This problem has led to the recognition of the importance of identifying risk factors.

IV. Risk Factors for Malnutrition

Malnutrition can have profound effects on behavior as well as on the musculoskeletal, pulmonary, or cardiovascular systems; it can cause poor wound healing, increased susceptibility to infections, and delayed convalescence. Its complications can include lack of coordination and increased risk of fractured femur, hypothermia, or pressure sores. The incidence of mortality is also increased with malnutrition.

If the potential risks of malnutrition in elderly people can be identified, preventative action can be taken rather than crisis treatment. When risk factors such as lack of sanitation or no food supplies become overwhelming, precedence may have to be given to high-level emergency action; however, in more settled conditions, the priority is to target the individuals and the groups likely to be at risk.

In most studies of elderly individuals, the incidence of malnutrition and the causes of nutritional deficiencies have been itemized under two headings: social–environmental factors and physical–mental disorders. The two are often interrelated. Table I indicates the elderly groups found to be most at risk in the U.K. study on nutrition and health in old age.

The housebound had already been identified as the largest single group at nutritional risk, with disability proving more crucial than advancing age. Interestingly, the bedfast were found to be less at risk because they, of necessity, receive more visits. Ta-

TABLE I Incidence of Undernutrition Associated with Different Medical and Social Conditions in Elderly People Living at Home

Group at risk	Men n^a	Men $\%^b$	Women n	Women $\%$
Whole sample	169	8	196	6
Living alone	41	15	103	7
No regular cooked meals	5	60***	5	40*
Supplementary benefit	47	19**	87	13**
Social classes IV and V	56	13	56	16***
Mental test score <13	48	17*	83	11*
Depression	7	14	9	22
Chronic bronchitis and emphysema	40	18*	22	27***
Gastrectomy	13	31*	7	14
Edentulous^c	17	29**	7	14
Difficulty in swallowing	10	20	6	50***
Housebound	13	23	33	15*
Smokers	80	11	36	6
Alcoholism	12	17	1	0

Reference: DHSS (1979).
^{a} Number in group.
^{b} % malnourished.
^{c} No dentures, or dentures never used for eating.
* $p < 0.05$; ** $p < 0.01$; *** $p < 0.001$.
[Reproduced with permission, from the controller of Her Majesty's Stationery office (1979). DHSS Report on Heath and Social Subjects.]

TABLE II Causes of Nutritional Deficiency

Primary	Secondary
Lack of knowledge	Impaired appetite
Social isolation	Masticatory inefficiency
Physical disability	Malabsorption
Mental disturbance	Alcoholism
Iatrogenic	Drugs
Poverty	Increased requirements

Reference: Exton-Smith. A. N. (1971). Nutrition of the elderly. *Br. J. Hosp. Med.* **5,** 639–45.

ble II itemizes some causes of nutritional deficiency.

Social isolation is a well-established cause of nutritional risk in many cultures, but its manifestation may be country-specific (e.g., low food intake while family works in the fields during day; temporarily left alone while family takes long school holidays out of town; alone in a high-rise flat, unable to shop for food at the street hawker stands). Poverty is cited in many studies as a major cause of malnutrition. It has, however, been pointed out that the re-

TABLE III The A–Z of Possible Nutritional Risk Factors in Old People's Homes[a]

A Weekly cyclic menu *or* monotony of menu.
B Difficulty with supper meal menus.[b]
C Supper meal at or before 5 P.M.[c]
D Lack of rapport between matron and cook *or* cook resists and resents suggestions.
E Residents' suggestions (e.g., for recipes) unheeded; residents' needs for modified diets ignored; inadequate committee contact.[d]
F Residents not allowed choice of portion size *or* poor portion control *or* no second helpings.
G No heed taken of food wastage.
H Little home-style cooking.[e]
I No special occasions for food treats from local community or from the home, apart from Christmas.
J For active residents: poor or no facilities for independence in providing food and drink.
K Hot foods served lukewarm *or* poor flavoring.
L Poor presentation of food, including table setting and appearance of dining room.
M Unfriendly or undignified waitress service; meal too rushed.
N No observation of weight changes.[f]
O No help in feeding very frail residents. No measures taken to protect other residents from those with offensive eating habits.
P Matron and cook lacking basic nutritional or catering knowledge; isolation from possible help.
Q Lengthy period between preparation, cooking, and serving. Time lag between staff meals and residents' meals, especially affecting vegetables.
R Lack of foods containing ascorbic acid *or* risk of unnecessary destruction of ascorbic acid.[g]
S Few vitamin D foods used, combined with lack of exposure of residents to sunlight.[g]
T Low-fiber diet and evidence of constipation.[g]
U Possible low intake of other nutrients (e.g., iron, folate, vitamin B12).[g]
V Preponderance of convenience foods of poor nutritional content.
W Disproportionate cost among (1) animal protein, (2) fruit and vegetables, and (3) energy foods.[g]
X Obvious food perks to staff to detriment of residents' meals; high proportion of food served to others rather than to the residents.
Y Conditions conducive to food poisoning; lack of cleanliness.
Z Recommendations may not be implemented.

Reference: L. Davies and M. D. Holdsworth, 1980. An at-risk concept used in homes for the elderly in the UK. *JADA* **76(3),** 264–267.
[a] Each factor to be assessed with degree of risk to the residents (i.e., high, moderate, or low).
[b] This highlighted lack of experience in menu planning and recipe ideas and frequently affected cost.
[c] This frequently occurs in the United Kingdom, mainly because of staffing difficulties. Cookies often must be supplied later in the evening because some residents become hungry before bedtime.
[d] Most homes in the voluntary sector have a committee of interested local citizens who are consulted on administrative matters.
[e] Residents frequently expressed a desire for the familiar foods they had been used to eating rather than institutional type catering. Our observation on factors E and V gave the necessary confirmation.
[f] Significant changes in weight can be used as an early diagnostic tool for illness, depression, or other conditions that can easily affect nutritional status.
[g] The nutrients chosen for special investigation in factors R, S, T, U, and W were among those highlighted in previous research.

moval of financial constraints will not necessarily improve nutritional status: when other risk factors such as depression or physical handicap are present, malnourishment can occur even in those with high income.

At this stage, it is essential to make the point that risk factors do not necessarily lead to alterations in nutrient intakes: different individuals react differently to the same potential stresses. For instance, some people positively revel in being on their own and stubbornly resist invitations from friends to join them for quiet meals or Christmas dinner. In some, depression after a bereavement swiftly results in anorexia; in others, it might lead to the overeating of "comfort foods"; whereas others still may make no changes in their food intake even though they are mourning a loved one.

Thus, it is not so much the risk factor as the reaction to it that needs to be noticed. For this reason, a list of easily recognizable warning signals has been suggested, including a recent, unplanned gain or loss in weight (± 3 kg), difficulties with food shopping or preparation, lack of sunlight, missed meals/snacks/fluids, food wastage/rejection, insufficient food stores at home, and mental confusion affecting eating. These are signals that can be noticed, and reported, by neighbors, relatives, friends, home aides, health visitors, district nurses, and other professional and lay carers.

This interaction of risks, warning signals and needs has led to the development in some areas of assessment techniques for use by the social services. The aim of such assessments is to evaluate the risks for each individual to offer a total package of care, whether it be for delivered meals or for some other service more in keeping with the needs and wishes of the individual.

Innovative measures have been introduced in some areas to improve the acceptability and nutritional content of delivered meals and luncheon club meals. Wider publicity and further trials are needed.

But what are the nutritional risks when an individual is placed into care, either in a residential home or in a hospital or nursing home? Surely, if all food is provided, the only nutritional risk is that it may prove inadequate in nutrient content—not so! Enjoyment of food can be affected not only by health but also by loss of control over food choices and portion size, unattractive surroundings, inappropriate food timing and food temperature, need for assisted feeding, monotony of menu, and many other examples of seeming lack of care, affection, or expertise. Studies in old peoples' residential homes have suggested an A–Z list of at least 26 potential risk factors (see Table III).

It is necessary to ask the following. (1) Does the risk exist? (2) With what severity? (3) What is the reason? (4) What are the possible solutions? (5) How can the solution be implemented?

V. Practical Preventative Measures

Practical nutrition for old people has been defined as seeking to influence or endorse food choice, thus helping them to achieve and maintain health and well-being. In successful counseling, research findings have been turned into practical action for the "young old," the "old old," and the frail old. For the young old, the implementation of national nutritional guidelines—linked with campaigns for increased activity—has taken its rightful place in the forefront of preventative medicine.

In a number of countries, group guidelines (e.g., to reduce total fats, sugar, and salt; to increase complex carbohydrates, including dietary fiber foods; to moderate alcohol consumption) are recommended. Evidence indicates that well-publicized group campaigns for lowering fat intake have helped to achieve dramatic successes in reducing the incidence of coronary heart disease.

It has been found that retirement from work is a time of high motivation for making changes in lifestyle and food choice. In some firms, preretirement counseling has been introduced, but the need for follow-up counseling as circumstances change is recognized.

It is, however, generally agreed that, once elderly individuals have survived to become old old, dietary contributions to the risk of chronic disease are reduced. The question at this stage is not so much "can we alter food patterns," but rather "should we"? Each nutritional guideline needs to be separately queried: Would it prove unnecessarily restrictive for this individual?

Recent work endorses the importance of encouraging and helping the old old to keep to their well-established patterns of food choice, to include plenty of dietary fiber foods and fluids, to eat in company if they wish, and to take as much exercise as possible.

With the onset of frailty at any age, the aim is to ensure maximum nourishment with minimum effort. Those people with reduced energy intake are also likely to have reduced intakes of essential nutrients, so that with the smaller appetite the foods chosen need to be particularly nourishing (i.e., nutrient-dense). The immense value of the patience and skills needed to feed stroke patients and frail or demented old people has been demonstrated in studies showing increased intakes with assistant feeding.

Current nutrition education, in the developed and developing world, is focused on spreading knowledge not only to elderly people themselves but also to those caring and providing for them. This includes physicians who generally have little nutritional content in their training but whose instructions are trusted. It also includes those who write for the media; food producers, manufacturers, and shopkeepers; large-scale caterers and meals-on-wheels organizers; and other paid or voluntary social and health workers, all of whom wield influence over food availability and choice.

What practical tools are available to encourage better nutrition? They include specialized nutrition–cook books, videos, films, TV and radio shows; over-60s and retirement cookery classes; discussions and peer-counseling held in day centers or clubs, including those for the deaf, blind, or otherwise handicapped; and leaflets, booklets, and informative mail shots (mailers), with advice on budgeting and menu planning, distributed with the pension or delivered meal. Equally important is the increased availability of assistant feeding utensils, adaptations for the kitchen or home, and small portion nourishing foods available in markets, shops, and supermarkets.

With these practical measures at hand, the type of action called for at a grass roots level is generally simple and basic: help with food shopping and food preparation, an occasional accompanied visit to shops, a shared meal or an easy recipe idea, or even the provision of something non-nutritional such as a shopping trolley or a social visit. This "first-aid" approach in no way disregards the more structured care of the social services. On the contrary, the lay person's observations may highlight the need for medical or paramedical care or the urgency of an introduction to meal programs or other community services.

Bibliography

Davies, L. (1990). Socioeconomic, psychological and educational aspects of nutrition in old age. *Age Ageing* **19,** S37–42.

Department of Health & Social Security (1979). Nutrition and health in old age, Report on health and social subjects, No. 16. London, Her Majesty's Stationery Office (HMSO).

Exton-Smith, A. N., and Caird, F. I. (eds.) (1980). "Metabolic and Nutritional Disorders in the Elderly." John Wright & Sons Ltd., Bristol, United Kingdom.

Henderson, C. T. (1988). Nutrition and malnutrition in the elderly nursing home patient. *Clin. Geriatr. Med.* **4(3),** 527–547.

Horwitz, A., Macfadyen, D. M., Munro, H., Scrimshaw, N. S., Steen, B., and Williams, T. F. (eds.) (1989). "Nutrition in the Elderly." Published on behalf of the World Health Organization, Oxford University Press.

Lehmann, A. B. (1989). Review: Under nutrition in elderly people. *Age Ageing* **18,** 339–353.

Prinsley, D. M., and Sandstead, H. H. (eds.) (1989). "Nutrition and Aging." Alan R. Liss Inc., New York.

Marijuana and Cannabinoids

MONIQUE C. BRAUDE, *Director, Graduate Women in Science*
(previously with the National Institute on Drug Abuse)

Glossary

Amotivational syndrome Possible consequence of chronic *Cannabis* use, producing changes in the basic personality of users so that they become unattracted to work or to strive for success
Bradycardia Abnormal slowness of the heart beat; a rate usually less than 60 beats per minute
Cannabinoids Active components of the marijuana plant
Glaucoma Group of eye diseases characterized by an increase in intraocular pressure
Hashish Resin extracted from the tops of the flowering *Cannabis sativa* (marijuana) plant
Pharmacokinetics Study of the passage of a drug through the body; its absorption, distribution, localization in tissues, degradation, and elimination.
Sensemilla Crude preparation made from the flowering tops of female *Cannabis* plants that have not been pollinated (seedless marijuana)
Tachycardia Abnormally fast heart beat

MARIJUANA, the herbaceous plant, has a complex composition and more than 400 chemicals have already been isolated. The major class of active principles is called cannabinoids. Among them, Δ-9-THC (THC) is the compound generally recognized to be responsible for the psychoactivity of marijuana. Analytical determination of cannabinoids in the plant and in biological specimens (mostly urine) is now routine. Immunoassays have been developed for rapid identification, and gas chromatography–mass spectrometric methods are available for confirmation and quantitation of the compounds.

Short-term exposure produces little toxicity, but long-term use may lead to changes in many organs and biological systems such as the brain, the lungs, the heart, and the reproductive and immune systems. Although tolerance to the effects of marijuana is well recognized, dependence occurs rarely and only at higher doses upon repeated use. THC has been found to be effective in preventing the nausea associated with cancer chemotherapy and has been used for glaucoma treatment in a small number of patients.

I. The Plant and Its Components

The name *marijuana* refers to preparations containing primarily the leaves and flowering tops of the herbaceous plant, *Cannabis sativa* L. It is a complex mixture of more than 400 chemicals that can be classified into 17 categories.

The cannabinoids are the most studied and are believed to be responsible for most of the biological activities. As of 1974, only four were well identified, (i.e., Δ-9-THC (THC), Δ-8-THC, Cannabinol (CBN), and cannabidiol (CBD). Currently, 61 of them have been isolated. There are two major types of marijuana: the fiber type with a high CBD content and the drug type containing primarily THC. Depending on the method of cultivation or collection, the marijuana available on the "street" can vary in potency from 0.4% to more than 4% THC and the overall potency of marijuana has been steadily increasing over the years. Sensemilla may contain from 6 to 14% THC.

Other *Cannabis* preparations, such as hashish or "hash oil" (an extract from the plant) contain higher concentrations of THC—10–20%, respectively. Apart from cannabinoids, the other major categories of chemicals identified in the plant are terpenes, sugars, hydrocarbons, steroids, and nitrogeneous compounds.

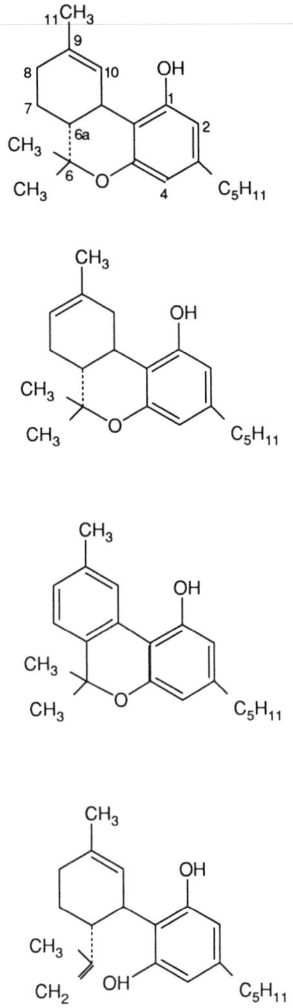

Δ^9-Tetrahydrocannabinol (Δ^9-THC)

Δ^8-Tetrahydrocannabinol (Δ^8-THC)

Cannabinol (CBN)

Cannabidiol (CBD)

FIGURE 1 Chemical structure of the primary cannabinoids.

II. Analytical Determination

The chemical structure of the four main cannabinoids is shown in Fig. 1. The numbering shown in this figure (i.e., the dibenzopyrane numbering) has been adopted by *Chemical Abstracts* and most U.S. authors. However, most non-U.S. scientists use the monoterpene system to identify cannabinoids. Thus, Δ-9-THC is often referred to as Δ-1-THC, and Δ-8-THC as Δ-1,6-THC in many publications. This can be very confusing for those new to the cannabinoid literature. THC, the major psychoactive cannabinoid is the (−)-*trans* isomer. Like all cannabinoids, it is very soluble in lipid and insoluble in water.

Methods for the determination of cannabinoids in plants by GC/MS have been available since the early 1970s. However, due to the high protein-binding properties of THC, it was not until the early 1980s that methods for detection of cannabinoids in biological fluids and tissues were developed. At present, a variety of methods are available including radioimmnoassays which allow detection of a terminal metabolite, the carboxy derivative of THC, in the urine. These rapid and less expensive methods have allowed widespread urine testing for evidence of prior marijuana use. [*See* RADIOIMMUNOASSAYS.]

As marijuana is usually taken by smoking, the chemistry of the smoke components delivered to the users is important. Marijuana cigarettes, popularly called "joints", are made of the leaves and flowering tops of the plant—finely chopped, rolled in cigarette paper, and smoked with deep inhalation. Many factors inherent to the smoking process decrease the overall potency of the drug's bioavailability. Approximately 20–30% of the THC content is destroyed by pyrolysis and as much as 40–50% can be lost in side stream smoke. Of concern was the finding that benzopyrene, a known carcinogen found in tobacco smoke, was 70% more abundant in marijuana smoke than in tobacco smoke. It should also be noted that passive inhalation of the sidestream smoke of marijuana does occur and may lead to mild intoxication. However, in order to get positive results in urine and blood tests the degree of exposure must be high.

III. Metabolism and Pharmacokinetics

When smoked, marijuana is rapidly absorbed into the bloodstream and metabolized by the liver and lungs to a number of metabolites. The most studied have been the 11-hydroxy-THC (11-OH-THC) and the terminal 9-carboxy-THC metabolites. In plasma, THC levels may rise rapidly to a peak of over 100ng/ml but decrease within 1 hr to levels around 10ng/ml and within 4–6 hr to below 1 ng/ml. However, recent studies in humans indicate that the THC plasma levels should be followed for longer periods of time. Concentration-versus-time curves reveal THC plasma half-lives of about 3 days when followed for 10 days and as much as 12.6 days when followed for 28 days.

A correlation between the elimination half-life of THC in plasma and of 9-carboxy-THC in urine is apparent, strongly indicating that the elimination of these compounds are rate-limited by the redistribution of THC from fat. The 9-carboxy metabolite is detectable in the blood within minutes and usually reaches a concentration equivalent to that of THC within 20 min. THC is very fat soluble and accumulates in fat tissue. Eighty to ninety percent of the dose of THC is excreted within 5 days, primarily in the feces. The urinary fraction consists primarily of acidic metabolites detectable for a long time (up to a week or more).

Depending on the route of administration, the contribution of the various metabolites to the biological effect of marijuana may be different. For example, the psychoactive metabolite, 11-OH-THC, probably contributes little to the overall effect of marijuana when taken by smoking or given iv as only 10% of the THC is degraded to this metabolite when given by these routes. However, after oral administration, the proportion of 11-OH-THC is equal to or may exceed that of THC itself and contributes substantially to the psychoactive effect of marijuana. It should be noted that positive results from a urine sample indicate only prior use and cannot be related directly to impairment since inactive metabolites (e.g., 9-carboxy-THC) persist in the urine for several days after smoking a single marijuana cigarette.

IV. Effects on Biological Systems

A. Mechanisms of Action

At the 2nd Meeting of the International *Cannabis* Study Group (ICSG) in 1989, important questions regarding the mechanism of action of cannabinoids were discussed. The search for a receptor responsible for cannabinoid effects has been the goal for research at the cellular level since the early 1970s. Unfortunately, the liposolubility of marijuana components and the lack of an available specific antagonist have prevented rapid advances in this field. However, in the last 5 yr, real progress has occurred using synthetic derivatives of cannabinoids such as CP-55,940. Binding sites for these compounds have been identified and recent claims of receptor(s) have been generally accepted.

It has also been proposed that the primary mode of action of psychoactive cannabinoids is by interacting with lipid components of cell membranes, possibly by increasing synaptic plasma membrane fluidity.

B. Patterns of Use

Although *Cannabis* has been used since antiquity for medicinal purposes and was abused by artists and writers in Europe in the 19th century (Club des Hashichins) it was not until the 1960s that its use became widespread in the United States especially among students and young adults. Epidemiological surveys show a significant increase in use from the late 1960s to the late 1970s followed by some decrease in recent years. Rate of use varies considerably and occasional use is still more frequent than daily use. However, age of use has decreased recently so that school-age children in their formative years are now smoking marijuana. Peer influence has been found to be the strongest predictor of future use.

C. Toxicity

The systemic toxicity of marijuana is low. So far no documented death in humans has been attributed to the acute use of marijuana alone. In animals, the acute lethal dose is greater than 1 g/kg by most routes except the intravenous route. Even after chronic administration, there is a very high safety ratio between pharmacologically effective and lethal doses. The major long-term chronic toxic effect of marijuana has been its effect on the lungs (see Section IV, F).

D. Behavioral Effects

As often seen, but especially significant for marijuana, the extrapolation of animal results to humans has been fraught with difficulties and the effects observed in animal species and humans can even been opposite. For instance, although tachycardia is a consequence of *Cannabis* use in humans, bradycardia is seen in such animals as rodents and dogs. These differences are primarily due to higher doses used in animal studies and different routes of administration. Oral, intraperitoneal, or subcutaneous routes have usually been used in animals, while smoking or intravenous routes are used in humans. The vehicle used in animal studies to dissolve or suspend cannabinoids can also influence absorption and distribution.

Marijuana has a multiplicity of behavioral effects resulting from alterations of central nervous system (CNS) function. The overt effects in humans are quite complex. Subjective effects include: excitement and dissociation of ideas, sensory enhancement, errors in judgement of space and time, delusions, irresistible impulses, illusions, and hallucinations. The so-called "amotivational syndrome" has also been attributed to *Cannabis* use by young people. At higher doses, decrements in psychomotor performance (e.g., driving and flying), attention span, memory, and a reduction in physical strength have also been observed. A variety of psychotic reactions have been reported after long-term frequent marijuana use, such as acute panic reactions, toxic delirium, acute paranoid states, or acute mania.

In animals, behavioral changes are characterized by a mixture of stimulant and depressant effects. There seem to be two types of cannabinoids, both of which predominantly produce CNS depression. The first group includes THC and is made up of cannabinoids that produce a behavioral depression accompanied by a stimulatory component. The second group, best exemplified by CBD and CBN, also produces CNS depression at high doses but without the stimulatory component. Corneal areflexia in rabbits, hindleg ataxia in dogs, and stimulus generalization studies (discrimination tests) in rats and monkeys have been widely used as predictors of the psychomimetic activity of cannabinoids.

E. Cardiovascular Effects

Tachycardia, orthostatic hypotension, and increased blood concentrations of carboxyhemoglobin (COHb) are produced by marijuana smoking. Tachycardia, primarily due to inhibition of vagal tone, could be deleterious in patients with heart disease due to arteriosclerosis of the coronary arteries or congestive heart failure. Smoking marijuana also increases myocardial oxygen demand and decreases myocardial oxygen delivery producing a 48% decrease in exercise time until angina appears. This risk is compounded by the fact that many patients also smoke tobacco cigarettes which are detrimental in case of angina.

F. Pulmonary Effects

As with cardiovascular effects, damage to the lungs produced by marijuana smoking is often complicated by use of tobacco in conjunction. A series of studies at the University of California has attempted to elucidate the damage produced by marijuana in chronic users versus those produced by tobacco. It was found that heavy marijuana smoking for 6–8 wk caused mild but significant airway obstruction, which may be reversible over time after cessation of use.

G. Effects on the Endocrine and Reproductive Systems

Changes in male (testosterone) and in female (LH) sex hormones have been reported with marijuana use. As the levels of these hormones vary during the day, and as tolerance develops to these effects, definitive answers have been hard to obtain. Decreased levels of testosterone have been associated with morphological abnormalities in sperm and with decreased sexual function. In two studies of chronic, male heavy-users of marijuana, abnormalities in count, structural characteristics, and motility of sperm were found. In one controlled study, a gradual return to normal function when marijuana use was discontinued was observed.

Data on effects of marijuana on the female reproductive system are sparse in humans. In animals, there is good evidence that marijuana and THC inhibit the secretion of the pituitary hormones, LH (luteinizing hormone) and FSH (follicle-stimulating hormone). Prolactin may be decreased or stimulated. These changes in pituitary hormone levels produce decreases in steroid sex hormones and cause disruption in menstrual cycles and ovulation. These effects are reversible when *Cannabis* use is discontinued or when tolerance develops. In view of these findings, the use of marijuana in pregnant women has been a cause for concern. Three studies have examined the effects of marijuana use during pregnancy on large samples of mother–child pairs. The general conclusions were that mothers who were marijuana users were more likely than nonusers to have an unplanned pregnancy and premature labor. Their children were more likely to have lower birth rates, congenital features compatible with those produced by fetal alcohol syndrome (FAS), or some other malformation. Investigators also point out that those mothers using marijuana during pregnancy were more likely than nonusers to smoke cigarettes, drink alcohol, or exhibit effects of malnutrition. These factors also have a deleterious effect on fetal growth and may be additive to marijuana effects. Interestingly, neurological disturbances noted at birth in babies born to regular marijuana users did not affect subsequent cognitive and motor performance when they were retested at $1\frac{1}{2}$ and 2 years of age.

H. Effects on the Immune System

In animals, there is consistent evidence that cannabinoids administered parenterally or by inhalation induce defects in both cell-mediated immunity (CMI) and humoral immune (HI) systems. THC suppresses the activity of natural killer (NK) cells which are important to defense against tumor cells and microbial infections. Cell culture studies have also shown that THC stimulates the maturation of key immune system cells called monocytes, but that this maturation is defective. These cells do not exhibit a mature monocytic morphology and produce aberrant proteins that seem to block transition to the more advanced states of maturation. Given these data, it would be reasonable to expect similar effects in man and early clinical studies reported decreases in human immune system function. However, later studies did not confirm these observations and there is at present no conclusive evidence that consumption of cannabinoids predisposes man to immune dysfunction. What is needed to resolve this and other controversial findings are long term epidemiological studies on large samples of marijuana users. [*See* NATURAL KILLER AND OTHER EFFECTOR CELLS.]

V. Tolerance and Dependence

Tolerance (i.e., a diminished response to the repeated administration of a given dose) to marijuana and cannabinoids has been demonstrated in many animal species and in humans. Dependence, however, with accompanying withdrawal symptoms has been difficult to document. Up to the mid 1980s, marijuana dependence in humans was not of concern in the United States. In contrast to other countries using the more potent forms of *Cannabis* such as hashish or ganga, the THC content of the marijuana grown in the United States varied from 1 to 4%. In the 1980s, more potent forms of the plant, such as sensemilla, became available on the "street" with THC content of 6–14%. These potent preparations as well as hashish have emerged as the usual forms of self-administration. In chronic marijuana users, withdrawal signs and symptoms have been noted after abrupt discontinuation of use. The abstinence phenomena resemble the restlessness, irritability, mild agitation, insomnia, and autonomic disturbance that follow discontinuation of low doses of alcohol or sedative drugs given over a period of time. Two clinical forms of marijuana dependence have been defined. Type I is represented by an individual who will self-administer marijuana several times per day, who may complain to a physician that his or her daily dosage has escalated and he or she is unable to cease use without medical assistance. This patient may or may not report mental impairment related to memory, motivation, timekeeping, abnormal thoughts, and work or school performance. The relapse after withdrawal from marijuana is high. Type II marijuana dependence is primarily identified as a result of mandatory urine screening and treatment referral in the workplace. The patient is usually self-administering marijuana every 24–36 hours and may have done so for several years. As in Type I, reported impairment relative to memory, motivation, timekeeping, and job performance is variable. In contrast to Type I, the patient may report few if any symptoms of withdrawal upon abrupt cessation. Relapse, however, is common. In both types, chronic and heavy use are the prerequisites for the development of dependence.

VI. Therapeutic Aspects

For centuries (and still now in many parts of the world) marijuana has been used as a folk medicine. It has been reported to be effective as an analgesic and anticonvulsant and has been used to treat diseases ranging from dysmenorrhea, postpartum psychoses, senile depression, and gonorrhea to opium or chloral addiction. In the 1970s, THC and the synthetic analogue, Nabilone, have been reported as useful antiemics for patients receiving cancer chemotherapy. In over 14 controlled clinical trials, THC efficacy was compared to that of other antiemetics such as prochlorperazine and was found to be at least as useful or even superior to marketed products. THC in capsule form, is now available by prescription but, in view of its psychoomimetic side effects, the amount that can be dispensed has been limited to that necessary for a single cycle of chemotherapy. When it was found in clinical trials that marijuana and THC given intravenously lower intraocular pressure by 30–50%, testing was undertaken for possible treatment of glaucoma. Unfortunately, tolerance develops to this beneficial effect of cannabinoids.

Cannabidiol, one of the major cannabinoids present in the plant but devoid of psychoactivity, has been shown to be effective as an anticonvulsant in animal and human studies. As it has a low toxicity, its use in the treatment of epileptic disorders appears promising. For many years, marijuana and THC were reported to have some analgesic activity, but the results of studies reported by different laboratories were often contradictory. The differences seem to depend on the route of administration used. Intravenous THC is 45 times more potent as an analgesic than subcutaneous THC and, given iv, THC is 3 times more potent than morphine given by the same route. It is intriguing that reasonable doses of THC iv, cause an effect identical to that of morphine, but that high doses given by different routes have no effect. The major problem in the search for therapeutic agents in the cannabinoid series has been the inability to synthethize a cannabinoid with therapeutic effects at doses below those producing psychomimetic side effects.

VII. Conclusions

This article has attempted to summarize the present state of knowledge regarding marijuana and the cannabinoids without dwelling on past controversies. Marijuana and its major components, the cannabinoids, have been the subject of extensive research studies in the 1970s and the 1980s. Much is now known about their pharmacologic effects and mechanism of action and some of the early fears regarding their systemic toxicity have been alleviated. However, the development and use of more potent forms of marijuana is a cause for concern and may lead to increased dependence on the drug. The tolerance development and reversibility of effects af-

ter discontinuation of use have been the reasons for many conflicting data published in the literature.

One must remember, however, that although the use of marijuana has decreased in recent years in the population as a whole, it has been increasingly abused by younger individuals in their developmental years when it could be most damaging. Furthermore, cannabinoids are often used concomitantly with alcohol or smoking which potentiate their effects. Long-term epidemiological studies will be the only way to obtain definitive answers to the long-term effects of marijuana and the cannabinoids on the population.

Bibliography

Agurell, S., Dewey, W., and Willette, R. E. (eds) (1984). "The Cannabinoids: Chemical, Pharmacologic and Therapeutic Aspects." Academic Press, New York.

Braude, M. C., and Szara, S. (eds.) (1976). "Pharmacology of Marijuana," Vols. 1 and 2, Raven, New York.

Braude, M. C., and Ludford, J. P. (eds.) (1984). Marijuana effects on the endocrine and reproductive systems. *NIDA Res. Monograph* **44.**

Chesher, G., Consroe, P., and Musty, R. (1987). Marijuana: an international research report. *Monograph series* **7.** Australian Government Publishing Service, Canberra.

Dewey, W. L. (1986). Cannabinoid pharmacology. *Pharmacological Rev.* **38** (2) 151.

Fehr, K., and Kalant, H. (1983). *Cannabis* and health hazards. Addiction Research Foundation, Toronto.

Harvey, D. J. (1985). "Marijuana 84." IRL Press, Oxford.

Hawks, R. L. (1982). The analysis of cannabinoids in biological fluids. *NIDA Res. Monograph* **42.**

Hutchings, D. E., *et al.* (1987). Δ-9-THC during pregnancy in the rat. I. Differential effects on maternal nutrition, embryotoxicity and growth in the offspring. *Neurotoxicol. and Teratol.* **9,** 39.

Johansson, E., Agurell, S., Hollister, L. E., and Halldin, M. M. (1988). Prolonged apparent half-life of Δ-1-tetrahydrocannabinol in plasma of chronic marijuana users. *J. Pharm. Pharmacol.* **40,** 374.

Maykut, M. O. (1984). "Health Consequences of Acute and Chronic marijuana Use." Pergamon Press, Oxford.

Petersen, R. C., ed. (1980). Marijuana research findings: 1980. *NIDA Research Monograph,* **31.**

Pinkert, T. M. (1985) Current research on the consequences of marijuana use. *NIDA Res. Monograph* **59.**

Szara, S. (1989). Marijuana *in* "Clinical Management of Poisoning and Drug Overdose" L. M. Haddad and J. F. Winchester, eds.)

Tennant, F. S. (1986). The clinical syndrome of marijuana dependence. *Psych. Annals* **16**(4), 225.

Zuckerman, B., *et al.* (1989). Effects of maternal marijuana and cocaine use on fetal growth. *New Engl. J. Med.* **320** (12), 762.

Maternal Mortality

ALLAN ROSENFIELD, *Columbia University School of Public Health*

I. Developed World: 1700–Present
II. Developing World
III. Conclusion

Glossary

Cesarean section Delivery of an infant by an abdominal surgical procedure, necessitated by an abnormal maternal problem or where signs of serious distress are evident in the fetus

Eclampsia End stage of pregnancy-induced hypertension in which the woman convulses repeatedly; without treatment, it almost always leads to death

Ectopic pregnancy Pregnancy in a fallopian tube instead of in the uterus; most ectopic pregnancies will rupture and lead to death unless surgery is performed

Hemorrhage Excessive amount of bleeding from a variety of causes and one of the most common causes of pregnancy-related deaths

Maternal mortality rate Number of pregnancy-related deaths per 100,000 women of reproductive age (15–45 yr old); because the denominator is difficult to obtain, it is rarely used

Maternal mortality ratio (rate) Number of pregnancy-related deaths per 100,000 livebirths; in the literature, although correctly this should be termed a ratio, it is most commonly referred to as a rate

Obstructed labor For a variety of reasons, normal vaginal delivery is impossible and a cesarean section is necessary to save the life of both the infant and the mother

Placenta previa Abnormal placement of the placenta (the afterbirth) such that it blocks the opening of the womb; with this condition vaginal delivery is impossible without massive hemorrhage and cesarean section is required

Pregnancy-induced hypertension (toxemia) Previously called toxemia of pregnancy, this condition results in excessive swelling of the feet, ankles, and, in later stages, hands and face; elevation of blood pressure, and protein in the urine

THE DEATH OF WOMEN from pregnancy-related causes has been an important, if not the most important, cause of mortality for women of reproductive age (age 15–45 or 50) throughout history. Only since the early 1900s have maternal mortality ratios in the developed world fallen to <100/100,000 livebirths, reaching the current low levels of 8–15/100,000 livebirths in most of the western world. In most developing countries of Asia, Africa, Latin America, and the Middle East, however, maternal mortality ratios remain extremely high. Unfortunately, it is only recently that public health and obstetrical attention has been given to this tragic developing-world health problem, in which the World Health Organization (WHO) estimates that as many as 500,000 women die each year from essentially preventable and/or treatable causes. As was the case in the West, the major causes vary in different countries but generally include obstructed labor (often leading to rupture of the uterus), hemorrhage (most commonly after the delivery of the baby), pregnancy-induced hypertension, infection, and complications of an illegal and/or unsafe abortion procedure. Any of these complications can be exacerbated by such problems as malaria, diabetes, high blood pressure, and heart disease. Finally, many women die early in pregnancy from a ruptured tubal (or ectopic) pregnancy. In rural areas, many deaths from this cause go unrecognized, because often it is not known that the woman is pregnant and death is often sudden.

I. Developed World: 1700–Present

Some of the most relevant historical information on mortality comes from Sweden, where mortality levels were reviewed from 1751 to 1980, with a primary focus on changes that took place prior to 1900. In 1751, the estimated level of mortality was >800/100,000 livebirths, a level not very different from that estimated to exist in many developing countries today. During the period under assessment, utilizing unusually reliable Swedish data, a rather dramatic fall in mortality levels was noted, particularly in the period 1861–1900, long before the introduction of modern obstetrical interventions.

Interestingly, the Swedish Board of Health in 1751 stated that "out of 651 women dying in childbirth, at least 400 could have been saved if only there were enough midwives." By 1861, the ratio is estimated to have fallen to a level of about 567/100,000 livebirths. In the following 40-yr period ending in 1900, the ratio fell further to an estimated level of 227/100,000 livebirths. Two interventions are described as having made the major impact on these falling ratios before the advent of modern obstetrics. The first was aseptic techniques, introduced in the late 1800s, which had a direct impact in reducing infection rates. This was thought to be of particular importance because ratios of sepsis in the new maternity hospitals created in Sweden in the late eighteenth century were extraordinarily high, often resulting in the death of the mother, as well as of the infant. In 1864–1880, 1 out of 35 women delivering in the hospital died of sepsis, as compared to 1 out of 1,124 women delivering at home during that time (maternal mortality ratios of 2,700/100,000 and 89/100,000, respectively).

The problem of hospital-related sepsis, seen in many other hospitals established to provide maternity care during that period, accounted for what is probably one of the most dramatic demonstrations of iatrogenic (medically induced) disease and mortality in medical history. As aseptic techniques were introduced into Swedish hospitals, however, these rates fell by 1895 to 96/100,000. Another important innovation was introduced in 1881, when a law was passed that required health personnel involved in caring for patients to use carbolic acid to disinfect their hands and instruments. During this same period, mortality due to sepsis also fell in home deliveries to a level of 33/100,000, with no apparent reason being noted.

Although Holmes (in 1843) and Semmelweiss (in 1844) are generally credited with the introduction of

the concept of puerperal sepsis, in 1768 a physician named Denman wrote that a disease is carried by a physician or midwife from one parturient to another, and White, in 1772, wrote about the discharges of septic women being contagious. The most definitive work was by Gordon who, in 1795, wrote that "this disease seized women only as were visited or delivered by a practitioner or taken care of by a nurse who had previously attended patients affected with the disease."

Despite these farsighted writings in the 1700s and the work of Holmes and Semmelweiss in the mid-1800s, it was not until the late 1800s that Swedish medicine and law took note of this knowledge. Even Swedish recognition of the importance of asepsis preceded attention to the issue in the U.K. and U.S. In 1900, for example, when Sweden's maternal mortality ratio was approximately 227/100,000 livebirths, the ratios in the U.K. and U.S. were 427 and 600, respectively.

The second intervention contributing to a reduction in mortality rates during this period was the introduction of trained and licensed midwives, who practiced widely throughout Sweden. By the early 1800s, the midwives were taught to use forceps for delivery and sharp hooks and perforators to remove an already dead baby in cases of severe obstructed labors. Although the available data suggests that the midwives did not use these techniques often, in some instances their use surely resulted in saving the life of a woman who otherwise would have died.

Based on Swedish experience, also noted elsewhere, the control of sepsis and the expanded training and utilization of midwives were the most important early innovations, although other public health changes, including better sanitation, improved nutrition and cleaner water supplies, were also of great importance.

A review of records from the late sixteenth century in an English parish suggests maternal mortality ratios as high as 2,300/100,000 livebirths, and similar ratios were reported in France. These ratios must be rather gross estimates, as universal reporting of deaths was unlikely, a problem that hinders research in developing countries today. But, due to the influences of the clergy and others, unusually complete, village-based data systems were established in some parts of England and Europe during the seventeenth and eighteenth centuries. Review of the available data reveals an excess mortality among women 15–49 yr old as compared with men of the same age group, and the difference was related to pregnancy-related deaths. Similar findings have been noted more recently in Bangladesh, India, Pakistan, and Sri Lanka.

What is striking in reviewing the past is that conditions found in much of the developing world today existed as recently as 50–60 yr ago in the U.S. and Europe. Even as recently as 1946, Klein and Clahr described a maternal mortality ratio of 116/100,000 livebirths in New York City, a ratio that fell over the next 10 yr to 40/100,000 livebirths, primarily through reductions in mortality from infection, hemorrhage, heart disease, and anesthesia. Although infection-related mortality played perhaps the single most important role in the decline, as late as 1956 it remained a leading cause of mortality because of the sepsis related to illegal abortions.

The careful studies by the Committee on Maternal Mortality of the Massachusetts Medical Society, initiated in 1941, traced the decline in mortality in Massachusetts from 50/100,000 livebirths in 1954 to 10 /100,000 livebirths in 1985. The leading causes from 1954 to 1957 were infection, cardiac disease, pregnancy-induced hypertension, and hemorrhage. From 1982 to 1985, the leading cause was pulmonary embolus, plus nonpregnancy-related trauma. The decreases were due to both legislative actions (licensing of maternity services, blood banks, and, probably of greatest importance, the legalization of abortion) and improvements in medical practice (antibiotic therapy, aggressive treatment of toxemia, and prompt replacement of blood in cases of hemorrhage).

One factor that is sometimes taken for granted in analyses of maternal mortality in the West is that almost 100% of pregnant women are either delivered in an institutional setting or have rapid access to such a facility in the case of complications in a home delivery. Recent data from Sweden, for example, demonstrates that the declines in maternal mortality ratios noted in the nineteenth century were even more dramatic in the twentieth century, reaching a level of 6/100,000 livebirths in 1980.

II. Developing World

A. Introduction

Professionals involved in health care in developing countries have devoted most of their attention in the past to the eradication of various communicable, infectious, and parasitic diseases and to the improvement of sanitation and clean water supplies. More recently, a great deal of attention has been given to the "child-survival" movement, which focuses on interventions aimed at reducing rates of infant and childhood mortality. As many as 14 million deaths are estimated to occur each year among infants and young children under the age of 5 yr from essentially preventable or treatable conditions. These high-priority, child-survival interventions include the management of diarrheal diseases and immunizations against the most common of the communicable diseases. Indeed, WHO and the Children's Fund (UNICEF), together with ministries of health throughout the world, have set a goal of "Health for All by the Year 2000," with a heavy emphasis on the delivery of simple primary health care services to underserved populations.

One neglected health problem in developing countries, however, remains: the health status of women of reproductive age. This is discouraging because for many years maternal mortality ratios have been recognized as extremely high in rural areas, as well as in the urban slums, of most developing countries.

What makes the lack of attention to maternal health so surprising and sad is the fact that WHO estimates that approximately 500,000 women die each year from pregnancy-related causes; over 98% of these deaths occur in developing countries. While ratios of maternal mortality in the U.S. and other developed countries today are in the range of 8–15 deaths per 100,000 livebirths, the ratios in many areas of the developing world are in the range of 200–1,000 or more per 100,000 livebirths. This differential between ratios in the developed versus the developing world is even higher than that seen with infant mortality rates (10–15 vs. 100–200/1,000, respectively).

Recently, the World Bank and WHO have given more attention to this problem, initially by sponsoring the first "Safe Motherhood" meeting in 1987 in Nairobi, Kenya. At that meeting, Halfdan Mahler (until very recently the Director General of WHO) gave a keynote address, which clearly summarized the tragic neglect that this problem has received up to now and stressed the importance of finding solutions that will lower rates of mortality.

B. Case Studies

To provide a clearer understanding of the many factors that contribute to maternal deaths in developing countries, a few case reports follow—the names have been changed, but the deaths are actual ones reported to WHO.

1. Obstructed Labor, Nigeria

Bola, aged 17, and her husband were farmers living in a remote village in Nigeria. They had no formal education and their house had no electricity, no water supply, and no modern sanitation facilities. Bola was married at age 13. Her first child was born dead after 4 days of a difficult obstructed labor, with a resultant vesicovaginal fistula, a permanent hole between her bladder and vagina. The consequences, incontinence of urine and constant odor, almost ended her marriage, but she was fortunate to have reconstructive surgery at a hospital far from her home. Two years later she again became pregnant, but living very far from any maternity care facility she had no prenatal care. After early rupture of her membranes, she eventually went into labor but did not progress for 3 days. Finally, the family took her the long distance to the nearest hospital, a trip of several hours. The baby had already died, and she was febrile and acutely ill with infection and a ruptured uterus. Despite heroic efforts at the hospital, she died.

2. Postpartum Hemorrhage, Yemen

Fatima, 27 yr old, lived in a remote village on the top of a steep hill. Her husband Mohammed worked as a laborer in Saudi Arabia. Fatima had five girls, whom she had delivered safely in the village with no prenatal care, because the nearest health station was a 2-hr walk. She had a very short labor, delivering in less than a half hour after her membranes had ruptured. The afterbirth (placenta) did not deliver as it had in previous pregnancies. Initially, there was little bleeding, but it soon became very heavy and the family became concerned, particularly as Fatima became increasingly weaker. Yemen culture required a husband's permission to go to the hospital, but because Mohammed was out of the country another male relative was sought to give the permission; in about 2 hr, her uncle was found in a neighboring village, and he gave his permission. A stretcher was made, and several men began the long trip down the hill and then 15 km along a rough road to a main road, where they hoped to flag down a vehicle to take them the last 40 km to the hospital. When they got down they noticed the stretcher covered with blood and Fatima lying very quietly, apparently sleeping. Soon, however, they realized that she had died; the delays were simply too long.

3. Illegal Abortion, Colombia

Esperanza, age 30, had already had five children, and neither she nor her husband wanted another child. They were upset when she again became pregnant. She had received health care at a nearby health center, but they did not offer either family planning advice or services. Esperanza did not know about the private family planning clinic in Bogota. She had recently found a job as a housemaid in the city, and the wages she earned were badly needed by the family. She decided on her own to obtain an illegal abortion. She did not know exactly what the abortionist had done, but it was painful and resulted in much bleeding. After 3 days, the bleeding decreased, but she developed a fever and abdominal pain and began to vomit. Her husband took her to the hospital where she received antibiotics for 2 days and then was discharged home. The pain returned, but Esperanza was afraid she would lose her job if she took more time off. However, her symptoms worsened, and she was readmitted to the hospital. Here she soon went into irreversible shock and, despite surgery, she died.

These three cases are typical examples of pregnancy-related deaths, tragically common in many parts of today's developing countries, particularly in the rural areas, where a majority of people live and where modern medical services and personnel are generally not available. It is unfortunate that the vast majority of physicians, trained midwives, nurses, and hospitals are located in the larger urban areas, with few such resources available in rural areas. In many countries, the capital city has a large modern tertiary care hospital, accessible to a very small percentage of the country's population, but using an inordinate percentage of the country's total health budget. The inequitable distribution of health resources is one of the greatest problems in many developing countries.

C. Current Status

In discussing the problem of maternal mortality in the developing world, one must note the absence of good data. In most developing countries, the vital registration system is poorly developed. Reports of deaths from rural areas are grossly underreported, and the cause of death is often absent because these reports usually originate from untrained local village headmen or other reporters. Unfortunately, few community-based studies exist because of the expense and difficulty in conducting such studies. Therefore, most published reports about maternal deaths come from hospital studies, and these obviously are limited to data on those patients that come to the hospital. However, the majority of women who die in many countries, particularly those who live in rural areas, probably never reach a hospital. Depending on the country, anywhere from 40% (in many Latin American countries) to >80% (in sub-Saharan African and Indian subcontinent countries) live in underserved rural areas. And the majority of women living in rural areas of these countries (and some of those living in the growing urban slum areas) deliver their babies, without any modern prenatal care, in their home with a traditional birth attendant (TBA), relative, or, in some cases, no one in attendance.

A detailed community survey (one of the few of its kind) of all women aged 13–49 yr in 32,215 households in Addis Ababa, Ethiopia, was conducted to obtain pregnancy outcome data over a 2-yr period (1981–1983). There were 45 pregnancy-related deaths among the 9,315 pregnancies identified during the 2-yr period, resulting in a maternal mortality ratio of 566/100,000 livebirths. Mortality in this urban study was highest among the unmarried nullipara, students, and women employed as housemaids. In this setting, complications of illegal abortions were the most common cause (54%), a not uncommon finding in studies of urban populations in the developing world, as compared with rural areas where the majority of pregnancy-related deaths are most likely to occur among high-parity, married women.

In a hospital-based study in Lagos, Nigeria, 51% of maternal deaths in a predominantly urban population were also due to abortion complications. In a large study of illegal abortions in 60 developing countries, somewhere between 70,000 and 100,000 maternal deaths each year were associated with complications of abortion, an extraordinary toll on predominantly poor women.

Similarly, primarily urban hospital-based studies in Zimbabwe, Ghana, Zambia, Kenya, and Bangladesh have demonstrated the tragic importance of illegal abortion as a major cause of maternal mortality, particularly in the rapidly growing urban slum areas. One large hospital in Nairobi, Kenya, allegedly admits daily 30–40 women suffering from complications of poorly performed illegal abortions. In Latin America, septic abortion reportedly accounts for a disproportionate share of the funds spent on transfusions and operating room costs, as well as total bed-nights.

Evidence, albeit primarily anecdotal, indicates that educated, and middle- and upper-class women generally have access to safe abortions, regardless of the legality, whereas predominantly poor women must resort to highly unsafe illegal procedures.

Overall, the series of hospital-based studies on maternal mortality paint a dismal picture. Information from such studies, as well as from those reported earlier, suggest maternal mortality rates of between 500 and 1,000 per 100,000 livebirths. In addition, as data from a number of both developed and developing countries have demonstrated, as much as 50% of maternal deaths go unreported. Even with such underreporting, in poor countries apparently as many as 21–46% of deaths among women aged 15–49 yr are pregnancy related, as compared to about 1% in the U.S.

In a rural study in Gambia, a mortality ratio of >2000/100,000 livebirths was noted, with postpartum hemorrhage and infection being the leading causes of death in this country. The groups at highest risk were found to be women <20 yr old, >40 yr old, and with multiple pregnancies. A detailed study in Zaria, Nigeria, noted that 219 maternal deaths occurred among 7,654 women seen for the first time in labor, as compared with 19 deaths among 15,020 women who had prenatal care through the hospital in Zaria, still a large number but dramatically lower than that seen among patients who were "unbooked." In a 1980s U.S. study of a religious group, which refused any modern medical care, a maternal mortality ratio of 872/100,000 livebirths was noted, a ratio lower than the highest ratio seen in the developing world, but comparable with overall figures for many developing countries.

There are many life-threatening complications of pregnancy that do not result in death but rather in serious, often chronic, conditions. Perhaps the most severe is a complication of obstructed labor in which a hole develops due to trauma between the bladder and the vagina, a vesicovaginal fistula. This complication often results in the ostracism of the woman, who is not acceptable to either her husband or her family because of the constant urine odor. In many hospitals in Africa, waiting lists for surgery are long, and many women suffering from this complication never have access to essential surgical repair. The surgical procedure is a difficult one and carries a relatively high rate of failure. A study from Ethiopia reported approximately 1,000 cases per year, while in the Khartoum Hospital in the Sudan, there were 122 cases in a 20-mo period, accounting for 16% of all gynecological admissions.

D. Possible Interventions

A major lesson from the experiences of the developed world is that aseptic technique plays a key role in the reduction of maternal mortality. This is important in both hospital and home deliveries. TBAs in many countries have undergone training programs, and clearly a major focus should be to instill an understanding and appreciation of the importance of asepsis. Less well recognized, but of equally great importance, is the use of careful sterile technique in hospitals, as is clearly evident from the experience in Sweden in the 1800s.

A second lesson from the past is that the training of midwives can also have an effect. To date, however, little evidence indicates that the training of TBAs has had an impact on maternal mortality in rural areas of developing countries. Nonetheless, because TBAs still deliver a large percentage of pregnant women in many of these countries, attention must continue to be given to attempts to work with these personnel.

Similarly, much greater attention should be given to the provision of health education at the community level so that women and their families have a better understanding of the various warning signs that mean assistance should be sought. In hospital reports on maternal deaths, there are repeated comments on the long delays between the onset of signs of problems and the actual decision to seek help at a hospital. Unfortunately, the reasons for the delays often are more related to distance, difficult access, or fears about the costs than they are to a lack of understanding that a problem exists.

An overwhelming problem in many rural areas is simply that maternity care services are much too far away to be truly accessible to the majority of women. To teach TBAs or the women themselves the signs that warrant referral when such referral is almost impossible creates major problems. For example, if a woman of parity greater than four is high risk, what should the woman do if she lives 2 or more hr from the nearest hospital or maternity care facility. Once labor has begun, it generally is too late for the woman to move because higher parity labors are usually shorter in duration than lower parity ones.

In a few countries, there are maternity waiting homes, adjacent to a hospital, where a woman can come in the last weeks of her pregnancy to await the onset of labor, assuming that she has been able to have someone else take care of her other children. The maternity home concept, where facilities and personnel are limited, is attractive, at least as an intervention worth testing. Experiments are currently underway to assess the effectiveness of such an approach.

Some African women at high risk of obstructed labor (e.g., very short stature or obvious bony pelvic abnormality), who are unlikely to be delivered in an institutionalized setting, have had surgical procedures performed to separate the anterior pelvic bones (called a symphysiotomy). This procedure is carried out early in pregnancy to facilitate a safer home delivery. This does not appear to be an active practice at present.

In the long run, in the opinion of the author, there is no question that the ultimate solution to the problem of maternal deaths lies in the development of networks of facilities that have the capability of providing blood transfusions and performing emergency cesarean sections. WHO has prepared a report on first-referral-level facilities, suggesting the steps needed to increase access to such care. Because a majority of women in many countries today do not have access to such simple facilities, upgrading, while expensive, is of the highest priority so as to increase access. A key component of such a plan, however, is not simply the building of such facilities but the provision of sufficient budget to pay for the necessary personnel, equipment, and supplies on a continuous basis. Facilities are available in some rural areas, but they have insufficient equipment and supplies, not to mention improperly trained personnel. Eventually, all women should be able to deliver in a well-equipped facility or, at least, have ready access to such a facility if low-risk women, after appropriate screening, continue to deliver at home with, however, well-trained personnel.

A significant problem at present is insufficient physicians, in general, and obstetricians, specifically, to staff the first-referral-level facilities in rural areas. Where general physicians are available, they should be trained in simple operations such as appendectomy, herniorrhaphy, and cesarean section. Where such physicians are not available, one should consider training other personnel in these techniques. One report from East Africa demonstrated that only approximately 10% of women needing a cesarean section for a maternal indication received one, the majority of those not having the procedure resulting in severe morbidity and a high rate of mortality. More appropriate use of personnel, other than physicians, for a range of procedures generally not allowed to be performed by such personnel at present has been suggested.

Two rural hospitals in Zaire have reported a very important experience. The two hospitals serve as referral centers for the surrounding rural communities and provide obstetrical services for women who often arrive very late for care. Because of severe understaffing of physicians, a pioneer missionary physician initiated a training program for local nurses to serve as emergency surgeons able to perform cesarean sections. In an 18-mo period in 1985–1986, trained nurses at the two hospitals performed 310 emergency cesarean sections, primarily for obstructed labor, inactive labor, previous cesarean section, and placenta previa. Only 3 of these 310 women died, a remarkable record, particularly given the long delays in arrival at the hospital and, as a result, the poor condition of many of the women. It is fair to say that without these procedures a high proportion of these women would have died. This experience in Zaire is a dramatic demonstration of what is possible where one does not follow conventional approaches when such approaches are simply impossible, although this is not likely to be emulated in many other settings.

In the short run, steps need to be taken that perhaps can have an impact prior to the establishment of a network of first-referral facilities. In the child-survival movement, strategies have been developed for village-level interventions such as immunization programs, oral rehydration fluids, etc. For maternal mortality, such community interventions are more difficult because of the nature of the conditions needing treatment. One most important exception to this is the role of family planning in helping to reduce maternal mortality rates. Women at young ages (<17 yr) and older ages (>35 yr), and, perhaps most importantly, women having their fifth or higher child are all at increased risk of death during a pregnancy. Therefore, effective family planning services, which can be delivered at village level, can have a direct impact on maternal mortality rates

in this group, not to mention on the large numbers of women who have stated in fertility surveys that they do not want any more children.

In reviewing the five major causes of deaths during pregnancy (obstructed labor and ruptured uterus, postpartum hemorrhage, pregnancy-induced hypertension, postpartum infection, and septic abortion), not a great deal can be done for most of these complications other than referral to a first-level facility, if such a facility is readily available. The following are some possible community-level strategies, most of which have not yet been carefully assessed.

1. Obstructed Labor
At village level, TBAs and women themselves need to understand the signs of obstructed labor and that, no matter how far, there often is time to get to the nearest referral facility if the decision to move is made before 24 hr of labor has passed.

2. Postpartum Hemorrhage
This complication is most common in high-parity women and, unfortunately, usually cannot be predicted in advance. When it does occur, the bleeding can be very extensive in a short time; very few women with this complication in a rural setting will survive a trip to a distant facility. TBAs and others can be taught to carry out effective uterine massage, the means to remove a retained placenta, and possibly the use of drugs that will make the uterus contract.

3. Pregnancy-Induced Hypertension
Little can be done for this complication at the village level other than to recommend bedrest and possibly the use of sedatives. Once severe preeclampsia or convulsions have occurred, the woman will almost certainly die unless she can be moved to a first-referral-level facility. [See HYPERTENSION.]

4. Postpartum Infection
The use of careful sterile technique can do much to decrease the incidence of puerperal sepsis, and TBAs in many countries have been so trained. While as yet not studied at village level, TBAs might possibly be trained to use a broad spectrum antibiotic for women who have had ruptured membranes for longer than 12 or 18 hr. The chance of

infection increases the longer the membranes are ruptured before delivery. Introduction of an antibiotic early in this course could have an impact; it is an intervention that needs to be studied.

5. Septic Abortion

The issue of legalized abortion is the most controversial of all issues within the field of reproductive health. Clearly the prevention of the need for an abortion through the practice of family planning should be encouraged. However, as the experience of the U.S. has shown, even widespread availability of contraceptive services does not do away with the demand for abortions. TBAs and other untrained practitioners perform abortions through one approach or another in villages, as well as in urban slum areas. Even where abortion is illegal, first-level referral centers should have suction equipment, together with personnel trained to use it (including midwives and/or nurses) for the management of incomplete abortion, hopefully before they have become infected.

III. Conclusion

This paper has reviewed past Western experiences, described the current situation in the developing world, and suggested some possible interventions for the short and long term. The death of some 500,000 women in the prime of their lives from a pregnancy-related cause that could have been either prevented or treated with simple existing technologies is a true tragedy.

Bibliography

Harrison, K. (1985). Child-bearing, health and social priorities: A survey of 22,774 consecutive hospital births in Zaria, Northern Nigeria. *Brit. J. Obstet. Gynaec.* **92** (Suppl. 5), 1–119.

Hogberg, U., and Wall, S. (1986). Secular trends in maternal mortality in Sweden from 1750 to 1980. *Bull. WHO* **64(1),** 79–84.

Kwast, B. E., Rochat, R. W., and Kidane-Mariam, W. (1986). Maternal mortality in Addis Ababa, Ethiopia. *Stud. Fam. Plann.* **17,** 288–301.

Lettenmaier, C., Liskin, L., Church, C., Harris, J. (1988). ''Mother's lives matter; Maternal health in the community'' *Pop. Reports, Series L, No. 7;* 1–31.

Mahler, H. (1987). The safe motherhood initiative: A call to action. *Lancet* **i,** 668–670.

Measham, A. R., and Herz, B. (1987). ''The Safe Motherhood Initiative: Proposals for Action.'' World Bank Background Paper for the Safe Brotherhood Conference, Nairobi, Kenya, Feb.

Rochat, R. W., Jaheen, S., Rosenberg, M. J., Measham, A. R., *et al.* (1981). Maternal and abortion related death in Bangladesh. *Int. J. Gynec. Obstet.* **19,** 155–164.

Rochat, R. W., Kramer, D., Senanayake, G., and Howell, C. (1980). Induced abortion and health problems in developing countries. *Lancet* **ii,** 484.

Rosenfield, A., and Maine, D. (1985). Maternal mortality: A neglected tragedy. *Lancet* **ii,** 83–85.

Rosenfield, A., (1989). Maternal mortality in developing countries: An ongoing but neglected epidemic.'' *J. Amer. Med. Assoc.* **262,** 376–379.

Royston, E., and Ferguson, J. (1985). The coverage of maternity care: A critical review of available information. *Wld. Hlth. Statist. Quart.* **38,** 267–288.

White, S. M., Thorpe, R. G., and Maine, D. (1987). Emergency obstetric surgery performed by nurses in Zaire. *Lancet* **ii,** 612–613.

Mathematical Modeling of Drug Resistance in Cancer Chemotherapy

JAMES H. GOLDIE, ANDREW J. COLDMAN, *Cancer Control Agency of British Columbia*

Glossary

Cross-resistance Property seen in cancer cells wherein a cell that is resistant to one drug is also resistant to a second, different drug

Doubling time Time, usually in hours or days, that it takes for a population of tumor cells to double their number

Drug resistance Phenomenon whereby a cell or organism expresses a reduction in its sensitivity to a particular toxic drug, as compared with its initial degree of sensitivity prior to drug exposure

Fluctuation test Experimental method used to determine whether mutations arise spontaneously or are induced by some environmental agent

Isogenic Strains of animals that have (virtually) identical genetic makeup, as in identical twins

Log kill Amount of cell killing that is achieved by chemotherapeutic drugs and radiation, as measured on a logarithmic scale

Noncross-resistance Situation where resistance to one drug is not associated with resistance to a second, different drug

Stem cell Cell that is capable of giving rise to an unlimited number of daughter cells

Transplantable leukemia Specifically developed strain of leukemia cells (usually derived from mice) that can be transferred from one inbred mouse to another

Tumor burden Quantitative extent of cancer present in a patient or experimental animal, usually expressed by cell numbers in logarithms to the base 10

DURING TREATMENT of microbial or neoplastic diseases, a progressive change in susceptibility to chemotherapy by the target cell population is commonly observed. The microorganisms or cancer cells show increasingly diminished susceptibility to drug-induced killing and may reach a stage where their growth appears to be completely unimpeded by therapeutic agents that were initially highly effective. Substitution of different drugs may result in renewed susceptibility by the offending population, but this maneuver is by no means always successful. This is especially true in the case of cancer chemotherapy, where secondary responses to new treatment are difficult to achieve and are usually of a very transient nature.

This phenomenon of change in susceptibility to a therapeutic agent has been referred to as acquired resistance and may be distinguished, at least in an operational sense, from intrinsic resistance, which implies initial refractoriness to treatment without any intervening stage of susceptibility.

Because the goal of chemotherapy in treating either infectious or neoplastic diseases is to either eradicate the offending population or to significantly impair its growth, allowing host defense mechanisms to come into effect, the phenomenon of drug resistance poses a major obstacle to achieving satisfactory therapeutic results.

In this article, we will discuss how some of the processes involved in the development of acquired resistance can be studied by appropriate mathematical techniques, with particular reference to their application to the specific problem of cancer chemotherapy.

921

I. Skipper–Schabel Model of Cancer Chemotherapy Effects

In their now classic studies of the growth kinetics of the murine leukemias, Skipper, Schabel, and co-workers established the basic quantitative model of cell growth and antitumor drug effect that has been the basis for much of our understanding of the effect of cancer chemotherapy at the clinical level. Researchers employed a transplanted mouse leukemia line (L1210 leukemia) that could be serially propagated in an appropriate strain of isogenic laboratory mice. Initially derived from a spontaneous mouse leukemia, after a number of transplant generations the behavior of such a tumor system becomes very regular and predictable. When injected into a suitable recipient host, these tumor cells grow with a doubling time of approximately 10 hr and in a nearly exponential fashion. When the total body burden of leukemic cells in the host animal approximates 1 billion (10^9 cells), the animals die because of the effects of widespread malignancy. A single viable leukemic cell transplanted into the host animal was shown to be sufficient to produce a fatal leukemia approximately 2 wk postinoculation. [See LEUKEMIA.]

If treatment with chemotherapeutic agents was commenced at some time prior to the fatal termination of the disease, significant reductions in the body burden of leukemic cells could be observed and the life span of the treated animals would be prolonged in proportion to the extent to which their leukemic cell population has been reduced (see Fig. 1).

The same dose of chemotherapeutic agent always reduced the tumor population by about the same fraction and not by the same absolute number of cells. If the dose of drug was increased or reduced, the proportion of cells that were killed by the application of treatment was concomitantly increased or reduced. For instance, if a given dose of drug was given to animals bearing 10^6 leukemic cells, it might be observed that immediately following treatment the residual population is reduced to 10^4 cells. This means that 99% of the cells undergoing treatment have been killed by the application of therapy. A further application of treatment at the time when the population was 10^4 would result in a further 99% reduction in the cell population, leaving a residual tumor size of 10^2. This relationship between applied dosage of agent and the quantitative impact on cell killing had been previously observed for agents

such as antiseptics and ionizing radiation. A 99% killing effect implied that any individual cell in the target population has a 0.99 probability of being killed by the drug dose under consideration. The chance of any individual cells surviving was, accordingly, 0.01. The negative logarithm of the probability of survival, in this case equal to 2, is referred to as the log kill value for the particular treatment. The probabilistic nature of this log kill function is consistent with the process as being fundamentally chemical in nature; therefore, one would predict that the magnitude of the log kill can be increased by (1) increasing the concentration of drug, (2) increasing the duration of the exposure to drug, or (3) raising the overall temperature of the tumor–drug system. This latter maneuver is not generally feasible in clinical treatment, but the first two options are commonly used to maximize therapeutic kill within the limits of normal tissue tolerance to the drug.

If repetitive doses of drug could be given at a time when the initial tumor burden was relatively low (i.e., $<10^6$ cells), the cure of the tumor-bearing animals was possible if the treatment was carried on past the point when the average surviving tumor burden had fallen below one cell.

However, if treatment was commenced when the tumor burden was high (i.e., 10^8 cells), it was commonly observed that there would be initial response and then recurrence of tumor despite repeat applications of the therapeutic agent. If the recurrent tumor cells were harvested from the animals who had died of advanced leukemia and reinoculated into similar hosts, then the new tumors so formed continued to be insensitive to the drug. This confirmed that the resistance was a property that resided within the tumor cells and was not due to induced changes in the initial host. Much work has been done over time to characterize the biochemical changes that have been observed in the cells displaying acquired resistance, and it can be generally stated that tumor cells have the capacity to display resistance toward all of the therapeutic agents currently available for the drug treatment of cancer.

II. Luria–Delbruck Model of Acquired Resistance

The phenomenon of acquired resistance has been seen earlier in the behavior of bacterial populations in their response to either antimicrobial drugs or biological agents such as bacteriophages. At the

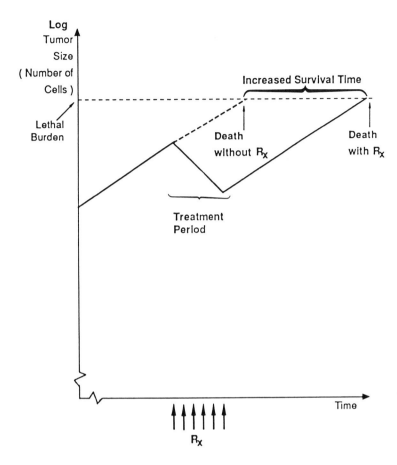

FIGURE 1 Diagram displays growth of experimental leukemia transplanted into a mouse host. Log number of cells is plotted against time when tumor burden reaches critical size ($\approx 10^9$ cells) and animals die (lethal burden). If chemotherapy (R_x) is commenced prior to the leukemia reaching the lethal burden level, there is a logarithmic reduction in leukemia cell numbers. With no further treatment, the leukemia commences regrowth and eventually kills the host, but at a point in time shifted to the right and proportional to the extent of drug-induced cell kill.

time of its discovery, debate centered around the question as to whether the resistance had in some way been specifically induced by the toxic agent or the toxin functioned by selecting for already-existing variants that had spontaneously arisen within the treated population.

In 1943, Luria and Delbruck published their paper in which they proposed an experimental and mathematical technique for distinguishing between the two hypotheses. Resistant phenotypes arising through spontaneous mutations would produce a situation where the variances in the number of resistant organisms in a series of parallel subclonal populations grown without the toxic agent and then tested for resistance would be much greater than the mean. If the process were being directly induced, then the distribution of resistant variants should follow a Poisson distribution, and hence the mean number of mutants would equal its variance.

Experimental studies carried out with populations of bacteria susceptible to a particular bacteriophage consistently yielded results in which the variance, or fluctuation, in numbers of resistant bacteria were much greater than the mean consistent with a random and stochastic origin of these phenotypes prior to the introduction of the bacteriophage. In turn, this suggested that they were arising through spontaneous mutations and that the bacteriophage acted by killing off sensitive organisms, leaving behind those that were resistant by virtue of spontaneous genetic change.

It is important to note that a key element in the initial Luria–Delbruck model is the assumption of identical growth rates by sensitive and resistant organisms. If the resistant organisms grow either more rapidly or more slowly than the sensitive counterparts, the estimates of the mean number of resistant cells per culture will be affected, and, in turn, the expected degree of variance from culture to culture will also be affected.

The Luria–Delbruck model not only provides a basis for distinguishing between an induced versus selected origin of resistant phenotypes but allows one to explicitly calculate the mutation rate and to estimate the frequency distribution of resistant forms in a particular population.

III. Model for a Single Class of Drug-Resistant Cells in a Tumor Population

Provided that a number of assumptions are made, it is possible to extend the basic Luria–Delbruck model to consideration of the behavior of a population of neoplastic cells. In the initial minimal model, it is desirable to make the conditions as close as feasible to the behavior of a population of growing bacterial cells. This is probably not too much of an oversimplification when it comes to dealing with the behavior of well-established transplanted tumor lines, although significant refinements of the model would be required in attempts to apply it to complex clinical situations.

Examination of large tumors show them frequently to be composed of necrotic regions. Cell culture experiments also show that not all apparently viable cells retain unlimited growth potential; therefore, these phenomena must be included in models of drug resistance. This is most easily accomplished by incorporating drug resistance within a model where stem cell divisions can result in either the development of two stem cells or the production of two cells of limited or zero growth potential. If we only keep track of the stem cells, because by definition they are the only cells capable of causing eventual treatment failure, the two types of division appear, respectively, as the birth and death of a stem cell.

If we assume that the rate of stem cell division is constant and that new stem cells are created at a rate b and that stem cells mature losing their stem cell capacity at a rate d, then the number of stem cells at time t is given by the simple birth and death process with parameters b and d. Stem cells will be assumed to be in one of two states with respect to the drug: sensitive or resistant. Resistance does not necessarily need to be assumed as absolute, but resistant stem cells are, by definition, less susceptible to drug-induced death than a corresponding sensi-

tive stem cell. Spontaneous evolution to form resistant cells can be incorporated by assuming that at each successful stem cell division, there is a probability α that one of the stem cell progeny will be resistant.

Then, if $R_0(t)$ designates the number of sensitive stem cells at time t, and $R_1(t)$ the number of resistant cells at time t (which are both random quantities), a mathematical model of resistance requires that we explicitly calculate these functions. Further development of the theory is most easily facilitated by considering the probability generating function $\phi(s_0,s_1;t)$ of the process

$$\phi(s_0,s_1;t) = \sum_{i=0}^{\infty} \sum_{j=0}^{\infty} (s_0)^i (s_1)^j P_{ij}(t),$$

where s_0 and s_1 are dimensionless parameters, α is the mutation rate to resistance for the drug, and $P_{ij}(t)$ is the probability that there are i sensitive cells and j resistant cells at time t. Now it is a straightforward matter to show (using the Kolmogorov forward equations) that ϕ must satisfy the partial differential equation:

$$\frac{\partial \phi}{\partial t} = [(bs_0 - d)(s_0 - 1) + \alpha bs_0(s_1 - s_0)] \frac{\partial \phi}{\partial s_0} + (bs_1 - d)(s_1 - 1) \frac{\partial \phi}{\partial s_1}. \quad (1)$$

The general solution can be calculated but does not take a simple form. Both the mean and variance of the number of resistant cells can be calculated from ϕ, but these generally are not overly informative.

Of some interest is the probability that there are no resistant cells, which is given by $P_0 = \phi(1,0;t)$, because this indicates the likely resistant status of the tumor. In simple cases where $d = 0$, the P_0 is given by $P_0 = e^{-\alpha(N-1)}$ where N is the number of cells in the tumor. If this function is plotted for a particular value of α (Fig. 2) against increasing tumor burden, it produces a characteristic sigmoid curve indicating the steepness with which the probability of zero resistant mutants drops as the tumor population increases.

If resistance is effectively absolute and we make the assumption that the condition of having zero resistant cells present is the minimum condition for

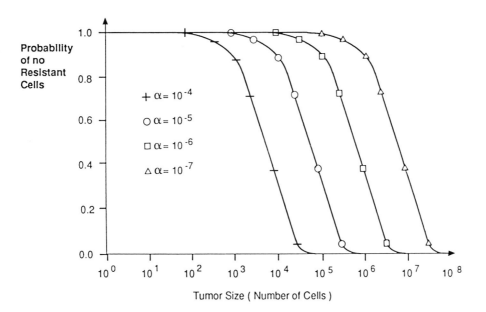

Tumor Size (Number of Cells)

FIGURE 2 Plot of the function $P_0 = e^{-\alpha(N-1)}$, where P_0 is the probability of zero resistant cells, α is the mutation rate per cell generation, N is the tumor size in cell numbers, and e is the base of natural logarithms. The plot is of probability of zero resistant cells against log tumor burden, for different values of α (10^{-4}–10^{-7}). All curves have the same form, but lower values of α are shifted to the right. If P_0 is considered approximately equivalent to probability of cure, then the relationship between tumor burden and curability becomes apparent.

drug-induced cure to be achieved, and moreover that we give enough courses of therapy to rapidly eliminate all of the sensitive cells in the tumor population, then the probability of zero resistant cells can be considered approximately equivalent to the maximum likelihood of drug-induced cure. However, this quantity cannot be calculated without consideration of how chemotherapy is applied. The calculation of this probability thus requires the specification of chemotherapy effects within the mathematical model.

Incorporating the effects of chemotherapy may be simply done if we assume that the chemotherapy induces an instant or near-instant kill. If it is assumed that each sensitive cell has a probability $\pi_0(D)$ of surviving the chemotherapy given at dose D, and each resistant cell a similar probability $\pi_1(D)$, where by definition $\pi_1(D) > \pi_0(D)$, then

$$\phi(s_0, s_1; t) = \phi[\xi_0(s_0), \xi_1(s_1); t^-], \qquad (2)$$

where $\xi_i(s_i) = 1 - \pi_i(D) + \pi_i(D)s_i$ and t represents the time of chemotherapy, and t^- an instant before.

Use of Equation (2) and the solution to Equation (1) permit the calculation of the probability generat-

ing function for any combination of treatments. In particular, it permits the calculation of the probability of cure P_c, defined as the probability that the stem cells have been eradicated by the regimen. P_c is given by

$$P_c = \phi(d/b, d/b; t_J),$$

where t_J is the time of the last treatment. This quantity P_c is deserving of some comment: At the time of completion of treatment, each stem cell has a probability equal to d/b that it will become extinct (i.e., at some future time, all of its progeny will be nonstem cells); thus, P_c is the probability that no stem cells are present plus the probability that any present will become extinct.

We can then proceed to test the model's ability to predict cure rates in tumor-bearing animals in which the tumor burden is accurately estimated and who were given treatment with a single antineoplastic agent. Such an analysis has been carried out by Skipper, and an example of the concordance between theoretic predictions and the observed data is given in Table I. It can be seen that the simple minimal model can accurately predict treatment outcome in at least some tumor systems where the behavior of the tumor is highly reproducible and where a single effective chemotherapeutic agent is employed. At least two important therapeutic inferences can be made from considerations of this basic minimal model. The first involves an inverse relationship between probability of cure of a neoplasm and the extent of tumor burden at the time treat-

TABLE I Observed and Predicted Rates of Cure for Intravenously Innoculated L1210 Leukemia Treated with Repetitive Courses of Arabinosylcytosine, Where $\alpha = 1.8 \times 10^{-7}$ [a]

Dose (mg/kg)	Schedule	Size at treatment	Observed cure rate	Predicted cure rate
15	q 3 hr (×8) 1 course	8×10^6	0.00	0.00
		8×10^5	0.03	0.02
		8×10^4	0.55	0.68
	q 3 hr (×8) 2 courses	8×10^7	0.00	0.00
		8×10^6	0.08	0.22
		8×10^5	0.58	0.86
	q 3 hr (×8) 3 courses	8×10^6	0.33	0.22
		8×10^5	0.83	0.86
	q 3 hr (×8) 4 courses	8×10^7	0.00	0.00
		8×10^6	0.31	0.22
		8×10^5	0.87	0.86
		8×10^4	1.00	0.99

[a] Actual experimental data of Skipper *et al.* showing observed relationship between leukemic cell burden and cure rate. The last column shows predicted cure rate using formula $P_0 = e^{-\alpha(N-1)}$ (see text and Fig. 2). This confirms the ability of the mathematical model to predict outcomes in well-defined experimental systems.

ment is initiated. The second (which is somewhat more subtle) involves combinations of noncross-resistant antitumor agents that have an enhanced capacity of cure in a neoplasm for any given tumor burden. This effect comes about because the combination of noncross-resistant agents can be considered to have the same mathematical effect as lowering the mutation rate to resistance; i.e., in those circumstances where acquisition of resistance to more than one drug is not closely linked, then the probability of a cell expressing more than one level of resistance simultaneously will be less than that of it expressing resistance to a single drug. This would imply that the major impact of combination chemotherapy would be one generating cures as opposed to simply producing measurable regressions of disease.

In principle, the previous theory may be expanded to allow for two or more drugs, although the increase in mathematical complexity with the addition of each drug is considerable. In the case of two drugs, A and B, a cell may occupy any of four states: (1) sensitive to both drugs (R_0); (2) sensitive to B and resistant to A (R_A); (3) sensitive to A and resistant to B (R_B); and (4) resistant to both (R_{AB}).

Proceeding as for a single drug, we may calculate the distribution of the various resistant subtypes using the same theoretical model. Because several resistant subtypes now exist, the number of parameters that refer to the likelihood of spontaneous transitions among the states correspondingly increases. Although the numbers of parameters to be specified may always be reduced by giving them biologically plausible values, the necessity of specifying them increases the difficulty in the analysis of multidrug models. Once the parameters have been specified, we may then calculate a probability generating function $\phi(s_0, s_1, s_2, s_3; t)$ and, thus, the probability of cure is given by $P_c = \phi(d/b, d/b, d/b, d/b; t)$. Numeric calculations must generally be made by computer, and the effect of varying the timing, dosage, and sequencing on the probability of cure P_c may be explored. Consideration of the nature of the acquisition of multidrug resistance indicates the kind of strategies that are likely to be successful in the control of this phenomenon. If the frequency of mutations from sensitivity to double-drug resistance is high, then the cells that are resistant to two drugs will be common; use of both drugs in such situations is unlikely to be advantageous. In situations where direct mutations from sensitivity to double-drug resistance are low, then doubly resistant cells will arise via a stepwise pattern, acquiring resistance to a single drug at each mutational step. In this case, control of the growth of the singly resistant cells is critical to limit the likelihood of doubly resistant cells. Prolonged therapy with a single drug will generally not be optimal because its use rapidly selects for a subgroup that is resistant to it; these cells continue to proliferate and may thus acquire resistance to the second drug. After such therapy, introduction of the second drug is ineffective because resistance to it is already widespread. It is advisable to expose the tumor to both drugs, possibly simultaneously, early in the treatment regimen. The "best" thing for the introduction of each drug will, of course, depend on its characteristics.

Special cases exist where optimal strategies may be identified without explicit knowledge of the parameter values. In particular, if the two drugs are equivalent (i.e., each parameter for drug A has the same value as for drug B), then the best strategy is always to alternate the sequence of delivery. Although interesting, this finding may be of little practical significance in attempts to optimize the delivery of therapy because few, if any, drugs are equivalent. Numeric calculation of arbitrary drug combinations is, in principal, simple but is made difficult in practice by the lack of detailed information on the parameters that must be specified. For example, even such basic information as the tumor growth rate is only known to an order of magnitude and varies considerably among individual patients.

For passaged experimental tumor systems, the requirement of this model for the numeric specification of parameters is burdensome but not impossible. For human disease, where each individual tumor arises *de novo,* each individual case is unique and has its own parameter values. The collection of tumors into groups based on site of origin, extent of disease, histologic appearance, etc. does not necessarily provide a group that is homogeneous for the model parameter values. For example, measured growth rates show a wide variation in most groups that have been examined. Log-kill and mutation rates may also display a similar variation between human tumors. However, unlike the case for passaged human tumors, assessing mutation rates in individual tumors is currently impossible, so that accurate modeling of human disease is made very difficult if not impossible. We may attempt to simulate the effect of variation by replacing individual parameter values with statistical distributions. By changing the variance (spread) of these distributions, the effect of variations in mutation rates, etc. among individuals may be assessed. Consideration of the relationships involved indicates that the qualitative predictions previously discussed will be unchanged because these predictions do not depend on individual parameter values. More precisely, the mathematical relationships are based on functions that are monotonic in the appropriate parameters, so changes in them do not change their nature; however, the quantitative predictions will be changed. In general, the details of any change must be calculated by computer; however, it is possible to arrive at a general idea of the likely behavior by considering the following example.

Consider a tumor treated using strategy C_1, which achieves a cure rate of 50%, for instance. A second strategy is proposed, C_2, and calculations are made, assuming fixed values for all the parameters, which predicts a cure rate of 70%. Thus, if the model is correct and the parameter values are as assumed, then there will be an increase in the cure rate of 20% by changing from C_1 to C_2. A second investigator using the same information contends that the patient population is not homogeneous but consists of two groups: one with very favorable

model parameters and the other with very poor parameters. These two groups are approximately equal in number so that the overall cure rate for s_1 (50%) is based on a 100% cure rate in the favorable group and a 0% cure rate in the unfavorable group. The favorable group's cure rate (100%) cannot be increased by the new strategy, s_2, while the unfavorable group has such poor parameters that its survival is unaffected in any significant way. Thus, there is no real increase in cure rate despite the theoretical superiority of s_2 over s_1, even in cases where the model holds exactly and completely (we may think of the favorable group as consisting of tumors with only sensitive cells, whereas the unfavorable group consists only of resistant ones). This may be summarized by stating that, in general, tumor inhomogeneity serves to diminish therapeutic gains, which would be realized if such homogeneity were not present.

IV. Conclusion

The mutation theory of drug resistance is a powerful model for explaining much of the behavior seen in the drug treatment of originally selected experimental systems. The general behavior of human cancer during chemotherapy is also in accord with many of the predictions of this model; however, the wide diversity of response is seen, undoubtedly, due to the interaction of a large number of different factors. Any model of the treatment of human cancers is limited by the relative paucity of detailed information and, thus, it is unlikely that any model will be successful without the increased availability of biologically relevant functions on the individual. The ability of this model to explain experimental results suggests that the availability of accurate human data will permit a better understanding of the effects of cancer chemotherapy in humans.

Bibliography

Coldman, A. J., and Goldie, J. H. (1983). A model for the resistance of tumor cells to cancer chemotherapeutic agents. *Math. Biosci.* **65,** 291–307.
Goldie, J. H., and Coldman, A. J. (1979). A mathematical model relating the drug sensitivity of tumors to their spontaneous mutation rate. *Cancer Treatment Rep.* **63,** 1727–1733.

Law, L. W. (1952). Origin of resistance of leukemic cells to folic acid antagonists. *Nature* **169,** 628–629.

Ling, V. (1982). Genetic basis of drug resistance in mammalian cells. *In* "Drug and Hormone Resistance in Neoplasia," Vol. 1 (N. Bruchovsky and J. H. Goldie, eds.). CRC Press, Boca Raton, Florida.

Luria, S. E., and Delbruck, M. (1943). Mutations of bacteria from virus sensitivity to virus resistance. *Genetics* **28,** 491–511.

Mackillop, J., Ciampi, A., Till, J. E., and Buick, R. N. (1983). Stem cell model of human tumor growth: Implications for tumor cell clonogenic assays. *J. Natl. Cancer Inst.* **70,** 9–16.

Skipper, H. S., Schabel, F. M., and Lloyd, H. M. (1979). Dose response and tumor cell repopulation rate in chemotherapeutic trials. *In* "Advances in Cancer Chemotherapy," Vol. 1 (A. Rozowski, ed.), pp. 205–253. Marcel Dekker, New York.

Mating Behavior

RICHARD P. MICHAEL AND D. ZUMPE, *Emory University School of Medicine*

Glossary

Ejaculation Expulsion of semen from the penis

Gonad Gland producing germ cells and hormones (i.e., the testis or the ovary)

Gonadal hormones Steroid hormones produced by the gonads, including estradiol and progesterone by the ovary and testosterone by the testis

Gonadotropins Gonad-stimulating peptide hormones, such as follicle-stimulating hormone and luteinizing hormone produced by the anterior lobe of the pituitary gland

Intromission Penetration of the female vagina by the male erect penis; may or may not be accompanied by pelvic thrusting by either partner

Ovulation Escape of the egg or ovum from the ripe ovarian follicle

MATING BEHAVIOR, also known as copulatory behavior, is the activity of a male and female by which the male inserts the penis into the female's genital tract (the vagina in mammals) to introduce sperm for the internal fertilization of eggs. This occurs in most vertebrates adapted to life on land, but external fertilization can occur in fishes and amphibia and is also called mating. There are many patterns of behavior associated with and leading up to mating that constitute courtship activity. The term *mating behavior* is sometimes employed more broadly to include courtship behavior, but for present purposes the narrower meaning will be used.

I. Introduction

Reproduction is of two kinds, asexual and sexual. The former occurs in simple forms of life such as viruses and bacteria and consists of a replication of identical copies of the parental type by a process of division, which in bacteria and unicellular protozoa is called *binary fission*. Sexual reproduction occurs in more complex forms, namely, plants and animals, and consists of a fusion between the genetic material derived from two different parental types—the male and the female. In animals, the germ cells or gametes derived from the parents, sperm from males and ova from females, then fuse to produce a zygote, which therefore contains genes inherited from both sides. Apart from spontaneous mutations, it is sexual reproduction that offers the possibility of genetic variation owing to recombinations that provide the basis on which natural selection can mold the process of evolution. Sexual reproduction in animals (i.e., the fusion of genetic material from two dissimilar parents) involves some type of motor activity by which the male and female elements, no matter how simple, come to approach each other and unite. In primarily aquatic forms such as fishes and amphibia, fertilization is generally external, the male liberating vast numbers of sperm into the water in the immediate vicinity of the female as she releases her eggs. In reptiles, birds, and mammals, which are adapted to life on land, fertilization is internal and, in mammals, involves the insertion of a penis into a vagina. The penis is the erectile organ of males that conducts sperm into the female and urine to the outside. Both males and females have a phallus (i.e., the penis in males and an homologous structure, the clitoris, in females that can also be erectile but does not function as a conduit). Of interest, only the virgin human female has a hymen. For the purposes of this article, the act of insertion is called *mating*, but another widely used term is *copulation*, a word derived from the Latin *copulatus*, past participle of *copulare*, meaning to unite or couple. In vertebrates, there are many patterns of behavior associated with and leading up to mating, and these behaviors constitute courtship activity. Courtship is so varied and so species-specific that it cannot be considered here. Internal fertilization usually occurs shortly after mating, in the minutes or hours required for the ovum to enter the oviduct and for the sperm to travel up the female genital tract to meet it. However, fertilization can be long delayed (e.g., bats mate in the fall; sperm are stored within the female and remain viable within the female until fertilization occurs the following spring when food is abundant). The enormous variations in reproductive processes reflect the richness of life itself.

Most of us are accustomed to thinking of males and females as two very different types, and in the human, distinctions are emphasized by dress and convention; although in the West, these tend to have been somewhat deemphasized in recent years. In a biological perspective, differences between the sexes are much less profound than appear at first sight. In some invertebrates such as the earthworm, each individual is bisexual and has both male and female gonads. Sperm are exchanged between individuals during mating, but self-fertilization can occur. Bisexuality may also be expressed sequentially: in some teleost fishes, individuals start adult life as females and become functional males later on in response to environmental signals. In other fish, the bisexual potential is evident because embryonic sex is determined by very small differences in water temperature. In amphibia such as the *Ranidae*, the addition of one part in 500 million of testosterone to the aquarium water can produce complete, permanent transformation of genetic females into functional males; but in some species, pretreatment with gonadotropin is necessary. The effect of estradiol in male amphibia is less certain, but hormonally induced changes to a functional ovary are well-documented. Among birds, the chick and duckling have been most studied, and matters are complicated by a lateral asymmetry: the left ovary and genital tract are large, whereas the right ovary and tract are rudimentary. Spontaneous reversal of the sex of the rudimentary right ovary may follow removal or disease of the dominant left ovary so that the hen becomes masculinized. All this refers to the induction of sex reversals with the gonads themselves, but such transformations are not so readily produced in mammals. Most experimental work has been with marsupials because they are accessible at early stages in development, and the embryonic testis of the Virginia opossum can be transformed into an ovotestis or ovary by the administration of estradiol from birth. The bisexual potential of the mammal is perhaps most strikingly displayed, however, in the rare but natural freemartin condition of cattle, in which the female of a pair of dizygotic twins is masculinized by the male twin as a result of anastomoses between the two placental circulations. One should note that the homozygous condition (ZZ) of amphibia, reptiles, and birds is male whereas that of mammals is female (XX). The fact that the basic somatotype is female in mammals has important consequences for sexual differentiation and behavior in the human.

To place human mating behavior in a wider biological perspective, emphasis is given to the comparative aspects by adopting a phylogenetic approach, one that has been used by such noted observers as Konrad Lorenz and Frank Beach. The mating behavior of men and women causes emotional reactions in us all, but we take the view here that a moralistic stance contributes little that is useful to our understanding of causation.

II. Comparative Aspects of Mating

From an evolutionary viewpoint, a critical consideration is that mating should result in successful fertilization. Reproductive success is predicated on successful fertilization, and this has enormous survival value for an individual's genes and for the species as a whole. Mechanisms that ensure successful fertilization have much selective advantage; a mating that does not result in fertilization simply means wasted effort and wasted gametes. The mating activity of the male and female are therefore carefully synchronized, and the appropriate behavior must itself occur at a time when ripe ova are available for fertilization. This may all seem rather self-evident, but it is critical for the many species with highly circumscribed breeding seasons and restricted periods of fertility.

Mating periodicity is the rule in most fishes, amphibia, reptiles, and birds, as well as in the majority of mammals, although the human may be an exception. Seasonal influences, called exteroceptive factors by the Cambridge physiologist F. H. A. Marshall, include changes in temperature, humidity, and rainfall and, very importantly, in the photoperiod (day length). These environmental influences impinge on the brain via the various distance receptors (e.g., the retina), which in turn influence the activity of the endocrine system, and the pathways have been well worked out for some species. Mammals that breed seasonally do so in spring or fall in response respectively either to increasing or to decreasing day lengths. The quantity of light reaching optic pathways changes the activity of the suprachiasmatic nuclei within the brain via retinohypothalamic projections to the most ventral parts of these nuclei, which then influence the secretion of gonadotropin-releasing hormones (GnRH) from other nuclei situated more posteriorly in the medial basal hypothalamus. These decapeptides, which can be synthesized in the laboratory, are liberated from nerve endings into a tiny vascular plexus at the base of the brain, which conveys them to the anterior lobe of the pituitary gland. This vascular plexus, the link between brain and gland, is called *the hypothalamic-hypophyseal-portal system.* The anterior lobe of the pituitary releases follicle-stimulating hormone, luteinizing hormone, and prolactin (among others) into the general circulation, and all are important for controlling the activity of the ovaries and testes. The whole complex is called *the hypothalamic-pituitary-gonadal axis*; this controls the majority of vertebrate reproductive functions and, albeit less directly, mating behavior itself in many species as well as its synchronization between the sexes. We have alluded to the long-term, often annual, rhythms of seasonal breeders, but there are shorter rhythms with daily or monthly periodicities. Twenty-four hour, or circadian, cycles affect a wide range of processes, among them the timing of ovulation, and it is thought that this is under the control of the pineal gland. This structure is regarded as an endocrine transducer, which secretes melatonin in response to stimulation by norepinephrine. Melatonin mostly has inhibitory effects on the gonad; its secretion occurs during darkness and ceases during daylight. But this rhythm persists in blinded or dark-housed animals although losing the strict 24-hour periodicity, and light cycles appear to entrain the rhythm.

The red deer of Scotland provide an excellent example of seasonal breeding. The testes are relatively quiescent between March and June when there is rapid antler growth. In August, as plasma testosterone increases, the velvet of the antlers is cast and males come into ''rut'' for a 6–8-week period in September–October. This is in response to greatly increased blood testosterone levels: stags start to roar, herd hinds, masturbate, thrash the undergrowth with their antlers, and show greatly increased aggression toward other males, both mature and immature. Mating activity becomes intense, and they have little time for anything else including feeding so they lose weight. During this time in the fall, social behavior changes, they leave the all-male herds and become locally territorial as they round up hinds. The hinds have 18-day ovarian cycles from October to the following March, and the first ovulation appears to come shortly after the onset of male rut. It is possible that an odor cue may be a timing factor here because rutting stags urinate extensively over their front legs and a very strong odor is produced. Testosterone secretion is reduced by increasing day lengths in early spring, and this decline causes the antlers to be cast. Early this century, Walter Heape distinguished heat in the female from rut in the male, and adopted the term ''estrus'' to describe that period when, and only when, the female is willing to receive the male. The word *estrus* comes from the Greek root meaning a gadfly or ''frenzy,'' alluding to the condition of great agitation in cattle when pestered by the fly's egg-laying. Later on, the morphological and endocrine changes of the estrous cycle were worked out, and the hormones involved were both isolated and identified. Many female mammals, those that are seasonal and those that are not, are polyestrous, showing a sequence of estrous cycles that are interrupted by mating when either pseudopregnancy or pregnancy ensues. During only a short time during the estrous cycle is the female receptive, and then she is said to be ''in estrus'' or ''in heat.'' This lasts only a few hours in mice and rats or a few days in a species like the domestic cat. The estrous phase is preceded by a proestrous phase during which the female becomes very attractive to the male and there is much courtship activity, but when proestrous the female is not yet receptive and will not permit copulation, namely, the insertion of the penis into the vagina. Let us consider this in more detail.

All male mammals, except the great apes and human, mount females from behind and mate in the dorsoventral position. For the majority of female quadrupeds, it is essential that certain postural reflexes are activated to make it possible for the male to achieve intromission. These reflexes are hormone-dependent, but other activating stimuli are involved, including the sight, vocalizations, and odor of the male; coital contact itself is an important sensory stimulus. The most notable postural reflex is the lordosis response, which is typical of many female rodents and carnivores. In lordosis, the lumbar spine is hyperextended so that the back is curved to bring the perineum and vulva into an exposed, upturned position; unless this posture is adopted and the female stands still (estrous sows, for example, have an immobilization reflex that permits them to sustain the weight of the boar), intromission cannot occur. During other phases of the estrous cycle, when the hormonal milieu is not appropriate, this reflex is not evoked, the posture is not adopted, the male cannot intromit, and the female may react aggressively instead of cooperatively. When a Graafian follicle containing an ovum is ready to ovulate, estradiol is being maximally produced by the granulosa and interstitial cells of the ovary; this activates the brain mechanisms responsible for these reflexes (but in some species progesterone is also needed), the female's motivation to mate is increased, the vaginal epithelium cornifies, and lubricating vaginal secretions are produced. Some species ovulate spontaneously every few days, in the human female every 28 days, whereas others ovulate reflexly in response to the male's intromission (rabbit, ferret, and cat). Estrogen is responsible for estrous behavior in all female mammals, but the effects of progesterone vary widely. In some species, progesterone from the corpus luteum acts synergistically with estradiol to enhance heat behavior; in others it appears to antagonize heat, and in yet others, progesterone has little effect either way. In a species such as the rat, the male makes several mounts and intromissions before ejaculating, whereas in others (cat, dog) there is a single intromission with ejaculation. In cats, intromission is brief, lasting a few seconds, whereas in dogs it may last for half an hour because the pair lock. The cat's penis is a few millimeters long and covered with spines which stimulate the vaginal walls, whereas the dog's penis is proportionally large and the glans swells, providing a different stimulus to the vagina while tying the pair together until detumescence occurs. In many male mammals and all male primates, ejaculation follows deep, piston-like thrusts and is signaled to the observer by involuntary, spasmodic contractions of the musculature. Detumescence then occurs, and thereafter the male is refractory to sexual stimulation for a variable period of time. The movements associated with ejaculation are thought to signal orgasm, which is the climax of the sexual act in the male and, by intuition, is regarded as pleasurable and rewarding. The signs of sexual climax in female mammals are sparse, but the female cat displays an "after-reaction" immediately after intromission consisting of violent rolling and squirming movements, and this may represent an external sign of sexual climax in this species. These much abbreviated descriptions of mating cycles in males and females provide us with a background against which to view anthropoid primates and the human.

III. Mating in Nonhuman Primates

We must jump over the Prosimians (*Prosimii*) and New World species (platyrrhine monkeys or *Ceboidea*) and direct our attention to the mostly ground-living macaques of Asia and baboons of Africa. These are Old World or catarrhine monkeys belonging to the *Cercopithecoidea*. The apes and the humans belong to the superfamily *Hominoidea*. Although the females of many mammalian species have an estrous cycle, and this refers to the periodic ripening of Graafian follicles within the ovary, it is not generally realized that only the Old World monkeys, the great apes, and the humans have a menstrual cycle, and this refers to the periodic breakdown of the uterine endometrium, which results in vaginal bleeding. Menstruation, which occurs about every 28 days in both rhesus monkey (*Macaca mulatta*) and the human, comprises 2–5 days of vaginal bleeding caused by the periodic breakdown of the uterine endometrium. Ovulation occurs midway between two menstruations, and midcycle is the most fertile time. There is well-marked seasonality in the mating of rhesus monkeys (September–January) that come from fairly northern latitudes in India, but in species living nearer the equator, where changes in photoperiod are minor, mating occurs throughout the year. This is the situation in the great apes (orangutan, chimpanzee, gorilla). In contrast, the mating activity of the pair does vary with the phase of the female's cycle in many primates, strikingly so in some species [*See* PRIMATES.]

A. Old World Monkeys

The rhesus monkey has been most studied. There is a well-marked rhythm in the ejaculatory activity of males, which increases during the first half (follicular phase) of the female's menstrual cycle and declines, often to very low levels, during the second half of the cycle (luteal phase). The sexual interactions of the pair are most intense near the time of ovulation. Baboons possess a specialized area of skin in the perineal region that becomes red, hot, and edematous under the influence of ovarian estradiol, and the swollen sexual skin visually stimulates the male. It collapses abruptly after ovulation, when progesterone is secreted by the corpus luteum, and mating activity soon ceases also. In these primates, then, there is a clear synchronization between mating activity and ovulation, but they differ greatly from lower mammals in an important respect. We have already seen that females in many species of nonprimate mammal show circumscribed periods of heat and at other times reject the male. This is not so in Old World monkeys and apes because females can accept males throughout the menstrual cycle, and they do not show the specialized, hormone-dependent postures that characterize heat; so it is questionable whether the term *estrus*, although used, really applies. Nevertheless, as we have seen, females certainly exhibit periods of heightened receptivity and sexual excitation near midcycle. In these primates, mating itself is different: it is less reflexive and stereotyped, and females simply adopt a quadrupedal "presentation" posture, directing their hindquarters toward the male. This is not a reflex akin to the lordosis posture of lower mammals but appears to be more volitional. Furthermore, it is used for purposes other than those directly associated with sexual motivation. In primates, mating is complicated by the dominance and potential for aggression of the male. Females may accept the male's coital attempts when not fully receptive as an act of submission to prevent male aggression, and the presentation posture is used to diminish or divert male threats and thereby gain some social advantages (e.g., access to food). In addition to the ubiquitous presentation posture of anthropoid primates, there are more subtle gestures (small, brief movements of the head or arm) that also act as female sexual invitations. The communication systems between males and females are quite complex, and both vocal and odor cues may supplement these sexual displays.

The male rhesus monkey mounts the female from the rear (dorsoventral position) by grasping the backs of her ankles with his feet and her haunches with his hands, so that the female supports his entire weight (often twice that of the female). Thus mounted, the male's penis can be directed into the vagina, usually with the help of some assisting movements and postural adjustments by the female. After intromission, a rhesus male makes some 6–12 pelvic thrusts and dismounts. There follows a period of quiet when the animals may groom each other. There follows another mount with more thrusting, and the sequence continues for several mounts, the final mount of the series terminating in ejaculation after a larger number of deeper thrusts. At ejaculation, the male may grimace characteristically and utter a vocalization, and there are spasmodic muscle contractions of the trunk, legs, and tail. During all but the final mount, the female quietly cooperates and maintains her posture while mounted, but as the male ejaculates, the female often turns her head around to look into the male's face, even touching it with her lips, lipsmacks vigorously, and at the same time reaches back to clutch the hair of the male's head, trunk, or leg. This reaction of females of several macaque species is thought by some to be a concomitant of the female's orgasm. After a lapse of some 15–30 minutes, another mounting series is initiated, also terminating in ejaculation, and these behaviors can persist for several days while the pair maintains a consortship, foraging and traveling about together. Although the rhesus has a pattern of multiple mounts leading to ejaculation, the closely related bonnet monkey (*M. radiata*) characteristically makes only a single mount with some 30 intromitted thrusts, whereas the mating pattern of the cynomolgus monkey (*M. fascicularis*) is intermediate, with some ejaculations occurring in a single mount but others after several mounts.

More-dominant males appear to form consortships during the middle part of the menstrual cycle, but lower-ranking males may mate with the same female both earlier and later in her cycle; so females may have several male partners during a single cycle. This polygamy is related to the social organization of the species. In macaques and baboons there are large multi-male, multi-female troops, but in certain species (barbary macaque, hamadryas baboon), troops are made up of subunits having a "harem" organization, with copulations occurring only between a single male and his small group of females. Among primates, a multi-male, single-female social organization has not been described. There is also the matter of individual partner preferences to consider, which play only a minor role in lower mammals, the dog perhaps being an exception. However, in rhesus monkeys, as in other primates, a clear preference for particular individuals as sexual partners is shown by males and females alike, but its basis is not fully understood: age, genealogy, familiarity, dominance rank, and hormonal variables all play a role, and certain pairs are more compatible than others.

B. The Great Apes

The orangutan, chimpanzee, and gorilla all have menstrual cycles with durations of between 30 and 36 days. Orangutans, probably the least studied of the great apes, apparently have no external signs of ovulation, and the female may be mated at any time of the cycle. Chimpanzees have a conspicuous sexual skin swelling lasting some 3–10 days; 80% of matings are restricted to the period of swelling. The gorilla has no sexual skin swelling comparable to that of the chimpanzee, but there is a visible labial turgescence lasting 1–4 days, the period to which copulations seem to be confined. The great apes ejaculate with a single intromission, and several ejaculations can occur in an hour. Orangutans are semisolitary: adult males defend territories against other males and copulate with females in their territory. Fertile matings appear to result from temporary consortships formed by a female near midcycle and an adult male; such matings are gentle, mostly female initiated, and generally involve ventro-ventral postures and intromissions of up to 30 minutes. Mouth–genital contacts have been observed in both males and females. Female orangutans are also mated more violently outside consortships by subadult males: these copulations are opportunistic, nonfertile, and resemble "rape" because the male aggressively wrestles the resisting, screaming female to the ground, and this unusual type of copulation is also in a ventro-ventral position. Chimpanzees live in loose-knit social groups, and although copulations can occur at any time of the cycle, the majority takes place during sexual skin-swelling when a male and female may form a consortship for several days. Both males and females initiate matings with a variety of behaviors, but females initiate more copulations than do males during turgescence, and males initiate more at other times. Intromissions are brief (i.e., about 10 seconds) and usually occur during dorsoventral mounts, although ventroventral intromissions have been observed. Gorillas live in groups containing one fully adult silver-backed male and several adult females together with their offspring at varying stages of maturity.

Only the silver-backed male mates these females: most copulations are initiated by the female with a rich variety of gestures (e.g., an outstretched hand), facial expressions, postures, and behavior patterns (e.g., backing the male into a corner with her hindquarters). Most copulations (80%) occur in the dorsoventral position, with intromission lasting about 60 seconds. However, ventro-ventral copulations are frequently observed; the female may squat in the male's lap, and mutual mouth–genital contacts in the "soixant-neuf" position have been described as a prelude to mating in captive gorillas. Thus, there are major differences between the great apes in social organization and mating patterns, but all show more variability than monkeys both in the mating positions themselves and in the forms of their sexual initiations.

IV. Mating in Humans

The act of mating has had enormous implications for all human institutions throughout recorded history and in every society that has been studied. It impinges on myth, religion, morality, education, medicine, hygiene, law, and the acceptable social norms of every culture including those determining the distribution and inheritance of wealth. We must take a narrow and restricted approach to these multifaceted issues.

The thoughtful reader will have noted that as we have proceeded from lower (simpler) to higher (more complex) mammals, mating behavior loses its stereotyped, ritualized quality, and the conditions under which it takes place, as well as the patterns of behavior themselves, show increased variation; this is more evident in the great apes and reaches its apogee in the human. To explain this trend, it has been hypothesized that hormonal factors become less important but that learning, itself associated with increased neocortical development, becomes more important as we approach the human more closely. For example, a female kitten at 6 weeks of age will show the typical adult estrous postures and reflexes when hormones are administered; however, a primate has a prolonged period of infantile dependence, and there is a period of maturation before adult mating responses are developed and integrated. In anthropoid primates, opportunities for "cognitive rehearsals" are essential for the full development of normal adult sexual behavior. These occur in a social context during the play of infants and juveniles, and if individuals are deprived of these experiences by social isolation, normal adult sexuality is totally disrupted. In adult rodents, when the testes or ovaries are removed, mating generally ceases in a few days or weeks, but in nonhuman primates and humans it may continue, although somewhat modified, for several years; mating in these latter forms appears to be less hormone-dependent. Nevertheless, there is strong evidence that hormones continue to play an important role in modulating the interactions between male and female primates including humans. This is illustrated in Fig. 1, in which the copulatory activities of rhesus monkey and human are compared in relation to the female's menstrual cycle: data from the two species are not strictly comparable but illustrate the modulating effect of the female. [*See* HORMONAL INFLUENCES ON BEHAVIOR.]

Most diurnal mammals mate during daylight, whereas nocturnal species mate at night. The human is different: although diurnal, humans mate mostly at night. This might be an artifact of our social organization, but whereas all other animals mate in "public" and show no embarrassment at doing so, with very few exceptions, in all human societies the mating pair seeks privacy. Both courtship (heavy petting) and copulation are vulnerable to even minor outside interferences. Many parts of the body are sensitive to stimulation by the opposite sex but most important are the "erogenous zones": lips, inside of the mouth, nipple regions, lower abdomen, inner thighs, and perineum. Most sensitive are the penile glans and shaft, and the labia minora and clitoris themselves. The vaginal mucosa is less sensitive to touch (except perhaps the anterior vaginal wall) but more sensitive to the displacement caused by deep thrusting when the penis is in the posterior fornix. The most frequently used position (70–80%) is face-to-face contact (ventro-ventral position) as in the orangutan, and not the dorsoventral, rear mounting of other species. It has been noted that the gorilla uses both positions and other variations such as squatting, and so does the human. The most frequent alternative to the man lying on the woman is the posture in which the woman sits or lies on the man in a face-to-face position. Anthropological studies have shown that any posture that facilitates penile entry and provides freedom of movement can be employed, but in certain societies there are definite postural traditions (e.g., the Trukese in some circumstances mate standing up). The extent to which sexual initiatives change between the male and female during mating is also influenced by cultural norms as well as by individual temperament. Different societies at different times have variously emphasized and deemphasized women's breasts: in some cultures they are fully exposed, in others scrupulously concealed. In no other species, including the great apes, are nonlactating mammary glands consistently conspicuous from puberty on. Anthropologically, prominent breasts signify fertility and fecundity, and ethologically they signal sexual maturity and sometimes promote intense sexual arousal. In the human female, it appears that the breast has evolved specific functions in courtship and lovemaking that are independent of nutrition and feeding. [*See* SEXUAL BEHAVIOR, HUMAN.]

The length of the flaccid human penis measured from pubic bone to tip is about 8–10 cm but is highly variable. The erect penis is about 16–18 cm with an approximate circumference of 10 cm, and the erect dimensions are more constant between individuals and relatively independent of general body stature; there are minor racial variations. These days, only about 10% of males worldwide between 20 to 40 years are circumcised, but fashions change. (Currently, over 90% of male children in the U.S. are circumcised shortly after birth.) The two dorsal corpora cavernosa and the midline, ventral corpus spongiosus make up the penile shaft and receive

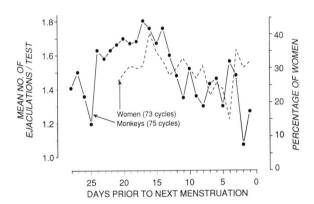

FIGURE 1 Comparison of mating activity of rhesus monkeys and humans in relation to the menstrual cycle. ●—●, Mean numbers of ejaculation per 1-hour behavior test (32 pairs of rhesus monkeys); -----, percentage of women reporting sexual intercourse (40 women) [From Udry, J. R., and Morris, N. M. (1968). Distribution of coitus in the menstrual cycle. *Nature* **220,** 593]. [From Michael, R. P., and Zumpe, D. (1970). Rhythmic changes in the copulatory frequency of rhesus monkeys (*Macaca mulatta*) in relation to the menstrual cycle and a comparison with the human cycle. *J. Reprod. Fert.* **21,** 199, with permission.]

blood from the internal pudendal arteries. Erection results from splanchnic nerve stimulation originating in the low lumbar–upper sacral segments of the spinal cord, which causes dilation of the arterioles and constriction of a valve-like system in the venous return, which drains into the deep and superficial dorsal penile veins; but the mechanisms of erection are more complex than this description. Arteriolar constriction is followed rapidly by detumescence. Erections occur both spontaneously and with stimulation from early infancy to great old age, and there are episodes of erection, termed *nocturnal penile tumescence*, during normal sleep, which may or may not be associated with erotic imagery. Erections on awakening in the morning are common, particularly in young adults. In relation to body size, the clitoris is proportionally small in women, generally about 0.6 cm long, and it is considerably smaller than in nonhuman primates in which both clitoris and vulva are more posteriorly placed. The anatomical differences may be related to postural differences because, in women, pressure on the clitoris by the male's pubis is an important source of stimulation, and this does not occur in the dorsoventral position. Ejaculation usually occurs during a single intromission lasting a few minutes, as in the apes, generally with from 5 to 50 pelvic thrusts. Mating frequency declines from about 4 per week at 20 years to once per week after 50 years. [*See* REPRODUCTIVE SYSTEM, ANATOMY.]

In association with mating, four stages of human sexual arousal have been described: excitement, plateau, orgasm, and resolution. In males, immediately before orgasm, a small amount of clear lubricating secretion from Cowper's glands exudes from the penile meatus, and the testes themselves increase in size and are elevated into a high position within the scrotum. Ejaculation results from rhythmic muscular contractions (0.8/sec) along the length of the urethral sphincter, from contractions of both ischio-cavernosus and ischio-spongiosus muscles and of other muscles in the pelvic floor; there are also contractions involving the vas deferens, seminal vesicles, and prostate gland, and the pelvic sensations are pleasurable. This is mostly an involuntary experience, but sensations are augmented by simultaneous contractions of somatic muscles. After ejaculation, detumescence occurs shortly, and there is often an urge to urinate. In females, within a few seconds of sexual stimulation, vaginal lubrication occurs as a direct transudate through the vagi-

nal walls. The secretions of cervical and Bartholin's glands are not now thought to play a role in lubrication. In the plateau phase, the breasts and areolae become engorged and increase in size, and the nipples become erect; these changes are less marked in those who have breast-fed. At orgasm, a highly variable erythematous flush or rash develops over the upper abdominal region and spreads to the chest, breasts, neck, and face, which rapidly disappears after orgasm. During orgasm in both males and females, there are increases in systolic blood pressure (+50 to 70 mm) and in pulse rate (+50/minute), and there may be hyperventilation, heavy perspiration, and, particularly in women, facial grimacing, clutching, carpopedal spasm, and occasionally vocalizations. The following changes occur in the female genital structures. Because of vascular engorgement, the labia minora double in size and the glans clitoris enlarges, but during orgasm itself the clitoris actually seems to retract. As plateau proceeds to orgasm, the vagina lengthens by 2–3 cm, the uterus elevates in the pelvis to produce a "tenting effect" in the upper vagina, and vascular engorgement with myotonia (increased muscle tone) of the vagina's lower third results in the formation of the orgasmic platform where contractions (also at 0.8/sec) occur during orgasm: these are thought to assist in triggering the male's ejaculation. Many women (~30%) do not experience orgasm during either masturbation or mating, although most men (99%) experience ejaculation; ejaculation can occur without erection and without the subjective experience of pleasure. Ejaculatory activity appears to decline gradually toward middle life, whereas the woman's capacity for orgasm appears gradually to increase somewhat. Orgasm in men is briefer than in women, and men are refractory for a period thereafter. In contrast, some women have several orgasms in succession with little additional stimulation. Because of these physiological differences, and because in the experienced pair mating is more within voluntary control, it is thought that greater sexual satisfaction can be obtained when the beginning of orgasm in the woman somewhat precedes that in the man. Orgasm in lactating women may be accompanied by forceful milk ejection caused by the release of oxytocin from the neural lobe of the pituitary gland.

Many sweeping generalizations have been made in the foregoing, and in a précis of this type it is impossible to avoid superficiality, so please read further.

Bibliography

Dewsbury, D. A., and Pierce, J. D., Jr. (1989). Copulatory patterns of primates as viewed in broad mammalian perspective. *Am. J. Primatol.* **17,** 51.

Michael, R. P., and Zumpe, D. (1988). Some long-term rhythms in the sexual and aggressive behavior of nonhuman primates and man. *In* ''Biorhythms and Stress in the Physiopathology of Reproduction'' (P. Pancheri and L. Zichella, eds.). Hemisphere, New York.

Nadler, R. D., Herndon, J. G., and Wallis, J. (1986). Adult sexual behavior: Hormones and reproduction. *In* ''Comparative Primate Biology: 2A. Behavior, Conservation, and Ecology'' (G. Mitchell and J. Erwin, eds.). Liss, New York.

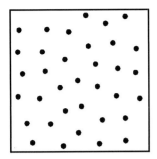

Measles

FRANCIS L. BLACK, *Yale University*

Glossary

Herd immunity Protection of a susceptible minority within a population by virtue of the fact that individual immunity in the majority prevents propagation of an infectious agent

Negative-sense RNA RNA strand carrying a sequence of bases that is complementary to the sequence recognized by the protein-synthesizing complex. In the negative sense adenine replaces guanine, cytosine replaces uracil, and vice versa

RNA polymerase Enzyme capable of synthesizing RNA from triphosphonucleotides according to a sequence complementary to a preexisting strand

Syncytium Unit of tissue formed from up to several hundred fused cells which retains the nuclei and other organelles of the original cells, but no internal membrane partitions

Virion Complete virus particle, as distinguished from vegetative virus, the parts of which may be dispersed throughout a cell

MEASLES is a sometimes dangerous acute virus infection to which all humans are susceptible unless protected by vaccine. Everyone becomes at risk once he or she loses, at a few months of age, passive immunity acquired from the mother. The virus is so infectious that all unvaccinated persons, except some who live only a few years or live in extreme isolation, will become infected. After infection there is an unusually uniform incubation period, in children, of 10–14 days.

Following this silent period, overt disease passes through two distinct phases. In the first, "prodromal," period virus continues to replicate and reaches levels that cause direct damage to circulating white blood cells and membrane surfaces throughout the body. The prodrome is difficult to distinguish from a severe cold, except that it is accompanied by fever. Large amounts of virus are excreted in the nasal discharge, tears, and urine and smaller amounts circulate in the blood.

The appearance of rash marks the development of the second phase. More fundamentally, this phase is distinguished by activation of an immune response. Virus is quickly cleared from the circulation and respiratory secretions. Clearing virus from the tissues, however, requires the destruction of infected cells, and this process raises the fever to new highs. The rash reflects damage done to infected cells in the deeper layers of the skin. In the majority of cases, the second phase lasts only a few days, and recovery is complete without serious lasting damage. However, where malnutrition, crowding, and intense exposure to secondary infections play roles, the death rate can be high, and blindness or deafness may follow. Some adverse outcomes are direct effects of the virus, but many are caused by other pathogens that spread because of virus-caused damage to the immune system. Occasionally, a third phase, encephalitis, follows as a result of sensitization of the patient's immune system to his or her own brain tissue. Once a person has had measles, immunity lasts for life.

I. Definition

This article covers only the disease otherwise referred to as "red measles," "7-day measles," "morbilli," or "rubeola." Common usage has sometimes been to use the word "measles," with some qualifying adjective, such as "German," to refer to other infectious diseases characterized by rash, but these—rubella, scarlet fever, roseola, exanthem subitum, and parvovirus infection or erythema infectiosum—are so varied in cause and significance that the name has little value used in this way. Neither of the terms "morbilli" or "rubeola" has the same recognition value as "measles" in the English language, and the latter term has the disadvantage of specifically meaning "rubella" in Spanish and Portuguese. As used most generally, and exclusively here, "measles" refers to a single well-defined disease.

II. Cause

Measles is caused by a virus. Environmental factors may modify the severity of the disease, but it always occurs as a consequence of infection by this one virus. The virus is structurally and metabolically similar to other members of the family Paramyxoviridae, a group which includes mumps virus and prominent causes of croup and bronchitis (i.e., the paramyxoviruses). Together with the viruses that cause canine distemper in dogs and rinderpest in cattle, it composes the genus *Morbillivirus*. All Paramyxoviridae are formed of a single strand of negative-sense RNA and six proteins—three internal and three associated with a lipid membrane—termed the "envelope." The RNA is large, relative to other viruses, with nearly 14,000 bases. [*See* VIROLOGY, MEDICAL.]

The negative sense of the measles genome prevents it from functioning as a messenger in coding protein synthesis, but its size, too, would make it inefficient in that role. Therefore, it cannot directly direct the synthesis of the proteins required for its own replication. Instead, measles virus inserts its three internal proteins, together with its RNA genome, into a host cell to function as RNA-primed RNA polymerase. In the cell this enzyme system makes shorter complementary-sequence copies of the RNA, which serve as messages capable of instructing the synthesis of all viral proteins (mRNAs). A few full-length copies of the complementary sequence (i.e., positive-sense strands) are also made and, from the latter, negative-sense strands for progeny genomes. Newly made virus mRNA usurps a place on cell ribosomes, and synthesis of new virus protein then begins.

During viral replication in cells, the amino acid chains of the newly synthesized H and F virus membrane proteins are modified by cell enzymes by the addition of carbohydrate side chains. These glycoproteins are inserted in the cell's outer membrane, displacing normal cell proteins from discrete sections of membrane. A smaller M protein attaches to the inner surface of this membrane section. Newly made negative-sense nucleic acid is wrapped in the NP protein to form a long loose helix. This, in turn, binds to the M protein and the complex bows outward, forming a blister on the cell surface. The enzymic P and L proteins bind to the complex, and the blister pinches off to form a new virus particle. As well as the carbohydrate specified by the host cell, the new virion carries, as an integral part of its structure, the lipid portion of the original cell membrane.

The measles virion is unstable. About one-half of it dies every 2 hours at body temperature. An acid environment or proteolytic enzymes accelerate this rate of destruction. If the virus dries on a surface, it is totally inactivated. Apparently, this effect is caused by the crushing force of surface tension as the last water evaporates, but not by the loss of water per se; if virus is suspended in fine droplets and dried in the air, it survives well and the usual rate of decay is slowed. Because of these characteristics, measles is not easily transmitted through food or water and is never spread by dry contaminated objects such as clothing or books. However, it spreads readily if it is sneezed or coughed into the air, dries there, and remains suspended until breathed in by a susceptible person.

For measles virus to infect a new cell, it must first attach to the cell by forming a bond between the H protein and a cellular receptor protein. The receptor on the cell is different from that used by other Paramyxoviridae and is found only on certain types of cells in species of the order Primates. Presumably, it has some normal function, but we do not know what that might be. Its limited distribution is responsible both for the fact that measles affects only humans, apes, and some monkeys and for the specific types of cell affected within the body. Thus, the virus attachment mechanism plays a large role

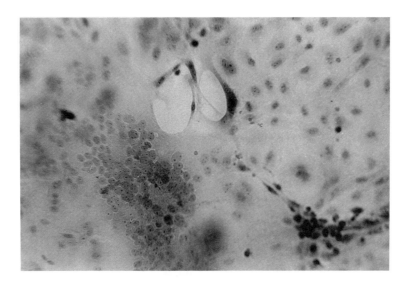

FIGURE 1   A measles-induced syncytium in tissue culture cells, Giemsa stained. Normal cells are upper right and a syncytium fills the left half of the field. The darkly stained cells at lower right might also be virus infected, but this is not certain from their appearance.

in determining the distinctive nature of measles as a disease. While the virus is bound to a cell by the H protein, the F protein goes to work to join the virus envelope to the cell membrane. The merged membrane segment then opens up like two bubbles fusing, and the virus nucleic acid, with its essential enzyme, is dumped into the cell cytoplasm to begin the cycle again. It takes about 24 hours to complete the cycle, but once infection is established, one cell can go on producing virus for a few days. [See CELL RECEPTORS.]

Often, Paramyxoviridae use a shortcut in this process, to speed replication and to avoid exposure to the host's immune defenses. If the point on the cell membrane where newly synthesized viral proteins are inserted happens to touch the membrane of an adjacent cell with its own receptor, the viral proteins can react with the second cell as if they were part of a mature virus particle. Fusion occurs between the membranes of the infected and uninfected cells, and the virus-replicating complex diffuses into a new cell without going through the packaging process or exposing itself to circulating antibody or white blood cells. This cell fusion process can be very extensive, giving rise to an unpartitioned unit, called a syncytium, that includes the nuclei and the cytoplasm of several hundred cells,

and has a diameter of more than 1 mm (Fig. 1). Ultimately, the central area of the syncytium becomes exhausted, ruptures, and sloughs off. These changes are characteristic of the damage caused by all paramyxoviruses, but especially measles virus, and are seen both in tissue culture and in infected membranes.

The packaging of measles virus is a sloppy process, and incomplete genomes as well as pieces of complementary RNA, are often wrapped in nucleocapsid protein and packaged in an envelope. Because of this, the size of the virions varies widely, but the minimum diameter is about 150 nm. Moreover, single-stranded RNA offers a less consistently accurate system of transmitting genetic information than does double-stranded DNA, and mutations occur at a high rate. Many apparently intact virus particles are incapable of independent replication because some genes are missing, but if their RNA is introduced into a cell together with a whole virus genome that can provide the functions of the missing genes, the short strands replicate faster than the whole ones and lead to mostly defective progeny. In spite of this, measles virus, as encountered in nature, is an extraordinarily consistent entity. Minor variations occur in the virus, but they have no significant effect on pathogenicity or resistance to immune defenses. The process of attachment to the specific receptor and subsequent replication must be intolerant of variation, and the ecological niche in which this virus propagates must be well separated by less efficient interactions from any alternative pathway. Measles is an essentially uniform disease.

III. Pathogenesis

A. Virus Penetration and the Incubation Period

Measles virus infects through the mucous membranes of the eyes, nose, throat, and especially the lungs. Only the finest droplets can pass into the lungs without being trapped along the way, but a dried aerosol can do so. Two days later, time enough for the virus to cycle through two or three layers of cells, virus appears in the bloodstream. Direct injection of virus into the bloodstream bypasses this phase and proportionally shortens the time to development of the disease.

Cells lining the local blood vessels are probably among the first internal tissues to be infected, but the virus also gets into and destroys circulating lymphocytes. In this regard measles virus behaves like human immunodeficiency virus (HIV), the virus of the acquired immunodeficiency syndrome (AIDS); the lymphocytes serve as a vehicle by which virus is rapidly carried to nearly all parts of the body, and lymphocyte destruction damages a key host defense. Lymphocytes are needed both for the production of antibody and for cell-mediated destruction of infected tissue cells. Measles virus is not as selective as HIV in destroying only the lymphocytes responsible for amplifying the immune response, nor is it as thorough in destroying this class of cells. Nevertheless, the short-term effects are similar and, as in AIDS, the patient is made more vulnerable to tuberculosis, localized bacterial infections, and a wide variety of childhood infections. A reduction in the number of circulating lymphocytes is the first clear sign of the disease, becoming evident about 5 days after infection. The virus spreads widely through the body. Syncytia form in the membrane surfaces of the conjunctiva and the respiratory, intestinal, and urinary tracts. [*See* LYM-PHOCYTES.]

B. Prodromal Period

By about 10 days after initiation, the areas of virus infection are extensive enough to cause coldlike symptoms of tearing, nasal discharge, and dry cough. During this period virus becomes established at many focal points in the lining of small blood vessels of the skin, kidneys, and intestines, but, except on the lining of the cheeks, these foci are not visible without biopsy and microscopic examination. Inside the cheeks the syncytia might be deep enough to be seen, without visual aid, as small white patches surrounded by reddened tissue. These lesions are the first distinctive clinical characteristic of measles infection and are referred to as Koplik spots, for the man who first described them.

Meanwhile, in the blood the infection spreads to various types of white cells besides the lymphocytes, and they, too, become reduced in number. Most severely affected are the eosinophils, cells that normally help defend against parasitic infection and reduce damage done by hypersensitivity reactions. Low levels of virus are present in the blood at this time, but most of it quickly finds and enters susceptible cells. There is enough cell destruction to release substantial pyrogen, and fever develops.

Large amounts of virus are produced during this phase and it is spread through the environment by coughing and sneezing induced by irritation of the respiratory membrane, as well as in the urine.

C. Florid Period

About 14 days after initial infection, both humoral and surviving cellular components of the immune system are ready to fight the infection. Free virus in the blood is quickly neutralized by antibody and cleared away. The new antibody also reacts with viral glycoproteins on infected cell surfaces, thus targeting the cells for destruction by newly sensitized lymphocytes. The lymphocytes attract macrophages, which release toxic factors, adding to the destruction and killing any actively infected cell with its internal virus. The most visible site of this reaction is the skin. At each focus of virus replication, the capillary vessels are damaged and distended, resulting in a red spot. This induration raises the spots slightly and smoothly above the normal skin surface. Each spot has a diameter of several millimeters, but if there has been much virus growth, they might be confluent. If poor nutrition has delayed the immune response or weakened the tissues, the capillaries might rupture. In this case there is bleeding beneath the skin and the rash is termed "hemorrhagic." Similar processes must go on wherever the virus has become established in deeper tissues, and this tissue destruction causes a high fever. [*See* IMMUNE SURVEILLANCE.]

D. Secondary Effects

In the normal course of events, virus and virus-infected tissues are rapidly eliminated and recovery begins within a few days. Often, however, all of the dead tissue on exposed surfaces, coupled with the damage to the immune system, permits bacteria to become established and to cause ear infections, pneumonia, and other secondary damage. Tuberculosis might reactivate. The immunodeficiency lasts for 1 month or more, during which the reaction to tuberculin is suppressed, and autoimmune kidney disease, if present, is actually ameliorated. Much of the mortality associated with measles occurs not during the acute disease, but in the month following.

E. Encephalitic Disease

Occasionally, especially in older children and adults, the process of tissue damage exposes proteins of the nervous system that are ordinarily hidden from the immune system, and hence not recognized as self. An immune response might then be mounted against the patient's own central nervous system. This process involves a whole new cycle of immunological stimulation, and commonly another week passes before the reaction is manifest as encephalitis. Measles encephalitis is not usually fatal, but permanent damage can be done.

The blood–brain barrier, a permeability limit that prevents leukocytes from wandering freely through the brain, keeps infected lymphocytes out of this organ, and the brain is not usually directly infected. In perhaps one in 100,000 cases, but relatively more often in young boys, the virus does get established in glial cells of the brain to cause the progressive disease subacute sclerosing panencephalitis. The virus does not grow well in glial cells, but there it is out of reach of immune surveillance. The infection persists, producing mainly defective progeny and spreading slowly through adjacent cells. After a few years the spreading lesions become big enough to cause noticeable brain damage. The course is inexorable and leads to death in 1–2 years. [*See* BLOOD–BRAIN BARRIER.]

IV. Vaccines

There is no effective treatment for measles, although secondary invaders might be controlled with antibiotics, and basic nursing care reduces mortality. Control of the disease is dependent on prevention by immunization. Temporary immunization can be achieved by the passive transfer of immunoglobulin from naturally immunized persons, but more lasting active immunization requires the introduction into the bloodstream of components of the infecting agent without the virulent organisms. For this the organism had to be isolated so that its antigens could be obtained free of other disease agents. This was done by John Enders and Thomas Peebles at Harvard University in 1954, when they demonstrated the now-familiar measles giant cells in cultures of cells grown from human foreskins collected at circumcision. Foreskin cultures are difficult to work with, but the virus was soon adapted to grow in lines of human cells that could be propagated in perpetuity.

Two routes to a vaccine were now open, and both were followed. The virus could be grown in quantity, killed with formalin, and injected into children to induce antibody directly. This killed virus vaccine proved disastrous, making the recipients allergic to measles virus, but not immune to infection.

Alternatively, the virus could be attenuated so that it would grow in susceptible persons only to a limited extent. The attenuation was accomplished by Enders and associates in 1957. They adapted the virus to grow in cultures of cells from chick embryos, stopping at a point before it lost the ability to grow in humans. This virus still causes all of the usual effects of wild measles—leukopenia, syncytia, fever, and rash—but with greatly reduced intensity. Like wild virus, it induces lifetime immunity after a single exposure. Unlike the attenuated poliovirus vaccines, it is not excreted to infect others secondarily. After some adjustments to obtain a level of attenuation that would immunize without causing too much disease, this vaccine has been tremendously successful.

Technical problems remained that were troublesome in less developed parts of the world. As already noted, measles virus is labile, and the live virus vaccine must be kept cold to remain effective. "Cold chains," a series of freezers and insulated containers, had to be established along the supply lines to keep the vaccine viable. Essentially, every child acquires enough measles antibody from his or her mother to prevent virus growth for several months. Vaccine virus given during this period cannot multiply, and the amount injected is inadequate to have any direct effect. The durability of passively acquired antibody varies from one child to another both within a community and, in its average time, between communities. Generally, it is lost more quickly in children of economically disadvantaged countries.

The vaccine may be wasted if it is given before 15 months of age in the United States, but if it were withheld that long in many other countries, half of the deaths that might have been prevented would already have occurred. Nine months is commonly recommended in these countries, but even this is too late for many. Always, some children are susceptible to disease, and responsive to vaccine, before others. There is no perfect age for vaccination, although the choice of the best age is crucial in Third World urban centers, where the virus circulates actively. A second dose of vaccine can be given, if the country can afford it, without harm to those who have already been actively immunized, but its capacity to immunize is reduced where there is a history of earlier unsuccessful vaccination. Recently, some strains of vaccine have been found to grow better than the standard strains, in the presence of small amounts of passive antibody. By overlapping vaccine induced with passive immunity in some children, it is possible to reduce the number of children left without either protection.

V. Epidemiology

A. Prevaccine Era in Developed Countries

Our traditional view of measles as a minor, but inevitable, event of childhood was shaped by the epidemiology of this disease in northern countries before it was modified by the vaccine. This infection most often occurred in elementary school-age children. If an older child or adult became infected, it seemed a reversion to childishness and hence vaguely humorous. Often, these older victims had escaped disease earlier because they came from a rural background, and thus the aspect of the "country bumpkin" was associated with older patients.

Epidemics occurred in the larger cities in alternate years with considerable regularity. Each epidemic built in intensity through the school year and peaked in April or May. Less frequently, but nevertheless with regularity, the epidemics spilled into smaller communities. This pattern has been subject to thorough mathematical analysis, which built our knowledge of the process of spread of infectious diseases. [*See* INFECTIOUS DISEASE, PEDIATRIC.]

It was found that the number of cases continued to increase as long as the average case gave rise to infection of more than one secondary case. The probability of this happening depended on the proportion of susceptible persons in the population, and this fraction oscillated around 10% in the cities. As an epidemic spread, more and more susceptible people were infected, passed through the disease, and became immune. Thus, the proportion of susceptible people declined progressively. Ultimately, the chance that an infected person would come into contact with a susceptible individual fell below 1.0, and the epidemic started to wane. The chance for such contacts was substantially greater in school than at home, because so many susceptible persons were gathered there, but younger siblings were often infected by children who brought the disease home from school. When schools let out for the summer, the already declining rate of spread fell precipitously, and the epidemic terminated.

The next year there were too many immune persons in the school to permit the virus to spread efficiently, but in the larger cities the chain of infection did not break. After 2 years new school entrants made up a sufficient part of the total enrollment, so that any child carrying the virus had a high probability of giving rise to more than one new case and starting a new epidemic.

Through the years, as rural schools were consolidated, this pattern became progressively more strongly established. In the Civil War measles was a major threat to the military; it probably affected 10% of all recruits, killing 0.6%. In World War II only 5 of every 1000 soldiers were affected each year and very few died. In the Vietnam War it was scarcely a problem, because nearly all military-age persons were immune.

School-age children might be the group best able to tolerate measles without serious harm. Nevertheless, in the United States in the prevaccine era, about 400 people died each year directly from measles. Many others, particularly babies living in the poorer sections of the country, were sufficiently weakened that they died shortly afterward of secondary causes. Compared to some other diseases, this death rate did not seem high, because essentially all cases were apparent and the denominator was large. Poliomyelitis killed about the same number of persons at that time, but less than 1% of the poliovirus infections were recognized, so when a child contracted an identifiable disease, the danger from poliomyelitis seemed much greater. Poliomyelitis also caused concern because of its conspicuous paralysis, whereas the damage to the brain from measles encephalitis, and the hearing and visual impairment caused by measles-associated infections, were more difficult to recognize.

B. Less Developed Countries

If we look back in time to early and preindustrial Europe and North America, or over to modern less developed countries, we see much more serious consequences of measles. Measles death rates rarely equaled those of smallpox, but now that smallpox is eliminated, measles is probably the most important single cause of death in childhood. Schools do not play a big role in measles distribution in these countries. Even if schooling is available, most children are infected with measles before they reach school age. In urban areas it is common for one-half of the children to be infected before they are 24 months old. This occurs in spite of the fact that, here too, every child is born with maternal antibody, and infection is rare during the first 6 months. The reason is the domestic crowding, with extended families living in a few rooms and many extended families crowded into a small area.

Although the school year does not play an important role in these countries, measles epidemics may still be strongly seasonal. During winter people tend to stay within buildings, and the virus is less likely to be inactivated by ultraviolet light or diluted by great volumes of air than when people are outside. The epidemics in northern climes, therefore, always occurred most often in winter. Conversely, in tropical countries travel between villages may be difficult during the rainy season, and measles spreads poorly when the people are isolated from one another. When the dry season arrives, major festivals are often scheduled and epidemics of measles virus break out.

The death rate in less developed countries is manyfold higher than in the United States. Measles can kill over 10% of the children it infects, and, overall, it is probable that 1% of all deaths in less developed countries are due to measles. Partly, this is due to the fact that young children are inherently at greater risk than older people; they are more likely to get dehydrated, and their smaller respiratory passages are more easily blocked. Another reason is that nutrition is often poor. The most frequently infected age group is often that which is being weaned, and therefore, their nutrition may be especially bad. Vitamin A seems to be particularly important, but the child with measles will lose weight, and, if already undernourished, this might tip him or her over the edge of survival. Third, it is because these areas are heavily contaminated with a great variety of pathogenic organisms, and the chance of secondary infection during the period of immunosuppression is high. Finally, part of the effect seems to be due to the fact that children in these crowded living conditions often live in close contact with the index case and receive large infecting doses; this shortens the time it takes the virus to reach harmful levels and allows it to do more damage before immune defenses can be mustered. Many measles-associated deaths occur in the acute phase of the disease, but many more follow in the next 1–2 months, as the effects of these several factors are compounded.

C. Epidemics in Isolated Areas

If we look farther back in time, before the development of agriculturally based civilization, or out to small surviving island or forest populations, we find another pattern of measles epidemiology. It has been noted that measles virus circulation persists at a low level through interepidemic periods in large cities. This is not true of cities with fewer than 300,000 inhabitants. There, and in most rural parts of the world, the virus fails, at some point during the interepidemic period, to find a new victim and, unable to persist outside a human host, becomes locally extinct. The initiation of the next cycle must wait until someone incubating the virus comes from outside the community. If the community is very isolated, many years can go by before this happens, and a large part of the population grows up without experiencing the disease. When the virus is ultimately reintroduced, it finds nearly everyone susceptible, and one initial case can give rise to hundreds of secondary cases. Much of the population, including all ages except perhaps the very oldest, become sick at once. Few people are left to perform simple necessities such as stoking fires and hauling water. Basic nursing care breaks down and mortality goes up. Such was the situation in the Faroe Islands in 1846, in Tahiti in 1951, and among the Yanomama Indians of Venezuela in 1968.

D. Vaccine Era

The measles vaccine was first licensed in the United States in 1963, but it was not put into general use until 1967 (Fig. 2). When this was done there was an

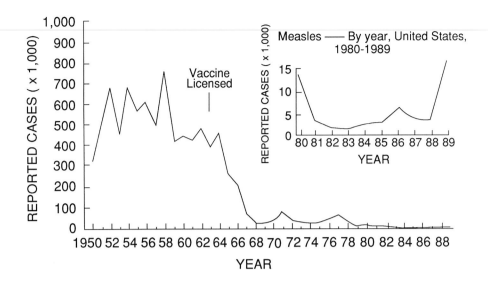

FIGURE 2 Reported cases of measles in the United States 1950–1989. Data from Centers for Disease Control (1989). *Morbidity Mortality Weekly Reports,* **37,** 54, 28; (1990) **38,** 888.

almost immediate reduction in the number of reported cases. Whereas there must have been about 4 million cases per year in the prevaccine era, there have only been a few thousand per year in the last decade—a 1000-fold reduction. Measles deaths and cases of subacute sclerosing panencephalitis have almost disappeared. Quantitatively, this has been a great achievement, but the disease has not been eliminated, and a recrudescence in 1989 demonstrates the need for unremitting control efforts.

Eastern European countries have had nearly as good a record, but there has been resistance to vaccination in Western Europe and Japan. The old idea that measles is a normal part of childhood persists, and progress there has been slower. Much less impact has been made on measles in less developed parts of the world. Vaccination programs are difficult to mount in these areas for the reasons described above and because mothers often lose interest in bringing babies to child care centers before they reach an appropriate age for vaccination. The World Health Organization has made a major effort to improve measles vaccine coverage in these areas, and progress is now being made. The disease is still, however, a major killer in many tropical countries.

In North America and Western Europe most continuing cases occur in the upper school grades. There is no indication that this reflects waning vaccine-induced immunity, but, rather, people gather in progressively larger groups as they advance through the educational system. No matter how strenuous the effort to vaccinate children in infancy, it is difficult to be sure that as much as 2% are not missed or ineffectively vaccinated. If there are only two susceptible children among 100 in an elementary school and one catches measles, the disease may not pass to the other, and if it does, the chain will probably end there. If there are 200 susceptible students among 10,000 in a college, the chance that one of them will catch measles is much greater and when that happens the probability of subsequent spread is also much greater. A program to catch those who remain susceptible with a second round of vaccinations in the upper grades is gaining support.

There is another age group that currently has a substantial attack rate: infants between 10 and 16 months of age. Fifteen months was set as the optimal age for immunization when most mothers themselves had had measles. Now most mothers have vaccine-derived immunity, and their antibody titers, while adequate for their own protection, are lower and fall sooner to inadequate levels in the child. Earlier vaccination in those population segments at greatest risk is now being advocated. Although there has been a relative increase in the proportion of cases in these two age strata, it is important to recognize that the absolute number of continuing cases has been greatly reduced in all age groups.

VI. Origin of the Virus

If measles cannot persist in populations of fewer than 300,000 people, it is apparent that it could not have existed prior to the appearance of the first cities in the Middle East about 4500 years ago. In fact, although it is one of the most distinctive of diseases, there is no mention of it in the writings of Hippocrates or other authors of the classic period. There is no classic Greek name for the disease, and the Latin term "morbilli" only appeared in the Middle Ages. Our first written record of measles is by Abu Becr, who lived in Baghdad during the 10th century. The disease was commonly recognized then, however, and he quotes a description from the seventh century by Al Yehudi, "The Jew."

Measles doubtless arose by mutation from either canine distemper or rinderpest. Its protein sequences are most like rinderpest, but rinderpest may be dependent on herds of domestic cattle for its perpetuation. Only canine distemper is known to have a mechanism that permits its persistence in small natural populations. The first forms of measles may not have been identical to what we know now, and its continued adaptation to human hosts would have been paralleled by a much slower evolution of humans to resistance to the new infection. Nevertheless, the whole world was susceptible, and epidemics that killed a large part of the human population seem to have occurred. Writers of the time were unfamiliar with modern diagnostic criteria and could not describe the disease precisely. There were epidemics in the Roman Empire in 165 and 251 AD and in China in 161 and 310 that could have been measles and smallpox successively.

If measles only appeared in the late classic period, the major continents and islands of the world already had human populations, and many had been cut off from the Afro-Eurasian center by rising sea levels. The populations of Australia, the Americas, and the Pacific probably had no experience with measles until it was brought to them by European explorers after the 16th century. When this occurred in continental areas, measles and other newly introduced diseases spread so rapidly that the epidemics got far ahead of the explorers, and often no record was retained. Vivid descriptions of the havoc wreaked by measles were recorded in Mexico, Hawaii, Fiji, and Tierra del Fuego. We know that up to one-half of these populations may have died. Yet now, the successor populations show no unusual susceptibility, and when vaccine virus has been used to test the responsiveness of the world's few remaining "virgin" populations, they showed only a little evidence of greater susceptibility. The main reason for the high initial mortality seems to have been the destruction of social support and the concurrence of measles with other new diseases introduced by the same contact, not lack of selection for resistance.

VII. Prospects for Elimination

The eradication of smallpox is one of the greatest achievements of modern medicine. Measles shares with smallpox the chief characteristics that made this feat possible: the absence of a reservoir for virus maintenance between outbreaks, lasting immunity, and a good vaccine. In fact, smallpox virus, unlike measles virus, can occasionally persist outside the infected host, and the measles vaccine, both with respect to the safety and durability of its effect, is superior to the smallpox vaccine. It has been a great disappointment, therefore, that measles has not been eliminated from any part of the globe. Of course, until it is eliminated from the whole world, elimination from one country would not mean much, because it could be reintroduced at any time. The problem in the United States, however, is not merely one of frequent introductions, but a large proportion of the continuing cases are acquired by transmission within the country.

All theories indicate that when the proportion of the population that is susceptible is reduced below a certain level, the average new case will come in contact with fewer than 1.0 susceptible person, and the chain of transmission will break. The prevaccine era population maintained a steady state, with about 10% susceptible to measles, and it was assumed that if this proportion could be reduced substantially, herd immunity would be established, transmission would stop, and the virus would die out. Intensive efforts to get good vaccine coverage have resulted in immunization of at least 98% of the population of the United States. The problem is that the original 10% were mostly semisequestered in their homes as preschoolers, whereas the 2% have grown, without acquiring individual immunity, to enter high schools and colleges, where their chances of acquiring and transmitting infection are much greater.

The analogy to smallpox breaks down on the fact that measles is much more infectious than smallpox, and the proportion susceptible must be reduced to a much lower level before herd immunity becomes effective. When our schools and military were free of measles, they included fewer than 1% susceptible people. The level of population immunity must be raised this extra notch before indigenous transmission will stop. Some of those who continue to be susceptible do so because of medical or religious reasons; some think they have been immunized, but the record is erroneous or the vaccination failed for some technical reason. Probably the majority, however, have never been vaccinated because they lacked good medical care. Many of these people are recent immigrants or members of underprivileged classes. It is they who must be reached to establish effective herd immunity. Children between 9 and 16 months of age are also an increasing problem because of their decreasing levels of passive protection, and because they are now more often brought together in child-care centers.

The United States is close to stopping measles virus transmission, but the task will become more difficult in the next decades, not easier. The older age cohorts were exposed to repeated epidemics of measles in the past. They have high levels of antibody prevalence and have served to dampen the spread of measles in the whole population. Now, however, they are dying off, and we will become entirely dependent on vaccine-induced immunity.

Bibliography

Behrman, R. E., and Vaughan, V. C. (eds.) (1987). "Nelson Textbook of Pediatrics," 13th ed., pp. 665–678. Saunders, Philadelphia, Pennsylvania.

Kingsbury, D. W. (ed.) (1990). The Paramyxoviruses. The Viruses.

Norrby, E. and Oxman, M. N. (1990). Measles Virus. *In* "Virology 2nd Edition." (B. N. Fields and D. M. Knipe, eds.) pp. 1013–1044. Raven, New York.

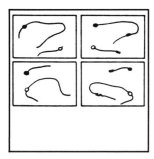

Meiosis

PETER B. MOENS, *York University*

Glossary

Chromosome Compacted strand of DNA with associated proteins that carries hereditary information as well as sequences of undetermined function. Specialized regions of the chromosome include the centromeric region (functional in the movement of chromosomes), the telomeres (specialized and sequences required for replication and integrity), and heterochromatin (highly repeated sequences or inactivated sequences).

Diploid Cells or individuals with a double genome (two sets of chromosomes)

Gamete Mature reproductive cell capable of fusing with a similar cell from the opposite sex to produce a zygote

Haploid Cells or individuals with a single genome (a single set of chromosomes)

Heredity Transmission of biological determinants from parent to offspring. This is the genetic material, usually contained in the chromosomes

Meiosis Two successive nuclear divisions preceding the formation of gametes. The first division reduces the diploid state to the haploid state. The daughter cells have one set of chromosomes each instead of the two sets in the original cell.

Ovum (pl: Ova) Mature reproductive cell of the female

Recombination Process that gives rise to new DNA combinations from (usually) two parental DNA strands

Sexual reproduction Regular alternation between the haploid and diploid phase of the life cycle

Spermatozoon (pl: Spermatozoa) Mature reproductive cell of a male animal

Synapsis Pairing of homologous chromosomes at prophase of meiosis

SEXUAL REPRODUCTION in humans, as well as in most animals and plants, involves the fertilization of an egg from the female by a sperm from the male. The product of fertilization is a zygote that develops to become an embryo, then a fetus, and then a newborn child. Thus the child has one set of hereditary information from the mother, the maternal genome, and one set from the father, the paternal genome (Fig. 1). Although this individual has two genomes, at maturity it will make reproductive cells (i.e., eggs or sperm) that contain only one genome. *Meiosis* is the name of the cellular and nuclear processes that reduce the diploid (two genomes) to the haploid (one genome) by way of two successive cell divisions. Conceptually there are several ways in which this task may be accomplished, but in practice a single mechanism is found in most sexually reproducing organisms. Unusual forms turn out to be derived from the same basic mechanism.

I. Gametogenesis

Although all gametes are similar in that they contain a single genome, the cellular processes that are associated with meiosis give vastly different cells in the male and in the female. Whereas the egg is a huge repository of structural, enzymatic, and energetic components dedicated to the early development of the zygote, the sperm is a much smaller transport vehicle for one male genome. Also, the ovary contains a total of about 100,000 eggs, but the

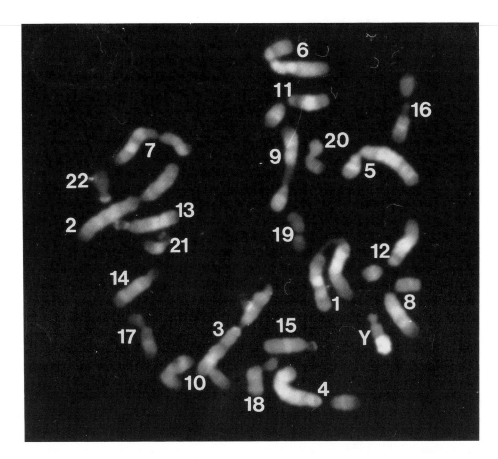

testes produces that number of spermatozoa every day. The cellular processes that surround meiosis differ greatly in the male and the female even though the genetic purpose (i.e., the reduction of diploidy to haploidy) is the same in both. The complexities of shaping an egg or a sperm tend to obscure the study of fundamental aspects of meiosis. The latter have often been learned from a variety of organisms, often simpler, single-celled organisms, where one aspect or another is particularly accessible.

A. Male Spermatogenesis

The stem cells that eventually give rise to the spermatozoa lie against the inside walls of the testicular tubules. When a stem cell becomes active, eight consecutive cell divisions take place. Unlike divisions in other tissues, the products of each division remain connected to each other by cell bridges. The resulting syncytium has less than the expected 256 cells because of occasional cell death and failure to divide. As the syncytium grows, it moves away from the periphery of the tubule toward the center

FIGURE 1 A single human genome consists of 22 chromosomes and an X or a Y chromosome. A single genome can only be found in the ovum (the unfertilized egg cell) or in the spermatozoa. Most cells of the body have two genomes as a result of the fusion of an egg cell with a spermatozoa. (Some cells such as liver cells have more than two genomes as a result of endoreplication.) The set of human chromosomes derives from a human sperm nucleus. In the figure, each of the chromosomes of a Y-bearing sperm is identified. The longest chromosome is #1; the shortest is #22. [From Martin, R. H. (1983). *Cytogenet. Cell Genet.* **35**.]

while new syncytia start forming in the vacated spaces. The approximately 200 primary spermatocytes divide at meiosis I to give twice that many (400) haploid secondary spermatocytes, and these in turn undergo meiosis II to produce about 800 spermatids. These develop into spermatozoa, which accumulate in the central lumen of the tubule and are subsequently transported out of the testis. [*See* SPERM.]

B. Female Oogenesis

Before birth the ovary contains about 100,000 egg cells (i.e., oocytes), which undergo the preliminary stages of meiosis at 7 months of gestation. The cells then arrest just before the first meiotic division. The oocytes remain in the arrested state for 12 to 50 years. Gradually, several oocytes per menstrual cycle are recruited for further development. Meiosis resumes as the released oocyte travels through the fallopian tube. After the first division one haploid nucleus remains with the egg while the other moves to the outside of the oocyte where it becomes a polar body. After the second meiotic division, one nucleus is again expelled from the egg where it joins the divided polar nuclei. There are now one egg nucleus and three polar nuclei. The sperm nucleus joins the egg nucleus on fertilization. [*See* OOGENESIS.]

C. Concepts in Gametogenesis

The discovery in the late 1800s that the reproductive cells of animals originate in a specific organ, the gonad, spelled the end of pangenesis, the widely held belief that all parts of the body contributed their characteristics to the reproductive cell. Experimental confirmation was obtained by transplanting the ovary of a black-haired guinea pig into a white-haired female who then produced black-haired offspring even though mated to a white-haired male. Clearly the body of the adult does not impart genetic information to the germ cells. This evidence further argued against Lamarck's theory of biological evolution through the inheritance of acquired characteristics.

Although the function of meiosis is to form haploid cells from diploid cells, there seems to be endless variation in the details of the process and of the function of haploid cells. For example, the haploid cells may first have a life of their own, multiply and form multicellular forms as can happen in fungi and ferns. In those cases the haploid cells are not called gametes until they function in fertilization. In hymenopterous insects such as ants, bees, and wasps the female can prevent some of her eggs from being fertilized. The unfertilized eggs become males (these males themselves have an unusual meiosis because they have only one set of chromosomes instead of the usual two sets). In the females of a number of species, one of the polar bodies reenters the egg and a quasi-fertilization takes place (parthenogenesis). In this method females produce females in the absence of males. Some species of frogs, salamanders, and insects have only females and they have three genomes instead of the usual two. Be-

cause of an extra DNA duplication during meiosis they produce eggs that have three genomes and that become females without fertilization. The conclusion is that the meiotic system found in humans is one of many possible systems, each system an adaptation to the demands of the life of the species.

II. Chromosome Synapsis at Meiosis

When a human germ cell enters meiosis it has a nucleus that contains 46 chromosomes: 23 individually distinct chromosomes that came from the mother and a similar set of 23 that came from the father (Fig. 1). The extraordinary event in the meiotic nucleus is that chromosome #1 from the mother finds and pairs lengthwise with chromosome #1 from the father. The same happens for all other chromosome pairs until there are not 46 chromosomes but 23 bivalents (i.e., pairs of homologous chromosomes). Before pairing, all chromosomes had duplicated so that each chromosome consists of two sister chromatids. As a consequence, the bivalent consists of two pairs of sister chromatids (four chromatids in total). The pairing of the homologous chromosomes is mediated by the protein cores that form inside the length of the chromosomes at early meiotic prophase. The cores of pairs of homologous chromosomes align in parallel to form what is called the synaptonemal complex (SC, Fig. 2). Thus there are 23 SCs in human meiotic prophase cells (Fig. 2). This pairing stage and the SC appear remarkably similar for nearly all organisms that undergo meiosis. [*See* CHROMOSOMES.]

Where there are structural or numerical rearrangements of the chromosomes, the structure of the SC can manifest this rearrangement. In a human with an extra chromosome #21 (Down's syndrome) there are three #21 cores, which partially pair two-by-two to give a trivalent; or alternatively, two cores pair completely, leaving the third by itself as a univalent. Occasionally the third core is aligned with the two paired cores (Fig. 3). If a piece of chromosome #1 has become translocated to chromosome #2, the core of the chromosome with the translocated piece will form a partial SC with the normal #2, then switch over so that the piece from #1 can pair with the corresponding section of the normal #1 (Fig. 4). Similar structural effects occur when there is an inverted section in the chromosome or a deletion or duplication. [*See* CHROMOSOME ANOMALIES.]

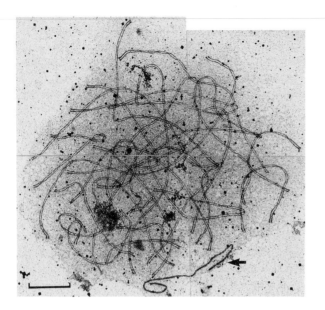

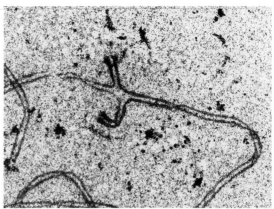

FIGURE 4 Occasionally, two chromosomes have exchanged ends. Such a translocation can be detected by the type of cross (*arrow*) shown in the figure. [From Speed, R. M. (1988). *Hum. Genet.* **78**.]

FIGURE 2 In preparation for the meiotic divisions, each of the 46 chromosomes forms a core, and the cores of homologous chromosomes pair with each other by parallel alignment. As a result, there are 23 synaptonemal complexes in the oocyte shown. Only the cores/SCs are visible in this electron micrograph. The chromatin portion of the chromosomes is not condensed enough at this stage to be visible in a light microscope picture. The *arrow* marks an abnormal extra chromosome of unknown origin in this oocyte. Scale bar = 5 μm. [From Speed, R. M. (1988). *Hum. Genet.* **78**.]

III. Recombination

While the chromosomes are held together by the SCs, a further unusual event takes place. Numerous breaks are formed in the chromatids, either in one

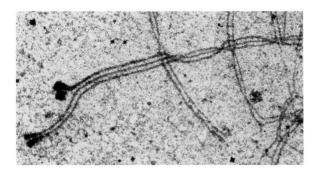

FIGURE 3 Trisomy is the chromosomal defect in which there is an extra copy of one of the chromosomes. A trisomy can be detected by looking for an association of three cores (instead of the usual two) similar to the three-paired structure shown in this figure. [From Speed, R. M. (1988). *Hum. Genet.* **78**.]

or both DNA strands (Fig. 5a). Such free DNA ends tend to invade the DNA of any one of the other three chromatids where the DNA sequence is the same as that of the broken end (Fig. 5b–d). After the crossed strands in Fig. 5h are cut, the nonsister chromatids (heavy and light lines) become continuous. The molecular basis of the reciprocal exchange guarantees that normally there is absolutely no gain or loss of genetic material in either of the participating chromatids. If the event occurs between sister chromatids (i.e., copies of the same chromosome), the genetic consequences are difficult to detect. If, however, nonsister chromatids (copies of the maternally and paternally derived chromosomes) are involved, the event can be detected in a genetic experiment. Suppose the original female parent has the a_1, b_1, c_1, and d_1 forms of genes a, b, c, and d on a given chromosome. The father contributed forms a_2, b_2, c_2, and d_2 on his chromosome. These chromosomes meet when the cells undergo meiosis in the offspring. If nonsister chromatids have an exchange event between genes b and c, the new chromatids are a_1, b_1, c_2, d_2, and a_2, b_2, c_1, d_1. The new combination can be detected in the offspring.

The frequency with which an exchange occurs between, in this case, genes b and c is regulated to the extend that in different repeats of the same experiment the same percent recombinant offspring are found. The percent is called the genetic distance from b to c. This is an operational definition of the distance (determined by an experiment) and the numerical value bears only a vague relation to the physical distance between genes. If the distance be-

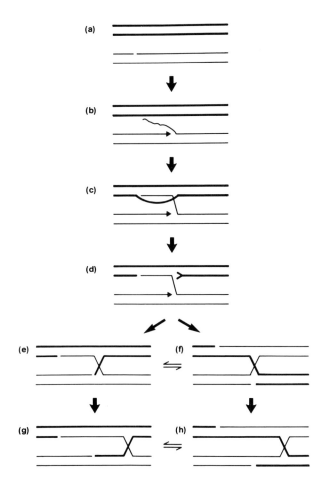

FIGURE 5 The Aviemore model of reciprocal recombination and gene conversion: **a:** Only two participating nonsister chromatids of a set of paired homologous chromosomes are shown. The DNA helix is drawn as parallel lines, heavy for one chromatid, light for the other. During meiotic prophase, occasional single-stranded breaks are introduced in the chromatids (*gap*). **b:** The broken strand unwinds and is replaced by newly synthesized DNA (*arrow*). **c:** The free strand invades the nonsister chromatid where the nucleotide sequences are homologous. **d:** The displaced strand is digested. **e, f:** The top and the base of the arrows join together. In three dimensions this structure, termed a "Holliday configuration," can exist in two physically similar states. **g, h:** The cross can move back and forth until it is cut. In the case of h, one now has a reciprocal recombination such that the chromatid denoted by the heavy lines has become continuous with the chromatid with the light lines and vice versa. This is the type of event that is detected in classical inheritance studies. The effect of such a crossover is shown in Fig. 6. At the site of the recombination itself, there are now three light lines and one heavy line instead of the normal two of each. This is referred to as gene conversion and requires special techniques to be detected. When the cross in g is cut, all that remains is a conversion without a reciprocal recombination. [From Hastings, P. J. (1987). "Meiosis" (P. B. Moens, ed.) Academic Press, Orlando, Florida.]

tween b and c is 5 map units and that between c and d is found be 10, there will be about 15% recombinant offspring for b and d. In other words, the distances are additive. The simplest, by now well-supported, explanation is that the genes are linked in a linear order along the chromosome. For numerous organisms the gene maps of the chromosomes have been prepared in this manner. Human families, however, have too few children to do meiotic linkage studies, and alternative methods have been used.

It should be noted that the genetic organization of the parental chromosomes comes to an end at meiosis because of this process of recombination. The implication is that what is passed on essentially unaltered from generation to generation is the gene. The ultimate effect of natural selection is which genes will appear in the next generation. Neither the individual complement of genes nor the individual chromosomes are passed on intact. These are dissolved at meiosis.

IV. Gene Conversion

The result of a reciprocal crossover is detected genetically because the chromosome has a new gene combination, which may be expressed in the offspring. The process of recombination itself occurred somewhere between the genes, and because of the topology of the DNA strand, the chances of observing the event are small. Suppose the DNA of a given chromosome is 4 mm long. To appreciate the topography, magnify the strand by 10^6, so that it is 4 km long and $\frac{1}{2}$ mm thick. At this magnification, a gene could be 1 m long. A recombination event would take up a few centimeters and could occur anywhere along the 4 km. The chances that it happens in a given gene are small indeed. With special techniques that can detect rare events, such as the survival only of recombinants in a given gene or the appearance, in experiments with some species of fungi, of a black spore in thousands of pale spores, it is possible to study the process of meiotic recombination itself. In intragenic recombination it appears that as often as one gets the balanced reciprocal recombination described above, one obtains a nonreciprocal gene conversion event. As before, the one DNA strand invades that of a nonsister chromatid and inserts its sequence in place of the existing one (Fig. 5c). The displaced strand is then digested (Fig. 5d). There are now three light strands and one heavy instead of the usual two of each (Fig. 5e–h). Because it seems as if one form converted the other, this is called *gene conversion*. By this process, a small part of a gene can be altered. The present structure of the several human genes bears evidence that they are the products of one or more conversion events.

V. Meiosis I Division

When pairing and crossing-over are completed (Fig. 6c,d), the paired chromosomes start to repel each other except at the positions of the crossovers (chiasmata) (Fig. 6e). At those points a chromatid of one chromosome has become continuous with a chromatid of the other chromosome, and because the sister chromatids are still locked together, the chromosomes cannot pull apart. It is at this stage that oocytes (i.e., cells that can develop into eggs) enter prolonged arrest. The arrest does not only occur in human females, but also in other vertebrate females, as well as in plants where the buds overwinter. In males there is no such arrest.

The repelled chromosomes shorten by condensation of the chromatin (i.e., single strand of DNA associated with protein). The condensed bivalents move to the equatorial plane of the nucleus (Fig. 6f). Here the specialized part of the chromosome devoted to chromosome movement, the centromere, attaches to microtubules that lie between the top and bottom pole of the nucleus. While connected to the tubules, one homologue of each pair moves up to the top pole, and the other member moves down to the bottom pole (apparently the cohesion between the sister chromatids is lost at this point) (Fig. 6g). The basic purpose of meiosis is now achieved. At the top pole there are 23 chromosomes, only one of each kind, one #1, one #2, etc. The other set of 23 chromosomes is at the bottom pole. Thus the diploid organism has produced two haploid cells. The composition of the 23 chromosomes is unique. Although there is one of each kind, one #1, one #2, etc., the parental origin is unpredictable. This means that as result of this random assortment of chromosomes at meiosis I there are 2^{23} chromosome combinations possible. The number of combinations becomes infinitely large if the exchanges (crossovers) are taken into account. Even casual observation of groups of people bears out the existence of this unending amount of variation. Genetic variation creates the richness of diversity and talents that defies generalizations of unique group and race characteristics.

VI. Meiosis II Division

At the end of meiosis I, the male produces two haploid cells (secondary spermatocytes), each with one set of 23 chromosomes. The already duplicated chromosomes (consisting of two chromatids each) soon move to the equatorial plane of the nucleus. The centromeres become attached to the microtubules and at anaphase II, the chromatids pull apart (Fig. 6h). Each pole thus has one of each chromosome (Fig. 6i). In total, one spermatocyte produces four haploid spermatids. In the female, one oocyte produces one haploid egg nucleus and three haploid polar bodies.

Textbooks commonly show meiosis II as an ordinary mitotic cell division such as occurs in cells of the body whereby the two daughter cells are identical genetic copies of the mother cell. This gives no credit to the specialized nature of meiosis, and it is inaccurate. Meiosis II is, for example, the only division in the body that is not directly preceded by a duplication of chromosomes. In the case of meiosis, duplication took place before prophase of meiosis I (Fig. 6b), and at the first division duplicated chromosomes went to the poles. Therefore no duplication is needed for the second division. Furthermore, and more significantly, because of the crossovers between nonsister chromatids, the chromatids of meiosis II chromosomes are nonidentical (Fig. 6g). In mitosis the chromatids of each chromosome are identical because they are copies of the same chromosome. The effect of this is that the products of a mitotic division are genetically identical, whereas the two spermatids that come from the secondary spermatocyte are not. (As always, there are rare exceptions to this mechanism in some organisms).

In the female, there are 23 pairs of matched chromosomes at meiosis. In the male, however, there are 22 such pairs plus one partially matched pair. This pair consists of a medium-sized X chromosome partially paired to a small Y chromosome. Because of their different distribution in females (XX) and males (XY), these are referred to as the sex chromosomes. In humans the Y plays a significant, be it indirect, role in determining maleness. As a result of the first meiotic division, half the spermatozoa carry an X chromosome, the other half a Y chromosome. Fertilization of the X-bearing

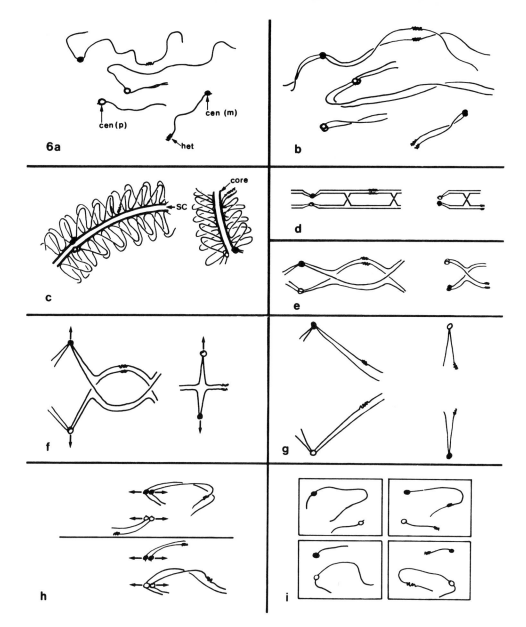

FIGURE 6 Diagramatic representation of meiosis. Only two of the 23 pairs of chromosomes are shown. **a:** When the cell enters the meiotic pathway, there are 46 chromosomes in the nucleus, 23 of maternal origin (m) and 23 of paternal origin (p). In the drawing, only two pairs of chromosomes are shown, and their origin is marked with a solid centromere (cen) for maternal origin and open centromeres for those of paternal origin. To illustrate the process of segregation, assortment, and recombination, the pairs of chromosomes have been drawn differently for hetero-chromatic blocks (het). The chromosomes are drawn as simple lines, but they are, in fact, intricately constructed protein–DNA structures. The basic structure, however, is a continuous double-stranded DNA helix, indicated by the split end of one of the chromosomes. **b:** Chromosome duplication. Each chromosome now consists of two sister chromatids. **c:** During meiotic pro-phase, the sister chromatids become attached to a core (core), and the cores of homologous chromosomes pair to form the syn-aptonemal complex (SC). This chromosome behavior and the SCs are exclusive to meiotic nuclei. **d:** Simplification of Fig. 6c. The chromosomes are paired and the reciprocal recombination events shown in Fig. 5 have been completed. The heterochro-matic blocks indicate that the sister chromatids are still together. **e:** At the diplotene stage of meiotic prophase, the chromosomes repel each other, but the adhesion between the sister chromatids

prevents separation. **f:** At metaphase I, the bivalents all lie on the equatorial plane of the nucleus. The centromeres have become attached to microtubules that function in the movement of the chromosomes to the poles of the nucleus (*arrows*). **g:** At ana-phase I, the essential function of meiosis is performed. The top and bottom half of the nucleus each has only one set of 23 chro-mosomes. Whether the maternal centromere (*solid*) or the pater-nal (*open*) goes to the top is a matter of chance. As a result, a large number of combinations are possible (2^{23}). As a conse-quence of the reciprocal recombination events, the chromatids are partially of maternal and paternal origin. This is illustrated by the instances in which the paternal centromere (*open*) is con-nected to a heterochromatic block of maternal origin. Recombi-nation further increases the number of possible gene combina-tions in the newly formed nuclei. **h:** After the first meiotic division, the chromosomes in the sister nuclei soon line up on the equatorial plane at metaphase II, and the centromeres become attached to the microtubules. The direction of chromatid move-ment is indicated by the arrows. **i:** After the second meiotic division, there are four nucli, each with one set of chromosomes. Thus the purpose of meiosis to make nuclei with single genomes from nuclei with two genomes is accomplished. An indication of the chromosomal variation in terms of maternal and paternal origins is evident from the figure.

egg by an X-bearing sperm will result in an XX female, whereas fertilization with a Y-bearing sperm will produce an XY male. The prediction would be that at conception there are equal numbers of XX and XY zygotes. In fact, the ratio appears to be 1:1.5, a large excess of male zygotes. Because of a biased loss of male zygotes during pregnancy, the ratio is 100:102 at birth. At sexual maturity the ratio approaches a perfect 1:1, an admirable case of biological fine-tuning that produces the right ratio at the right time.

VII. Defects During Meiosis

Like all other biological processes, meiosis is subject to errors. Unlike other cellular processes, however, the consequences of an error at meiosis may be extremely serious because an entire individual may be the victim if the defective gamete partakes in fertilization. Gross defects such as the loss or addition of an entire chromosome or an accidental extra set of chromosomes usually results in a loss of the embryo at an early stage of development. Surprisingly, variations in the numbers of sex chromosomes seem to be tolerable, but the carriers suffer a variety of effects. Variations in numbers of X chromosomes are probably tolerated because numbers in excess of one are partially inactivated. Thus cells of a normal XX female have one active and one inactive X chromosome. The Y chromosome appears to have few essential genes, and its absence is tolerated in X0 individuals (0 indicating the absence of Y).

In the oocytes, the two paired X chromosomes sometimes do not go to separate poles at anaphase I, but stay together (nondisjunction). As a result, the egg nucleus or the polar body has two X chromosomes. If an egg nucleus has two X chromosomes as a result of nondisjunction, and it is fertilized by a Y-bearing sperm, the zygote has the usual 22 pairs of chromosomes and the unusual XXY sex chromosomes. The abnormalities that are associated with the chromosome abnomality are described by Klinefelter's syndrome. By the similar nondisjunction errors at meiosis I or II or both, such forms as X0 (Turner's syndrome), XXX, and XXXXY can be produced. A nondisjunction event in any other chromosome can similarly produce a trisomy such as the well-known trisomy 21, Down's syndrome. [*See* DOWN'S SYNDROME, MOLECULAR GENETICS.]

Bibliography

Brennan, J. R. (1985). "Patterns of Human Heredity." Prentice-Hall, Englewood Cliffs, New Jersey.
John, B. (1990). Meiosis. *In* "Developmental and Cell Biology." (D. W. Barlow, P. Bray, P. B. Green, and J. W. Slack, eds.) Cambridge University Press, Cambridge.
Moens, P. B., ed. (1987). "Meiosis." Academic Press, Orlando, Florida.
Riley, R., Bennett, M. D., and Flavell, R. B., eds. (1977). A discussion on the meiotic process. *Phil. Trans. R. Soc. Lond. B.*

Melanoma Antigens and Antibodies

B. M. MUELLER, R. A. REISFELD, *Research Institute of Scripps Clinic*

Glossary

Ganglioside Complex glycolipids, containing one or more sialic acid residue, abundant in the plasma membrane of cells of neuroectodermal origin

Melanoma-associated antigens Antigens expressed on melanoma cells but not on normal skin melanocytes, including antigens that are unique for melanoma and others that are typical for all tumors of neuroectodermal origin

Monoclonal antibody Antibody derived from a clone of B-lymphocytes, usually a B-lymphocyte hybridoma; they are homogeneous in structure

Proteoglycans Structural component of the extracellular matrix consisting of glycosaminoglycans covalently bound to a core polypeptide chain; carbohydrate moiety can be up to 95% of the weight of a proteoglycan

MELANOMA-ASSOCIATED antigens (MAAs) and monoclonal antibodies (Mabs) that define some of their biochemical, immunological, and functional characteristics have become of increasing interest for determining biological and pathophysiological parameters of human melanoma. A variety of these molecules are discussed here, including cell-substrate interacting glycoproteins, ion-transport and -binding proteins, gangliosides, receptors for growth factors, and class II HLA antigens. Emphasis has also been placed on antigens that are expressed during melanoma tumor progression. The major focus is on those MAAs that are defined by Mabs that can potentially serve as molecular probes to gain a better understanding of tumor invasion and metastasis.

I. Introduction

There has been an increasing interest in the biology and pathophysiology of human cutaneous melanoma during the last two decades, in part because of the rapidly mounting incidence of this disease. Consequently, human cutaneous melanoma has become both clinically and experimentally one of the best-studied solid human tumors. The biological foundation of much of this research has come from studies on tumor progression based on histopathological studies of tumor lesions at six different steps (Clark levels). Briefly, this starts with the acquired melanocyte nevus, which lacks any architectural or cytological atypia. In step two of this progression, the melanocyte nevus acquires architectural but no cytological atypia. Step three features the dysplastic nevus with both architectural and cytological atypia. Both melanocyte and dysplastic nevi are considered precursor lesions of melanoma, although they usually differentiate and disappear. In clear distinction, step four of this tumor progression is represented by the radial growth phase (RGP) of primary melanoma, which lacks competence for metastasis, and its surgical excision usually leads to the cure of cutaneous melanoma. However, during the RGP, focal lesions may arise that ultimately lead to the vertical growth phase (VGP) of primary melanoma (i.e., step five, with acquired competence for metastasis). The incidence of melanoma metastasis is directly related to the thickness of the VGP lesion (Breslow index). Finally, the sixth step of tumor is characterized by the development of extensive metastasis. [*See* METASTASIS.]

Although the 27,000 annual cases of malignant melanoma account for only about 2.5% of cancer in the United States and approximately 7% of all skin cancer, they cause approximately 65% of all skin cancer deaths. The fatal nature of this highly disseminated disease is caused by its poor response to conventional radiation and chemotherapy. Partly because of this rather dismal clinical outlook and possibly due to melanoma being among the tumors with the most rapidly mounting incidence at 7% annually, an intense interest has developed in MAAs as a potential means for diagnosis and therapy. This interest is also based, in part, on prior, well-documented findings indicating that changes in cell-surface antigens were associated with neoplastic transformation. During the last 10 years, a rapid expansion of this type of biomedical research was triggered by two major technical developments: Mabs and modern molecular biology. These technologies actually catalyzed recent progress in the characterization of antigens expressed by melanoma cells and identified by Mabs. In turn, these antibodies together with their well-defined antigen targets were used as tools to develop new approaches for the diagnosis and therapy of melanoma. Equally important, these Mabs directed to human MAAs constitute effective molecular probes that may ultimately prove highly useful in delineating basic mechanisms of tumor invasion and metastasis. [*See* MONOCLONAL ANTIBODY TECHNOLOGY.]

The intent of this brief chapter is to focus mainly on those MAAs that were defined by Mabs that can potentially serve as molecular probes to gain a better understanding of tumor invasion and metastasis. Particular emphasis has been placed on the biochemical and immunological characterization of such melanoma tumor antigens and a delineation of their possible biological functions in normal and neoplastic cells.

II. Biochemical, Immunological, and Functional Characterization of Melanoma-Associated Antigens

The use of Mab directed against MAAs as molecular probes has led to the identification of several antigenic structures with distinct biochemical and immunological characteristics. Based on these properties and on their established as well as putative biological functions, these antigens and the Mabs recognizing them are ordered here into several major categories, and their biochemical and immunological characteristics are summarized. These categories can be extended to those antigens that are expressed on the cell surface and/or shed into the circulation. Thus, some extracellular matrix proteins such as fibronectin are not expressed on the surface of melanoma cells, although they are shed in relatively large quantities. Gangliosides and cell substrate-interacting proteins such as the MAA–chondroitin sulfate proteoglycan (CSP) are both expressed on the cell surface and shed. A number of cell-surface receptors for various growth factors, as well as several ion-transport and -binding proteins are concerned with the intracellular transport of biologically relevant material. HLA antigens are involved in immune recognition and are both expressed and shed by melanoma cells. Finally, some pigmentation antigens are located in the cytoplasm and can be found in the spent medium of pigmented melanoma cells.

A. Cell Substrate-Interacting Glycoproteins

Several molecules of relatively high molecu weight, which are uniformly expressed and she little heterogeneity within each tumor when an lyzed either on cultured cells or tumor tissue se tions, belong to the cell substrate-interacting glyco protein of MAAs.

1. Chondroitin Sulfate Proteoglycans

a. Biochemical and Immunological Characterization The gene encoding melanoma-associated CSP was mapped to chromosome 15, and this major MAA has produced several murine Mabs because of its high expression on human melanoma cells and its immunogenicity in mice. Thus, a unique glycoprotein–proteoglycan complex associated with human melanoma cells has been defined by Mab 9.2.27. Biosynthetic studies indicated this molecule to consist of an N-linked glycoprotein of 250 kDa and a proteoglycan component of >450 kDa. The 250-kDa glycoprotein has endo-β-N-acetylglucosaminidase (Endo-A)-sensitive precursors that are N-asparagine-linked, unprocessed, high-mannose oligosaccharides. The 250-kDa molecule is the core glycoprotein of the proteoglycan and various tests show that it is a CSP. Initial biosynthetic analysis indicated that the cationic ionophore monensin blocked biosynthesis of the proteoglycan and that its appearance was linked kinetically to biosynthetic functions of the Golgi apparatus, the proposed site of glycosyltransferases involved in proteoglycan biosynthesis. These data suggest that the addition of N-linked oligosaccharides is an early biosynthetic event in CSP core protein biosynthesis and that the chondroitin sulfate chains are added, in a series of steps, onto the 250-kDa core glycoprotein to form the >450-kDa proteoglycan. [*See* Proteoglycans.]

Another Mab, 155.8 (IgGl), produced against purified membranes of human melanoma cells also was shown to react with a CSP expressed on such cells; however, binding inhibition studies revealed that Mab 155.8 reacts with a different epitope than that recognized by Mab 9.2.27 on the same proteoglycan molecule. Mab 155.8 differs further from Mab 9.2.27 in that it binds melanoma cells to a lesser extent and immunodepletion experiments indicated that 155.8 determinants are present only on a subgroup of those molecules bearing the 9.2.27 epitope. Nevertheless, the molecules recognized by Mab 155.8 are proteoglycan in nature and contain chondroitin sulfate type A and/or C. Similar to Mab 9.2.27, Mab 155.8 recognizes determinants on the 250-kDa core glycoprotein and on the intact CSP. Also, following chondroitinase digestion of proteoglycans immunoprecipitated by Mab 155.8, the CSP disappeared with a concomitant increase in the concentration of the 250-kDa core glycoprotein, thus proving that this glycoprotein was included in the CSP and was indeed its core protein.

Several other Mabs have been reported that recognize antigens designated melanoma-association high-molecular weight antigens (MAA-HMW). Considerable evidence from several laboratories indicates that these Mabs also recognize epitopes on a melanoma-associated proteoglycan. The most intensely investigated ones among these reagents is Mab 225.28S (IgGl). Comparison of this antibody with Mab 9.2.27 in a competition antibody-binding assay revealed a striking difference in reactivity. Also, although both these antibodies bind to the core glycoprotein of CSP, they clearly recognize different epitopes. Moreover, only antibody 225.28S reacts slightly with the fibrosarcoma cell line HT 1080, leading to the speculation that only the epitope recognized by 225.28S is on proteoglycans found on such cells. This would support the

argument for structural heterogeneity of CSP. This certainly seems to be a possibility in view of the very complex structure of proteoglycan aggregates, which can include isomeric, sulfate-substituted carbohydrates and differential chain length of oligosaccharides such as hyaluronic acid. This type of molecular complexity could easily explain why a whole variety of antigen epitopes are recognized by different Mabs. In conclusion, Mab 225.285 recognizes a different epitope on the same antigen structure recognized by Mab 9.2.27.

b. Characterization of Functional Properties
Proteoglycans have been implicated in growth control, involvement in adhesion in cell–substratum interactions, cell–cell contacts, and other functional properties of potential relevance. The potential of CSP as a functional molecule is brought into focus by its presence on >90% of melanoma cultures, where 80–100% of cells express this molecule at densities ranging from 1×10^5 to 6×10^6 binding sites per cell. The CSP is expressed on the melanoma cell surface on microspikes (i.e., 1–2 μm structures on the upper cell surface that range up to 20 μm at the cell periphery). Peripheral CSP microspikes are involved in cell–cell contacts and also form complex footpads that make contact with the substratum. The CSP antigen is also very well expressed on adhesion plaques that are deposited along cell membranes.

A possible functional role of CSPs in human melanoma was demonstrated in their interaction with basement membrane components. Thus, Mab 9.2.27 can block the early events of melanoma cell spreading on endothelial basement membranes, while only slightly inhibiting cell adhesion, suggesting a possible role for CSPs in stabilizing cell–substratum interactions in this *in vitro* model for metastatic invasion. Another potential functional role for CSP was indicated when Mab 9.2.27 specifically inhibited anchorage-independent growth of human melanoma cells in soft agar. The precise mechanism underlying this phenomenon is not entirely clear, and it has been proposed that it relates to the physicochemical nature and location of CSP recognized by Mab 9.2.27 on the microspikes of the melanoma cell surface.

Another functional property of melanoma-associated proteoglycans was suggested when Mab 9.2.27, employed as a molecular and functional probe, demonstrated that the core protein of the CSP recognized by this antibody is present in two forms on the cell surface, either free or modified by the addition of chondroitin sulfate chains. These findings suggest that the addition of glycosaminoglycan chains may not be a prerequisite for cell-surface expression of the proteoglycan core protein. Thus, it is unlikely that such a modification serves as a marker to segregate molecules on the cell surface. Signals involved in targeting CSP or its core protein to the cell surface are more likely to reside in the N- and/or O-linked oligosaccharides that are added to the core protein prior to addition of glycosaminoglycan chains.

2. Melanoma-Associated Cell-Adhesion Molecules

Several Mabs have been reported that immunoprecipitate from melanoma cell extracts (under nonreducing conditions molecules that range from 130 to 150 kDa and from 90 to 105 kDa. Under reducing conditions, these molecular species are of 116–120, 95, 29, and 25–26 kDa. Among these Mabs are Nu4B, which binds to most cultured melanoma and glioma cells and to most melanomas *in situ*. Another Mab, DA3, was reported to have a similar binding pattern but higher binding affinity. Mabs DA3 and AMF7 inhibit attachment of melanoma cells to different substrates, and the antigen recognized by Mab AMF7 has been designated melanoma cell-adhesion molecule-1 (MECAM-1), as it apparently plays a role in motility and migration of melanoma cells.

Among the various Mabs recognizing melanoma-associated cell-adhesion molecules, Mab LM609 is probably best defined by its recognition of the biochemically well-characterized vitronectin receptor ($\alpha_v\beta_3$) expressed on human endothelial cells and on the cultured human melanoma cell line M21. The adhesion receptor complex recognized by LM609, which consists of two unequal length subunits, was isolated from umbilical vein endothelial cells and purified on an affinity matrix consisting of an Arg-Gly-Asp (RGD) containing heptapeptide. The receptor consists of noncovalently associated α- and β-subunits, which under reducing conditions have molecular masses of 135 and 115 kDa, respectively. The α-subunit also contains a disulfide bridge-linked 27-kDa light chain.

Additional Mabs have been reported that recognize melanoma-associated cellular-adhesion molecules. Among these is Mab CL203.4, which is secreted from hybridomas produced by immunizing BALB/c mice with human melanoma cells COLO

38 treated with gamma interferon (IFN-γ. This antibody recognizes a single-chain glycoprotein of 96 kDa with a polypeptide moiety of 50 kDa. This antigen is identical in biochemical and immunological properties of the well-characterized intercellular adhesion molecule (ICAM-1).

Mab P3.58 and MUC18 are two other murine Mabs raised against fresh human melanoma tumor tissue that bind to cell-surface molecules involved in cellular adhesion. They were first reported to bind to membrane glycoproteins of 89 and 113 kDa, respectively. The antigen detected by P3.58 is again homologous to ICAM-1, while the integral membrane glycoprotein detected by MUC18 shows extensive sequence similarity with neural cell-adhesion molecules (N-CAM).

3. Functional Properties of MAA Cell-Adhesion Molecules

A functional property of the adhesion receptor complex identified by Mab LM609 is well delineated by this antibody. Thus, it inhibits the attachment of human endothelial cells and of M21 melanoma cells to such extracellular matrix proteins as fibrinogen, von Willebrand factor, and vitronectin but has no effect on the attachment of these cells to fibronectin, collagen, or laminin. Mab LM609 also inhibits attachment of these cells to an immobilized peptide containing the RGD sequence. This same antibody was also used to determine for the first time that platelets contain the vitronectin receptor complex (VNR) in addition to glycoprotein IIb/IIIa, its prominent RGD-binding adhesion receptor complex. It has been postulated that RGD recognition may play a significant role in biological events associated in vascular proliferation and in motility and migration of melanoma cells that are key elements of the metastatic process.

The ICAM-1 antigen recognized by Mab CL203.4 is highly susceptible to modulation by IFN-γ, interleukin-1 (IL-1), and tumor necrosis factor α (TNF-α). Mab CL203.4 has a markedly lower reactivity with benign than with malignant lesions and stains a higher percentage of metastatic than primary lesions. Melanoma cells isolated from primary lesion acquire reactivity with Mab CL203.4 following a 24-hr *in vitro* incubation with IFN-γ (100 U/ml), suggesting that such cells are able to synthesize and express ICAM-1. The lack of staining of primary lesions with CL203.4 is thought to reflect insufficient concentration of the cytokines that modulate this antigen. Reactivity with CL203.4 of primary lesions removed from patients with stage I melanoma showed a highly significant correlation with course of disease. Thus, the disease-free interval of patients without detectable reactivity of their primary lesion with Mab CL203.4 was significantly longer ($p = 0.004$) than that of patients whose primary lesions stained with this antibody. Similar observations were made for the antigens defined by P3.58 and MUC18. Expression of both adhesion molecules seems to correlate with increasing vertical thickness (i.e., malignancy of melanoma biopsies); therefore, these antigens have been proposed to be directly involved in tumor progression.

B. Ion Transport and Binding Proteins

1. Melanotransferrin (p97)

a. Biochemical and Immunological Characterization A family of antigens of 97 kDa, designated p97, is preferentially expressed on human melanoma cells and tissues. Similar to CSP, the family of p97 glycoproteins is highly immunogenic when human melanoma cells are injected into mice. In initial studies, Mab 4.1 (IgG1) was found to bind significantly to 90% of melanomas tested as well as to 55% of a variety of other tumor cells analyzed. It recognizes a glycoprotein of 97 kDa. Although p97 is not melanoma-specific, normal tissues and nonmelanoma tumors contain p97 in far lower amounts than melanoma tissues. Antigen p97 was also observed in a number of fetal tissues such as colon, umbilical cord, lung, and benign nevi.

Several Mabs recognized different epitopes on p97. Thus, Mabs 4.1 and 96.5 reacted with the same antigen epitope p97a, whereas Mabs 118.1 and 8.2 defined epitopes p97b and p97c, respectively. The relatively high frequency of hybridomas secreting antibodies to p97 suggested this antigen to be either particularly immunogenic and/or expressed in larger amounts than other antigens on the melanoma cell lines (SK-MEL-28 and K-2) used for immunization. Quantitation by antibody-binding analysis indicated that SK-MEL-28 cells express at least 4×10^5 molecules of p97 per cell (i.e., roughly the same number of HLA-ABC determinants identified by anti-HLA Mab W6/32 on these cells). Melanoma cell lines bound $3.0-3.8 \times 10^5$ molecules of antibodies defining epitopes p97a, b, and c per cell, whereas lung carcinoma cell lines bound $4.0-6.2 \times 10^3$ molecules per cell and B- and T-lymphoid cell lines bound only 250–1,400 molecules per cell.

The initial chemical characterization of p97 indicated sialylated carbohydrate chains and a single polypeptide chain with some intrachain disulfide bridges. The p97 antigen is functionally related to transferrin and may have evolved from a common ancestor. In fact, there is partial amino-terminal amino acid sequence homology between these two molecules. Moreover, p97 binds iron, and an antiserum made to denatured p97 cross-reacts with both denatured transferrin and lactotransferrin. Under reducing conditions, p97 has a similar molecular weight as the transferrin receptor, but the transferrin receptor forms a dimer of 200 kDa, whereas p97 does not. Mab OKT9, which is specific for transferrin receptors, does not immunoprecipitate p97 from radiolabeled human melanoma cells.

Antigen p97 is one of the relatively few human tumor-associated antigens of which the gene has been cloned and sequenced. The overall structure based on amino acid sequence shows a mature p97 molecule with two extracellular domains of 342 and 352 residues. A possible membrane anchor for p97 is provided by a 25-residue C-terminal stretch of mainly hydrophobic amino acid residues. There is considerable sequence conservation, because each of the extracellular domains contain 14 cystine residues that form seven intradomain disulfide bridges. The p97 molecule has relatively little glycosylation, because the extracellular domain contains only one or two potential N-glycosylation sites. The three antigenic determinants of p97 are protein in nature and are situated on the N-terminal domain. These domains show 46% sequence homology to each other and 39% homology to the corresponding domains of human transferrin. The conservation of the disulfide bridges is thought to create iron-binding pockets for p97. Based on the structural homology between p97 and the three members of the transferrin superfamily, it was suggested that p97 may have diverged from serum transferrin more than 300 million years ago, when mammalian and avian lineages diverged.

b. Functional Properties Although p97 has been named melanotransferrin, it binds far less iron than transferrin, and its role played in cellular iron metabolism may differ considerably from that played by circulating serum transferrin and its cellular receptors. The anti-p97 Mab is not effective in mediating antibody-dependent cellular cytotoxicity (ADCC) and cannot mediate complement-dependent cytotoxicity (CDC) with human complement.

To increase the immune response of the tumor-bearing host by inducing strong T-cell help, the gene encoding for p97 was introduced into the vaccinia virus genome to produce an antitumor vaccine vp97. When this vaccine is introduced into mice bearing murine tumors, they produced an immune response against p97. The effectiveness of this vaccine in humans remains to be tested.

2. Calcium-Binding S-100 Protein

a. Biochemical Characterization and Functional Properties

The S-100 proteins, named for their solubility in saturated ammonium sulfate, are a group of abundant, low-molecular weight (10–12 kDa) acid proteins that are highly enriched in nerve tissue. The S-100 fraction contains a mixture of hetero- and homodimers of α- and β-subunits with different amino acid compositions. These proteins contain two Ca^{++} binding regions known as EF hands. Upon calcium binding, several properties of S-100 protein are altered, such as the α-helix and β-sheet structures, absorption and circular dichroism spectra, reflecting a structural rearrangement. No enzymatic functions have been shown for S-100 proteins, and they are thought to exert their biological effects by binding to and modulating effector proteins in a manner similar to calmodulin. In fact, the Ca^{++}-dependent changes in S-100 protein and in calmodulin are similar. In all vertebrates studied, the primary amino acid sequences have been >95% conserved, suggesting that the S-100 proteins perform important biological function. The various proposed and demonstrated effect of S-100 proteins in cell-cycle progression, cell-type differentiation, signal transduction, and morphological differentiation are thought to be mediated by the same basic process. Inhibition of cytoskeletal-associated protein phosphorylation by the S-100 proteins is considered key in controlling all of these biological processes.

The S-100 proteins are found in addition to neural crest-derived tissues, also in chondrocytes, adipocytes, myoepithelial cells, dendritic cells of lymphoid tissue, Langerhans cells, and T lymphocytes. They are present in a high proportion of malignant melanomas and neurocytic skin nevi. On melanoma cells, the S-100 molecules are a particularly good diagnostic marker for secondary, amelanotic melanoma, whose tumor cells lack typical melanoma morphology. In identifying such melanomas, antiserum to S-100 proteins is often used in combination with a Mab to cellular keratin and to common leukocyte antigens. Malignant melanoma expresses S-100 proteins, but not the latter two antigens, and can be distinguished from carcinoma, which expresses keratins but not S-100 proteins, and from lymphomas, which usually express common leukocyte antigens but no S-100 proteins. This approach also serves to detect single melanoma cells in lymph nodes regional to melanoma when conventional histology fails to identify them.

C. Ganglioside Antigens

1. Biochemical and Immunological Characterization

Numerous studies have characterized Mabs directed to the carbohydrate portion of glycolipids, particularly those containing one or more sialic acid residues (i.e., gangliosides). Figure 1 is a schematic representation of ganglioside structures designated according to the nomenclature of Svennerholm. It is possible to characterize Mabs directed against glycolipids mainly by immunostaining individual glycolipid determinants separated on a thin-layer chromatography plate. [See GANGLIOSIDES.]

One advantage of Mabs directed to carbohydrate antigens is their potential to establish a structure–function relationship for these determinants on the tumor cell surface, because Mabs can recognize a carbohydrate determinant with known sugar composition and anomeric linkages. By using Mabs to define oligosaccharide structures on the tumor cell surface, one can pose questions regarding their functional properties as they relate to the malignant or metastatic phenotype. This is much more difficult to do with Mabs directed to protein or glycoprotein antigens whose epitopes are in most cases structurally ill defined, because they depend on unpredictable conformation and three-dimensional structure. Moreover, the complete primary amino acid sequence frequently is not available.

Human melanoma, which is derived from the neuroectoderm, is rich in ganglioside content. A number of reports have described murine Mabs directed to the disialoganglioside GD3, which is similar in structure to GD2 but lacks a terminal N-acetylgalactosamine. Mabs R-24 (IgG$_3$) and 4.2 (IgM) recognize GD3 on the surfaces of human melanoma cells. The ceramide portion of tumor GD3, however, manifests a predominance of longer-chain fatty acids than brain GD3. Because ceramide chain length affects exposure of the antigen epitope, it has

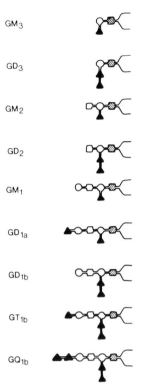

FIGURE 1 Schematic representations of ganglioside structures according to the nomenclature of Svennerholm. SA (▲) sialic acid; Gal (○) galactose; GalNAc (□) N-acetylgalactosamine; Glc (▨) glucose; Ceramide (⊂) lipid portion of the ganglioside molecule.

been suggested that GD3 in melanoma is more antigenic than GD3 in normal cell membranes.

A number of studies have implicated gangliosides expressed on the surfaces of melanoma cells as potentially relevant antigens in humans. In this regard, the "AH antigen" was detected during the initial analysis of the humoral immune response of melanoma patients to autologous melanoma cells. Serum from patient AH reacted with a cell-surface antigen ultimately identified as the GD2 ganglioside. In other studies, peripheral blood lymphocytes from melanoma patients transformed into B-lymphoblastoid cells by Epstein–Barr virus (which produce Mabs) were used in vitro to generate monospecific human IgM antibodies of two kinds, one specifically reactive with GD2 (OFA-I-2) and another directed to the monosialyated version, i.e. GM2 (OFA-I-1). Attempts to produce human Mabs with melanoma cells as immunogen produced essentially only antibodies to gangliosides. The rationale for making human Mabs was to avoid human anti-mouse antibody (HAMA) responses that usually occur when pa-

tients are injected with murine Mabs and to ascertain whether the human immune system in general can recognize an antigen repertoire other than its murine counterpart. In contrast to murine Mabs, all human Mabs to ganglioside that were produced thus far are of IgM isotype. Such antibodies are negative in ADCC-type interactions with human effector cells but effectively fix human complement. Some human Mabs such as AbFCM1 (IgM) and AbHJM1 (IgM) that were more recently produced have a diverse reactivity with gangliosides. Thus, AbFCM1 recognizes mainly GM3 but also several epitopes on GD3, GD2, GD1a, and GT1b. Human Mab AbHJM1 recognizes mainly GD3, but in contrast to murine, anti-GD3 Mab R24, it also recognizes epitopes of GD2 and GD1b. At least at this time, no clearly documented advantage of human over murine Mabs is evident, with the exception of less HAMA responses in the very limited number of patients studied thus far with human Mabs.

Murine Mabs produced in several laboratories have defined at least five melanoma-associated gangliosides: GD2, GD3, 9-O-acetylated GD3, GM2, and GM3. Gangliosides GD2 and GD3 are also shed by melanoma cells but usually form micelles in the circulation and have to be extracted with chloroform-methanol before they can be recognized by antiganglioside Mabs. GD2 thus determined in the serum extracts ranges from 25 to 85 ng/ml of serum, while GD3 ranges from 10 to 30 ng/ml serum.

The use of anti-GD3 Mab R24 in an indirect immunoperoxidase-staining technique on 175 cryopreserved, unfixed human tissue sections established a striking specificity for malignant melanoma. Thus, all of the primary and malignant melanoma tissue sections contained GD3. Among normal tissues, 9 of 11 nevi were GD3 positive, and melanocytes in the basal layer of the skin stained weakly. Another Mab (D1.1) that was originally generated against an undifferentiated rat neuronal tumor cell line reacts specifically with a 9-O-acetylated form of the disialoganglioside GD3. The expression of this antigen is even more restricted to human melanoma cells than that of GD3, because Mab D1.1 does not react with normal melanocytes, fetal or adult brain tissue, or normal skin.

Mabs directed to ganglioside GD2, such as 126-4 (IgM), 14.18 (IgG3), 14G2a (IgG2a), and 3F8 (IgG3), generally react well by standard ELISA technique with most cultured human melanoma and neuroblastoma cells, as well as with several glioma and small cell lung cancer cell lines. These antibodies generally fail to react in this assay with most cell lines derived from other tumors.

A most comprehensive profile of ganglioside distribution in human melanoma tissues was provided by extensive studies of 80 melanoma specimens, including 52 surgical specimens and 28 cultured cell lines. In this case, ganglioside fractions were isolated and purified from each of these tissues and the amount of each component ganglioside was assessed by thin-layer chromatography. The most commonly expressed gangliosides on these specimens were GM3, GD3, GM2, and GD2. The distribution of each ganglioside was found to be widely heterogeneous, and the total amount of ganglioside ranged from 33 to 302 μg/g net weight tissue. GM2 and GD2 were relatively minor components of biopsied melanomas but often became major components of cultured melanoma cells. Alkali-labile gangliosides, such as 9-O-acetylated GD3, were usually more strongly expressed on biopsied melanomas than on cultured melanoma cells.

2. Functional Properties of Gangliosides

It has been proposed that glycolipids (particularly gangliosides) are involved in cell–substratum interactions. In fact, anti-GD3 Mab MB3.6 and anti-GD2 Mab 126 localized each of these gangliosides in focal adhesion plaques and on the surfaces of human melanoma cells in microprocesses. These structures emanate from the plasma membrane and, in some cases, appear to make contact with the substrate to which the cells are attached. It has also been established that GD2 and GD3 are involved in a generalized mechanism of melanoma cell attachment. Anti-GD2 as well as anti-GD3 was capable of inhibiting the attachment of human melanoma and neuroblastoma cells to a variety of immobilized extracellular matrix components including fibronectin, vitronectin, collagen, and laminin.

Pretreatment of human melanoma cells with anti-GD2 antibody resulted in 92% inhibition of attachment to a matrix of bovine endothelial cells laid down on tissue culture plastic. Ganglioside GD2 appears to be involved in the initial phases of the cell attachment process. Moreover, anti-GD2 Mabs induced cell rounding and loss of attachment by cells preadhered and spread on a fibronectin substrate. This observation was also made with anti-GD3 Mab R24 (IgG3). In contrast, Mabs directed to other determinants on the cell surface fail to alter the cells' adhesive properties, indicating that the inhibitory effects observed were not simply due to Mab bind-

ing to any antigen on the cell surface. Ultrastructural analysis of human melanoma cell attachment with both scanning and transmission immunoelectron microscopy demonstrated that GD2 gangliosides were preferentially expressed on cell membrane-associated microprocesses in direct contact with the substrate upon which the cells were attached. In fact, an actual gradient of antigen expression could be observed where ganglioside expression increased on the microprocesses as they neared the substrate. Additional evidence suggests that both GD2 and GD3 can interact directly with melanoma cell-surface glycoprotein receptors for fibronectin or vitronectin; thus, this interaction may potentiate the appropriate receptor configuration leading to optimal cell attachment.

A number of anti-GD2 and GD3 murine Mabs of either IgG3 or IgG2a isotype were shown to effectively mediate ADCC with a variety of effector cells and also mediate complement-dependent cytotoxicity with human complement. Some of these same antibodies (i.e., anti-GD3 Mabs R24 [IgG3] and ME36.1 [IgG2a]) produced regressions of melanoma tumors in phase I clinical trials, and anti-GD2 Mabs 3F8 (IgG3) and 14G2a (IgG2a) did the same in clinical trials with neuroblastoma patients. A mechanism of cellular immunity was proposed to explain this effect, because the tumor regression took many weeks and because subpopulations of peripheral T-lymphocytes expressed GD3. Such T cells could be activated *in vitro* by Mab R24 or its (Fab')$_2$ fragment to proliferate and to express IL-2 receptors. Moreover, tumor regression achieved by Mab R24 was effective only at lower doses of antibody, suggesting that larger doses kill GD3-bearing T cells by complement-mediated lysis. Another factor is that GD2 was expressed on tumor-infiltrating lymphocytes, which had been shown to be important in melanoma tumor regression when activated *in vitro* by IL-2. Whatever the mechanism(s), it is important that murine Mabs directed to gangliosides GD2 and GD3 are thus far the only ones that could achieve measurable tumor regression in melanoma and neuroblastoma patients.

There seems to be little doubt that gangliosides react more frequently than any other Mab-deficient MAAs with the immune system. In fact, human anti-GD2 antibodies have been isolated from sera of tumor-bearing patients, and human Mabs to GD2, GM2, and GM3 have been produced by lymphocytes obtained from melanoma patients. Also, after immunizations with melanoma cells, patients developed antibodies to GD2 and GM2. It is also certain that gangliosides play a potentially important role in growth regulation and differentiation of melanoma cells, similar to the stimulatory effect of gangliosides on the proliferation of astroglial cells.

D. Receptors for Growth Factors

Growth factor receptors have been found on cultured human tumor cell lines and human tumor tissues. Even though these receptors are not specific for malignant cells, their number can be orders of magnitude higher on tumor cells than on normal cells and they can mediate a mitogenic effect. Some growth factor receptors are homologous to the products of viral oncogenes (e.g., the epidermal growth factor receptor [EGF receptor] that is coded for by a protooncogene corresponding to the v-erb-B2 oncogene). Therefore, there may be a link between the abundance of these receptor molecules on tumor cells and neoplastic transformation and cell proliferation. Human malignant melanoma cells have been shown to express EGF receptors as well as receptors for nerve growth factor (NGF receptor) and for α-melanocyte-stimulating hormone (α-MSH-receptor). [*See* TISSUE REPAIR AND GROWTH FACTORS.]

EGF receptor is a transmembrane mitogenic glycoprotein of 170 kDa. Its natural ligands are EGF and TGF-α. Among Mabs to EGF receptor, some inhibit the binding of EGF and some mimic the mitogenic effect of EGF. Normal human melanocytes do not express EGF receptor, and its expression on melanoma cell lines is related to the expression of phenotypic markers of cell differentiation. Less-differentiated, unpigmented melanoma cell lines of epithelioid morphology often express EGF receptor, whereas more highly differentiated, pigmented cell lines with a dendritic morphology do not. EGF is mitogenic for freshly isolated metastatic melanoma cells, but their responsiveness tends to disappear in long-term tissue culture. In patients, the presence of EGF receptor has been correlated with malignancy, exhibiting its highest expression on advanced metastatic lesions.

NGF is of importance for regulating the development of neurons, and NGF receptor is also expressed by some tumors, such as melanomas, that are derived from the neural crest. Mabs against NGF receptor have been raised in mice using human melanoma cell lines such as A875 and WM245 as immunogens. Such antibodies made it possible to detect NGF receptor in the cytoplasm and on the plasma membrane of human melanoma cells, from where it could be immunoprecipitated as a 75-kDa glycoprotein. Normal melanocytes fail to express NGF receptors, but they are expressed on common and dysplastic nevi, as well as on primary and metastatic melanoma cells. [*See* NERVE GROWTH FACTOR.]

Human melanoma cell lines also produce a number of cytokines and growth factors, such as TGF-α and TGF-β), platelet-derived growth factor, fibroblast growth factor, melanoma growth stimulating activity, IL-1. An autocrine function has been proposed for most of these factors and is thought to influence tumor growth with positive and negative signals. However, an appropriate surface receptor for these growth factors has not yet been described on human melanoma cells.

E. HLA Class II Antigens

In contrast to HLA class I antigens, which are expressed on virtually all nucleated cells, HLA class II antigens are mainly found on lymphocytes and antigen-presenting cells. They cannot be detected on melanocytes in normal skin or in cultures. In culture, expression of HLA class II (i.e., HLA-DR, DQ, and DP), however, can readily be induced with IFN. In contrast, approximately 70% of human melanoma cell lines express HLA class II antigens, which is also expressed in surgically removed benign and malignant melanoma lesions. This expression is frequently observed in primary and metastatic lesions. Whether this expression reflects the stage of differentiation of the melanocytes from which these tumor cells have originated or is induced by the malignant process is unclear. The level of expression of HLA class II antigens in primary and metastatic melanoma lesions was suggested to correlate with the clinical course of the disease.

III. Antigen Expression during Melanoma Tumor Progression

A major objective of tumor immunology has been to identify "tumor-associated" antigens, which might be expressed only on malignant tumors or during particular stages of tumor progression. Such antigens would be useful for tumor diagnosis, and the identification of their biochemical nature could elu-

TABLE I Antigen Expression and Shedding during Melanoma Progression

Antigen	Expression[a]				Shedding[b]			
	Melanocyte	Nevus	Primary melanoma in VGP	Metastatic melanoma	Melanocyte	Nevus	Primary melanoma in VGP	Metastatic melanoma
p97	++	++	+++	++++	−	−	+/−	+/−
Proteoglycan	+	++	+++	++++	−	−	+/−	+
GD3	+/−	+/−	++	++++	+	++	+++	++++
9-O-acetyl GD3	+/−	+/−	+	++	+	+	++	+++
GD2	−	−	+++	++++	−	−	++	+++
HLA-DR	−	+/−	++	++	−	−	+/−	+
EGF-R	−	+/−	++	++				
NGF-R	+	++	+++	+++				
ICAM-1	−	−	++	+++				
N-CAM	−	−	+	++				

Data combined from M. Herlyn *et al.,* 1987, *Lab. Inv.* **56,** 461; D. E. Elder *et al.,* 1987, *Cancer Res.* **49,** 5091; and B. Holzmann, *et al.,* 1987, *Int. J. Cancer* **39,** 466.

[a] Expression detected with monoclonal antibodies by indirect immunofluorescence, mixed hemabsorption assay, or immunocytochemistry.

[b] Shedding determined after incubation of cells in serum-free media and testing supernatants for binding to monoclonal antibodies in solid phase radioimmunoassay.

cidate mechanisms of tumor progression or be attributed to a neoplastic phenotype.

In Table I, the expression of MAA discussed above is summarized with regard to melanoma tumor progression. It is evident that none of the various Mabs against MAA is exclusively reactive with cells from advanced malignant lesions. Most antigens can be detected on at least a few melanocytic cells in benign lesions or on normal melanocytes. Among these are the p97 antigen, CSP, and the ganglioside GD3. Although these antigens are not restricted to malignant cells, their expression increases during the malignant process. The literature regarding the expression of these MAAs is somewhat inconsistent. This may be due to the use of Mabs that differ in affinity and avidity, to technical differences, or to a variation in antigen expression in the patient versus cultures.

Mabs to other MAAs discriminate benign from malignant neoplastic melanocytes. The ganglioside antigen GD2, HLA-DR, and EGF receptor, and the adhesion molecules ICAM-1 and N-CAM are found in only a low percentage of nevi and primary melanoma <0.75 mm thick. They appear to be mainly restricted to the two latest stages of melanoma progression, VGP primary melanoma, and metastatic melanoma. Differences in antigen expression during tumor progression appear to be quantitative rather than qualitative.

As also shown in Table I, some MAA are shed from the tumor cells. Shedding increases with the amount of antigen expressed on the cell surface and varies for different antigens. Thus, MAA proteoglycan and the ganglioside antigens are shed to a larger extent than melanotransferrin or EGF receptor. This observation may have some significance for tumor diagnosis using MAAs as tumor markers or as probes for tumor imaging, which have opposite requirements (i.e., either antigen shed into the bloodstream or preferentially not shed to a large extent). Also, for tumor therapy, Mabs directed against antigens that are not shed may be preferred to obtain better tumor targeting. The practical usefulness of Mabs against MAAs is determined by the expression of these antigens in tumor progression. Whereas antigens expressed strongly on many or all tumor cells in the advanced stages of melanoma progression should be considered for antibody-directed tumor therapy, expression of ICAM-1 and HLA-DR may be used as a prognostic marker in primary lesions because it correlates with early occurrence of metastatic disease.

In summary, MAAs and the Mabs directed against them have provided thus far a basis for an extensive investigation of molecular and cellular mechanisms of tumor invasion and progression that will likely continue well into the next decade. Hopefully, this approach will yield basic information that will lead to advances in diagnosis and therapy of cancer.

Bibliography

Cheresh, D. A., and Spiro, R. C. (1987). Biosynthetic and functional properties of an Arg-Gly-Asp-directed receptor involved in human melanoma cell attachment to vitronectin, fibrinogen, and von Willebrand factor. *J. Biol. Chem.* **262,** 17703.

Graf, Jr., L. H., and Ferrone, S. (1989). Human melanoma-associated antigens. *In* "Human Immunogenetics" (S. D. Litwin, ed.). Marcel Dekker, Inc., New York.

Hellstrom, K. E., and Hellstrom, I. (1989). Immunological approaches to tumor therapy: Monoclonal antibodies, tumor vaccines, and anti-idiotypes. *In* "Covalently Modified Antigens and Antibodies in Diagnosis and Therapy" (Quash and Rodwell, eds.). Marcel Dekker, Inc., New York.

Herlyn, M., Clark, W. H., Rodeck, U., Mancianti, M. L., Jambrosic, J., and Koprowski, H. (1987). Biology of tumor progression in human melanocytes. *Lab. Inv.* **56,** 461.

Houghton, A. N., Cordon-Cardo, C., and Eisinger, M. (1986). Differentiation antigens of melanoma and melanocytes. *Int. Rev. Exp. Pathol.* **28,** 217.

Johnson, J. P., Stade, B. G., Holzmann, B., Schwable, W., and Riethmuller, G. (1989). De novo expression of intercellular-adhesion molecule 1 in melanoma correlates with increased risk of metastasis. *Proc. Natl. Acad. Sci. USA* **86,** 641.

Johnson, J. P., Stade, B. G., Rothbacher, U., Lehmann, J. M., and Riethmuller, G. (1990). Melanoma antigens and their association with malignant transformation and tumor progression. *In* "Human Melanoma" (S. Ferrone, ed.). Springer, Berlin.

Kligman, D., and Hilt, D. C. (1988). The S-100 protein family. *Trends Biochem. Sci.* **11,** 437.

Natali, P., Bigotti, A., Cavaliere, R., Ruiter, D. J., and Ferrone, S. (1987). HLA class II antigens synthesized by melanoma cells. *Cancer Rev.* **9,** 1.

Reisfeld, R. A., and Cheresh, D. A. (1987). Human tumor antigens. *In* "Advances in Immunology" (F. J. Dixon, ed.). Academic Press, New York.

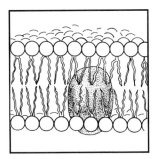

Membranes, Biological

THEODORE L. STECK, *The University of Chicago*

I. Membrane Lipids
II. Membrane Proteins
III. Membrane Transport
IV. Membrane Transduction
V. Membrane Connections
VI. Membrane Biogenesis

Glossary

Amphipathy Tendency of molecules containing substantial polar and nonpolar centers to seek two environments and hence to orient at interfaces

Bilayer Bimolecular leaflet of amphipathic lipids that confers continuity on biological membranes

***Cis* configuration** Alkyl chain (e.g., a fatty acid) that is bent or kinked by an unsaturation (a double bond between carbon atoms)

Eukaryotes Life forms characterized by complex cells with a true (membrane-enclosed) nucleus

Fusion When two vesicles unite by coalescence at the point of membrane contact. The reverse process is called *fission*

Gradient Difference in the concentration of a solute between two points, which drives its flow, as illustrated by the movement of ethanol across the plasma membrane into cells

Hydrophobic effect Thermodynamic mechanism that drives nonpolar moieties from the aqueous phase, resulting from an unfavorable balance of the enthalpy and entropy of hydration

Kinase Enzyme that transfers the terminal phosphate group of adenosine triphosphate (ATP) to an acceptor molecule. For example, protein kinases phosphorylate specific serine, threonine, or tyrosine amino acid residues in proteins

Lysosome Digestive organelle of eukaryotic cells

Monosaccharide Carbohydrate containing one sugar group

Oligosaccaride Carbohydrate containing a few sugar groups

Oxidative phosphorylation Trapping of the energy of cellular respiration (i.e., redox reactions) in the synthesis of ATP by a specific assembly of membrane-spanning proteins

pH "Power of hydrogen": $pH = -\log_{10}[H^+]$. The greater the acidity, the lower the pH

Photosynthetic organisms Those that can trap the energy absorbed from photons of light in the synthesis of ATP and other energy-rich compounds by a specific assembly of membrane-spanning proteins

Polar group Part of a complex molecule that possesses a higher affinity for electrons than its neighbors (e.g., the oxygen-containing carboxyl in a fatty acid)

Prokaryotes Life forms characterized by simple cells without a true (membrane-enclosed) nucleus

Protons Hydrogen cations

Redox reaction Chemical process in which electrons pass from a compound of low electron affinity (the reductant) to one of high electron affinity (the oxidant)

Saturated fatty acids Those with no double bond between carbon atoms

Transduction Conversion of stimuli, signals, or information received at the cell surface to a form readable by the cytoplasm

Unsaturated fatty acids Those with one or more double bonds between carbon atoms

MEMBRANES are the structures by which cells and organelles are demarcated into discrete aqueous compartments. Biological membranes also perform numerous specific tasks for the cell; for example, they provide the mechanisms by which materials and information are conveyed from one compartment to another. The molecular organization of biological membranes can be understood in terms of the physical and chemical properties of their constituent lipid and protein molecules. Similarly, membrane functions follow general rules best illustrated by particularly favorable systems.

I. Membrane Lipids

Biological membranes are composed primarily of lipids, proteins, and carbohydrates—the latter attached to lipids to form glycolipids, or to proteins to form glycoproteins. That the lipids provide the basis for membrane structure is demonstrated simply by dispersing purified phospholipids in water; they spontaneously assemble into *bilayers*, sheets two molecules thick, which resemble natural membranes to a striking degree. Membrane assembly is therefore said to be driven by a thermodynamic mechanism: the minimization of free energy.

A. Lipid Classes

The three major classes of membrane lipids are glycerolipids, sphingolipids, and sterols. *Glycerolipids* are the most universal of membrane constituents. Esterified to the first and second carbon atoms of the glycerol backbone are fatty acids. These have chain lengths in the range of 14 to 24 methylene groups, the most common being 16 or 18. The fatty acid linked to the first carbon atom of the glycerol is usually saturated. The second fatty acid is usually unsaturated with one to three or more double bonds in the *cis* configuration. The third carbon atom on the glycerol backbone bears a polar group, typically, phosphate or an oligosaccharide. The phosphate group is usually also esterified to a polar alcohol such as choline, ethanolamine, serine, or inositol.

Sphingolipids resemble and often substitute for glycerolipids. However, they are built on sphingosine, an amine-alcohol that serves both as a backbone and as the bearer of one of the two long aliphatic chains of these lipids. The other alkyl chain derives from a fatty acid linked through an amide bond to the primary amino group. As with the glycerolipids, the polar head groups of sphingolipids are either oligosaccharides or phosphodiesters (e.g., phosphorylcholine in sphingomyelin).

Sterols, like cholesterol, are polycyclic hydrocarbons, devoid of polar centers except for the oxygen atom in the hydroxyl group at carbon three. They are designed by nature to intercalate into phospholipid membranes.

Membrane lipid composition varies characteristically among the phyla for reasons we do not understand. Prokaryotes generally use glycerolipids containing fatty acids with no more than one unsaturation. Gram-positive bacteria and plants are rich in glycoglycerolipids, whereas animals use glycosphingolipids. The most common polar head group in animals, phosphorylcholine, is absent in prokaryotes. Sterols are ubiquitous in eukaryotes but rare in prokaryotes; they, like sphingolipids, are specifically enriched in plasma membranes. A complex membrane lipid, called *diphosphatidylglycerol* or *cardiolipin*, is characteristic of gram-negative bacteria, mitochondrial inner membranes, and the thylakoid membranes of chloroplasts, perhaps reflecting an evolutionary relation. [*See* LIPIDS.]

B. Bilayer Membranes

Membrane lipids share a special feature: they are *amphipaths*. Although their polar moieties seek association with water, their nonpolar moieties are driven from the aqueous phase by the *hydrophobic effect* to form a nonaqueous phase of their own. Membrane lipids are shaped so that their aggregates take the form of two apposed monomolecular layers associated through the ends of their buried apolar chains. These planar bimolecular leaflets, called *bilayers*, are 4–5 nm thick.

The hydrophobic edges of planar bilayers are not stable as long as they contact water and will add lipid indefinitely. This thermodynamic restlessness is resolved when the bilayer closes into a spherical shell, generically, a vesicle. It is this configuration—the closed sac—in which all biological membranes are found and which enables them to create sealed cellular compartments of any size and shape.

C. The Fluid Bilayer

Lipid bilayers pass from a solid state (gel or crystalline) to a fluid state (liquid crystalline) at a critical temperature. Such behavior reflects the cooperative disordering of the assembled fatty acyl chains from a relatively close-packed, linear configuration to a more expanded array of thermally disordered chains. Cells adjust the melting point of their membranes biosynthetically. They can introduce phospholipids with long-chain fatty acids and a melting temperature greater than 50°C or fatty acids with multiple unsaturations and a melting temperature below 0°C. (*Cis* double bonds thwart close-packing by kinking alkyl chains acutely.)

When microbes are shifted to a new growth temperature, they make new membrane lipids that melt near that temperature. Their goal cannot be merely to maintain the membrane in a fluid state, for the organism could assure this simply by synthesizing at all times lipids with short-chain, polyunsaturated tails. Instead, they maintain the bilayer as a dynamic mosaic of solid and fluid domains. What purpose this serves we do not know.

The *viscosity* (the inverse of *fluidity*) in fluid bilayers is commonly a few hundred times that of water. The simple bilayer thus has a resistance to flow comparable to that of a vegetable oil. In such a medium, a phospholipid molecule can diffuse laterally at a rate of about 1 μm/sec, taking just a few seconds to migrate around an organelle or a bacterial cell.

D. Transverse Asymmetry

Although the lateral motion of polar membrane lipids (i.e., along the plane of the bilayer) is generally unrestricted, diffusion across the plane of the bilayer is exceedingly slow; typically, the half-time is measured in hours or days. Presumably, the hydration of the polar head groups of phospholipids and glycolipids impedes their entry into the core of the bilayer. In contrast, sterols, which lack a large polar head group, diffuse across the membrane on the time sale of seconds or less.

Given the low rate of transverse diffusion in bilayers, it is not surprising that naturally occurring membranes maintain a high degree of lipid asymmetry. In the human erythrocyte, the outer leaflet is rich in neutral lipids bearing phosphocholine and sugars; the inner leaflet is rich in anionic phospholipids bearing phosphate esters of ethanolamine, serine, glycerol, and inositol. Transverse asymmetry is apparently generated during the original assembly of the bilayer, after which the redistribution of the lipids is slow. In addition, there seems to be an ATP-using enzyme in plasma membranes that catalyzes the pumping of negatively charged phospholipids from the outer to the cytoplasmic leaflet.

E. Membrane Fusion and Fission

The lipid bilayers of membrane vesicles, including cells, can merge by *fusion* or divide by *fission* without the loss of either membrane integrity or vesicle contents. Fusion allows secretory vesicles in the cytoplasm to discharge their contents (e.g., peptide hormones, neurotransmitters, serum lipoproteins, and antibodies) through the plasma membrane. Fission underlies phagocytosis, pinocytosis, the budding of enveloped viruses, and cell division.

II. Membrane Proteins

Proteins carry out most membrane activities (e.g., solute transport, information transduction, and energy conversion). Each membrane and each function has its own specific proteins; no membrane protein is universal. Some membranes of limited function contain less than 25% protein by mass; myelin, the membranous insulator wrapped around nerve axons, is an example. Busy membranes, such as the mitochondrial inner membrane, are comprised of 50–75% protein by weight. Myelin and viral envelopes have only a few types of proteins; the inner mitochondrial membrane and that of *Escherichia coli* have dozens or even hundreds.

A. Associations of Proteins with Membranes

Proteins are associated with membranes in three general ways. (1) *Integral* or *intrinsic* proteins are dissolved in the core of the bilayer by means of their hydrophobic segment. (2) Several kinds of proteins are covalently bound to membrane lipids. Some of these proteins are also integrated into the bilayer, but others are simply tethered through the lipid link. (3) *Peripheral* or *extrinsic* proteins bind stereochemically (i.e., with a defined stoichiometry and affinity) to specific membrane sites such as integral proteins.

There are simple functional tests to determine the mode of association of a protein with a membrane. A protein solubilized from the membrane without disrupting the bilayer is probably peripheral. Integral and lipid-linked proteins are released by solubilizing the bilayer. For example, nonionic *detergents* and *bile salts* carry away the membrane lipids as soluble mixed *micelles* and take their place at the hydrophobic contact surface of the solubilized proteins, usually without altering native protein conformation. Lipid-linked proteins may sometimes be solubilized by the action of the appropriate lipolytic enzyme.

B. Composition of Membrane Proteins

Membrane proteins contain the same amino acids as water-soluble proteins; membrane glycoproteins are conjugated with the same oligosaccharides as are soluble glycoproteins. Most membrane proteins tend to be acidic in charge. (The mutual repulsion of their like charges may forestall their aggregation, which might otherwise occur at the high concentrations found within membranes.) Peripheral proteins usually resemble water-soluble proteins in amino acid composition. Integral proteins may be enriched in hydrophobic residues if their contact surface with the core of the bilayer is extensive.

C. Disposition of Membrane Proteins

Peripheral proteins may be globular or filamentous, monomeric or polymeric, just like their nonmembrane counterparts. Integral proteins are amphipathic, making contact with water and lipid polar head groups as well as with apolar fatty acyl chains. Some integral proteins span the membrane, thereby communicating with three compartments. The oligosaccharides of glycoproteins are almost always found on the membrane surface opposite the cytoplasm; in phosphoproteins, the phosphate groups are usually confined to the cytoplasmic aspect of the protein. Integral proteins may be anchored in the bilayer by a single strand of amino acids spanning the membrane, a patch of hydrophobic residues, or an extensive hydrophobic surface. In some cases, the bulk of the protein is submerged in the lipid.

The amino acid sequences of integral proteins often provide insights into their spatial organization. Water-soluble and membrane-intercalated domains may lie in separate segments of the primary structure of the protein. These domains may fold independently and may be cleaved apart experimentally by limited proteolysis if they are joined by an exposed hinge region.

A common feature of the primary structure of integral proteins are sequences approximately 25 residues in length composed almost entirely of nonpolar residues. If alpha-helical in secondary structure, these hydrophobic segments would be approx-

imately 4 nm in length, just long enough to span the apolar core of the bilayer. The rarity of hydrophobic stretches longer than 30 residues suggests that such segments are essentially linear and perpendicular to the plane of the membrane; more complex intramembrane organization is as yet virtually unknown. When several such hydrophobic segments occur in tandem, they are typically separated by two to 10 or more polar residues and lace back and forth across the bilayer in an antiparallel bundle.

Integral membrane proteins are frequently oligomeric: dimers and tetramers compounded of either identical, similar, or dissimilar subunits. Their axis of symmetry typically lies normal to the plane of the membrane. Some integral membrane proteins are tightly packed in patches containing dozens or hundreds of units in hexagonal array. (Two examples are *gap junctions* (see next section) in vertebrate cells and the purple patches of *bacterial rhodopsin*.) Integral proteins often form heterocomplexes with peripheral proteins at the membrane surface.

Integral membrane proteins may interact specifically with their lipid environment, selectively binding or shunning various lipids. Their hydrophobic contact surfaces may perturb bilayer organization and lipid motion. The activity of transport proteins (discussed below) and enzymes embedded in the membrane often responds profoundly to the neighboring lipid head groups and/or fatty acyl chains. The activity of some integral proteins may also be sensitive to the state of fluidity of the bilayer and even to bilayer thickness, which presumably should just match the span of the hydrophobic contact surface of the protein.

Some integral proteins display a diffusional mobility commensurate with the fluidity of the lipids in which they are dissolved. In many cases, however, lateral motion is constrained by protein–protein associations (e.g., the formation of large patches or the tethering of integral proteins to cytoplasmic filaments). Integral proteins may also be sequestered within the interstices of submembrane protein networks or may be kept in one portion of a cell surface by proteinaceous barriers (e.g., the tight junctions linking epithelial cells).

Finally, membrane proteins experience an overwhelming energy barrier to translocation across the plane of the bilayer, even when the protein spans the membrane. This thermodynamic mechanism maintains the absolute asymmetry or "sidedness" of the proteins at two faces of a biological membrane, every copy of a given protein having the same orientation.

III. Membrane Transport

The movement of solutes (substances dissolved in water) across membranes is driven by two general kinds of energy. First, solutes may move from a distribution of high transmembrane potential energy to one of lower transmembrane potential energy. This potential energy gradient derives from both the difference in solute concentration and, if the solute is charged, from the electrical field across the membrane; it is therefore referred to as an *electrochemical gradient*. For example, in the case of the potassium ion, either a solute concentration difference of 10-fold or a membrane potential of 59 mV would impose a driving force of approximately 1.4 kcal per mole of solute. [*See* CELL MEMBRANE TRANSPORT.]

The second source of energy is one external to the membrane. For example, "active transport" may be fueled directly by adenosine triphosphate (ATP), respiration, or light. In such cases, the dissipation of the external energy is *coupled* to solute movement in that the transport protein can catalyze both processes only through a single conformational cycle. [*See* ADENOSINE TRIPHOSPHATE (ATP).]

A. Simple Diffusion in Bilayers

A primordial form of transport is the unaided diffusion of lipid-soluble molecules through the bilayer. This is the way we get our oxygen and our alcohol. Lipoidal hormones (e.g., steroids and prostaglandins) can also take this route.

B. Transmembrane Channels

Some cells have protein pores that create aqueous continuity between the two water compartments adjoining their plasma membrane, permitting the transmembrane diffusion of polar solutes. Because there is no barrier other than sieving, this is the most rapid mode of transport known: some channels conduct more than 10^7 ions per second through a pore less than 1 nm wide.

The outer envelopes of gram-negative bacteria are perforated by multiple channels that allow passage of polar solutes less than 700 daltons in molecular weight. These channels are created by the asso-

ciation of three 58-residue amphipathic helical peptides (called *porins*) to form a hydrophobic coil with a central polar pore 1.3 nm in diameter. The outer membranes of mitochondria have comparable structures performing a similar function: the rapid and nonselective movement of microsolutes.

Many tissue cells are coupled by *gap junctions*. These are patches in the apposed plasma membranes of adjacent cells that contain multiple copies of a pore protein. These proteins create aqueous channels about 1.2 nm in diameter between the adjoining cytoplasmic compartments. Cells coupled in this way share their metabolites and cytoplasmic signal molecules. [*See* CELL JUNCTIONS.]

Channels make the plasma membranes of nerves and muscles excitable. Neurotransmitters like acetylcholine bind to integral receptor proteins of postsynaptic membranes. The binding event presumably drives a conformational change in the receptor, rapidly opening its small cation-selective pore. These receptors are called *chemically gated channels*, because the opening or closing of the pore is chemically controlled. Ions immediately flow across the membrane down their concentration gradients. Within a few milliseconds, a second, slower conformational change inactivates the pore. In this way, the conductance of the channel is rapidly terminated, sparing the cell the deleterious effects of prolonged stimulation. Later, the acetylcholine dissociates from the receptor, and the channel regains its resting conformation and excitability.

Scattered through the outlying plasma membrane of these excitable cells are other channels selectively permeable to Na^+ or K^+. They are *electrically gated*; they open by undergoing a conformational change in response to the change in membrane electrical potential, which was initially triggered by the flow of ions through the chemically gated channels at the synapse. Because these voltage-gated channels add to the change in membrane electrical potential that opened them, they cause the opening of similar nearby channels, propagating a wave of electrical activity (the *action potential*) along the length of a nerve or muscle cell membrane. [*See* ION PUMPS.]

C. Reciprocating Transporters

Several kinds of transport proteins bind solutes and move them across the membrane; they must then complete the cycle by returning through a reciprocal step. Presumably, a cyclic change in protein conformation is required. If the transporter is unable to change conformation without transferring a solute molecule, transport will be strictly one-for-one, and the transporter is called an *exchanger*. This is essentially the case for the anion transporter in the erythrocyte, which rapidly and randomly swaps bicarbonate and chloride ions to facilitate the movement of carbon dioxide from the tissues to the lungs. A second example is the sodium/proton exchanger, which adjusts the pH of the cytoplasm in many cells. In contrast to these one-for-one systems, reciprocating transporters for solutes such as glucose can cycle either with their binding site filled or empty. They can therefore mediate the net transfer of nutrients down their concentration gradients by crossing filled and returning empty.

D. Coupled Transport

Some reciprocating transporters have separate binding sites for two different solutes (e.g., Na^+ and glucose). Because the transporter cannot cycle unless both solute species are bound simultaneously, the transport of the two solutes must be coupled. Therefore, the solute with the higher potential energy gradient will drive the other "uphill" until the two gradient energies balance. In the plasma membranes of kidney and intestinal epithelium, for example, the movement of Na^+ down its gradient drives the uptake of glucose against its gradient, concentrating it in the cells. Microbes use proton gradients to assimilate nutrients in a similar way.

Depending on the system, two coupled solutes may be required to bind on the same or on opposite sides of the membrane for their transporter to cycle. In the case of the Na^+/glucose transporter mentioned above, the two solutes must come from the same side. In another system, Na^+ gradients drive the movement of Ca^{2+} from the opposite side of the plasma membrane; in yet another, the efflux of H^+ from secretory vesicles causes neurotransmitter to be concentrated inside.

E. Transport ATPases

Cation pumps, powered by the hydrolysis of ATP, and therefore called ATPases, transport Na^+, K^+, H^+, and Ca^{2+} across plasma membranes. Some of these pumps trade one cation for another and therefore do not change the electrical potential of the membrane; the H^+/K^+ ATPase, which acidifies the stomach, is an example. The Na^+/K^+ ATPase, how-

ever, expels 3 Na^+ from the cell for every 2 K^+ it takes up and is therefore said to be *electrogenic*.

The mechanism by which these transporters function also appears to involve a reciprocating conformational cycle. However, in this case, the cycle is tightly coupled to the hydrolysis of ATP. As a result, ATP is cleaved as the appropriate cations, bound to their transporter sites, cross the membrane in the mandated direction. In this way, the free energy of ATP hydrolysis is converted into concentration gradient potentials.

F. Redox-Driven Transport

Many chemical reactions can be coupled to the movement of solutes across membranes against their electrochemical gradient by appropriately disposed membrane proteins. For example, the free energy of redox reactions is used by mitochondria to transport electrons across the membrane; this generates an electrical gradient, which, in turn, creates a transmembrane proton gradient. The proton gradient is as universal an energy currency as is ATP; it can be tapped to power many other processes, including the synthesis of ATP in *oxidative phosphorylation*. Similarly, a primary consequence of the absorption of light by photosynthetic organisms is the generation of membrane electrical and proton gradients that are subsequently harvested in the form of energy-storing chemical compounds. [*See* ATP SYNTHESIS BY OXIDATIVE PHOSPHORYLATION IN MITOCHONDRIA.]

IV. Membrane Transduction

Just as plasma membranes mediate the selective passage of solutes, so must they also transmit information from outside the cell to the cytoplasm.

A. Stimuli and Receptors

Unlike steroid hormones, which dissolve in and diffuse through membranes en route to their intracellular receptors, peptide hormones, neurotransmitters, chemoattractants, antigens, and growth factors act at the outer surface of the plasma membrane. Specific to each of these *stimuli* is a *receptor*: an integral (glyco)protein that asymmetrically spans the plasma membrane. Transduction occurs when a receptor binds its *ligand* (i.e., the stimulus molecule) at the extracellular surface and manifests a change at the cytoplasmic surface, as discussed below.

B. Mechanisms of Signal Transduction

We can describe three general mechanisms by which the binding of a chemical stimulus at the cell surface is communicated to the cytoplasm. First, there is the direct alteration of membrane permeability by receptors that are also chemically gated channels. The resulting flow of cations can change the membrane potential or the level of an intracellular messenger such as Ca^{2+}, generating a cytoplasmic signal.

Other stimuli act by inducing the clustering of their mobile receptors in the fluid membrane. The resulting multimeric receptor–ligand complex triggers a cytoplasmic response, perhaps through a change in membrane cation permeability or, as in the case of insulin and epidermal growth factor, by activating a protein kinase built into the cytoplasmic domain of the receptor. The phosphorylation of specific tyrosine residues of target proteins in the cytoplasm constitutes the signal.

In the third category, ligand binding to the receptor imposes a transmembrane conformational change, which activates a specific cytoplasmic cascade. Thus, the receptor for the hormone epinephrine, when unfilled, holds inactive at its cytoplasmic domain a detachable protein complexed with guanosine diphosphate (GDP). Hormone binding causes the dissociation of this so-called *G protein*, whereupon it releases GDP and takes up guanosine triphosphate (GTP). It is this form of the G protein that stimulates adenylate cyclase to synthesize 3',5'-cyclic adenosine monophosphate (cyclic AMP), a *second messenger* that regulates multiple cellular activities. Stimulation is terminated when the G protein cleaves its own GTP back to GDP. The inactive G protein then returns to its receptor, completing the cycle. Several types of regulatory G proteins are now known; there is one in the retina, for example, that transduces light into a chemical message. [*See* G PROTEINS; SIGNAL TRANSDUCTION AT THE CELL SURFACE.]

C. Receptor Regulation

The level of activity of membrane receptors is physiologically regulated in several ways. Some receptors, such as that for acetylcholine, automatically shut off soon after activation, even though the stimulus is still bound. Because the release of a highly avid ligand may be slow, this mechanism spares the cell from enduring prolonged responses to transient stimuli. Sensitivity is regained after the ligand departs. Some receptors are temporarily inactivated (*desensitized*) in response to regulatory signals through the phosphorylation of their cytoplasmic domain. A slower mode of regulation uses the internalization of receptors by *endocytosis* (the pinching off of vesicles from the plasma membrane into the cytoplasm). Internalized receptors later return to the cell surface by *exocytosis*. This cycle controls the fraction of total receptors positioned at the cell surface. Ligands internalized on endocytosed receptors may be returned to the cell surface or may be disposed of by selective transfer to lysosomes, whereas free receptors are returned to the cell surface. Receptors enjoy dozens of circuits before they too are degraded in the lysosomes.

V. Membrane Connections

Membranes and organelles are typically coupled to filamentous matrices both within the cell and without. The function of these meshworks is generally mechanical: to strengthen or protect the thin and fragile membranes (i.e., to bear and distribute tension); to determine membrane contour; and to direct cell and organelle movement.

A. The Cytoskeleton

When eukaryotic cells are treated with a nondenaturing detergent such as Triton X-100, their bilayers are dissolved, and their water-soluble contents are washed away. What remains are the water-insoluble, filamentous matrices of the extracellular, cytoplasmic, and nuclear compartments. The filaments constituting the *cytoskeleton* are polymers of actin, tubulin, and intermediate filament proteins, bound to each other and to various membranes by a battery of accessory proteins. The filaments of actin and tubulin mediate the change of position of cells and organelles.

B. The Extracellular Matrix

Living cells of all kinds are imbedded in various supporting matrices that make specific associations with the cell surface. *Fibronectin*, for example, is a large filamentous glycoprotein, which binds both to special receptors on tissue cell surfaces and to intercellular fibrils of collagen, fibrin, and proteoglycans. These associations couple the cell to its environment. Fibronectin filaments are often gathered on the cell surface into patches that parallel arrays of actin on the cytoplasmic side of the plasma membrane. Extracellular and intracellular matrices are thus linked and coordinated by such *focal adhesions*. [*See* EXTRACELLULAR MATRIX.]

VI. Membrane Biogenesis

A. Proteins

Membranes are elaborated by the assimilation of new molecular components into preexisting structures that act both as templates and catalysts for their own assembly. Consider first the integral proteins. Their polypeptide chains are synthesized by ribosomes in the cytoplasm until a sequence of approximately 25 nonpolar residues emerges. This hydrophobic signal becomes inserted across the bilayer of the rough endoplasmic reticulum membrane by a special recognition mechanism. As further synthesis proceeds, the nascent protein passes into and through this membrane, in some cases crossing the membrane several times. A variety of maturational events occurs even as the synthesis of the polypeptide is proceeding: the amino-terminal hydrophobic signal sequence is often cleaved from the nascent protein at the lumenal surface of the rough endoplasmic reticulum by a *signal peptidase*; the protein remains anchored to the membrane by a hydrophobic *anchor sequence*. The membrane-bound polypeptide becomes folded into its proper tertiary and quaternary (oligomeric) structure, often within the membrane; certain pairs of cysteine residues are oxidized to create disulfide bonds, and glycosylation at asparagine residues is effected. The newly completed integral proteins are thought to diffuse along the endoplasmic reticulum, ultimately congregating in small membrane-bound transport vesicles that carry them to the Golgi apparatus. A similar process conveys the appropriate proteins along the Golgi stack where further maturation occurs, with reshuffling and extension of glycosylation. Finally, the newly fashioned proteins are carried by small vesicles to their final destination, in the membranes of the cell surface, the lysosomes, and other organelles.

Secreted proteins begin in much the same way as the integral proteins, except that the entire polypeptide chain is threaded through the membrane of the rough endoplasmic reticulum and is deposited in its lumen in a water-soluble form. Secretory proteins, such as peptide hormones or immunoglobulins, are packaged inside transport vesicles for release from the cell by fusion with the plasma membrane.

The routing of newly synthesized proteins from the lumen of the endoplasmic reticulum to the lumen of the appropriate organelle is determined by specific signals present in the proteins. For example, lysosomal hydrolases carry as a temporary signal phosphate groups bound to terminal mannose residues on their oligosaccharides. Binding proteins for this unusual *mannose-6-phosphate* moiety are found within the Golgi apparatus and on other membranes of the cell; presumably these help to guide the hydrolases to the lysosomes. Once there, the signal phosphate groups are excised, and the enzymes are retained.

Other membrane proteins are synthesized entirely on ribosomes free in the cytoplasm rather than being bound to the endoplasmic reticulum. Some of these bind to specific sites on the cytoplasmic surface of cellular membranes as peripheral proteins. Others integrate into the bilayers of organelle membranes. Still others are imported to the interior of peroxisomes, chloroplasts, and mitochondria. Indeed, even though the latter two organelles have their own genes and ribosomes, 90% of their protein is coded for by nuclear genes and is synthesized on ribosomes free in the cytoplasm.

B. Lipids

Most membrane lipids are initiated if not entirely synthesized in the endoplasmic reticulum and are then delivered to outlying membranes. The pathway of transport of membrane lipids to their destination is largely uncharted but could entail, as is the case with integral membrane proteins, diffusion along the endoplasmic reticulum, packaging in small transport vesicles, and delivery through fusion of these vesicles with a succession of target membranes. There are also soluble lipid binding proteins in the cytoplasm that could serve as carriers. Membrane lipids are frequently turned over (broken down and resynthesized) or otherwise modified or completed at distal sites. Mitochondria and chloroplasts participate in the synthesis of the phospholipids of other organelles; they also have the capacity to synthesize some of their own phospholipids, diphosphatidylglycerol (cardiolipin) in particular.

C. Complex Carbohydrates

Oligosaccharides are added to membrane glycoproteins and glycolipids in two ways. Monosaccharide residues may be transferred stepwise from sugar nucleotides to growing oligosaccharide chains by specific glycosyltransferases in the endoplasmic reticulum and Golgi apparatus. Alternatively, oligosaccharide cores are presynthesized on dolichol phosphate carriers in the endoplasmic reticulum and then are transferred to designated asparagine residues on nascent proteins as they emerge into the lumen of the rough endoplasmic reticulum. The trimming and extension of these oligosaccharides accompany, and perhaps guide, the transit of these glycoproteins through the endoplasmic reticulum, Golgi apparatus, and transport vesicles to their various destinations.

Bibliography

Alberts, B., Bray, D., Lewis, J., Raff, M., Roberts, K., and Watson, J. D. (1989). "Molecular Biology of the Cell." Garland Publishing, New York.

Bishop, W. R., and Bell, R. M. (1988). Assembly of phospholipids into cellular membranes: Biosynthesis, transmembrane movement, and intracellular translocation. *Annu. Rev. Cell Biol.* **4**, 580–610.

Burridge, K., Fath, K., Kelly, T., Nuckolls, G., and Turner, C. (1988). Focal adhesions: Transmembrane junctions between the extracellular matrix and the cytoskeleton. *Annu. Rev. Cell Biol.* **4**, 487–525.

Catterall, W. A. (1988). Structure and function of voltage-sensitive ion channels. *Science* **242**, 50–61.

Gennis, R. B. (1989). "Biomembranes: Molecular Structure and Function." Springer-Verlag, New York.

Gilman, A. G. (1987). G proteins: transducers of receptor-generated signals. *Annu. Rev. Biochem.* **56**, 615–649.

Harold, F. M. (1986). "The Vital Force: A Study of Bioenergetics." W. H. Freeman, New York.

Jain, M. K. (1988). "Introduction to Biological Membranes," 2nd ed. Wiley, New York.

Verner, K., and Schatz, G. (1988). Protein translocation across membranes. *Science* **241**, 1307–1313.

Yeagle, P. (1987). "The Membranes of Cells." Academic Press, New York.

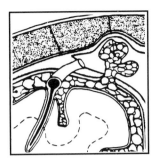

Meninges

J. E. BRUNI, *The University of Manitoba*

Glossary

Arachnoid mater   Middle of the three meninges enclosing the brain and spinal cord; it is a thin, translucent membrane closely adherent to the dura mater; trabeculae extend from its inner surface across the subarachnoid space to the pia

Arachnoid villi   Microscopic tufts of pia-arachnoid mater that pierce the dura and open into venous sinuses, lacunae, or extradural veins; they permit unidirectional flow of cerebrospinal fluid from the subarachnoid space to the venous system

Calvaria   Dome of the skull formed by the superior parts of the frontal, parietal, and occipital bones that roofs over and protects the brain.

Dura mater   Tough, fibrous, outermost covering of the brain and spinal cord

Dural septae   Folds of dura mater that reflect off the surface of the skull and incompletely divide the cranial cavity into compartments

Leptomeninges   Term used for the two delicate inner membranes (pia and arachnoid mater) that invest the brain and spinal cord

Meningiomas   Benign brain tumors of arachnoid origin

Pachymeninx   Term used synonymously with dura mater

Pia mater   Delicate vascular membrane that is closely adherent to the surface of the brain and spinal cord; it is the innermost of the three meningeal investments

Subarachnoid cisternae   Enlarged subarachnoid spaces between the pia and arachnoid mater filled with cerebrospinal fluid

Venous sinuses   Venous channels wholly enclosed within the dura mater that convey blood from the brain, meninges, and skull to the internal jugular veins

THE MENINGES ARE membranes that envelope the brain and spinal cord, providing support and protection. They consist of three components: the dura, arachnoid, and pia mater. They are composed of interlacing bundles of connective tissue fibers and cells. The outermost cranial membrane, the dura, gives rise to a number of partitions, which incompletely divide the cranial cavity into compartments. In certain locations, it also encloses venous channels or sinuses, which convey blood from intracranial sources to the internal jugular veins. The two innermost membranes, the arachnoid and pia mater, enclose the subarachnoid space. This space is of variable dimension, is continuous, and contains cerebrospinal fluid. It surrounds both the brain and spinal cord and is in direct communication with the cerebral ventricles. The arachnoid gives rise to villi, which allow the unidirectional flow of cerebrospinal fluid from the subarachnoid space to the venous channels. The spinal meninges, although similar to the cranial meninges, exhibit some important regional differences. Meningiomas, a common form of benign brain tumor, arise from the pia-arachnoid and express features reflecting their embryonic origin from the mesoderm and neural crest.

I. Cranial Meninges

A. Cranial Dura Mater

The cranial dura mater (L. *durus,* hard or strong; *mater,* mother) is a thick, tough fibrous membrane that envelops the brain. At the margin of the foramen magnum, which connects the skull with the spinal cord spaces, the adherent cranial dura becomes continuous with the spinal dura. The dura mater is also referred to as the pachymeninx (Gr. *pachys,* thick; *meninx,* membrane). [*See* BRAIN.]

The cranial dura consists of two layers: an outer or endosteal (endocranial) layer and an inner or meningeal layer. The outer layer is attached to the inner surface of the skull and functions as the periosteum of the bone. This layer of dura is well vascularized and richly innervated. It is continuous through the foramina of the skull with the periosteum outside of the skull (pericranium). The more fibrous inner layer of the dura is apposed to the brain and considered to be the true dura.

These two layers are closely adherent and together form the dura mater (Fig. 1). They are separated only in certain locations, where they enclose venous channels called sinuses (see Section I,A,3) and where they form folds, or septae (see Section I,A,2), which divide the cranial cavity into compartments.

The dura is composed mainly of interlacing bundles of collagenous connective tissue fibers arranged in layers. Interspersed among them are elastic fibers and flattened dural cells. The dural cells from which the connective tissue fibers are derived resemble fibroblasts or fibrocytes. They have a dense nucleus with clumped chromatin. The cell body may contain abundant organelles, including a well-developed rough endoplasmic reticulum, and is drawn out into long branching processes.

The inner and outer layers of the dura can be differentiated histologically. As the periosteum of the skull, the outer layer of dura is a looser connective tissue than the dura proper. It is also richer in cells and contains more blood vessels than the inner meningeal layer. The connective tissue fibers in the latter layer form an almost continuous sheet. [*See* CONNECTIVE TISSUE.]

The innermost margin of the dura has been identified as a specialized layer of cells and designated the dural border cell layer. In this location, the dura consists of one or more distinct layers of flattened cells, which are tightly bound to form thin overlapping sheets. The overlapping and interdigitating processes of these cells are attached to one another by junctional complexes of the desmosomal variety, but tight junctions are conspicuously absent. Many irregularly dilated spaces are present between the tightly packed cells of this layer, but capillaries and connective tissue fibers are absent. This interface layer of dural cells has been variously referred to as the elastic endothelial membrane, subdural neurothelium, dural mesothelium, or the layer of dural border cells.

1. Subdural and Epidural Spaces

There is some dispute regarding the precise location of the dura-arachnoid interface. The border between the dura and subjacent arachnoid mater is identified microscopically by the contrast between the dense flattened dural cells and the more lucid arachnoid cells. The commonly held view is that a gap, called the subdural space, containing a film of fluid is present between these two meningeal layers. This fluid-filled space, which surrounds both the brain and spinal cord has been thought to act like a bursa, preventing movement of the dura from being transmitted to the central nervous system.

Studies of the dura-arachnoid interface, however, do not appear to substantiate the classic concept of the subdural space. No space is said to normally exist between the two layers. They are attached to each other by two distinct layers of cells: the dural border cells and the subjacent arachnoid barrier cells (see Section I,B). The surfaces of the dura and arachnoid may under certain circumstances separate, however, to form a narrow artifactual space, the subdural space. The cleavage that forms this subdural space occurs intradurally, i.e., entirely within the layer of dural border cells and not between dural border cells and the arachnoid layer. The significance of the dura-arachnoid interface is related to the proneness of venous channels in this area to rupture and bleed into the subdural space, resulting in profound neurological consequences, and is discussed in Section III,A.

The cranial epidural space (L. *epi,* above) is also a potential space but, in contrast to the subdural space, resides outside the periosteal layer of the dura. Extensive bleeding may also occur at this location with certain kinds of injuries and is detailed in Section III,B.

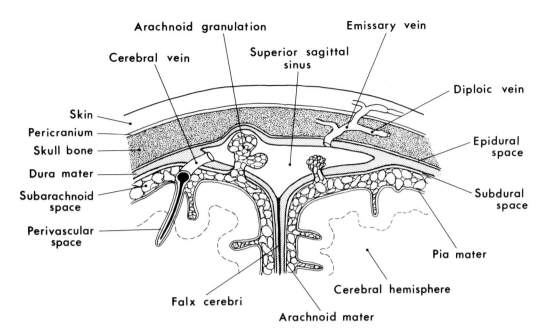

Arachnoid granulation

Cerebral vein

Superior sagittal sinus

Emissary vein

Diploic vein

Skin

Pericranium

Skull bone

Dura mater

Subarachnoid space

Perivascular space

Epidural space

Subdural space

Pia mater

Falx cerebri

Arachnoid mater

Cerebral hemisphere

FIGURE 1 Drawing of coronal section through the apex of the skull showing the cranial meninges and their relationship to the brain, skull, and vascular channels. [Copyright 1981 Ciba-Geigy Corporation. Adapted with permission from *Clinical Symposia* by Frank H. Netter, M.D. All rights reserved.]

2. Dural Septae

Dural septae are duplications of the meningeal layer of dura, which reflect off the surface of the skull and divide the cranial cavity into compartments (Figs. 1 and 2). The following are the commonly recognized dural septae.

a. Falx Cerebri   The falx cerebri (L. *falx,* sickle-shaped) is a midline septum that extends vertically between the two cerebral hemispheres (Figs. 1 and 2C). It is sickle-shaped and is attached anteriorly to bone; the crista galli of the ethmoid bone and the frontal crest of the frontal bone. Its upper margin is attached to the inner surface of the calvarium in the midline and extends posteriorly to the internal occipital protuberance. The lower surface of the falx cerebri joins posteriorly with the upper surface of the tentorium cerebelli. The falx cerebri encloses in its attached upper margin the superior sagittal sinus. In its inferior free margin, the inferior sagittal sinus is located.

b. Tentorium Cerebelli   The tentorium cerebelli is a tent-shaped horizontal partition that roofs over the posterior cranial fossa and separates the occipi-

tal lobes of the brain from the cerebellum (Fig. 2B). This partition is used to conveniently divide the cranial cavity into a supra- and infratentorial compartment. It is attached to the occipital bone posteriorly and along the petrous temporal bone anteriorly. Its free anterior midline border forms the tentorial notch or incisure through which the brain stem passes. The superior petrosal sinus is enclosed along the attached margin of the tentorium to the petrous temporal bone. Along its attachment to the falx cerebri, it encloses the straight sinus. Along its attachment to the occipital bone posterolaterally, it encloses the transverse sinuses (Fig. 2B).

c. Falx Cerebelli   The falx cerebelli is a small vertical septum located in the posterior midline (Fig. 2A). It partially separates the two cerebellar hemispheres. It is attached to the occipital bone and to the under surface of the tentorium. It encloses the occipital sinus along its attachment to the bone.

d. Diaphragma Sellae   The diaphragma sellae is a small dural partition that extends from the anterior to posterior clinoid processes, forming a roof over the sella turcica of the sphenoid bone (Fig. 2A). This horizontal partition separates the pituitary gland (hypophysis) from the brain and is perforated by the stalk (infundibular) of the gland.

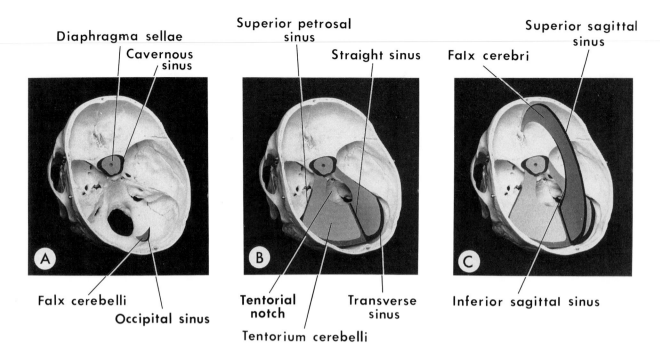

Diaphragma sellae
Cavernous sinus
Superior petrosal sinus
Straight sinus
Superior sagittal sinus
Falx cerebri

Falx cerebelli
Occipital sinus
Tentorial notch
Tentorium cerebelli
Transverse sinus
Inferior sagittal sinus

FIGURE 2 Interior views of the base of the skull showing the location of the dural septae and dural venous sinuses. A. The diaphragma sellae roofing over the sella turcica and the cavernous sinuses adjoining the sphenoid bone are shown. Also shown is the falx cerebelli and occipital sinus in its attached margin. B. The tentorium cerebelli with the transverse and superior petrosal sinuses along its attached margin and the straight sinus in the midline have been added. C. The falx cerebri with the superior sagittal sinus at the attached margin and the inferior sagittal sinus at its free margin have been added. [Adapted, with permission, from Montemurro and Bruni, 1988, "The Human Brain in Dissection"; courtesy of Oxford University Press.]

3. Dural Venous Sinuses

In certain locations, the meningeal dura and endosteum separate to enclose dural venous sinuses. Most of the venous sinuses are located along the margins of dural attachment to bone (Figs. 1, 2, and 3). Exceptions, however, are the inferior sagittal (Fig. 2C) and the straight sinus (Fig. 2B and C), which are wholly enclosed within the reduplicated folds or septae of the dura. The venous sinuses convey blood from the brain via cerebral veins, from the meninges via meningeal veins (see Section I,A,3,c) and from the skull via diploic veins to the internal jugular veins (Fig. 1). Several of them also communicate with veins outside the skull via emissary veins. This route of communication is of important clinical significance because it may serve as a route for the spread of infection and thrombosis. The venous sinuses are lined by endothelium, are without valves, and are held open against an intracranial pressure gradient by the taut dura mater.

Diploic veins are endothelial-lined channels located between the inner and outer tables of the skull bones. They drain the calvaria into cerebral veins and venous sinuses and communicate freely with meningeal and emissary veins (Fig. 1).

The dural venous sinuses can be conveniently divided into two groups: paired and unpaired sinuses.

a. Unpaired Sinuses

i. Superior Sagittal Sinus The superior sagittal sinus lies along the attached border of the falx cerebri to the midline under surface of the calvarium (Figs. 1 and 2C). It arises at the foramen caecum and extends backward to the internal occipital protuberance, where it usually continues to the right as the right transverse sinus. It is a quite small triangular-shaped channel anteriorly but it enlarges as it runs posteriorly. In the posterior midline, at the internal occipital protuberance, it contributes to the formation of the confluens sinuum or torcular herophili. The confluens sinuum is the point where the superior sagittal, straight, right and left transverse, and occipital sinuses are confluent.

The superior sagittal sinus receives venous blood from the upper and posterior parts of the medial and

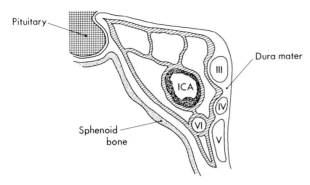

Pituitary

Dura mater

ICA

III

IV

VI

V

Sphenoid
bone

FIGURE 3 Cross section of the cavernous sinus adjoining the body of sphenoid bone showing its labyrinthine network of endothelial lined (cross-hatched) venous channels. Its lateral wall is formed by the dura mater, and embedded in it are the oculomotor (III), trochlear (IV) nerves, and the ophthalmic and maxillary divisions of the trigeminal nerve (V). The internal carotid artery (ICA) and abducens nerve (VI) run forward through it.

lateral surfaces of both cerebral hemispheres. Superior cerebral veins from the lateral convex surface of the hemisphere empty into superior sagittal sinus. They run obliquely forward, penetrating the sinus against the flow of blood within.

The sinus receives diploic veins, emissary veins passing through the foramen caecum, which communicate with veins of nose, and emissary veins, which communicate with veins of the face and scalp. These connections with veins outside the cranial cavity can allow infections such as from the scalp or nose to be transmitted to the sinus.

Laterally along its course and communicating with it are irregular expansions within the dura called venous lacunae. Lacunae are the largest and most constant in the posterior parietal region. The venous lacunae receive projecting arachnoid villi or granulations (Fig. 1). As detailed elsewhere (see Section I,B,2), arachnoid granulations protrude into the lacunae and are the means by which cerebrospinal fluid is returned to the venous circulation.

ii. Inferior Sagittal Sinus The inferior sagittal sinus lies within the free inferior margin of the falx cerebri (Fig. 2C). It drains the lower medial surfaces of the cerebral hemispheres. At the junction of falx cerebri and the tentorium cerebelli, the inferior sagittal sinus joins the great cerebral vein (of Galen) to form the straight sinus.

iii. Straight Sinus The straight sinus, or sinus rectus, is located within the folds of the dura along the attached margin of the falx cerebri to the tento-

rium cerebelli (Fig. 2B and C). It receives the inferior sagittal sinus and the great cerebral vein of Galen anteriorly and posteriorly ends at the internal occipital protuberance, usually by turning into the left transverse sinus.

iv. Occipital Sinus The occipital sinus is a small vertical channel that runs downward in the attached margin of the falx cerebelli (Fig. 2A). It extends from the internal occipital protuberance to the margin of the foramen magnum, where it communicates with the internal vertebral venous plexus. It joins the confluens sinuum and drains into the sigmoid sinus.

b. Paired Sinuses
i. Transverse Sinus The transverse sinus is a bilateral structure located along the attachment of the tentorium cerebelli to the occipital bone (Fig. 2B and C). It extends laterally and forward on each side from the internal occipital protuberance to the junction of the petrous and mastoid temporal bones. At this point, it joins the sigmoid sinus. It receives the superior petrosal sinus and venous tributaries from the nearby cerebral and cerebellar hemispheres, the pons and medulla oblongata. It also communicates with extracranial occipital and vertebral veins by way of emissary veins.

ii. Sigmoid Sinus The sigmoid sinus is a continuation of the transverse sinus bilaterally. It curves downward and forward along the mastoid process of each temporal bone and opens through the jugular foramen into the bulb of the internal jugular vein. It receives the superior petrosal sinus at its upper end and the occipital sinus at its lower end. It communicates with cerebellar veins and veins outside the skull via emissary veins.

iii. Superior and Inferior Petrosal Sinuses The superior petrosal sinuses arise posteriorly from the cavernous sinus on either side of sella turcica. They run within the attached margins of the tentorium cerebelli to the petrous part of temporal bone (Fig. 2B and C). They empty into the sigmoid sinuses, where they are joined by the transverse sinus.

From the cavernous sinus, the inferior petrosal sinus courses backward between the petrous and occipital bones to the jugular foramen, where it empties into the bulb of the internal jugular vein. It receives venous tributaries from the lower cerebellum and brain stem.

iv. Basilar Sinus The basilar sinus is a plexus of veins located in the dura overlying the body of the sphenoid bone and basilar part (clivus) of the occipital bone. It is a continuation of the anterior internal plexus of vertebral veins, which communicates with the cavernous, inferior petrosal and occipital sinuses. It also receives venous tributaries from the pons and medulla oblongata.

v. Sphenoparietal Sinuses The sphenoparietal sinuses are located below the lesser wings of the sphenoid bone. They arise from veins of the dura mater, diploic veins, and middle meningeal veins and empty into the cavernous sinuses.

vi. Cavernous Sinus The cavernous sinuses are located on either side of the body of sphenoid bone (Fig. 2A–C). They consist of a labyrinth of endothelial-lined channels extending from the superior orbital fissure to the apex of the petrous temporal bones. In its lateral wall are embedded the oculomotor, trochlear nerves, and the ophthalmic and maxillary divisions of the trigeminal nerve. The internal carotid artery and abducens nerve run forward through it (Fig. 3).

Each cavernous sinus receives the sphenoparietal sinuses anteriorly. Posteriorly they drain via the superior and inferior petrosal sinuses into the internal jugular vein.

In addition to receiving veins from the frontal and temporal lobes of the brain, they also communicate with veins outside the skull. This provides a route by which inflammation may spread to the cavernous sinuses from extracranial sites. This is of particular clinical significance because if infected it may involve important structures within the sinus or in its walls.

Blood from the face, cheek, jaw, and pharynx reach the cavernous sinuses via the ophthalmic veins and emissary veins, which communicate with the pharyngeal and pterygoid plexuses. Frequent infection arises through their connections with the pharynx, particularly following tonsillectomy.

vii. Anterior and Posterior Intercavernous Sinuses Anterior and posterior intercavernous sinuses connect the cavernous sinuses on each side, in front and behind the stalk of the pituitary gland

(Fig. 2A–C). The whole arrangement is sometimes referred to as the circular sinus.

c. Flow of Blood in the Venous Sinuses Almost all of the blood from the intracranial compartment leaves via the transverse, sigmoid sinuses and ultimately the internal jugular veins. The dural venous sinuses, meningeal, diploic, emissary veins, and veins of the scalp freely communicate and are without valves (Fig. 1).

The superficial cerebral veins drain into the superior sagittal, cavernous, petrosal, or transverse sinuses. They enter the superior sagittal and transverse sinuses opposite to the direction of blood flow in them. Blood from the superior sagittal sinus usually flows in turn into the right transverse sinus.

The deep veins of the brain drain via the great vein of Galen into the straight sinus, which in turn flows into the left transverse sinus.

The veins of the posterior cranial fossa drain the cerebellum and brain stem into the great vein of Galen, straight sinus, and sigmoid, transverse, and petrosal sinuses.

At the level of the internal jugular veins, blood initially derived from the two internal carotid arteries is incompletely mixed. Approximately two-thirds of the venous blood comes from the ipsilateral side of the brain, whereas only one-third derives from the contralateral side. Furthermore, about one-fifth of arterial blood from the internal carotids finds its way into the external jugular veins, whereas only 3% of blood reaching the internal jugular veins is derived from extracerebral sources. The emissary veins that perforate the skull and connect intracranial channels with veins of the scalp are mainly responsible for the outward flow of blood (Fig. 1).

B. Cranial Arachnoid Mater

The arachnoid mater (Gr. *arachni* + *eidos,* spiderweblike; L. *mater,* mother) is a thin translucent membrane located between the outer dura mater and the innermost pia mater (Fig. 1). The arachnoid is adherent to the overlying inner surface of the dura and follows its contours bridging over the gyri and sulci of the brain. Between the arachnoid and dural membranes is a potential space known as the subdural space, through which the delicate cerebral veins pass on their way to the dural sinuses. At this point they are vulnerable to trauma, which may result in extravasation of blood into the subdural space (see Section III,A).

Between the arachnoid and the pia mater at the surface of the brain is the subarachnoid space, which contains cerebrospinal fluid and which is in direct communication with the fourth cerebral ventricle. The inner surface of the arachnoid is irregular and forms many trabeculae, which span the subarachnoid space and merge with the pia (Fig. 1). Their cells have a similar appearance, and because of this association the arachnoid and pia are sometimes referred to collectively as the pia-arachnoid or leptomeninges (Gr. *leptos,* slender or delicate; *meninx,* membrane). While the arachnoid membrane is itself avascular, all superficial cerebral blood vessels are located in the subarachnoid space and are covered by arachnoid.

The arachnoid mater is several layers thick and is composed of overlapping arachnoid cells intermixed with bundles of connective tissue fibers. Between arachnoid cells are fixed macrophages with pale cytoplasm and many lysosomes capable of engulfing foreign bodies. The connective tissue fibers, primarily collagenous intermixed with some elastic and reticular fibers are found within the extracellular tunnels interspersed among the arachnoid cells. The arachnoid cells resemble fibroblasts and are flattened cells oriented in the plane of the meninges. They have elongated nuclei and their cytoplasm contains many mitochondria, prominent Golgi apparatus, ribosomal rosettes, and dispersed microfibrils. Lysosomes and lipid droplets are also present within cytoplasm.

Two distinct cell layers have been distinguished in the arachnoid: an outer barrier layer next to the dura and an inner layer, the arachnoid proper, next to the subarachnoid space.

The barrier layer, three to six cells thick, is distinguished by its compact arrangement of flattened lucent cells and absence of connective tissue fibers. Unlike the dural border cell component of this interface (see Section I,A), tight junctions are typically present between the arachnoid barrier cells. These junctions allow the arachnoid mater to be an effective physiological barrier preventing the intercellular diffusion of large molecules from the dura into or out of the subarachnoid space. The inner aspect of the arachnoid barrier layer of cells is covered with a discontinuous basal lamina.

In contrast, the innermost layer of arachnoid cells are more dense and loosely arranged. Between the

cells are bundles of connective tissue fibers. Between the inner arachnoid cells, gap junctions and desmosomes are found.

Similar cells form the arachnoid trabeculae and merge with the pia. A loose network of these branching arachnoid cells are interwoven with bundles of collagen fibrils and together form arachnoid trabeculae of variable sizes.

1. Subarachnoid Cisterns

There is variable separation between the innermost pia mater and the arachnoid at different regions of the brain; therefore, the subarachnoid space is larger in certain regions than that in others. It is always larger over a sulcus than over a gyrus. Along the base of brain it is particularly large in regions and contains large amounts of cerebrospinal fluid. In these regions, the large cerebrospinal fluid-containing compartments are known as subarachnoid cisterns (Fig. 4). All are interconnected and in direct communication with the fourth ventricle via the foramina of Magendie and Luschka.

The largest of the subarachnoid cisterns is situated in the cerebello-medullary angle and is called the cerebello-medullary cistern, or the cisterna magna. Into this cistern, the midline foramen of Magendie opens. The pontine cistern is located on the anterior surface of the pons and is continuous posteriorly with the cisterna magna and anteriorly with the cisterna basalis. Into this cistern, the paired lateral foramina of Luschka open. The cisterna basalis consists of the interpeduncular cistern, which surrounds the midbrain anteriorly and laterally as well as the more rostral chiasmatic cistern, which surrounds the optic chiasma. Communicating with the chiasmatic cistern rostrally is the cistern of the lamina terminalis. On the superior convex surface of the corpus callosum is the cistern of the corpus callosum. The superior cistern, or cisterna ambiens, surrounds the superior and lateral surface of the midbrain. The cistern of the lateral fissure is located on the lateral convex surface of the hemisphere where the arachnoid bridges over the lateral fissure from the frontal and parietal to the temporal lobe.

2. Arachnoid

Arachnoid villi are microscopic tufts of pia-arachnoid mater that pierce the inner meningeal layer of the dura mater and extend into the venous sinuses, adjoining lateral lacunae and extradural veins of the spinal cord (Fig. 1). Villi that are macroscopically

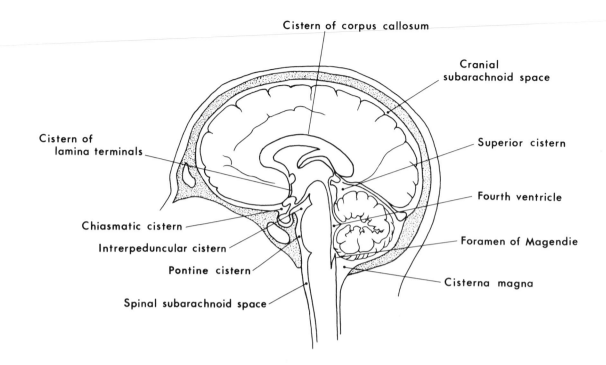

FIGURE 4 Sagittal section of the head with the brain in situ showing the subarachnoid space and its enlargements known as the subarachnoid cisterns. All are interconnected and in direct communication with the fourth ventricle via the foramena of Magendie and Luschka. [Redrawn from Carpenter, Core text of neuroanatomy (1985); from Carpenter and Sutin, "Human Neuroanatomy," (1983); courtesy of Williams and Wilkins Co., Baltimore.]

visible but similar otherwise are referred to as Pacchionian bodies, or arachnoid granulations. They are most numerous within the superior sagittal sinus and adjacent venous lacunae as well as within the transverse sinuses. Arachnoid villi are the site of unidirectional transfer of cerebrospinal fluid from the subarachnoid space to the venous system.

They increase in number and size until adult life, often becoming calcified and sometimes forming the site of origin of neoplasms. The calcified villi often thin the dura and indent the skull bone, leaving their impressions on the inner table of the skull bones.

Each villus consists of a body located in the lumen of a blood vessel and a neck encircled by the dura mater through which the villus protrudes. The body is capped by several layers of flattened arachnoid cells and overlying connective tissue fibers reflected from the dura. Some researchers, however, contend that a complete and uninterrupted covering of endothelial cells reflected from the vessel wall cap the villi. The core of the villus contains a stroma of arachnoid cells, fibroblasts, and collagen fibers interspersed with interstitial spaces of variable size.

The precise nature of the communication between the subarachnoid space, villus spaces, and the venous system is still debated and, in addition, there may be species differences. Reports suggest that the villus core contains tubules or other open tortuous channels that pass through the stroma of

the villus core opening into the sinus at one end and into the subarachnoid space at the other end.

Villi are normally turgid and their tubules open because the hydrostatic pressure of the cerebrospinal fluid usually exceeds that of venous blood within the dural sinuses. Accordingly, the villi act as valves allowing the unidirectional flow of cerebrospinal fluid from the subarachnoid space to the venous blood along the pressure gradient.

C. Cranial Pia Mater

The pia mater (L. *pia*, tender; *mater*, mother) is the innermost of the three meningeal investments (Fig. 1). It is a fine delicate membrane that is closely apposed to the sulci and gyri of the brain. The pia contains numerous blood vessels that supply underlying nervous tissue.

The pia consists of one or more laminae that may in areas be interrupted or only one cell thick. Two layers have been traditionally distinguished; an inner membranous layer adherent to the nervous tissue called the intima pia (of Key and Retzius) and an outer epipial layer. The intima pia is more fibrous and is composed of elastic and fine reticular fibers. The outer epipial layer consists of a loose meshwork of collagenous fibers that is continuous with the arachnoid trabeculae spanning the subarachnoid space.

Pial cells are flat connective tissue cells that cannot be distinguished from those in the arachnoid. The pial cells and their slender elongated processes are interspersed with bundles of collagen fibers. In addition to the pial fibroblast, some cells of the pia and, indeed, throughout the leptomeninges have features of macrophages. Macrophages are distinguished from pial cells by their lack of long cytoplasmic processes and their cytoplasmic content of membrane-bound inclusions and vacuoles. Evidence suggests that all cells lining the subarachnoid space may be potential macrophages that may actively phagocytose foreign protein entering the subarachnoid space.

On its inner surface, the pia mater is usually separated from and anchored to the basal lamina of the external glia membrane by collagen fibers. The external glia membrane surrounds the neural elements of the brain and is formed by foot processes of subjacent astrocytes. In places where the pia is interrupted, the basal lamina of the external glia limitans is exposed directly to subarachnoid space.

1. Perivascular Space

Blood vessels traversing the subarachnoid space are ensheathed by the squamous cells of the leptomeninges. As blood vessels invaginate into the neural tissue, the pia-arachnoid and the subarachnoid space follow suite (Fig. 1). The arachnoid and pia form the outer and inner investments, respectively, of the invaginated subarachnoid space now called the perivascular (or Virchow-Robin) space. Blood vessels are so ensheathed until they become capillaries. Some reports, however, suggest that the perivascular compartment does not communicate with the subarachnoid space. The arachnoid reflects over blood vessels traversing the subarachnoid space, whereas the pia accompanies them. As blood vessels enter the neural tissue, the pia accompanies them and splits to enclose a funnellike space. This pial space merges with the perivascular space but is sealed off from the subarachnoid space.

II. Innervation of the Cranial Meninges

The dura and pia mater are supplied with nerve fibers that are frequently associated with blood vessels. These nerves belong to the sympathetic nervous system and originate from the carotid and vertebral plexuses and are sympathetic vasomotor in function. Some nerve fibers, however, form sensory endings within the meninges and around the blood vessels.

The dura mater receives its innervation from two main sources. The supratentorial dura is innervated by branches of the fifth cranial nerve or trigeminal nerve. The infratentorial dura is supplied by branches from the upper cervical spinal nerves as well as from the tenth and twelfth cranial nerves (the vagus and hypoglossal nerves). The dura is largely insensitive except in areas where blood vessels and nerves traverse it, and pain is the only sensation that may be elicited.

III. Blood Supply of the Cranial Meninges

The blood supply to the cranial dura is provided mainly by the middle meningeal artery, a branch of the internal maxillary artery, which arises from the external carotid artery. The middle meningeal artery enters the skull through the foramen spinosum. An accessory meningeal artery comes off the middle meningeal and usually enters the skull through the foramen oval to assist in supplying the dura.

The dura covering the anterior cranial fossa or compartment receives its blood supply from the anterior meningeal artery, which arises from the internal carotid artery via the ophthalmic artery.

The dura of the posterior cranial fossa is supplied by the posterior meningeal artery, a branch of the vertebral and occipital arteries that enters the skull through the jugular foramen.

All of these vessels are located between the dura and the cranium, which they also supply. Veins, which are similarly located, communicate with diploic veins, emissary veins, and superficial cerebral veins and terminate in the venous sinuses.

A. Subdural Hematoma

Cerebral veins must pass from the arachnoid to the dura on their way from the cerebral cortex to the venous sinuses. At the point where they enter the tough dura, they have little support and are vulnerable to injury (Fig. 1). Trauma to the head, which displaces the brain, or any other pathological condition may rupture the delicate thin-walled veins as they traverse the arachnoid–dura interface. Extravasation of blood at this location separates the dura and arachnoid along the artificial subdural space and results in a subdural hematoma. If untreated, the hematoma enlarges, and the brain is compressed and may herniate resulting in profound neurological impairment and even death.

B. Epidural Hematoma

Head injuries may also produce extensive hemorrhaging into the cranial epidural space between the periosteal layer of dura and the inner table of the skull bone (Fig. 1). Epidural hematomas are present in 1–3% of major head injuries. Skull fractures often tear underlying veins, dural sinuses, or arteries such as the middle meningeal, which supplies blood to the meninges. The loose attachment of the periosteal layer of dura to the inner table of bone permits extensive epidural hemorrhaging. Extradural bleeding from injury to middle meningeal vessels is 10 times less frequent than a subdural hematoma.

With hematomas the brain must be rapidly decompressed, the clot removed surgically, and the bleeding vessel coagulated to prevent ensuing coma and death.

IV. Spinal Meninges

A. Spinal Dura Mater

The spinal dura mater is a fibrous tubular sheath that invests the spinal cord in the vertebral canal (Fig. 5). It extends from the foramen magnum, where it is attached to the occipital bone, to a level just below the second sacral vertebrae. The spinal dura corresponds to the meningeal layer of the cranial dura and is continuous with it at the foramen magnum. Beyond the second sacral vertebrae, the dural tube tapers abruptly to form the slender coccygeal ligament, which descends the sacral canal and attaches to the coccyx or sacrum. The dural tube is also anchored by connective tissue to the ligaments and periosteum of the vertebrae at various levels of the vertebral column, particularly in the cervical and lumbar regions. The spinal dura is also attached to the margins of the intervertebral

foraminae, where it forms a sleeve over the roots of all spinal nerves. By virtue of these attachments, the dura and its contents are anchored in place within the vertebral canal. [See SPINAL CORD.]

The spinal dura differs from the cranial dura in that it consists of a single meningeal layer only and, unlike its cranial counterpart, is without septae and does not enclose venous sinuses. The collagenous connective tissue fibers of the spinal dura run in a longitudinal direction paralleling the cord. Elastic fibers are less prominent in the spinal than in the cranial dura.

1. Epidural Spaces

Unlike the cranial dura, the spinal dura is separated from the bony and ligamentous walls of the vertebral canal by a large epidural space (Fig. 5). The cranial epidural space, it will be recalled, is merely a potential space outside of the periosteal layer of the dura (Fig. 1). The spinal epidural space, in contrast, is occupied by loose connective tissue, fat, and a major plexus of veins. It is neither continuous with or homologous with its cranial counterpart.

2. Subdural Space

The inner surface of the spinal dura is separated from the underlying arachnoid mater by a potential subdural space equivalent to that found intracranially.

B. Spinal Arachnoid Mater

The arachnoid mater of the spinal cord is continuous with and not dissimilar from the arachnoid investing the brain. The leptomeninges of the brain and spinal cord are structurally similar. The spinal arachnoid is closely adherent to the dural sleeve accompanying it to the level just below the second sacral vertebrae. Like its counterpart in the cranial cavity, it differs from the dura by its thinness and transparency.

The spinal arachnoid mater forms the outer membranous boundary of the cerebrospinal fluid containing subarachnoid space (Fig. 5). The subarachnoid space is bridged by many delicate trabeculae of pia-arachnoid composition, which condense in the posterior midline to form a subarachnoid septum. The subarachnoid space is incompletely separated on either side of the spinal cord by the denticulate ligaments, which pierce the arachnoid membrane and attach to the dura (see Section IV,C,1).

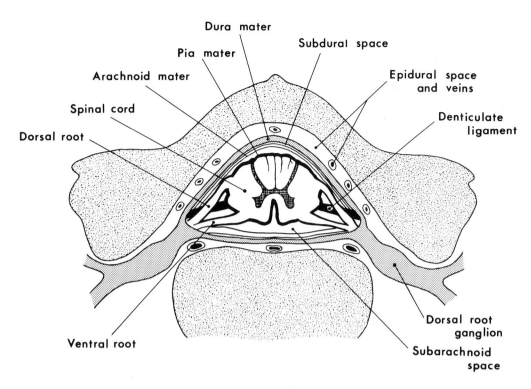

Dura mater
Pia mater
Subdural space
Arachnoid mater
Epidural space and veins
Spinal cord
Denticulate ligament
Dorsal root
Ventral root
Dorsal root ganglion
Subarachnoid space

FIGURE 5 Drawing of a transverse section through a typical vertebra showing the spinal cord *in situ* and its relation to its meningeal investments. [Adapted from Carpenter, Core text of neuroanatomy (1985); from Carpenter and Sutin, "Human Neuroanatomy," (1983); as modified from Corning, '22; courtesy of Williams and Wilkins.]

1. Subarachnoid Space and Lumbar Cistern

The spinal subarachnoid space occupies the interval between the arachnoid and pia mater and is continuous with the cranial subarachnoid space (Fig. 5). The spinal subarachnoid space is largest beyond the level of termination of the spinal cord where it forms a cul-de-sac containing the filum terminale and the lumbosacral nerve roots called the cauda equina (L. horse tail). This cul-de-sac containing cerebrospinal fluid is known as the lumbar cistern. It exists because the spinal cord in the adult terminates at a higher vertebral level (between the first and second lumbar vertebrae) than its tubular dural-arachnoid sleeve, which terminates below the level of the second sacral vertebral segment. Clinical advantage is taken of this enlarged subarachnoid space to gain access to the cerebrospinal fluid without risk of injury to the spinal cord as described below.

2. Lumbar Puncture

A lumbar puncture is performed with the patient sitting or lying on one side and with the lumbar spine flexed. A line drawn across the back joining the highest points on the iliac bones usually passes between the spinous processes of the third and fourth lumbar vertebrae. Puncture of the lumbar cistern is usually done at this point or between the spinous processes of the fourth and fifth lumbar vertebrae. A needle is inserted into the intervertebral space between the spinous processes of the vertebrae to a depth of about 5 cm penetrating the dura and entering the subarachnoid space. This procedure is made possible because the laminae of lumbar vertebrae, unlike other vertebrae are broad and short and do not overlap and because there is no risk of injury to the spinal cord.

Lumbar puncture, or spinal tap, is commonly used for cytological and chemical investigations to introduce therapeutic agents or to measure cerebrospinal fluid pressure. It is also used to relieve intracranial pressure and to introduce air or opaque media into the subarachnoid space for radiographic imaging. As an alternative, when this procedure is contraindicated, puncture of the cerebral ventricle or cisterna magna can be performed instead.

C. Spinal Pia Mater

The spinal pia is a delicate membrane that adheres to the spinal cord and ensheaths blood vessels supplying the spinal cord (Fig. 5). The spinal pia extends beyond the tapering end of the spinal cord (conus medullaris) as a threadlike median filament among the lumbosacral nerve roots. It is about 15 cm long and is known as the filum terminale. It extends to the level of the second sacral vertebrae, at which point it terminates by inserting into the dura-arachnoid tube. Beyond this point it continues as the coccygeal ligament.

The spinal pia consists of the same two layers, the inner membranous intima pia and the outer epipial layer as the cranial pia. Blood vessels of the spinal cord lie within the epipial layer. The spinal pia, unlike the cranial pia, is thicker due to the presence of an outer longitudinal layer of connective tissue fibers. An extension of this layer of pia on either side of the spinal cord forms the denticulate ligaments.

1. Denticulate Ligaments

The denticulate ligaments are thin collagenous bands covered by pia-arachnoid cells, which arise from the outer longitudinal fibers of the epipia. They extend laterally on each side of the spinal cord midway between the dorsal and ventral roots (Fig. 5). The lateral margin is serrated, and it impales the arachnoid to the dura at intervals along the length of the cord. There are about 21 points of attachment on each side, which alternate with the emergence of nerve roots from the cord. The first point of attachment is at the foramen magnum; the last one is at the level of emergence of the first lumbar nerve roots. The denticulate ligaments suspend the cord within the cerebrospinal fluid-filled subarachnoid space. This effectively anchors the cord, preventing it from being displaced with movement of the dura.

V. Function of the Meninges

The meninges primarily provide support and protection for the delicate brain and spinal cord. The buoyancy provided by the fluid environment within the subarachnoid space allows movement while maintaining shape and protecting the nervous system from distorting forces. In the spinal cord, the meningeal septae and ligaments support and stabilize the spinal cord during movement of the spine.

In addition to providing for cerebrospinal fluid drainage via the arachnoid villi, the perivascular and perineural arrangement of the meninges may facilitate exchange of cerebrospinal fluid and/or metabolites between brain extracellular spaces and the subarachnoid space as well as provide an alternate route of drainage to extracranial lymphatic channels outside the dura.

The leptomeninges by virtue of their cell-to-cell contacts may function as a barrier, preventing the free diffusion of certain substances to or from the central nervous system. They may additionally accumulate and inactivate various substances protecting the nervous system from substances such as neurotransmitters.

VI. Tumors of the Meninges

Brain tumors, both primary and metastatic, occur with a frequency of 0.02% in the general population. Meningiomas are the most common of the benign brain tumors, accounting for 13–21% of all primary intracranial tumors. They are rarely found in children; only about 6% occur within the first 10 yr of life. With age, however, the incidence increases. Sixty-five percent of meningiomas occur from 50 to 70 yr of age, with women twice as likely as men to develop them.

A. Clinical Manifestations

Meningiomas are leptomeningeal in origin, arising from arachnoid cells. They are slow-growing and almost always benign, compressing nervous tissue without invading it. Accordingly, the symptoms and signs of neurologic dysfunction produced by meningiomas depend on their size, location, growth rate, and rate at which intracranial pressure is increased. Symptoms exhibited are usually related to dysfunction of the part of the brain being compromised and may include headache, confusion, weakness, seizures, paralysis, incoordination, double vision, facial numbness, drowsiness, loss of consciousness, and personality changes.

B. Classification

Classification of meningiomas is made on the basis of histologic makeup and location of the tumor within the nervous system.

1. Histology

Meningiomas are highly cellular, slow-growing, encapsulated masses. They have an irregularly lobulated or flattened shape and may become quite large. They may be soft and hemorrhagic or firm and rubbery in consistency and may calcify. They have a characteristic vasculature, which consists of fine vessels and no arteriovenous shunts. Because of their location, they may attach to the dura and invade the overlying skull, producing increased vascularity and stimulating bony proliferation (hyperostosis). Their microscopic appearance is variable because the neoplastic cells can transform into a variety of connective tissue types or exhibit epithelial features. They are usually classified histologically according to their most prominent tissue elements.

2. Location

Meningiomas are found preferentially in certain intracranial locations often in the vicinity of the arachnoid granulations. The majority are concentrated over the cerebral convexities, in the region around the superior sagittal sinus and along the lesser wing of the sphenoid bone. Others may be found in the olfactory region, around the tuberculum sellae, and in the cerebellopontine angle.

C. Prognosis and Treatment

Prognosis and treatment depends on both the tumor's histologic composition and its location. Because they are most often a benign encapsulated mass, excision is the preferred treatment. Local radiation may be required for tumors that have been incompletely removed. Because meningiomas are rarely malignant their removal is usually curative, although long-term prognosis depends on how completely they have been resected.

Bibliography

Carpenter, M. B. (1985). "Core Text of Neuroanatomy," 3rd ed. Williams and Wilkins, Baltimore.

Freidberg, S. R. (1986). Tumors of the brain. *Clin. Symp.* **38(4),** 2–32.

Friedman, W. A. (1984). Head injuries. *Clin. Symp.* **36(1),** 2–32.

Krisch, B., Leonhardt, H., and Oksche, A. (1984). Compartments and perivascular arrangement of the meninges covering the cerebral cortex of the rat. *Cell Tissue Res.* **238,** 459–474.

Lopes, C. A. S., and Mair, W. G. P. (1974). Ultrastructure of the arachnoid membrane in man. *Acta Neuropath.* **28,** 167–173.

Lopes, C. A. S., and Mair, W. G. P. (1974). Ultrastructure of the outer cortex and the pia mater in man. *Acta Neuropath.* **28,** 79–86.

Montemurro, D. G., and Bruni, J. E. (1988). "The Human Brain in Dissection," 2nd ed. Oxford University Press, New York.

Nicholas, D. S., and Weller, R. O. (1988). The fine anatomy of the human spinal meninges. *J. Neurosurg.* **69,** 276–282.

Peters, A., Palay, S., and Webster, H. deF. (1976). The meninges. *In* "The Fine Structure of the Nervous System: The Neurons and Supporting Cells," chap. XIII, pp. 332–343. W. B. Saunders Co., Philadelphia.

Schachenmayr, W., and Friede, R. L. (1978). The origin of subdural neomembranes I. Fine structure of the dura-arachnoid interface in man. *Am. J. Pathol.* **92,** 53–68.

Sheldon, J. J. (1981). Blood vessels of the scalp and brain. *Clin. Symp.* **33(5),** 3–36.

Mental Disorders

CHARLES L. BOWDEN, *The University of Texas*

I. Indicators of Mental Disorders
II. Diagnosis of Mental Disorders
III. Diagnostic Categories

Glossary

Criterion variance Variability and divergence in the diagnosis selected by mental health clinicians derived from the usage of differing criteria for a disorder. An objective of the DSM-III-R is to establish criteria that can reliably be applied by all mental health clinicians throughout the United States and in many other countries.

CT scans Computerized axial tomography is a radiographic technique that allows quantitative determination of the size and density of discrete brain regions. The resulting pictures provide a two-dimensional ''slice'' of brain at any selected level. Evidence of tumor, atrophy, or enlargement of fluid containing areas can be determined.

DSM-III-R The American Psychiatric Association's *Diagnostic and Statistical Manual of Mental Disorders* (the current version is better known as DSM-III-R) is a listing of emotional disorders that is widely used by psychiatrists, other mental health clinicians, and researchers throughout the world. It is an atheoretical, noninferential, descriptive approach to diagnosis, which will be revised as new information warrants.

Research Diagnostic Criteria A set of diagnostic criteria for a subset of major psychiatric disorders. They differ from DSM-III-R in that they have been held constant over time; thus, the criteria for a particular disorder have not changed. This facilitates comparison between studies conducted at different times.

MENTAL DISORDERS are conditions in which behavior, mood, or thinking processes are dysfunc-

tional. In most instances the dysfunction causes distress to the person, but for some disorders the dysfunction may be only recognized by or interfere with others. An additional requirement for a mental disorder is that it must last at least a minimal period, generally most days for one or two weeks. By this standard, a person momentarily anxious during a robbery would not be considered to have a mental disorder, despite having a broad range of symptoms that are seen in anxious disorders. This example suggests another difference between mood disorders and ordinary responses to daily life occurrences. For the latter, the responses are appropriate to the stimulus; for the former, they may be unrelated to the stimulus, inappropriate to it, or abnormal in magnitude.

I. Indicators of Mental Disorders

The diagnosis of a mental disorder is based on the presence of a relatively small range of observable or inferable phenomena. One reason for this is the limited range of central nervous system–mediated responses to widely varying etiologic factors. For example, slowed thinking and impaired concentration may be seen in dementias, substance abuse, schizophrenia, depression, and anxiety disorders. Clinicians organize these phenomena into categories to facilitate systematic collection of data and diagnostic decision making. The major categories of these phenomena are summarized here.

Observable physical characteristics include disturbances of sleep, appearance, weight, appetite, and physical activity. Restlessness, purposeless fidgeting, pacing, grimacing, and similar manifestations of increased motor activity occur in a wide range of disorders. The exact form of such restlessness can often be of importance in terms of diagnostic specificity. Tourette's syndrome, a movement disorder first appearing in childhood, is characterized by dramatic grimacing. Aimless repetitive playing with buttons or other parts of apparel is observed in some organic mental disorders. Calmness at rest replaced by shakiness when using one's hands, such as holding a cup, characterizes a side effect to the psychotropic drug lithium and a familial type of tremor. General appearance usually provides nonspecific information, such as care of hair and dress, but sometimes provides relatively specific information such as ruddy cheeks or characteristic breath in alcoholism.

Mood—the sustained feeling tone of a person—is described in terms depression or elation, anxiety, and angry tone. Each of these may vary in terms of intensity (slight elation to florid mania), stability (fixed or labile), reactivity to events, periodicity, duration and congruence with thought content (consistent or inconsistent).

Disturbances in thought are seen in nearly all mental disorders. Thought may be slowed or speeded up. The thread of associations of a person may be hard to follow. The person may be preoccupied with a single theme, or have idiosyncratic thoughts that are also considered bizarre by the observer. Some forms of thought disturbance are evident to lay as well as professional persons. The person who believes firmly that a foreign government is relaying messages to him through a radio implanted in dental fillings would be so considered. This is an example of delusional thought, or beliefs held firmly despite contrary rational evidence. A related disturbance is that of perception, in which the person either distorts information actually coming through sense organs, or imagines input in the absence of any external input. Thus hallucinations are perceptions in the absence of any corresponding sensory stimulus. They are strongly associated with delusions. Even with a phenomenon as frankly abnormal as hallucinations generally are, some types of hallucinations can occur in healthy individuals. Hypnagogic hallucinations are those vivid visual images that occur just at the moment of falling asleep, for example.

Disturbances of sensorium, or consciousness, occur in some but not all mental disorders. These are evident as impaired alertness or disorientation as to the time, place, or who one is. Disorientation is largely limited to organic mental disorders, some severe psychoses, and dissociative states.

Disturbances of memory or intellectual function also occur in some but not all mental disorders. In some organic mental disorders there are differences between impaired short-term memory (what was eaten at breakfast) and retained long-term memory (where one was born). Memory disturbance may arise from structural lesions in the brain, from drug effects, or from inattention and disinterest. [*See* LEARNING AND MEMORY.]

The above summary of a few of the major ways in which mental disorders may manifest themselves allows several points to be made. In most instances a degree of subjectivity exists. Even if we observe a patient apparently asleep, we cannot ignore his later stating that he slept restlessly and without satisfaction. The content in which signs are observed is quite important. Lastly, the constellation of all signs, rather than consideration of one single characteristic, is essential for diagnostic utilization of this information.

II. Diagnosis of Mental Disorders

The diagnosis of mental disorders is based on the presence of a characteristic set of signs and symptoms and a characteristic course. In most instances accurate diagnosis is aided by historical features (age of onset, number of episodes), family history, and evidence of response to specific treatments. Laboratory test information, and information from special tests such as electroencephalograms, are also used for some conditions.

New approaches to diagnosis constitute one of the major changes in psychiatry over the past 20 years. Beginning in the late 1950s, a revolutionary development of specifically effective therapies, most of which are psychopharmacological, has occurred. Previously, treatment had tended to be relatively nonspecific, with an emphasis on the use of custodial approaches, generally sedative drugs, and psychotherapy. With the expansion of the range of specific treatments, the importance of accurate diagnosis to link treatment to diagnosis increased. Concurrently, scientists found that the greatest discrepancy in diagnosis between different psychiatrists could be attributed to low reliability and criterion variance. That is, two psychiatrists would proficiently and consistently obtain the same clinical information, yet each might interpret the information in a theoretically driven, somewhat subjective way, and thereby arrive at a different diagnosis. Feighner and other psychiatrists at Washington University, along with Spitzer and Endicott at Columbia, and Katz and others in the major NIMH Collaborative Programs on the Psychobiology of Depression set in motion a diagnostic system relying on observable, rather than inferential information. The *Research Diagnostic Criteria* and *Diagnostic and Statistical Manual* (DSM) which emerged from this work have the great advantage that a person diagnosed as having a manic disorder in Dubuque would be highly likely to be similarly diagnosed in Dallas. This development has spurred epidemiological studies that have reliably established the incidence and prevalence of major psychiatric disorders in the United States and several other countries. An illustration of the value of such work is the surprising finding that the severe, often disabling obsessive-compulsive disorder is present in over 2% of adult Americans. It has been assumed to be much less common, since less than one in ten persons with the illness seeks treatment. The ability to define the symptomatic boundaries of the disorder has also contributed to other important developments. For example, distinguishing obsessive-compulsive disorder from similar appearing conditions has uncovered evidence of biochemical dysfunction of serotonergic activity (that is, a possible functional deficiency of the neurotransmitter serotonin) in obsessive-compulsive disorders. Concurrently, drugs with specific effects on serotonergic transmission have been developed and appear to be effective in the treatment of obsessive-compulsive disorder.

The development of evidence for a biochemical component or other physically based abnormality has been one of the most important themes in psychiatric disorders, as it lays the foundation both for more specific therapies, and more useful discrimination of diagnostic entities with differing prognosis, and causal factors.

The current diagnostic system has limitations. Biological data, such as from *computed tomography* (CT) *scans* and laboratory tests, are not included as criteria. The approach of a committee-based revision of the so-called DSM nomenclature has resulted in a certain politicization of the process. This has led to establishment of diagnostic labels (such as deviant expressive writing disorder) which, even if reliably identified, many authorities doubt represent bonafide medical entities.

Mental disorders have been difficult to investigate and treat because of the frequency of phenocopies. That is, the constellation of major symptoms of two persons may be similar if not identical, yet the underlying disease process, and by implication

the treatment needed, may be quite different. This is illustrated by depression. One of the major advances of the criterion-based approach to diagnosis has been the division of patients with serious depressive conditions into those with only depressive episodes and those (bipolar depressed) who also have manic episodes. The two groups show differences in heritability, age of onset, course of illness, and treatment response, all of which have important practical implications. Undoubtedly, scientists have yet to disentangle many of these "look-alike" disorders, but the means to study such issues has made this one of the most active areas in psychiatric investigation. [*See* DEPRESSION.]

One of the major barriers to the study of mental disorders has been lack of access to the tissue involved in the brain. Not only is the brain encased in a highly protective bony structure, there are also special barriers to the passage into the brain of many chemical substances that diffuse freely through the rest of the body. Furthermore, autopsy study of the brain frequently does not identify any visible abnormality. This inability to see gross pathological changes led in the past to considering many major, chronic mental disorders as "functional." New techniques involving biochemical analysis of very small concentrations of metabolites of chemicals active in the brain, and measurement of brain electrical activity and brain structure, now allow the identification and quantitation of key dimensions of some mental disorders. Imaging techniques in particular have had a major influence on clinical diagnosis. Computerized tomography (CT) is a widely available technique that allows identification of atrophic changes seen in some forms of dementia and schizophrenia. Nuclear magnetic resonance imagery is another recently developed technique that offers a high degree of brain structure resolution. To date it has been of greater utility in neurological than in psychiatric disorders. A more expensive and complex procedure, positron emission tomography, allows the study of regional brain energy metabolism and of the number and localized distribution of specific receptors to which neurotransmitters in the brain bind. This procedure may be particularly helpful in determining whether certain receptor changes and altered patterns of brain blood flow are specific to, or simply nonspecific factors in, complex disorders such as schizophrenia. [*See* BLOOD–BRAIN BARRIER; MAGNETIC RESONANCE IMAGING.]

Although the above-described new approaches are important additions to psychiatric diagnosis and investigation, the fundamental approach to diagnosis of mental disorders remains a detailed history and careful observation. Unlike a condition such as hypertension, which is established by a blood pressure determination, mental disorders are by definition behavioral. Although in most instances signs of a psychiatric disorder may be evident to the lay person and the patient is able to recognize and summarize major symptoms, this is often not the case. Persons with manic disorders may function quite effectively for substantial periods of time, and may be able to rationalize away some of the excesses of their behavior, such as spending sprees or temper outbursts. A relatively lengthy period of evaluation, observation by professional staff in a hospital, or information obtained from family members thus may be needed for diagnostic certainty. Certain symptoms of mental disorders are not always distressing. The manic person may find increased energy an asset in his work. Focused inquiry is also needed because a patient may not have linked all distressing behavior to one underlying condition. Persons who wake frequently in the night and have difficulty returning to sleep do not always recognize this as a major feature of depressive disorder, for example.

The above considerations also influence one factor that distinguishes mental disorders from most other medical disorders. Whereas denial of illness is unfortunately common across all of medicine, it is for the most part not immediately self-injurious or harmful to others. In mental disorders, both from the perspective of suicidal behavior and aggressive-hostile behavior, it may be harmful. Therefore, due-process procedures for involuntary evaluation are still required. It is difficult for some persons to understand this seeming paradox between the major advances in the understanding of mental disorders and the need for such restrictive procedures. In fact the need for such involuntary commitment is now much less frequent, as a consequence of improved treatments. Nevertheless, major impairment of insight, judgment, or intellectual function, and key aspects of certain untreated mental disorders make it nearly certain that some form of involuntary evaluation and treatment (at least over the short term) will continue to be needed.

Severity plays a complex role in many mental disorders. Alcoholism is a convenient example, as most persons in Western countries engage in some social drinking. Whereas no one would have difficulty with the diagnosis of alcoholism in a person with severe alcohol withdrawal syndrome and physical sequelae of prolonged, excessive alcohol consumption, there is a gray zone in which even authorities have difficulty in saying whether alcohol abuse exists. Viewed differently, for some conditions, a continuum, as opposed to a discrete presence or absence of indicators of disease, may be present in some disorders.

Certain mental disorders have the characteristics of causing severe impairment and distress in the absence of observable disturbance. Some severely, even suicidally depressed persons are able to go about routine daily responsibilities with no more than slight diminution in efficiency, and maintain a generally responsive, appropriate demeanor, albeit not without great effort on their part.

III. Diagnostic Categories

The following section reviews the major diagnostic categories of mental disorders and includes common synonyms when applicable. Brief descriptions of the disorders follow.

A. Developmental Disorders

These usually first appear during childhood or adolescence. The essential feature is a disturbance in the acquisition of information processing, language, motor, or social skills. They may persist throughout life, as in the case of mental retardation, or may resolve with maturation or treatment, as with some eating disorders and tics.

Mental retardation is characterized by impaired learning, general adaptive function, knowledge of common facts, and conceptualizing capacity. The causes of most cases of mental retardation are unknown. Autistic disorder is a rare, severe impairment in the ability to relate to other persons, with attendant learning difficulties. Stereotyped behaviors and impaired language use are common. Attention-deficit hyperactivity disorder is marked by restlessness and inability to maintain attention. Much more common in boys, it is often benefited by several stimulant-type medications. Conduct disorders are marked by repetitive socially miscreant behaviors such as lying, property destruction, and fighting. There is some evidence for two types of conduct disorder, those with one type having apparently good affiliative relationships, those with the other type being socially distant. Several anxiety disorders are identified in children, including one in which worry about harm from separation from parental figures is the key element. Eating disorders are another large subclass. Anorexia nervosa entails refusal to maintain a normal body weight and intense fear of becoming fat. Bulimia nervosa involves recurrent binge eating and self-induced vomiting or other measures to prevent weight gain. Both anorexia and bulimia are seen principally in girls. Gender identity disorders involve persistent, intense distress about and aversion to one's biological sex. Transsexualism involves a sustained discomfort with one's biological sex and a wish to acquire the physical sex characteristics of the other sex. Of several stereotyped movement disorders, *Tourette's* is noteworthy because of its positive response to drug treatment. It consists of multiple recurrent motor or vocal tics. [*See* AUTISM; EATING DISORDERS.]

B. Organic Mental Disorders

These conditions entail a dysfunction of the brain, usually identifiable on examination through specific memory or intellectual deficits or through special tests such as CT scans. They are generally related to aging, or to substance abuse, or as complications and manifestations of other medical disorders. Alzheimer's dementia has an insidious, progressively worsening loss in a broad range of intellectual functions. Over 2% of the population over the age of 65 are estimated to have this type dementia. Although certain pathological changes (such as tangled scar-like fibers) are present in brain at autopsy, no specific cause or treatment for Alzheimer's has been found. [*See* ALZHEIMER'S DISEASE; DEMENTIA IN THE ELDERLY.]

C. Substance Abuse and Dependence

In these disorders the disturbed behavior is caused by the ingestion of chemical substances that alter brain function. The range of such substances is very large. Among the most common are alcohol, narcotics, and stimulants such as amphetamines and cocaine. The concept of dependence is defined to mean that the person must take an increased amount of the substance to obtain the same psychological effect, and that abrupt cessation of the substance will result in a withdrawal reaction charac-

teristic for the particular group of drugs. [*See* COCAINE AND STIMULANT ADDICTION.]

D. Schizophrenia

Schizophrenic disorders involve some psychotic symptomatology, such as delusions, hallucinations, or bizarre thought pattern. Schizophrenia is often chronic and severe. Despite improvements in treatment, schizophrenia remains one of the least satisfactorily treated of the major psychiatric disorders. Persons with psychotic symptoms of brief duration are classified as having brief reactive psychosis or schizophreniform disorders. Persons with fixed, nonbizarre delusions of situations that might occur in real life—such as being poisoned—are classified as having *delusional,* or *paranoid disorder.* [*See* SCHIZOPHRENIC DISORDERS.]

E. Mood Disorders

These involve a persistent disturbance in mood, expressed either as depression or mania. Major depression and dysthymia represent the two forms of strictly depressive disorders. A range of other symptoms related to diminished energy and interest also characterize depressive disorders. The diagnosis of bipolar disorder requires evidence of a current or prior manic episode. The current symptoms of a bipolar disorder may present as bipolar depression, bipolar mania, or a mixed disorder with current symptoms of both mania and depression. Manic episodes require a persistently elevated, expansive, or variable mood. Cyclothymia is a chronic milder cycling of episodes with both manic and depressed features. [*See* MOOD DISORDERS.]

F. Anxiety Disorders

These disorders are characterized by symptoms of anxiety and avoident behavior. They are subcategorized, based on predominant symptoms, into panic disorder, phobias, generalized anxiety disorders, and obsessive-compulsive disorder. The latter is characterized by recurrent obsessions or compulsions that cause distress and that persist although the person knows that they are unreasonable.

G. Somatoform Disorders

In these disorders the patient's concern is about bodily dysfunction, although there are no physical findings sufficient to explain the complaints. A synonymous vernacular term is hypochondriasis.

H. Dissociative Disorders

Such disorders are characterized by sudden, temporary loss of a complex bodily function, such as memory or personal identity. *Amnesia* is included in this group.

I. Sexual Disorders

These disorders involve discomfort with one's anatomical sex, or paraphilias, in which the person is sexually aroused by circumstances not usually associated with arousal, or they may entail inhibited sexual desire or response. [*See* SEXUAL BEHAVIOR, HUMAN.]

J. Sleep Disorders

These involve either an abnormality in the amount, quality or timing of sleep, or abnormal events occurring during sleep. An example of the former category is insomnia, of the latter sleepwalking. [*See* SLEEP DISORDERS.]

K. Impulse Control Disorders

In these disorders the person is unable to resist an impulse to act in a way that would be harmful to himself or others, such as intermittent explosive disorder, characterized by discrete episodes of hostility and aggression not accounted for by other mental disorders. Other examples in this category include kleptomania and pathological gambling.

L. Adjustment Disorders

In these conditions behavioral symptoms arise in response to identifiable psychosocial stressors, and do so within a time-limited way.

M. Personality Disorders

Personality disorders encompass a large group of conditions in which enduring personality traits contribute to the person relating to others in a relatively inflexible, maladaptive way. These include very diverse patterns, such as paranoid, antisocial, and borderline personality disorders. [*See* PERSONALITY DISORDERS (PSYCHIATRY).]

Bibliography

American Psychiatric Association (1987). ''Diagnostic and Statistical Manual of Mental Disorders, Third Edition, Revised.'' American Psychiatric Association, Washington, D.C.

Kocsis, J. H., and Frances, A. J. (1987). A critical discussion of DSM-III dysthymic disorder, *American Journal of Psychiatry* **144**, 1534–42.

Leon, R. L., Bowden, C. L., and Faber, R. (1989). Psychiatric interview, history, and mental status, *in* ''Comprehensive Textbook of Psychiatry, Vol. 1,'' 5th ed. (H. I. Kaplan and B. J. Sadvck, eds.). Williams & Wilkins, Baltimore.

Spitzer, R. L., Endicott, V., and Robins, E. (1978). Research diagnostic criteria, *Archives of General Psychiatry* **35**, 773–82.

Yager, J. (1989). Clinical manifestations of psychiatric disorders, *in* ''Comprehensive Textbook of Psychiatry, Vol. 1,'' 5th ed. (H. I. Kaplan and B. J. Sadock, eds.). Williams & Wilkins, Baltimore, pp. 553–82.

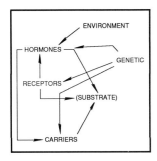

Metabolic Regulation

CAROLYN D. BERDANIER, *The University of Georgia*

Glossary

Anabolism Reactions or reaction sequences that result in the synthesis of macromolecules needed by the body

Catabolism Reactions or reaction sequences that result in degradation of macromolecules to their smaller component molecules

Cell components Parts of the cell; each part or organelle has a specific role in cellular metabolism

Citric acid cycle (Krebs cycle) Cyclic series of reactions that enables cells to oxidize metabolites and that results in the production of reducing equivalents for use by the respiratory chain and citrate, which can be transported to the cytosol for hydrolysis to oxalacetate and acetyl CoA. This cycle takes place in the mitochondria

Cytosol Also known as the cell sap. This is the compartment, surrounded by the plasma membrane, in which the organelles are suspended

Gluconeogenesis Synthesis of glucose from two and three carbon metabolic intermediates. Gluconeogenesis is stimulated when the intake of glucose is deficient or when glucose is not being metabolized by cells. This occurs in the disease called *diabetes mellitus*. Gluconeogenesis takes place in the cytosol and shares many of its reactions with glycolysis

Glycogenesis When more glucose is provided to the body than can be immediately used, some of this glucose is stored in the form of glycogen through a series of reactions beginning with glucose-6-phosphate and ending with a branched glucose polymer. Glucoses are joined by linkages between carbon 1 and carbons 4 or 6

Glycogenolysis When the body is in need of glucose it raids its glycogen stores through glycogen hydrolysis

Glycolysis The main pathway for the use of glucose by the cell. This use begins with the phosphorylation of glucose and, through a series of enzymatic steps in the cytosol, results in the production of the three carbon molecule pyruvate. If oxygen is in short supply, pyruvate is converted to lactate. If not converted to lactate, pyruvate enters the mitochondria for further use by the citric acid cycle

Lipogenesis Synthesis of fatty acids from acetyl CoA and the esterification of these fatty acids to glycerol to make triglycerides. Lipogenesis occurs in the cytosol

Lipolysis Cleavage of fatty acids from glycerol usually followed by the oxidation of these fatty acids and the reuse of glycerol. Lipolysis occurs in the cytosol, whereas fatty acid oxidation occurs in the mitochondria

Membrane Lipid bilayer consisting of phospholipids and cholesterol. Proteins, which serve as receptors, carriers, or enzymes, are embedded in this bilayer

Mitochondria Organelle responsible for energy transformation; sometimes called the powerhouse of the cell. The citric acid cycle, respiratory chain, oxidative phosphorylation, and fatty acid oxidation occur in this compartment

Nucleus Organelle of the cell that contains the genetic material, DNA. Gene codes for almost all the proteins synthesized by the cell are in this cell component

Pentose phosphate shunt When glucose is consumed in excess of need, some of the glucose is metabolized by way of this shunt. This series of reactions yields two reducing equivalents per molecule of glucose used. These reducing equivalents are used in the *de novo* synthesis of fatty acids. The shunt also produces ribose-5-phosphate, an important constituent of ribonucleic acid, and three carbon intermediates, which can enter the glycolytic sequence. The shunt takes place in the cytosol

Respiration/oxidative phosphorylation Series of reactions that results in the joining of oxygen to hydrogen to make water. Respiration produces energy, which is trapped (under closely regulated conditions) in the high energy bond of ATP; ATP in turn transfers this energy to the synthetic processes of the cell

Ureogenesis Cyclic series of reactions for the synthesis of urea from ammonia

METABOLIC REGULATION is defined as the sum of all those processes and reactions designed to provide a continuous supply of substrates for the maintenance of cell life. It includes both catabolic and anabolic processes; both energy-producing and energy-using pathways are included. Living cells must interdigitate these pathways so that they are able to survive times of energy deficit *and* energy surfeit. They must be able to replace their functional parts as they are used up or worn out or damaged. As simple one-cell organisms evolved into complex entities consisting of specialized cells organized into tissues, organs, and systems, the need for this interdigitation of pathways became even more important as animals had to adapt to changing environments and fluctuations in the food supply. Only those animals that could adapt by having more versatile metabolic control mechanisms survived. Those that could not adapt did not survive.

I. Levels of Control

The regulation of metabolism can be viewed either as the controls exerted on a single reaction in the cell or the control of an animal's response to a change in its environment. Neither view is wholly correct because regulation can and does occur at many levels in the body as illustrated in Fig. 1. The

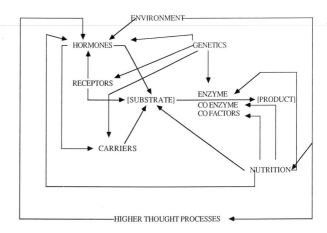

FIGURE 1 Metabolic regulation encompasses all factors that influence the flux of nutrients and metabolic substrates into, out of, and between subcellular compartments, cells, tissues, organs, and organisms. Levels of control illustrated here are various and complex.

simplest level of control is that which is exerted over a single enzymatic reaction. This reaction is controlled by the amount of available substrate, the amount and activity of the enzyme that catalyzes the reaction, the presence of appropriate amounts of required cofactors and/or coenzymes, and the accumulated product.

II. Control of Enzymatic Reactions

The amount and activity of the enzyme is controlled by the rate at which the enzyme is synthesized in the cell, the rate at which it is activated, and the rate at which it is destroyed. Enzymes that are in short supply in the cell and that are the least active in a reaction sequence are called *rate-limiting enzymes*. In glycolysis, for instance, the rate-limiting enzymes are glucokinase and phosphofructokinase. Table 1 lists the different pathways of intermediary metabolism and their rate-limiting steps. Figures 2 and 3 show how pathways for the metabolism of glucose and fatty acids interdigitate. Shown are glycolysis, gluconeogenesis, citric acid cycle, lipolysis, glycogenesis, and glycogenolysis. Those reactions having a letter designation are rate-limiting reactions. Carrier-mediated metabolite exchange across the mitochondrial membrane is designated by a circle. Much of the literature in metabolism has concerned itself with the factors that control these enzymes and how they function in limiting the activity of the pathways to which they are connected.

TABLE I Metabolic Pathways and Their Control

Pathway	Factors that affect the pathway
Glycolysis	a) Transport of glucose into the cell (mobile glucose transporter b) Glucokinase c) Phosphofructokinase d) α-glycerophosphate shuttle e) Redox state, phosphorylation state
Pentose phosphate shunt	a) Glucose-6-phosphate dehydrogenase b) 6-phosphogluconate dehydrogenase
Glycogenesis	a) Stimulated by insulin and glucose b) High-phosphorylation state (ratio of ATP : ADP)
Glycogenolysis	a) Low-phosphorylation state b) Stimulated by catecholamines
Lipogenesis	a) Stimulated by insulin b) Acetyl-CoA carboxylase c) High-phosphorylation state d) Malate citrate shuttle
Gluconeogenesis	a) Stimulated by epinephrine b) Malate aspartate shuttle c) Redox state d) Phosphoenopyruvate carboxykinase e) Pyruvate kinase
Cholesterogenesis	a) HMG CoA reductase
Ureogenesis	a) Carbamyl phosphate synthesis b) [ATP]
Citric acid cycle	a) All three shuttles b) Phosphorylation state
Lipolysis	a) Lipoprotein lipase
Respiration	a) ADP influx into the mitochondria b) Ca^{2+} flux c) Shuttle activities d) Substrate transporters
Oxidative phosphorylation	a) ADP : ATP exchange b) Ca^{2+} ion

For example, one of the first enzymes to be isolated and studied was the first rate-limiting step in the pentose phosphate shunt, glucose-6-phosphate dehydrogenase. This step and the one that follows it in the shunt provide about half the reducing equivalents needed for *de novo* fatty acid synthesis. Thus, the study of the rate-limiting steps of the shunt also is relevant to the study of fatty acid synthesis. [*See* GLYCOLYSIS.]

Numerous publications exist reporting in detail how the various enzymes in metabolism work *in vitro*. However, one must appreciate the fact that most *in vitro* studies of enzyme activity are conducted with idealized amounts of substrate and cofactors and with little subsequent product removal. Usually the enzyme is studied at saturation[1], a condition that seldom occurs in the living organism. *In vivo*, enzymes are rarely saturated; they usually are working so rapidly that 10–20% saturation is more likely. This allows for greater control of the flux of metabolites through the system by factors other than the amount and activity of the enzyme *per se*. Thus, in cells having 50% less of a given enzyme, as measured *in vitro*, compared with normal cells, the flux of substrate through the pathway *in vivo* would not necessarily be slower. In fact, the inheritance of a number of the genetic diseases characterized by an absence of activity of a certain enzyme can be tracked within a given family by the activity of that enzyme in unaffected homozygotes and heterozygotes. Heterozygotes will have reduced enzyme activity yet be normal with respect to their metabolism. Table 2 lists some of the genetic diseases that are due to a spontaneous mutation in the gene code for a given enzyme and that are characterized by an absence or loss of activity of an enzyme. The genetic codes dictate the sequence of amino acids that constitute the proteins that serve as receptors, carriers, and enzymes. If the code is aberrant, an amino acid substitution might occur. If the substitution occurs in a part of the molecule away from the active site and is a substitution of a structurally related amino acid, the activity of the carrier or enzyme may not be affected. However, if the substitution occurs with a dissimilar amino acid (e.g., glycine substituted for cysteine or methionine), the activity of the protein could be so drastically affected as to render that protein nonfunctional. Examples of this kind of substitution may be found in the medical literature. In the disease phenylketonuria, an amino acid substitution caused by a mutation in the gene code for the enzyme phenylalanine hydroxylase can result in a nonfunctional enzyme. This enzyme cannot catalyze the conversion of phenylalanine, an essential amino acid, to tyrosine. The dietary phenylalanine must be metabolized, and so alternative reactions are stimulated. When this occurs, normal minor metabolites of phenyl-

1. Saturation—when substrate is in great excess, all the enzyme is used to catalyze the reaction.

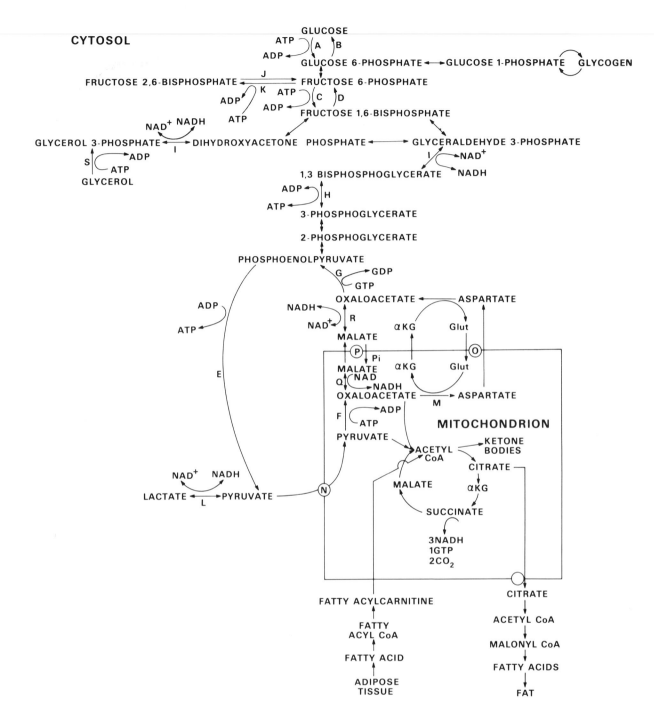

FIGURE 2 Abbreviated metabolic map showing shared pathways of glycolysis and gluconeogenesis, glycogenesis and glycogenolysis, lipolysis, citric acid cycle, fatty acid oxidation, and fatty acid synthesis. Not shown is pentose shunt, which contributes reducing equivalents to fatty acid synthesis. Rate-limiting enzymatic steps and carriers are by letters.

alanine accumulate because the activity of further reaction sequences is not sufficient to metabolize them. These normal metabolites then are in abnormally large amounts and are thought to cause the severe mental retardation characteristic of the disease. What makes this gene-directed amino acid substitution in the enzyme protein so devastating is the fact that the affected enzyme is in a major reac-

tion sequence in the processing of dietary amino acids. If a spontaneous mutation in the gene code

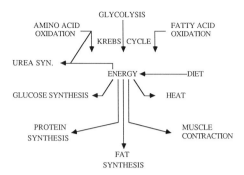

FIGURE 3 Energy and substrates supplied by diet and by intermediary metabolism are used to support protein synthesis, muscle contraction, fat synthesis, heat production, glucose synthesis, and urea synthesis.

occurs for an enzyme of minor consequence, the resultant disease might be inconsequential indeed. Here, an example might be the error in the code for the enzyme necessary for the reduction of L-xylulose to xylitol. In this instance, individuals having this gene code error will excrete xylulose in the urine when xylulose is consumed. Because only a few foods (plums, cherries) contain this sugar, this disease is inconsequential. [*See* GENETIC DISEASES; PHENYLKETONURIA, MOLECULAR GENETICS.]

III. Carrier/Receptor Activity

The amount of substrate can be controlled by factors that influence the entry of the substrate into the cellular compartment in which the reaction occurs.

This represents the next level of control of that metabolic reaction. Substrate entry can be regulated by the availability of carriers (if required) and the mobility of these carriers within the membrane surrounding the cellular compartment. The membrane lipid fluidity may affect carrier mobility if the carrier must change its shape as it functions in substrate transport. Membrane fluidity is determined by the saturation of the phospholipid fatty acids and the amount of cholesterol. Membranes containing higher than normal amounts of saturated fatty acids and cholesterol are also less fluid than those containing more unsaturated fatty acids and less cholesterol. Both diet and hormonal status can influence the ratio of saturated to unsaturated fatty acid and fatty acids to cholesterol. Both the carrier's affinity and mobility and the amount and activity of the enzyme for the reaction in question may be under genetic control. [*See* CHOLESTEROL.]

Just as genetic errors in the amino acid sequence of an enzyme in a key pathway can be devastating, genetic errors in proteins active in higher levels of metabolic control can also have serious consequences. As mentioned, metabolism can be controlled by the activity and availability of carriers that make a substrate available for further processing by enzymes. These carriers in turn may be activated by a variety of factors that comprise the next level of control—a level of control that exists as part of the membrane that surrounds the cell. Membranes form the boundaries of cells and subcellular organelles. Contrary to what was thought for a number of years, they are not inert barriers but

TABLE II Some Diseases Caused by a Spontaneous Mutation in a Gene Code for an Enzyme or Carrier or Hormones or Receptor Important in Metabolic Regulation

Disease	Site of error	Main symptom
Phenylketonuria	Phenylalanine hydroxylase	Mental retardation
Hartnup's disease	Defective epithelial cell transport of neutral amino acids	Tryptophane deficiency
Sickle cell anemia	Hemoglobin: valine substituted for glutamate	Decreased oxygen-carrying capacity of hemoglobin
Diabetes mellitus	Insulin Insulin receptor Intermediary metabolism	Hyperglycemia; abnormal glucose tolerance
Type I lipemia	Adipocyte lipoprotein lipase	High blood levels of triglycerides and cholesterol
Type II lipemia	β-lipoprotein receptor	High blood levels of triglycerides and cholesterol, cardiovascular disease
Hemolytic anemia	Glucose-6-phosphate dehydrogenase	Fragile RBCs
	Pyruvate kinase	Fragile RBCs

function as highly selective permeable structures. Numerous ions and metabolites that are necessary participants of cellular processes traverse them. In addition, proteins, which function either as protein carriers for these molecules or as enzymes for their reactions, are integral parts of the membrane structure. This structure is shown in Fig. 4. Membrane structure consists of localized regions of hydrophobicity and hydrophilicity. Thus, metabolites, nutrients, ions, and other factors important to metabolism but differing in solubility can traverse these barriers. As a consequence, membranes function as controllers and integrators of the various metabolic pathways within the cell, the tissues, and the organs of the body.

The mitochondrial membrane plays a particularly important role in the regulation of cellular metabolism. Catabolism of carbohydrates, proteins, and lipids is initiated in the cytosol but terminated in the mitochondrion. Anabolic processes such as fatty acid synthesis and gluconeogenesis may be initiated in the mitochondrion and finished in the cytosol. If the communication between mitochondrion and cytosol that is necessary in these processes is altered, regulation of intermediary metabolism is changed.

To understand how this communication can be affected, knowledge of the mitochondrial membrane's composition and structure has been obtained. It is known that these membranes are 27% lipid, most of which is phospholipid. A small percentage (1–2%) is cholesterol. The remainder is protein. The activity of the membrane-bound proteins is influenced by their environment (i.e., the kinds of lipids that surround them). For instance, the fatty acids of the phospholipids of the membranes can be modulated by diet. When rats are fed

highly saturated fats, their membrane phospholipid fatty acids reflect this diet and are highly saturated as well. Similar changes in phospholipid fatty acids have been shown with diets containing unusual fatty acids or large amounts of unsaturated fatty acids. Membranes from rats fed these diets have these fatty acids in them. Changes in the membrane lipid in turn can affect metabolism by affecting the activity of the membrane-bound carriers, receptors, and enzymes.

The control that is exerted by the proteins that serve as hormone receptors or receptors for other carrier proteins is genetically dictated. Again, the gene codes dictate whether these proteins are active and functional. The Nobel Prize winning work of Brown and Goldstein showed that if an individual possesses the code for an aberrant receptor for the triglyceride and cholesterol carrying blood protein (β lipoproteins, LDLs), their blood lipid levels, particularly blood cholesterol levels, will rise as will their risk of severe coronary vessel disease. Similarly, if the gene code for the plasma membrane insulin receptor is aberrant, insulin will not adequately bind to it and the entry of glucose into the cell will not be facilitated. Thus, a condition of high levels of blood glucose will develop, and one of the diseases known as diabetes mellitus (a generic disease name that refers to an abnormal insulin–glucose relation) will be diagnosed. Thus, membranes possess metabolic control properties.

IV. Tissue Interactions

Interactions between tissues can exert control on metabolism. For example, muscles at heavy work produce lactate as well as ketones, which are contributed to the circulation and are transported to the liver. These metabolites exert control on hepatocyte metabolism by perturbing the redox state (i.e., the ratios of the reduced to oxidized substrates and coenzymes) of the cell. This perturbation creates a demand for oxygen and increases the activity of the citric acid cycle, mitochondrial respiration, ATP synthesis, and glycogenolysis. These processes are turned on not so much by the energy demand created by the muscles as they contract and relax but by the demand created by the need to metabolize the end products of muscle metabolism because the muscle itself cannot take the oxidation of glucose and fatty acids to completion. The liver must "help out."

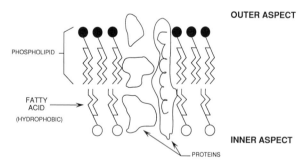

FIGURE 4 *Schematic representation of biological membrane. Some proteins are on the surface, whereas others extend through the lipid bilayer. Some of those extending through membrane are carriers.*

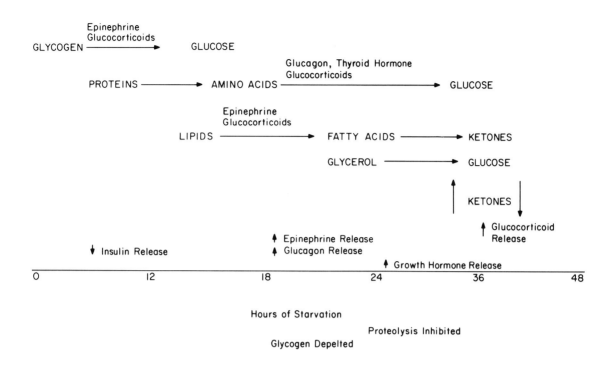

FIGURE 5 Time sequence of events in mobilization of energy and proteins in the body in the first 48 hours of starvation.

V. Hormones

A further integrator of tissue interdependence is the hormones. By definition, a hormone is a substance released by one tissue, carried in the blood, and having an effect on another tissue. Typical of the hormones is insulin. It is released by the β cells of the islets of Langerhans in the pancreas, carried in the blood, and made active when bound to the plasma membrane receptor of the target cell. On binding it exerts its influence on glucose oxidation, fatty acid synthesis, amino acid incorporation to protein, and a whole host of major and minor metabolic events. This integration of metabolic processes occurs in several different cell types and tissues throughout the body, and these metabolic hormone effects can be of both short- and long-term. Substrate flux between tissues is important to survival during starvation. Energy stored as glycogen in the liver or as fat in the peripheral fat cells is released in measured, carefully controlled amounts. This control is orchestrated by the endocrine system. Shown in Fig. 5 is the time course of fuel mobilization in humans as we adapt to the lack of food. Normally, a 70-kg human can survive 30–60 days on the body's fuel reserves. Shown in Fig. 6 are

estimates of the source of energy in this average human. [*See* INSULIN AND GLUCAGON.]

VI. Hormone Interactions

In addition to the control of metabolism by hormones, there is a higher level of control exerted on these hormones that regulates their synthesis and release. For example, the pituitary synthesizes and releases several hormones whose target tissues are

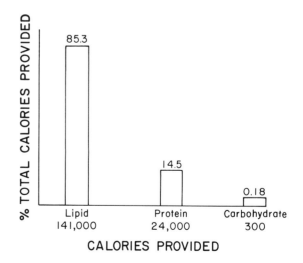

FIGURE 6 Estimates of energy reserves available in an average 70-kg man.

the endocrine glands. Thyroid-stimulating hormone (TSH) is released by the pituitary and stimulates thyroxine release by the thyroid gland and the enzymatic conversion of thyroxine to triiodothyronine by 5' deiodinase. Triiodothyronine is the active hormone form and stimulates respiration, substrate use, oxidative phosphorylation, lipogenesis, lipolysis, gluconeogenesis, glycogenolysis, fatty acid oxidation, and the synthesis of a number of key enzymes in intermediary metabolism. Some of these thyroid hormone effects on metabolism are direct, whereas some are indirect (i.e., the indirect control exerted by the hormone via enzyme or carrier protein synthesis through its effect on DNA translation and/or transcription).

VII. Nutrition

Nutritional state can affect metabolic controls by altering the amounts and kinds of substrates offered to the system for metabolism. In the absence of food, the metabolic pathways are adjusted such that life can be sustained. In starvation, glycolysis, lipogenesis, glycogenesis, and ureogenesis are minimized while lipolysis, fatty acid oxidation, gluconeogenesis, and glycogenolysis are enhanced. A glucose-rich diet has the reverse effect, whereas a fat-rich diet can mimic, to some extent, the metabolic patterns observed in the starved animal. Feeding a fat-rich diet results in minimal glycolytic activity and lipogenic activity as well as minimal glycolysis.

VIII. Higher Thought Processes

Coupled with the effect of nutritional state on metabolic activity is the effect of higher thought processes, which stimulate eating and cessation of eating. These processes also dictate the type of food consumed. The desire for a high-fat or a high-protein or a high-carbohydrate food is initiated in the brain by as yet unknown messages. Some of these messages are rhythmic. That is, they follow a set pattern from day to day, month to month, year to year. These rhythms are circadian. As such, they impose a rhythm on metabolism such that pathways can vary from day to day or within the day in response to changes in light intensity or other geophysical factors. [See CIRCADIAN RHYTHMS AND PERIODIC PROCESSES.]

IX. Environment

Lastly, metabolic control can be perturbed by environmental factors such as temperature, light, and geophysical rhythms. External stress that is unexpected or uncontrolled can also exert control that indeed may influence the continuance of life of the individual. Such is the case in the individual who becomes traumatized by massive burns, multiple injuries, bone fractures, or overwhelming infections. These factors can exert a powerful influence on metabolism, which, if it cannot respond, means the death of the individual. Indeed, death occurs when metabolism is uncontrolled, for the cell can no longer regulate its energy input and outflow.

In summary, metabolic control mechanisms exist so that both coarse and fine adjustments in energy balance and substrate flux can be made. Energy from the food and energy used by the body to sustain life are carefully balanced. So too are amino acids, vitamins, and minerals. Together these materials are used to create and sustain life. Without functioning metabolic controls, life cannot be sustained.

Bibliography

Denton, R. M., and Pogson, C. I. (1976). ''Metabolic Regulation.'' John Wiley & Sons, New York.

Devlin, T. M. (1982). ''Textbook of Biochemistry with Clinical Correlations.'' John Wiley & Sons, New York.

Finean, J. B., Coleman, R., and Michell, R. H. (1978). ''Membranes and Their Cellular Functions.'' Blackwell Scientific Publications, London.

Harold, F. M. (1986). ''A Study of Bioenergetics.'' W. H. Freeman, New York.

Saier, M. H., Jr. (1987). ''Enzymes in Metabolic Pathways.'' Harper & Row, New York.

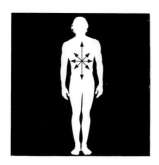

Metastasis

BEVERLY TAYLOR SHER

Glossary

Metastasis Process of formation of secondary tumors; also used as a synonym for a secondary tumor

Metastatic potential Capacity of a population of cancer cells to form new tumors when introduced into an animal

Micrometastasis Secondary tumor that is microscopic in size

Preneoplastic cell Cell that has a genomic change(s), rendering it more likely to become transformed at a later time

Seed/soil hypothesis Explanation of why different tumor types preferentially metastasize to different sites; this preference is hypothesized to be due to the different microenvironmental requirements of different kinds of tumor cells for growth

Tumor progression Development of a tumor from a single transformed cell into a full-blown, aggressive tumor; characterized by a number of successive genetic and phenotypic changes

METASTASIS IS THE PROCESS by which cancer cells spread from a primary tumor to other, often very distant, parts of the body and form secondary tumors, or metastases. The characteristics of this complex phenomenon are different for different tumor types. Some tumors, such as melanomas, which arise from melanocytes in the skin, metastasize with relative ease; others, such as basal cell carcinomas, which arise from epithelial cells in the basal layer of the skin, can grow for years before metastasis occurs. In addition, different tumors show different site preferences for metastasis formation. Metastasis is also a problem of great clinical importance. In most cases, cancer is already a systemic disease with micrometastases present at the time a patient begins treatment, and the available treatments are not very effective against widely disseminated metastatic disease. The problem of metastasis is being intensively studied by many laboratories throughout the world. Some key questions commonly addressed by these studies include the following.

1. What tumor cell characteristics (cell-surface markers, nuclear DNA changes, etc.) are associated with the metastatic phenotype?
2. Why does a given type of tumor show preferential metastasis to particular sites in the body rather than others?
3. What treatments block metastasis and why?

In this article, I will first discuss the major experimental systems used to study metastasis, and then discuss the metastatic process itself.

I. Experimental Approaches to the Problem of Metastasis

A. Establishing Correlations between Phenotypic Characteristics of Cells and Metastatic Behavior

As metastasis is a clinically relevant problem, studies of metastasis using human tumor cells are most desirable. Obvious ethical problems exist with most direct forms of experimentation, however, so until recently the only method for studying metastasis of human tumors involved determining correlations between various tumor cell properties and the stage of the disease in which they made themselves manifest. Once a tumor cell attribute had been defined as interesting by virtue of such a correlation, it could then be studied in a more direct fashion, either in vitro or in an appropriate animal model system.

B. Classical Experimental Metastasis

"Experimental metastasis" is a term that refers to the process of injecting a single-cell suspension into the bloodstream of a syngeneic (genetically identical) test animal, allowing the cells to establish metastases in the animal, and then counting the tumor colonies in the different organs of the animal after an appropriate time. Cell suspensions that result in the formation of large numbers of metastases are said to have a high metastatic potential, whereas those that form few or no tumor colonies have a low metastatic potential. Comparison of the number of metastases found in different organs of the animal also can reveal whether the test cells have a discernible organ preference for metastasis formation.

Experimental metastasis has been a highly useful and informative technique. It allows the direct comparison of the behaviors of cell suspensions subjected to different treatments or of cells known to differ in only one phenotypic attribute; it can also be used to test the effects of a particular pretreatment of the host or therapy performed during tumor development. Some limitations on the system, however, should be mentioned.

First, if normal animals are used as tumor hosts, only syngeneic test cells can be used, because the host's immune system will attack allogeneic (genetically distinct but from the same species) or xenogeneic (from a different species, i.e., human) cells.

Second, the introduction of relatively large numbers of tumor cells in a single-cell suspension directly into the bloodstream is a nonphysiological model for metastasis. Different doses of tumor cells will often produce different patterns of metastasis. In addition, the usual method involves the injection of the cells into the tail vein of the mouse or rat. Thus, the first organ that they enter is the heart, and the first capillary bed that they encounter is in the lungs. Injection of test cells into different parts of the circulation or through specific, noncirculatory routes will often produce different results. Also, pretreatment of the test cells before injection (i.e., homogenization of tumor tissue in buffer to make a single-cell suspension) could theoretically alter the properties of those cells.

Variations on the technique of classical experimental metastasis include the injection of test cells into different parts of the host's body, injection of different numbers of cells into different hosts, examination of test animals at different times after injection, and so on.

C. Experimental Metastasis in Immunodeficient Hosts

An important variation on the technique of classical experimental metastasis involves the use of immunodeficient host animals. The most frequently used immunodeficient host is the nude mouse. Nude mice are homozygous for the nu allele. This mutated gene renders the mice congenitally athymic, and thus the T-cell compartment of the immune system fails to develop normally. As a result, these mice cannot mount a normal cellular immune response and cannot effectively reject most allogeneic and xenogeneic cells. Thus, human tumors can be studied readily in these mice. Although nude mice have some residual cellular immunity, this can be abrogated by treatment with immunosuppressive drugs. A potential limitation to the use of this system is the possibility that mouse and human tissue microenvironments might not be identical; this has not stopped some very productive studies in this system. Other immunodeficient hosts, such as beige mice, which lack natural killer (NK) cell activity (see below), and nude rats, can also be used.

D. Transgenic Mice

Another fairly recent development in the study of metastasis is the use of various transgenic strains of mice. Transgenic mice are made by injecting specific DNA constructs into fertilized mouse eggs, which take up the foreign DNA and integrate it into their genomes. The injected eggs are implanted into foster mothers and develop into mice as usual. They are then used to establish new inbred strains of mice, which express the foreign gene. Strains of mice bearing oncogene and growth factor constructs often develop tumors in a predictable fashion. Such strains can be used to study tumor development and metastasis.

II. Tumor Formation and Growth

A. Introduction

The process of metastasis can best be understood in the context of the life history of the tumor. The process by which a single normal cell becomes preneoplastic, loses normal growth control, and grows, divides, and undergoes further changes to achieve the final stage of malignant, widely disseminated metastatic cancer is known as tumor progression. Tumor progression, with emphasis on the features of this process that are most important for metastasis formation, is discussed in the following section. [*See* NEOPLASMS, ETIOLOGY.]

B. Genomic Change

A wide variety of experiments clearly shows that the vast majority of tumors are monoclonal in origin. Thus, the process of tumor progression is generally thought to begin with a genomic change in a single normal cell. This alteration can be the result of a mutation caused by a chemical mutagen or radiation in some form, or it could be the result of the introduction of viral oncogenes subsequent to the infection of the cell by a tumor virus. Although direct genetic change is the major force driving tumor progression, nonmutational genomic change can also be involved. For example, a tumor-promoting chemical can induce a change in the differentiated state of a normal cell that results in the cell's increased susceptibility to the effects of some carcinogenic substance.

The initial genomic alteration in the normal cell can have subtle or dramatic effects. Relatively subtle changes result in preneoplastic cells. Such cells are often identifiable by differences in morphology, rate of growth, pattern of growth, or loss of expression of differentiated functions when compared to surrounding normal cells. These cells can grow and divide to form preneoplastic lesions. An example of a preneoplastic lesion is the colon polyp: these growths are not cancerous but can become so if more genetic changes occur. Not all tumor types have identifiable precancerous states, however. Renal carcinoma, some forms of lung carcinoma, and a group of tumors categorized as "unknown primary tumors," which show up only as widely disseminated disease, seem to compress the process of tumor progression into a very short time and have no identifiable precancerous lesions.

The process of tumor progression continues with the accumulation of further genomic changes, which can result in the gradual loss of normal control over cell division and alterations of differentiated function. The next major step in tumor progression is the fully transformed, or neoplastic, cell. A neoplastic cell is a cell that has lost normal growth control and grows and divides autono-

mously to form a neoplasm (a new growth). Many different cellular genes, which, when mutationally altered, can result in neoplastic transformation of their host cells, have been identified. Their functions, when known, have fallen into several interesting groups: growth factors, growth factor receptors, other signal-transducing factors, and nuclear proteins involved in the regulation of other genes. These proteins can be involved in growth regulation either during development, during the course of mature cellular function, or both. These genes are known as cellular protooncogenes.

Fully neoplastic cells are also referred to as transformed cells. The concept of cellular transformation was initially defined in tissue culture. Transformed cells have several unique properties:

1. They have lost the ability to stop dividing when they are in contact with other cells, a property known as contact inhibition.
2. They generally have a higher growth rate than normal cells.
3. They divide indefinitely in culture, unlike normal cells, which die after a set number of generations (immortalization).
4. They form tumors when injected into syngeneic animals.

The progression from the normal cell stage to the fully transformed neoplastic cell stage can require the accumulation of many genomic changes or can be quite rapid, depending on the nature of the genomic changes and the differentiated state of the normal parental cell in question. This point is best illustrated by the example of inherited retinoblastoma, a tumor of the eye that develops only in children under the age of 5 yr. Genetically susceptible individuals are heterozygous for the RB1 gene: they have one functional copy and one nonfunctional copy. If, during the growth of the retina (which is over by the age of 5 yr), a retinal cell in a susceptible individual has its normal Rb gene mutationally inactivated, that cell will grow and divide to form a retinoblastoma. Thus, in this case a single genetic change can result in the formation of a fully malignant tumor. Although functional RB1 alleles are undoubtedly mutationally inactivated in other cell types during a susceptible person's lifetime, their susceptibility to tumors arising from other tissues, while higher in some cases than that of an RB1 homozygote, is not nearly as great as their susceptibility to retinoblastoma. Comparison of people affected by bilateral retinoblastoma (two independent retinoblastoma tumors arise, one in each eye) with people afflicted with unilateral retinoblastoma has shown that all cases of bilateral retinoblastoma arise in people who are heterozygous for the RB1 gene; people suffering from unilateral retinoblastoma are frequently homozygous for the functional RB1 gene and develop retinoblastoma due to the sequential mutational inactivation of both functional RB1 alleles later in development. Thus, cells in the developing retina, by virtue of their differentiated state, are uniquely susceptible to the mutational inactivation of a single gene. [*See* RETINOBLASTOMA, MOLECULAR GENETICS.]

C. Tumor Heterogeneity

The process of accumulation of genetic and phenotypic changes does not stop after the formation of the initial transformed cell. As the tumor cells grow and divide, individual tumor cells continue to accumulate changes, and, thus, tumor cell subpopulations with different characteristics arise. This process is accelerated by two different forces: genetic instability and phenotypic selection within the tumor. Compared with normal cells, tumor cells are genetically unstable, particularly in more advanced tumors. They often have aberrant DNA content, having either lost or amplified various stretches of chromosome. In some types of tumors, the loss or amplification of particular stretches of chromosome can be characteristic of (and perhaps causative for) the stage of tumor progression arrived at by the tumor. Colon carcinomas, for example, typically suffer mutations involving the RAS protooncogene early in their development, whereas mutations involving the gene encoding the transformation-associated protein P53 are associated with the transition from the benign to the malignant state. As the tumor grows, different microenvironments within the tumor arise, due to differences in such things as the availability of nutrients. A subpopulation of cells that has a selective advantage over the other tumor cells in one of these microenvironments, or in the tumor as a whole, can come to dominate the tumor.

The presence of multiple phenotypically and genetically distinct tumor cell subpopulations within a tumor is known as tumor heterogeneity. It is important to the oncologist because different subpopulations have disparate responses to different therapies, and the processes that drive tumor heterogeneity also drive the creation of therapy-resistant tumor cell subpopulations. Tumor heterogeneity is also important for the process of metastasis forma-

tion because, in most primary tumors, only a small subpopulation of the cells is actually metastasis-competent. Through the processes that create tumor heterogeneity, subpopulations acquire additional characteristics and do more than evade normal growth control and divide to form a tumor.

D. Angiogenesis

The initial growth of a tumor is limited by the tumor cells' supplies of nutrients and oxygen. As the tumor becomes larger, the cells in the middle become farther away from the blood vessels surrounding the tumor and often die because diffusion limits their nutrient supply to the point of starvation. The center of a tumor is thus often filled with areas of dead, dying, and dormant cells, known as necrotic areas. Until the tumor becomes capable of stimulating the growth of blood vessels into itself, a phenomenon known as neovascularization, its diameter will be limited to a few millimeters. Therefore, angiogenesis, the formation of new blood vessels, is a critical process for tumor progression.

The neovascularization of the tumor begins when some of its cells acquire the ability to secrete one or more angiogenic factors. These factors stimulate changes in the behavior of the endothelial cells lining blood vessels adjacent to the tumor, which begin to divide and migrate toward the tumor. The endothelial cells also acquire the ability to break down the different components of the extracellular matrix surrounding the tumor. Eventually, they form themselves into functional capillaries within the tumor tissue.

A number of different angiogenic factors, secreted by both tumor cells and normal cells (during processes such as wound healing) have been identified. Some of these factors are proteins, including acidic fibroblast growth factor, basic fibroblast growth factor, angiogenin, and transforming growth factor. Other substances, including some lipids and prostaglandins, also have angiogenic activity. [*See* TRANSFORMING GROWTH FACTOR-α; TRANSFORMING GROWTH FACTOR-β.]

A particularly useful model system for the study of angiogenesis and its role in tumor progression should be noted here. In a strain of transgenic mice, which carries the SV40 large T antigen oncogene under the control of the rat insulin promoter regulatory region, the beta cells of the islets of Langerhans (the cells that produce insulin) reproducibly proliferate in an unusual manner. With a predictable frequency, these areas of hyperplasia turn malignant and begin to invade surrounding tissue. This important step in tumor progression has been shown to be preceded by the production of angiogenic factors by some, but not all, of the hyperplastic islets. All of these areas of focal hyperplasia that turn malignant have undergone neovascularization. Thus, in this system, angiogenesis is crucial for tumor progression toward malignancy.

Angiogenesis is also important for the process of metastasis. In order to move to distant sites in the body, tumor cells need to get into the circulation. Increasing the proximity and number of available capillaries increases the likelihood of metastasis.

III. What Makes a Metastatic Cell?

A. Introduction

The previous section focused on properties of the tumor as a whole that are relevant for metastasis formation. The next steps in the metastatic process, however, depend on the properties of individual tumor cells. In order for distant metastases to form, individual tumor cells or small clusters of tumor cells must perform the following steps:

1. They must leave the site of the primary tumor.
2. They must enter the circulatory system.
3. They must survive in the circulatory system long enough to reach a new part of the body.
4. They must become arrested in the new site, either by mechanical trapping or by active adhesion to blood vessel endothelial walls.
5. They (often) must pass through blood vessel walls, a process known as extravasation.
6. They must invade surrounding tissue.
7. They must grow and divide to form a new tumor mass.

Although not all of these steps are obligatory for metastasis formation in every situation, they are all important to the process as a whole and will be discussed in detail in this section.

As discussed above, in most primary tumors, a few tumor cell subpopulations are actually capable of metastasis. These subpopulations develop by forces that create tumor heterogeneity, namely genetic change and phenotypic selection. The ease with which subpopulations of metastasis-competent cells arise within the primary tumor depends to a large extent on the cell type of the initial transformed cell. If the original cell already possessed some of the biochemical properties necessary for metastasis formation and these properties were retained by the cells of the tumor, metastasis will occur more readily because these properties will not have to be acquired by chance. In accordance with this idea, lymphomas, which arise from transformed lymphocytes, often readily enter the bloodstream and rapidly accumulate in all of the organs of the host that are accessible to normal recirculating lymphocytes. Basal cell carcinoma cells, on the other hand, can grow for a long time before some of the cells in the tumor acquire the ability to enter the bloodstream; their parental normal cells, found in the bottom layer of the epidermis, lack motility and the ability to enter the bloodstream as differentiated functions. [*See* LYMPHOCYTES; LYMPHOMA.]

B. Entry into the Circulation

The formation of a secondary tumor begins when a tumor cell, or a clump of tumor cells, leaves the site of the primary tumor. In some cases, this occurs as the result of simple mechanical forces. Tumor cells that grow into the wall of nearby blood vessels can bud into the lumen of those vessels and eventually be swept away as emboli by the pressure of the circulation. Tumor cells can also enter the circulation as a result of damage to the primary tumor (i.e., caused by a surgeon's scalpel). Also, some evidence indicates that some tumor cells adhere less well to each other than do cells in normal tissue. This is particularly true for the cells in the necrotic areas in the centers of tumors. Neoangiogenesis of these areas can therefore result in the entry of tumor cells into the bloodstream. Tumor cells can also actively enter the circulation. Tumor cells that have acquired an invasive phenotype, i.e., the ability to autonomously move through host tissue, can leave the site of the primary tumor with ease. (Invasiveness will be discussed in detail in Section F.) An actively invasive tumor cell that is motile and can chew its way through surrounding host tissue can either grow and divide to form a new tumor not far from the parental tumor or go farther away to initiate a more distant metastasis. If the tumor cell chews its way into a lymphatic vessel, it will end up in the lymph node, draining the region of the primary tumor. It can either lodge and form a new tumor (providing it escapes the attention of the immune system [see Section D], or it can eventually end up in the thoracic duct lymph, which is emptied directly into the bloodstream. A less complicated way to enter the bloodstream and seed distant sites is, however, direct entry through a blood vessel wall, a process known as intravasation.

C. Survival in the Bloodstream

For a tumor cell, life in the bloodstream is dramatically different from life in the tumor or in the tissue surrounding it. Hydrodynamic pressures reduce the ability of the cell to grow and divide and can destroy large, fragile cells. The vast majority of tumor cells that enter the bloodstream die before they can leave it. Resistance to hydrodynamic stress is, thus, a potentially important part of the metastatic phenotype. The effects of hydrodynamic stress on different cell types have been measured in vitro. Highly metastatic B16 melanoma cells were as resistant to hydrodynamically induced cell lysis as human red blood cells, much more resistant to such stress than were mouse hybridoma cells (which are relatively large cells) and human white blood cells. Hydrodynamic stress is also survived better by clumps of tumor cells than by individual cells. Tumor cells that form emboli by adhering to each other (homotypic adhesion) or to host platelets or lymphocytes (heterotypic adhesion) often have higher metastatic potential than cells that do not form emboli. Similarly, cells that form thrombi by interacting with the host's clotting system can show enhanced metastatic potential in some systems.

D. Interactions with the Host's Immune System

In order to form a metastasis, a tumor cell must escape the attention of the host's immune system. The cellular branch of the immune system is thought to have immune surveillance (the search for and destruction of transformed cells) as one of its primary functions. This is accomplished in different ways by different immune system components. Circulating host macrophages, as well as resident tissue macrophages such as the Kupffer cells in the liver, are able to destroy tumor cells. Natural killer cells recognize transformed cells and destroy them. Natural killer cells can kill both syngeneic and allogeneic transformed target cells. T cells recognize specific tumor cell-surface molecules (tumor antigens) in the context of host major histocompatibility molecules (MHC antigens: HLA in human, H-2 in mouse) and kill the tumor cells bearing the appropriate combination of self-MHC molecules and tumor antigens. T cells will also kill target cells that

bear allogeneic MHC molecules. Other, less well-defined types of cells may be involved in tumoricidal activity as well. [See IMMUNE SURVEILLANCE; MACROPHAGES; NATURAL KILLER AND OTHER EFFECTOR CELLS.]

A potentially metastatic tumor cell can avoid immune action in several ways. Tumor cells that do not display appreciable amounts of MHC antigens on their surfaces can avoid the attention of killer T cells, and indeed low expression of MHC molecules has been correlated with increased metastatic potential in vivo. Similarly, tumor cells that display lower amounts of tumor-specific antigens also are less likely to be recognized and killed. The reduced expression of such cell-surface molecules is known as antigenic modulation. Tumor cells can also be selected in vitro for resistance to NK cells and macrophage-induced lysis; this presumably can also happen in vivo.

The parental tumor can produce factors that render host immune cells incapable of action. Although advanced tumors usually contain T cells (known as TIL cells, or tumor-infiltrating lymphocytes), these cells are not activated and do not kill the tumor cells. Upon removal from the tumor and activation with such substances as interleukin 2 or interferon, TIL cells will readily kill the tumor, so the tumor is only rendering them incapable of responding, not actively deleting the set that is capable of killing it. The presence of the primary tumor in the body can therefore protect metastatic cells as well, by virtue of its general suppression of antitumor action.

E. Arrest of the Blood-Borne Tumor Cell

The next step in metastasis formation is the arrest of the blood-borne tumor cell(s) in the vasculature. Tumor cell emboli and large tumor cells can become trapped in the microcirculation, often stopping in the very first capillary bed they encounter. Small tumor cells, on the other hand, can circulate through a number of capillary beds before becoming arrested by virtue of their adhesion to the wall of the blood vessel. A number of biochemically distinct adhesion interactions have been implicated in this process. These include, but are not limited to: receptors for various basement membrane components; cell-adhesion molecules, usually involved in adhesive interactions during development; the liver asialoglycoprotein receptor; certain glycolipids; cell-surface molecules involved in the recirculation of normal lymphocytes, such as lymphocyte homing receptors and LFA-1; a variety of lectins (macromolecules which bind specific sugar groups) found on both the tumor cell and the endothelial cell surface; and other endothelial cell-surface and tumor cell-surface molecules. It is worth noting that all of these adhesive interactions are probably involved in normal host functions; tumor cells that acquire the expression of these molecules at adequate levels can use them to adhere to host blood vessel walls.

Some of these adhesive interactions can occur in almost any tissue that a tumor cell ends up in. Increased expression of laminin receptors has been correlated with increased metastatic potential in some experimental systems. Laminin is a common component of vascular endothelial basement membranes (although the amount of laminin in different basement membranes varies), and thus tumor cell laminin receptors could facilitate tumor entry into a number of different host tissues.

Other adhesion mechanisms are tissue-specific and could thus contribute to the site-specificity shown by some types of tumors. For example, it is known that different endothelia bear tissue-specific cell-surface molecules. Tumor cells selected for metastasis to a particular site will often adhere preferentially to endothelia from that tissue. This suggests that some of the tissue-specific endothelial cell-surface molecules are involved in tumor cell adhesion to endothelia. One situation in which this is almost certainly the case involves the molecules known as lymphocyte homing receptors. Normal lymphocytes enter peripheral lymph nodes and Peyer's patches from the blood by means of adhesive interactions between lymphocyte homing receptors on their surfaces and specific endothelial cell-surface molecules found only in specialized blood vessels known as high endothelial venules. In mice, lymphoma cells bearing either the peripheral node or Peyer's patch lymphocyte homing receptor (or both) form metastases in all peripheral lymph nodes, as well as in other organs, such as liver and spleen. Lymphomas that do not express homing receptors form solid tumors when injected subcutaneously and do not form widespread metastases.

F. Invasion of Surrounding Host Tissue

Once a tumor cell has become arrested, either by becoming trapped in a capillary bed or by adhering to a blood vessel wall, the next step in metastasis formation is its entry into surrounding tissue and the eventual establishment of a new tumor. This can occur in two different ways. Arrested tumor cells or emboli can grow and divide in the microcirculation and eventually break through the blood vessel wall, or they can actively pass through it by stimulating the retraction of endothelial cells and moving through the endothelial wall's basement membrane into the surrounding tissue. This second alternative requires a constellation of biochemical attributes that together constitute the invasive phenotype. A given tumor cell does not need to express every single invasive attribute in order to enter a host tissue; rather, different tumor cells can have only a subset of invasive properties. These properties include the capacity for autonomous or chemotactic movement; alterations in the ability to bind to host extracellular matrix components or to cell-surface molecules of parenchymal cells in the host's tissue; and the ability to degrade the extracellular matrix of the host's tissue, or to induce host cells to do so. These properties will now be discussed in detail. [See EXTRACELLULAR MATRIX.]

1. Tumor Cell Motility

One major component of the invasive phenotype is the ability of the tumor cell to move within the host tissue. The cell can do this in two different ways. The first is to acquire the ability to produce autocrine mobility factors, molecules that bind to specific tumor cell-surface receptors and cause the cell to begin to move in random directions. Alternatively, the tumor cell can be stimulated to migrate in a directed chemotactic fashion by interactions between specific receptors on its surface and elements in the extracellular matrix surrounding the tumor, including Type IV collagen, laminin, and fibronectin. This stimulation, which is mediated by specific cell-surface receptors for these proteins, has been demonstrated by the use of in vitro assays of the mobility of tumor cells on purified substrates and has been shown to be correlated with metastatic potential in a variety of experimental systems. Synthetic peptides corresponding to fragments of these extracellular matrix elements have been shown to block migration along artificial substrates in vitro and experimental metastasis in vivo (although the in vivo mechanism is probably more complicated than this, as adhesive interactions between the tumor cells and blood vessel endothelial walls can also be blocked by such treatments).

2. Altered Adhesive Properties

Alterations in the adhesive properties of tumor cells have been correlated with increases in invasiveness in vitro and in metastatic potential in vivo. Increases in the ability of tumor cells to adhere to cells in the target tissue (increased heterotypic adhesiveness) often correlate with increased metastatic potential, perhaps because a tumor cell that can more effectively bind to tissue parenchymal cells can more effectively invade its target tissue. Alterations in the ability of a tumor cell to bind to various extracellular matrix components have also been correlated with increased invasiveness in vitro and increased metastatic potential in vivo. Perhaps the best example of this involves the fibronectin receptor. Fibronectin is a protein component of the extracellular matrix. Many tumor cells display reduced synthesis of fibronectin as well as lower numbers of fibronectin receptors on their surfaces and less cell-surface fibronectin. The fibronectin receptor is a member of a family known as the integrins, which are all cell-surface proteins that link extracellular elements to the cytoskeleton and are thus critical to adhesion, maintenance of cell shape, and cell movement. Decreases in the transcription of the fibronectin receptor genes and, thus, in its synthesis, have been shown to correlate with increased invasiveness in vitro and increased metastatic potential in vivo.

3. Degradation of the Host Tissue's Extracellular Matrix

The production of enzymes that degrade different components of the extracellular matrix has been correlated with increased metastatic potential in many different systems. The breakdown of the matrix, both by the potentially metastatic cell and by cells surrounding it, facilitates the tumor cell's movement through normal host tissue. Some of the degradative enzymes produced by tumor cells of high metastatic potential include heparan sulfatases, which degrade heparan sulfate, a major component of the extracellular matrix; Type IV collagenases, which degrade Type IV collagen, the major form of collagen found in the extracellular matrix; and such general-utility proteases as trypsin and cathepsin B. Interestingly enough, although some tumor cells produce collagenases specific for other types of collagen, the expression of these enzymes does not always correlate with increased metastatic potential.

The ability of a tumor cell to invade host tissue can be facilitated by the effects it has on surrounding host cells. Fibroblasts, the cellular elements of connective tissue, can be induced to produce collagenases and other degradative enzymes by factors secreted either by tumor cells or by host cells responding to the presence of tumor cells. These factors, normally produced during such processes as tissue growth and remodeling and wound healing, include platelet-derived growth factor, epidermal growth factor, beta-transforming growth factor, epithelial cytokine, interleukin 1, and tumor necrosis factor. In addition, the various angiogenic factors (see above) will make endothelial cells invasive and, therefore, contribute indirectly to the destruction of the host tissue's extracellular matrix and thus to increased tumor cell spread.

G. Growth in the New Location: The Seed/Soil Hypothesis

Once a tumor cell has entered a new host tissue, it has to grow and divide to form a new tumor for the process of metastasis to be complete. Whether it is able to do so depends on both its own properties and the properties of the tissue microenvironment in which it finds itself. If it is capable of responding to locally produced growth factors, or of stimulating its own growth in an autocrine fashion, or of stimulating surrounding cells to produce growth factors to which it can respond, it will be able to grow and divide rapidly. Likewise, it must be insensitive to locally produced substances involved in the limitation of cell growth and division in order to create a new tumor. This idea, first formulated by Stephen Paget in the nineteenth century, is known as the seed/soil hypothesis. It is supported by numerous experimental observations, which include the following.

1. Tumor cells can be subcloned, which are then capable of growth in one organ in vivo but not in others; appropriate selection of tumor cells from the same parental line can produce subclones with different organ preferences.
2. In vitro, such subcloned tumor cell lines proliferate in response to soluble factors from their target organ, but not from others.
3. Tumor cells selected for growth in a particular organ will grow better in vitro on extracellular matrix extracted from that organ than on extracellular matrix extracted from other organs.

4. Negative effects on tumor cell growth of soluble factors from various organs can also be demonstrated; thus, tumor cell growth can be actively inhibited in some microenvironments.

Thus, the microenvironment in which a tumor cell finds itself is probably critically important for metastasis formation; therefore, seed/soil interactions are thought to be a major source of the target organ preference evinced by many tumor types.

As the secondary tumor grows in the new site, cells within it continue to diversify. Factors in the new microenvironment can induce phenotypic changes in appropriately responsive cells, which can either enhance or reduce their ability to grow and divide, and mutations continue to accumulate as the new tumor grows. Late in tumor progression, though, tumor heterogeneity is often greatly reduced, as metastases and even the primary tumor can become dominated by particularly successful highly metastatic cell populations. These cells have gradually lost their susceptibility to normal growth regulation and increased their ability to autonomously grow and divide. As a result, these cells can grow practically anywhere. At the death of the host, frequently every organ system has some tumor burden, as host tissues have been invaded by cells originating in both the primary tumor and by cells originating in metastases.

IV. Summary

The process of metastasis is highly complex and is driven by genetic change and phenotypic selection within the primary tumor as well as later in tumor progression. The timing of metastasis formation and the pattern of metastatic spread depend on both properties of the tumor cells and properties of the host tissues in which they find themselves. Many different biochemical properties of normal host cells can be used by metastatic cells during the process of metastasis formation. Each potentially metastatic cell needs only a subset of the possible attributes that could contribute to its success to eventually create a metastasis. Thus, there is no single metastatic phenotype, but rather a whole collection of possible metastatic phenotypes. Similarly, the site preferences evinced by particular types of tumors are only relative; given enough time, almost any host tissue will have some metastatic tumor burden. These features of the metastatic process will continue to pose challenges for both researchers and oncologists.

Bibliography

Folkman, J., and Klagsbrun, M. (1987). Angiogenic factors. *Science* **235,** 442–447.

Hood, L. E., Weissman, I. L., Wood, W. B., and Wilson, J. H. (1984). "Immunology," 2nd ed. The Benjamin Cummings Publishing Co., Inc. Menlo Park, CA.

Kerbel, R. S., Waghorne, C., Korczak, B., Lagarde, A., and Breitman, M. L. (1988). Clonal dominance of primary tumors by metastatic cells: Genetic analysis and biological implications. *Cancer Surveys* **I(4),** 597–629.

Nicolson, G. L. (1988). Cancer metastasis: Tumor cell and host organ properties important in metastasis to specific secondary sites. *Biochim. Biophys. Acta* **948,** 175–224.

Sher, B. T., Bargatze, R., Holzmann, B., Gallatin, W. M., Matthews, D., Wu, N., Picker, L., Butcher, E. C., and Weissman, I. L. (1988). Homing receptors and metastasis. *Adv. Canc. Res.* **51,** 361–390.